The Book of Wood Names

Buch der Holznamen
Les Noms des Bois
Los Nombres de las Maderas

by
Dr. Hans Meyer

LINDEN PUBLISHING
Fresno, CA

Originally published in 1936 by M. & H. Schaper,
Hannover.
First edition printed in Germany.

123456789

THE BOOK OF WOOD NAMES

Library of Congress Cataloging-in-Publication Data
Meyer, Hans, 1885-1935.
  (Buch der Holznamen. English)
  The book of wood names : vernacular and botanical nomencla-
ture of worlds woods / Hans Meyer.
    p. cm.
  ISBN 0-941936-62-7 (pbk.)
  1. Botany--Nomenclature. 2. Botany--Nomenclature (Popular) 3.
Wood. 4. Timber. I. Title.

QK96 .M413 2000
582.16'01'4--dc21

LINDEN PUBLISHING

*The Woodworker's
Library*

Linden Publishing Inc.
336 West Bedford, Ste 107
Fresno, CA 93711 USA
tel 800-345-4447
www.lindenpub.com

Vorwort.

Das vorliegende Buch soll einem Mangel abhelfen, der sowohl in wissenschaftlichen als auch in praktischen Kreisen immer fühlbarer wird durch das Fehlen eines umfassenden schnellen Nachschlagewerkes für die den Holzbezeichnungen nach ihrer Herkunft zugrunde liegenden Stammpflanzen.

Von wissenschaftlicher Seite aus ist dieser Frage in gewisser Weise stets Beachtung geschenkt. In den wissenschaftlichen Veröffentlichungen sind die Vulgär- (Volks-), Handels- und Eingeborenenbezeichnungen der Nutzpflanzen, unter denen die Hölzer eine große Rolle spielen, wohl erwähnt, jedoch sind diese meistenteils nicht alphabetisch geordnet oder in Sachregistern zusammengestellt. Die Namen sind deshalb in der Literatur schwer aufzufinden und können infolgedessen in der Praxis zur Bestimmung eines Holzes nur mit großem Zeitaufwand ausgewertet werden. In den beiden letzten Jahrzehnten hat besonders das Ausland (d. h. außerhalb Deutschlands) die Bedeutung der Holznamen erkannt und auf diesem Gebiete eine rege Tätigkeit entfaltet. Es hat zahlreiche Zusammenstellungen derselben veröffentlicht, doch beziehen sich diese entweder nur auf die eingeführten Hölzer eines Einzelstaates oder beschränken sich nur auf begrenzte Herkunftsgebiete.

Das vorliegende Buch versucht, diese Mängel zu beseitigen einerseits durch alphabetische Anordnung der Kennwörter, die ein schnelles Auffinden gewährleistet, andererseits durch Erfassung möglichst sämtlicher Erzeugungs- und Verbrauchsgebiete. Es sind die zugrunde liegende Stammpflanze, Pflanzenfamilie sowie die Herkunft des Holzes oder des betreffenden Namens, soweit bekannt, sofort ersichtlich.

Das Buch verdankt seine Entstehung einer 20-jährigen Untersuchungs- und Auskunftstätigkeit auf dem Gebiete der Nutzhölzer, während welcher Zeit die Namen aus der einschlägigen Literatur gesammelt wurden. Das Interesse in den Holzeinfuhr-, Holzhandels- und Holzindustriekreisen für die den in- und ausländischen Nutzholzsorten bezw. -namen zugrunde liegenden Stammpflanzen wächst zusehends, wie die ständig zunehmende Zahl der am „Institut für angewandte Botanik zu Hamburg" einlaufenden Anfragen erkennen läßt. Ferner hat die Feststellung der jeweiligen Pflanzenart eine besonders hohe Bedeutung für verschiedene Behörden, wie die Zollbehörden, Eisenbahn, Hafenbehörden, Handelskammern u. a. m. Beispielsweise teilt der deutsche Zolltarif die eingeführten Nutzhölzer in folgende Gruppen auf: 1) Zedernhölzer; 2) Weichhölzer, unter diesen bis vor kurzem als besonders begünstigt Tannen- (Abies), Fichten- (Picea) und Lärchen- (Larix-)hölzer; 3) Harthölzer, von denen die Obstbaumhölzer und einige andere tariflich begünstigt sind; 4) Edelhölzer. Diese Gruppen fallen unter verschiedene Tarifnummern und erfahren damit eine verschiedene zolltarifliche Bewertung. Auch der deutsche Eisenbahngütertarif bewertet die Nutzhölzer hinsichtlich der Beförderungssätze nach verschiedenen Klassen. Die Einstufung bezw. die Einreihung in die jeweilige Gruppe ist aber meist nur mit Kenntnis der Stammpflanze möglich. Aenderungen der Tarife oder die Einführung solcher Begünstigungen sind natürlich jederzeit gegeben.

Die Zusammenstellung gestaltete sich insofern schwierig, als die vorgefundenen Angaben vielfach mangelhaft und ungenau sind. So wurden im Interesse der Vollständigkeit Ergänzungen notwendig, die nach Möglichkeit eingefügt wurden. Sie beziehen sich vornehmlich auf den Autor des botanischen Namens und auf die Pflanzenfamilie. Sofern eine Klärung nicht möglich war, mußten die Literaturangaben als solche wiedergegeben werden, d. h. mit den hier oft vorhandenen Fehlern und Unvollständigkeiten.

Da die Literaturangaben auf verschiedenen Pflanzensystemen basieren, so ist naturgemäß auch die vorliegende Zusammenstellung davon betroffen. Die Ergänzungen erfolgten meistenteils nach dem „Kew-Index", der auf dem Gebiete der Synonymbezeichnungen zu Einwendungen Anlaß geben kann, teils nach dem „Engler-Prantl" und, wo diese beiden Systeme, speziell bei neuen Arten, versagten, nach anderweitig vorgefundenen Literaturangaben.

Alle Nadelhölzer wurden einheitlich als „Coniferae" aufgeführt, ferner sind die „Leguminosae" in ihre Unterfamilien der „Mimosaceae", „Caesalpiniaceae" und „Papilionaceae" aufgeteilt. Weiter ist zu betonen, daß entsprechend der zoologischen Systematik alle Artbezeichnungen durchgehend mit kleinen Anfangsbuchstaben wiedergegeben sind. Dieses geschah, da auf diesem Gebiete in der Literatur ein außergewöhnlicher Wirrwarr herrscht und die gleiche Artbezeichnung bald mit großem, bald mit kleinem Anfangsbuchstaben vorgefunden wird. Eine Vereinheitlichung im vorliegenden Werk schien aber dringend erwünscht.

Die alphabetisch geordneten Kennwörter sind, soweit dieses erkennbar war, nach der Grund- oder Hauptbezeichnung eingereiht, jede adjektive Bezeichnung nachgestellt. Dadurch wird das Ziel erreicht, zusammenfassende Übersichten über ähnlich klingende und deshalb leicht zu Verwechslungen führende Namen zu schaffen. Ferner tritt die Verschiedenheit der den gleich oder ähnlich klingenden Holzbezeichnungen je nach Herkunft zugrunde liegenden Stammpflanzen besonders klar hervor, wie z. B. bei den verschiedenen Mahagonis, Teaks, Eichen, Walnußhölzern u. s. w. In gleicher Weise sind die Namen angeordnet, welche mit den Wörtern „Bois", „Cây", „Kajoe", „Palo", „Pao" oder „Pau" etc. verbunden sind. Obgleich diese „Holz" oder „Baum" bedeutenden Wörter keine eigentlichen Kennwörter darstellen, mußten sie trotzdem aus praktischen Gründen der Übersichtlichkeit als solche behandelt werden.

Ferner ist darauf hinzuweisen, daß, falls für ein Kennwort mehrere Arten derselben Gattung in Frage kommen, diese aus drucktechnischen Gründen nicht einzeln aufgeführt werden konnten, sondern unter Nennung der Gattung mit der nachstehenden Bezeichnung „spp" zusammengefaßt sind.

Dem Wissenschaftler soll das Werk, besonders wenn Herbarmaterial unvollständig vorliegt, eine annähernde Bestimmungsmöglichkeit geben; dem Forschungsreisenden die Erkennung und Auswahl des Sammelmaterials erleichtern; die Behörden bei der Einreihung fraglicher Nutzholzarten in ihre Tarife unterstützen; den Handels- und Verbrauchskreisen für neu eingeführte Hölzer durch die Angabe der Stammpflanze ihre eventuelle Verwendungsmöglichkeit und sonstige Aufschlüsse durch Vergleich mit bekannten Arten derselben Gattung an die Hand geben und diesen Kreisen den Weiterbezug genau desselben Holzes sichern.

Da die Holzgewächse in großem Umfange auch Rohstoffe für die Farb- und Gerbstoffindustrie, Ölindustrie, Papier- und Textilindustrie, ferner medizinische und technische Drogen liefern, die dann meist den gleichen Namen wie die Pflanzen führen, hat das vorliegende Buch auch für die Handels- und Industriekreise dieser Rohstoffe hohe Bedeutung und kann ihnen als wichtiges und nützliches Nachschlagewerk dienlich sein.

Auf Vollständigkeit kann das Buch als Werk eines Einzelnen selbstverständlich keinen Anspruch erheben. Es enthält, wie schon erwähnt, vielfach noch die in der Literatur vorhandenen Unvollständigkeiten. Das Buch gibt aber eine für Weiterarbeit geeignete erste Grundlage, an der sich sämtliche interessierten Kreise beteiligen möchten. Durch Kritik und Ergänzungsvorschläge werden wesentliche Verbesserungen gemacht werden können. Zweckmäßig wäre es, wenn dann durch spätere internationale Vereinbarung für jede Bezeichnung nur eine Art festgelegt und für die anderen in Frage kommenden Arten diese Bezeichnung geändert würde, sodaß künftig die durch die gleich oder ähnlich lautenden Namen entstehenden Verwechslungen vermieden werden.

Falls die Vereinheitlichung bezw. Aenderung der gleich oder ähnlich klingenden Namen durch internationales Übereinkommen erreicht werden kann, und falls weiterhin durch Kritik und Ergänzungen aus allen interessierten Kreisen dieses Buch zu einem in jeder Hinsicht einwandfreien und anerkannten Nachschlagewerk erweitert werden kann, so würden zwei von den hauptsächlich beabsichtigten Zwecken erreicht sein.

Der Autor.

Preface.

This book is intended to supply a much-felt want, not only in scientific but also in practical circles, which has become more and more perceptible every day owing to the lack of a comprehensive and handy book of reference showing the species which correspond to the commercial, common and vernacular wood-denominations according to their origin.

Science, to some degree, has always paid attention to this question. It is true that the commercial, common and vernacular names of the useful plants, of which timber plays an important part, have been mentioned in scientific publications, but in most cases they are neither alphabetically arranged nor indexed. These names, therefore, are very difficult to find in literature and in consequence can only be made use of in practice with great loss of time, when the wood has to be determined by means of such a given name. During the last two decades the importance of these wood-names has been recognized abroad, that is to say, outside of Germany, where brisk activity in this field has developed. Numerous lists of wood-names have been published abroad, but these lists refer only to either timber imported by a single country or confine themselves to certain limited territories of origin.

This book tries to do away with these deficiencies, on the one hand through alphabetical order which effects a speedy finding of the desired name, and on the other hand by possibly including all domains of origin and consumption. The botanical species in question, the family as well as the origin of the wood or its respective name can, as far as are known, be found at once.

This work is the result of 20 years examination and information activity in regard to timber, during which period the names have been obtained from the corresponding literature. The interest shown by commercial and industrial timber importing circles in scientific species of home- and foreign-grown woods, or their names respectively, shows a noticeable increase; this is proved by the continuously increasing number of enquiries received at the "Institut für angewandte Botanik" (Institute for Applied Botany) in Hamburg. Furthermore establishing the respective species is highly important for such different authorities, as the customs, railway, port officials, boards of commerce and so forth. As an example, the German customs-tariff divides timber imported into the following groups: 1) Cedar-woods; 2) Softwoods, including as until recently having been specially privileged: fir (Abies), spruce (Picea) and larch (Larix); 3) Hardwoods with the woods of fruit trees and some others being specially privileged; 4) specially designated Fancywoods. These groups are classified into different tariff-numbers and are therefore subject to different assessment. The German railway-freight-tariff also classifies timbers differently. Classification under respective groups is in most cases only possible when the botanical species is known. Amendments to the tariffs or the introduction of such privileges are of course possible at any time.

Compilation has been rather difficult insofar as the literary data obtained were very often faulty and incomplete. In the interest of accuracy, therefore, supplementary data became necessary, and have been added wherever possible. These supplements mostly refer to the author of the botanical name and the family. Where a clearing-up has not been possible, the literary data have been given as such, that means, with existing faults and deficiencies.

The literary data being based upon different plant-systems, the compilation is naturally influenced accordingly. Supplements for the greater part have been taken from the "Kew-Index", which as regards synonyms may give cause for objection, and partly from "Engler-Prantl", and, whenever these two were found wanting, especially as to new species, from other literature at the author's disposal.

All conifers are listed uniformly as "Coniferae", furthermore "Leguminosae" are divided up into their sub-families "Mimosaceae", "Caesalpiniaceae" and "Papilionaceae". Moreover, attention has to be called to the fact that according to zoological system all species-denominations were written throughout with small initials. This was done as on this point extraordinary confusion prevails in the literature insofar as the same species is often found written with a capital letter and again and again with a small initial. Uniformity in the way of writing seems to be highly desirable in this book.

Alphabetically arranged catch-words are inserted, as far as it was recognizable, according to the main-denomination, the adjectives being inserted behind. It has thus been possible to attain comprehensive synoptical tables of similar names which often caused confusion and mistakes. Furthermore the diversity of the botanical species, falling under the same or similar name, is thus clearly shown, as is the case with the various mahoganies, teaks, oaks, walnuts etc. Arranged in the same way are the denominations connected with the respective words "bois", "cây", "kajoe", "Palo", "Pao" or "Pau" etc. Though these words, meaning "wood" or "tree", do not signify any main-denomination they have nevertheless had to be treated thus for synoptical reasons.

It must also be pointed out, that wherever one catch-word was given for several species of the same genus in the literature, all the species could not be listed for technical printing reasons, and are taken collectively under the genus heading with the designation "spp" thereafter.

The book is intended to give scientists, especially when only incomplete herbar-material is at their disposal, a more or less approximate possibility of determining the species in question; to facilitate for explorers the recognition and choice of the material to be collected; to assist authorities in regard to classification in their tariffs; to enable commerce and industry, by stating the botanical species, to eventually ascertain the best utilization for newly imported woods and to give other information in comparison with species of the same genus already known so that exactly the same wood may be procured again.

As the wood-plants, to a large extent, also furnish raw materials for the dyeing, tanning, oil, paper and textile industries and in addition also provide medicinal and technical drugs, which in most cases bear the same name as that of the plant, this book is of great importance to commercial and industrial circles by whom it will be found to be a very useful and valuable book of reference.

As the work of a single person the book cannot of course claim to be wholly complete. It very often gives, as already stated, the deficiencies still contained in the literature used. The book however will certainly present an appropriate primary basis for further work, in which all those interested are kindly asked to participate. Through criticism and supplementary proposals considerable improvements and corrections could be made. Finally, it would be highly desirable if later on through international agreement only one species could be fixed for each catch-word, simultaneously altering all other species in question, so that in future any confusion and mistake, caused through similar names, would be eliminated.

Should it be possible to attain a simplification or alteration of similar names through international agreement, and if by means of criticism and supplements, received from all those persons interested, this work could be enlarged into a book of reference generally recognized and in every respect unassailable, two of the main purposes, chiefly intended, would, in addition to the ones above stated, be accomplished.

The Author.

Préface.

Le présent ouvrage est destiné à remédier à une lacune qui devient chaque jour plus sensible, aussi bien pour les milieux scientifiques que pour les praticiens, faute d'un lexique suffisamment complet et facile à consulter des dénominations des diverses sortes de bois dont on peut avoir à s'occuper.

Les savants ont toujours attribué un certain intérêt à cette question. Toutefois, si dans les publications scientifiques, on mentionne bien les noms vulgaires, commerciaux ou vernaculaires des plantes utiles, parmi lesquelles les bois jouent un grand rôle, ces noms ne sont généralement pas classés alphabétiquement ou relevés dans des registres. C'est pourquoi ils sont difficiles à trouver dans la bibliographie, et ils ne peuvent, par suite, en pratique, être utilisés pour déterminer une sorte de bois qu'avec une grande perte de temps. Au cours des deux dernières décades, on a, tout particulièrement hors d'Allemagne, reconnu l'importance qu'il convient d'attribuer aux noms des bois, et une vive activité a été déployée dans ce domaine. On a publié de nombreuses listes de ces noms, mais celles-ci n'ont trait qu'aux bois importés par le pays dans lequel la publication a été faite, ou bien elles ne se limitent qu'à des domaines d'origine restreints.

L'ouvrage actuel veut combler cette lacune, d'une part par une classification alphabétique permettant de trouver rapidement le nom désiré, d'autre part en faisant entrer dans cette classification les bois de tous ou presque tous les territoires de production et de consommation. On trouve de suite l'espèce d'arbre qui a fourni le bois, ainsi que l'origine du bois ou du nom en question, lorsque toutefois ces renseignements sont connus.

Cette oeuvre est le résultat d'une activité de vingt années d'investigations et d'informations dans le domaine des bois d'oeuvre, période durant laquelle les noms ont été puisés dans les publications ayant trait à la question et rassemblés.

L'intérêt dans les milieux s'occupant de l'importation, du commerce et de l'industrie des bois pour les sortes de bois d'oeuvre allemands et étrangers et pour les noms des espèces botaniques qui fournissent ces bois, s'accroît très rapidement, ainsi que l'atteste le nombre constamment croissant des demandes adressées à l'„Institut für angewandte Botanik" (Institut de Botanique appliquée) de Hambourg. En outre, la détermination de l'espèce des plantes présente une extrême importance pour diverses administrations telles que la douane, le chemin de fer, les autorités portuaires, les chambres de commerce etc. Par exemple, le tarif douanier allemand répartit les bois d'oeuvre importés selon les groupes suivants: 1) les bois de cèdre; 2) les bois tendres, parmi lesquels les essences suivantes ont joui, il y a peu de temps, d'un traitement particulièrement favorable: sapins (Abies), pins (Picea) et mélèzes (Larix); 3) les bois durs, parmi lesquels les bois d'arbres fruitiers et quelques autres jouissent d'un tarif privilégié; 4) les bois supérieurs. Ces groupes tombent sous divers numéros du tarif douanier et ils sont par suite soumis à des taxes douanières différentes. Le tarif de transport de marchandises du Chemin de Fer allemand applique aux bois d'oeuvre des taux de transport différents selon les classes dans lesquelles il les a groupés. Mais la classification des différents bois dans les groupes n'est le plus souvent possible que si l'on connaît les espèces auxquelles ils appartiennent. Des modifications des tarifs ou l'introduction des taxes privilégiées sont bien possible à tout temps.

La classification des différents noms a été difficile, parce que les indications trouvées étaient souvent imparfaites ou inexactes. Et c'est pourquoi pour obtenir un travail aussi complet que possible il a fallu procéder à des additions complémentaires. Ces additions ont trait surtout à l'auteur du nom botanique et à la famille de la plante. Lorsqu'un éclaircissement n'a pas été possible, les indications d'ordre bibliographique ont été reproduites sans changement, c'est-à-dire avec les inexactitudes et les imperfections qu'elles renferment souvent.

Comme les indications fournies par la bibliographie sont basées sur des systèmes de plantes différents, le groupement auquel nous avons procédé s'en ressent naturellement. Les additions faites ont eu lieu le plus souvent d'après le „Kew-Index" qui, dans le domaine des synonymes, peut donner lieu à des objections. Elles ont été faites également, en partie, d'après le „Engler-Prantl" et, dans le cas où les deux systèmes n'ont pas suffi, surtout en ce qui concerne les sortes nouvelles, d'après des indications recueillies ailleurs.

Tous les conifères ont été classés uniformément sous la rubrique „Coniferae"; en outre, les „Leguminosae" ont été réparties dans leurs sous-familles, à savoir les „Mimosaceae", les „Caesalpiniaceae" et les „Papilionaceae". En outre, il faut préciser que, conformément à la systématologie zoologique, toutes les dénominations d'espèces commencent par une lettre minuscule. On a procédé ainsi parce qu'une extrême confusion règne dans la bibliographie relative à cette question et que la dénomination d'espèces a lieu tantôt avec une lettre minuscule, tantôt avec une lettre majuscule. Une uniformité nous a semblé vivement désirable.

Les mots classés alphabétiquement ont été, lorsque cela a pu être reconnu, groupés selon leur signification principale ou fondamentale, chaque signification accessoire étant indiquée ensuite. De cette manière, on est parvenu à donner des aperçus d'ensemble sur des noms à consonnance semblable pouvant conduire facilement à des erreurs. En outre, la diversité des dénominations de bois à même consonnance ou à consonnance semblable a été particulièrement mise en évidence en faisant ressortir les noms des plantes qui fournissent ces divers bois, comme par exemple pour les divers acajous, teaks, chênes, noyers etc. De la même manière sont arrangés les noms composés avec les mots „Bois", „Cây", „Kajoe", „Palo", „Pao" ou „Pau" etc. Quoique ces mots, qui désignent „bois" ou „arbre", ne représentent pas une dénomination fondamentale ou principale, ils ont dû, quandmême, pour des raisons synoptiques, être traités comme telles.

En outre, il faut signaler que dans le cas où il y a plusieurs espèces pour un seul nom, ces espèces n'ont pu être nommées séparément en raison des difficultés d'impression et qu'elles ont été réunies en nommant le genre suivi de la marque „spp".

Le présent ouvrage doit permettre aux savants, en particulier lorsqu'ils disposent d'un herbier insuffisant, de trouver une détermination des bois aussi approximative que possible; de faciliter à l'explorateur le choix et la reconnaissance de son herbier; d'aider les autorités lors de la classification des sortes de bois d'oeuvre dans leurs tarifs; de donner aux milieux commerciaux et aux consommateurs des renseignements sur des bois nouveaux importés en indiquant les plantes qui fournissent ces bois et par suite les possibilités d'emploi éventuelles et d'autres renseignements par la comparaison avec des sortes connues de la même espèce, enfin d'assurer à ces milieux, dans le cas d'achats ultérieurs, l'importation de bois identiques aux premiers.

Comme les plantes ligneuses fournissent en grand nombre les matières premières pour la teinturerie et tannerie, ainsi que pour les industries travaillant l'huile, le papier, les textiles, et qu'elles fournissent en outre des produits pharmaceutiques et onguents techniques qui pour la plupart portent le même nom que la plante dont ils proviennent, il en résulte que cet ouvrage suscite également l'intérêt des milieux commerciaux et industriels auxquels il peut servir de lexique par son importance et son utilité.

Le présent livre, qui est l'oeuvre d'une seule personne, n'a pas naturellement la prétention d'être complet. Il comprend, ainsi que cela est déjà mentionné plus haut, très souvent encore les imperfections contenues dans les ouvrages précédents relatifs à la question. Mais il fournit la première base permettant de continuer une étude à laquelle tous les milieux intéressés sont priés de s'associer. Par la critique et les observations des lecteurs compétents, des améliorations notables seront possibles. Il serait utile par une entente internationale future de fixer une seule sorte pour chaque dénomination et de modifier cette dénomination pour toutes les autres sortes auxquelles elle s'applique également, en vue d'éviter, à l'avenir, les confusions qui peuvent provenir de noms identiques ou semblables.

Si l'uniformisation ou la modification de noms identiques ou semblables peut être obtenue et si, en outre, par la critique et les observations de tous les milieux intéressés, cet ouvrage peut être élargi et devenir un lexique reconnu et irréprochable à tous les points de vue, deux des buts principaux poursuivis se trouveront atteints.

L'Auteur.

Preámbulo.

El presente libro tiene por objeto remediar una falta que no solo en el mundo científico, sino que también en la práctica se hace sentir cada vez más, á causa de no existir una obra de referencia, comprensiva y fácilmente utilizable, que dé las especies botánicas con sus respectivas designaciones comerciales, comunes ó indígenas.

La ciencia, en cierta forma, siempre se ha ocupado de este problema. Verdad es que en las publicaciones científicas estas designaciones de las plantas útiles, entre las cuales á las maderas toca un rol de gran importancia, ya se mencionan, pero en la mayor parte no están ordenadas alfabéticamente ó reunidas en índices. Por consecuencia, es muy difícil encontrarlas en la literatura, y por este motivo se puede aprovecharlas en la práctica solo á costa de una gran pérdida de tiempo, cuando se desea establecer una determinada madera. Durante los últimos 20 años, especialmente en el extranjero, es decir fuera de Alemania, se ha reconocido la importancia de tales designaciones y se ha desarrollado una actividad intensa en este terreno. Allá se ha publicado un gran número de índices con tales nombres, pero estos se refieren solamente á las manderas importadas en un país particular ó están limitadas á ciertos territorios de procedencia.

El presente libro trata de subsanar estos defectos: 1) dando, por una parte, las designaciones en orden alfabético, con lo que fácil y rápidamente se las encuentra y, 2) incluyendo, por otra parte, dentro de lo posible, todos los distritos de producción y de consumo. Asi inmediatamente se puede encontrar la especie botánica, la familia y el orígen de la madera ó de la denominación respectiva en cuestión, dentro de lo que se conoce.

Este libro representa el resultado de una labor de investigación é información durante los últimos 20 años, con respecto á las maderas. En el transcurso de este tiempo los nombres en cuestión se han sacado de la literatura correspondiente. El interés demostrado por los círculos de importación, del comercio y de la industria en las especies botánicas, las cuales basan en las diversas clases de las maderas, tanto indígenas como extranjeras, ó en las denominaciones en cuestión, crece constantemente, como lo demuestra el número siempre creciente de los pedidos de información recibido en el „Institut für angewandte Botanik" (Instituto de Botánica aplicada) de Hamburgo. Además, la fijación de la respectiva especie botánica es de la más alta importancia para las diversas administraciones: p. e. la aduana, los ferrocarriles, el puerto, las cámaras de comercio etc. etc. El arancel de aduana alemán p. e. clasifica las maderas en los grupos siguientes: 1) cedros; 2) maderas blandas, entre éstas como especialmente favorecidas hasta hace poco tiempo, maderas de abeto (Abies), de pino (Picea) y de alerce (Larix); 3) maderas duras, entre éstas como favorecidas las de árboles frutales y algunas otras; 4) maderas nobles. Estos grupos están clasificados bajo diferentes números del arancel y están sujetos por consiguiente á derechos distintos. También la tarifa de fletes de ferrocarriles alemanes clasifica las maderas en varios grupos. La clasificación en los grupos correspondientes solamente es posible conociéndose la especie botánica. Alteraciones en los aranceles ó la introducción de tales preferencias son posibles naturalmente en todo momento.

Esta compilación ha resultado bastante difícil, puesto que los datos encontrados en la literatura son frecuentemente imperfectos é inexactos. Por eso, y en pro de una obra completa, se han hecho necesarios los suplementos, los cuales han sido insertados dentro de lo posible. Estos datos suplementarios se refieren principalmente al autor del nombre botánico y á la familia de la madera. Cuando una aclaración no ha sido posible, se han reproducido los datos de la literatura tal cual estaban, es decir con las faltas é inexactitudes adherentes.

Ya que los datos literarios tienen por base varios sistemas, esta circunstancia ha ejercido su influencia sobre el libro. Los suplementos se han formado en su mayoría de acuerdo con el „Kew-Indice", el cual, en cuanto á sinónimos, puede dar causa á objeciones, por otra parte según el sistema „Engler-Prantl" y, cuando estos dos sistemas fallaban especialmente en cuanto á especies nuevas, conforme á los datos literarios encontrados en otras fuentes. Todas las coníferas están registradas solamente como „Coníferae", además las „Leguminosae" se dividen en sus sub-familias „Mimosaceae", „Caesalpiniaceae" y „Papilionaceae". Fuera de eso, es preciso mencionar que correspondiente al sistema zoológico, todas las designaciones de especie son dadas enteramente con letra minúscula. Esto se ha hecho porque en este dominio reina una confusión excesiva y se encuentra la misma designación de especie, escrita ya con mayúscula, ya con minúscula. Una uniformidad en esta obra parecía ser en alto grado deseable.

Los nombres en cuestión de las maderas, puestos en órden alfabético, han sido arreglados, siempre dentro de lo que se ha podido constatar, según la significación principal, cada signatura adjetiva colocada detrás. En esta forma se obtiene el objeto perseguido, es decir, crear sumarios sinópticos de denominaciones consonantes y asonantes, las cuales pueden fácilmente dar motivo á equivocaciones. Además, la diferencia entre las especies botánicas, en las cuales basan las denominaciones consonantes y asonantes, está demostrada de una manera muy clara, como p. e. entre las diversas Caobas, Teaks, Robles, Nogales etc. De la misma manera están arreglados los nombres que se componen con las palabras „Bois", „Cây", „Kajoe", „Palo", „Pao" ó „Pau" etc. Apesar de que estas palabras, que corresponden á „madera" ó „árbol", no represénten una significación principal, tuvieron que ser tratadas como tales para mayor claridad.

Además es menester hacer resaltar que en caso de tratarse de varias especies del mismo género para una sola denominación, estas varias especies no son mencionadas cada una por separado, por razones técnicas de impresión, sino que van unidas bajo la signatura del género con un „spp" puesto detrás.

La obra tiene por objeto: dar al erudito una posibilidad toda lo más aproximada posible de determinar la planta en cuestión, especialmente cuando el material herbario á su disposición sea incompleto: facilitar al viajero-explorador el reconocimiento y la selección del material á seleccionar: ayudar á las administraciones, por fijación de la especie respectiva, á clasificar las diversas maderas en sus aranceles: señalar al comercio y consumo de maderas, mediante el dato de la especie botánica, la posibilidad de establecer como se puede emplear en la mejor forma, maderas importadas por primera vez y dar ulteriores aclaraciones para la comparación con especies ya conocidas del mismo género, y fuera de esto posibilitar á los círculos comerciales mencionados la nueva obtención de la misma especie exacta de madera ya recibida.

Como las plantas maderables, en su mayoría, también proporcionan las materias primas para las industrias de los colorantes y de los materiales para curtiembre, para las industrias de los aceites y la fabricación de papeles y de textiles, como que también dan drogas medicinales y técnicas, materiales que por lo general son conocidos bajo la misma denominación que la planta respectiva, el presente libro es también de la más alta importancia para los círculos comerciales é industriales mencionados, y por consiguiente puede prestarles valiosos servicios como vademécum importante y útil.

Este libro, como obra de una sola persona, naturalmente no puede tener pretensiones de absoluta perfección. Contiene aún, como queda dicho más arriba, muchas veces los defectos contenidos en la literatura utilizada. El libro empero da, en todo caso, una primera base para trabajo ulterior, á participar en el cual, se suplica á todos los círculos interesados. Por medio de la crítica y de propuestas suplementarias, se pueden realizar importantes rectificaciones y mejoras. Sería muy ventajoso, si entonces, por medio de un acuerdo internacional, para cada denominación se pudiera establecer una sola especie, modificando esta denominación para todas las otras especies respectivas, de manera que en el futuro se evitaran las equivocaciones causadas á consecuencia de los nombres consonantes y asonantes.

Para el caso de que la uniformidad ó la modificación de estos últimos nombres se pudiesen obtener por medio de un acuerdo tal, y si además por medio de la crítica y de suplementos de parte de todos los círculos interesados, este libro pudiese ser ampliado á un vademécum sin objeción y generalmente apreciado en todo sentido, dos de los fines principales buscados habrían encontrado su realización.

El Autor.

Abkürzungen
Abbreviations
Abréviations
Abreviaciones

I: hinter dem Kennwort, eingeklammert ( ), bedeuten den Handels- oder Volks-
namen des betreffenden Landes, bei einem (H) den in dem a n g e g e b e n e n
H e r k u n f t s g e b i e t üblichen bezw. vorgeschlagenen Handelsnamen.

behind the catch-word, in parenthesis ( ), signify the commercial or the common
name in the respective country, with the exception that a capital (H) signifies
the catch-word as in use or proposed as a commercial name in the c o u n t r y
o f o r i g i n i n d i c a t e d.

après le nom en question, en parenthèses ( ), signifient le nom commercial
ou commun du pays respectif, à l'exception de la majuscule (H), signifiant
le nom en question comme nom commercial usé ou proposé dans le p a y s
d ' o r i g i n e i n d i q u é.

detrás del nombre en cuestión, en paréntesis ( ), significan el nombre comercial
ó popular del país respectivo, á excepción de la letra mayúscula (H)
que significa el nombre en cuestión como el nombre comercial usado ó propuesto
en el p a í s d e o r i g e n i n d i c a d o.

dä = Dänemark,	Denmark,	Danemark,	Dinamarca.
d = Deutschland,	Germany,	Allemagne,	Alemania.
e = England,	England,	Angleterre,	Inglaterra.
eu = europäischer Handel,	european commerce,	commerce européen,	comercio europeo.
f = Frankreich,	France,	France,	Francia.
h = Holland,	Holland,	Hollande,	Holanda.
i = Italien,	Italy,	Italie,	Italia.
p = Portugal,	Portugal,	Portugal,	Portugal.
sch = Schweden,	Sweden,	Suède,	Suecia.
s = Spanien,	Spain,	Espagne,	España.
u = Vereinigte Staaten,	United States,	Etats Unis,	Estados Unidos.
H = wie oben,	as above,	comme ci-dessus,	como arriba.

II: der Herkunftsgebiete,
of the origin,
de l'origine,
del origen.

a)

Af . . .	Afrika Africa Afrique Africa	Amaz . .	Amazonas Amazon Amazones Amazonas
Afg . . .	Afghanistan Afghanistan Afghanistan Afgánistan	Am . . .	Amerika America Amérique América
Alas . .	Alaska Alaska Alaska Alasca	Amb . .	Amboina Amboyna Ile Amboine Isla Amboino

And . . .	Andamanen	
	Andamans	
	Iles Andaman	
	Islas Andamanas	
Ang . . .	Angola	
	Angola	
	Angolares	
	Angola	
Ant . . .	Antillen	
	Antilles	
	Antilles	
	Antillas	
Aeq . . .	Aequatorial	
	equatorial	
	équatorial	
	ecuatorial	
Arab . .	Arabien	
	Arabia	
	Arabie	
	Arabía	
Arg . . .	Argentinien	
	Argentina	
	Argentine	
	República Argentina	
As . . .	Asien	
	Asia	
	Asie	
	Asia	
Au . . .	Australien	
	Australia	
	Australie	
	Australia	
Bah . . .	Bahamas	
	Bahama Islands	
	Iles Bahama	
	Islas de Bahama	
Barb . .	Barbados	
	Barbados Island	
	la Barbade	
	la Barbada	
B.-K. . .	Belgisch Kongo	
	Belgian Congo	
	Congo Belge	
	Congo Bélgico	
Beng . .	Bengalen	
	Bengal	
	Bengale	
	Bengala	
Bl. B . .	Blaue Berge (Au)	
	Blue Mountains (Au)	
	Montagnes bleues (Au)	
	Montañas azules (Au)	
Bol . . .	Bolivien	
	Bolivia	
	Bolivie	
	Bolivia	
Bor . . .	Borneo	
	Borneo	
	Bornéo	
	Bórneo	

Bras . .	Brasilien	
	Brazil	
	Brésil	
	Brasil	
br . . .	britisch	
	british	
	britannique	
	británico	
Cay . . .	Cayenne	
	Cayenne	
	Cayenne	
	Cayena	
Cel . . .	Celebes	
	Celebes	
	Célèbes	
	Célebes	
C . . .	Central	
	Central	
	Central, - - e	
	Central	
Cey . . .	Ceylon	
	Ceylon	
	Ceylan	
	Ceilán	
Ch . . .	China	
	China	
	Chine	
	China	
Co.-Ch . .	Cochinchina	
	Cochinchina	
	Cochinchine	
	Cochinchina	
Cord . .	Cordoba (Arg)	
	Cordoba	
	Cordoba	
	Córdoba	
Corr . .	Corrientes (Arg)	
	Corrientes	
	Corrientes	
	Corrientes	
C.-R. . . .	Costa Rica	
	Costa Rica	
	Costa Rica	
	Costa Rica	
Dah . . .	Dahomey	
	Dahomey	
	Dahomey	
	Dahomey, Dahomé	
Ek . . .	Ekuador	
	Ecuador	
	Equateur	
	Ecuador	
Elf . . .	Elfenbeinküste	
	Ivory Coast	
	Côte d'Ivoire	
	Costa del Marfil	
Eu . . .	Europa	
	Europe	
	Europe	
	Europa	

Fid . . .	Fidschiinseln	H.-Ind. .	Hinterindien
	Fiji Islands		Further India
	Iles de Fidji		Indochine
	Islas Vití, Islas Fidjí		Indo-China
Fl . . .	Florida	Hind . .	Hindustan
	Florida		Hindustan
	Florida		Hindoustan
	Florida		Indostán
Form . .	Formosa	Holl . .	Holland
	Formosa		Holland, Netherlands
	Ile Formose		Hollande, Pays-Bas
	Isla Formosa		Holanda, Países Bajos
Fr . . .	Frankreich	holl . . .	holländisch
	France		dutch
	France		hollandais
	Francia		holandés
fr . . .	französisch	Hond . .	Honduras
	french		Honduras
	français, - - e		Honduras
	francés		Honduras
Gab . . .	Gabun	Ind . . .	Indien
	Gabon		India
	Gabon		Inde
	Gabón		India
Go . . .	Goldküste	Ind.-Ch .	Indochina
	Gold Coast		Indochina
	Côte d'Or		Indochine
	Costa de Oro		Indochina
Gren . .	Grenada (Ant)	It . . .	Italien
	Grenada Island		Italy
	Ile Grenade		Italie
	Isla Grenada		Italia
Griech . .	Griechenland	it . . .	italienisch
	Greece		italian
	Grèce		italien
	Grecia		italiano
Guad . .	Guadeloupe (Ant)	Jam . .	Jamaika
	Guadeloupe		Jamaica
	Guadeloupe		Jamaïque
	Isla Guadalupe		Jamaica
Guat . .	Guatemala	Jap . . .	Japan
	Guatemala		Japan
	Guatemala		Japon
	Guatemala		Japón
Gu . . .	Guayana	Kalif . .	Kalifornien
	Guiana		California
	Guyane		Californie
	Guyana		California
Guin . .	Guinea	Kamb . .	Kambodscha
	Guinea		Cambodia
	Guinée		Cambodge
	Guinea		Camboja, Camboya
Hem . .	Hemisphäre	Kam . .	Kamerun
	Hemisphere		Cameroons
	Hémisphère		Cameroun
	Hemisferio		Camerón
Himal . .	Himalaja	Kan . . .	Kanada
	Himalaya		Canada
	Himalaya		Canada
	Himalaya		Canadá

Kanar . .	Kanarische Inseln Canary Islands Iles Canaries Islas Canarias	Maur . .	Mauritius Maurice Island Ile Maurice Isla Mauricio
Kap . . .	Kap der Guten Hoffnung Cape of Good Hope Cape de Bonne Espérance Cabo de Buena Esperanza	Mex . .	Mexiko Mexico Mexique Mejico
Karp . .	Karpathen Carpathian Mountains Carpathes Carpatos	Mind . .	Mindoroinsel (Phil) Mindoro Island Ile Mindoro Isla Mindoro
Kauk . .	Kaukasus Caucasus Caucase Cáucaso	Mis . . .	Misiones (Arg) Misiones Misiones Misiones
Kl.-As . .	Kleinasien Asia Minor Asie Mineure Asia Menor	Mitt.-Eu .	Mitteleuropa Central Europe Europe centrale Europa central
Kol . . .	Kolumbien Colombia Colombie Colombia	m . . .	mittel, mittlerer middle moyen mediano
Leew . .	Lewardinseln oder Inseln unter dem Winde (Ant) Leeward Islands (Ant) Iles-sous-le-Vent (Ant) Islas de Sotavento (Ant)	M.-M. . .	Mittelmeergebiet Mediterranean territory Région Méditerranée Territorio Mediterráneo
Lib . . .	Liberia Liberia Libéria Liberia	Moluk . .	Molukken Moluccas Iles Moluques Islas Molucas
Luz . . .	Luzon, Phil. Luzon, P. I. Luzon, Luzon, Fil.	Monts . .	Montserrat (Ant) Montserrat Island Ile Montserrat Isla Montserrat
Mad . .	Madagaskar Madagascar Madagascar Madagascar	N.-Fundl .	Neufundland New Foundland Terre Neuve Terranova
Mal . . .	Malabar Malabar Malabar Malabar	N.-Kal . .	Neukaledonien New Caledonia Nouvelle Calédonie Nova Caledonia
Malak . .	Malakka Malacca Malacca Malaca	N.-Seel .	Neuseeland New Zealand Nouvelle Zélande Nova Zelandia
Marok . .	Marokko Morocco Maroc Marruecos	N.-S.-W. .	Neusüdwales New Southwales Nouvelle Galles du Sud Nueva Gales del Sur
Mart . .	Martinique (Ant) Martinique Martinique Isla Martinique	Nic . . .	Nicaragua Nicaragua Nicaragua Nicaragua
Mask . .	Maskareneninseln Mascarene Islands Iles Mascareignes Islas Mascarenas	Nied . .	Niederländisch Dutch Hollandais, Néerlandais Holandés, Neerlandés

Nig . . .	Nigerien	Rhod . .	Rhodesien	
	Nigeria		Rhodesia	
	Nigéria		Rhodésia	
	Nigeria		Rhodesia	
N . . .	Nord	Salv . . .	San Salvador	
	North		San Salvador	
	du Nord		San Salvador	
	del Norte		San Salvador	
ob . . .	oberer	Sans . .	Sansibar	
	upper		Zanzibar	
	haut		Zanzibar	
	superior		Isla Zanzibar	
Or . . .	Orient	S. Fé . .	Santa Fé (Arg)	
	Orient		Santa Fé	
	Orient		Santa Fé	
	Oriente		Santa Fé	
O . . .	Ost	Sao P . .	Sao Paulo (Bras)	
	East		Sao Paulo	
	de l'Est		Sao Paulo	
	de Este		Sao Paulo	
Pan . . .	Panama	Schwed .	Schweden	
	Panama		Sweden	
	Panama		Suède	
	Panama		Suecia	
Par . . .	Paraguay	Senl . . .	Senegal	
	Paraguay		Senegal	
	Paraguay		Sénégal	
	Paraguay		Senegal	
Pers . . .	Persien	Senb . .	Senegambien	
	Persia		Senegambia	
	Perse		Sénégambie	
	Persia		Senegambia	
Phil . . .	Philippinen	Sey . . .	Seychellen	
	Philippine Islands		Seychelle Islands	
	Iles Philippines		Iles Seychelles	
	Filipinas		Islas Seychelas	
P.-R. . .	Porto Rico	Sib . . .	Sibirien	
	Porto Rico		Siberia	
	Porto Rico		Sibérie	
	Porto Rico		Siberia	
Port . . .	Portugal	S.-L . .	Sierra Leone	
	Portugal		Sierra Leone	
	Portugal		Sierra Leone	
	Portugal		Sierra Leone	
port . . .	portugiesisch	Sp . . .	Spanien	
	portuguese		Spain	
	portugais		Espagne	
	portugués		España	
Pyr . . .	Pyrenäen	sp . . .	spanisch	
	the Pyrenees		spanish	
	Pyrénées		espagnol	
	Pirineos		español	
Queensl .	Queensland	S	Süd	
	Queensland		South	
	Queensland (Au)		du Sud	
	Queensland (Au)		del Sur	
Réu . . .	Réunion	Südsee . .	Südsee	
	Réunion Island		Oceania	
	Réunion		Océanie	
	Isla Reunión		Oceánia	

Sum . . .	Sumatra	Tr . . .	Tropen
	Sumatra		Tropics
	Sumatra		Tropiques
	Sumatra		Trópicos
Sunda . .	Sundainseln	tr . . .	tropisch
	Sunda-Islands		tropical
	Iles de la Sonde		tropical
	Islas de la Sonda		tropical
Sur . . .	Surinam	Türk . .	Türkei
	Surinam		Turkey
	Surinam		Turquie
	Surinam		Turquia
Syr . . .	Syrien	Urug . .	Uruguay
	Syria		Uruguay
	Syrie		Uruguay
	Siria		Uruguay
Tasm . .	Tasmanien	Ven . . .	Venezuela
	Tasmania		Venezuela
	Tasmanie		Vénézuéla
	Tasmania		Venezuela
Tener . .	Teneriffa	Vict . . .	Victoria (Au)
	Teneriffe		Victoria
	Ténérife		Victoria
	Tenerife		Victoria
Tob . . .	Tobago (Ant)	V.-Ind . .	Vorderindien
	Tobago Island		East India
	Ile Tobago		Inde Anglaise
	Isla Tobago		Indias Orientales
Tonk . .	Tonkin	W.-Ind .	Westindien
	Tongking		West India
	Tonkin		Indes Occidentales
	Tonquín, Tonkin		Indias Occidentales
Transkauk	Transkaukasien	Windw .	Windwardinseln oder Inseln
	Transcaucasia		über . dem Winde, Ant.
	Transcaucase		Windward Islands (Ant)
	Transcáucasia		Iles sur le Vent (Ant)
			Islas de Barlovento (Ant)
Trin . .	Trinidad	Zamb . .	Zamboanga (Phil)
	Trinidad		Zamboanga
	Trinidad		Zamboanga
	Trinidad		Zamboanga

b)

n =	nördlich	northern	septentrional, du Nord	septentrional, del Norte,
s =	südlich	southern	méridional, du Sud	meridional, del Sur
ö =	östlich	eastern	oriental, d'Est	oriental, de Este
w =	westlich	western	occidental, d'Ouest	occidental, al Oeste

c) Beispiele für Zusammensetzungen,
Examples for combinations,
Des exemples pour des combinaisons,
Ejemplos para las combinaciones.

N.-Am	C.-Am	S.-Am
=	=	=
Nordamerika	Centralamerika	Südamerika
Northamerica	Centralamerica	Southamerica
Amérique du Nord	Amérique centrale	Amérique du Sud
América del Norte	América central	América del Sur

sö. N.-Am
=
südöstliches Nordamerika
southeastern Northamerica
Amérique du Nord sud-est
América del Norte sudeste
etc.

sö. C.-Am
=
südöstliches Centralamerika
southeastern Centralamerica
Amérique centrale sud-est
América central sudeste
etc.

sö. S.-Am
=
südöstliches Südamerika
southeastern Southamerica
Amérique du Sud sud-est
América del Sur sudeste
etc.

Anordnung der nachstehenden Liste.

Kennwort, durch Doppelpunkt (:) von der Pflanzenart getrennt.
Pflanzenart, durch Semikolon (;) von der Pflanzenfamilie getrennt,
    2 Spezies durch Komma (,) von einander getrennt.
Pflanzenfamilie, durch Semikolon (;) von der Herkunft getrennt.
Herkunft, fast stets gekürzt laut vorstehenden Übersichten der verschiedenen Abkürzungen.

Beispiel:

Kennwort: Pflanzenart   ; Pflanzenfamilie; Herkunft.

oder

Kennwort: Pflanzenart, Pflanzenart; Pflanzenfamilie; Herkunft.

Arrangement of the following list.

Catch-word, separated from the Species by colon (:).
Species, separated from the Family by semicolon (;),
    2 Species separated from one another by comma (,).
Family, separated from the Origin by semicolon (;).
Origin, nearly always abbreviated as per the preceding list of the various abbreviations.

Example:

Catch-word: Species   ; Family; Origin.

or

Catch-word: Species, Species; Family; Origin.

Arrangement de la liste suivante.

Nom en question, séparé de l'Espèce par deux-points (:).
Espèce, séparée de la Famille par point-virgule (;),
    2 Espèces séparées par virgule (,) l'une de l'autre.
Famille, séparée de l'Origine par point-virgule (;).
Origine, presque toujours abrégée selon les tableaux précédents des différentes abréviations.

Exemple:

Nom en question: Espèce   ; Famille; Origine.

ou

Nom en question: Espèce, Espèce; Famille; Origine.

Orden de la lista siguiente.

Nombre en cuestión, separado de la Especie por dos puntos (:).
Especie, separada de la Familia por punto y coma (;),
    2 Especies separadas por coma (,) una de la otra.
Familia, separada del Origen por punto y coma (;).
Origen, casí siempre abreviado según los sumarios precedentes de las diferentes abreviaciones.

Ejemplo:

Nombre en cuestión: Especie   ; Familia; Origen.

ó

Nombre en cuestión: Especie, Especie; Familia; Origen.

A

Aach-a-yong: *Cestrum panamense* Standl.; *Solanac.;* br. Hond.
Aagatsura: *Toona sinensis* Roem.; *Meliac.;* Jap.
Aapiesdoorn: *Acacia burkei* Bth. et Hook.; *Mimosac.;* S.-Af.
Aba: *Bombacopsis sp.; Bombacac.;* Nic.
Ababele: *Chrysobalanus ellipticus* Sol.; *Rosac.;* W.-Af.
Abachi (d): *Tiplochiton scleroxylon* K. Schum.; *Sterculiac.;* W.-Af.
Abaij: *Tessmannia parvifolia* Harms; *Caesalpiniac.;* Kam.
Abaku: *Mimusops sp.; Sapotac.;* Go.
„ baku: *Mimusops sp.; Sapotac.;* Go.
Abal: *Spondias sp.; Anacardiac.;* Mex.
Abale: *Blighia sapida* Koenig; *Sapindac.;* Kam.
Abalé (f): *Petersia viridiflora* Chev.; *Myrtac.;* Elf., Gab.
Abalu: *Plectronia didyma* Kurz; *Rubiac.;* tr. As.
Abam: *Chrysophyllum autranianum* Chev.; *Sapotac.;* Kam., Gab.
Abamm: *Chrysophyllum lacourtianum* de Wild.; *Sapotac.;* Gab.
Aban: *Chlorophora excelsa* Bth. et Hook.; *Morac.;* Gab.
Abang: *Chlorophora excelsa* Bth. et Hook.; *Morac.;* Kam.
„ : *Chlorophora regia* Chev.; *Morac.;* sp. Guin.
„ -ábang: *Oroxylum indicum* Vent.; *Bignoniac.;* Phil.
Abangboua: *Eugenia sp.; Myrtac.;* Gab.
? Abgang mekoué: *Calpocalyx dinklagei* Harms; *Mimosac.;* Kam.
Aban Héli: *Chlorophora excelsa* Bth. et Hook.; *Morac.;* Gab.
Abaracaatinga: *Mimosa bracaatinga* Hoehne; *Mimosac.;* Bras.
„ : *Mimosa bracaatinga* Hoehne var. *aspericarpa Hoehne; Mimosac.;* Bras.
Abari: *Anopyxis ealaensis* Spr.; *Rhizophorac.;* Go.
Abati timbary: *Hymenaea courbaril* L.; *Caesalpiniac.;* Par.
Abayuelo: *Colubrina colubrina* Millsp.; *Rhamnac.;* P.-R.
Abazi: *Buchholzia macrophylla* Pax; *Capparidac.;* Elf.
Abbíhal: *Erythrophloeum densiflorum* Merr.; *Caesalpiniac.;* Phil.
Abé: *Homalium africanum* Bth.; *Flacourtiac.;* Kam.
Abe: *Cola acuminata* Sch. et Engl.; *Sterculiac.;* Kam.
Abé: *Canarium schweinfurthi* Engl.; *Burserac..* Gab., Kam.
Abé: ? *Piptadenia africana* Hook f.; *Mimosac.;* Elf.
Abe, white: *Homalium africanum* Bth.; *Flacourtiac.;* Kam.
Abedul: *Betula verrucosa* Ehrh.; *Betulac.;* Sp
„ : *Alnus sp.; Betulac.;* Mex.
Abeille: *Mimusops balata* L.; *Sapotac.;* Gu.
Abejuelo: *Colubrina ferruginosa* Brongn.; *Rhamnac.;* P.-R.
Abel: *Cola ballayi* Cornu; *Sterculiac.;* Gab.
„ : *Canarium velutinum* Guill.; *Burserac.;* Gab.
„ : *Canarium schweinfurthi* Engl.; *Burserac.,* Kam.
Abele: *Populus alba* L.; *Salicac.;* ö. N.-Am.
Abell: *Canarium schweinfurthi* Engl.; *Burserac.;* Kam.
Abelluello: *Colubrina ferruginosa* Brongn.; *Rhamnac.;* P.-R.
Abem: *Berlinia bracteosa* Bth., *Macrolobium sp.; Caesalpiniac.;* Kam.
Abemaki: *Quercus variabilis* Bl., *Q. serrata* Thunb.; *Fagac.;* Jap.
Abenbélé-njambokka: *Pouteria guianensis* Aubl.; *Sapotac.;* Sur.
Aberemou: ? *Perebea sp.; Morac.;* Gu.
Abeto: *Pseudotsuga macrocarpa* Mayr, *P. taxifolia* Bütt.; *Conifer.;* Mex.
Abeubeu: *Khaya ivorensis* Chev.; *Meliac.;* Aeq. Af.
Abey: *Jacaranda sagreana* DC.; *Bignoniac.;* Cuba.
„ hembra: *Peltophorum adnatum* Gris.; *Caesalpiniac.;* Cuba.
„ macho: *Jacaranda sagreana* DC.; *Bignoniac.;* Cuba.
Abezguen: *Salvadora persica* L.; *Salvadorac.;* Sahara.

Abfan: *Panda oleosa* Pierre; *Pandac.*; Kam.
Abi: *Cola acuminata* Sch. et Endl.; *Sterculiac.*; Kam.
Abiang: *Livistona rotundifolia* Mart. var. *luzonensis* Becc.; *Palmac.*; Phil.
  „ : *Livistona merrilli* Becc.; *Palmac.*; Phil.
Abianguar: *Rhaptopetalum soyauxi* Oliv.; *Olacac.*; Gab.
  „ : *Brazzeia spp.*; *Tiliac.*; Gab.
Abiarana: *Lucuma sp.*; *Sapotac.*; Bras.
Abieiro: *Lucuma sp.*; *Sapotac.*; Bras.
Abies: *Xylopia hypolampra* Mildbr.; *Anonac.*; Kam.
Abigón: *Pterocymbium tinctorium* Merr.; *Sterculiac.*; Phil.
Abil: *Cola acuminata* Sch. et Endl.; *Sterculiac.*; Kam.
Abilo: *Garuga spp.*; *Burserac.*; Phil.
Abimba: *Macrolobium dewevrei* de Wild.; *Caesalpiniac.*; B.-K
Abimpé: *Petersia viridiflora* Chev.; *Myrtac.*; Elf.
Abin: *Rhizophora racemosa* E. Mey.; *Rhizophorac.*; Go.
  „ : *Laguncularia racemosa* Gaertn. f.; *Combretac.*; Go.
Abine: *Petersia viridiflora* Chev.; *Myrtac.*; Kam.
Abing: *Petersia minor* Ndz.; *Myrtac.*; Kam.
  „ : *Mimusops djave* Engl.; *Sapotac.*; Kam.
Abinioro: *Allexis cauliflora* Pierre; *Violac.*; Gab.
Abio: *Lucuma cainito* R. et P.; *Sapotac.*; n. Bras. (Amaz.)
Abiorana gutta: ? *Lucuma piriry* Ducke; *Sapotac.*; Bras.
Abisoa: *Vitex cienkowskii* K. et Payer; *Verbenac.*; Go.
Abiu: *Lucuma sp.*; *Sapotac.*; Bras.
Abiurana: *Lucuma sp., Ecclinusa balata* Ducke; *Sapotac.*; Bras.
Abnoos: *Diospyros ebenum* Koenig; *Ebenac.*; s. Ind., Cey.
Abo: *Macaranga heudeloti* Baill.; *Euphorbiac.*; Elf.
  „ , Abö: *Cola acuminata* Sch. et Endl.; *Sterculiac.*; Kam.
Abö: *Canarium schweinfurthi* Engl.; *Burserac.*; W.-Af.
Aboc: *Omphalocarpum procerum* Beauv.; *Sapotac.*; Kam.
Aboé: *Alchornea cordata* Bth.; *Euphorbiac.*; Kam.
Aboel: *Canarium schweinfurthi* Engl.; *Burserac.*; Gab.
Abo-idofun: *Parinarium curatellifolium* Planch.; *Rosac.*; C.-Af.
Abomé: *Musanga smithi* R. Br.; *Morac.*; Elf.
Abondi: *Tylostemon sp.*; *Laurac.*; Kam.
Abondog: *Mimusops sp.*; *Sapotac.*; Kam.
Abonko: *Sarcocephalus trillesi* Pierre; *Rubiac.*; Kam.
Abono: *Ficus goliath* Chev.; *Morac.*; Elf.
Abonsannua: *Afrormosia laxiflora* Harms; *Papilionac.*; Go.
Aboonkini: *Inga alba* Willd., *I. sertalifera* DC.; *Mimosac.*; Sur.
  „ : *Pithecolobium pedicellare* Bth.; *Mimosac.*; Sur.
Aborkpor: *Diospyros crassiflora* Hiern; *Ebenac.*; Nig.
Aborobié: *Pachystela cinerea* Pierre; *Sapotac.*; Elf.
Abot: *Ochrocarpus africanus* Oliv.; *Guttifer.*; Kam.
Abotnsok: *Ochrocarpus africanus* Oliv.; *Guttifer.*; Kam.
Abot zoc: *Mammea klaineana* Pierre; *Guttifer.*; Kam.
Aboudikro: *Entandrophragma cylindricum* Spr.; *Meliac.*; Elf.
  „ : *Entandrophragma tomentosum* Chev.; *Meliac.*; Elf.
Aboué: *Macaranga heudeloti* Baill.; *Euphorbiac.*; Elf.
Aboundou: *Barteria sp.*; *Passiflorac.*; Gab.
Aboya: *Trichilia prieuriana* A. Juss.; *Meliac.*; Go.
Abracouélé: *Lannea acidissima* Chev.; *Anacardiac.*; Elf.
Abran de Costa: *Bunchosia nitida* Juss.; *Malpighiac.*; Cuba.
Abriboca: *Grabowskya sp.*; *Solanac.*; S.-Am.
Abriboquillo del campo: *Ximenia americana* L.; *Olacac.*; Arg.
Abricó do Pará: *Mammea americana* L.; *Guttifer.*; Bras.
Abricot d'Afrique: *Mammea klaineana* Pierre; *Guttifer.*; Gab.
  „ des bois: *Sideroxylon mastichodendron* Jacq.; *Sapotac.*; fr. W.-Ind.
  „ de singe: *Couroupita guianensis* Aubl.; *Lecythidac.*; fr. Gu.
  „ du pays: *Mammea americana* L.; *Guttifer.*; fr. W.-Ind.
  „ jaune d'oeuf: *Lucuma sp.*; *Sapotac.*; St.-D.
Abricotier: *Prunus armeniaca* L.; *Rosac.*; Fr.

Abricotier d'Amérique: *Mammea americana* L.; *Guttifer.;* Guat., fr. W.-Ind.
„   des Antilles: *Mammea americana* L.; *Guttifer.;* W.-Ind.
„   de St. Domingue: *Mammea americana* L.; *Guttifer.;* W.-Ind.
„   sauvage: *Mammea americana* L.; *Guttifer.;* W.-Ind.
Abrojo: *Genista hispanica* L.; *Papilionac.;* Sp.
Abrotano hembra: *Santolina chamaecyparissus* L.; *Composit.;* Sp.
Abru Koyin: *Hura crepitans* L.; *Euphorbiac.;* Go.
Abser: *Acacia tortilis* Hayne; *Mimosac.;* Sahara.
Absinthe: *Canariellum oleiferum* Engl.; *Burserac.;* N.-Kal.
Abua: *Albizzia fastigiata* Oliv.; *Mimosac.;* B.-K.
Abubungon: *Shorea polysperma* Merr.; *Dipterocarpac.;* Phil.
Abura: *Mitragyne macrophylla* Hiern; *Rubiac.;* W.-Af.
Aburachan: *Lindera praecox* Blume; *Laurac.;* Jap.
Aburagiri: *Aleurites cordata* Steud.; *Euphorbiac.;* Jap.
Aburakiri: *Aleurites cordata* Steud.; *Euphorbiac.;* Jap.
Abura-sugi: *Keteleeria davidiana* Beissn.; *Conifer.;* Jap., Form., Ch.
Aburésé baka: *Grumilea venosa* Hiern; *Rubiac.;* Elf.
Aburuhi: *Scottellia coriacea* Chev.; *Flacourtiac.;* Elf.
Aburuki: *Scottellia coriacea* Chev.; *Flacourtiac.:* Elf.
Abut: *Trichoscypha oddoni* de Wild.; *Anacardiac.;* Kam.
Abutra: *Anamirta paniculata* Colebr.; *Menispermac.;* Jap.
Abuwe: *Lovoa klaineana* Pierre; *Meliac.;* W.-Af.
Abúyo: *Celtis spp.; Ulmac.;* Phil.
Acacia (i): *Robinia pseudacacia* L.; *Papilionac.;* N.-Am.
Acacia: *Acacia sp.; Mimosac.;* Ven.
„   : *Acacia vera* Willd.; *Mimosac.;* Hond.
„   : *Acacia farnesiana* Willd.; *Mimosac.;* Jap.
„   : *Acacia confusa* Merr.; *Mimosac.;* Form.
„   : *Piptadenia buchanani* Baker; *Mimosac.;* O.-Af.
„   : *Leucaena glauca* Bth.; *Mimosac.;* N.-Kal.
„   amarilla: *Albizzia lebbek* Bth.; *Mimosac.;* P.-R.
„   bella rosa: *Robinia hispida* L.; *Papilionac.;* N.-Am.
„   black: *Robinia pseudacacia* L.; *Papilionac.;* N.-Am.
„   blanca: *Robinia pseudacacia* L.; *Papilionac.;* N.-Am.
„   Bullhorn: *Acacia hindsi* Bth.; *Mimosac.;* Hond.
„   centenario: *Acacia decurrens* Willd. *var. mollis* Lind.; *Mimosac.;* Arg.
„   de Catarina: *Prosopis juliflora* DC.; *Mimosac.;* Nic.
„   de Constantinopla: *Albizzia julibrissin* Durazz.; *Mimosac.;* Arg.
„   del Japón (s): *Sophora japonica* L.; *Papilionac.;* ö. As., Arg.
„   de Rusia (s.): *Caragana arborescens* Lam.; *Papilionac.;* Sib., Mand.
„   du Gabon: *Distemonanthus benthamianus* Baill.; *Caesalpiniac.;* Gab.
„   falsa (i): *Robinia pseudacacia* L.; *Papilionac.;* N.-Am.
„   false (e): *Robinia pseudacacia* L.; *Papilionac.;* N.-Am.
„   franc (f.): *Enterolobium schomburgki* Bth.; *Mimosac.;* fr. Gu.
„   francesca: *Acacia decurrens* Willd. *var. dealbata* F. v. M.; *Mimosac.;* Arg.
„   jam scented: *Acacia acuminata* Bth.; *Mimosac.;* W.-Au.
„   mâle (f.): *Parkia pendula* Bth.; *Mimosac.;* fr. Gu.
„   Natal: *Acacia decurrens* Willd. *var. mollis* Lindl.; *Mimosac.;* Arg.
„   negra: *Gleditschia triacanthos* L.; *Caesalpiniac.;* Arg.
„   raspberry scented: *Acacia acuminata* Bth.; *Mimosac.;* W.-Au.
„   rose (e): *Robinia hispida* L.; *Papilionac.;* N.-Am.
„   Southamerican: *Calliandra saman* Griseb.; *Mimosac.;* Jam.
„   three-thorned: *Gleditschia triacanthos* L.; *Caesalpiniac.;* ö. N.-Am.
„   willow: *Acacia salicina* Lindl.; *Mimosac.;* Au.
Acagiu (i.): *Swietenia spp.; Meliac.;* tr. Am.
Acahuite: *Pseudotsuga taxifolia* Britt., *P.macrocarpa* Mayr; *Conifer.;* Mex.
Acajou: *Swietenia spp.; Meliac.;* tr. Am.
„   : *Khaya spp.; Meliac.;* W.-Af.
„   (f.) : *Treculia africana* Decne.; *Morac.;* Elf.
„   : *Eucalyptus spp.; Myrtac.;* Au.
„   : *Calophyllum burmanni* Wight, *C. spectabile* Willd.; *Guttifer.;* Fid.
„   : *Anacardium occidentale* L.; *Anacardiac.;* S.-Am.

1*

Acajou Aboudikro: *Entandrophragma cylindricum* Sp.; *Meliac.;* Elf.
„ „ : *Entandrophragma tomentosum* Chev.; *Meliac.;* Elf.
„ à fruits: *Anacardium occidentale* L.; *Anacardiac.;* tr. Am.
„ à grandes folioles: *Khaya grandifoliola* Thompson; *Meliac.;* Elf.
„ amer: *Cedrela odorata* L.; *Meliac.;* W.-Ind.
„ à meubles (f.): *Swietenia mahagoni* Jacq.; *Meliac.;* W.-Ind.
„ à planches (f.): *Cedrela odorata* L.; *Meliac.;* tr. Am.
„ à pommes: *Anacardium occidentale* L.; *Anacardiac.;* fr. Gu.
„ Assié (f.): *Entandrophragma utile* Spr.; *Meliac.;* Kam.
„ bàtard: *Curatella americana* L.; *Dilleniac.;* fr. Gu.
„ „ de Cayenne: *Curatella americana* L.; *Dilleniac.;* fr. Gu.
„ baum (d.): *Anacardium occidentale* L.; *Anacardiac.;* S.-Am.
„ blanc: *Khaya anthotheca* C. DC.; *Meliac.;* Elf.
„ „ : *Khaya agboensis* Chev.; *Meliac.;* Elf.
„ „ : *Entandrophragma macrophylla* Chev.; *Meliac.;* W.-Af.
„ „ : *Cedrela odorata* L.; *Meliac.;* fr. Gu.
„ „ : *Simaruba amara* Aubl.; *Simarubac.;* Guad.
„ „ : *Semecarpus atra* Vieill.; *Anacardiac.;* N.-Kal.
„ Krala (f.): *Khaya anthotheca* C. DC.; *Meliac.;* Elf.
„ Bossé (f.): *Guarea cedrata* Pell.; *Meliac.;* Elf.
„ caïl-cédrat (f.): *Khaya senegalensis* Juss.; *Meliac.;* fr. Guin.
„ cédrel: *Cedrela odorata* L.; *Meliac.;* W.-Ind.
„ d'Afrique (f): *Khaya ivorensis* Chev.; *Meliac.;* W.-Af.
„ „ (f): *Khaya senegalensis* Juss.; *Meliac.;* Senl.
„ „ : *Zizyphus vulgaris* Lamk.; *Rhamnac.;* Tunis.
„ d'Amérique: *Swietenia* spp.; *Meliac.;* tr. Am.
„ d'Annam: *Aglaia gigantea* Pell.; *Meliac.;* O.-As.
„ de Bassam: *Khaya ivorensis* Chev.; *Meliac.;* Elf.
„ de Cayenne: *Cedrela guianensis* A. Juss.; *Meliac.;* fr. Gu.
„ de Chine (f): *Toona sinensis* Roem.; *Meliac.;* O.-As.
„ de Cuba: *Swietenia mahagoni* Jacq.; *Meliac.;* W.-Ind.
„ de Guadeloupe: *Anacardium occidentale* L.; *Anacardiac.;* S.-Am.
„ de Haiti: *Swietenia mahagoni* Jacq.; *Meliac.;* W.-Ind.
„ de la Guyane: *Cedrela guianensis* A. Juss.; *Meliac.;* Gu.
„ de Madagascar (f): *Khaya madagascariensis* Jum. et Perr.; *Meliac.;* Mad.
„ de St. Domingue: *Swietenia mahagoni* Jacq.; *Meliac.;* St.-D.
„ d'Europe: *Juglans regia* L. var. ?; *Juglandac.;* Fr.
„ d'Indochine: *Melanorrhoea laccifera* Pierre; *Anacardiac.;* O.-As.
„ du Cambodge: *Tarrietia cochinchinensis* Pierre; *Sterculiac.;* Kamb.
„ du Gabon: *Khaya ivorensis* Chev.; *Meliac.;* Gab.
„ du Honduras: *Swietenia macrophylla* King; *Meliac.;* Hond.
„ du Mexique: *Swietenia humilis* Zucc.; *Meliac.;* Mex.
„ du pays: *Cedrela odorata* L.; *Meliac.;* W.-Ind.
„ du Sénégal: *Khaya senegalensis* Juss.; *Meliac.;* Senl.
„ du Vénézuela: *Swietenia candollei* Pittier; *Meliac.;* Ven.
„ du Yucatan: *Swietenia* spp.; *Meliac.;* Mex.
„ faux (f): *Cedrela odorata* L.; *Meliac.;* tr. Am.
„ femelle: *Cedrela odorata* L.; *Meliac.;* tr. Am., Ind.-Ch.
„ „ (f): *Cedrela guianensis* A. Juss.; *Meliac.;* Gu.
„ franc: *Cedrela* sp.; *Meliac.;* Gu.
„ frisé: *Entandrophragma septentrionalis* Chev.; *Meliac.;* W.-Af.
„ Grand-Bassam: *Khaya ivorensis* Chev.; *Meliac.;* Elf.
„ Kosipo: *Entandrophragma candollei* Harms; *Meliac.;* Elf.
„ Krala: *Khaya anthotheca* C. DC.; *Meliac.;* Elf.
„ „ : *Entandrophragma* sp.; *Meliac.;* W.-Af.
„ Lahou: *Khaya ivorensis* Chev.; *Meliac.;* W.-Af.
„ magona: *Lannea welwitschi* Engl.; *Anacardiac.;* Kam.
„ mahogani: *Swietenia mahagoni* Jacq.; *Meliac.;* fr. Gu.
„ Mahagon (f): *Swietenia mahagoni* Jacq.; *Meliac.;* fr. W.-Ind.
„ Mangona (f): *Khaya anthotheca* C. DC.; *Meliac.;* Kam.
„ marron: *Myristica* sp.; *Myristicac.;* Trin.
„ Mebrou: *Entandrophragma utile* Spr.; *Meliac.;* Elf.

Acajou N'dola: *Khaya ivorensis* Chev.; *Meliac.;* Gab.
„ noir (f): *Lovoa klaineana* Pierre; *Meliac.;* W.-Af.
„ pâle: *Khaya anthotheca* C. DC.; *Meliac.;* Kam.
„ Pommier: *Anacardium occidentale* L.; *Anacardiac.;* Senl.
„ résineux (f): *Tarrietia utilis* Spr.; *Sterculiac.;* W.-Af.
„ „ : *Cola proteiformis* Chev.; *Sterculiac.;* Elf.
„ ronceux: *Swietenia mahagoni* Jacq.; *Meliac.;* Haiti.
„ rosé: *Guarea cedrata* Pell.; *Meliac.;* Elf.
„ rosé d'Afrique (f): *Guarea cedrata* Pell.; *Meliàc.;* Elf.
„ rose du Tonkin: *Melia azedarach* L.; *Meliac.;* Ind.-Ch.
„ rouge (f): *Khaya ivorensis* Chev.; *Meliac.;* Kam.
„ „ : *Cedrela odorata* L.; *Meliac.;* Guad.
„ sapeli: *Entandrophragma cylindricum* Spr., *E. utile* Spr.; *Meliac.;* Nig.
„ Sapelli: *Entandrophragma cylindricum* Spr.; *Meliac.;* Kam.
„ „ : *Entandrophragma utile* Spr.; *Meliac.;* Kam.
„ Sipo: *Entandrophragma utile* Spr.; *Meliac.;* Elf.
„ Tabasco: *Swietenia macrophylla* King; *Meliac.;* Mex.
„ Tiama: *Entandrophragma spp.; Meliac.;* W.-Af.
„ wood: *Cedrela fissilis* Vell.; *Meliac.;* Bras.
Acaju (i, p): *Swietenia mahagoni* Jacq.; *Meliac.;* W.-Ind.
„ : *Cedrela odorata* L.; *Meliac.;* Bras.
Acalocahuite: *Pinus sp.; Coniferae;* Mex.
Acana (hell): *Bassia albescens* Griseb.; *Sapotac.;* Cuba.
„ : *Mimusops wrightiana* Bth.; *Sapotac.;* Cuba.
Acána: *Lucuma sp.; Sapotac.;* P.-R.
Acanita: *Pinus spp.; Coniferae;* Mex.
Acapro: *? Tecoma sp.; Bignoniac.;* Ven.
Acapú: *Andira aubleti* Bth.; *Papilionac.;* Bras.
Acapu: *Vouacapoua americana* Aubl.; *Papilionac.;* Bras.
Acapulco: *Diospyros obtusiflora* Willd.; *Ebenac.;* Mex.
Acaricara branca: *? Minquartia guianensis* Aubl.; *Olacac.;* Bras.
Acaricuára: *Minquartia guianensis* Aubl.; *Olacac.;* Bras.
Acariquára: *Minquartia guianensis* Aubl.; *Olacac.;* tr. Am.
„ da varzea: *Minquartia guianensis* Aubl.; *Olacac.;* Bras.
„ do igapó: *Minquartia guianensis* Aubl.; *Olacac.;* Bras.
Acariuba: *Minquartia guianensis* Aubl.; *Olacac.;* Bras.
Acary: *Minquartia guianensis* Aubl.; *Olacac.;* Bras.
„ coara: *Minquartia guianensis* Aubl.; *Olacac.;* Bras.
„ uba: *Minquartia guianensis* Aubl.; *Olacac.;* Bras.
Acatta: *Leucaena glauca* Bth.; *Mimosac.;* Bras.
Accyon tendre: *Mimosa sp.; Mimosac.;* Gren.
Acebo: *Ilex aquifolium* L.; *Aquifoliac.;* Sp.
Acebuche: *Olea europaea* L.; *Oleac.;* Sp.
Acecincle: *Acer spp.; Acerac.;* Mex.
Aceite de canime: *Copaifera spp.; Caesalpiniac.;* Kol.
„ de copaiba: *Copaifera spp.; Caesalpiniac.;* Kol.
„ de María: *Calophyllum calaba* Jacq.; *Guttifer.;* Kol., W.-Ind., Gu.
Aceitillo: *Simaruba tulac* Urb.; *Simarubac.;* Bras.
„ : *Gymnanthes lucida* Sw.; *Euphorbiac.;* Cuba.
„ : *Zanthoxylum martinicensis* Lam.; *Rutac.;* P.-R.
Aceito de cavalho: *Luehea divaricata* M. et Z.; *Tiliac.;* s. Bras., Arg.
„ de María: *Calophyllum calaba* Jacq.; *Guttifer.;* Trin.
Aceituna: *Symplocos martinicensis* Jacq.; *Symplocac.;* W.-Ind.
„ silvestre: *? Symplocos sp.; Symplocac.;* Ven.
Aceitunillo: *Lucuma salicifolia* H. B. K.; *Sapotac.;* C.-Am.
„ : *Terminalia sp.; Combretac.;* Trin.
Aceituno: *Simaruba glauca* DC.; *Simarubac.;* Hond.
„ : *Vitex colombiensis* Pitt.; *Verbenac.;* Kol.
„ : *? Calophyllum longifolium* Wall.; *Guttifer.;* Ven.
„ amarillo: *Ocotea sp.; Laurac.;* Ven.
„ montes: *Platymiscium polystachyum* Bth.; *Papilionac.;* Salv.
Acero: *Krugiodendron ferreum* Urb.; *Rhamnac.;* Cuba.

Acero riccio: *Acer platanoides* L.; *Acerac.;* It.
Acevino: *Ilex canariensis* Poir.; *Aquifoliac.;* Tener.
Acezuintle: *Acer spp.; Acerac.;* Mex.
Acha maram: *Hardwickia binata* Roxb.; *Caesalpiniac.;* Ind.
Achara: *Picea morinda* Link; *Conifer.;* n. Ind.
Achata: *Symplocos paniculata* Miq.; *Symplocac.;* Jap.
Achin: *Blighia sapida* Koen.; *Sapindac.;* Go.
Achiote: *Bixa orellana* L.; *Bixac.;* tr. Am.
   „ : *Oncoba laurina* Warb.; *Flacourtiac.;* Guat.
Achiotillo: *Colubrina ferruginosa* Brongn.; *Rhamnac.;* P.-R.
Achira-mourou: *Cordia sp.; Borraginac.;* fr. Gu. (Galibis)
Ac ho: *Randia longiflora* Lamk.; *Rubiac.;* Annam.
Achote: *Bixa orellana* L.; *Bixac.;* Pan., Kol.
Achotillo: *Vismia ferruginea* H. B. K.; *Guttifer.;* Guat.
Ach sat: *Brownlowia emarginata* Pierre; *Tiliac.;* Ind-Ch.
   „ „ : *Brownlowia tabularis* Pierre; *Tiliac.;* Ind.-Ch.
Achuete: *Bixa orellana* L.; *Bixac.;* Phil.
Acibuche: *Celtis sp.; Ulmac.;* Mex.
Acietuno negrito: *Simaruba sp.; Simarubac.;* Nic.
Acioua: *Couepia spp.; Rosac.;* fr. Gu.
Acle: *Albizzia acle* Merr.; *Mimosac.;* Phil.
   „ : *Xylia xylocarpa* Taub.; *Mimosac.;* Phil.
Acleng-párang: *Albizzia procera* Bth.; *Mimosac.;* Phil.
Açoita cavallo: *Luehea divaricata* Mart.; *Tiliac.;* Bras. (Sao P.)
   „   „ : *Luehea grandiflora* Mart.; *Tiliac.;* Bras.
   „   „ branco: *Luehea divaricata* Mart.; *Tiliac.;* Bras.
Acolé: *Acolea ?moniliformis* Pierre, *A.missionis* Pierre; *Sapotac.;* Gab.
Acolla: *Brazzeia spp.; Tiliac.;* Gab.
   „ : *Erythropyxis scandens* Pierre; *Styracac.;* Gab.
Acoma: *Sideroxylon foetidissimum* Jacq.; *Sapotac.;* Trin.
   „ : *Sideroxylon quadriloculare* Pierre; *Sapotac.;* Trin., Tob.
   „ : *Sideroxylon mastichodendron* Jacq.; *Sapotac.;* Trin.
   „ batard: *Sideroxylon mastichodendron* Jacq.; *Sapotac.;* Guad.
   „ blanc: *Homalium sp.; Flacourtiac.;* Guad.
   „ sauvage: *Homalium sp.; Flacourtiac.;* Guad.
Acomas à épis: *Homalium sp.; Flacourtiac.;* fr. Gu.
   „ à grappes: *Homalium sp.; Flacourtiac.;* fr. Gu.
Acomat (f): *Sideroxylon mastichodendron* Jacq.; *Sapotac.;* Mart.
   „ : *Homalium sp.; Flacourtiac.;* Guad.
   „ bastard: *Dipholis salicifolia* A. DC.; *Sapotac.;* fr. W.-Ind.
   „ franc: *Sideroxylon mastichodendron* Jacq.; *Sapotac.;* Guad.
   „ rouge: *Dipholis salicifolia* A. DC.; *Sapotac.;* Haiti.
Acouchini: *Icica sp.; Burserac.;* br. Gu.
Acoui: *Xylopia aethiopica* A. Rich.; *Anonac.;* Kam.
Acouma: *Homalium sp.; Flacourtiac.;* Mart.
   „ jaune: *Sideroxylon mastichodendron* Jacq.; *Sapotac.;* fr. W.-Ind.
Acoumé: *Aucoumea klaineana* Pierre; *Burserac.;* Gab.
Acouri: *Andira sp.,* ? *A.coriacea* Pulle; *Papilionac.;* fr. Gu.
Açouta-cavallo: *Luehea divaricata* Mart.; *Tiliac.;* Bras.
   „ cavallos: *Luehea grandiflora* Mart.; *Tiliac.;* Bras.
Acoute Cavallo: *Lueha divaricata* Mart.; *Tiliac.;* Bras.
Acuapar: *Hura crepitans* L.; *Euphorbiac.;* Kol.
Açucena branco: *Randia formosa* Schum.; *Rubiac.;* Bras.
Acundum: *Sarcocephalus sp.; S.sambucinus* K. Schum.; *Rubiac.;* Kam.
Acurel: *Trichilia oblanceolata* Rusby; *Meliac.;* Trin.
Acuyuru: *Astrocaryum aculeatum* G. F. W. Mey.; *Palmac.;* fr. Gu.
Adaán: *Albizzia procera* Bth.; *Mimosac.;* Phil.
Adabo: *Ceiba pentandra* Gaertn.; *Bombacac.;* Elf.
   „ : *Bombax buonopozense* Beauv.; *Bombacac.;* Elf.
Ada boumbia: *Terminalia ivorensis* Chev.; *Combretac.;* Elf.
Adada: *Cylicodiscus gabunensis* Harms; *Mimosac.;* Go.
Adadawa: *Cylicodiscus gabunensis* Harms; *Mimosac.;* Go.

Adadua: *Cylicodiscus gabunensis* Harms; *Mimosac.; Go.
Adagwe-matjapau: ? *Swartzia triphylla* Willd.; *Caesalpiniac.;* Sur.
Adamsapfel: *Mimusops kauki* L.; *Sapotac.;* O.-As.
Adan: *Mimusops micrantha* Chev.; *Sapotac.;* Elf.
Adáng: *Eugenia calubcob* C. B. Rob.; *Myrtac.;* Phil.
Adánga: *Homalium villarianum* Vid.; *Flacourtiac.;* Phil.
Adatsigo: *Spathodia campanulata* Beauv.; *Bignoniac.;* Go.
Adda: *Ferolia guianensis* Aubl.; *Rosac.;* Sur.
Addá: *Brosimum paraense* Huber; *Morac.;* Sur.
Adeka: *Irvingia gabonensis* Baill.; *Simarubac.;* W.-Af.
Adelfa: *Nerium oleander* L.; *Apocynac.;* Sp.
Adelfilla: *Daphne laureola* L.; *Thymelaeac.;* Sp.
Adelopo: *Klainedoxa ovalifolia* Verm.; *Simarubac.;* B.-K.
Adema: *Berlinia heudelotiana* Baill.; *Caesalpiniac.;* Togo.
Aderno: *Ardisia excelsa* Ait.; *Myrsinac.;* Tener.
„ : ? *Roupala brasiliensis* Klotzsch; *Proteac.;* Bras.
„ : ? *Astronium commune* Jacq.; *Anacardiac.;* Bras.
Aderno preto: *Astronium spp.; Anacardiac.;* Bras.
Adgánon: *Vitex parviflora* Juss.; *Verbenac.;* Phil.
Adiakoua: *Funtumia africana* Stapf; *Apocynac.;* Elf.
Adiana: *Adina cordifolia* Hook. f.; *Rubiac.;* Ind.
Adiáñgau: *Agathis alba* Foxw.; *Conifer.;* Phil.
Adidi: *Elaeophorbia drupifera* Stapf; *Euphorbiac.;* Go.
Adido: *Adansonia digitata* L.; *Bombacac.;* Go., Togo.
Adios: *Tristania spp.; Myrtac.;* Phil.
Adioumkoué: *Klainedoxa sp.; Simarubac.;* Elf.
Adivi nimma: *Atalantia monophylla* Correa; *Rutac.;* O.-Ind.
Adjab: *Mimusops djave* Engl.; *Sapotac.;* Kam.
Adjansé: *Cicca discoïdea* Baill.; *Euphorbiac.;* Elf.
Adjansi (f): *Cicca discoïdea* Baill.; *Euphorbiac.;* Elf.
Adjap: *Mimusops djave* Engl.; *Sapotac.;* W.-Af., Kam.
Adjar: *Maerua crassifolia* Forsk.; *Capparidac.;* Sahara.
Adjawakie: *Inga heterophylla* Willd., *I.lateriflora* Miq.; *Mimosac.;* Sur.
Adjensi: *Cicca discoidea* Baill.; *Euphorbiac.;* Elf.
Adjip: *Lavalleopsis densivenia* Engl.; *Olacac.;* Kam.
Adjondé: *Dracaena perrotteti* Baker; *Liliac.;* Elf.
Adjop: *Mimusops djave* Engl.; *Sapotac.;* Kam., Gab.
„ : *Mimusops pierreana* Engl.; *Sapotac.;* Kam., Gab.
Adjouaba (f): *Haematostaphis barteri* Hook. f.; *Anacardiac.;* Elf., Gab.
Adjouga: *Erytrophloeum guineense* G. Don; *Caesalpiniac.;* Elf.
Adlerholz (d): *Aquilaria malaccensis* Lam.; *Thymelaeac.;* H.-Ind., Malay.
Adlerholz: *Excoecaria agallocha* L.; *Euphorbiac.;* Kam.
Adoek: ? *Polyalthia subcordata* Blume; *Anonac.;* N.-Kal.
Adokbé-matjapau: ? *Swartzia triphylla* Willd.; *Caesalpiniac.;* Sur.
Adolumbi: ? *Klainedoxa ovalifolia* Verm.; *Simarubac.;* B.-K.
Adonisi-do-o: *Simaruba amara* Aubl.; *Simarubac.;* Sur.
Adonisidoro: *Simaruba amara* Aubl.; *Simarubac.;* Sur.
Adoonsidero: *Simaruba amara* Aubl.; *Simarubac.;* Sur.
Adoubrouya: *Lannea acidissima* Chev.; *Anacardiac.;* Elf.
Adoum: *Cylicodiscus gabunensis* Harms; *Mimosac.;* Kam.
„ : *Chlorophora excelsa* Bth. et Hook.; *Morac.;* Kam.
Adrakof: *Funtumia africana* Stapf; *Apocynac.;* Elf.
Adrakoi: *Funtumia africana* Stapf; *Apocynac.;* Elf.
Adria: *Pycnanthus kombo* Warb.; *Myristicac.;* Elf.
Aduaba: *Psidium guajava* L.; *Myrtac.;* Go.
Aduás: *Dracontomelum cumingianum* Baill.; *Anacardiac.;* Phil.
„ : *Dracontomelum edule* Skeels; *Anacardiac.;* Phil.
Adum: *Cylicodiscus gabunensis* Harms; *Caesalpiniac.;* Kam.
Adumói: *Shorea polysperma* Merr.; *Dipterocarpac.;* Phil.
Adupar: *Dysoxylum turczaninowi* C. DC.; *Meliac.;* Phil.
Adupong: *Sterculia spp.; Sterculiac.;* Phil.
Aduwa: *Balanites aegyptiaca* Delile; *Simarubac.;* n. Nig.

Adwindwera: *Lecaniodiscus cupanioides* Planch.; *Sapindac.;* Go.
Adwuba: *Dialium guineense* Willd.; *Caesalpiniac.;* Go.
Adwunkobi: *Hymenostegia afzelii* Oliv.; *Caesalpiniac.;* Go.
Adyanya peso: *Trichilia emetica* Vahl; *Meliac.;* tr. Af.
Adza: *Mimusops djave* Engl.; *Sapotac.;* Kam., Gab.
„ : *Mimusops pierreana* Engl.; *Sapotac.;* Kam., Gab.
„ : *Blighia sapida* Koenig; *Sapindac;* Go., Togo.
Adzadze: *Phyllanthus discoideus* Muell. Arg.; *Euphorbiac.;* Go.
Adzap: *Mimusops djave* Engl.; *Sapotac.;* Kam., Gab.
„ : *Mimusops pierreana* Engl.; *Sapotac.;* Kam., Gab.
Adzo: *Mimusops djave* Engl.; *Sapotac.;* Gab.
„ : *Mimusops pierreana* Engl.; *Sapotac.;* Gab.
Adzom: *Pachylobus edulis* G. Don; *Burserac.;* Gab.
? Adzuro: *Rhizophora racemosa* E. Mey.; *Rhizophorac.;* Go.
Aeng: *Dipterocarpus turbinatus* Gaertn.; *Dipterocarpac.;* Birma.
Aengdah: *Dipterocarpus laevis* Hamilt.; *Dipterocarpac.;* Birma.
Aeta: *Mauritia flexuosa* L. f.; *Palmac.;* br. Gu.
„ -balli: *Vochysia melinoni* Beckm.; *Vochysiac.;* Sur.
Aetan pandak: *Ctenolophon parvifolius* Oliv.; *Olacac.;* Malak.
Afam: *Panda oleosa* Pierre; *Pandac.;* Gab.
Afame: *Panda oleosa* Pierre; *Pandac.;* Gab.
Afamo: *Carapa sp.; Meliac.;* Bras.
Afan: *Panda oleosa* Pierre; *Pandac.;* Kam., Gab.
Afane (f): *Panda oleosa* Pierre; *Pandac;* Gab., Kam.
„ : *Porphyranthus zenkeri* Engl.; *Burserac.;* Gab.
Afann: *Panda oleosa* Pierre; *Pandac.;* Gab.
Afanu: *Macrolobium limba* Scott Ell.; *Caesalpiniac.;* Go.
Afara: *Terminalia superba* Engl. et Diels; *Combretac.;* Nig.
Afârez: *Quercus castaneifolia* C. A. Mey.; *Fagac.;* Kl.-As.
Afata blanca: *Heliocarpus americanus* L.; *Tiliac.;* Arg.
„ colorada: *Trema micrantha* Blume; *Ulmac.;* Arg.
„ grande: *Heliocarpus americana* L.; *Tiliac.;* Arg.
„ „ : *Cordia gerascanthus* L.; *Borraginac.;* Arg.
Afendek: *Grewia oligoneura* Spr.; *Tiliac.;* Kam.
Afetewa: *Vitex cienkowskii* K. et Payer; *Verbenac.;* Go.
Affenbrotbaum: *Adansonia digitata* L.; *Bombacac.;* Tr.
Affromouondou: *Funtumia africana* Stapf; *Apocynac.;* Elf.
Afia-fia: *Carpolobia lutea* G. Don; *Polygalac.;* Go.
Afiafy: *Avicennia officinalis* L.; *Verbenac.;* Mad.
Afiti: *Parkia africana* R. Br.; *Mimosac.;* tr. Af.
Aflangaloubé: *Strombosia pustulata* Oliv.; *Olacac.;* Elf.
Afo: *Poga oleosa* Pierre; *Rhizophorac.;* Gab.
Afollado: *Viburnum rugosum* Pers.; *Caprifoliac.;* Tener.
Afom: *Cleistopholis patens* Engl. et Diels; *Anonac.;* Kam.
Afop zam: *Mitragyne macrophylla* Hiern; *Rubiac.;* Kam.
? Afouebi Kouendé: *Berlinia sp.; Caesalpiniac.;* Kam.
Afraa: *Terminalia superba* Engl. et Diels; *Combretac.;* Go.
Afraba: ? *Malacantha warneckeana* Engl.; *Sapotac.;* Go.
African bark: *Crossopteryx kotschyana* Engl.; *Rubiac.;* Go.
Afrokouanda: *Albizzia sp.; Mimosac.;* Elf.
Afromouondou: *Funtumia africana* Stapf; *Apocynac.;* Elf.
Afu: *Anisoptera brunnea* Foxw.; *Dipterocarpac.;* Phil.
Afú: *Dipterocarpus vernicifluus* Blanco; *Dipterocarpac.;* Phil.
Afuma: *Microdesmis puberula* Hook. f.; *Euphorbiac.;* Go.
Afwom: *Cleistopholis patens* Engl. et Diels; *Anonac.;* Kam.
Afzeliaholz (d): *Afzelia africana* Smith; *Caesalpiniac.;* tr. Af.
Agag: *Commiphora africana* Engl.; *Burserac.;* it. Somali.
Agai: *Aglaia laevigata* Merr.; *Meliac.;* Phil.
Agajo: *Adelia triloba* Hemsl.; *Euphorbiac.;* Hond.
Agalla: *Bourreria calophylla* Griseb.; *Borraginac.;* Cuba.
Agallo: *Caesalpinia coriaria* Willd.; *Caesalpiniac.;* Pan.
Agallocha black: *Aquilaria agallocha* Roxb.; *Thymelaeac.;* Ind.

Agallocha black: *Aquilaria malaccensis* Lam.; *Thymelaeac.;* Ind.
Agallochum: *Aquilaria agallocha* Roxb.; *Thymelaeac.;* Ind.
„ : *Aquilaria malaccensis* Lam.; *Thymelaeac.;* Ind.
Agando: *Sideroxylon triflorum* Vahl; *Rubiac.;* Sur.
Aganokwi: *Mimusops lacera* Bak.; *Sapotac.;* W.-Af.
Agar-agar: *Aquilaria agallocha* Roxb.; *Thymelaeac.;* Ind.
Agarobilla: *Caesalpinia brevifolia* Baill.; *Caesalpiniac.;* tr. S.-Am.
Agarrobo morado: *Prosopis nigra* Hiern; *Mimosac.;* Arg.
Agáru: *Dysoxylum decandrum* Merr., *D. turczaninowi* C. DC.; *Meliac.;* Phil.
„ : *Terminalia comintana* Merr.; *Combretac.;* Phil.
Agarwood: *Aquilaria agallocha* Roxb.; *Thymelaeac.;* Ind.
„ real: *Aquilaria agallocha* Roxb.; *Thymelaeac.;* Assam.
Agás: *Semecarpus spp.; Anacardiac.;* Phil.
Agati: *Adenanthera pavonina* L.; *Mimosac.;* Sey.
Agatsura: *Toona sinensis* Roem.; *Meliac.;* Jap.
Agáu: *Neonauclea spp.; Rubiac.;* Phil.
Agbaia: *Haematostaphis barteri* Hook. f.; *Anacardiac.;* Elf.
Agbana: *Uvaria chamae* Beauv.; *Anonac.;* Go.
Agbomé: *Musanga smithi* R. Br.; *Morac.;* Elf.
Agboui: *Turraeanthus africana* Pell.; *Meliac.;* Elf.
Aggai: *Dillenia pentagyna* Roxb.; *Dilleniac.;* O.-Ind.
Aggan-Egbo: *Diospyros monbuttensis* Gürke; *Ebenac.;* W.-Af.
Agilawood: *Aquilaria agallocha* Roxb.; *Thymelaeac.;* Ind.
„ : *Aquilaria malaccensis* Lam.; *Thymelaeac.;* Ind.
Agipau: *Ormosia melanocarpa* Kleinh.; *Papilionac.;* Sur.
„ : *Ormosia coccinea* Jacks.; *Papilionac.;* Sur.
Aglaia: *Aglaia sp.; Meliac.;* Phil.
Aglay-marom: *Chukrasia tabularis* A. Juss.; *Meliac.;* tr. As.
Agliko: *Spondias lutea* L.; *Anacardiac.;* Togo.
Agni (f): *Petersia viridiflora* Chev.; *Myrtac.;* Elf.
Agnieré: *Myrianthus arboreus* Beauv.; *Morac.;* Elf.
Agnon: *Myrianthus arboreus* Beauv.; *Morac.;* tr. Af.
Agnûhé: *Pentadesma butyracea* Sabine; *Guttifer.;* Gab.
Agó: *Casuarina equisetifolia* Forst.; *Casuarinac.;* Phil.
Agofa: *Mitragyne macrophylla* Hiern; *Rubiac.;* Elf.
Agóho: *Casuarina equisetifolia* Forst.; *Casuarinac.;* Phil.
Agoniada: *Plumiera sp.; Apocynac.;* Bras.
Agóo: *Casuarina equisetifolia* Forst.; *Casuarinac.;* Phil.
Agóso: *Casuarina equisetifolia* Forst.; *Casuarinac.;* Phil.
Agosthyo: *Dillenia scabrella* Roxb.; *Dilleniac.;* Beng.
Agotót: *Cordia subcordata* Lam.; *Borraginac.;* Phil.
Agouelé: *Strombosia pustulata* Oliv.; *Olacac.;* Elf.
Agoumi: *Musanga smithi* R. Br.; *Morac.;* Elf.
Agouty-treva: *Anona sp.; Anonac.;* Bras.
Agqui: *Ricinodendron africanum* Muell. Arg.; *Euphorbiac.;* Elf.
Agracejo: *Phillyrea latifolia* L., *P. media* L.; *Oleac.;* Sp.
„ : *Ardisia sp.; Myrsinac.;* Cuba.
„ de Monte: *Casearia eriophora* Wright; *Flacourtiac.;* Cuba.
„ de sabana: *Ardisia sp.; Myrsinac.;* Cuba.
Agrifoglio (i): *Ilex aquifolium* L.; *Aquifoliac.;* Eu.
Agrifolia (p): *Ilex aquifolium* L.; *Aquifoliac.;* Eu.
Aguabola: *Maytenus spp.; Celastrac.;* Mex.
Aguacate: *Persea americana* Mill.; *Laurac.;* tr. Am.
„ cimaron: ? *Persea sp.; Laurac.;* Ven.
„ cimorrón: *Beilschmiedia pendula* Bth. et Hook. f.; *Laurac.;* P.-R.
„ dulce: *Persea gratissima* Gaertn.; *Laurac.;* Ven.
Aguacatillo: *Misanteca capitata* Cham. et Schlecht.; *Laurac.;* Hond.
„ : *Nectandra globosa* Mez; *Laurac.;* Hond.
„ : *Persea amplifolia* Mez et Sm.; *Laurac.;* Hond.
„ : *Ocotea cernua* Mez; *Laurac.;* Hond.
„ : *Nectandra laurel* Kl. et Karst.; *Laurac.;* Nic.
„ : *Nectandra glabrescens* Bth.; *Laurac.;* Guat., Hond.

Aguacatillo: *Phoebe ambigens* Blake; *Laurac.;* Guat., Hond.
„   : *Hernandia guianensis* Aubl.; *Hernandiac.;* Guat.
„   : *Persea coerulea* Mez; *Laurac.;* Ven.
„   : *Oreodaphne leucoxylon* Nees; *Laurac.;* Cuba.
„   : *Meliosma herberti* Rolfe; *Sabiac.;* P.-R.
Aguacatire: *Sickingia erythroxylon* Willd.; *Rubiac.;* Ven.
Agua-egbua: *Linociera manni* Solered.; *Oleac.;* Elf.
Agúaí: *Chrysophyllum lucumifolium* Gris.; *Sapotac.;* Arg.
„ -guazú: *Lucuma laurifolia* A. DC.; *Sapotac.;* Arg.
„ -guazú: *Citharexylum barbinerve* Cham.; *Verbenac.;* Arg.
„ -saiyú: *Labatia glomerata* Radlk.; *Sapotac.;* Arg.
Aguameque: *Robinsonella pilosa* Rose; *Malvac.;* Hond.
Aguano: *?* *Swietenia macrophylla* King; *Meliac.;* Peru.
Aguaribay: *Schinus molle* L.; *Anacardiac.;* Arg.
Aguatire: *Sickingia erythroxylon* Willd.; *Rubiac.;* Ven.
Aguatle: *Quercus sp.; Fagac.;* Mex.
Aguatope: *Inga sp.; Mimosac.;* Mex.
Aguay: *Chrysophyllum lucumifolium* Griseb.; *Sapotac.;* Arg.
„   : *Chrysophyllum cainito* L.; *Sapotac.;* Arg.
„   : *Pouteria suavis* Hemsl.; *Sapotac.;* Arg.
„ amarillo: *Labatia glomerata* Radlk.; *Sapotac.;* Arg.
„   „   : *Chrysophyllum cainito* L.; *Sapotac.;* Arg.
„ -guasu: *Lucuma neriifolia* Hook. et Arn.; *Sapotac.;* tr. Am.
„ -guazu: *Citharexylum barbinerve* Cham.; *Verbenac.;* n. Arg.
„ -sayyú: *Chrysophyllum cainito* L.; *Sapotac.;* Arg.
Aguaya: *Alchornea sp.; Euphorbiac.;* Elf.
„ gueia: *Omphalocarpum anocentrum* Pierre; *Sapotac.;* Elf.
Aguaytarán: *Colubrina ferruginosa* Brongn.; *Rhamnac.;* P.-R.
Agudesi: *Bombax buonopozense* Beauv.; *Bombacac.;* Go.
Aguë: *Ceiba pentandra* Gaertn.; *Bombacac.;* Elf.
Agueghé: *Ceiba pentandra* Gaertn.; *Bombacac.;* W.-Af.
Agui: *Chlorophora excelsa* Bth. et Hook.; *Morac.;* Elf.
Aguillon: *Cananga odorata* Hook. f. et Thoms.; *Anonac.;* O.-As.
Agulasing: *Aglaia diffusa* Merr.; *Meliac.;* Phil.
Agulu: *Autranella congolensis* Chev.; *Sapotac.;* Kongo.
Agupañgá: *Pometia pinnata* Forst.; *Sapindac.;* Phil.
Agutúb: *Cordia subcordata* Lam.; *Borraginac.;* Phil. (Cebú)
Agúu: *Pinus merkusi* Jungh. et de Vr.; *Conifer.;* Phil.
Agyama: *Musanga smithi* R. Br.; *Morac.;* Go.
Agyiapa: *Pentadesma butyracea* Sabine; *Guttifer.;* Go.
Ahamman: *Tamarix gallica* L.; *Tamaricac.;* n. Marok.
Ahanta: *Afzelia africana* Smith; *Caesalpiniac.;* Go.
Ahate: *Anona spp.; Anonac.;* Mex.
Ahenkyen: *Fagara macrophylla* Engl.; *Rutac.;* Go.
Ahetes: *Acacia albida* Delile; *Mimosac.;* Sahara.
Ahia: *Mimusops djave* Engl.; *Sapotac.;* Kam., Gab.
„   : *Mimusops pierreana* Engl.; *Sapotac.;* Kam., Gab.
Ahiani: *Stephegyne parvifolia* Korth.; *Rubiac.;* Ind.-Ch.
Ahinebé (f): *Anthocleista nobilis* G. Don; *Loganiac.;* Elf., Gab.
Ahinébé (f): *Anthocleista zenkeri* Gilg; *Loganiac.;* Kam.
Ahlbeere, gemeine: *Ribes nigrum* L.; *Saxifragac.;* Eu.
Ahoaquahuitl: *Quercus sp.; Fagac.;* Mex.
Ahoatl: *Quercus sp.; Fagac.;* Mex.
Ahorn (d): *Acer spp.; Acerac.;* Eu.
„ Berg: *Acer pseudoplatanus* L.; *Acerac.;* Eu.
„ eschenblättriger: *Negundo aceroides* Moench; *Acerac.;* N.-Am.
„ Feld: *Acer campestre* L.; *Acerac.;* Eu.
„ französischer: *Acer monspessulanum* L.; *Acerac.;* s. Eu.
„ Spitz: *Acer platanoides* L.; *Acerac.;* Deutschld.
„ Trauben: *Acer pseudoplatanus* L.; *Acerac.;* Deutschld.
„ Vogelaugen (d): *Acer saccharum* Marsh.; *Acerac.;* ö. N.-Am.
„ Weißer: *Acer pseudoplatanus* L.; *Acerac.;* Deutschld.

Ahorn Zucker- (d): *Acer saccharum* Marsh.; *Acerac.;* N.-Am.
Ah-pill: *Erythrophloeum laboucheri* F. v. M.; *Caesalpiniac.;* Queensl.
Ahuacata: *Persea americana* Mill.; *Laurac.;* Mex., C.-Am.
Ahualtzocotlque: *Malpighia urens* L.; *Malpighiac.;* Mex.
Ahuehuete: *Taxodium mucronatum* Ten.; *Conifer.;* Mex.
Ahuehuetl: *Taxodium mucronatum* Ten.; *Conifer.;* Mex.
Ahuejote: *Salix sp.; Salicac.;* Mex.
    „   : *Erythrina sp.; Papilionac.;* Salv.
Ahuijote: *Erythrina sp.; Papilionac.;* Salv.
Aiaúma: *Couroupita peruviana* Miers; *Lecythidac.;* Peru.
Aiauman: *Couroupita guianensis* Aubl.; *Lecythidac.;* n. Bras.
Aiélé (f): *Canarium schweinfurthi* Engl.; *Burserac.;* W.-Af.
Aigle montagne: *Moronobea spp.; Guttifer.;* fr. Gu.
Aigrefoux (f): *Ilex aquifolium* L.; *Aquifoliac.;* Eu.
Ailante de Malabar (f): *Ailanthus malabarica* DC.; *Simarubac.;* Ind.
    „ glanduleux (f): *Ailanthus glandulosa* Desf.; *Simarubac.;* O.-As.
Ailanthus: *Ailanthus glandulosa* Desf.; *Simarubac.;* N.-Am.
Ailanto: *Ailanthus glandulosa* Desf.; *Simarubac.;* Moluk., Jap.
Ailé (f): *Canarium occidentale* Chev.; *Burserac.;* Elf.
Ailon: *Magnolia sp.; Magnoliac.;* s. USA.
Aimpen: ? *Piratinera sp.; Morac.;* Bras. (Bahia)
Ainee: *Artocarpus hirsuta* Lamk.; *Morac.;* Ind.
Aing-Kamgien-zée: *Dipterocarpus alatus* Roxb.; *Dipterocarpac.;* Birma.
Aini: *Artocarpus hirsuta* Lamk.; *Morac.;* Ind.
Ainyeran: *Afrormosia sp.; Papilionac.;* W.-Af.
Aiquiʒ: *Thevetia gaumeri* Hemsl.; *Apocynac.;* br. Hond.
Aité: *Gymnanthes lucida* Sw.; *Euphorbiac.;* Cuba.
Aiti: *Moquilea tomentosa* Bth.; *Rosac.;* Bras.
Aiuí-pará: *Ocotea diospyrifolia* Mez; *Laurac.;* Arg.
Aiuy-moroti: *Nectandra tweedi* Mez; *Laurac.;* Arg.
Aja-Igi: *Cylicodiscus gabunensis* Harms; *Mimosac.;* Nig.
Ajama: *Musanga smithi* R. Br.; *Morac.;* Go.
Ajar: *Chrysophyllum sp.; Sapotac.;* Go.
Ajawa: *Protium spp.; Burserac.;* Sur.
Aja zoc: *Mimusops djave* Engl.; *Sapotac.;* Kam.
    „   : *Mimusops congolensis* Russ. et Héd.; *Sapotac.;* Kam.
Ajeersi: *Ferolia guianensis* Aubl.; *Morac.;* Sur.
Ajhar: *Lagerstroemia flos-reginae* Reʒ.; *Lythrac.;* Assam.
Ajicillo: *Erythroxylon argentinum* E. Sch.; *Erythroxylac.;* Arg.
    „   : *Scutia buxifolia* Reiss.; *Rhamnac.;* Arg.
    „   : ? *Zanthoxylum sp.; Rutac.;* Arg.
    „   : *Heisteria macrophylla* Oerst.; *Olacac.;* Pan.
Ajicito: *Capparis pachaca* H. B. K.; *Capparidac.;* Ven.
Ajoewa: *Parkia nitida* Miq., *P. microcephala* Kleinh.; *Mimosac.;* Sur.
Ajoewi: *Acrodiclidium canella* Mez, *A. chrysophyllum* Meissn.; *Laurac.;* Sur.
Ajonc (f): *Ulex europaeus* L.; *Papilionac.;* Eu.
Ajone: *Ulex europaeus* L.; *Papilionac.;* Fr.
Ajubo: *Ajouea tenella* Nees; *Laurac.;* Bras.
Ajunjurrah: *Lecaniodiscus cupanioides* Planch.; *Sapindac.;* Go.
Ajuru: *Hirtella sp.; Rosac.;* Bras.
Ajuss: *Triplochiton scleroxylon* K. Schum.; *Sterculiac.;* Kam.
Aka: *Duboscia macrocarpa* Bocq.; *Tiliac.;* Gab.
    „ : *Desplaʒia caudata* Pierre; *Tiliac.;* Gab.
Akäba: *Prosopis oblonga* Bth.; *Mimosac.;* W.-Af.
Akaboa: *Cordia irvingi* Baker; *Borraginac.;* Go.
Aka-eso-matsu: *Picea ajanensis* Fisch.; *Conifer.;* n. Jap.
Akaezo-matsu: *Picea glehni* Mast.; *Conifer.;* Jap.
Akafekafei: *Macaranga rowlandi* Prain; *Euphorbiac.;* Go.
Akagashi: *Quercus acuta* Thunb.; *Fagac.;* Jap.
Akagi: *Bischoffia javanica* Blume; *Euphorbiac.;* Form.
Akajoeran: *Dimorphandra latifolia* Tul.; *Caesalpiniac.;* Sur.
Akak: *Diplanthemum viridiflorum* Hutch.; *Tiliac.;* Kam.

Akak: *Duboscia macrocarpa* Bocq., *Desplaţia caudata* Pierre; *Tiliac.;* Gab.
Akäkä: *Prosopis oblonga* Bth.; *Mimosac.;* Togo.
Akakarie: *Eschweilera longipes* Miers; *Lecythidac.;* Sur.
„ : *Eschweilera subglandulosa* Miers, *Lecythis sp.; Lecythidac.;* Sur.
Aka-kashi: *Quercus acuta* Thunb.; *Fagac.;* Jap.
? Akali: *Dialium guineense* Willd.; *Caesalpiniac.;* Go.
Akamamenoki: *Ormosia formosana* Kanehira; *Papilionac.;* Form.
Akamatsu: *Pinus densiflora* S. et. Z.; *Conifer.;* Jap.
Akamba: *Heisteria trillesiana* Pierre; *Olacac.;* Gab.
Akame-gashiwa: *Mallotus japonicus* Muell. Arg.; *Euphorbiac.;* Jap.
Akamekashiwa: *Mallotus japonicus* Muell. Arg.; *Euphorbiac.;* Jap.
Akara: *Tapirira sp. aff. guianensis* Aubl.; *Anacardiac.;* Sur.
Akarak: *Hymenostegia aff. afzelii* Harms; *Caesalpiniac.;* Kam.
Akark: *Duboscia macrocarpa* Bocq.; *Tiliac.;* Gab.
Akar lakak: *Dalbergia parviflora* Roxb.; *Papilionac.;* ö. As.
Akasa: *Chrysophyllum albidum* G. Don; *Sapotac.;* Go.
Akashide: *Carpinus laxiflora* Blume; *Corylac.;* Jap.
Aka-si: *Glyphaea laterifolia* Hutch.; *Tiliac.;* Kam.
Akata: *Bombax buonopozense* Beauv.; *Bombacac.;* Go.
Akatakunkuntuni: *Bombax buonopozense* Beauv.; *Bombacac.;* Go.
Akátan: *Palaquium spp.; Sapotac.;* Phil.
Akayanagi: *Salix urbaniana* Seem.; *Salicac.;* Jap.
Akayezo: *Picea glehni* Mayr; *Conifer.;* Jap.
Akazia: *Robinia pseudacacia* L.; *Papilionac.;* Holl.
Akazie unechte: *Robinia pseudacacia* L.; *Papilionac.;* N.-Am.
Akbaij: *Cola sp.; Sterculiac.;* Kam.
Akcam: *Pinus nigra* Arnold *var. pallasiana* Schneid.; *Conifer.;* Türk.
Ake: *Pausinystalia spp., Pseudocinchona spp.; Rubiac.;* Kam.
„ : *Pseudocinchona pachyceras* K. Schum.; *Rubiac.;* Kam.
„ : *Morinda lucida* Bth.; *Rubiac.;* Go.
„ : *Fagara xanthoxyloides* Lam.; *Rutac.;* Go.
„ : *Dodonaea viscosa* Jacq.; *Sapindac.;* N.-Seel.
Akeake: *Dodonaea viscosa* Jacq.; *Sapindac.;* N.-Seel.
Aké-Atok-Evila: *Maba sp.; Ebenac.;* Gab.
Akédé: *Chlorophora excelsa* Bth. et Hook.; *Morac.;* Elf.
„ : *Antiaris africana* Engl.; *Morac.;* Elf.
Akee: *Blighia sapida* Koenig; *Sapindac.;* W.-Ind., C.-Am.
Akee wild: *Guarea sp.; Meliac.;* br. W.-Ind.
Akélé: *Lophira procera* Chev.; *Ochnac.;* Kam.
Akelet: *Pausinystalia spp.; Rubiac.;* Kam.
„ : *Pseudocinchona spp., P.pachyceras* K. Schum.; *Rubiac.;* Kam.
Akelläng: *Corynanthe pachyceras* K. Schum.; *Rubiac.;* Kam.
Akemereibli: *Macrolobium limba* Scott Ell.; *Caesalpiniac.;* Go.
Akenden (f): *Grewia coriacea* Mast.; *Tiliac.;* Kam.
Akenneng: *Grewia coriacea* Mast.; *Tiliac.;* Kam.
Akeul: *Corynanthe gabonensis* Chev.; *Rubiac.;* Gab.
Akhor: *Juglans regia* L.; *Juglandac.;* Hind.
Akhrot: *Juglans regia* L.; *Juglandac.;* Ind.
Aki-Aki: ? *Dodonaea viscosa* Jacq.; *Sapindac.;* N.-Seel.
Akian: *Morinda citrifolia* L.; *Rubiac.;* Gab.
Aki Rantangi: ? *Dodonaea viscosa* Jacq.; *Sapindac.;* N.-Seel.
Aki-gumi: *Elaeagnus pungens* Thunb.; *Elaeagnac.;* Jap.
Akibaum: *Metrosideros scandens* Banks et Sol.; *Myrtac.;* N.-Seel.
Akiko: *Spondias lutea* L.; *Anacardiac.;* Togo.
Akinire: *Ulmus parvifolia* Jacq.; *Ulmac.;* Jap.
Akion: *Coula edulis* Baill.; *Olacac.;* Elf.
Akiou: *Coula edulis* Baill.; *Olacac.;* Elf.
Akisika: *Sçottellia kamerunensis* Gilg; *Flacourtiac.;* Elf.
Akkarandan: *Paypayrola guianensis* Aubl.; *Violac.;* Sur.
Akkèja: *Tecoma sp., Tabebuia sp.; Bignoniac.;* Sur.
Akkekèja: *Tecoma sp., Tabebuia sp.; Bignoniac.;* Sur.
Akkojaarie: *Cedrela odorata* L.; *Meliac.;* Sur.

Akleng-Parang: *Albizzia procera* Bth.; *Mimosac.;* Phil.
Akli: *Albizzia acle* Merr.; *Mimosac.;* Phil.
Aklike: *Dichrostachys glomerata* Hutch. et Dalz.; *Mimosac.;* Go.
Aknedum: *Sarcocephalus sp.; Rubiac.;* Kam.
Ako: *Ficus wightiana* Wall. *var. japonica* Miq.; *Morac.;* Jap.
  „ (f): *Antiaris toxicaria* Lesch.; *Morac.;* Elf.
  „ (f): *Antiaris africana* Engl.; *Morac.;* Gab.
Akö: *Corynanthe dolichocarpa* Brandt; *Rubiac.;* Kam.
Akodiombi: *Linociera manni* Solered.; *Oleac.;* Elf.
Akodo: *Ochrocarpus africanus* Oliv.; *Guttifer.;* Elf.
Akoejalli: *Cedrela odorata* L.; *Meliac.;* Sur.
Akoeli tjèrèrè: *Andira coriacea* Pulle; *Papilionac.;* Sur.
Akoelie-kiererie: *Andira coriacea* Pulle; *Papilionac.;* Sur.
Akoema: *Couma guianensis* Aubl.; *Apocynac.;* Sur.
Akoga: *Lophira procera* Chev.; *Ochnac.;* sp. Guin.
Akogha: *Lophira procera* Chev.; *Ochnac.;* Elf., Gab.
Akoïma: *Homalium africanum* Bth.; *Flacourtiac.;* Elf.
Akoïssou: *Homalium africanum* Bth.; *Flacourtiac.;* Elf.
Akök: *Tylostemon cf. crassifolius* Engl.; *Laurac.;* Kam.
Akokamot: *Alstonia congensis* Engl.; *Apocynac.;* Kam.
Akoko: *Firmiana barteri* K. Schum.; *Sterculiac.;* W.-Af. (Niger)
Akokotona: *Linociera manni* Solered.; *Oleac.;* Elf.
Akola: *Alangium lamarcki* Thw.; *Cornac.;* Ind.
Akola: *Xylopia aethiopica* A. Rich.; *Anonac.;* Gab.
Akom: *Terminalia altissima* Chev.; *Combretac.;* Kam.
  „ : *Terminalia superba* Engl. et Diels; *Combretac.;* Kam.
Akondoc: *Sarcocephalus trillesi* Pierre, *S. pobeguini* Pob.; *Rubiac.;* Kam.
Akondok osoé: *Sarcocephalus sp.; Rubiac.;* Kam.
Akondug: *Sarcocephalus cf. trillesi* Pierre; *Rubiac.;* Kam.
Akonibia: *Homalium africanum* Bth.; *Flacourtiac.;* Elf.
Akonkore: *Bombax buonopozense* Beauv.; *Bombacac.;* Go.
Akor: *Albizzia zygia* Macb.; *Mimosac.;* Go.
Akora: *Sclerosperma walkeri* Chev.; *Palmac.;* Gab.
Akoriüé: *Linociera manni* Solered.; *Oleac.;* Elf.
Akossika (f): *Scottellia kamerunensis* Gilg; *Flacourtiac.;* Elf.
Akotombo: *Buchholzia macrophylla* Pax; *Capparidac.;* Elf.
Akotombo-atoin: *Conopharyngia crassa* Stapf; *Apocynac.;* Elf.
Akou: *Ficus exasperata* Vahl; *Morac.;* Gab.
Akoul: *Ficus exasperata* Vahl; *Morac.;* Gab.
Akoura: *Lophira procera* Chev.; *Ochnac.;* Elf., Gab.
Akouya: *Irvingia sp.; Simarubac.;* Elf.
Akowa: ? *Irvingia klainei* Pierre; *Simarubac.;* Gab.
Akpa: *Tetrapleura thonningi* Bth.; *Mimosac.;* Kam.
Akpakossa: *Mitragyne africana* Korth.; *Rubiac.;* Togo.
Akré-akoué: *Pynaertia occidentalis* Chev.; *Meliac.;* Elf.
Akrot: *Juglans regia* L.; *Juglandac.;* Pers.
Aks: *Turraeanthus africana* Pellegr.; *Meliac.;* W.-Af.
Akuama: *Cistanthera papaverifera* Chev.; *Tiliac.;* Go.
Akuamma: *Cistanthera papaverifera* Chev.; *Tiliac.;* Go.
  „ : *Pentaclethra macrophylla* Bth.; *Mimosac.;* Go.
Akuana: *Porphyranthus zenkeri* Engl.; *Burserac.;* Elf.
Akudoni: *Bombax buonopozense* Beauv.; *Bombacac.;* Go.
Akuédo: *Eugenia rowlandi* Spr.; *Myrtac.;* Elf.
Akuisi-amba: *Lonchocarpus sericeus* H. B. K.; *Papilionac.;* Elf.
Akumadua: *Cistanthera papaverifera* Chev.; *Tiliac.;* Go.
Akumasé: *Allanblackia parviflora* Chev.; *Guttifer.;* Elf.
Akumassé: *Carapa velutina* C. DC.; *Meliac.;* Elf.
Akun: *Uapaca staudti* Pax; *Euphorbiac.;* Nig.
Akunduk: *Sarcocephalus cf. trillesi* Pierre; *Rubiac.;* Kam.
Akvankusuma: *Porphyranthus zenkeri* Engl.; *Burserac.;* Elf.
Akwabu: ? *Irvingia sp.; Simarubac.;* Elf.
Akwan siba: *Swartzia tomentosa* DC.; *Caesalpiniac.;* Sur.

Akwantanuro: *Lovoa klaineana* Pierre; *Meliac.*; Go.
Akye: *Phialodiscus unijugatus* Radlk.; *Sapindac.*; Go.
Akyere: *Phialodiscus unijugatus* Radlk.; *Sapindac.*; Go.
Alaän-kopie: *Erisma uncinatum* Warm.; *Vochysiac.*; Sur.
Alaän kopie: *Vochysia tomentosa* DC.; *Vochysiac.*; Sur.
Alaba: *Uapaca bingervillensis* Beille; *Euphorbiac.*; Elf.
Alabanunu: *Pentadesma leucantha* Chev.; *Guttifer.*; Elf.
Alabenun: *Allanblackia parviflora* Chev.; *Guttifer.*; Elf.
Alabo: *Uapaca bingervillensis* Beille; *Euphorbiac.*; Elf.
Alacayuela: *Helianthemum ocymoides* Pers.; *Cistac.*; Sp.
Alagba: *Adansonia digitata* L.; *Bombacac.*; Go.
Alahorré: *Tecoma leucoxylon* Mart.; *Bignoniac.*; Sur.
   „   : *Tecoma araliacea* P. DC., *Tabebuia sp.*; *Bignoniac.*; Sur.
Alahua: *Uapaca bingervillensis* Beille; *Euphorbiac.*; Elf.
Alai: *Bruguiera eriopetala* Lam.; *Rhizophorac.*; Phil.
Alaká: *Palaquium sp.*; *Sapotac.*; Phil.
Alakáak: *Palaquium sp.*, *P. gigantifolium* Merr.; *Sapotac.*; Phil.
Alalangád: *Albizzia procera* Bth.; *Mimosac.*; Phil.
Alám: *Shorea squamata* Dyer; *Dipterocarpac.*; Phil.
Alámag: *Aglaia clarki* Merr.; *Meliac.*; Phil.
Alamo: *Populus nigra* L.; *Salicac.*; Sp., Arg.
   „   : *Pentapanax angelicifolius* Griseb.; *Araliac.*; Arg.
   „   : *Ficus religiosa* L.; *Morac.*; Cuba.
   „   blanco: *Populus alba* L.; *Salicac.*; Sp.
   „   „   : *Platanus spp.*; *Platanac.*; Mex.
   „   de Italia: *Populus nigra* L., *P. pyramidalis* Salisb.; *Salicac.*; Arg.
   „   de la Carolina: *Populus angulata* James; *Salicac.*; Arg.
   „   llamado de Italia: *Populus nigra* L.; *Salicac.*; Arg.
   „   negro: *Populus nigra* L.; *Salicac.*; Sp.
   „   „   : *Ulmus campestris* Smith; *Ulmac.*; Sp.
   „   plateado: *Populus alba* L.; *Salicac.*; Arg.
Alan: *Hylodendron gabunense* Taub.; *Caesalpiniac.*; Kam.
Alañgigan: *Canangium odoratum* Baill.; *Anonac.*; Phil.
Alanguillan: *Cananga odorata* Hook. f. et Th.; *Anonac.*; tr. O.-As.
Alanígi: *Myristica spp.*; *Myristicac.*; Phil.
Alaoelama: *Sclerolobium paniculatum* Vogel; *Caesalpiniac.*; Sur.
Ala-oné: *Tecoma leucoxylon* Mart. var. *pentaphylla* Juss.; *Bignoniac.*; Sur.
   „   „ : *Tecoma araliacea* P. DC., *Tabebuia pentaphylla* Hemsl.; *Bignoniac.*; Sur.
Ala-onni: *Tecoma leucoxylon* Mart.; *Bignoniac.*; Sur.
   „   „ : *Tecoma araliacea* P. DC., *Tabebuia pentaphylla* Hemsl.; *Bignoniac.*; Sur.
Alasa Pegrecou: *Xylopia frutescens* Aubl.; *Anonac.*; Caraïb.
Alasema: *Chrysophyllum aff. africanum* A. DC.; *Sapotac.*; Go.
Alastanhout: *Hymenaea sp.*; *Caesalpiniac.*; Sur.
Alas-waboe: *Tabebuia sp.*; *Bignoniac.*; Sur.
Alato: *Dillenia sp.*; *Dilleniac.*; Phil. (n. Luzon)
Alatrique: *Cordia sp.*; *Borraginac.*; Trin.
Alauíhau: *Dracontomelum cumingianum* Merr.; *Anacardiac.*; Phil. (s. Luzon)
   „   : *Dracontomelum edule* Skeels; *Anacardiac.*; Phil.
   „   : *Pometia pinnata* Forst.; *Sapindac.*; Phil.
Alauna: *Licania macrophylla* Bth.; *Rosac.*; Sur.
Alawata poesoeloekoeloe: *Inga spp.*; *Mimosac.*; Sur.
Alawatta-moelerie: *Pithecolobium pedicellare* Bth.; *Mimosac.*; Sur.
Alazano: *Calycophyllum candidissimum* DC.; *Rubiac.*; Pan.
Albahaquilla de Chile (s): *Psoralea glandulosa* L.; *Papilionac.*; Chile.
Albarco: *Cariniana pyriformis* Miers; *Lecythidac.*; Kol.
Albaricoquillo: *Ximenia americana* L.; *Olacac.*; Arg.
Albarillo: *Ximenia americana* L.; *Olacac.*; Arg.
   „   del campo: *Ximenia americana* L.; *Olacac.*; Arg.
Albero di Acajou (i): *Swietenia mahagoni* Jacq.; *Meliac.*; tr. Am.
   „   „   „   (i): *Swietenia macrophylla* King; *Meliac.*; tr. Am.
Albero di paradiso (i): *Ailanthus glandulosa* Desf.; *Simarubac.*; N.-Ch.
Albiche: *Pentapanax angelicifolius* Griseb.; *Araliac.*; Arg.

Alcabé: *Zanthoxylum sp.; Rutac.;* Pan.
Alcabu: *Zanthoxylum panamense* P. Wils.; *Rutac.;* Pan.
Alcanfor sacha: *Zanthoxylum sp.; Rutac.;* Peru.
Alcarreto: *Sickingia sp.; Rubiac.;* Pan.
Alcornoque: *Quercus suber* L.; *Fagac.;* Sp.
„   : *Licania sp.; Rosac.;* C.-R.
„   : *Ormosia panamensis* Bth.; *Papilionac.;* Pan.
„   : *Dimorphandra oleifera* Triana; *Caesalpiniac.;* Pan.
„   : *Bowdichia virgilioides* H. B. K.; *Papilionac.;* Ven., Kol.
Alder (e): *Alnus spp.; Betulac.;* Eu.
„   : *Alnus incana* Willd. *var. sibirica* Spach; *Betulac.;* Jap.
„ black: *Alnus glutinosa* Gaertn.; *Betulac.;* Eu.
„   „ : *Rhamnus frangula* L.; *Rhamnac.;* Eu., As., Af.
„   „ : *Alnus glauca* Michx.; *Betulac.;* N.-Am.
„ brown: *Ackama quadrivalvis* White; *Saxifragac.;* O.-Au.
„ California (u): *Alnus rhombifolia* Nutt.; *Betulac.;* w. N.-Am.
„ Cherry: *Eugenia luehmanni* F. v. M.; *Myrtac.;* Au.
„ common: *Alnus glutinosa* Gaertn.; *Betulac.;* Eu.
„   „ : *Alnus serrulata* Ait.; *Betulac.;* N.-Am.
„ european: *Alnus glutinosa* Gaertn.; *Betulac.;* ö. USA.
„   black: *Alnus vulgaris* Hill.; *Betulac.;* N.-Am.
„ hoary: *Alnus incana* Moench; *Betulac.;* N.-Am.
„ indian: *Alnus spp.; Betulac.;* Ind.
„ Mountain: *Alnus rhombifolia* Nutt., *A. tenuifolia* Nutt.; *Betulac.;* w. N.-Am.
„ Oregon (u): *Alnus oregona* Nutt.; *Betulac.;* nw. N.-Am.
„ pink: *Quintinia sieberi* A. DC.; *Saxifragac.;* O.-Au.
„ red: *Alnus rubra* Brongn.; *Betulac.;* N.-Am.
„ rose: *Ackama paniculata* Engl.; *Cunoniac.;* Au.
„ sea-side: *Alnus maritima* Nutt.; *Betulac.;* ö. USA.
„ Sitka: *Alnus sitchensis* Sarg.; *Betulac.;* w. Kan.
„ smooth: *Alnus rugosa* Spreng.; *Betulac.;* nö. USA.
„ speckled: *Alnus incana* Moench; *Betulac.;* ö. Kan.
„ western: *Alnus rubra* Brongn.; *Betulac.;* br. Kol.
„ white (u): *Alnus rhombifolia* Nutt.; *Betulac.;* w. N.-Am.
„   „ : *Platylophus trifoliatus* D. Don; *Saxifragac.;* Kap.
Alébié: *Uapaca bingervillensis* Beille; *Euphorbiac.;* Elf.
Alecrim: *Holocalyx balansae* Micheli; *Caesalpiniac.;* Bras. (Sao P.), Arg.
Alecrin: *Holocalyx balansae* Micheli; *Caesalpiniac.;* Arg.
Alef: *Desbordesia glaucescens* Pierre; *Simarubac.;* Kam.
Alelí: *Plumiera spp.; Apocynac.;* P.-R.
Aleli: *Melia azedarach* L.; *Meliac.;* Ven.
Alelí cimarrón: *Plumiera spp.; Apocynac.;* P.-R.
? Alen: ? *Amphimas sp.; Papilionac.;* Kam.
Alèn: *Elaeis guineensis* Jacq., *E. nigrescens* Chev.; *Palmac.;* Gab.
Alèn-Bingom: *Elaeis guineensis* Jacq., *E. virescens* Chev.; *Palmac.;* Gab.
Alenda: *Ephedra alata* Dcne.; *Ephedrac.;* Sahara.
Alèn-Ntang'a: *Elaeis guineensis* Jacq., *E. virescens* Chev.; *Palmac.;* Gab.
Alen okpoué: *Dracaena fragrans* Gawl.; *Liliac.;* Kam.
Aleo: *Olea laperrini* Batt. et Trab.; *Oleac.;* Sahara.
Alèp (f): *Desbordesia insignis* Pierre; *Simarubac.;* Gab.
Alep (f): *Irvingia oblonga* Chev.; *Simarubac.;* Kam., Gab.
Alerce: *Fitzroya patagonica* Hook. f.; *Conifer.;* Chile.
Alerzeholz (d): *Fitzroya patagonica* Hook. f.; *Conifer.;* Chile.
Alfaje: *Trichilia tuberculata* C. DC.; *Meliac.;* Pan.
Alfajeo: *Trichilia propinqua* C. DC.; *Meliac.;* Pan.
„ colorado: *Trichilia tuberculata* C. DC.; *Meliac.;* Pan.
Alfalfa arbórea: *Medicago arborea* L.; *Papilionac.;* Arg.
„ tree (e): *Medicago arborea* L.; *Papilionac.;* Pers., Syr., s. Eu.
Afiler: *Bougainvillea stipitata* Griseb.; *Nyctaginac.;* Arg.
Algarob: *Calliandra saman* Griseb.; *Mimosac.;* Jam.
Algaroba (u): *Prosopis juliflora* DC.; *Mimosac.;* tr. Am.
Algarobilla: *Prosopis panta* Hieron.; *Mimosac.;* tr. Am.

Algarobilla: *Caesalpinia brevifolia* Baill.; *Caesalpiniac.;* Chile.
Algarroba: *Prosopis chilensis* Stuntz; *Mimosac.;* Hond.
„ carretera: *Enterolobium cyclocarpum* Gris.; *Mimosac.;* Cuba.
„ de Olor: *Albizzia lebbek* Bth.; *Mimosac.;* Cuba.
„ dulce: *Prosopis nigra* Hieron.; *Mimosac.;* Arg.
„ negro: *Prosopis nigra* Hieron.; *Mimosac.;* Arg.
Algarrobe blanco: *Prosopis alba* Griseb.; *Mimosac.;* tr. Am.
Algarrobilla: *Prosopis algarrobilla* Griseb.; *Mimosac.;* Arg. (B. Air.)
Algarrobillo: *Acacia moniliformis* Griseb.; *Mimosac.;* Arg. (S. Fé)
„ : *Prosopis ñandubay* Lorentz; *Mimosac.;* Arg. (Cord.)
Algarrobo: *Hymenaea courbaril* L.; *Caesalpiniac.;* tr. Am.
„ : *Pithecolobium saman* Bth.; *Mimosac.;* Cuba.
„ : *Peltogyne sp.; Caesalpiniac.;* Pan.
„ : *Prosopis juliflora* DC.; *Mimosac.;* S.-Am.
„ : *Prosopis hassleri* Harms; *Mimosac.;* Arg. (Form.)
„ : *Prosopis nigra* Hieron., *P. alba* Gris.; *Mimosac.;* Arg., Urug.
„ : *Caesalpinia brevifolia* Baill.; *Caesalpiniac.;* Chile.
„ amarillo: *Prosopis sp.; Mimosac.;* Arg.
„ blanco: *Prosopis vinalillo* Stuckert; *Mimosac.;* Arg.
„ „ : *Prosopis alba* Griseb.; *Mimosac.;* Arg.
„ „ : *Prosopis ruscifolia* Griseb.; *Mimosac.;* Arg.
„ colorado: *Prosopis juliflora* DC.; *Mimosac.;* Arg. (Corr.)
„ cupesi: *Prosopis sp.; Mimosac.;* Bol.
„ del pais: *Enterolobium saman* Prain; *Mimosac.;* Cuba.
„ de miel: *Gleditschia triacanthos* L.; *Caesalpiniac.;* C.-USA.
„ europea: *Ceratonia siliqua* L.; *Caesalpiniac.;* Arg.
„ ñandubay: *Prosopis sp.; Mimosac.;* Arg.
„ negro: *Prosopis nigra* Hieron.; *Mimosac.;* Arg.
„ „ del Chaco: *Prosopis hassleri* Harms; *Mimosac.;* Arg. (Form.)
„ panta: *Prosopis panta* Hieron.; *Mimosac.;* Arg.
Algodoeiro do praia: *Hibiscus tiliaceus* L.; *Malvac.;* Bras.
Algodón: *Chorisia speciosa* St. Hil.; *Bombacac.;* Arg.
Algodoncillo: *Luehea sp.; Tiliac.;* Mex.
„ : *Helicteres baruensis* Jacq.; *Sterculiac.;* Mex.
„ : *Hibiscus tiliaceus* L.; *Malvac.;* Pan., Ven.
Algodonero: *Chorisia speciosa* St. Hil.; *Bombacac.;* Arg.
Alho: *Desbordesia spp.; Simarubac.;* Gab.
Aliaga: *Calycotome spinosa* Link; *Papilionac.;* Sp.
„ : *Genista scorpius* DC.; *Papilionac.;* Sp.
„ : *Ulex europaeus* L.; *Papilionac.;* Arg.
Aliana-oeu: *Swartzia spp.; Caesalpiniac.;* Sur.
Alibabái: *Allaeanthus glaber* Warb.; *Morac.;* Phil.
Alibágon: *Cratoxylon sp.; Guttifer.;* Phil.
Aliboufier benjoin (f): *Styrax tonkinense* Pierre; *Styracac.;* Ind.-Ch.
„ „ „ : *Styrax macrothyrsus* Perk.; *Styracac.;* Ind.-Ch.
„ „ „ : *Anthostyrax tonkinense* Pierre; *Styracac.;* Ind.-Ch.
Alieskie-ie: *Jacaranda filicifolia* D. Don; *Bignoniac.;* Sur.
Alieskie-ie-wéwé: *Jacaranda filicifolia* D. Don; *Bignoniac.;* Sur.
Aligámen: *Zizyphus sp.; Rhamnac.;* Phil.
Aligógon: *Cratoxylon sp.; Guttifer.;* Phil.
Aligustre: *Ligustrum vulgare* L.; *Oleac.;* Sp.
Aliki: *Pithecolobium latifolium* Bth.; *Mimosac.;* br. Gu.
Aliku: *Pithecolobium cauliflorum* Mart.; *Mimosac.;* br. Gu.
„ : *Pithecolobium latifolium* Bth.; *Mimosac.;* br. Gu.
Alilaila: *Melia azedarach* L.; *Meliac.;* P.-R.
Alilem: *Terminalia oocarpa* Merr.; *Combretac.;* Phil.
Aliliba: *Cordia abyssinica* R. Br.; *Borraginac.;* n. Nig.
Alinau: *Cyathocalyx globosus* Merr.; *Anonac.;* Phil.
Alináu: *Diospyros currani* Merr.; *Ebenac.;* Phil.
Alínau: *Zizyphus sp.; Rhamnac.;* Phil.
Alintatáu: *Neonauclea sp.; Rubiac.;* Phil.
Alintau: *Neonauclea sp.; Rubiac.;* Phil.

Alipata: *Excoecaria agallocha* L.; *Euphorbiac.*; Phil.
Alipáuen: *Alstonia scholaris* R. Br.; *Apocynac.*; Phil.
Alisalañga: *Zizyphus sp.*; *Rhamnac.*; Phil.
Alisier blanc: *Sorbus aria* Cranß; *Rosac.*; Fr.
„ torminal: *Sorbus torminalis* Cranß; *Rosac.*; Fr.
Aliso: *Alnus glutinosa* Gaertn.; *Betulac.*; Sp.
„ : *Alnus sp.*; *Betulac.*; Mex.
„ : *Tessaria integrifolia* R. et P.; *Composit.*; Arg.
„ : *Eupatorium sp.*; *Composit.*; Arg.
„ del cerro: *Alnus spachi* Regel; *Betulac.*; Arg.
„ del río: *Tessaria sp.*; *Composit.*; Arg., Par.
Alitagtág: *Allaeanthus glaber* Warb.; *Morac.*; Phil.
„ : *Buchanania sp.*; *Anacardiac.*; Phil.
Aljaba: *Fuchsia macrostemma* R. et P.; *Oenotherac.*; Arg. (Patag.), Chile.
Alla: *Carapa sp.*; *Meliac.*; Elf.
Allabahnunu: *Pentadesma leucantha* Chev.; *Guttifer.*; Elf.
Allaïa: *Carapa sp.*; *Meliac.*; Elf.
Allen: *Cynometra aff. lujaei* de Wild.; *Caesalpiniac.*; Kam.
Allerheiligenholz (d): *Haematoxylon campechianum* L.; *Caesalpiniac.*; tr. Am.
Alligatorwood: *Liquidambar styraciflua* L.; *Hamamelidac.*; ö. N.-Am.
„ : *Guarea gomma* Pulle; *Meliac.*; Sur.
Allo: *Irvingia klainei* Pierre; *Simarubac.*; Gab.
„ : *Desbordesia spp.*; *Simarubac.*; Gab.
„ : *Fillaeopsis discophora* Harms; *Mimosac.*; Gab.
Allom: *Irvingia klainei* Pierre; *Simarubac.*; Gab.
Alloro spinosa (i): *Ilex aquifolium* L.; *Aquifoliac.*; Eu.
Allouchier: *Pirus aria* Cranß; *Rosac.*; Eu., W.-As.
Allspice tree: *Pimenta officinalis* Lindl.; *Myrtac.*; br. Hond., W.-Ind.
Almáciga (s): *Agathis alba* Foxw.; *Conifer.*; Phil.
„ : *Bursera gummifera* Jacq.; *Burserac.*; Ven.
Almacigo: *Bursera gummifera* Jacq.; *Burserac.*; Cuba.
„ : *Bursera simaruba* Sarg.; *Burserac.*; tr. Am.
Almácigo amarillo: *Bursera simaruba* Sarg.; *Burserac.*; Cuba.
„ blanco: *Bursera simaruba* Sarg.; *Burserac.*; Cuba.
„ colorado: *Bursera simaruba* Sarg.; *Burserac.*; Cuba.
„ encarnado: *Bursera simaruba* Sarg.; *Burserac.*; P.-R.
Almandier: *Laplacea curtyana* A. Rich.; *Theaceae*; Cuba.
Almecega: *Protium icicariba* March.; *Burserac.*; Bras.
Almecegueira: *Protium icicariba* March.; *Burserac.*; Bras.
Almendrillo: *Prunus occidentalis* Sw.; *Rosac.*; P.-R.
„ : *Rhamnidium revolutum* Griseb.; *Rhamnac.*; Cuba.
„ : *Saurauia sp.*; *Dilleniac.*; Mex.
„ : *Terminalia catappa* L.; *Combretac.*; tr. Am.
Almendro: *Amygdalus communis* L.; *Rosac.*; Sp.
„ : *Laplacea curtyana* A. Rich.; *Theaceae*; Cuba.
„ : *Dipteryx panamensis* Pittier; *Papilionac.*; Pan.
„ : *Lonchocarpus sp.*; *Papilionac.*; br. Hond.
„ : *Andira inermis* H. B. K.; *Papilionac.*; C.-Am.
„ : *Terminalia obovata* Eichl.; *Combretac.*; Hond.
„ : *Bertholletia excelsa* Humb. et Bonpl., *B. nobilis* Miers; *Lecythidac.*; Kol.
„ : *Prunus sphaerocarpa* Sw.; *Rosac.*; Ven.
„ : *Geoffraea superba* Humb. et Bonpl.; *Papilionac.*; Ven.
„ : *Prunus amygdalus* Stokes; *Rosac.*; Arg. (B. Air.)
„ del río: *Andira inermis* H. B. K.; *Papilionac.*; Salv.
„ macho: *Andira inermis* H. B. K.; *Papilionac.*; Salv.
„ montes: *Andira inermis* H. B. K.; *Papilionac.*; Salv.
„ sylvestre: *Dipholis salicifolia* A. DC.; *Sapotac.*; Fl., Ant.
Almendron: *Caryocar sp.*; *Caryocarac.*; Kol.
Almendrón: *Terminalia catappa* L.; *Combretac.*; tr. Am.
„ : *Prunus occidentalis* Sw.; *Rosac.*; P.-R.
„ : *Dipholis salicifolia* A. DC.; *Sapotac.*; P.-R.
„ de playa: *Terminalia sp.*; *Combretac.*; Trin.

Almesca ussú: *Icica sp.; Burserac.;* Bras. (Bahia)
Almescega: *Icica sp.; Burserac.;* Bras.
Almez: *Celtis australis* L.; *Ulmac.;* Sp.
„ americano: *Celtis occidentalis* L.; *Ulmac.;* ö. N.-Am.
Almique (hell): *Labourdonnaisia albescens* Bth.; *Sapotac.;* Cuba.
Almiquí: *Labourdonnaisia albescens* Bth.; *Sapotac.;* Cuba.
Almon: *Shorea eximia* Scheff.; *Dipterocarpac.;* Phil.
Almond: *Tamarindus indica* L., *T. catappa* L.; *Caesalpiniac.;* St.-D.
„ Dog: *Andira inermis* H. B. K.; *Papilionac.;* tr. Am.
„ Horse: *Sterculia foetida* L.; *Sterculiac.;* N.-Au.
„ Indian: *Terminalia catappa* L.; *Combretac.;* Tr.
„ Red: *Alphitonia petriei* Braid et C. T. White; *Rhamnac.;* Au.
„ Rose: *Owenia venosa* F. v. M.; *Meliac.;* Au.
„ Tropical: *Terminalia catappa* L.; *Combretac.;* Malay, Mad.
„ Wild: *Brabejum stellatifolium* L.; *Proteac.;* Kap.
Almonlauan: *Shorea furfuracea* Miqu.; *Dipterocarpac.;* Phil.
Almrausch: *Rhododendron hirsutum* L.; *Ericac.;* Alpen.
Almug: *Pterocarpus santalinus* L. f.; *Papilionac.;* O.-As., Java.
Alo: *Desbordesia insignis* Pierre; *Simarubac.;* Gab.
Aloa: *Antiaris cf. welwitschi* Engl.; *Morac.;* Kam.
Aloang bo loang: *Baccaurea annamensis* Gagnep.; *Euphorbiac.;* Ind.-Ch.
„ canaysa biêt: *Alangium costatum* Valeton; *Cornac.;* Ind.-Ch.
„ dông do: *Endospermum chinense* Bth.; *Euphorbiac.;* Ind.-Ch.
„ lay sa lang: *Mallotus philippinensis* Muell. Arg.; *Euphorbiac.;* Ind.-Ch.
„ mân bâu: *Macaranga henricorum* Hemsl.; *Euphorbiac.;* Ind.-Ch.
„ tang: *Trevesia palmata* Vis.; *Araliac.;* Ind.-Ch.
Alobo: *Uapaca bingervillensis* Beille; *Euphorbiac.;* Elf.
Alocotimon: *Malacantha robusta* Chev.; *Sapotac.;* Elf.
Aloès: *Michelia champaca* L.; *Magnoliac.;* Nied.-Ind.
Aloes lign: *Aquilaria agallocha* Roxb.; *Thymelaeac.;* Ind.
Aloeswood: *Aquilaria ovata* Cav., *A. agallocha* Roxb.; *Thymelaeac.;* tr. As.
Aloewa-oe: *Protium spp.; Burserac.;* Sur.
Aloewau-oe: *Hedwigia balsamifera* Swartz; *Burserac.;* Sur.
„ „ : *Protium sagotianum* March., *P. polybotryum* Engl.; *Burserac.;* Sur.
„ „ : *? Crepidospermum rhoïfolium* Tr. et Pl.; *Burserac.;* Sur.
Aloewood: *Cordia sp.; Borraginac.;* br. W.-Ind.
Aloího: *Pometia pinnata* Forst.; *Sapindac.;* Phil.
Alokoba: *Uapaca bingervillensis* Beille; *Euphorbiac.;* Elf.
Alokon: *Allaeanthus glaber* Warb.; *Morac.;* Phil.
Alokotimon: *Malacantha robusta* Chev.; *Sapotac.;* Elf.
Alokwstumon: *Malacantha robusta* Chev.; *Sapotac.;* Elf.
Alologpo: *Dracaena fragrans* Ker.-Gawl.; *Liliac.;* Kam.
Alom: *Diospyros mespiliformis* Hochst.; *Ebenac.;* Senl.
„ : *Lannea zenkeri* Engl. et Krause; *Anacardiac.;* Gab.
Aloma: *Sarcocephalus trillesi* Pierre; *Rubiac.;* W.-Af., Gab.
Alombo: *? Irvingia klainei* Pierre; *Simarubac.;* Gab.
Alona wood (u): *Lovoa klaineana* Pierre; *Meliac.;* W.-Af.
Alone: *Lovoa klaineana* Pierre; *Meliac.;* Gab.
Alongua: *Morinda citrifolia* L.; *Rubiac.;* Elf.
Aloo: *Adansonia digitata* L.; *Bombacac.;* Senl.
Alop: *Lovoa klaineana* Pierre; *Meliac.;* Kam.
„ : *Gossweilerodendron sp.; Caesalpiniac.;* Kam.
Alou: *Desbordesia sp.; Simarubac.;* Gab.
„ : *Psychotria sp.; Rubiac.;* Kam.
Alpear: *Grewia laevigata* Vahl; *Tiliac.;* Ind.
Alpenheckenkirsche: *Lonicera alpigena* L.; *Caprifoliac.;* Eu.
Alpenrose: *Rhododendron spp.; Ericac.;* Alpen.
„ behaarte: *Rhododendron hirsutum* L.; *Ericac.;* Mitt.-Eu.
„ rostblättrige: *Rhododendron ferrugineum* L.; *Ericac.;* Mitt.-Eu.
Alquifoux: *Ilex aquifolium* L.; *Aquifoliac.;* Fr.
Alu: *Pseudocedrela kotschyi* Harms; *Meliac.;* Go.
Aluan drau: *Tritaxis gaudichaudi* Baill.; *Euphorbiac.;* Ind.-Ch.

Aluan sa coi: *Baccaurea oxycarpa* Gagnep.; *Euphorbiac.;* Ind.-Ch.
„ te he: *Microdesmis caseariaefolia* Planch.; *Euphorbiac.;* Ind.-Ch.
Aludel: *Artocarpus nobilis* Thw.; *Morac.;* Cey.
Alukó: *Garcinia sp.; Guttifer.;* Phil.
? Alum: *Alnus sp.; Betulac.;* C.-R.
Alumbi: ? *Klainedoxa ovalifolia* Verm.; *Simarubac.;* B.-K.
Alupag: *Euphoria cinerea* Radlk.; *Sapindac.;* Phil.
„ : *Euphoria didyma* Blco.; *Sapindac.;* Phil.
„ -amó: *Litchi philippinensis* Radlk.; *Sapindac.;* Phil.
„ -maching: *Santiria nitida* Merr.; *Burserac.;* Phil.
Alupaí: *Euphoria cinerea* Radlk.; *Sapindac.;* Phil.
Alupák: *Euphoria cinerea* Radlk.; *Sapindac.;* Phil.
Alupí: *Terminalia edulis* Blanco; *Combretac.;* Phil.
Alvarillo del campo: *Ximenia americana* L.; *Olacac.;* S.-Am.
Alvier: *Pinus cembra* L.; *Conifer.;* w. Sib., Karp., Alpen.
Am: *Mangifera indica* L.; *Anacardiac.;* tr. & subtr. As.
Ama-apa: *Couma guianensis* Aubl.; *Apocynac.;* Sur.
Amacei: *Copaifera spp.; Caesalpiniac.;* St.-D.
Amáciga: *Bursera gummifera* Jacq.; *Burserac.;* Ven.
Amaco: *Ixora nicaraguensis* Wernh.; *Rubiac.;* Guat.
Amaga: *Diospyros discolor* Willd., *D. multiflora* Blanco; *Ebenac.;* Phil.
Amajouva: *Ajouea brasiliensis* Meissn.; *Laurac.;* Bras.
Amamáhi: *Vitex aherniana* Merr.; *Verbenac.;* Phil.
Amamáhit: *Vitex aherniana* Merr.; *Verbenac.;* Phil.
Amamanít: *Eucalyptus naudiniana* F. v. Muell.; *Myrtac.;* Phil.
Amamba Pombolo: *Dialium connaroïdes* Harms; *Caesalpiniac.;* Gab.
Aman: *Afzelia africana* Smith; *Caesalpiniac.;* Kam.
Amancayo: *Plumiera spp.; Apocynac.;* Kol.
Amandelboom: *Brabeium stellatifolium* L.; *Proteac.;* S.-Af.
Amandier: *Amygdalus communis* L.; *Rosac.;* Fr.
„ : *Prunus occidentalis* Swartz; *Rosac.;* St.-D.
„ du bord du mer: *Terminalia sp.; Combretac.;* Trin.
„ du pays: *Terminalia sp.; Combretac.;* Trin.
Amanga: *Sclerosperma walkeri* Chev.; *Palmac.;* Gab.
Amangkás: *Sideroxylon luzoniense* Merr.; *Sapotac.;* Phil.
Amanguila: *Khaya klainei* Pierre; *Meliac.;* Gab.
Amanseidua: *Cassia occidentalis* L.; *Caesalpiniac.;* Go.
Amanza mujer: *Prioria copaifera* Gris.; *Caesalpiniac.;* Pan.
Amapa: *Tecoma pentaphylla* Juss.; *Bignoniac.;* Mex.
Amapá: *Tabebuia palmeri* Rose; *Bignoniac.;* Mex.
„ : *Hancornia amapa* Huber; *Apocynac.;* Bras. (Amaz.)
„ blanca: *Cordia sonorae* Rose; *Borraginac.;* Mex.
„ prieta: *Tecoma palmeri* Rose; *Bignoniac.;* Mex.
„ prieto: *Tabebuia palmeri* Rose; *Bignoniac.;* Mex.
„ rosa: *Tecoma pentaphylla* Juss.; *Bignoniac.;* Mex.
Amaparian: *Couma guianensis* Aubl.; *Apocynac.;* Sur.
Amape: *Tecoma pentaphylla* Juss.; *Bignoniac.;* Mex.
Amapola: *Bombax ellipticum* H. B. K.; *Bombacac.;* Mex.
„ : *Pereskia amapola* Web.; *Cactac.;* Bras.
„ : *Tecoma pentaphylla* Juss.; *Bignoniac.;* Mex.
„ blanca: *Bombax ellipticum* H. B. K.; *Bombacac.;* Mex.
„ colorada: *Bombax ellipticum* H. B. K.; *Bombacac.;* Mex.
Amaracciole (i): *Cytisus scoparius* Link-Bailey; *Papilionac.;* Eu.
Amarant (d): *Peltogyne spp.; Caesalpiniac.;* Gu.
Amarante: *Copaifera ?bracteata var. pubiflora* Bth.; *Caesalpiniac.;* tr. Am.
„ : *Peltogyne spp.; Caesalpiniac.;* tr. Am.
„ rouge: *Martiusia excelsa* Bth.; *Caesalpiniac.;* Sur.
„ „ : *Martiusia parvifolia* Bth.; *Caesalpiniac.;* Sur.
Amaranth (e): *Peltogyne spp.; Caesalpiniac.;* tr. Am.
Amarantholz (d): *Peltogyne spp.; Caesalpiniac.;* tr. Am.
Amaranthout: *Peltogyne spp.; Caesalpiniac.;* tr. S.-Am.
Amarellinho: *Symplocos sp.; Symplocac.;* Bras.

Amarello: *Pithecolobium vinhatico* Record; *Mimosac.;* Bras.
„ : *Plathymenia reticulata* Bth.; *Mimosac.;* Bras.
„ Araribá: *Centrolobium tomentosum* Guill.; *Papilionac.;* Bras. (Sao P.)
„ Ipé: *Tecoma sp.; Bignoniac.;* Bras. (Sao P.)
Amargo: ? *Simaruba glauca* DC.; *Simarubac.;* Ven.
Amargoso: *Andira sp.; Papilionac.;* Hond.
„ : *Aspidosperma sp.; Apocynac.;* Ven.
Amarguillo: *Capparis speciosa* Griseb.; *Capparidac.;* Arg.
Amarilho: *Terminalia sp.; Combretac.;* Bras.
Amarilla yema de huevo: *Tabebuia pentaphylla* Hemsl.; *Bignoniac.;* W.-Ind., Bras., Pan.
„ „ „ „ : *Aspidosperma vargasi* A. DC.; *Apocynac.;* Ven.
„ „ „ huore: *Aspidosperma vargasi* A. DC.; *Apocynac.;* Ven.
Amarillo: *Bucida buceras* L.; *Combretac.;* Pan.
„ : *Aspidosperma vargasi* A. DC.; *Apocynac.;* Ven.
„ : *Terminalia hilariana* Steud.; *Combretac.;* Ven.
„ : *Zanthoxylum cumanense* ?; *Rutac.;* Ven.
„ : *Chlorophora tinctoria* Gaudich.; *Morac.;* Bol.
„ : *Terminalia australis* Camb.; *Combretac.;* Arg.
„ : *Ptilochaeta nudipes* Griseb.; *Malpighiac.;* Arg. (Salta)
„ : *Eugenia ligustrina* Willd.; *Myrtac.;* Arg. (Entre Rios)
„ blanco: *Ficus spp.; Morac.;* Mex.
„ boj: *Aspidosperma vargasi* A. DC.; *Apocynac.;* Ven.
„ de Guayaquil: *Centrolobium ochroxylon* Rose; *Papilionac.;* Ek.
„ „ „ : *Centrolobium patinense* Pitt.; *Papilionac.;* Pan.
„ del rio: *Terminalia australis* Camb.; *Combretac.;* Arg. (Corr.)
„ de Puteña: *Centrolobium sp.; Papilionac.;* Pan.
„ Espino: *Pithecolobium tortum* Mart.; *Mimosac.;* Pan.
„ fruta: *Lafoensia punicifolia* DC.; *Lythrac.;* Pan.
„ prieto: *Ficus spp.; Morac.;* Mex.
„ real: *Terminalia obovata* Eich.; *Combretac.;* Pan.
Amarra jabón: *Neomillspaughia paniculata* Blake; *Polygonac.;* Hond.
Amarro „ : *Neomillspaughia paniculata* Blake; *Polygonac.;* Hond.
Amate: *Ficus glabrata* H. B. K.; *Morac.;* C.-Am.
„ : *Ficus oerstedtiana* Miq., *F.radula* Willd.; *Morac.;* br. Hond.
Amate amarillo: *Ficus spp.; Morac.;* Mex.
Amateltel: *Cocculus pendulus* Forst.; *Menispermac.;* Sahara.
Amatillo: *Ficus spp.; Morac.;* Mex.
Amatito de montana: *Ficus spp.; Morac.;* Mex.
Amato de hijo grande: *Ficus spp.; Morac.;* Salv.
Amava-buruhi: *Erythroxylon areolatum* L.; *Erythroxylac.;* Ind.
Ambabálud: *Neonauclea sp.; Rubiac.;* Phil.
Ambafolo: *Homalium africanum* Bth.; *Flacourtiac.;* Kam.
Ambai: *Cecropia adenopus* Mart.; *Morac.;* Arg.
„ guazú: *Didymopanax morototoni* Dcne. et Pl.; *Araliac.;* Arg.
Ambaiba: *Cecropia adenopus* Mart.; *Morac.;* Bras.
Ambal: *Phyllanthus emblica* L.; *Euphorbiac.;* O.-As.
Ambalam: *Spondias mangifera* Pers.; *Anacardiac.;* Ind., Burma.
Ambaloe: *Dysoxylum acutangulum* Miq.; *Meliac.;* Nied.-Ind.
Ambara: *Spondias mangifera* Pers.; *Anacardiac.;* Ind., Burma.
Ambarwood (u): *Liquidambar styraciflua* L.; *Hamamelidac.;* sö. USA., C.-Am.
Ambatschholz (d): *Aeschynomene elaphroxylon* Taub.; *Papilionac.;* O.-Af.
Ambauva brava: *Pourouma bicolor* Mart.; *Morac.;* Bras. (Amaz.)
„ de vinho: *Pourouma tomentosa* Mart.; *Morac.;* Bras. (Amaz.)
„ mirim de vinho: *Pourouma acuminata* Mart.; *Morac.;* Bras.; (Amaz.)
Ambaville: *Senecio ambavilla* Pers.; *Composit.;* Réunion.
„ : *Hypericum lanceolatum* Lamk.; *Hyperic.;* Réunion.
Ambay: *Cecropia peltata* L.; *Morac.;* tr. Am.
Ambaÿ: *Cecropia adenopus* Mart.; *Morac.;* Arg.
Ambeng chés: *Diospyros filipendula* Pierre; *Ebenac.;* Ind.-Ch.
„ thngé: *Heritiera littoralis* Dry.; *Sterculiac.;* Ind.-Ch.
Amberbaum (d): *Liquidambar styraciflua* L.; *Hamamelidac.;* s. N.-Am.
Ambit: *Elaeocarpus angustifolius* Blume; *Tiliac.;* Nied.-Ind.

Amboinamaser (d): *Pterocarpus sp.; Papilionac.;* Moluk.
Amboine (f): *Pterocarpus sp.; Papilionac.;* Moluk.
Amboine: *Flindersia amboinensis* Poir.; *Rutac.;* Moluk.
Ambonhout: *Pterocarpus sp.; Papilionac.;* Moluk.
Ambora: *Tambourissa thouvenoti* Danguy; *Monimiac.;* Mad.
Ambovitsika: *Pittosporum stenopetalum* Bak.; *Pittosporac.;* Mad.
Amboyna (u): *Pterocarpus sp.; Papilionac.;* Moluk.
Ambug: *Cecropia adenopus* Mart.; *Morac.;* Bras.
Ambúgis: *Koordersiodendron pinnatum* Merr.; *Anacardiac.;* Phil.
Ambuláuan: *Vitex parviflora* Juss.; *Verbenac.;* Phil. (Sulu-Arch.)
Amdali: *Drimycarpus racemosa* Hook. f.; *Anacardiac.;* ö. Himal.
Ameixa: *Ximenia americana* L.; *Olacac.;* Bras.
   „ de espinha: *Ximenia americana* L.; *Olacac.;* Bras.
Ameixero: *Ximenia americana* L.; *Olacac.;* Bras.
Amélanchier du Canada: *Amelanchier canadensis* Medic.; *Rosac.;* ö. N.-Am.
Amelliki: *Sarcocephalus esculentus* Afzel.; *Rubiac.;* S.-L.
Amemmai: *Tamarix gallica* L.; *Tamaricac.;* N.-Af.
Amendoeira: *Terminalia catappa* L.; *Combretac.;* Bras.
Amendoim: *Pterogyne nitens* Tul.; *Caesalpiniac.;* Bras. (Sao P.)
Amendro: *Terminalia catappa* L.; *Combretac.;* P.-R.
Ameneiro: *Alnus glutinosa* Gaertn.; *Betulac.;* Sp.
Ameraé-amoelaoe: *Mouriria anomala* Pulle; *Melastomatac.;* Sur.
   „   „   : *Mouriria plaschaerti* Pulle; *Melastomatac.;* Sur.
Amerau: *Mouriria anomala* Pulle, *M. plaschaerti* Pulle; *Melastomatac.;* Sur.
Améréré: *Erythrophloeum guineense* G. Don; *Caesalpiniac.;* Elf.
Amhidja: *Terminalia ivorensis* Chev.; *Combretac.;* Elf.
Amhio: *Cola cordifolia* H. Baill,; *Sterculiac.;* Elf.
Amia: *Calpocalyx klainei* Pierre; *Mimosac.;* Kam.
Ami-Gashi: *Lithocarpus amygdalifolia* Rehder; *Fagac.;* Form.
Amislág: *Securinega flexuosa* Muell. Arg.; *Euphorbiac.;* Phil.
Amizi: *Buchholzia coriacea* Engl.; *Capparidac.;* Elf.
Amlabaum (d): *Phyllanthus emblica* Gaertn.; *Euphorbiac.;* O.-As., Sunda.
Amla ka: *Phyllanthus emblica* Gaertn.; *Euphorbiac.;* O.-As.
Amli: *Tamarindus indica* L.; *Caesalpiniac.;* Ind.
   „   : *Bauhinia malabarica* Roxb.; *Caesalpiniac.;* O.-Ind.
Amlika: *Tamarindus indica* L.; *Caesalpiniac.;* Tr.
Amlika jhar: *Tamarindus indica* L.; *Caesalpiniac.;* Tr.
Amlilèce: *Rhamnus alaternus* L.; *Rhamnac.;* Arabien.
Amluk: *Diospyros lotus* L.; *Ebenac.;* Jap.
Amo: *Albizzia welwitschi* Oliv.; *Mimosac.;* Kam.
Amoelau: *Mouriria anomala* Pulle, *M. plaschaerti* Pulle; *Melastomatac.;* Sur.
Amoerau-balli: *Mouriria acutiflora* Naud.; *Melastomatac.;* Sur.
Amoguis: *Koordersiodendron pinnatum* Merr.; *Anacardiac.;* Phil.
Amoora: *Amoora cucullata* Roxb.; *Meliac.;* Beng.
Amonilla: *Berrya ammonilla* Roxb.; *Tiliac.;* s. Ind.
Amopola: *Bombax ellipticum* H. B. K.; *Bombacac.;* Mex.
Amor platonico: *Albizzia lebbek* Bth.; *Mimosac.;* Tr.
   „ seco: *Alchornea glandulosa* Poepp. et Endl.; *Euphorbiac.;* Arg. (Mis.)
   „   „ : *Heliocarpus americanus* L.; *Tiliac.;* Arg.
   „   „ guazú: *Heliocarpus americanus* L.; *Tiliac.;* Arg.
Amora branca: *Piratinera sp.; Morac.;* Bras. (Bahia)
Amoreira: *Chlorophora excelsa* Bth. et Hook.; *Morac.;* Angola.
   „   : *Chlorophora tinctoria* Gaudich.; *Morac.;* Bras.
   „ de espinho: *Chlorophora tinctoria* Gaudich.; *Morac.;* Bras.
Amorethout: *Piratinera guianensis* Aubl.; *Morac.;* Sur.
Amorpha, shrubby (e): *Amorpha fruticosa* L.; *Papilionac.;* N.-Am.
Amouk: *Cynometra aff. lujaei* de Wild.; *Caesalpiniac.;* Kam.
Amourette: *Piratinera guianensis* Aubl.; *Morac.;* fr. Gu.
   „   : *Drepanocarpus lunatus* G. F. W. Mey.; *Papilionac.;* fr. Gu.
Ampang: *Ouratea calophylla* Engl.; *Ochnac.;* Kam.
Ampelas: *Ficus ampelos* Burm.; *Morac.;* Java.
Ampil: *Tamarindus indica* L.; *Caesalpiniac.;* Ind.-Ch.

Ampil tuc: *Pithecolobium dulce* Bth.; *Mimosac.;* Ind.-Ch.
„ „ : *Inga dulcis* Willd.; *Mimosac.;* Ind.-Ch.
Ampong prahok: *Nothaphoebe umbelliflora* Blume; *Laurac.;* Ind.-Ch.
Amra: *Spondias mangifera* Pers.; *Anacardiac.;* Ind., Burma.
Amrataca: *Spondias mangifera* Pers.; *Anacardiac.;* Ind., Burma.
Amuan: *Saccoglottis gabonensis* Urb.; *Humiriac.;* Elf.
Amuáuan: *Vitex parviflora* Juss.; *Verbenac.;* Phil.
Amugán: *Pygeum sp.; Rosac.;* Phil.
Amugáuan: *Vitex parviflora* Juss.; *Verbenac.;* Phil.
Amúgis: *Bassia ramiflora* Merr.; *Sapotac.;* Phil.
„ : *Buchanania sp.; Anacardiac.;* Phil.
„ : *Garuga sp.; Burserac.;* Phil.
„ : *Koordersiodendron pinnatum* Merr.; *Anacardiac.;* Phil.
„ corriente: *Palaquium sp.; Sapotac.;* Phil.
Amuláuon: *Vitex pentaphylla* Merr.; *Verbenac.;* Phil. (Zamb.)
Amuñgiáng: *Pygeum sp.; Rosac.;* Phil.
Amur: *Amoora cucullata* Roxb.; *Meliac.;* Ind.
Amuráuon: *Vitex parviflora* Juss.; *Verbenac.;* Phil.
Amutsi: *Rhizophora racemosa* E. Mey.; *Rhizophorac.;* Go.
„ : *Avicennia nitida* Jacq.; *Verbenac.;* Go.
Amuyáuon: *Vitex parviflora* Juss.; *Verbenac.;* Phil.
Amúyong: *Ormosia calavensis* Azaola; *Papilionac.;* Phil.
Amvim: *Melocarpidium lepidotum* ?; *Anonac.;* Kam.
Amyris legitimo: *Amyris balsamifera* L.; *Rutac.;* Ven.
Amyriswood (hell): *Amyris balsamifera* L.; *Rutac.;* tr. Am.
Ana: *Leptadenia pyrotechnica* Dcne.; *Asclepiadac.;* Sahara.
Ana-akara: *Tapirira aff. guianensis* Aubl.; *Anacardiac.;* Sur.
Anáau: *Livistona rotundifolia* Mart.; *Palmac.;* Phil.
Anabíong: *Trema amboinensis* Blume; *Ulmac.;* Phil.
Anablíng: *Artocarpus rubrovenia* Warb.; *Morac.;* Phil.
Anabó: *Allaeanthus luzonicus* Vid.; *Morac.;* Phil.
Anabovahatra: *Rhizophora mucronata* Lam.; *Rhizophorac.;* Mad.
Anacagüite: *Sterculia sp.; Sterculiac.;* P.-R.
Anacahuitewood: *Cordia boissieri* A. DC.; *Borraginac.;* Mex.
Anacahuitholz: *Cordia boissieri* A. DC..; *Borraginac.;* Mex.
Anacardo da Americo: *Swietenia spp.; Meliac.;* tr. Am.
Anaco: *Erythrina sp.; Papilionac.;* Kol.
Anacoco: *Swartzia tomentosa* DC.; *Caesalpiniac.;* Gu.
„ wanebala: *Swartzia tomentosa* DC.; *Caesalpiniac.;* Bras.
Anagá: *Dehaasia triandra* Merr.; *Laurac.;* Phil.
Anagáp: *Nothaphoebe malabonga* Merr.; *Laurac.;* Phil.
Anagap: *Pithecolobium scutiferum* Bth.; *Mimosac.;* Phil.
Anagás: *Buchanania sp., Semecarpus sp.; Anacardiac.;* Phil.
Anagau: *Dillenia sp.; Dilleniac.;* Phil.
Anagep: *Litsea sp., Beilschmiedia cairocan* Vid.; *Laurac.;* Phil.
„ : *Albizzia acle* Merr.; *Mimosac.;* Phil.
Anagép: *Terminalia nitens* Presl; *Combretac.;* Phil.
Anagó-ñgisí: *Beilschmiedia cairocan* Vid.; *Laurac.;* Phil.
Anago-switie: *Tabebuia sp.; Bignoniac.;* Sur.
Anagós: *Litsea sp.; Laurac.;* Phil.
Anahau: *Livistona rotundifolia* Mart. *var. luzonensis* Becc.; *Palmac.;* Phil.
„ : *Livistona merrilli* Becc.; *Palmac.;* Phil.
Anaháuon: *Dipterocarpus spp.; Dipterocarpac.;* Phil.
Anaingüéri: *Malacantha robusta* Chev.; *Sapotac.;* Elf.
Anai-puliya-koy: *Adansonia digitata* L.; *Bombacac.;* Ind.
Anakara: *Inga spp.; Mimosac.;* Sur.
Anakin: ? *Virola surinamensis* Warb.; *Myristicac.;* br. Gu.
Anakoko: *Ormosia spp.; Papilionac.;* Sur.
Anakué: *Pycnanthus kombo* Warb.; *Myristicac.;* Elf.
Anám: *Buchanania sp.; Anacardiac.;* Phil.
Anamomilla: *Lovoa klaineana* Pierre; *Meliac.;* Nig.
Anamúchil: *Pithecolobium dulce* Bth.; *Mimosac.;* Mex.

Anan: *Fagraea fragrans* Roxb.; *Loganiac.;* Ind.
Anán: *Buchanania sp.; Anacardiac.;* Phil.
Ananbo: *Crypteronia pubescens* Bl.; *Lythrac.;* Burma.
Anandjo: *Chrysophyllum obovatum* Sabine; *Sapotac.;* Elf.
Anang: *Diospyros spp.; Ebenac.;* Phil.
Anánggi: *Canarium villosum* F. Vill.; *Burserac.;* Phil.
Anangilan: *Canangium odoratum* Baill.; *Anonac.;* Phil.
Anañgíran: *Canangium odoratum* Baill.; *Anonac.;* Phil.
Ananguéri: *Chrysophyllum obovatum* Sabine; *Sapotac.;* Elf.
Anani: *Citrus aurantium* L.; *Rubiac.;* Tahiti.
Ananiwana: *Hieronymia laxiflora* Muell. Arg.; *Euphorbiac.;* Sur..
Ananma: *Fagraea fragrans* Roxb.; *Loganiac.;* Burma.
Ananta: *Cynometra ananta* Hutch. et Dalz.; *Caesalpiniac.;* Go.
Anany: *Symphonia globulifera* L. f.; *Guttifer.;* Bras.
Ananyo: *Chrysophyllum obovatum* Sabine; *Sapotac.;* Elf.
Ananzo: *Albizzia fastigiata* Oliv.; *Mimosac.;* Elf.
Anaoura: *Licania spp.; Rosac.;* Sur.
Anaplá: *Albizzia procera* Bth.; *Mimosac.;* Phil.
Anatto: *Bixa orellana* L.; *Bixac.;* Hond.
Anau: *Livistona rotundifolia* Mart. var. *luzonensis* Becc.; *Palmac.;* Phil.
Anauerá: *Licania hypoleuca* Bth.; *Rosac.;* Bras.
Anaura: *Licania spp.; Rosac.;* Sur.
„ : *Couepia surinamensis* Kleinh.; *Rosac.;* Sur.
Anavinga: *Casearia sp.; Flacourtiac.;* Bras.
Anchico: *Piptadenia rigida* Bth.; *Mimosac.;* Arg.
„ blanca: *Cassia brasiliensis* Vog.; *Caesalpiniac.;* tr. Am.
„ blanco: *Pithecolobium hassleri* Chod.; *Mimosac.;* Arg.
„ „ : *Piptadenia nitida* Bth.; *Mimosac.;* Arg. (Mis. & Corr.)
„ colorado: *Acacia angico* Mart.; *Mimosac.;* C.-Am.
„ „ : *Piptadenia rigida* Bth.; *Mimosac.;* Arg.
Anchóan: *Cassia javanica* L.; *Caesalpiniac.;* Phil.
Ancoche: *Vallesia cymbaefolia* Ortega; *Apocynac.;* S.-Am.
Anda-Assu: *Joannesia princeps* Vell.; *Euphorbiac.;* Bras.
„ leboe: ? *Canarium sp.; Burserac.;* s. Sumatra.
Andaia Uichi: *Andira racemosa* Lam.; *Papilionac.;* Bras. (Amaz.)
Andara-Uichi: *Andira amazonum* Mart.; *Papilionac.;* Bras. (Amaz.)
Andaráian: *Alstonia scholaris* R. Br.; *Apocynac.;* Phil.
Andingding: *Mitragyne sp.; Rubiac.;* Kam.
Andira uchy: *Drepanocarpus inundatus* Mart.; *Papilionac.;* Bras.
Andiroba: *Carapa guianensis* Aubl.; *Meliac.;* Bras.
„ : *Carapa procera* DC., *C. surinamensis* Miq.; *Meliac.;* Sur.
„ Bordado: *Carapa guianensis* Aubl.; *Meliac.;* Bras.
„ branca: *Carapa guianensis* Aubl.; *Meliac.;* Bras.
„ carapa: *Carapa guianensis* Aubl.; *Meliac.;* fr. Gu.
„ saruba: *Carapa guianensis* Aubl.; *Meliac.;* Bras.
Andirova: *Carapa guianensis* Aubl.; *Meliac.;* Bras.
Andjec: *Ongokea klaineana* Pierre; *Olacac.;* Kam.
Andjejang: *Ricinodendron africanum* Muell. Arg.; *Euphorbiac.;* Kam.
Andjenda: *Cephaelis manni* Bth.; *Rubiac.;* Kam.
Andoc: *Irvingia gabonensis* Baill.; *Simarubac.;* Kam.
Andog: *Irvingia gabonensis* Baill.; *Simarubac.;* Kam.
Andogh: *Irvingia gabonensis* Baill.; *Simarubac.;* Gab.
Andok: *Irvingia gabonensis* Baill.; *Simarubac.;* Gab.
Andom: *Ficus cf. recurvata* de Wild.; *Morac.;* B.-K., Fernando Po.
Andoo: *Irvingia gabonensis* Baill.; *Simarubac.;* Kam.
Ando ongoué: *Desbordesia insignis* Pierre; *Simarubac.;* Kam.
Andoung (f): *Berlinia sp.; Caesalpiniac.;* Gab.
Andrèze: *Celtis madagascariensis* Boj.; *Ulmac.;* Réunion
Andubon: *Cornus nuttalli* Aud.; *Cornac.;* USA.
Anduli: *Diospyros chloroxylon* Roxb.; *Ebenac.;* O.-Ind.
Andum: *Berlinia sp.; Caesalpiniac.;* Kam.
Andys: *Juniperus drupacea* Labill.; *Conifer.;* Kl.-As.

Aneisi wiwiri: *Piper sp.; Piperac.;* fr. Gu.
Anemene: *Cassia timorensis* DC.; *Caesalpiniac.;* tr. Au., Cey.
Angan: *Fraxinus floribunda* Wall.; *Oleac.;* O.-Ind.
„ : *Dialium connaroïdes* Harms; *Caesalpiniac.;* Gab.
Angana: *Timonius jambosella* Thw.; *Rubiac.;* Malay.
Añgátuan: *Alangium meyeri* Merr.; *Cornac.;* Phil.
Angeco vermelho: *Pithecolobium spp.; Mimosac.;* Bras.
Angèk: *Ongokea klaineana* Pierre; *Olacac.;* Gab.
Angeleen: *Andira inermis* H. B. K.; *Papilionac.;* Trin.
Angelica: *Andira inermis* H. B. K.; *Papilionac.;* Mart.
„ do Pará: *Dicorynia paraënsis* Bth.; *Caesalpiniac.;* Bras.
Angelim: *Andira anthelmintica* Bth., *A. inermis* H. B. K.; *Papilionac.;* tr. Am
„ amargosa: ? *Andira vermifuga* Mart.; *Papilionac.;* Bras.
„ araroba: *Andira sp.; Papilionac.;* Bras.
„ Bark of: *Andira inermis* H. B. K.; *Papilionac.;* tr. Am.
„ de Para: *Andira amazonum* Mart.; *Papilionac.;* Bras. (Amaz.)
„ doce: *Andira fraxinifolia* Bth.; *Papilionac.;* Bras.
„ dos campos: ? *Andira vermifuga* Mart.; *Papilionac.;* Bras.
„ pedra: *Hymenolobium petraeum* Ducke; *Papilionac.;* Bras.
„ „ : *Andira spectabilis* Sald.; *Papilionac.;* Bras.
? „ preto: *Andira inermis* H. B. K.; *Papilionac.;* Bras.
„ rajado: *Pithecolobium racemiflorum* Ducke; *Mimosac.;* Bras.
„ „ : *Stryphnodendron flammatum* Kleinh.; *Mimosac.;* Bras.
„ rosa: *Platycyamus regnelli* Bth.; *Papilionac.;* Bras.
? „ vermelho: *Andira inermis* H. B. K.; *Papilionac.;* Bras.
Angelin: *Vouacapoua americana* Aubl.; *Papilionac.;* Sur.
„ : *Andira inermis* H. B. K.; *Papilionac.;* Trin., Tobago, Mart.
Angelino: ? *Nectandra discolor* Nees; *Laurac.;* Ven.
„ : *Homalium sp.; Flacourtiac.;* Ven.
„ : *Andira inermis* H. B. K.; *Papilionac.;* Kol.
„ laurel: *Homalium sp.; Flacourtiac.;* Ven.
Angélique: *Ocotea spp.; Laurac.;* Guad., Mart.
„ (f): *Dicorynia paraënsis* Bth.; *Caesalpiniac.;* Gu.
„ bâtard: *Dicorynia paraënsis* Bth.; *Caesalpiniac.;* fr. Gu.
„ franc: *Dicorynia paraënsis* Bth.; *Caesalpiniac.;* fr. Gu.
„ gris: *Dicorynia paraënsis* Bth.; *Caesalpiniac.;* fr. Gu.
„ rouge: *Dicorynia paraënsis* Bth.; *Caesalpiniac.;* fr. Gu.
Angel Sisa: *Caesalpinia pulcherrima* Sw.; *Caesalpiniac.;* Peru.
Angeuk: *Ongokea klaineana* Pierre; *Olacac.;* Gab.
Anggrit: *Nauclea lanceolata* Blume; *Rubiac.;* Java.
Angica do Sertao: *Piptadenia rigida* Bth.; *Mimosac.;* Bras.
„ vermelho: *Piptadenia rigida* Bth.; *Mimosac.;* Bras.
Angico: *Piptadenia spp.; Mimosac.;* Bras.
„ : *Peltophorum vogelianum* Bth.; *Caesalpiniac.;* Bras.
„ amarello: *Piptadenia sp.; Mimosac.;* Bras.
„ blanco: *Piptadenia rigida* Bth.; *Mimosac.;* Arg.
„ branco: *Piptadenia sp.; Mimosac.;* Bras.
„ coco: *Piptadenia sp.; Mimosac.;* Bras.
„ colorado: *Piptadenia rigida* Bth.; *Mimosac.;* Arg.
„ oscuro: *Piptadenia sp.; Mimosac.;* Bras.
„ preto: *Piptadenia rigida* Bth.; *Mimosac.;* Bras.
„ verdadeiro: *Piptadenia rigida* Bth.; *Mimosac.;* Bras.
„ vermelho: *Piptadenia rigida* Bth.; *Mimosac.;* Bras.
Angik: *Piptadenia rigida* Bth.; *Mimosac.;* Bras.
Angika: *Piptadenia rigida* Bth.; *Mimosac.;* Bras.
Angiroba: *Carapa guianensis* Aubl.; *Meliac.;* Bras.
Ang kanh: *Cassia siamea* Lam.; *Caesalpiniac.;* Ind.-Ch.
„ kea dey: *Sesbania grandiflora* Pers.; *Papilionac.;* Ind.-Ch.
„ kel: *Cassia siamea* Lam.; *Caesalpiniac.;* Ind.-Ch.
Angkol: *Eugenia formosa* Wall., ? *Careya globosa* ?; *Myrtac.;* Kamb.
Angkot khmau: *Diospyros ebenum* Koenig; *Ebenac.;* Ind.-Ch.
Angoc: *Ongokea gore* Engl.; *Olacac.;* Kam.

Angök: *Ongokea klaineana* Pierre; *Olacac.;* Kam.
Angokolo: *Sclerosperma walkeri* Chev.; *Palmac.;* Gab.
Angokoum: *Barteria dewevrei* de Wild. et Dur.; *Passiflorac.;* Gab.
Angolaholz (d): *Baphia nitida* Afzel.; *Papilionac.;* W.-Af.
Angona: *Vitex pachyphylla* Baker; *Verbenac.;* Gab.
Angongi: *Fillaeopsis discophora* Harms; *Mimosac.;* W.-Af.
Angongo: *Antrocaryon klaineanum* Pierre; *Anacardiac.;* Kam.
Angossa: *Markhamia tomentosa* K. Schum.; *Bignoniac.;* Kam.
Angouann: *Randia cladantha* K. Schum.; *Rubiac.;* Kam.
Angouaran: *Oldfieldia africana* Bth.; *Euphorbiac.;* Elf.
Angouma: *Aucoumea klaineana* Pierre; *Burserac.;* Gab.
Angoumea: *Aucoumea klaineana* Pierre; *Burserac.;* Gab.
Angoüssa: *Markhamia tomentosa* K. Schum.; *Bignoniac.;* Kam.
Angsana: *Pterocarpus indicus* Willd.; *Papilionac.;* Malay.
Angsanah: *Pterocarpus indicus* Willd.; *Papilionac.;* Burma, And.
Angsóan: *Cassia javanica* L.; *Caesalpiniac.;* Phil.
Angsóban: *Radermachera sp.; Bignoniac.;* Phil.
Angú: *Pterocarpus spp.; Papilionac.;* Bras.
Angueuk (f): *Ongokea klaineana* Pierre; *Olacac.;* Elf., Kam., Gab.
Angumá: *Aucoumea klaineana* Pierre; *Burserac.;* sp. Guin.
Anhauiana: *Cryptocarya densiflora* Nees; *Laurac.;* Bras.
Ani: *Pangium edule* Reinw.; *Bixac.;* Amboina.
Anibé: *Ochrocarpus africanus* Oliv.; *Guttifer.;* Elf.
„ : *Mammea africana* G. Don; *Guttifer.;* Elf.
Aníbong: *Oncosperma filamentosum* Blume; *Palmac.;* Phil.
Anibony: *Caryota urens* L.; *Palmac.;* Ind.
Anigá: *Vatica mangachapoi* Blanco; *Dipterocarpac.;* Phil.
Anigád: *Tristania sp.; Myrtac.;* Phil.
Anílai: *Lumnitzera littorea* Voigt; *Combretac.;* Phil.
Anil-assú: *Eupatorium laeve* DC.; *Composit.;* Bras.
Animbé: *Ochrocarpus africanus* Oliv.; *Guttifer.;* Elf.
„ : *Mammea africana* G. Don; *Guttifer.;* Elf.
Animebaum: *Hymenaea courbaril* L.; *Caesalpiniac.;* W.-Ind.
Anime blanca: *Protium guianense* March.; *Burserac.;* Kol.
„ blanco: *Protium heptaphyllum* March.; *Burserac.;* Kol.
„ comino: *Icica sp.; Burserac.;* Kol.
Aninaplá: *Albizzia procera* Bth.; *Mimosac.;* Phil.
Aningá: *Agathis alba* Foxw.; *Conifer.;* Phil.
Aningát: *Parinarium sp.; Rosac.;* Phil.
Aninggát: *Vatica mangachapoi* Blanco; *Dipterocarpac.;* Phil.
Aningnai: *Litchi philippinensis* Radlk.; *Sapindac.;* Phil.
Aning-ñgai: *Pygeum sp.; Rosac.;* Phil. (Zamboanga).
Aninguéri (f): *Malacantha robusta* Chev.; *Sapotac.;* Elf.
Aniokouéty: *Pachypodanthium staudti* Engl. et Diels; *Anonac.;* Elf.
Anionkéti: *Pachypodanthium staudti* Engl. et Diels; *Anonac.;* Elf.
Anioukéti (f): *Pachypodanthium staudti* Engl. et Diels; *Anonac.;* Elf.
Añipla: *Toona calantas* Merr. et Rolfe; *Meliac.;* Phil.
Anis étoilé (f): *Illicium verum* Hook.; *Magnoliac.;* Ind.-Ch.
Anislág: *Securinega flexuosa* Muell. Arg.; *Euphorbiac.;* Phil.
Anjan: *Hardwickia binata* Roxb.; *Caesalpiniac.;* Ind.
Anjananjana: *Leptochlaena multiflora* Dup. Thou.; *Chlaenac.;* Mad.
Anjek: *Ongokea kamerunensis* Engl.; *Olacac.;* Kam.
„ : *Ongokea klaineana* Pierre; *Olacac.;* Kam.
Anjera: *Enterolobium cyclocarpum* Gris.; *Caesalpiniac.;* Kol.
An-koe: *Buxus stenophylla* Hance; *Buxac.;* Ch.
Ankye Biri: *Phialodiscus unijugatus* Radlk.; *Sapindac.;* Go.
Ankyewa: *Phialodiscus unijugatus* Radlk.; *Sapindac.;* Go.
Ankyi: *Anopyxis ealaensis* Sprague; *Rhizophorac.;* Go.
Annchioin: *Cola dasysperma* Pierre; *Sterculiac.;* Gab.
Anne: *Alnus glutinosa* Gaertn.; *Betulac.;* Fr.
Anneb: *Ziziphus vulgaris* Lamk.; *Rhamnac.;* Tunis.
Anngokoum: *Barteria dewevrei* de Wild. et Dur.; *Passiflorac.;* Gab.

Anobilíng: *Artocarpus rubrovenia* Warb.; *Morac.;* Phil.
Anoeh: *Pterygota macrocarpa* K. Schum.; *Sterculiac.;* Kam.
Anoema latti: *Tapirira aff. guianensis* Aubl.; *Anacardiac.;* Sur.
Anohwere: ? *Oxytenanthera abyssinica* Munro; *Gramineae;* Go.
Anokué: *Guarea cedrata* Pell.; *Meliac.;* Elf.
Anón: *Anona glabra* L.; *Anonac.;* P.-R.
„ : *Rollinia mucosa* Baill.; *Anonac.;* P.-R.
Anona: *Anona glabra* L.; *Anonac.;* Mex.
„ : *Anona sp., Rollina jimenezi* Safford; *Anonac.;* C.-Am.
„ blanca: *Anona spp.; Anonac.;* Salv.
„ cincuya: *Anona sp.; Anonac.;* Guat.
„ colorada: *Anona spp.; Anonac.;* Salv.
„ de montaña: *Anona purpurea* Moç. et Ses.; *Anonac.;* Guat.
„ de redecilla: *Anona reticulata* L.; *Anonac.;* Hond.
„ poshte: *Anona spp.; Anonac.;* Salv.
Anonang: *Cordia myxa* L.; *Borraginac.;* O.-Ind.
Anonde escamas: *Anona spp.; Anonac.;* P.-R.
Anone: *Anona glabra* L.; *Anonac.;* Fl., W.-Ind.
„ á peau dure: *Anona scleroderma* Safford; *Anonac.;* Guat.
„ de montagne: *Anona testudinea* ?; *Anonac.;* Guat.
„ pourpre: *Anona purpurea* Moç. et Ses.; *Anonac.;* Guat.
„ reticulé: *Anona reticulata* L.; *Anonac.;* Guat.
„ squameuse: *Anona squamosa* L.; *Anonac.;* tr. Am.
Anonillo: *Anona glabra* L.; *Anonac.;* Guat., Hond.
„ : *Desmopsis panamensis* Safford; *Anonac.;* Pan.
Anora: *Pachira aquatica* Aubl.; *Bombacac.;* fr. Gu.
Anosép: *Carallia integerrima* DC.; *Rhizophorac.;* Phil.
„ : *Mimusops elengi* L.; *Sapotac.;* Phil.
An'Reyjan: *Canarium laxum* A. W. Benn.; *Burserac.;* Malakka.
Ansall sroû: *Elaeocarpus madopelatus* Pierre; *Tiliac.;* Ind.-Ch.
Anso-huma: *Ostryoderris leucobotrya* Dunn; *Papilionac.;* Go.
Ansu-dua: *Ostryoderris leucobotrya* Dunn; *Papilionac.;* Go.
Antaga: *Pachylobus edulis* G. Don; *Burserac.;* Gab.
Antagan: *Pterocarpus sbp.; Papilionac.;* Phil.
Antám: *Shorea guiso* Blume; *Dipterocarpac.;* Phil.
Antáng: *Canarium luzonicum* A. Gray; *Burserac.;* Phil.
Anté: *Desbordesia sp.; Simarubac.;* Kam.
Anténg: *Agathis alba* Foxw.; *Conifer.;* Phil.
Anteng: *Buchanania sp.; Anacardiac.;* Phil.
Anténg: *Canarium luzonicum* A. Gray, *C.villosum* F. Vill.; *Burserac.;* Phil.
Antipólo· *Artocarpus communis* Forst., *A.treculiana* Elm.; *Morac.;* Phil.
Ants' wood (u): *Bumelia angustifolia* Nutt.; *Sapotac.;* ö. Fl., Bah., Cuba.
Ant tree: *Triplaris caracasana* Cham.; *Polygonac.;* n. S.-Am., C.-Am.
Anubíng: *Artocarpus spp.; Morac.;* Phil.
Anubing-kadiós: *Gymnartocarpus woodi* Merr.; *Morac.;* Phil.
Anubing na nangká: *Gymnartocarpus woodi* Merr.; *Morac.;* Phil.
Anublíng: *Artocarpus cumingiana* Tréc., *A.rubrovenia* Warb.; *Morac.;* Phil.
Anula: *Phyllanthus emblica* L.; *Euphorbiac.;* O.-As., Ind.
Anwama: *Ricinodendron africanum* Muell. Arg.; *Euphorbiac.;* Go.
Anyeran: *Afrormosia elata* Harms, *A.laxiflora* Harms; *Papilionac.;* Nig.
Anyico colorado: *Piptadenia rigida* Bth.; *Mimosac.;* Arg. (Corr. & Mis.)
Anyoe: *Allanblackia klainei* Pierre; *Guttifer.;* Kam.
Anzala: *Mimusops congolensis* de Wild.; *Sapotac.;* Gab.
Anzeuk: *Ongokea klaineana* Pierre; *Olacac.;* Gab.
Aoatl: *Quercus sp.; Fagac.;* Mex.
Ao giri: *Sterculia platanifolia* L. f.; *Sterculiac.;* Jap.
Aohada: *Ilex macropoda* Miq.; *Aquifoliac.;* Jap.
Aoka: *Tecoma chrysantha* Jacq.; *Bignoniac.;* Nic.
Aokago: *Actinodaphne acuminata* Meissn.; *Laurac.;* Jap.
Aokiba: *Aucuba japonica* Thunb.; *Cornac.;* Jap.
Ao-giri: *Sterculia platanifolia* L.; *Sterculiac.;* Jap.
Aokiri: *Sterculia platanifolia* L.; *Sterculiac.;* Jap.

Aomer: *Phoenix dactylifera* L.; *Palmac.*; M.-M.
Aomoritodomatsu: *Abies mariesi* Mast.; *Conifer.*; Jap.
Aotago: *Fraxinus longicuspis* S. et Z.; *Oleac.*; Jap.
Aouamé: *Malacantha robusta* Chev.; *Sapotac.*; Elf.
Aouara: *Astrocaryum vulgare* Mart.; *Palmac.*; fr. Gu.
Aouara d'Afrique: *Elaeis guineensis* Jacq.; *Palmac.*; fr. Gu.
Aouassé: *Cola vera* K. Schum.; *Sterculiac.*; Elf.
Aoud ech-chouk: *Ilex aquifolium* L.; *Aquifoliac.*; Arabien.
Aoula: *Phyllanthus emblica* L.; *Euphorbiac.*; Ind.
Apa: *Eperua falcata* Aubl.; *Caesalpiniac.*; Bras.
„ : *Afzelia africana* Smith; *Caesalpiniac.*; Togo, sw. Nig.
„ akaniran: *Parkia pendula* Bth.; *Mimosac.*; Sur.
Apakanierian: *Stryphnodendron flammatum* Kleinh.; *Mimosac.*; Sur.
„ : *Stryphnodendron guianense* Bth.; *Mimosac.*; Sur.
Apa-kanilan: *Parkia pendula* Bth.; *Mimosac.*; Sur.
Apaka-paká: *Palaquium sp.*; *Sapotac.*; Phil.
Apakwie-ie: *Platonia insignis* Mart.; *Guttifer.*; Sur.
Apakyisie: *Bridelia micrantha* Baill.; *Euphorbiac.*; Go.
Apalachentee: *Ilex vomitoria* Ait.; *Aquifoliac.*; s. USA.
Apalan: *Couma guianensis* Aubl.; *Apocynac.*; Sur.
Apálang: *Bischofia javanica* Blume, *Cyclostemon sp.*; *Euphorbiac.*; Phil.
„ : *Planchonia spectabilis* Merr.; *Lecythidac.*; Phil.
Apálit: *Pahudia rhomboidea* Prain; *Caesalpiniac.*; Phil.
„ : *Pterocarpus spp.*; *Papilionac.*; Phil.
Apálong: *Euphoria cinerea* Radlk.; *Sapindac.*; Phil.
Apam: *Parinarium tenuifolium* Chev.; *Rosac.*; Go.
Apamate: *Tecoma pentaphylla* Juss.; *Bignoniac.*; tr. Am.
Ap ánh: *Celtis australis* L.; *Ulmac.*; Ind.-Ch.
Apaoewa: *Copaifera guianensis* Desf.; *Caesalpiniac.*; Sur.
Apapaya: *Phyllanthus discoideus* Muell. Arg.; *Euphorbiac.*; Go.
Aparahiú: *Mimusops spp.*; *Sapotac.*; Bras.
Apareiba: *Rhizophora mangle* L.; *Rhizophorac.*; Bras.
Apas: *Bridelia speciosa* Muell. Arg.; *Euphorbiac.*; Elf.
Apatoe-hout: *Swartzia tomentosa* DC.; *Caesalpiniac.*; Sur.
Apaya: *Turraeanthus sp.*; *Meliac.*; Go.
Apazeiro: *Eperua falcata* Aubl.; *Caesalpiniac.*; Bras.
Apé: *Anona spp.*; *Anonac*; Bras.
„ : *Pterygota cordifolia* Chev.; *Sterculiac.*; Elf.
Apehiripe: *Pterocarpus santalinoides* L'Hérit.; *Papilionac.*; Go.
Apenewa: *Khaya grandifolia* C. DC.; *Meliac.*; Go.
Apenkam: ? *Apeiba aspera* Aubl.; *Tiliac.*; fr. Gu.
Apepú: *Citrus aurantium* L.; *Rutac.*; Arg.
Apfelbaum: *Pirus malus* L.; *Rosac.*; Eu.
Apfel wild: *Pirus malus* L.; *Rosac.*; Eu., w. As.
Aphobesam: *Mitragyne macrophylla* Hiern; *Rubiac.*; Kam.
Api-api: *Avicennia officinalis* L.; *Verbenac.*; Phil.
Apicsie-ie: *Acrodiclidium canella* Mez; *Laurac.*; Sur.
„ „ : *Acrodiclidium chrysophyllum* Meissn.; *Laurac.*; Sur.
Apiesie-ie: *Nectandra sp.*, *Ocotea sp.*; *Laurac.*; Sur.
Apiesie-ie a blaakawan: *Nectandra sp.*, *Ocotea sp.*; *Laurac.*; Sur.
? Apihirape: *Pterocarpus santalinoides* L'Hérit.; *Papilionac.*; Go.
Apikara: *Campsiandra laurifolia* Bth.; *Caesalpiniac.*; br. Gu.
Apiníg: *Eugenia xanthophylla* C. B. Rob.; *Myrtac.*; Phil.
Apiranga: *Mouriria sp.*; *Melastomatac.*; Bras.
Apiri: *Dodonaea viscosa* Jacq.; *Sapindac.*; Tahiti.
Apisie-ie: *Nectandra sp.*, *Ocotea sp.*; *Laurac.*; Sur.
Apítan: *Alangium longiflorum* Merr.; *Cornac.*; Phil.
Apitáng: *Pygeum sp.*; *Rosac.*; Phil.
Apítong: *Dipterocarpus spp.*; *Dipterocarpac.*; Phil.
Apiy: *Ficus maximiliana* Mart.; *Morac.*; Bras.
Apnít: *Anisoptera thurifera* Blume; *Dipterocarpac.*; Phil.
„ : *Parashorea plicata* Brand, *P. malaanonan* Merr.; *Dipterocarpac.*; Phil.

Apnít: *Pentacme contorta* Merr. et Rolfe; *Dipterocarpac.;* Phil.
„ : *Shorea malaanonan* Blume; *Dipterocarpac.;* Phil.
Apo: *Cola vera* K. Schum.; *Sterculiac.;* Elf.
Apobo: *Lovoa klaineana* Pierre; *Meliac.;* Nig.
Apoekoetja: *Aspidosperma nitidum* Bth., *A. excelsum* Bth.; *Apocynac.;* Sur.
Apoeroekonnie: *Inga alba* Willd., *I. sertulifera* DC.; *Mimosac.;* Sur.
Apoetoe: *Swartzia tomentosa* DC.; *Caesalpiniac.;* Sur.
Apog: *Pterygota macrocarpa* K. Schum.; *Sterculiac.;* Kam.
Apohita: *Aspidosperma spp.; Apocynac.;* Sur.
Apokoita: *Aspidosperma nitidum* Bth., *A. excelsum* Bth.; *Apocynac.;* Sur.
Apollimerie: *Piratinera guianensis* Aubl.; *Morac.;* Sur.
Apomé (f): *Cynometra ananta* Hutch. et Dalz.; *Caesalpiniac.;* Elf.
Apompo: *Pachira macrocarpa* Walp.; *Bombacac.;* Mex.
Apopo: *Lovoa klaineana* Pierre; *Meliac.;* Nig.
Aporosa: *Aporosa oblonga* Muell. Arg.; *Euphorbiac.;* Assam.
Apose: *Cnestis ferruginea* DC.; *Connarac.;* Go.
Apóstola: *Cassia javanica* L.; *Caesalpiniac.;* Phil.
Apoto ajawa: *Protium spp.; Burserac.;* Sur.
Apoupé: *Pachylobus balsamifera* Guill.; *Burserac.;* Aeq.-Af.
Apple: *Pirus malus* L.; *Rosac.;* gemäß. Zonen, Subtr.
„ : *Angophora intermedia* DC.; *Myrtac.;* Au.
„ : *Angophora subvelutina* F. v. M.; *Myrtac.;* Au.
„ african: *Ochrocarpus africanus* Oliv.; *Guttifer.;* Lib.
„ Akee: *Blighia sapida* Koenig; *Sapindac.;* Go., Lib.
„ Alligator: *Anona spp.; Anonac.;* Fl., br. W.-Ind.
„ black: *Sideroxylon australe* Bth. et Hook. f.; *Sapotac.;* Queensl.
„ broad-leaved: *Angophora subvelutina* F. v. M.; *Myrtac.;* sö. Au.
„ Bull: *Sideroxylon sp.; Sapotac.;* br. W.-Ind.
„ Bush: *Heinsia pulchella* K. Schum.; *Rubiac.;* Lib.
„ Crab: *Schizomeria ovata* D. Don; *Cunoniac.;* Au.
„ „ : *Pirus coronaria* L., *P. angustifolia* Ait.; *Rosac.;* USA.
„ „ American: *Pirus coronaria* L.; *Rosac.;* USA. (Ontario)
„ „ Oregon: *Pirus rivularis* Dougl.; *Rosac.;* USA. (Washington.)
„ „ Western: *Pirus rivularis* Dougl.; *Rosac.;* Brit. Columbia.
„ Crow's: *Owenia venosa* F. v. M.; *Meliac.;* Au.
„ Custard: *Asimina triloba* Dun.; *Anonac.;* ö. N.-Am.
„ „ wild: *Anona squamosa* L., *A. glabra* L.; *Anonac.;* br. Hond., Fl., W.-Ind.
„ Elephant: *Feronia elephantum* Correa; *Rutac.;* O.-Ind., Cey.
„ Emu: *Owenia acidula* F. v. M., *O. venosa* F. v. M.; *Meliac.;* Au.
„ „ : *Petalostigma quadriloculare* F. v. M.; *Euphorbiac.;* Au.
„ Mammee, african: *Ochrocarpus africanus* Oliv.; *Guttifer.;* Go.
„ Monkey: *Anona spp.; Anonac.;* Fl., br. W.-Ind.
„ „ : *Posoqueria latifolia* Roem. et Schult.; *Rubiac.;* Pan.
„ „ : *Anisophyllea laurina* R. Br.; *Rhizophorac.;* S.-L.
„ Mooley: *Owenia acidula* F. v. M.; *Meliac.;* Queensl.
„ Mountain: *Eugenia malaccensis* L.; *Myrtac.;* Pan.
„ Narrow-leaved: *Angophora intermedia* DC.; *Myrtac.;* ö. Au.
„ Pitch: *Clusia rosea* Jacq.; *Guttifer.;* br. W.-Ind.
„ Pond: *Anona glabra* L.; *Anonac.;* Fl., W.-Ind.
„ Red: *Eugenia brachyandra* Maid. et Betche; *Myrtac.;* Au.
„ Rose: *Eugenia jambos* L.; *Myrtac.;* Hond.
„ smooth-barked: *Angophora lanceolata* Cav.; *Myrtac.;* N.-S.-W.
„ Spurious: *Angophora intermedia* DC.; *Myrtac.;* Au.
„ Star: *Chrysophyllum roxburghi* G. Don; *Sapotac.;* Ind., Sunda.
„ „ : *Chrysophyllum cainito* L.; *Sapotac.;* br. Hond.
„ „ common: *Chrysophyllum cainito* L.; *Sapotac.;* C.-Am.
„ „ white: *Chrysophyllum cainito* L.; *Sapotac.;* br. Hond.
„ „ wild: *Chrysophyllum mexicanum* Brandeg.; *Sapotac.;* br. Hond.
„ „ „ : *Chrysophyllum argenteum* Jacq.; *Sapotac.;* Pan.
„ Sugar, wild: *Rollinia multiflora* Splt.; *Anonac.;* Fl., Trin., Tob.
„ „ „ : *Rollinia mucosa* Baill.; *Anonac.;* Fl., Trin., Tob.
„ Water: *Anona spp.; Anonac.* Fl., br. W.-Ind.

Apple White: *Eugenia grandis* Wight; *Myrtac.*; Queensl.
„ Wild: *Pirus malus* L.; *Rosac.*; Eu., w. As., Himal.
„ Wood-: *Feronia elephantum* Correa; *Rutac.*; O.-Ind., Cey.
Apraiú: *Mimusops spp.*; *Sapotac.*; Bras.
Apraua: *Mimusops spp.*; *Sapotac.*; Bras.
Apricot African: *Ochrocarpus africanus* Oliv.; *Guttifer.*; Lib.
„ Wild: *Mammea americana* L.; *Guttifer.*; W.-Ind.
Aprikosenbaum: *Prunus armeniaca* L.; *Rosac.*; Eu.
„ amerikanischer: *Mammea americana* L.; *Guttifer.*; W.-Ind.
Aprikose von St. Domingo: *Mammea americana* L.; *Guttifer.*; W.-Ind., tr. Am.
Aprokonie: *Inga alba* Willd., *I. sertulifera* DC.; *Mimosac.*; Sur.
Aprono: *Mansonia altissima* Chev.; *Sterculiac.*; Go.
Apuhy grande: *Coussapoa nitida* Miq.; *Morac.*; Bras.
Apukur: *Conopharyngia crassa* Stapf; *Apocynac.*; Elf.
Apuñgá: *Terminalia comintana* Merr.; *Combretac.*; Phil.
Apurokuma: *Panda oleosa* Pierre; *Pandac.*; Go.
Aquaba: ? *Klainedoxa sp.*; *Simarubac.*; Elf.
Aquapou: *Irvingia sp.*; *Simarubac.*; Elf.
Aquatapana: *Chytroma idatimon* Miers; *Lecythidac.*; Trin.
Aque: *Corynanthe brachythyrsus* K. Schum.; *Rubiac.*; Kam.
Aquiche: *Guazuma ulmifolia* Lam.; *Sterculiac.*; Mex.
Aquilon: *Laugeria resinosa* Vahl; *Rubiac.*; P.-R.
Aqulugin: *Aquilaria agallocha* Roxb.; *Thymelaeac.*; O.-Ind.
„ : *Aquilaria malaccensis* Lam.; *Thymelaeac.*; O.-Ind.
Ara: *Ficus spp.*; *Morac.*; Malay.
Ara-ara-koereroe: *Xylopia spp.*; *Anonac.*; Sur.
Ara atroeka: *Cordia tetrandra* Aubl.; *Borraginac.*; Sur.
Arában: *Eugenia benthami* A. Gray; *Myrtac.*; Phil.
„ : *Litsea sp.*; *Laurac.*; Phil.
Arabmétu: *Aphanocalyx sp.*; *Caesalpiniac.*; Elf.
Arabo: *Erythroxylon areolatum* L.; *Erythroxylac.*; Cuba.
„ colorado: *Erythroxylon obovatum* L.; *Erythroxylac.*; Cuba.
Arabutan: *Caesalpinia echinata* Lam.; *Caesalpiniac.*; C.-Am.
Araçá: *Psidium araça* Raddi; *Myrtac.*; Bras.
„ d'agua: *Terminalia aff. januarensis* DC.; *Combretac.*; Bras.
„ de igapo: *Eugenia sp.*; *Myrtac.*; Bras.
„ do campo: *Psidium araça* Raddi; *Myrtac.*; Bras. (Amaz.)
„ do matto: *Psidium sp.*; *Myrtac.*; Bras. (Amaz.)
„ -guazú: *Psidium laurifolium* R. Br.; *Myrtac.*; Arg. (Mis.)
„ piranga: *Psidium acutangulum* DC.; *Myrtac.*; Bras. (Sao P.)
„ „ : *Eugenia bagensis* Berg; *Myrtac.*; Bras. (Sao P.)
„ preto: *Psidium sp.*; *Myrtac.*; Bras.
Arachichú: *Anona spp.*; *Anonac.*; Arg.
„ -guazú: *Anona spinescens* Mart.; *Anonac.*; Arg. (Mis.)
Aracuhy: ? *Andira vermifuga* Mart.; *Papilionac.*; Bras.
Aracui: ? *Andira vermifuga* Mart.; *Papilionac.*; Bras.
Aragashi: *Quercus glauca* Thunb.; *Fagac.*; Jap.
Araguaney: *Tecoma chrysantha* DC.; *Bignoniac.*; Ven.
Araguato: *Calycophyllum candidissimum* DC.; *Rubiac.*; Ven.
Arahara: *Phylloxylon perrieri* Drake; *Papilionac.*; Mad.
Arahoni: *Tecoma spp.*; *Bignoniac.*; fr. Gu.
Araká: *Palaquium sp.*; *Sapotac.*; Phil.
Arakachi: *Quercus acuta* Thunb.; *Fagac.*; Jap.
Arakadak: *Byrsonima ceranthera* Bth., *B. rugosa* Bth.; *Malpighiac.*; br. Gu.
Arakadako: *Byrsonima rugosa* Bth.; *Malpighiac.*; br. Gu.
Arakashi: *Quercus glauca* Thunb.; *Fagac.*; Jap., Form.
Aralee: *Clusia rosea* Jacq.; *Guttifer.*; Trin.
Arali matepalo: *Clusia rosea* Jacq.; *Guttifer.*; Trin.
Aralu: *Terminalia chebula* Retz.; *Combretac.*; Ind.
Aramata: *Clathrotropis brachypetalum* Kleinh.; *Papilionac.*; Sur.
Aramatta: *Diplotropis brachypetalum* Tul.; *Papilionac.*; br. Gu.
„ : *Diplotropis sp.*; *Papilionac.*; br. Gu.

Aramby: *Eugenia montana* Wight; *Myrtac.;* Ind.
Aramillo: ? *Lysiloma sp.; Mimosac.;* Pan.
Aramo (f): *Mammea africana* G. Don; *Guttifer.;* Elf.
„ „ : *Parinarium robustum* Oliv.; *Rosac.;* Elf.
„ „ : *Mitragyne macrophylla* Hiern; *Rubiac.;* Elf.
Aramon: *Parinarium robustum* Oliv.; *Rosac.;* Elf.
Arandano: *Vaccinium myrtillus* L.; *Ericac.;* Sp.
Arang: *Diospyros spp.*, Maba spp., *Ebenac.;* Malay.
„ : *Eugenia similis* Merr.; *Myrtac.;* Phil.
Aranga: *Homalium luzoniense* F. Vill.; *Flacourtiac.;* Phil.
Aráñgan: *Homalium bracteatum* Bth., *H. luzoniense* F. Vill.; *Flacourtiac.;* Phil.
Aráñgas: *Buchanania sp.; Anacardiac.;* Phil.
Aráñgen: *Aglaia diffusa* Merr., *A. turczaninowi* C. DC.; *Meliac.;* Phil.
Arangen: *Ganophyllum sp.; Burserac.;* Phil.
Aráñges: *Buchanania sp.; Anacardiac.;* Phil.
Arang-mill: *Eucalyptus terminalis* F. v. M.; *Myrtac.;* sö. Au.
Aranígan: *Canangium odoratum* Baill.; *Anonac.;* Phil.
Arapary vermelho: *Elizabetha paraënsis* Ducke; *Caesalpiniac.;* Bras.
Arapawané: *Eschweilera sp.*, *Lecythis sp.; Lecythidac.;* Sur.
Arapiraca: *Piptadenia sp.; Mimosac.;* Bras.
Arapoca: *Raputia magnifica* Engl.; *Rutac.;* Bras. (Sao P.)
„ amarella: *Raputia magnifica* Engl.; *Rutac.;* Bras.
„ „ : *Galipea dichotoma* Allem.; *Rutac.;* Bras.
„ branca: *Raputia alba* Engl.; *Rutac.;* Bras.
Araque: *Iriartea fusca* Drude; *Palmac.;* Ven.
Arâr: *Callitris quadrivalvis* Vent.; *Conifer.;* Marok.
Arara: *Joannesia heveoides* Ducke; *Euphorbiac.;* Bras.
„ : *Bocagea virgata* Bth. et Hook.; *Anonac.;* Sur.
„ : *Uvaria sp.*, *Anaxagorea sp.; Anonac.;* Demerara, Jam.
„ : *Unonopsis sp.*, *Duguetia sp.; Anonac.;* Demerara, Jam.
Ararabi: *Boswellia dalzieli* Hutch.; *Burserac.;* n. Nig.
Araracanga: *Aspidosperma aff. desmanthum* Muell. Arg.; *Apocynac.;* Bras. (Para)
Araragi: *Taxus cuspidata* S. et Z.; *Conifer.;* Jap.
Araragni: *Ilex latifolia* Thunb.; *Aquifoliac.;* Jap.
Arara-koereroe: *Xylopia spp.; Anonac.;* Sur.
Ararama: *Sclerolobium sp.; Caesalpiniac.;* Sur.
Arareua: *Sickingia sp.; Rubiac.;* Bras.
Arargé: *Taxus cuspidata* S. et Z.; *Conifer.;* Jap.
Araríba: *Sickingia sp.; Rubiac.;* Bras.
Araribá: *Centrolobium robustum* Mart.; *Papilionac.;* Bras. (Sao P.)
„ amarella: *Sickingia oliveri* K. Schum.; *Rubiac.;* Bras. (R. J.)
„ amarello: *Centrolobium robustum* Mart.; *Papilionac.;* Bras. (Sao P.)
„ „ : *Centrolobium tomentosum* Guill.; *Papilionac.;* Bras. (Sao P.)
„ branca: *Sickingia viridifolia* K. Schum.; *Rubiac.;* Bras. (R. J.)
„ „ : *Centrolobium sp.; Papilionac.;* Bras. (R. J., Sao P.)
„ rosa: *Sickingia tinctoria* K. Schum.; *Rubiac.;* Bras. (Amaz.)
„ „ : *Centrolobium tomentosum* Guill.; *Papilionac.;* Bras.
„ „ : *Centrolobium robustum* Mart.; *Papilionac.;* fr. Gu.
„ roxo: *Sickingia sp.; Rubiac.;* Bras.
Arariba vermelha: *Sickingia rubescens* K. Schum.; *Rubiac.;* Bras. (R. J.)
„ vermelho: *Centrolobium sp.; Papilionac.;* fr. Gu.
Araroba: *Andira araroba* Aguiar; *Papilionac.;* Bras.
„ : *Centrolobium robustum* Mart.; *Papilionac.;* Bras.
Araroeira: *Astronium urundeuva* Engl.; *Anacardiac.;* Bras. (Sao P.)
Arasaru: *Andira sp.; Papilionac.;* br. Gu.
Arasay: *Psidium guajava* L. *var. piriferum* Raddi; *Myrtac.;* S.-Am.
Arassou: *Calligonum comosum* L'Hérit.; *Polygonac.;* Sahara.
Arata: *Minquartia guianensis* Aubl.; *Olacac.;* Sur.
Aratahoedoe: *Minquartia guianensis* Aubl.; *Olacac.;* Sur.
Araticú: *Cordia salicifolia* Cham.; *Borraginac.;* Arg. (Mis.)
„ : *Anona spp.; Anonac.* Bras.
„ apé: *Anona sp.; Anonac.;* Bras.

Araticú d'agua: *Anona sp.; Anonac.;* Bras.
„ de espino: *Anona sp.; Anonac.;* Bras.
„ do brejo: *Anona palustris* L.; *Anonac.;* Bras.
„ „ mangue: *Anona sp.; Anonac.;* Bras.
„ „ matto: *Anona sp.; Anonac.;* Bras.
„ dulce: *Anona sp.; Anonac.;* Arg.
„ grande: *Anona sp.; Anonac.;* Bras.
„ guazú: *Cordia salicifolia* Cham.; *Borraginac.;* Arg. (Corr.)
„ „ : *Cordia sp.; Borraginac.;* Arg.
„ „ : *Anona silvestris* St. Hil.; *Anonac.;* Par.
„ mi: *Anona sp.; Anonac.;* Par.
„ nu: *Anona sp.; Anonac.;* Par.
„ panan: *Anona sp.; Anonac.;* Bras.
„ pedra: *Anona sp.; Anonac.;* Bras.
„ ponbê: *Anona sp.; Anonac.;* Bras.
Araticum do brejo: *Anona palustris* L.; *Anonac.;* Bras.
„ fruta de páo: *Talauma sp.; Magnoliac.;* Bras.
Aratikú: *Rollinia emarginata* Schlecht.; *Anonac.;* Arg.
Aratroeka: *Cordia tetrandra* Aubl.; *Borraginac.;* Sur.
Aratta: *Minquartia guianensis* Aubl.; *Olacac.;* Sur.
Aratte: *Minquartia guianensis* Aubl.; *Olacac.;* fr. Gu.
Araucaria: *Araucaria imbricata* Pav.; *Conifer.;* Chile.
Araukarienholz (d): *Araucaria spp.; Conifer.;* S.-Am., Au.
Araumatta: *Diplotropis sp.; Papilionac.;* br. Gu.
Araurama: *Sclerolobium paniculatum* Vogel; *Caesalpiniac.;* Sur.
Araúva: *Centrolobium sp.; Papilionac.;* Bras. (R. J., Sao P.)
Arawata-mo-èrèrè: *Pithecolobium pedicellare* Bth.; *Mimosac.;* Sur.
Arazá: *Eugenia guabijú* Berg; *Myrtac.;* Arg. (Mis., Corr.)
Araza: *Psidium cattleyanum* Sabine; *Myrtac.;* Urug.
„ guazú: *Myrcia sp.; Myrtac.;* Arg.
„ -hay: *Psidium arasa-hu* Pdi.; *Myrtac.;* Arg. (Form., Chaco)
„ -hu: *Psidium arasa-hu* Pdi.; *Myrtac.;* Arg. (Form., Chaco)
„ ñuati: *Machaonia spinosa* Cham, et Schlecht.; *Rubiac.;* Arg. (Chaco)
„ puitá: *Psidium guajava* Raddi; *Myrtac.;* Arg. (Mis.)
„ -saiyú: *Psidium cattleyanum* Sabine; *Myrtac.;* Arg. (Mis., Corr.)
Arbo del ajo: *Cordia alliodora* Cham.; *Borraginac.;* Peru, Bras.
Arbol blanco: *Prosopis sp.; Mimosac.;* Arg.
„ de agi: *Drimys winteri* Forst.; *Magnoliac.;* Ven.
„ de corcho: *Anona glabra* L.; *Anonac.;* Mex.
„ de diablo: *Hura sp.; Euphorbiac.;* Mex.
„ de Judas: *Cercis siliquastrum* L.; *Caesalpiniac.;* Arg.
„ de Judea: *Cercis siliquastrum* L.; *Caesalpiniac.;* Arg.
„ del aceita de Maria: *Calophyllum calaba* Jacq.; *Guttifer.;* Cuba.
„ „ algodon: *Ceiba pentandra* Gaertn.; *Bombacac.;* Mex.
„ de la muerte: *Hippomane mancinella* L.; *Euphorbiac.;* Mex.
„ de la pimienta: *Schinus molle* L.; *Anacardiac.;* Arg.
„ de las orejas: *Enterolobium cyclocarpum* Gris.; *Mimosac.;* Cuba.
„ del balsamo: *Toluifera pereira* Baill.; *Papilionac.;* Peru.
„ „ leche: *Sapium aucuparium* Jacq. *var. stenophyllum* Griseb.; *Euphorbiac.;* S.-Am.
„ de lluvia: *Lonchocarpus albiflorus* Hassl.; *Papilionac.;* Par., Arg.
„ del mate: *Ilex paraguariensis* St. Hil.; *Aquifoliac.;* S.-Am.
„ del Peru: *Schinus molle* L.; *Anacardiac.;* Mex.
„ de maría: *Calophyllum sp.; Guttifer.;* Kol.
„ de orejas: *Enterolobium cyclocarpum* Gris.; *Mimosac.;* Salv.
„ de sal: *Avicennia nitida* Jacq.; *Verbenac.;* Salv.
„ „ sangre de drago: *Croton succirubrus* Parodi; *Euphorbiac.;* S.-Am.
„ „ sombra: *Phytolacca dioica* L.; *Phytolaccac.;* Tener.
„ „ ule: *Castilla sp.; Morac.;* Mex.
„ madre: *Erythrina corallodendron* L.; *Papilionac.;* tr. Am.
„ negro: *Prosopis nigra* Hieron.; *Mimosac.;* Arg. (San Juan)
„ paraiso: *Melia azedarach* L.; *Meliac.;* Mex.
„ santo: ? *Guaiacum palmeri* Vail.; *Zygophyllac.;* Mex.

Arboldelpan: *Artocarpus communis* Forst.; *Morac.;* Malay.
Arbolito: *Phyllanthus acidus* Skeels; *Euphorbiac.;* Kol.
Arboloco: *Montanoa lehmanni* Blake; *Composit.;* Kol.
„ : *Montanoa morißiana* Sch. Bip.; *Composit.;* Kol.
Arbor del diablo: *Morisonia americana* L.; *Capparidac.;* Mex.
Arbor vitae: *Biota orientalis* Endl.; *Conifer.;* Jap.
„ „ : *Thuja occidentalis* L.; *Conifer.;* N.-Am.
„ „ giant: *Thuja plicata* Donn; *Conifer.;* w. Kan.
„ „ Pacific: *Thuja gigantea* Nutt.; *Conifer.;* nw. N.-Am.
Arbousier: *Arbutus unedo* L.; *Ericac.;* Fr.
„ de Menzies: *Arbutus menziesi* Pursh; *Ericac.;* w. N.-Am.
Arbre à ail: *Hua gaboni* Pierre; *Sterculiac.;* Gab.
„ „ „ : *Scorodophloeus zenkeri* Harms; *Caesalpiniac.;* Kam., Gab.
„ „ baume: *Canarium commune* L.; *Burserac.;* O.-As.
„ „ „ : *Bursera gummifera* Jacq.; *Burserac.;* Ant.
„ „ beurre (f): *Pentadesma butyracea* Sabine; *Guttifer.;* Gab.
„ „ „ : *Caryocar glabrum* Pers.; *Caryocarac.;* fr. Gu.
„ „ „ : *Caryocar tomentosum* Willd.; *Caryocarac.;* fr. Gu.
„ „ bombes: *Couroupita guianensis* Aubl.; *Lecythidac.;* Trin.
„ „ brai: *Humiria spp.; Humiriac.;* fr. Gu.
„ „ caoutchouc (f): *Funtumia elastica* Stapf; *Apocynac.;* Elf.
„ „ chapelets: *Melia azedarach* L.; *Meliac.;* Guad.
„ „ chou: *Andira inermis* H. B. K.; *Papilionac.;* tr. Am.
„ „ cire: *Gardenia aubryi* Vieill.; *Rubiac.;* N.-Kal.
„ „ concombres (f): *Magnolia spp.; Magnoliac.;* USA.
„ „ coton (f): *Ceiba pentandra* Gaertn.; *Bombacac.;* Tr.
„ „ encens (f): *Boswellia serrata* Roxb.; *Burserac.;* Ind.
„ „ fourmis: *Barteria spp.; Passiflorac.;* Gab.
„ „ graisse (f): *Mimusops djave* Engl.; *Sapotac.;* Gab.
„ „ huile de la Guyane: *Carapa guianensis* Aubl.; *Meliac.;* fr. Gu.
„ „ la fièvre: *Vismia guianensis* Pers.; *Hypericac.;* Gu.
„ „ lait: *Tabernaemontana utilis* Wight et Arn.; *Apocynac.;* fr. Gu.
„ „ laque: *Melanorrhoea laccifera* Pierre; *Anacardiac.;* O.-As.
„ „ l'oselle: *Andromeda arborea* L.; *Ericac.;* N.-Am.
„ „ noix longues: *Juglans cinerea* L.; *Juglandac.;* ö. Kan.
„ „ pagaies: *Staudtia gabunensis* Warb.; *Myristicac.;* Gab.
„ „ pain (f): *Treculia africana* Decne.; *Morac.;* Elf.
„ „ „ : *Artocarpus incisa* Forst.; *Morac.;* Elf.
„ „ „ : *Artocarpus communis* Forst.; *Morac.;* Tr.
„ „ „ à graines: *Artocarpus integrifolia* L.; *Morac.;* fr. Gu.
„ „ „ d'Afrique: *Treculia africana* Decne.; *Morac.;* Gab.
„ „ papier du Tonkin: *Rhamnoneuron balansae* Gilg; *Thymelaeac.;* Tonk.
„ „ savon: *Sapindus saponaria* L.; *Sapindac.;* Ant.
„ „ singe (f): *Myrianthus arboreus* Beauv.; *Morac.;* Gab.
„ „ suif (f): *Sapium sebiferum* Roxb.; *Euphorbiac.;* s. USA, W.-Ind., Ch., Ind.
„ „ „ : *Virola sp.; Myristicac.;* Gu.
„ „ „ du Gabon (f): *Pycnanthus kombo* Warb.; *Myristicac.;* Gab.
„ „ tabous: ? *Pycnandra benthami* Baill.; *Sapotac.;* N.-Kal.
„ „ Toung: *Aleurites fordi* Hemsl.; *Euphorbiac.;* Ch.
„ aux Epices (f): *Xylopia frutescens* Aubl.; *Anonac.;* tr. S.-Am.
„ „ palabres: *Hibiscus sterculiaefolius* Steud.; *Malvac.;* W.-Af.
„ -candélabre: *Pterochrosia vieillardi* Baill.; *Apocynac.;* N.-Kal.
„ -carotte: *Myodocarpus fraxinifolius* Brongn. et Gris.; *Araliac.;* N.-Kal.
„ cathédrale: *Irvingia sp.; Simarubac.;* Elf.
„ de Judée: *Cercis siliquastrum* L.; *Caesalpiniac.;* s. Eu., sw. As.
„ de la liberté: *Uvaria longifolia* R. et P.; *Anonac.;* Ind.
„ „ „ „ : *Uvaria odorata* Lam.; *Anonac.;* Ind.
„ des Upas: *Antiaris toxicaria* Lesch.; *Morac.;* Ind., Malay, Gab.
„ du diable: *Morisonia americana* L.; *Capparidac.;* Guad.
„ „ Castor: *Magnolia glauca* L.; *Magnoliac.;* sö. N.-Am.
„ orseille: *Montrouziera sphaeroidea* Planch.; *Guttifer.;* N.-Kal.
„ puant: *Olax zeylanica* L.; *Olacac.;* Cey.

Arbre saint: *Melia azedarach* L.; *Meliac.;* Guad.
Arbutus: *Arbutus menziesi* Pursh; *Ericac.;* w. Kan.
Arca: *Acacia visite* Griseb.; *Mimosac.;* Arg.
Arce: *Acer monspessulanum* L., *A. campestre* L.; *Acerac.;* Sp.
„ : *Acer negundo* L.; *Acerac.;* Arg. (B. Air)
„ blanco: *Acer pseudoplatanus* L.; *Acerac.;* Sp.
Arceira: *Hura crepitans* L.; *Euphorbiac.;* Bras.
Arceta: *Sapindus trifoliatus* L.; *Sapindac.;* s. As.
Arco de pipa: *Coccoloba sp.; Polygonac.;* Bras.
„ „ pipi: *Erythroxylon pulchrum* St. Hil.; *Erythroxylac.;* Bras.
Arditsch: ? *Juniperus excelsa* Bieb.; *Conifer.;* Kaukasus.
Arègè: *Poga oleosa* Pierre; *Rhizophorac.;* Gab.
Areira negra: *Schinus terebinthifolius* Radlk. var. *pobliana* Engl.; *Anacardiac.;* S.-Am.
Arèkè: *Poga oleosa* Pierre; *Rhizophorac.;* Gab.
Arellano: *Caesalpinia platyloba* Wats.; *Caesalpiniac.;* Mex.
Arendj: *Citrus bigaradia* Duham.; *Rutac.;* Arabien.
Arenillero: *Hura crepitans* L.; *Euphorbiac.;* Kol.
Arenillo: *Andira inermis* H. B. K.; *Papilionac.;* Pan.
„ : *Hura crepitans* L.; *Euphorbiac.;* Kol.
Arere: *Triplochiton spp.; Sterculiac.;* Nig.
Argan: *Argania sideroxylon* Roem. et Schult.; *Sapotac.;* Marok.
Argaña: *Calluna vulgaris* Salisb.; *Ericac.;* Sp.
Arganier: *Argania sideroxylon* Roem. et Schult.; *Sapotac.;* Marok.
Arhoné: *Erythrophloeum guineense* G. Don; *Caesalpiniac.;* Elf.
Aricurana: *Hieronymia alchorneoides* Allem.; *Euphorbiac.;* Bras. (ob. Amaz.)
Aridda: *Campnosperma zeylanicum* Thw.; *Anacardiac.;* Cey.
Arik: *Duabanga moluccana* Blume; *Lythrac.;* Phil.
Arilla: *Lothospermum fruticosum* L.; *Borraginac.;* Sp.
Arínkubál: *Terminalia nitens* Presl; *Combretac.;* Phil.
Aripawana: *Gustavia angusta* L.; *Lecythidac.;* Sur.
„ énnékan: *Gustavia pterocarpa* Poit.; *Lecythidac.;* Sur.
Arisauru: ? *Vatairea guianensis* Aubl.; *Papilionac.;* br. Gu.
Arisaurú: *Andira sp.; Papilionac.;* br. Gu.
Arishta: *Sapindus emarginata* Vahl; *Sapindac.;* Ind.
Arisoeroe: *Vatairea guianensis* Aubl.; *Papilionac.;* Sur.
Arisower: *Andira sp.; Papilionac.;* Sur.
Aritongtóng: *Terminalia pellucida* Presl; *Combretac.;* Phil,
Arjun: *Terminalia arjuna* Bedd.; *Combretac.;* C.-Ind.
Arlo: *Berberis vulgaris* L.; *Berberidac.;* Sp.
Armata: *Clathrotropis brachypetalum* Kleinh.; *Papilionac.;* Sur.
Arnato: *Bixa orellana* L.; *Bixac.;* Mex.
Arnatto: *Bixa orellana* L.; *Bixac.;* Bras.
Aroba: *Parinarium robustum* Oliv.; *Rosac.;* Elf.
„ : *Erythroxylon spp.; Erythroxylac.;* Cuba.
„ colorada: *Erythroxylon spp.; Erythroxylac.;* Cuba.
Arocurana: *Hieronymia alchorneoides* Allem.; *Euphorbiac.;* Bras.
Aroeira: *Schinus terebinthifolius* Raddi; *Anacardiac.;* Bras. (Sao P.)
„ : *Astronium fraxinifolium* Schott; *Anacardiac.;* Bras. (Sao P.)
„ blanca: *Lithraea molleoides* Engl., *Schinus sp.; Anacardiac.;* Arg.
„ colorada: *Schinus weinmanniaefolius* Engl.; *Anacardiac.;* Arg.
„ do campo: *Astronium fraxinifolium* Schott; *Anacardiac.;* Bras.
„ „ „ : *Astronium urundeuva* Engl., *Schinus sp.; Anacardiac.;* Bras.
„ „ sertão: *Astronium urundeuva* Engl.; *Anacardiac.;* Bras.
„ molle: *Schinus molle* L.; *Anacardiac.;* Bras.
„ negra: *Lithraea chichita* Speg.; *Anacardiac.;* Arg. (Mis., Corr.)
„ preta: *Astronium urundeuva* Engl.; *Anacardiac.;* Bras.
„ vermelha: *Schinus terebinthifolius* Raddi var. *rhoifolius* ?; *Anacardiac.;* Bras.
Aroematta: *Clathrotropis brachypetalum* Kleinh.; *Papilionac.;* Sur.
Aróho: *Casuarina equisetifolia* Forst.; *Casuarinac.;* Phil.
Arolla: *Pinus cembra* L.; *Conifer.;* w. Sib., Karp., Alpen.
Arom: *Azara microphylla* Hook. f.; *Flacourtiac.;* Chile.
Aroma: *Prosopis juliflora* DC.; *Mimosac.;* Phil.

Aroma: *Acacia spp.; Mimosac.;* Arg.
„ Francesa: *Albizzia lebbek* Bth.; *Mimosac.;* Cuba.
Aromatie: *Clathrotropis brachypetalum* Kleinh.; *Papilionac.;* Sur.
Aromo: *Azara microphylla* Hook. f.; *Bixac.;* Chile.
„ : *Acacia praecox* Griseb.; *Mimosac.;* Arg.
„ : *Acacia farnesiana* Willd.; *Mimosac.;* Hond.
„ : *Calliandra tonduzi* Britt. et Rose; *Mimosac.;* Pan.
„ negro: *Acacia melanoxylon* R. Br.; *Mimosac.;* Arg. (B. Air.)
Aromói: *Shorea guiso* Blume; *Dipterocarpac.;* Phil. (Zambonaga)
Aróo: *Casuarina equisetifolia* Forst.; *Casuarinac.;* Phil.
Arouaou: *Icica sp.; Burserac.;* fr. Gu. (Galibis)
Arowonè: *Tecoma leucoxylon* Mart., *T. araliacea* P. DC.; *Bignoniac.;* Sur.
Arquané: *Symphonia globulifera* L.; *Guttifer.;* Elf.
Arraclan: *Rhamnus frangula* L.; *Rhamnac.;* Sp.
Arraicán: *Eugenia guaviyú* Berg; *Myrtac.;* Arg. (Mis.)
Arraiyan: *Eugenia sp.; Myrtac.;* Arg.
Arrangen: *Canophyllum falcatum* Blume; *Sapindac.;* Phil.
Arra-óné: *Tecoma sp.; Bignoniac.;* Sur.
Arrata: *Minquartia guianensis* Aubl.; *Olacac.;* Sur.
Arratawerie: *Minquartia guianensis* Aubl.; *Olacac.;* Sur.
Arrayan: *Myrtus communis* L.; *Myrtac.;* Sp.
„ : *Blepharocalyx cisplatensis* Gris.; *Myrtac.;* Ind.-Ch.
„ : *Myrceugenia apiculata* Ndz.; *Myrtac.;* S.-Am. (Anden)
„ : *Eugenia patagonica* Phil.; *Myrtac.;* Arg. (Patagonien)
Arrayán: *Psidium sartorianum* Ndz.; *Myrtac.;* Mex.
? Arrewewa: *Oxandra sp.; Anonac.;* br. Gu.
Arrhonée: *Tecoma spp.; Bignoniac.;* fr. Gu. (Galibis)
Arrisouroo: *Andira sp.; Papilionac.;* br. Gu.
Arromatta: *Clathrotropis brachypetalum* Kleinh.; *Papilionac.;* Sur.
Arrowwood: *Viburnum dentatum* L.; *Caprifoliac.;* N.-Am.
„ : *Sapium sp.; Euphorbiac.;* C.-Am.
Arroyo: *Meliosma obtusifolia* Krug et Urb.; *Sabiac.;* P.-R.
Ars: *Cedrus libani* Barr.; *Conifer.;* Kleinasien.
Arta: *Calligonum comosum* L'Hérit.; *Polygonac.;* Sahara.
Artos: *Rhamnus infectoria* L.; *Rhamnac.;* Sp.
Artscha: *Juniperus excelsa* Bieb.; *Conifer.;* Kauk.
Aruané: *Symphonia globulifera* L.; *Guttifer.;* Elf.
Aruás: *Shorea negrosensis* Foxw.; *Dipterocarpac.;* Phil.
Aruba: *Simaruba amara* Aubl., *S. versicolor* St. Hil.; *Simarubac.;* Sur.
Aruera: *Schinus molle* L.; *Anacardiac.;* Urug.
Arupág: *Aglaia diffusa* Merr.; *Meliac.;* Phil.
„ : *Euphoria cinerea* Radlk.; *Sapindac.;* Phil.
Arve: *Pinus cembra* L.; *Conifer.;* Eu.
Arvore da morte: *Hippomane mancinella* L.; *Euphorbiac.;* Bras.
„ „ preguiça: *Cecropia sp.; Morac.;* Bras.
„ de cuia: *Crescentia cujete* L.; *Bignoniac.;* Bras.
„ „ sebo: *Virola bicuhyba* Warb.; *Myristicac.;* Bras.
„ „ umbella: *Cordia sp.; Borraginac.;* Bras.
Asada: *Ostrya japonica* Sarg.; *Betulac.;* Jap.
„ Carne: *Cordia ferruginea* Roem. et Schult.; *Borraginac.;* Hond.
Asafra: *Eugenia rowlandi* Spr.; *Myrtac.;* Elf.
Asagara: *Halesia corymbosa* Bth. et Hook.; *Styracac.;* Jap.
Asa Gidi: *Bridelia micrantha* Muell. Arg.; *Euphorbiac.;* W.-Af.
Asaina: *Parkia agboensis* Chev.; *Mimosac.;* Elf.
„ : *Terminalia tomentosa* Bedd.; *Combretac.;* Ind.
Asam: *Uapaca sp.; Euphorbiac.;* Gab.
„ kumbang: *Parinarium rubiginosum* Ridl.; *Rosac.;* Malay.
„ „ : *Angelisia splendens* Korth.; *Rosac.;* Malay.
Asama: *Parkia agboensis* Chev.; *Mimosac.;* Elf.
„ -tsuge: *Buxus sempervirens* L.; *Buxac.;* Jap.
Asamoiaké: *Trichilia sp.; Meliac.;* Elf.
Asan: *Terminalia tomentosa* Bedd.; *Combretac.;* Ind.

Asaná: *Pterocarpus spp.; Papilionac.;* Phil.
Asanama: *Dialium guineense* Willd.; *Caesalpiniac.;* Go.
Asanamba: *Dialium guineense* Willd.; *Caesalpiniac.;* Go.
Asanohakaede: *Acer argutum* Maxim.; *Acerac.;* Jap.
Asante: *Pterocarpus santalinoides* L'Hérit.; *Papilionac.;* Go.
Asar: *Acer opulifolium* Vill.; *Acerac.;* Sp.
„ -quiro: *Hedyosmum sp.; Chloranthac.;* Peru.
Asas (f): *Bridelia speciosa* Muell. Arg.; *Euphorbiac.;* Elf., Gab.
„ (f): *Bridelia grandis* Pierre et Hutch.; *Euphorbiac.;* Gab.
Asasalá: *Litsea sp.; Laurac.; Phil.*
Asat: *Aglaia laevigata* Merr.; *Meliac.;* Phil.
Asau: *Pithecolobium spp.; Mimosac.;* Sur.
Asebi: *Andromeda japonica* Thunb.; *Ericac.;* Jap.
Asémigniri: *Afzelia microcarpa* Chev.; *Caesalpiniac.;* Elf.
Asenã: *Dialium guineense* Willd.; *Caesalpiniac.;* Go.
Asenama: *Dialium guineense* Willd.; *Caesalpiniac.;* Go.
Asép: *Vatica mangachapoi* Blco.; *Dipterocarpac.;* Phil.
Asep: *Sapium ellipticum* Pax; *Euphorbiac.;* Gab.
Asé-poekoe: *Pouteria guianensis* Aubl.; *Sapotac.;* Sur.
Asepoko: *Pouteria guianensis* Aubl.; *Sapotac.;* Sur.
Ash (e): *Fraxinus excelsior* L.; *Oleac.;* Eu.
„ : *Pithecolobium muellerianum* Maid. et Bak.; *Mimosac.;* Au.
„ : *Eucalyptus regnans* F. v. M.; *Myrtac.;* Tasm.
„ : *Flindersia australis* R Br.; *Rutac.;* N.-S.-W.
„ Alpine: *Eucalyptus delegatensis* Bak.; *Myrtac.;* S.-Au.
„ American: *Fraxinus americana* L.; *Oleac.;* ö. Kan.
„ Arkansas: *Fraxinus platycarpa* Michx.; *Oleac.;* N.-Am.
„ Basket: *Fraxinus nigra* Marsh.; *Oleac.;* ö. N.-Am.
„ Bennett's: *Flindersia bennettiana* F. v. M.; *Rutac.;* Au.
„ Biltmore: *Fraxinus biltmoreana* Beadle; *Oleac.;* ö. USA.
„ Bitter: *Quassia amara* L. f.; *Simarubac.;* W.-Ind.
„ „ : *Picraena excelsa* Lindl.; *Simarubac.;* Mart.
„ Black: *Fraxinus spp.; Oleac.;* N.-Am.
„ „ : *Acer negundo* L.; *Acerac.;* N.-Am.
„ Blue: *Fraxinus quadrangulata* Michx.; *Oleac.;* USA.
„ „ : *Aphanopetalum resinosum* Endl.; *Saxifragac.;* N.-S.-W.
„ „ Berry: *Elaeocarpus obovatus* G. Don; *Elaeocarpac.;* Au.
„ Brown: *Fraxinus nigra* Marsh.; *Oleac.;* N.-Am.
„ Brush: *Acronychia baueri* Schott; *Rutac.;* sö. Au.
„ Bumpy: *Flindersia schottiana* F. v. M.; *Rutac.;* Au.
„ Cape: *Ekebergia capensis* Sparrm.; *Meliac.;* S.-Af.
„ Carolina: *Fraxinus platycarpa* Michx.; *Oleac.;* USA.
„ Common: *Fraxinus excelsior* L.; *Oleac.;* Eu.
„ Crow's: *Flindersia australis* R. Br.; *Rutac.;* Au.
„ „ Bastard: *Pentaceras australis* Hook. f.: *Rutac.;* Au.
„ „ „ : *Flindersia collina* F. M. B.; *Rutac.;* Au.
„ Darlington: *Fraxinus darlingtoni* Britton; *Oleac.;* ö. USA.
„ Elderberry: *Panax sambucifolius* Sieber; *Araliac.;* N.- & O.-Au.
„ English: *Fraxinus excelsior* L.; *Oleac.;* Eu.
„ European: *Fraxinus excelsior* L.; *Oleac.;* USA.
„ Flowering: *Fraxinus dipetala* Hook. et Arn.; *Oleac.;* w. USA.
„ „ : *Fraxinus ornus* L.; *Oleac.;* It.
„ „ Californian: *Fraxinus dipetala* Hook. et Arn.; *Oleac.;* USA.
„ Fringe-flower: *Fraxinus dipetala* Hook. et Arn.; *Oleac.;* w. USA.
„ Gray: *Fraxinus pubescens* Lam., *F.pennsylvanica* Marsh.; *Oleac.;* ö. N.-Am.
„ Green: *Fraxinus americana* L., *F.viridis* Michx.; *Oleac.;* N.-Am.
„ Ground: *Fraxinus americana* L., *F.sambucifolia* Lam.; *Oleac.;* ö. Kan.
„ Gully: *Eucalyptus badjensis* de Beuz. et Welch; *Myrtac.;* s. N.-S.-W.
„ Himalayan: *Fraxinus floribunda* Wall.; *Oleac.;* Ind.
„ Hoop: *Fraxinus sambucifolia* Lam.; *Oleac.;* ö. N.-Am.
„ Hungarian: *Fraxinus excelsior* L.; *Oleac.;* Eu.
„ Japanese: *Fraxinus mandschurica* Rupr.; *Oleac.;* O.-As.

Ash Japanese: *Fraxinus sieboldiana* Blume *var. serrata* Nakai; *Oleac.;* Jap.
„ Leather-leaf: *Fraxinus velutina* Torr.; *Oleac.;* sw. N.-Am.
„ Leopard: *Flindersia collina* F. M. B.; *Rutac.;* Au.
„ Manna: *Fraxinus ornus* L.; *Oleac.;* It.
„ Moreton Bay: *Eucalyptus tessellaris* F. v. M.; *Myrtac.;* Queensl.
„ Mountain: *Pirus aucuparia* Gaertn.; *Rosac.;* Engl.
„ „ : *Pirus americana* DC.; *Rosac.;* N.-Am.
„ „ : *Fraxinus americana* L.; *Oleac.;* N.-Fundl.
„ „ : *Eucalyptus microcorys* F. v. M.; *Myrtac.;* O.-Au.
„ „ : *Eucalyptus pilularis* Smith; *Myrtac.;* O.-Au.
„ „ : *Eucalyptus regnans* F. v. M.; *Myrtac.;* Au., Tasm.
„ „ : *Eucalyptus gigantea* Hook. f.; *Myrtac.;* Au., Tasm.
„ „ : *Eucalyptus amygdalina* Lab.; *Myrtac.;* N.-S.-W.
„ „ : *Eucalyptus sieberiana* F. v. M.; *Myrtac.;* N.-S.-W.
„ „ : *Alphitonia excelsa* Reiss.; *Rhamnac.;* Queensl.
„ „ : *Eucalyptus delegatensis* Bak.; *Myrtac.;* S.-Au.
„ „ American: *Pirus americana* DC.; *Rosac.;* USA.
„ „ Austral: *Elaeocarpus grandis* F. v. M.; *Tiliac.;* Au.
„ „ Europ.: *Pirus americana* DC.; *Rosac.;* ö. N.-Am.
„ „ Red: *Eucalyptus gigantea* Hook. f.; *Myrtac.;* Au., Tasm.
„ „ Western: *Pirus sitchensis* Piper; *Rosac.;* N.-Am.
„ „ White: *Eucalyptus regnans* F. v. M.; *Myrtac.;* O.-Au.
„ Nova Scotia: *Fraxinus sambucifolia* Lam.; *Oleac.;* N.-Am.
„ Oregon: *Fraxinus oregona* Nutt.; *Oleac.;* w. N.-Am.
„ Phul: *Euphorbia longana* Lamk.; *Euphorbiac.;* Beng.
„ Pigeonberry: *Olea paniculata* R. Br.; *Oleac.;* Au.
„ „ : *Cryptocarya erythroxylon* Maid. et Betche; *Laurac.;* Au.
„ Prickly: *Fagara schinifolia* Engl.; *Rutac.;* Jap.
„ „ : *Zanthoxylum fraxineum* Willd.; *Rutac.;* N.-Am.
„ „ : *Zanthoxylum carolinianum* Lam.; *Rutac.;* USA.
„ „ : *Orites excelsa* R. Br.; *Proteac.;* N.-S.-W.
„ Pumkin: *Fraxinus profunda* Bush; *Oleac.;* m. USA.
„ Quebec: *Fraxinus americana* L.; *Oleac.;* N.-Am.
„ Red: *Fraxinus americana* L., *F. tomentosa* Michx.; *Oleac.;* N.-Am.
„ „ : *Alphitonia excelsa* Reiss.; *Rhamnac.;* N.- & O.-Au.
„ „ : *Alphitonia petriei* Braid et C. T. White; *Rhamnac.;* N.- & O.-Au.
„ „ : *Tarrietia argyrodendron* Bth.; *Sterculiac.;* N.- & O.-Au.
„ „ : *Eucalyptus gigantea* Hook. f.; *Myrtac.;* Au.
„ „ : *Pomaderris sp.;* *Rhamnac.;* S.-Au.
„ Rim: *Fraxinus pennsylvanica* Marsh.; *Oleac.;* Kan.
„ River: *Fraxinus pubescens* Lamk.; *Oleac.;* Ontario (USA.)
„ „ White: *Fraxinus americana* L.; *Oleac.;* Ontario (USA).
„ Rock: *Rhus thunbergi* Hook.; *Anacardiac.;* Kap.
„ Satin Rose: *Eugenia francisi* F. M. B.; *Myrtac.;* Au.
„ Sea: *Zanthoxylum sp.;* *Rutac.;* USA.
„ Silky: *Ehretia acuminata* R. Br.; *Borraginac.;* Au.
„ Silver: *Flindersia schottiana* F. v. M.; *Rutac.;* Au.
„ Soft: *Fraxinus pennsylvanica* Marsh.; *Oleac.;* ö. Kan.
„ Swamp: *Fraxinus sambucifolia* Lam.; *Oleac.;* N.-Am.
„ „ : *Fraxinus nigra* Marsh.; *Oleac.;* Kan.
„ Water: *Fraxinus platycarpa* Michx.; *Oleac.;* Florida.
„ „ : *Fraxinus nigra* Marsh.; *Oleac.;* USA.
„ White: *Fraxinus pennsylvanica* Marsh.; *Oleac.;* N.-Am.
„ „ : *Fraxinus americana* L.; *Oleac.;* N.-Am.
„ „ : *Alphitonia petriei* Braid et C. T. White; *Rhamnac.;* Au.
„ Yellow: *Emmenosperma alphitonioides* F. v. M.; *Rhamnac.;* Au.
„ Yellowwood: *Flindersia oxleyana* F. v. M.; *Rutac.;* Au.
Asia: *Pachylobus büttneri* Engl.; *Burserac.;* Gab.
Asiminier (f): *Asimina triloba* Dun.; *Anonac.;* s. USA.
Asipoko: *Lucuma sp.;* *Sapotac.;* br. Gu.
Asisimansa: *Newbouldia laevis* Seem.; *Bignoniac.;* Go.
Asna: *Terminalia tomentosa* Bedd.; *Combretac.;* Ind.

Asni: *Thujopsis dolabrata* S. et Z.; *Conifer.*; Jap.
Asogomon: *Mallotus subulatus* Muell. Arg.; *Sapotac.*; Gab.
Asohin: *Anthostema aubryanum* Baill.; *Euphorbiac.*; Gab.
Asok: *Polyalthia longifolia* Bth.; *Anonac.*; O.-Ind., Cey.
Asokolo: *Conocarpus erectus* Jacq.; *Combretac.*; Go.
Asokpolo: *Avicennia nitida* Jacq.; *Verbenac.*; Go.
Asoma: *Ricinodendron africanum* Muell. Arg.; *Euphorbiac.*; Go.
Asong'o: ? *Irvingia klainei* Pierre; *Simarubac.*; Gab.
Asonná: *Anthostema aubryanum* Baill.; *Euphorbiac.*; sp. Guin.
Asopolo: *Conocarpus erectus* Jacq.; *Combretac.*; Go.
Asoporo: *Avicennia nitida* Jacq.; *Verbenac.*; Go.
? Asopro: *Rhizophora racemosa* E. Mey.; *Rhizophorac.*; Go.
Asoré: *Entandrophragma candollei* Harms; *Meliac.*; Nig.
Asoroa: *Carapa procera* DC.; *Meliac.*; Go.
Asorowa: *Carapa procera* DC.; *Meliac.*; Go.
Asourémon: *Croton sp.*; *Euphorbiac.*; Gab.
Asp: *Populus tremula* L.; *Salicac.*; Eu.
Asp, Quaking: *Populus tremuloides* Michx.; *Salicac.*; N.-Am.
? Aspaï: ? *Quararibea sp.*; *Bombacac.*; Ven.
Aspavé: *Anacardium rhinocarpum* DC.; *Anacardiac.*; C.-Am.
Aspen: *Populus tremuloides* Michx.; *Salicac.*; Eu., N.-Am.
„ American: *Populus tremuloides* Michx.; *Salicac.*; USA.
„ „ : *Populus grandidentata* Michx.; *Salicac.*; USA.
„ largo-tooth: *Populus grandidentata* Michx.; *Salicac.*; N.-Am.
„ Quaking: *Populus tremuloides* Michx.; *Salicac.*; N.-Am.
„ Small toothed: *Populus tremuloides* Michx.; *Salicac.*; N.-Am.
„ Trembling: *Populus tremuloides* Michx.; *Salicac.*; N.-Am.
Assa: *Pachylobus edulis* G. Don; *Burserac.*; Gab.
Assa-bogué: *Maesobotrya stapfiana* Beille; *Euphorbiac.*; Elf.
Assachi: *Rheedia madruno* Pl. et Tr., *R. virens* Pl. et Tr.; *Guttifer.*; br. Gu.
Assacú: *Hura crepitans* L., *H. brasiliensis* Willd.; *Euphorbiac.*; Bras.
„ : *Erythrina sp.*; *Papilionac.*; Bras.
Assacurâna: *Erythrina glauca* Willd.; *Papilionac.*; Bras.
Assagay-boom: *Curtisia faginea* Ait.; *Cornac.*; S.-Af.
Assagaywood: *Curtisia faginea* Ait.; *Cornac.*; Kap.
Assam: *Uapaca staudti* Pax; *Euphorbiac.*; Kam.
Assama: *Parkia agboënsis* Chev.; *Mimosac.*; Elf.
Assan: *Coula edulis* Baill.; *Olacac.*; Elf.
„ : *Musanga smithi* R. Br.; *Morac.*; Kam., Gab.
Assana-baké: *Maesobotrya stapfiana* Beille; *Euphorbiac.*; Elf.
Assang: *Musanga smithi* R. Br.; *Morac.*; Fernando Po, Gab.
Assapoko: *Lucuma sp.*; *Sapotac.*; br. Gu.
„ : *Pouteria guianensis* Aubl.; *Sapotac.*; Sur.
Assapookoo: *Mimusops spp.*; *Sapotac.*; fr. Gu.
Assas: *Bridelia grandis* Pierre et Hutch.; *Euphorbiac.*; Gab.
„ (f): *Bridelia speciosa* Muell. Arg.; *Euphorbiac.*; Kam., Gab.
„ : *Pachylobus edulis* G. Don; *Burserac.*; Kam.
„ mingou: *Pachylobus edulis* G. Don *var. silvestris* Chev.; *Burserac.*; Kam.
„ onone: *Pachylobus edulis* G. Don *var. silvestris* Chev.; *Burserac.*; Kam.
Assasodau: *Sterculia aff. oblonga* Mart.; *Sterculiac.*; Elf.
Asse: *Entandrophragma cf. cylindricum* Spr.; *Meliac.*; Kam.
Assegaai: *Curtisia faginea* Ait.; *Cornac.*; S.-Af.
Assegai: *Curtisia faginea* Ait.; *Cornac.*; S.-Af.
Assem: *Tamarindus indica* L.; *Caesalpiniac.*; Java.
Asseng: *Musanga smithi* R. B.; *Morac.*; Gab.
Assia: *Pachylobus büttneri* Engl.; *Burserac.*; Gab.
Assié (f): *Entandrophragma utile* Spr.; *Meliac.*; Kam.
„ false: *Entandrophragma aff. rhederi* Harms; *Meliac.*; Kam.
Assiga: *Petersia viridiflora* Chev.; *Myrtac.*; Elf.
Assihaoteau: *Pterocarpus esculentus* Schum.; *Papilionac.*; Elf.
Assikum: *Geophila leucocarpa* Krause; *Rubiac.*; Kam.
Assita: *Parinarium excelsum* Sabine; *Rosac.*; Kam.

Asso: *Lophira procera* Chev.; *Ochnac.;* Elf.
Assol: *Lophira procera* Chev.; *Ochnac.;* Elf.
Assolotin: *Irvingia sp.; Simarubac.;* Dah.
Assomah: *Ricinodendron africanum* Muell. Arg.; *Euphorbiac.;* W.-Af.
Assongha: *Anthostema aubryanum* Baill.; *Euphorbiac.;* Gab.
Assonha: *Anthostema aubryanum* Baill.; *Euphorbiac.;* Gab.
Assoré: *Entandrophragma candollei* Harms; *Meliac.;* W.-Af.
Assoséka: *Diospyros sanza-minika* Chev.; *Ebenac.;* Elf.
Assou: *Coula edulis* Baill.; *Olacac.;* Elf.
Assoua: *Anthostema aubryanum* Baill.; *Euphorbiac.;* Gab.
Assoun: *Anthostema aubryanum* Baill.; *Euphorbiac.;* Gab.
Assumséka: *Diospyros sanza-minika* Chev.; *Ebenac.;* Elf.
Asugi: *Taiwania cryptomerioides* Hayata; *Conifer.;* Jap.
Asul: *Tamarix articulata* Vahl; *Tamaricac.;* Arabien.
Asumpa: *Anonidium manni* Engl. et Diels; *Anonac.;* Go.
Asunaro: *Thujopsis dolabrata* S. et Z.; *Conifer.;* Jap.
Asun-kruma: *Homalium sp.; Flacourtiac.;* Go.
Asya: *Pachylobus büttneri* Engl.; *Burserac.;* W.-Af.
Ata: *Pentaclethra macrophylla* Bth.; *Mimosac.;* Go.
„ : *Anona spp.; Anonac.;* fr. Gu.
Ataa: *Pentaclethra macrophylla* Bth.; *Mimosac.;* Go.
Ata-apa: *Macrolobium chrysostachyum* Bth.; *Caesalpiniac.;* Sur.
„ „ : *Macrolobium bifolium* Pers.; *Caesalpiniac.;* Sur.
„ -apiririe: *Tapirira aff. guianensis* Aubl.; *Anacardiac.;* Sur.
Ata-áta: *Diospyros spp.; Ebenac.;* Phil.
Ata-ata: *Esenbeckia atata* Pittier; *Rutac.;* Ven.
Ataba: *Bridelia speciosa* Muell. Arg.; *Euphorbiac.;* Elf.
„ : *Macrolobium chrysostachyum* Bth.; *Caesalpiniac.;* Sur.
„ : *Macrolobium bifolium* Pers.; *Caesalpiniac.;* Sur.
Atabaca: *Inula viscosa* Ait.; *Composit.;* Tener.
Atabene: *Chrysophyllum aff. africanum* A. DC.; *Sapotac.;* Go.
Atadijo: *Trema micrantha* Blume; *Ulmac.;* Peru.
Atagua: *Ficus spp.; Morac.;* Trin.
Atakamara: *Pouteria guianensis* Aubl.; *Sapotac.;* Sur.
Atambú-assú: *Podocarpus lamberti* Klotzsch; *Conifer.;* Bras.
Atamisque: *Atamisquea emarginata* Miers; *Capparidac.;* Arg.
Atanga: *Pachylobus edulis* G. Don; *Burserac.;* Gab.
Atapa: *Macrolobium chrysostachyum* Bth.; *Caesalpiniac.;* Sur.
„ : *Macrolobium bifolium* Pers.; *Caesalpiniac.;* Sur.
Atapiririe: *Tapirira aff. guianensis* Aubl.; *Anacardiac.* Sur.
Atata: *Esenbeckia atata* Pittier; *Rutac.;* Ven.
„ : *Ticorea sp.; Rutac.;* Ven.
Ataúba: *Guarea spp.; Meliac.;* Bras.
Atawa: *Pentaclethra macrophylla* Bth.; *Mimosac.;* Go.
Atay-Atay: *Eranthemum bicolor* Schrank; *Acanthac.;* Phil.
Atcha: *Coula edulis* Baill.; *Olacac.;* Elf.
Atchang: *Morinda citrifolia* L.; *Rubiac.;* Kam.
Atchek: *Morinda sp.; Rubiac.;* Kam.
Atchoourgo: *Melaleuca leucadendron* L.; *Myrtac.;* Queensl.
Atchuipon: *Mitragyne macrophylla* Hiern; *Rutac.;* Elf.
Até: *Buchholzia macrophylla* Pax; *Capparidac.;* Elf.
Ate: *Anona spp.; Anonac.;* fr. Gu.
Atéang: *Hopea ferrea* Pierre; *Dipterocarpac.;* Ind.-Ch.
Atégué: *Spondianthus preussi* Engl.; *Euphorbiac.;* Gab.
Ateje: *Bourreria havanensis* Miers; *Borraginac.;* Cuba.
„ : *Cordia collococca* L.; *Borraginac.;* Cuba.
„ amarillo: *Cordia sp.; Borraginac.;* Cuba.
„ macho: *Cordia sp.; Borraginac.;* Cuba.
Atejillo: *Cordia nitida* L.; *Borraginac.;* Cuba.
Atenli: *Pycnanthus kombo* Warb.; *Myristicac.;* Go.
Atewa: *Calliandra sp.; Mimosac.;* Go.
Ati: *Nauclea grandifolia* Bl.; *Rubiac.;* Malay.

Ati: *Hannoa klaineana* Pierre et Engl.; *Simarubac.;* Elf.
„ : *Calophyllum inophyllum* L.; *Guttifer.;* Tahiti.
Atia: *Anacardium occidentale* L.; *Anacardiac.;* Go.
Atiati: *Morinda lucida* Bth.; *Rubiac.;* Go.
Atich: *Cyanodaphne cuneata* Blume; *Laurac.;* Ind.-Ch.
Atiéma: *Erythrophloeum guineense* G. Don; *Caesalpiniac.;* Elf.
Atikóko: *Vitex longifolia* Merr.; *Verbenac.;* Phil.
Atilmá: *Diospyros nitida* Merr.; *Ebenac.;* Phil.
Atiouya: *Erythrophloeum guineense* G. Don; *Caesalpiniac.;* Elf.
Atitoe: *Dialium guineense* Willd.; *Caesalpiniac.;* Go.
Atjarie kjanari: *Acrodiclidium canella* Mez; *Laurac.;* Sur.
„ „ : *Acrodiclidium chrysophyllum* Meissn.; *Laurac.;* Sur.
Atjek: *Corynanthe sp.; Rubiac.;* Kam.
Atjeng: *Morinda citrifolia* L.; *Rubiac.;* Kam.
Atlasbeerbaum: *Sorbus torminalis* Crantz; *Rosac.;* Mitt.- & S.-Eu.
Atlasholz (d): *Fagara flava* Krug et Urb.; *Rutac.;* W.-Ind.
„ „ : *Ferolia guianensis* Aubl.; *Rosac.;* Bras.
Atocire: *Anona spp.; Anonac.;* fr. Gu.
Atoel lahoet: *Heritiera littoralis* Dryand.; *Sterculiac.;* Malay.
Atoeng-laoet: *Heritiera littoralis* Dryand.; *Sterculiac.;* Moluk.
„ -makan: *Heritiera sp.; Sterculiac.;* Moluk.
Atolaïe: *Myrianthus arboreus* Beauv.; *Morac.;* Elf.
Atom: *Trichoscypha sp.; Anacardiac.;* Kam.
Atome: ? *Pachypodantium confine* Engl. et Diels; *Anonac.;* Gab.
Atoto-ito: *Eschweilera sp.; Lecythidac.;* Sur.
Atou-krouya: *Lannea acidissima* Chev.; *Anacardiac.;* Elf.
Atoul: *Parkia klainei* Pierre; *Mimosac.;* Kam.
Atoyaxocotl: *Spondias spp.; Anacardiac.;* Mex.
Atpái: *Vatica mangachapoi* Blco.; *Dipterocarpac.;* Phil.
Atpúi: *Dillenia sp.; Dilleniac.;* Phil. (Palawan)
Atra: *Rhizophora racemosa* E. Mey.; *Rhizophorac.;* Go.
Atrati: *Rhizophora racemosa* E. Mey.; *Rhizophorac.;* Go.
Atsa: *Blighia sapida* Koenig; *Sapindac.;* Go.
Atsan: *Coula edulis* Baill.; *Olacac.;* Elf.
Atsel: *Tamarix articulata* Vahl; *Tamaricac.;* Arabien.
Atsitoe: *Dialium guineense* Willd.; *Caesalpiniac.;* Go.
Atsui: *Haronga madagascariensis* Choisy; *Hypericac.;* Gab.
Atta: *Anona spp.; Anonac.;* Bras.
„ : *Bixa orellana* L.; *Bixac.;* br. Hond.
Attabini: *Tarrietia utilis* Spr.; *Sterculiac.;* Go.
Atta-bini: *Pycnanthus kombo* Warb.; *Myristicac.;* Go.
Attajam: *Olea dioica* Roxb.; *Oleac.;* O.-Ind.
Atta vanji: *Anthocephalus cadamba* Miq.; *Rubiac.;* O.-Ind.
Attawood wild: *Sloanea faginea* Standl.; *Elaeocarpac.;* br. Hond.
Atte: *Anona spp.; Anonac.;* fr. Gu.
Attéang: *Hopea ferrea* Pierre; *Dipterocarpac.;* Ind.-Ch.
Attia: *Coula edulis* Baill.; *Olacac.;* Elf.
Attich: *Sambucus ebulus* L.; *Caprifoliac.;* Eu.
Attier: *Anona squamosa* L.; *Anonac.;* tr. Am.
Attum: *Olea chrysophylla* Lam.; *Oleac.;* O.-Af., Maur.
Atui: *Piptadenia africana* Hook. f.; *Mimosac.;* Kam.
Atuij: *Piptadenia africana* Hook. f.; *Mimosac.;* Kam.
Atuel: *Tamarix articulata* Vahl; *Tamaricac.;* N.- & C.-Af.
Aturonsu: *Newbouldia laevis* Seem.; *Bignoniac.;* Go.
Aturu: *Ficus guineensis* Stapf; *Morac.;* Elf.
Atzapotl: *Lucuma salicifolia* H. B. K.; *Sapotac.;* C.-Am.
Aubry: *Irvingia gabonensis* Baill.; *Simarubac.;* Kam.
Auhi: *Cordia abyssinica* R. Br.; *Borraginac.;* Abessinien.
Aukchinza: *Dysoxylum binectariferum* Hook. f.; *Meliac.;* Burma.
Aulaga: *Genista hirsuta* Vahl, *G. anglica* L.; *Papilionac.;* Sp.
„ : *Ulex australis* Clem.; *Papilionac.;* Sp.
„ : *Ulex europaeus* L.; *Papilionac.;* Eu.

Aulage fina: *Genista tourneti* Spach; *Papilionac.;* Sp.
„ morisca: *Genista triacanthos* Brot.; *Papilionac.;* Sp.
Aulanche: *Diospyros chloroxylon* Roxb.; *Ebenac.;* O.-Ind.
Auleh: *Olea chrysophylla* Lam.; *Oleac.;* O.-Af., Maur.
Auloc oulek: *Hymenodictyon excelsum* Wall.; *Rubiac.;* Ind.-Ch.
Aune baumier: *Populus balsamifera* L.; *Salicac.;* Kan.
„ -buis: *Acer negundo* L.; *Acerac.;* Mittel-Kan.
„ commun: *Alnus incana* Moench; *Betulac.;* ö. Kan.
„ de l'Orégon: *Alnus oregona* Nutt.; *Betulac.;* w. Kan.
„ noir (f): *Rhamnus frangula* L.; *Rhamnac.;* Eu.
Aura: *Tapura cubensis* Gris.; *Dichapetalac.;* Cuba.
Au sú: *Myrcia pagani* Krug et Urb.; *Myrtac.;* P.-R.
Ausubo: *Sideroxylon foetidissimum* Jacq.; *Sapotac.;* P.-R.
„ : *Mimusops spp., Sideroxylon mastichodendron* Jacq.; *Sapotac.;* P.-R.
Auzerole: *Acer campestre* L.; *Acerac.;* Mittel-Eu.
Avalo: *Bumelia guatemalensis* Standl.; *Sapotac.;* C.-Am.
Avane de Magellan: *Caesalpinia pectinata* Cav.; *Caesalpiniac.;* Peru, Chile, Bol.
Avara: *Astrocaryum segregatum* Drude; *Palmac.;* br. Gu.
Avaremotemo: *Pithecolobium spp.; Mimosac.;* Bras.
Avarempotimbó: *Pithecolobium cauliflorum* Mart.; *Mimosac.;* Arg.
Avati: *Hymenaea courbaril* L.; *Caesalpiniac.;* Par.
„ tymbati: *Casearia silvestris* Sw.; *Flacourtiac.;* Arg. (Mis.)
Avellaneiro: *Corylus avellana* L.; *Corylac.;* Sp.
Avellanillo: *Lomatia dentata* R. Br.; *Proteac.;* S.-Am.
Avellano: *Corylus avellana* L.; *Corylac.;* Sp.
„ : *Guevina avellana* Mol.; *Proteac.;* Chile.
Avenja: *Vernonia senegalensis* Less.; *Composit.;* W.-Af.
Aveup: *Ricinodendron africanum* Muell. Arg.; *Euphorbiac.;* Kam.
Avia: *Newbouldia laevis* Seem.; *Bignoniac.;* Go.
Avichuri: *Brosimum utile* Pittier; *Morac.;* Kol.
Avispa: *Hibiscus rosa-sinensis* L.; *Malvac.;* Nic.
Avispillo: *Phoebe montana* Griseb.; *Laurac.;* P.-R.
Avoa: *Cola cordifolia* R. Br.; *Sterculiac.;* Elf.
Avoané: *Psychotria klaineana* Pierre; *Rubiac.;* Gab.
Avoane: *Sterculia tragacantha* Lindl.; *Sterculiac.;* W.-Af.
Avocado: *Persea americana* Mill.; *Laurac.;* tr. Am.
„ Pear: *Persea gratissima* Gaertn.; *Laurac.;* W.-Ind.
Avocat: *Ruprechtia sp.; Polygonac.;* Trin.
Avokate: *Persea americana* Mill.; *Laurac.;* Mex.
Avocatier: *Persea americana* Mill.; *Laurac.;* tr. Am.
Avodiré (f): *Turraeanthus africana* Pell.; *Meliac.;* Elf.
Avoira: *Astrocaryum vulgare* Mart.; *Palmac.;* fr. Gu.
Avom: *Cleistopholis patens* Engl. et Diels; *Anonac.;* Kam.
Avome: *Cleistopholis patens* Engl. et Diels; *Anonac.;* Gab.
Avomé: *Malacantha robusta* Chev.; *Sapotac.;* Elf.
Avorniello (i): *Laburnum anagyroides* Medic; *Papilionac.;* Eu.
Avouana: *Psychotria klaineana* Pierre; *Rubiac.;* Gab.
Avoum: *Pentadesma butyracea* Sabine; *Guttifer.;* Kam.
Avuli: *Dialium guineense* Willd.; *Caesalpiniac.;* Go.
Awaakoko: *Ormosia melanocarpa* Kleinh.; *Papilionac.;* Sur.
Awabuki: *Meliosma myriantha* S. et Z.; *Sapindac.;* Jap.
Awamé: *Malacantha robusta* Chev.; *Sapotac.;* Elf.
Awapa: *Eperua falcata* Aubl.; *Caesalpiniac.;* Sur.
Awapau: *Sideroxylon guianense* A. DC.; *Sapotac.;* Sur.
Awari: *Pterygota macrocarpa* K. Schum.; *Sterculiac.;* Go.
Awari-Kokori: *Cola cordifolia* R. Br.; *Sterculiac.;* Go.
Awarra: *Astrocaryum spp.; Palmac.;* Sur.
„ Afrikaansche: *Elaeis guineensis* Jacq.; *Palmac.;* fr. Gu.
Awasakuli: *Tovomita guianensis* Aubl., *T. obovata* Engl.; *Guttifer.;* br. Gu.
„ : *Tovomita schomburgki* Tr. et Pl.; *Guttifer.;* br. Gu.
Awassé: *Cola vera* K. Schum.; *Sterculiac.;* Elf.
Awati: *Inga spp.; Mimosac.;* br. Gu.

Awatie: *Maprounea guianensis* Aubl.; *Euphorbiac.;* Sur.
Aweb: *Holoptelea grandis* Mildbr.; *Ulmac.;* Kam.
Awep: *Holoptelea grandis* Mildbr.; *Ulmac.;* Kam .
Awési: *Klainedoxa gabonensis* Pierre; *Simarubac.;* Gab.
Awiejoeu: *Xylopia spp.; Anonac.;* Sur.
Awiem-fosemena: *Albizzia ferruginea* Bth.; *Mimosac.;* Go.
Awie-öe: *Xylopia spp.; Anonac.;* Sur.
Awo giri: *Sterculia platanifolia* L. f.; *Sterculiac.;* Jap.
Awo guiri: *Sterculia platanifolia* L. f.; *Sterculiac.;* Jap.
Awogye: *Bridelia micrantha* Baill.; *Euphorbiac.;* Go.
Awole: *Phialodiscus unijugatus* Radlk.; *Sapindac.;* Go.
Awonia: *Cynometra ananta* Hutch. et Dalz.; *Caesalpiniac.;* Go.
Awraw: *Sterculia rhinopetala* K. Schum.; *Sterculiac.;* S.-Kam., Nig.
Axe-breaker: *Notelaea longifolia* Vent.; *Oleac.;* N.-S.-W.
Axehandlewood: *Aphananthe philippinensis* Planch.; *Ulmac.;* Au.
    „   : *Pseudomorus brunoniana* Bur.; *Morac.;* Au.
Axemaster: *Krugiodendron ferreum* Urb.; *Rhamnac.;* br. Hond.
    „   : *Schinopsis spp.; Anacardiac.;* S.-Am.
    „   bastard: *Allophylus longeracemosus* Standl.; *Sapindac.;* br. Hond.
Axle wood: *Anogeissus latifolia* Wall.; *Combretac.;* Ind.
Aya: *Platanus spp.; Platanac.;* Mex.
Ayacahuite: *Pinus spp.; Conifer.;* Mex.
    „   colorado: *Pinus spp.; Conifer.;* Mex.
Ayal: *Crescentia sp.; Bignoniac.;* Mex.
Ayale: *Crescentia sp.; Bignoniac.;* Mex.
Ayalid: *Balanites aegyptiaca* Delile; *Simarubac.;* Senegal.
Ayapana: *Eupatorium ayapana* Vent.; *Composit.;* Réunion.
Ayawa: *Icica sp.; Burserac.;* fr. Gu.
Ayedru: *Thespesia populnea* Correa; *Malvac.;* Go.
Ayele: *Crescentia alata* H. B. K.; *Bignoniac.;* Mex.
Ayenda: *Anthocleista zenkeri* Gilg; *Loganiac.;* Kam.
Ayési: *Klainedoxa gabonensis* Pierre; *Simarubac.;* Gab.
Ayigbe atia: *Blighia sapida* Koenig; *Sapindac.;* Go.
    „   ogbedei: *Ceiba pentandra* Gaertn. *var. ?; Bombacac.;* Go.
Ayikui: *Monodora myristica* Dunal; *Anonac.;* Go.
Ayin: *Anogeissus leiocarpa* Guill. et Perr.; *Combretac.;* Af.
Ayinebe: *Anthocleista nobilis* G. Don; *Loganiac.;* Kam., Gab.
Ayinre: *Albizzia browni* Walp.; *Mimosac.;* W.-Af.
Ayirewamba: *Monodora myristica* Dunal; *Anonac.;* Go.
Aylanthe: *Ailanthus glandulosa* Desf.; *Simarubac.;* O.-As.
Aylapia: *Gordonia sp.; Theaceae;* Amboina.
Ayle: *Alnus sp.; Betulac.;* Mex.
Aynre oga: *Albizzia fastigiata* Oliv.; *Mimosac.;* tr. Af.
Ayos: *Triplochiton scleroxylon* K. Schum.; *Sterculiac.;* Kam.
Ayous (f): *Triplochiton scleroxylon* K. Schum.; *Sterculiac.;* Kam.
Ayua: *Zanthoxylum sp.; Rutac.;* Cuba.
    „   amarilla: *Zanthoxylum sp.; Rutac.;* Cuba.
    „   blanca: *Zanthoxylum sp.; Rutac.;* Cuba.
Ayuda: *Zanthoxylum sp.; Rutac.;* Cuba.
    „   amarilla: *Zanthoxylum lanceolatum* Poir.; *Rutac.;* Cuba.
    „   blanca: *Zanthoxylum juglandifolium* Willd.; *Rutac.;* Cuba.
    „   hembra: *Zanthoxylum juglandifolium* Willd.; *Rutac.;* Cuba.
    „   macho: *Zanthoxylum lanceolatum* Poir.; *Rutac.;* Cuba.
    „   varía: *Zanthoxylum sp.; Rutac.;* Cuba.
Ayuelo: *Erythroxylon sp.; Erythroxylac.;* Kol.
Ayui-hu: *Ocotea spectabilis* Mez; *Laurac.;* Arg. (Mis.)
Ayuí-morotí: *Ocotea sp.; Laurac.;* Arg.
    „   -pará: *Phoebe sp.; Laurac.;* Arg.
    „   -pichaí: *Phoebe sp.; Laurac.;* Arg.
    „   -saiyú: *Ocotea puberula* Nees, *O. lanceolata* Nees; *Laurac.;* Arg.
Ayunandy: *Persea sp.; Laurac.;* Par.
Ayuní: *Bischofia javanica* Blume; *Euphorbiac.;* Phil.

Ayupag: *Euphoria cinerea* Radlk.; *Sapindac.;* Phil.
Ayuy hú: *Phoebe porphyria* Mez; *Laurac.;* Arg.
„ -pichai: *Phoebe vesiculosa* Mez; *Laurac.;* Arg.
Aza: *Mimusops djave* Engl., *M. pierreana* Engl.; *Sapotac.;* Kam.. Gab.
Azabar de monte: *Clusia sp.; Guttifer.;* C.-R.
Azabon: *Citrus decumana* Willd.; *Rutac.;* Jap.
Azad: *Ulmus sp.; Ulmac.;* Klein-As., Pers.
Azahar de la India: *Murraya exotica* L.; *Rutac.;* Ven.
Azaharillo: *Capparis sp.; Capparidac.;* Hond.
Azaharito: *Tabernaemontana sp.; Apocynac.;* Ven. (Maracaibo)
Azala: *Croton oligandrus* Pierre; *Euphorbiac.;* Kam.
Azang: *Mimusops sp.; Sapotac.;* Kam.
Azar-rá: *Capparis cyanophallophora* Griseb.; *Capparidac.;* Arg. (Form.)
Azédarach (f): *Melia azedàrach* L.; *Meliac.;* O.-As.
„ du Japon (f): *Melia japonica* G. Don; *Meliac.;* Jap.
Azeitona do matto: *Rapanea sp.; Myrsinac.;* Bras.
Azevinho (p): *Ilex aquifolium* L.; *Aquifoliac.;* Eu., N.-Af.
Azobé (f): *Lophira procera* Chev., *L. alata* Banks; *Ochnac.;* Elf., Kam., Gab.
Azobe kyere pone: *Lophira alata* Banks; *Ochnac.;* Go.
Azodo: *Sterculia oblonga* Mart.; *Sterculiac.;* Elf.
Azoite cavallo: *Luehea grandiflora* Mart.; *Tiliac.;* S.-Am.
Azongue vegetal: *Ficus cystopoda* Miq.; *Morac.;* Bras.
Azota caballo: *Luehea divaricata* Mart.; *Tiliac.;* Arg.
Azote: *Hampea panamensis* Standl.; *Bombacac.;* Pan.
„ de Caballos: *Luehea divaricata* Mart.; *Tiliac.;* W.-Ind.
Azou: *Chrysophyllum balansae* Baill.; *Sapotac.;* N.-Kal.
„ noir: *Chrysophyllum wakere* Panch. et Sébert; *Sapotac.;* N.-Kal.
Azrelei: *Tamarix articulata* Vahl; *Tamaricac.;* Arabien.
Azucarero: *Hedwigia balsamifera* Swartz; *Burserac.;* Cuba.
„ del monte: *Hedwigia balsamifera* Swartz; *Burserac.;* Kol.
„ de montaña: *Hedwigia balsamifera* Swartz; *Burserac.;* Cuba.
Azucena del monte: *Capparis sp.; Capparidac.;* Arg.
Azuceno: *Plumiera sp.; Apocynac.;* Kol.
Azufaifo: *Zizyphus mauritiana* Lam.; *Rhamnac.;* Ven.
Azukinashi: *Micromeles alnifolia* Koehne; *Rosac.;* Jap.
Azulillo: *Haematoxylon brasiletto* Karst.; *Caesalpiniac.;* Mex. (Oaxaca)

B

Ba: *Pentaclethra macrophylla* Bth.; *Mimosac.;* Kam.
Baaka-apiesi-ie: *Nectandra sp., Ocotea sp.; Laurac.;* Sur.
Baäkalaka: *Eschweilera spp.; Lecythidac.;* Sur.
Baaka moonba: *Virola mycetis* Pulle; *Myristicac.;* Sur.
Baäka-tjabisie: *Diplotropis guianensis* Bth.; *Papilionac.;* Sur.
„ „ : *Diplotropis leptophylla* Kleinh.; *Papilionac.;* Sur.
Baäklaka: *Eschweilera spp.; Lecythidac.;* Sur.
Baäsa monbé: *Tapirira aff. guianensis* Aubl.; *Anacardiac.;* Sur.
Baäsa mopé: *Tapirira aff. guianensis* Aubl.; *Anacardiac.;* Sur.
Baäsra-kokonie-oedoe: *Pouteria sp.; Sapotac.;* Sur.
Babacón: *Rhytidophyllum auriculatum* Hook.; *Gesneriac.;* P.-R.
Babaï: *Maxwellia lepidota* Baill.; *Sterculiac.;* N.-Kal.
Babáian: *Allaeanthus luzonicus* Vid.; *Morac.;* Phil.
Babaisákan: *Quercus sp.; Fagac.;* Phil.
Babam: *Anthocleista nobilis* G. Don; *Loganiac.;* Kam.
Babangánon: *Shorea negrosensis* Foxw.; *Dipterocarpac.;* Phil.
Babassu: *Orbignia martiana* Barb. Rod.; *Palmac.;* Bras.
Babdah: *Lagerstroemia hypoleuca* Kurz; *Lythrac.;* And.
Babela: *Terminalia belerica* Roxb.; *Combretac.;* Ind.
Ba bep: *Commersonia echinata* Forst. *var. platyphylla* Andr.; *Sterculiac.;* Ind.-Ch.
Ba bêt: *Alangium chinense* Lour.; *Cornac.;* Ind.-Ch.
Babie Koeroes: *Eurycoma longifolia* Jack; *Simarubac.;* Nied.-Ind.

Bá binh (H): *Eurycoma longifolia* Jack; *Simarubac.*; Ind.-Ch.
Babla: *Acacia arabica* Willd.; *Mimosac.*; n. Ind.
Baboen: *Myristica surinamensis* Warb.; *Myristicac.*; Sur.
Baboen-banjie: *Pithecolobium pedicellare* Bth.; *Mimosac.*; Sur.
Baboenhoedoe: *Virola surinamensis* Warb.; *Myristicac.*; Sur.
Baboenhout: *Virola surinamensis* Warb.; *Myristicac.*; Sur.
„ Bastard: *Virola mycetis* Pulle; *Myristicac.*; Sur.
Baboennoot: *Omphalea diandra* L.; *Euphorbiac.*; Sur.
Baboen trie: *Myristica sebifera* Sw.; *Myristicac.*; fr. Gu.
„ walaba: *Eperua schomburgkiana* Bth.; *Caesalpiniac.*; Sur.
Babosa branca: *Cordia sp.*; *Borraginac.*; Bras.
Babou: *Dumoria heckeli* Chev.; *Sapotac.*; Elf.
Baboul: *Salvadora persica* L.; *Salvadorac.*; Sahara.
Baboun houdou: *Virola sp.*; *Myristicac.*; Gu.
Babubali: *Duguetia guianensis* DC.; *Anonac.*; fr. Gu.
Babul: *Acacia arabica* Willd.; *Mimosac.*; Ind.
Babun: *Virola prob. surinamensis* Warb.; *Myristicac.*; Sur.
„ hudu: *Virola sp.*; *Myristicac.*; Gu.
Babusano: *Phoebe barbusano* Webb; *Laurac.*; Tenerife.
Bác (H): *Kurrimia robusta* Kurz var. *tonkinensis* Kurg.; *Celastrac.*; Ind.-Ch.
Bac: *Anisoptera cochinchinensis* Pierre; *Dipterocarpac.*; Ind.-Ch.
„ : *Araucaria excelsa* R. Br.; *Conifer.*; Ind.-Ch.
Bacabuay: *Curatella americana* L.; *Dilleniac.*; Cuba.
Bacauan: *Rhizophora mucronata* Lam.; *Rhizophorac.*; tr. As., Au.
„ : *Rhizophora conjugata* L.; *Rhizophorac.*; Phil.
„ babáe: *Rhizophora sp.*; *Rhizophorac.*; Phil.
„ gubat: *Carallia integerrima* DC.; *Rhizophorac.*; Phil.
„ laláki: *Rhizophora sp.*; *Rhizophorac.*; Phil.
Bacellina (i): *Genista tinctoria* L.; *Papilionac.*; Eu., w. As.
Bách (H): *Cupressus funebris* Endl.; *Conifer.*; Ind.-Ch.
Bach-dan: *Dysoxylum loureiri* Pierre; *Meliac.*; Ind.-Ch.
„ dau: *Dysoxylum loureiri* Pierre; *Meliac.*; Ind.-Ch.
„ diép: *Cupressus funebris* Endl.; *Conifer.*; Ind.-Ch.
„ duóng: *Santalum album* L.; *Santalac.*; Ind.-Ch.
Ba chia: *Aglaia merostela* Pellegr.; *Meliac.*; Ind.-Ch.
„. chó: *Ilex rotunda* Thib.; *Aquifoliac.*; Ind.-Ch.
Bách tán: *Araucaria excelsa* R. Br.; *Conifer.*; Ind.-Ch.
„ túng: *Podocarpus cupressina* R. Br.; *Conifer.*; Ind.-Ch.
Bac ken: *Aesculus chinensis* Bunge; *Hippocastanac.*; Ind.-Ch.
Bac lá: *Croton argyratus* Blume; *Euphorbiac.*; Ind.-Ch.
Baclo: *Pterospermum suberifolium* Lam.; *Sterculiac.*; O.-Ind.
Baco: *Gustavia spp.*; *Lecythidac.*; Kol.
Bacon wood: *Pithecolobium lovellae* Bailey; *Mimosac.*; Au.
Bacú: *Cariniana pyriformis* Miers; *Lecythidac.*; Kol., Ven.
Bacuri: *Platonia insignis* Mart.; *Guttifer.*; Bras.
„ grande: *Platonia insignis* Mart.; *Guttifer.*; fr. Gu.
Bacurubú: *Schizolobium excelsum* Vog.; *Caesalpiniac.*; Bras. (Sao P.)
Bacury (H): *Platonia insignis* Mart.; *Guttifer.*; Bras.
„ -pary: *Rheedia macrophylla* Pl. et Tr.; *Guttifer.*; Bras. (Amaz.)
Badam: *Terminalia catappa* L.; *Combretac.*; And.
„ : *Terminalia procera* Roxb.; *Combretac.*; Ind.
Badamier (f): *Terminalia catappa* L.; *Combretac.*; Tr.
Badi (f): *Sarcocephalus trillesi* Pierre; *Rubiac.*; Elf., Gab.
„ : *Parinarium robustum* Oliv.; *Rosac.*; Elf.
„ : *Bombax buonopozense* Beauv.; *Bombacac.*; Elf.
„ : *Bridelia micrantha* Baill.; *Euphorbiac.*; Go.
Badiane (f): *Illicium anisatum* L.; *Magnoliac.*; Jap.
. „ du Cambodge: *Illicium cambodgianum* Hance; *Magnoliac.*; Kamb.
Badiángau: *Agathis alba* Foxw.; *Conifer.*; Phil.
Badjibri: *Strombosia pustulata* Oliv.; *Olacac.*; Elf.
Badjoer: *Pterospermum sp.*; *Sterculiac.*; Bangka.
Badjong: *Acacia microbotrya* Bth.; *Mimosac.*; W.-Au.

Badláu: *Radermachera sp.; Bignoniac.;* Phil.
Bado: *Schleichera trijuga* Willd.; *Sapindac.;* Macassar.
Badomier: *Terminalia catappa* L.; *Combretac.;* Senegal.
Badula: *Ardisia spp.; Myrsinac.;* P.-R.
Baeg: *Allaeanthus glaber* Warb.; *Morac.;* Phil.
Bael: *Aegle marmelos* Correa; *Rutac.;* Ind.
Baerentraube, Kalifornische: *Arctostaphylos pungens* H. B. K.; *Ericac.;* USA.
Baga: *Anona glabra* L.; *Anonac.;* Cuba.
Bàga: *Erythrophloeum guineense* G. Don; *Caesalpiniac.;* Gab.
Bagabág: *Eugenia mananquil* Blco.; *Myrtac.;* Phil.
Bagac (u): *Dipterocarpus grandiflorus* Blco.; *Dipterocarpac.;* Phil.
Bagaceira: *Bagassa guianensis* Aubl.; *Morac.;* Bras.
Bagákai: *Schizostachyum lumampao* Merr.; *Gramineae;* Phil.
Bagalayáu: *Pahudia rhomboidea* Prain; *Caesalpiniac.;* Phil.
Bagamani: *Tapirira myriantha* Triana et Pl.; *Anacardiac.;* Pan. (Colon)
Bagana: *Acacia arabica* Willd.; *Mimosac.;* Sudan.
Bagangsúsu: *Vatica mangachapoi* Blco.; *Dipterocarpac.;* Phil.
Baganitó: *Diospyros currani* Merr., *D. pilosanthera* Blco.; *Ebenac.;* Phil.
Bagarilao: *Anthocephalus cadamba* Miq.; *Rubiac.;* Phil.
Bagariláu: *Litsea sp.; Laurac.;* Phil.
„   : *Neonauclea sp.; Rubiac.;* Phil.
Bagaroua: *Acacia arabica* Willd.; *Mimosac.;* Sudan.
Bagaruwa: *Acacia arabica* Willd.; *Mimosac.;* n. Nig.
„   namiji: *Acacia nilotica* Delile; *Mimosac.;* n. Nig.
„   ta mata: *Acacia arabica* Willd.; *Mimosac.;* n. Nig.
Bagaruwar makka: *Moringa pterygosperma* Gaertn.; *Moringac.,* n. Nig.
Bagasse: *Bagassa guianensis* Aubl.; *Morac.;* fr. Gu.
„   : *Icica sp.; Burserac.;* fr. Gu.
„   jaune: *Bagassa guianensis* Aubl.; *Morac.;* fr. Gu.
Bagasusu: *Vatica mangachapoi* Blco.; *Dipterocarpac.;* Phil.
Bagatal: *Kayea paniculata* Merr.; *Guttifer.;* Phil.
Bagatáru: *Cyclostemon sp.; Euphorbiac.;* Phil.
Bagbalógo: *Kingiodendron alternifolium* Merr.; *Caesalpiniac.;* Phil.
Bage: *Albizzia lebbek* Bth.; *Mimosac.;* Ind.
Baggie-baggie: *Minquartia guianensis* Aubl.; *Olacac.;* Sur.
Bagilumbáng: *Aleurites trisperma* Blco.; *Euphorbiac.;* Phil.
Bagin: *Parkia timoriana* Merr.; *Mimosac.;* Phil.
Bagiróro: *Adenanthera intermedia* Merr.; *Mimosac.;* Phil.
„   : *Cassia javanica* L.; *Caesalpiniac.;* Phil.
Bagná: *Bischofia javanica* Blume; *Euphorbiac.;* Phil.
Bagodiláu: *Neonauclea sp.; Rubiac.;* Phil.
Bagosantól: *Sandoricum vidali* Merr.; *Meliac.;* Phil.
Bagot: *Cynometra notzmaniana* ?; *Caesalpiniac.;* fr. Gu.
„   : *Copaifera pubiflora* Heyne; *Caesalpiniac.;* fr. Gu.
Bagotambis: *Eugenia sp.; Myrtac.;* Phil.
Bagre: *Trichilia triphylla* Blake; *Meliac.;* Kol.
Bagsang: *Livistona rotundifolia* Mart. *var. luzonensis* Becc.; *Palmac.;* Phil.
Bagtican: *Parashorea malaanonan* Merr.; *Dipterocarpac.;* Phil.
Bagtikan: *Parashorea malaanonan* Merr.; *Dipterocarpac.;* Phil.
„   Bird's-eye: *Shorea eximia* Scheff.; *Dipterocarpac.;* Phil.
Baguasú: *Ilex paraguariensis* St. Hil.; *Aquifoliac.;* Bras.
Baguenaudier (f): *Colutea arborescens* L.; *Papilionac.;* M.-M.
Bagulañgog: *Terminalia quadrialata* Merr.; *Combretac.;* Phil.
Bagulibás: *Buchanania sp.; Anacardiac.;* Phil.
„   : *Dysoxylum decandrum* Merr.; *Meliac.;* Phil.
„   : *Garuga sp.; Burserac.;* Phil.
Bagunárem: *Lagerstroemia piriformis* Koehne; *Lythrac.;* Phil.
Bagunáum: *Lagerstroemia piriformis* Koehne; *Lythrac.;* Phil. (Zamb.)
Bah: *Scaphopetalum amoenum* Chev.; *Sterculiac.;* Lib.
Báhai: *Adenanthera intermedia* Merr.; *Mimosac.;* Phil. (Zamb.)
„   : *Kingiodendron alternifolium* Merr.; *Caesalpiniac.;* Phil. (Zamb.)
„   : *Ormosia calavensis* Azaola; *Papilionac.;* Phil. (Zamb.)

Bahan: *Populus euphratica* Oliv.; *Salicac.;* Ind. (Bombay)
Bahê: *Fagara macrophylla* Engl.; *Rutac.;* Elf.
Bahera: *Terminalia belerica* Roxb.; *Combretac.;* Ind.
Bahi: *Livistona rotundifolia* Mart. *var.luzonensis* Becc.; *Palmac.;* Phil.
Bahia (f): *Mitragyne macrophylla* Hiern; *Rubiac.;* W.-Af., Gab.
„   „ : *Coula edulis* Baill.; *Olacac.;* Elf.
Bahiawood (H): *Caesalpinia echinata* Lam.; *Caesalpiniac.;* Bras.
Bahn: *Ochrocarpus africanus* Oliv.; *Guttifer.;* Lib.
Báho: *Terminalia edulis* Blco.; *Combretac.;* Phil.
Baibya: *Cratoxylon neriifolium* Kurz; *Guttifer.;* H.-Ind.
Bai cánh: *Sterculia populifolia* Roxb.; *Sterculiac.;* Ind.-Ch.
Baid: *Sideroxylon macranthum* Merr.; *Sapotac.;* Phil.
Bailador: *Guarea spp.; Meliac.;* Kol.
Baing: *Tetrameles nudiflora* R. Br.; *Datiscac.;* Ind.
Baiñha de Espada: *Eperua falcata* Aubl.; *Caesalpiniac.;* Gu., Bras.
Bainha de espado: *Sahagunia strepitans* Liebw.; *Morac.;* Bras. (R. J.)
Baïra: *Cynometra alexandri* Wright; *Caesalpiniac.;* B.-K.
Bairadah: *Rhizophora mucronata* Lam.; *Rhizophorac.;* And.
Baít: *Euphoria cinerea* Radlk.; *Sapindac.;* Phil.
Bai thura (H): *Sterculia spp.; Sterculiac.;* Ind.-Ch.
Baitoa: *Phyllostylon brasiliensis* Capanema; *Ulmac.;* St.-D.
Baiukan: *Parashorea malaanonan* Merr.; *Dipterocarpac.;* Phil.
„   : *Shorea squamata* Dyer; *Dipterocarpac.;* Phil.
Baiúkan: *Quercus sp.; Fagac.;* Phil.
„   -pulá: *Parashorea malaanonan* Merr.; *Dipterocarpac.;* Phil.
Baja: *Cynometra alexandri* Wright; *Caesalpiniac.;* B.-K.
Bajamhout: *Afzelia bijuga* A. Gray; *Caesalpiniac.;* Cel.
Bajoer: *Pterospermum blumeanum* Korth.; *Sterculiac.;* Sunda.
„   sipoeloet: *Pterospermum sp.; Sterculiac.;* Sum.
Bajoor Kalang: *Pterospermum sp.; Sterculiac.;* Sum.
Bajor: *Pterospermum blumeanum* Korth.; *Sterculiac.;* Malay.
Bak: *Anisoptera cochinchinensis* Pierre; *Dipterocarpac.;* Ind.-Ch.
Bakabe-ie: *Humiria floribunda* Mart., *H.balsamifera* Aubl.; *Humiriac.;* Sur.
Baka-Biringui: *Entandrophragma septentrionale* Chev.; *Meliac.;* Elf.
„   -kounini: *Cynometra ananta* Hutch. et Dalz.; *Caesalpiniac.;* Elf.
Bakalao: *Euphoria cinerea* Radlk.; *Sapindac.;* Phil.
Bakaláu: *Euphoria cinerea* Radlk.; *Sapindac.;* Phil.
„   : *Nephelium mutabile* Blume; *Sapindac.;* Phil.
„   : *Quercus sp.; Fagac.;* Phil. (Zamboanga)
„   : *Aglaia bicolor* Merr.; *Meliac.;* Phil.
Bakam: *Caesalpinia sappan* L.; *Caesalpiniac.;* Ind.
Bâkân: *Rhizophora sp.; Rhizophorac.;* Malay.
Bakân: *Litsea philippinensis* Merr.; *Laurac.;* Phil.
„   : *Canarium luzonicum* A. Gray; *Burserac.;* Phil.
„   -kalaánan: *Litsea sp.; Laurac.;* Phil.
Bakan Hak: *Thujopsis dolobrata* S. et Z.; *Conifer.;* Jap.
Bakana: *Chlorophora excelsa* Bth. et Hook.; *Morac.;* Elf.
Bakapu: *Caesalpinia sappan* L.; *Caesalpiniac.;* O.-Ind., Malay.
Bakau: *Rhizophora spp.; Rhizophorac.;* Phil., Burma.
Bakáu: *Bruguiera gymnorrhiza* Lam.; *Rhizophorac.;* Phil.
Bakau Kurap: *Rhizophora mucronata* Lam.; *Rhizophorac.;* Malay.
Bakáuan: *Bruguiera spp., Ceriops roxburghiana* Arn.; *Rhizophorac.;* Phil.
„   -babáe: *Rhizophora mucronata* Lam.; *Rhizophorac.;* Phil.
„   -baláki: *Rhizophora sp.; Rhizophorac.;* Phil.
Bakáyau: *Parinarium sp.; Rosac.;* Phil.
Bakaza: *Scottellia coriacea* Chev.; *Flacourtiac.;* Elf.
Bakbaboi: *Albizzia fastigiata* Oliv.; *Mimosac.;* tr. Af.
Bakeles: *Euphoria cinerea* Radlk.; *Sapindac.;* Phil.
Bakhau: *Rhizophora sp.; Rhizophorac.;* Phil.
Ba khia: *Lophopetalum wightianum* Arn. *var. macrocarpum* Arn.; *Celastrac.;* Ind.-Ch.
Bakin makarfo: *Burkea africana* Hook.; *Caesalpiniac.;* n. Nig.
Bakkaláu: *Euphoria cinerea* Radlk.; *Sapindac.;* Phil.

Bakkie-bakki: *Conceveiba guianensis* Aubl., *C. hostmanni* Bth.; *Euphorbiac.;* Sur.
Bakli: *Anogeissus latifolia* Wall.; *Combretac.;* Ind.
„ : *Lagerstroemia parvifolia* Roxb. *var. majuscula* Clarke; *Lythrac.;* nw. Ind.
Bakmi: *Sarcocephalus cordatus* Miqu.; *Rubiac.;* Cey.
Baknítan: *Shorea polysperma* Merr.; *Dipterocarpac.;* Phil.
Bako: *Rhizophora conjugata* L.; *Rhizophorac.;* Java.
Bakoerie: *Platonia insignis* Mart.; *Guttifer.;* Sur.
Bakoly: *Aleurites fordi* Hemsl., *A. triloba* Forst.; *Euphorbiac.;* Mad.
Bakondo: *Pycnanthus kombo* Warb.; *Myristicac.;* Kam.
Bakóog: *Canarium luzonicum* A. Gray; *Burserac.;* Phil.
Bâkor: *Shorea vulgaris* Pierre; *Dipterocarpac.;* Ind.-Ch.
Bakororo: *Gymnosporia senegalensis* Loes.; *Celastrac.;* n. Nig.
Bakota: *Endospermum malaccense* Bth.; *Euphorbiac.;* Ind.
Bakting: *Lumnitzera littorea* Voigt; *Combretac.;* Phil.
Baku: *Mimusops sp.; Sapotac.;* Go.
Bakuchinoki: *Prunus macrophylla* S. et Z.; *Rosac.;* Jap.
Bakul: *Mimusops elengi* L.; *Sapotac.;* Ind.
Bakula: *Mimusops elengi* L.; *Sapotac.;* Ind., Cey.
Bakunin: *Alstonia congensis* Engl.; *Apocynac.;* Go.
Baku tjolan: ? *Aglaia odorata* Lour.; *Meliac.;* Java.
Bala: *Hibiscus tiliaceus* L.; *Malvac.;* Ind.
„ de cañon: *Couroupita guianensis* Aubl.; *Lecythidac.;* Cuba.
Balábak: *Shorea squamata* Dyer; *Dipterocarpac.;* Phil.
Balabían: *Intsia acuminata* Merr.; *Caesalpiniac.;* Phil.
Balácat: *Zizyphus talanai* Merr.; *Rhamnac.;* Phil.
Balagáyan: *Shorea polysperma* Merr.; *Dipterocarpac.;* Phil.
Balagbág: *Pentacme contorta* Merr. et Rolfe; *Dipterocarpac.;* Phil.
Balag Tiong: *Shorea glauca* King; *Dipterocarpac.;* Malay.
Balahiáu: *Pahudia rhomboidea* Prain; *Caesalpiniac.;* Phil.
Balai: *Diospyros melanoxylon* Roxb.; *Ebenac.;* s. Ind.
„ des bas: *Erythroxylon hypericifolium* Lamk.; *Erythroxylac.;* Réu.
„ „ bois: *Erythroxylon hypericifolium* Lamk.; *Erythroxylac.;* Réu.
Bálai-nák: *Oroxylum indicum* Vent.; *Bignoniac.;* Phil. (Zamb.)
Balaire: *Desmoncus polyacanthos* Mart.; *Palmac.;* Hond.
Balais: *Quivisia laciniata* Balf. f.; *Meliac.;* Maur.
Balák: *Pentacme contorta* Merr. et Rolfe; *Dipterocarpac.;* Phil.
Balak: *Livistona rotundifolia* Mart. *var. luzonensis* Becc.; *Palmac.;* Phil.
Balakbák: *Eugenia xanthophylla* C. B. Rob.; *Myrtac.;* Phil. (Zamb.)
„ : *Pentacme contorta* Merr. et Rolfe; *Dipterocarpac.;* Phil.
Balakbákan: *Parashorea malaanonan* Merr.; *Dipterocarpac.;* Phil.
„ : *Shorea spp.; Dipterocarpac.;* Phil.
Balak Tiong: *Shorea sp.; Dipterocarpac.;* Malay.
Balalaboué: *Couratari guianensis* Aubl.; *Lecythidac.;* fr. Gu.
Balambanan: *Pometia pinnata* Forst.; *Sapindac.;* Phil.
Balambu: *Averrhoa bilimbi* L.; *Oxalidac.;* Guad.
Balamo: *Citharexylum fruticosum* L.; *Verbenac.;* P.-R.
Balam Teroeng: *Palaquium hexandrum* Endl.; *Sapotac.;* Java.
Balang: *Shorea balangeran* Dyer; *Dipterocarpac.;* Phil. (Zamb.)
Baláñga: *Buchanania sp.; Anacardiac.;* Phil.
Balangáñan: *Litsea sp.; Laurac.;* Phil .
Balangeran: *Shorea balangeran* Burck; *Dipterocarpac.;* Phil., Borneo.
Balangód: *Litsea sp.; Laurac.;* Phil.
Balangra: *Flacourtia ramontchi* L'Hérit.; *Flacourtiac.;* O.-Ind.
Balantí: *Aglaia llanosiana* C. DC.; *Meliac.;* Phil.
Balao: *Dipterocarpus vernicifluus* Blume; *Dipterocarpac.;* w. Borneo.
„ : *Dipterocarpus grandiflorus* Blco.; *Dipterocarpac.;* Phil.
Balata: *Manilkara bidentata* Chev.; *Sapotac.;* Gu.
„ : *Mimusops balata* Gaertn.; *Sapotac.;* Gu.
„ : *Micropholis melinoniana* Pierre; *Sapotac.;* fr. Gu.
„ : *Bumelia retusa* Sw.; *Sapotac.;* St.-D.
„ : *Mimusops globosa* Gaertn.; *Sapotac.;* Trin.
„ : *Achras sapota* L.; *Sapotac.;* Monts.

Balata: *Mimusops darienensis* Pittier; *Sapotac.;* Pan.
„ : *Mimusops elata* Allem.; *Sapotac.;* Ven.
„ : *Ecclinusa balata* Ducke; *Sapotac.;* Bras. (Amaz.)
Balataballi: *Calocarpum mammosum* Pierre; *Sapotac.;* br. Gu.
Balata blanc: *Plumiera spp.; Apocynac.;* fr. Gu.
„ „ : *Micropholis melinoniana* Pierre; *Sapotac.;* Gu.
„ „ Soms: *Couratari guianensis* Aubl.; *Myrtac.;* fr. Gu.
„ Bloed: *Mimusops spp., Manilkara balata* Pierre; *Sapotac.;* Gu.
„ „ : *Ecclinusa sanguinolenta* Pierre; *Sapotac.;* Gu.
„ boom: *Mimusops spp.; Sapotac.;* Sur.
„ des galibis: *Mimusops balata* Gaertn.; *Sapotac.;* Gu.
„ Dooier: *Lucuma rivicoa* Gaertn.; *Sapotac.;* fr. Gu.
„ franc (f): *Manilkara balata* Pierre; *Sapotac.;* Gu.
„ „ (f): *Mimusops spp.; Sapotac.;* Sur.
„ Indianen: *Labatia macrocarpa* Mart.; *Sapotac.;* fr. Gu.
„ indien: *Labatia macrocarpa* Mart.; *Sapotac.;* fr. Gu.
„ „ (f): *Sideroxylon guianense* A. DC.; *Sapotac.;* Sur.
„ jaune d'oeuf: *Lucuma rivicoa* Gaertn.; *Sapotac.;* fr. Gu.
„ „ „ : *Labatia macrocarpa* Mart.; *Sapotac.;* fr. Gu.
„ „ „ (f): *Sideroxylon guianense* A. DC.; *Sapotac.;* Sur.
„ moiré (f): *Sideroxylon guianense* A. DC.; *Sapotac.;* Sur.
„ pommier: *Ecclinusa sanguinolenta* Pierre; *Sapotac.;* Gu.
„ red: *Mimusops sp.; Sapotac.;* fr. Gu.
„ roode: *Mimusops spp., Manilkara balata* Pierre; *Sapotac.;* Gu.
„ „ : *Ecclinusa sanguinolenta* Pierre; *Sapotac.;* Gu.
„ rosada: *Sideroxylon cyrtobotryum* Miq.; *Sapotac.;* Bras.
„ „ : *Sideroxylon resiniferum* Ducke; *Sapotac.;* Bras.
„ rouge (f): *Manilkara balata* Pierre; *Sapotac.;* Gu.
„ „ „ : *Mimusops spp.; Sapotac.;* Gu.
„ „ : *Ecclinusa sanguinolenta* Pierre; *Sapotac.;* Gu.
„ „ : *Oxytheca hahnianum* Pierre; *Sapotac.;* Guad.
„ saignant: *Mimusops spp., Manilkara balata* Pierre; *Sapotac.;* Gu.
„ „ : *Ecclinusa sanguinolenta* Pierre; *Sapotac.;* Gu.
„ „ à lait rouge: *Ecclinusa sessiliflora* Pierre; *Sapotac.;* fr. Gu.
„ singe rouge: *Labatia macrocarpa* Mart.; *Sapotac.;* fr. Gu.
„ white: *Micropholis melinoniana* Pierre; *Sapotac.;* fr. Gu.
„ witte: *Micropholis melinoniana* Pierre; *Sapotac.;* Gu.
„ „ : *Plumiera spp.; Apocynac.;* Gu.
Balatinán: *Diospyros pilosanthera* Blco.; *Ebenac.;* Phil. (n. Luzon)
Balat-usín: *Planchonia spectabilis* Merr.; *Lecythidac.;* Phil.
Bálau: *Anisoptera curtisi* Dyer; *Dipterocarpac.;* Phil.
„ : *Dipterocarpus spp.; Dipterocarpac.;* Phil.
„ : *Pentacme contorta* Merr. et Rolfe; *Dipterocarpac.;* Phil.
„ : *Parinarium oblongifolium* Hook. f.; *Rosac.;* Malay.
„ : *Shorea materialis* Ridl., *S. collina* Ridl.; *Dipterocarpac.;* Malay.
„ bukit: *Shorea collina* Ridl.; *Dipterocarpac.;* Malay.
„ Giam: *Vatica sp.; Dipterocarpac.;* Borneo, Sum.
„ „ : *Shorea sp., Hopea sp., Isoptera sp.; Dipterocarpac.;* Malay.
„ pipit: *Shorea collina* Ridl.; *Dipterocarpac.;* Malay.
„ ulat: *Parinarium rubiginosum* Ridl.; *Rosac.;* Malay.
„ „ : *Angelisia splendens* Korth.; *Rosac.;* Malay.
Balaustre: *Centrolobium paraënse* Tul.; *Papilionac.;* Ven.
„ : *Centrolobium robustum* Mart.; *Papilionac.;* Ven. (Maracaibo)
Balawo-Kudu: *Afzelia africana* Smith; *Caesalpiniac.;* B.-K.
Balayóhot: *Buchanania sp.; Anacardiac.;* Phil.
Baláyong: *Cassia javanica* L.; *Caesalpiniac.;* Phil.
Baláyung: *Pahudia rhomboidea* Prain; *Caesalpiniac.;* Phil.
„ : *Sindora supa* Merr.; *Caesalpiniac.;* Phil.
„ : *Wallaceodendron celebicum* Koord.; *Mimosac.;* Phil.
Balché: *Lonchocarpus hondurensis* Bth.; *Papilionac.;* br. Hond.
„ : *Lonchocarpus longistylus* Pittier; *Papilionac.;* Mex.
Balche-cebi: *Lonchocarpus yucatanensis* Pittier; *Papilionac.;* Mex.

Balchechi: *Lonchocarpus yucatanensis* Pittier; *Papilionac.*; Mex.
Balel: *Coriaria nepalensis* Wall.; *Malpighiac.*; Ind. (Punjab)
Balí: *Nephelium mutabile* Blume; *Sapindac.*; Phil. (Zamboanga)
Bali: *Virola sp.*; *Myristicac.*; Gu.
,. : *Pentaclethra macrophylla* Bth.; *Mimosac.*; Kam.
Baliaro: *Schizostachyum diffusum* Merr.; *Gramin.*; Phil. (Iloilo)
Baliboup: *Uvaria parviflora* A. Rich.; *Anonac.*; w. Senegamb.
Balibud: *Buchanania sp.*; *Anacardiac.*; Phil.
Baliembieng bissie: *Averrhoa bilimbi* L.; *Oxalidac.*; Sunda.
Bali'gáng: *Eugenia sp.*; *Myrtac.*; Phil. (s. Luzon)
Baligóhot: *Buchanania sp.*; *Anacardiac.*; Phil.
Balik: *Euphoria cinerea* Radlk.; *Sapindac.*; Phil. (Zamb.)
Balikau: *Schizostachyum diffusum* Merr.; *Gramin.*; Phil. (Iloilo)
Balikbíkan: *Cyclostemon sp.*; *Euphorbiac.*; Phil.
Baliknít: *Zizyphus sp.*; *Rhamnac.*; Phil.
Balílang-uák: *Oroxylum indicum* Vent.; *Bignoniac.*; Phil.
Balillo: *Sonchus leptocephalus* Cass.; *Composit.*; Tener.
Balimbíngan: *Nephelium mutabile* Blume; *Sapindac.*; Phil.
Balínad: *Sterculia sp.*; *Sterculiac.*; Phil.
Baling-agtá: *Diospyros discolor* Willd., *D. pilosanthera* Blco.; *Ebenac.*; Phil.
Balinghásay: *Buchanania sp.*; *Anacardiac.*; Phil.
Balinsaráyan: *Bruguiera eriopetala* Lam.; *Rhizophorac.*; Phil.
Balinsiagáu: *Aglaia harmsiana* Perk.; *Meliac.*; Phil.
Balióhod: *Buchanania sp.*; *Anacardiac.*; Phil.
Balisáyin: *Terminalia edulis* Blco.; *Combretac.*; Phil. (Mind.)
Balít: *Euphoria cinerea* Radlk.; *Sapindac.*; Phil.
„ : *Litchi philippinensis* Radlk.; *Sapindac.*; Phil.
Balitagtág: *Allaeanthus glaber* Warb.; *Morac.*; Phil.
Balitangtáng: *Buchanania sp.*; *Anacardiac.*; Phil.
Baliwiswís: *Shorea malaanonan* Blume; *Dipterocarpac.*; Phil.
Balla: *Livistona rotundifolia* Mart. var. *luzonensis* Becc.; *Palmac.*; Phil.
Ballang: *Livistona rotundifolia* Mart. var. *luzonensis* Becc.; *Palmac.*; Phil.
Ballatináu: *Diospyros discolor* Willd.; *Ebenac.*; Phil. (n. Luzon)
Ballet-tree: *Mimusops spp.*; *Sapotac.*; Sur.
Balluk: *Eucalyptus globulus* Lab.; *Myrtac.*; N.-S.-W., Tasm.
Balm: *Populus balsamifera* L.; *Salicac.*; Kan.
„ of Gilead: *Populus balsamifera* L.; *Salicac.*; N.-Am.
„ „ heaven: *Oreodaphne californica* Nees; *Laurac.*; USA.
Balo: *Plocama pendula* Ait.; *Rubiac.*; Tenerife.
Baloc baloc: *Pongamia glabra* Vent.; *Papilionac.*; Phil. (Manila)
Baloen adoek: ? *Polyalthia subcordata* Blume; *Anonac.*; N.-Kal.
„ anjoek: ? *Polyalthia subcordata* Blume; *Anonac.*; N.-Kal.
Balógo: *Albizzia saponaria* Blume; *Mimosac.*; Phil.
Báloi: *Pterospermum sp.*; *Sterculiac.*; Phil.
Balong ayam: *Tarrietia spp.*; *Sterculiac.*; Malay.
Balsa: *Ochroma lagopus* Swartz; *Bombacac.*; C.-R., Bras. (ob. Amaz.)
„ : *Ochroma grandiflora* Rowlee; *Bombacac.*; Ek.
„ : *Ochroma bicolor* Rowlee; *Bombacac.*; br. Hond.
„ : *Ochroma limonensis* Rowlee, *O. velutina* Rowlee; *Bombacac.*; Pan.
„ : *Ochroma obtusa* Rowlee; *Bombacac.*; Kol.
„ : *Ochroma boliviana* Rowlee; *Bombacac.*; Bol., Peru.
„ : *Heliocarpus appendiculatus* Turcz.; *Tiliac.*; Nic.
„ : *Heisteria parvifolia* Smith; *Olacac.*; Lib.
Balsa-hout: *Ochroma lagopus* Swartz; *Bombacac.*; Sur.
Balsam: *Abies grandis* L.; *Conifer.*; N.-Am.
„ : *Abies balsamea* Marsh.; *Conifer.*; USA.
„ : *Populus balsamifera* L.; *Salicac.*; N.-Am.
„ : *Myroxylon pereirae* Klotzsch; *Papilionac.*; br. Hond.
„ : *Myroxylon toluiferum* H. B. K.; *Papilionac.*; br. Hond.
„ : *Mimusops globosa* Gaertn.; *Sapotac.*; Trin., Tobago.
„ : *Copaifera officinalis* Willd.; *Caesalpiniac.*; Trin., Tob.
„ of capivitree: *Geijera salicifolia* Schott; *Rutac.*; N.-S.-W.

Balsamtree: *Clusia rosea* Jacq.; *Guttifer.;* br. W.-Ind.
„ : *Clusia insignis* Mart.; *Guttifer.;* Gu.
„ Copaiba: *Daniella oliveri* Hutch. et Dalz.; *Caesalpiniac.;* Go.
Balsam Western: *Abies subalpina* Engelm.; *Conifer.;* Kan.
„ Yellow-flowered: *Clusia flava* Jacq.; *Guttifer.;* USA.
Balsamaria: *Calophyllum sp.; Guttifer.;* Bol.
Balsamillo: *Elaphrium jaquinianum* H. B. K.; *Burserac.;* Ven.
Balsamito: *Myroxylon toluiferum* H. B. K.; *Papilionac.;* C.-R.
Balsamo: *Dalbergia sp.; Papilionac.;* s. Mex.
Bálsamo: *Myroxylon pereirae* Kloßsch; *Papilionac.;* Hond.
Balsamo: *Myroxylon toluiferum* H. B. K.; *Papilionac.;* tr. S.-Am.
Balsámo: *Bursera˟tomentosa* Tr. et Pl.; *Burserac.;* Ven.
Balsamo: *Myrospermum erythroxylon* Allem.; *Papilionac.;* Bras.
Bálsamo: *Schinus molle* L.; *Anacardiac.;* Arg.
Balsamo de tolú: *Myroxylon toluiferum* H. B. K.; *Papilionac.;* Kol.
Balsían: *Shorea polysperma* Merr.; *Dipterocarpac.;* Phil.
Balso: *Ochroma lagopus* Swarß; *Bombacac.;* Ven., Kol.
Baltík: *Agathis alba* Foxw.; *Conifer.;* Phil. (Palawan)
Balú: *Erythrina sp.; Papilionac.;* Kol.
Balu: *Macrolobium dewevrei* de Wild.; *Caesalpiniac.;* B.-K.
Bálu: *Cordia subcordata* Lam.; *Borraginac.;* Phil.
Baluch: *Lonchocarpus sp.; Papilionac.;* br. Hond.
Bálug: *Neonauclea sp.; Rubiac.;* Phil.
Balúi: *Aglaia clarkei* Merr.; *Meliac.;* Phil.
Balukanád: *Aleurites trisperma* Blco.; *Euphorbiac.;* Phil.
Balukanág: *Aleurites trisperma* Blco.; *Euphorbiac.;* Phil.
„ : *Aphanomyxis cumingiana* Harms; *Meliac.;* Phil.
Balukáui: *Dinochloa scandens* O. Kße.; *Gramin.;* Phil. (Mind., Cebu)
Baluknít: *Lagerstroemia piriformis* Koehne; *Lythrac.;* Phil.
Balukúk: *Garcinia sp.; Guttifer.;* Phil.
Balungáu: *Neonauclea sp.; Rubiac.;* Phil.
Balungkauái: *Cordia subcordata* Lam.; *Borraginac.;* Phil.
Balúno: *Cumingia philippinensis* Vid.; *Bombacac.;* Phil. (Zamb.)
Bamanga: *Macrolobium coeruleoides* de Wild.; *Caesalpiniac.;* B.-K.
Bamanta: *Xylia evansi* Hutch.; *Mimosac.;* Go.
Bamba (f): *Croton oligandrum* Pierre; *Euphorbiac.;* Gab. (Loango)
„ : *Triplochiton scleroxylon* K. Schum.; *Sterculiac.;* Elf.
„ -apiesi-ie: *Nectandra sp., Ocotea sp.; Laurac.;* Sur.
„ -apisieaie: *Acrodiclidium canella* Mez; *Laurac.;* Sur.
„ „ : *Acrodiclidium chrysophyllum* Meissn.; *Laurac.;* Sur.
Bambaï: *Discostigma vitiensis* A. Gay; *Guttifer.;* N.-Kal.
Bamboo: *Guadua aculeata* Rupr.; *Gramin.;* br. Hond.
„ Bindura: *Oxytenanthera abyssinica* Munro; *Gramin.;* Rhod.
„ of Bengal, Common: *Bambusa tulda* Roxb.; *Gramin.;* O.-Ind.
„ Female: *Bambusa balcooa* Roxb.; *Gramin.;* O.-Ind.
„ Kayin: *Melocanna bambusoides* Trin.; *Gramin.;* Ind.
„ Male: *Dendrocalamus strictus* Nees; *Gramin.;* O.-Ind.
„ Spiny: *Bambusa arundinacea* Reß.; *Gramin.;* O.-Ind.
Bamboutouli: *Chrysophyllum lacourtianum* de Wild.; *Sapotac.;* Gab.
Bambus: *Oxytenanthera abyssinica* Munro; *Gramin.;* Rhod.
„ bunt: *Phyllostachys puberula* Munro *forma nigropunctata* Makino; *Gramin.;* O-As.
Bambusrohr: *Phyllostachys puberula* Munro; *Gramin.;* O.-As.
Bambus Flecken-: *Phyllostachys puberula* Munro *forma nigropunctata* Makino;
[*Gramin.;* O.-As.
Bambway Nee: *Planchonia littoralis* Van Houtte; *Myrtac.;* Ind.
Bampoo: ? *Oxytenanthera abyssinica* Munro; *Gramin.;* Go.
Bá mu: *Aglaïa gigantea* Pellegrin; *Meliac.;* Ind.-Ch.
Bân (H): *Sonneratia acida* L.; *Punicac.;* Ind.-Ch.
„ : *Pithecolobium clypearia var. acuminatum* Bth.; *Mimosac.;* Ind.-Ch.
Ban: *Bauhinia variegata* L.; *Caesalpiniac.;* Ind.-Ch.
„ : *Lophira alata* Banks; *Ochnac.;* Kam.
Bana: *Xylopia rubescens* Oliv.; *Anonac.;* Kam.

Banaási: *Murraya exotica* L.; *Rutac.;* Phil.
Banaba: *Lagerstroemia speciosa* Pers., *L. flos-reginae* Reß.; *Lythrac.;* Phil.
Banabáng-bugtúng: *Lagerstroemia piriformis* Koehne; *Lythrac.;* Phil.
Banábang-dinglás: *Lagerstroemia piriformis* Koehne; *Lythrac.;* Phil.
Banabáng-tináan: *Lagerstroemia piriformis* Koehne; *Lythrac.;* Phil.
Banágau: *Cordia subcordata* Lam.; *Borraginac.;* Phil.
Banágo: *Cordia subcordata* Lam.; *Borraginac.;* Phil.
Banaibánai: *Albizzia saponaria* Blume; *Mimosac.;* Phil.
Banai-bánai: *Radermachera pinnata* Seem.; *Bignoniac.;* Phil.
Banak: *Virola merendonis* Pittier, *V. panamensis* Warb.; *Myristicac.;* C.-Am.
Banálo: *Thespesia populnea* Correa; *Malvac.;* Phil.
„ : *Cordia subcordata* Lam.; *Borraginac.;* Phil.
Banan: *Ceiba pentandra* Gaertn. *var. dehiscens* Ulb.; *Bombacac.;* Sudan.
Banana, false: *Asimina triloba* Dunal; *Anonac.;* ö. N.-Am.
Banáog: *Trichadenia philippinensis* Merr.; *Flacourtiac.;* Phil.
Banarish: *Fraxinus floribunda* Wall.; *Oleac.;* O.-As.
Banási: *Murraya exotica* L.; *Rutac.;* Phil.
Banati: *Murraya exotica* L.; *Rutac.;* Phil.
„ : *Lophopetalum wightianum* Arn.; *Celastrac.;* Ind.
Banáu: *Trichadenia philippinensis* Merr.; *Flacourtiac.;* Phil.
Banáui: *Cyclostemon spp.; Euphorbiac.;* Phil.
„ : *Homalium oblongifolium* Merr.; *Flacourtiac.;* Phil.
Banaye: *Trichilia heudeloti* Planch.; *Meliac.;* Elf.
„ : *Dialium guineense* Willd.; *Caesalpiniac.;* Elf.
„ pubescent: *Trichilia zenkeri* Harms; *Meliac.;* Elf.
Banba-apisieaie: *Acrodiclidium canella* Mez; *Laurac.;* Sur.
„ „ : *Acrodiclidium chrysophyllum* Meissn.; *Laurac.;* Sur.
? Banborla: *Litsea sebifera* Pers.; *Laurac.;* O.-Ind.
Bancal: *Sarcocephalus cordatus* Miq.; *Rubiac.;* Phil.
Bancoulier: *Aleurites triloba* Forst.; *Euphorbiac.;* Tr.
Banda: *Crossopteryx kotschyana* Fenzl; *Rubiac.;* Togo.
„ : *Cola proteïformis* Chev.; *Sterculiac.;* Elf.
„ : *Tarrietia utilis* Sprague; *Sterculiac.;* Elf.
Bandara: *Lagerstroemia parviflora* Roxb.; *Lythrac.;* Ind.
Bandeng lai: *Bruguiera eriopetala* Bth.; *Rhizophorac.;* Ind.-Ch.
Bandolinewood, Chinese: *Machilus thunbergi* S. et Z.; *Laurac.;* Ch., Jap.
Bandongué: *Rauwolfia macrophylla* Stapf; *Apocynac.;* Kam.
Bandoor-pala: *Amoora rohituka* Wr. et Arn.; *Meliac.;* Beng.
Bandorhulla: *Duabanga sonneratioides* Buch.-Ham.; *Lythrac.;* Ind.
Bané-canida: *Licania microcarpa* Hook. f.; *Rosac.;* Bras.
Bang: *Chlorophora excelsa* Bth. et Hook.; *Morac.;* Kam., Elf.
Báng: *Microdesmis caseariaefolia* Planch.; *Euphorbiac.;* Ind.-Ch.
„ (H): *Terminalia catappa* L., *T. procera* Roxb.; *Combretac.;* Ind.-Ch.
Banga: *Mammea klaineana* Pierre; *Guttifer.;* Gab. (Loango)
„ : *Hymenostegia aff. afzelii* Harms; *Caesalpiniac.;* Kam.
Bangalay: *Eucalyptus botryoides* Smith; *Myrtac.;* ö. Au.
Bangal bulala: *Sarcocephalus cordatus* Miq.; *Rubiac.;* tr. As.
Bang all: *Acronychia laurifolia* Blume; *Rutac.;* Ind.-Ch.
Banga Ouiba: *Irvingia gabonensis* Baill.; *Simarubac.;* Kam.
Bañgát: *Pterocymbium tinctorium* Merr.; *Sterculiac.;* Phil. (Zamb.)
Bañgayás: *Terminalia comintana* Merr.; *Combretac.;* Phil.
Bang-bang: *Lagerstroemia spp.; Lythrac.;* Ind.-Ch.
Bangbang: *Dipterocarpus sp.; Dipterocarpac.;* Sum.
Bangbaye: *Albizzia fastigiata* Oliv.; *Mimosac.;* Elf.
Bang bi (H): *Barringtonia speciosa* Forst.; *Myrtac.;* Ind.-Ch.
„ cou: *Aglaïa pirifera* Hance; *Meliac.;* Ind.-Ch.
Banggái: *Quercus sp.; Fagac.;* Phil.
Bang hao: *Schleichera trijuga* Willd.; *Sapindac.;* Ind.-Ch.
Bangkahási: *Koordersiodendron pinnatum* Merr.; *Anacardiac.;* Phil.
Bangkál: *Neonauclea sp., Nauclea orientalis* L.; *Rubiac.;* Phil.
Bangkal: *Sarcocephalus cordatus* Miq.; *Rubiac.;* Phil.
Bangkalaguán: *Terminalia calamansanai* Rolfe; *Combretac.;* Phil.

Bangkálan: *Pometia pinnata* Forst.; *Sapindac.;* Phil.
Bangkalandi: *Sideroxylon duclitan* Blume; *Sapotac.;* Phil.
Bangkalári: *Koordersiodendron pinnatum* Merr.; *Anacardiac.;* Phil.
Bangkaláuag: *Terminalia calamansanai* Rolfe, *T. oocarpa* Merr.; *Combretac.;* Phil.
Bangkaláuang: *Buchanania sp.; Anacardiac.;* Phil.
Bangkeou: *Xanthophyllum sp.; Rutac.;* Kamb.
Bang khuôi: *Dipterocarpus insularis* Hance; *Dipterocarpac.;* Ind.-Ch.
„ lam (H): *Lagerstroemia thoreli* Gagnep.; *Lythrac.;* Ind.-Ch.
Banglang (f): *Lagerstroemia angustifolia* Pierre; *Lythrac.;* Ind.-Ch.
„ „ : *Lagerstroemia flos-reginae* Retz.; *Lythrac.;* Ind.-Ch.
Bang lang ban: *Duabanga sonneratioïdes* Buch.-Ham.; *Punicac.;* Ind.-Ch.
„ „ chéo (H): *Lagerstroemia duperreana* Pierre; *Lythrac.;* Ind.-Ch.
„ „ cum: *Lagerstroemia sp.; Lythrac.;* Ind.-Ch.
„ „ cuôm (H): *Lagerstroemia angustifolia* Pierre; *Lythrac.;* Ind.-Ch.
„ „ muôc (H): *Lagerstroemia floribunda* Jack; *Lythrac.;* Ind.-Ch.
„ „ nuôc (H): *Lagerstroemia flos-reginae* Retz.; *Lythrac.;* Ind.-Ch.
„ „ oi (H): *Lagerstroemia crispa* Pierre; *Lythrac.;* Ind.-Ch.
„ „ tia (H): *Lagerstroemia hirsuta* Willd.; *Lythrac.;* Ind.-Ch.
„ „ trang: *Lagerstroemia sp.; Lythrac.;* Ind.-Ch.
Bañglís: *Terminalia comintana* Merr.; *Combretac.;* Phil.
Báng nhoc: *Terminalia procera* Roxb.; *Combretac.;* Annam.
Báng nhuôm: *Terminalia tomentosa* Bedd.; *Combretac.;* Ind.-Ch.
„ nuóc (H): *Terminalia procera* Roxb.; *Combretac.;* Ind.-Ch.
Bañgóran: *Hopea acuminata* Merr.; *Dipterocarpac.;* Phil.
Bang ró: *Schleichera trijuga* Willd.; *Sapindac.;* Ind.-Ch.
Bangue: *Landolphia kirki* Dyer; *Apocynac.;* O.-Af.
Bañgulo: *Dehaasia triandra* Merr.; *Myristicac.;* Phil. (Mindoro)
„ : *Litsea garciae* Vid.; *Myristicac.;* Phil. (Mindoro)
Bang xe (H): *Albizzia lucida* Bth.; *Mimosac.;* Ind.-Ch.
Bánh lanh (H): *Lagerstroemia duperreana* Pierre; *Lythrac.;* Ind.-Ch.
Bania: *Swartzia spp.; Caesalpiniac.;* br. Gu.
Baniakáu: *Hopea acuminata* Merr.; *Dipterocarpac.;* Phil.
Bani-bani: *Beilschmiedia sp.; Laurac.;* Phil. (Agusan Prov.)
Banigna: *Acacia arabica* Willd.; *Mimosac.;* Sudan.
Banik: *Vatica mangachapoi* Blco.; *Dipterocarpac.;* Phil.
Banikad: *Sterculia sp.; Sterculiac.;* Phil. (Mindoro)
Banílad: *Sterculia philippinensis* Merr.; *Sterculiac.;* Phil.
Banítan: *Mangifera altissima* Blco.; *Anacardiac.;* Phil.
Baníti: *Bassia ramiflora* Merr., *Palaquium sp.; Sapotac.;* Phil. (Bataan Pr.)
„ : *Wrightia sp.; Apocynac.;* Phil. (Camarines Prov.)
Banítis: *Bassia betis* Merr.; *Sapotac.;* Phil.
Bankalla: *Pterospermum lancaefolium* Roxb.; *Sterculiac.;* O.-Ind.
Bankhor: *Aesculus indica* Colebr.; *Hippocastanac.;* Ind.
Banknor: *Aesculus indica* Colebr.; *Hippocastanac.;* nw. Himal.
Banksia: *Banksia littoralis* R. Br.; *Proteac.;* Au.
Banksia, River: *Banksia verticillata* R. Br.; *Proteac.;* w. Au.
Bankul: *Aleurites triloba* Forst.; *Euphorbiac.;* Tr.
Banlang (f): *Lagerstroemia spp.; Lythrac.;* Ind.-Ch.
Ban muôc: *Hymenodictyon excelsum* Wall.; *Rubiac.;* Ind.-Ch.
Banógan: *Phoebe sterculioides* Merr.; *Laurac.;* Phil.
Bân oi: *Sonneratia acida* L. f.; *Punicac.;* Ind.-Ch.
Banokbók: *Sideroxylon luzoniense* Merr.; *Sapotac.;* Phil.
Bansalagın: *Mimusops parvifolia* R. Br.; *Sapotac.;* Phil.
Bansalágin: *Planchonia spectabilis* Merr.; *Lecythidac.;* Phil.
Bansalágon: *Mimusops elengi* L.; *Sapotac.;* Phil.
Bansilai: *Cratoxylon sp.; Guttifer.;* Phil.
Bansilaían: *Cratoxylon sp.; Guttifer.;* Phil.
Bantam: *Ceiba pentandra* Gaertn.; *Bombacac.;* W.-Af.
Bantan: *Ceiba pentandra* Gaertn. var. *dehiscens* Ulb.; *Bombacac.;* Dah.
„ : *Ceiba pentandra* Gaertn.; *Bombacac.;* Gab.
Bantang: *Ceiba pentandra* Gaertn.; *Bombacac.;* W.-Af.
Bantaógan: *Calophyllum cumingi* Pl. et Tr.; *Guttifer.;* Phil. (Mind.)

Bantigné: *Ceiba pentandra* Gaertn. *var. dehiscens* Ulb.; *Bombacac.*; Senl.
Bantínen: *Toona calantas* Merr. et Rolfe; *Meliac.*; Phil.
Banting: *Lumnitzera littorea* Voigt; *Combretac.*; Phil.
Bantolinao: *Diospyros pilosanthera* Blco.; *Ebenac.*; Phil.
Bân trang: *Sonneratia alba* Smith; *Punicac.*; Ind.-Ch.
Bantulináu: *Maba buxifolia* Pers.; *Ebenac.*; Phil.
   „   : *Diospyros alvarezi* Merr.; *Ebenac.*; Phil.
Banukalád: *Aleurites trisperma* Blco.; *Euphorbiac.*; Phil.
Banútan: *Hopea plagata* Vid.; *Dipterocarpac.*; Phil.
Banúyo: *Albizzia acle* Merr.; *Mimosac.*; Phil.
   „   : *Phoebe sterculioides* Merr.; *Laurac.*; Phil.
   „   : *Wallaceodendron celebicum* Koord.; *Mimosac.*; Phil.
Ban váng: *Barringtonia sp.*; *Myrtac.*; Ind.-Ch.
Banyan: *Ficus bengalensis* L.; *Morac.*; Ind.
Bao: *Garcinia tonkinensis* Vesque; *Guttifer.*; Ind.-Ch.
Baobab (f): *Adansonia digitata* L.; *Bombacac.*; Tr.
Baoélébako: *Ongokea gore* Engl.; *Olacac.*; Kam.
Bapa: *Shorea selanica* Blume; *Dipterocarpac.*; Nied.-Ind.
   „   : *Cynometra alexandri* Wright; *Caesalpiniac.*; B.-K.
Bapeba: *Lucuma sp.*; *Sapotac.*; Bras.
   „   assú: *Lucuma sp.*; *Sapotac.*; Bras.
Bapi: *Maesobotrya stapfiana* Beille; *Euphorbiac.*; Elf.
Bapu: *Michelia champaca* L.; *Magnoliac.*; Cey.
Bara: *Adina microcephala* Hiern; *Rubiac.*; W.-Af., Togo.
Bara-bara: *Diospyros guianensis* Gürke; *Ebenac.*; br. Gu.
Barabás: *Vitex kuyleni* Standl.; *Verbenac.*; Guat.
Barabú: *Peltogyne spp.*; *Caesalpiniac.*; Bras.
Baracarra: *Swartzia tomentosa* DC.; *Caesalpiniac.*; br. Gu.
Bara champ: *Michelia excelsa* Blume; *Magnoliac.*; Ind.
Barada-balli: *Icica sp.*; *Burserac.*; br. Gu.
Baraja: *Cassia reticulata* Willd.; *Caesalpiniac.*; Hond.
Barajo: *Cassia reticulata* Willd.; *Caesalpiniac.*; Guat.
Barakara: *Ormosia spp.*; *Papilionac.*; Sur.
Bara-kara: *Ormosia coccinea* Jacks.; *Papilionac.*; br. Gu.
Barakaroe: *Ormosia spp.*; *Papilionac.*; Sur.
Barakaroeballi: *Dicorynia paraënsis* Bth.; *Caesalpiniac.*; Sur.
Barakaro fieroberoe: *Ormosia fastigiata* Tul.; *Papilionac.*; Sur.
   „   ibikoro: *Ormosia coccinea* Jacks.; *Papilionac.*; Sur.
   „   korero ibibero iwi: *Ormosia melanocarpa* Kleinh.; *Papilionac.*; Sur.
Barakbak: *Myristica sp.*; *Myristicac.*; Phil.
Barakbák: *Eugenia calubcob* C. B. Rob., *E. xanthophylla* C. B. Rob.; *Myrtac.*: Phil.
Baramalli: *Catostemma fragrans* Bth.; *Malvac.*; br. Gu.
Baramomi: *Picea polita* Carr.; *Conifer.*; Jap.
Baráñgau: *Oroxylum indicum* Vent.; *Bignoniac.*; Phil.
Barata: *Lucuma sp.*; *Sapotac.*; br. Gu.
Baratara: *Bambusa aculeata* Hitchc.; *Gramin.*; Nic.
Barauna: *Melanoxylon brauna* Schott; *Caesalpiniac.*; Bras.
   „   : *Schinopsis sp.*; *Anacardiac.*; Bras.
Baraus: *Terminalia edulis* Blco.; *Combretac.*; Phil.
Baráwisan: *Octomeles sumatrana* Miq.; *Datiscac.*; Phil.
Barawiswísan: *Octomeles sumatrana* Miq.; *Datiscac.*; Phil.
Baráyung: *Pahudia rhomboidea* Prain; *Caesalpiniac.*; Phil.
Barba de boi: *Cordia sp.*; *Borraginac.*; Bras.
   „   „   chivo: *Caesalpinia gilliesi* Wall.; *Caesalpiniac.*; Arg.
   „   „   Jupiter: *Anthyllis barba-jovis* L.; *Papilionac.*; Arg.
   „   „   tigre: *Colletia ferox* Gill.; *Rhamnac.*; Arg.
   „   „   „   : *Prosopis kuntzei* Harms; *Mimosac.*; Arg.
   „   die Jonkemann: *Albizzia lebbek* Bth.; *Mimosac.*; Nied.-W.-Ind.
   „   jolote: *Cassia aff. emarginata* L.; *Caesalpiniac.*; br. Hond.
   „   „   : *Pithecolobium arboreum* Urb.; *Mimosac.*; br. Hond.
Barbados-kers: *Malpighia punicifolia* L.; *Malpighiac.*; fr. Gu.
Barbás: *Vitex kuyleni* Standl., *V. longeracemosa* Pitt.; *Verbenac.*; Guat.

Barbas de macho: *Phlomis crinita* Cav.; *Labiat.;* Sp.
Barbasco: *Jacquinia barbasco* Mez; *Myrsinac.;* P.-R.
„ : *Canella winterana* Gaertn.; *Canellac.;* P.-R.
„ : *Jacquinia armillaris* Jacq.; *Myrsinac.;* Peru, Trin.
„ : *Jacquinia revoluta* Jacq.; *Myrsinac.;* Ven.
„ : *Piscidia erythrina* L.; *Papilionac.;* Ven.
„ de árbol: *Ichthyomethia piscipula* Hitch.; *Papilionac.;* Kol.
Barbatimão: *Stryphnodendron barbatimão* Mart.; *Mimosac.;* P.-R.
Barbatuco: *Erythrina sp.; Papilionac.;* Kol.
Barbed wood: *Anona glabra* L.; *Anonac.;* br. Hond.
Barberry: *Rhamnus purshiana* DC.; *Rhamnac.;* N.-Am.
„ Common: *Berberis vulgaris* L.; *Berberidac.;* England.
Bardaguera: *Salix cinerea* L., *S. pedicellata* Desf.; *Salicac.;* Sp.
„ blanca: *Salix oleaefolia* Vill.; *Salicac.;* Sp.
Bardottier: *Mimusops imbricaria* Willd.; *Sapotac.;* Maur.
Barenillo: *Croton glabellus* L.; *Euphorbiac.;* Hond.
Bärentraube, gemeine: *Arctostaphylos uva-ursi* Sprengel; *Ericac.;* n. Hem.
„ Alpen-: *Arctostaphylos alpina* Sprengel; *Ericac.;* Alp., Karp., Pyr.
„ kalifornische: *Arctostaphylos manzanita* Parry; *Ericac.;* Kalif.
Baria: *Cordia gerascanthoides* H. B. K.; *Borraginac.;* Cuba.
Baría: *Luehea sp.; Tiliac.;* Cuba.
„ : *Cordia gerascanthus* L.; *Borraginac.;* Mex.
„ : *Calophyllum calaba* Jacq.; *Guttifer.;* St.-D.
Bariaco: *Krugiodendron ferreum* Urb.; *Rhamnac.;* P.-R.
Baridikutshi: *Rinorea sp.; Uiolac.;* br. Gu.
Barigudo: *Chorisia ventricosa* Nees; *Bombacac.;* Bras.
Barillo: *Calophyllum sp.; Guttifer.;* Salv.
„ : *Symphonia globulifera* L. f.; *Guttifer.;* Guat., Pan.
Baringkukúrung: *Cratoxylon sp.; Guttifer.;* Phil.
Bárit: *Heritiera littoralis* Dry.; *Sterculiac.;* Phil. (Zamb.)
„ : *Parinarium sp.; Rosac.;* Phil.
Bariwiswís: *Anisoptera thurifera* Blume; *Dipterocarpac.;* Phil.
Barkclothtree: *Antiaris africana* Engl.; *Morac.;* Go.
Barklak: *Lecythis ollaria* L.; *Lecythidac.;* Sur.
„ : *Eschweilera sp.; Lecythidac.;* Sur.
„ Man-: *Eschweilera spp.; Lecythidac.;* Sur.
Bark of Angelim: *Andira inermis* H. B. K.; *Papilionac.;* tr. Am.
Barkraki: *Eschweilera sp., Lecythis ollaria* L.; *Lecythidac.;* Sur.
Barktree, Cosmetic: *Murraya paniculata* Jack; *Rutac.;* Ind., Sunda.
Barkwood, Cabbage: *Andira inermis* H. B. K.; *Papilionac.;* W.-Ind.
Barl: *Cordia gerascanthoides* H. B. K.; *Borraginac.;* Mex.
Barma: *Taxus baccata* L.; *Conifer.;* Eu., N.-Äf., As.
Barna: *Crataeva religiosa* Forst.; *Capparidac.;* tr. Af.
Barniz falso del Japon (s): *Ailanthus glandulosa* Desf.; *Simarubac.;* O.-As.
Baro: *Hibiscus tiliaceus* L.; *Malvac.;* Mad.
Baroe: *Hibiscus tiliaceus* L.; *Malvac.;* Nied.-Ind.
Bároi: *Pterospermum sp.; Sterculiac.;* Phil.
Barokaro-koereroe-ibi-bero-iewi: *Ormosia melanocarpa* Kleinh.; *Papilionac.;* Sur.
Baromalli: *Tabebuia sp.; Bignoniac.;* br. Gu.
Baromé: *Mimusops globosa* Gaertn.; *Sapotac.;* br. Gu., W.-Ind.
Baros: *Magnolia blumei* Prantl; *Magnoliac.;* Java.
Barosingsíng: *Hopea acuminata* Merr.; *Dipterocarpac.;* Phil.
Barranduna: *Trochocarpa laurina* R. Br.; *Epacridac.;* N.-S.-W.
Barredero: *Trichilia sp.; Meliac.;* Salv.
Barrehorno: *Trichilia havanensis* Jacq.; *Meliac.;* Hond., Salv.
Barren Privet (e): *Rhamnus alaternus* L.; *Rhamnac.;* w. M.-M.
Barreta: *Helietta parviflora* Bth.; *Rutac.;* Mex.
Barreto: *Uernonia spp.; Composit.;* Salv.
Barrigón: *Copernicia excelsa* León; *Palmac.;* Cuba.
Barriguda: *Cavanillesia arborea* K. Schum.; *Bombacac.;* Bras.
„ : *Chorisia ventricosa* Nees; *Bombacac.;* Bras.
Barrigudo: *Ceiba sp.; Bombacac.;* Arg.

Barruch: *Brosimum discolor* Schott; *Morac.;* Bras.
Barsino: ? *Khaya anthotheca* C. DC.; *Meliac.;* Mex.
Bartaballi: *Lucuma bonplandia* H. B. K.; *Sapotac.;* Gu.
? Bartara silver balli: *Bellucia grossularioides* Triana; *Melastomatac.;* br. Gu.
Bartu: *Hymenodictyon excelsum* Wall.; *Rubiac.;* w. Himalaya.
Baruch: *Mimusops spp.;* Sapotac.; br. Gu.
„ : *Mimusops globosa* Gaertn.; *Sapotac.;* br. Gu., W-.Ind.
Barungi: *Quercus dilatata* Lindl.; *Fagac.;* nw. Himalaya, Afg.
Barúto: *Sideroxylon macranthum* Merr.; *Sapotac.;* Phil.
Barwood: *Pterocarpus santalinoides* L'Hérit.; *Papilionac.;* W.-Af.
„ : *Pterocarpus soyauxi* Taub.; *Papilionac.;* W.-Af.
„ : *Baphia nitida* Lodd.; *Papilionac.;* W.-Af.
Basaä-botie-ie: ? *Chaetocarpus schomburgkiana* Pax et Hoffm.; *Euphorbiac.:* Sur.
Basákan: *Buchanania sp.; Anacardiac.;* Phil. (Mindoro)
„ : *Quercus sp.; Fagac.;* Phil.
Basakanda: *Ormosia spp.; Papilionac.;* Sur.
Basal: *Mimusops elengi* L.; *Sapotac.;* Phil. (Mindoro)
Basamu: *Boswellia dalzieli* Hutch., *B. odorata* Hutch.; *Burserac.;* n. Nig.
Basán: *Garcinia sp.; Guttifer.;* Phil. (Mindoro)
Basí: *Terminalia edulis* Blco.; *Combretac.;* Phil.
Basikálang: *Alstonia macrophylla* Wall.; *Apocynac.;* Phil.
Basikárang: *Alstonia macrophylla* Wall.; *Apocynac.;* Phil.
Basílan: *Shorea polysperma* Merr.; *Dipterocarpac.;* Phil.
Basiláyan: *Litsea sp.; Laurac.;* Phil.
Basilikon: *Juglans regia* L.; *Juglandac.;* w. As., S.-Eu.
Basinau: *Aglaia bicolor* Merr.; *Meliac.;* Phil.
Basipi: *Mimusops lacera* Bak.; *Sapotac.;* Kam.
Basít: *Lagerstroemia piriformis* Koehne; *Lythrac.;* Phil. (Zamb.)
Basláyan: *Dehaasia triandra* Merr.; *Laurac.;* Phil. (Mindoro)
Basóg: *Palaquium sp.; Sapotac.;* Phil.
Ba soi: *Macaranga denticulata* Muell. Arg.; *Euphorbiac.;* Ind.-Ch.
Basra-Ajoewa: *Parkia microcephala* Kleinh.; *Mimosac.;* Sur.
Basra locus: *Dicorynia paraënsis* Bth.; *Caesalpiniac.;* Bras. (Amaz.)
Basra-loksi: *Dicorynia paraënsis* Bth.; *Caesalpiniac.;* Sur.
Basra-Tonto-awha: *Parkia microcephala* Kleinh.; *Mimosac.;* Sur.
Bass (e): *Tilia parvifolia* Ehrh.; *Tiliac.;* Eu.
Bassèc: *Terminalia ruspoli* Engl. et Diels; *Combretac.;* Somali.
Basswood: *Tilia americana* L., *T. glabra* Vent.; *Tiliac.;* N.-Am.
„ , Michaux: *Tilia michauxi* Nutt.; *Tiliac.;* ö. N.-Am.
„ , Silver: *Panax elegans* Moore et Muell.; *Araliac.;* Au.
„ , White: *Tilia heterophylla* Vent.; *Tiliac.;* USA.
Bast-tree (e): *Tilia parvifolia* Ehrh.; *Tiliac.;* Eu.
Bát: *Strychnos nux-vomica* L.; *Loganiac.;* Ind.-Ch.
Bataan: *Shorea polysperma* Merr.; *Dipterocarpac.;* Phil.
Batákan: *Bambusa blumeana* Schultes f.; *Gramin.;* Phil.
Batakirilea: *Erythroxylon lucidum* Moon; *Erythroxylac.;* Cey.
Batalha: *Nectandra robusta* Loefg. et Evet.; *Laurac.;* Bras. (Sao P.)
Batam: *Ceiba pentandra* Gaertn. *var. clausa* Ulb.; *Bombacac.;* Dah.
Bat and Ball: *Endiandra virens* F. v. M.; *Laurac.;* N.-S.-W.
Batañgáli: *Pometia pinnata* Forst.; *Sapindac.;* Phil.
Batárau: *Calophyllum inophyllum* L.; *Guttifer.;* Phil.
Bâtard: *Curatella americana* L.; *Dilleniac.;* Cay.
Batavi-neboo: *Citrus decumana* Willd.; *Rutac.;* Ind.
Batete: *Kingiodendron alternifolium* Merr.; *Caesalpiniac.;* Phil.
Ba thu'a (H): *Sterculia thoreli* Pierre; *Sterculiac.;* Ind.-Ch.
Baticulín: *Litsea obtusata* F. Vill., ? *Phoebe sp.; Laurac.;* Phil.
Batidos: *Quararibea fieldi* Miller; *Bombacac.;* br. Hond.
Bá tiên: *Pterospermum diversifolium* Blume; *Sterculiac.;* Ind.-Ch.
Bátik: *Erythrophloeum densiflorum* Merr.; *Caesalpiniac.;* Phil.
„ : *Hopea plagata* Vid.; *Dipterocarpac.;* Phil.
Batikalág: *Lagerstroemia piriformis* Koehne; *Lythrac.;* Phil.
„ : *Alstonia macrophylla* Wall.; *Apocynac.;* Phil.

Batikaláng: *Alstonia macrophylla* Wall.; *Apocynac.;* Phil.
Batikulíng: *Litsea sp., Phoebe sterculioides* Merr.; *Laurac.;* Phil.
„ : *Paralstonia clusiacea* Baill.; *Apocynac.;* Phil.
Batíno: *Alstonia macrophylla* Wall., *A. paucinervia* Merr.; *Apocynac.;* Phil.
Batínong-dágat: *Cyclostemon sp.; Euphorbiac.;* Phil.
Batitinan: *Lagerstroemia piriformis* Koehne; *Lythrac.;* Phil.
Batitínan-babáe: *Terminalia comintana* Merr.; *Combretac.;* Phil.
Batlatináu: *Maba buxifolia* Pers.; *Ebenac.;* Phil.
Bato-bató: *Cyclostemon sp.; Euphorbiac.;* Phil.
Batopoua: *Sterculia tragacantha* Lindl.; *Sterculiac.;* Elf.
Bat seed: *Andira inermis* H. B. K.; *Papilionac.;* br. Gu.
Bats souari: *Caryocar glabrum* Pers.; *Caryocarac.;* br. Gu.
Batteo: ? *Arariba sp.; Papilionac.;* Pan.
? Batteo: *Carapa guianensis* Aubl.; *Meliac.;* Pan.
Batúan: *Dracontomelum dao* Merr. et Rolfe; *Anacardiac.;* Phil.
„ : *Garcinia sp.; Guttifer.;* Phil.
Batukanág: *Aglaia spp.; Meliac.;* Phil.
Batulinau: *Diospyros sp., Maba buxifolia* Pers.; *Ebenac.;* Phil.
Batun: *Sideroxylon macranthum* Merr.; *Sapotac.;* Phil. (Mind.)
Batwi: *Pterocarpus esculentus* Schum. et Thom.; *Papilionac.;* W.-Af.
Batwiri: ? *Khaya anthotheca* C. DC.; *Meliac.;* Kam.
Báu chó: *Pometia pinnata* Forst.; *Sapindac.;* Ind.-Ch.
Baúgin: *Bambusa blumeana* Schultes f.; *Gramin.;* Phil
Bauji: *Acacia sieberiana* DC.; *Mimosac.;* n. Nig.
Baûka Mapa: *Couma guianensis* Aubl.; *Apocynac.;* Sur.
Baume à cochon: *Hedwigia balsamifera* Swartz; *Burserac.;* Jam.
Baumhasel: *Corylus columna* L.; *Betulac.;* sö. Eu., As.
Baumheide: *Erica arborea* L.; *Ericac.;* M.-M.
Baumier: *Populus balsamifera* L.; *Salicac.;* n. N.-Am.
Baumier de la Jamaïque (f): *Amyris balsamifera* L.; *Rutac.;* W.-Ind.
Baumwollbaum: *Ceiba pentandra* Gaertn.; *Bombacac.;* Tr.
Baumwollholz, afrikanisch: *Ceiba pentandra* Gaertn.; *Bombacac.;* W.-Af.
„ „ : *Bombax buonopozense* Beauv.; *Bombacac.;* W.-Af.
Bau nâu (H): *Aegle marmelos* Correa; *Rutac.;* Ind.-Ch.
Baure: *Ficus gnaphalocarpa* A. Rich.; *Morac.;* n. Nig.
Baurlo: *Cordia macleodi* Hook. f. et Thoms.; *Borraginac.;* n. Ind.
Baushi: *Terminalia avicennioides* Guill. et Perr., *T. glaucescens* Pl.; *Combretac.;* n. Nig.
Baví: *Casearia sp.; Flacourtiac.;* Arg.
Bawa: *Garcinia speciosa* Wall.; *Guttifer.;* Burma.
Bawang utan: *Scorodocarpus borneensis* Becc.; *Olacac.;* Malay.
Bay: *Swietenia macrophylla* King; *Meliac.;* C.-Am.
Bây: *Canarium nigrum* Engl.; *Burserac.;* Ind.-Ch.
Bay bay: *Croton joufra* Roxb.; *Euphorbiac.;* Ind.-Ch.
Bay, Bull-: *Magnolia grandiflora* L.; *Magnoliac.;* USA.
„ Rose-: *Rhododendron maximum* L.; *Ericac.;* ö. USA.
„ Dwarf rose: *Rhododendron maximum* L.; *Ericac.;* USA.
„ Laurel-: *Magnolia grandiflora* L.; *Magnoliac.;* USA.
„ Loblolly-: *Gordonia lasianthus* L.; *Theac.;* sö. N.-Am.
„ Moreton-: *Araucaria cunninghami* Ait.; *Conifer.;* Queensl.
„ Red-: *Laurus carolinensis* L.; *Laurac.;* USA.
„ „ : *Persea borbonica* Spreng.; *Laurac.;* USA.
„ Swamp: *Magnolia virginiana* L., *M. glauca* L.; *Magnoliac.;* N.-Am.
„ „ : *Persea pubescens* Sarg.; *Laurac.;* ö. USA.
„ „ red: *Persea carolinensis* Nees *var. palustris* Ch.; *Laurac.;* sö. N.-Am.
„ Sweet: *Magnolia virginiana* L., *M. glauca* L.; *Magnoliac.;* USA.
„ Tan-: *Gordonia lasianthus* L.; *Theac.;* sö. N.-Am., tr. Am.
„ White: *Magnolia glauca* L.; *Magnoliac.;* N.-Am.
Baytree (u): *Umbellularia californica* Nutt.; *Laurac.;* nw. USA.
„ : *Magnolia glauca* L.; *Magnoliac.;* USA.
Baywood: *Swietenia macrophylla* King; *Meliac.;* C.-Am.
Baya: *Mitragyne macrophylla* Hiern; *Rubiac.;* Go.
Bayábas: *Psidium guajava* L.; *Myrtac.;* Phil.

Bayábas: *Shorea balangeran* Dyer; *Dipterocarpac.;* Phil.
Bayábo: *Terminalia calamansanai* Rolfe; *Combretac.;* Phil.
Bayádgung: *Pahudia rhomboidea* Prain; *Caesalpiniac.;* Phil.
Bayag-kabáyo: *Heritiera littoralis* Dry.; *Sterculiac.;* Phil. (Manila)
Bayag-usá: *Paralstonia clusiacea* Baill.; *Apocynac.;* Phil. (Mind.)
Bayahonda: *Prosopis juliflora* DC.; *Mimosac.;* Haiti.
Bayakbák: *Eugenia costulata* C. B. Rob.; *Myrtac.;* Phil.
Bayama: *Swartzia madagascariensis* Desv.; *Caesalpiniac.;* n. Nig.
Bayam badak: *Strombosia javanica* Blume; *Olacac.;* Malay.
Bayantí: *Aglaia harmsiana* Perk., *A.llanosiana* C. DC.; *Meliac.;* Phil.
Bayaó: *Pterocymbium tinctorium* Merr.; *Sterculiac.;* Phil.
Bayátis: *Palaquium sp.; Sapotac.;* Phil.
Bayauas: *Shorea balangeran* Dyer; *Dipterocarpac.;* Phil.
Bayayat: *Sterculia sp.; Sterculiac.;* Phil.
Bayberries: *Myrica rubra* S. et Z.; *Myricac.;* Jap.
Bayberry: *Myrica cerifera* L.; *Myricac.;* Hond.
Bay-doo: *Pycnanthus dinklagei* Warb.; *Myristicac.;* Lib.
Baye: *Terminalia superba* Engl. et Diels; *Combretac.;* Lib.
Bayít: *Euphoria cinerea* Radlk.; *Sapindac.;* Phil.
Bayito: *Haenianthus salicifolium* Griseb.; *Oleac.;* Cuba.
Bayleaf tree: *Pimenta acris* Kostel.; *Myrtac.;* St.-Lucia.
Bayóg: *Aglaia harmsiana* Perk.; *Meliac.;* Phil.
  „ : *Bambusa blumeana* Schultes f.; *Gramin.;* Phil.
  „ : *Pterospermum sp.; Sterculiac.;* Phil.
Bayog-bayóg: *Pterospermum sp.; Sterculiac.;* Phil. (Zamb.)
Bayók: *Pterospermum sp.; Sterculiac.;* Phil.
Bayokbayókan: *Pterospermum niveum* Vid.; *Sterculiac.;* Phil.
Bayóto: *Ormosia calavensis* Azaola; *Papilionac.;* Phil.
  „ : *Pometia pinnata* Forst.; *Sapindac.;* Phil.
Bayr: *Ziziphus jujuba* Lamk.; *Rhamnac.;* Ind. (Hindustan)
Báyu: *Dipterocarpus affinis* Brandis; *Dipterocarpac.;* Phil.
Bayúa: *Zanthoxylum sp.; Rutac.;* Cuba.
Bayúa lisa: *Zanthoxylum sp.; Rutac.;* Cuba.
Bayuda: *Zanthoxylum sp.; Rutac.;* Cuba.
Bayúg: *Bambusa blumeana* Schultes f.; *Gramin.;* Phil.
Bayuk: *Pterospermum sp.; Sterculiac.;* Phil.
Bayúkan: *Pentacme contorta* Merr. et Rolfe; *Dipterocarpac.;* Phil. (Zamb.)
Bayúko: *Artocarpus cumingiana* Tréc.; *Morac.;* Phil.
  „ : *Gymnartocarpus woodi* Merr.; *Morac.;* Phil.
Báyung: *Pahudia rhomboidea* Prain; *Caesalpiniac.;* Phil.
Bayur: *Pterospermum blumeanum* Korth.; *Sterculiac.;* Java.
Baza (f): *Blighia sapida* Koenig; *Sapindac.;* Elf.
Bazoa: *Strombosiopsis tetrandra* Engl.; *Olacac.;* Kam.
B'Béré: *Pentadesma sp.; Guttifer.;* Elf.
Beach, White: *Gmelina fasciculiflora* Bth.; *Uerbenac.;* Au.
Bead tree: *Melia azedarach* L.; *Meliac.;* Ind., s. USA.
  „ „ : *Melia japonica* G. Don; *Meliac.;* Jap.
  „ „ : *Adenanthera pavonina* L.; *Mimosac.;* Go.
Beag thuge: *Pterospermum acerifolium* Willd.; *Sterculiac.;* Co.-Ch.
Bean, Atta-: *Pentaclethra macrophylla* Bth.; *Mimosac.;* Go.
  „ Black: *Castanospermum australe* A. Cunn.; *Papilionac.;* Au.
  „ Carob: *Ceratonia siliqua* L.; *Caesalpiniac.;* M.-M.
  „ Coffee: *Gymnocladus dioica* C. Koch; *Caesalpiniac.;* Kan.
  „ Indian: *Catalpa speciosa* Warder; *Bignoniac.;* ö. N.-Am.
  „ Red: *Dysoxylum muelleri* Bth.; *Meliac.;* sö. Au.
  „ Tonka: *Dipteryx odorata* Willd.; *Papilionac.;* Sur.
  „ Tonkin: *Dipteryx odorata* Willd.; *Papilionac.;* br. Gu.
  „ Tonquina: *Dipteryx odorata* Willd.; *Papilionac.;* Sur.
Bean-tree (e): *Laburnum anagyroides* Medic.; *Papilionac.;* Eu.
  „ : *Castanospermum australe* A. Cunn.; *Papilionac.;* Au.
  „ Locust: *Parkia filicoidea* Welw.; *Mimosac.;* n. Nig.
  „ Oil-: *Pentaclethra macrophylla* Bth.; *Mimosac.;* Go.

Bean Walnut-: *Endiandra palmerstoni* C. T. White; *Laurac.;* Au.
Bean, Whistling: *Albizzia lebbek* Bth.; *Mimosac.;* Bah.
„ White: *Ailanthus malabarica* A. DC.; *Simarubac.;* Au.
Bearberry: *Rhamnus purshiana* DC.; *Rhamnac.;* N.-Am.
Beari: *Chrysophyllum ellipticum* Chev.; *Sapotac.;* Lib.
Bearwood: *Rhamnus purshiana* DC.; *Rhamnac.;* N.-Am.
Beati: *Cassia siamea* Lamk.; *Caesalpiniac.;* Ind.
Beaume de San Tomé: *Pachylobus balsamifera* Guill.; *Burserac.;* Gab.
Beautree: *Catalpa catalpa* Karst.; *Bignoniac.;* sö. USA.
Beaver tree: *Magnolia virginiana* L., *M.glauca* L.; *Magnoliac.;* ö. N.-Am.
Be-ay: *Tylostemon manni* Stapf; *Laurac.;* Lib.
Bebam: *Chrysophyllum africanum* A. DC.; *Sapotac.;* Fernando Po.
? Bébamme: *Calpocalyx klainei* Pierre; *Mimosac.;* Kam.
Bébé: *Pterocarpus draco* L., *P.rhori* Vahl; *Papilionac.;* Sur.
Bebehoedoe: *Pterocarpus draco* L., *P.rhori* Vahl; *Papilionac.;* Sur.
Bébé, Hoogland: *Pterocarpus rhori* Vahl; *Papilionac.;* Sur.
Bébéhout, Laaglandsch: *Pterocarpus draco* L.; *Papilionac.;* Gu.
Bebeeree: *Nectandra rodioei* Hook.; *Laurac.;* br. Gu.
Bebeeren: *Nectandra rodioei* Hook.; *Laurac.;* br. Gu.
Bebeeru: *Nectandra rodioei* Hook.; *Laurac.;* br. Gu.
Beberimi-itu: ? *Treculia africana* Decne.; *Morac.;* Go.
Beberubaum (d): *Nectandra rodioei* Hook.; *Laurac.;* br. Gu.
Beberuboom (h): *Nectandra rodioei* Hook.; *Laurac.;* br. Gu.
Bébèti: *Funtumia elastica* Stapf; *Apocynac.;* Elf.
Bebuhago: *Pachylobus edulis* G. Don; *Burserac.;* W.-Af.
Bebvo: *Saccoglottis gabonensis* Urb.; *Humiriac.;* Kam.
Beb vouo: *Placodiscus pseudostipularis* Radlk.; *Sapindac.;* Kam.
Becuiba: *Virola bicuhyba* Warb.; *Myristicac.;* Bras.
Becuiba assú: *Virola bicuhyba* Warb.; *Myristicac.;* Bras.
„ branca: *Virola bicuhyba* Warb.; *Myristicac.;* Bras.
„ mirim: *Virola bicuhyba* Warb.; *Myristicac.;* Bras.
„ vermelha: *Virola bicuhyba* Warb.; *Myristicac.;* Bras.
Becya: *Pachylobus edulis* G. Don; *Burserac.;* Gab.
Bedara hutan: *Parinarium rubiginosum* Ridl.; *Rosac.;* Malay.
„ „ : *Angelisia splendens* Korth.; *Rosac.;* Malay.
„ poeti: *Eurycoma longifolia* Jacq.; *Simarubac.;* Nied.-Ind.
Bedaroh: *Ziziphus sp.;* *Rhamnac.;* Nied.-Ind.
„ : *Nephelium eriopetalum* Miq.; *Sapindac.;* Sum.
Bedaru: *Cantleya johorica* Ridley; *Icacinac.;* Borneo.
„ : *Urandra corniculata* Foxw.; *Icacinac.;* Borneo.
Bediwunua: *Canarium schweinfurthi* Engl.; *Burserac.;* Go.
Bédjabi: *Mimusops pierreana* Engl.; *Sapotac.;* Gab.
Bedo: *Sarcocephalus trillesi* Pierre; *Rubiac.;* Elf.
Bedoll: *Betula verrucosa* Ehrh.; *Betulac.;* Sp.
Bedou Angouna: *Placodiscus pseudostipularis* Radlk.; *Sapindac.;* Kam.
Bedunua: *Canarium schweinfurthi* Engl.; *Burserac.;* Go.
Beeberoe: *Nectandra rodioei* Hook.; *Laurac.;* Sur.
Beech (e): *Fagus silvatica* L.; *Fagac.;* Eu.
„ : *Fagus grandifolia* Ehrh.; *Fagac.;* N.-Am.
„ : *Flindersia australis* R. Br.; *Rutac.;* N.-S.-W.
„ : *Trochocarpa laurina* R. Br.; *Epacridac.;* N.-S.-W.
„ : *Nothofagus cunninghami* Oerst.; *Fagac.;* Tasm.
„ American: *Fagus grandifolia* Ehrh.; *Fagac.;* N.-Am.
„ Antarctic: *Fagus moorei* F. v. M.; *Fagac.;* s. S.-Am., Au.
„ Australisch: *Gmelina leichhardti* Bth.; *Verbenac.;* N.-S.-W.
„ Black: *Nothofagus solandri* Oerst.; *Fagac.;* N.-Seel.
„ „ : *Weinmannia racemosa* L. f.; *Saxifragac.;* N.-Seel.
„ Blue (u): *Carpinus caroliniana* Walt.; *Corylac.;* ö. N.-Am.
„ Bolly: *Litsea reticulata* Bth.; *Laurac.;* Au.
„ Brown: *Pennantia cunninghami* Miers; *Icacinac.;* Au.
„ „ : *Cryptocarya patentinervis* F. v. M.; *Laurac.;* Au.
„ „ : *Litsea reticulata* Bth.; *Laurac.;* Au.

Beech Brown Bolly: *Litsea reticulata* Bth.; *Laurac.*; Au.
„ Cape: *Myrsine melanophleos* R. Br.; *Myrsinac.*; S.-Af.
„ Chilian: *Fagus antarctica* Forst.; *Fagac.*; S.-Am.
„ Clinker: *Nothofagus truncata* Cockayne; *Fagac.*; N.-Seel.
„ Entire-leaved: *Nothofagus solandri* Oerst.; *Fagac.*; N.-Seel.
„ European (u): *Fagus silvatica* L.; *Fagac.*; Eu.
„ Evergreen: *Nothofagus cunninghami* Oerst.; *Fagac.*; O.-Au.
„ Hard: *Nothofagus truncata* Cockayne; *Fagac.*; N.-Seel.
„ Indian: *Pongamia glabra* Vent.; *Papilionac.*; Ind., tr. Au.
„ Leaved: *Curtisia faginea* Ait.; *Cornac.*; S.-Af.
„ Mountain: *Nothofagus cliffortioides* Oerst.; *Fagac.*; N.-Seel.
„ Negro-head: *Nothofagus cunninghami* Oerst.; *Fagac.*; Vict., Tasm.
„ New Zealand: *Nothofagus menziesi* Oerst., *N. fusca* Oerst.; *Fagac.*; Au.
„ „ „ : *Nothofagus solandri* Oerst.; *Fagac.*; N.-Seel.
„ Queensland: *Gmelina leichhardti* F. v. M.; *Verbenac.*; Queensl.
„ Red: *Fagus grandifolia* Ehrh.; *Fagac.*; N.-Am.
„ „ : *Nothofagus fusca* Oerst.; *Fagac.*; N.-Seel.
„ „ : *Flindersia brayleyana* F. v. M.; *Rutac.*; Queensl.
„ „ : *Flindersia pimenteliana* F. v. M.; *Rutac.*; Queensl.
„ She: *Cryptocarya glaucescens* R. Br., *C. obovata* R. Br.; *Laurac.*; nö. Au.
„ Silky: *Villaresia moorei* F. v. M.; *Icacinac.*; Au.
„ Silver: *Nothofagus menziesi* Oerst.; *Fagac.*; N.-Seel.
„ Southland: *Nothofagus menziesi* Oerst.; *Fagac.*; N.-Seel.
„ Tasmanian: *Nothofagus cunninghami* Oerst.; *Fagac.*; Au.
„ Tooth-leaved: *Nothofagus fusca* Oerst.; *Fagac.*; N.-Seel.
„ Water- (u): *Carpinus carolineana* Walt.; *Corylac.*; ö. N.-Am.
„ White: *Fagus silvatica* L.; *Fagac.*; N.-Am.
„ „ : *Phyllanthus ferdinandi* Muell. Arg.; *Euphorbiac.*; Au.
„ „ : *Elaeocarpus kirtoni* F. v. M.; *Elaeocarpac.*; O.-Au.
„ „ : *Gmelina leichhardti* F. v. M.; *Verbenac.*; Au.
„ „ of Bunya Mountains: *Elaeocarpus kirtoni* F. v. M.; *Elaeocarpac.*; Au.
Beef-food: *Cassia grandis* L.; *Caesalpiniac.*; br. Hond.
Beefwood (u): *Casuarina equisetifolia* Forst.; *Casuarinac.*; Tr.
„ : *Mimusops surinamensis* Miq., *M. balata* Gaertn.; *Sapotac.*; Sur.
„ : *Swartzia tomentosa* DC.; *Caesalpiniac.*; Gu.
„ : *Mimusops globosa* Gaertn.; *Sapotac.*; Trin.
„ : *Roupala montana* Aubl.; *Proteac.*; Trin.
„ : *Pisonia obtusata* Jacq.; *Nyctaginac.*; tr. Am.
„ : *Casuarina* spp.; *Casuarinac.*; Au.
„ : *Macadamia praealta* Bailey; *Proteac.*; Au.
„ : *Stenocarpus salignus* R. Br.; *Proteac.*; N.-S.-W.
„ : *Grevillea striata* R. Br.; *Proteac.*; Queensl.
„ , Scrub: *Stenocarpus salignus* R. Br.; *Proteac.*; Au.
„ White: *Stenocarpus sinuatus* Endl.; *Proteac.*; Au.
„ „ : *Orites excelsa* R. Br.; *Proteac.*; Au.
Beeja-poora: *Citrus medica* L.; *Rutac.*; Ind.
„ „ : *Citrus limonum* Risso; *Rutac.*; Ind.
Beereegah: *Careya sphaerica* Roxb.; *Myrtac.*; Andamanen.
Bee tree (u): *Tilia americana* L.; *Tiliac.*; N.-Am.
Beg-poora: *Citrus limonum* Risso; *Rutac.*; Beng.
Béguan: *Berlinia acuminata* Soland.; *Caesalpiniac.*; Elf.
Beh: *Chrysophyllum obovatum* G. Don; *Sapotac.*; Lib.
Beh-eck: *Bourreria mollis* Standl.; *Borraginac.*; br. Hond.
Bèhèrada: *Parinarium campestre* Aubl.; *Rosac.*; Sur.
Bèhèrie kotton: ? *Andira retusa* H. B. K.; *Papilionac.*; Sur.
Behlehketehbeh: *Echinocarpus sigun* Blume; *Tiliac.*; Sunda.
Behlo poetih: *Cupania sideroxylon* Cambess.; *Sapindac.*; Nied.-Ind.
Behn: ? *Chrysophyllum ellipticum* Chev.; *Sapotac.*; Lib.
Behoerada: *Parinarium campestre* Aubl.; *Rosac.*; Sur.
Behra: *Chloroxylon swietenia* DC.; *Meliac.*; Ind.
Beinholz (d): *Lonicera xylosteum* L.; *Caprifoliac.*; Eu.
Beira: *Cynometra alexandri* Wright; *Caesalpiniac.*; B.-K.

Bei són lóc: *Tarrietia javanica* Blume; *Sterculiac.;* Co.-Ch.
Bejoura: *Citrus medica* L.; *Rutac.;* Ind.
Bejuco cadena: *Bauhinia heterophylla* Kunth; *Caesalpiniac.;* Kol.
„ de agua: *Vitis sp.; Vitac.;* br. Hond.
„ „ lucá: *Gustavia superba* Berg; *Lecythidac.;* Pan.
„ negro: *Cordia ferruginea* Roem. et Schult.; *Borraginac.;* Hond.
Beko: *Calocarpum mammosum* Pierre; *Sapotac.;* C.-R.
„ : *Klainedoxa ovalifolia* Verm.; *Simarubac.;* B.-K.
Békolo: *Tieghemella africana* Pierre; *Sapotac.;* Gab.
Be-kwai: *Qualea dinizi* Ducke; *Vochysiac.;* Sur.
Be kwa-ie: *Vochysia melinoni* Beckm.; *Vochysiac.;* Sur.
Bela indien (f): *Aegle marmelos* Correa; *Rutac.;* Ind.
Belapoa: *Diospyros spp.; Ebenac.;* Mad.
Belar: *Casuarina lepidophloia* F. v. M.; *Casuarinac.;* sö. Au.
Bélarbinikelé: *Scottellia spp.; Bixac.;* Gab.
Belarbre: *Dimorphandra mora* Bth. et Hook.; *Caesalpiniac.;* Mart.
Belbél: *Pinus insularis* Endl.; *Conifer.;* Phil.
Belbil: *Villaresia moorei* F. v. M.; *Icacinac.;* Au., N.-S.-W.
Bélèbé: *Desbordesia insignis* Pierre; *Simarubac.;* Gab.
Belehketehpeh: *Echinocarpus sigun* Blume; *Tiliac.;* Sunda.
Belekoro: *Piratinera guianensis* Aubl.; *Morac.;* Sur.
Be leng: *Zollingeria dongnaiensis* Pierre; *Sapindac.;* Ind.-Ch.
Beletoné: *Pentadesma leucantha* Chev.; *Guttifer.;* Elf.
Beleyleh: *Terminalia belerica* Roxb.; *Combretac.;* Ind., Cey., Burma.
Belfou: *Tetrapleura thonningi* Bth.; *Mimosac.;* Elf.
Belfruittree: *Aegle marmelos* Correa; *Rutac.;* O.-Ind.
Belgaum: *Aleurites moluccana* Willd.; *Euphorbiac.;* Tr.
Belian: *Payena utilis* Ridl., *Madhuca spp.; Sapotac.;* Malay.
„ : *Palaquium stellatum* King et Gamble; *Sapotac.;* Malay.
„ : *Eusideroxylon zwageri* Teijsm. et Binn.; *Laurac.;* Borneo.
Bel indien (f): *Aegle marmelos* Correa; *Rutac.;* Ind.
Belis: *Canarium luzonicum* A. Gray; *Burserac.;* Phil.
Bellfruit tree: *Codonocarpus australis* A. Cunn.; *Phytolaccac.;* Au.
Belli patta: *Hibiscus tiliaceus* L., *H. elatus* Swartz; *Malvac.;* Ind.
Bellota: *Quercus sp.; Fagac.;* Salv.
„ : *Sterculia sp.; Sterculiac.;* Mex.
„ : *Bellota miersi* C. Gay; *Laurac.;* Chile.
Bellotto: *Bellotta miersi* C. Gay; *Laurac.;* Chile.
Be loi (H): *Litsea vang* H. Lec.; *Laurac.;* Ind.-Ch.
Belon: *Ceiba pentandra* Gaertn. *var. dehiscens* Ulb.; *Bombacac.;* Sudan.
Belookar: *Garcinia andersoni* Hook. f.; *Guttifer.;* Kamb.
Belukap: *Rhizophora mucronata* Lam.; *Rhizophorac.;* tr. As. & Au., Sans., Mad., Sey.
Belvi: *Enantia chlorantha* Oliv.; *Anonac.;* Lib.
Bemba: *Pentaclethra macrophylla* Bth.; *Mimosac.;* Kam.
Bémbada: *Pentaclethra macrophylla* Bth.; *Mimosac.;* Gab.
Bembé: *Odina sp.; Anacardiac.;* fr. Guin.
Be-moonba: *Iryanthera sagotiana* Warb., *I. hostmanni* Warb.; *Myristicac.;* Sur.
Bên: *Pithecolobium balansae* Oliv.; *Mimosac.;* Ind.-Ch.
Ben: *Heisteria trillesiana* Pierre; *Olacac.;* Gab.
Benak: *Balanocarpus heimi* King; *Dipterocarpac.;* Malay.
Benang: *Cotoneaster bacillaris* Wall.; *Rosac.;* Himalaya.
Benaroon: *Eucalyptus pilularis* Smith; *Myrtac.;* ö. Au.
Benda: *Pithecolobium trapezifolium* Bth.; *Mimosac.;* br. Gu.
Bende Bende: *Monodora myristica* Dun.; *Anonac.;* B.-K.
Bendi: *Thespesia populnea* Correa; *Malvac.;* Tr.
Béndjèkè: *Ongokea klaineana* Pierre; *Olacac.;* Gab.
Beng: *Pahudia cochinchinensis* Pierre; *Caesalpiniac.;* Ind.-Ch.
„ : *Chlorophora excelsa* Bth. et Hook.; *Morac.;* Kam.
Bengang: *Neesia altissima* Blume; *Bombacac.;* Sunda.
Beng Kheou: *Aglaïa sp.; Meliac.;* Ind.-Ch.
Béngomba: *Coula edulis* Baill.; *Olacac.;* Gab.
Bengouma: *Aucoumea klaineana* Pierre; *Burserac.;* Gab.

Benibrun: *Mangifera foetida* Lour.; *Anacardiac.*; Java.
Beni-Byakushin: *Juniperus japonica* Carrier; *Conifer.*; Jap.
„   -byakusin: *Juniperus nipponica* Maxw.; *Conifer.*; Jap.
Beni-hi: *Chamaecyparis formosensis* Matsum.; *Conifer.*; Form., Jap.
Benin wood: *Khaya ivorensis* Chev.; *Meliac.*; Nig.
Benisugi: *Cryptomeria japonica* D. Don; *Conifer.*; Ch., Jap.
Benjoin: *Terminalia benzoin* L.; *Combretac.*; Mauritius.
„   : *Terminalia mauritiana* Lam.; *Combretac.*; Réunion.
Benkyi: *Diospyros pissatoria* Gürke; *Ebenac.*; Go.
„   : *Phyllanthus discoideus* Muell. Arg.; *Euphorbiac.*; Go.
Ben nam (H): *Gymnosporia mekongensis* Pierre; *Celastrac.*; Ind.-Ch.
Ben nay: *Grewia microcos* L.; *Tiliac.*; Ind.-Ch.
Ben oil: *Moringa pterygosperma* Gaertn.; *Moringac.*; Sudan.
Bentan Habou: *Ceiba pentandra* Gaertn. var. *dehiscens* Ulb.; *Bombacac.*; Dah.
Benten: *Ceiba pentandra* Gaertn.; *Bombacac.*; W.-Af.
Bentenier: *Ceiba pentandra* Gaertn.; *Bombacac.*; W.-Af.
Bent Pod: *Pithecolobium altissimum* Oliv.; *Mimosac.*; O.-Af.
Bénzèng: *Odyendea gabonensis* Pierre; *Simarubac.*; Gab.
Benzoin: *Croton benzoe* Murr.; *Euphorbiac.*; Mauritius.
Bépanda: *Panda oleosa* Pierre; *Pandac.*; Gab.
Bepaulétoe: *Piratinera guianensis* Aubl.; *Morac.*; Sur.
Bépobo: *Poga oleosa* Pierre; *Rhizophorac.*; Gab.
Ber: *Zizyphus jujuba* Lamk.; *Rhamnac.*; Ind.
Bera: *Bulnesia arborea* Engl.; *Zygophyllac.*; Ven.
Beradie hohoradikora: *Nectandra sp.*, *Ocotea sp.*; *Laurac.*; Sur.
Berangan: *Castanopsis sp.*; *Fagac.*; Malay.
„   babi: *Pasania sp.*, *Quercus sp.*; *Fagac.*; Malay.
Berbá: *Heliocostylis latifolia* Pittier; *Morac.*; Pan.
Berberine, Bevat-: *Zanthoxylum caribaeum* Lam.; *Rutac.*; fr. Gu.
„   „   : *Zanthoxylum clava-herculis* L.; *Rutac.*; fr. Gu.
Berberiʒe (d): *Berberis vulgaris* L.; *Berberidac.*; Eu.
Berekoro: *Heliocostylis sp.*; *Morac.*; Eu.
Beresklet: *Evonymus europaea* L.; *Celastrac.*; Russland.
Beretué: *Omphalocarpum anocentrum* Pierre; *Sapotac.*; Elf.
Bereza: *Betula alba* L.; *Betulac.*; N.-Eu., As.
Beri: *Zizyphus jujuba* Lamk.; *Rhamnac.*; Ind.
„   : *Canarium schweinfurthi* Engl.; *Burserac.*; Lib.
Beriba: *Oxandra lanceolata* Baill.; *Anonac.*; fr. Gu.
Berkhout: *Vouacapoua americana* Aubl.; *Papilionac.*; Sur.
„   : *Lecythis ollaria* L., *L. corrugata* Poit.; *Lecythidac.*; Sur.
Berombong: *Adina rubescens* Hemsl.; *Rubiac.*; Malay.
Berra: *Bulnesia arborea* Engl.; *Zygophyllac.*; Ven.
Berraco: *Celtis sp.*; *Ulmac.*; Kol.
Berus: *Bruguiera caryophyllaeoides* Blume; *Rhizophorac.*; Malay.
Besenginster: *Cytisus scoparius* Link; *Papilionac.*; Eu.
Besenheide: *Calluna vulgaris* Salisb.; *Ericac.*; Eu.
Besenpfriemen, gemeiner (d): *Cytisus scoparius* Link; *Papilionac.*; Eu.
Besí: *Terminalia edulis* Blco.; *Combretac.*; Phil.
Besi: *Afzelia palembanica* Baker; *Caesalpiniac.*; Nied.-Ind.
Bét bét: *Alangium chinense* Lour.; *Cornac.*; Ind.-Ch.
Bethabara (u): *Tecoma leucoxylon* Mart.; *Bignoniac.*; Nied.-Gu.
Betis: *Payena utilis* Ridl.; *Sapotac.*; Malay.
„   : *Madhuca spp.*, *Palaquium stellatum* King et Gamble; *Sapotac.*; Malay.
„   : *Bassia betis* Merr.; *Sapotac.*; Phil.
„   : *Dehaasia triandra* Merr.; *Laurac.*; Phil.
Bétik: *Shorea guiso* Blume, *S. teysmanniana* Dyer; *Dipterocarpac.*; Phil.
Betoum: *Pistacia atlantica* Desf.; *Anacardiac.*; Sahara.
Betún: *Calycophyllum candidissimum* DC.; *Rubiac.*; Ven.
Beu (f): *Symphonia globulifera* L.; *Guttifer.*; Elf.
Beudding: *Acronychia baueri* Schott; *Rutac.*; N.-S.-W.
Beujabi: ? *Dumoria heckeli* Chev.; *Sapotac.*; Kam.
Beukenhout: *Myrsine melanophleos* R. Br.; *Myrsinac.*; S.-Af.

Beukenhout Transvaal-: *Faurea saligna* Harv.; *Proteac.;* S.-Af.
Beurata: *Parinarium campestre* Aubl.; *Rosac.;* Sur.
Bew- boc: *Bruguiera gymnorrhiza* Lam.; *Rhizophorac.;* And.
Béwinda: *Klainedoxa gabonensis* Pierre; *Simarubac.;* Gab.
Bewmah: *Rhizophora mucronata* Lam.; *Rhizophorac.;* Burma.
Beya: *Mitragyne macrophylla* Hiern; *Rubiac.;* Go.
Beymadah: *Albizzia elata* Bth.; *Mimosac.,* And.
Beyor: *Uapaca guineensis* Muell. Arg.; *Euphorbiac.;* Lib.
Bey sanh: *Tarrietia cochinchinensis* Pierre; *Sterculiac.;* Ind.-Ch.
„   sanlek: *Tarrietia cochinchinensis* Pierre; *Sterculiac.;* Ind.-Ch.
„   son loc: *Tarrietia cochinchinensis* Pierre; *Sterculiac.;* Ind.-Ch.
Bhadrat: *Elaeocarpus lancaefolius* Roxb.; *Elaeocarpac.;* ö. Himal.
Bhadroi: *Phoebe lanceolata* Nees; *Laurac.;* O.-Ind.
Bhalkua: *Bambusa balcooa* Roxb.; *Gramin.;* O.-Ind.
Bharjapatri: *Betula bhojpattra* Wall.; *Betulac.;* N.-Ind.
Bhel: *Aegle marmelos* Correa; *Rutac.;* Ind.
Bhoso: *Humiria floribunda* Mart., *H. balsamifera* Aubl.; *Humiriac.;* Sur.
Bhujpattra: *Betula bhojpattra* Wall.; *Betulac.;* Ind.
Bhurjama: *Betula bhojpattra* Wall.; *Betulac.;* N.-Ind.
Bia: *Dinochloa scandens* O. Ktʒe.; *Gramin.;* Phil.
Biaá: *Zizyphus sp.; Rhamnac.;* Phil.
Biacca Contó: *Grimmeodendron jamaicense* Urb.; *Euphorbiac.;* Jam.
Biakushin: *Juniperus chinensis* L.; *Conifer.;* Jap.
Bialé: *Garcinia klaineana* Pierre; *Guttifer.;* Gab.
Biálung: *Pahudia rhomboidea* Prain; *Caesalpiniac.;* Phil.
Biarúng: *Pahudia rhomboidea* Prain; *Caesalpiniac.;* Phil.
Biasbías: *Polyscias nodosa* Seem.; *Araliac.;* Phil.
Biáu: *Aleurites moluccana* Willd.; *Euphorbiac.;* Phil.
Biaxbi: *Spondias sp.; Anacardiac.;* Mex.
Biba: *Irvingia gabonensis* Baill.; *Simarubac.;* W.-Af.
Bibacier (f): *Eriobotrya japonica* Lindl.; *Rosac.;* Ch., Jap.
Bi bai: *Evodia spp.; Rutac.;* Ind.-Ch.
„   „   (H): *Elaeodendron glaucum* Pers.; *Celastrac.;* Ind.-Ch.
Biberoe sokoné: *Piratinera guianensis* Aubl.; *Morac.;* Sur.
Biberoo: *Nectandra rodioei* Hook.; *Laurac.;* Sur.
Bibi-abé: *Carapa velutina* C. DC.; *Meliac.;* Elf.
Bibíli: *Cordia myxa* L., *C. subcordata* Lam.; *Borraginac.;* Phil.
Bibindi: *Diospyros atropurpurea* Gürke; *Ebenac.;* Kam.
Bibir: *Nectandra rodioei* Hook.; *Laurac.;* br. Gu.
Bibira: *Nectandra rodioei* Hook.; *Laurac.;* br. Gu.
Bibiri: *Nectandra rodioei* Hook.; *Laurac.;* br. Gu.
Bibiritim: ? *Treculia africana* Dcne.; *Morac.;* Go.
Bibiroo: *Nectandra rodioei* Hook.; *Laurac.;* br. Gu.
Bibiru: *Nectandra rodioei* Hook.; *Laurac.;* br. Gu.
Bibít: *Vatica mangachapoi* Blco.; *Dipterocarpac.;* Phil.
Bibla: *Pterocarpus marsupium* Roxb.; *Papilionac.;* C.- & S.-Ind.
Bibo: *Rhaphia monbuttorum* Drude; *Palmac.;* O.-Af. (Uganda)
Bibolo (f): *Lovoa klaineana* Pierre; *Meliac.;* Kam.
Bibzar: *Murraya exotica* L.; *Rutac.;* Ind.
Bichet: *Bixa orellana* L.; *Bixac.;* fr. Gu.
Bico de papagaio: *Machaerium sp.; Papilionac.;* Bras. (Sao P.)
„   „   pato: *Machaerium acutifolium* Vog.; *Papilionac.;* Bras.
„   „   „   : *Machaerium angustifolium* Vog.; *Papilionac.;* Bras. (Sao P.)
Bicona: *Vitex grandifolia* Gürke; *Verbenac.;* Gab.
Bienhiba: *Virola bicuhyba* Warb.; *Myristicac.;* Bras.
Bicuiba: *Virola bicuhyba* Warb.; *Myristicac.;* Bras. (Bahia)
„   : *Myristica officinalis* Mart.; *Myristicac.;* Bras.
„   crespa: *Virola bicuhyba* Warb.; *Myristicac.;* Bras. (Sao P.)
Bidara: *Zizyphus jujuba* Lamk.; *Rhamnac.;* Malay.
Bidbid: *Eugenia mananquil* Blco.; *Myrtac.;* Phil.
Bideiro: *Betula verrucosa* Ehrh.; *Betulac.;* Sp.
Bidóso: *Pometia pinnata* Forst.; *Sapindac.;* Phil. (Mind.)

Bidou: *Placodiscus pseudostipularis* Radlk.; *Sapindac.*; Kam.
„ : *Oxystigma manni* Harms; *Caesalpiniac.*; Kam.
„ : *Saccoglottis gabonensis* Urb.; *Humiriac.*; Kam.
Bidu: *Saccoglottis gabonensis* Urb.; *Humiriac.*; Kam.
Bié: *Sarcocephalus trillesi* Pierre; *Rubiac.*; Kam.
Biedjietan: *Lansium domesticum* Jack; *Meliac.*; Java.
Bie-ie oedoe: *Eperua jenmanni* Oliv.; *Caesalpiniac.*; Sur.
Bienédora: *Sindora klaineana* Pierre; *Caesalpiniac.*; Gab.
Bientangoor: *Calophyllum inophyllum* L.; *Guttifer.*; Sum.
Bientienoh: *Melochia odorata* Forst.; *Sterculiac.*; Nied.-Ind.
Bientinoe: *Melochia odorata* Forst.; *Sterculiac.*; Java.
Bier: *Zizyphus jujuba* Lamk.; *Rhamnac.*; Ind.
Biércol: *Erica vagans* L.; *Ericac.*; Sp.
Bierie hoedoe: *Eperua sp.*; *Caesalpiniac.*; Sur.
Biesietan: *Lansium domesticum* Jack; *Meliac.*; Java.
Bieslen: *Evonymus europaea* L.; *Celastrac.*; Eu. (Böhmen)
Biga: *Gymnartocarpus woodi* Merr.; *Morac.*; Phil.
Bigá: *Zizyphus sp.*; *Rhamnac.*; Phil.
Bigaá: *Zizyphus sp.*; *Rhamnac.*; Phil.
Bigaignon: *Psiloxylon mauritianum* Baill.; *Flacourtiac.*; Maur.
„ Batard: *Antidesma madagascariense* Lam.; *Euphorbiac.*; Maur.
Bigaradier (f): *Citrus bigaradia* Duham.; *Rutac.*; Tr.
Bigbé: *Pterocarpus draco* L., *P.rhori* Vahl; *Papilionac.*; Sur.
Big-tree: *Sequoia gigantea* Dcne.; *Conifer.*; USA. (Kalif.)
Bihambi: *Uapaca staudti* Pax; *Euphorbiac.*; tr. Af.
Bija: *Bixa orellana* L.; *Bixac.*; Cuba.
Bijaguara: *Colubrina reclinata* Brongn.; *Rhamnac.*; s. Fl., W.-Ind.
„ : *Colubrina ferruginosa* Brongn.; *Rhamnac.*; Cuba.
Bija sal: *Pterocarpus marsupium* Roxb.; *Papilionac.*; Ind.
Bijlhout: *Eperua falcata* Aubl.; *Caesalpiniac.*; Sur.
„ echte: *Eperua falcata* Aubl.; *Caesalpiniac.*; Sur.
„ Roode: *Eperua falcata* Aubl.; *Caesalpiniac.*; Sur.
„ Water: *Macrolobium chrysostachyum* Bth.; *Caesalpiniac.*; Sur.
„ „ : *Macrolobium bifolium* Pers.; *Caesalpiniac.*; Sur.
„ Wit: *Eperua schomburgkiana* Bth.; *Caesalpiniac.*; Sur.
Bijouree: *Citrus medica* L.; *Rutac.*; Ind.
Bikal: *Schizostachyum dielsianum* Merr.; *Gramin.*; Phil.
„ -bábi: *Schizostachyum dielsianum* Merr.; *Gramin.*; Phil.
„ -bábui: *Schizostachyum dielsianum* Merr.; *Gramin.*; Phil.
„ -machin: *Schizostachyum dielsianum* Merr.; *Gramin.*; Phil.
Bikiti: *Qualea albiflora* Warm.; *Uochysiac.*; Sur.
Bikuwe: *Xylopia aethiopica* A. Rich.; *Anonac.*; B.-K.
Bilabila: *Cupania asperula* Standl.; *Sapindac.*; Nic.
Bilge-watertree: *Andira inermis* H.B.K.; *Papilionac.*; tr. Am.
Bili-bili: *Guarea spp.*; *Meliac.*; Kol.
„ -bíli: *Dracontomelum cumingianum* Baill.; *Anacardiac.*; Phil.
„ „ : *Dracontomelum edule* Skeels; *Anacardiac.*; Phil.
Biligra: *Albizzia sp.*; *Mimosac.*; Elf.
Bili hoedoe: *Eperua falcata* Aubl.; *Caesalpiniac.*; Sur.
Bilimbi: *Averrhoa bilimbi* L.; *Oxalidac.*; Bras.
Bilin: *Feronia elephantum* Correa; *Rutac.*; O.-Ind., Cey., Java.
Bilinga (d): *Sarcocephalus trillesi* Pierre; *Rubiac.*; Kam., Gab.
„ des marais: *Sarcocephalus trillesi* Pierre var. *paludosus* Chev.; *Rubiac.*; Elf.
Bilis: *Garcinia sp.*; *Guttifer.*; Phil.
Billa: *Casuarina glauca* Sieb.; *Casuarinac.*; Queensl.
Billan-billan: *Olea paniculata* R. Br.; *Oleac.*; Queensl.
Billbird patter: *Ouratea pyramidalis* Riley; *Ochnac.*; br. Hond.
Billian: *Eusideroxylon zwageri* Teijsm. et Binn.; *Laurac.*; Malay., Borneo.
Billoo: *Chloroxylon swietenia* DC.; *Meliac.*; Ind.
Billu: *Chloroxylon swietenia* DC.; *Meliac.*; Ind.
Billy Webb: *Sweetia panamensis* Bth.; *Papilionac.*; br. Hond.
„ „ Bastard: *Caesalpinia yucatanensis* Greenm.; *Caesalpiniac.*; br. Hond.

Bilolo: *Khaya ivorensis* Chev.; *Meliac.*; Gab.
Bilólo: *Eugenia sp.*; *Myrtac.*; Phil.
Bilorkbingkélé: *Scottellia kamerunensis* Gilg; *Bixac.*; Gab.
Bilreiro: *Guarea spp.*; *Meliac.*; Bras.
Bilsted: *Liquidambar styraciflua* L.; *Hamamelidac.*; sö. USA.
Biluáng: *Endospermum peltatum* Merr.; *Euphorbiac.*; Phil.
„ : *Octomeles sumatrana* Miq.; *Datiscac.*; Phil.
Bilúkau: *Carallia integerrima* DC.; *Rhizophorac.*; Phil.
„ : *Garcinia sp.*; *Guttifer.*; Phil.
Bilwara: *Albizzia odoratissima* Bth.; *Mimosac.*; O.-Ind.
Bimba: *Aeschynomene elaphroxylon* Taub.; *Papilionac.*; tr. Af.
Bimbi: *Macrolobium aff. limba* Scott Ell.; *Caesalpiniac.*; Kam
Bimiti: ? *Eperua schomburgkiana* Bth.; *Caesalpiniac.*; br. Gu.
„ -walaba: *Eperua schomburgkiana* Bth.; *Caesalpiniac.*; Sur.
Binalíuan: *Parashorea plicata* Brandis; *Dipterocarpac.*; Phil.
Binburra: *Gmelina leichhardti* F. v. M.; *Verbenac.*; sö. Au.
Bine: *Petersia viridiflora* Chev.; *Myrtac.*; Kam.
Bing: *Chlorophora excelsa* Bth. et Hook.; *Morac.*; Kam.
Binga: *Stephegyne diversifolia* Hook. f.; *Rubiac.*; Burma.
Biñgas: *Terminalia comintana* Merr.; *Combretac.*; Phil.
Binggás: *Terminalia comintana* Merr.; *Combretac,*; Phil.
„ : *Parinarium sp.*; *Rosac.*; Phil.
Binggáu: *Parinarium sp.*; *Rosac.*; Phil.
Binh bát (H): *Anona reticulata* L.; *Anonac.*; Ind.-Ch.
„ linh (H): *Vitex pubescens* Vahl; *Verbenac.*; Ind.-Ch.
„ tchnai: *Schoutenia hypoleuca* Pierre; *Tiliac.*; Ind.-Ch.
„ „ : *Crypteronia paniculata* Kurz; *Crypteroniac.*; Ind.-Ch.
Binkey: *Sterculia rupestris* Bth.; *Sterculiac.*; Queensl.
Bini: *Pycnanthus kombo* Warb.; *Myristicac.*; Go.
Binnak: *Eucalyptus botryoides* Smith; *Myrtac.*; ö. Au.
Binólo: *Eugenia sp.*; *Myrtac.*; Phil.
Binolóan: *Eugenia saligna* C. B. Rob.; *Myrtac.*; Phil.
Bintaforo: *Ceiba pentandra* Gaertn.; *Bombacac.*; Senl.
Bin tagou: *Calophyllum inophyllum* L.; *Guttifer.*; Singapore.
Bintangoer boenga: *Calophyllum sp.*; *Guttifer.*; Nied.-Ind.
„ boenoet: *Calophyllum sp.*; *Guttifer.*; Nied.-Ind.
„ djalai: *Calophyllum sp.*; *Guttifer.*; Nied.-Ind.
„ djangkai: *Calophyllum sp.*; *Guttifer.*; Nied.-Ind.
„ mera: *Calophyllum sp.*; *Guttifer.*; Nied.-Ind.
„ oendjam: *Calophyllum sp.*; *Guttifer.*; Nied.-Ind.
„ poetih: *Calophyllum sp.*; *Guttifer.*; Nied.-Ind.
Bintangoor batoe: *Calophyllum lanigerum* Miq.; *Guttifer.*; Sunda.
„ priët: *Calophyllum plicipes* Miq.; *Guttifer.*; Sunda.
Bintangor: *Calophyllum spp.*; *Guttifer.*; Malay.
„ bunga: *Calophyllum inophyllum* L.; *Guttifer.*; Malay.
„ tawah: *Calophyllum sp.*; *Guttifer.*; Nied.-Ind.
Bintangore: *Calophyllum inophyllum* L.; *Guttifer.*; Malay.
Bintanoe: *Melochia odorata* Forst.; *Sterculiac.*; Nied.-Ind.
Bintégnié: *Ceiba pentandra* Gaertn. *var. dehiscens* Ulb.; *Bombacac.*; Senl.
Bintinoe: *Melochia odorata* Forst.; *Sterculiac.*; Nied.-Ind.
Bintoela: *Sclerolobium paniculatum* Vogel; *Caesalpiniac.*; Sur.
Binuáng: *Octomeles sumatrana* Miq.; *Datiscac.*; Phil.
„ : *Duabanga moluccana* Blume; *Lythrac.*; Phil.
„ : *Endospermum peltatum* Merr.; *Euphorbiac.*; Phil.
Binúkau: *Garcinia binucao* Choisy; *Guttifer.*; Phil.
Binúnga: *Endospermum peltatum* Merr., *Macaranga sp.*; *Euphorbiac.*; Phil.
Bió: *Dracontomelum cumingianum* Baill., *D. edule* Skeels; *Anacardiac.*; Phil.
„ : *Garuga sp.*; *Burserac.*; Phil.
Bióso: *Pometia pinnata* Forst.; *Sapindac.*; Phil.
Biot: *Sandoricum vidali* Merr.; *Meliac.*; Phil.
Bir: *Zizyphus jujuba* Lamk.; *Rhamnac.*; Ind.
Birapita mini: *Daphnopsis sp.*; *Thymelaeac.*; Arg. (Corr.)

Birbogh: *Juglans regia* L.; *Juglandac.*; S.-Eu., W.-As.
Birch: *Betula alba* L.; *Betulac.*; England.
„ , Alaska: *Betula alaskana* Sarg.; *Betulac.*; Alas., w. Kan.
„ American: *Betula lenta* L.; *Betulac.*; N.-Am.
„ Black: *Nothofagus fusca* Oerst., *N. solandri* Oerst.; *Fagac.*; N.-Seel.
„ „ : *Betula occidentalis* Hook., *B. lenta* L.; *Betulac.*; N.-Am.
„ „ : *Betula nigra* L.; *Betulac.*; N.-Am.
„ „ : ? *Betula lutea* Michx.; *Betulac.*; Neu-Schottl.
„ Blue: *Betula coerulea* Blanchard; *Betulac.*; ö. USA.
„ Brown: *Nothofagus fusca* Oerst.; *Fagac.*; N.-Seel.
„ Canadian: *Betula lutea* Michx.; *Betulac.*; N.-Am.
„ Canoe: *Betula papyrifera* Marsh.; *Betulac.*; N.-Am.
„ Cherry: *Betula lenta* L.; *Betulac.*; USA.
„ Curly: *Betula lutea* Michx.; *Betulac.*; Kan.
„ Dwarf: *Betula lenta* L.; *Betulac.*; N.-Am.
„ European (u): *Betula alba* L.; *Betulac.*; Eu.
„ Gold-: *Betula lutea* Michx.; *Betulac.*; ö. N.-Am.
„ Grey-: *Betula lutea* Michx., *B. populifolia* Marsh.; *Betulac.*; N.-Am.
„ „ : *Bridelia exaltata* F. v. M.; *Euphorbiac.*; Au.
„ Hard: *Betula lutea* Michx.; *Betulac.*; ö. N.-Am.
„ Indian: *Betula bhojpattra* Wall.; *Betulac.*; C.-As., O.-As.
„ Jamaica-: *Bursera simaruba* Sarg.; *Burserac.*; Jam.
„ Mahogany- (u): *Betula lenta* L.; *Betulac.*; ö. N.-Am.
„ Mountain-: *Betula fontinalis* Sarg.; *Betulac.*; Kan.
„ Native: *Dodonaea viscosa* Jacq.; *Sapindac.*; Tasm.
„ New Zealand-: *Nothofagus solandri* Oerst.; *Fagac.*; N.-Seel.
„ Old-field- (u): *Betula populifolia* Marsh.; *Betulac.*; ö. N.-Am.
„ Paper-: *Betula papyrifera* Marsh.; *Betulac.*; N.-Am.
„ „ European: *Betula alba* L.; *Betulac.*; ö. N.-Am.
„ „ Indian: *Betula bhojpattra* Wall.; *Betulac.*; C.-As., O.-As.
„ Poplar-leaved: *Betula populifolia* Marsh.; *Betulac.*; N.-Am.
„ Poverty-: *Betula populifolia* Marsh.; *Betulac.*; ö. N.-Am.
„ Puget-Sound-: *Betula occidentalis* Hook.; *Betulac.*; nw. N.-Am.
„ Red: *Betula lenta* L., *B. nigra* L.; *Betulac.*; USA.
„ „ : *Nothofagus fusca* Oerst.; *Fagac.*; N.-Seel.
„ River-: *Betula nigra* L.; *Betulac.*; USA.
„ Russian: *Betula alba* L.; *Betulac.*; N.-Eu., As., Am.
„ Silver-: *Betula papyrifera* Marsh.; *Betulac.*; N.-Am.
„ „ : *Nothofagus menziesi* Oerst.; *Fagac.*; N.-Seel.
„ Southland-: *Weinmannia racemosa* L. f.; *Cunoniac.*; N.-Seel.
„ Sweet: *Betula lenta* L.; *Betulac.*; N.-Am.
„ Tall: *Betula lutea* Michx.; *Betulac.*; ö. N.-Am.
„ Water-: *Betula nigra* L.; *Betulac.*; ö. USA.
„ Western: *Betula occidentalis* Hook.; *Betulac.*; nw. N.-Am.
„ West Indian (u): *Bursera gummifera* Jacq.; *Burserac.*; W.-Ind., tr. Am.
„ White (e): *Betula alba* L.; *Betulac.*; Eu.
„ „ : *Betula papyrifera* Marsh.; *Betulac.*; N.-Am.
„ „ (u): *Betula populifolia* Marsh.; *Betulac.*; ö. N.-Am.
„ „ Small: *Betula populifolia* Marsh.; *Betulac.*; ö. N.-Am.
„ „ : *Nothofagus solandri* Oerst.; *Fagac.*; N.-Seel. (Southland)
„ Wire-: *Betula populifolia* Marsh.; *Betulac.*; ö. Kan.
„ Yellow: *Betula lutea* Michx.; *Betulac.*; N.-Am.
„ „ : *Betula lenta* L.; *Betulac.*; Prinz Eduard Ins.
„ wood: *Bursera gummifera* L.; *Burserac.*; br. Hond.
Biribá: *Oxandra lanceolata* Baill.; *Anonac.*; Bras.
Biriba: *Eschweilera luschnathi* Miers; *Lecythidac.*; ö. Bras.
Biribi: *Eschweilera luschnathi* Miers; *Lecythidac.*; ö. Bras.
Biribu: *Khaya ivorensis* Chev.; *Meliac.*; Elf.
Biri-hoedoe: *Eperua jenmanni* Oliv.; *Caesalpiniac.*; Sur.
„ „ Wittie: *Eperua schomburgkiana* Bth.; *Caesalpiniac.*; Sur.
Birjagua: *Colubrina ferruginosa* Brongn.; *Rhamnac.*; Cuba.
Birke, amerikanische: *Betula spp.*; *Betulac.*; N.-Am.

Birke, Gelb-: *Betula lutea* Michx.; *Betulac.;* N.-Am.
„ Gemeine (d): *Betula verrucosa* Ehrh.; *Betulac.;* Eu.
„ Haar- (d): *Betula pubescens* Ehrh.; *Betulac.;* N.-Eu.
„ Hain- (d): *Betula lenta* L.; *Betulac.;* N.-Am.
„ Moor- (d): *Betula pubescens* Ehrh.; *Betulac.;* N.-Eu.
„ Nachen- (d): *Betula papyrifera* Marsh.; *Betulac.;* N.-Am.
„ Rau- (d): *Betula verrucosa* Ehrh.; *Betulac.;* Eu.
„ Ruch- (d): *Betula pubescens* Ehrh.; *Betulac.;* N.-Eu.
„ Weiss- (d): *Betula verrucosa* Ehrh.; *Betulac.;* Eu.
„ „ nordische: *Betula pubescens* Ehrh.; *Betulac.;* N.-Eu.
Birma: *Calophyllum antillanum* Britton; *Guttifer.;* br. Hond.
„ (u): *Myristica aff. panamensis* Hemsl.; *Myristicac.;* br. Hond.
Birmah (u): *Calophyllum calaba* Jacq.; *Guttifer.;* br. Hond.
Birnbaum: *Pirus communis* L.; *Rosac.;* Eu.
„ afrikanisch (d): *Mimusops spp., Dumoria sp.; Sapotac.;* W.-Af.
„ herber: *Pirus acerba* DC.; *Rosac.;* Eu.
Birne, Holz-: *Pirus communis* L.; *Rosac.;* Eu., W.-As.
Bíro: *Cratoxylon sp.; Guttifer.;* Jap.
Bironji: *Areca catechu* L.; *Palmac.;* Jap.
Birringo: *Rhus integrifolia* Bth. et H. f., *R. juglandifolia* H. B. K.; *Anacardiac.;* n. S.-Am.
Birumanemu: *Albizzia lebbek* Bth.; *Mimosac.;* Jap.
Bisál: *Terminalia calamansanai* Rolfe; *Combretac.;* Phil.
„ : *Terminalia edulis* Blco., *T. nitens* Presl; *Combretac.;* Phil.
Bisamholz, Australisches (d): *Olearia argophylla* F. v. M.; *Composit.;* Au.
Biscoyal: *Bactris horrida* Oerst.; *Palmac.;* C.-R.
Biscoyol: *Bactris spp.; Palmac.;* Hond.
Bishop wood: *Bischofia javanica* Blume; *Euphorbiac.;* Ind.
Biskan: *Dillenia sp.; Dilleniac.;* Phil.
Bislót: *Eugenia xanthophylla* C. B. Rob.; *Myrtac.;* Phil.
Bisong: *Sterculia sp.; Sterculiac.;* Phil.
Bisquite: *Acacia paniculata* Willd.; *Mimosac.;* Hond.
Bissaboko: *Piptadenia chevalieri* Harms; *Mimosac.;* Elf.
Bissé: *Daniella klainei* Pierre; *Caesalpiniac.;* Gab.
Bisseye: *Daniella klainei* Pierre; *Caesalpiniac.;* Gab.
Bitá: *Alstonia scholaris* R. Br.; *Apocynac.;* Phil.
Bita hoedoe: *Potalia amara* Aubl.; *Loganiac.;* fr. Gu.
„ „ : *Homalium racoubea* Sw.; *Flacourtiac.;* fr. Gu.
Bitalí: *Pterocarpus spp.; Papilionac.;* Phil. (Zamb.)
Bitamók: *Calophyllum whitfordi* Merr.; *Guttifer.;* Phil. (Mind.)
Bitañgól: *Calophyllum spp., Garcinia sp.; Guttifer.;* Phil.
„ : *Kingiodendron alternifolium* Merr.; *Caesalpiniac.;* Phil.
Bitanhól: *Calophyllum blancoi* Pl. et Tr.; *Guttifer.;* Phil.
Bitáog: *Calophyllum spp.; Guttifer.;* Phil.
Bitáoi-bákil: *Calophyllum blancoi* Pl. et Tr.; *Guttifer.;* Phil.
„ „ : *Calophyllum cumingi* Pl. et Tr.; *Guttifer.;* Phil.
Bitáong: *Calophyllum blancoi* Pl. et Tr.; *Guttifer.;* Phil.
Bitchwood: *Ichthyomethia sp.; Papilionac.;* C.-Am., W.-Ind.
Bitik: *Shorea guiso* Blume; *Dipterocarpac.;* Phil.
Bitok: *Pausinystalia spp.; Rubiac.;* Kam.
Bitók-gúbat: *Calophyllum cumingi* Pl. et Tr.; *Guttifer.;* Phil. (Zamb.)
Bittáog: *Calophyllum inophyllum* L.; *Guttifer.;* Phil.
Bitterbark: *Petalostigma quadriloculare* F. v. M.; *Euphorbiac.;* O.-Au.
„ : *Alstonia constricta* F. v. M.; *Apocynac.;* sö. Au.
Bitter dan: *Simaruba sp.; Simarubac.;* Jam.
Bitterholz: *Quassia amara* L.; *Simarubac.;* tr. Am.
„ , Jamaica- (d): *Picraena excelsa* Lindl.; *Simarubac.;* W.-Ind.
Bitterhout: *Quassia amara* L.; *Simarubac.;* Gu.
„ : *Acrodiclidium camara* Schomb.; *Laurac.;* Gu.
„ : *Aspidosperma sp.; Apocynac.;* Gu.
Bitters (e): *Colubrina ferruginosa* Brongn.; *Rhamnac.;* W.-Ind.
Bittersweet (u): *Celastrus scandens* L.; *Celastrac.;* tr. Am.
Bitterwood (u): *Simaruba glauca* DC.; *Simarubac.;* Fl., tr. Am., W.-Ind.

Bitterwood: *Picrasma ailanthoïdes* Planch.; *Simarubac.*; Jap.
  „   : *Picrasma quassioides* Benn.; *Simarubac.*; Jap.
  „   : *Quassia amara* L.; *Simarubac.*; tr. Am.
  „   : *Picraena excelsa* Lindl.; *Simarubac.*; tr. Am.
  „ , Florida-: *Simaruba glauca* DC.; *Simarubac.*; Fl.
  „ South-America- (H): *Quassia amara* L.; *Simarubac.*; tr. Am.
  „ Westindian-: *Picraena excelsa* Lindl.; *Simarubac.*; W.-Ind.
Biʒe: *Inga sp.; Mimosac.*; Mex.
Biung: *Grewia oppositifolia* Roxb.; *Tiliac.*; nw. Himalaya.
Biús: *Bruguiera caryophyllaeoides* Blume; *Rhizophorac.*; Phil.
Biwa: *Eriobotrya japonica* Lindl.; *Rosac.*; Jap.
Bixa: *Bixa orellana* L.; *Bixac.*; Cuba.
Björk: *Betula odorata* Bechst., *B. verrucosa* Ehrh.; *Betulac.*; Schwed.
Blackbark: *Royena lucida* L.; *Ebenac.*; Kap.
Blackberry: *Callicarpa acuminata* H. B. K.; *Verbenac.*; Pan.
Black boy: *Xanthorrhoea preissi* Endl.; *Liliac.*; O.-Au.
Black bread (e): *Pithecolobium unguis-cati* Bth.; *Mimosac.*; Fl., br. W.-Ind.
Blackbutt: *Eucalyptus pilularis* Smith, *E. viminalis* Lab.; *Myrtac.*; ö. Au.
  „   : *Eucalyptus patens* Bth.; *Myrtac.*; w. Au.
Blackheart: *Vouacapoua americana* Aubl.; *Papilionac.*; br. Gu.
  „   : *Dissiliaria baloghioides* F. v. M.; *Euphorbiac.*; Au.
  „   : *Nothofagus solandri* Oerst.; *Fagac.*; N.-Seel. (Westland)
Blacktree: *Avicennia nitida* Jacq.; *Verbenac.*; Fl.
Blackwood: *Avicennia nitida* Jacq.; *Verbenac.*; Fl.
  „   (e): *Haematoxylon campechianum* L.; *Caesalpiniac.*; tr. Am.
  „   : *Dalbergia latifolia* Roxb.; *Papilionac.*; Ind.
  „   : *Celastrus sp.; Celastrac.*; Kap.
  „   : *Royena nitida* Thunb.; *Ebenac.*; S.-Af. (Natal)
  „   : *Acacia melanoxylon* R. Br.; *Mimosac.*; Au., Tasm.
  „ African: *Dalbergia saxatilis* Hook. f.; *Papilionac.*; Nig.
  „   „ : *Dalbergia melanoxylon* Guill. et Perr.; *Papilionac.*; tr. Af.
  „ Bombay-: *Dalbergia latifolia* Roxb.; *Papilionac.*; Ind.
  „ Burma-: *Dalbergia cultrata* R. Grah.; *Papilionac.*; Ind.
  „ Malabar-: *Dalbergia latifolia* Roxb.; *Papilionac.*; Ind.
  „ Nilghiri-: *Dalbergia latifolia* Roxb. var. *sissoides* R. Grah.; *Papilionac.*; Ind.
Bladder-pod tree (e): *Diphysa carthagenensis* Jacq.; *Papilionac.*; C.-Am., S.-Am.
Blahn: *Stenanthera bakuana* Chev.; *Anonac.*; Lib.
Blakka kabbisi: *Vouacapoua americana* Aubl.; *Papilionac.*; Sur.
  „ kabisie: *Diplotropis guianensis* Bth.; *Papilionac.*; Sur.
  „   „ : *Diplotropis leptophylla* Kleinh.; *Papilionac.*; Sur.
  „ parihoedoe: *Swarʒia acuminata* Willd.; *Caesalpiniac.*; Sur.
  „ tjabisie: *Diplotropis guianensis* Bth.; *Papilionac.*; Sur.
  „   „ : *Diplotropis leptophylla* Kleinh.; *Papilionac.*; Sur.
Blancito: *Dichapetalum donnell-smithi* Engl.; *Dichapetalac.*; Pan.
Blankholz (d): *Haematoxylon campechianum* L.; *Caesalpiniac.*; tr. Am.
Blanquillo: *Sebastiana kloʒschiana* Muell. Arg.; *Euphorbiac.*; tr. Am.
  „   : *Excoecaria marginata* Griseb.; *Euphorbiac.*; Arg.
  „   : *Sapium marginatum* Muell. Arg.; *Euphorbiac.*; Arg. (Entre Rios)
  „ blanco: *Pithecolobium hassleri* Chod.; *Mimosac.*; Arg. (Mis.)
  „ colorado: *Chrysophyllum grisebachi* Mez; *Sapotac.*; Arg. (Mis.)
  „   „ : *Chrysophyllum maytenoides* Mart.; *Sapotac.*; Arg. (Mis.)
  „   „ : *Sideroxylon ligustrinum* Speg.; *Sapotac.*; Arg.
Blar-jee: *Randia genipaeflora* DC.; *Rubiac.*; Lib.
Blasenschote (d): *Diphysa carthagenensis* Jacq.; *Papilionac.*; C.-Am., S.-Am.
Blasenstrauch, hoher (d): *Colutea arborescens* L.; *Papilionac.*; M.-M.
Blauholz (d): *Haematoxylon campechianum* L.; *Caesalpiniac.*; Mex.
Blayhu: *Pentaclethra macrophylla* Bth.; *Mimosac.*; Lib.
Bléblendon: *Treculia africana* Baill.; *Morac.*; Elf.
Bleeding heart, Native-: *Homalanthus populifolius* R. Grah.; *Euphorbiac.*; Au.
Bleeding-heart tree: *Euonymus atropurpureus* Jacq.; *Celastrac.*; ö. USA.
Bleh: *Scaphopetalum amoenum* Chev.; *Sterculiac.*; Lib.
Bleketembi: *Echinocarpus sigun* Blume; *Tiliac.*; Sunda.

Blékoré: *Maesobotrya stapfiana* Baill.; *Euphorbiac.*; Elf.
Blendreng: *Hymenodictyon excelsum* Wall.; *Rubiac.*; w. Himalaya.
Blétiné: *Ochrocarpus africanus* Oliv.; *Guttifer.*; Elf.
Blétuné: *Ochrocarpus africanus* Oliv.; *Guttifer.*; Elf.
Bliembieng: *Averrhoa bilimbi* L.; *Oxalidac.*; Malay.
  „ oeloe: *Averrhoa bilimbi* L.; *Oxalidac.*; Malay.
Blimah: *Trichoscypha spp.*, *Sorindeia longifolia* Oliv.; *Anacardiac.*; Lib.
  „ -pu: *Turraeanthus sp.*; *Meliac.*; Lib.
Blimbing: *Averrhoa bilimbi* L.; *Oxalidac.*; Java.
Blimbling: *Averrhoa bilimbi* L.; *Oxalidac.*; Java.
Blinding tree: *Excoecaria agallocha* L.; *Euphorbiac.*; Phil.
Blindwood: *Ceiba pentandra* Gaertn.; *Bombacac.*; Tr.
Blo: *Ceiba pentandra* Gaertn. *var. dehiscens* Ulb.; *Bombacac.*; Sudan.
Blockwood (e): *Haematoxylon campechianum* L.; *Caesalpiniac.*; tr. Am.
Bloe: *Eugenia whytei* Sprague; *Myrtac.*; Lib.
Blolly: *Pisonia obtusata* Jacq.; *Nyctaginac.*; Fl., br. W.-Ind.
Blom-poe: *Cola caricifolia* K. Schum.; *Sterculiac.*; Lib.
Bloodwood: *Pterocarpus spp.*; *Papilionac.*; C.-Am.
  „   : *Laplacea haematoxylon* G. Don; *Theac.*; Jam.
  „   : *Vismia macrophylla* H. B. K.; *Guttifer.*; br. Gu.
  „   : *Vismia cayennensis* Pers.; *Guttifer.*; br. Gu .
  „   : *Lagerstroemia flos-reginae* Retz.; *Lythrac.*; Ind.
  „   : *Pterocarpus angolensis* DC.; *Papilionac.*; O.-Af.
  „   : *Eucalyptus corymbosa* Sm.; *Myrtac.*; O.-Au.
  „ Mountain-: *Eucalyptus eximia* Schauer; *Myrtac.*; O.-Au. (Bl. B.)
  „ Red: *Eucalyptus corymbosa* Sm.; *Myrtac.*; Queensl.
  „ Scrub-: *Baloghia lucida* Endl.; *Euphorbiac.*; O.-Au.
  „ Smoothbar-ked: *Eucalyptus eximia* Schauer; *Myrtac.*; O.-Au. (Bl. B.)
  „ Swamp-: *Pterocarpus spp.*; *Papilionac.*; Tr.
  „ White: *Eucalyptus trachyphloia* F. v. M.; *Myrtac.*; Queensl.
  „ Yellow: *Eucalyptus eximia* Schauer; *Myrtac.*; O.-Au. (Bl. B.)
Blorh: ? *Ficus vogeliana* Miq.; *Morac.*; Lib.
  „ -feh: *Albizzia zygia* Macbr.; *Mimosac.*; Lib.
Blossomberry: *Eugenia axillaris* Willd.; *Myrtac.*; br. Hond.
Blossom, Blue-: *Vitex gaumeri* Greenm.; *Verbenac.*; br. Hond.
Blu: *Lonchocarpus cyanescens* Bth.; *Papilionac.*; Lib.
Blu-chu: *Maba cooperi* Hutch. et Dalz.; *Ebenac.*; Lib.
Blu-koh: *Alchornea cordifolia* Muell. Arg.; *Euphorbiac.*; Lib.
Blutholzbaum (d): *Haematoxylon campechianum* L.; *Caesalpiniac.*; tr. Am., W.-Ind.
  „   : *Eucalyptus corymbosa* Sm., *E. terminalis* F. v. M.; *Myrtac.*; Au.
Bo (H): *Sterculia nobilis* Sm.; *Sterculiac.*; Ind.-Ch.
Bo: *Hibiscus praeclarus* Gagnep.; *Malvac.*; Annam.
Bô: *Barringtonia acutangula* Gaertn., *B. annamica* Gagnep.; *Myrtac.*; Ind.-Ch.
  „ : *Mallotus cochinchinensis* Lour.; *Euphorbiac.*; Ind.-Ch.
Boa: *Euphoria longana* Lamk.; *Sapindac.*; Co.-Ch.
Boabab: *Adansonia digitata* L.; *Bombacac.*; Af.
Bo-ah: *Cola lateritia* K. Schum.; *Sterculiac.*; Lib.
  „ : *Neoboutonia glabrescens* Prain; *Euphorbiac.*; Lib.
Boandjo: *Avicennia nitida* Jacq.; *Verbenac.*; Kam.
Boango (f): ? *Avicennia nitida* Jacq.; *Verbenac.*; Kam.
Boa Pow: *Mangifera longipes* Griff.; *Anacardiac.*; Ind.-Ch.
Boar wood (H): *Symphonia globulifera* L.; *Guttifer.*; br. W.-Ind.
Boasa: *Panda oleosa* Pierre; *Pandac.*; Kam.
Boba: *Tessaria integrifolia* R. et P.; *Composit.*; Arg. (Salta, Jujuy)
Bobai (f): *Afzelia africana* Smith; *Caesalpiniac.*; Kam.
Bobaï (f): *Albizzia welwitschi* Oliv.; *Mimosac.*; Kam.
Bobam: *Chrysophyllum africanum* A. DC.; *Sapotac.*; Fernando-Po.
Bobata: *Millettia versicolor* Welw.; *Papilionac.*; Angola.
Bobáue: *Dysoxylum decandrum* Merr.; *Meliac.*; Phil. (Mind.)
Bobbi: *Calophyllum inophyllum* L.; *Guttifer.*; Ind.
Bobé: *Staudtia gabonensis* Warb.; *Myristicac.*; Gab.
Bobe ba nduku: ? *Entandrophragma macrophyllum* Chev.; *Meliac.*; Kam.

Bobimbi: *Scorodophloeus zenkeri* Harms; *Caesalpiniac.;* Kam.
Bobinga: *Brachystegia sp.; Caesalpiniac.;* Gab.
Bobi- waäta: *Sideroxylon guianense* A. DC.; *Sapotac.;* Sur.
Bobo: *Tessaria integrifolia* R. et P.; *Composit.;* Arg. (Salta, Jujuy)
„ : ? *Hopea odorata* Roxb.; *Dipterocarpac.;* Co.-Ch.
Bobó: *Sterculia sp.; Sterculiac.;* Phil.
Bo bo (H): *Shorea sp.; Dipterocarpac.;* Ind.-Ch.
Boboa: *Euphoria longana* Lamk.; *Sapindac.;* Co.-Ch.
Bobóg: *Sterculia sp.; Sterculiac.;* Phil.
Bobolu: *Irvingia wombolu* Verm.; *Simarubac.;* B.-K.
Bobom: *Xylopia rubescens* Oliv.; *Anonac.;* Kam.
Bobota: *Millettia versicolor* Welw.; *Papilionac.;* B.-K.
Boboy: *Sterculia sp.; Sterculiac.;* Phil.
Bobwood: *Anona glabra* L.; *Anonac.;* br.Hond.
Bob wood (u): *Ochroma spp.; Bombacac.;* tr. Am.
Bô cap dông: *Cassia javanica* L.; *Caesalpiniac.;* Ind.-Ch.
„ „ nurôc: *Cassia floribunda* Cav.; *Caesalpiniac.;* Ind.-Ch.
Boccoholz (d): *Inocarpus edulis* Aubl.; *Papilionac.;* Gu.
Boco (f): *Bocoa prouacensis* Aubl.; *Papilionac.;* fr. Gu.
Boco: *Swartzia spp.; Caesalpiniac.;* Sur.
Bocoma: *Parinarium robustum* Oliv.; *Rosac.;* Elf.
Bocote: *Cordia gerascanthoides* H.B.K.; *Borraginac.;* Mex.
Bodaiju: *Tilia miquelina* Maxim.; *Tiliac.;* Jap.
Bodé (f): *Styrax tonkinensis* Pierre, *S.macrothyrsus* Perk.; *Styracac.;* Ind.-Ch.
„ „ : *Anthostyrax tonkinensis* Pierre; *Styracac.;* Ind.-Ch.
Bô dè tiá: *Styrax sp.; Styracac.;* Ind.-Ch.
„ „ tráng: *Styrax tonkinensis* Pierre; *Styracac.;* Ind.-Ch.
Bodia (f): *Pynaertia occidentalis* Chev.; *Meliac.;* Elf.
Bodioa (f): *Anopyxis occidentalis* Chev.; *Rhizophorac.;* Elf.
„ „ : *Anopyxis ealaensis* Sprague; *Rhizophorac.;* Elf.
„ „ : *Cleistopholis patens* Engl. et Diels; *Anonac.;* Elf.
Bodjabé: *Mimusops pierreana* Engl.; *Sapotac.;* Gab.
Bodjembo: *Afzelia africana* Smith; *Caesalpiniac.;* B.-K.
Bodo: *Mitragyne macrophylla* Hiern; *Rubiac.;* Elf.
Bodoua: *Saccoglottis gabonensis* Urb.; *Humiriac.;* Kam.
Boe: *Parkia bicolor* Chev.; *Mimosac.;* Lib.
Boea nona: *Anona sp.; Anonac.;* fr.Gu.
Boegoe-boegoe: *Swartzia acuminata* Willd.; *Caesalpiniac.;* Sur.
Boegroe makka: *Guilielma speciosa* Mart.; *Palmac.;* fr.Gu.
Boeirata: *Parinarium campestre* Aubl.; *Rosac.;* Sur.
Boelewé: *Mimusops spp.; Sapotac.;* Sur.
Boelikoro: *Helicostylis poeppigiana* Trec.; *Morac.;* Gu.
Boeloekoro: *Piratinera guianensis* Aubl.; *Morac.;* Sur.
Boeloemé balli: ? *Chaetocarpus schomburgkiana* Pax et Hoffm.; *Euphorbiac.:* Sur.
Boeloewé: *Mimusops spp.; Sapotac.;* Sur.
Boendah: *Artocarpus elastica* Reinw.; *Morac.;* Java.
Boengang: *Neesia altissima* Blume; *Bombacac.;* Java.
Boengoer: *Lagerstroemia flos-reginae* Retz.; *Lythrac.;* Ind.
Boerbone: *Schotia latifolia* Jacq.; *Caesalpiniac.;* Kap.
Boerboom, Bosch-: *Schotia latifolia* Jacq.; *Caesalpiniac.;* S.-Af.
Boeroewé: *Mimusops spp.; Sapotac.;* Sur.
Boesie kasjoe: *Dimorphandra latifolia* Tul.; *Caesalpiniac.;* Sur.
Boesie-mahonie: *Loxopterygium sagoti* Hook. f.; *Anacardiac.;* Sur.
Boesie tamalin: *Stryphnodendron guianense* Bth.; *Mimosac.;* Sur.
Boesi Kakam: ? *Apeiba aspera* Aubl.; *Tiliac.;* fr.Gu.
„ papaja: *Cecropia peltata* L.; *Morac.;* fr.Gu.
„ Soersakkar: *Anona sp.; Anonac.;* fr.Gu.
Boeton galeh: *Irina glabra* Blume; *Sapindac.;* Nied.-Ind.
Boewah Raoe: *Dracontomelum dao* Merr. et Rolfe; *Anacardiac.;* Cel.
Bofaka: *Copaifera demeusei* Harms; *Caesalpiniac.;* B.-K.
Bofale: *Parinarium glabrum* Oliv.; *Rosac.;* B.-K.
Bofali: *Parinarium glabrum* Oliv.; *Rosac.;* B.-K.

69

Bofonghe: *Bosquiea welwitschi* Engl.; *Morac.;* B.-K.
Bofote: *Platysepalum chevalieri* Harms; *Papilionac.;* B.-K.
Bofu: *Cleistopholis patens* Engl. et Diels; *Anonac.;* Elf.
Bofwe: *Allanblackia leucantha* Hutch. et Dalz.; *Guttifer.;* Go.
Bogabog-u-veto: *Xylosma monospora* Harv.; *Bixac.;* S.-Af.
Bogamani: *Virola spp.; Myristicac.;* Pan.
„ verde: *Dialyanthera otoba* Warb.; *Myristicac.;* Pan.
Bo-gar: *Conopharyngia durissima* Stapf; *Apocynac.;* Lib.
Bogarrier bâtard: *Myginda sp.; Celastrac.;* Mart.
„ sauvage: *Myginda sp.; Celastrac.;* Mart.
Bogenholz: *Maclura aurantiaca* Nutt.; *Morac.;* N.-Am.
Bogenhout: *Faurea speciosa* Welw., *F. usambarensis* Engl.; *Morac.;* Af.
Boggi-lobbi: *Cordia tetrandra* Aubl.; *Borraginac.;* Sur.
Bógo: *Garuga sp., Santiria nitida* Merr.; *Burserac.;* Phil.
Bog-onion: *Dysoxylum fraseranum* Bth.; *Meliac.;* Au.
Bogo zage: *Swartzia madagascariensis* Desv.; *Caesalpiniac.;* n. Nig.
Bogué: *Coula edulis* Baill.; *Olacac.;* Elf.
Bogüiè: *Coula edulis* Baill.; *Olacac.;* Elf.
Bogum: *Symphonia globulifera* L. f.; *Guttifer.;* Pan.
„ -Bogum: *Flindersia bennettiana* F. v. M.; *Meliac.;* Queensl.
Boh: *Mitragyne stipulosa* O. Ktze.; *Rubiac.;* Lib.
Bohamamua: *Mimusops clitandrifolia* Chev.; *Sapotac.;* Elf.
Bohambo: *Piptadenia africana* Hook. f.; *Mimosac.;* Kam.
? Bohari: *Cordia myxa* L.; *Borraginac.;* Aegypten, tr. Au.
Bohingo: *Pachypodanthium confine* Engl. et Diels; *Anonac.;* Kam.
Bohnenbaum (d): *Laburnum vulgare* Griseb.; *Papilionac.;* Eu.
„ Alpen- (d): *Laburnum alpinum* Griseb.; *Papilionac.;* Eu.
„ Gemeiner (d): *Laburnum anagyroides* Medic.; *Papilionac.;* Eu.
Bohnenstrauch: *Laburnum vulgare* Griseb.; *Papilionac.;* Eu.
Bohoi: *Shorea curtisi* Dyer; *Dipterocarpac.;* Malay.
Bohókan: *Cyathocalyx globosus* Merr.; *Anonac.;* Phil.
„ : *Eugenia saligna* C. B. Rob.; *Myrtac.;* Phil.
Bôhôkô: *Ochocoa gaboni* Pierre; *Myristicac.;* Gab.
Bohom: *Cordia gerascanthus* L.; *Borraginac.;* Mex.
Bô hón (H): *Sapindus mokorossi* Gaertn.; *Sapindac.;* Ind.-Ch.
Bohun: *Cordia alliodora* Cham.; *Borraginac.;* br. Hond.
„ -ché: *Cordia gerascanthus* L.; *Borraginac.;* br. Hond.
Boia: *Sterculia sp.; Sterculiac.;* Bras.
Boilam: *Swintonia schwenki* Kurz; *Anacardiac.;* Malay.
Bôi lôi (H): *Litsea vàng* H. Lec.; *Laurac.;* Ind.-Ch.
„ „ „ : *Tetranthera monopetala* Roxb.; *Laurac.;* Ind.-Ch.
„ „ „ : *Tritaxis gaudichaudi* Baill.; *Euphorbiac.;* Ind.-Ch.
„ „ : *Casearia membranacea* Hance; *Samydac.;* Ind.-Ch.
„ „ (H): *Nothaphoebe kingiana* Gamble; *Laurac.;* Ind.-Ch.
„ „ lông bên: *Litsea váng* St. Hil. *var. lobata* H. Lec.; *Laurac.;* Ind.-Ch.
„ „ nhúrt: *Litsea sp.; Laurac.;* Ind.-Ch.
„ „ tiá: *Litsea pierrei* H. Lec.; *Laurac.;* Ind.-Ch.
„ „ (tim): *Tritaxis gaudichaudi* Baill.; *Euphorbiac.;* Ind.-Ch.
„ „ tráng: *Litsea sp.; Laurac.;* Ind.-Ch.
„ „ váng: *Nothaphoebe umbelliflora* Blume; *Laurac.;* Ind.-Ch.
Bo-in-dah: *Guarea thompsoni* Spr. et Hutch.; *Meliac.;* Lib.
„ „ „ : *Trichilia heudeloti* Planch.; *Meliac.;* Lib.
Boiong: *Tarrietia argyrodendron* Bth.; *Sterculiac.;* O.-Au.
„ : *Tarrietia trifoliata* F. v. M.; *Sterculiac.;* O.-Au.
Bois Abeille (f): *Manilkara balata* Pierre; *Sapotac.;* fr. Gu.
„ agouti: *Minquartia guianensis* Aubl.; *Olacac.;* Sur.
„ amadou: ? *Hernandia guianensis* Aubl.; *Laurac.;* Gu.
„ „ : *Apeiba glabra* Aubl.; *Tiliac.;* Gu.
„ amer: *Quassia amara* L.; *Simarubac.;* Gu., Ant.
„ „ : *Acrodiclidium camara* Schomb.; *Laurac.;* Gu., Ant.
„ „ : *Picraena excelsa* Lindl.; *Simarubac.;* Guad.
„ „ blanc: *Trichilia spondioïdes* Sw.; *Meliac.;* Guad

Bois Anguilles: *Grangeria buxifolia* Sm.; *Rosac.;* Maur.
„ ara: *Pithecolobium pedicellare* Bth.; *Mimosac.;* fr. Gu.
„ babtiste: *Vismia sp.; Guttifer.;* fr. Gu.
„ bagasse: *Bagassa guianensis* Aubl.; *Morac.;* fr. Gu.
? „ bagot: *Peltogyne spp.; Caesalpiniac.;* fr. Gu.
„ balle: *Trichilia guarea* L.; *Meliac.;* Gu., fr. W.-Ind.
„ „ : *Guarea aubleti* A. Juss.; *Meliac.;* Gu., fr. W.-Ind.
„ banane: *Apeiba sp.; Tiliac.;* fr. Gu.
„ bande: *Roupala montana* Aubl.; *Proteac.;* Trin., Tob.
„ bandé: *Richeria grandis* Vahl; *Euphorbiac.;* Guad.
„ Baptiste: *Vismia cayennensis* Pers.; *Guttifer.;* Gu.
„ „ : *Hypericum sessilifolium* Aubl.; *Guttifer.;* Gu.
„ baroit: *Brosimum paraense* Huber; *Morac.;* fr. Gu.
„ barré: *Acer pennsylvanicum* L.; *Acerac.;* ö. N.-Am.
„ batard-canon: *Didymopanax morototoni* Dcne. et Pl.; *Araliac.;* Trin.
„ beni: *Buxus sempervirens* L.; *Buxac.;* S.-Eu., N.-Af., N.- & W.-As.
„ benjoin: *Terminalia mauritiana* Blco.; *Combretac.;* Maur.
„ blanc: *Tilia americana* L.; *Tiliac.;* ö. N.-Am.
„ „ : *Bumelia sp.; Sapotac.;* Haiti.
„ „ : *Phyllostylon brasiliensis* Capanema; *Ulmac.;* St.-D., Haiti.
„ „ : *Simaruba amara* Aubl.; *Simarubac.;* Sur., fr. W.-Ind.
„ „ : *Simaruba versicolor* A. St. Hil.; *Simarubac.;* Sur., fr. W.-Ind.
„ „ : *Hernandia ovigera* L.; *Hernandiac.;* Réunion.
„ „ : *Hernandia peltata* Meissn.; *Hernandiac.;* Sey.
„ „ du Nord: *Picea excelsa* Link; *Conifer.;* Eu.
„ blanchet: *Couralia fluviatilis* Splitg.; *Bignoniac.;* fr. Gu. (St. Laur.)
„ bleu: *Haematoxylon campechianum* L.; *Caesalpiniac.;* tr. Am.
„ cabri: *Aegiphila martinicensis* L.; *Verbenac.;* Bras.
„ Cabrit: *Zanthoxylum tragodes* Jacq.; *Rutac.;* Guad., Mart.
„ caca: *Gustavia spp.; Lecythidac.;* Gu.
„ „ : *Sterculia sp.; Sterculiac;* fr. Gu.
„ „ : *Capparis ferruginea* L.; *Capparidac.;* Ant.
„ cachiment: *Talauma plumieri* Swartz; *Magnoliac.;* Ant.
„ caconnier rouge: *Ormosia dasycarpa* Jacks.; *Papilionac.;* Guad.
„ caille: *Carapa guianensis* Aubl.; *Meliac.;* fr. Gu.
„ calalou: *Apeiba sp.; Tiliac.;* fr. Gu.
„ canari: *Hirtella sp.; Rosac.;* Trin.
„ canelle: *Licaria guianensis* Aubl.; *Laurac.;* fr. Gu.
„ „ : *Acrodiclidium canella* Mez; *Laurac.;* Sur.
„ canne: *Byrsonima sp.; Malpighiac.;* Sur., Guad.
„ cannelle: *Licaria guianensis* Aubl.; *Laurac.;* Gu.
„ canon: *Cecropia peltata* L.; *Morac.;* fr. Gu., Trin.
„ „ batard: *Didymopanax morototoni* Dcne. et Pl.; *Araliac.;* fr. Gu.
„ Capitaine: *Malpighia urens* L.; *Malpighiac.;* Guad.
„ capucin: *Northea seychellana* Hook. f.; *Sapotac.;* Sey.
„ carré: *Evonymus europaea* L.; *Celastrac.;* Fr.
„ cassant: *Claoxylon sp.; Euphorbiac.;* Ind.
„ „ : *Psathura borbonica* Gmel.; *Rubiac.;* Réunion.
„ casse: *Cassia apoucouita* Aubl.; *Caesalpiniac.;* Gu.
„ cerf (f): *Pithecolobium pedicellare* Bth.; *Mimosac.;* fr. Gu.
„ „ : *Olea chrysophylla* Lam.; *Oleac.;* Mauritius,
„ chaine: *Catalpa longissima* Jacq.; *Bignoniac.;* Haiti.
„ chaire: *Tecoma leucoxylon* Mart.; *Bignoniac.;* Ant., S.-Am.
„ chandelle: *Amyris balsamifera* L.; *Rutac.;* Guad.
„ „ blanc: *Amyris silvatica* Jacq.; *Rutac.;* Guad.
„ chatousieux (f): *Pterocarpus draco* L., *P. rhori* Vahl; *Papilionac.;* Sur., Guad.
„ chauve souris: *Fernelia buxifolia* Lam.; *Rubiac.;* Maur.
„ chêne: *Catalpa longissima* Jacq.; *Bignoniac.;* Haiti.
„ clou: *Eugenia cotinifolia* Jacq.; *Myrtac.;* Maur.
„ cochon: *Ecclinusa sanguinolenta* Pierre; *Sapotac.;* fr. Gu.
„ „ : *Hedwigia balsamifera* Swartz; *Burserac.;* Ant., fr. Gu.
„ „ : *Symphonia globulifera* L. f.; *Guttifer.;* Gu.

Bois conleuvre: *Colubrina ferruginosa* Brongn.; *Rhamnac.;* Mart.
„ corail: *Pterocarpus soyauxi* Taub.; *Papilionac.;* Kam., Gab.
„ corbeau: ? *Swartҙia sp.; Mimosac.;* Gu.
„ cossais: *Vismia guianensis* Pers.; *Guttifer.;* Gu.
„ costière: *Colubrina reclinata* Brongn.; *Rhamnac.;* Guad., Trin.
„ côtelet: *Casearia parviflora* Willd.; *Flacourtiac.;* Guad.
„ côte noir: *Tapura guianensis* Aubl.; *Dichapetalac.;* Guad.
„ coton: *Ceiba pentandra* Gaertn.; *Bombacac.;* fr. Gu.
„ couleuvre: *Colubrina ferruginea* Brongn.; *Rhamnac.;* Guad.
„ crapaud: ? *Swartҙia sp.; Caesalpiniac.;* fr. Gu.
„ „ : *Micropholis melinoniana* Pierre; *Sapotac.;* fr. Gu.
„ creusot: *Vochysia guianensis* Aubl.; *Vochysiac.;* Gu.
„ creuzot: *Vochysia guianensis* Aubl.; *Vochysiac.;* fr. Gu.
„ cruzeau: *Vochysia guianensis* Aubl.; *Vochysiac.;* Gu.
„ dard: *Swartҙia sp.; Caesalpiniac.;* fr. Gu.
„ dartre: *Pterocarpus guianensis* Aubl.; *Papilionac.;* Gu.
„ „ : *Cassia alata* L.; *Caesalpiniac.;* Gu.
„ „ : *Hypericum sessilifolium* Aubl.; *Guttifer.;* Gu.
„ doux: *Phoebe elongata* Nees, *Persea sp.; Laurac.;* Guad.
„ „ : *Craterispermum microdon* Baker; *Rubiac.;* Sey.
„ „ jaune: *Ceanothus chloroxylon* Nees; *Rhamnac.;* Jam.
„ „ maron: *Inga ingoides* Willd.; *Mimosac.;* St.-D.
„ „ noir: *Ocotea sp.; Laurac.;* Guad., Mart.
„ dur: *Ostrya virginiana* C. Koch; *Corylac.;* ö. N.-Am.
„ dysentérique: *Byrsonima spicata* DC.; *Malpighiac.;* fr. Gu., Mart.
„ épineux: *Ceiba pentandra* Gaertn.; *Bombacac.;* Guad.
„ „ blanc (f): *Zanthoxylum fraxineum* Willd.; *Rutac.;* USA., Mex.
„ „ jaune: *Zanthoxylum clava-herculis* L.; *Rutac.;* Ant.
„ fidèle: *Citharexylum sp.; Verbenac.;* W.-Ind.
„ flambeau: *Toulicia guianensis* Aubl.; *Sapindac.;* Gu.
„ flèche: *Cornus florida* L.; *Cornac.;* N.-Am.
„ flot: *Ochroma lagopus* Swartҙ; *Bombacac.;* Ant.
„ fourmis (f): *Triplaris surinamensis* Cham.; *Polygonac.;* Sur.
„ fourmi: *Triplaris sp.; Polygonac.;* Mart.
„ franc (f): *Ilex aquifolium* L.; *Aquifoliac.;* Eu.
„ fromage: *Enterolobium schomburgki* Bth.; *Mimosac.;* fr. Gu.
„ galeux: *Dombeya populnea* Cav.; *Sterculiac.;* Réunion.
„ gaulettes: *Melicocca diversifolia* Juss.; *Sapindac.;* Réu., Maur.
„ „ : *Cupania alternifolia* Pers.; *Sapindac.;* Réu., Maur.
„ „ rouge (f): *Licania heteromorpha* Bth.; *Rosac.;* Sur.
„ gommier: *Icica sp.; Burserac.;* Mart.
„ gonte: *Vitex sp.; Verbenac.;* Guad.
„ grage: *Apeiba aspera* Aubl.; *Tiliac.;* Gu.
„ graine bleue: *Symplocos martinicensis* Jacq.; *Symplocac.;* St.-D.
„ grège: *Apeiba sp.; Tiliac.;* fr. Gu.
„ gris: *Licania sp.; Rosac.;* Guad., Trin.
„ handelle: *Amyris sp.; Rutac.;* Guad.
„ Hinselin: *Malpighia urens* L.; *Malpighiac.;* Guad.
„ immortelle: *Erythrina sp.; Papilionac.;* Trin.
„ incorruptible: *Homalium sp.; Flacourtiac.;* fr. Gu.
„ ivoire: *Sideroxylon eburneum* Chev.; *Sapotac.;* Ind.-Ch.
„ ivrant de la Jamaique: *Ichthyomethia piscipula* Hitch.; *Papilionac.;* Jam.
„ jaune: *Chlorophora tinctoria* Gaudich.; *Morac.;* C.-Am., S.-Am.
„ „ : *Ceanothus chloroxylon* Nees; *Rhamnac.;* Jam.
„ „ : *Zanthoxylum sp.; Rutac.;* fr. W.-Ind.
„ „ : *Aniba bracteata* Mez; *Laurac.;* Guad.
„ „ : *Ocotea sp.; Laurac.;* Guad., Mart.
„ „ : ? *Aniba panurensis* Mez; *Laurac.;* fr. Gu.
„ „ : *Ochrosia borbonica* Gmel.; *Apocynac.;* Réunion.
„ „ de l'Australie: *Flindersia oxleyana* F. v. M.; *Rutac.;* Au.
„ „ „ Tampico: *Chlorophora tinctoria* Gaudich.; *Morac.;* Mex.
„ „ „ Brésil: *Chlorophora tinctoria* Gaudich.; *Morac.;* Bras.

Bois jaune de Cuba: *Chlorophora tinctoria* Gaudich.; *Morac.*; Cuba.
„ „ des Antilles (f): *Zanthoxylum clava-herculis* L.; *Rutac.*; Ant.
„ Jonquille: *Melanorrhoea laccifera* Pierre; *Anacardiac.*; Ind.-Ch.
„ la morue: *Pithecolobium pedicellare* Bth.; *Mimosac.*; fr. Gu.
„ „ „ : ? *Vitex sp.; Verbenac.*; fr. Gu.
„ Lamoussé noir: ? *Swartzia sp.; Caesalpiniac.*; Gu.
„ l'étang: *Pterocarpus spp.; Papilionac.*; Trin.
„ lèzard: *Piratinera guianensis* Aubl.; *Morac.*; Leew., W.-Ind.
„ „ : *Vitex sp.; Verbenac.*; Mart., Guad.
„ „ : *Vitex capitata* Vahl, *V. divaricata* Sw.; *Verbenac.*; Trin.
„ lizard: ? *Brosimum aubleti* Poepp. et Endl.; *Morac.*; St.-D.
„ loustau: *Evonymus europaea* L.; *Celastrac.*; Fr.
„ „ : *Antirrhoea verticillata* DC.; *Rubiac.*; Maur.
„ Mabou: *Morisonia americana* L.; *Capparidac.*; Mart.
„ macaque: *Enterolobium schomburgki* Bth.; *Mimosac.*; fr. Gu.
„ „ gris: *Enterolobium schomburgki* Bth.; *Mimosac.*; fr. Gu
„ „ rouge: *Pithecolobium pedicellare* Bth.; *Mimosac.*; fr. Gu., Cay.
„ madre: *Gymnanthes lucida* Sw.; *Euphorbiac.*; s. Fl., W.-Ind.
„ maigre: *Nuxia verticillata* Lam.; *Loganiac.*; Maur., Réu.
„ major: *Erythroxylon areolatum* L.; *Erythroxylac.*; tr. Am., Ind.
„ mamzelle: *Ochna mauritiana* Lamk.; *Ochnac.*; Maur.
„ manahé: *Securinega durissima* Gmel.; *Euphorbiac.*; Maur.
„ manche-houe: *Zanthoxylum clava-herculis* L.; *Rutac.*; Guad.
„ manglier rouge: *Bruguiera gymnorrhiza* Lam.; *Rhizophorac.*; Sey.
„ marbré: *Richeria grandis* Vahl; *Euphorbiac.*; Guad.
„ „ : *Brosimum paraënse* Huber; *Morac.*; fr. Gu., Bras.
„ maré: *Barringtonia speciosa* Forst.; *Lecythidac.*; Sey.
„ Marguerite: *Cordia sp.; Borraginac.*; fr. Gu.
„ Marie: *Caryocar spp.; Caryocarac.*; Sur.
„ „ : *Calophyllum calaba* Jacq.; *Guttifer.*; St.-D., Trin.
„ Mary: *Caryocar sp.; Caryocarac.*; fr. Gu.
„ mêche: *Apeiba glabra* Aubl.; *Tiliac.*; fr. Gu.
„ merle: *Aphloia integrifolia* Benn.; *Flacourtiac.*; Sey.
„ moucheté: *Dysoxylum lessertianum* Bth.; *Meliac.*; N.-Kal.
„ mozambique: *Ludia sessiliflora* Lam.; *Flacourtiac.*; Maur.
„ mulâtre: *Pentaclethra filamentosa* Bth.; *Mimosac.*; Sur.
? „ negre: *Cordia sp.; Borraginac.*; Trin.
„ négresse: *Beilschmiedia pendula* Hemsl.; *Laurac.*; Guad.
„ „ : *Nectandra coriacea* Gris.; *Laurac.*; Guad.
„ néphrétique (f): *Guaiacum officinale* L.; *Zygophyllac.*; tr. Am.
„ noir: *Acer pennsylvanicum* L.; *Acerac.*; ö. N.-Am.
„ „ : *Haematoxylon campechianum* L.; *Caesalpiniac.*; tr. Am.
„ „ : *Albizzia lebbek* Bth.; *Mimosac.*; Tr.
„ pagaies: *Swartzia acuminata* Willd.; *Caesalpiniac.*; Sur.
„ pagode: *Pithecolobium pedicellare* Bth., *Inga sp.; Mimosac.*; fr. Gu.
„ palmiste: *Andira inermis* H. B. K.; *Papilionac.*; fr. Gu.
„ parasol: *Cordia sp.; Borraginac.*; fr. Gu.
„ pelé: *Colubrina ferruginosa* Brongn.; *Rhamnac.*; Haiti.
„ perdrix (f): *Vouacapoua americana* Aubl.; *Papilionac.*; Gu.
„ „ „ : *Cassia siamea* Lamk.; *Caesalpiniac.*; Ind.-Ch.
„ pétrole: ? *Pycnandra benthami* Baill.; *Sapotac.*; N.-Kal.
„ pian: *Zanthoxylum pterota* H. B. K.; *Rutac.*; Mart.
„ piant: *Gustavia angusta* L.; *Lecythidac.*; Sur.
„ pigeon: *Mallotus integrifolius* Muell. Arg.; *Euphorbiac.*; Maur.
„ pin: *Talauma plumieri* Swartz; *Magnoliac.*; Ant.
„ piquant: *Zanthoxylum clava-herculis* L.; *Rutac.*; fr. Gu., fr. W.-Ind.
„ „ : *Zanthoxylum caribaeum* Lam.; *Rutac.*; fr. Gu., fr. W.-Ind.
„ pistolet: *Guarea spp.; Meliac.*; Gu., Ant.
„ plan: ? *Jacaranda copaia* D. Don; *Bignoniac.*; fr. Gu.
„ pois: *Swartzia pinnata* Willd.; *Caesalpiniac.*; Trin., Tobago.
„ porche: *Cordia subcordata* Lam.; *Borraginac.*; Sey.
„ pourpre: *Peltogyne spp.; Caesalpiniac.*; fr. Gu.

Bois puant: *Capparis ferruginea* L.; *Capparidac.;* Ant.
  „   „   : *Grias aubletiana* Miers, *Gustavia spp.; Lecythidac.;* Gu.
  „   „   : *Sterculia sp.; Sterculiac.;* Gu.
  „   „   : *Sterculia foetida* L.; *Sterculiac.;* tr. Am., Ind.
  „   „   : *Foetida borbonica* Gmel.; *Lecythidac.;* Réunion.
  „ quinquina des savanes: *Byrsonima crassifolia* DC.; *Malpighiac.;* Guad.
  „ quivi: *Quivisia mauritiana* Baker; *Meliac.;* Mauritius.
  „ ramier: *Muntingia calabura* L.; *Tiliac.;* Guad.
  „ Ramon: *Sapındus saponaria* L.; *Sapindac.;* Ant.
  „ résolu: *Chimarrhis cymosa* Jacq.; *Rubiac.;* Guat.
  „ roi: *Clusia rosea* Jacq.; *Guttifer.;* fr. Gu.
  „ ross: *Gaertnera vaginata* Lam.; *Loganiac.;* Maur.
  „ rouge: *Mimusops spp.; Sapotac.;* fr. Gu.
  „   „   : *Humiria balsamifera* Aubl.; *Humiriac.;* Gu.
  „   „   : *Coccoloba barbadensis* Jacq.; *Polygonac.;* Guad.
  „   „   : *Erythroxylon squamatum* Vahl; *Erythroxylac.;* Guad.
  „   „   : *Pterocarpus soyauxi* Taub.; *Papilionac.;* W.-Af., Gab.
  „   „   : *Elaeodendron orientale* Jacq.; *Celastrac.;* Réu.
  „   „   : *? Tetracera sp.; Dilleniac.;* Sey.
  „   „   : *Wormia ferruginea* Baill.; *Dilleniac.;* Sey.
  „ „ carapat: *Carapa guianensis* Aubl.; *Meliac.;* Guad.
  „ „ de St. Domingue: *Guarea trichilioides* L.; *Meliac.;* Gu., Ant.
  „ „ du Nord: *Pinus silvestris* L.; *Conifer.;* N.-Eu., N.-As., N.-Am.
  „ „ montagne: *Coccoloba sp.; Polygonac.;* fr. Gu., fr. W.-Ind.
  „ „ tisane: *Licania heteromorpha* Bth.; *Rosac.;* Sur.
  „ „   „   : *Humiria spp.; Humiriac.;* fr. Gu.
  „ sabre: *Eperua falcata* Aubl.; *Caesalpiniac.;* Sur.
  „ sagaie: *Cupania laevis* Pers.; *Sapindac.;* Maur.
  „ saint: *Guaiacum officinale* L.; *Zygophyllac.;* tr. Am.
  „   „   : *Dipteryx odorata* Willd.; *Papilionac.;* tr. Am.
  „ sandal: *Carissa seychellensis* Baker; *Apocynac.;* Sey.
  „ sanglant: *Haematoxylon campechianum* L.; *Caesalpiniac.;* tr. Am.
  „   „   : *Vismia guianensis* Pers.; *Guttifer.;* Gu.
  „ sans écorce: *Ludia heterophylla* Lamk.; *Bixac.;* Réunion.
  „ satin (f): *Chloroxylon swietenia* DC.; *Meliac.;* Ind.
  „ satiné: *Ferolia guianensis* Aubl.; *Morac.;* Gu.
  „ „ d'Amérique: *Fagara flava* Kr. et Urb.; *Rutac.;* Ant.
  „ „ de l'Inde: *Chloroxylon swietenia* DC.; *Meliac.;* Ind.
  „ savonneux: *Sapindus saponaria* L.; *Sapindac.;* Ant.
  „ serpent: *Clusia sp.; Guttifer.;* fr. Gu.
  „ „ (f): *Stryphnodendron flammatum* Kleinh.; *Mimosac.;* Gu.
  „ „ „ : *Pithecolobium racemiflorum* Ducke; *Mimosac.;* Gu.
  „ Shavanon: *Catalpa speziosa* Warder; *Bignoniac.;* s. USA.
  „ tabac: *Exostemma floribundum* Roem.; *Rubiac.;* W.-Ind.
  „ tan: *Byrsonima sp.; Malpighiac.;* fr. Gu.
  „ „ : *Byrsonima altissima* DC.; *Malpighiac.;* Guad.
  „ „ : *Byrsonima spicata* DC.; *Malpighiac.;* Mart.
  „ tapiré: *Tapirira guianensis* Aubl.; *Anacardiac.;* fr. Gu.
  „ „ : *? Tapirira aff. guianensis* Aubl.; *Anacardiac.;* Sur.
  „ tressé: *Acacia sp.; Mimosac.;* Au.
  „ trompette: *Cecropia peltata* L.; *Morac.;* Gu.
  „ verdoyant: *Ceanothus chloroxylon* Nees; *Rhamnac.;* Jam.
  „ vert: *Gymnanthes lucida* Sw.; *Euphorbiac.;* Guad.
  „ violet (f): *Peltogyne spp.; Caesalpiniac.;* Gu.
  „ „ „ : *Copaifera bracteata* Bth.; *Caesalpiniac.;* Gu.
  „ zébra: *Stryphnodendron flammatum* Kleinh.; *Mimosac.;* Sur.
  „ „ : *Pithecolobium racemiflorum* Ducke; *Mimosac.;* Sur.
  „ zébré (f): *Cynometra aff. lujaei* de Wild.; *Caesalpiniac.;* Kam., Gab.
  „ à baguettes: *Coccoloba uvifera* L.; *Polygonac.;* Ant.
  „ „ balais: *Erythroxylon hypericifolium* Lamk.; *Erythroxylac.;* Réu.
  „ „ barrique: *Hedwigia balsamifera* Swartz; *Burserac.;* Ant.
  „ „ cochon: *Protium heptaphyllum* March.; *Burserac.;* Gu.

Bois à dartres: *Vismia guianensis* Pers.; *Guttifer.;* Gu.
„ „ énivrer: *Ichthyomethia piscipula* Hitch.; *Papilionac.;* Guad.
„ „ flambeau: *Humiria spp.; Humiriac.;* fr. Gu.
„ „ flambeaux: *Toulicia guianensis* Aubl.; *Sapindac.;* Gu.
„ „ „ : *Hedwigia balsamifera* Swartz; *Burserac.;* Ant., fr. Gu.
„ „ „ : *Erythroxylon laurifolium* Lamk.; *Erythroxylac.;* Réu.
„ „ „ : *Erythroxylon longifolium* Lamk.; *Erythroxylac.;* Réu.
„ „ flèches: *Swartzia sp.; Caesalpiniac.;* fr. Gu.
„ „ pian: *Chlorophora tinctoria* Gaudich.; *Morac.;* tr. Am.
„ „ „ : *Jacaranda copaia* D. Don; *Bignoniac.;* Sur.
„ „ „ : *Zanthoxylum tragodes* Jacq.; *Rutac.;* Guad., Mart.
„ d'absinthe: *Quassia amara* L.; *Simarubac.;* Ant.
„ „ : *Canariellum oleiferum* Engl.; *Burserac.;* N.-Kal.
„ d'acajou à planches: *Cedrela spp.; Meliac.;* Barb.
„ d'Acossais: *Vismia guianensis* Pers.; *Guttifer.;* Gu.
„ d'acouma: *Homalium sp.; Flacourtiac;* fr. Gu.
„ d'aigle (f): *Aquilaria agallocha* Roxb.; *Thymelaeac.;* Ind.-Ch.
„ d'alès du Mexique (f): *Bursera delpechiana* J. Poiss.; *Burserac.;* Mex.
„ d'Aloès, Faux: *Michelia spp.; Magnoliac.;* Nied.-Ind.
„ d'Amarante: *Peltogyne spp.; Caesalpiniac.;* Gu.
„ „ : *Copaifera bracteata* Bth.; *Caesalpiniac.;* Gu.
„ „ rouge: *Martiusia excelsa* Bth.; *Caesalpiniac.;* Sur.
„ „ : *Martiusia parvifolia* Bth.; *Caesalpiniac.;* Sur.
„ d'amaranthe: ? *Swietenia sp.; Meliac.;* Cay.
„ d'Amboine (f): *Flindersia amboinensis* Poir.; *Rutac.;* Moluk.
„ d'amourette moucheté (f): *Piratinera guianensis* Aubl.; *Morac.;* Sur.
„ d'Angélique: *Dicorynia paraënsis* Bth.; *Caesalpiniac.;* fr. Gu.
„ d'Anis: *Ocotea cymbarum* H. B. K.; *Laurac.;* Ven.
„ „ : *Mesua ferrea* L.; *Guttifer.;* Co.-Ch.
„ „ : *Limonia madagascariensis* Lamk.; *Rutac.;* Mad.
„ d'anisette: *Piper sp.; Piperac.;* fr. Gu.
„ „ : *Ocotea cymbarum* H. B. K.; *Laurac.;* Ven.
„ „ : *Limonia madagascariensis* Lamk.; *Rutac.;* Mad.
„ d'arc: *Oxandra lanceolata* Baill.; *Anonac.;* Gu., W.-Ind.
„ d'Arrada: *Byrsonima lucida* DC.; *Malpighiac.;* Guad.
„ d'artre: *Vismia sp.; Guttifer.;* fr. Gu.
„ d'Atlas (e): *Chloroxylon swietenia* DC.; *Meliac.;* Ind.
„ d'eau de la Guadeloupe: *Ocotea sp.; Laurac.;* Guad., Mart.
„ d'ébène: *Diospyros melanida* Poir.; *Ebenac.;* Réunion.
„ „ verte: *Tecoma spp.; Bignoniac.;* fr. Gu.
„ d'encens: *Hedwigia balsamifera* Swartz; *Burserac.;* Ant.
„ „ : *Humiria balsamifera* Aubl.; *Humiriac.;* Gu.
„ „ : *Protium heptaphyllum* March.; *Burserac.;* Gu.
„ d'encre: *Exotea paniculata* Radlk.; *Sapindac.;* Fl., Ant.
„ d'huile: ? *Tapirira guianensis* Aubl.; *Anacardiac.;* fr. Gu.
„ „ : *Erythroxylon hypericifolium* Lamk;. *Erythroxylac.;* Réu.
„ d'Inde: *Amomis caryophyllata* Kr. et Urb.; *Myrtac.;* Guad.
„ d'ivoire: *Sideroxylon eburneum* Chev.; *Sapotac.;* Ind.-Ch.
„ d'oiseau: *Claoxylon sp.; Euphorbiac.;* Réunion.
„ d'olive: *Capparis sp.; Capparidac.;* Trin.
„ „ : *Elaeodendron orientale* Jacq.; *Celastrac.;* Maur.
„ „ : *Olea lancea* Lam.; *Oleac.;* Réunion.
„ d'or: ? *Pithecolobium vinhatico* Record; *Mimosac.;* Bras.
„ „ du Cap (f): *Elaeodendron croceum* DC.; *Celastrac.;* Kap.
„ „ „ „ : *Cassine crocea* O. Ktze.; *Celastrac.;* Kap.
„ d'orange: *Chlorophora tinctoria* Gaudich.; *Morac.;* Trin.
„ d'Orme (f): *Guazuma ulmifolia* Lamk.; *Sterculiac.;* Trin.
„ „ d'Amérique: *Guazuma tomentosa* H. B. K.; *Sterculiac.;* Guad.
„ d'ortie: *Obetia sp.; Morac.;* Réunion.
„ de Bassin: *Blackwellia paniculata* Lam.; *Flacourtiac.;* Réu.
„ „ bouis: *Bumelia sp.; Sapotac.;* fr. W.-Ind.
„ „ Bourg épine (f): *Rhamnus alaternus* L.; *Rhamnac.;* M.-M.

Bois de Brésil: *Caesalpinia crista* L.; *Caesalpiniac.;* Gu.
„ „ Caju-bélo: *Schmidelia pinnata* DC.; *Sapindac.;* Südsee.
„ „ campêche: *Haematoxylon campechianum* L.; *Caesalpiniac.;* Gu.
„ „ Canne: *Melicocca diversifolia* Juss.; *Sapindac.;* Réu., Maur.
„ „ cannelle: *Ocotea sp.;* *Laurac.;* fr. Gu.
„ „ capahu: ? *Aniba panurensis* Mez; *Laurac.;* fr. Gu.
„ „ Capitaine: *Malpighia glabra* L.; *Malpighiac.;* Haiti.
„ „ casse-tête: *Dodonaea dioica* Roxb.; *Sapindac.;* N.-Kal.
„ „ Cavalam (f): *Sterculia foetida* L.; *Sterculiac.;* tr. As.
„ „ Cavalone: *Sterculia sp.;* *Sterculiac.;* fr. Gu.
„ „ Cayan (f): *Simaruba amara* Aubl.; *Simarubac.;* fr. W.-Ind.
„ „ Cayenne: *Brosimum paraënse* Huber; *Morac.;* fr. Gu.
„ „ cedra: *Cedrela spp.;* *Meliac.;* fr. Gu.
„ „ Cédrel (f): *Cedrela guianensis* A. Juss.; *Meliac.;* fr. Gu.
„ „ Cerisier: *Malpighia glabra* L.; *Malpighiac.;* Ant.
„ „ chenille: *Psiadia sp.;* *Composit.;* Réunion.
„ „ chinchin: *Azara microphylla* Phil.; *Bixac.;* Chile.
„ „ Chine: *Murraya exotica* L.; *Rutac.;* Réunion.
„ „ Chittagong (e): *Chukrasia tabularis* A. Juss.; *Meliac.;* Ind.
„ „ „ „ : *Toona ciliata* Roem.; *Meliac.;* Ind.
„ „ citron (f): *Amyris silvatica* Jacq.; *Rutac.;* Ant.
„ „ „ : *Zanthoxylum horridum* Welw.; *Rutac.;* port. Af.
„ „ „ de Cayenne: ? *Aniba panurensis* Mez; *Laurac.;* fr. Gu.
„ „ coeur pourpre: *Peltogyne spp.;* *Caesalpiniac.;* fr. Gu.
„ „ Colophane: *Canarium paniculatum* Bth.; *Burserac.;* Maur.
„ „ „ franc: *Canarium paniculatum* Bth.; *Burserac.;* Maur.
„ „ „ bâtard: *Protium obtusifolium* March.; *Burserac.;* Mask.
„ „ Compagnie: *Protium obtusifolium* March.; *Burserac.;* Mask.
„ „ corail tendre: *Pterocarpus spp.;* *Papilionac.;* Guad.
„ „ corne fétide: *Capparis ferruginea* L.; *Capparidac.;* Ant.
„ „ „ „ : *Sterculia sp.;* *Sterculiac.;* fr. Gu.
„ „ cotelet: *Citharexylum spp.;* *Verbenac.;* tr. Am.
„ „ coumarouna: *Dipteryx odorata* Willd.; *Papilionac.;* Mart.
„ „ courbaril: *Hymenaea courbaril* L.; *Caesalpiniac.;* fr. Gu.
„ „ crave: *Dicypellium caryophyllatum* Nees; *Laurac.;* fr. Gu.
„ „ cype: *Cordia gerascanthus* L.; *Borraginac.;* Mart.
„ „ Cypre: *Cordia gerascanthus* L.; *Borraginac.;* tr. Am., Mart.
„ „ fee: *Bumelia angustifolia* Nutt.; *Sapotac.;* Mart.
„ „ fer: *Ostrya virginiana* C. Koch; *Corylac.;* ö. N.-Am.
„ „ „ : *Krugiodendron ferreum* Urb.; *Rhamnac.;* C.-Am., W.-Ind.
„ „ „ : *Ixora ferrea* Bth.; *Rubiac.;* Guad.
„ „ „ : *Colubrina ferruginosa* Brongn.; *Rhamnac.;* Ant.
„ „ „ : *Colubrina reclinata* Brongn.; *Rhamnac.;* Ant.
„ „ „ : *Zanthoxylum pterota* H. B. K.; *Rutac.;* Jam.
„ „ „ : *Sideroxylon sp.;* *Sapotac.;* Gu.
„ „ „ : *Mouriria spp.;* *Melastomatac.;* Gu.
„ „ „ (f): *Swartjia spp., Swartjia tomentosa* DC.; *Caesalpiniac.;* Sur.
„ „ „ : *Tecoma sp.;* *Bignoniac.;* Sur.
„ „ „ (f): *Mesua ferrea* L.; *Guttifer.;* Co.-Ch.
„ „ „ „ : *Lophira procera* Chev.; *Ochnac.;* W.-Af., Gab.
„ „ „ : *Sideroxylon borbonicum* A. DC.; *Sapotac.;* Réu.
„ „ „ . *Stadmannia sideroxylon* DC.; *Sapindac.;* Maur.
„ „ „ : *Vateria seychellarum* Dyer; *Dipterocarpac.;* Sey.
„ „ „ : *Tarrietia argyrodendron* Bth.; *Sterculiac.;* Queensl.
„ „ „ : *Casuarina equisetifolia* Forst.; *Casuarinac.;* N.-Kal.
„ „ „ : *Casuarina cunninghamia* Miq.; *Casuarinac.;* N.-Kal.
„ „ „ blanc: *Bumelia sp.;* *Sapotac.;* fr. W.-Ind.
„ „ „ de Judas: *Cossignia borbonica* DC.; *Sapindac.;* Réu.
„ „ „ de la Réunion (f): *Cupania sideroxylon* Cambess.; *Sapindac;.* Réu.
„ „ „ de montagne: *Casuarina poissoniana* Schltr.; *Casuarinac.;* N.-Kal.
„ „ Féroles: *Ferolia guianensis* Aubl.; *Morac.;* Gu.
„ „ Filao: *Casuarina equisetifolia* Forst.; *Casuarinac.;* Senl.

Bois de flot: *Hibiscus tiliaceus* L.; *Malvac.;* N.-Kal.
„ „ forêt: *Aglaia elaeagnoidea* Bth.; *Meliac.;* N.-Kal.
„ „ frêne: *Quassia amara* L.; *Simarubac.;* fr. W.-Ind.
„ „ Fresne: *Quassia amara* L.; *Simarubac.;* Ant.
„ „ gaiac: *Guaiacum officinale* L.; *Zygophyllac.;* tr. Am.
„ „ Gaillard: *Melicocca diversifolia* Juss.; *Sapindac.;* Réu., Maur.
„ „ gaulette: *Hirtella sp.; Rosac.;* fr. Gu.
„ „ gayac: *Guaiacum officinale* L.; *Zygophyllac.;* tr. Am.
„ „ Gommier blanc: *Bursera gummifera* L.; *Burserac.;* Ant.
„ „ „ rouge: *Hedwigia balsamifera* Swarƺ; *Burserac.;* fr. W.-Ind., fr. Gu.
„ „ „ „ : *Protium spp.; Burserac.;* fr. W.-Ind., fr. Gu.
„ „ guiatre: *Citharexylum fruticosum* L.; *Verbenac.;* Fl., Ant.
„ „ Jasmin: *Ochna mauritiana* Lamk.; *Ochnac.;* Maur.
„ „ joli coeur: *Celastrus undulatus* Lamk.; *Celastrac.;* Kap., Mask.
„ „ Judas: *Cossignia borbonica* DC.; *Sapindac.;* Réunion.
„ „ la fièvre: *Vismia sp.; Guttifer.;* fr. Gu.
„ „ „ Jamaique: *Haematoxylon campechianum* L.; *Caesalpiniac.;* tr. Am.
„ „ lait: *Plumiera alba* L.; *Apocynac.;* fr. W.-Ind.
„ „ „ : *Tabernaemontana persicariaefolia* Jacq.; *Apocynac.;* Maur.
„ „ lance: *Oxandra lanceolata* Baill.; *Anonac.;* tr. Am.
„ „ lardoire: *Evonymus europaea* L.; *Celastrac.;* Fr.
„ „ lettre: *Machaerium schomburgki* Bth.; *Papilionac.;* Gu.
„ „ „ gris: *Piratinera guianensis* Aubl.; *Morac.;* fr. Gu.
„ „ „ moucheté: *Piratinera guianensis* Aubl.; *Morac.;* fr. Gu.
„ „ „ rouge: *Amanoa guianensis* Aubl.; *Euphorbiac.;* fr. Gu.
„ „ „ „ : *Brosimum paraënse* Huber; *Morac.;* fr. Gu.
„ „ „ tigré: *Machaerium schomburgki* Bth.; *Papilionac.;* fr. Gu.
„ „ lettres: *Piratinera guianensis* Aubl.; *Morac.;* fr. Gu.
„ „ „ de Chine: *Piratinera guianensis* Aubl.; *Morac.;* tr. Am.
„ „ „ marbré: *Machaerium schomburgki* Bth.; *Papilionac.;* fr. Gu.
„ „ „ moucheté: *Machaerium schomburgki* Bth.; *Papilionac.;* fr. Gu.
„ „ licari: ? *Aniba panurensis* Mez; *Laurac.;* fr. Gu.
„ „ liège: *Hibiscus elatus* Swarƺ; *Malvac.;* Barb.
„ „ „ (f): *Hibiscus tiliaceus* L.; *Malvac.;* Ant.
„ „ lilas: *Syringa vulgaris* L.; *Oleac.;* sö. Eu.
„ „ lima: *Haematoxylon brasiletto* Karst.; *Caesalpiniac.;* tr. Am.
„ „ Mahot: *Dombeya angulata* Cav.; *Sterculiac.;* Réunion.
„ „ Mai: *Didymopanax morototoni* Dcne. et Pl.; *Araliac.;* fr. Gu.
„ „ mapou: *Andromeda pyrifolia* Thou.; *Ericac.;* Réu.
„ „ Marigni: *Protium obtusifolium* March.; *Burserac.;* Mask.
„ „ mèche: *Avicennia nitida* Jacq.; *Verbenac.;* Guad.
„ „ merde: *Sterculia sp.; Sterculiac.;* fr. Gu.
„ „ merle: *Sapindus saponaria* L.; *Sapindac.;* Ant.
„ „ „ : *Schmidelia sp.; Sapindac.;* Réunion.
„ „ „ : *Celastrus undulatus* Lamk.; *Celastrac.;* Kap., Mask.
„ „ Mora: *Dimorphandra mora* Bth.; *Caesalpiniac.;* br. Gu., Ven., C.-Am.
„ „ natte: *Mimusops spp.; Sapotac.;* fr. Gu.
„ „ „ : *Mimusops commersoni* Engl.; *Sapotac.;* Mad.
„ „ nèfle: *Jossinia mespiloides* DC.; *Myrtac.;* Réu.
„ „ Nerprun (f): *Rhamnus alaternus* L.; *Rhamnac.;* M.-M.
„ „ Nicaragua: *Haematoxylon campechianum* L.; *Caesalpiniac.;* tr. Am.
„ „ Noirprun (f): *Rhamnus alaternus* L.; *Rhamnac.;* M.-M.
„ „ pagaie blanc: *Swarƺia tomentosa* DC.; *Caesalpiniac.;* fr. Gu.
„ „ Parcouri: *Clusia insignis* Mart.; *Guttifer.;* Gu.
„ „ pêche: *Sizygium paniculatum* Gaertn.; *Myrtac.;* Réu.
„ „ perdrix: *Swarƺia tomentosa* DC.; *Caesalpiniac.;* fr. Gu.
„ „ „ : *Heisteria coccinea* Jacq.; *Olacac.;* Guad.
„ „ perroquet: *Fissilia psittacorum* Vahl; *Olacac.;* Réu., Maur.
„ „ petit Frêne: *Quassia amara* L.; *Simarubac.;* Ant.
„ „ pieux: *Schmidelia pinnata* DC.; *Sapindac.;* Südsee.
„ „ poivre: ? *Aniba panurensis* Mez; *Laurac.;* fr. Gu.
„ „ pomme: *Eugenia glomerata* Lam ; *Myrtac.;* Réu., Maur.

Bois de Quassie: *Quassia amara* L.; *Simarubac.*; Ant.
„ „ quassia de la Jamaique: *Picraena sp.*; *Simarubac.*; fr. W.-Ind.
„ „ rat: *Erythroxylon hypericifolium* Lamk.; *Erythroxylac.*; Réu.
„ „ Rainette: *Dodonaea salicifolia* DC.; *Sapindac.*; O.-Ind., Réu.
„ „ Reinette: *Dodonaea salicifolia* DC.; *Sapindac.*; O.-Ind., Réu.
„ „ rempart: *Agauria spp.*; *Ericac.*; Réunion.
„ „ „ : *Monimia myrtifolia* ?; *Monimiac.*; Réu.
„ „ Requin: *Melicocca diversifolia* Juss.; *Sapindac.*; Réu., Maur.
„ „ resonnance: *Picea excelsa* Link; *Conifer.*; Eu.
„ „ Rhodes: *Cordia gerascanthus* L.; *Borraginac.*; Guad.
„ „ „ : *Zanthoxylum emarginatum* Sw.; *Rutac.*; Guad.
„ „ „ des Parfumeurs: *Convolvulus spp.*; *Convolvulac.*; Kanar.
„ „ ronce: *Toddalia asiatica* Baill.; *Rutac.*; Mask.
„ „ Ronde: *Erythroxylon laurifolium* Lamk.; *Erythroxylac.*; Maur., Réu.
„ „ rose: *Machaerium sp.*; *Papilionac.*; Ant.
„ „ „ : *Cordia gerascanthus* L.; *Borraginac.*; W.-Ind.
„ „ „ : *Zanthoxylum emarginatum* Sw.; *Rutac.*; Guad.
„ „ „ : *Aniba panurensis* Mez; *Laurac.*; Sur.
„ „ „ : *Dalbergia nigra* Allem.; *Papilionac.*; Bras.
„ „ „ : *Aniba parviflora* Mez; *Laurac.*; Bras. (Amaz.)
„ „ „ : *Didelotia africana* Pierre; *Caesalpiniac.*; W.-Af.
„ „ „ : ? *Brachystegia sp.*; *Caesalpiniac.*; W.-Af.
„ „ „ : *Thespesia populnea* Correa; *Malvac.*; N.-Kal.
„ „ „ de Cayenne: ? *Aniba panurensis* Mez; *Laurac.*; fr. Gu.
„ „ „ „ l'Océanie (f): *Thespesia populnea* Correa; *Malvac.*; Südsee.
„ „ „ , faux: *Didelotia africana* Pierre; *Caesalpiniac.*; Gab.
„ „ „ „ : *Thespesia populnea* Correa; *Malvac.*; Südsee.
„ „ „ femelle: *Aniba panurensis* Mez; *Laurac.*; fr. Gu.
„ „ „ „ : *Icica sp.*; *Burserac.*; fr. Gu.
„ „ „ male: ? *Aniba panurensis* Mez; *Laurac.*; fr. Gu.
„ „ Sagaie: *Melicocca diversifolia* Juss.; *Sapindac.*; Réu., Maur.
„ „ sagai rouge: *Doratoxylon mauritianum* Thou.; *Hippocastanac.*; Maur.
„ „ sagaye: *Doratoxylon mauritianum* Thou.; *Hippocastanac.*; Maur., Réu.
„ „ Sagouer: *Garcinia picrorrhiza* Miq.; *Guttifer.*; Moluk.
„ „ Saint-Jean: *Didymopanax morototoni* Dcne. et Pl.; *Araliac.*; fr. Gu.
„ „ Saint-Martin: *Picraena excelsa* Lindl.; *Simarubac.*; fr. W.-Ind.
„ „ sang: *Vismia sp.*; *Guttifer.*; fr. Gu.
„ „ „ : *Haematoxylon campechianum* L.; *Caesalpiniac.*; tr. Am.
„ „ santé (f): *Guaiacum officinale* L.; *Zygophyllac.*; tr. Am.
„ „ Sappon: *Caesalpinia sappan* L.; *Caesalpiniac.*; Ind.
„ „ satin: *Zanthoxylum brachyacanthum* F. v. M.; *Rutac.*; Queensl.
„ „ savon: *Sapindus saponaria* L.; *Sapindac.*; Ant.
„ „ savonette: *Sapindus saponaria* L.; *Sapindac.*; Ant.
„ „ senteur: *Dombeya populnea* Cav.; *Sterculiac.*; Réu.
„ „ „ galet: *Olea cernua* Vahl; *Oleac.*; Réu.
„ „ serpent: *Colubrina ferruginosa* Brongn.; *Rhamnac.*; Mart.
„ „ simire: *Hymenaea courbaril* L.; *Caesalpiniac.*; W.-Ind.
„ „ soie: *Muntingia calabura* L.; *Tiliac.*; Guad.
„ „ Spa: *Aesculus hippocastanum* L.; *Hippocastanac.*; Eu., As.
„ „ sucrier: *Hedwigia balsamifera* Swartz; *Burserac.*; Ant.
„ „ Surinam: *Quassia amara* L.; *Simarubac.*; Gu., fr. W.-Ind.
„ „ tabac: *Psiadia sp.*; *Composit.*; Réunion
„ „ table *Heritiera littoralis* Dryand.; *Sterculiac.*; Sey.
„ „ Tam: *Byrsonima spicata* DC.; *Malpighiac.*; Mart.
„ „ tambour: *Tambourissa quadrifida* Sonner.; *Monimiac.*; Mask.
„ „ tan: *Weinmannia macrostachya* DC.; *Cunoniac.*; Réu.
„ „ tani: *Byrsonima sp.*; *Malpighiac.*; fr. Gu.
„ „ Tatajuba (f): *Caryocar tomentosum* Willd.; *Caryocarac.*; Gu.
„ „ Tatayouba (f): *Caryocar tomentosum* Willd.; *Caryocarac.*; Gu.
„ „ taxalm: ? *Aniba panurensis* Mez; *Laurac.*; fr. Gu.
„ „ Toon: *Toona ciliata* Roem.; *Meliac.*; As.
„ „ Trinque male (f): *Berrya ammonilla* Roxb.; *Tiliac.*; Ind.

Bois de vie: *Guaiacum sanctum* L.; *Zygophyllac.;* W.-Ind.
„ „ vouacapou: *Uouacapoua americana* Aubl.; *Papilionac.;* Sur.
„ „ zèbre: *Connarus guianensis* Lamb.; *Connarac.;* fr. Gu.
„ des dames: *Erythroxylon hypericifolium* Lamk.; *Erythroxylac.;* Réu.
„ du Merisier d'or: *Malpighia spicata* R.; *Malpighiac.;* Ant., Gu.
Boisima: *Sarcocephalus trillesi* Pierre; *Rubiac.;* Elf.
Boissima: *Sarcocephalus trillesi* Pierre; *Rubiac.;* Elf.
Boisulu: *Pterocarpus soyauxi* Taub.; *Papilionac.;* B.-K.
Boj: *Buxus sempervirens* L.; *Buxac.;* Eu., As., N.-Af.
Boja: *Xylia dolabriformis* Bth.; *Mimosac.;* Ind.
Boje: *Maytenus buxifolia* Griseb.; *Celastrac.;* Cuba.
Bojon: *Cordia gerascanthus* L.; *Borraginac.;* Mex.
Bok: *Fagus silvatica* L.; *Fagac.;* Schweden.
Bokaara-gass: *Gomphia angustifolia* Vahl; *Ochnac.;* Cey.
Bokáue: *Schizostachyum lumampao* Merr.; *Gramin.;* Phil.
Bokbók: *Cyclostemon sp.; Euphorbiac.;* Phil.
„ : *Palaquium sp.; Sapotac.;* Phil.
„ : *Phoebe sterculioides* Merr.; *Laurac.;* Phil.
Bokerah: *Ochna wightiana* Wall.; *Ochnac.;* Ind.
Bô kêt: *Gleditschia sinensis* L.; *Caesalpiniac.;* Ind.-Ch.
„ „ dai: *Leucaena glauca* Bth.; *Mimosac.;* Ind.-Ch.
Boking: *Afzelia bipindensis* Harms; *Caesalpiniac.;* Kam.
Bokke-Droll: *Plectronia spinosa* Klotzsch; *Rubiac.;* Kap.
Bokkenoot: *Caryocar tomentosum* Willd.; *Caryocarac.;* Sur.
„ Onechte-: *Bertholletia excelsa* Humb.; *Lecythidac.;* fr. Gu.
Bokkobokkoton: *Hirtella spp.; Rosac.;* Sur.
Bokó: *Gigantochloa levis* Merr.; *Gramin.;* Phil.
Bokobi: *Mitragyne macrophylla* Hiern; *Rubiac.;* Kam.
Bokobokoton: *Hirtella spp.; Rosac.;* Sur.
Bokoka: *Lophira procera* Chev.; *Ochnac.;* Kam.
Bokol: *Panda oleosa* Pierre; *Pandac.;* Kam.
Bokolo: *Tieghemella africana* Pierre; *Sapotac.;* Gab.
Bokolopolo: *Anthocleista zenkeri* Gilg; *Loganiac.;* Kam.
Bokoma: *Parinarium robustum* Oliv.; *Rosac.;* Elf.
Bokombo: *Musanga smithi* R. Br.; *Morac.;* W.-Af.
Bokombolo (f): *Piptadenia africana* Hook. f.; *Mimosac.;* Kam.
Bokome: *Terminalia superba* Engl. et Diels; *Combretac.;* W.-Af.
Bokondo: *Pycnanthus kombo* Warb.; *Myristicac.;* Kam.
Bokonghe: *Millettia laurenti* de Wild.; *Papilionac.;* B.-K.
Bokongo: *Copaifera demeusei* Harms; *Caesalpiniac.;* B.-K.
Bokouda: *Avicennia nitida* Jacq.; *Uerbenac.;* Kam.
Bokouka: *Alstonia congensis* Engl.; *Apocynac.;* Kam.
„ -bambalé: *Alstonia congensis* Engl.; *Apocynac.;* Kam.
Boksboom (dä, h): *Buxus sempervirens* L.; *Buxac.;* Eu., As., N.-Af.
Bokuka: *Alstonia congensis* Engl.; *Apocynac.;* Kam.
Bokukulu: *Dialium yambataense* Vermoesen; *Caesalpiniac.;* B.-K.
Bokumia: *Coula edulis* Baill.; *Oleac.;* Kam.
Bola: *Hibiscus tiliaceus* L.; *Malvac.;* Ind.
Bolador: *Terminalia obovata* Eichl.; *Combretac.;* Guat.
Boladora: *Terminalia sp.; Combretac.;* Guat., Hond.
Bolafa: *Macrolobium dewevrei* de Wild.; *Caesalpiniac.;* B.-K.
Bolagu: *Pterospermum suberifolium* Lam.; *Sterculiac.;* O.-Ind.
Bolaka: *Symphonia gabonensis* Pierre; *Guttifer.;* B.-K.
Bolda: *Boldoa chilensis* Juss.; *Monimiac.;* Chile.
Boldack: *Schima wallichi* Choisy; *Theac.;* Ind. (Himalaya)
Boldo: *Peumus boldus* Mol.; *Monimiac.;* Chile.
Boldu: *Peumus boldus* Mol.; *Monimiac.;* Chile.
Boleko: *Ongokea klaineana* Pierre; *Oleac.;* B.-K.
Bolengu: *Afzelia africana* Smith; *Caesalpiniac.;* B.-K.
Bolina: *Genista umbellata* Poir.; *Papilionac.;* Sp.
Polinda: *Polyalthia suaveolens* Engl. et Diels; *Anonac.;* B.-K.
Bollén: *Kageneckia oblonga* R. et Pav.; *Rosac.;* Chile.

Bolletree, Basra-: *Humiria floribunda* Mart.; *Humiriac.*; Sur.
„   „   : *Humiria balsamifera* Aubl.; *Humiriac.*; Sur.
Bolletrie: *Mimusops balata* Gaertn.; *Sapotac.*; Sur.
„   : *Lucuma mammosa* Gaertn.; *Sapotac.*; holl. Gu.
„   : *Mimusops spp.*; *Sapotac.*; holl. Gu.
„   Bastaard-: *Humiria floribunda* Mart.; *Humiriac.*; Sur.
„   „   : *Humiria balsamifera* Aubl.; *Humiriac.*; Sur.
„   Witte: *Dipholis salicifolia* A. DC.; *Sapotac.*; fr. Gu.
Bollet tree: *Mimusops balata* Gaertn.; *Sapotac.*; Gu.
Bollitree: *Mimusops globosa* Gaertn.; *Sapotac.*; br. Gu., W.-Ind.
Bolltrie: *Swartѯia tomentosa* DC.; *Caesalpiniac.*; n. S.-Am.
Bollywood, Brown: *Litsea ferruginea* Bth. et Hook.; *Laurac.*; Au.
„   „   : *Litsea reticulata* Bth.; *Laurac.*; Au.
Bolo: *Pachylobus trimerus* Guill.; *Burserac.*; tr. Af.
? Bolo: *Hopea odorata* Roxb.; *Dipterocarpac.*; ö. As.
Boló: *Gigantochloa levis* Merr.; *Gramin.*; Phil.
Boloagnita: *Diospyros pilosanthera* Blco.; *Ebenac.*; Phil.
Bolondo: *Piptadenia africana* Hook. f.; *Mimosac.*; tr. W.-Af.
„   : *Erythrophloeum guineense* G. Don; *Caesalpiniac.*; Kam.
„   : *Parkia sp.*; *Mimosac.*; Kam.
„   : *Dialium sp.*; *Caesalpiniac.*; Kam.
„   : *Chlorophora excelsa* Bth. et Hook.; *Morac.*; B.-K.
Belongaeta: *Diospyros pilosanthera* Blco.; *Ebenac.*; Phil.
Bolong-eta: *Diospyros pilosanthera* Blco.; *Ebenac.*; Phil.
Bolongo: *Symphonia gabonensis* Pierre; *Guttifer.*; B.-K.
Bólong-sína: *Dendrocalamus latiflorus* Munro; *Gramin.*; Phil.
Bolundu: *Chlorophora excelsa* Bth. et Hook.; *Morac.*; B.-K.
Bolungu: *Symphonia gabonensis* Pierre; *Guttifer.*; B.-K.
Bomba: *Cordia irvingi* Baker; *Borraginac.*; Kam.
Bombaba (f): *Dialium macranthum* Chev.; *Caesalpiniac.*; Kam.
Bombaha: *Pentaclethra macrophylla* Bth.; *Mimosac.*; Gab.
Bombala (f): *Dialium guineense* Willd.; *Caesalpiniac.*; Elf., Kam., Gab.
Bombali: *Macrolobium dewevrei* de Wild.; *Caesalpiniac.*; B.-K.
Bombax (f): *Bombax buonopozense* Beauv.; *Bombacac.*; Elf., Gab.
„   „: *Carapa velutina* C. DC.; *Meliac.*; Elf.
Bombaya: *Xylopia striata* Engl.; *Anonac.*; Kam.
Bombi: *Anonidium manni* Engl.; *Anonac.*; Kam.
Bombôlo: *Melia bambolo* Welw.; *Meliac.*; port. Af.
Bombol: *Melia dubia* Cav.; *Meliac.*; O.-Ind.
Bombolo ia n'puto: *Melia azedarach* L.; *Meliac.*; port. Af.
Bombway, Red: *Planchonia andamanica* King; *Lecythidac.*; Ind.
„   White: *Terminalia procera* Roxb.; *Combretac.*; Ind.
Bombwe, Red: *Planchonia andamanica* King; *Lecythidac.*; And.
„   White: *Careya arborea* Roxb.; *Myrtac.*; And.
Bo-mi: *Litsea sebifera* Pers.; *Laurac.*; Cey.
Bomoku: *Mammea africana* G. Don; *Guttifer.*; Elf.
Bompegya: *Ochrocarpus africanus* Oliv.; *Guttifer.*; Go.
Bomsamdua: *Distemonanthus benthamianus* Baill.; *Caesalpiniac.*; Go.
Bom vâng (H): *Abroma augusta* L.; *Sterculiac.*; Ind.-Ch.
? Bon: *Calpocalyx klainei* Pierre; *Mimosac.*; Kam.
Bona: *Terminalia ivorensis* Chev.; *Combretac.*; Elf.
Bo nang (H): *Chisocheton globosus* Pierre; *Meliac.*; Ind.-Ch.
Bóndjengi: *Odyendea gabonensis* Engl.; *Simarubac.*; Gab.
Bonete: *Luehea sp.*; *Tiliac.*; Salv.
„   : *Pileus heptaphyllus* Ramirez; *Passiflorac.*; Mex.
Bonetero (s): *Evonymus europaea* L.; *Celastrac.*; Eu., W.-As., N.-Af.
Bonewood: *Emmenosperma alphitonioides* F. v. M.; *Rhamnac.*; ö. Au.
„   : *Medicosma cunninghami* Hook. f.; *Rutac.*; Au.
Bông bac (H): *Vernonia arborea* Buch.-Ham.; *Composit.*; Ind.-Ch.
Bongbóng: *Schizostachyum diffusum* Merr.; *Gramin.*; Phil. (Cebu)
Bôngèké: *Ongokea klaineana* Pierre; *Olacac.*; Gab.
Bongelé: *Sterculia oblonga* Mast.; *Sterculiac.*; Kam.

Bongheli: *Macrolobium coeruleoides* de Wild.; *Caesalpiniac.;* B.-K.
Boñglín: *Polyscias nodosa* Seem.; *Araliac.;* Phil.
Bongo: *Cavanillesia platanifolia* H. B. K.; *Bombacac.;* Pan., Kol.
Boñgóg: *Parinarium sp.; Rosac.;* Phil.
„ : *Vitex pentaphylla* Merr.; *Verbenac.;* Phil.
Boñgógon: *Vitex turczaninowi* Merr.; *Verbenac.;* Phil.
Boñgon: *Allaeanthus glaber* Warb.; *Morac.;* Phil.
Bongongi: *Antrocaryon klaineanum* Pierre; *Anacardiac.;* Kam.
„ : *Staudtia kamerunensis* Warb.; *Myristicac.;* Kam.
„ : *Millettia sp.; Papilionac.;* Kam.
„ : *Hylodendron gabunense* Taub.; *Caesalpiniac.;* Kam.
„ : *Fillaeopsis discophora* Harms; *Mimosac.;* Kam.
Boñgóog: *Vitex pentaphylla* Merr.; *Verbenac.;* Phil.
Bongor: *Lagerstroemia spp.; Lythrac.;* Malay.
Bongosi (d): *Lophira procera* Chev.; *Ochnac.;* W.-Af.
Bongossi: *Lophira procera* Chev.; *Ochnac.;* Kam.
Bongro: *Licania pachystachya* Kleinh.; *Rosac.;* Sur.
„ : *Parinarium campestre* Aubl.; *Rosac.;* Sur.
Bông su (H): *Duabanga sonneratioïdes* Buch.-Ham.; *Punicac.;* Ind.-Ch.
Boniato amarillo: *Nectandra exaltata* Griseb.; *Laurac.;* Cuba.
„ blanco: *Phoebe sp.; Laurac.;* Cuba.
„ cigua: *Ocotea sp.; Laurac.;* Cuba.
„ laurel: *Ocotea sp.; Laurac.;* Cuba.
Bonjabi: *Mimusops djave* Engl.; *Sapotac.;* Kam.
Bon jasanga: *Ricinodendron africanum* Muell. Arg.; *Euphorbiac.;* W.-Af.
Bonkankangu: *Sarcocephalus diederichi* de Wild.; *Rubiac.;* B.-K.
Bonkeka: *Conopharyngia smithi* Stapf; *Apocynac.;* B.-K.
Bon nang: *Litsea vang* H. Lec.; *Laurac.;* Ind.-Ch.
Bonmeza: *Albizzia stipulata* Boiv.; *Mimosac.;* tr. & subtr. As.
Bonnet carré (f): *Evonymus europaea* L.; *Celastrac.;* S.-Eu., W.-As., N.-Af.
„ de prêtre (f): *Evonymus europaea* L.; *Celastrac.;* S.-Eu., W.-As., N.-Af.
Bonni: *Acacia farnesiana* Willd.; *Mimosac.;* S.-Am.
Bonnie-bonnie-hoedoe: *Maprounea guianensis* Aubl.; *Euphorbiac.;* Sur.
Bon njabi: *Mimusops congolensis* Russel et Hédin; *Sapotac.;* Kam.
Bono: *Mitragyne macrophylla* Hiern; *Rubiac.;* Elf.
Bonsamdua: *Afrormosia laxiflora* Harms; *Papilionac.;* Go.
Bonsum: *Phoebe hainesiana* Brandis; *Laurac.;* Ind.
Bontolei: *Cleistopholis pynaerti* de Wild.; *Anonac.;* B.-K.
Bontòna: *Adansonia madagascariensis* Baill.; *Bombacac.;* Mad.
Bonuku bololo: *Enantia chlorantha* Oliv.; *Anonac.;* Kam.
Bonwula: *Coula edulis* Baill.; *Olacac.;* Kam.
Bonyuromé: *Alchornea sp.; Euphorbiac.;* Elf.
Bonzo: *Chlorophora excelsa* Bth. et Hook.; *Morac.;* Elf.
Boo: *Sorindeia juglandifolia* Planch.; *Anacardiac.;* Kam.
Boo-bóo: *Pinus insularis* Endl.; *Conifer.;* Phil.
Boobooraballi: ? *Chaetocarpus schomburgkiana* Pax et Hoffm.; *Euphorbiac.;* Sur.
Bood Joong: *Callistemon salignus* DC.; *Myrtac.;* N.-S.-W.
Boohoorada: *Parinarium campestre* Aubl.; *Rosac.;* Sur.
Bookoot: *Cassia grandis* L. f.; *Caesalpiniac.;* br. Hond.
Bookut: *Cassia grandis* L. f.; *Caesalpiniac.;* br. Hond.
Boola: *Ceratopetalum apetalum* D. Don; *Saxifragac.;* N.-S.-W.
Boolerchu: *Tristania suaveolens* Sm.; *Myrtac.;* Au.
Boona: *Eucalyptus corymbosa* Sm.; *Myrtac.;* ö. Au.
Boone wood: *Myristica surinamensis* Warb.; *Myristicac.;* holl. Gu.
Booni: *Cola lateritia* K. Schum.; *Sterculiac.;* Lib.
Booral: *Persoonia falcata* R. Br.; *Proteac.;* Au.
Boorooch-gaha: *Chloroxylon swietenia* DC.; *Meliac.;* Ind.
Booroota-gass: *Chloroxylon swietenia* DC.; *Meliac.;* Ind.
Booscuru: *Premna tomentosa* Bl.; *Verbenac.;* Cey.
Booyong: *Tarrietia argyrodendron* Bth.; *Sterculiac.;* Au.
Bôp (H): *Cinnamomum sp.; Laurac.;* Ind.
Bopala: *Irvingia gabonensis* Baill.; *Simarubac.;* Kam.

Bopalo: *Afzelia sp.; Caesalpiniac.;* Kam.
„ : *Enantia chlorantha* Oliv.; *Anonac.;* Kam.
Bopande: *Uvaria büsgeni* Diels; *Anonac.;* Kam.
Bope: *Uvaria büsgeni* Diels; *Anonac.;* Kam.
„ : *Staudtia gabonensis* Warb.; *Myristicac.;* Kam.
„ : *Irvingia barteri* Hook. f.; *Simarubac.;* Kam.
Bopé: *Mitragyne macrophylla* Hiern; *Rubiac.;* Kam.
„ bambale: *Staudtia gabonensis* Warb.; *Myristicac.;* Kam.
„ „ : *Staudtia kamerunensis* Warb.; *Myristicac.;* Kam.
„ „ : *Carapa microcarpa* Chev.; *Meliac.;* Kam.
Bopolo-polo: *Anthocleistus nobilis* G. Don; *Loganiac.;* Kam.
Bôp váng: *Cinnamomum sp.; Laurac.;* Ind.-Ch.
Bôp xanh: *Cinnamomum sp.; Laurac.;* Ind.-Ch.
Borbor: *Irvingia gabonensis* Baill.; *Simarubac.;* W.-Af.
Borbur: *Ficus religiosa* L.; *Morac.;* Ind.
Boree: *Acacia pendula* A. Cunn.; *Mimosac.;* ö. Au.
Boré-porè: *Lannea acidissima* Chev.; *Anacardiac.;* Elf.
Borgonli: *Casearia glomerata* Roxb.; *Flacourtiac.;* O.-Ind.
Borlas de obispo: *Calliandra parvifolia* Speg.; *Mimosac.;* Arg.
Boróan: *Buchanania sp.; Anacardiac.;* Phil.
Boromé: *Mimusops balata* Gaertn.; *Sapotac.;* Gu.
Borowé: *Mimusops spp.; Sapotac.;* Sur.
Borrachero: *Datura arborea* L.; *Solanac.;* Kol.
Borrachin: *Arbutus unedo* L.; *Ericac.;* Sp.
Borracho: *Ichthyomethia piscipula* Hitch.; *Papilionac.;* Ven.
Bosambi: *Uapaca staudti* Pax; *Euphorbiac.;* Kam.
Bosanglói: *Ceiba pentandra* Gaertn.; *Bombacac.;* Phil.
Bosáo: *Pachylobus edulis* G. Don; *Burserac.;* W.-Af.
Bosch kasjoe: *Dimorphandra latifolia* Tul.; *Caesalpiniac.;* Sur.
Boschkers: *Eugenia pumila* Gaertn.; *Myrtac.;* fr. Gu.
Bosch marmel doos: *Duroia eriophila* L.; *Rubiac.;* Sur.
Boschtom: *Buxus sempervirens* L.; *Buxac.;* Transkauk.
Boschzuurzak: *Anona sp.; Anonac.;* Sur.
Bosé: ? *Xylopia striata* Engl.; *Anonac.;* Kam.
Bosenge: *Musanga smithi* R. Br.; *Morac.;* tr. W.-Af.
Bosengué: *Musanga smithi* R. Br.; *Morac.;* Kam.
Bosganna: *Rhus laevigata* L.; *Anacardiac.;* Kap.
Bosi: *Guarea thompsoni* Sprague et Hutch.; *Meliac.;* Go.
Bosipi: ? *Oxystigma manni* Harms; *Caesalpiniac.;* Kam., B.-K.
Bosmoué: *Bussea occidentalis* Hutch.; *Caesalpiniac.;* Elf.
Bosneak: *Mesua ferrea* L.; *Guttifer.;* Kamb.
Bosoho: *Eschweilera corrugata* Miers; *Lecythidac.;* Sur.
Bosong: *Distemonanthus benthamianus* Baill.; *Caesalpiniac.;* W.-Af.
Bosongé: *Musanga smithi* R. Br.; *Morac.;* Kam.
Bosow: *Pseudospondias microcarpa* Engl.; *Anacardiac.;* B.-K.
Bos phnom: *Hibiscus macrophyllus* Roxb.; *Malvac.;* Ind.-Ch.
Bossanghe: *Xylopia aethiopica* A. Rich.; *Anonac.;* B.-K.
Bossanghi: *Xylopia aethiopica* A. Rich.; *Anonac.;* B.-K.
Bossé (f): *Guarea cedrata* Pellegr.; *Meliac.;* Elf.
„ d'acajou: *Entandrophragma aff. E. utile* Sprague; *Meliac.;* Elf.
„ „ : *Entandrophragma aff. E. cylindricum* Spr.; *Meliac.;* Elf.
„ blanc: *Guarea cedrata* Pellegr.; *Meliac.;* Elf.
„ Kisoko: *Trichilia kisoko* de Wild.; *Meliac.;* Gab.
„ rouge (f): *Entandrophragma aff. cylindricum* Spr.; *Meliac.;* Elf.
„ „ „ : *Entandrophragma aff. utile* Sprague; *Meliac.;* Elf.
Bossenghe: *Uapaca guineensis* Muell. Arg.; *Euphorbiac.;* B.-K.
Bossengue: *Musanga smithi* R. Br.; *Morac.;* Kam.
Bossipi: *Oxystigma manni* Harms; *Caesalpiniac.;* Kam.
Bosso (i): *Buxus sempervirens* L.; *Buxac.;* S.-Eu., W.-As., N.-Af.
„ comun (i): *Buxus sempervirens* L.; *Buxac.;* S.-Eu., W.-As., N.-Af.
„ verde (i): *Buxus sempervirens* L.; *Buxac.;* S.-Eu., W.-As., N.-Af.
Bossolo (i): *Buxus sempervirens* L.; *Buxac.;* S.-Eu., W.-As., N.-Af.

Bossolo di Constantinopoli (i): *Buxus sempervirens* L.; *Buxac.*; S.-Eu., W.-As., N.-Af.
„ gentile (i): *Buxus balearica* Lamarck; *Buxac.*; w. M.-M.
Bosúa: *Zanthoxylum ochroxylum* DC.; *Rutac.*; Ven.
Bosuga: *Zanthoxylum ochroxylum* DC.; *Rutac.*; Ven.
Bosulu: *Pterocarpus soyauxi* Taub.; *Papilionac.*; B.-K.
Bot: *Cordia macleodi* Hook. f.; *Borraginac.*; w. subtr. Himal.
Bota: *Millettia versicolor* Welw.; *Papilionac.*; B.-K.
Botábon: *Parinarium sp.*; *Rosac.*; Phil.
Botan: *Sabal excelsa* Mart.; *Palmac.*; br. Hond.
Botan-Geyaki: *Planera japonica* Miq.; *Ulmac.*; Jap.
Botanyholz, schwarzes (d): *Dalbergia latifolia* Roxb.; *Papilionac.*; O.-Ind.
Boterboom: *Caryocar tomentosum* Willd.; *Caryocarac.*; Gu.
Boternoot: *Caryocar butyrosum* Willd.; *Caryocarac.*; Sur.
Botgó: *Sideroxylon macranthum* Merr.; *Sapotac.*; Phil.
Bô thê deng: *Quercus sp.*; *Fagac.*; Ind.-Ch.
Boti: *Terminalia ivorensis* Chev.; *Combretac.*; Elf.
Botie-ie: *Mimusops spp.*; *Sapotac.*; Sur.
Botón: *Quararibea guianensis* Aubl.; *Bombacac.*; Kol.
„ blanco: *Melanthera aspera* Rich.; *Composit.*; Nic.
Botoncillo: *Borreria capitata* DC.; *Rubiac.*; Ven.
„ : *Conocarpus erectus* L.; *Combretac.*; Ven.
Botóng: *Bambusa vulgaris* Schrad.; *Gramin.*; Phil.
„ : *Dendrocalamus latiflorus* Munro; *Gramin.*; Phil.
„ : *Gigantochloa levis* Merr.; *Gramin.*; Phil.
„ : *Barringtonia asiatica* Kurz; *Lecythidac.*; Phil.
Botong-bótong: *Barringtonia asiatica* Kurz; *Lecythidac.*; Phil.
Botopia: *Sterculia tragacantha* Lindl.; *Sterculiac.*; Elf.
Botopoua: *Sterculia tragacantha* Lindl.; *Sterculiac.*; Elf.
Botrie: *Mimusops spp.*; *Sapotac.*; Sur.
„ , Basra-: *Humiria balsamifera* Mart.; *Humiriac.*; Sur.
„ „ : *Humiria balsamifera* Aubl.; *Humiriac.*; Sur.
Botro-hoedoe: *Gustavia pterocarpa* Poit.; *Lecythidac.*; Sur.
Bottlebrush, Red: *Callistemon lanceolatus* DC.; *Myrtac.*; Queensl.
Bottletree: *Sterculia rupestris* Bth.; *Sterculiac.*; Queensl.
„ : *Sterculia trichosiphon* Bth.; *Sterculiac.*; Queensl.
„ narrow-leaved: *Brachychiton rupestris* K. Schum.; *Sterculiac.*; Au.
„ Scrub-: *Brachychiton discolor* F. v. M.; *Sterculiac.*; Au.
„ Victorian: *Sterculia diversifolia* G. Don; *Sterculiac.*; Au.
Botuba: *Irvingia barteri* Hook. f.; *Simarubac.*; Kam.
Botukisemto: *Ceiba pentandra* Gaertn. *var. clausa* Ulb.; *Bombacac.*; Togo.
Botukocholemotu: *Ceiba pentandra* Gaertn. *var. clausa* Ulb.; *Bombacac.*; Togo.
Bouandjo (f): *Allanblackia parviflora* Chev.; *Guttifer.*; Gab.
„ „ : *Allanblackia floribunda* Oliv.; *Guttifer.*; Gab.
Bouba: *Myrianthus arboreus* Beauv.; *Morac.*; Gab.
Boubambou: *Chrysophyllum lacourtianum* de Wild.; *Sapotac.*; Gab.
Bouboussou: *Entandrophragma aff. cylindricum* Spr.; *Meliac.*; Elf.
Boucara: *Swartzia tomentosa* DC.; *Caesalpiniac.*; Sur.
Bouday: *Ceiba pentandra* Gaertn. *var. dehiscens* Ulb.; *Bombacac.*; Senl.
Boudinga linga: *Mimusops lacera* Bak. *var. newtoni* Engl.; *Sapotac.*; Kam.
Bouelébandjo: *Garcinia punctata* Oliv.; *Guttifer.*; Kam.
Bougouni: *Inga spp.*; *Mimosac.*; Sur.
„ pet. feuilles: *Pithecolobium pedicellare* Bth.; *Mimosac.*; fr. Gu.
Bouï: *Citrus decumana* L.; *Rutac.*; Annam.
Bouillard: *Populus nigra* L.; *Salicac.*; Eu.
Bouis: *Chrysophyllum cainito* L.; *Sapotac.*; fr. W.-Ind.
Boujabi: *Dumoria heckeli* Chev.; *Sapotac.*; Elf.
Boukouissou: *Ternstroemia japonica* Thunb.; *Theac.*; Jap.
Bouleau (f): *Betula odorata* Bechst., *B. verrucosa* Ehrh.; *Betulac.*; Eu.
„ acajou: *Betula lenta* L.; *Betulac.*; ö. N.-Am.
„ blanc: *Betula papyrifera* Marsh.; *Betulac.*; N.-Am.
„ gris: *Betula populifolia* Marsh.; *Betulac.*; ö. N.-Am.
„ jaune: *Betula lutea* Michx.; *Betulac.*; N.-Am.

Bouleau merisier: *Betula lenta* L.; *Betulac.*; ö. N.-Am.
„ occidental: *Betula occidentalis* Hook.; *Betulac.*; w. N.-Am.
„ rouge: *Betula populifolia* Marsh.; *Betulac.*; ö. N.-Am.
„ à capot: *Betula papyrifera* Marsh.; *Betulac.*; N.-Am.
„ „ feuilles de peuplier: *Betula populifolia* Marsh.; *Betulac.*; ö. N.-Am.
„ „ papier: *Betula papyrifera* Marsh.; *Betulac.*; N.-Am.
„ „ sucre: *Betula lenta* L.; *Betulac.*; ö. N.-Am.
Boulet de canon: *Couroupita guianensis* Aubl.; *Lecythidac.*; fr. Gu.
Bouliboba: *Dracaena fragrans* Ker-Gawl.; *Liliac.*; Kam.
Bouma: *Ceiba pentandra* Gaertn.; *Bombacac.*; W.-Af.
Boume boumirí: *Humiria* spp.; *Humiriac.*; fr. Gu.
Boumou: *Bombax buonopozense* Beauv.; *Bombacac.*; fr. Guin.
Bouna: *Terminalia ivorensis* Chev.; *Combretac.*; Elf.
Boung: *Erythrina indica* Lam.; *Papilionac.*; Ind.-Ch.
„ rep: *Parkia streptocarpa* Hance; *Mimosac.*; Ind.-Ch.
„ to: *Mitrephora bousigoniana* Pierre; *Anonac.*; Ind.-Ch.
Boura-coura: *Brosimum aubleti* Poepp. et Endl.; *Morac.*; fr. Gu.
Bourao (f): *Hibiscus tiliaceus* L.; *Malvac.*; Tr.
Bourdaine: *Rhamnus frangula* L.; *Rhamnac.*; Eu.
Bouré: *Gardenia* sp.; *Rubiac.*; Sudan.
Bourgène (f): *Rhamnus frangula* L.; *Rhamnac.*; Eu.
Bourgoni: *Inga bourgoni* DC.; *Mimosac.*; fr. Gu.
Bourracourra: *Piratinera guianensis* Aubl.; *Morac.*; br. Gu.
Bourrao: *Couratari guianensis* Aubl.; *Lecythidac.*; fr. Gu.
Bourrayero gourroo: *Eucalyptus corymbosa* Sm.; *Myrtac.*; N.-S.-W.
Bourreria, havanische: *Bourreria havanensis* Miers; *Borraginac.*; Fl., W.-Ind.
Bourrerier de Havana: *Bourreria havanensis* Miers; *Borraginac.*; Fl., W.-Ind.
Bousaïra: *Ceiba pentandra* Gaertn. *var. dehiscens* Ulb.; *Bombacac.*; sw. Senl.
Bousi soursakka: *Anona* sp.; *Anonac.*; Sur.
„ tamarin: *Pithecolobium racemiflorum* Ducke; *Mimosac.*; Sur.
„ „ : *Stryphnodendron flammatum* Kleinh.; *Mimosac.*; Sur.
Boussana: *Ceiba pentandra* Gaertn.; *Bombacac.*; fr. Guin.
Bout: *Garcinia vilersiana* Pierre; *Guttifer.*; Annam.
Boutous: *Piratinera guianensis* Aubl.; *Morac.*; br. Gu.
Bouvinga: *Brachystegia* sp.; *Caesalpiniac.*; Gab.
Bouyouvi: *Barteria nigritiana* Hook. f.; *Passiflorac.*; Kam.
Bouzu: *Chlorophora excelsa* Bth. et Hook.; *Morac.*; Elf.
Bôvanda: *Panda oleosa* Pierre; *Pandac.*; Gab.
Bovitia: *Osbeckia aspera* Blume; *Melastomatac.*; Cey.
Bovo: *Poga oleosa* Oliv.; *Rhizophorac.*; Gab.
Bowhanti: *Vouacapoua americana* Aubl.; *Papilionac.*; Sur.
Bow-pigeon: *Coccoloba* sp.; *Polygonac.*; br. W.-Ind.
Bowwood (u): *Maclura aurantiaca* Nutt.; *Morac.*; USA.
„ : *Amanoa guianensis* Aubl.; *Euphorbiac.*; Gu.
„ : *Parinarium* sp. (? *guianensis* Aubl.); *Rosac.*; Gu., Guad.
„ : *Tecoma leucoxylon* Mart. *var. pentaphylla* Juss.; *Bignoniac.*; Sur.
„ : *Tecoma araliacea* P. DC.; *Bignoniac.*; Sur.
„ Andaman: *Sageraea elliptica* Hook. f. et Thoms.; *Anonac.*; Ind.
Bowae: *Ficus thonningi* Blume; *Morac.*; Lib.
Box (e): *Buxus sempervirens* L.; *Buxac.*; S.-Eu., W.-As., N.-Af.
„ (e, u): *Cornus florida* L.; *Cornac.*; sö. N.-Am.
„ : *Bumelia* sp.; *Sapotac.*; Windw. (Ant.)
„ : *Vitex umbrosa* Sw.; *Verbenac.*; Bah.
„ : *Jacaranda coerulea* Griseb.; *Bignoniac.*; Bah.
„ : *Casearia praecox* Griseb.; *Flacourtiac.*; Ven.
„ : *Aspidosperma eburneum* Allem.; *Apocynac.*; S.-Bras.
„ : *Bursaria spinosa* Cav.; *Pittosporac.*; Tasm.
„ , African: *Buxella mac-owani* Tiegh.; *Buxac.*; S.-Af.
„ American (e, u): *Cornus florida* L.; *Cornac.*; sö. N.-Am.
„ Anatolian: *Buxus sempervirens* L.; *Buxac.*; Eu., As.
„ Balearic (e): *Buxus balearica* Lamarck; *Buxac.*; w. M.-M.
„ Bastard: *Eucalyptus goniocalyx* F. v. M.; *Myrtac.*; Au. (Vict.)

Box Bastard: *Tristania conferta* R. Br.; *Myrtac.*; N.-S.-W.
„ Bembil: *Eucalyptus populifolia* Hook.; *Myrtac.*; nö. Au.
„ Black: *Eucalyptus bicolor* Cunn.; *Myrtac.*; sö. Au.
„ Brazilian: *Euxylophora paraënsis* Huber; *Rutac.*; Bras.
„ Brisbane: *Tristania conferta* R. Br.; *Myrtac.*; ö. Au.
„ Broad-leaved: *Eucalyptus acmenioides* Schauer; *Myrtac.*; sö. Au.
„ Brown: *Eucalyptus polyanthemos* Schauer; *Myrtac.*; sö. Au.
„ Brush-: *Tristania conferta* R. Br.; *Myrtac.*; O.-Au.
„ Burmese: *Murraya exotica* L.; *Rutac.*; O.-Ind.
„ Cambodscha-: *Wrightia annamensis* Eberh. et Dubard; *Apocynac.*; Ind.-Ch.
„ Cape-: *Buxella mac-owani* Tiegh.; *Buxac.*; S.-Af.
„   „ : *Celastrus buxifolius* L.; *Celastrac.*; S.-Af.
„   „ : *Gonioma kamassi* E. Mey.; *Apocynac.*; S.-Af.
„ Ceylon-: *Canthium didymum* Gaertn.; *Rubiac.*; Ind.
„ Chinese: *Murraya exotica* L.; *Rutac.*; Ind.
„ Circassian: *Buxus sempervirens* L.; *Buxac.*; Eu., As.
„ Colonial: *Buxella mac-owani* Tiegh.; *Buxac.*; S.-Af.
„ Common (e): *Buxus sempervirens* L.; *Buxac.*; S.-Eu., W.-As., N.-Af.
„ Cooburu: *Eucalyptus largiflorens* F. v. M.; *Myrtac.*; O.-Au.
„ Cornel: *Cornus florida* L.; *Cornac.*; USA.
„ Corsican: *Buxus sempervirens* L.; *Buxac.*; Eu., As.
„ Dog- (e, u): *Cornus florida* L.; *Cornac.*; sö. N.-Am.
„ Dwarf-: *Eucalyptus microtheca* F. v. M.; *Myrtac.*; Au.
„ Eastafrican: *Buxella mac-owani* Tiegh.; *Buxac.*; S.-Af.
„ East London: *Gonioma kamassi* E. Mey.; *Apocynac.*; S.-Af.
„ European (e): *Buxus sempervirens* L.; *Buxac.*; S.-Eu., W.-As., N.-Af.
„ False (e, u): *Cornus florida* L.; *Cornac.*; sö. N.-Am.
„   „ : *Gyminda grisebachi* Sarg.; *Celastrac.*; Fl., Cuba, P.-R.
„   „ : *Gonioma kamassi* E. Mey.; *Apocynac.*; tr. Af.
„ Flooded: *Eucalyptus microtheca* F. v. M.; *Myrtac.*; Au.
„ Florida-: *Schaefferia frutescens* Jacq.; *Celastrac.*; tr. & subtr. Am.
„ Grey: *Eucalyptus hemiphloia* F. v. M.; *Myrtac.*; sö. Au.
„   „ : *Eucalyptus polyanthemos* Schauer; *Myrtac.*; sö. Au.
„   „ : *Hemicyclia australasica* Muell. Arg.; *Euphorbiac.*; Au.
„   „ , Bairnsdale-: *Eucalyptus bosistoana* F. v. M.; *Myrtac.*; Au. (Vict.)
„   „ Gippsland-: *Eucalyptus bosistoana* F. v. M.; *Myrtac.*; sö. Au.
„ Gum-topped: *Eucalyptus hemiphloia* F. v. M.; *Myrtac.*; sö. Queensl. (Au.)
„ India (eu): *Casearia praecox* Griseb.; *Flacourtiac.*; W.-Ind.
„   „ (e): *Aspidosperma vargasi* A. DC.; *Apocynac.*; Ven.
„   „ : *Gardenia latifolia* Ait.; *Rubiac.*; Ind.
„ Iron-, Black: *Eucalyptus raveretiana* F. v. M.; *Myrtac.*; Queensl.
„ Ironbark-: *Eucalyptus obliqua* L'Hérit.; *Myrtac.*; sö. Au.
„ Jamaica- (u): *Schaefferia frutescens* Jacq.; *Celastrac.*; Jam.
„   „ : *Tecoma pentaphylla* Juss.; *Bignoniac.*; Bras., Ven., W.-Ind.
„ Kamassi-: *Gonioma kamassi* E. Mey.; *Apocynac.*; S.-Af.
„ Kanuka-: *Tristania laurina* R. Br.; *Myrtac.*; Au.
„ Knysna-: *Gonioma kamassi* E. Mey.; *Apocynac.*; Kap.
„ Malabar-: *Hemicyclia spp.*; *Euphorbiac.*; Ind.
„ Maracaibo-: *Casearia praecox* Griseb.; *Flacourtiac.*; W.-Ind.
„ Minorca (e): *Buxus balearica* Lamarck; *Buxac.*; w. M.-M.
„ Narrow-leaved: *Eucalyptus microtheca* F. v. M.; *Myrtac.*; Au.
„ Native-: *Bursaria spinosa* Cav.; *Pittosporac.*; S.-Au.
„ New Zealand-: *Sapota costata* A. DC.; *Sapotac.*; N.-Seel.
„ Persian: *Buxus sempervirens* L.; *Buxac.*; S.-Eu., W.-As., N.-Af.
„ Poplar-: *Eucalyptus populifolia* Hook.; *Myrtac.*; ö. Au.
„ Prickly: *Bursaria spinosa* Cav.; *Pittosporac.*; sö. Au.
„ Red: *Eucalyptus polyanthemos* Schauer; *Myrtac.*; Au.
„   „ : *Tristania conferta* R. Br.; *Myrtac.*; N.-S.-W.
„ Sand- (u): *Hura crepitans* L.; *Euphorbiac.*; tr. Am.
„   „ , White: *Hura crepitans* L.; *Euphorbiac.*; Trin.
„   „ Yellow: *Hura crepitans* L.; *Euphorbiac.*; Trin.
„ Satin-: *Phebalium billardieri* A. Juss.; *Rutac.*; sö. Au.

Box Scrub-: *Tristania conferta* R. Br.; *Myrtac.;* Au.
„ Siamese: *Gardenia sp.; Rubiac.;* Siam.
„ Soap-: *Villaresia moorei* F. v. M.; *Icacinac.;* Au.
„ Southafrican: *Gonioma kamassi* E. Mey.; *Apocynac.;* S.-Af.
„ „ : *Buxella mac-owani* Tiegh.; *Buxac.;* S.-Af.
„ Stanthorpe-: *Eucalyptus stuartiana* F. v. M.; *Myrtac.;* O.-Au., Tasm.
„ St. Domingo-: *Phyllostylon brasiliensis* Capanema; *Ulmac.;* St.-D.
„ Thozet's: *Eucalyptus raveretiana* F. v. M.; *Myrtac.;* Queensl.
„ True: *Eucalyptus corymbosa* Sm.; *Myrtac.;* N.-S.-W.
„ Turkish (u): *Buxus sempervirens* L.; *Buxac.;* S.-Eu., W.-As., N.-Af.
„ Venezuelan: *Casearia praecox* Griseb.; *Flacourtiac.;* tr. Am.
„ Westafrican (e): *Gonioma kamassi* E. Meyer; *Apocynac.;* S.-Af.
„ Westindian: *Casearia praecox* Griseb.; *Flacourtiac.;* W.-Ind.
„ „ : *Phyllostylon brasiliensis* Capanema; *Ulmac.;* tr. Am.
„ „ : *Tabebuia pentaphylla* Hemsl.; *Bignoniac.;* W.-Ind.
„ „ (e) : *Aspidosperma vargasi* A. DC.; *Apocynac.;* Ven.
„ White: *Eucalyptus hemiphloia* F. v. M.; *Myrtac.;* N.-S.-W.
„ Yellow: *Eucalyptus melliodora* A. Cunn.; *Myrtac.;* sö. Au.
„ „ : *Eucalyptus corymbosa* Sm.; *Myrtac.;* N.-S.-W.
„ „ : *Sideroxylon pohlmanianum* Bth. et Hook.; *Sapotac.;* Au.
„ tree: *Bursaria spinosa* Cav.; *Pittosporac.;* Au.
Boye: *Pycnanthus kombo* Warb.; *Myristicac.;* Lib.
Boy job: *Matayba apetala* Radlk.; *Sapindac.;* br. Hond.
Boyung: *Tarrietia argyrodendron* Bth.; *Sterculiac.;* nö. Au.
Bpak-pei: *Albizzia zygia* Macbr.; *Mimosac.;* Lib.
Bracaatinga: *Mimosa bracaatinga* Hoehne; *Mimosac.;* Bras.
Bracuhy de pedra: *Copaifera langsdorfi* Desf.; *Caesalpiniac.;* Bras. (Sao P)
Bradi lifi: *Coccoloba sp.; Polygonac.;* Sur.
Brahmani:. *Evonymus hamiltoniana* Wall.; *Celastrac.;* N.-Ind., C.-As., Jap.
Brakassa: *Cicca discoïdea* Baill.; *Euphorbiac.;* Elf.
Braloc: *Ocotea rubra* Mez; *Laurac.;* Bras.
Branquilho: *Gymnanthes marginata* Baill.; *Euphorbiac.;* Bras.
Brasil: *Haematoxylon campechianum* L.; *Caesalpiniac.;* tr. Am.
„ : *Haematoxylon brasiletto* Karst.; *Caesalpiniac.;* Nic., Kol., Ven.
„ : *Calderonia salvadorensis* Standl.; *Rubiac.;* Salv., Guat., Hond.
„ : *Exandra rhodoclada* Standl.; *Rubiac.;* Salv.
„ cojones de gato: *Caesalpinia crista* L.; *Caesalpiniac.;* Mex.
Brasilete: *Haematoxylon brasiletto* Karst.; *Caesalpiniac.;* Ven.
„ colorado: *Caesalpinia crista* L.; *Caesalpiniac.;* Cuba.
Brasiletto: *Haematoxylon brasiletto* Karst.; *Caesalpiniac.;* Kol.
Brasilienholz: *Caesalpinia echinata* Lam.; *Caesalpiniac.;* Bras.
„ : *Chlorophora tinctoria* Gaudich.; *Morac.;* Bras.
Brateco: *Cordia goeldiana* Huber; *Borraginac.;* Bras.
Brauna: *Melanoxylon brauna* Schott; *Caesalpiniac.;*. Bras. (Sao P)
„ : *Schinopsis brasiliensis* Engl.; *Anacardiac.;* Bras. (Sao P)
„ parda: *Melanoxylon brauna* Schott; *Caesalpiniac.;* Bras.
Braunherz (d): *Vouacapoua americana* Aubl.; *Papilionac.;* Sur.
Brazil: *Caesalpinia echinata* Lam.; *Caesalpiniac.;* Bras. (Bahia)
„ : *Caesalpinia tinctoria* Domb.; *Caesalpiniac.;* Ven.
„ : *Caesalpinia insignis* Steud.; *Caesalpiniac.;* Ven.
Brazilette: *Haematoxylon brasiletto* Karst.; *Caesalpiniac.;* tr. Am.
„ : *Sweetia sp.; Papilionac.;* Bras.
Braziletto: *Caesalpinia platyloba* S. Watson; *Caesalpiniac.;* br. Hond.
„ Bastard-: *Weinmannia sp.; Cunoniac.;* Jam.
„ Narrowleaved: *Caesalpinia sappan* L.; *Caesalpiniac.;* tr. O.-As.
„ Wild: *Weinmannia sp.; Cunoniac.;* Jam.
Brazo del fuego: *Chlorophora tinctoria* Gaudich.; *Morac.;* Arg.
Brea: *Cercidium sp.; Caesalpiniac.;* Arg.
„ : *Caesalpinia praecox* R. et P.; *Caesalpiniac.;* Arg.
Bréa: *Canarium villosum* F. Vill.; *Burserac.;* Phil. (Zamb.)
Breabri: *Inga spectabilis* Willd.; *Mimosac.;* br. Hond.
Bread and cheeses: *Pithęcolobium unguis-cati* Bth.; *Mimosac.;* Fl., br. W.-Ind.

Breadfruit: *Artocarpus communis* Forst.; *Morac.;* Hond.
„ : *Artocarpus incisa* Forst.; *Morac.;* Sey.
„ African: ? *Treculia africana* Dcne.; *Morac.;* Go.
Breadtree, Monkey-: *Adansonia digitata* L.; *Bombacac.;* Tr.
Break-axe: *Sloanea jamaicensis* Hook. f.; *Tiliac.;* Jam.
„ „ : *Schinopsis spp.; Anacardiac.;* S.-Am.
Breakbill: *Bumelia sp.; Sapotac.;* Windw. (Ant.)
Breiapfel: *Achras sapota* L.; *Sapotac.;* tr. Am.
Brésillet: *Caesalpinia tinctoria* Domb.; *Caesalpiniac.;* Chile.
„ : *Caesalpinia granadillo* Pittier; *Caesalpiniac.;* Ven.
„ : *Caesalpinia crista* L.; *Caesalpiniac.;* Ant.
„ : *Caesalpinia bahamensis* Lamk.; *Caesalpiniac.;* Bah.
„ : *Caesalpinia sappan* L.; *Caesalpiniac.;* Ind.
Bressilet franc: *Comocladia sp.; Anacardiac.;* St.-D.
Brettbaum: *Heritiera fomes* Buch.; *Sterculiac.;* Beng., Malay. Bor.
Breu branco: ? *Protium giganteum* Engl.; *Burserac.;* Bras. (Amaz.)
„ „ : *Protium heptaphyllum* March.; *Burserac.;* Bras. (Amaz.)
„ jauaricia: *Icica sp.; Burserac.;* Bras.
„ preto: *Icica sp.; Burserac.;* Bras.
„ sucuriu: *Icica sp.; Burserac.;* Bras. (Amaz.)
Brezo: *Erica arborea* L.; *Ericac.;* Tener.
„ : *Erica vagans* L.; *Ericac.;* Sp.
„ : *Calluna vulgaris* Salisb.; *Ericac.;* Sp.
„ blanco: *Erica arborea* L.; *Ericac.;* Sp.
„ de escobas: *Erica scoparia* L.; *Ericac.;* Sp.
„ negro: *Erica aragonensis* Willk.; *Ericac.;* Sp.
„ rubio: *Erica australis* L.; *Ericac.;* Sp.
Briar: *Erica arborea* L.; *Ericac.;* Af.
„ American: *Kalmia sp., Rhododendron sp.; Ericac.;* sö. N.-Am.
„ rood: *Cytisus scoparius* Link; *Papilionac.;* S.-Eu.
Bri-bri: *Inga spp.; Mimosac.;* C.-Am.
„ „ Bastard: *Inga recordi* Britt. et Rose; *Mimosac.;* br. Hond.
„ „ macho: *Inga recordi* Britt. et Rose; *Mimosac.;* br. Hond.
Brigalow: *Acacia harpophylla* F. v. M.; *Mimosac.;* Au.
„ Mountain-: *Acacia glaucescens* Willd.; *Mimosac.;* sö. Au.
Brigbau: *Albizzia sp.; Mimosac.;* Elf.
Brigelow: *Acacia lasiophylla* Bth.; *Mimosac.;* sö. Au.
Brimstonetree: *Morinda citrifolia* L.; *Rubiac.;* Ind., Sunda, Au.
Brin d'amour: *Malpighia urens* L.; *Malpighiac.;* Guad.
Broad-leaf: *Ocotea sp.; Laurac.;* Jam.
„ „ : *Terminalia latifolia* Sw.; *Combretac.;* Jam.
„ „ : *Vismia ferruginea* H. B. K.; *Guttifer.;* br. Hond.
„ „ : *Griselinia littoralis* Raoul; *Cornac.;* N.-Seel.
Broinharti: *Vouacapoua americana* Aubl.; *Papilionac.;* Sur.
Bro-kpar: *Homalium spp.; Flacourtiac.;* Lib.
Bromabine: *Pentadesma butyracea* Sabine; *Guttifer.;* Go.
Broodboom: *Artocarpus incisa* L.; *Morac.;* fr. Gu.
Broom, Scotch (e): *Cytisus scoparius* Link; *Papilionac.;* Eu.
„ Spanish (e): *Spartium junceum* L.; *Papilionac.;* M.-M., Kanar., Arg.
„ „ White (e): *Cytisus multiflorus* Sweet; *Papilionac.;* w. M.-M.
Broomwood: *Carpinus caroliniana* Walt.; *Corylac.;* N.-Am.
Brorogery: *Callitris rhomboidea* R. Br.; *Conifer.;* sö. Au.
Brosse à dents: *Grevillea heterochroma* Brongn. et Gris.;*Proteac.;* N.-Kal.
Brotfruchtbaum (d): *Artocarpus integrifolia* L.; *Morac.;* Tr.
Brownheart: *Vouacapoua americana* Aubl.; *Papilionac.;* Gu.
Bruco: *Erica vagans* L.; *Ericac.;* Sp.
Bruinhart: *Vouacapoua americana* Aubl.; *Papilionac.;* Sur.
Brush: *Trochocarpa laurina* R. Br.; *Epacridac.;* sö. Au.
Bruyère (f): *Erica arborea* L.; *Ericac.;* M.-M.
Bruyères: *Philippia sp.; Ericac.;* Mad.
Bsa: *Buxus sempervirens* L.; *Buxac.;* S.-Rußland.
Búa (H): *Garcinia oliveri* Pierre; *Guttifer.;* Ind.-Ch.

Búa (H): *Garcinia schomburgkiana* Pierre; *Guttifer.;* Ind.-Ch.
Bua: *Fagraea berteriana* A. Gray; *Loganiac.;* Fid.
Buabúa: *Eugenia mananquil* Blco.; *Myrtac.;* Phil. (Mindoro)
Bua-bua: *Guettarda speciosa* L.; *Rubiac.;* Fid.
Buah keras: *Aleurites triloba* Forst.; *Euphorbiac.;* Malay.
Bu-ah-vohn-doo: *Hugonia planchoni* Hook. f.; *Linac.;* Lib.
Búa lúeur: *Garcinia fusca* Pierre; *Guttifer.;* Annam.
„ mói: *Garcinia harmandi* Pierre; *Guttifer.;* Annam.
„ nha: *Garcinia sp.; Guttifer.;* Ind.-Ch.
Buamúchil: *Pithecolobium dulce* Bth.; *Mimosac.;* Mex.
Buá nui: *Garcinia oliveri* Pierre; *Guttifer.;* Annam.
Bua ring: *Garcinia sp.; Guttifer.;* Ind.-Ch.
Buá ruñg: *Garcinia oliveri* Pierre; *Guttifer.;* Annam.
Bu-ay-boh: *Funtumia africana* Stapf; *Apocynac.;* Lib.
Bu-ay-wreh: *Maesopsis emini* Engl.; *Rhamnac.;* Lib.
Bubai: ? *Khaya anthotheca* C. DC.; *Meliac.;* Kam.
Bubinga (f): *Didelotia africana* Pierre; *Caesalpiniac.;* Kam., Gab.
„ : ? *Brachystegia sp.; Caesalpiniac.;* W.-Af.
Bubingo: *Didelotia africana* Pierre; *Caesalpiniac.;* Gab.
Bubóg: *Sterculia sp.; Sterculiac.;* Phil.
Búboi: *Ceiba pentandra* Gaertn.; *Bombacac.;* Phil.
Buboi: *Funtumia africana* Stapf; *Apocynac.;* Lib.
Búboi-gúlat: *Bombax malabaricum* DC.; *Bombacac.;* Phil.
Bubrome: *Guazuma ulmifolia* Lamk.; *Sterculiac.;* Ant.
Bubúa: *Aglaia everetti* Merr.; *Meliac.;* Phil.
Bucabally: *Cedrela guianensis* A. Juss.; *Meliac.;* Gu.
Bucago: *Erythrina sp.; Papilionac.;* P.-R., Cuba.
Bucare: *Erythrina glauca* Willd.; *Papilionac.;* Ven.
Búcare espinoso: *Erythrina sp.; Papilionac.;* Cuba.
Búcaro: *Erythrina sp.; Papilionac.;* Kol.
Bucco: *Rhizophora sp.; Rhizophorac.;* Malay.
Buche (d): *Fagus silvatica* L.; *Fagac.;* Eu.
„ amerikanische (d): *Fagus grandifolia* Ehrh.; *Fagac.;* N.-Am.
„ Hage-: *Carpinus betulus* L.; *Corylac.;* Mitt.-Eu.
„ Hain-: *Carpinus betulus* L.; *Corylac.;* Eu.
„ Hopfen-: *Ostrya vulgaris* Willd.; *Corylac.;* S.-Eu., Kl.-As.
„ „ virginische: *Ostrya virginica* Willd.; *Corylac.;* ö. N.-Am.
„ Horn-: *Carpinus betulus* L.; *Corylac.;* Mitt.-Eu.
„ Rot-: *Fagus silvatica* L.; *Fagac.;* Eu.
„ „ amerikanische: *Fagus grandifolia* Ehrh.; *Fagac.;* ö. N.-Am.
„ Schwarz-: *Ostrya vulgaris* Willd.; *Corylac.;* S.-Eu., Kl.-As.
„ Weiß-, Gemeine: *Carpinus betulus* L.; *Corylac.;* Eu.
„ „ , Amerikanische: *Carpinus americana* Michx.; *Corylac.;* N.-Am.
Bucheira: *Aspidosperma duckei* Huber; *Apocynac.;* Bras. (Amaz.)
Buchstabenholz (d): *Piratinera guianensis* Aubl.; *Morac.;* Sur.
Buchs: *Buxus sempervirens* L.; *Buxac.;* S.-Eu., W.-As., N.-Af.
„ : *Celastrus buxifolius* L.; *Celastrac.;* S.-Af.
„ : *Gonioma kamassi* E. Meyer; *Apocynac.;* S.-Af.
„ : *Heterodendrum oleaefolium* Desf.; *Sapindac.;* Queensl.
„ : *Wrightia annamensis* Eberh. et Dubard; *Apocynac.;* Annam.
„ afrikanisch: *Buxus mac-owani* Oliv.; *Buxac.;* S.-Af.
„ Australisch: *Pittosporum undulatum* Vent.; *Pittosporac.;* Au.
„ Balearisch (d): *Buxus balearica* Lamk.; *Buxac.;* w. M.-M.
„ Echtes: *Buxus sempervirens* L.; *Buxac.;* S.-Eu., W.-As., N.-Af.
„ Gemeiner: *Buxus sempervirens* L.; *Buxac.;* S.-Eu., W.-As., N.-Af.
„ Himalaya-: *Gardenia latifolia* Ait.; *Rubiac.;* Ind.
„ Japanisch: *Buxus sempervirens* L. var. *japonica* Muell. Arg.; *Buxac.;* Jap.
„ Kaukasisch (d): *Buxus sempervirens* L.; *Buxac.;* Kauk.
„ Levantinisch: *Buxus sempervirens* L.; *Buxac.;* Türkei.
„ Orientalisch: *Buxus sempervirens* L.; *Buxac.;* Kl.-As.
„ Usambara-: *Schefflerodendron usambarense* Harms; *Araliac.;* O.-Af.
„ Westindisches: *Casearia praecox* Griseb.; *Flacourtiac.;* W.-Ind.

Buchs Westindisches: *Tecoma pentaphylla* Juss.; *Bignoniac.*; n. S.-Am.
„   „   : *Aspidosperma vargasi* A. DC.; *Apocynac.*; Ven.
Buckeye: *Aesculus californica* Nutt.; *Hippocastanac.*; USA. (Cal.)
„   : *Aesculus octandra* Marsh., *A. glabra* Willd.; *Hippocastanac.*; USA. (Ohio)
„   (e): *Aesculus turbinata* Blume; *Hippocastanac.*; Jap.
„   Fetid-: *Aesculus glabra* Willd.; *Hippocastanac.*; USA.
„   Large: *Aesculus flava* Ait.; *Hippocastanac.*; USA.
„   Ohio-: *Aesculus glabra* Willd.; *Hippocastanac.*; USA.
„   Sweet: *Aesculus octandra* Marsh.; *Hippocastanac.*; ö. USA.
„   „   : *Aesculus flava* Ait.; *Hippocastanac.*; N.-Am.
„   Yellow: *Aesculus octandra* Marsh.; *Hippocastanac.*; ö. USA.
Buckthorn, Canadian: *Rhamnus purshiana* DC.; *Rhamnac.*; N.-Am.
„   Sacred-bark-: *Rhamnus purshiana* DC.; *Rhamnac.*; N.-Am.
Buck-was tree: *Symphonia globulifera* L.; *Guttifer.*; br. Gu.
Buck-wheat tree: *Cliftonia ligustrina* Banks; *Cyrillac.*; USA.
Buda: *Coula edulis* Baill.; *Olacac.*; Gab.
Budgó: *Shorea teysmanniana* Dyer; *Dipterocarpac.*; Phil.
Budree: *Zizyphus jujuba* Lamk.; *Rhamnac.*; Beng.
Budrunga: *Zanthoxylum budrunga* DC.; *Rutac.*; Beng.
Bu-eh: *Deinbollia polypus* Stapf; *Sapindac.*; Lib.
Bu-ehn-waye: *Spondianthus ugandensis* Hutch.; *Euphorbiac.*; Lib.
Buele ba nyou: *Entandrophragma rederi* Harms; *Meliac.*; Kam.
Buen amigo: *Pisonia aculeata* L.; *Nyctaginac.*; Kol.
Buenas noches: *Datura metel* L.; *Solanac.*; Tener.
Buessé: *Cola vera* K. Schum.; *Sterculiac.*; Elf.
Buey: *Diospyros spp.*, *Maba spp.*; *Ebenac.*; Malay.
Bufalaga: *Thymelaea tartonraira* All.; *Thymelaeac.*; Sp.
Buffalo berry: *Shepherdia argentea* Nutt.; *Elaeagnac.*; USA.
Büffelholz (d): *Burchellia bubalina* R. Br.; *Rubiac.*; Kap.
Buffelsbal: *Gardenia thunbergi* L.; *Rubiac.*; Kap.
Bugalót: *Garcinia sp.*; *Guttifer.*; Phil.
Bug'áom: *Lagerstroemia piriformis* Koehne; *Lythrac.*; Phil.
Bug'árom: *Lagerstroemia piriformis* Koehne; *Lythrac.*; Phil.
„   : *Lagerstroemia speciosa* Pers.; *Lythrac.*; Phil.
Bugás: *Neonauclea sp.*; *Rubiac.*; Phil.
Bugáyong: *Ormosia calavensis* Azaola; *Papilionac.*; Phil.
Bugayong-china: *Adenanthera intermedia* Merr.; *Mimosac.*; Phil.
Buge: *Albizzia lebbek* Bth.; *Mimosoac.*; tr. As., Af.
Búgis: *Pentacme contorta* Merr. et Rolfe; *Dipterocarpac.*; Phil.
Búgo: *Garuga sp.*; *Burserac.*; Phil.
„   : *Phoebe sterculioides* Merr.; *Laurac.*; Phil.
Bugre: *Albizzia lebbek* Bth.; *Mimosac.*; tr. As., Af.
Búgu: *Garuga sp.*; *Burserac*; Phil.
Buguarom: *Lagerstroemia piriformis* Koehne; *Lythrac.*; Phil.
Buhían: *Buchanania sp.*; *Anacardiac.*; Phil.
Buho: *Schizostachyum lumampao* Merr.; *Gramin.*; Phil.
Buhoorada: *Parinarium campestre* Aubl.; *Rosac.*; br. Gu.
Buhs: *Buxus sempervirens* L.; *Buxac.*; S.-Eu., W.-As., N.-Af.
Buhsboum: *Buxus sempervirens* L.; *Buxac.*; S.-Eu., W.-As., N.-Af.
Buhúkan: *Planchonia spectabilis* Merr.; *Lecythidac.*; Phil.
Buhurada: *Parinarium campestre* Aubl.; *Rosac.*; br. Gu.
Bui: *Canarium nigrum* Engl., *C. commune* L.; *Burserac.*; Ind.-Ch.
Buig-mij-niet: *Buxella mac-owani* Tiegh.; *Buxac.*; S.-Af.
Buig-my-nie: *Buxella mac-owani* Tiegh.; *Buxac.*; S.-Af.
Buirata: *Parinarium campestre* Aubl.; *Rosac.*; Sur.
Buis (f): *Buxus sempervirens* L.; *Buxac.*; S.-Eu., W.-As., N.-Af.
„   bâtard: *Eucalyptus hemiphloia* F. v. M., *E. globulus* Lab.; *Myrtac.*; Au.
„   bei (f): *Buxus sempervirens* L.; *Buxac.*; S.-Eu., W.-As., N.-Af.
„   commun (f): *Buxus sempervirens* L.; *Buxac.*; S.-Eu., W.-As., N.-Af.
„   Faux: *Gyminda latifolia* Urb.; *Celastrac.*; Fl., Cuba, P.-R.
„   Ibéro: *Picralima nitida* Pierre, *P. umbellata* Stapf; *Apocynac.*; W.-Af.
„   Kamassi (f): *Gonioma kamassi* E. Meyer; *Apocynac.*; Kap.

Buis toujours vert (f): *Buxus sempervirens* L.; *Buxac.;* S.-Eu., W.-As., N.-Af.
„ de Wallich: *Buxus wallichiana* Baill.; *Buxac.;* Ind.
„ d'Amérique (f): *Casearia praecox* Griseb.; *Flacourtiac.;* W.-Ind. *
„ d'Annam (f): *Sideroxylon eburneum* Chev.; *Sapotac.;* Ind.-Ch.
„ de Brésil: *Euxylophora paraënsis* Huber; *Rutac.;* Bras.
„ de Ceylan (f): *Canthium didymum* Roxb.; *Rubiac.;* Ind.-Ch.
„ „ Chine: *Murraya exotica* L.; *Rutac.;* Ind.-Ch.
„ „ l'Inde: *Murraya exotica* L.; *Rutac.;* Réunion.
„ „ Mahon: *Buxus balearica* Lamk.; *Buxac.;* w. M.-M.
„ „ Minorque (f): *Buxus balearica* Lamk.; *Buxac.;* w. M.-M.
„ „ Siam: *Gardenia spp.;* *Rubiac.;* Ind.-Ch.
„ des Antilles (f): *Casearia praecox* Griseb.; *Flacourtiac.;* W.-Ind.
„ „ „ : *Tecoma pentaphylla* Juss.; *Bignoniac.;* Ant.
„ „ Baléares: *Buxus balearica* Lamk.; *Buxac.;* w. M.-M.
„ du Cap (f): *Gonioma kamassi* E. Meyer; *Apocynac.;* Kap.
Buje (s): *Buxus sempervirens* L.; *Buxac.;* S.-Eu., W.-As., N.-Af.
Buk: *Quercus lamellosa* Smith; *Fagac.;* ö. Himal.
Búkag: *Duabanga moluccana* Blume; *Lythrac.;* Phil.
Bukain: *Melia azedarach* L.; *Meliac.;* Ind.
? Bukal: *Mimusops elengi* L.; *Sapotac.;* w. O.-Ind., Cey.
Bukampadaruka: *Cordia myxa* L.; *Borraginac.;* Kl.-As., Ind.
Bukáu: *Dinochloa scandens* O. Kᵗ ͻe.; *Gramin.;* Phil.
Bukáui: *Dinochloa scandens* O. Kᵗ ͻe.; *Gramin.;* Phil.
Bukkaláu: *Euphoria cinerea* Radlk.; *Sapindac.;* Phil.
Bukkum-wood: *Caesalpinia sappan* L.; *Caesalpiniac.;* tr. O.-As.
Bukome: *Terminalia superba* Engl. et Diels; *Combretac.;* tr. W.-Af.
Buksboom (dä): *Buxus sempervirens* L.; *Buxac.;* S.-Eu., W.-As., N.-Af.
Buku-buku: *Hirtella hirsuta* Lam.; *Rosac.;* br. Gu.
Bukyu-Akwa: *Cistanthera papaverifera* Chev.; *Tiliac.;* Go.
Bulá': *Shorea eximia* Scheff.; *Dipterocarpac.;* Phil.
Bulábog: *Parishia malabog* Merr.; *Anacardiac.;* Phil.
Bulagsóg: *Eugenia claviflora* Roxb.; *Myrtac.;* Phil.
Búlak: *Ceiba pentandra* Gaertn.; *Bombacac.;* Phil.
Bulakán: *Sterculia sp.;* *Sterculiac.;* Phil.
Bulála: *Litchi philippinensis* Radlk.; *Sapindac.;* Phil.
„ : *Nephelium mutabile* Blume; *Sapindac.;* Phil.
„ : *Nauclea sp.,* ? *Neonauclea sp.;* *Rubiac.;* Phil.
„ -na-tolón: *Wrightia sp.;* *Apocynac.;* Phil.
Bulan-bulán (f): *Palaquium sp.;* *Sapotac.;* Phil.
Bulatináu: *Maba buxifolia* Pers.; *Ebenac.;* Phil.
Buláu: *Canarium luzonicum* A. Gray; *Burserac.;* Phil.
Buláuen: *Vitex parviflora* Juss.; *Verbenac.;* Phil.
Bulbúl: *Pinus insularis* Endl.; *Conifer.;* Phil.
Bulínau: *Bambusa vulgaris* Schrad.; *Gramin.;* Phil.
Bulines: *Guazuma ulmifolia* Lamk.; *Sterculiac.;* Mex.
Bulinga: *Copaifera sp.;* *Caesalpiniac.;* Kam.
Bulkokra: *Chaetocarpus castanocarpus* Thwait.; *Euphorbiac.;* O.-Ind.
Bull Bay: *Magnolia grandiflora* L.; *Magnoliac.;* USA.
Bullet, Bastard-: *Humiria floribunda* Mart.; *Humiriac.;* Sur.
„ „ : *Humiria balsamifera* Aubl.; *Humiriac.;* Sur.
Bulletree: *Manilkara bidentata* Chev.; *Sapotac.;* Gu.
Bullet-tree (e): *Manilkara bidentata* Chev.; *Sapotac.;* Gu.
„ „ : *Mimusops globosa* Gaertn.; *Sapotac.;* Trin.
„ „ : *Bumelia retusa* Sw.; *Sapotac.;* St.-D.
„ „ : *Sapota sideroxylon* Griseb.; *Sapotac.;* W.-Ind.
„ „ : *Bucida buceras* L.; *Combretac.;* br. Hond.
„ „ : *Guarea trichilioides* L.; *Meliac.;* Hond.
„ „ : *Robinia panacoco* Aubl.; *Papilionac.;* tr. Am.
„ „ male: *Lucuma sp.;* *Sapotac.;* br. Gu.
„ „ Naseberry: *Mimusops sp.;* *Sapotac.;* Jam.
„ „ Red: ? *Dipholis nigra* Gris.; *Sapotac.;* Jam.
„ „ sapodilla: *Labourdonnaisia albescens* Bth.; *Sapotac.;* Jam.

Bulletrie: *Manilkara bidentata* Chev.; *Sapotac.;* Gu.
Bulletwood: *Mimusops elengi* L., *M. littoralis* Kurz; *Sapotac.,* Ind.
„ Bastard-: *Humiria spp.; Humiriac.;* br. Gu.
„ Yellow: *Sideroxylon australe* Bth. et Hook.; *Sapotac.;* Au.
Bullhoof: *Drypetes browni* Standl.; *Euphorbiac.;* Hond.
„ : *Celtis hottlei* Standl.; *Ulmac.;* br. Hond.
„ , macho: *Drypetes browni* Standl.; *Euphorbiac.;* br. Hond.
Bulltree, White: *Dipholis salicifolia* A. DC.; *Sapotac.;* W.-Ind.
Bully, Bastard-: *Humiria spp.; Humiriac.;* br. Gu.
Bully-tree: *Mimusops globosa* Gaertn.; *Sapotac.;* br. Gu., W.-Ind.
„ „ : *Achras sapota* L.; *Sapotac.;* Ant., C.-Am.
„ „ : *Bucida buceras* L.; *Combretac.;* br. Hond.
„ „ : *Mimusops spectabilis* Pittier; *Sapotac.;* C.-R.
„ „ : ? *Hieronymia alchorneoides* Allem.; *Euphorbiac.;* Pan.
„ „ macho: *Eugenia sp.; Myrtac.;* br. Hond.
„ „ Naseberry-: *Sapota sideroxylon* Griseb.; *Sapotac.;* Jam.
Bulno: *Livistona rotundifolia* Mart. var. *luzonensis* Becc.; *Palmac.;* Phil.
Bulóg: *Aglaia spp.; Meliac.;* Phil.
Bulokbúlok: *Lumnitzera littorea* Voigt; *Combretac.;* Phil.
Bultók: *Quercus sp.; Fagac.;* Phil.
Bulu: *Terminalia belerica* Roxb.; *Combretac.;* Ind.
Buluán: *Bassia ramiflora* Merr.; *Sapotac.;* Phil. (Zamb.)
Buluáng: *Endospermum peltatum* Merr.; *Euphorbiac.;* Phil.
„ : *Terminalia edulis* Blco.; *Combretac.;* Phil.
Bulubangkál: *Nauclea sp.; Rubiac.;* Phil.
Bulubitóon: *Nauclea sp.; Rubiac.;* Phil.
Búlus: *Litsea spp.; Laurac.;* Phil.
Buma (d): *Ceiba pentandra* Gaertn.; *Bombacac.;* Kam.
Bumwood: *Metopium toxiferum* K. et U.; *Anacardiac.;* Fl., br. W.-Ind.
Buna: *Fagus silvatica* L. var. *sieboldi* Maxim.; *Fagac.;* Jap.
„ : *Terminalia ivorensis* Chev.; *Combretac.;* Elf.
Bunég: *Garcinia sp.; Guttifer.;* Phil.
Bung: *Niebuhria sp., Crataeva sp.; Capparidac.;* Ind.-Ch.
Búñga: *Areca catechu* L.; *Palmac.;* Phil.
„ : *Sterculia sp.; Sterculiac.;* Phil.
Buñgá: *Artocarpus rubrovenia* Warb.; *Morac.;* Phil.
„ : *Pentacme contorta* Merr. et Rolfe; *Dipterocarpac.;* Phil.
„ : *Shorea spp.; Dipterocarpac.;* Phil.
Bunga: *Lagerstroemia spp.; Lythrac.;* Malay.
Buñgálon: *Avicennia officinalis* L.; *Verbenac.;* Phil.
„ : *Cumingia philippinensis* Vid.; *Bombacac.;* Phil.
„ : *Sonneratia caseolaris* Engl.; *Lythrac.;* Phil.
Bungang: *Neesia altissima* Blume; *Bombacac.;* Java.
Buñgat: *Sterculia sp.; Sterculiac.;* Phil.
Bunggás: *Terminalia comintana* Merr.; *Combretac.;* Phil.
Bunggáson-tugás: *Terminalia comintana* Merr.; *Combretac.;* Phil.
Bungkúlan: *Eugenia mananquil* Blco.; *Myrtac.;* Phil.
Buñglás: *Alangium longiflorum* Merr.; *Cornac.;* Phil.
„ : *Terminalia comintana* Merr.; *Combretac.;* Phil.
Buñglín: *Polyscias nodosa* Seem.; *Araliac.;* Phil.
Bunglíu: *Aphanomyxis cumingiana* Harms; *Meliac.;* Phil.
Bungló: *Tristania sp.; Myrtac.;* Phil.
Bunglói: *Dysoxylum turczaninowi* C. DC.; *Meliac.;* Phil. (Iloilo)
Bungor: *Lagerstroemia spp.; Lythrac.;* Malay.
Bungrás: *Terminalia comintana* Merr.; *Combretac.;* Phil.
Buñguás: *Aglaia everetti* Merr.; *Meliac.;* Phil. (Cebu)
Buñgúg: *Vitex pentaphylla* Merr.; *Verbenac.;* Phil.
Buni: *Antidesma bunius* Wall.; *Euphorbiac.;* Java.
Bun mukh: *Schima wallichi* Choisy; *Theac.;* Ind.
Bunec-walwal: *Evodia accedens* Blume; *Rutac.;* O.-Au.
Bunog: *Garcinia sp.; Guttifer.;* Phil.
Bunsóg: *Agathis alba* Foxw.; *Conifer.;* Phil.

Buntan: *Citrus decumana* L.; *Rutac.*; Jap.
Buntúgon: *Dysoxylum decandrum* Merr.; *Meliac.*; Phil.
Bunúg: *Garcinia sp.*; *Guttifer.*; Phil.
Bunut: *Calophyllum spp.*; *Guttifer.*; Malay.
Bunya bunya: *Araucaria bidwilli* Hook.; *Conifer.*; Au.
Buobi: *Diospyros sanza-minika* Chev.; *Ebenac.*; Elf.
Buoi: *Citrus decumana* L.; *Rutac.*; Annam.
Buoi-bong: *Acronychia laurifolia* Blume; *Rutac.*; Tonkin.
Buraem: *Pradosia latescens* Radlk.; *Sapotac.*; Bras.
Buragít: *Artocarpus subrotundifolia* Elmer; *Morac.*; Phil.
Buragrís: *Garcinia sp.*; *Guttifer.*; Phil.
Buragu: *Ceiba pentandra* Gaertn.; *Bombacac.*; Ind.
Burahem: *Pradosia latescens* Radlk.; *Sapotac.*; Bras.
Buranhé: *Pradosia latescens* Radlk.; *Sapotac.*; Bras.
Buranhem: *Pradosia latescens* Radlk.; *Sapotac.*; Bras. (Bahia)
Buranhen: *Pradosia latescens* Radlk.; *Sapotac.*; Bras.
Burao: *Hibiscus tiliaceus* L.; *Malvac.*; Tahiti.
Bura-rutha: *Sapindus emarginatus* Vahl; *Sapindac.*; Ind.
Buratu: *Gymnartocarpus woodi* Merr.; *Morac.*; Phil.
Burau: *Hibiscus tiliaceus* L.; *Malvac.*; Tahiti.
Buráwis: *Terminalia calamansanai* Rolfe; *Combretac.*; Phil.
Bureria: *Bourreria havanensis* Miers; *Borraginac.*; Fl., W.-Ind.
Burgan: *Kunzea peduncularis* F. v. M.; *Myrtac.*; sö. Au.
Burhoorada: *Parinarium campestre* Aubl.; *Rosac.*; br. Gu.
Burhuda: *Endlicheria multiflora* Mez; *Laurac.*; br. Gu.
Buri: *Mollia lepidota* Spruce; *Tiliac.*; br. Gu.
Burillo: ? *Apeiba aspera* Aubl.; *Tiliac.*; Nic.
   „   : *Heliocarpus sp.*; *Tiliac.*; C.-R.
   „ blanco: *Heliocarpus sp.*; *Tiliac.*; C.-R.
   „ falso: *Heliocarpus sp.*; *Tiliac.*; Nic.
Burío: *Heliocarpus sp.*; *Tiliac.*; C.-R.
Burio: *Hampea panamensis* Standl.; *Bombacac.*; Pan.
Buriogre: *Hampea integerrima* Schlechter *var. appendiculata* ?; *Bombacac.*; C.-R.
Burírau: *Bambusa vulgaris* Schrad.; *Gramin.*; Phil.
Buritisa: *Mauritia vinifera* Mart.; *Palmac.*; Bras.
Burito: *Souroubea guianensis* Aubl.; *Marcgraviac.*; Nic.
Burletta (u): *Acer sp.*; *Acerac.*; USA.
Burning bush: *Evonymus atropurpurea* Jacq.; *Celastrac.*; N.-Am.
Burnúd: *Myristica sp.*; *Myristicac.*; Phil. (Zamb.)
Buro-buro: *Anthocleista nobilis* G. Don; *Loganiac.*; Elf.
Burocossa: *Mitragyne macrophylla* Hiern; *Rubiac.*; W.-Af.
Buro-koro: *Piratinera guianensis* Aubl.; *Morac.*; br. Gu.
Burr: *Triumfetta lappula* L.; *Tiliac.*; br. Hond.
Burracura: *Piratinera guianensis* Aubl.; *Morac.*; br. Gu.
Burro: *Capparis sp.*; *Capparidac.*; P.-R.
   „ blanco: *Capparis sp.*; *Capparidac.*; P.-R.
   „ , Cajon de: *Saccoglottis amazonica* Mart.; *Humiriac.*; Trin.
Burrunedura: *Cupania anacardioides* A. Rich.; *Sapindac.*; N.-S.-W.
Bursera, Gum-bearing: *Bursera simaruba* Sarg.; *Burserac.*; br. W.-Ind.
Bursere, Gummitragende (d): *Bursera simaruba* Sarg.; *Burserac.*; tr. Am.
Buruch: *Manilkara bidentata* Chev.; *Sapotac.*; br. Gu.
Burué: *Mimusops spp.*; *Sapotac.*; Sur.
Buruea: *Mimusops spp.*; *Sapotac.*; br. Gu.
Buruta: *Chloroxylon swietenia* DC.; *Meliac.*; Ind.
Burutu: *Chloroxylon swietenia* DC.; *Meliac.*; Cey.
Bus (h): *Buxus sempervirens* L.; *Buxac.*; S.-Eu., W.-As., N.-Af.
Buság: *Tristania sp.*; *Myrtac.*; Phil.
Busáhin: *Alangium longiflorum* Merr.; *Cornac.*; Phil.
Busáin: *Bruguiera spp.*; *Rhizophorac.*; Phil.
Busáing: *Bruguiera eriopetala* Lam.; *Rhizophorac.*; Phil.
Busenlói: *Aphanomyxis cumingiana* Harms; *Meliac.*; Phil.
Bushrope: *Ancistrophyllum secundiflorum* G. Mann et H. Wendl.; *Palmac.*; tr. Af.

Busi: *Guarea thompsoni* Spr. et Hutch.; *Meliac.;* Go.
Busíli: *Neonauclea sp.; Rubiac.;* Phil.
„   : *Terminalia calamansanai* Rolfe; *Combretac.;* Phil.
Busserole de Californie: *Arctostaphylos pungens* H. B. K.; *Ericac.;* USA.
Bussipi: *Oxystigma manni* Harms; *Caesalpiniac.;* Kam.
Busso (i): *Buxus sempervirens* L.; *Buxac.;* S.-Eu., W.-As., N.-Af.
Bussu: *Manicaria saccifera* Gaertn.; *Palmac.;* P.-R.
Bustic (u): *Dipholis salicifolia* A. DC.; *Sapotac.;* s. Fl., Ant.
Buta-buta: *Excoecaria agallocha* L.; *Euphorbiac.;* Phil.
Butarik: *Adenanthera intermedia* Merr.; *Mimosac.;* Phil.
Butigáu: *Cyclostemon sp.; Euphorbiac.;* Phil.
Butíg-bábui: *Cyclostemon sp.; Euphorbiac.;* Phil.
Butong-manúk: *Cyclostemon microphyllus* Merr.; *Euphorbiac.;* Phil.
Butter-bough (u): *Exothea paniculata* Radlk.; *Sapindac.;* Fl., Ant.
Butterfly tree: *Erblichia odorata* Seem.; *Turnerac.;* br. Hond.
„   wings: *Copaifera mopane* J. Kirk; *Caesalpiniac.;* O.-Af., C.-Af.
Butternut: *Juglans cinerea* L.; *Juglandac.;* N.-Am.
Buttertree: *Illipe latifolia* Engl.; *Sapotac.;* Ind.
„   : *Pentandesma butyraceum* Don; *Guttifer.;* Go.
Butterwood: *Platanus occidentalis* L.; *Platanac.;* N.-Am.
„   : *Callicoma serratifolia* Andr.; *Saxifragac.;* O.-Au.
Button-ball: *Platanus occidentalis* L.; *Platanac.;* N.-Am.
„   -bush: *Cephalanthus occidentalis* L.; *Rubiac.;* USA.
„   -wood: *Platanus occidentalis* L.; *Platanac.;* USA.
„   „   California-: *Platanus racemosa* Nutt.; *Platanac.;* USA.
„   „   Bastard-: *Laguncularia racemosa* Gaertn. f.; *Combretac.;* Bah.
„   „   False: *Laguncularia racemosa* Gaertn. f.; *Combretac.;* tr. Am.
„   „   Florida-: *Conocarpus erectus* L.; *Combretac.;* tr. Am., W.-Af.
„   „   White (u): *Laguncularia racemosa* Gaertn. f.; *Combretac.;* tr. Am., W.-Af.
Butubutú: *Buchanania sp.; Anacardiac.;* Phil. (Cebu)
Butún: *Dendrocalamus latiflorus* Munro; *Gramin.;* Phil. (Cebu)
Butúng: *Bambusa vulgaris* Schrad.; *Gramin.;* Phil.
Butusu: *Dumoria heckeli* Chev.; *Sapotac.;* Elf.
Bux (dä, sch): *Buxus sempervirens* L.; *Buxac.;* S.-Eu., W.-As., N.-Af.
Buxbom (dä, sch): *Buxus sempervirens* L.; *Buxac.;* S.-Eu., W.-As., N.-Af.
Buxbomsträ (sch): *Buxus sempervirens* L.; *Buxac.;* S.-Eu., W.-As., N.-Af.
Buxo (p): *Buxus sempervirens* L.; *Buxac.;* S.-Eu., W.-As., N.-Af.
Buyúkan: *Duabanga moluccana* Blume; *Lythrac.;* Phil.
Buza pundi: *Balanites mayumbensis* Exell.; *Simarubac.;* Angola.
Buzzard-head-tree: *Swietenia sp.; Meliac.;* Mex.
Bwaegyee: *Adenanthera pavonina* L.; *Mimosac.;* And.
Bwamba: *Klainedoxa gabonensis* Pierre; *Simarubac.;* Gab.
Bwayè: *Garcinia klaineana* Pierre; *Guttifer.;* Gab.
Bwélébako: *Ongokea gore* Engl.; *Olacac.;* Kam.
Bwiba: *Irvingia gabonensis* Baill.; *Simarubac.;* Kam.
„   : *Irvingia barteri* Hook. f.; *Simarubac.;* Kam.
„   bambale: ? *Irvingia gabonensis* Baill.; *Simarubac.;* Kam.
„   „   : *Irvingia barteri* Hook. f.; *Simarubac.;* W.-Af.
„   banjo: ? *Irvingia oblonga* Chev.; *Simarubac.;* Kam.
„   „   : *Klainedoxa latifolia* Pierre; *Simarubac.;* Kam.
Bwibanjoc: *Desbordesia insignis* Pierre; *Simarubac.;* Kam.
Byakushin: *Juniperus chinensis* L.; *Conifer.;* Jap.
Byala: *Garcinia klaineana* Pierre; *Guttifer.;* Gab.
Bylhout: *Eperua falcata* Aubl.; *Caesalpiniac.;* Sur.

C

Ca: *Antidesma glaesembilla* Gaertn.; *Euphorbiac.;* Ind.-Ch.
„   chac: *Shorea obtusa* Wall.; *Dipterocarpac.;* Ind.-Ch.
„   chit: *Shorea obtusa* Wall.; *Dipterocarpac.;* Ind.-Ch.
„   duôi (H): *Cyanodaphne cuneata* Blume; *Laurac.;* Ind.-Ch.

Ca gan (H): *Terminalia tomentosa* Bedd.; *Combretac.;* Ind.-Ch.
„ giam (H): *Stephegyne parvifolia* Korth.; *Rubiac.;* Ind.-Ch.
„ lau (H): *Pleurostylia cochinchinensis* Pierre; *Celastrac.;* Ind.-Ch.
„ lich: *Terminalia tomentosa* Bedd.; *Combretac.;* Ind.-Ch.
„ -lo-gnauh: *Crataeva religiosa* Forst.; *Capparidac.;* Co.-Ch.
„ ma (H):ꞏ *Buxus cochinchinensis* Pierre; *Buxac.;* Ind.-Ch.
„ na: *Canarium album* Engl.; *Burserac.;* Ind.-Ch.
„ „ : *Garcinia gaudichaudi* Pl. et Tr.; *Guttifer.;* Ind.-Ch.
„ „ : *Elaeocarpus madopetalus* Pierre; *Elaeocarpac.;* Ind.-Ch.
„ nan: *Phoebe lanceolata* Nees; *Laurac.;* Ind.-Ch.
„ nhum: *Dalbergia cochinchinensis* Pierre; *Papilionac.;* Ind.-Ch.
„ ôi (H): *Castanopsis tribuloïdes* A. DC.; *Fagac.;* Ind.-Ch.
„ vi: *Acronychia laurifolia* Blume; *Rutac.;* Ind.-Ch.
Caá-apoam: *Anona sp.; Anonac.;* Bras.
„ -ingá: *Pithecolobium sp.; Mimosac.;* Bras.
„ jussará: *Duroia saccifera* Hook. f.; *Rubiac.;* Bras.
„ -nambi: *Gleditschia amorphoides* Taub.; *Caesalpiniac.;* Arg.
„ -o-veti-guazú: *Luehea divaricata* Mart.; *Tiliac.;* Arg.
„ -tiguá: *Trichilia catigua* A. Juss.; *Meliac.;* Par.
Caabera: *Myrsine sp.; Myrsinac.;* tr. S.-Am.
Caalang: *Doryphora sassafras* Endl.; *Monimiac.;* N.-S.-W.
Cababa: *Khaya anthotheca* C. DC.; *Meliac.;* tr. Af.
Cabacalli: *Goupia glabra* Aubl.; *Celastrac.;* Gu.
Cabaceira: *Crescentia cujete* L.; *Bignoniac.;* Bras.
Caballitos: *Jacaranda sp.; Bignoniac.;* Kol.
Cabanholz, Afrikanisch: *Baphia nitida* Afzel.; *Papilionac.;* W.-Af.
Cabareiba: *Myrocarpus fastigiatus* Allem.; *Papilionac.;* Bras.
Cabbage: *Oreodoxa oleracea* Morris; *Palmac.;* br. Hond.
„ Bastard: *Andira inermis* H. B. K.; *Papilionac.;* Jam.
„ Mountain-: *Oreodoxa oleracea* Morris; *Palmac.;* br. Hond.
„ Palmetto-: ꞏ*Sabal palmetto* Lodd.; *Palmac.;* Fl.
Cabbage-bark: *Andira inermis* H. B. K.; *Papilionac.;* br. Hond.
„ „ Bastard: *Andira sp.; Papilionac.;* Trin.
„ „ Black: *Lonchocarpus rugosus* Bth.; *Papilionac.;* br. Hond.
„ „ Red: *Andira coriacea* Pulle; *Papilionac.;* Sur.
Cabbagetree: *Andira inermis* H. B. K.; *Papilionac.;* W.-Ind.
„ : *Chamaerops palmetto* Michx.; *Palmac.;* USA.
„ : *Sabal palmetto* Lodd.; *Palmac.;* tr. Am.
„ : *Heisteria coccinea* Jacq.; *Olacac.;* Hond.
„ : *Vernonia conferta* Bth.; *Composit.;* Lib.
„ Red: *Andira coriacea* Pulle; *Papilionac.;* Sur.
abela: *Xylopia aethiopica* A. Rich.; *Anonac.;* St.-Thomé.
Cabellito: *Aphelandra tetragona* Nees; *Acanthac.;* Kol.
Cabello de angel: *Myginda oxyphylla* Blake; *Celastrac.;* br. Hond.
Cabellos de Angel: *Albizzia lebbek* Bth.; *Mimosac.;* Cuba.
„ „ „ : *Ceiba pentandra* Gaertn.; *Bombacac.;* Mex.
„ „ „ : *Bombax ellipticum* H. B. K.; *Bombacac.;* s. Mex., Guat.
Cabende: *Celtis soyauxi* Engl.; *Ulmac.;* Angola.
Cabeti: *Luehea divaricata* Mart.; *Tiliac.;* Arg.
? Cabeza culebra: *Tanaecium jaroba* Sw.; *Bignoniac.;* Nic.
„ viejo: *Cereus sp.; Cactac.;* Mex.
„ de ilama: *Anona sp.; Anonac.;* Mex.
„ „ loro: *Eugenia sp.; Myrtac.;* Kol.
„ „ negra: *Anona sp.; Anonac.;* Mex.
„ „ negro: *Apeiba tibourbou* Aubl.; *Tiliac.;* Ven.
Cabiche: *Heisteria coccinea* Jacq.; *Olacac.;* Hond.
Cabil caspi: *Acacia cebil* Griseb.; *Mimosac.;* Arg.
Cabima: *Guarea trichilioides* L.; *Meliac.;* St.-D.
„ : *Copaifera spp.; Caesalpiniac.;* Ven.
Cabimba: *Copaifera spp.; Caesalpiniac.;* Ven.
Cabimbo: *Icica sp.; Burserac.;* Ven.
Cabimo: *Copaifera langsdorfi* Desf.; *Caesalpiniac.;* Ven.

Cabismo: *Copaifera chiriquensis* Pitt.; *Caesalpiniac.*; Pan.
Cablote: *Guazuma ulmifolia* Lamk.; *Sterculiac.*; Guat.
Cabiúna: *Dalbergia nigra* Allem.; *Papilionac.*; Bras.
Cabo de hacha: *Trichilia spondioïdes* Sw.; *Meliac.*; sp. Am.
„   „   „   : *Trichilia sp.*; *Meliac.*; P.-R.
„   „   „   : *Lonchocarpus lanceolatus* Bth.; *Papilionac.*; Mex.
„   „   lanza: *Bumelia obtusifolia* Roem. et Schult.; *Sapotac.*; Arg.
Caboré: *Myrocarpus frondosus* Allem.; *Papilionac.*; Bras.
Caboucalli: *Couepia guianensis* Aubl.; *Rosac.*; fr. Gu.
Cabra: *Schaefferia frutescens* Jacq.; *Celastrac.*; St.-D.
Cabrahigo: *Ficus carica* L.; *Morac.*; Sp.
Cabrahosca: *Zschokkea armata* Pittier; *Apocynac.*; Ven.
Cabreúba: *Myrocarpus frondosus* Allem.; *Papilionac.*; Arg.
„   : *Myrocarpus fastigiatus* Allem.; *Papilionac.*; Arg.
Cabreúva: *Myrocarpus frondosus* Allem.; *Papilionac.*; Bras. (Sao P)
„   : *Myroxylon pereirae* Klotzsch; *Papilionac.*; Bras.
„   : *Myroxylon toluiferum* H. B. K.; *Papilionac.*; Bras.
„   amarella: *Myrocarpus frondosus* Allem.; *Papilionac.*; Bras.
„   do campo: *Myrocarpus fastigiatus* Allem.; *Papilionac.*; Bras.
„   vermelho: *Myroxylon toluiferum* H. B. K.; *Papilionac.*; Bras. (Sao P)
Cabrewa: *Myrocarpus fastigiatus* Allem.; *Papilionac.*; Bras.
Cabrinova: *Myrocarpus frondosus* Allem.; *Papilionac.*; Bras.
Cabrito-negro: *Coutarea hexandra* Schum.; *Rubiac.*; Ven.
Cabritón: *Ruprechtia hamani* Blake; *Polygonac.*; Ven.
Cabriuba: *Myrocarpus frondosus* Allem.; *Papilionac.*; Arg.
Cabriuva: *Myrocarpus frondosus* Allem.; *Papilionac.*; Bras.
„   : *Myrocarpus fastigiatus* Allem.; *Papilionac.*; Bras.
Cabriwa: *Myrocarpus frondosus* Allem.; *Papilionac.*; S.-Am.
Cabui: *Enterolobium lutescens* Allem.; *Mimosac.*; Bras.
Cabuina: *Dalbergia miscolobium* Bth.; *Papilionac.*; Bras.
Cac heo (H): *Parkia streptocarpa* Hance; *Mimosac.*; Ind.-Ch.
„   si sièt: *Shorea cochinchinensis* Pierre; *Dipterocarpac.*; Ind.-Ch.
Caca de niño: *Licania sp.*; *Rosac.*; Mex.
Cacachien: *Hymenaea courbaril* L.; *Caesalpiniac.*; fr. Gu.
Cacagua: *Gliricidia sepium* Steud.; *Papilionac.*; Hond.
Cacaguillo: *Zizyphus melastomoides* Pitt.; *Rhamnac.*; Ven.
„   : *Sterculia carthagenensis* Cav.; *Sterculiac.*; Ven.
Cacagüito: *Sterculia carthagenensis* Cav.; *Sterculiac.*; Ven.
Cacahoananche: *Licania sp.*; *Rosac.*; Mex.
Cacahoanantzin: *Licania sp.*; *Rosac.*; Mex.
Cacahuanche: *Licania sp.*; *Rosac.*; Mex.
Cacahuate: *Licania sp.*; *Rosac.*; Mex.
Cacahuil: *Anacardium occidentale* L.; *Anacardiac.*; P.-R.
Cacaillo: *Ocotea sp., Nectandra sp.*; *Laurac.*; P.-R.
Cacaillo: *Sloanea sp.*; *Elaeocarpac.*; P.-R.
Cacaíto: *Sterculia sp.*; *Sterculiac.*; Ven.
Cacao: *Theobroma cacao* L.; *Sterculiac.*; tr. Am.
„   Bosch-: *Carolinea princeps* L. f.: *Bombacac.*; fr. Gu.
„   cimarrón: *Theobroma angustifolium* DC.; *Sterculiac.*; Pan.
„   grand bois: *Humiria floribunda* Mart.; *Humiriac.*; fr. Gu.
„   Madre-: *Lonchocarpus sp.*; *Papilionac.*; br. Hond.
„   „   : *Gliricidia sepium* Steud.; *Papilionac.*; C.-Am.
„   mani: *Theobroma purpureum* Pitt.; *Sterculiac.*; Pan.
„   motillo: *Sloanea berteriana* Choisy; *Elaeocarpac.*; P.-R.
„   otillo: ? *Sloanea berteriana* Choisy; *Elaeocarpac.*; P.-R.
„   roseto: *Sloanea sp.*; *Elaeocarpac.*; P.-R.
„   sauvage: *Humiria floribunda* Mart.; *Humiriac.*; fr. Gu.
„   Valsche: *Carolinea princeps* L. f.; *Bombacac.*; fr. Gu.
„   Wild: *Theobroma purpureum* Pitt.; *Sterculiac.*; Pan.
„   „   : *Humiria floribunda* Mart.; *Humiriac.*; fr. Gu.
„   del monte: *Theobroma bernouilli* Pitt.; *Sterculiac.*; Pan.
Cacaoeiro: *Theobroma cacao* L.; *Sterculiac.*; port. Af.

Cacaotier (f): *Theobroma cacao* L.; *Sterculiac.;* tr. Am.
Cacaoyer (f): *Theobroma cacao* L.; *Sterculiac.;* tr. Am.
„ Faux: *Carolinea princeps* L. f.; *Bombacac.;* fr. Gu.
„ à feuilles d'Orme: *Guazuma ulmifolia* Lamk.; *Sterculiac.;* fr. Gu., Ant.
Cacapoule: *Malpighia sp.; Malpighiac.;* Trin.
Cacatier: *Zanthoxylum sp.; Rutac.;* fr. Gu. (Galibis)
Cacáu: *Theobroma cacao* L.; *Sterculiac.;* Bras.
Cachaporra do gentio: *Terminalia sp.; Combretac.;* Bras.
Cacheta: *Tabebuia leucoxyla* DC.; *Bignoniac.;* Bras. (Sao P)
Cachibou: *Bursera gummifera* Jacq.; *Burserac.;* Gu.
Cachibu: *Bursera simaruba* Sarg.; *Burserac.;* Cuba.
Cachiman: *Anona reticulata* L.; *Anonac.;* Ant.
„   : *Anona squamosa* L.; *Anonac.;* W.-Ind.
„   : *Anona spp.; Anonac.;* fr. Gu.
„ épineux: *Anona muricata* L.; *Anonac.;* Guat.
„ morveux: *Anona mucosa* Jacq.; *Anonac.;* fr. Ant.
„ de montagne: *Talauma plumieri* Swartz; *Magnoliac.;* Ant.
Cachimbo: *Platymiscium sp.; Papilionac.;* Guat.
„   : *Erythrina sp.; Papilionac.;* Kol.
Cachimentier: *Anona sp.; Anonac.;* fr. Gu.
Cachito: *Posoqueria latifolia* Roem. et Schult.; *Rubiac.;* Hond.
„ de aromo: *Acacia farnesiana* Willd.; *Mimosac.;* Hond.
Cacho de toro: *Acacia sp.; Mimosac.;* Guat.
„ venado: *Mouriria parvifolia* Bth.; *Melastomatac.;* br. Hond.
Cacicuto: *Bixa orellana* L.; *Bixac.;* Cuba.
Cacique: *Brosimum paraënse* Huber; *Morac.;* Pan.
„   : *Diphysa carthagenensis* Jacq.; *Papilionac.;* Pan.
„ Bastard-: ? *Prunus annularis* Koehne; *Rosac.;* Pan.
„ Bloodwood-: *Brosimum conduru* Allem.; *Morac.;* Pan.
„   „   : *Brosimum caloxylon* Standl.; *Morac.;* Pan.
„ White: *Eugenia cricamolensis* Standl.; *Myrtac.;* Pan.
Caco: *Jacaranda sp.; Bignoniac.;* Kol.
Caconier: *Ormosia dasycarpa* Jacks.; *Papilionac.;* St.-D.
Cacoo, Bosch-: *Carolinea princeps* L. f.; *Bombacac.;* fr. Gu.
Cactus, Giant-: *Cereus giganteus* Engl.; *Cactac.;* s. USA.
„ Mission-: *Opuntia tuna* Mill.; *Cactac.;* S.-Am., W.-Ind.
Cadeno: *Albizzia longipedata* Britt. et Rose; *Mimosac.;* Guat.
Cadillo: *Triumfetta lappula* L.; *Tiliac.;* Kol.
Cadol: *Careya sphaerica* Roxb.; *Lecythidac.;* Kamb.
Caesar-wood: *Zanthoxylum sp.; Rutac.;* Jam.
Café: *Coffea arabica* L.; *Rubiac.;* Hond.
„ blanco: *Guarea sp.; Meliac.;* Bras.
„ cimarron: *Faramea odoratissima* DC.; *Rubiac.;* Cuba.
„   „   : *Sesbania marginata* Bth.; *Papilionac.;* Arg.
„ de monte: *Palicourea spp.; Rubiac.;* Ven.
Cafecillo: *Rinorea guatemalensis* Bartlett; *Violac.;* br. Hond.
„ de danta: *Faramea odoratissima* DC.; *Rubiac.;* Ven.
Cafecito: *Inga englesingi* Standl.; *Mimosac.;* Nic.
Cafeillo: *Casearia sp.; Flacourtiac.;* P.-R.
„ cimarrón: *Casearia sp.; Flacourtiac.;* P.-R.
Cafetillo: *Casearia sp.; Flacourtiac.;* P.-R.
Cafézinho: *Mouriria pseudo-geminata* Pitt.; *Melastomatac.;* Bras.
Cagalera: *Ximenia americana* L.; *Olacac.;* Hond.
„ comestible: *Celtis sp.; Ulmac.;* Nic.
Cagalero: *Randia armata* DC.; *Rubiac.;* Hond.
„ negro: *Pisonia sp.; Nyctaginac.;* Salv.
Caguairán: *Pseudocopaiva hymenaefolia* Britton et Wilson; *Caesalpiniac.;* Cuba.
Caguani: *Sideroxylon mastichodendron* Jacq.; *Sapotac.;* Cuba.
Cahualagua: *Heliocarpus americana* L.; *Tiliac.;* Mex.
Cahuite: *Pseudotsuga douglasi* Carr.; *Conifer.;* Mex.
Cai saug: *Horsfieldia amygdalina* Warb.; *Myristicac.;* Ind.-Ch.
Cail: *Khaya senegalensis* A. Juss.; *Meliac.;* Senl.

Caïl-cédra: *Khaya senegalensis* A. Juss.; *Meliac.;* W.-Af.
Cailcédra: *Cedrela odorata* L.; *Meliac.;* Cuba.
Caïl-cédrat: *Khaya senegalensis* A. Juss.; *Meliac.;* W.-Af.
Caimbahiba: *Curatella americana* L.; *Dilleniac.;* Bras.
Caïmitier: *Chrysophyllum cainito* L.; *Sapotac.;* fr. W.-Ind.
Caimitillo: *Chrysophyllum olivaeforme* Lamk.; *Sapotac.;* Cuba.
Caimito: *Chrysophyllum mexicanum* Brandeg.; *Sapotac.;* Salv.
   „   : *Chrysophyllum cainito* L.; *Sapotac.;* tr. Am.
   „   blanco: *Chrysophyllum sp.;* *Sapotac.;* Cuba.
   „   cimarrón: *Chrysophyllum argenteum* Jacq.; *Sapotac.;* C.-R.
   „   morado: *Chrysophyllum sp.;* *Sapotac.;* P.-R.
   „   verde: *Chrysophyllum sp.;* *Sapotac.;* P.-R.
   „   de montaña: *Rheedia edulis* Tr. et Pl.; *Guttifer.;* Hond.
   „   de perro: *Chrysophyllum sp.;* *Sapotac.;* P.-R.
Caimo: *Lucuma sp.;* *Sapotac.;* Kol.
Caindawood (eu): *Parkia africana* R. Br.; *Mimosac.;* tr. Af .
Cainga: *Aspidosperma sp.;* *Apocynac.;* Bras.
Caïnitier: *Chrysophyllum cainito* L.; *Sapotac.;* fr. Gu.
Caixeta: *Simaruba versicolor* St. Hil.; *Simarubac.;* Bras.
   „   : *Vochysia sp.;* *Vochysiac.;* Bras.
   „   branca: *Simaruba sp.;* *Simarubac.;* Bras.
Caixycahen: *Ilex ebenacea* Reiss.; *Aquifoliac.;* Bras.
Cajá comun: *Spondias sp.;* *Anacardiac.;* Bras.
   „   manga: *Spondias dulcis* Forst.; *Anacardiac.;* Bras.
   „   mirim: *Spondias lutea* L.; *Anacardiac.;* Bras.
   „   do sertão: *Spondias sp.;* *Anacardiac.;* Bras.
Cajarana: *Cabralea multijuga* C. DC.; *Meliac.;* Arg.
Cajaseiro: *Spondias dulcis* Forst.; *Anacardiac.;* ö. Bras.
Cajaty: *Cryptocarya mandioccana* Meissn.; *Laurac.;* Bras.
Cajazuero: *Spondias sp.;* *Anacardiac.;* Bras.
Cajeput tree: *Oreodaphne californica* Nees; *Laurac.;* USA.
   „   „   : *Melaleuca leucadendron* L.; *Myrtac.;* H.-Ind., Malay., Au.
Cajetillo: *Hirtella sp.;* *Rosac.;* Mex.
Cajeto: *Ochroma sp.;* *Bombacac.;* Guat.
Cajou senti: *Cedrela spp.;* *Meliac.;* Guad.
Cajú: *Anacardium occidentale* L.; *Anacardiac.;* Bras.
   „   -açú: *Anacardium spruceanum* Bth.; *Anacardiac.;* Bras. (Amaz.)
   „   -assú: *Anacardium spruceanum* Bth.; *Anacardiac.;* Bras. (Amaz.)
Caju Baroedan: *Flindersia amboinensis* Poir.; *Rutac.;* Amb.
   „   Malta Buta: *Excoecaria agallocha* L.; *Euphorbiac.;* S.-As., Au.
   „   Ticcos major: *Albizzia montana* Bth.; *Mimosac.;* Java, N.-Kal., Au.
Cajú do matto: *Anacardium giganteum* Hance; *Anacardiac.;* Bras. (Amaz.)
Cajuca: *Virola surinamensis* Warb.; *Myristicac.;* Trin., Tob.
Cajueiro: *Anacardium occidentale* L.; *Anacardiac.;* Bras.
   „   bravo: *Curatella americana* L.; *Dilleniac.;* Bras.
   „   do campo: *Anacardium occidentale* L.; *Anacardiac.;* Bras.
Cajuela: *Hieronymia alchorneoides* Allem.; *Euphorbiac.;* Cuba.
Cajuil: *Anacardium occidentale* L.; *Anacardiac.;* P.-R.
Cajurana: *Anacardium nanum* St. Hil.; *Anacardiac.;* Bras.
Calaba: *Calophyllum calaba* Jacq.; *Guttifer.;* fr. Ant.
   „   à fruits allongés (f): *Calophyllum calaba* Jacq.; *Guttifer.;* Ant.
   „   „   „   ronds (f): *Calophyllum inophyllum* L.; *Guttifer.;* Tr.
Calabash: *Crescentia cujete* L.; *Bignoniac.;* tr. Am.
   „   black: *Enallagma cucurbitina* Baill.; *Bignoniac.;* Hond., sö. USA.
   „   wild: *Enallagma latifolia* Small; *Bignoniac.;* Pan.
   „   „   : *Parmentiera macrophylla* Standl.; *Bignoniac.;* Pan.
   „   vine: ? *Drymonia spectabilis* Mart.; *Gesneriac.;* Pan.
Calabasse: *Crescentia cujete* L.; *Bignoniac.;* Guad.
   „   Schwarze: *Enallagma cucurbitina* Baill.; *Bignoniac.;* sö. USA.
Calabazo: *Crescentia cujete* L.; *Bignoniac.;* Kol.
Calabori: *Zanthoxylum sp.;* *Rutac.;* Ven. (Curaçao)
Calabure (f): *Muntingia calabura* L.; *Tiliac.;* tr. Am.

Calagua: *Heliocarpus sp.; Tiliac.;* Salv.
Calagual: *Heliocarpus sp.; Tiliac.;* Salv.
Calagüe: *Heliocarpus sp.; Tiliac.;* Salv.
Calamander: *Diospyros quaesita* Thw.; *Ebenac.;* Cey.
Calamansanay: *Nauclea sp.; Rubiac.;* Phil.
Calambreña: *Coccoloba sp.; Polygonac.;* P.-R.
Calambreñas: *Coccoloba sp.; Polygonac.;* P.-R.
Calambrujo: *Rosa tomentosa* Sm.; *Rosac.;* Sp.
Caland: *Mesua ferrea* L.; *Guttifer.;* Kamb.
Calantas: *Toona calantas* Merr. et Rolfe; *Meliac.;* Phil.
  ,, Bird's eye-: *Azadirachta integrifolia* Merr.; *Meliac.;* O.-As.
  ,, Curly: *Azadirachta integrifolia* Merr.; *Meliac.;* Phil.
Caldén: *Prosopis algarrobilla* Griseb.; *Mimosac.;* Arg.
Calebasse colin: *Couroupita guianensis* Aubl.; *Lecythidac.;* fr. Gu.
Calebassier: *Crescentia cujete* L.; *Bignoniac.;* fr. Gu.
Caliaturholz (d): *Pterocarpus santalinus* L. f.; *Papilionac.;* As.
Calicedra: *Cedrela odorata* L.; *Meliac.;* Mex.
Calico du Pape: *Tecoma leucoxylon* Mart.; *Bignoniac.;* Sey.
Calla de St. Domingue: *Sideroxylon sp.; Sapotac.;* St.-D.
Callcedrawood: *Flindersia australis* R. Br.; *Rutac.;* O.-Au.
Callhoon: *Elaeocarpus grandis* F. v. M.; *Elaeocarpac.;* Queensl.
Calumpit: *Terminalia edulis* Blco.; *Combretac.;* Phil.
Calshum: *Elaeocarpus grandis* F. v. M.; *Elaeocarpac.;* Queensl.
Cam (H): *Parinarium annamense* Hance; *Rosac.;* Annam.
  ,, dran: *Albizzia lebbekoïdes* Bth.; *Mimosac.;* Ind.-Ch.
  ,, lai (H): *Dalbergia spp.; Papilionac.;* Ind.-Ch.
  ,, ,, bông: *Dalbergia oliveri* Gamble; *Papilionac.;* Ind.-Ch.
  ,, lang (H): *Barringtonia longipes* Gagnep.; *Euphorbiac.;* Ind.-Ch.
  ,, liên (H): *Pentacme siamensis* Kurz; *Dipterocarpac.;* Ind.-Ch.
  ,, moro: *Acronychia laurifolia* Blume; *Rutac.;* Ind.-Ch.
  ,, ne: *Xylia dolabriformis* Bth.; *Mimosac.;* Burma.
  ,, non: *Citrus limonum* Risso; *Rutac.;* Annam.
  ,, nuóc: *Glycosmis cochinchinensis* Pierre; *Rutac.;* Ind.-Ch.
  ,, thi (H): *Diospyros siamensis* Hochr.; *Ebenac.;* Ind.-Ch.
  ,, ,, : *Diospyros rubra* H. Lec.; *Ebenac.;* Ind.-Ch.
  ,, tien: *Citrus aurantium* L.; *Rutac.;* Annam.
  ,, tóng huong: *Ailanthus fauveliana* Pierre; *Simarubac.;* Ind.-Ch.
Cama tala: *Prosopis sp.; Mimosac.;* Arg.
  ,, tale: *Prosopis panta* Hieron.; *Mimosac.;* tr. & subtr. Am.
Camaca: *Ardisia compressa* H. B. K.; *Myrsinac.;* Hond.
Camachile: *Pithecolobium dulce* Bth.; *Mimosac.;* Guam. (Phil.)
Camaco: *Parathesis serrulata* Mez; *Myrsinac.;* Guat.
Camagon: *Diospyros discolor* Willd.; *Ebenac.;* Phil.
Camagoon: *Diospyros pilosanthera* Blco.; *Ebenac.;* Phil.
Camagua: *Wallenia laurifolia* Sw.; *Myrsinac.;* Cuba.
Camajonduro: *Sterculia sp.; Sterculiac.;* Kol.
Camajurú: *Sterculia sp.; Sterculiac.;* Kol.
Camará de bilro: *Platycyamus regnelli* Bth.; *Papilionac.;* Bras.
Camari: *Dipteryx odorata* Willd.; *Papilionac.;* n. S.-Am.
Camarón: *Calycophyllum candidissimum* DC.; *Rubiac.;* Mex.
Camaroncillo: *Hirtella triandra* Swartz; *Rosac.;* Pan.
Camasey: *Miconia sp.; Melastomatac.;* tr. S.-Am.
Camay-cuy: *Quercus sp.; Fagac.;* Mex.
Camba-acá: *Guazuma ulmifolia* Lamk.; *Sterculiac.;* Arg.
Cambá-nambi: *Gleditschia amorphoides* Taub.; *Caesalpiniac.;* S.-Am.
Cambaiba: *Curatella americana* L.; *Dilleniac.;* Bras.
Cambalholz (d): *Baphia nitida* Lodd.; *Papilionac.;* W.-Af.
Cambará: *Tecomía pentaphylla* Juss.; *Bignoniac.;* Bras. (Sao P)
  ,, : *Zanthoxylum erytropappa* ?; *Rutac.;* Bras.
  ,, : *Moquinia polymorpha* DC.: *Composit.;* Arg.
  ,, : *Vernonia crotonoides* Schultz-Bip.; *Composit.;* S.-Am.
Cambay: *Sesbania marginata* Bth.; *Papilionac.;* Arg.

Cambeza de negro: *Guazuma ulmifolia* Lamk.; *Sterculiac.*; Arg.
Camboata: *Guarea excelsa* H. B. K.; *Meliac.*; Bras.
„ : *Guarea tuberculata* Vell.; *Meliac.*; Bras. (Paraná)
„ : *Guarea trichilioides* L.; *Meliac.*; Arg.
„ : *Talisia esculenta* Radlk.; *Sapindac.*; Arg.
„ : *Cupania uruguensis* Hook. et Arn.; *Sapindac.*; Arg.
„ blanca: *Guarea trichilioides* L.; *Meliac.*; Arg.
Cambogala: *Aucoumea klaineana* Pierre; *Burserac.*; W.-Af.
Cambrará: *Moquinia polymorpha* DC.; *Composit.*; Bras.
Cambrión: *Genista barnadesi* Graells; *Papilionac.*; Sp.
Cambui: *Eugenia vulgaris* Baill.; *Myrtac.*; tr. As.
Cambulo: *Erythrina sp.*; *Papilionac.*; Kol.
Camdeboo: *Celtis rhamnifolia* Presl; *Ulmac.*; S.-Af.
„ : *Celtis kraussiana* Bernh.; *Ulmac.*; S.-Af.
Camdeboom: *Rhamnus celtifolia* Thunb.; *Rhamnac.*; S.-Af.
Camellia: *Thea japonica* Nois.; *Theac.*; Jap.
Camerunga: *Averrhoa carambola* L.; *Oxalidac.*; Ang.
Camesito: *Rapanea sp.*; *Myrsinac.*; Peru.
Camfine: *Trichilia tuberculata* C. DC.; *Meliac.*; Pan.
Camiba: *Copaifera sp.*; *Caesalpiniac.*; Pan.
? Camibar: *Copaifera sp.*; *Caesalpiniac.*; Pan.
Camichín: *Ficus sp.*; *Morac.*; Salv.
? Camimba: *Copaifera sp.*; *Caesalpiniac.*; Pan.
? Camimbar: *Copaifera sp.*; *Caesalpiniac.*; Pan.
Camoruco: *Sterculia apetala* Karst.; *Sterculiac.*; Ven.
Campanilla: *Coutarea spp.*; *Rubiac.*; Ven.
Campaño: *Samanea saman* Merr.; *Mimosac.*; Kol.
Campeachy wood: *Haematoxylon campechianum* L.; *Caesalpiniac.*; tr. Am.
Campeche: *Haematoxylon campechianum* L.; *Caesalpiniac.*; tr. Am.
Campéche: *Leucaena glauca* Bth.; *Mimosac.*; P.-R.
Campeche: *Calderonia salvadorensis* Standl.; *Rubiac.*; Salv., Guat.
Campechy wood: *Haematoxylon campechianum* L.; *Caesalpiniac.*; tr. Am.
Campeggio (i): *Haematoxylon campechianum* L.; *Caesalpiniac.*; tr. Am.
Campeschenholz (d): *Haematoxylon campechianum* L.; *Caesalpiniac.*; tr. Am.
Camphorwood: *Cinnamomum camphora* Nees et Eberm.; *Laurac.*; O.-As.
„ : *Cinnamomum kanechirai* Hayata; *Laurac.*; Form.
„ : *Protium copal* Engl.; *Burserac.*; br. Hond.
„ : *Laurus sp.*; *Laurac.*; Hond.
„ : *Cinnamomum oliveri* F. M. B.; *Laurac.*; Au.
„ Borneo-: *Dryobalanops aromatica* Gaertn. f.; *Dipterocarpac.*; Malay.
„ Eastafrican-: *Ocotea usambarensis* Engl.; *Laurac.*; O.-Af.
„ Nepal-: *Cinnamomum glanduliferum* Meissn.; *Laurac.*; Ind.
Camphre: *Cinnamomum camphora* Nees et Eberm.; *Laurac.*; O.-As.
Camphrier: *Cinnamomum camphora* Nees et Eberm.; *Laurac.*; O.-As.
„ Faux: *Melia azedarach* L.; *Meliac.*; Co.-Ch.
„ „ : *Hernandia voyroni* H. Jum.; *Laurac.*; Mad.
„ de Bornéo: *Dryobalanops aromatica* Gaertn. f.; *Dipterocarpac.*; Malay.
Camwood: *Baphia nitida* Lodd.; *Papilionac.*; W.-Af.
„ : *Baphia pubescens* Hook. f.; *Papilionac.*; Go.
Camxé: *Xylia dolabriformis* Bth.; *Mimosac.*; Ind.-Ch.
Can dai: *Randia oxyodonta* Drake; *Rubiac.*; Ind.-Ch.
„ luong: *Adina sessilifolia* Hook.; *Rubiac.*; Ind.-Ch.
„ than: *Feronia lucida* Teijsm. et Binn.; *Rutac.*; Ind.-Ch.
„ thang (H): *Feronia lucida* Teijsm. et Binn.; *Rutac.*; Ind.-Ch.
„ „ „ : *Feronia elephantum* Correa; *Rutac.*; Ind.-Ch.
„ thau: *Feronia elephantum* Correa; *Rutac.*; Co.-Ch.
„ xú: *Citrus aurantium* L.; *Rutac.*; Annam.
Caña-boho: *Schizostachyum lumampao* Merr.; *Gramin.*; Phil.
„ -bojo: *Schizostachyum lumampao* Merr.; *Gramin.*; Phil.
„ caijino: *Bactris sp.*; *Palmac.*; Pan.
„ danta: *Geonoma sp.*; *Palmac.*; Nic.
„ dulce: *Licania sp.*; *Rosac.*; Mex.

Caña-espina: *Bambusa blumeana* Schultes f.; *Gramin.*; Phil.
„ -fistola: *Cassia brasiliana* Lam.; *Caesalpiniac.*; Ven.
„ „ Marimari: *Cassia bonplandiana* DC.; *Caesalpiniac.*; Ven.
„ „ du bemana santa: ? *Cassia brasiliana* Lam.; *Caesalpiniac.*; Ven.
„ -fistula: *Peltophorum vogelianum* Walp.; *Caesalpiniac.*; P.-R.
„ „ : *Cassia grandis* L.; *Caesalpiniac.*; Pan.
„ „ : *Cassia fistula* L.; *Caesalpiniac.*; br. Hond.
„ „ : *Peltophorum vogelianum* Walp.; *Caesalpiniac.*; Arg.
„ „ : *Cassia javanica* L.; *Caesalpiniac.*; Phil.
„ molinillo: *Geonoma pinnatifrons* Willd.; *Palmac.*; Ven.
Cañaguate: *Tecoma spectabilis* Pl. et Linden; *Bignoniac.*; Kol.
Canalete: *Cordia sebestena* L.; *Borraginac.*; Kol., Ven.
Canalón guazú: *Rapanea sp.*; *Myrsinac.*; Arg.
Canang: *Cananga odorata* Hook. f. et Thoms.; *Anonac.*; Mart.
Canari macaque: *Lecythis grandiflora* Aubl.; *Lecythidac.*; fr. Gu.
„ vulgaire (f): *Canarium commune* L.; *Burserac.*; sö. As.
Canarywood: *Liriodendron tulipifera* L.; *Magnoliac.*; ö. N.-Am.
„ : *Centrolobium robustum* Mart.; *Papilionac.*; Bras.
„ : *Sarcocephalus cordatus* Miq.; *Rubiac.*; tr. As., tr. Au.
Canayan: *Bambusa blumeana* Schultes f.; *Gramin.*; Phil.
Cancertree: *Jacaranda coerulea* Griseb.; *Bignoniac.*; Bah.
Cancharana: *Cabralea cangerana* Sald.; *Meliac.*; Arg.
Canchorena: *Cabralea sp.*; *Meliac.*; Arg.
Canchorono: *Cabralea sp.*; *Meliac.*; Arg.
Cancorosa: *Maytenus boaria* Mol.; *Celastrac.*; Arg. (Patag.)
Candeboo: *Celtis rhamnifolia* Presl; *Ulmac.*; Kap.
Candeia: *Lychnophora ericoides* Mart.; *Composit.*; Bras.
„ : *Piptocarpha rotundifolia* Baker; *Composit.*; Bras.
Candela: *Antirrhoea trichantha* Hemsl.; *Rubiac.*; Pan.
„ de breyo: *Machaerium brasiliense* Vog.; *Papilionac.*; Arg.
Candelero: *Cordia sp.*; *Borraginac.*; Mex.
Candelilla: *Euphorbia antisyphilitica* Zucc.; *Euphorbiac.*; Mex.
„ : *Pedilanthus tithymaloides* Poit.; *Euphorbiac.*; Mex.
„ : *Pedilanthus pavonis* Boiss.; *Euphorbiac.*; Mex.
Candelillo: *Cassia spectabilis* DC.; *Caesalpiniac.*; Hond.
„ : *Rondeletia deami* Standl.; *Rubiac.*; Hond.
Candelo: *Gyranthera caribensis* Pitt.; *Bombacac.*; Ven.
Candelón: *Rhizophora mangle* L.; *Rhizophorac.*; Mex.
Candil: *Amyris sp.*; *Rutac.*; Ven.
Candil de montaña: *Amyris sp.*; *Rutac.*; Ven.
„ „ playa: *Amyris simplicifolia* Karst.; *Rutac.*; Ven.
Candilera: *Phlomis lychnitis* L.; *Labiat.*; Sp.
Candle berry: *Byrsonima sp.*; *Malpighiac.*; br. W.-Ind.
Candlenut: *Aleurites triloba* Forst., *A. moluccana* Willd.; *Euphorbiac.*; Malay.
Candletree: *Myrica inodora* Bartr.; *Myricac.*; USA.
Candlewood: *Amyris balsamifera* L.; *Rutac.*; tr. Am.
„ : *Toulicia guianensis* Aubl.; *Sapindac.*; br. Gu.
„ : *Dracaena americana* Donn. Sm.; *Liliac.*; br. Hond.
„ black: *Ocotea sp., Nectandra sp.*; *Laurac.*; Jam.
„ false: *Esenbeckia pentaphylla* Griseb.; *Rutac.*; Guat.
„ red: *Sideroxylon sp.*; *Sapotac.*; br. Hond.
Cane, Tsinglee-: *Bambusa sp.*; *Gramin.*; Ind.?
Canela: *Ocotea sp., Nectandra spp.*; *Laurac.*; Bras.
„ : *Persea sp.*; *Laurac.*; P.-R.
„ blanca: *Canella winterana* Gaertn.; *Canellac.*; Cuba.
„ -guaicá: *Ocotea puberula* Nees; *Laurac.*; Arg.
„ -guaiká: *Nectandra tweediei* Mez; *Laurac.*; Arg. (Mis.)
„ moena: *Aniba canelilla* Mez; *Laurac.*; Peru.
„ muena: *Aniba canelilla* Mez; *Laurac.*; Peru.
„ preta: *Phoebe porphyria* Mez; *Laurac.*; Arg.
„ de benado: *Myrsine floribunda* R. Br.; *Myrsinac.*; Arg.
Canelillo: *Ocotea veraguensis* Mez; *Laurac.*; C.-R.

Canelito: *Alchornea latifolia* Swartz; *Euphorbiac.;* Hond.
Canella: *Nectandra sp.; Laurac.;* Bras.
„ açafrão: *Ocotea sp., Nectandra sp.; Laurac.;* S.-Bras.
„ alba: *Nectandra spp., Ocotea spp.; Laurac.;* Bras.
„ amarella: *Nectandra nitidula* Nees et Mart.; *Laurac.;* Bras.
„ amarga: *Drimys winteri* Forst.; *Magnoliac.;* Bras.
„ bastarda: *Ocotea sp., Nectandra sp.; Laurac.;* S.-Bras.
„ batalha: *Nectandra robusta* Loef., *N. rigida* Nees; *Laurac.;* Bras.
„ blanca: *Canella winterana* Gaertn.; *Canellac.;* tr. Am.
„ branca: *Endlicheria hirsuta* Nees; *Laurac.;* Bras.
„ „ : *Nectandra leucothyrsus* Meissn.; *Laurac.;* Bras.
„ burra: *Nectandra myriantha* Meissn.; *Laurac.;* S.-Bras.
„ capitão-mór: *Ocotea sp., Nectandra sp.; Laurac.;* S.-Bras.
„ cedro: *Ocotea sp., Nectandra sp.; Laurac.;* S.-Bras.
„ foreta: *Nectandra mollis* Nees; *Laurac.;* Bras.
„ imbuia: *Phoebe porosa* Mez; *Laurac.;* Bras.
„ „ clara: *Phoebe porosa* Mez; *Laurac.;* Bras.
„ „ escura: *Phoebe porosa* Mez; *Laurac.;* Bras.
„ legitima: *Ocotea sp., Nectandra sp.; Laurac.;* S.-Bras.
„ limão: *Ocotea sp., Nectandra sp.; Laurac.;* S.-Bras.
„ louro: *Ocotea blancheti* Mez; *Laurac.;* S.-Bras.
„ mandóca: *Ocotea sp., Nectandra sp.; Laurac.;* S.-Bras.
„ parda: *Ocotea sp., Nectandra sp.; Laurac.;* S.-Bras.
„ pimenta: *Ocotea sp., Nectandra sp.; Laurac.;* S.-Bras.
„ prego: *Ocotea sp., Nectandra sp.; Laurac.;* S.-Bras.
„ preta: *Nectandra mollis* Nees; *Laurac.;* Bras. (Sao P)
„ „ : *Ocotea spectabilis* Mez; *Laurac.;* Bras. (Sao P)
„ puante: *Ocotea sp., Nectandra sp.; Laurac.;* S.-Bras.
„ rapaduro: *Licania utilis* Fritsch; *Rosac.;* Bras.
„ ruiva: *Ocotea sp., Nectandra sp.; Laurac.;* S.-Bras.
„ Samambaia: *Sickingia sampaioana* Standl.; *Rubiac.;* Bras.
„ santa: *Ocotea sp., Nectandra sp.; Laurac.;* S.-Bras.
„ sassafraz: *Ocotea pretiosa* Bth. et Hook.; *Laurac.;* Bras.
„ „ amarella: *Ocotea sp., Nectandra sp.; Laurac.;* S.-Bras.
„ „ preta: *Ocotea sp., Nectandra sp.; Laurac.;* S.-Bras.
„ sebo: *Ocotea sp., Nectandra sp.; Laurac.;* S.-Bras.
„ seibo: *Nectandra rigida* Nees; *Laurac.;* S.-Bras.
„ vermelha: *Ocotea sp., Nectandra sp.; Laurac.;* S.-Bras.
„ de cheiro: *Mespilodaphne opifera* Meissn.; *Laurac.;* Bras.
„ „ cotia: *Esenbeckia grandiflora* Mart.; *Rutac.;* Bras.
„ „ paramo: *Drimys winteri* Forst.; *Magnoliac.;* Bras.
„ „ Sebo: *Nectandra angustifolia* Nees et Mart.; *Laurac.;* Bras.
„ do brejo: *Talauma sp.; Magnoliac.;* Bras.
„ „ venado: *Helietta cuspidata* Chod. et Hassl.; *Rutac.;* Arg.
„ „ viado: *Helietta cuspidata* Chod. et Hassl.; *Rutac.;* Arg.
Canellão: *Nectandra myriantha* Meissn.; *Laurac.;* S.-Bras.
Canellilo: *Trichilia sp.; Meliac.;* Salv.
Canellinha: *Nectandra tweediei* Mez; *Laurac.;* S.-Bras.
„ : *Ocotea pulchella* Mart., *O. pretiosa* Bth. et Hook.; *Laurac.;* S.-Bras.
„ : *Nectandra linhearia* Meissn.; *Laurac.;* S.-Bras.
„ embuia: *Ocotea sp., Nectandra sp.; Laurac.;* S.-Bras.
„ marçanahyba: *Ocotea sp., Nectandra sp.; Laurac.;* S.-Bras.
„ puante: *Ocotea sp., Nectandra sp.; Laurac.;* S.-Bras.
Canelo: *Drimys winteri* Forst.; *Magnoliac.;* tr. Am.
„ de páramo: *Drimys winteri* Forst.; *Magnoliac.;* Kol.
Canelón: *Ocotea wrighti* Mez; *Laurac.;* P.-R.
„ : *Rapanea laetevirens* Mez; *Myrsinac.;* Arg.
„ blanco: *Myrsine sp.; Myrsinac.;* Arg.
„ capororoca: *Rapanea sp.; Myrsinac.;* Arg.
Cáng: *Albizzia stipulata* Boivin; *Mimosac.;* Ind.-Ch.
„ giáo: *Adina cordifolia* Hook. f.; *Rubiac.;* Ind.-Ch.
„ hom thom: *Ailanthus fauveliana* Pierre; *Simarubac.;* Ind.-Ch.

Cáng son: *Canthium parvifolium* Roxb.; *Rubiac.*; Ind.-Ch.
„ trau: *Randia dumetorum* Lamk.; *Rubiac.;* Ind.-Ch.
Cangerana: *Cabralea cangerana* Sald.; *Meliac.;* Bras. (Sao P)
„ -mirim: *Cabralea sp.; Meliac.;* Bras.
Cangorosa: *Maytenus sp.; Celastrac.;* S.-Am.
Cangica: *Piratinera guianensis* Aubl.; *Morac.;* br. Gu.
Canida: *Licania microcarpa* Hook. f.; *Rosac.;* Bras.
„ de benado: *Myrsine floribunda* R. Br.; *Myrsinac.;* Arg.
„ „ mula: *Triplaris sp.; Polygonac.;* Salv.
„ „ „ : *Licania sp.; Rosac.;* Salv.
„ „ venado: *Bredemeyera floribunda* Willd.; *Polygalac.;* Ven.
Canillo de venado: *Rondeletia purdiei* Hook. f.; *Rubiac.;* Ven.
Canique gris: *Caesalpinia crista* L.; *Caesalpiniac.;* Guad.
Canime: *Copaifera officinalis* Willd.; *Caesalpiniac.;* Kol.
Canistel: *Lucuma nervosa* Griseb.; *Sapotac.;* Cuba.
Caniva: *Copaifera sp.; Caesalpiniac.;* Pan.
Canjarana: *Cabralea sp.; Meliac.;* Bras.
Canjerana: *Cabralea cangerana* Sald.; *Meliac.;* Bras.
Canjuro: *Trichilia sp.; Meliac.;* Salv.
Cannariboom: *Hymenaea courbaril* L.; *Caesalpiniac.;* tr. Am.
Cannelle: *Cinnamomum zeylanicum* Nees; *Laurac.;* Ind.-Ch.
„ : *Cinnamomum obtusifolium* Nees; *Laurac.;* Ind.-Ch.
Cannellier royal: *Cinnamomum cassia* Blume; *Laurac.;* Ind.-Ch.
Cannes marronnes: *Cordyline spp.; Liliac.;* Réunion.
Cannon-balltree: *Xylocarpus granatum* Koen.; *Meliac.;* Ind.
„ „ : *Couroupita parviflora* Standl.; *Lecythidac.;* Pan.
„ „ : *Couroupita cutteri* Morton Skutch; *Lecythidac.;* Pan.
Cañoeto: *Platypodium maxonianum* Pitt.; *Papilionac.;* Kol.
Canomai: *Diospyros multiflora* Blco.; *Ebenac.;* Phil.
Canomoi: *Diospyros multiflora* Blco.; *Ebenac.;* Phil.
Cañotillo: *Piper smilacifolium* H. B. K.; *Piperac.;* Pan.
Canudo: *Carpotroche brasiliensis* Endl.; *Flacourtiac.;* S.-Bras.
Canxarana: *Cabralea cangerana* Sald.; *Meliac.;* Arg.
Cao: *Aleurites cordata* Steud.; *Euphorbiac.;* Ind.-Ch.
„ ly vong: *Murraya exotica* L.; *Rutac.;* Ind.-Ch.
„ ly yong: *Murraya exotica* L.; *Rutac.;* Ind.-Ch.
„ sanh: *Feronia lucida* Teijsm. et Binn.; *Rutac.;* Ind.-Ch.
„ xio: *Cryptocarya guianensis* Meissn.; *Laurac.;* Bras.
Caoba (s): *Swietenia spp.; Meliac.;* W.-Ind.
„ : *Swietenia mahagoni* Jacq.; *Meliac.;* Cuba.
„ : *Swietenia macrophylla* King; *Meliac.;* C.-Am.
„ : *Swietenia humilis* Zucc.; *Meliac.;* Salv.
„ : *Swietenia candollei* Pitt.; *Meliac.;* Ven.
„ blanca: *Swietenia sp.; Meliac.;* Pan.
„ falsa: *Bauhinia candicans* Bth.; *Caesalpiniac.;* Arg.
„ monde: *Swietenia sp.; Meliac.;* Kol.
„ roja: *Swietenia macrophylla* King; *Meliac.;* Pan.
„ , Venadillo-: *Swietenia cirrhata* Blake; *Meliac.;* Mex.
Caóbano: *Guarea spp.; Meliac.;* Ven.
Caobillo: *Tapirira mexicana* March.; *Anacardiac.;* Mex.
Caomao: *Wallenia laurifolia* Sw.; *Myrsinac.;* Cuba.
Caona: *Agonandra excelsa* Griseb.; *Olacac.;* Arg.
Caopiá: *Vismia sp.; Guttifer.;* Bras.
„ de Pison: *Vismia guianensis* Pers.; *Guttifer.;* Gu.
„ „ Marcgraff: *Vismia guianensis* Pers.; *Guttifer.;* Gu.
Caouroubara: *Couratari guianensis* Aubl.; *Lecythidac.;* fr. Gu.
Caoutchouc-boom: *Hevea guianensis* Aubl.; *Euphorbiac.;* fr. Gu.
Capá: *Cordia gerascanthus* L.; *Borraginac.;* P.-R.
„ : *Petitia domingensis* Jacq.; *Verbenac.;* P.-R.
„ amarillo: *Petitia domingensis* Jacq.; *Verbenac.;* P.-R.
„ de sábana: *Petitia domingensis* Jacq.; *Verbenac.;* P.-R., St.-D.
„ blanco: *Petitia domingensis* Jacq.; *Verbenac.,;* P.-R.

Capá blanco: *Cordia alliodora* Cham.; *Borraginac.*; P.-R.
„ sabanero: *Petitia domingensis* Jacq.; *Verbenac.*; P.-R.
„ de sábana: *Petitia domingensis* Jacq.; *Verbenac.*; P.-R.
? Caparo: *Cordia sp.*; *Borraginac.*; Pan.
Capay-yè-wood: *Vochysia guianensis* Aubl.; *Vochysiac.*; Gu.
Cap-berry: *Ocotea sp.*, *Nectandra sp.*; *Laurac.*; Jam.
Cape: *Clusia sp.*; *Guttifer.*; Kol.
„ grande: *Clusia rosea* Jacq.; *Guttifer.*; Kol.
Capecillo oloroso: *Clusia sp.*; *Guttifer.*; Kol.
Caper, Florida-: *Capparis sp.*; *Capparidac.*; Fl.
Capertree: *Capparis sp.*; *Capparidac.*; Fl.
Capire: *Sideroxylon capiri* Pitt.; *Sapotac.*; Mex.
Capiri: *Sideroxylon capiri* Pitt.; *Sapotac.*; Mex.
Capirona: *Calycophyllum spruceanum* Bth.; *Rubiac.*; Peru.
Capirote: *Bellucia costaricensis* Cogn.; *Melastomatac.*; Nic.
Capiroto: *Conostegia xalapensis* D. Don; *Melastomatac.*; Hond.
Capoeira: *Jacaranda sp.*; *Bignoniac.*; Bras. (Sao P.)
Capokier: *Ceiba pentandra* Gaertn.; *Bombacac.*; W.-Af.
Capolín: *Prunus capuli* Cav.; *Rosac.*; C.-Am., Ek., Peru.
Capomo: *Brosimum sp.*; *Morac.*; Mex.
Capororoca: *Myrsine sp.*; *Myrsinac.*; S.-Am.
„ -assú: *Rapanea ferruginea* Mez; *Myrsinac.*; Bras.
„ colorada: *Rapanea sp.*; *Myrsinac.*; Arg.
Capote: *Sterculia pruriens* Schum.; *Sterculiac.*; Bras.
Capour Barros: *Dryobalanops aromatica* Gaertn.; *Dipterocarpac.*; Bor.
Câprier commun (f): *Capparis spinosa* L.; *Capparidac.*; M.-M.
„ épineux (f): *Capparis spinosa* L.; *Capparidac.*; M.-M.
„ ferrugineux (f): *Capparis ferruginea* L.; *Capparidac.*; Ant.
Capriwa: *Myrocarpus frondosus* Allem.; *Papilionac.*; Bras.
Capuallcacao: *Spondias sp.*; *Anacardiac.*; Mex.
Capueiro: *Cordia sp.*; *Borraginac.*; Bras.
Capul: *Celtis sp.*; *Ulmac.*; Mex.
Capulí: *Prunus capuli* Cav.; *Rosac.*; C.-Am., Ek., Peru.
Capulamate: *Ficus padifolia* H. B. K.; *Morac.*; Guat.
Capulin: *Muntingia calabura* L.; *Tiliac.*; C.-Am.
„ : *Prunus capuli* Cav.; *Rosac.*; Mex.
„ : *Belotia campbelli* Sprague; *Tiliac.*; Hond.
„ : *Trema micrantha* Blume; *Ulmac.*; Hond.
„ amate: *Ficus sp.*; *Morac.*; Guat.
„ grande: *Ficus sp.*; *Morac.*; Mex.
„ macho: *Trema micrantha* Blume; *Ulmac.*; Pan.
„ montes: *Trema micrantha* Blume; *Ulmac.*; Salv.
„ negro: *Trema micrantha* Blume; *Ulmac.*; Hond.
„ savanero: *Belotia reticulata* Sprague; *Tiliac.*; Nic.
„ de corona: *Calliandra winzerlingi* Standl.; *Mimosac.*; br. Hond.
Capulincillo: *Casearia sp.*; *Flacourtiac.*; Mex.
„ : *Trema micrantha* Blume; *Ulmac.*; Salv.
Capur Barros: *Dryobalanops aromatica* Gaertn.; *Dipterocarpac.*; Bor.
Caput mortui: *Couroupita peruviana* Miers; *Lecythidac.*; Peru.
Caraba: *Carapa spp.*; *Meliac.*; br. Gu.
Carabeen, Yellow: *Sloanea woollsi* F. v. M.; *Elaeocarpac.*; N.-S.-W.
Caracoli: *Anacardium rhinocarpus* DC.; *Anacardiac.*; Ven., Kol.
Caracollilo: *Casearia sp.*; *Flacourtiac.*; P.-R.
„ : *Trichilia sp.*; *Meliac.*; P.-R.
Caracollillo: *Casearia sp.*; *Flacourtiac.*; P.-R.
Caragérou: *Bignonia tinctoria* Arruda; *Bignoniac.*; fr. Gu.
Caragne blanche: *Icica sp.*; *Burserac.*; fr. Gu.
Carahiba: *Simaruba versicolor* St. Hil.; *Simarubac.*; Bras.
Carahyba: *Cordia calocephala* Cham.; *Borraginac.*; Bras.
Caraipé: *Licania microcarpa* Hook. f.; *Rosac.*; Bras.
„ : *Moquilia guianensis* Aubl.; *Rosac.*; Bras.
Caraipé rana: *Parinarium barbatum* Ducke; *Rosac.*; Bras.

Caraipé verdadeiro: *Licania sclerophylla* Mart., *L. utilis* Fritsch; *Rosac.;* Bras.
„ das aguas: *Licania turiuva* Cham. et Schlecht.; *Rosac.;* Bras.
Caraipérana: *Licania micranthea* Miq.; *Rosac.;* Bras.
Caraiperana: *Licania turiuva* Cham. et Schlecht.; *Rosac.;* Bras.
Carallia: *Carallia integerrima* DC.; *Rhizophorac.;* Ind.
Caramayo: *Caesalpinia recordi* Britt. et Rose; *Caesalpiniac.;* br. Hond.
Caramba: *Averrhoa carambola* L.; *Oxalidac.;* Bras.
Carambola: *Averrhoa carambola* L.; *Oxalidac.;* Mal.
Carambold (u): *Averrhoa carambola* L.; *Oxalidac.;* Bras.
Carambole: *Averrhoa carambola* L.; *Oxalidac.;* Guad.
Caramboleira: *Averrhoa carambola* L.; *Oxalidac.;* Bras.
Carambolero (s): *Averrhoa carambola* L.; *Oxalidac.;* Bras.
Carambolier (f): *Averrhoa carambola* L.; *Oxalidac.;* Bras.
„ Blimbing (f): *Averrhoa bilimbi* L.; *Oxalidac.;* O.-Ind., Bras., Ant.
„ vrai (f): *Averrhoa carambola* L.; *Oxalidac.;* tr. As.
Caramura: *Humiria sp.; Humiriac.;* fr. Gu.
Carana: *Mauritia carana* Wallace; *Palmac.;* Bras.
Caraña: *Protium carana* March.; *Burserac.;* Ven.
Carana branca: *Protium altissimum* March.; *Burserac.;* Bras.
„ do Rio Negro: *Bactris cuspidata* Mart.; *Palmac.;* Bras.
Caranahy: *Mauritia limnophila* Barb. Rodr.; *Palmac.;* Bras.
„ do matto: *Lepidocaryum tenue* Mart.; *Palmac.;* Bras.
Caranda: *Copernicia australis* Becc.; *Palmac.;* Bras.
Carandá: *Copernicia cerifera* Mart.; *Palmac.;* Par.
Carandahy: *Trithrinax brasiliensis* Mart.; *Palmac.;* Bras.
Caranguda: *Caesalpinia acinaciformis* Mart.; *Caesalpiniac.;* Bras.
Caraño: *Bursera simaruba* Sarg.; *Burserac.;* tr. Am.
„ : *Protium asperum* Standl.; *Burserac.;* C.-Am.
„ : *Casearia nitida* Jacq.; *Flacourtiac.;* Pan.
Carano: *Bursera gummifera* Jacq.; *Burserac.;* Trin.
Carao: *Cassia grandis* L. f.; *Caesalpiniac.;* C.-Am.
Carap: *Carapa guianensis* Aubl.; *Meliac.;* Trin.
Carapa: *Carapa guianensis* Aubl.; *Meliac.;* tr. Am.
„ blanc: *Carapa guianensis* Aubl.; *Meliac.;* fr. Gu.
„ jaune: *Carapa guianensis* Aubl.; *Meliac.;* fr. Gu.
„ Roode: *Carapa guianensis* Aubl.; *Meliac.;* Gu.
„ rouge: *Carapa procera* DC., *C. surinamensis* Miq.; *Meliac.;* Sur.
„ „ : *Carapa guianensis* Aubl.; *Meliac.;* fr. Gu.
„ Witte: *Carapa guianensis* Aubl.; *Meliac.;* Gu.
Carapanauba: *Aspidosperma excelsum* Bth.; *Apocynac.;* Bras.
Carapeirana: *Licania turiuva* Cham. et Schlecht.; *Rosac.;* Bras.
Carapo: *Carapa guianensis* Aubl.; *Meliac.;* Trin.
Carapun: *Chrysophyllum cainito* L.; *Sapotac.;* Arg.
Carareba: *Myrocarpus fastigiatus* Allem.; *Papilionac.;* Bras.
Carat: *Sabal sp.; Palmac.;* Trin.
Carate: *Bursera sp.; Burserac.;* Kol.
Caratero: *Bursera sp.; Burserac.;* Kol.
Carba suella: *Terminalia sp.; Combretac.;* Pan.
Carballo: *Quercus pedunculata* Ehrh.; *Fagac.;* Sp.
Carbón: *Andira inermis* H. B. K.; *Papilionac.;* br. Hond.
„ : ? *Guarea chichon* C. DC.; *Meliac.;* br. Hond.
„ : *Myginda eucymosa* Loes. et Pitt.; *Celastrac.;* br. Hond.
„ : *Tetragastris stevensoni* Standl.; *Burserac.;* br. Hond.
„ : *Guarea longipetiola* C. DC.; *Meliac.;* Hond.
„ : *Matayba glaberrima* Radlk.; *Sapindac.;* Hond.
„ : *Trichilia izabalana* Blake; *Meliac.;* Guat.
„ : *Cordia decandra* Hook. et Arn.; *Borraginac.;* S.-Am. (Anden)
„ colorado: *Cupania macrophylla* A. Rich.; *Sapindac.;* Guat.
Carboncillo: *Cupania guatemalensis* Radlk.; *Sapindac.;* Guat.
„ : *Trichilia izabalana* Blake; *Meliac.;* Guat.
? Carbonera: *Pithecolobium brevifolium* Bth.; *Mimosac.;* Mex.
Carbonero: *Licania sp.; Rosac.;* Kol.

Carbonero: *Piptadenia pittieri* Harms; *Mimosac.;* Ven.
Carcaño: *Caesalpinia affinis* Hemsl.; *Caesalpiniac.;* Guat.
Carcuera: *Platypodium maxonianum* Pitt.; *Papilionac.;* Pan.
Cardboard: *Pycnanthus kombo* Warb.; *Myristicac.;* W.-Af.
Cardinalwood: *Brosimum paraënse* Huber; *Morac.;* Bras. (Amaz.)
Cardo cabrero: *Kentrophyllum arborescens* Hook.; *Composit.;* Sp.
Cardón: *Cephalocereus russelianus* Rose; *Cactac.;* Kol.
„   : *Cereus spp.; Cactac.;* Ven.
„ de brava: *Cereus sp.; Cactac.;* Ven.
„ „ dato: *Cereus sp.; Cactac.;* Ven.
„ „ guanarije: *Cereus sp.; Cactac.;* Ven.
„ del Paraná: *Cereus sp.; Cactac.;* Arg.
Careicillo: *Curatella americana* L.; *Dilleniac.;* Cuba.
Carey de costa: *Krugiodendron ferreum* Urb.; *Rhamnac.;* W.-Ind., C.-Am.
Cargalera: *Pisonia aculeata* L.; *Nyctaginac.;* Hond.
Caria quillo: *Lantana camara* L.; *Verbenac.;* P.-R.
Caribeen, Yellow: *Sloanea woollsi* F. v. M.; *Elaeocarpac.;* Au.
Carica: *Bambusa aculeata* Hitchc.; *Gramin.;* Nic.
Caricari: *Icica sp.; Burserac.;* Ven.
Caricarito: *Bursera simaruba* Sarg.; *Burserac.;* Ven. (Curaçao)
Caricaro: *Icica sp.; Burserac.;* Ven.
Caricolí: *Anacardium rhinocarpus* DC.; *Anacardiac.;* Kol.
? Carinquita negra: *Cordia sp.; Borraginac.;* Trin.
Caripé: *Moquilea sp.; Rosac.;* Bras.
„ rana: *Licania sp.; Rosac.;* Bras.
„ verdadeiro: *Licania utilis* Fritsch; *Rosac.;* Bras.
Caripérana de folha larga: *Licania micrantha* Miq.; *Rosac.;* Bras.
Carisiri: ? *Oxandra lanceolata* Baill.; *Anonac.;* br. Gu.
Carita: ? *Acacia sarmentosa* Dcne.; *Mimosac.;* Kol.
Carito: *Enterolobium cyclocarpum* Griseb.; *Mimosac.;* Kol.
Caritiva: *Prosopis sp.; Mimosac.;* Ven.
Carnauba: *Copernicia cerifera* Mart.; *Palmac.;* Bras.
„ do Matta: *Jacaranda copaia* D. Don; *Bignoniac.;* Bras.
? Carnaveira: *Sweetia sp.; Papilionac.;* Bras.
Carne asada: *Cordia ferruginea* Roem. et Schult.; *Borraginac.;* Hond.
„ d'anta: *Maytenus obtusifolia* Mart.; *Celastrac.;* Bras. (Bahia)
„ „ : *Drimys winteri* Forst.; *Magnoliac.;* Bras.
„ de doncella: *Byrsonima lucida* DC.; *Malpighiac.;* Cuba.
„ „ vaca: *Roupala sp.; Proteac.;* Bras. (Sao P.)
„ „ „ : *Chrysophyllum cainito* L.; *Sapotac.;* Arg.
„ „ „ : *Styrax leprosum* Hook. et Arn.; *Styracac.;* Arg.
Carnero: *Coussapoa rekoi* Standl.; *Morac.;* Mex. (Oaxaca)
„ : *Coccoloba sp.; Polygonac.;* Mex.
Caro: *Enterolobium cyclocarpum* Griseb.; *Mimosac.;* Ven.
„ hembra: *Enterolobium cyclocarpum* Griseb.; *Mimosac.;* Salv.
„ huesco de pescado: *Samanea polycephala* Pitt.; *Mimosac;.* Ven.
Carob: *Ceratonia siliqua* L.; *Caesalpiniac.;* Cypern.
Caroba: *Jacaranda copaia* D. Don; *Bignoniac.;* Bras.
„ : *Jacaranda caroba* DC.; *Bignoniac.;* Arg.
„ do matto: *Jacaranda copaia* D. Don; *Bignoniac.;* Bras.
Carobucu: *Jacaranda copaia* D. Don; *Bignoniac.;* Bras.
Carocolillo: *Homalium sp.; Flacourtiac.;* P.-R.
Caroubier: *Ceratonia siliqua* L.; *Caesalpiniac.;* M.-M.
„ de la Guyane: *Hymenaea courbaril* L.; *Caesalpiniac.;* fr. Gu.
Caroukouva-marom: *Ziziphus jujuba* Lamk.; *Rhamnac.;* S.-Ind.
Carparo: *Cistus hirsutus* Lam.; *Cistac.;* Sp.
„ : *Halimium occidentale* Willk.; *Cistac.;* Sp.
Carquesia: *Genista sagitalis* L.; *Papilionac.;* Sp.
Carquexia: *Genista sagitalis* L.; *Papilionac.;* Sp.
Carra-seri: ? *Oxandra lanceolata* Baill.; *Anonac.;* br. Gu.
Carrabean: *Sloanea woollsi* F. v. M.; *Elaeocarpac.;* Au.
Carrabeen, Red: *Geissois benthami* F. v. M.; *Cunoniac.;* Au.

Carrapeteiro: *Guarea trichilioides* L.; *Meliac.*; Bras.
Carrasca: *Quercus ilex* L.; *Fagac.*; Sp.
Carrascina: *Erica cinerea* L.; *Ericac.*; Sp.
Carrasco: *Comocladia glabra* Spreng.; *Anacardiac.*; P.-R.
Carreto: *Ruprechtia deami* Robinson; *Polygonac.*; Guat.
„   : *Samanea saman* Merr.; *Mimosac.*; Salv.
„   : *Aspidosperma elliptica* Rusby; *Apocynac.*; Ven., Kol.
Carribin: *Sloanea woollsi* F. v. M.; *Elaeocarpac.*; Au.
Carrobean, Blush-: *Sloanea australis* F. v. M.; *Elaeocarpac.*; Au.
„   Grey: *Elaeocarpus obovatus* G. Don; *Elaeocarpac.*; Au.
„   „   : *Sloanea woollsi* F. v. M.; *Elaeocarpac.*; Au.
„   Red: *Geissois benthami* F. v. M.; *Cunoniac.*; Au.
Carropeta: *Guarea sp.*; *Meliac.*; Bras.
Carrot-wood: *Protium australasicum* Sprague; *Burserac.*; Au.
Cartán: *Centrolobium paraënse* Tul.; *Papilionac.*; Ven.
Carua: *Juglans regia* L.; *Juglandac.*; Griech.
Carubio: *Zanthoxylum sp.*; *Rutac.*; P.-R.
Caruto: *Genipa spp.*; *Rubiac.*; Ven.
Carvalho: ? *Roupala brasiliensis* Klotzsch; *Proteac.*; Bras.
„   nacional: *Roupala brasiliensis* Klotzsch; *Proteac.*; Bras. (Sao P)
Carvoeiro (p): *Miconia sp.*; *Melastomatac.*; Bras.
Carya amer: *Carya cordiformis* C. Koch; *Juglandac.*; Kan.
„   blanc: *Carya ovata* C. Koch; *Juglandac.*; Kan.
„   glabre: *Carya glabra* Spach; *Juglandac.*; Kan.
„   des pourceaux: *Carya glabra* Spach; *Juglandac.*; Kan.
„   tomenteux: *Carya alba* Nutt.; *Juglandac.*; Kan.
Cás: *Psidium friedrichsthalianum* Ndz.; *Myrtac.*; C.-R.
Casabe: *Pisonia sp.*; *Nyctaginac.*; Ven.
Casada: *Psychotria chiapensis* Standl.; *Rubiac.*; br. Hond.
Casadawood: *Ilex sp.*; *Aquifoliac.*; br. Gu.
Casca doce: *Pradosia latescens* Radlk.; *Sapotac.*; Bras.
„   preciosa: *Aniba canelilla* Mez; *Laurac.*; Bras. (Amaz.)
„   d'anta: *Drimys winteri* Forst.; *Magnoliac.*; Bras.
„   de anta: *Drimys winteri* Forst.; *Magnoliac.*; Bras.
„   de Winter: *Drimys winteri* Forst.; *Magnoliac.*; Bras.
Cascabel sonaja: *Enterolobium cyclocarpum* Griseb.; *Mimosac.*; Mex.
Cascalote: *Caesalpinia coriaria* Willd.; *Caesalpiniac.*; Cuba.
„   : *Caesalpinia coriaria* Willd.; *Caesalpiniac.*; Mex.
Cascara: *Rhamnus purshiana* DC.; *Rhamnac.*; USA. (Oregon)
„   amarga: *Sweetia panamensis* Bth.; *Papilionac.*; C.-Am., Mex.
Cascarilián: *Croton glabellus* L.; *Euphorbiac.*; Hond.
Cascarilla: *Pogonopus febrifugus* Bth. et Hook.; *Rubiac.*; tr. Am.
„   : *Coutarea hexandra* K. Schum.; *Rubiac.*; Arg.
Cascarilla Bark: *Croton insularis* Baill.; *Euphorbiac.*; Queensl.
Cascaron: *Cascaronia astragalina* Griseb.; *Papilionac.*; tr. Am.
Casco de venado: *Bauhinia divaricata* L.; *Caesalpiniac.*; Hond.
Case: *Licania sp.*; *Rosac.*; Trin.
Cashaw: *Prosopis juliflora* DC.; *Mimosac.*; Jam.
Cashew: *Anacardium occidentale* L.; *Anacardiac.*; tr. Am.
„   , Bosh-: *Curatella americana* L.; *Dilleniac.*; Gu.
„   Wild: *Curatella americana* L.; *Dilleniac.*; Gu.
„   „   : *Anacardium giganteum* Hancock; *Anacardiac.*; br. Gu.
„   „   : *Anacardium rhinocarpus* DC.; *Anacardiac.*; Ven.
Cashew-nut: *Anacardium occidentale* L.; *Anacardiac.*; Tr.
Cashewtree, Giant: *Anacardium rhinocarpus* DC.; *Anacardiac.*; tr. Am.
Cashoema: *Anona sp.*; *Anonac.*; Gu.
Casique care: ? *Piratinera guianensis* Aubl.; *Morac.*; Pan.
Casita: *Sapindus divaricatus* St. Hil.; *Sapindac.*; Arg.
Casmagua: *Wallenia laurifolia* Sw.; *Myrsinac.*; Cuba.
Caspi: *Rhus juglandifolia* H. B. K.; *Anacardiac.*; Kol.
Cassada: *Dipholis salicifolia* A. DC.; *Sapotac.*; s. Fl., Bah.
Cassavawood: *Alchornea triplinervia* Muell. Arg.; *Euphorbiac.*; br. Gu.

Casse: *Cassia spectabilis* DC.; *Caesalpiniac.*; Trin.
  „ haches: *Gymnanthes sp.*; *Euphorbiac.*; Mart.
Cassena: *Ilex vomitoria* Ait.; *Anacardiac.*; sö. USA.
Cassia of Brazil, Clove-: *Licaria guianensis* Aubl.; *Laurac.*; fr. Gu.
Cassia lignea: *Cinnamomum tamala* Nees et Eberm.; *Laurac.*; Ind.
Cassie: *Acacia farnesiana* Willd.; *Mimosac.*; S.-Am.
Cassier: *Cassia apoucouita* Aubl.; *Caesalpiniac.*; Gu.
Castagno d'India (i): *Aesculus hippocastanum* L.; *Hippocastanac.*; It.
Castañeto: *Hura crepitans* L.; *Euphorbiac.*; Kol.
Castanha: *Bertholletia excelsa* H. B. K.; *Lecythidac.*; Bras. (Amaz.)
  „ de arara: *Joannesia heveoides* Ducke; *Euphorbiac.*; Bras.
  „ „ macaco: *Curupita sp.*; *Lecythidac.*; Bras.
  „ „ Maranhão: *Pachira aquatica* Aubl.; *Bombacac.*; Bras.
  „ do Para: *Bertholletia excelsa* H. B. K.; *Lecythidac.*; N.-Bras.
Castanheira: *Bertholletia excelsa* H. B. K.; *Lecythidac.*; Bras.
  „ das aguas: *Eschweilera sp.*, *Jugastrum sp.*; *Lecythidac.*; Bras.
Castanheiro: *Bertholletia excelsa* H. B. K.; *Lecythidac.*; Bras.
  „ : *Bertholletia nobilis* Miers; *Lecythidac.*; Bras.
  „ da India (p): *Aesculus hippocastanum* L.; *Hippocastanac.*; Or., Ind.
  „ „ Para: *Bertholletia excelsa* H. B. K.; *Lecythidac.*; Bras.
Castaño: *Castanea vulgaris* L.; *Fagac.*; Sp.
  „ : *Sterculia apetala* Karst.; *Sterculiac.*; Hond.
  „ silvestre: *Pachira aquatica* Aubl.; *Bombacac.*; Cuba.
  „ de Indias (s): *Aesculus hippocastanum* L.; *Hippocastanac.*; Or., Ind.
Castor-bean tree: *Ricinus communis* L.; *Euphorbiac.*; tr. As., tr. Af.
Casuarina: *Casuarina sp.*; *Casuarinac.*; Bras. (Sao P)
Casuarine (f): *Casuarina equisetifolia* Forst.; *Casuarinac.*; W.-Ind., s. Fl.
Casuarine, Berg- (d): *Casuarina montana* Lesch.; *Casuarinac.*; Malay.
Catagoá: *Trichilia catigua* A. Juss.; *Meliac.*; Bras.
Cataia: *Drimys winteri* Forst.; *Magnoliac.*; Bras.
Catalonazno: *Calatola mollis* Standl.; *Icacinac.*; Mex.
Catalpa: *Catalpa bignonioides* Walt.; *Bignoniac.*; USA.
  „ : *Catalpa longissima* Jacq.; *Bignoniac.*; Haiti.
  „ (f): *Thespesia populnea* Correa; *Malvac.*; Ant.
Catappa (f): *Thespesia populnea* Correa; *Malvac.*; Ant.
Cataúba: *Erythroxylon sp.*; *Erythroxylac.*; Bras.
Catechu: *Acacia catechu* Willd.; *Mimosac.*; Ind.
Catena: *Heliocarpus sp.*; *Tiliac.*; Mex.
Catéping: *Parkinsonia aculeata* L.; *Caesalpiniac.*; W.-Ind., s. Fl.
Catésima: *Parinarium tenuifolium* Chev.; *Rosac.*; Elf.
Catiguá: *Trichilia catigua* A. Juss.; *Meliac.*; Bras., Arg.
  „ blanca: *Trichilia hieronymi* Griseb.; *Meliac.*; Arg. (Mis.)
  „ colorado: *Trichilia catigua* A. Juss.; *Meliac.*; Arg.
  „ graúdo: *Trichilia sp.*; *Meliac.*; Bras.
  „ miúdo: *Trichilia sp.*; *Meliac.*; Bras.
  „ -obŷ: *Casearia sp.*; *Flacourtiac.*; Arg.
  „ puíta: *Trichilia sp.*; *Meliac.*; Arg.
  „ verde: *Casearia sp.*; *Flacourtiac.*; Arg.
Cativo: *Prioria copaifera* Griseb.; *Caesalpiniac.*; tr. Am.
Cat-Korundoo: *Atalantia monophylla* Correa; *Rutac.*; Ind.
Cat's-claw: *Pithecolobium unguis-cati* Bth.; *Mimosac.*; s. Fl., br. W.-Ind.
Catmón: *Dillenia philippinensis* Rolfe; *Dilleniac.*; Phil.
  „ -calabáu: *Dillenia reifferscheidia* F. Villar; *Dilleniac.*; Phil.
Cattle-bush: *Atalaya hemiglauca* F. v. M.; *Sapindac.*; O.-Au.
Catuaba: *Erythroxylon sp.*; *Erythroxylac.*; Bras.
Catucanha: *Roupala sp.*; *Proteac.*; Bras. (Sao P.)
Catucanhem: *Roupala sp.*; *Proteac.*; Bras.
Cauache: *Trichilia sp.*; *Meliac.*; Mex.
Cauassú: *Coccoloba sp.*; *Polygonac.*; Bras.
Cauayan: *Bambusa blumeana* F. Schultes f.; *Gramin.*; Phil.
Cauba: *Bauhinia candicans* Bth.; *Caesalpiniac.*; Arg.
Cauchillo: *Lecythis sp.*; *Lecythidac.*; Kol.

Caucho: *Castilla fallax* O. F. Cook, *C. panamensis* O. F. Cook; *Morac.; Pan.*
„ : *Castilla ulei* Warb.; *Morac.;* Bras.
„ centro-americano: *Castilla sp.; Morac.;* Ven.
„ macho: *Castilla fallax* O. F. Cook; *Morac.;* C.-R.
„ -rana: *Perebea calophylla* Bth. et Hook.; *Morac.;* Bras. (Amaz.)
Caujara: *Cordia sp.; Borraginac.;* Ven.
Caujaro: *Cordia alba* Roem. et Schult.; *Borraginac.;* Kol.
Caukalou: *Casuarina sp.; Casuarinac.;* Fid.
Caukuru: *Casuarina sp.; Casuarinac.;* Fid.
Caulote: *Guazuma ulmifolia* Lamk.; *Sterculiac.;* tr. Am.
„ : *Luehea seemanni* Tr. et Pl.; *Tiliac.;* Hond.
„ blanco: *Luehea candida* Mart.; *Tiliac.;* Hond.
Caumao: *Wallenia laurifolia* Sw.; *Myrsinac.;* Cuba.
Cauré: *Uapaca bingervillensis* Beille; *Euphorbiac.;* Elf.
Caüri: *Terminalia ivorensis* Chev.; *Combretac.;* Elf.
Cautivo: *Prioria copaifera* Griseb.; *Caesalpiniac.;* Pan.
Cauto: *Hirtella silicea* Griseb.; *Rosac.;* Trin.
Cauturo: *Parinarium campestre* Aubl.; *Rosac.;* Trin.
Cavalonga: *Thevetia neriifolia* Juss.; *Apocynac.;* P.-R.
Cavana Tabua: *Podocarpus cupressina* R. Br.; *Conifer.;* Fid.
Caviuna: *Dalbergia nigra* Allem.; *Papilionac.;* S.-Bras.
„ preta: *Machaerium sp.; Papilionac.;* Bras. (Sao P.)
Caxarana: *Cabralea sp.; Meliac.;* Arg.
Caxeta: *Tabebuia leucoxyla* DC.; *Bignoniac.;* S.-Bras.
Caxinguba: *Ficus anthelmintica* Mart.; *Morac.;* Bras. (Amaz.)
Cây (H): *Irvingia oliveri* Pierre; *Simarubac.;* Ind.-Ch.
„ ac ho: *Randia longiflora* Lamk.; *Rubiac.;* Annam.
„ bai: *Litchi chinensis* Sonn.; *Sapindac.;* Annam.
„ bang: *Terminalia catappa* L.; *Combretac.;* Ind.-Ch.
„ bây thúa: *Sterculia thoreli* Pierre, *S. bicolor* Mast.; *Sterculiac.;* Annam.
„ bo: *Hibiscus praeclarus* Gagnep.; *Malvac.;* Annam.
„ bui: *Canarium nigrum* Engl., *C. album* Raeusch; *Burserac.;* Ind.-Ch.
„ bung: *Niebuhria sp., Crataeva sp.; Capparidac.;* Ind.-Ch.
„ cá đťác: *Pentacme siamensis* Kurz; *Dipterocarpac.;* Annam.
„ -cây: *Irvingia malayana* Oliv., *I. oliveri* Pierre; *Simarubac.;* Annam.
„ choi moi: *Alangium salviifolium* Wangerin; *Cornac.;* Ind.-Ch.
„ Co gan: *Randia dumetorum* Lamk.; *Rubiac.;* Ind.-Ch.
„ Công: *Calophyllum pulcherrimum* Wall.; *Guttifer.;* Annam.
„ Daikloai: *Randia oxyodonta* Drake; *Rubiac.;* Ind.-Ch.
„ dâu rái: *Dipterocarpus laevis* Hamilt.; *Dipterocarpac.;* Annam.
„ -dau-tra-ben: *Dipterocarpus artocarpifolius* Pierre; *Dipterocarpac.;* Annam.
„ Dau: *Dipterocarpus spp.; Dipterocarpac.;* Annam.
„ Doc: *Garcinia tonkinensis* Vesque; *Guttifer.;* Tonkin.
„ dzau núóc: *Dipterocarpus alatus* Roxb.; *Dipterocarpac.;* Annam.
„ Gio: *Rhamnoneuron balansae* Gilg; *Thymelaeac.;* Ind.-Ch.
„ Gio niet: *Wickstroemia viridiflora* Meissn.; *Thymelaeac.;* Ind.-Ch.
„ Giuong: *Broussonetia papyrifera* L.; *Morac.;* Ind.-Ch.
„ go: *Sindora cochinchinensis* Baill.; *Caesalpiniac.;* Co.-Ch.
„ gôn: *Ceiba pentandra* Gaertn.; *Bombacac.;* Annam.
„ -hom-tom: *Ailanthus fauveliana* Pierre; *Simarubac.;* Co.-Ch.
„ houm ván: *Dipterocarpus duperreanus* Pierre; *Dipterocarpac.;* Annam.
„ Kêú: *Sterculia bicolor* Mast.; *Sterculiac.;* Annam.;
„ la Buong: *Corypha lecomtei* ?; *Palmac.;* Ind.-Ch.
„ Lang: *Randia tomentosa* Bl.; *Rubiac.;* Ind.-Ch.
„ lo bo: *Brownlowia tabularis* Pierre; *Tiliac.;* Annam.
„ lo bo la long: *Brownlowia denysiana* Pierre; *Tiliac.;* Annam.
„ loman: *Pterospermum diversifolium* Blume; *Sterculiac.;* Annam.
„ -lom-vang: *Ailanthus malabarica* A. DC.; *Simarubac.;* Co.-Ch.
„ long mang: *Pterospermum diversifolium* Blume; *Sterculiac.;* Annam.
„ Long Muc: *Wrightia annamensis* Eberh. et Dubard; *Apocynac.;* Annam.
„ lú úói: *Sterculia lynchnophora* Hance; *Sterculiac.;* Annam.
„ main: *Capparis grandis* Heyn.; *Capparidac.;* Co.-Ch.

Cây mây: *Ochrocarpus siamensis* T. Anders.; *Guttifer.;* Annam.
  „ -mu-cuon: *Flacourtia cataphracta* Roxb.; *Flacourtiac.;* Co.-Ch.
  „ -Mun: *Diospyros vera* Chev.; *Ebenac.;* Ind.-Ch.
  „ mun: *Calophyllum inophyllum* L.; *Guttifer.;* Ind.-Ch.
  „ -muong-troung: *Zanthoxylum clava-herculis* L.; *Rutac.;* Co.-Ch.
  „ Ngau: *Aglaia odorata* Lour.; *Meliac.;* Co.-Ch.
  „ Nguyet: *Murraya exotica* L.; *Rutac.;* Ind.-Ch.
  „ -nhaoc: *Unona silvatica* Dun.; *Anonac.;* Co.-Ch.
  „ nhoc quich: *Unona corticosa* Pierre; *Anonac.;* Annam.
  „ -Nhon: *Euphoria longana* Lamk.; *Sapindac.;* Annam.
  „ ói: *Holoptelea integrifolia* Planch.; *Ulmac.;* Ind.-Ch.
  „ Ruoi: *Streblus asper* Lour.; *Morac.;* Ind.-Ch.
  „ tao: *Ximenia americana* L.; *Olacac.;* Annam.
  „ Thung luc: *Randia pycnantha* Drake; *Rubiac.;* Ind.-Ch.
  „ tia: *Thespesia populnea* Correa; *Malvac.;* Co.-Ch.
  „ trâ: *Kleinhovia hospita* L.; *Sterculiac.;* Annam.
  „ tra budhe: *Thespesia populnea* Correa; *Malvac.;* Annam.
  „ trau trau: *Ochrocarpus siamensis* T. Anders.; *Guttifer.;* Annam.
  „ umain: *Capparis grandis* Heyn.; *Capparidac.;* Co.-Ch.
Caya (u): *Sideroxylon sp.; Sapotac.;* St.-D.
Cayampa: *Capparis sp.; Capparidac.;* Arg.
Cayaté: *Omphalea diandra* L.; *Euphorbiac.;* Bras.
Cayennehout: *Brosimum paraënse* Huber; *Morac.;* tr. S.-Am.
Cayuca: *Virola sp.; Myristicac.;* Trin.
Cayul: *Anona sp.; Anonac.;* P.-R.
Cayumito: *Chrysophyllum cainito* L.; *Sapotac.;* Mex. (Yucatan)
Cayur: *Anona glabra* L.; *Anonac.;* P.-R.
Cayures: *Anona sp.; Anonac.;* P.-R.
Cazuela: *Hieronymia clusioides* Griseb.; *Euphorbiac.;* Cuba.

Cé: *Butyrospermum parki* Kotschy; *Sapotac.;* Sudan.
„ : *Haematostaphis barteri* Hook. f.; *Anacardiac.;* Elf.
Cebil: *Piptadenia macrocarpa* Bth.; *Mimosac.;* Arg.
„ blanco· *Piptadenia macrocarpa* Bth.; *Mimosac.;* Arg.
„ colorado: *Piptadenia macrocarpa* Bth.; *Mimosac.;* Arg.
„ moro: *Piptadenia communis* Bth.; *Mimosac.;* Arg.
Cebo burro: *Hernandia guianensis* Aubl.; *Hernandiac.;* Pan.
„ macho: *Hernandia guianensis* Aubl.; *Hernandiac.;* Pan.
Cebolino: *Pisonia sp.; Nyctaginac.;* Bras.
Cécérüi: *Morus mesozygia* Stapf; *Morac.;* Elf.
Cedar: *Chamaecyparis thyoides* B. S. P.; *Conifer.;* USA.
„ : *Thuja occidentalis* L.; *Conifer.;* N.-Am.
„ : *Juniperus virginiana* L.; *Conifer.;* USA.
„ (e): *Cedrela odorata* L.; *Meliac.;* W.-Ind.
„ : *Cedrela mexicana* Roem.; *Meliac.;* C.-Am., Trin., Tob.
„ : *Juniperus flaccida* Schlechtendal; *Conifer.;* Mex.
-, : *Protium spp.; Burserac.;* Gu.
„ : *Cedrela guianensis* A. Juss.; *Meliac.;* fr. Gu.
„ : *Cedrela fissilis* Vell.; *Meliac.;* S.-Bras.
„ : *Cedrus atlantica* Manetti; *Conifer.;* Marok.
„ : *Guarea cedrata* Pellegr.; *Meliac.;* Elf.
„ : *Guarea thompsoni* Sprague et Hutch.; *Meliac.;* Nig.
„ : *Widdringtonia juniperoides* Endl.; *Conifer.;* Kap.
„ : *Callitris arborea* Schrad.; *Conifer.;* Kap.
„ : *Juniperus procera* Hochst.; *Conifer.;* O.-Af.
„ : *Widdringtonia whytei* Rendle; *Conifer.;* O.-Af. (Nyassa).
„ : *Toona australis* Harms; *Meliac.;* N.-S.-W.
„ : *Libocedrus bidwilli* Hook. f.; *Conifer.;* N.-Seel.
„ : *Athrotaxis selaginoides* D. Don; *Conifer.;* Tasm.
„ : *Chamaecyparis obtusa* S. et Z., *C. pisifera* Endl.; *Conifer.;* Jap.
„ : *Thuja japonica* Maxim.; *Conifer.;* Jap.
„ : *Juniperus rigida* S. et Z.; *Conifer.;* Jap.
„ , Acacia-: *Albizzia toona* F. M. Bailey; *Mimosac.;* Queensl.
„ , african: *Guarea cedrata* Pellegr.; *Meliac.;* Elf.
„ , „ : *Entandrophragma spp.; Meliac.;* Go.
„ , „ : *Entandrophragma septentrionalis* Chev.; *Meliac.;* Elf.
„ , Alaska-: *Chamaecyparis nutkaënsis* Spach; *Conifer.;* Kan.
„ , Bass-: *Trema micrantha* Blume; *Ulmac.;* Jam.
„ , Bastar- (e): *Guazuma tomentosa* H. B. K.; *Sterculiac.;* Ind.
„ , Bastard- (e): *Cedrela odorata* L.; *Meliac.;* tr. Am.
„ , „ : *Trichilia sp.; Meliac.;* br. Hond.
„ , „ : *Guazuma ulmifolia* Lam.; *Sterculiac.;* Pan.
„ , „ : *Protium heptaphyllum* March.; *Burserac.;* br. Gu.
„ , „ : *Protium sagotianum* March.; *Burserac.;* Sur.
„ , „ : *Protium heptaphyllum* March.; *Burserac.;* Sur.
„ , „ : *Protium puberulum* Engl., *P. aracouchini* Mart.; *Burserac.;* Sur.
„ , „ : *Protium altissimum* March.; *Burserac.;* Sur.
„ , „ (e): *Chukrasia tabularis* A. Juss.; *Meliac.;* tr. As.
„ , „ : *Melia azedarach* L.; *Meliac.;* Ind.
„ , „ (e): *Soymida febrifuga* A. Juss.; *Meliac.;* Ind., Cey.
„ , Bay-: *Suriana maritima* L.; *Simarubac.;* Bah.
„ , „ : *Guazuma ulmifolia* Lam.; *Sterculiac.;* br. Hond.
„ , Bermuda-: *Juniperus bermudiana* L.; *Conifer.;* Berm.
„ , black: *Nectandra sp., Ocotea sp.; Laurac.;* Sur.
„ , „ : *Nectandra pisi* Miq.; *Laurac.;* fr. Gu.
„ , Borneo-: *Shorea spp.; Dipterocarpac.;* n. Bor.

Cedar, Britisch Columbia-: *Thuja plicata* J. Donn; *Conifer.;* Kan.
,, , Britisch Honduras-: *Cedrela mexicana* Roem.; *Meliac.;* C.-Am.
,, , brown: *Ehretia acuminata* R. Br.; *Borraginac.;* ö. Au.
,, , Canoe-: *Thuja gigantea* Nutt.; *Conifer.;* w. N.-Am.
,, , Ceylon-: *Melia dubia* Cav.; *Meliac.;* O.-Ind.
,, , Cigar-box-: *Cedrela odorata* L.; *Meliac.;* W.-Ind., Sur.
,, , Clanwilliam-: *Callitris arborea* Schrad.; *Conifer.;* O.-Af.
,, , ,, : *Widdringtonia juniperoides* Endl.; *Conifer.;* Kap.
,, , Cuba- (e): *Cedrela odorata* L.; *Meliac.;* W.-Ind.
,, , Deodar-: *Cedrus deodara* Loud.; *Conifer.;* Ind.
,, , Dry zone-: *Pseudocedrela kotschyi* Harms; *Meliac.;* Go.
,, , East African- (e): *Juniperus procera* Hochst.; *Conifer.;* O.-Af.
,, , Eastern: *Thuja occidentalis* L.; *Conifer.;* Kan.
,, , giant: *Thuja gigantea* Nutt.; *Conifer.;* nw. N.-Am.
,, , ,, : *Thuja plicata* J. Donn; *Conifer.;* Kan.
,, , gray: *Qualea rosea* Aubl.; *Vochysiac.;* fr. Gu.
,, , Haiti-: *Juniperus gracilior* Pilger; *Conifer.;* Haiti.
,, , Himalayan-: *Cedrus deodara* Loud.; *Conifer.;* Ind.
,, , Incense: *Libocedrus decurrens* Torr.; *Conifer.;* N.-Am.
,, , Indian: *Cedrus deodara* Loud.; *Conifer.;* Ind.
,, , Jamaican: *Cedrela sp.; Meliac.;* Jam.
,, , Japanese: *Cryptomeria japonica* D. Don; *Conifer.;* Jap.
,, , Lebanon: *Cedrus libani* Loud.; *Conifer.;* Palest.
,, , Mackay-: *Albizzia toona* F. M. Bailey; *Mimosac.;* Queensl.
,, , Mexican: *Toona sinensis* M. Roemer; *Meliac.;* Jap.
,, , Mlanje-: *Widdringtonia whytei* Rendle; *Conifer.;* O.-Af. (Nyassa).
,, , Moulmain- (e): *Toona serrata* M. Roemer; *Meliac.;* Ind.
,, , ,, - (e): *Toona ciliata* M. Roemer; *Meliac.;* Ind., Burma.
,, , New Zealand-: *Dysoxylum spectabile* Hook. f.; *Meliac.;* N.-Seel.
,, , ,, ,, : *Libocedrus bidwilli* Hook. f.; *Conifer.;* N.-Seel.
,, , Oregon-: *Chamaecyparis lawsoniana* Parl.; *Conifer.;* N.-Am.
,, , Pencil- (e): *Juniperus barbadensis* L.; *Conifer.;* sö. USA., W.-Ind.
,, , ,, : *Juniperus virginiana* L.; *Conifer.;* Hond., W.-Ind.
,, , ,, : *Juniperus procera* Hochst.; *Conifer.;* O.-Af.
,, , ,, : *Phyllanthus ferdinandi* Muell. Arg.; *Euphorbiac.;* Au.
,, , ,, : *Dysoxylum muelleri* Bth., *D. fraserianum* Bth.; *Meliac.;* Queensl.
,, , ,, : *Synoum glandulosum* A. Juss.; *Meliac.;* N.-S.-W.
,, , ,, : *Ackama muelleri* Bth., *A. quadrivalvis* C. T. White; *Cunoniac.;* O.-Au.
,, , ,, , african: *Juniperus procera* Hochst.; *Conifer.;* O.-Af.
,, , ,, , black: *Panax elegans* Moore et F. v. M.; *Araliac.;* Au.
,, , ,, , east-african: *Juniperus procera* Hochst.; *Conifer.;* O.-Af. (Kenya).
,, , ,, , himalayan: *Juniperus excelsa* M. Bieb.; *Conifer.;* Ind.
,, , ,, , scentless: *Synoum lardneri* ?; *Meliac.;* N.-S.-W.
,, , pink: *Acrocarpus fraxinifolius* Wight; *Caesalpiniac.;* O.-Ind.
,, , ,, , african: *Guarea cedrata* Pellegr.; *Meliac.;* Elf.
,, , Port Orford-: *Chamaecyparis lawsoniana* Parl.; *Conifer.;* USA. (Kalif.)
,, , Post-: *Libocedrus decurrens* Torr.; *Conifer.;* w. USA.
,, , Queensland-: *Pentaceras australis* Hook. f.; *Rutac.;* Queensl.
,, , Red (e): *Juniperus virginiana* L.; *Conifer.;* sö. USA., W.-Ind.
,, , ,, ,, : *Juniperus barbadensis* L.; *Conifer.;* sö. USA., W.-Ind.
,, , ,, : *Cedrela odorata* L.; *Meliac.;* Sur.
,, , ,, : *Protium altissimum* March.; *Burserac.;* br. Gu.
,, , ,, : *Cedrela mexicana* Roem.; *Meliac.;* br. Gu.
,, , ,, : *Tarrietia utilis* Sprague; *Sterculiac.;* Lib.
,, , ,, : *Uapaca guineensis* Muell. Arg.; *Euphorbiac.;* Lib.
,, , ,, : *Cunonia capensis* L.; *Cunoniac.;* S.-Af. (Knysna).
,, , ,, : *Acrocarpus fraxinifolius* Wight; *Caesalpiniac.;* O.-Ind.
,, , ,, : *Toona ciliata* M. Roem.; *Meliac.;* Phil.
,, , ,, : *Toona australis* Harms, *T. febrifuga* M. Roem.; *Meliac.;* Au.
,, , ,, , Britisch Columbia-: *Thuja plicata* J. Donn; *Conifer.;* Kan.
,, , ,, , Eastern: *Juniperus virginiana* L.; *Conifer.;* N.-Am.
,, , ,, , northwestern: *Thuja gigantea* Nutt.; *Conifer.;* nw. N.-Am.

Cedar, Red, southern: *Juniperus virginiana* L.; *Conifer.*; s. USA.
 „ , „ , „ : *Juniperus barbadensis* L.; *Conifer.*; sö. USA., W.-Ind.
 „ , „ , „ : *Juniperus lucayana* Britton; *Conifer.*; USA.
 „ , „ , western: *Thuja plicata* J. Donn; *Conifer.*; N.-Am.
 „ , „ , „ : *Juniperus occidentalis* Hook.; *Conifer.*; nw. N.-Am.
 „ , „ , of the West: *Thuja gigantea* Nutt.; *Conifer.*; w. N.-Am.
 „ , Rock-: *Juniperus sabinoides* Sarg.; *Conifer.*; USA. (Texas), Mex.
 „ , soft: ? *Virola surinamensis* Warb.; *Myristicac.*; Trin.
 „ , spanish: *Cedrela odorata* L.; *Meliac.*; W.-Ind.
 „ , „ : *Cedrela mexicana* M. Roem.; *Meliac.*; C.-Am.
 „ , „ : *Cedrela sp.*; *Meliac.*; Bras. (Bahia).
 „ , spicy: *Tylostemon manni* Stapf; *Laurac.*; Lib.
 „ , stinking: *Torreya taxifolia* Arn.; *Conifer.*; N.-Am.
 „ , sweet: *Guarea thompsoni* Sprague et Hutch.; *Meliac.*; Lib.
 „ , Tulip-: *Melia azedarach* L. var. *australasica* C. DC.; *Meliac.*; Au.
 „ , western: *Thuja plicata* J. Donn; *Conifer.*; Kan.
 „ , West Indian-: *Cedrela odorata* L.; *Meliac.*; W.-Ind.
 „ , white: *Cupressus thyoides* L.; *Conifer.*; USA.
 „ , „ : *Thuja occidentalis* L.; *Conifer.*; N.-Am.
 „ , „ : *Libocedrus decurrens* Torr.; *Conifer.*; w. USA.
 „ , „ : *Tabebuia pentaphylla* Hemsl.; *Bignoniac.*; W.-Ind.
 „ , „ : *Dialyanthera otoba* Warb.; *Myristicac.*; Pan.
 „ , „ : *Tabebuia longipes* Baker; *Bignoniac.*; br. Gu.
 „ , „ : *Pycnanthus kombo* Warb.; *Myristicac.*; Lib.
 „ , „ : *Chukrasia tabularis* A. Juss.; *Meliac.*; tr. As.
 „ , „ : *Dysoxylum glandulosum* Talbot; *Meliac.*; Ind.
 „ , „ (H): *Dysoxylum malabaricum* Bedd.; *Meliac.*; Ind.
 „ , „ : *Melia dubia* Cav.; *Meliac.*; Ind., Queensl.
 „ , „ : *Melia azedarach* L. var. *australasica* C. DC.; *Meliac.*; Au.
 „ , „ : *Melia composita* Willd.; *Meliac.*; Au.
 „ , „ : *Pentaceras australis* Hook. f.; *Rutac.*; Queensl.
 „ , „ , coast-: *Chamaecyparis thyoides* B. S. P.; *Conifer.*; ö. USA.
 „ , „ , northern: *Thuja occidentalis* L.; *Conifer.*; USA.
 „ , „ , southern: *Chamaecyparis thyoides* B. S. P.; *Conifer.*; USA.
 „ , wild: *Bombacopsis sp.*; *Bombacac.*; Nic.
 „ , yellow: *Juniperus occidentalis* Hook.; *Conifer.*; nw. N.-Am.
 „ , „ : *Chamaecyparis* nutkaënsis Spach; *Conifer.*; w. N-Am.
 „ , „ : *Rhodosphaera rhodanthema* Engl.; *Anacardiac.*; Au.
Ceddre: *Cedrela sp.*; *Meliac.*; Sur.
Ceder (h): *Cedrela odorata* L.; *Meliac.*; Sur.
 „ , Java-: *Toona febrifuga* M. Roem., *T. sinensis* M. Roem.; *Meliac.*; Java.
 „ , witte: *Protium spp.*; *Burserac.*; Sur.
 „ , Zilverblad-: *Ocotea guianensis* Aubl.; *Laurac.*; Sur.
Cederboom: *Callitris juniperoides* Eichler; *Conifer.*; Kap.
Cederhout (h): *Cedrela spp.*; *Meliac.*; tr. Am.
 „ , bruin: *Cedrela odorata* L.; *Meliac.*; Sur.
 „ , rood: *Cedrela odorata* L.; *Meliac.*; Sur.
Ceder-pesi: *Aniba guianensis* Aubl.; *Laurac.*; Sur.
Cedoe: *Cedrela odorata* L.; *Meliac.*; Sur.
Cédra (f): *Cedrela odorata* L.; *Meliac.*; W.-Ind.
Cedra: *Ziziphus lotus* Lamb.; *Rhamnac.*; N.-Af.
Cedra (f): *Guarea cedrata* Pellegr.; *Meliac.*; Elf.
 „ d'Afrique: *Entandrophragma cylindricum* Spr.; *Meliac.*; W.-Af.
Cedrão: *Guarea sp.*; *Meliac.*; Bras.
Cédrat (f): *Cedrela guianensis* A. Juss.; *Meliac.*; fr. Gu.
 „ : *Entandrophragma rufa* Chev.; *Meliac.*; Elf.
Cédratier (f): *Citrus medica* L.: *Rutac.*; M.-M., As.
Cèdre: *Cedrela spp.*; *Meliac.*; tr. Am.
Cèdre: *Cedrus atlantica* Manetti; *Conifer.*; Marok.
 „ : *Guarea cedrata* Pellegr.; *Meliac.*; Elf.
Cèdre-acajou (f): *Cedrela guianensis* A. Juss.; *Meliac.*; fr. Gu.
 „ „ „ : *Cedrela odorata* L.; *Meliac.*; Sur.

Cèdre bagasse: *Protium altissimum* March.; *Burserac.*; fr. Gu.
  ,, bâtard (f): *Toona ciliata* M. Roem.; *Meliac.*; Ind.
  ,,   ,,   ,, : *Chukrasia tabularis* A. Juss.; *Meliac.*; Ind.
  ,, blanc: *Thuja occidentalis* L.; *Conifer.*; Kan.
  ,,   ,, : *Protium altissimum* March.; *Burserac.*; fr. Gu.
  ,,   ,, : *Couralia fluviatilis* Splitg.; *Bignoniac.*; fr. Gu.
  ,, canelle: *Ocotea sp., Nectandra sp.; Laurac.;* Guad., Mart.
  ,, gris: *Nectandra leucantha* Nees; *Laurac.*; fr. Gu.
  ,,   ,, : *Qualea rosea* Aubl., *Q. coerulea* Aubl.; *Vochysiac.*; fr. Gu.
  ,,   ,, , Couari-: *Qualea coerulea* Aubl.; *Vochysiac.*; Sur.
  ,,   ,, ,   ,, : *Qualea albiflora* Warm., *Q. rosea* Aubl.; *Vochysiac.*; Sur.
  ,, , Iciquier-: *Protium altissimum* March.; *Burserac.*; Guad.
  ,, jaune: *Nectandra sp., Aniba guianensis* Aubl.; *Laurac.*; fr. Gu.
  ,, noir (f): *Nectandra sp., Ocotea sp.; Laurac.;* Sur.
  ,,   ,, de montagne: *Nectandra pisi* Miq.; *Laurac.*; fr. Gu.
  ,, odorant: *Cedrela sp.; Meliac.;* fr. Gu.
  ,, rose d'Afrique (f): *Guarea cedrata* Pellegr.; *Meliac.*; Elf.
  :, rouge: *Juniperus virginiana* L.; *Conifer.*; Kan.
  ,,   ,, : *Protium altissimum* March.; *Burserac.*; fr. Gu.
  ,,   ,, (f): *Toona serrata* M. Roem.; *Meliac.*; Ind.
  ,, à feuilles d'argent: *Ocotea guianensis* Aubl.; *Laurac.*; fr. Gu.
  ,, d'Afrique (f): *Guarea cedrata* Pellegr.; *Meliac.*; Elf.
  ,, d'argent: *Pithecolobium pedicellare* Bth.; *Mimosac.*; fr. Gu.
  ,, de Batna: *Cedrus atlantica* Manetti; *Conifer.*; Algier (Atlas).
  ,,   ,, Cuba: *Cedrela odorata* L.; *Meliac.*; Ant.
  ,,   ,, Singapore (f): *Toona serrata* M. Roem.; *Meliac.*; Ind.
  ,,   ,, la Barbade: *Cedrela odorata* L.; *Meliac.*; Ant.
  ,,   ,, ,, Jamaique: *Cedrela odorata* L.; *Meliac.*; Ant.
  ,,   ,, ,,   ,, : *Guazuma ulmifolia* Lamk.; *Sterculiac.*; Réu.
  ,, des Antilles (f): *Swietenia mahagoni* Jacq.; *Meliac.*; W.-Ind.
Cédrel: *Cedrela sp.; Meliac.;* fr. Gu.
  ,, odorant (f): *Cedrela odorata* L.; *Meliac.*; W.-Ind.
  ,, rouge (f): *Toona serrata* M. Roem.; *Meliac.*; tr. As.
Cedrela: *Cedrela fissilis* Vell.; *Meliac.*; Arg.
Cedrela: *Toona sinensis* M. Roem.; *Meliac.*; Ch.
Cedrela-hout (h): *Cedrela odorata* L.; *Meliac.*; Sur.
  ,, odorant (f): *Cedrela odorata* L.; *Meliac.*; Sur.
Cedrillo: *Trichilia sp.; Meliac.;* br. Hond.
  ,,   ,, : *Zanthoxylum procerum* Donn. Sm.; *Rutac.*; br. Hond.
  ,,   ,, : *Guarea excelsa* H. B. K.; *Meliac.*; Guat.
  ,,   ,, : *Virola merendonis* Pittier; *Myristicac.*; Guat.
  ,,   ,, : *Trichilia spondioides* Jacq.; *Meliac.*; Ven.
  ,,   ,, : *Guarea spiciflora* A. Juss.; *Meliac.*; Arg.
  ,,   ,, : *Cupania sp.; Sapindac.;* Arg.
  ,, blanco: *Guarea sp.; Meliac.;* Arg.
  ,, cimarrón: *Guarea sp.; Meliac.;* Mex.
Cedro (i): *Citrus medica* L.; *Rutac.*; M.-M.
  ,, : *Cedrela odorata* L.; *Meliac.*; W.-Ind.
  ,, : *Cedrela occidentalis* Rose; *Meliac.*; w. C.-Am.
  ,, : *Cedrela mexicana* M. Roem., *C. odorata* L.; *Meliac.*; br. Hond.
? ,, : *Podocarpus coriacea* Rich.; *Conifer.*; Hond.
  ,, : *Cedrela mexicana* M. Roem.; *Meliac.*; Guat.
  ,, : *Cedrela fissilis* Vell.; *Meliac.*; S.-Am.
  ,, : *Cedrela odorata* L.; *Meliac.*; Sur.
  ,, : *Cedrela huberi* Ducke; *Meliac.*; Bras. (Para).
  ,, : *Protium altissimum* March.; *Burserac.*; Bras.
  ,, : *Cedrela tubiflora* Bertoni; *Meliac.*; Arg.
  ,, africano: *Sorindeia acutifolia* Engl.; *Anacardiac.*; W.-Af. (St. Thomé, Principe).
  ,, amarello: *Cedrela fissilis* Vell.; *Meliac.*; Bras.
  ,, amargo: *Cedrela sp.; Meliac.;* C.-Am.
  ,,   ,, : *Cedrela glaziovi* C. DC.; *Meliac.*; Ven.
  ,,   ,, : *Simaba cedron* Planch.; *Simarubac.*; Ven.

Cedro amarillo: *Cupressus sp.; Conifer.;* Mex.
„ aromático: *Cedrela sp.; Meliac.;* Bras.
„ balata: *Cedrela fissilis* Vell.; *Meliac.;* S.-Am.
„ „ : *Cedrela fissilis* Vell. *var.* ?; *Meliac.;* Bras.
„ batteo: *Carapa slateri* Standl.; *Meliac.;* Pan.
„ blanco: *Cupressus sp.; Conifer.;* Mex.
„ „ : *Cedrela fissilis* Vell.; *Meliac.;* Par.
„ bordado: *Andriapetalum sp.; Proteac.;* Bras.
„ branco: *Cedrela fissilis* Vell.; *Meliac.;* Bras.
„ caoba: *Cedrela sp.; Meliac.;* Kol.
„ caopiuva: *Cedrela fissilis* Vell.; *Meliac.;* Bras.
„ carmesi: *Swietenia sp.; Meliac.;* Kol.
„ cebello: *Swietenia sp.; Meliac.;* Pan.
„ cebolla: *Cedrela mexicana* M. Roem.; *Meliac.;* Pan.
„ cheiroso: *Cedrela sp.; Meliac.;* Bras.
„ chino: *Cedrela sp.; Meliac.;* Mex.
„ cimarrón: *Calophyllum rekoi* Standley; *Guttifer.;* Mex. (Oaxaca).
„ colorado: *Juniperus sp.; Conifer.;* Mex.
„ „ : *Cedrela sp.; Meliac.;* Mex., Ek.
„ „ : *Cedrela fissilis* Vell., *C. tubiflora* Bertoni; *Meliac.;* Par.
„ crespo: *Cedrela fissilis* Vell.; *Meliac.;* Arg.
„ dulce: *Cedrela montana* Turcz.; *Meliac.;* Ven.
„ „ : *Bombax mompoxense* H. B. K.; *Bombacac.;* Ven.
„ español: *Cedrela sp.; Meliac.;* tr. S.-Am.
„ espino: *Zanthoxylum kellermani* P. Wilson; *Rutac.;* Hond.
„ „ : *Bombacopsis fendleri* Pittier; *Bombacac.;* Hond.
„ espinoso: *Bombacopsis fendleri* Pittier; *Bombacac.;* C.-Am., Pan.
„ fino: *Cedrela sp.; Meliac.;* Mex.
„ gaspeado: *Cedrela sp.; Meliac.;* Arg.
„ Grenadine: *Cedrela fissilis* Vell.; *Meliac.;* tr. Am.
„ „ : *Cedrela sp.; Meliac.;* C.-R.
„ hembra: *Cedrela odorata* L.; *Meliac.;* Cuba.
„ jaspeado: *Cedrela fissilis* Vell.; *Meliac.;* Arg.
„ liso: *Cedrela sp.; Meliac.;* Mex.
„ macho: *Cedrela odorata* L.; *Meliac.;* Mex.
„ „ : *Cedrela fissilis* Vell.; *Meliac.;* Salv.
„ „ : *Carapa nicaraguensis* C. DC.; *Meliac.;* Nic.
„ „ : *Bombacopsis sp.; Bombacac.;* Nic.
„ „ : *Carapa slateri* Standley; *Meliac.;* Pan., C.-R.
„ „ : *Cedrela fissilis* Vell.; *Meliac.;* Par., Arg.
„ „ : *Cabralea multijuga* C. DC.; *Meliac.;* Arg.
„ negro: *Juglans sp.; Juglandac.;* Hond.
„ obscuro: *Cedrela sp.; Meliac.;* Arg.
„ oloroso: *Cedrela sp.; Meliac.;* Kol., Mex.
„ prieto: *Metopium toxiferum* Krug et Urban; *Anacardiac.;* P.-R.
„ -rá: *Cabralea spp.; Meliac.;* Arg.
„ real: *Cedrela fissilis* Vell.; *Meliac.;* Pan.
„ rojo: *Cedrela spp.; Meliac.;* Arg.
„ rosa: *Cedrela glaziovi* C. DC.; *Meliac.;* Bras., (R.-J., Sao P.).
„ vermelho: *Cedrela fissilis* Vell., *C. odorata* L.; *Meliac.;* Bras.
„ „ : *Cedrela velloziana* M. Roem.; *Meliac.;* Bras. (R.-J.)
„ „ : *Cedrela glaziovi* C. DC.; *Meliac.;* Bras. (R.-J., Sao P.)
„ „ : *Juniperus virginiana* L.; *Conifer.;* Bras.
„ da Bahia: *Cedrela fissilis* Vell.; *Meliac.;* Bras.
„ de Bogotá: *Cedrela spp.; Meliac.;* Kol.
„ „ ramazón: *Cedrela sp.; Meliac.;* Cuba.
„ „ la Habana: *Cedrela odorata* L.; *Meliac.;* Mex.
„ „ „ sierra: *Cupressus sp.; Conifer.;* Mex.
„ do Amazonas: *Cedrela sp.; Meliac.;* Bras.
„ „ Brazil: *Cedrela fissilis* Vell.; *Meliac.;* Bras.
Cedrohy: *Guarea guara* P. Wilson; *Meliac.;* Bras.
„ : *Tapirira guianensis* Aubl.; *Anacardiac.;* Bras.

Cedron macho: *Beilschmiedia pendula* Hemsl.; *Laurac.;* P.-R.
Cego maschado: *Physocalymma scaberrimum* Pohl; *Lythrac.;* Bras.
Ceiba: *Ceiba pentandra* Gaertn.; *Bombacac.;* tr. Am.
„   : *Hura crepitans* L.; *Euphorbiac.;* Kol.
„   amarilla: *Hura crepitans* L.; *Euphorbiac.;* Kol.
„   colorado: *Bombacopsis sp.; Bombacac.;* Kol., Ven.
„   pochote: *Ceiba pentandra* Gaertn.; *Bombacac.;* Mex.
„   de agua: *Bombax cumanense* H. B. K.; *Bombacac.;* Kol.
„   „   leche: *Hura crepitans* L.; *Euphorbiac.;* Kol.
Ceibillo: *Zanthoxylum kellermani* P. Wilson; *Rutac.;* Guat.
Ceibo: *Ceiba pentandra* Gaertn.; *Bombacac.;* Ven., Ek.
„   : *Erythrina falcata* Bth., *E. crista-galli* L.; *Papilionac.;* Arg.
„   jabillo: *Ceiba pentandra* Gaertn.; *Bombacac.;* Ven.
„   macho: *Erythrina crista-galli* L.; *Papilionac.;* Arg.
Ceibón: *Ceiba pentandra* Gaertn.; *Bombacac.;* Nic.
„   de agua: *Pachira aquatica* Aubl.; *Bombacac.;* Cuba.
„   „   arroyo: *Pachira aquatica* Aubl.; *Bombacac.;* Cuba.
Ceil de chat: *Caesalpinia crista* L.; *Caesalpiniac.;* Guad.
Celery-wood: *Panax elegans* Moore et F. v. M.; *Araliac.;* Au.
Cempoalchuatl: *Ulmus mexicana* Planch.; *Ulmac.;* Mex.
Cenicero: *Pithecolobium saman* Bth.; *Mimosac.;* Guat.
Ceniza: *Ulmus mexicana* Planch.; *Ulmac.;* Pan.
Cenizero: *Enterolobium cyclocarpum* Griseb.; *Mimosac.;* C.-R.
Cenizo: *Miconia argentea* DC.; *Melastomatac.;* Hond.
Centro: *Carapa sp.; Meliac.;* Elf.
Cera vegetal: *Myrica mexicana* Willd.; *Myricac.;* Hond.
„   „   : *Lacistema aggregatum* Rusby; *Lacistemac.;* Hond.
Cerbatana: *Dracaena americana* Donn. Sm.; *Liliac.;* br. Hond.
Cerecillo: *Lonicera xylosteum* L.; *Caprifoliac.;* Sp.
Ceregeira: *Myrcianthes edulis* Berg; *Myrtac.;* Bras.
Cereipo: *Myrospermum frutescens* Jacq.; *Papilionac.;* Ven.
Cereja: *Eugenia retusa* Arechav.; *Myrtac.;* s. Bras.
Cerejeira: *Torresea cearensis* Allem.; *Papilionac.;* Bras.
„   : *Myrcia laevigata* Berg, *Phyllocalyx laevigatus* Berg; *Myrtac.;* s. Bras.
Cerel: *Inga leptoloba* Schld.; *Mimosac.;* Guat.
Cerelha: *Eugenia retusa* Arechav.; *Myrtac.;* Arg.
Cerelillo: *Inga leptoloba* Schld.; *Mimosac.;* Guat.
Ceresa: *Eugenia retusa* Arechav.; *Myrtac.;* Arg.
„   del monte: *Byrsonima sp.; Malpighiac.;* Trin.
Cereza: *Eugenia retusa* Arechav.; *Myrtac.;* Arg.
„   cimarróna: *Cordia sp.; Borraginac.;* P.-R.
Cerezo: *Homalium sp.; Flacourtiac.;* P.-R.
„   : *Malpighia punicifolia* L.; *Malpighiac.;* Ven.
„   : *Bunchosia glauca* H. B. K.; *Malpighiac.;* Ven.
„   -aliso: *Prunus padus* L.; *Rosac.;* Sp.
„   macho: *Trichilia trifoliata* Jacq.; *Meliac.;* Ven.
„   silvestre: *Prunus prostrata* Labill.; *Rosac.;* Sp.
„   de Santa Lucia: *Prunus mahaleb* L.; *Rosac.;* Sp.
Cerillo: *Exostemma caribaeum* Roem. et Schult.; *Rubiac.;* Cuba.
„   : *Ceanothus chloroxylon* Nees; *Rhamnac.;* Jam.
Cerillos: *Casearia sp.; Flacourtiac.;* Nic.
Cérise du pays: *Eugenia uniflora* L.; *Myrtac.;* fr. Gu.
Cerisier: *Malpighia glabra* L.; *Malpighiac.;* Ant.
„   : *Malpighia punicifolia* L.; *Malpighiac.;* Mart.
„   : *Elaeocarpus persicifolius* Brongn. et Gris.; *Elaeocarpac.;* N.-Kal.
„   amer: *Prunus emarginata* Walp.; *Rosac.;* Kan.
„   Capitaine: *Malpighia urens* L.; *Malpighiac.;* Guad.
„   merisier: *Prunus avium* L.; *Rosac.;* Fr.
„   noir: *Prunus serotina* Ehrh.; *Rosac.;* Kan.
„   sauvage: *Prunus avium* L.; *Rosac.;* Fr.
„   „   : *Prunus virginiana* L.; *Rosac.;* Kan.
„   „   de l'Ouest: *Prunus serotina* Ehrh.; *Rosac.;* Kan.

Cerisier tardif: *Prunus serotina* Ehrh.; *Rosac.;* Kan.
„ à grappes: *Prunus virginiana* L.; *Rosac.;* Kan.
„ d'automme: *Prunus serotina* Ehrh.; *Rosac.;* Kan.
„ de Courwith: *Malpighia urens* L.; *Malpighiac.;* Guad.
„ „ virginie: *Prunus virginiana* L.; *Rosac.;* Kan.
Ceriúba: *Avicennia nitida* Jacq.; *Verbenac.;* Bras.
Cero: *Astronium conzatti* Blake; *Anacardiac.;* Pan.
„ : *Rheedia edulis* Triana et Pl.; *Guttifer.;* Pan.
„ : *Symphonia globulifera* L. f.; *Guttifer.;* Pan.
Cerón: *Phyllostylon brasiliensis* Capanema; *Ulmac.;* Mex.
Cerote: *Ficus spp.; Morac.;* Kol.
Cerro: *Quercus cerris* L.; *Fagac.;* It.
Cevil: *Piptadenia excelsa* Lillo; *Mimosac.;* Arg.
„ colorado: *Piptadenia macrocarpa* Bth.; *Mimosac.;* Arg.
Cha chia: *Podocarpus latifolia* Wall.; *Conifer.;* Ind.-Ch.
„ chou: *Cunninghamia sinensis* R. Br.; *Conifer.;* Ind.-Ch.
„ dam: *Sapium cochinchinensis* O. Ktze.; *Euphorbiac.;* Ind.-Ch.
„ pa: *Alangium chinense* Lour.; *Cornac.;* Ind.-Ch.
„ phay: *Duabonga sonneratioides* Buch-Ham.; *Sonneratiac.;* Ind.-Ch.
Chac-anal: *Aphelandra deppeana* Schl. et Cham.; *Acanthac.;* br. Hond.
„ khe (H): *Dysoxylum binectariferum* Hook.; *Meliac.;* Ind.-Ch.
„ „ „ : *Dysoxylum translucidum* Hook.; *Meliac.;* Ind.-Ch.
„ khé: *Dysoxylum tonkinense* ?; *Meliac.;* Tonkin.
„ -mol-ché: *Erythrina rubrinervia* H. B. K.; *Papilionac.;* br. Hond.
„ -toc: *Hamelia erecta* Jacq.; *Rubiac.;* br. Hond.
Chaca: *Bursera gummifera* L.; *Burserac.;* Mex.
Chacafruto: *Erythrina sp.; Papilionac.;* Kol.
Chacah: *Bursera gummifera* L.; *Burserac.;* br. Hond.
Chacal haaz: *Calocarpum mammosum* Pierre; *Sapotac.;* Mex.
Chacanúai: *Apurimacia incarum* Harms; *Leguminos.;* Peru.
Chacay: *Colletia cruzerillo* Bert., *C. doniana* Clos; *Rhamnac.;* Arg.
Chaccanhuai: *Apurimacia incarum* Harms; *Leguminos.;* Peru.
Cha-cha: *Albizzia lebbek* Bth.; *Mimosac.;* Haiti.
Chachaca: *Drimys winteri* Forst.; *Magnoliac.;* Mex.
„ : *Prosopis juliflora* DC.; *Mimosac.;* Mex.
Chachacoma: *Escallonia myrtilloides* L. f.; *Saxifragac.;* Ek.
Chachacomo: *Escallonia resinosa* Pers.; *Saxifragac.;* Peru.
Chachacuma: *Escallonia resinosa* Pers.; *Saxifragac.;* Peru.
Chachalaco: *Ruprechtia occidentalis* Standl.; *Polygonac.;* Mex.
„ : *Cordia diversifolia* Pavón; *Borraginac.;* Hond.
Cháchiga: *Dipholis salicifolia* A. DC.; *Sapotac.;* br. Hond.
Chacmaloché: *Erythrina sp.; Papilionac.;* Mex.
Chadacula: *Boswellia serrata* Roxb.; *Burserac.;* Ind.
Chaddec: *Citrus decumana* L.; *Rutac.;* Guad.
Chadek (f): *Citrus decumana* L.; *Rutac.;* Tr.
Châgne: *Quercus pedunculata* Ehrh.; *Fagac.;* Eu.
Chagualo: *Clusia sp., Calophyllum sp.; Guttifer.;* Kol.
Chahan: *Rhododendron arboreum* Sm.; *Ericac.;* nw. O.-Ind. (Punjab).
Châhrr tuk: *Dipterocarpus alatus* Roxb.; *Dipterocarpac.;* Kamb.
Chai: *Shorea vulgaris* Pierre; *Dipterocarpac.;* Ind.-Ch. (Annam).
„ : *Sageraea elliptica* Hook. f. et Thoms.; *Anonac.;* And.
„ xanh: *Shorea vulgaris* Pierre; *Dipterocarpac.;* Ind.-Ch.
Chaili: *Morinda tinctoria* Roxb.; *Rubiac.;* ö. Beng.
Chain, golden (e): *Cytisus laburnum* L.; *Papilionac.;* Eu.
Chakma: *Anogeissus acuminata* Wall.; *Combretac.;* Ind.
„ : *Nauclea sessilifolia* Roxb.; *Rubiac.;* Burma (Chitt.)
Chakota: *Schleichera trijuga* Willd.; *Sapindac.;* w. O.-Ind. (Bombay).
Châkrâleck: *Dipterocarpus sp.; Dipterocarpac.;* Kamb.
Chakrey: *Lagerstroemia parviflora* Roxb.; *Lythrac.;* C.-O.-Ind.
Chakua: *Albizzia stipulata* Boivin; *Mimosac.;* Beng.
Chakwa: *Anogeissus acuminata* Wall.; *Combretac.;* Beng.
Chal: *Anogeissus latifolia* Wall.; *Combretac.;* nw. O.-Ind. (Punjab).

Chalata: *Dillenia indica* L.; *Dilleniac.*; Beng.
Chalcha: *Butea frondosa* Roxb.; *Papilionac.*; C.-Ind. (Bandelkhand).
Chal-Chal: *Schmidelia edulis* St. Hil.; *Sapindac.*; Arg.
  „ „ : *Chrysophyllum grisebachi* Mez; *Sapotac*; Arg.
Chalchal de gallina: *Acnistus parviflorus* Griseb.; *Solanac.*; Arg.
  „ „ „ : *Allophylus edulis* Radlk.; *Sapindac.*; Arg.
Chaldua: *Erythrina suberosa* Roxb.; *Papilionac.*; nö. O.-Ind. (Orissa).
Chaljuti: *Murraya exotica* L.; *Rutac.*; w. O.-Ind. (Bombay).
Challane: *Dipterocarpus indicus* Bedd.; *Dipterocarpac.*; sw. O.-Ind.
Challe: *Zizyphus xylocarpus* Willd.; *Rhamnac.*; w. O.-Ind. (Bombay).
Chalni: *Populus ciliata* Wall.; *Salicac.*; n. O.-Ind. (Garwhal).
Chalta: *Dillenia indica* L.; *Dilleniac.*; Beng.
Chalter: *Dillenia indica* L.; *Dilleniac.*; tr. As.
Chalun: *Populus ciliata* Wall.; *Salicac.*; n. O.-Ind. (Punjab).
Cham: *Artocarpus chaplasha* Roxb.; *Morac.*; Assam, Burma.
  „ : *Albizzia stipulata* Boivin; *Mimosac.*; Ind.-Ch.
  „ (H): *Entada scandens* Bth.; *Mimosac.*; Ind.-Ch.
  „ „ : *Canarium spp.*; *Burserac.*; Ind.-Ch.
  „ : *Melaleuca leucadendron* L.; *Myrtac.*; Ind.-Ch.
  „ ba: *Ixonanthes cochinchinensis* Pierre; *Linac.*; Ind.-Ch.
  „ bâc: *Irvingia oliveri* Pierre; *Simarubac.*; Kamb.
  „ „ parang: *Elaeocarpus lacunosus* Wall., *E. madopetalus* Pierre; *Elaeocarpac.*;
  „ „ prang: *Elaeocarpus lacunosus* Wall.; *Elaeocarpac.*; Ind.-Ch [Ind.-Ch.
  „ „ trang: *Elaeocarpus lacunosus* Wall.; *Elaeocarpac.*; Ind.-Ch.
  „ -bao: *Hydnocarpus anthelmintica* Pierre; *Flacourtiac.*; Co.-Ch.
  „ „ : *Terminalia corticosa* Pierre; *Combretac.*; Ind.-Ch.
  „ bok barang: *Terminalia catappa* L.; *Combretac.*; Ind.-Ch.
  „ chim: *Canarium tonkinense* Engl.; *Burserac.*; Ind.-Ch.
  „ den: *Canarium nigrum* Engl.; *Burserac.*; Ind.-Ch.
  „ hông: *Holoptelea integrifolia* Planch.; *Ulmac.*; Ind.-Ch.
  „ ôi (H): *Daphniphyllum pierrei* Hance; *Euphorbiac.*; Ind.-Ch.
  „ ruong: *Atalantia monophylla* Correa; *Rutac.*; Ind.-Ch.
  „ trang: *Canarium copaliferum* Chev.; *Burserac.*; Ind.-Ch.
Chama: *Artocarpus lakoocha* Roxb; *Morac.*; O.-Ind. (Assam).
Chamaggai: *Dillenia aurea* Sm.; *Dilleniac.*; O.-As.
Chamar karhar: *Gardenia turgida* Roxb.; *Rubiac.*; C.-O.-Ind.
Chamb: *Alnus nitida* Endl.; *Betulac.*; n. O.-Ind. (Kasch.).
Chamba: *Artocarpus lakoocha* Roxb.: *Morac.*; O.-Ind. (Assam).
Chambagam: *Michelia champaca* L.; *Magnoliac.*; w. O.-Ind. (Deccan).
Châmbâk: *Irvingia oliveri* Pierre; *Simarubac.*; Kamb.
Chamiari: *Prunus puddum* Roxb.; *Rosac.*; nw. O.-Ind. (Punjab).
Champ: *Michelia champaca* L.; *Magnoliac.*; Ind.
Champa: *Michelia champaca* L.; *Magnoliac.*; Beng.
Champac (f): *Michelia champaca* L.; *Magnoliac.*; tr. As.
  „ prang: *Canarium copaliferum* Chev.; *Burserac.*; Ind.-Ch.
Champaca: *Michelia champaca* L.; *Magnoliac.*; Burma.
Champacam: *Michelia champaca* L.; *Magnoliac.*; w. Ind. (s. Deccan).
Champak (H): *Michelia champaca* L.; *Magnoliac.*; Ind.
Champakan: *Michelia champaca* L.; *Magnoliac.*; sw. O.-Ind.
Chamror: *Ehretia laevis* Roxb.; *Borraginac.*; C.-O.-Ind.
Chan: *Machilus odoratissima* Nees; *Laurac.*; nw. O.-Ind. (Punjab).
  „ (H): *Wendlandia glabrata* DC.; *Rubiac.*; Ind.-Ch.
  „ chim (H): *Vitex heterophylla* Roxb.; *Verbenac.*; Ind.-Ch.
  „ „ núi: *Schefflera tonkinensis* R. Vig.; *Araliac.*; Ind.-Ch.
  „ hiông: *Gordonia sp.*; *Theac.*; Annam.
  „ khao: *Santalum album* L.; *Santalac.*; Ind.-Ch.
  „ krasna: *Aquilaria crassna* Pierre; *Thymelaeac.*; Ind.-Ch.
  „ sar: *Santalum album* L.; *Santalac.*; Ind.-Ch.
  „ tampeang: *Randia densiflora* Bth.; *Rubiac.*; Ind.-Ch.
  „ tonea: *Randia densiflora* Bth.; *Rubiac.*; Ind.-Ch.
  „ vit (H): *Acer campbelli* Hook. f.; *Acerac.*; Ind.-Ch.
Chañar: *Gourliea decorticans* Gill.; *Papilionac.*; Chile, Arg.

Chañar breda: *Gourliea decorticans* Gill.; *Papilionac.;* Arg.
Chancán: *Pithecolobium dulce* Bth.; *Mimosac.;* Kol.
Chancanguarica: *Bixa orellana* L.; *Bixac.;* Mex.
Chanchin: *Toona sinensis* M. Roem.; *Meliac.;* Ch., Jap.
Chanchito: *Tabernaemontana citrifolia* L.; *Apocynac.;* Hond.
  „ de flores blancas: *Tabernaemontana citrifolia* L.; *Apocynac.;* Hond.
Chanchorana (H): *Trichilia canjerana* Vell.; *Meliac.;* Arg.
Chandal: *Santalum album* L.; *Santalac.;* C.-O.-Ind.
Chandan: *Santalum album* L.; *Santalac.;* O.-Ind., Malay.
  „ : *Juniperus macropoda* Boiss.; *Conifer.;* n. O.-Ind. (Nepal).
Chandana: *Santalum album* L.; *Santalac.;* Ind.
Chandanam: *Pterocarpus santalinus* L. f.; *Papilionac.;* sö. O.-Ind.
Chandelle: *Amyris balsamifera* L.; *Rutac.;* Guad.
Chandmara: *Heritiera littoralis* Dryander; *Sterculiac.;* w. O.-Ind.
Chandna: *Litsea sebifera* Pers.; *Laurac.;* O.-Ind. (Dun).
Chang: *Rhizophora mucronata* Lam.; *Rhizophorac.;* Ind.-Ch.
  „ bá: *Ixonanthes hancei* Pierre; *Linac.;* Co.-Ch.
  „ lá: *Styrax tonkinense* Craib; *Styracac.;* Ind.-Ch.
  „ ma: *Carallia lucida* Kurz; *Rhizophorac.;* Ind.-Ch.
Changal feyrak: *Shorea leprolusa* Miq.; *Dipterocarpac.;* Malak.
Change écorce: *Prockia theaeformis* Willd.; *Flacourtiac.;* Réu.
  „   „ : *Ludia heterophylla* Lamk.; *Flacourtiac.;* Réu.
Changugo: *Byrsonima crassifolia* H. B. K.; *Malpighiac.;* Mex.
Chanh chanh: *Kandelia rheedi* Kurz; *Rhizophorac.;* Ind.-Ch.
  „ nún: *Citrus limonum* Risso; *Rutac.;* Annam.
  „ ôc (H): *Microdesmis caseariaefolia* Planch.; *Euphorbiac.;* Ind.-Ch.
  „ vet (H): *Kandelia rheedi* Kurz; *Rhizophorac.;* Ind.-Ch.
Chann mou: *Melia azedarach* L.; *Meliac.;* Annam.
Chanté: *Tecoma stans* H. B. K.; *Bignoniac.;* Guat.
Cháo: *Engelhardtia chrysolepis* Hance; *Juglandac.;* Ind.-Ch.
Chao: *Liquidambar formosana* Hance; *Hamamelidac.;* Ind.-Ch.
Chaouari: *Caryocar glabrum* Pers.; *Caryocarac.;* fr. Gu.
Chap choa (H): *Beilschmiedia sphaerocarpa* H. Winkler; *Laurac.;* Ind.-Ch.
Chaparro: *Quercus ilex* L., *Q. coccifera* L.; *Fagac.;* Sp.
  „ : *Curatella americana* L.; *Dilleniac.;* tr. Am.
  „ : *Davilla kunthi* St. Hil.; *Dilleniac.;* br. Hond.
  „ bobo: *Palicourea rigida* H. B. K.; *Rubiac.;* Ven.
  „ colorado: *Curatella americana* L.; *Dilleniac.;* Ven.
  „ manteco: *Byrsonima crassifolia* DC.; *Malpighiac.;* Ven.
Chaparrón: *Rheedia edulis* Triana et Pl.; *Guttifer.;* Salv.
Chapel: *Lonchocarpus guatemalensis* Bth.; *Papilionac.;* Hond.
Chapernillo: *Inga sp.;* *Mimosac.;* Salv.
Chaperno: *Aspidosperma megalocarpon* Muell. Arg.; *Apocynac.;* Hond.
  „ : *Lonchocarpus atropurpureus* Bth.; *Papilionac.;* Hond.
  „ : *Andira inermis* H. B. K.; *Papilionac.;* br. Hond.
  „ : *Lonchocarpus hondurensis* Bth.; *Papilionac.;* Guat.
  „ : *Lonchocarpus lucidus* Pittier; *Papilionac.;* Pan.
Chaplach (H): *Artocarpus chaplasha* Roxb.; *Morac.;* Beng.
Chaplis: *Artocarpus chaplasha* Roxb.; *Morac.;* Beng.
Chapot-siris: *Dalbergia lanceolaria* L.; *Papilionac.;* n. O.-Ind.
Chapote: *Diospyros texana* Scheele; *Ebenac.;* s. USA., Mex.
Chapui: *Shorea polysperma* Merr.; *Dipterocarpac.;* Phil.
Chapulaltapa: *Pithecolobium discolor* Britton; *Mimosac.;* tr. Am.
  „ : *Schizolobium excelsum* Vogel; *Caesalpiniac.;* tr. Am.
  „ : *Machaerium latifolium* Rusby; *Papilionac.;* tr. Am.
  „ : *Lonchocarpus rugosus* Bth.; *Papilionac.;* Salv.
Chapupo: *Tabernaemontana citrifolia* L.; *Apocynac.;* Guat.
Chaputa: *Afzelia quanzensis* Welw.; *Caesalpiniac.;* O.-Af. (Moz.).
Chaquera: *Podocarpus sp.;* *Conifer.;* Kol.
Chaquira: *Colubrina ferruginosa* Brongn.; *Rhamnac.;* Bol.
Chaquiro: *Colubrina ferruginosa* Brongn.; *Rhamnac.;* Bol.
Char: *Buchanania latifolia* Roxb.; *Anacardiac.;* sw. O.-Ind.

Char: *Stereospermum chelonioides* DC.; *Bignoniac.;* Ind.-Ch.
Charachi: *Grewia tiliaefolia* Vahl; *Tiliac.;* w. O.-Ind. (n. Bombay).
Charaiguria: *Vitex peduncularis* Wall.; *Verbenac.;* sw. Beng. (Orissa).
Charan: *Homalium griffithianum* Kurz; *Flacourtiac.;* Ind.-Ch.
Charanj: *Castanopsis indica* A. DC., *Castanea indica* Roxb.; *Fagac.;* nö. O.-Ind.
Charcoal tree of India: *Trema orientalis* Blume; *Ulmac.;* ö. Au.
Charha: *Holoptelea integrifolia* Planch.; *Ulmac.;* ö. Beng.
Charlijal: *Salvadora persica* L.; *Salvadorac.;* Ind., N.- & O.-Af.
Charmaghz: *Juglans regia* L.; *Juglandac.;* Pers.
Charme: *Carpinus betulus* L.; *Betulac.;* Fr.
„ commun: *Carpinus betulus* L.; *Betulac.;* Eu.
„ d'Amérique: *Carpinus caroliniana* Walt.; *Betulac.;* Kan.
Charo: *Byrsonima coriacea* DC.; *Malpighiac.;* Ven.
Charoli: *Buchanania latifolia* Roxb.; *Anacardiac.;* sw. O.-Ind.
Charrasquillo: *Quercus sp.; Fagac.;* Mex.
Charu: *Buchanania latifolia* Roxb.; *Anacardiac.;* sw. Beng. (Orissa).
Charungi: *Sonneratia apetala* Ham.; *Sonneratiac.;* sw. O.-Ind.
Charwari: *Buchanania latifolia* Roxb.; *Anacardiac.;* O.-Ind. (Haid.).
Chasedue: *Erythrina suberosa* Roxb.; *Papilionac.;* sw. Beng. (Orissa).
Chashing: *Symplocos theaefolia* Ham.; *Symplocac.;* O.-Ind.
Chat: *Gironniera chinensis* Bth.; *Ulmac.;* Ind.-Ch.
Chataigne, wild: *Pachira insignis* DC.; *Bombacac.;* Trin.
„ du Brésil (f): *Bertholletia excelsa* H. B. K.; *Lecythidac.;* fr. Gu.
Châtaignier: *Castanea vesca* Gaertn.; *Fagac.;* M.-M.
„ : *Castanea dentata* Borkh.; *Fagac.;* Kan.
„ : *Sloanea sp.; Elaeocarpac.;* Mart.
„ coco: *Sloanea massoni* Swarꞩ; *Elaeocarpac.;* Mart.
„ franc: *Sloanea sp.; Elaeocarpac.;* Guad.
„ grandes feuilles: *Sloanea massoni* Swarꞩ; *Elaeocarpac.;* Guad.
„ petit coco: *Sloanea sp.; Elaeocarpac.;* Guad.
„ d'Amérique: *Castanea dentata* Borkh.; *Fagac.;* Kan.
„ de montagne: *Sloanea massoni* Swarꞩ; *Elaeocarpac.;* Guad.
„ de la Martinique: *Sloanea sinemariensis* Aubl.; *Elaeocarpac.;* Mart.
Chatim: *Alstonia scholaris* R. Br.; *Apocynac.;* tr. As.
Chatinn: *Alstonia scholaris* R. Br.; *Apocynac.;* Beng.
Chatiun: *Aistonia scholaris* R. Br.; *Apocynac.;* n. O.-Ind.
Chatiyan (H): *Alstonia scholaris* R. Br.; *Apocynac.;* n. O.-Ind. (Nepal).
Chatni: *Alstonia scholaris* R. Br.; *Apocynac.;* ö. Beng.
Chatterbox tree: *Albizzia lebbek* Bth.; *Mimosac.;* Guad.
Chatwan: *Alstonia scholaris* R. Br.; *Apocynac.;* Beng.
Cháu (H): *Carya tonkinensis* H. Lec.; *Juglandac.;* Ind.-Ch.
„ : *Aleurites montana* Pierre; *Euphorbiac.;* Ind.-Ch.
„ kram: *Xylia dolabriformis* Bth.; *Mimosac.;* Ind.-Ch.
„ krasna: *Aquilaria agallocha* Roxb.; *Thymelaeac.;* Ind.-Ch.
Châu vit: *Acer campbelli* Hook. f.; *Acerac.;* Ind.-Ch. (Tonkin).
Chauá: *Mimusops elata* Allem.; *Sapotac.;* Bras. (Sao P).
„ : *Lucuma procera* Mart.; *Sapotac.;* Bras.
Chaulai: *Wendlandia exserta* DC.; *Rubiac.;* Himal.
Chaulmugri: *Gynocardia odorata* R. Br.; *Flacourtiac.;* Beng.
Chavandalai: *Berrya ammonilla* Roxb.; *Tiliac.;* w. O.-Ind. (s. Deccan).
Chavandi: *Ehretia laevis* Roxb.; *Borraginac.;* tr. As.
Chavarie: *Caryocar spp.; Caryocarac.;* Sur.
Chavoundel-marom: *Berrya ammonilla* Roxb.; *Tiliac.;* s. Ind.
Chavuku: *Casuarina equisetifolia* Forst.; *Casuarinac.;* w. H.-Ind.
Chawari: *Caryocar glabrum* Pers.; *Caryocarac.;* Gu.
Chay: *Gluta tavoyana* Wall.; *Anacardiac.;* Burma.
„ : *Artocarpus tonkinensis* Chev.; *Morac.;* Ind.-Ch.
„ (H): *Palaquium obovatum* Engl.; *Sapotac.;* Ind.-Ch.
„ dai: *Acronychia laurifolia* Blume; *Rutac.;* Ind.-Ch.
Chaya-kaya: *Amoora rohituka* Wight et Arn.; *Meliac.;* Burma.
Che phai: *Liquidambar formosana* Hance; *Hamamelidac.;* Ind.-Ch.
Ché quay: *Homalium fagifolium* Bth.; *Flacourtiac.;* Ind.-Ch.

Chea: *Butea superba* Roxb.; *Papilionac.;* Ind.-Ch.
Chebulic: *Terminalia chebula* Roxb.; *Combretac.;* Ind.
Chechém: *Metopium browni* Urban; *Anacardiac.;* br. Hond.
„ de caballo: *Cameraria belizensis* Standl.; *Apocynac.;* br. Hond.
Chediya: *Ficus thonningi* Blume; *Morac.;* n. Nig.
Chee (H): *Barringtonia acutangula* L.; *Lecythidac.;* Ind.
Cheesewood: *Pittosporum tobira* Ait.; *Pittosporac.;* Jap.
„ : *Pittosporum undulatum* Vent.; *Pittosporac.;* Au.
„ , Victorian: *Pittosporum bicolor* Hook.; *Pittosporac.;* s. Au.
Chein: *Melia azedarach* L.; *Meliac.;* nw. O.-Ind. (Punjab).
Chek tuôm: *Cinnamomum sp.; Laurac.;* Ind.-Ch.
Chelele: *Inga sp.; Mimosac.;* Mex.
Chella: *Cordia myxa* L.; *Borraginac.;* w. O.-Ind. (Bombay).
Chem: *Tarrietia cochinchinensis* Pierre; *Sterculiac.;* Ind.-Ch.
Chemmaram: *Amoora rohituka* Wight et Arn.; *Meliac.;* sw. O.-Ind.
Chemnen: *Afzelia quanzensis* Welw.; *Caesalpiniac.;* S.-Af.
Chemona: *Pynaertia occidentalis* Chev.; *Meliac.;* Elf.
Chemouan: *Pynaertia occidentalis* Chev.; *Meliac.;* Elf.
Chena: *Celtis australis* L.; *Ulmac.;* n. O.-Ind. (Dehra Dun).
Chenalla: *Myristica attenuata* Wall.; *Myristicac.;* ö. O.-Ind.
Chendra: *Mallotus philippinensis* Muell. Arg.; *Euphorbiac.;* w. O.-Ind.
Chêne (f)· *Quercus spp.; Fagac.;* Eu., Am., As.
„ bicolore: *Quercus bicolor* Willd.; *Fagac.;* Kan.
„ blanc: *Quercus pedunculata* Ehrh., *Q. sessiliflora* Salisb.; *Fagac.;* Eu.
„ „ : *Quercus alba* L.; *Fagac.;* Kan.
„ „ : *Flindersia fournieri* Panch. et Sébert; *Meliac.;* N.-Kal.
„ „ frisé: *Quercus macrocarpa* Michx.; *Fagac.;* Kan.
„ bleu: *Quercus bicolor* Willd.; *Fagac.;* Kan.
„ chevelu: *Quercus cerris* L.; *Fagac.;* s. Eu.
„ chinquapin: *Quercus muehlenbergi* Engelm.; *Fagac.;* Kan.
„ commun: *Quercus pedunculata* Ehrh.; *Fagac.;* Eu.
„ „ de bourgogne: *Quercus sessiliflora* Salisb.; *Fagac.;* Eu.
„ drillard: *Quercus sessiliflora* Salisb.; *Fagac.;* Eu.
„ drille: *Quercus sessiliflora* Salisb.; *Fagac.;* Eu.
„ durelin: *Quercus sessiliflora* Salisb.; *Fagac.;* Eu.
„ écarlate: *Quercus coccinea* Wangenh.; *Fagac.;* Kan.
„ écousé: *Quercus ilex* L.; *Fagac.;* M.-M.
„ étoile: *Quercus stellata* Wang.; *Fagac.;* Kan.
„ , faux: *Syzygium wagapense* Vieill.; *Myrtac.;* N.-Kal.
„ femelle: *Quercus pedunculata* Ehrh.; *Fagac.;* Eu.
„ française des Antilles: *Bucida buceras* L.; *Combretac.;* tr. Am.
„ -gomme: *Spermolepis gummifera* Brongn. et Gris.; *Myrtac.;* N.-Kal.
„ gris: *Pancheria seberti* Guill.; *Cunoniac.;* N.-Kal.
„ Limbo (f): *Terminalia superba* Engl. et Diels; *Combretac.;* Kam., Gab., B.-K.
„ lombard: *Quercus cerris* L.; *Fagac.;* Eu.
„ noir: *Quercus velutina* Lam.; *Fagac.;* Kan.
„ prin: *Quercus prinus* L.; *Fagac.;* Kan.
„ rouge: *Quercus rubra* L.; *Fagac.;* Kan.
„ „ : *Codia floribunda* Brongn. et Gris.; *Cunoniac.;* N.-Kal.
„ rouvre: *Quercus pedunculata* Ehrh., *Q. sessiliflora* Salisb.; *Fagac.;* Eu.
„ tigré: ? *Dysoxylum sp.; Meliac.;* N.-Kal.
„ vert: *Quercus ilex* L.; *Fagac.;* M.-M.
„ yeuse: *Quercus ilex* L.; *Fagac.;* M.-M.
„ zeen: *Quercus mirbecki* Dur.; *Fagac.;* Marok.
„ à grappe: *Quercus pedunculata* Ehrh.; *Fagac.;* Eu.
„ à gros glands: *Quercus macrocarpa* Michx.; *Fagac.;* Kan.
„ à lobes obtus: *Quercus stellata* Wang.; *Fagac.;* Kan.
„ à trochet: *Quercus sessiliflora* Salisb.; *Fagac.;* Eu.
„ d'Afrique (f): *Chlorophora excelsa* Bth. et Hook.; *Morac.;* W.-Af.
„ „ : *Uapaca benguelensis* Muell. Arg.; *Euphorbiac.;* Elf.
„ de Bourgogne: *Quercus cerris* L.; *Fagac.;* Eu.
„ „ chevelu: *Quercus cerris* L.; *Fagac.;* S.-Eu.

Chêne de Garry: *Quercus garryana* Dougl.; *Fagac.;* Kan.
„ „ marais: *Quercus palustris* Muenchh.; *Fagac.;* Kan.
„ des Antilles: *Catalpa longissima* Jacq.; *Bignoniac.;* Mart.
„ „ teinturiers: *Quercus velutina* Lam.; *Fagac.;* Kan.
„ du Cambodge (f): *Dipterocarpus spp.; Dipterocarpac.;* As.
„ „ „ „ : *Manglietia glauca* Blume; *Magnoliac.;* Ind.-Ch.
„ „ „ „ : *Manglietia fordiana* Oliv.; *Magnoliac.;* Ind.-Ch.
Chengai: *Balanocarpus heimi* King; *Dipterocarpac.;* Malay.
Chengal: *Balanocarpus heimi* King; *Dipterocarpac.;* Malay.
„ : *Parashorea stellata* Kurz; *Dipterocarpac.;* H.-Ind., Malay.
„ pasir: *Hopea spp., Pachychlamys hemsleyanus* Ridl.; *Dipterocarpac.;* Malay.
Chéo: *Engelhardtia chrysolepis* Hance; *Juglandac.;* Ind.-Ch.
„ den: *Engelhardtia chrysolepis* Hance; *Juglandac.;* Ind.-Ch.
„ tía: *Engelhardtia chrysolepis* Hance; *Juglandac.;* Ind.-Ch.
„ trang: *Engelhardtia chrysolepis* Hance; *Juglandac.;* Ind.-Ch.
Cheola: *Butea frondosa* Roxb.; *Papilionac.;* C.-O.-Ind.
Cheongbu: *Taxus baccata* L.; *Conifer.;* n. O.-Ind. (ö. Nepal).
Chequen: *Eugenia chequen* Mol.; *Myrtac.;* S.-Am. (Anden).
Chera: *Holigarna arnottiana* Hook. f.; *Anacardiac.;* sw. O.-Ind. (Mal.)
Cheribon: *Tectona grandis* L. f.; *Verbenac.;* Java.
Cherimole: *Anona cherimolia* Mill.; *Anonac.;* Guat.
Cherimolia: *Anona sp.; Anonac.;* P.-R.
Chérimolier (f): *Anona cherimolia* Mill.; *Anonac.;* tr. Am.
Cherimoya: *Anona cherimolia* Mill.; *Anonac.;* tr. Am.
Cherry: *Pseudolmedia oxyphyllaria* Donn. Sm., *P. spuria* Gris.; *Morac.;* br. Hond.
„ : *Saccoglottis gabonensis* Urban; *Humiriac.;* Lib.
„ (e): *Prunus spp.; Rosac.;* Jap.
„ , african: *Kayea sp.; Guttifer.;* Gab.
„ , „ : *Pterocelastrus variabilis* Sond.; *Celastrac.;* S.-Af. (Knysna).
„ , amarelles: *Prunus cerasus* L.; *Rosac.;* USA.
„ , bird-: *Prunus padus* L.; *Rosac.;* Eu.
„ , „ : *Prunus avium* L.; *Rosac.;* ö. N.-Am.
„ , „ : *Prunus pennsylvanica* L. f.; *Rosac.;* USA.
„ , bitter: *Prunus emarginata* Walp.; *Rosac.;* Kan.
„ , black: *Prunus serotina* Ehrh.; *Rosac.;* ö. N.-Am.
„ , „ : ? *Parathesis serrulata* Mez; *Myrsinac.;* Pan.
„ , Broad-leaved: *Exocarpus latifolia* R. Br.; *Santalac.;* Au.
„ , Brush-: *Trochocarpa laurina* R. Br.; *Epacridac.;* N.-S.-W.
„ , Cabinet-: *Prunus serotina* Ehrh.; *Rosac.;* USA.
„ , Choke-: *Prunus virginiana* L.; *Rosac.;* N.-Am.
„ , „ , western: *Prunus demissa* Walp.; *Rosac.;* Kan.
„ , Clammy-: ? *Cordia sp.; Borraginac.;* Jam.
„ , Cypress-: *Exocarpus cupressiformis* Labill.; *Santalac.;* ö. Au.
„ , english: *Prunus avium* L.; *Rosac.;* Eu., C.-As.
„ , Evergreen-: *Prunus ilicifolia* Walp.; *Rosac.;* w. USA. (Kalif.)
„ , Fire-: *Prunus pennsylvanica* L. f.; *Rosac.;* USA.
„ , holly-leaved: *Prunus ilicifolia* Walp.; *Rosac.;* USA. (Kalif.).
„ , Hottentot-: *Cassine capensis* L.; *Celastrac.;* Kap.
„ , „ : *Maurocenia frangularia* Mill.; *Celastrac.;* Kap.
„ , Laurel-: *Prunus caroliniana* Ait.; *Rosac.;* sö. USA.
„ , Mexican: *Sequoia sempervirens* Endl.; *Conifer.;* USA.
„ , Native: *Exocarpus cupressiformis* Labill.; *Santalac.;* Au. (Vict.), Tasm.
„ , Ox-heart-: *Prunus avium* L.; *Rosac.;* Eu., C.-As.
„ , Pigeon-: *Prunus pennsylvanica* L. f.; *Rosac.;* Kan.
„ , Pin-: *Prunus pennsylvanica* L. f.; *Rosac.;* Kan.
„ , Red: *Prunus pennsylvanica* L. f.; *Rosac.;* N.-Am.
„ , Rum: *Prunus serotina* Ehrh.; *Rosac.;* Kan.
„ , Scrub: *Eugenia australis* Wendl.; *Myrtac.;* Au.
„ , „ : *Eugenia myrtifolia* Sims; *Myrtac.;* ö. Au.
„ , Sour: *Eugenia corynantha* F. v. M.; *Myrtac.;* Au.
„ , Tartarian, black: *Prunus avium* L.; *Rosac.;* USA.
„ , West-Indian-: *Prunus myrtifolia* Urban; *Rosac.;* ö. USA., W.-Ind., S.-Am.

Cherry, White: *Schizomeria ovata* D. Don; *Cunoniac.;* ö. Au.
„ , Wild: *Prunus emarginata* Walp.; *Rosac.;* Kan.
„ , „ : *Prunus ilicifolia* Walp.; *Rosac.;* USA. (Kalif.)
„ , „ : *Rhamnus purshiana* DC.; *Rhamnac.;* w. USA.
„ , „ : *Prunus virginiana* L.; *Rosac.;* USA.
„ , „ : *Byrsonima crassifolia* DC.; *Malpighiac.;* Pan.
„ , „ : *Citharexylum caudatum* L.; *Verbenac.;* Pan.
„ , „ , black: *Prunus serotina* Ehrh.; *Rosac.;* N.-Am.
„ , „ , red: *Prunus pennsylvanica* L. f.; *Rosac.;* Kan.
„ , „ , western: *Prunus emarginata* Walp.; *Rosac.;* w. Kan. (br. Columbia).
„ , Woolly-leaf, bitter: *Prunus mollis* Walp.; *Rosac.;* w. N.-Am.
Cherrywood (e): *Pterocelastrus variabilis* Meissn.; *Celastrac.;* S.-Af.
„ : *Prunus pseudocerasus* Lindl.; *Rosac.;* Jap.
Cherumali: *Zizyphus jujuba* Lamk.; *Rhamnac.;* sw. O.-Ind.
Cherupinnay: *Calophyllum wightianum* Wall.; *Guttifer.;* w. O.-Ind. (s. Deccan).
Chestnut: *Castanea vesca* Gaertn.; *Fagac.;* S.-Eu., N.-Af.
„ : *Castanea vulgaris* Lam. var. *americana* Michx.; *Fagac.;* USA. (Ontario).
„ : *Castanea dentata* Borkh.; *Fagac.;* nö. N.-Am.
„ : *Castanopsis spp.; Fagac.;* Malay.
„ : *Castanea pubinervis* Schneid.; *Fagac.;* Jap.
„ , American: *Castanea vesca* L.; *Fagac.;* USA.
„ , Cape-: *Calodendron capense* Thunb.; *Rutac.;* S.-Af.
„ , Dwarf-: *Castanea alnifolia* Nutt.; *Fagac.;* USA.
„ , Evergreen-: *Castanopsis chrysophylla* A. DC.; *Fagac.;* USA. (Oregon).
„ , Horse-: *Aesculus hippocastanum* L.; *Hippocastanac.;* Eu., As.
„ , „ : *Aesculus turbinata* Blume; *Hippocastanac.;* Jap.
„ , „ , Indian: *Aesculus spp.; Hippocastanac.;* Ind.
„ , Indian (H): *Castanopsis indica* A. DC.; *Fagac.;* Ind.
„ , „ , „ : *Castanea indica* Roxb.; *Fagac.;* Ind.
„ , Moreton Bay-: *Castanospermum australe* A. Cunn.; *Papilionac.;* Au.
„ , Rose-, indian: *Mesua ferrea* L.; *Guttifer.;* O.-Ind.
„ , Spanish: *Castanea vulgaris* Lam.; *Fagac.;* E.
„ , Sweet: *Castanea sativa* Mill.; *Fagac.;* N.-Am.
„ , „ : *Castanea dentata* Borkh.; *Fagac.;* Kan.
„ , „ , indian: *Castanopsis indica* A. DC.; *Fagac.;* Ind.
„ , Wild: *Calodendron capense* Thunb.; *Rutac.;* S.-Af.
Chew: *Mareya spicata* Baill.; *Euphorbiac.;* Lib.
Chewing-gum tree: *Achras zapote* L.; *Sapotac.;* s. Mex., C.-Am.
Chewstick: *Symphonia globulifera* L. f. *Guttifer.;* br. Hond.
„ : *Anogeissus leiocarpus* Guill. et Perr.; *Combretac.;* Togo.
„ : *Garcinia manni* Oliv.; *Guttifer.;* Lib.
„ , Waika-: *Symphonia globulifera* L. f.; *Guttifer.;* br. Hond.
„ , Whykee-: *Symphonia globulifera* L. f.; *Guttifer.;* br. Hond.
Chhatiana: *Alstonia scholaris* R. Br.; *Apocynac.;* sw. Beng. (Orissa).
Chheu dây khla: *Wrightia annamensis* Ehrh. et Dub.; *Apocynac.;* Kamb.
Chhlic: *Erythrophloeum sp.; Caesalpiniac.;* Kamb.
Chhlik: *Terminalia tomentosa* Bedd.; *Combretac.;* Ind.-Ch.
„ sneng: *Terminalia tomentosa* Bedd.; *Combretac.;* Ind.-Ch.
Chhnar: *Dipterocarpus dyeri* Pierre; *Dipterocarpac.;* Ind.-Ch.
Chhoeu khlong: *Dipterocarpus tuberculatus* Gaertn.; *Dipterocarpac.;* Kamb.
„ teal: *Dipterocarpus alatus* Roxb., *D. dyeri* Pierre; *Dipterocarpac.;* Kamb.
„ „ tuc: *Dipterocarpus alatus* Roxb.; *Dipterocarpac.;* Kamb.
Chhota dundhera: *Hardwickia binata* Roxb.; *Caesalpiniac.;* C.-O.-Ind.
Chi: *Malpighia glabra* L.; *Malpighiac.;* Mex.
„ : *Byrsonima crassifolia* DC.; *Malpighiac.;* br. Hond.
„ : *Litchi chinensis* Radlk.; *Sapindac.;* Ind.-Ch.
„ bân: *Nephelium lappaceum* L.; *Sapindac.;* Ind.-Ch.
„ chéon: *Pasania dealbata* Oerst.; *Fagac.;* Ind.-Ch.
„ -mou: *Castanopsis tribuloides* A. DC.; *Fagac.;* Ind.-Ch.
Chiabal: *Spondias sp.; Anacardiac.;* Mex.
Chian: *Butea frondosa* Roxb.; *Papilionac.;* Ind.-Ch.
Chianchin: *Toona sinensis* M. Roem.; *Meliac.;* Jap.

Chibatan: *Astronium fraxinifolium* Schott; *Anacardiac.;* Bras.
Chibatão: *Astronium gracile* Engl.; *Anacardiac.;* Bras. (Sao P).
„ pedra: *Astronium sp.; Anacardiac.;* Bras. (Sao P).
„ vidrado: *Astronium sp.; Anacardiac.;* Bras. (Sao P).
Chibou: *Bursera gummifera* L.; *Burserac.;* W.-Ind., Gu.
Chibra: *Randia nilotica* Stapf; *Rubiac.;* n. Nig.
Chica: *Lundia chica* Seem.; *Bignoniac.;* Ven.
Chicalá: *Tecoma spectabilis* Planch. et Linden; *Bignoniac.;* Kol.
Chicarrón: *Comocladia sp.; Anacardiac.;* St.-D.
„ : *Guazuma ulmifolia* Lam.; *Sterculiac.;* Salv.
Chiceh: *Chrysophyllum viride* L.; *Sapotac.;* Mex. (Yucatan), br. Hond.
Chich: *Shorea obtusa* Wall.; *Dipterocarpac.;* Ind.-Ch.
„ dông: *Vatica astrotricha* Hance; *Dipterocarpac.;* Ind.-Ch.
Chicha: *Sterculia chicha* St. Hil.; *Sterculiac.;* Bras.
Chicharia-caspi: *Lippia virgata* Steud.; *Verbenac.;* Peru.
Chicharro: *Terminalia obovata* Steud.; *Combretac.;* Ven.
Chicharrón: *Terminalia chicharronia* Wright; *Combretac.;* Cuba.
„ : *Hirtella triandra* Swartz; *Rosac.;* Pan.
„ prieto: *Chuncoa obovata* Pers.; *Combretac.;* Cuba.
Chichiboa: *Zizyphus melastomoides* Pittier; *Rhamnac.;* Ven.
Chichicaste: *Jatropha urens* L.; *Euphorbiac.;* C.-Am.
„ : *Urera spp.; Urticac.;* Hond.
„ : *Wigandia caracasana* H. B. K.; *Hydrophyllac.;* Hond.
Chichicastillo: *Myriocarpa yzabalensis* Killip; *Urticac.;* Hond., Nic.
Chichilegua: *Solanum hirtum* Vahl; *Solanac.;* Mex.
Chichimeca: *Mosquitoxylum jamaicense* Krug et Urb.; *Anacardiac.;* br. Hond.
Chichimi: *Hasseltia mexicana* Standl.; *Flacourtiac.;* Guat.
Chichipate: *Sweetia panamensis* Bth.; *Papilionac.;* C.-Am.
Chichita mononi: *Schinus sp.; Anacardiac.;* Arg.
Chichiwa: *Maerua angolensis* DC.; *Capparidac.;* n. Nig.
Chichola: *Albizzia lebbek* Bth.; *Mimosac.;* Ind.
Chichra: *Butea frondosa* Roxb.; *Papilionac.;* n. O.-Ind.
Chicipate: *Sweetia panamensis* Bth.; *Papilionac.;* Salv.
Chickalda: *Albizzia odoratissima* Bth.; *Mimosac.;* C.-O.-Ind.
Chickdi: *Buxus wallichiana* Baill.; *Buxac.;* nw. O.-Ind. (Kasch.).
Chickla: *Albizzia stipulata* Boivin; *Mimosac.;* Beng.
Chickrassi (H): *Chukrasia tabularis* A. Juss.; *Meliac.;* Ind.
Chickrasy: *Chukrasia tabularis* A. Juss.; *Meliac.;* Beng.
Chickri: *Buxus wallichiana* Baill.; *Buxac.;* nw. O.-Ind. (Kasch.).
Chickwa: *Albizzia odoratissima* Bth.; *Mimosac.;* C.-O.-Ind.
Chicle: *Achras zapota* L.; *Sapotac.;* br. Hond.
„ macho: *Achras chicle* Pittier; *Sapotac.;* br. Hond.
Chico: *Achras zapota* L.; *Sapotac.;* tr. Am.
Chicochuchi: *Bombax ellipticum* H. B. K.; *Bombacac.;* s. Mex., Guat.
Chicochutl: *Bombax ellipticum* H. B. K.; *Bombacac.;* s. Mex., Guat.
Chicosapote: *Lucuma hypoglauca* Standl.; *Sapotac.;* C.-Am.
Chicot du Canada: *Gymnocladus dioica* C. Koch; *Caesalpiniac.;* Kan.
Chicote de niño: *Calliandra bicolor* Bth.; *Mimosac.;* Arg.
Chicozapote: *Achras zapota* L.; *Sapotac.;* Mex.
Chiculte: *Swietenia macrophylla* King; *Meliac.;* br. Hond.
Chiéli: *Placodiscus pseudostipularis* Radlk.; *Sapindac.;* Kam.
Chiêu: *Nephelium bassacense* Pierre; *Sapindac.;* Ind.-Ch.
„ liêu (H): *Terminalia chebula* Retz.; *Combretac.;* Ind.-Ch.
„ „ „ : *Terminalia nigrovenulosa* Pierre; *Combretac.;* Ind.-Ch.
„ „ dong: *Terminalia nigrovenulosa* Pierre; *Combretac.;* Ind.-Ch.
„ „ miêt: *Terminalia darfeuillana* Pierre; *Combretac.;* Ind.-Ch.
„ „ nôc: *Terminalia papilio* Hance; *Combretac.;* Ind.-Ch.
„ „ núi: *Terminalia corticosa* Pierre; *Combretac.;* Ind.-Ch.
„ „ núoc: *Terminalia papilio* Hance; *Combretac.;* Ind.-Ch.
„ „ xanh: *Terminalia chebula* Retz.; *Combretac.;* Ind.-Ch.
Chifle do vaca: *Avicennia nitida* Jacq.; *Verbenac.;* P.-R.
Chii: *Pasania cuspidata* Oerst.; *Fagac.;* Jap.

Chiire dochi: *Aesculus turbinata* Blume *var.* ?; *Hippocastanac.;* Jap.
Chijira: *Aesculus turbinata* Blume; *Hippocastanac.;* Jap.
Chijire-dochi: *Aesculus turbinata* Blume; *Hippocastanac.;* Jap.
Chijol: *Ichthyomethia piscipula* Hitch., *Erythrina sp.; Papilionac.;* Mex.
Chik: *Shorea obtusa* Wall.; *Dipterocarpac.;* Ind.-Ch.
„ -bevu: *Melia azedarach* L.; *Meliac.;* w. O.-Ind. (Bombay).
Chiké: *Chrysophyllum mexicanum* Brandeg.; *Sapotac.;* br. Hond.
Chikrassi: *Chukrasia tabularis* A. Juss.; *Meliac.;* Beng.
Chikri: *Buxus wallichiana* Baill.; *Buxac.;* nw. O.-Ind. (Kasch.).
Chiküé: *Bridelia speciosa* Muell Arg.; *Euphorbiac.;* Elf.
Chikwani: *Piptadenia buchanani* Baker; *Mimosac.;* s. C.-Af.
„ : *Albizzia gummifera* C. A. Smith; *Mimosac.;* s. C.-Af.
Chil (H): *Pinus longifolia* Roxb.; *Conifer.;* nw. O.-Ind. (Punjab).
„ : *Pinus excelsa* Wall.; *Conifer.;* n. O.-Ind. (Jaunsar).
Chila: *Pinus excelsa* Wall.; *Conifer.;* n. O.-Ind. (Garhwal).
Chilamate: *Sapium sp.; Euphorbiac.;* Salv.
Chilauni: *Nyssa sessiliflora* Hook. f.; *Cornac.;* Beng.
„ : *Schima wallichi* Choisy; *Theac.;* Beng., Assam.
Chilauwe: *Schima wallichi* Choisy; *Theac.;* Himal., H.-Ind.
Chilbil: *Holoptelea integrifolia* Planch.; *Ulmac.;* ö. Beng.
Chilca: *Thevetia peruviana* Merr.; *Apocynac.;* Hond.
„ : *Podocarpus coriacea* Rich.; *Conifer.;* Hond.
„ : *Baccharis polyantha* H. B. K.; *Composit.;* Ek.
Chileholz: *Laurelia aromatica* Juss.; *Monimiac.;* Chile.
Chilgoza: *Pinus gerardiana* Wall.; *Conifer.;* nw. Himal.
Chilicote: *Erythrina sp.; Papilionac.;* Mex.
Chilicuate: *Sapium pleiostachys* K. Schum.; *Euphorbiac.;* Guat.
Chili-gidda: *Strychnos potatorum* L. f.; *Loganiac.;* w. O.-Ind. (s. Bombay).
Chilillo: *Casearia praecox* Griseb.; *Flacourtiac.;* Mex.
„ : *Drimys winteri* Forst.; *Magnoliac.;* Mex.
„ : *Zygia latifolia* St. Hil., ? *Pithecolobium sp.; Mimosac.;* br. Hond.
„ : *Picraena excelsa* Lindl.; *Simarubac.;* br. Hond.
„ : *Amyris elemifera* L.; *Rutac.;* Hond.
Chilio de perro: *Hamelia axillaris* Sw.; *Rubiac.;* Nic.
Chilka duduga: *Polyalthia cerasoides* Bth. et Hook. f.; *Anonac.;* w. O.-Ind. (n. Bombay).
Chilla: *Casearia tomentosa* Roxb.; *Flacourtiac.;* tr. As.
„ : *Excoecaria agallocha* L.; *Euphorbiac.;* ö. O.-Ind. (n. Madras).
„ : *Holoptelea integrifolia* Planch.; *Ulmac.;* C.-O.-Ind.
„ : *Strychnos potatorum* L. f.; *Loganiac.;* ö. O.-Ind. (n. Madras).
Chillu: *Strychnos potatorum* L. f.; *Loganiac.;* w. O.-Ind. (s. Bombay).
Chilwal: *Holoptelea integrifolia* Planch.; *Ulmac.;* C.-O.-Ind.
Chim chim: *Schefflera octophylla* Harms; *Araliac.;* Ind.-Ch.
„ „ rung: *Sterculia foetida* L.; *Sterculiac.;* Co.-Ch.
Chiman-sag: *Gmelina arborea* L.; *Verbenac.;* C.-O.-Ind.
Chimas: *Chaetocarpus castanocarpus* Thw.; *Euphorbiac.;* Ind.-Ch.
Chimbó: *Enterolobium timbouva* Mart.; *Mimosac.;* Bras. (Sao P.)
Chimiche: *Parathesis rekoi* Standl.; *Myrsinac.;* Guat.
Chimidida: *Hymenaea courbaril* L.; *Caesalpiniac.;* Sur.
Chimkani: *Cassia fistula* L.; *Caesalpiniac.;* nw. O.-Ind. (Bombay).
? Chimpaka mearah outan: *Pterospermum diversifolium* Blume; *Sterculiac.;* Malak.
Chimpampa: *Monotes spp.; Dipterocarpac.;* n. Rhod.
Chimu: *Morus serrata* Roxb.; *Morac.;* Himal.
China Berry: *Melia azedarach* L.; *Meliac.;* C.-Am.
Chinacahuite: *Bursera gummifera* L.; *Burserac.;* Hond.
Chinacuite: *Bursera gummifera* L.; *Burserac.;* Hond.
Chinamulli: *Schinus pearcei* Engl.; *Anacardiac.;* Peru.
Cinangi: *Lagerstroemia parviflora* Roxb.; *Lythrac.;* sö. O.-Ind.
Chinarindenbaum: *Cinchona spp.; Rubiac.;* Tr.
China tree: *Melia azedarach* L.; *Meliac.;* s. Eu., Pers.
„ „ , wild: *Sapindus marginatus* Willd.; *Sapindac.;* s. C.-USA.
Chinbyit: *Bauhinia malabarica* Roxb.; *Caesalpiniac.;* Burma.
Chincapin: *Castanea pumila* Mill.; *Fagac.;* USA.

Chinchinholz: *Azara microphylla* Hook. f.; *Flacourtiac.;* Chile.
Chincho: *Zanthoxylum fagara* Sarg.; *Rutac.;* Hond.
Chinchola: *Albizzia lebbek* Bth.; *Mimosac.;* C.-O.-Ind.
Chinchula: *Albizzia lebbek* Bth.; *Mimosac.;* sw. O.-Ind. (Bombay).
Chincua: *Anona sp.; Anonac.;* Mex.
Chincuya: *Anona purpurea* Moç. et Sesse; *Anonac.;* Guat.
Chindarey: *Grewia paniculata* Roxb.; *Tiliac.;* Malak.
Chindaryeh: *Grewia paniculata* Roxb.; *Tiliac.;* Malak.
Chindia: *Acer pictum* Thunb.; *Acerac.;* O.-Ind.
Chinduga: *Albizzia odoratissima* Bth.; *Mimosac.;* w. O.-Ind. (s. Deccan).
Chingalé: ? *Acanthorrhiza sp.; Palmac.;* Kol.
Chingali: *Jacaranda filicifolia* D. Don; *Bignoniac.;* Kol.
Chinh dong (H): *Bridelia minutiflora* Hook. f.; *Euphorbiac.;* Ind.-Ch.
Chinh hoi: *Claoxylon indicum* Hassk.; *Euphorbiac.;* Ind.-Ch.
    „ oi: *Terminalia corticosa* Pierre; *Combretac.;* Ind.-Ch.
Chini: *Tetrameles nudiflora* R. Br.; *Datiscac.;* sw. O.-Ind.
Chinil-té: *Comocladia sp.; Anacardiac.;* Mex.
Chinna kadambu: *Mitragyne parvifolia* Korth.; *Rubiac.;* tr. As.
    „ -kalinga: *Dillenia pentagyna* Roxb.; *Dilleniac.;* w. O.-Ind. (n. Bombay).
Chino: *Bursera gummifera* L.; *Burserac.;* Hond.
    „ : *Lagerstroemia parviflora* Roxb.; *Lythrac.;* Beng.
Chinor: *Ehretia laevis* Roxb.; *Borraginac.;* C.-O.-Ind.
Chinquapin: *Castanea pumila* Mill.; *Fagac.;* N.-Am.
    „ : *Castanopsis chrysophylla* A. DC.; *Fagac.;* USA. (Oregon).
    „ : *Castanopsis kawakamii* Hayata; *Fagac.;* Form.
Chinquepin: *Castanea pumila* Mill.; *Fagac.;* N.-Am.
Chinta: *Cordia vestita* Hook. f. et Thoms.; *Borraginac.;* n. O.-Ind.
    „ mola: *Gomphia sumatrana* Jack; *Ochnac.;* Mola.
Chintah: *Erythroxylon cuneatum* Kurz; *Erythroxylac.;* Malay, Phil.
Chintonrol: *Posoqueria latifolia* Roem. et Schult.; *Rubiac.;* Guat., br. Hond.
Chinyok: *Garuga pinnata* Roxb.; *Burserac.;* Burma.
Chioukoue: *Bridelia speciosa* Muell. Arg.; *Euphorbiac.;* Elf.
Chipa: *Protium sp.; Burserac.;* Gu.
Chip-ché: *Jatropha gaumeri* Greenm.; *Euphorbiac.;* br. Hond.
Chiquichiqui: *Attalea funifera* Mart.; *Palmac.;* Ven. ˙
Chir (H): *Pinus longifolia* Roxb.; *Conifer.;* nw. O.-Ind. (Punjab).
Chiragni: *Ilex aquifolium* L.; *Aquifoliac.;* Eu.
Chirara: *Litsea umbrosa* Nees; *Laurac.;* n. O.-Ind.
Chirauli: *Buchanania latifolia* Roxb.; *Anacardiac.;* Ind.
Chirco. *Thevetia nitida* A. DC.; *Apocynac.;* Pan.
Chiréon: *Pasania dealbata* Oerst.; *Fagac.;* Ind.-Ch.
Chirhol: *Holoptelea integrifolia* Planch.; *Ulmac.;* C.-O.-Ind.
Chiriman: *Anogeissus latifolia* Wall.; *Combretac.;* w. O.-Ind. (n. Bombay).
Chirimolia: *Anona sp.; Anonac.;* Chile.
Chirimoya: *Coussapoa rekoi* Standley; *Morac.;* Mex. (Oaxaca).
    „ : *Anona sp.; Anonac.;* Peru.
    „ : *Anona cherimolia* Mill.; *Anonac.;* Ek.
Chirimuyo: *Anona sp.; Anonac.;* Salv.
Chirindi: *Litsea umbrosa* Nees; *Laurac.;* nw. O.-Ind. (Punjab).
Chiriri: *Combretum leucanthum* Heurck et Muell. Arg.; *Combretac.;* Nig.
    „ : *Combretum lecananthum* Engl. et Diels; *Combretac.;* n. Nig.
Chiroli: *Buchanania latifolia* Roxb.; *Anacardiac.;* O.-Ind.
Chiromas: *Schima crenata* Korth.; *Theac.;* Kamb.
Chironji: *Buchanania latifolia* Roxb.; *Anacardiac.;* C.-O.-Ind.
Chirta mola: ? *Gomphia sumatrana* Jack; *Ochnac.;* Malak.
Chisanoki: *Styrax japonica* S. et Z.; *Styracac.;* Jap.
Chicha-No-Ki: *Ehretia acuminata* R. Br.; *Borraginac.;* Jap.
Chita: *Tamarindus indicus* L.; *Caesalpiniac.;* tr. Af.
    „ mota: *Gardenia gummifera* L.; *Rubiac.;* ö. O.-Ind. (Madras).
Chitea: *Baccharis sp.; Composit.;* S.-Am.
Chitreka: *Bursera serrata* Colebr.; *Burserac.;* w. O.-Ind. (n. Bombay).
Chitta bikki: *Gardenia gummifera* L.; *Rubiac.;* w. O.-Ind. (Bombay).

Chittagong wood (e): *Chukrasia tabularis* A. Juss.; *Meliac.;* tr. As.
Chittam wood: *Cotinus americana* Nutt.; *Anacardiac.;* USA.
Chittim: *Dalbergia sissoo* Roxb.; *Papilionac.;* Ind.
Chittin wood (u): *Bumelia lanuginosa* Pers.; *Sapotac.;* sw. USA.
Chitulia: *Acer pictum* Thunb.; *Acerac.;* O.-Ind. (Dotial).
Chiu: *Rhododendron arboreum* Sm.; *Ericac.;* nw. O.-Ind. (Punjab).
„ : *Juglans sieboldiana* Maxim.; *Juglandac.;* O.-As.
Chiukoué: ? *Bridelia speciosa* Muell. Arg.; *Euphorbiac.;* Elf.
Chiur: *Litsea sebifera* Pers.; *Laurac.;* ö. Beng.
Chiviri: *Pterocarpus stevensoni* Burtt Davy; *Papilionac.;* s. C.-Af.
„ : ? *Baphia ovata* T. R. Sim; *Papilionac.;* s. C.-Af.
Chlague: *Elaeodendron glaucum* Pers.; *Celastrac.;* Ind.-Ch.
Chliêu lai bon doum: *Dalbergia oliveri* Gamble; *Papilionac.;* Ind.-Ch.
Chloeu teal: *Dipterocarpus sp.; Dipterocarpac.;* Kamb.
Chlôr: *Palaquium obovatum* Engl.; *Sapotac.;* Ind.-Ch.
Chnar: *Dipterocarpus dyeri* Pierre; *Dipterocarpac.;* Kamb.
Chó: *Randia dumetorum* Lamk.; *Rubiac.;* Ind.-Ch.
„ : *Dipterocarpus tonkinensis* Chev.; *Dipterocarpac.;* Ind.-Ch.
„ : ? *Aglaia sp.; Meliac.;* Annam.
„ (H): *Bassia latifolia* Roxb.; *Sapotac.;* Ind.-Ch.
„ chai: *Hopea recopeï* Pierre; *Dipterocarpac.;* Ind.-Ch.
„ chi: *Dipterocarpus tonkinensis* Chev.; *Dipterocarpac.;* Ind.-Ch.
„ „ : *Parashorea stellata* Kurz; *Dipterocarpac.;* Ind.-Ch.
„ -„ : *Engelhardtia chrysolepis* Hance; *Juglandac.;* Ind.-Ch.
„ -dô: *Azadirachta indica* A. Juss.; *Meliac.;* Co.-Ch.
„ dong: *Shorea sp.; Dipterocarpac.;* Ind.-Ch.
„ miêt: *Shorea sp.; Dipterocarpac.;* Ind.-Ch.
„ nâu (H): *Elaeocarpus balansa* A. DC.; *Elaeocarpac.;* Ind.-Ch.
„ nen: *Shorea sp.; Dipterocarpac.;* Ind.-Ch.
„ nhai: *Anogeissus rivularis* Gagnep.; *Combretac.;* Ind.-Ch.
„ núi: *Shorea sp.; Dipterocarpac.;* Ind.-Ch.
„ tack ket: *Glochidion obliquum* Dcne.; *Euphorbiac.;* Ind.-Ch.
„ thoi: *Sterculia hypochra* Pierre; *Sterculiac.;* Ind.-Ch.
„ voy: *Shorea thoreli* Pierre; *Dipterocarpac.;* Ind.-Ch.
„ xanh: *Euphoria longana* Lamk.; *Sapindac.;* Annam.
Choben-ché: *Trichilia sp.; Meliac.;* Mex.
Choc ché: *Zygia latifolia* St. Hil.; *Mimosac.;* br. Hond.
Chóc móc: *Sterculia alata* Roxb.; *Sterculiac.;* Annam.
„ môt (H): *Mallotus furetianus* Muell. Arg.; *Euphorbiac.;* Ind.-Ch.
Choch: *Lucuma hypoglauca* Standl.; *Sapotac.;* Mex. (Yucatan).
Chocho: *Erythrina sp.; Papilionac.;* Kol.
„ colorado: *Erythrina sp.; Papilionac.;* Kol.
Chocolatillo: *Acacia polyphylla* DC.; *Mimosac.;* Kol.
Chocolín: *Erythrina sp.; Papilionac.;* Mex.
Chocomico: *Ximenia americana* L.; *Olacac.;* Hond.
Choeu teal: *Dipterocarpus sp.; Dipterocarpac.;* Ind.-Ch.
Choeung chap: *Cyanodaphne cuneata* Blume; *Laurac.;* Ind.-Ch.
Chogar maddi: *Morinda tinctoria* Roxb.; *Rubiac.;* ö. O.-Ind. (Madras).
Choha: *Conopharyngia crassa* Stapf; *Apocynac.;* Elf.
Chói (H): *Poinciana racemosa* ?; *Caesalpiniac.;* Ind.-Ch.
„ dá (H): *Elaeocarpus bonii* Gagnep.; *Elaeocarpac.;* Ind.-Ch.
„ mói (H): *Antidesma bunius* Spreng., *A. japonicum* Bth.; *Euphorbiac.;* Ind.-Ch.
„ „ : *Zanthoxylum budrunga* DC.; *Rutac.;* Annam.
„ „ nep: *Mallotus hookerianus* Muell. Arg.; *Euphorbiac.;* Ind.-Ch.
„ xê (H): *Baeckea frutescens* L.; *Myrtac.;* Ind.-Ch.
Chokri: *Cordia myxa* L.; *Borraginac.;* sw. O.-Ind. (Bombay).
Chom cha: *Toona febrifuga* M. Roem.; *Meliac.;* Ind.-Ch.
„ -chom: *Nephelium lappaceum* L.; *Sapindac.;* Annam.
„ sar: *Santalum album* L.; *Santalac.;* Ind.-Ch.
Chomnay Povéang: *Aglaia gigantea* Pellegr.; *Meliac.;* Kamb.
„ „ : *Amoora gigantea* Pierre; *Meliac.;* Kamb.
Chon Chon: *Apodytes tonkinensis* Gagnep.; *Icacinac.;* Ind.-Ch.

Chong: *Shorea vulgaris* Pierre; *Dipterocarpac.;* Ind.-Ch.
„ kong au dâeut: *Crypteronia paniculata* Blume; *Crypteroniac.;* Ind.-Ch.
„ „ dâh: *Antidesma coriaceum* Bl.; *Euphorbiac.;* Ind.-Ch.
„ riêt: *Albizzia lebbekoides* Bth.; *Mimosac.;* Ind.-Ch.
Chonta: *Bactris sp.; Palmac.;* Peru
„ ruru: *Guilelma speciosa* Mart.; *Palmac.;* Ek.
Chooi: *Sagraea elliptica* Hook. f. et Thoms.; *Anonac.;* And.
Choomuntri: *Heritiera littoralis* Dryander; *Sterculiac.;* w. O.-Ind. (s. Deccan).
Choonoo: *Croton argyratus* Blume; *Euphorbiac.;* Ind.
Chop moi: *Antidesma ghesaembilla* Gaertn.; *Euphorbiac.;* Ind.-Ch.
Chope: *Gustavia sp.; Lecythidac.;* Peru.
Chopera: *Rhamnus pumila* L., *R. frangula* L.; *Rhamnac.;* Sp.
Chope: *Populus nigra* L.; *Salicac.;* Sp., Arg.
„ blanco: *Populus alba* L.; *Salicac.;* Sp.
„ tremblon: *Populus tremula* L.; *Salicac.;* Sp.
Chor chang: *Shorea vulgaris* Pierre; *Dipterocarpac.;* Ind.-Ch.
„ chong: *Shorea vulgaris* Pierre; *Dipterocarpac.;* Ind.-Ch.
„ ny: *Palaquium obovatus* Engl.; *Sapotac.;* Ind.-Ch.
„ tue: *Dipterocarpus laevis* Hamilt.; *Dipterocarpac.;* Kamb.
Chora: *Butea frondosa* Roxb.; *Papilionac.;* C.-O.-Ind. (Merwara).
Chorão: *Salix sp.; Salicac.;* Bras.
„ : *Cariniana uahupensis* Miers; *Lecythidac.;* n. Bras.
Choro: *Cariniana uahupensis* Miers; *Lecythidac.;* n. Bras.
Chosen-hakuyo: *Populus koreana* Rehder; *Salicac.;* Jap.
„ -matsu: *Pinus koreiensis* S. et Z.; *Conifer.;* Jap.
Chosenmomı: *Abies holophylla* Maxim.; *Conifer.;* n. Korea.
Chosen-toneriko: *Fraxinus rhynchophylla* Hance; *Oleac.;* Korea.
Chosı: *Rhus vernicifera* DC.; *Anacardiac.;* Jap.
Chota padar: *Stereospermum chelonoides* DC.; *Bignoniac.;* C.-O.-Ind.
„ palang: *Stereospermum chelonoides* DC.; *Bignoniac.;* C.-O.-Ind.
Choti karandi: *Schrebera swietenioides* Roxb.; *Oleac.;* C.-O.-Ind.
Chottza: *Erythrina sp.; Papilionac.;* Mex.
Chou: *Sterculia dongnaiensis* Pierre; *Sterculiac.;* Ind.-Ch.
Chouk: *Casuarina equisetifolia* Forst.; *Casuarinac.;* s. O.-Ind.
Choupultea: *Kydia calycina* Roxb.; *Malvac.;* n. O.-Ind.
Chow: *Casuarina equisetifolia* Forst.; *Casuarinac.;* Bor.
Chozo: *Licania hypoleuca* Bth.; *Rosac.;* Hond., br. Hond., Guat.
Chram: *Artocarpus chaplasha* Roxb.; *Morac.;* Assam, Beng., Burma.
Chramas: *Vatica cochinchinensis* Pierre; *Dipterocarpac.;* Ind.-Ch.
Chrès: *Pentacme siamensis* Kurz; *Dipterocarpac.;* Ind.-Ch.
„ : *Albizzia lebbek* Bth.; *Mimosac.;* Kamb.
Christmas-berry: *Heteromeles arbutifolia* M. Roem.; *Rosac.;* USA. (Kalif.).
„ -tree: *Nuytsia floribunda* R. Br.; *Loranthac.;* ? Au.
Chroui: *Casearia grewiaefolia* Vent.; *Flacourtiac.;* Ind.-Ch.
Chu (H): *Sterculia dongnaiensis* Pierre; *Sterculiac.;* Ind.-Ch.
Chú: *Dracontomelum mangiferum* Blume; *Anacardiac.;* Ind.-Ch.
Chu me: *Albizzia stipulata* Boivin; *Mimosac.;* Ind.-Ch.
Chua (H): *Albizzia stipulata* Boivin; *Mimosac.;* Ind.-Ch.
„ khét: *Chukrasia sp.; Meliac.;* Ind.-Ch.
„ me: *Albizzia stipulata* Boivin; *Mimosac.;* Ind.-Ch.
„ moi: *Antidesma ghesaembilla* Gaertn.; *Euphorbiac.;* Ind.-Ch.
„ nao: *Chukrasia sp.; Meliac.;* Ind.-Ch.
„ nga: *Mallotus hookerianus* Muell. Arg.; *Euphorbiac.;* Ind.-Ch.
Chucarea: *Porliera lorenẓi* Engl.; *Zygophyllac.;* Arg.
Chúcata: *Prosopis juliflora* DC.; *Mimosac.;* Mex.
Chucchemuch: *Calderonia salvadorensis* Standl.; *Rubiac.;* br. Hond.
Chuchi am: *Mangifera silvatica* Roxb.; *Anacardiac.;* nö. O.-Ind. (Nepal).
Chuchum: *Pithecolobium albicans* Bth.; *Mimosac.;* Mex. (Yucatan), br. Hond.
„ : ? *Acacia riparia* H. B. K.; *Mimosac.;* br. Hond.
Chuchupate: *Guarea longipetiola* C. DC.; *Meliac.;* Pan.
Chuchupi: *Porliera lorenẓi* Engl.; *Zygophyllac.;* Gu.
Chuckem: ? *Pithecolobium albicans* Bth.; *Mimosac.;* br. Hond.

Chuckem: ? *Acacia riparia* H. B. K.; *Mimosac.*; br. Hond.
Chucul has: *Calocarpum mammosum* Pierre; *Sapotac.*; br. Hond.
Chucupi: *Porliera hygrometrica* R. et P.; *Zygophyllac.*; Arg.
  ,,   : *Porliera lorenzi* Engl.; *Zygophyllac.*; Arg.
Chuglam: *Terminalia bialata* Wall.; *Combretac.*; And.
  ,,   : *Myristica irya* Gaertn.; *Myristicac.*; And.
  ,, , black: *Terminalia mani* King; *Combretac.*; And.
  ,, , ,, : *Myristica irya* Gaertn.; *Myristicac.*; And.
  ,, , Kala-: *Terminalia mani* King; *Combretac.*; And.
  ,, , white (H): *Terminalia bialata* Wall.; *Combretac.*; And.
Chuichona: *Exostemma floribundum* Roem. et Schult.; *Rubiac.*; Cuba.
Chukrassi: *Chukrasia tabularis* A. Juss.; *Meliac.*; Beng.
Chukupi: *Porliera lorenzi* Engl.; *Zygophyllac.*; Arg.
Chul: *Calocarpum mammosum* Pierre; *Sapotac.*; Guat.
  ,, -ul: *Calocarpum mammosum* Pierre; *Sapotac.*; Guat.
Chule: *Prunus padus* L.; *Rosac.*; nw. O.-Ind. (Punjab).
Chum bac: *Kurrimia robusta* Kurz; *Celastrac.*; Ind.-Ch.
  ,, bao: *Terminalia corticosa* Pierre; *Combretac.*; Ind.-Ch.
  ,, moi: *Antidesma ghesaembilla* Gaertn.; *Euphorbiac.*; Ind.-Ch.
  ,, -pich: *Calliandra confusa* Spr. et Riley; *Mimosac.*; br. Hond.
Chumbinha roxo: *Lantana camara* L.; *Verbenac.*; Bras.
Chumcintoc: *Bulnesia arborea* Engl.; *Zygophyllac.*; Mex.
Chumico: *Curatella americana* L.; *Dilleniac.*; Pan.
  ,, palo: *Curatella americana* L.; *Dilleniac.*; Pan.
  ,, de palo: *Curatella americana* L.; *Dilleniac.*; C.-R.
Chúng bao (H): *Hydnocarpus anthelmintica* Pierre; *Flacourtiac.*; Ind.-Ch.
  ,, ,, : *Crataeva macrocarpa* Kurz; *Capparidac.*; Ind.-Ch.
  ,, ,, lón: *Hydnocarpus anthelmintica* Pierre; *Flacourtiac.*; Ind.-Ch.
  ,, ,, ,, : *Taractogenos kursi* King; *Flacourtiac.*; Ind.-Ch.
  ,, ,, nho: *Taractogenos subintegra* Pierre; *Flacourtiac.*; Ind.-Ch.
  ,, nôm: *Archytaea vahli* Choisy; *Theac.*; Annam.
Chungi: *Lagerstroemia parviflora* Roxb.; *Lythrac.*; O.-Ind. (Haid.)
Chunup: *Clusia sp.*; *Guttifer.*; Mex.
Chuoc: *Markhamia stipulata* Seem.; *Bignoniac.*; Ind.-Ch.
  ,, bung (H): *Elaeocarpus griffithi* Mast.; *Elaeocarpac.*; Ind.-Ch.
Chuông (H): ? *Laurus camphorata* Ham.; *Laurac.*; Ind.-Ch.
Chupa: *Licania platypus* Fritsch; *Rosac.*; Kol.
Chupamiel: *Hamelia rovirosae* Wernham; *Rubiac.*; Nic.
Chupandilla: *Spondias sp.*; *Anacardiac.*; Mex.
Chupón: *Chrysophyllum aulacocarpum* ?; *Sapotac.*; Ven.
  ,, : ? *Gustavia fastuosa* Willd.; *Lecythidac.*; Ven.
  ,, : *Gustavia yaracuyensis* Pittier; *Lecythidac.*; Ven.
  ,, colorado: *Sideroxylon amygdalicarpum* Pittier; *Sapotac.*; Ven
  ,, ,, : *Gustavia speciosa* DC., *Lecythidac.*; Ven.
  ,, ventoso: *Gustavia eximia* Pittier; *Lecythidac.*; Ven.
Churee-chenz: *Adansonia digitata* L.; *Bombacac.*; Ind.
Churla: *Holoptelea integrifolia* Planch.; *Ulmac.*; sw. O.-Ind.
Churnwood: *Villaresia moorei* F. v. M.; *Icacinac.*; Au.
  ,, : *Ehretia acuminata* R. Br.; *Borraginac.*; n. Queensl.
Churqui: *Prosopis ferox* Griseb.; *Mimosac.*; Arg.
  , : *Acacia farnesiana* Willd.; *Mimosac.*; Arg.
  ,, blanco: *Prosopis ferox* Griseb.; *Mimosac.*; S.-Am.
  ,, negro: *Acacia cavenia* Hook. et Arn.; *Mimosac.*; S.-Am.
  ,, mola: *Gomphis sp.*; *Ochnac.*; Malak.
Churu: *Goeldinia spp.*; *Lecythidac.*; Bras. (Pará).
Chu-say-dor-kohn: *Carapa procera* DC.; *Meliac.*; Lib.
Chusquitala: *Celtis tala* Gill.; *Ulmac.*; Arg.
Chutama: *Bursera sp.*; *Burserac.*; Mex.
Chutéea Pongkoui: *Dipterocarpus alatus* Roxb.; *Dipterocarpac.*; Ind.-Ch.
Chutra: *Protium sp.*; *Burserac.*; Pan.
Chutras: *Protium sessiliflorum* Standl.; *Burserac.*; Pan.
Chyahkya: *Phyllanthus emblica* L.; *Euphorbiac.*; O.-Ind.

Chyamka: *Michelia champaca* L.; *Magnoliac.;* O.-Ind.
Ci: *Butyrospermum parki* Kotschy; *Sapotac.;* Sudan.
Ciavardello: *Sorbus torminalis* Crantz; *Rosac.;* It.
Cibi cibi: *Pterocarpus indicus* Willd.; *Papilionac.;* Fid.
Cibuero: *Coussapoa sp.; Morac.;* Bras.
Cicahuite: *Lysiloma aurita* Bth.; *Mimosac.;* Pan., Salv.
Cicerolas: *Ribes alpinum* L.; *Saxifragac.;* Sp.
Ciceru: *Pithecolobium micradenium* Bth.; *Mimosac.;* St.-D.
Ciclamor: *Cercis siliquastrum* L.; *Caesalpiniac.;* Arg.
Cider tree: *Eucalyptus gunni* Hook. f.; *Myrtac.;* Au. (Vict.).
Cidra: *Citrus medica* L.; *Rutac.;* Hond.
   „   : *Citrus decumana* Willd.; *Rutac.;* Arg.
Cidreira do matto: *Siparuna sp.; Monimiac.;* Bras.
Cieneguillo: *Daphnopsis philippiana* Krug et Urb.; *Thymelaeac.;* P.-R.
Cierito: *Mouriria parvifolia* Bth.; *Melastomatac.;* Pan.
Ciesta de gallo: *Digitalis canariensis* L.; *Scrophulariac.;* Tener.
Cigarboxwood (e): *Cedrela odorata* L.; *Meliac.;* W.-Ind.
   „   „ : *Toona sinensis* M. Roem.; *Meliac.;* Ch.
Cigar tree: *Catalpa speciosa* Warder; *Bignoniac.;* USA.
   „   : *Catalpa bignonioides* Walt.; *Bignoniac.;* w. Fl.
Cigarrenholz: *Cedrela spp.; Meliac.;* tr. Am.
Cigarrenkistenholz: *Cedrela spp.; Meliac.;* tr. Am.
Cigua: *Ocotea sp., Nectandra sp.; Laurac.;* Cuba.
Ciliegio salvatico: *Prunus avium* L.; *Rosac.;* It.
Cimarrón: *Anona glabra* L.; *Anonac.;* s. Fl., W.-Ind.
   „   : *Juglans australis* Griseb.; *Juglandac.;* ? Arg.
Cimiri: *Hymenaea courbaril* L.; *Caesalpiniac.;* br. Gu.
Cimmaron, Grenada-: *Eugenia axillaris* Willd.; *Myrtac.;* br. Hond.
Cinacina: *Parkinsonia aculeata* L.; *Caesalpiniac.;* Arg.
Cinamomo: *Hippophaë rhamnoides* L.; *Elaeagnac.;* Sp.
   „   : *Elaeagnus angustifolia* L.; *Elaeagnac.;* Sp.
   „   : *Melia azedarach* L.; *Meliac.;* Bras.
Cincho: *Lonchocarpus guatemalensis* Bth.; *Papilionac.;* br. Hond.
Cinco dedos: *Quararibea asterolepis* Pittier; *Bombacac.;* Pan.
Cinnamon (H): *Cinnamomum tavoyanum* Meissn.; *Laurac.;* s. Burma.
   „   : *Cinnamomum tamala* Nees et Eberm.; *Laurac.;* Ind.
   „   : *Cinnamomum pedunculatum* J. S. Presl; *Laurac.;* Jap.
   „ , black: ? *Pimenta acris* Kostel.; *Myrtac.;* St.-D.
   „ , Mindanao-: *Cinnamomum mindanaënse* Elmer; *Laurac.;* Phil.
   „ , Native: *Cinnamomum tamala* Nees et Eberm.; *Laurac.;* Queensl.
   „ , wild: *Canella winterana* Gaertn.; *Canellac.;* Fl., Bah.
   „ , „ : *Phoebe montana* Griseb.; *Laurac.;* Jam.
   „ , „ : *Croton glabellus* L.; *Euphorbiac.;* br. Hond.
   „ , „ (H): *Cinnamomum zeylanicum* Breyn; *Laurac.;* Ind.
Cinnamon bark: *Canella winterana* Gaertn.; *Canellac.;* Fl., Bah.
Cinnamon tree: *Cinnamomum zeylanicum* Breyn; *Laurac.;* Sey.
   „ wood: *Cinnamomum spp.; Laurac.;* Ind.
Cinzeiro: *Terminalia tanibouca* Rich.; *Combretac.;* Bras.
   „   : *Vochysia guatemalensis* J. D. Smith; *Vochysiac.;* Bras.
Cipiri: *Ocotea rodioei* Mez; *Laurac.;* br. Gu.
Cipó-cururú: *Echites cururu* Mart.; *Apocynac.;* n. Bras.
Cipo mata-páo: *Ficus sp.; Morac.;* Bras.
Cipó de Jatobá: *Anisosperma passiflora* Manso; *Cucurbitac.;* Bras. (Amaz.)
Cipo de sapo: *Relbunium hypocarpium* Hemsl.; *Rubiac.;* Bras., Ven.
Cipre: *Cordia gerascanthus* L.; *Borraginac.;* Guad.
   „ balanic: *Cordia sp.; Borraginac.;* Guad.
   „ oranger: *Cordia sp.; Borraginac.;* Guad.
Ciprés: *Cupressus benthami* Endl.; *Conifer.;* Hond., Guat.
Cipres: *Libocedrus tetragona* Endl.; *Conifer.;* s. Chile.
   „ de Guaitecas: *Libocedrus tetragona* Endl.; *Conifer.;* Chile.
Ciprès de Mexico: *Cupressus sp.; Conifer.;* Mex.
Ciprés de Montezuma: *Taxodium mucronatum* Tenare; *Conifer.;* Mex.

Cipreso: *Juniperus flaccida* Schld.; *Conifer.;* Mex.
Cipresso: *Cupressus sempervirens* L.; *Conifer.;* It.
Circuelillo: *Embothrium coccineum* Forst.; *Proteac.;* S.-Am.
Circ: *Gardenia aubryi* Vieill.; *Rubiac.;* N.-Kal.
Cirimo: *Tilia sp.; Tiliac.;* Mex.
Ciriúba: *Avicennia nitida* Jacq.; *Verbenac.;* Bras.
Cirolier (f): *Rheedia lateriflora* L.; *Guttifer.;* W.-Ind.
Cironballi, brown: *Nectandra sp.; Laurac.;* br. Gu.
    „ , yellow: *Nectandra pisi* Miq.; *Laurac.;* br. Gu.
Ciroua-Balli: *Nectandra sp.; Laurac.;* br. Gu.
Ciruela: *Spondias cirouella* Tussac ; *Anacardiac.;* P.-R.
    „   : *Astronium conzattii* Blake; *Anacardiac.;* Guat.
    „   : *Spondias purpurea* L.; *Anacardiac.;* Hond.
    „ agria: *Spondias sp.; Anacardiac.;* Cuba, Mex.
    „ amarilla: *Spondias sp.; Anacardiac.;* Mex.
    „ calentana: *Spondias sp.; Anacardiac.;* Kol.
    „ campechana: *Spondias sp.; Anacardiac.;* Cuba.
    „ colorada: *Spondias sp.; Anacardiac.;* Mex.
    „ loca: *Spondias sp.; Anacardiac.;* Cuba.
    „ obo: *Spondias sp.; Anacardiac.;* Mex.
    „ rojo: *Spondias sp.; Anacardiac.;* Mex.
    „ de Mexico: *Spondias sp.; Anacardiac.;* Mex.
    „ del pais: *Spondias sp.; Anacardiac.;* P.-R., Mex.
Ciruelillo: *Ximenia americana* L.; *Olacac.;* Cuba.
    „   : *Behaimia cubensis* Griseb.; *Papilionac.;* Cuba.
    „   : *Astronium graveolens* Jacq.; *Anacardiac.;* Hond.
    „   : *Phyllanthus conami* Sw.; *Euphorbiac.;* br. Hond.
    „   : *Embothrium coccineum* Forst.; *Proteac.;* Chile, Arg.
Ciruello: *Astronium sp.; Anacardiac.;* Hond.
Ciruelo: *Spondias purpurea* L.; *Anacardiac.;* br. Hond., Ek.
    „   : *Spondias lutea* L.; *Anacardiac.;* Ek., Cuba.
    „ cimarrón: *Ximenia americana* L.; *Olacac.;* Cuba.
    „ de hueso: *Spondias purpurea* L.; *Anacardiac.;* Ven.
Cirujano: *Bursera sp.; Burserac.;* Mex.
Citroenboomhout: *Chlorophora tinctoria* Gaudich.; *Morac.;* tr. Am.
Citroenhout: *Chlorophora tinctoria* Gaudich.; *Morac.;* tr. Am.
Citron: *Citrus medica* L.; *Rutac.;* Tr., Sub-Tr.
    „   : *Chlorophora tinctoria* Gaudich.; *Morac.;* tr. Am.
Citrón: *Cashalia panamensis* Standl.; *Leguminos.;* Pan.
Citron de mer: *Ximenia americana* L.; *Olacac.;* Gab.
Citronenholz, afrikanisch: *Distemonanthus benthamianus* Baill.; *Caesalpiniac.;* W.-Af.
Citronera: *Murraya exotica* L.; *Rutac.;* Trin.
Citronier: *Fagara flava* Krug et Urb.; *Rutac.;* W.-Ind.
    „   : *Chloroxylon swietenia* DC.; *Rutac.;* O.-Ind.
Citronnier (f): *Citrus limonum* Risso; *Rutac.;* M.-M.
    „   : *Chlorophora tinctoria* Gaudich.; *Morac.;* tr. Am.
    „ blanc: *Ilex sideroxyloides* Griseb.; *Aquifoliac.;* Guad.
    „ doux: *Citrus decumana* L.; *Rutac.;* Réu.
    „ vrai: *Citrus medica* L.; *Rutac.;* Tr., Sub-Tr.
    „ de Ceylon: *Chloroxylon swietenia* DC.; *Rutac.;* Ind.-Ch.
    „ des juifs: *Citrus medica* L.; *Rutac.;* Tr., Sub-Tr.
Ckara-huilca: *Piptadenia sp.; Mimosac.;* Peru.
Clarahiba: *Cordia sp.; Borraginac.;* Bras.
Claraiba: *Cordia sp.; Borraginac.;* s. Bras.
    „ das algoas: *Cordia sp.; Borraginac.;* Bras.
Clavalier (f): *Zanthoxylum fraxineum* Willd.; *Rutac.;* USA., Mex.
    „ des Antilles (f): *Zanthoxylum clava-herculis* L.; *Rutac.;* W.-Ind.
Clavellina: *Bombax ellipticum* H. B. K.; *Bombacac.;* Mex., Guat.
    „   : *Tapetes erecta* L.; *Composit.;* Pan.
Clavellina: *Calliandra spp.; Mimosac.;* Ven.
    „ colorado: *Caesalpinia pulcherrima* Sw.; *Caesalpiniac.;* Ven.
    „ de la barranca: *Bombax palmeri* S. Watson; *Bombacac.;* Mex.

Clavillo: *Hamelia rovirosae* Wernham; *Rubiac.;* Hond.
Clavito: *Hamelia erecta* Jacq.; *Rubiac.;* Guat.
,, : *Erythroxylon sp.; Erythroxylac.;* Ven.
Clavo: *Chomelia recordi* Standl.; *Rubiac.;* Guat.
,, : *Pisonia macranthocarpa* Donn. Smith; *Nyctaginac.;* Guat.·
Claw berries: *Phyllanthus nobilis* Muell. Arg.; *Euphorbiac.;* br. Hond.
Clemón: *Thespesia populnea* Correa; *Malvac.;* Ven.
Clou: *Eugenia cotinifolia* Jacq.; *Celastrac.;* Maur.
,, à Maqui: *Aristotelia macqui* L'Hérit.; *Elaeocarpac.;* Chile.
Clubwood: *Swartzia tomentosa* DC.; *Caesalpiniac.;* br. Gu.
Clusier rose: *Clusia rosea* L.; *Guttifer.;* tr. Am.
Cô: *Quercus spp.; Fagac.;* Ind.-Ch.
,, : *Pterospermum grewiaefolium* Pierre; *Sterculiac.;* Ind.-Ch.
,, bay: *Canarium tonkinense* Engl.; *Burserac.;* Ind.-Ch.
,, cà nan: *Phoebe lanceolata* Nees; *Laurac.;* Ind.-Ch.
,, cai san: *Horsfieldia amygdalina* Warb.; *Myristicac.;* Ind.-Ch.
,, cha ba heng: *Chisocheton cochinchinensis* Pierre; *Meliac.;* Ind.-Ch.
,, ,, chiâ: *Podocarpus latifolia* Wall.; *Conifer.;* Ind.-Ch.
,, ,, pá: *Alangium chinense* Rehder; *Cornac.;* Ind.-Ch.
,, ,, ,, : *Marlea begoniaefolia* Roxb.; *Cornac.;* Ind.-Ch.
,, chi: *Strychnos nux-vomica* L.; *Loganiac.;* Ind.-Ch.
,, deng: *Quercus chevalieri* Hick. et A. Camus; *Fagac.;* Ind.-Ch.
,, giáy: *Bocagea philastreana* Pierre; *Anonac.;* Co.-Ch.
,, gié: *Mitrephora bousigoniana* Pierre; *Anonac.;* Annam.
,, ,, nui: *Mitrephora thoreli* Pierre; *Anonac.;* Kamb.
,, ham hoc: *Ficus sp.; Morac.;* Ind.-Ch.
,, hang quay: *Polyalthia jucunda* Finet et Gagnep.; *Anonac.;* Ind.-Ch.
,, ke (H): *Grewia paniculata* Roxb.; *Tiliac.;* Annam.
,, khet: *Pithecolobium clypearia* Bth. var. *acuminatum* Gagnep.; *Mimosac.;* Ind.-Ch.
,, khe mu: *Quercus sp.; Fagac.;* Ind.-Ch.
,, ky: *Hopea odorata* Roxb.; *Dipterocarpac.;* Kamb.
,, lan: *Lagerstroemia tomentosa* Presl; *Lythrac.;* Ind.-Ch.
,, lim: *Erythrophloeum fordi* Oliv.; *Caesalpiniac.;* Ind.-Ch.
,, mac cang: *Albizzia stipulata* Boivin; *Mimosac.;* Ind.-Ch.
,, muc: *Wrightia ovata* A. DC.; *Apocynac.;* Ind.-Ch.
,, -nhap: *Columbia thoreli* Gagnep.; *Tiliac.;* Ind.-Ch.
,, -nhôm-nhôm: *Elaeocarpus dubius* A. DC.; *Elaeocarpac.;* Ind.-Ch.
,, pâng: *Sapium baccatum* Roxb.; *Euphorbiac.;* Ind.-Ch.
,, pen: *Sapium sp.,* *Mallotus albus* Muell. Arg.; *Euphorbiac.;* Ind.-Ch.
., phan: *Eugenia resinosa* Gagnep.; *Myrtac.;* Ind.-Ch.
,, phuong: *Averrhoa carambola* L.; *Oxalidac.;* Ind.-Ch.
,, pô: *Hibiscus praeclarus* Gagnep.; *Malvac.;* Ind.-Ch.
,, san: *Dillenia indica* L.; *Dilleniac.;* Ind.-Ch.
,, sen: *Pterospermum grewiaefolium* Pierre; *Sterculiac.;* Ind.-Ch.
,, song: *Wendlandia paniculata* DC.; *Rubiac.;* Ind.-Ch.
Coa obety: *Luehea grandiflora* Mart.; *Tiliac.;* Par.
Coach wood: *Ceratopetalum apetalum* D. Don; *Cunoniac.;* ö. Au.
,, ,, : *Callicoma serratifolia* Andr.; *Cunoniac.;* ö. Au.
Coacha machalli: *Pithecolobium dulce* Bth.; *Mimosac.;* Mex.
Coache: *Picraena sp.; Simarubac.;* fr. W.-Ind.
Coachi: *Quassia amara* L.; *Simarubac.;* fr. W.-Ind.
Coajin-guva: *Ficus anthelmintica* Mart.; *Morac.;* Bras. (Amaz.)
Coaopía: *Vismia sp.; Guttifer.;* Bras.
Coapinol: *Hymenaea courbaril* L.; *Caesalpiniac.;* Mex.
Coapma: *Erythrina rubrinervia* H. B. K.; *Papilionac.;* br. Hond.
Coariuba: *Vochysia sp.; Vochysiac.;* Bras.
Coataquiçaua: *Peltogyne paradoxa* Ducke; *Caesalpiniac.;* Bras. (Pará).
Coatari: *Couratari guianensis* Aubl.; *Myrtac.;* fr. Gu.
,, rouge: *Lecythis sp.; Lecythidac.;* fr. Gu. (Cayenne).
Coate: *Eysenhardtia polystachia* Sarg.; *Papilionac.;* Mex.
Coatindiva: *Trema micrantha* Blume; *Ulmac.;* Bras.
Coatl: *Eysenhardtia polystachia* Sarg.; *Papilionac.;* Mex.

Coatli: *Eysenhardtia polystachia* Sarg.; *Papilionac.;* Mex.
Cobana negra: *Stahlia monosperma* Urban; *Caesalpiniac.;* P.-R.
Cobano: *Swietenia sp.,* ? *S. macrophylla* King; *Meliac.;* Mex.
  „ : *Cedrela sp.,* ? *C. mexicana* M. Roemer; *Meliac.;* Mex.
  „ : *Stahlia maritima* Bello; *Caesalpiniac.;* P.-R.
Cobola: *Podocarpus sp.; Conifer.;* C.-R.
Côc (H): *Lumnitzera coccinea* Wight et Arn., *L. racemosa* Willd.; *Combretac.;* Ind.-Ch.
  „ : *Spondias lutea* L.; *Anacardiac.;* Ind.-Ch.
  „ : *Sterculia foetida* L.; *Sterculiac.;* Annam.
  „ rúng: *Spondias mangifera* Willd.; *Anacardiac.;* Ind.-Ch.
Coca: *Erythroxylon coca* Lam.; *Erythroxylac.;* Peru.
  „ del monte: *Erythroxylon sp.,* ? *E. coca* Lam.; *Erythroxylac.;* Arg.
Coccalobaholz (d): *Coccoloba laurifolia* Jacq.; *Polygonac.;* USA.
Cocco (i): *Cocos nucifera* L.; *Palmac.;* Tr.
Coccolobaholz (d): *Coccoloba uvifera* Jacq.; *Polygonac.;* tr. Am.
Coche: *Heliocarpus sp.; Tiliac.;* Mex.
Cochiman de montagne: *Talauma sp.; Magnoliac.;* Mart.
Cochimbo de jaboty: *Erisma sp.; Vochysiac.;* Bras. (Amaz.)
Cochinillo: *Metopium browni* Urban; *Anacardiac.;* St.-D.
Cochinito: *Escallonia floribunda* H. B. K.; *Saxifragac.;* Ven.
Cochitzapotl: *Casimiroa edulis* Llave et Lexarza; *Rutac.;* Mex.
Cochizquilitl: *Erythrina sp.; Papilionac.;* Mex.
Cochli: *Zanthoxylum rhetsa* DC.; *Rutac.;* sw. O.-Ind. (Bombay).
Cochucho: *Zanthoxylum coco* Gill.; *Rutac.;* Arg.
Cockspur: *Crataegus crus-galli* L.; *Rosac.;* USA.
  „ : *Acacia cooki* Safford; *Mimosac.;* br. Hond.
  „ : *Erythrina crista-galli* L.; *Papilionac.;* Arg.
Coco: *Cocos nucifera* L.; *Palmac.;* Hond.
  „ : *Inocarpus edulis* Aubl.; *Papilionac.;* Gu.
  „ : *Cocos romanzoffiana* Cham.; *Palmac.;* Arg.
  „ : *Zanthoxylum coco* Gill.; *Rutac.;* Arg.
  „ mamá: *Quararibea fieldi* Millsp.; *Bombacac.;* Hond.
  „ zapote: *Couroupita darienensis* Pittier; *Lecythidac.;* Pan.
  „ de cordoba: *Zanthoxylum sp.; Rutac.;* Arg.
  „ „ mer: *Lodoicea sechellarum* Labill.; *Palmac.;* Sey.
  „ „ mono: *Jugastrum christi* Pittier; *Lecythidac.;* Ven.
  „ „ „ : *Lecythis sp.; Lecythidac.;* Kol.
  „ „ purga: *Joannesia princeps* Vell.; *Euphorbiac.;* Bras.
  „ do airy: *Cocos sp.; Palmac.;* Bras.
Cocoa tree: *Theobroma cacao* L.; *Sterculiac.;* Tr.
  „ , Wild: *Pachira aquatica* Aubl.; *Bombacac.;* br. Gu.
  „ , „ : *Warszewiczia coccinea* Klotzsch; *Rubiac.;* Trin., Tob.
Cocobolito: *Psychotria chiapensis* Standl.; *Rubiac.;* Pan.
Cocobolo: *Dalbergia retusa* Hemsl.; *Papilionac.;* tr, Am.
  „ : *Dalbergia granadillo* Pittier; *Papilionac.;* Mex.
  „ : *Dalbergia hypoleuca* Pittier; *Papilionac.;* C.-R.
  „ : *Lecythis costaricensis* Pittier; *Lecythidac.;* C.-R.
  „ : *Dalbergia stevensoni* Standl., *D. laevigata* Standl.; *Papilionac.;* br. Hond.
  „ : *Dalbergia aff. lineata* Pittier; *Papilionac.;* Hond.
  „ : *Dalbergia retusa* Hemsl.; *Papilionac.;* Pan.
  „ : *Psychotria brachiata* Sw.; *Rubiac.;* Pan.
  „ : *Coccoloba nutans* Kunth; *Polygonac.;* Peru.
  „ : *Dalbergia cultrata* Graham; *Papilionac.;* Burma.
  „ , African (u): *Dalbergia obtusa* Lecomte; *Papilionac.;* Mad.
  „ , „ „ : *Dalbergia greveana* Baill.; *Papilionac.;* Mad.
  „ , Bastard-: *Platymiscium trifoliatum* Bth.; *Papilionac.;* Nic.
  „ , Indian (H): *Dalbergia cultrata* Graham; *Papilionac.;* Ind.
  „ ñambar: *Dalbergia retusa* Hemsl.; *Papilionac.;* C.-R.
  „ negro: *Dalbergia retusa* Hemsl.; *Papilionac.;* C.-R.
  „ prieto: *Dalbergia retusa* Hemsl.; *Papilionac.;* Pan.
  „ yama (New York): *Platymiscium sp.; Papilionac.;* Nic.
Cococho: ? *Zanthoxylum sp.; Rutac.;* Arg.

9*

Cocoholz: *Aporosa dioica* Muell. Arg.; *Euphorbiac.;* tr. As.
Coconut: *Cocos nucifera* L.; *Palmac.;* tr. Am.
Cocorite: *Maximiliana insignis* Mart.; *Palmac.;* Trin.
Cocos-Pflaume: *Chrysobalanus icaco* L.; *Rosac.;* Fl.
Cocotero: *Cocos nucifera* L.; *Palmac.;* Mex.
Cocotier (f): *Cocos nucifera* L.; *Palmac.;* Tr.
Cocso: *Tetrapleura thonningi* Bth.; *Mimosac.;* Gab.
Cocú: *Andira* ? *inermis* H. B. K.; *Papilionac.;* Pan.
Cocucho: *Zanthoxylum coco* Griseb.; *Rutac.;* Arg.
Cocuile: *Ichthyomethia sp.; Papilionac.;* Mex.
Cocuite: *Ichtyomethia piscipula* Hitch.; *Papilionac.;* Mex.
Cocus: *Brya ebenus* DC.; *Papilionac.;* W.-Ind.
Cocuswood: *Brya ebenus* DC.; *Papilionac.;* Jam.
Cocuyo: *Bumelia cuneata* Swartz, *B. nigra* Sw.; *Sapotac.;* Cuba.
Codeso: *Cytisus laburnum* L.; *Papilionac.;* Arg.
   „   de los Alpes: *Laburnum alpinum* Griseb.; *Papilionac.;* Arg.
Coemari: *Coccoloba sp.; Polygonac.;* holl. W.-Ind.
Coenatepi: *Platymiscium sp.; Papilionac.;* holl. Gu.
Coentrilho: *Zanthoxylum hyemale* St. Hil.; *Rutac.;* Bras.
Coerana: *Solanum auriculatum* Ait.; *Solanac.;* Bras. (Sao P).
   „   : *Ipomoea bona-nox* L.; *Convolvulac.;* Bras.
Coeur dehors: *Diplotropis guianensis* Bth., *D. leptophylla* Kleinh.; *Papilionac.;* Sur.
   „   rouge (f): *Haematoxylon campechianum* L.; *Caesalpiniac.;* tr. Am.
   „   vert (f): *Ocotea rodioei* Mez; *Laurac.;* Sur.
   „   de boeuf: *Anona reticulata* L.; *Anonac.;* Guat.
   „   „   cachiman: *Talauma sp.; Magnoliac.;* Guad.
Cofa: *Triplochiton scleroxylon* K. Schum.; *Sterculiac.;* Elf.
Coffee: *Coffea arabica* L.; *Rubiac.;* Tr.
   „   mortar: *Terminalia sp.; Combretac.;* br. Gu.
   „   , wild: *Rinorea hummelli* Sprague; *Violac.;* br. Hond.
   „   ,   „   : *Colubrina panamensis* Standl.; *Rhamnac.;* Pan.
   „   ,   „   : *Compsoneura costaricensis* Warb.; *Myristicac.;* Pan.
   „   ,   „   : *Posoqueria latifolia* Roem. et Schult.; *Rubiac.;* Pan.
   „   ,   „   : *Coussarea paniculata* Standl.; *Rubiac.;* Trin., Tob.
   „   ,   „   : *Faramea occidentalis* A. Rich.; *Rubiac.;* Trin., Tob.
   „   -nut: *Gymnocladus dioica* C. Koch; *Caesalpiniac.;* ö. USA.
   „   -tree: *Gymnocladus dioica* C. Koch; *Caesalpiniac.;* USA.
   „   - „   , Kentucky-: *Gymnocladus dioica* C. Koch; *Caesalpiniac.;* ö N.-Am.
Cognassier: *Cydonia oblonga* Mill.; *Rosac.;* Guat.
   „   du Bengale (f): *Aegle marmelos* Correa; *Rutac.;* Ind.
Cogotone: *Tabernaemontana citrifolia* L.; *Apocynac.;* br. Hond.
Cogwood: *Zizyphus chloroxylon* Oliv.; *Rhamnac.;* Jam.
   „   : *Tecoma pentaphylla* Juss.; *Bignoniac.;* Tob.
   „   , White: *Homalium sp.; Flacourtiac.;* Jam.
Cohigüe: *Nothofagus sp.; Fagac.;* Arg.
Cohune: *Orbygnia cohune* Dahlgren; *Palmac.;* br. Hond.
Coi (H): *Pterocarya stenoptera* A. DC.; *Juglandac.;* Ind.-Ch.
Coigüe: *Nothofagus dombeyi* Mirb.; *Fagac.;* Chile.
Coihue: *Nothofagus dombeyi* Mirb.; *Fagac.;* Chile.
Coirana: *Solanum auriculatum* Ait.; *Solanac.;* Bras. (Sao P).
Coji: *Prosopis sp.; Mimosac.;* Curacao.
Cojoba: *Piptadenia peregrina* Bth.; *Mimosac.;* P.-R.
   „   : *Pithecolobium arboreum* Urban; *Mimosac.;* P.-R.
Cójobana: *Pithecolobium arboreum* Urban; *Mimosac.;* P.-R.
   „   : *Piptadenia peregrina* Bth.; *Mimosac.;* P.-R.
Cojobilla: *Piptadenia peregrina* Bth.; *Mimosac.;* P.-R.
Cojobo: *Piptadenia sp.; Mimosac.;* P.-R.
Cojón: *Tabernaemontana citrifolia* L.; *Apocynac.;* Guat.
   „   de burro: *Stemmadenia donnell-smithi* Woodson; *Apocynac.;* Hond.
   „   „   cabrito: *Tabernaemontana sp.; Apocynac.;* Bras.
   „   „   fraille: *Bunchosia glandulosa* DC.; *Malpighiac.;* br. Hond.
   „   „   mico: *Stemmadenia donnell-smithi* Woodson; *Apocynac.;* br. Hond.

Cojón de mico: *Tabernaemontana amygdalifolia* Jacq.; *Apocynac.*; Guat.
„   „   „ : *Tabernaemontana grandiflora* Jacq.; *Apocynac.*; Pan.
„   „ venado: *Sapranthus nicaraguensis* Standl.; *Anonac.*; Guat.
Cojotón: *Stemmadenia donnell-smithi* Woodson; *Apocynac.*; br. Hond.
„   : *Tabernaemontana spp.*; *Apocynac.*; br. Hond.
Cokewood: *Chlorophora excelsa* Bth. et Hook.; *Morac.*; Elf.
Cola de marano: *Pithecolobium arboreum* Urban; *Mimosac.*; Guat., Hond.
„   „ mico: *Pithecolobium arboreum* Urban; *Mimosac.*; Guat.
„   „ pava: *Cupania guatemalensis* Radlk.; *Sapindac.*; Guat.
„   „   „ : *Cupania asperula* Standl.; *Sapindac.*; Nic.
„   „ pavo: *Trichilia izabalana* Blake; *Meliac.*; Guat.
„   „   „ : *Cupania glabra* Sw.; *Sapindac.*; Hond.
Colabazo: *Crescentia cucurbitina* L.; *Bignoniac.*; tr. Am.
Colas: *Gymnanthes sp.*; *Euphorbiac.*; Mart.
Colatier (f): *Cola vera* K. Schum.; *Sterculiac.*; Elf.
„   : *Cola ballayi* Cornu; *Sterculiac.*; Kam.
Colca: *Miconia pastoënsis* Triana; *Melastomatac.*; Ek.
Colebasse: *Crescentia cucurbitina* L.; *Bignoniac.*; tr. Am.
Coleira: *Cola acuminata* R. Br.; *Sterculiac.*; port. O.-Af.
Coletui: *Coronilla emerus* L.; *Papilionac.*; Sp.
Coletuy: *Coronilla emerus* L.; *Papilionac.*; Arg.
Colima: *Zanthoxylum sp.*; *Rutac.*; Mex.
Colita: *Cordia sp.*; *Borraginac.*; Arg.
Coloc: *Talisia floresi* Standl.; *Sapindac.*; Mex. (Yucatan).
Colon: *Adina cordifolia* Hook. f.; *Rubiac.*; Cey.
Colonial deal: *Podocarpus totara* G. Benn.; *Conifer.*; N.-S.-W.
Colophane: *Canarium colophania* Baker; *Burserac.*; Au., Maur.
Colophanholz: *Canarium paniculatum* Bth.; *Burserac.*; Maur.
Coloradillo: *Grislea secunda* Loefl.; *Lythrac.*; Hond.
„   : *Hamelia spp.*; *Rubiac.*; Hond.
Colorado: *Aspidosperma megalocarpon* Muell. Arg.; *Apocynac.*; s. Mex.
Coloridito: *Tovomitopsis multiflora* Standl.; *Guttifer.*; Pan.
Colorín: *Erythrina rubrinervia* H. B. K.; *Papilionac.*; br. Hond.
„ negro: *Erythrina sp.*; *Papilionac.*; Mex.
„ de peces: *Ichthyomethia sp.*; *Papilionac.*; Mex.
Colpachi: *Croton glabellus* L.; *Euphorbiac.*; Pan.
Colutea (i): *Colutea arborescens* L.; *Papilionac.*; Mitt.-Eu.
Côm (H): *Xantonnea quocensis* Pierre; *Rubiac.*; Ind.-Ch.
„ : *Elaeocarpus dubius* A. DC.; *Elaeocarpac.*; Ind.-Ch.
„ nép: *Aporosa microcalyx* Hassk.; *Euphorbiac.*; Ind.-Ch.
„ nguôi (H): *Popowia aberrans* Pierre; *Anonac.*; Ind.-Ch.
Coma: *Bumelia spiniflora* A. DC.; *Sapotac.*; Mex.
Comandatubá: *Hirtella sp.*; *Rosac.*; Bras.
Comarre: *Dipteryx odorata* Willd.; *Papilionac.*; Sur.
Comasuche: *Maximilianea vitifolia* Krug et Urban; *Bixac.*; Guat.
Comayagua: *Cassia biflora* L.; *Caesalpiniac.*; Hond.
Combeningo: *Sarcocephalus sp.*; *Rubiac.*; Gab.
Combo-combo: *Musanga smithi* R. Br.; *Morac.*; W.-Af.
„ - „ : *Aucoumea klaineana* Pierre; *Burserac.*; Gab.
Come negro: *Lonchocarpus latifolius* H. B. K.; *Papilionac.*; Pan.
„   „ : *Dialium divaricatum* Vahl; *Caesalpiniac.*; Nic.
Comida de culebra: *Casearia nitida* Jacq.; *Flacourtiac.*; Hond.
„ de loro: *Casearia nitida* Jacq.; *Flacourtiac.*; Pan.
„ del mono: *Protium sessiliflorum* Standl.; *Burserac.*; Pan.
Comino liso: *Aniba perutilis* Hemsl.; *Laurac.*; Kol.
„ wood (u): *Aniba perutilis* Hemsl.; *Laurac.*; Kol.
Como: *Bumelia sp.*; *Sapotac.*; Haiti.
„ -como: *Barteria dewevrei* de Wild. et Dur.; *Flacourtiac.*; Gab.
Comou: *Oenocarpus bacaba* Mart.; *Palmac.*; fr. Gu.
Comuchín: *Ficus sp.*; *Morac.*; Mex.
Con: *Perymenium strigillosum* Greenm.; *Composit.*; Hond.
Côn côn: *Elaeocarpus dongnaiensis* Pierre; *Elaeocarpac.*; Ind.-Ch.

Côn hen ti tey: *Mitrephora edwardsi* Pierre; *Anonac.;* Ind.-Ch.
Conacaste (u): *Enterolobium cyclocarpum* Griseb.; *Mimosac.;* Guat.
Concha de huevo: *Morinda panamensis* Seem.; *Rubiac.;* Hond.
„ de indio: *Naucleopsis naga* Pittier; *Morac.;* Hond.
Conchido: *Pithecolobium arboreum* Urban; *Mimosac.;* C.-R.
Conchudo: *Pithecolobium filicifolium* Bth.; *Mimosac.;* W.-Ind., C.-Am.
Condessa: *Anona spp.; Anonac.;* n. Bras. (Pará).
Condori wood: *Adenanthera pavonia* L.; *Mimosac.;* tr. As.
Condurù: *Ferolia guianensis* Aubl.; *Rosac.;* n. Bras.
Condurú: *Brosimum conduru* Allem.; *Morac.;* n. Bras. (Pará).
„ de sangue: *Brosimum paraënse* Huber; *Morac.;* Bras.
Conejo colorado: *Trichilia hirta* L.; *Meliac.;* Pan.
Conessi bark: *Holarrhena antidysenterica* Wall.; *Apocynac.;* Ind.
Confite: *Zizyphus sonorensis* Watson; *Rhamnac.;* Mex.
Cong: *Antiaris toxicaria* Leschen.; *Morac.;* Ind.-Ch.
„ : *Calophyllum saïgonense* Pierre; *Guttifer.;* Ind.-Ch.
„ : *Calophyllum dryobalanoïdes* Pierre; *Guttifer.;* Ind.-Ch.
„ : *Endospermum chinense* Bth.; *Euphorbiac.;* Ind.-Ch.
„ -gass: *Schleichera trijuga* Willd.; *Sapindac.;* Ind.
„ giay: *Calophyllum sp.; Guttifer.;* Annam.
„ mùu: *Calophyllum thoreli* Pierre; *Guttifer.;* Annam.
„ nuï: *Calophyllum dryobalanoïdes* Pierre; *Guttifer.;* Annam.
„ nuóc: *Calophyllum dongnaiensis* Pierre; *Guttifer.;* Ind.-Ch.
„ tau lau: *Calophyllum spectabile* Willd.; *Guttifer.;* Annam.
„ tia: *Calophyllum saïgonense* Pierre; *Guttifer.;* Annam.
„ trang: *Calophyllum dryobalanoïdes* Pierre; *Guttifer.;* Annam.
„ „ : *Calophyllum spectabile* Willd.; *Guttifer.;* Annam.
„ vay ôc: *Calophyllum pulcherrimum* Wall.; *Guttifer.;* Ind.-Ch.
Conghas: *Schleichera trijuga* Willd.; *Sapindac.;* Ind.
Congo: *Andira inermis* H. B. K.; *Papilionac.;* Kol.
„ : *Shorea tumbuggaia* Roxb.; *Dipterocarpac.;* sö. O.-Ind.
„ pump: *Cecropia peltata* L.; *Morac.;* br. Gu.
Congonha: *Villaresia mucronata* R. et P.; *Icacinac.;* Bras. (Sao P).
„ do campo: *Vochysia sp.; Vochysiac.;* Bras.
Congenheiro: *Vochysia sp.; Vochysiac.;* Bras.
Congowood (u): *Lovoa klaineana* Pierre; *Meliac.;* W.-Af.
Congu: *Stephegyne parvifolia* Korth.; *Rubiac.;* w. O.-Ind. (n. Bombay).
Conguérecou: *Xylopia frutescens* Aubl.; *Anonac.;* Gu.
Conh ranh (H): *Erismanthus indochinensis* Gagnep.; *Euphorbiac.;* Ind.-Ch.
„ so ca lui: *Mallotus furetianus* Muell. Arg.; *Euphorbiac.;* Ind.-Ch.
Conjerana-assú: *Cabralea sp.; Meliac.;* Bras.
Conocarpe droit: *Conocarpus erecta* L.; *Combretac.;* tr. Am.
Contrevent: *Lucuma multiflora* A. DC.; *Sapotac.;* tr. Am.
Coobiaby: *Sarcocephalus cordatus* Miq.; *Rubiac.;* Queensl.
Coobo: *Ficus sp.; Morac.;* Mex.
Coochin-coochin: *Eugenia smithi* Poir.; *Myrtac.;* Queensl.
Coolibah: *Eucalyptus microtheca* F. v. M.; *Myrtac.;* w. Au.
Coolibar: *Eucalyptus microtheca* F. v. M.; *Myrtac.;* Queensl.
Cooloon: *Elaeocarpus grandis* F. v. M.; *Elaeocarpac.;* Au.
Coomeroo-coomeroo: *Pipturus argenteus* Wedd.; *Urticac.;* Au.
Coonam: *Araucaria cunninghami* Ait.; *Conifer.;* ö. Au.
Coondoo: *Sideroxylon richardi* F. v. M.; *Sapotac.;* Au.
„ , Blush-: *Sideroxylon richardi* F. v. M.; *Sapotac.;* Au.
Cooperhood: *Brownea latifolia* Jacq.; *Caesalpiniac.;* Trin., Tob.
Coopers wood: *Pomaderris apetala* Labill.; *Rhamnac.;* s. Au.
„ „ : *Alphitonia excelsa* Reisseck; *Rhamnac.;* N.-S.-W.
Coorong: *Callitris robusta* R. Br.; *Conifer.;* N.-S.-W.
Cop smoul: *Walsura villosa* Wall.; *Meliac.;* Ind.-Ch.
Copá: *Protium panamensis* Rose; *Burserac.;* Pan.
Copahiba: *Copaifera langsdorffi* Desf.; *Caesalpiniac.;* Bras.
„ angelim: *Copaifera multijuga* Hayne; *Caesalpiniac.;* Bras.
„ cuiarana: *Copaifera glycycarpa* Ducke; *Caesalpiniac.;* Bras. (Amaz.)

Copahiba jutahy: *Copaifera reticulata* Ducke, *C. marti* Hayne; *Caesalpiniac.*; Bras.
„ marimary: *Copaifera multijuga* Hayne; *Caesalpiniac.*; Bras. (Amaz.)   [(Amaz.)
„   „   : *Copaifera reticulata* Ducke; *Caesalpiniac.*; Bras. (Amaz.)
„ preta: *Copaifera glycycarpa* Ducke; *Caesalpiniac.*; Bras. (Amaz.)
„ rana: *Eperua purpurea* Bth.; *Caesalpiniac.*; Bras. (Amaz.)
Copahibeira: *Copaiba spp.*; *Caesalpinia.*; Bras. (Amaz.)
Copahú: *Copaifera sp.*; *Caesalpiniac.*; fr. W.-Ind.
Copahúva: *Copaifera langsdorffi* Desf.; *Caesalpiniac.*; Bras. (Sao P).
Copahy: *Copaifera langsdorffi* Desf.; *Caesalpiniac.*; Bras. (Sao P).
Copahyba: *Copaifera officinalis* L., *C. langsdorffi* Desf.; *Caesalpiniac.*; Bras.
„ parda: *Copaifera sp.*; *Caesalpiniac.*; Bras. (Sao P).
„ vermelha: *Copaifera sp.*; *Caesalpiniac.*; Bras. (Sao P).
Copaïa: *Jacaranda copaia* D. Don; *Bignoniac.*; fr. Gu.
Copaiba: *Copaifera sp.*; *Caesalpiniac.*; Ven.
Copaiva: *Copaifera guianensis* Desf.; *Caesalpiniac.*; Sur.
„   : *Copaifera sp.*; *Caesalpiniac.*; Peru.
Copaiyé: *Vochysia guianensis* Aubl.; *Vochysiac.*; br. Gu.
Copal: *Protium guianense* March.; *Burserac.*; Cuba.
„   : *Protium sessiliflorum* Standl.; *Burserac.*; Hond.
„   : *Hedwigia balsamifera* Swartz; *Burserac.*; Salv.
„   : *Hymenaea courbaril* L.; *Caesalpiniac.*; Ek.
„   : *Daniella similis* Craib; *Caesalpiniac.*; Lib.
„ amargo: · *Bursera sp.*; *Burserac.*; Mex.
„ amargoso: *Bursera sp.*; *Burserac.*; Mex.
„ blanco: *Bursera sp.*; *Burserac.*; Mex.
„ chino: *Bursera sp.*; *Burserac.*; Mex.
„ , Indian: *Vateria indica* L.; *Dipterocarpac.*; Ind.
„ macho: *Protium sessiliflorum* Standl.; *Burserac.*; br. Hond.
„ quahuitl: *Schinus molle* L.; *Anacardiac.*; Peru.
„ , red: *Cupania guatemalensis* Radlk.; *Sapindac.*; br. Hond.
„ santo: *Bursera sp.*; *Burserac.*; Mex.
„ de penca: *Bursera sp.*; *Burserac.*; Mex.
Copalier: *Copaïfera spp.*; *Caesalpiniac.*; W.-Af.
„ d'Afrique: *Copaifera sp.*; *Caesalpiniac.*; Gab.
„ d'Amérique: *Hymenaea courbaril* L.; *Caesalpiniac.*; fr. Gu.
Copalillo: *Eupatorium daleoides* Hemsl.; *Composit.*; Guat.
Copalm: *Liquidambar styraciflua* L.; *Hamamelidac.*; ö. USA.
Copalquahuitl: *Bursera sp.*; *Burserac.*; Mex.
Copalquin: *Coutarea pterosperma* Standl.; *Rubiac.*; Mex.
Copalxochitl: *Bursera sp.*; *Burserac.*; Mex.
Copaúba: *Copaifera sp.*; *Caesalpiniac.*; Bras.
Copay-yè: *Vochysia guianensis* Aubl.; *Vochysiac.*; br. Gu.
Copé: *Clusia odorata* Seem.; *Guttifer.*; Pan.
Cope grande: *Clusia rosea* Jacq.; *Guttifer.*; Pan.
Copey: *Clusia rosea* Jacq.; *Guttifer.*; tr. Am.
Copeysillo: *Clusia flava* Jacq.; *Guttifer.*; Cuba.
Copi: *Goupia glabra* Aubl.; *Celastrac.*; Gu.
Copié: *Goupia glabra* Aubl.; *Celastrac.*; Gu.
Copie, red: *Goupia glabra* Aubl.; *Celastrac.*; holl. Gu.
Copinol: *Hymenaea courbaril* L.; *Bignoniac.*; Salv.
Copinole: *Hymenaea courbaril* L.; *Caesalpiniac.*; Mex.
Copón: *Bursera gummifera* L.; *Burserac.*; Hond.
Copté: *Cordia sp.*; *Borraginac.*; Mex.
Copuda: *Couepia sp.*; *Rosac.*; Bras.
Coqi: *Styrax leprosum* Hook. et Arn.; *Styracac.*; Arg.
Coquillo: *Lecythis currani* Pittier; *Lecythidac.*; Kol
Coquirana: *Ecclinusa balata* Ducke; *Sapotac.*; Bras. (Amaz.)
Coquito: *Bombax ellipticum* H. B. K.; *Bombacac.*; Mex.
Corabina: *Plumiera sp.*; *Apocynac.*; Bras.
Coracão de boi: *Anona sp.*; *Anonac.*; Bras.
„ „ negro: *Swartzia ingaefolia* Ducke; *Caesalpiniac.*; Bras. (Amaz.)
„ „ „ : *Cassia apoucouita* Aubl.; *Caesalpiniac.*; Bras. (Bahia).

Corail: *Pterocarpus soyauxi* Taub.; *Papiliónac.;* Gab.
„ : *Pterocarpus sp.; Papilionac.;* As.
Coral: *Hamelia patens* Jacq.; *Rubiac.;* Hond.
„ tree: *Erythrina crista-galli* L.; *Papilionac.;* Arg.
„ „ : *Erythrina indica* Lam.; *Papilionac.;* Queensl.
„ „ , Bat's-wing-: *Erythrina vespertilio* Bth.; *Papilionac.;* Au.
„ „ , common: *Erythrina crista-galli* L.; *Papilionac.;* Arg.
„ wood: *Adenanthera pavonina* L.; *Mimosac.;* tr. As., And.
Coralillo: *Inga rufescens* Bth.; *Mimosac.;* Pan.
„ : ? *Inga spuria* Humb. et Bonpl.; *Mimosac.;* Pan.
„ : *Hamelia patens* Jacq.; *Rubiac.;* Hond., Nic.
„ : *Hamelia spp.; Rubiac.;* Ven.
Coralleira: *Isertia hypoleuca* Bth.; *Rubiac.;* Bras.
Corallero: *Coccoloba ramosissima* Lindl.; *Polygonac.;* Kol.
„ : *Randia aculeata* L.; *Rubiac.;* Kol.
Coralmeca: *Schlegelia nicaraguensis* Standl.; *Bignoniac.;* Nic.
Corana: *Solanum auriculatum* Ait.; *Solanac.;* Bras. (Sao P).
Coratu: *Enterolobium cyclocarpum* Griseb.; *Mimosac.;* Pan.
Corazón: *Anona glabra* L.; *Anonac.;* s. Fl., W.-Ind.
„ : *Anona reticulata* L.; *Anonac.;* P.-R.
„ : *Anona sp.; Anonac.;* Trin.
„ cimarrón: *Anona glabra* L.; *Anonac.;* P.-R.
„ de bugre: *Lithraea lorentziana* Hieron.; *Anacardiac.;* S.-Am.
„ „ paloma: *Doerpfeldia cubensis* Urban; *Rhamnac.;* Cuba.
Corbán: *Symphonia globulifera* L. f.; *Guttifer.;* br. Hond.
Corcho: *Anona palustris* L.; *Anonac.;* tr. Am.
„ -Cajete: *Ochroma lagopus* Swartz; *Bombacac.;* tr. Am.
Corcolén: *Azara integrifolia* R. et P.; *Flacourtiac.;* S.-Am. (Anden).
Corcorron: *Elaeodendron sp.; Celastrac.;* P.-R.
Cordiawood (u): *Cordia goeldiana* Huber; *Borraginac.;* Bras.
Cordoncillo: *Trichilia havanensis* Jacq.; *Meliac.;* br. Hond.
„ : *Piper sp.; Piperac.;* Nic.
Cordovan: *Didymopanax morototoni* Dcne. et Planch.; *Araliac.;* Cuba.
Cordovancillo: *Miconia rangeliana* Wright; *Melastomatac.;* Cuba.
Corduroy: *Sarcopteryx stipitata* Radlk.; *Sapindac.;* Au.
Corindiba: *Celtis brasiliensis* Planch.; *Ulmac.;* Bras. (R.-J.)
„ : *Trema micrantha* Blume; *Ulmac.;* Bras.
Corindiuba: *Celtis brasiliensis* Planch.; *Ulmac.;* Bras. (R.-J.)
Cork, spanish: *Thespesia populnea* Correa; *Malvac.;* Bah.
„ tree: *Quercus suber* L.; *Fagac.;* s. M.-M.
„ „ : *Erythrina vespertilio* Bth.; *Papilionac.;* Au.
„ „ , indian: *Millingtonia hortensis* L. f.; *Bignoniac.;* Tr.
Corkwood: *Leitneria floridana* Chapmann; *Leitneriac.;* N.-Am.
„ : *Anona palustris* L.; *Anonac.;* tr. Am.
„ : *Hibiscus tiliaceus* L.; *Malvac.;* Ant.
„ : *Ochroma lagopus* Swartz; *Bombacac.;* tr. Am., Ant.
„ : *Anona glabra* L.; *Anonac.;* br. Hond.
„ : *Pterocarpus draco* L., *P. rhori* Vahl; *Papilionac.;* Gu.
„ : *Musanga smithi* R. Br.; *Morac.;* W.-Af.
„ : *Bombax malabaricum* DC.; *Bombacac.;* tr. As.
„ : *Duboisia myoporoides* R. Br.; *Solanac.;* Au.
„ : *Ackama paniculata* Engl., *A. muelleri* Bth.; *Cunoniac.;* ö. Au.
„ : *Quintinia sieberi* A. DC.; *Saxifragac.;* ö. Au.
„ : *Weinmannia rubifolia* Bth.; *Cunoniac.;* N.-S.-W.
„ : *Entelea arborescens* R. Br.; *Tiliac.;* N.-Seel.
„ , grey: *Erythrina vespertilio* Bth.; *Papilionac.;* Au.
Corla: *Sequoia sempervirens* Endl.; *Conifer.;* USA. (Kalif.)
Cormier: *Pirus americana* DC.; *Rosac.;* Kan.
Cornel (e, u): *Cornus florida* L.; *Cornac.;* ö. N.-Am.
„ (e, u): *Cornus mas* L., *C. sanguinea* L.; *Cornac.;* Eu.
„ : *Cornus controversa* Hemsl.; *Cornac.;* Jap.
„ , Flowering (e, u): *Cornus florida* L.; *Cornac.;* ö. N.-Am.

Cornelwood, red: *Alphitonia excelsa* Reisseck; *Rhamnac.;* Queensl.
Cornelianwood (e, u): *Cornus florida* L.; *Cornac.;* ö. N.-Am.
Cornezuelo: *Acacia donnelliana* Safford; *Mimosac.;* Hond.
Cornicabra: *Pistacia terebinthus* L.; *Anacardiac.;* Sp.
Cornical: *Periploca laevigata* Ait.; *Asclepiadac.;* Tener.
Cornichon: *Averrhoa bilimbi* L.; *Geraniac.;* Guad.
Cornizo: *Cornus sanguinea* Forsk.; *Cornac.;* Sp.
Cornizuelo: *Acacia sp.; Mimosac.;* Nic.
Cornouiller de la Floride: *Cornus florida* L.; *Cornac.;* Kan.
Cornudo: *Swartzia sp.; Caesalpiniac.;* Pan.
Cornwood: *Andira inermis* H. B. K,; *Papilionac.;* br. Hond.
Corobore: *Hymenaea courbaril* L.; *Caesalpiniac.;* Ven.
Corojo: *Cocos crispa* H. B. K.; *Palmac.;* Cuba.
Coromandelholz: *Diospyros hirsuta* L. f.; *Ebenac.;* Ind., Cey.
    „   : *Diospyros melanoxylon* Roxb.; *Ebenac.;* Ind., Cey.
Coromandelhout (h): *Diospyros melanoxylon* Roxb.; *Ebenac.;* Nied.-Ind.
Corona de Cristo: *Koeberlinia spinosa* Zuccarini; *Koeberliniac.,* s. USA., Mex.
    „ „ espinas: *Gleditschia amorphoides* Taub.; *Caesalpiniac.;* Arg.
Coronel: *Krugiodendron ferreum* Urban; *Rhamnac.;* W.-Ind., C.-Am.
Coronilla: *Scutia buxifolia* Reisseck; *Rhamnac.;* Arg.
    „ colorado: *Scutia buxifolia* Reisseck; *Rhamnac.;* W.-Ind.
    „ de Rey: *Globularia alypum* L.; *Globulariac.;* Sp.
Coronillo: *Scutia buxifolia* Reisseck; *Rhamnac.;* Arg.
    „   : *Gleditschia amorphoides* Taub.; *Caesalpiniac.;* Arg.
Corosol: *Anona sp.; Anonac.;* Trin.
Corosolier: *Anona muricata* L.; *Anonac.;* Guat.
Corossol: *Anona spp.; Anonac.;* fr. Gu.
    „ cachiman sauvage: *Anona muscosa* Aubl.; *Anonac.;* fr. Gu.
    „ commun: *Anona muricata* L.; *Anonac.;* fr. Gu.
    „ écailleux: *Anona squamosa* L.; *Anonac.;* W.-Ind.
    „ Grand-: *Anona muricata* L.; *Anonac.;* C.-Am.
    „ Petit-: *Anona reticulata* L.; *Anonac.;* Ant.
    „ sauvage: *Anona palludosa* Aubl.; *Anonac.;* fr. Gu.
    „ des chiens: *Anona sp.; Anonac.;* Guad., Mart.
Corossolier: *Anona muricata* L.; *Anonac.;* C.-Am.
Corozo: *Orbignya cohune* Dahlgren; *Palmac.;* n. C.-Am.
    „ : *Acrocomia sclerocarpa* Mart.; *Palmac.;* Ven.
    „ gallinazo: *Attalea gomphococca* Mart.; *Palmac.;* C.-R.
Corrimiente: *Citharexylum cooperi* Standl.; *Verbenac.;* Pan.
Corta lengua: *Casearia silvestris* Sw.; *Flacourtiac.;* Pan.
Cortes: *Tecoma guayacan* Seem.; *Bignoniac.;* Mex.
    „ : *Tecoma sp.; Bignoniac.;* Guat.
Cortés: *Tabebuia chrysantha* Nicholson; *Bignoniac.;* Hond., Nic.
    „ colorado: *Tabebuia palmeri* Rose; *Bignoniac.;* Guat.
    „ negro: *Tabebuia chrysantha* Nicholson; *Bignoniac.;* Guat.
Cortez: *Tecoma sp.; Bignoniac.;* br. Hond.
    „ : *Tabebuia donnell-smithi* Rose; *Bignoniac.;* Guat.
    „ : *Apeiba tibourbou* Aubl.; *Tiliac.;* Pan.
    „ amarillo: *Tecoma sp.; Bignoniac.;* Salv.
    „ blanco: *Tabebuia donnell-smithi* Rose; *Bignoniac.;* Salv.
    „ negro: *Tecoma sp.; Bignoniac.;* Salv.
    „ prieto: *Tabebuia chrysantha* Nicholson; *Bignoniac.;* Salv.
Corteza: *Tecoma sp.; Bignoniac.;* Guat.
    „ : *Apeiba sp.; Tiliac.;* Kol.
    „ amarilla: *Tecoma sp.; Bignoniac.;* C.-R.
Cortica: *Anona sp.; Anonac.;* Bras.
    „ : *Apeiba sp.; Tiliac.;* Bras.
Corticeira: *Anona sp.; Anonac.;* Bras.
    „ : *Tabebuia cassinoides* P. DC.; *Bignoniac.;* Bras.
    „ : *Erythrina crista-galli* L.; *Papilionac.;* Arg.
Cortissa: *Anona palustris* L.; *Anonac.;* Ant., s. Bras.
Coruba: *Attalea speciosa* Mart.; *Palmac.;* Ven.

Corusi-caá: *Calycophyllum sp.; Rubiac.;* Ek.
Cosahuico: *Sideroxylon capiri* Pittier; *Sapotac.;* Mex.
Coscoja: *Quercus coccifera* L.; *Fagac.;* Sp.
Coscorron: *Elaeodendron sp.; Celastrac.;* P.-R.
Cosomon: *Uapaca benguelensis* Muell. Arg.; *Euphorbiac.;* Elf.
Costilla de danto: *Acalypha diversifolia* Jacq.; *Euphorbiac.;* Hond.
Costixocotl: *Spondias sp.; Anacardiac.;* Mex.
Costus afer: *Bambusa nana* Roxb.; *Gramin.;* Maur.
Cotésima: *Parinarium tenuifolium* Chev.; *Rosac.;* Elf.
Cotinillo: *Metopium browni* Urban; *Anacardiac.;* St.-D.
Coton: *Ceiba pentandra* Gaertn.; *Bombacac.;* fr. Gu. (Sinnamary).
  „ perruche: *Ceiba pentandra* Gaertn.; *Bombacac.;* fr. Gu. (Oyapock).
Cotonnier: *Populus deltoides* Marsh.; *Salicac.;* Kan.
  „ -arbre: *Ceiba pentandra* Gaertn.; *Bombacac.;* Tr.
Cotonnier faux: *Ceiba pentandra* Gaertn.; *Bombacac.;* W.-Af.
  „ grand bois: *Ceiba pentandra* Gaertn.; *Bombacac.;* Sur.
Cotonron: *Luehea seemanni* Triana et Planch.; *Tiliac.;* Guat.
Cotoperis: *Talisia olivaeformis* Radlk.; *Sapindac.;* Ven.
  „ : *Myrtus erythroxyloides* H. B. K.; *Myrtac.;* Ven.
Cotoperiz: *Talisia olivaeformis* Radlk.; *Sapindac.;* Ven.
Cotoprís: *Talisia olivaeformis* Eadlk.; *Sapindac.;* Kol.
Cotorrerillo verde: *Casearia sp.; Flacourtiac.;* P.-R.
Cotorrón: *Luehea sp.; Tiliac.;* Salv.
Cotton: *Gossypium mexicanum* Tod.; *Malvac.;* br. Hond.
  „ tree: *Ceiba pentandra* Gaertn.; *Bombacac.;* br. Hond.
  „ „ : *Hibiscus tiliaceus* L.; *Malvac.;* Queensl.
  „ „ , Silk-: *Ceiba pentandra* Gaertn.; *Bombacac.;* Tr.
  „ „ , „ : *Bombax malabaricum* DC.; *Bombacac.;* Ind.
Cotton-varay: *Albizzia lebbek* Bth.; *Mimosac.;* sö. O.-Ind.
  „ varray: *Albizzia julibrissin* Durazzini; *Mimosac.;* Jap.
Cottonwood: *Populus deltoides* Marsh.; *Salicac.;* N.-Am.
  „ (H): *Bombax malabaricum* DC.; *Bombacac.;* Ind.
  „ , Balm-: *Populus trichocarpa* Torr. et Gray; *Salicac.;* Kan.
  „ , „ , western: *Populus trichocarpa* Torr. et Gray; *Salicac.;* Kan.
  „ , black: *Populus trichocarpa* Torr. et Gray; *Salicac.;* w. N.-Am.
  „ , common: *Populus deltoides* Marsh.; *Salicac.;* Kan.
  „ , eastern: *Populus deltoides* Marsh.; *Salicac.;* Kan.
  „ , Fremont-: *Populus fremonti* Watson; *Salicac.;* N.-Am.
  „ , Lanceleaf-: *Populus acuminata* Ryd.; *Salicac.;* Kan.
  „ , Narrowleaf-: *Populus angustifolia* James; *Salicac.;* Kan.
  „ , River-: *Populus heterophylla* L.; *Salicac.;* N.-Am.
  „ , Swamp-: *Populus heterophylla* L.; *Salicac.;* N.-Am.
  „ , white: *Populus fremonti* Watson; *Salicac.;* N.-Am.
Cotumo: *Crescentia cujete* L.; *Bignoniac.;* P.-R.
Cou kirp: *Cryptocarya ochracea* Lecomte; *Laurac.;* Ind.-Ch.
Couaie: *Myristica sp.; Myristicac.;* fr. Gu. (Tollinche).
  „ : *Qualea coerulea* Aubl.; *Vochysiac.;* fr. Gu.
  „ : *Virola surinamensis* Warb.; *Myristicac.;* Sur.
? Couari: *Qualea rosea* Aubl.; *Vochysiac.;* Sur.
Couatari: *Couratari guianensis* Aubl.; *Lecythidac.;* fr. Gu.
Coudrier: *Corylus avellana* L.; *Betulac.;* Eu.
Couepi: *Couepia sp.; Rosac.;* fr. Gu.
  „ : *Goupia glabra* Aubl.; *Celastrac.;* Gu.
Couhaya: *Malpighia urens* L.; *Malpighiac.;* Guad.
Coula (f): *Sarcocephalus pobeguini* Pob.; *Rubiac.;* Elf.
  „ „ : *Coula edulis* Baill.; *Olacac.;* W.-Af.
Coulequin: *Cecropia peltata* L.; *Artocarpac.;* Gu.
Couma: *Couma guianensis* Aubl.; *Apocynac.;* fr. Gu.
Coumarounaholz: *Coumarouna odorata* Aubl.; *Papilionac.;* n. Bras.
  „ : *Coumarouna oppositifolia* Taub.; *Papilionac.;* n. Bras.
Coumarou-rana: *Dipteryx sp.; Papilionac.;* fr. Gu.
Coumaté: *Pterocarpus guianensis* Aubl.; *Papilionac.;* fr. Gu.

Counana: *Astrocaryum acaule* Mart.; *Palmac.;* fr. Gu.
Couninou: *Coula edulis* Baill.; *Olacac.;* Gab.
Country timber: *Cornus florida* L.; *Cornac.;* ö. N.-Am.
Coupaia des chantiers: *Jacaranda copaia* D. Don; *Bignoniac.;* tr S.-Am.
Coupi: *Couepia guianensis* Aubl.; *Rosac.;* fr. Gu.
Couranira: *Humiria floribunda* Mart.; *Humiriac.;* Gu.
Couranoura: *Humiria floribunda* Mart.; *Humiriac.;* Gu.
Couratari: *Couratari guianensis* Aubl.; *Lecythidac.;* fr. Gu.
Courbaril: *Hymenaea courbaril* L.; *Caesalpiniac.;* fr. Gu.
  „ de Montagne: *Hymenaea courbaril* L.; *Caesalpiniac.;* Sur.
  „ „ savanna: *Hymenaea courbaril* L.; *Caesalpiniac.;* Sur.
Courida: *Avicennia nitida* Jacq.; *Verbenac.;* br. Gu.
Couroucay: *Protium heptaphyllum* March.; *Burserac.;* Trin.
Courouttay: *Atalantia monophylla* Correa; *Rutac.;* s. Ind.
Courridjah: *Banksia integrifolia* L. f.; *Proteac.;* N.-S.-W.
Coutaballi: *Licania sp.; Rosac.;* br. Gu.
Covalam: *Aegle marmelos* Correa; *Rutac.;* Ind.
Cowdie: *Agathis australis* Steud.; *Conifer.;* N.-Seel.
Cowce: *Pterocarpus belizensis* Standl.; *Papilionac.;* Hond.
Cow-itch: *Urera baccifera* Gaudich.; *Urticac.;* br. Hond.
  „ „ : *Myriocarpa yzabalensis* Killip; *Urticac.;* Pan.
  „ -okra: *Parmentiera edulis* DC.; *Bignoniac.;* br. Hond.
Cowrie: *Agathis australis* Steud.; *Conifer.;* N.-Seel.
Cow tree: *Brosimum utile* Pittier; *Morac.;* Ven.
  „ „ : *Mimusops elata* Allem.; *Sapotac.;* n. Bras. (Pará).
  „ „ : *Couma sapida* Pittier; *Apocynac.;* Ven.
  „ „ : *Couma guatemalensis* Standl.; *Apocynac.;* Guat.
  „ „ , Humboldt-: *Brosimum utile* Pittier; *Morac.;* Pan., Ven.
Coxa de frango: *Vochysia sp.; Vochysiac.;* Bras.
Coyán: *Nothofagus sp.; Fagac.;* Arg.
Coyghue: *Nothofagus dombeyi* Mirb.; *Fagac.;* Chile.
Coyol: *Acrocomia mexicana* Karw.; *Palmac.;* Hond.
Coyolillo: *Matayba glaberrima* Radlk.; *Sapindac.;* Guat.
Cozquelite: *Erythrina sp.; Papilionac.;* Mex.
Cozticxocotl: *Spondias sp.; Anacardiac.;* Mex.
Cozticzapotl: *Lucuma salicifolia* H. B. K.; *Sapotac.;* Mex.
Coztilxocotl: *Spondias sp.; Anacardiac.;* Mex.
Crab, sweet-scented: *Pirus coronaria* L.; *Rosac.;* ö. USA.
  „ , wild: *Pirus coronaria* L.; *Rosac.;* ö. USA.
Crabbaum (d): *Carapa guianensis* Aubl.; *Meliac.;* Gu., Bras. (Amaz.)
Craboo: *Byrsonima crassifolia* H. B. K.; *Malpighiac.;* br. Hond.
Crab tree: *Petalostigma quadriloculare* F. v. M.; *Euphorbiac.;* ö. Au.
Crabwood: *Gymnanthes lucida* Sw.; *Euphorbiac.;* sö. USA.
  „ : *Carapa guianensis* Aubl.; *Meliac.;* tr. Am.
  „ : *Sebastiania lucida* Muell. Arg.; *Euphorbiac.;* tr. Am.
  „ : *Ardisia compressa* H. B. K.; *Myrsinac.;* Pan.
  „ : *Carapa procera* DC.; *Meliac.;* Go.
  „ : *Carapa surinamensis* Miq.; *Meliac.;* Sur.
  „ , Highland-: *Carapa guianensis* Aubl.; *Meliac.;* br. Gu.
  „ , Lowland-: *Carapa guianensis* Aubl.; *Meliac.;* br. Gu.
  „ , red: *Carapa sp.; Meliac.;* br. Gu.
  „ , white: *Carapa guianensis* Aubl.; *Meliac.;* br. Gu.
Cramantee: *Guarea excelsa* H. B. K.; *Meliac.;* br. Hond.
Crapa: *Carapa guianensis* Aubl.; *Meliac.;* Sur.
Crapaud: *Carapa guianensis* Aubl.; *Meliac.;* Trin.
Crapo: *Carapa guianensis* Aubl.; *Meliac.;* fr. Gu.
Crapoo: *Byrsonima crassifolia* DC.; *Malpighiac.;* br. Hond.
Crappa: *Carapa guianensis* Aubl.; *Meliac.;* Sur.
Crappo: *Carapa guianensis* Aubl.; *Meliac.;* Trin.
Crasan: *Feronia elephantum* Correa; *Rutac.;* Ind.-Ch.
  „ : *Feronia lucida* Teijsm. et Binn.; *Rutac.;* Ind.-Ch.
Crassé: *Guarea cedrata* Pellegr.; *Meliac.;* Elf.

Cravo do maranhão: *Dicypellium caryophyllatum* Nees.; *Laurac.;* Bras. (Amaz.)
Craw-wood: *Carapa guianensis* Aubl.; *Meliac.;* fr. Gu.
Cream tree: *Sideroxylon gaumeri* Pittier; *Sapotac.;* br. Hond.
„   -nut tree: *Lecythis ollaria* L.; *Lecythidac.;* Bras. (Bahia).
Cream of Tartar-tree: *Adansonia gregori* F. v. M.; *Bombacac.;* n. Au.
Crementillo: *Casearia sp.; Flacourtiac.;* Mex.
Crémon: *Thespesia populnea* Correa; *Malvac.;* Ven.
Cremor tartboom: *Adansonia digitata* L.; *Bombacac.;* sw. Af.
Creona: *Premna taïtensis* Schauer; *Verbenac.;* Fid.
Cresciuma: *Chusquea ramosissima* Pilg.; *Gramin.;* Arg.
Cresta de gallo: *Erythrina crista-galli* L.; *Papilionac.;* Arg.
Crête de coq: *Erythrina crista-galli* L.; *Papilionac.;* Arg.
Crindeuva: *Trema micrantha* Blume; *Ulmac.;* Bras.
Crindiuva: *Trema micrantha* Blume; *Ulmac.;* Bras
Criollo: *Minquartia guianensis* Aubl.; *Olacac.;* Pan.
Crispin: *Bursera sp.; Burserac.;* Kol.
Cristobal: *Platymiscium polystachyum* Bth.; *Papilionac.;* C-.R.
Criuba: *Clusia sp.; Guttifer.;* Bras.
Cro-cro: *Zanthoxylum sp.; Rutac.;* Jam.
Croc: *Ximenia americana* L.; *Olacac.;* St.-D.
Croma: *Klainedoxa sp.; Simarubac.;* Elf.
Croton: ? *Croton coriaceus* H. B. K.; *Euphorbiac.;* Ven.
Crowsfoot: *Acacia costaricensis* Schenck; *Mimosac.;* br. Hond.
Crow wood, John-: *Cespedesia macrophylla* Seem.; *Ochnac.;* Pan.
Cruceta: *Randia spinosa* Karst.; *Rubiac.;* Ven.
„   negra: *Randia spinosa* Karst.; *Rubiac.;* Ven.
„   real: *Randia spinosa* Karst.; *Rubiac.;* Ven.
„   de sabana: *Rondeletia purdici* Hook. f.; *Rubiac.;* Ven.
Crucetilla: *Randia armata* DC.; *Rubiac.;* Hond.
Cruceto: *Gymnopodium antigonoides* Blake; *Polygonac.;* br. Hond.
„   : *Guettarda ambigua* DC.; *Rubiac.;* Ven.
Crucito: *Pisonia sp.; Nyctaginac.;* Salv.
Crugia: *Digitalis obscura* L.; *Scrophulariac.;* Sp.
Cruz de espina: *Achatocarpus bicornutus* Schinz; *Phytolaccac.;* Arg.
Cruzeiro: *Eupatorium odoratum* L.; *Composit.;* Bras.
Cryptocarya, Red wooded: *Cryptocarya erythroxylon* M. et B.; *Laurac.;* Au.
„   , Small-leaved: *Cryptocarya foveolata* C. T. White et Francis; *Laurac.;* Au.
Cu (H): *Thespesia populnea* Correa *var. macrocarpa* Bl.; *Malvac.;* Ind.-Ch.
„   chi: *Strychnos nux-vomica* L.; *Loganiac.;* Ind.-Ch.
„   den (H): *Maesa indica* Wall.; *Myrsinac.;* Ind.-Ch.
„   „   : *Croton poilanei* Gagnep.; *Euphorbiac.;* Ind.-Ch.
„   giai: *Alangium chinense* Rehder; *Cornac.;* Ind.-Ch.
Cuaba: *Amyris balsamifera* L.; *Rutac.;* Cuba.
„   amarilla: *Amyris sp.; Rutac.;* Cuba.
„   „   de costa: *Amyris sp.; Rutac.;* Cuba.
„   blanca: *Amyris silvatica* Jacq.; *Rutac.;* Cuba.
„   de sabana: *Amyris sp.; Rutac.;* Cuba.
„   del monte: *Amyris sp.; Rutac.;* Cuba.
Cuabilla: *Amyris sp.; Rutac.;* Cuba.
„   : *Suriana maritima* L.; *Simarubac.;* Cuba.
Cuachepil: *Diphysa carthaginensis* Jacq.; *Papilionac.;* Mex.
Cuahualagua: *Heliocarpus sp.; Tiliac.;* Mex.
Cuahulote: *Guazuma ulmifolia* Lam.; *Sterculiac.;* Mex.
Cuajada: *Vitex cooperi* Standl.; *Verbenac.;* Pan.
Cuajani: *Prunus occidentalis* Swartz; *Rosac.;* Cuba.
Cuajilote: *Bombax palmeri* S. Watson; *Bombacac.;* Mex.
Cuajinicuil: *Inga rodrigueziana* Pittier; *Mimosac.;* Guat., Hond.
Cuajiniquil: *Inga punctata* Willd.; *Mimosac.;* Hond.
Cuajiote: *Bursera sp.; Burserac.;* Mex.
„   amarillo: *Bursera sp.; Burserac.;* Mex.
„   chino: *Bursera sp.; Burserac.;* Mex.
„   colorado: *Bursera sp.; Burserac.;* Mex.

Cuajiote verde: *Bursera sp.; Burserac.;* Mex.
Cuajo: *Virola venezuelensis* Warb.; *Myristicac.;* Ven.
Cuamara: *Dipteryx odorata* Willd.; *Papilionac.;* Gu.
„ Kumara: *Dipteryx odorata* Willd.; *Papilionac.;* Sur.
Cuamóchil: *Pithecolobium dulce* Bth.; *Mimosac.;* Mex.
Cuamúchil: *Pithecolobium dulce* Bth.; *Mimosac.;* Mex.
Cuanacaztle: *Enterolobium cyclocarpum* Griseb.; *Mimosac.;* Mex.
Cuapinol: *Hymenaea courbaril* L.; *Caesalpiniac.;* Mex.
Cuapinole: *Hymenaea courbaril* L.; *Caesalpiniac.;* Mex.
Cuapinoli: *Hymenaea courbaril* L.; *Caesalpiniac.;* Mex.
Cuaré: *Guarea spiciflora* A. Juss.; *Meliac.;* Arg.
Cuartololoti: *? Andira inermis* H. B. K.; *Papilionac.;* Mex.
Cuatatachi: *Hura sp.; Euphorbiac.;* Mex.
Cuate: *Eysenhardtia polystachya* Sarg.; *Papilionac.;* Mex.
Cuaulahuac: *Heliocarpus sp.; Tiliac.;* Mex.
Cuaulote: *Heliocarpus sp.; Tiliac.;* Mex.
„ : *Guazuma ulmifolia* Lam.; *Sterculiac.;* Mex.
Cuau-nacaztli: *Enterolobium cyclocarpum* Griseb.; *Mimosac.;* Mex.
Cuaxinguba: *Ficus sp.; Morac.;* Bras.
Cuayabí: *Patagonula americana* L.; *Borraginac.;* Arg.
Cuayavillo: *Caesalpinia mexicana* A. Gray; *Caesalpiniac.;* Mex.
Cubaholz: *Chlorophora tinctoria* Gaudich.; *Morac.;* tr. Am.
Cube: *Lonchocarpus nicou* DC.; *Caesalpiniac.;* n. Bras.
„ : *? Lonchocarpus sp.; Caesalpiniac.,* Peru.
„ : *? Derris negrensis* Bth.; *Papilionac.;* Peru.
Cubi: *Beilschmiedia cairocan* Vid.; *Laurac.;* Phil.
Cúc môc: *Xanthophyllum excelsum* Blume; *Polygalac.;* Ind.-Ch.
„ môt (H): *Zollingeria dongnaiensis* Pierre; *Sapindac.;* Ind.-Ch.
Cuca: *Erythroxylon coca* Lam.; *Erythroxylac.;* Peru.
Cucaracho: *Cornutia grandifolia* Schauer; *Verbenac.;* Hond.
Cuchape: *Coccoloba sp.; Polygonac.;* Trin.
Cuchara: *? Oreocallis grandiflora* R. Br.; *Proteac.;* Ven.
Cucharera: *Porliera lorenţi* Engl.; *Zygophyllac.;* Arg. (Tucuman).
Cucharillo: *Trichilia sp.; Meliac.;* Mex.
Cucharo: *Oreocallis grandiflora* R. Br.; *Proteac.,* Ven.
Cucharón: *Gyranthera caribensis* Pittier; *Bombacac.;* Ven.
Cucharrero: *Porliera hygrometrica* R. et P.; *Zygophyllac.;* Arg.
Cuchilleja: *Bupleurum spinosum* L.; *Umbellifer.;* Sp.
Cucu: *Theobroma cacao* L.; *Sterculiac.;* br. Hond.
Cucubano: *Coccoloba laurifolia* Jacq.; *Polygonac.;* USA.
Cucumber tree: *Magnolia acuminata* L.; *Magnoliac.;* USA.
„ „ , Earleaf-: *Magnolia fraseri* Wall.; *Magnoliac.;* USA.
„ „ , Longleaf-: *Magnolia macrophylla* Michx.; *Magnoliac.;* USA.
Cucumber, wild: *Carica dclichaula* D. Sm.; *Caricac.;* Pan.
„ , „ : *Cyphomandra caudata* Standl.; *Solanac.;* Pan.
Cucurito: *Maximiliana regia* Mart.; *Palmac.;* Ven.
Cucuyo: *Bumelia sp.; Sapotac.;* Cuba.
„ de sabana: *Bumelia sp.; Sapotac.;* Cuba.
Cucuyul: *Ardisia compressa* H. B. K.; *Myrsinac.;* Hond.
Cudgeree: *Sloanea australis* F. v. M.; *Elaeocarpac.;* Au.
Cudgerie: *Euroschinus falcatus* Hook. f.; *Anacardiac.;* Au.
„ : *Flindersia schottiana* F. v. M.; *Rutac.;* Au.
„ : *Litsea reticulata* Bth.; *Laurac.;* Queensl.
„ : *Hernandia bivalvis* Bth.; *Hernandiac.;* Queensl.
„ , Blush-: *Euroschinus falcatus* Hook. f.; *Anacardiac.;* Au.
„ , Brown: *Protium australasicum* Sprague; *Burserac.;* Au.
Cuentrillo: *Fagara hiemalis* Engl.; *Rutac.;* Arg.
Cueramo: *Cordia sp.; Borraginac.;* Mex.
Cuernava: *? Diospyros ebenaster* Retz.; *Ebenac.;* tr. As.
Cuero de sapo: *Exostemma caribaeum* Roem. et Schult.; *Rubiac.;* Mex., C.-Am.
Cugerie: *Flindersia australis* R. Br.; *Rutac.;* N.-S.-W.
Cúi: *Heritiera littoralis* Dryander; *Sterculiac.;* Annam.

Cui-Cui: *Heritiera littoralis* Dryander; *Sterculiac.;* Annam.
Cuia de macaco: *Couroupita guianensis* Aubl.; *Lecythidac.;* Bras. (Amaz.)
Cuieira: *Crescentia cujete* L.; *Bignoniac.;* Bras.
Cuieté: *Crescentia cujete* L.; *Bignoniac.;* Bras.
Cuilimbuca: ? *Andira inermis* H. B. K.; *Papilionac.;* Mex.
Cuirana: *Terminalia sp.; Combretac.;* Bras.
Cuité: *Crescentia cujete* L.; *Bignoniac.;* Bras.
Cuius-Cuius: *Taxotrophis ilicifolia* Vid.; *Morac.;* Phil.
Cujete: *Crescentia cujete* L.; *Bignoniac.;* Mex., C.-Am.
Cuji: *Inga cinerea* Humb. et Bonpl.; *Mimosac.;* Ven.
  „ : *Centrolobium orinocense* Pittier; *Papilionac.;* Ven.
  „ yaque: *Prosopis juliflora* DC.; *Mimosac.;* Ven.
Cujía: *Ardisia amplifolia* Standl.; *Myrsinac.;* Nic.
Cujinicuil: *Inga sp.; Mimosac.;* Salv.
Culé: *Psoralea glandulosa* L.; *Papilionac.;* Chile, Arg.
Culebra: *Astronium sp.; Anacardiac.;* Mex.
Culén: *Psoralea glandulosa* L.; *Papilionac.;* Chile, Arg.
Cullonen: *Gmelina leichhardti* F. v. M.; *Verbenac.;* ö. Au.
Culumata: *Avicennia nitida* Jacq.; *Verbenac.;* C.-Am.
Cuma: *Irvingia robur* Mildbr.; *Simarubac.;* Ang.
Cumã-assú: *Couma macrocarpa* Barb. Rodr.; *Apocynac.;* Bras. (Amaz.)
Cumacá-y: *Lophostoma calophylloides* Meissn.; *Thymelaeac.;* Bras. (Amaz.)
Cumandahy: *Securidaca volubilis* L.; *Polygalac.;* Bras.
Cumarica: *Machaerium melanophyllum* Standl.; *Papilionac.;* Ven.
Cumari-rana: *Dipteryx sp.; Papilionac.;* Bras.
Cumarú: *Dipteryx odorata* Willd.; *Papilionac.;* Bras.
  „ amarello: *Dipteryx odorata* Willd.; *Papilionac.;* Bras.
  „ rana: *Dipteryx sp.; Papilionac.;* Bras.
  „ do Amazonas: *Dipteryx odorata* Willd.; *Papilionac.;* Bras.
  „ do Ceará: *Torresea cearensis* Allem.; *Papilionac.;* Bras.
Cumaruco: *Sterculia sp.; Sterculiac.;* Ven.
Cumary (i): *Dipteryx odorata* Willd.; *Papilionac.;* n. S.-Am.
Cumbro: *Celtis sp.; Ulmac.;* Mex.
Cumburtu: *Araucaria cunninghami* Ait.; *Conifer.;* ö. Au.
Cumpang: *Myristica sp.; Myristicac.;* Borneo.
Cumquat: *Atalantia glauca* Hook.; *Rutac.;* Queensl.
Cunatuna: *Zanthoxylum naranjillo* Griseb.; *Rutac.;* Arg.
Cundurú: *Brosimum paraënse* Huber; *Morac.;* Bras.
Cung: *Machilus odoratissima* Nees; *Laurac.;* Ind.-Ch.
Cuom: *Canarium copaliferum* Chev.; *Burserac.;* Ind.-Ch.
Cuom-Thi: *Diospyros dodecandra* Lour.; *Ebenac.;* Ind.-Ch.
Cuong: *Dipterocarpus tuberculatus* Roxb.; *Dipterocarpac.;* Ind.-Ch.
Cupang: *Parkia roxburghi* G. Don; *Mimosac.;* Phil.
  „ : *Parkia timoriana* Merr.; *Mimosac.;* Phil.
Cupaúba-rana: *Eperua purpurea* Bth.; *Caesalpiniac.;* n. Bras.
Cupay: *Clusia rosea* Jacq.; *Guttifer.;* Ven.
Cupeillo: *Clusia krugiana* Urban; *Guttifer.;* P.-R.
Cupey: *Clusia rosea* Jacq.; *Guttifer.;* P.-R., Trin.
Cupia: *Bumelia sp.; Sapotac.;* Mex.
Cupiúba: *Goupia paraënsis* Huber; *Celastrac.:* Bras. (Pará).
  „ : *Goupia glabra* Aubl.; *Celastrac.;* Bras
Cupiúva: *Goupia glabra* Aubl.; *Celastrac.;* Bras
Cupu: *Spondias sp.; Anacardiac.;* Mex.
Cupuassú: *Theobroma grandiflorum* Schum.; *Sterculiac.;* Bras. (Amaz.)
Cupú-assúrana: *Matisia paraënsis* Huber; *Bombacac.;* Bras. (Amaz.)
Cupú-rana: *Matisia paraënsis* Huber; *Bombacac.;* Bras. (Amaz.)
Cutaçoa de negro: *Cassia apoucouita* Aubl.; *Caesalpiniac.;* Bras.
Curacy: *Warszewiczia coccinea* Klotzsch; *Rubiac.;* Bras.
Curanguay: *Schinus molle* L.; *Anacardiac.;* Arg.
Curari: *Tecoma serratifolia* G. Don; *Bignoniac.;* Ven.
Curarier: *Prosopis sp.; Mimosac.;* Ven.
Curarire: *Tecoma serratifolia* G. Don; *Bignoniac.;* Ven.

Curata: *Curatella americana* L.; *Dilleniac.;* Ven.
Curatahie: *Curatella americana* L.; *Dilleniac.;* Gu.
Curatelle d'Amérique: *Curatella americana* L.; *Dilleniac.;* tr. Am.
Curaturá: *Fagara stipitata* Engl.; *Rutac.;* Arg.
Curbana: *Canella winterana* Gaertn.; *Canellac.;* Cuba.
Curi-y: *Araucaria brasiliana* Lamb.; *Conifer.;* Par.
Currajong: *Sterculia diversifolia* G. Don; *Sterculiac.;* Au.
Currant, native: *Leptomeria billardieri* Sieber; *Santalac.;* Tasm.
„ , red: *Euclea sp.; Ebenac.;* Kap..
Currant tree: *Aristotelia racemosa* Hook. f.; *Elaeocarpac.;* N.-Seel.
Currijong: *Plagianthus pulchellus* Gray; *Malvac.;* Au. (Vict.)
Currungul: *Dissiliaria baloghioides* F. v. M.; *Euphorbiac.;* Queensl.
Curtidor: *Hieronyma alchorneoides* Allem.; *Euphorbiac.;* Hond.
„ : *Weinmannia glabra* L.; *Cunoniac.;* Ven.
„ montañero: *Eschweilera fendleriana* Miers; *Lecythidac.;* Ven.
Curtiscâo: *Erythrina crista-galli* L.; *Papilionac.;* Bras.
Curtiza: *Erythrina crista-galli* L.; *Papilionac.;* Arg.
Curtizera: *Erythrina crista-galli* L.; *Papilionac.;* Arg.
Curuberanda: *Picramnia sp.; Simarubac.;* br. Gu.
Curumo: *Capparis sp.; Capparidac.;* Salv.
Curupai: *Piptadenia cebil* Griseb.; *Mimosac.;* tr. S.-Am.
„ -nà: *Cassia brasiliensis* L.; *Caesalpiniac.;* tr. Am.
Curupa-y: *Piptadenia sp.; Mimosac.;* Bol.
Curupay: *Piptadenia rigida* Bth.; *Mimosac.;* Bras. (Sao P).
„ : *Piptadenia cebil* Griseb.; *Mimosac.;* Arg.
„ : *Piptadenia macrocarpa* Bth.; *Mimosac.;* Arg.
„ negro: *Piptadenia sp.; Mimosac.;* Arg.
„ -rá: *Piptadenia rigida* Bth.; *Mimosac.;* Arg.
Curupi: *Sapium aucuparicum* Jacq. *var. stenophyllum* Griseb.; *Euphorbiac.;* S.-Am.
Curupicayh: *Sapium longifolium* Boj.; *Euphorbiac.;* Arg.
Cury: *Araucaria brasiliana* Lamb.; *Conifer.;* Bras.
Cusombu: *Elaeis guineensis* Jacq. *var.* ?; *Palmac.;* Gab.
Cuspa: *Cusparia trifoliata* Engl.; *Rutac.;* Ven.
Cusú: *Cordia glabra* L.; *Borraginac.;* Kol.
Cút rê: *Raphiolepis indica* Lindl.; *Rosac.;* Ind.-Ch.
„ sat: *Acronychia laurifolia* Blume; *Rutac.;* Ind.-Ch.
Cútaro. *Swarţia sp.; Caesalpiniac.;* Pan.
Cutch tree: *Acacia catechu* Willd.; *Mimosac.;* tr. As., tr. Af.
Cutitiribá: *Lucuma sp.; Sapotac.;* Bras.
Cutucanhée: *Roupala sp.; Proteac.;* Bras.
Cutucanhem: *Roupala sp.; Proteac.;* Bras.
Cutufume: *Lippia kellermani* Greenm.; *Uerbenac.;* Hond.
Cutujume: *Lippia myriocephala* Schlecht. et Cham.; *Uerbenac.;* Guat.
Cuveraca: *Toona ciliata* M. Roemer; *Meliac.;* Ind.
Cuyá: *Dipholis salicifolia* A. DC.; *Sapotac.;* Cuba.
Cuya: *Ardisia paschalis* Donn. Smith; *Myrsinac.;* Hond.
Cuyapo: *Quararibea stenophylla* Pittier; *Bombacac.;* Guat.
Cuyas de macaco: *Lecythis sp.; Lecythidac.;* Bras.
Cyp: *Cordia gerascanthus* L.; *Borraginac.;* Trin.
Cypre: *Cordia gerascanthus* L.; *Borraginac.;* Trin.
„ : *Ocotea cernua* Mez; *Laurac.;* St.-D.
„ : *Cordia alliodora* Cham.; *Borraginac.;* Trin., Tob.
„ , doux: *Phoebe sp.; Laurac.;* Guad.
„ , Laurier-: *Ocotea glomerata* Bth. et Hook.; *Laurac.;* Trin., Tob.
„ , „ -: *Ocotea cernua* Mez; *Laurac.;* Trin., Tob., St.-D.
„ oranger: *Styrax glabrum* Sw.; *Styracac.;* Guad.
Cyprès: *Pinus banksiana* Lam.; *Conifer.;* Kan.
„ , faux, de Nootka: *Chamaecyparis nutkaensis* Spach; *Conifer.;* Kan.
„ jaune: *Chamaecyparis nutkaensis* Spach; *Conifer.;* Kan.
Cypress: *Cupressus sempervirens* L.; *Conifer.;* M.-M.
„ : *Pinus banksiana* Lam., *P. murrayana* Balf.; *Conifer.;* Kan.
„ : *Taxodium distichum* Rich.; *Conifer.;* s. USA.

Cypress: *Podocarpus coriacea* Rich.; *Conifer.;* br. Hond.
„ : *Podocarpus guatemalensis* Standl.; *Conifer.;* br. Hond.
„ : *Taxodium mucronatum* Tenore; *Conifer.;* Mex.
„ : *Callitris verrucosa* R. Br.; *Conifer.;* Au. (Vict.)
„ : *Libocedrus doniana* Endl.; *Conifer.;* N.-Seel.
„ , Alaska-: *Chamaecyparis nutkaensis* Spach; *Conifer.;* Kan.
„ , Arizona-: *Cupressus arizonica* Greene; *Conifer.;* s. USA.
„ , Atlas-: *Callitris quadrivalvis* Vent.; *Conifer.;* Marok.
„ , Bald-: *Taxodium distichum* Rich.; *Conifer.;* N.-Am.
„ , Black: *Taxodium distichum* Rich.; *Conifer.;* s. USA.
„ , „ : *Laurus sp.; Laurac.;* Tob.
„ , Congo-: *Bulnesia arborea* Engl.; *Zygophyllac.;* n. S.-Am.
„ , Deciduous: *Taxodium distichum* Rich.; *Conifer.;* USA.
„ , Gowen-: *Cupressus goveniana* Gord.; *Conifer.;* w. N.-Am. (s. Kalif.)
„ , Himalayan (H): *Cupressus torulosa* D. Don; *Conifer.;* Ind. (Himal.)
„ , Lawson-: *Chamaecyparis lawsoniana* Parl.; *Conifer.;* USA.
„ , Macnab-: *Cupressus macnabiana* Murr.; *Conifer.;* w. N.-Am. (Kalif.)
„ , Mlanje-: *Widdringtonia whytei* Rendle; *Conifer.;* O.-Af. (Nyassa).
„ , Nootka-: *Chamaecyparis nutkaensis* Spach; *Conifer.;* Kan.
„ , red: *Taxodium distichum* Rich.; *Conifer.;* ö. USA.
„ , Sitka-: *Chamaecyparis nutkaensis* Spach; *Conifer.;* nw. N.-Am.
„ , southern: *Taxodium distichum* Rich.; *Conifer.;* USA.
„ , Swamp-: *Taxodium distichum* Rich.; *Conifer.;* USA.
„ , „ : *Taxodium heterophyllum* Brongn.; *Conifer.;* Ch.
„ , true: *Cupressus sempervirens* L.; *Conifer.;* Paläst.
„ , white: *Taxodium distichum* Rich.; *Conifer.;* ö. USA.
„ , „ : *Cordia sp.; Borraginac.;* Tob.
„ , yellow: *Chamaecyparis nutkaensis* Spach; *Conifer.;* w. N.-Am.
Cyrouenne: *Melia azedarach* L.; *Meliac.;* s. Fr.
Cyroyer (f): *Rheedia lateriflora* L.; *Guttifer.;* W.-Ind.

D

Dá (H): *Ceriops candolleana* Arn., *Bruguiera sp.; Rhizophorac.;* Ind.-Ch.
„ : *Ficus religiosa* L.; *Morac.;* Ind.-Ch.
„ Dá: ? *Eusideroxylon sp.; Laurac.;* Ind.-Ch.
„ „ : *Xylia kerri* Craib et Hutch.; *Mimosac.;* Ind.-Ch.
„ „ trang: *Taractogenos serrata* Pierre; *Flacourtiac.;* Ind.-Ch.
„ dó: *Ceriops sp.; Rhizophorac.;* Ind.-Ch.
„ hop: *Talauma fistulosa* Finet et Gagnep.; *Magnoliac.;* Ind.-Ch.
„ „ : *Magnolia pumila* Andr.; *Magnoliac.;* Ind.-Ch.
„ „ ring: *Talauma spp.; Magnoliac.;* Ind.-Ch.
„ „ rung: *Talauma fistulosa* Finet et Gagnep.; *Magnoliac.;* Ind.-Ch.
„ -kom: *Adina cordifolia* Hook. f.; *Rubiac.;* Beng.
„ nuoe: *Ceriops candolleana* Arn.; *Rhizophorac.;* Ind.-Ch.
Daan: *Albizzia procera* Bth.; *Mimosac.;* Phil. (Benguet).
Dabarouida: *Pithecolobium unguis-cati* Bth.; *Mimosac.;* Curacao.
Dabdabbi: *Meliosma wallichi* Hook. f.; *Sabiac.;* ö. Himal.
„ : *Garuga pinnata* Roxb.; *Burserac.;* nö.-Ind. (Nepal).
Dabdabi: *Garuga pinnata* Roxb.; *Burserac.;* nö. O.-Ind.
Dabéma (f): *Piptadenia africana* Hook. f.; *Mimosac.;* W.-Af.
Dabkar: *Crataeva religiosa* Forst.; *Capparidac.;* tr. Af.
Dabo: *Carapa sp.; Meliac.;* Elf.
Dabu-Dabu: *Cola cordifolia* R. Br. *var. maclaudi* Chev.; *Sterculiac.;* Elf.
Dacamaballi: *Vouacapoua americana* Aubl.; *Papilionac.;* fr. Gu.
Dache: *Cleidion javanicum* Blume; *Euphorbiac.;* Ind.-Ch.
Dadap: *Erythrina indica* Lam.; *Papilionac.;* tr. As., tr. Af.
„ : *Erythrina lithosperma* Blume; *Papilionac.;* s. O.-Ind.
Dadar: *Cedrus deodara* Loudon; *Conifer.;* nw. O.-Ind. (Kasch.).
Dadiáñgau: *Agathis alba* Foxw.; *Conifer.;* Phil.
Dadie: *Adansonia digitata* L.; *Bombacac.;* Togo.
Dadsal: *Grewia tiliaefolia* Vahl; *Tiliac.;* w. O.-Ind. (Bombay).

Dagáaú: *Zizyphus sp.; Rhamnac.;* Phil. (Surigao).
Dágam: *Uatica mangachapoi* Blco.; *Dipterocarpac.;* Phil. (Cam.).
Dágame: *Calycophyllum candidissimum* DC.; *Rubiac.;* Cuba.
Dágang: *Anisoptera curtisi* Dyer; *Dipterocarpac.;* Phil.
„ : *Anisoptera thurifera* Blume; *Dipterocarpac.;* Phil.
Dagdagáan: *Knema glomerata* Merr.; *Myristicac.;* Phil. (Cag.).
Dagera: *Combretum lecananthum* Engl. et Diels; *Combretac.;* n. Nig.
Dagingdíñgan: *Hopea pierrei* Hance; *Dipterocarpac.;* Phil. (Zam.).
„ : *Shorea guiso* Blume; *Dipterocarpac.;* Phil. (Misamis).
Dagíñgiran: *Cyclostemon spp.; Euphorbiac.:* Phil. (Cag.).
Dagkálan: *Calophyllum inophyllum* L.; *Guttifer.;* Phil.
Dagpit: *Taxotrophis ilicifolia* Vid.; *Morac.;* Phil.
Daguilla: *Lagetta lintearia* Lam.; *Thymelaeac.;* W.-Ind.
Dagum: *Anisoptera thurifera* Blume; *Dipterocarpac.;* Phil. (Lag.).
Dah: *Cynometra ananta* Hutch. et Dalz.; *Caesalpiniac.;* Lib.
Daha: *Duabanga moluccana* Blume; *Sonneratiac.;* Phil. (Neg.)
Dahat (H): *Tectona hamiltoniana* Wall.; *Uerbenac.;* Burma.
Dahipalas: *Cordia macleodi* Hook. f. et Thoms.; *Borraginac.;* C.-O.-Ind.
Dahl Wah: *Casuarina suberosa* Otto et Dietr.; *Casuarinac.;* N.-S.-W.
Dahmun: *Eriolaena candollei* Wall.; *Sterculiac.;* w. O.-Ind. (Bombay).
Dahn-hay: *Bussea occidentalis* Hutch.; *Caesalpiniac.;* Lib.
Daho: *Carapa sp.; Meliac.;* Elf.
„ : *Artocarpus lakoocha* Roxb.; *Morac.;* C.-O.-Ind. (Nagpur).
Dahoe: *Dracontomelum mangiferum* Blume; *Anacardiac.;* Malay, Sunda.
Dahu: *Artocarpus lakoocha* Roxb.; *Morac.;* C.-O.-Ind. (Nagpur).
Dai: *Randia oxydonta* Drake; *Rubiac.;* Ind.-Ch.
„ hoi: *Illicium cambodgianum* Hance; *Magnoliac.;* Annam.
„ „ nui: *Illicium cambodgianum* Hance; *Magnoliac.;* Kamb.
„ khoai: *Randia oxyodonta* Drake; *Rubiac.;* Ind.-Ch.
„ kla: *Wrightia annamensis* Eberh. et Dub.; *Apocynac.;* Ind.-Ch.
„ ngua: *Sterculia cochinchinensis* Pierre; *Sterculiac.;* Annam.
„ -phong-tu: *Hydnocarpus anthelmintica* Pierre; *Flacourtiac.;* Co.-Ch.
„ vôi: *Eugenia operculata* Roxb.; *Myrtac.;* Ind.-Ch.
Daiamiras: *Aglaia harmsiana* Perk.; *Meliac.;* Phil. (Mind.).
Daïdaï: *Citrus aurantium* L. var. *amara* L.; *Rutac.;* Jap.
Daira: *Wrightia tomentosa* Roem. et Schultes; *Apocynac.;* n. O.-Ind.
Daiwas: *Cordia macleodi* Hook. f. et Thoms.; *Borraginac.;* w. O.-Ind.
Daiyar: *Cordia macleodi* Hook. f. et Thoms.; *Borraginac.;* O.-Ind.
Dakama: *Dimorphandra latifolia* Tul.; *Caesalpiniac.;* Gu.
Dakamaballi: *Vouacapoua americana* Aubl.; *Papilionac.;* Sur.
Dakartalada: *Calophyllum spectabile* Willd.; *Guttifer.;* And.
Dakhani babul: *Pithecolobium dulce* Bth.; *Mimosac.;* Ind.
Dakua: *Dammara vitiensis* Seem.; *Conifer.;* Fid.
„ salu salu: *Dacrydium elatum* Wall.; *Conifer.;* Fid.
Dakúg: *Eugenia clausa* C. B. Rob.; *Myrtac.;* Phil. (I.-S.).
Dakulau: *Shorea eximia* Scheff.; *Dipterocarpac.;* Phil. (Cam.)
Dakwora: *Acacia senegal* Willd.; *Mimosac.;* n. Nig.
Dala: *Hopea pierrei* Hance; *Dipterocarpac.;* Phil. (Cag.).
Dalákan: *Alstonia macrophylla* Wall.; *Apocynac.;* Phil. (I.-S.).
Dalby Myall: *Acacia stenophylla* A. Cunn.; *Mimosac.;* C.-Au.
Dalchini: *Cinnamomum zeylanicum* Breyn; *Laurac.;* sö. O.-Ind.
Dali bhimal: *Grewia laevigata* Vahl; *Tiliac.;* nw. O.-Ind. (Kumaon).
Dali dali: *Strombosia javanica* Blume; *Olacac.;* Malay.
Dalina: *Hieronyma alchorneoides* Allem.; *Euphorbiac.;* br. Gu.
Dalínas: *Cyathocalyx globosus* Merr.; *Anonac.;* Phil. (Bat.)
Dalingdíñgan: ? *Shorea philippinensis* Brandis; *Dipterocarpac.;* Phil. (Tay.).
„ : *Hopea acuminata* Merr.; *Dipterocarpac.;* Phil.
„ : *Hopea basilanica* Foxw.; *Dipterocarpac.;* Phil. (Bas.).
„ : *Hopea pierrei* Hance; *Dipterocarpac.;* Phil.
„ : *Hopea plagata* Vid.; *Dipterocarpac.;* Phil. (Bas.).
„ -isak: *Hopea pierrei* Hance; *Dipterocarpac.;* Phil.
Dalingsem: *Homalium tomentosum* Bth.; *Flacourtiac.;* tr. As.

Dalinsai: *Terminalia oocarpa* Merr.; *Combretac.;* Phil. (Cam.).
Dalinsi: *Terminalia pellucida* Presl; *Combretac.;* Phil. (Tay.).
„ : *Terminalia edulis* Blco.; *Combretac.;* Phil. (Lag., Tay.).
„ : *Terminalia nitens* Presl; *Combretac.;* Phil. (Tay.).
„ : *Terminalia oocarpa* Merr.; *Combretac.;* Phil. (Tay.).
Dalinsín: *Terminalia oocarpa* Merr.; *Combretac.;* Phil. (Alb.).
Dalipápa: *Vitex aherniana* Merr.; *Verbenac.;* Phil. (Cag.).
Dalipatsan: *Zizyphus spp.; Rhamnac.;* Phil. (Mas.).
Dalipáuen: *Alstonia scholaris* R. Br.; *Apocynac.;* Phil. (I.-N., Cag., Un., Abra).
Dalisai: *Terminalia catappa* L.; *Combretac.;* Phil.
Dalkaramcha: *Pongamia glabra* Vent.; *Papilionac.;* Beng.
Dalli: *Virola surinamensis* Warb.; *Myristicac.;* br. Gu.
Dallie: *Virola surinamensis* Warb.; *Myristicac.;* Gu.
Dalmara: *Chukrasia tabularis* A. Juss.; *Meliac.;* sw. O.-Ind.
Dalné katus: *Castanopsis hystrix* A. DC.; *Fagac.;* nö. O.-Ind. (Nepal).
Dalsingha: *Canthium didymum* Roxb.; *Rubiac.;* w. Beng.
Daluru: *Sonneratia pagatpat* Blco., *S. alba* Smith; *Sonneratiac.;* Phil.
„ -babáe: *Lumnitzera littorea* Voigt; *Combretac.;* Phil. (Tay.).
Dalutan: *Quercus spp.; Fagac.;* Phil. (I.-S.).
Dama de noche: *Cestrum panamense* Standl.; *Solanac.;* br. Hond.
Damadi: *Diospyros melanoxylon* Roxb.; *Ebenac.;* s. Ind.
Damajagua: *Hibiscus tiliaceus* L.; *Malvac.;* Peru.
Daman: *Grewia tiliaefolia* Vahl; *Tiliac.;* sw. O.-Ind.
Damanu: *Calophyllum burmanni* Wight; *Guttifer.;* Fid.
Damar: *Dipterocarpus spp.; Dipterocarpac.;* Malay.
„ Hitam: *Balanocarpus penangianus* King; *Dipterocarpac.;* Malay.
„ „ : *Balanocarpus curtisi* King; *Dipterocarpac.;* Malay.
„ „ : *Shorea sp.; Dipterocarpac.;* Malay.
„ itam: *Canarium rostratum* Zipp.; *Burserac.;* Nied.-Ind.
„ laoet: *Shorea sp., Hopea sp.; Dipterocarpac.;* Nied.-Ind.
„ Laut: *Agathis alba* Foxw.; *Conifer.;* Malay.
„ „ : *Parashorea stellata* Kurz; *Dipterocarpac.;* Malay.
„ „ : *Shorea glauca* King, *S. utilis* King; *Dipterocarpac.;* Malay.
„ „ Daun Kechil: *Shorea utilis* King; *Dipterocarpac.;* Malay.
„ „ merah: *Shorea sp.; Dipterocarpac.;* Malay.
„ „ numero satu: *Shorea utilis* King; *Dipterocarpac.;* Malay.
„ mata Koetjieng: *Hopea mengarawana* Miqu.; *Dipterocarpac.;* Nied.-Ind.
„ „ Kuching: *Hopea spp.; Dipterocarpac.;* Malay.
„ minyak: *Agathis alba* Foxw.; *Conifer.;* Malay.
„ Minyar: *Agathis alba* Foxw.; *Conifer.;* Malay.
„ noir: *Canarium rostratum* Zipp.; *Burserac.;* Nied.-Ind.
„ siput: *Hopea spp., Shorea sp.; Dipterocarpac.;* Malay.
Damarau: *Cyathocalyx globosus* Merr.; *Anonac.;* Phil. (Neg.).
Dambel: *Careya arborea* Roxb.; *Myrtac.;* Beng.
Damílang: *Shorea negrosensis* Foxw., *S. palosapis* Merr.; *Dipterocarpac.;* Phil
Daminne: *Grewia tiliaefolia* Vahl; *Tiliac.;* O.-Ind., Cey.
Dàmmal: *Acacia stenocarpa* Hochst.; *Mimosac.;* O.-Af. (Somali).
Dammar: *Agathis alba* Foxw.; *Conifer.;* Malay.
„ : *Dammara orientalis* Lamb.; *Conifer.;* Bor.
„ , black (H): *Dipterocarpus indicus* Bedd.; *Dipterocarpac.;* Ind.
„ blanc (f): *Vateria indica* L.; *Dipterocarpac.;* Ind.
„ -malaijoe: *Shorea selanica* Blume; *Dipterocarpac.;* Bor., Moluk.
„ -mata-koetjing: *Hopea dryobalanoides* Miq.; *Dipterocarpac.;* Sum.
„ -poetih: *Hopea dryobalanoides* Miq.; *Dipterocarpac.;* Sum.
„ -sila: *Shorea selanica* Blume; *Dipterocarpac.;* Bor., Moluk.
„ , Southern: *Agathis australis* Steud.; *Conifer.;* N.-Seel.
„ , White: *Vateria indica* L.; *Dipterocarpac.;* Ind.
„ de l'Inde: *Shorea robusta* Gaertn.; *Dipterocarpac.;* Beng.
Dammara: *Vatica rassak* Korth.; *Dipterocarpac.;* holl.-Bor.
Dammer: *Shorea robusta* Gaertn.; *Dipterocarpac.;* Ind.
Damnak: *Grewia tiliaefolia* Vahl; *Tiliac.;* Beng.
Damo: *Fraxinus mandschurica* Rupr.; *Oleac.;* Jap.

Damól: *Cyclostemon spp.; Euphorbiac.;* Phil. (Sam.).
Dampól: *Pygeum glandulosum* Merr., *P. presli* Merr.; *Rosac.;* Phil. (Bat.).
Damshing: *Viburnum erubescens* Wall.; *Caprifoliac.;* O.-Ind., Cey.
Damson, bitter: *Simaruba sp.; Simarubac.;* Jam.
 „ , Mountain-: *Simaruba sp.; Simarubac.;* Jam.
Dân: *Apodytes cambodiana* Pierre; *Icacinac.;* Annam.
Danala: *Elaeocarpus tuberculatus* Roxb.; *Elaeocarpac.;* sw. O.-Ind. (Mysore).
Dandoshi: *Dalbergia lanceolaria* L.; *Papilionac.;* sw. O.-Ind.
Dandous: *Dalbergia lanceolaria* L.; *Papilionac.;* sw. O.-Ind.
Dandua: *Anogeissus latifolia* Wall.; *Combretac.;* sw. O.-Ind.
Dandukit: *Gardenia turgida* Roxb.; *Rubiac.;* ö. Beng.
Dandúlit: *Cumingia philippinensis* Vid.; *Bombacac.;* Phil. (Zamb.)
Dang: *Schefflera tonkinensis* R. Vig.; *Araliac.;* Ind.-Ch.
 „ : *Schefflera octophylla* Harms; *Araliac.;* Ind.-Ch.
 „ : *Hopea odorata* Roxb.; *Dipterocarpac.;* Ind.-Ch.
 „ : *Rhizophora mucronata* Lam.; *Rhizophorac.;* Ind.-Ch.
 „ Dàng: *Spatholobus orientalis* Hassk.; *Papilionac.;* Ind.-Ch.
 „ dê: *Rubia nauclea* ?; *Rubiac.;* Ind.-Ch.
 „ „ : *Adina polycephala* Bth.; *Rubiac.;* Ind.-Ch.
 „ dinh: *Carapa obovata* Blume; *Meliac.;* Ind.-Ch.
 „ hung: *Schefflera tonkinensis* R. Vig.; *Araliac.;* Ind.-Ch.
 „ huong: *Pterocarpus cambodianus* Pierre; *Papilionac.;* Ind.-Ch.
 „ „ : *Pterocarpus pedatus* Pierre; *Papilionac.;* Ind.-Ch.
 „ „ : *Pterocarpus macrocarpus* Kurz; *Papilionac.;* Annam.
 „ kiêp kdam: *Antidesma ghesaembilla* Gaertn.; *Euphorbiac.;* Ind.-Ch.
 „ tu hu: *Randia dumetorum* Lamk.; *Rubiac.;* Ind.-Ch.
Dangdoer: *Bombax malabaricum* DC.; *Bombacac* Nied.-Ind.
 „ gédé: *Sterculia foetida* L.; *Sterculiac.;* Sunda.
Dangdur: *Bombax malabaricum* DC.; *Bombacac.;* Nied.-Ind.
 „ gedch: *Sterculia foetida* L.; *Sterculiac.;* Sunda.
Danggái: *Kingiodendron alternifolium* Merr.; *Papilionac.;* Phil.
Danggi: *Vatica mangachapoi* Blco.; *Dipterocarpac.;* Phil. (Riz.).
Danghkyam: *Holarrhena antidysenterica* Wall.; *Apocynac.;* O.-Ind.
 „ -kaba: *Holarrhena antidysenterica* Wall.; *Apocynac.;* O.-Ind.
 „ -kaji: *Wrightia tomentosa* Roem. et Schultes; *Apocynac.;* O.-Ind.
Dañgig: *Pentacme contorta* Merr. et Rolfe; *Dipterocarpac.;* Phil. (Ag.).
Dañgila: *Koordersiodendron pinnatum* Merr.; *Anacardiac.;* Phil.
Dañgióg: *Pentacme contorta* Merr. et Rolfe; *Dipterocarpac.;* Phil. (Sibu.).
 „ : *Parashorea malaanonan* Merr.; *Dipterocarpac.;* Phil. (Cap.).
Dañgiran: *Cyclostemon spp.; Euphorbiac.;* Phil. (Cag.).
Dangkálan: *Calophyllum inophyllum* L.; *Guttifer.;* Phil.
 „ : *Garcinia spp.; Guttifer.;* Phil. (Cam.).
Danglin: *Grewia rolfei* Merr.; *Tiliac.;* Phil.
Dañgúla: *Vitex aherniana* Merr.; *Verbenac.;* Phil. (Neg., Ilo.).
Danh ca: *Sesbania grandiflora* Pers.; *Papilionac.;* Ind.-Ch.
 „ ghét: *Rhaphiolepis indica* Lindl.; *Rosac.;* Ind.-Ch.
 „ „ : *Symplocos indica* ?; *Symplocac.;* Ind.-Ch.
 „ nganh: *Cratoxylon polyanthum* Korth.; *Guttifer.;* Ind.-Ch.
Dani: *Piptadenia africana* Hook. f.; *Mimosac.;* Go.
Daniella: *Daniella thurifera* J. J. Benn.; *Caesalpiniac.;* Kam., Gab.
 „ : ? *Daniella klainei* Pierre; *Caesalpiniac.;* Kam., Gab.
Daniggá: *Toona calantas* Merr. et Rolfe; *Meliac.;* Phil. (Cag.).
Daníri: *Shorea guiso* Blume; *Dipterocarpac.;* Phil. (Tay., Cam.).
Danlieba: *Tapirira aff. guianensis* Aubl.; *Anacardiac.;* Sur.
Danlíg: *Parashorea malaanonan* Merr.; *Dipterocarpac.;* Phil. (Tay.).
 „ : *Pentacme contorta* Merr. et Rolfe; *Dipterocarpac.;* Phil.
 „ : *Shorea eximia* Scheff.; *Dipterocarpac.;* Phil. (Tay.).
 „ : *Shorea malaanonan* Blume; *Dipterocarpac.;* Phil. (Tay.).
 „ : *Shorea palosapis* Merr.; *Dipterocarpac.;* Phil. (Tay.).
Danlóg: *Dipterocarpus grandiflorus* Blco.; *Dipterocarpac.;* Phil. (Cap.).
 „ : *Parashorea malaanonan* Merr.; *Dipterocarpac.;* Phil. (Cam.).
 „ : *Pentacme contorta* Merr. et Rolfe; *Dipterocarpac.;* Phil.

Danstan: *Peltogyne spp.*, *Copaifera bracteata* Bth.; *Caesalpiniac.*; Sur.
Danta: *Cistanthera papaverifera* Chev.; *Tiliac.*; Go.
Danupra: *Toona calantas* Merr. et Rolfe; *Meliac.*; Phil. (I.-N.).
Danya: *Sclerocarya birroea* Hochst.; *Anacardiac.*; n. Nig.
Daó: *Dracontomelum dao* Merr. et Rolfe; *Anacardiac.*; Phil.
Dao: *Artocarpus lakoocha* Roxb.; *Morac.*; Ind.
  „ lèo (H): *Tetrameles nudiflora* R. Br.; *Datiscac.*; Ind.-Ch.
Daoe: *Dracontomelum mangiferum* Blume; *Anacardiac.*; Malay., Sunda.
Daoen tjoetjoer atap: *Baeckea frutescens* L.; *Myrtac.*; Malay.
Daom: *Acer campbelli* Hook. f. et Thoms.; *Acerac.*; nö. O.-Ind. (ö. Nepal).
Daou: *Calophyllum inophyllum* L.; *Guttifer.*; Südsee (Mariannen).
Dapagan: *Palaquium spp.*; *Sapotac.*; Phil. (Un.).
Dapdap: *Erythrina indica* Lam.; *Papilionac.*; Phil.
Dapul: *Duabanga moluccana* Blume; *Sonneratiac.*; Phil. (Abra).
Daquilla: *Lagetta lintearia* Lam.; *Thymelaeac.*; Cuba.
Dar: *Boehmeria rugulosa* Wedd.; *Urticac.*; n. O.-Ind.(Nepal).
Darachik: *Melia azedarach* L.; *Meliac.*; O.-Ind. (Kurram).
Darbela: *Wrightia tomentosa* Roem. et Schultes; *Apocynac.*; n. O.-Ind. (Garhwal).
Dargu: *Ougeinia dalbergioides* Bth.; *Papilionac.*; w. O.-Ind. (n. Bombay).
Dari: *Virola sp.*; *Myristicac.*; Gu.
Darinaballi: *Stryphnodendron flammatum* Kleinh.; *Mimosac.*; Sur.
Darli: *Toona serrata* M. Roem.; *Meliac.*; n. O.-Ind. (Jaunsar).
Darloi: *Toona serrata* M. Roem.; *Meliac.*; n. O.-Ind. (Jaunsar).
Darlu: *Toona serrata* M. Roem.; *Meliac.*; n. O.-Ind. (Jaunsar).
Daroo-daroo: *Urandra sp.*; *Icacinac.*; br. Bor. (Sarawak).
Darroo: *Apodytes sp.*; *Olacac.*; Malay.
Dartrier: *Pterocarpus guianensis* Aubl.; *Papilionac.*; fr. Gu.
  „ : *Cassia alata* L.; *Papilionac.*; fr. Gu.
Daru: *Urandra sp.*; *Icacinac.*; br. Bor. (Sarawak).
  „ : *Sideroxylon malaccense* C. B. Clarke; *Sapotac.*; Malay.
  „ Daru: *Urandra corniculata* Foxw.; *Icacinac.*; Malay.
Dashi: *Balsamodendron africanum* A. Rich.; *Burserac.*; n. Nig.
Dastan: *Peltogyne spp.*, *Copaifera bracteata* Bth.; *Caesalpiniac.*; Sur.
Dât vân: *Ixonanthes cuneata* Miq.; *Linac.*; Annam.
Date, chinese: *Ziziphus jujuba* Lamk.; *Rhamnac.*; tr. Af., tr. As., tr. Au.
  „ , Desert-: *Balanites aegyptiaca* Delile; *Zygophyllac.*; n. Nig.
  „ , edible: *Phoenix dactylifera* L.; *Palmac.*; tr. Af., sw. As.
Dátil: *Phoenix dactylifera* L.; *Palmac.*; Hond.
Datrang: *Ehretia laevis* Roxb.; *Borraginac.*; w. O.-Ind.
Datranga: *Ehretia laevis* Roxb.; *Borraginac.*; C.-O.-Ind.
Datte-phal: *Barringtonia acutangula* Gaertn.; *Myrtac.*; w. O.-Ind.
Dattel, chinesische (d): *Ziziphus jujuba* Lamk.; *Rhamnac.*; tr. As., tr. Au., tr. Af.
Dattelfeige, mexikanische (d): *Diospyros texana* Scheele; *Ebenac.*; USA., Mex.
  „ , virginische: *Diospyros virginiana* L.; *Ebenac.*; sö. N.-Am.
Dattelpflaume, mexikanische (d): *Diospyros texana* Scheele; *Ebenac.*; s. USA., Mex.
  „ , schwarze: *Diospyros texana* Scheele; *Ebenac.*; s. USA., Mex.
  „ , virginische: *Diospyros virginiana* L.; *Ebenac.*; sö. N.-Am.
Dattier (f): *Phoenix dactylifera* L.; *Palmac.*; Af., sw. As.
  „ du désert (f): *Balanites aegyptiaca* Delile; *Zygophyllac.*; n. Af.
Dattock: *Detarium senegalense* Gmel.; *Caesalpiniac.*; W.-Af.
Dâu: *Dipterocarpus spp.*; *Dipterocarpac.*; Ind.-Ch.
  „ : *Apodytes cambodiana* Pierre; *Icacinac.*; Annam.
  „ : *Morus indica* L.; *Morac.*; Annam.
  „ : *Shorea thoreli* Pierre; *Dipterocarpac.*; Co.-Ch. (Saigon).
  „ : *Morinda citrifolia* L.; *Rubiac.*; Ind.-Ch.
  „ cal: *Dipterocarpus artocarpifolius* Pierre; *Dipterocarpac.*; Annam.
  „ „ : *Dipterocarpus insularis* Hance; *Dipterocarpac.*; Annam.
  „ cát: *Dipterocarpus artocarpifolius* Pierre; *Dipterocarpac.*; Ind.-Ch.
  „ „ : *Dipterocarpus insularis* Hance; *Dipterocarpac.*; Ind.-Ch.
  „ chai: *Dipterocarpus intricatus* Dyer; *Dipterocarpac.*; Annam.
? „ chi rui: *Diospyros nitidula* Lecomte; *Ebenac.*; s. Annam.
? „ „ rut: *Diospyros nitidula* Lecomte; *Ebenac.*; s. Annam.

Dâu con rai: *Dipterocarpus alatus* Roxb.; *Dipterocarpac.;* Annam.
„ „ „ nuoe: *Dipterocarpus jourdaini* Pierre; *Dipterocarpac.;* Ind.-Ch.
„ da: *Spondias lakonensis* Pierre; *Anacardiac.;* Ind.-Ch.
„ „ dàt: *Baccaurea sapida* Muell. Arg.; *Euphorbiac.;* Ind.-Ch.
„ dàt: *Baccaurea cauliflora* L., *B. annamensis* Gagnep.; *Euphorbiac.;* Ind.-Ch.
„ den: *Morus indica* L.; *Morac.;* Ind.-Ch.
„ do: *Dipterocarpus punctulatus* Pierre; *Dipterocarpac.;* Ind.-Ch.
„ du: *Diospyros nitidula* Lecomte; *Ebenac.;* s. Annam.
„ gia dát: *Baccaurea sapida* Muell. Arg.; *Euphorbiac.;* Ind.-Ch.
„ heo: *Garuga pinnata* Roxb.; *Burserac.;* Ind.-Ch.
„ la tat: *Cleidion javanicum* Blume; *Euphorbiac.;* Ind.-Ch.
„ lai: *Trewia nudiflora* L.; *Euphorbiac.;* Ind.-Ch.
„ lông: *Dipterocarpus duperreanus* Pierre; *Dipterocarpac.;* Annam.
„ lóng: *Dipterocarpus tuberculatus* Roxb.; *Dipterocarpac.;* Annam.
„ mit: *Dipterocarpus artocarpifolius* Pierre; *Dipterocarpac.;* Annam.
„ „ : *Dipterocarpus insularis* Hance; *Dipterocarpac.;* Annam.
„ ngo: *Dipterocarpus alatus* Roxb.; *Dipterocarpac.;* Annam.
„ nuoc: *Dipterocarpus alatus* Roxb.; *Dipterocarpac.;* Annam.
„ rai: *Canarium oleosum* Engl.; *Burserac.;* Ind.-Ch.
„ rùng: *Baccaurea sapida* Muell. Arg.; *Euphorbiac.;* Ind.-Ch.
„ sang nang: *Dipterocarpus dyeri* Pierre; *Dipterocarpac.;* Annam.
„ „ son: *Dipterocarpus tuberculatus* Roxb.; *Dipterocarpac.;* Annam.
„ son: *Dipterocarpus tuberculatus* Roxb.; *Dipterocarpac.;* Annam.
„ „ nang: *Dipterocarpus dyeri* Pierre; *Dipterocarpac.;* Annam.
„ song nang: *Dipterocarpus dyeri* Pierre; *Dipterocarpac.;* Ind.-Ch.
„ „ „ : *Dipterocarpus artocarpifolius* Pierre; *Dipterocarpac.;* Annam.
„ thiên: *Baccaurea sapida* Muell. Arg.; *Euphorbiac.;* Ind.-Ch.
„ tiên: *Baccaurea annamensis* Gagnep.; *Euphorbiac.;* Ind.-Ch.
„ tra beng: *Dipterocarpus obtusifolius* Teijsm.; *Dipterocarpac.;* Annam.
„ trai: *Dipterocarpus intricatus* Dyer; *Dipterocarpac.;* Annam.
„ truòng: *Schleichera trijuga* Willd.; *Sapindac.;* Ind.-Ch.
Dauél: *Wallaceodendron celebicum* Koord.; *Leguminos.;* Phil (Cag.).
Dauen: *Dipterocarpus grandiflorus* Blco.; *Dipterocarpac.;* Phil. (Cag.).
Dáueng: *Aglaia diffusa* Merr.; *Meliac.;* Phil. (Cag.).
Dauér: *Wallaceodendron celebicum* Koord.; *Leguminos.;* Phil. (Cag.).
Dauh: *Saccoglottis gabonensis* Urban; *Humiriac.;* Lib.
Daula: *Miliusa velutina* Hook. f. et Thoms.; *Anonac.;* nw. O.-Ind. (Kumeon).
Daun durian: *Boschia griffithi* Mast.; *Bombacac.;* Malak.
Daungsatpya: *Callicarpa arborea* Roxb.; *Verbenac.;* Burma.
Dauraenjo: *Pongamia glabra* Vent.; *Papilionac.;* sw. Beng. (Orissa).
Daurango: *Holoptelea integrifolia.* Planch.; *Ulmac.;* sw. Beng. (Orissa).
Dauri: *Toona serrata* Roem.; *Meliac.;* n. O.-Ind.
Dauta: *Cola cordifolia* R. Br.; *Sterculiac.;* Senl.
Dawa: *Pometia pinnata* Forst.; *Sapindac.;* Sunda., Südsee.
Dawata: *Carallia integerrima* DC.; *Rhizophorac.;* Cey.
Dawngding: *Dipterocarpus tuberculatus* Roxb.; *Dipterocarpac.;* O.-Ind.
Dawo: *Tetrapleura thonningi* Bth.; *Mimosac.;* Kam.
Dawoenbaroe: *Hibiscus tiliaceus* L.; *Malvac.;* Nied.-Ind.
Day khla: *Wrightia annamensis* Eberh. et Dubard; *Apocynac.;* Ind.-Ch.
„ mang: *Croton argyratus* Blume; *Euphorbiac.;* Ind.-Ch.
Dayamí: *Cereus* sp.; *Cactac.;* Arg.
Dáyap-amó: *Taxotrophis ilicifolia* Vid.; *Morac.;* Phil. (Pal.).
Dayapa: *Virola sp.; Myristicac.;* Gu.
Day-ne-waye: *Cola digitata* Mast.; *Sterculiac.;* Lib.
Ddjivica: ? *Maesobotrya stapfiana* Beille; *Euphorbiac.;* Elf.
Dê: *Diplospora singularis* Kurz; *Rubiac.;* Ind.-Ch.
Dea: *Timonius jambosella* Thw.; *Rubiac.;* Ind.-Ch.
Dead finish: *Albizzia basaltica* Bth.; *Mimosac.;* Queensl.
Dead man's bones: *Psychotria pinularis* Sessé et Moç.; *Rubiac.;* br. Hond.
Déaû rai: *Canarium oleosum* Engl.; *Burserac.;* Ind.-Ch.
Debima: *Uvaria* sp.; *Anonac.;* Go.
Deboo: *Anona muricata* L.; *Anonac.;* Go.

Debou: ? *Anona muricata* L.; *Anonac.;* Dah.
Dedadeda: *Digitalis canariensis* L.; *Scrophulariac.;* Tener.
Dedali: *Strombosia javanica* Blume; *Olacac.;* br. Malay.
Dedaru: *Urandra corniculata* Foxw.; *Icacinac.;* Malay.
Dedatera: *Digitalis canariensis* L.; *Scrophulariac.;* Tener.
Dedoari-janum: *Zizyphus jujuba* Lamb.; *Rhamnac.;* Beng. (Nagpur).
Dedule: *Protea bismarcki* Engl.; *Proteac.;* Togo.
Dedwar: *Cedrus deodara* Loudon; *Conifer.;* n. O.-Ind. (Kasch.).
Deehn: *Pycnanthus kombo* Warb.; *Myristicac.;* Lib.
Deer's horn: *Mouriria sp.; Melastomatac.;* br. Hond.
Deerwood: *Ostrya virginiana* C. Koch; *Betulac.;* nö. USA.
Dee-waye: *Acioa sp.; Rosac.;* Lib.
Defla: *Nerium oleander* L.; *Apocynac.;* Sahara.
Degame (u): *Calycophyllum candidissimum* DC.; *Rubiac.;* Cuba.
Degbo: *Celtis sp.; Ulmac.;* Dah.
Dego: ? *Parkia agboënsis* Chev.; *Mimosac.;* Elf.
Deh: *Omphalocarpum elatum* Miers; *Sapotac.;* Lib.
Dehakali: *Wrightia tinctoria* R. Br.; *Apocynac.;* C.-O.-Ind.
Dehua: *Artocarpus lakoocha* Roxb.; *Morac.;* Beng.
Dehwan: *Cordia macleodi* Hook. f. et Thoms.; *Borraginac.;* O.-Ind.
Deikna: *Melia azedarach* L.; *Meliac.;* n. O.-Ind.
Dek: *Melia azedarach* L.; *Meliac.;* n. O.-Ind.
Dekamali: *Gardenia lucida* Roxb.; *Rubiac.;* w. O.-Ind.
Dekie-hatti: *Maprounea guianensis* Aubl.; *Laurac.;* Sur.
Deknoi: *Melia azedarach* L.; *Meliac.;* n. O.-Ind.
Del: *Artocarpus nobilis* Thw.; *Morac.;* s. Ind.
Deleb: *Borassus flabellifer* L.; *Palmac.;* tr. As.
Delima hutan: *Gardenia tubifera* Wall.; *Rubiac.;* Malay, Sum., Bor.
Delimas: *Acrocarpus fraxinifolius* Wight et Arn.; *Caesalpiniac.;* Java.
Demb: *Diplanthemum viridiflorum* Hutch.; *Tiliac.;* Kam.
Dembudu: *Elaeis guineensis* Jacq. var. ?; *Palmac.;* Gab.
Demi-deuil (f): *Diospyros aggregata* Gürke; *Ebenac.;* Kam., Gab.
Demthy: *Erythrina sp.; Papilionac.;* Mex.
Dén: *Timonius jambosella* Thw.; *Rubiac.;* Ind.-Ch.
Den-be-haw: *Cola angustifolia* K. Schum.; *Sterculiac.;* Lib.
Dendé: *Borassus flabellifer* L. *var. aethiopum* Warb.; *Palmac.;* Elf.
Dendo: *Borassus flabellifer* L. *var. aethiopum* Warb.; *Palmac.;* Elf.
Dendurien: *Boschia griffithi* Mast.; *Bombacac.;* Malak.
Deng: *Xylia dolabriformis* Bth.; *Mimosac.;* Ind.-Ch.
Denkan: *Melia azedarach* L.; *Meliac.;* n. O.-Ind. (Garhwal).
Denya: *Cylicodiscus gabunensis* Harms; *Mimosac.;* Go.
Deo: *Xylopia aethiopica* A. Rich.; *Anonac.;* Lib.
Deodar (H): *Cedrus deodara* Loudon; *Conifer.;* n. O.-Ind.
  „ : *Cupressus torulosa* D. Don; *Conifer.;* n. O.-Ind. (Punjab).
De-orh: *Tarrietia utilis* Sprague; *Sterculiac.;* Lib.
Depapate: *Detarium senegalense* Gmel.; *Caesalpiniac.;* Togo.
Dephul: *Artocarpus lakoocha* Roxb.; *Morac.;* Beng.
Dera: *Callicarpa arborea* Roxb.; *Verbenac.;* n. O.-Ind.
Dervo: *Lophira procera* Chev.; *Ochnac.;* Kam., Gab.
Determa: *Bucida angustifolia* DC.; *Combretac.;* fr. Gu.
  „ (e): *Nectandra sp., Ocotea rubra* Mez; *Laurac.;* Sur.
  „   „ : ? *Terminalia buceras* Hook.; *Combretac.;* Sur.
Detʒe: *Prunus capuli* Cav.; *Rosac.;* Mex.
Devadarum: *Erythroxylon areolatum* L.; *Erythroxylac.;* s. Ind.
Deva-kanchan: *Bauhinia purpurea* L.; *Caesalpiniac.;* Beng.
Devdari: *Toona ciliata* M. Roem.; *Meliac.;* w. O.-Ind.
Devdaru: *Dysoxylum binectariferum* Hook. f.; *Meliac.;* sw. O.-Ind.
Devidar: *Cupressus torulosa* D. Don; *Conifer.;* n. O.-Ind. (Punjab).
Devidiar: *Cupressus torulosa* D. Don; *Conifer.;* n. O.-Ind.
Devil wood: *Olea americana* L.; *Oleac.;* USA.
Dewadari: *Erythroxylon areolatum* L.; *Erythroxylac.;* Ind.
Dewar: *Cedrus deodara* Loudon; *Conifer.;* n. O.-Ind.

Dewudar: *Erythroxylon areolatum* L.; *Erythroxylac.;* Ind.
Dhaia: *Callicarpa arborea* Roxb.; *Verbenac.;* n. O.-Ind. (Garhwal).
Dhaim: *Cordia macleodi* Hook. f. et Thoms.; *Borraginac.;* w. O.-Ind.
Dhaivan: *Cordia macleodi* Hook. f. et Thoms.; *Borraginac.;* w. O.-Ind.
Dhaiyan: *Cordia macleodi* Hook. f. et Thoms.; *Borraginac.;* O.-Ind.
Dhak: *Erythrina suberosa* Roxb.; *Papilionac.;* n. O.-Ind.
  „ : *Butea frondosa* Roxb.; *Papilionac.;* n. O.-Ind. (Punjab).
Dhal: *Cajanus indicus* Spreng.; *Papilionac.;* Tr.
Dhaman (H): *Grewia tiliaefolia* Vahl; *Tiliac.;* C.-O.-Ind.
  „ : *Grewia oppositifolia* Roxb.; *Tiliac.;* n. O.-Ind. (Punjab).
  „ : *Grewia elastica* Royle; *Tiliac.;* n. O.-Ind. (Punjab).
  „ : *Grewia laevigata* Vahl; *Tiliac.;* w. O.-Ind.
Dhamana: *Grewia tiliaefolia* Vahl; *Tiliac.;* O.-Ind.
Dhamani: *Grewia tiliaefolia* Vahl; *Tiliac.;* sw. O.-Ind.
  „ : *Grewia laevigata* Vahl; *Tiliac.;* nw. O.-Ind. (Punjab).
Dhamasi: *Dalbergia lanceolaria* L.; *Papilionac.;* C.-O.-Ind.
Dhamia: *Grewia elastica* Royle; *Tiliac.;* C.-O.-Ind. (Merwara).
Dhamin: *Grewia tiliaefolia* Vahl; *Tiliac.;* n. O.-Ind.
Dhamman: *Grewia oppositifolia* Roxb.; *Tiliac.;* nw. Himal.
Dhamni: *Grewia asiatica* L.; *Tiliac.;* Ind.
Dhamnoo: *Grewia elastica* Royle; *Tiliac.;* n. Ind.
Dhamora: *Anogeissus latifolia* Wall.; *Combretac.;* C.-O.-Ind.
Dhambabar: *Cassia fistula* L.; *Caesalpiniac.;* C.-O.-Ind.
Dhao: *Anogeissus latifolia* Wall.; *Combretac.;* nw. O.-Ind. (Punjab).
Dhaora: *Anogeissus latifolia* Wall.; *Combretac.;* C.-O.-Ind.
Dhaori: *Anogeissus latifolia* Wall.; *Combretac.;* sw. O.-Ind.
Dharauli: *Wrightia tomentosa* Roem. et Schultes; *Apocynac.;* n. O.-Ind.
Dharmara: *Stereospermum chelonioides* DC.; *Bignoniac.;* Beng.
Dhau: *Anogeissus pendula* Edgew.; *Combretac.;* O.-Ind. (Raip.).
  „ : *Anogeissus latifolia* Wall.; *Combretac.;* n. O.-Ind.
Dhaukra: *Anogeissus pendula* Edgew.; *Combretac.;* O.-Ind. (Raip.).
Dhaul: *Erythrina suberosa* Roxb.; *Papilionac.;* n. O.-Ind.
  „ -pedda: *Trewia nudiflora* L.; *Euphorbiac.;* O.-Ind. (Dehra-Dun).
Dhauli: *Hymenodictyon excelsum* Wall.; *Rubiac.;* n. O.-Ind.
Dhaulo: *Bridelia retusa* Spreng.; *Euphorbiac.;* n. O.-Ind. (Garhwal).
Dhaunda: *Anogeissus latifolia* Wall.; *Combretac.;* C.-O.-Ind.
Dhaura: *Lagerstroemia parviflora* Roxb.; *Lythrac.;* n. O.-Ind.
  „ : *Anogeissus latifolia* Wall.; *Combretac.;* w. O.-Ind.
Dhauri: *Lagerstroemia parviflora* Roxb.; *Lythrac.;* Ind.
Dhawa: *Anogeissus latifolia* Wall.; *Combretac.;* C.-O.-Ind.
Dhengau: *Cordia macleodi* Hook. f. et Thoms.; *Borraginac.;* O.-Ind.
Dheu: *Artocarpus lakoocha* Roxb.; *Morac.;* O.-Ind. (Dehra Dun).
Dhimeri: *Ficus glomerata* Roxb.; *Morac.;* nö. O.-Ind.
Dhobu: *Anogeissus latifolia* Wall.; *Combretac.;* w. Beng. (Orissa).
Dhorbiola: *Pterocarpus marsupium* Roxb.; *Papilionac.;* C.-O.-Ind.
Dhowda: *Holarrhena antidysenterica* Wall.; *Apocynac.;* O.-Ind. (Gujarat).
Dhum: *Aquilaria agallocha* Roxb.; *Thymelaeac.;* Assam.
Dhumnah: *Holoptelea integrifolia* Planch.; *Ulmac.;* n. O.-Ind.
Dhundol: *Carapa obovata* Blume; *Meliac.;* Beng.
Dhundul: *Carapa moluccensis* Lam.; *Meliac.;* Beng.
Dhup (H): *Canarium euphyllum* Kurz; *Burserac.;* And.
  „ : *Juniperus macropoda* Boiss.; *Conifer.;* n. O.-Ind. (Jaunsar).
  „ : *Pinus longifolia* Roxb.; *Conifer.;* n. O.-Ind. (Nepal, Beng.)
  „ : *Ailanthus malabarica* DC.; *Simarubac.;* n. O.-Ind., Cey.
  „ , black (H): *Canarium strictum* Roxb.; *Burserac.;* Ind.
  „ maram: *Vateria indica* L.; *Dipterocarpac.;* s. Ind.
  „ , red: *Parishia insignis* Hook. f.; *Anacardiac.;* And.
  „ , white (H): *Vateria indica* L.; *Dipterocarpac.;* O.-Ind.
  „ „ : *Canarium euphyllum* Kurz; *Burserac.;* And.
Dhupa: *Vateria indica* L.; *Dipterocarpac.;* O.-Ind.
  „ paini: *Vateria indica* L.; *Dipterocarpac.;* sw. O.-Ind.
Dhupi: *Juniperus macropoda* Boiss.; *Conifer.;* n. O.-Ind. (Nepal).

Dhupri: *Juniperus macropoda* Boiss.; *Conifer.*; n. O.-Ind. (Jaunsar).
Di: *Chlorophora excelsa* Bth. et Hook.; *Morac.*; Elf.
Diáan: *Zizyphus spp.*; *Rhamnac.*; Phil. (Pang.).
Diabé: *Acacia arabica* L.; *Mimosac.*; Sudan.
Diala: *Khaya senegalensis* A. Juss.; *Meliac.*; Senl., fr. Guin.
Dialamban: *Dalbergia melanoxylon* Guill. et Perr.; *Papilionac.*; Sudan.
Dialfélkété: *Hannoa undulata* Planch.; *Simarubac.*; fr. Guin.
Diam: *Stephegyne parvifolia* Korth.; *Rubiac.*; Annam.
Diamilikie: *Jacaranda filicifolia* D. Don; *Bignoniac.*; Sur.
Diancongué: *Myrianthus serratus* Bth. et Hook.; *Morac.*; Elf.
Diang: *Eugenia operculata* Roxb.; *Myrtac.*; Ind.-Ch.
Diangué: *Ficus sp.*; *Morac.*; Elf.
   „   : *Mimusops micrantha* Chev.; *Sapotac.*; Elf.
Diar: *Uvaria parviflora* Rich.; *Anonac.*; sw. Seng.
Diar: *Cedrus deodara* Loudon; *Conifer.*; n. O.-Ind. (Punjab).
Dibanga: *Cola acuminata* Schott et Endl.; *Sterculiac.*; Kam.
Dib-bah: *Diospyros spp.*; *Ebenac.*; Lib.
Dibétou (f): *Lovoa klaineana* Pierre; *Meliac.*; Elf.
Dibi: *Cola acuminata* Schott et Endl.; *Sterculiac.*; Kam.
   „ -Dibi: *Caesalpinia coriaria* Willd.; *Caesalpiniac.*; Cuba, St.-D.
Dibidou: *Cola acuminata* Schott et Endl.; *Sterculiac.*; Kam.
Dibindi: *Odyendea gabunensis* Engl.; *Simarubac.*; Gab.
Dibuál: *Pterospermum spp.*; *Sterculiac.*; Phil. (Bas.).
Dibwe-Mutshi: *Khaya wildemani* Ghesq.; *Meliac.*; B.-K.
Dickamali: *Gardenia gummifera* L.; *Rubiac.*; w. O.-Ind.
Dicky: *Gardenia gummifera* L.; *Rubiac.*; sw. O.-Ind.
Dididire: *Euphorbia poissoni* Pax; *Euphorbiac.*; Togo.
Didigkalín: *Uitex aherniana* Merr.; *Uerbenac.*; Phil. (Cam.).
Didôk: *Bombax insigne* Wall. var. *andamanica* Prain; *Bombacac.*; Burma.
   „   : *Bombax insigne* Wall. var. *wighti* Prain; *Bombacac.*; Burma.
Didu (H): *Bombax insigne* Wall.; *Bombacac.*; Ind., And.
   „   simal: *Bombax insigne* Wall.; *Bombacac.*; And.
Dien: *Pentace tonkinensis* Chev.; *Tiliac.*; Ind.-Ch.
Diên diên: *Aeschynomene aspera* L.; *Papilionac.*; Ind.-Ch.
Dieng: *Evodia fraxinifolia* Hook. f.; *Rutac.*; Assam.
? Dienya: *Afrormosia laxiflora* Bth.; *Papilionac.*; Go.
Diêp tây: *Poinciana regia* Bojer; *Caesalpiniac.*; Ind.-Ch.
   „   „ : *Hura crepitans* L.; *Euphorbiac.*; Ind.-Ch.
Diêu nhuôm: *Bixa orellana* L.; *Bixac.*; Ind.-Ch.
Digbere: *Hannoa undulata* Planch.; *Simarubac.*; Togo.
Digpapate: *Detarium microcarpum* Guill. et Perr.; *Caesalpiniac.*; Togo.
Dika: *Irvingia gabonensis* Baill.; *Simarubac.*; Gab.
Dikamali: *Gardenia lucida* Roxb.; *Rubiac.*; C.-O.-Ind.
Dikang: *Terminalia calamansanai* Rolfe; *Combretac.*; Phil. (Pamp.).
Dikassa-kassa: *Albizzia fastigiata* Oliv.; *Mimosac.*; B.-K. (Mayumba).
Diki: *Celtis integrifolia* Lam.; *Ulmac.*; Togo.
Dikláb: *Zizyphus spp.*; *Rhamnac.*; Phil. (I.).
Dikláp: *Zizyphus spp.*; *Rhamnac.*; Phil. (I.).
Dikotséké: *Sclerosperma walkeri* ?; *Palmac.*; Gab.
Dikusu: *Dialium laurenti* de Wild.; *Caesalpiniac.*; B.-K.
Diláan: *Buchanania spp.*; *Anacardiac.*; Phil. (Zam.).
Diladiláan: *Celtis spp.*; *Ulmac.*; Phil. (Riz.).
Diladíla-asu: *Celtis spp.*; *Ulmac.*; Phil. (Pamp.)
Dilang-butiki: *Knema glomerata* Merr.; *Myristicac.*; Phil. (Lag.).
   „   „   : *Myristica spp.*; *Myristicac.*; Phil. (Lag.).
Dílang-usá: *Carallia integerrima* DC.; *Rhizophorac.*; Phil. (Mind.).
Dileba: *Khaya sp.*; *Meliac.*; Gab.
Dilifu: *Trichilia prieuriana* Jussieu; *Meliac.*; W.-Af.
Dill: *Cedrus deodara* Loudon; *Conifer.*; Ind.
Dillenia (H): *Dillenia pentagyna* Roxb.; *Dilleniac.*; Ind.
Dilly, wild: *Mimusops parvifolia* Radlk.; *Sapotac.*; s. Fl., W.-Ind.
Dilo: *Calophyllum inophyllum* L.; *Guttifer.*; Fid.

Dilobidiba-Kouandzi: *Berlinia bracteosa* Bth.; *Caesalpiniac.;* Gab.
Dilobo: *Lovoa klaineana* Pierre; *Meliac.;* Gab.
Dilolo: *Khaya ivorensis* Chev.; *Meliac.;* Gab.
„ -di-benga: *Khaya sp.; Meliac.;* Gab.
„ „ -fiote: *Khaya sp.; Meliac.;* Gab.
„ „ -Mounaï: *Antrocaryon klaineanum* Pierre; *Anacardiac.;* Gab.
„ „ -Tangani: *Khaya sp.; Meliac.;* Gab.
Dilomba: *Pycnanthus kombo* Warb.; *Myristicac.;* Gab.
Dima: *Adansonia digitata* L.; *Bombacac.;* Af. (Abess.).
Dimagnımouirri: *Pierreodendron durissimum* Chev.; *Sapotac.;* Gab.
Dimb: *Cordyla senegalensis* ?; *Caesalpiniac.;* Senl.
Dimbalo: *Pentaclethra macrophylla* Bth.; *Mimosac.;* Gab.
Dimbilo: *Elaeis guineensis* Jacq. *var. nigrescens* Chev.; *Palmac.;* Gab.
Dimébo: ? *Entandrophragma sp.,* ? *Khaya spp.,* ? *Carapa spp.; Meliac.;* Kam.
Dimimbidiamamba: *Discoglypremna caloneura* Prain; *Euphorbiac.;* Gab.
Dim-petoi: *Dillenia aurea* Sm.; *Dilleniac.;* Kamb.
Dina: *Dialium aff. connaroides* Harms; *Caesalpiniac.;* Kam., Gab.
Dindal: *Anogeissus latifolia* Wall.; *Combretac.;* w. O.-Ind.
Dinde: *Chlorophora tinctoria* Gaudich.; *Morac.;* Kol.
„ : *Pithecolobium unguis-cati* Bth.; *Mimosac.;* Kol.
Dinduga: *Anogeissus latifolia* Wall.; *Combretac.;* w. O.-Ind.
Dingan: *Schima wallichi* Choisy; *Theac.;* Assam.
Dingdah: *Bucklandia populnea* R. Br.; *Hamamelidac.;* Assam.
Dingèko: *Ongokea klaineana* Pierre; *Olacac.;* Gab.
Dingi: *Rhizophora racemosa* G. F. W. Meyer; *Rhizophorac.;* Lib.
Diñgin: *Dillenia spp.; Dilleniac.;* Phil. (Zam.).
Dingkurlong: *Melia composita* Willd.; *Meliac.;* Assam.
Dinglás: *Lagerstroemia piriformis* Kochne; *Lythrac.;* Phil. (Tay.)
„ : *Terminalia comintana* Merr.; *Combretac.;* Phil. (Bat., Lag., Tay.).
„ : *Eucalyptus deglupta* Blume; *Myrtac.;* Phil. (Zamb., Cot.).
„ : *Tristania spp.; Myrtac.;* Phil.
Dingleen: *Betula alnoides* Ham.; *Betulac.;* Assam.
Dingo latoe: *Murraya sumatrana* Roxb.; *Rutac.;* Sum.
Dingrittang: *Quercus serrata* Thunb.; *Fagac.;* Assam.
Dingsa: *Pinus khasya* Royle; *Conifer.;* Assam.
Dingsableh: *Taxus baccata* L.; *Conifer.;* Assam.
„ : *Podocarpus neriifolia* D. Don; *Conifer.;* Assam.
Dinh: *Markhamia stipulata* Seem.; *Bignoniac.;* Ind.-Ch.
„ gan gá: *Markhamia stipulata* Seem.; *Bignoniac.;* Ind.-Ch.
„ khét: *Markhamia stipulata* Seem.; *Bignoniac.;* Ind.-Ch.
„ mât: *Markhamia stipulata* Seem.; *Bignoniac.;* Ind.-Ch.
Dinjóngoholz (d): *Kickxia elastica* Preuß; *Apocynac.;* Kam.
Dinya: *Vitex cienkowskyi* Kotschy et Peyr.; *Verbenac.;* n. Nig.
„ biri: *Vitex diversifolia* Baker; *Verbenac.;* n. Nig.
Diolosso: *Antiaris africana* Engl.; *Morac.;* Kam.
Diomate: *Astronium graveolens* Jacq.; *Anacardiac.;* Kol.
Diou: *Mitragyne africana* Korth.; *Rubiac.;* fr. Guin.
Di-peh: *Aporrhiza aff. talboti* Bak. f.; *Sapindac.;* Lib.
Dipérétou: *Picralima ellioti* Stapf; *Apocynac.;* Elf.
Diphole, weidenblättrige (d): *Dipholis salicifolia* A. DC.; *Sapotac.;* s. Fl., Bah., Ant.
Diphylle pois confiture (f): *Hymenaea courbaril* L.; *Caesalpiniac.;* fr. Gu.
Dipouta: *Coula edulis* Baill.; *Olacac.;* Gab.
Dir'án: *Zizyphus spp.; Rhamnac.;* Phil.
Diráan: *Quercus spp.; Fagac.;* Phil.
„ : *Zizyphus spp.; Rhamnac.;* Phil.
„ : *Pentacme contorta* Merr. et Rolfe; *Dipterocarpac.;* Phil. (Ben.).
Dirasana: *Albizzia lebbek* Bth.; *Mimosac.;* w. O.-Ind.
Dirasanam: *Albizzia lebbek* Bth.; *Mimosac.;* Ind.
Dirigkalín: *Terminalia edulis* Blco.; *Combretac.;* Phil. (Cam.).
Diritá: *Alstonia scholaris* R. Br.; *Apocynac.;* Phil. (Bat., Cam.).
Disciplina: *Bombax ellipticum* H. B. K.; *Bombacac.;* Mex.
„ de monja: *Caesalpinia gilliesi* Wall.; *Caesalpiniac.;* Arg.

Diseng: *Musanga smithi* R. Br.; *Morac.;* Kam.
Disèngo· *Odyendea gabunensis* Engl.; *Simarubac.;* Gab.
Disi: *Terminalia edulis* Blco.; *Combretac.;* Phil. (Nu.-V.).
Distel, assyrische (d): *Verbascum sp.; Scrophulariac.;* s. Eu.
Dita: *Detarium senegalense* Gmel.; *Caesalpiniac.;* Lib.
Ditá: *Alstonia scholaris* R. Br.; *Apocynac.;* Phil.
„ : *Paralstonia clusiacea* Baill.; *Apocynac.;* Phil. (Bat.).
Ditáa: *Alstonia scholaris* R Br.; *Apocynac.;* Phil.
Ditsende-Tsende: *Fagara macrophylla* Engl.; *Rutac.;* B.-K.
Divangwa: *Corynanthe paniculata* Welw.; *Rubiac.;* B.-K. (Mayumba).
Divavira: *Pachylobus balsamifera* Guill.; *Burserac.;* Gab.
Dividive: *Caesalpinia coriaria* Willd.; *Caesalpiniac.;* Ven.
„ de los Andes: *Caesalpinia tinctoria* Bth.; *Caesalpiniac.;* Ven.
Divi-Divi: *Caesalpinia coriaria* Willd.; *Caesalpiniac.;* Ant., C.-Am., Kol., Ven.
„ „ : *Prosopis chilensis* Stuntz; *Mimosac.;* Guat.
„ „ : *Caesalpinia digyna* Rottl.; *Caesalpiniac.;* Ind.
Dividivi: *Caesalpinia spinosa* Kuntze; *Caesalpiniac.;* Kol.
„ de los Andes: *Caesalpinia spinosa* Kuntze; *Caesalpiniac.;* Ven.
Diyapara: *Wormia triquestra* Rottb.; *Dilleniac.;* Cey.
Djakali: *Couma guianensis* Aubl.; *Apocynac.;* Sur.
Djakidja: *Sclerolobium sp.; Caesalpiniac.;* Sur.
Djakidji: *Sclerolobium sp.; Caesalpiniac.;* Sur.
Djalangadah: *Pterocarpus dalbergioides* Roxb.; *Papilionac.;* And.
Djalatrang: *Aromadendron elegans* Blume; *Magnoliac.;* Nied.-Ind.
Djalitri: *Wrightia javanica* A. DC.; *Apocynac.;* Java, Ind.
Djaloepang: *Columbia javanica* Blume; *Tiliac.;* Nied.-Ind.
Djambi: ? *Khaya anthotheca* C. DC.; *Meliac.;* Kam.
Djamboe: *Psidium pomiferum* L.; *Myrtac.;* W.-Ind.
Djamoedjoe: *Podocarpus imbricata* Blume; *Conifer.;* Java.
Djangkang: *Sterculia foetida* L.; *Sterculiac.;* Bor.
Djanglot: *Picrasma javanica* Blume; *Simarubac.;* Java.
Djanitrie: *Elaeocarpus angustifolius* Blume; *Elaeocarpac.;* Nied.-Ind.
Djansan· *Ricinodendron africanum* Muell. Arg.; *Euphorbiac.;* Kam.
Djati: *Tectona grandis* L.; *Verbenac.;* Java.
Djatti: *Tectona grandis* L.; *Verbenac.;* Gab.
„ -hollanda: *Guazuma tomentosa* H. B. K.; *Sterculiac.;* Malay, Sunda.
Djavé: *Mimusops djave* Engl., *M. Pierreana* Engl.; *Sapotac.;* Kam., Gab.
Djedoe: *Sclerolobium spp.; Caesalpiniac.;* Sur.
„ , roode: *Sclerolobium sp.; Caesalpiniac.;* Sur.
„ , witte: *Sclerolobium sp.; Caesalpiniac.;* Sur.
Djeloetoeng: *Dyera lowi* Hook. f.; *Apocynac.;* Bor.
Djempina: *Palaquium javense* Burck.; *Sapotac.;* Java.
„ : *Hopea fagifolia* Miq.; *Dipterocarpac.;* Java, Banka.
Djenar: ? *Murraya sp.; Rutac.;* Java.
Djengkol: *Pithecolobium bigeminum* Mart.; *Mimosac.;* O.-Ind.
Djerambong: *Evodia sp.; Rutac.;* Nied.-Ind.
Djeroek balie: *Citrus decumana* L.; *Rutac.;* Nied.-Ind.
„ matjang: *Citrus decumana* L.; *Rutac.;* Nied.-Ind.
Djib: *Strombosia grandifolia* Hook. f.; *Olacac.;* Kam.
Djiera Koewwanten: *Eurya hirsutula* Miq.; *Theac.;* Nied.-Ind.
Djika: *Mimusops heckeli* Lecomte; *Sapotac.;* Gab.
Djiladjila: ? *Dialium yambataense* Verm.; *Caesalpiniac.;* B.-K.
„ : ? *Petersia africana* Welw.; *Myrtac.;* B.-K.
Djilika: ? *Maesobotrya stapfiana* Beille; *Euphorbiac.;* Elf.
Djilo: *Pycnanthus kombo* Warb.; *Myristicac.;* Elf.
Djimbo: *Ochrocarpus africanus* Oliv.; *Guttifer.;* Elf.
Djinda: *Dialium laurenti* de Wild.; *Caesalpiniac.;* B.-K.
Djindjit: *Calophyllum sp.; Guttifer.;* Nied.-Ind.
Djirika: ? *Maesobotrya stapfiana* Beille; *Euphorbiac.;* Elf.
Djiuluton: *Dyera lowi* Hook. f.; *Apocynac.;* Bor.
Djo Arbi: *Rhaptopetalum tieghemi* Chev.; *Olacac.;* Elf.
Djoë y: *Bombax buonopozense* Beauv.; *Bombacac.;* tr. Af.

Djoebolletrie, witte: ? *Chaetocarpus schomburgkiana* Pax et Hoffm.; *Euphorbiac.; Sur.*
Djoebotrie, witte: ? *Chaetocarpus schomburgkiana* Pax et Hoffm.; *Euphorbiac.; Sur.*
Djoekabébé: *Pterocarpus draco* L., *P. rhori* Vahl; *Papilionac.; Sur.*
Djoemoe: *Couratari guianensis* Aubl.; *Lecythidac.; Sur.*
Djoengoeng: *Heritiera littoralis* Dryander; *Sterculiac.; Sunda.*
Djoenti: *Dillenia aurea* Sm.; *Dilleniac.; Java.*
Djoho: *Terminalia belerica* Roxb.; *Combretac.; Java.*
Djom: *Ceiba pentandra* Gaertn.; *Bombacac.; Kam.*
„ : *Cylicodiscus gabunensis* Harms; *Mimosac.; Kam.*
Djombé oua: *Pausinystalia spp.; Rubiac.; Kam.*
„ „ : *Corynanthe gabonensis* Chev.; *Rubiac.; Kam.*
„ „ : *Corynanthe johimbe* K. Schum.; *Rubiac.; Kam.*
D'jouga: *Humiria gabonensis* Baill.; *Humiriac.; Gab.*
Djoupie: *Ochrocarpus siamensis* T. Anders.; *Guttifer.;* w. Annam.
Djuna: *Musanga smithi* R. Br.; *Morac.;* Elf.
Djungu: *Dialium yambataense* Verm.; *Caesalpiniac.;* B.-K.
Dlingsem: *Homalium tomentosum* Bth.; *Flacourtiac.;* Java.
Dô: *Buchholzia macrophylla* Engl.; *Capparidac.;* Elf.
Do dot: *Mallotus eberhardti* Gagnep.; *Euphorbiac.;* Ind.-Ch.
„ „ : *Phyllanthus ruber* Spreng.; *Euphorbiac.;* Ind.-Ch.
„ giam: *Paradina hirsuta* Pitard; *Rubiac.;* Annam.
„ ngon: *Pithecolobium acuminatum* Bth.; *Mimosac.;* Ind.-Ch.
Doc: *Garcinia tonkinensis* Vesque; *Guttifer.;* Annam.
Doctor's club: *Zanthoxylum sp.; Rutac.;* Bah.
Dodankaha: *Memecylon capitellatum* Blume; *Melastomatac.;* Cey.
Dodo: *Adansonia digitata* L.; *Bombacac.;* Go.
Dodowa: *Cola caricifolia* K. Schum.; *Sterculiac.;* Go.
Doe: *Musanga smithi* R. Br.; *Morac.;* Lib.
Doe-doo: *Periploca nigrescens* Afz.; *Asclepiadac.;* Lib.
„ fiah: *Cola acuminata* Schott et Endl.; *Sterculiac.;* Lib.
„ pu: *Octoknema borealis* Hutch. et Dalz.; *Olacac.;* Lib.
„ yah: *Sarcocephalus esculentus* Afz.; *Rubiac.;* Lib.
Doekaliballi: *Platymiscium polystachyum* Bth.; *Papilionac.; Sur.*
Doekoe: *Lansium domesticum* Jack; *Meliac.;* Malay.
Doekoek: *Lumnitzera littorea* Voigt; *Combretac.;* tr. As., tr. Au., Südsee.
„ -ageng: *Lumnitzera littorea* Voigt; *Combretac.;* Java.
Doekoeli: *Couepia surinamensis* Kleinh.; *Rosac.; Sur.*
Doekoelia: *Ficus sp.; Morac.; Sur.*
„ : *Couepia surinamensis* Kleinh.; *Rosac.; Sur.*
Doengoen Kontal: *Heritiera littoralis* Dryander; *Sterculiac.;* Sunda.
Doeren: *Durio zibethinus* L.; *Malvac.;* Malay.
Doerian: *Durio zibethinus* L.; *Malvac.;* Malay.
Doeum-Chhoeung-Kô: *Bauhinia malabarica* Roxb.; *Caesalpiniac.;* Kamb.
Dogberry: *Ardisia sp.; Myrsinac.;* Fl.
Dogo: *Parkia agboënsis* Chev.; *Mimosac.;* Elf.
Dogwood: *Evonymus europaea* L.; *Celastrac.;* E.
„ : *Rhamnus frangula* L.; *Rhamnac.;* E.
„ (u, e): *Cornus florida* L.; *Cornac.;* sö. N.-Am.
„ : *Cornus nuttalli* Audubon; *Cornac.;* w. N.-Am.
„ : *Piscidia communis* Standl.; *Papilionac.;* br. Hond.
„ : ? *Lonchocarpus lucidus* Pittier; *Papilionac.;* Pan.
„ : *Cornus controversa* Hemsl.; *Cornac.;* Jap.
„ : *Jacksonia scoparia* R. Br.; *Papilionac.;* Queensl.
„ : *Pomaderris apetala* Lab.; *Rhamnac.;* Tasm.
„ : *Bedfordia salicina* DC.; *Composit.;* Tasm.
„ , alternate-leaved: *Cornus alternifolia* L.; *Cornac.;* USA.
„ , flowering (u, e): *Cornus florida* L.; *Cornac.;* sö. N.-Am.
„ , Jamaica- (u): *Piscidia erythrina* L.; *Papilionac.;* W.-Ind., C.-Am.
„ , Native-: *Bedfordia salicina* DC.; *Composit.;* Au. (Vict.)
„ , Pacific-: *Cornus nuttalli* Audubon; *Cornac.;* N.-Am.
„ , Poison-: *Rhus vernix* L.; *Anacardiac.;* ö. USA.
„ , „ : *Clusia cooperi* Standl.; *Guttifer.;* Pan.

Dogwood, Ridge-: *Lonchocarpus sp.; Papilionac.;* br. Hond.
  „ , Swamp-: *Lonchocarpus hondurensis* Bth.; *Papilionac.;* br. Hond.
  „ , „ : *Lonchocarpus latifolius* H. B. K.; *Papilionac.;* br. Hond.
  „ , wèstern: *Cornus nuttalli* Audubon; *Cornac.;* N.-Am.
Dohar: *Cordia myxa* L.; *Borraginac.;* O.-Ind.
Dohu: *Anogeissus latifolia* Wall.; *Combretac.;* sw. Beng. (Orissa).
Doi: *Eugenia jambos* L., *E. formosa* Wall.; *Myrtac.;* Annam.
  „ : *Eugenia formosa* Wall. *var. ternifolia* Roxb.; *Myrtac.;* Ind.-Ch.
  „ : *Alphitonia franguloides* A. Gray; *Rhamnac.;* Fid.
  „ -yanagi: *Salix doii* Hayata; *Salicac.;* Form.
Dok khao: *Randia longiflora* Lamk.; *Rubiac.;* Annam.
  „ koum: *Crataeva sp.; Capparidac.;* Ind.-Ch.
  „ mâ: *Saraca dives* Pierre; *Caesalpiniac.;* Ind.-Ch. (Laos).
  „ thong dine: *Paradina hirsuta* Pitard; *Rubiac.;* Annam.
Doka: *Isoberlinia doka* Craib et Stapf; *Caesalpiniac.;* n. Nig.
  „ : *Tapirira aff. guianensis* Aubl.; *Anacardiac.;* Sur.
  „ rafi: *Berlinia auriculata* Bth.; *Caesalpiniac.;* n. Nig.
Dokalli: *Couma guianensis* Aubl.; *Apocynac.;* Sur.
Dokhaa: *Tapirira aff. guianensis* Aubl.; *Anacardiac.;* Sur.
Dokka: *Tapirira aff. guianensis* Aubl.; *Anacardiac.;* Sur.
Dokke: *Ehretia laevis* Roxb.; *Borraginac.;* C.-O.-Ind.
Doldól: *Ceiba pentandra* Gaertn.; *Bombacac.;* Phil.
Dolley: *Ouratea aff. vogeli* Engl.; *Ochnac.;* Kam.
Dólo: *Fagraea fragrans* Roxb.; *Loganiac.;* Phil.
Doloko: *Cola sp.; Sterculiac.;* Elf.
Dom: *Acer campbelli* Hook. f. et Thoms.; *Acerac.;* Beng.
  „ : *Chukrasia tabularis* A. Juss.; *Meliac.;* Ind.-Ch.
  „ chem: *Tarrietia cochinchinensis* Pierre; *Sterculiac.;* Ind.-Ch.
?   „ chheu prus: *Garcinia sp.; Guttifer.;* Ind.-Ch.
  „ chhoeu con hen titey: *Mitrephora edwardsi* Pierre; *Anonac.;* Kamb.
  „ „ crai crohom: *Xylopia vielana* Pierre; *Anonac.;* Kamb.
  „ „ „ sâr: *Xylopia pierrei* Hance; *Anonac.;* Kamb.
  „ „ dpang: *Dipterocarpus duperreanus* Pierre; *Dipterocarpac.;* Kamb., Annam.
  „ „ klong: *Dipterocarpus tuberculatus* Roxb.; *Dipterocarpac.;* Kamb.
  „ „ Koeupai: *Miliusa bailloni* Pierre; *Anonac.;* Kamb.
  „ „ neang déng crohom: *Dipterocarpus insularis* Hance; *Dipterocarpac.;* Kamb.
  „ „ phaong k'ting: *Calophyllum thoreli* Pierre; *Guttifer.;* Kamb.
  „ „ phchoc: *Shorea obtusa* Wall.; *Dipterocarpac.;* Kamb.
  „ „ phdiec crohóm: *Shorea cambodiana* Pierre; *Dipterocarpac.;* Kamb.
  „ „ Pru: *Garcinia benthami* Pierre; *Guttifer.;* Kamb.
  „ „ prus: *Garcinia sp.; Guttifer.;* Ind.-Ch.
  „ „ rang pnhóm: *Pentacme siamensis* Kurz; *Dipterocarpac.;* Kamb.
  „ „ rát soul: *Dipterocarpus dyeri* Pierre; *Dipterocarpac.;* Kamb.
  „ „ rong puhom: *Pentacme siamensis* Kurz; *Dipterocarpac.;* Annam.
  „ „ roré: *Dillenia pentagyna* Roxb.; *Dilleniac.;* Kamb.
  „ „ rue: *Dillenia pentagyna* Roxb.; *Dilleniac.;* Kamb.
?   „ „ „ : *Dillenia bailloni* Pierre; *Dilleniac.;* Kamb., Annam.
  „ „ Teal: *Dipterocarpus spp.; Dipterocarpac.;* Kamb.
  „ „ teal dung: *Dipterocarpus jourdaini* Pierre; *Dipterocarpac.;* Kamb.
  „ „ „ neang dang phaoc: *Dipterocarpus artocarpifolius* Pierre; *Dipterocarpac.;*
  „ „ „ tom: *Dipterocarpus alatus* Roxb.; *Dipterocarpac.;* Kamb.   [Kamb.
  „ „ „ tuc: *Dipterocarpus alatus* Roxb.; *Dipterocarpac.;* Kamb.
  „ „ thbeng: *Dipterocarpus obtusifolius* Miq.; *Dipterocarpac.;* Kamb.
  „ „ trách: *Dipterocarpus intricatus* Dyer; *Dipterocarpac.;* Kamb.
  „ Dom: *Alchornea tiliaefolia* Muell. Arg.; *Euphorbiac.;* Ind.-Ch.
  „ kruol: *Melanorrhoea laccifera* Pierre; *Anacardiac.;* Ind.-Ch.
  „ om beng thugé: *Pterospermum diversifolium* Blume; *Sterculiac.;* Kamb.
  „ ong cól: *Garcinia lanessani* Pierre; *Guttifer.;* Kamb.
  „ „ „ : *Garcinia sp.; Guttifer.;* Ind.-Ch.
  „ pelou: *Dillenia ovata* Wall.; *Dilleniac.;* Kamb.
  „ phlong: *Ternstroemia penangiana* Choisy; *Theac.;* Kamb.
  „ prohút: *Garcinia vilersiana* Pierre; *Guttifer.;* Kamb.

Dom propel ma sau: *Shorea cochinchinensis* Pierre; *Dipterocarpac.;* Kamb.
„ roniduol: *Unona mesnyi* Pierre; *Anonac.;* Kamb.
„ -sal: *Miliusa velutina* Hook. f. et Thoms.; *Anonac.;* n. O.-Ind.
„ somrang: *Sterculia lychnophora* Hance; *Sterculiac.;* Kamb.
Domatia tree: *Endiandra discolor* Bth.; *Laurac.;* Au.
Domba: *Calophyllum inophyllum* L.; *Guttifer.;* Cey.
„ -gass: *Calophyllum inophyllum* L.; *Guttifer.;* Cey.
Dominguila: *Lovoa klaineana* Pierre; *Meliac.;* Gab.
Don chém: *Tarrietia cochinchinensis* Pierre; *Sterculiac.;* Ind.-Ch.
? „ chhoeu rué: *Dillenia bailloni* Pierre; *Dilleniac.;* Kamb., Annam.
„ sál: *Miliusa velutina* Hook. f. et Thoms.; *Anonac.;* Ind.
Dona: *Carapa microcarpa* Chev.; *Meliac.;* Elf.
Doncella: *Bombax ellipticum* H. B. K.; *Bombacac.;* Guat.
? „ : *Matayba oppositifolia* Britton; *Sapindac.;* P.-R.
? „ : *Byrsonima sp.; Malpighiac.;* W.-Ind.
„ : *Matayba apetala* Radlk.; *Sapindac.;* Jam.
Doncelo: *Hymenodictyon excelsum* Wall.; *Rubiac.;* w. O.-Ind.
Dondaguéenngo: *Odyendea gabunensis* Engl.; *Simarubac.;* Kam.
Don-doh: *Soyauxia grandifolia* Gilg et Stapf; *Flacourtiac.;* Lib.
Dondól: *Cciba pentandra* Gaertn.; *Bombacac.;* Phil.
Dondolino (i): *Coronilla emerus* L.; *Papilionac.;* Mitt.-& S.-Eu.
Dondru: *Hymenodictyon excelsum* Wall.; *Rubiac.;* w. O.-Ind.
Dong chau: *Alchornea tiliaefolia* Muell. Arg.; *Euphorbiac.;* Ind.-Ch.
Dongon: *Sterculia cymbiformis* Blco.; *Sterculiac.;* Phil.
Donhoedoe: *Triplaris surinamensis* Cham.; *Polygonac.;* Sur.
Don-oedoe: *Triplaris surinamensis* Cham.; *Polygonac.;* Sur.
Donoza: *Guarea cedrata* Pell.; *Meliac.;* Elf.
Donsella (u): *Labourdonnaisia albescens* Bth.; *Sapotac.;* Cuba.
Doo: *Bonamia cymosa* Hallier f.; *Convolvulac.;* Lib.
„ -vlehn: *Allophylus talboti* Bak. f.; *Sapindac.;* Lib.
Dooh: *Caloncoba echinata* Gilg; *Flacourtiac.;* Lib.
Dooka: ? *Tapirira guianensis* Aubl.; *Anacardiac.;* br. Gu.
Doon: *Doona zeylanica* Thw.; *Dipterocarpac.;* Cey.
„ , red: *Doona gardneri* Thw.; *Dipterocarpac.;* Cey.
Doopamara: *Canarium strictum* Roxb.; *Burserac.;* sw. O.-Ind.
Door: *Duabanga sonneratioides* Buch.-Ham.; *Sonneratiac.;* nö. O.-Ind. (ö. Nepal).
Dorita: *Guarea chiricana* Standl.; *Meliac.;* Pan.
Dorn, Christus-: *Ilex aquifolium* L.; *Aquifoliac.;* s. Eu.
„ , : *Gleditschia triacanthos* L.; *Caesalpiniac.;* N.-Am.
„ , Esch-, gevlekt, amerikanisch: *Acer saccharum* Marsh.; *Acerac.;* N.-Am.
„ , Juden- (d): *Zizyphus vulgaris* Lam.; *Rhamnac.;* M.-M., Or.
„ , Kreuz- (d): *Rhamnus alaterna* L.; *Rhamnac.;* s. Eu., n. Af.
„ , „ „ : *Rhamnus cathartica* L.; *Rhamnac.;* Eu.
„ , Sand-, gemeiner (d): *Hippophaë rhamnoides* L.; *Elaeagnac.;* Eu.
„ , Sauer-, gemeiner (d): *Berberis vulgaris* L.; *Berberidac.;* Eu.
„ , Schleh- (d): *Prunus spinosa* L.; *Rosac.;* Eu.
„ , Schoten-, borstiger (d): *Robinia hispida* L.; *Papilionac.;* N.-Am.
„ , „ , gemeiner (d): *Robinia pseudacacia* L.; *Papilionac.;* N.-Am.
„ , „ , klebriger (d): *Robinia viscosa* Vent.; *Papilionac.;* N.-Am.
„ , Schwarz- (d): *Prunus spinosa* L.; *Rosac.;* Eu.
„ , Weiss-, eingriffeliger (d): *Crataegus monogyna* Jacq.; *Rosac.;* Eu.
„ , „ , gemeiner (d): *Crataegus oxyacantha* L.; *Rosac.;* Eu.
„ , „ , zweigriffeliger (d): *Crataegus oxyacantha* L.; *Rosac.;* Eu.
Doro: *Apeiba sp.; Tiliac.;* br. Gu.
„ : *Populus balsamifera* L.; *Salicac.;* Jap.
„ -no-ki: *Populus suaveolens* Fisch.; *Salicac.;* Jap.
Doronoki: *Populus maximowiczi* Henry; *Salicac.;* Jap.
Dorosé populu: *Funtumia elastica* Stapf; *Apocynac.;* Elf.
Dorowa: *Parkia filicoidea* Welw.; *Mimosac.;* n. Nig.
Doti: *Pterocarpus erinaceus* Poir.; *Papilionac.;* Togo, Go.
Dottergelb (d): *Aspidosperma vargasi* A. DC.; *Apocynac.;* Ven.
Dou: *Chlorophora excelsa* Bth. et Hook.; *Morac.;* Elf.

158

Douégna: *Diospyros mespiliformis* Hochst.; *Ebenac.;* Sudan.
Douglas: *Pseudotsuga douglasi* Carr.; *Conifer.;* N.-Am.
Douglasie (d): *Pseudotsuga douglasi* Carr.; *Conifer.;* N.-Am.
Douglastanne, blaue (d): *Pseudotsuga glauca* Mayr; *Conifer.;* w. N.-Am.
„ , großfrüchtige (d): *Pseudotsuga macrocarpa* Mayr; *Conifer.;* w. N.-Am. (Kalif.).
„ , grüne (d): ? *Pseudotsuga douglasi* Carr.; *Conifer.;* w. N.-Am.
„ , japanische (d): *Pseudotsuga japonica* Shirasawa; *Conifer.;* Jap.
Dougonienguéla: *Lovoa klaineana* Pierre; *Meliac.;* Gab.
Dougoura: *Cordyla africana* Lour.; *Caesalpiniac.;* Sudan.
Douk: *Pterocarpus pedatus* Pierre; *Papilionac.;* Ind.-Ch.
Douka (f): *Mimusops africana* Lecomte, *M. heckeli* Lecomte; *Sapotac.;* Kam., Gab.
Doukaliebalie: *Ficus sp.; Morac.;* Sur.
Doukouma: *Khaya ivorensis* Chev.; *Meliac.;* Elf.
Doum: *Cylicodiscus gabunensis* Harms; *Mimosac.;* Kam.
„ : *Ceiba pentandra* Gaertn.; *Bombacac.;* Kam., Gab.
Doumier (f): *Hyphaene thebaica* Mart.; *Palmac.;* O.-Af.
Doumzou: *Cleistopholis patens* Bth.; *Anonac.;* Gab.
Doundaké: *Sarcocephalus esculentus* Afzel.; *Rubiac.;* Elf.
„ : *Sarcocephalus russeggeri* Kotschy; *Rubiac.;* Elf.
Doussié (f): *Afzelia sp.; Caesalpiniac.;* Elf.
Doux blanc: *Sterculia foetida* L.; *Sterculiac.;* Guad.
„ zabel: *Persea sp.; Laurac.;* Guad.
Dowa: *Artocarpus lakoocha* Roxb.; *Morac.;* Assam.
Dowla: *Holarrhena antidysenterica* Wall.; *Apocynac.;* w. O.-Ind.
Downtree: *Ochroma spp.; Bombacac.;* Sur.
Doy: *Sarcocephalus esculentus* Afzel.; *Rubiac.;* Elf.
„ : *Schefflera pes-avis* R. Viguier; *Araliac.;* Ind.-Ch.
Dozona: *Guarea cedrata* Pellegr.; *Meliac.;* Elf.
Dpon: *Sterculia alata* Roxb.; *Sterculiac.;* Ind.-Ch.
D'pou: *Sterculia alata* Roxb.; *Sterculiac.;* w. Annam.
Drago: *Pterocarpus sp.; Papilionac.;* Mex.
„ : *Calderonia salvadorensis* Standl.; *Rubiac.;* Salv., Guat.
„ : *Virola merendonis* Pittier; *Myristicac.;* Guat.
Dragonnier (f): *Ceiba pentandra* Gaertn.; *Bombacac.;* W.-Af.
Dragon tree: *Pterocarpus draco* L.; *Papilionac.;* Hond.
Drawa: *Toona serrata* M. Roem.; *Meliac.;* nw. O.-Ind. (Punjab).
Drawi: *Toona serrata* M. Roem.; *Meliac.;* nw. O.-Ind. (Punjab).
„ : *Toona ciliata* M. Roem.; *Meliac.;* nw. O.-Ind. (Punjab).
Dreb-bah: *Diospyros kamerunensis* Gürke; *Ebenac.;* Lib.
Drega quru quru: *Alstonia vitiensis* Seem.; *Apocynac.;* Fid.
Drehn: *Xylopia staudti* Engl.; *Anonac.;* Lib.
„ -gbar-doo: *Stephania dingklagei* Diels; *Menispermac.;* Lib.
Dreifi: *Coccoloba uvifera* L.; *Polygonac.;* holl. W.-Ind.
Drek: *Melia azedarach* L.; *Meliac.;* nw. O.-Ind. (Punjab).
Drewar: *Abies pindrow* Spach; *Conifer.;* nw. O.-Ind. (Kasch.).
Drolpeer: *Dombeya densiflora* Planch.; *Sterculiac.;* S.-Af.
Drowak: *Columbia javanica* Blume; *Tiliac.;* Nied.-Ind.
Druif: *Coccoloba uvifera* L.; *Polygonac.;* Sur.
Drum tree: *Cordia irvingi* Baker; *Borraginac.;* Go.
Dsato: *Daniella thurifera* Bth.; *Caesalpiniac.;* Togo.
Dthaaman: *Ficus rubiginosa* Desf.; *Morac.;* N.-S.-W.
Du-ah-dor: *Antidesma membranaceum* Muell. Arg.; *Euphorbiac.;* Lib.
Du-kpay: *Garcinia sp.; Guttifer.;* Lib.
Du tung deng: *Pterocarya stenoptera* A. DC.; *Juglandac.;* Ind.-Ch.
Dúa: *Ficus sp.; Morac.;* Ind.-Ch.
Dua abae: *Afrormosia laxiflora* Harms; *Papilionac.;* Go.
„ anyan: *Afrormosia laxiflora* Harms; *Papilionac.;* Go.
„ ben: *Haronga paniculata* Sprague; *Guttifer.;* Go.
„ kobin: *Afrormosia laxiflora* Harms; *Papilionac.;* Go.
Duabai: *Afrormosia laxiflora* Harms; *Papilionac.;* Go.
Duali (u): *Anisoptera thurifera* Blume; *Dipterocarpac.;* Phil.
Duana: *Guarea cedreta* Pellegr.; *Meliac.;* Elf.

Duang: *Altingia excelsa* Noronha; *Hamamelidac.;* Assam.
Duasika: *Uvaria sp.; Anonac.;* Go.
Duay-gray: ? *Drypetes afzelii* Hutch.; *Euphorbiac.;* Lib.
Dubdah: *Barringtonia speciosa* L.; *Lecythidac.;* And.
Dubibi: *Khaya ivorensis* Chev.; *Meliac.;* Elf.
Dubini: *Khaya grandifolia* C. DC., *K. ivorensis* Chev.; *Meliac.;* Go.
„   Blay: *Lovoa klaineana* Pierre; *Meliac.;* Go.
Dubir: *Khaya ivorensis* Chev.; *Meliac.;* Elf.
Dubiri: *Khaya ivorensis* Chev.; *Meliac.;* Elf.
Dubrefwo: *Mareya micrantha* Muell. Arg.; *Euphorbiac.;* Go.
Dubuma: *Uvaria sp.; Anonac.;* Go.
Dúc: *Pterocarpus pedatus* Pierre; *Papilionac.;* Ind.-Ch.
Dudagu: *Adina cordifolia* Hook. f.; *Rubiac.;* O.-Ind.
Dudcory: *Holarrhena antidysenterica* Wall.; *Apocynac.;* Assam.
Dudh-kainju: *Acer pictum* Thunb.; *Acerac.;* n. O.-Ind. (Jaunsar).
Dudhari: *Wrightia tomentosa* Roem. et Schultes; *Apocynac.;* C.-O.-Ind.
Dudhi: *Holarrhena antidysenterica* Wall.; *Apocynac.;* C.-O.-Ind.
Dudhi: *Wrightia tinctoria* R. Br.; *Apocynac.;* C.-O.-Ind.
„   : *Wrightia tomentosa* Roem. et Schultes; *Apocynac.;* C.-O.-Ind.
Dudhiari: *Holarrhena antidysenterica* Wall.; *Apocynac.;* nw. O.-Ind. (Bihar).
Dudhiya: *Wrightia tomentosa* Roem. et Schultes; *Apocynac.;* C.-O.-Ind.
Dudi maddi: *Terminalia tomentosa* Wight et Arn.; *Combretac.;* O.-Ind. (Haid.).
„   : *Bridelia retusa* Spreng.; *Euphorbiac.;* w. O.-Ind. (n. Bombay).
„   -yetta: *Hymenodictyon excelsum* Wall.; *Rubiac.;* w. O.-Ind. (n. Bombay).
Dudippa: *Hymenodictyon excelsum* Wall.; *Rubiac.;* w. O.-Ind. (n. Bombay).
Dudippi: *Careya arborea* Roxb.; *Myrtac.;* ö. O.-Ind. (Madras).
Dudiya: *Wrightia tinctoria* R. Br.; *Apocynac.;* C.-O.-Ind.
Dudla: *Prunus padus* L.; *Rosac.;* nw. O.-Ind. (Punjab).
Dudla: *Sapium insigne* Muell. Arg.; *Euphorbiac.;* O.-Ind.
Dudri: *Gardenia turgida* Roxb.; *Rubiac.;* nö. O.-Ind. (Orissa, Bihar).
Dudurede: *Protea bismarcki* Engl.; *Proteac.;* Togo.
Duég: *Bischofia javanica* Blume; *Euphorbiac.;* Phil. (I.-N., Tar.).
Duen: *Dipterocarpus grandiflorus* Blco.; *Dipterocarpac.;* Phil. (Cag.).
Dug-án: *Myristica spp.; Myristicac.;* Phil.
Dugarai: *Aphanomyxis cumingiana* Harms; *Meliac.;* Phil. (Pamp.).
Dugáum: *Lagerstroemia speciosa* Pers.; *Lythrac.;* Phil. (Ley.).
Dugbongbore: *Cola cordifolia* R. Br.; *Sterculiac.;* Togo.
Dugdugia: *Eugenia operculata* Roxb.; *Myrtac.;* C.-O.-Ind. (Oudh).
Dugían: *Bambusa blumeana* Schultes f.; *Gramin.;* Phil.
Dugkátan: *Cryptocarya bicolor* Merr.; *Laurac.;* Phil. (Lanao, Cot., Sulu).
Duguah: *Myristica philippinensis* Lam.; *Myristicac.;* Phil.
Dugúan: *Knema glomerata* Merr., *Myristica spp.; Myristicac.;* Phil.
„   : *Myristica philippinensis* Warb.; *Myristicac.;* Phil.
Dugura: *Khaya ivorensis* Chev.; *Meliac.;* Elf.
Dúhat: *Eugenia cumini* Merr.; *Myrtac.;* Phil.
Duhau: *Myristica spp.; Myristicac.;* Phil. (Zamb.)
Duho: *Dipterocarpus vernicifluus* Blco.; *Dipterocarpac.;* Phil.
Duidui: *Pterocymbium tinctorium* Merr.; *Sterculiac.;* Phil. (Tay.).
Duivenhout (h): *Rhizophora mangle* L.; *Rhizophorac.;* Sur.
Duizendbeen-boom (h): *Rhizophora mangle* L.; *Rhizophorac.;* Sur.
Duizenhout (h): *Rhizophora mangle* L.; *Rhizophorac.;* Sur.
Duka: *Tapirira guianensis* Aubl., *T. marchandi* Aubl.; *Anacardiac.;* br. Gu.
„   (f): *Mimusops africana* Lecomte, *M. heckeli* Lecomte; *Sapotac.;* Kam., Gab.
Duká: *Kingiodendron alternifolium* Merr.; *Papilionac.;* Phil. (Tab., Ley., w. Neg.).
Dukalaballi: *Ficus sp.; Morac.;* br. Gu.
„   : *Platymiscium polystachyum* Bth.; *Papilionac.;* Sur.
Dukalliballi: *Couma guianensis* Aubl.; *Apocynac.;* br. Gu.
Dukki. *Celtis integrifolia* Lam.; *Ulmac.;* n. Nig.
Duklàp: *Zizyphus spp.; Rhamnac.;* Phil. (Bat., Riz., Batg.).
Duklítan: *Sideroxylon duclitan* Blco.; *Sapotac.;* Phil. (Bat.).
Duko: *Lannea acidissima* Chev.; *Anacardiac.;* Elf.
Duku: *Lansium domesticum* Jack; *Meliac.;* Malay.

Dukuláb: *Zizyphus trinerva* DC., *Z. spp.; Rhamnac.;* Phil. (Bat.).
Dukuma: *Khaya ivorensis* Chev.; *Meliac.;* Elf.
„ : *Khaya grandifolia* C. DC.; *Meliac.;* Go.
„ : *Entandrophragma spp.; Meliac.;* Go.
Dukwa: *Lonchocarpus sericeus* H. B. K.; *Papilionac.;* Go.
Dukwa: *Platysepalum chevalieri* Harms; *Papilionac.;* B.-K. (Bangala).
Duláuen: *Cassia javanica* L.; *Caesalpiniac.;* Phil. (Bas.).
„ : *Neonauclea spp.; Rubiac.;* Phil. (Cag.).
„ : *Beilschmiedia cairocan* Vid., *Litsea sp.; Laurac.;* Phil. (Cag.).
„ : *Notaphoebe malabonga* Merr.; *Laurac.;* Phil. (Cag.).
„ : *Terminalia pellucida* Presl; *Combretac.;* Phil. (Cag.).
Dulia: *Dipterocarpus pilosus* Roxb.; *Dipterocarpac.;* Beng.
Dulit: *Canarium villosum* F. Vill.; *Burserac.;* Phil. (Pang., Zam.).
Dulítan: *Palaquium merrilli* Dubard, *P. spp.; Sapotac.;* Phil. (Lag., Tay., Cam.).
„ : *Eugenia benthami* A. Gray; *Myrtac.;* Phil. (Riz.).
„ : *Sideroxylon duclitan* Blco.; *Sapotac.;* Phil. (Lag.).
Dulokdúlok: *Lumnitzera littorea* Voigt; *Combretac.;* Phil. (Mas.).
Dulu: *Ficus vallis-choudae* Del.; *Morac.;* n. Nig.
Dúlu: *Fagraea fragrans* Roxb.; *Loganiac.;* Phil.
Dum hin: *Chukrasia tabularis* A. Juss.; *Meliac.;* Ind.-Ch.
„ kotokoi: *Callicarpa arborea* Roxb.; *Verbenac.;* ö. Beng.
„ -kurdu: *Gardenia latifolia* Ait.; *Rubiac.;* nw. O.-Ind. (Orissa).
Dumadára: *Knema glomerata* Merr.; *Myristicac.;* Phil. (Cag.).
Dumanami: *Khaya grandifolia* C. DC.; *Meliac.;* Go.
Dumar: *Ficus glomerata* Roxb.; *Morac.;* Beng.
Dumátc: *Lagerstroemia piriformis* Koehne; *Lythrac.;* Phil. (n. Luzon).
Dumboil: *Bombax insigne* Wall.; *Bombacac.;* Assam.
Dumitha: *Acer pictum* Thunb.; *Acerac.;* n. O.-Ind. (Garhwal).
Dumón: *Heritiera littoralis* Dryander; *Sterculiac.;* Phil. (Cag.).
Dumori: *Mimusops heckeli* Lecomte; *Sapotac.;* Elf.
Dumri: *Ficus glomerata* Roxb.; *Morac.;* n. O.-Ind. (Nepal).
Dumsa-gyaw: *Hymenodictyon excelsum* Wall.; *Rubiac.;* O.-Ind.
Dun: *Juglans regia* L.; *Juglandac.;* nw. O.-Ind. (Kasch.).
„ : *Doona zeylanica* Thw.; *Dipterocarpac.;* Cey.
Dundathu: *Agathis robusta* Salisb.; *Conifer.;* Queensl.
„ Kauri: *Agathis robusta* Salisb.; *Conifer.;* Au.
Dundi: *Barringtonia acutangula* Gaertn.; *Lecythidac.;* O.-Ind.
Dundra: *Bauhinia purpurea* L.; *Caesalpiniac.;* C.-O.-Ind.
Dundu: *Dichrostachys glomerata* Hutch. et Dalz.; *Mimosac.;* Go.
„ : *Dichrostachys nutans* Bth.; *Mimosac.;* n. Nig.
? Dundum: *Dichrostachys glomerata* Hutch. et Dalz.; *Mimosac.;* Go.
Dung: *Symplocos laurina* Wall.; *Symplocac.;* Annam.
„ mât: *Symplocos laurina* Wall.; *Symplocac.;* Ind.-Ch.
„ trang: *Symplocos laurina* Wall.; *Symplocac.;* Ind.-Ch.
„ xanh: *Symplocos laurina* Wall.; *Symplocac.;* Ind.-Ch.
Dunggói: *Litsea spp.; Laurac.;* Phil. (Riz.).
Dungohame: *Conopharyngia smithi* Stapf; *Apocynac.;* B.-K. (Bangala).
Dungohami: *Conopharyngia smithi* Stapf; *Apocynac.;* B.-K. (Bangala).
Dungomi: *Conopharyngia smithi* Stapf; *Apocynac.;* B.-K. (Bangala).
Dúngon: *Tarrietia silvatica* Merr.; *Sterculiac.;* Phil.
„ : *Heritiera littoralis* Dryander; *Sterculiac.;* Phil.
„ : *Planchonia spectabilis* Merr.; *Lecythidac.;* Phil. (Neg.).
„ : *Pterocarpus spp.; Papilionac.;* Phil. (I.-N., Cag.).
„ : *Shorea balangeran* Dyer; *Dipterocarpac.;* Phil. (Ag.).
„ -dungónan: *Parinarium spp.; Rosac.;* Phil. (Tay.).
„ -láte: *Heritiera littoralis* Dryander; *Sterculiac.;* Phil.
Duñgul: *Tarrietia silvatica* Merr.; *Sterculiac.;* Phil. (Cag.)
? Duñgúla: *Vitex aherniana* Merr.; *Verbenac.;* Phil. (Neg., Ilo.).
Dunlóg: *Parashorea malaanonan* Merr.; *Dipterocarpac.;* Phil. (Mas.).
Dunlúg: *Pentacme contorta* Merr. et Rolfe; *Dipterocarpac.;* Phil. (Bas.)
Dunya: *Vitex cienkowskyi* Kotschy et Peyr.; *Verbenac.;* n. Nig.
Dunyar biri: *Vitex diversifolia* Baker; *Verbenac.;* n. Nig.

Duoc: *Rhizophora conjugata* L.; *Rhizophorac.;* Ind.-Ch.
Duoc: *Bruguiera gymnorrhiza* Lamk.; *Rhizophorac.;* Ind.-Ch.
„ bôt:- *Rhizophora conjugata* L.; *Rhizophorac.;* Ind.-Ch.
„ do: *Ceriops candolleana* Arn.; *Rhizophorac.;* Ind.-Ch.
„ duàng: *Phyllanthus distichus* Muell-Arg.; *Euphorbiac.;* Ind.-Ch.
„ vàng: *Rhizophora conjugata* L.; *Rhizophorac.;* Ind.-Ch.
Duôi rùng: *Coelodiscus muricatus* Gagnep.; *Euphorbiac.;* Ind.-Ch.
„ tu: *Polyalthia jucunda* Finet et Gagnep.; *Anonac.;* Annam.
Duong: *Broussonetia papyrifera* Vent.; *Morac.;* Tonk.
„ lieu: *Casuarina equisetifolia* Forst.; *Casuarinac.;* Ind.-Ch.
Dupada telledamaru: *Vateria indica* L.; *Dipterocarpac.;* Ind.
Dupuin: *Khaya ivorensis* Chev.; *Meliac.;* Go.
Dur: *Tamarix aphylla* Lanza; *Tamaricac.;* Af. (it. Somali).
Duraznillo: *Celtis sp.; Ulmac.;* Salv.
„ colorado: *Ruprechtia sp.; Polygonac.;* Arg.
„ morado: *Coccoloba cordata* Cham.; *Polygonac.;* Arg.
Durazno: *Prunus persica* S. et Z.; *Rosac.;* Hond.
Durbi: *Ostryoderris chevalieri* Dunn; *Papilionac.;* n. Nig.
Durdi: *Gardenia turgida* Roxb.; *Rubiac.;* O.-Ind.
Durgo: *Lannea acidissima* Chev.; *Anacardiac.;* Elf.
Duri: *Toona serrata* M. Roem.; *Meliac.;* n. O.-Ind. (ö. Punjab).
Duriamaddi: *Bridelia retusa* Spreng.; *Euphorbiac.;* w. O.-Ind. (n. Bombay).
Durian: *Durio zibethinus* Murr.; *Bombacac.;* Malay.
„ api: *Durio sp.; Bombacac.;* Malay.
„ Badak: *Coelostegia griffithi* Bth.; *Bombacac.;* Malay.
„ burong: *Durio sp.; Bombacac.;* Malay.
„ daun: *Durio sp.; Bombacac.;* Malay.
„ , Doun-: *Boschia griffithi* Mast. *var.* ?; *Malvac.;* Malak.
„ , Mun-: *Boschia griffithi* Mast.; *Malvac.;* Malak.
„ tana: *Durio sp.; Bombacac.;* Malay.
„ tua: *Coelostegia griffithi* Bth.; *Bombacac.;* Malay.
„ , wild (H): *Cullenia excelsa* White; *Bombacac.;* sw. O.-Ind.
Durillo: *Viburnum tinus* L.; *Caprifoliac.;* Sp.
Durin: *Terminalia platyphylla* F. v. M.; *Combretac.;* Queensl.
Durioan: *Durio zibethinus* L.; *Bombacac.;* Malay.
Durion: *Durio zibethinus* L.; *Bombacac.;* Malay.
Duro: *Vatica mangachapoi* Blco.; *Dipterocarpac.;* Phil. (Sam., Ley.).
Durobby: *Eugenia moorei* F. v. M.; *Myrtac.;* Au.
Durog: *Vatica mangachapoi* Blco.; *Dipterocarpac.;* Phil. (Sam., Ley.).
Durote: *Centrolobium orinocense* Pittier; *Papilionac.;* Ven.
Durugú: *Knema glomerata* Merr.; *Myristicac.;* Phil. (Lag.).
„ : *Myristica spp.; Myristicac.;* Phil. (Bat.)
Duruko: *Lannea acidissima* Chev.; *Anacardiac.;* Elf.
Durumi: *Ficus polita* Vahl; *Morac.;* n. Nig.
Dushe: *Acacia seyal* Delile; *Mimosac.;* n. Nig.
Dussa: *Acacia seyal* Delile; *Mimosac.;* n. Nig.
Dututure: *Pseudocedrela kotschyi* Harms; *Meliac.;* Togo.
Dúung: *Anisoptera thurifera* Blume; *Dipterocarpac.;* Phil. (Ilocos).
Duyokdúyok: *Mimusops calophylloides* Merr.; *Sapotac.;* Phil. (Surigao).
Dúyong: *Cassia javanica* L.; *Caesalpiniac.;* Phil. (Cag.).
„ : *Vatica mangachapoi* Blco.; *Dipterocarpac.;* Phil. (Tay.).
Dúyung: *Anisoptera thurifera* Blume; *Dipterocarpac.;* Phil. (Ilocos).
Dwabok: *Kydia calycina* Roxb.; *Malvac.;* Burma.
Dwabote: *Kydia calycina* Roxb.; *Malvac.;* n. Ind.
Dwanee: *Eriolaena candollei* Wall.; *Sterculiac.;* w. O.-Ind.
Dwani: *Eriolaena candollei* Wall.; *Sterculiac.;* Burma.
Dwáyu: *Turraea robusta* Gürke; *Meliac.;* D.-O.-Af. (w. Usambara).
Dweh: *Anthocleista nobilis* G. Don; *Loganiac.;* Lib.
„ -vah: *Gaertnera cooperi* Hutch. et Moss; *Loganiac.;* Lib.
Dwenwere: *Lecaniodiscus cupanioides* Planch.; *Sapindac.;* Go.
Dweraba dua: *Crescentia cujete* L.; *Bignoniac.;* Go.
Dwuma: *Musanga smithi* R. Br.; *Morac.;* Go.

162

Dyrren-dyrren: *Podocarpus elata* R. Br.; *Conifer.;* N.-S.-W.
Dzalmuy: *Anona squamosa* L.; *Anonac.;* br. Hond.
Dzana: *Guarea cedrata* Pellegr.; *Meliac.;* Elf.
Dzao cát: *Dipterocarpus artocarpifolius* Pierre; *Dipterocarpac.;* Annam.
„ „ : *Dipterocarpus insularis* Hance; *Dipterocarpac.;* Annam.
„ con rai núóc: *Dipterocarpus jourdaini* Pierre; *Dipterocarpac.;* Annam.
„ do: *Dipterocarpus punctulatus* Pierre; *Dipterocarpac.;* Annam.
„ long: *Dipterocarpus tuberculatus* Roxb.; *Dipterocarpac.;* Annam.
„ „ : *Dipterocarpus alatus* Roxb.; *Dipterocarpac.;* Annam, Burma.
„ „ : *Dipterocarpus obtusifolius* Teijsm.; *Dipterocarpac.;* Annam, Burma.
„ mich: *Dipterocarpus insularis* Hance; *Dipterocarpac.;* Annam.
„ mit: *Dipterocarpus intricatus* Dyer; *Dipterocarpac.;* Annam.
„ ngho: *Dipterocarpus dyeri* Pierre; *Dipterocarpac.;* Annam.
„ núóc: *Dipterocarpus thoreli* Pierre; *Dipterocarpac.;* Annam.
„ truong: *Schleichera trijuga* Willd.; *Sapindac.;* Annam.
„ xam nau: *Dipterocarpus dyeri* Pierre; *Dipterocarpac.;* Annam.
„ xang neu: *Dipterocarpus dyeri* Pierre; *Dipterocarpac.;* Annam.
Dzau con raï traig: *Dipterocarpus alatus* Roxb.; *Dipterocarpac.;* Annam.
„ „ „ trâng: *Dipterocarpus alatus* Roxb.; *Dipterocarpac.;* Annam.
„ mich: *Dipterocarpus insularis* Hance; *Dipterocarpac.;* Annam.
„ ta beng: *Dipterocarpus obtusifolius* Miq.; *Dipterocarpac.;* Annam.
„ trai: *Dipterocarpus intricatus* Dyer; *Dipterocarpac.;* Annam.
Dzigne: *Hoplestigma klaineanum* Pierre; *Flacourtiac.;* Gab.
Dzin: *Monodora myristica* Dunal; *Anonac.;* Kam.
Dzogbedodo: *Terminalia ivorensis* Chev.; *Combretac.;* Go.
Dzoi: *Sideroxylon gaumeri* Pittier; *Sapotac.;* br. Hond.

E

Eaglewood: *Aquilaria malaccensis* Lam., *A. spp.; Thymelaeac.;* Ind., Malay.
„ : *Aquilaria ovata* Cav., *A. agallocha* Roxb.; *Thymelaeac.;* Malay.
„ , indian (H): *Aquilaria agallocha* Roxb.; *Thymelaeac.;* Assam, Beng., Burma.
Eba: *Pachylobus balsamifera* Oliv.; *Burserac.;* Gab.
Ebaij: *Cordia platythyrsa* Baker; *Borraginac.;* Kam.
Ebal: *Pentaclethra macrophylla* Bth.; *Mimosac.;* Kam.
Ebam: *Picralima klaineana* Pierre; *Apocynac.;* Kam.
„ : *Garcinia manni* Oliv.; *Guttifer.;* Kam.
„ -Bisana: *Chrysophyllum subnudum* Baker; *Sapotac.;* Gab.
Ebama: *Chrysophyllum subnudum* Baker; *Sapotac.;* Gab.
Ebamba: *Albizzia fastigiata* Oliv.; *Mimosac.;* B.-K. (Bangala).
Eban: *Canarium schweinfurthi* Engl.; *Burserac.;* Kam., Gab.
Ebana: ? *Copaifera sp.; Caesalpiniac.;* Gab.
Ebane: ? *Copaifera sp.; Caesalpiniac.;* Gab.
Ebangbemva: *Trichilia sp.; Meliac.;* Kam.
Ebano: *Caesalpinia sclerocarpa* Standl.; *Caesalpiniac.;* Mex. (Sinaloa).
„ : *Caesalpinia granadillo* Pittier; *Caesalpiniac.;* Mex. (Sinaloa), Ven.
„ : *Caesalpinia ebano* Karst.; *Caesalpiniac.;* Kol.
„ : *Caesalpinia punctata* Willd.; *Caesalpiniac.;* Ven.
„ : *Diospyros buxifolia* Hiern; *Ebenac.;* Phil. (Zamb., Pal.).
„ : *Maba buxifolia* Pers.; *Ebenac.;* Phil.
„ amarillo: *Sideroxylon mastichodendron* Jacq.; *Sapotac.;* Cuba.
„ carbonero: *Diospyros tetrasperma* Sw.; *Ebenac.;* Cuba.
„ falso: *Cytisus laburnum* L.; *Papilionac.;* Arg.
„ real: *Diospyros tetrasperma* Sw.; *Ebenac.;* Cuba.
Ebans: *Diospyros ebenum* Koenig; *Ebenac.;* s. O.-Ind.
Ebap: *Pachylobus balsamifera* Guill.; *Burserac.;* sp. Guin.
„ : *Sorindeia juglandifolia* Planch.; *Anacardiac.;* Kam.
Ebarmébéne: *Baphia laurifera* Baill.; *Papilionac.;* Gab.
Ebaya: *Cordia platythyrsa* Baker; *Borraginac.;* Kam.
Ebbenhout (h): *Diospyros spp., Maba spp.; Ebenac.;* As., Af.
„ „ : *Tecoma leucoxylon* Mart.; *Bignoniac.;* fr. Gu.

Ebbenhout: *Diospyros utilis* Koord. et Valet.; *Ebenac.;* Nied.-Ind.
„ , roodes (h): *Diospyros rubra* Gaertn.; *Ebenac.;* tr. Am.
„ , „ „ : *Brya ebenus* DC.; *Papilionac.;* tr. Am.
Ebé: *Pterocarpus tinctorius* Welw.; *Papilionac.;* sp. Guin.
„ : *Cordia platythyrsa* Baker; *Borraginac.;* Kam.
„ : *Macrolobium sp.; Caesalpiniac.;* Gab.
Ebè: *Pentaclethra macrophylla* Bth.; *Mimosac.;* Kam., Gab.
Ebébiliba: *Brachystegia klaineana* Pierre; *Caesalpiniac.;* Gab.
Ebel: *Pterocarpus soyauxi* Taub.; *Papilionac.;* Gab.
Ebéne: *Diospyros flavescens* Gürke; *Ebenac.;* W.-Af.
Ebène: *Diospyros spp., Maba spp.; Ebenac.;* Tr.
„ : *Diospyros evila* Pierre; *Ebenac.;* Gab.
„ grise: *Tecoma sp.; Bignoniac.;* fr. Gu.
„ marbré: *Diospyros melanida* Poir.; *Ebenac.;* Maur.
„ Mozambique-: *Diospyros sp.; Ebenac.;* Mad.
„ plaque minier: *Diospyros ebenaster* Retj.; *Ebenac.;* Guad.
„ verte: *Tecoma leucoxylon* Mart.; *Bignoniac.;* fr. Gu.
„ „ : *Tecoma leucoxylon* Mart. var. *pentaphylla* Juss.; *Bignoniac.;* Sur.
„ „ : *Tecoma araliacea* P. DC.; *Bignoniac.;* Sur.
„ „ brune: *Gymnanthes sp.; Euphorbiac.;* Mart.
„ d'Afrique: *Diospyros incarnata* Gürke; *Ebenac.;* Kam., Gab.
„ „ : *Diospyros crassiflora* Hiern; *Ebenac.;* Kam., Gab.
„ d'Ilkelemba: *Diospyros incarnata* Gürke; *Ebenac.;* tr. Af.
„ d'Indo-Chine: *Diospyros vera* Chev., *D. decandra* Lour.; *Ebenac.;* Ind.-Ch.
„ „ „ : *Diospyros dodecandra* Lour.; *Ebenac.;* Ind.-Ch.
„ du Cameroun: *Diospyros crassiflora* Hiern; *Ebenac.;* Kam.
„ „ Gabon: *Diospyros incarnata* Gürke; *Ebenac.;* Gab.
Ebenholz (d): *Diospyros spp., Maba spp.; Ebenac.;* Tr.
„ : *Bignonia leucoxylon* L.; *Bignoniac.;* Ant.
„ : *Dalbergia melanoxylon* Guill. et Perr.; *Papilionac.;* Senl., O.-Af.
„ : *Dombeya melanoxylon* Roxb.; *Sterculiac.;* St. Helena.
„ , abendländisches: *Buxus sempervirens* L.; *Buxac.;* s. Eu., w. As., n. Af.
„ , afrikanisches: *Diospyros dendo* Welw.; *Ebenac.;* tr. W.-Af.
„ , amerikanisches: *Brya ebenus* DC.; *Papilionac.;* Cuba, Jam.
„ , Berg-: *Bauhinia acuminata* L.; *Caesalpiniac.;* Ind., Ch.
„ , blaues: *Copaifera bracteata* Bth.; *Caesalpiniac.;* n. S.-Am.
„ , braunes: *Tecoma leucoxylon* Mart.; *Bignoniac.;* Ant., S.-Am.
„ , deutsches: *Taxus baccata* L.; *Conifer.;* Eu., w. As., n. Af.
„ , gelbes: *Tecoma leucoxylon* Mart.; *Bignoniac.;* Ant., S.-Am.
„ , gestreiftes: *Diospyros quaesita* Thw.; *Ebenac.;* Cey.
„ , grünes (d): *Tecoma leucoxylon* Mart.; *Bignoniac.;* fr. Gu.
„ , indisches: *Diospyros melanoxylon* Roxb.; *Ebenac.;* s. Ind.
„ , Madagaskar-: *Diospyros perrieri* H. Jum.; *Ebenac.;* Mad.
„ , nordisches (d): *Buxus sempervirens* L.; *Buxac.;* s. Eu, w. As., n. Af.
„ , rotes (d): *Brya ebenus* DC.; *Papilionac.;* tr. Am.
Ebénier: ? *Diospyros mespiliformis* Hochst.; *Ebenac.;* Sahara.
„ vrai d'Afrique: *Diospyros incarnata* Gürke; *Ebenac.;* Kam., Gab.
„ „ „ : *Diospyros crassifolia* Hiern; *Ebenac.;* Kam., Gab.
„ faux (f): *Cytisus laburnum* L.; *Papilionac.;* Eu.
„ du Sénégal: *Dalbergia melanoxylon* Guill. et Perr.; *Papilionac.;* Sudan.
Eberesche, gemeine: *Sorbus aucuparia* L.; *Rosac.;* Eu.
„ , skandinavische: *Sorbus suecica* Krock. et Almg.; *Rosac.;* n. N.-Eu.
„ , zahme: *Sorbus domestica* L.; *Rosac.;* s. Eu.
Eberrauten-Beifuß: *Artemisia abrotanum* L.; *Composit.;* s. Eu., Or.
Ebeul: *Pterocarpus soyauxi* Taub.; *Papilionac.;* Kam., Gab.
Ebey: *Lovoa klaineana* Pierre; *Meliac.;* Gab.
Ebiara (f): *Berlinia acuminata* Bth.; *Caesalpiniac.;* Elf., Gab.
„ : *Berlinia bracteosa* Bth.; *Caesalpiniac.;* Kam., Gab.
Ebien: *Alstonia congensis* Engl.; *Apocynac.;* Elf.
Ebine: *Croton sp.; Euphorbiac.;* Kam.
Ebiraro: *Pterogyne nitens* Tul.; *Caesalpiniac.;* Bras.
Ebo: *Dipteryx panamensis* Pittier; *Papilionac.;* Pan.

Ebo: *Pachylobus balsamifera* Oliv., *P. sp.; Burserac.;* Gab.
,, : *Pachylobus ebo* Pierre; *Burserac.;* Gab.
,, , scotch: *Hieronyma alchorneoides* Allem.; *Euphorbiac.;* Pan.
Ebodi m'lon: *Eremospatha cuspidata* G. Mann et H. Wendl.; *Palmac.;* Kam.
Eboe: *Dipteryx oleifera* Taub.; *Papilionac.;* Hond.
Ebom: *Anonidium manni* Engl. et Diels; *Anonac.;* Kam.
Ebomali: *Pachylobus ebo* Pierre; *Burserac.;* Gab.
Ebondso: *Pentadesma lecomteana* Pierre; *Guttifer.;* Gab.
Ebongagnagne: *Garcinia kola* Heckel; *Guttifer.;* Kam.
Ebongo adoi: *Grewia coriacea* Mast.; *Tiliac.;* Kam.
,, angoa: *Pentadesma leptonema* Pierre; *Guttifer.;* Kam.
,, eyidi: *Grewiopsis globosa* de Wild. et Dur.; *Tiliac.;* Kam.
,, maya: *Garcinia manni* Oliv.; *Guttifer.;* Kam.
,, mbonji: *Xylopia aethiopica* A. Rich.; *Anonac.;* Kam.
Ebony (e): *Diospyros spp., Maba spp.; Ebenac.;* Tr.
,, : *Brya ebenus* DC.; *Papilionac.;* Jam.
,, : *Krugiodendron ferreum* Urban; *Rhamnac.;* P.-R., Virg.-Ins.
,, : *Diospyros sp.; Ebenac.;* Simalu / (Sum.)
,, : *Diospyros virginalis* ?; *Ebenac.;* Leew., Monts.
,, : *Maba verae-crucis* Standl.; *Ebenac.;* br. Hond.
,, : *Swartzia sp.; Caesalpiniac.;* br. Gu.
,, : *Diospyros dendo* Welw.; *Ebenac.;* Ang.
,, : *Diospyros mespiliformis* Hochst.; *Ebenac.;* Go., n. Nig., B.-K.
,, : *Dombeya melanoxylon* Roxb.; *Sterculiac.;* St. Helena.
,, : *Diospyros perrieri* Jumelle; *Ebenac.;* Mad.
,, : *Diospyros reticulata* Willd., *D. tesselaria* Poir.; *Ebenac.;* Maur.
,, : *Diospyros mauritiana* A. DC.; *Ebenac.;* Maur.
,, : *Diospyros spp.; Ebenac.;* Ind.
,, (H): *Diospyros melanoxylon* Roxb., *Ebenac.;* Ind.
,, ,, : *Diospyros tomentosa* Roxb.; *Ebenac.;* Ind. (Punjab).
,, ,, : *Diospyros ebenum* Koenig; *Ebenac.;* Ind., Cey.
,, : *Diospyros ebenum* Koenig, *D. pilosula* Wall.; *Ebenac.;* And.
,, : *Maba buxifolia* Pers.; *Ebenac.;* Phil.
,, : *Diospyros sp.; Ebenac.;* Bor., Cel. (Makassar).
,, : *Diospyros discolor* Willd.; *Ebenac.;* Phil., Form.
,, , american (e): *Brya ebenus* DC.; *Papilionac.;* Cuba, Jam.
,, , Bastard-: *Diospyros ebenaster* Retz.; *Ebenac.;* Cey.
,, , black: *Diospyros tesselaria* Poir.; *Ebenac.;* Maur.
,, , British Guiana-: *Swartzia sp.; Caesalpiniac.;* br. Gu.
,, , brown (u): *Caesalpinia granadillo* Pittier; *Caesalpiniac.;* Ven.
,, , ,, : *Swartzia tomentosa* DC.; *Caesalpiniac.;* br. Gu.
,, , ,, (e): *Brya ebenus* DC.; *Papilionac.;* Cuba, Jam.
,, , ,, : *Swartzia minutiflora* Kleinh.; *Caesalpiniac.;* Sur.
,, , ,, : *Swartzia triphylla* Willd., *S. benthamianus* Miq.; *Caesalpiniac.;* Sur.
,, , Cape-: *Euclea pseudebenus* E. Meyer; *Ebenac.;* Kap.
,, , ,, : *Heywoodia lucens* Sim; *Euphorbiac.;* S.-Af.
,, , Coromandel-: *Diospyros quaesita* Thw.; *Ebenac.;* Cey.
,, , green: *Brya ebenus* DC.; *Papilionac.;* Bah., Cuba, Jam.
,, , ,, : *Tecoma arctiacea* P. DC.; *Bignoniac.;* fr. Gu.
,, , ,, : *Diospyros chloroxylon* Roxb.; *Ebenac.;* C.-O.-Ind., s. O.-Ind.
,, , Jamaican (e): *Brya ebenus* DC.; *Papilionac.;* Cuba, Jam.
,, , Macassar-: *Diospyros sp.; Ebenac.;* Cel.
,, , Mountain-: *Bauhinia acuminata* L.; *Caesalpiniac.;* tr. As.
,, , Myrtle-: *Diospyros pentamera* F. v. M.; *Ebenac.;* Au.
,, , Native-: *Maba humilis* R. Br.; *Ebenac.;* Au.
,, , Queensland-: *Bauhinia carroni* F. v. M.; *Caesalpiniac.;* n. Au.
,, , ,, : *Bauhinia hookeri* F. v. M.; *Caesalpiniac.;* n. Au.
,, , Senegal-: *Dalbergia melanoxylon* Guill. et Perr.; *Papilionac.;* Senl.
,, , speckled: *Diospyros spp.; Ebenac.;* Ind.
,, , streaked: *Diospyros spp.; Ebenac.;* Ind.
,, , West Indian- (e): *Brya ebenus* DC.; *Papilionac.;* Cuba, Jam.
,, , white: *Pterocarpus sp.; Papilionac.;* And.

Ebony, white: ? *Diospyros malacapai* A. DC.; *Ebenac.;* Phil.
Ebor: *Ochrocarpus africana* Oliv.; *Guttifer.;* Gab.
„ : *Mammea klaineana* Pierre; *Guttifer.;* Gab.
Eborha: *Berlinia acuminata* Sol.; *Caesalpiniac.:* Gab.
Eborodumuen: *Leptaulus daphnoides* Bth.; *Icacinac.;* Elf.
Eborro: *Mammea ebboro* Pierre; *Guttifer.;* Gab.
Ebot: *Mammea ebboro* Pierre; *Guttifer.;* Kam.
Ebotoc m'lon: *Eremospatha cuspidata* G. Mann et H. Wendl.; *Palmac.;* Kam.
Eboubouré: *Cola lateritia* K. Schum.; *Sterculiac.;* Gab.
Ebouiro: *Pentadesma butyraceum* Don; *Guttifer.;* Gab.
Ebruké: *Lannea sp.; Anacardiac.;* Elf.
Ebungo matatolo: *Allanblackia klainei* Pierre; *Guttifer.;* Kam.
Ebusok: *Sapium mannianum* Bth.; *Euphorbiac.;* Kam.
Ebvoughounzo: *Corynanthe johimbe* K. Schum.; *Rubiac.;* Gab.
Ecguèhié: *Khaya ivorensis* Chev.; *Meliac.;* Elf.
Echa-humo: *Tecoma serratifolia* G. Don; *Bignoniac.;* Ven.
Eche: *Fagara xanthoxyloides* Lam.; *Rutac.;* Togo.
Echeché: *Anogeissus leiocarpa* Guill. et Perr.; *Combretac.;* Togo.
Echirna: *Hasskarlia didymostemon* Baill.; *Euphorbiac.;* Elf.
Echirua: *Hasskarlia didymostemon* Baill.; *Euphorbiac.;* Elf.
Eck: *Bourreria oxyphylla* Standl.; *Borraginac.;* br. Hond.
Eckbié: *Khaya ivorensis* Chev.; *Meliac.;* Elf.
Eckon: *Alstonia congensis* Engl.; *Apocynac.;* Gab.
Eckou: *Alstonia congensis* Engl.; *Apocynac.;* Gab.
Eckoua: *Alstonia congensis* Engl.; *Apocynac.;* Gab.
Eco: *Macaranga heudeloti* Baill.; *Euphorbiac.;* Elf.
Ecobem: *Macrolobium aff. chrysophylloides* Hutch., *M. sp., Berlinia sp.; Caesalpiniac.;*
Ecorce blanc: *Olax wightiana* Wall.; *Olacac.;* Maur., Réu. [Kam.
„ de tapir: *Drimys winteri* Forst.; *Magnoliac.;* C.-Am., S.-Am.
Eda-balli: *Calophyllum sp.; Guttifer.;* br. Gu.
„ -kula: *Alstonia scholaris* R. Br.; *Apocynac.;* O.-Ind. (Madras).
Edan: *Uapaca bingervillensis* Beille; *Euphorbiac.;* Elf.
Edangkorna: *Stereospermum xylocarpum* Wight; *Bignoniac.;* O.-Ind.
Eddé: *Scottellia kamerunensis* Gilg, *S. chevalieri* Chipp; *Flacourtiac.;* Elf.
Eddi: *Strychnos nux-vomica* L.; *Loganiac.;* sw. O.-Ind.
Edel: *Bombax malabaricum* DC.; *Bombacac.;* ö. Beng.
Edid: *Chlorophora excelsa* Bth. et Hook.; *Morac.;* Go.
Edji Bari: *Uapaca bingervillensis* Beille; *Euphorbiac.;* Elf.
Edjin: *Sarcocephalus trillesi* Pierre; *Rubiac.;* Kam.
Edoku: *Mitragyne macrophylla* Hiern; *Rubiac.;* B.-K.
Edoucié (f): *Entandrophragma aff. choriandrum* Harms, *E. sp.; Meliac.;* Kam.
Edoué: *Saccoglottis gabonensis* Urban; *Humiriac.;* Kam.
Edoum: *Chlorophora excelsa* Bth. et Hook.; *Morac.;* Elf.
„ : *Cylicodiscus gabunensis* Harms; *Mimosac.;* Kam.
Edoup Banjoro: *Strombosiopsis tetandra* Engl.; *Olacac.;* Kam.
Edoup-Bazoa: *Lavalleopsis densivenia* Engl.; *Olacac.;* Kam.
„ -m'bazoa: *Lavalleopsis densivenia* Engl.; *Olacac.;* Kam.
Edoussié: *Entandrophragma sp.; Meliac.;* Kam.
Edua: *Pycnanthus kombo* Warb.; *Myristicac.;* Elf.
Edundu: *Piptadenia africana* Hook. f.; *Mimosac.;* tr. W.-Af.
Edwundwuna: *Phialodiscus unijugatus* Radlk.; *Sapindac.;* Go.
Edwunkobini: *Baphia pubescens* Hook. f.; *Papilionac.;* Go.
Eenndi: ? *Terminalia altissima* Chev.; *Combretac.;* Kam.
Efé: *Hannoa klaineana* Pierre et Engl.; *Simarubac.;* Elf.
Efeu (d): *Hedera helix* L.; *Araliac.;* Mitt.- & S.-Eu.
Effe (d): *Ulmus effusa* Willd.; *Ulmac.;* Mitt.-Eu.
„ : *Hannoa klaineana* Pierre et Engl.; *Simarubac.;* Elf.
Effoi: *Pycnanthus kombo* Warb.; *Myristicac.;* Elf.
Efok: *Cola heterophylla* Schott et Endl.; *Sterculiac.;* Kam.
Efombo: *Coula edulis* Baill.; *Olacac.;* B.-K. (Lulongo).
Efomu: *Xylopia aethiopica* A. Rich.; *Anonac.;* Elf.
Efonkyiya: *Entandrophragma spp.; Meliac.;* Go.

Efrobrodidwo: *Entandrophragma spp.; Meliac.;* Go.
Efuaba: *Monodora myristica* Dunal; *Anonac.;* Go.
Efuen: *Monodora myristica* Dunal; *Anonac.;* Elf.
Efueno: *Monodora myristica* Dunal; *Anonac.;* Elf.
Efuokonkonti: *Entandrophragma spp.; Meliac.;* Go.
Efurumundu: *Funtumia elastica* Stapf; *Apocynac.;* Elf.
Egg-fruit: *Lucuma sp.; Sapotac.;* Bah.
Eghessi-Wurzel (d): *Sarcocephalus esculentus* Afzel.; *Rubiac.;* Elf.
Egligh: *Balanites aegyptiaca* Delile; *Simarubac.;* Af. (Aeg.).
Egna: *Ceiba pentandra* Gaertn. *var. dehiscens* Ulb.; *Bombacac.;* Elf.
Egnougnouma: *Daniella soyauxi* Harms; *Caesalpiniac.;* Gab.
Egonoki: *Styrax japonica* S. et Z.; *Styracac.;* Jap.
Egoumé: *Didelotia africana* Pierre; *Caesalpiniac.;* Gab.
„ : *Coula edulis* Baill.; *Olacac.;* Gab.
Egouzi: *Chlorophora excelsa* Bth. et Hook.; *Morac.;* Elf.
Eguaba: *Psidium guajava* L.; *Myristicac.;* Go.
Egui: *Musanga smithi* R. Br.; *Morac.;* Elf., Go.
„ : *Funtumia africana* Stapf; *Apocynac.;* Kam.
Eguina: *Ceiba pentandra* Gaertn.; *Bombacac.;* Elf.
Eguna: *Ceiba pentandra* Gaertn.; *Bombacac.;* Elf.
Egunli: *Musanga smithi* R. Br.; *Morac.;* Go.
Egyan: *Antiaris africana* Engl.; *Morac.;* Go.
Egypt: *Lavalleopsis densiventa* Engl.; *Olacac.;* Gab.
Ehaletombo: *Drypetes paxi* Hutch.; *Euphorbiac.;* Kam.
Ehé: ? *Piptadenia africana* Hook. f.; *Mimosac.;* Elf.
Ehela: *Cassia fistula* L.; *Caesalpiniac.;* Cey.
Ehoronvian: *Afrormosia laxiflora* Harms; *Papilionac.;* Go.
Ehoumé: *Coula edulis* Baill.; *Olacac.;* Gab.
? Ehranbaum: *Acer pseudoplatanus* L.; *Acerac.;* De.
Ehurike: *Pentadesma butyraceum* Don; *Guttifer.;* Go.
Ehurufren: *Distemonanthus benthamianus* Baill.; *Caesalpiniac.;* Go.
Eibe (d): *Taxus baccata* L.; *Conifer.;* Eu., w. As., n. Af.
„ , gemeine: *Taxus baccata* L.; *Conifer.;* Eu., w. As., n. Af.
„ , japanische: *Taxus cuspidata* S. et Z.; *Conifer.;* Jap.
„ , pazifische: *Taxus brevifolia* Nutt.; *Conifer.;* w. N.-Am.
Eiche (d): *Quercus spp.; Fagac.;* n. Hem.
„ , adriatische: *Quercus cerris* L.; *Fagac.;* s. Eu.
„ , Afrikanisch- (d): *Chlorophora excelsa* Bth. et Hook.; *Morac.;* W.-Af., O.-Af.
„ , Antillen-: *Bignonia longissima* Swartz; *Bignoniac.;* Ant.
„ , brasilianische: *Posoqueria latifolia* Roem.; *Rubiac.;* Bras., Ven., Trin.
„ , Burgunder-· *Quercus cerris* L.; *Fagac.;* s. Eu., Kl.-As.
„ , Busch-: *Chlorophora excelsa* Bth. et Hook.; *Morac.;* W.-Af.
„ , dickfrüchtige: *Quercus conferta* Kit.; *Fagac.;* s. Eu.
„ , flaumhaarige: *Quercus pubescens* Willd.; *Fagac.;* s. Eu., Or.
„ , französische: *Quercus pubescens* Willd.; *Fagac.;* sö. Eu., Or.
„ , Gallapfel-: *Quercus infectoria* Oliv.; *Fagac.;* Griech., Kl.-As.
„ , gedrängtfrüchtige: *Quercus conferta* Kit.; *Fagac.;* s. Eu.
„ , gesägtblätterige: *Quercus serrata* Thunb.; *Fagac.;* O.-As.
„ , goldschuppige: *Quercus chrysolepis* Liebm.; *Fagac.;* USA.
„ , grossfrüchtige: *Quercus macrocarpa* Michx.; *Fagac.;* s. USA.
„ , Haselnuss-: *Quercus ballota* Desf.; *Fagac.;* M.-M.
„ , indische: *Tectona grandis* L. f.; *Verbenac.;* Ind., Malay.
„ , kastanienblätterige: *Quercus castaneaefolia* C. A. Meyer; *Fagac.;* Kl.-As.,
„ , keilblätterige: *Quercus digitata* Sudw.; *Fagac.;* USA.   [n. Pers.
„ , Ki- (d): *Castanea pumila* Mill.; *Fagac.;* USA.
„ , Kork- (d): *Quercus suber* L.; *Fagac.;* M.-M.
„ , langfrüchtige: *Quercus lobata* Née; *Fagac.;* USA. (w. Kalif.).
„ , Lebens- (d): *Quercus virens* Ait.; *Fagac.;* N.-Am.
„ , österreichische: *Quercus cerris* L.; *Fagac.;* s. Eu.
„ , Rot- (d): *Quercus rubra* L.; *Fagac.;* N.-Am.
„ , Sommer- (d): *Quercus pedunculata* Ehrh.; *Fagac.;* Eu.
„ , spanische: *Quercus digitata* Sudw.; *Fagac.;* USA.

Eiche, spitzblätterige: *Quercus acuta* Thunb.; *Fagac.;* s. Jap.
 „ , Stein (d): *Quercus sessiliflora* Salisb.; *Fagac.;* Eu.
 „ , Stiel- (d): *Quercus pedunculata* Ehrh.; *Fagac.;* Eu.
 „ , Trauben-: *Quercus sessiliflora* Salisb.; *Fagac.;* Eu.
 „ , türkische: *Quercus cerris* L.; *Fagac.;* s. Eu.
 „ , ungarische: *Quercus conferta* Kit.; *Fagac.;* ö. M.-M.
 „ , weichhaarige: *Quercus pubescens* Willd.; *Fagac.;* sö. Eu., As.
 „ , Weiss- (d): *Quercus alba* L.; *Fagac.;* N.-Am.
 „ , Winter-: *Quercus sessiliflora* Salisb.; *Fagac.;* Eu.
Eijan: *Thespesia populnea* Correa; *Malvac.;* Go.
E(i)jóng: *Sterculia oblonga* Mast.; *Sterculiac.;* Kam.
Eijserhout (h): *Eusideroxylon zwageri* Teijsm. et Binn.; *Laurac.;* Bor., Java, Banka.
 „ „ : *Metrosideros vera* Lindl.; *Myrtac.;* Amb., Java, Moluk., Ternate.
Eikmwe: *Lagerstroemia flos-reginae* Retz.; *Lythrac.;* Burma.
Einmwe: *Lagerstroemia flos-reginae* Retz.; *Lythrac.;* Burma.
Eisenholz (d): *Casuarina equisetifolia* Forst.; *Casuarinac.;* Au.
 „ „ : *Combretum imberbe* Wawra *var. petersi* Engl. et Diels; *Combretac.;*
 „ , Borneo- (d): *Eusideroxylon zwageri* Teijsm. et Binn.; *Laurac.;* Bor. [? M.-M.
 „ , Guayana (d): *Swartzia spp.; Caesalpiniac.;* Gu.
 „ , Jamaika-: *Fagara pterota* H. B. K.; *Rutac.;* W.-Ind., C.-Am., Kol.
 „ , ostindisches: *Mesua ferrea* L.; *Guttifer.;* O.-Ind.
 „ , rotes: *Reynosia latifolia* Griseb.; *Rhamnac.;* s. Fl., W.-Ind.
 „ , Südsee-: *Calophyllum inophyllum* L.; *Guttifer.;* Südsee.
 „ , weisses: *Hypelate trifoliata* Sw.; *Sapindac.;* s. Fl., W.-Ind.
 „ , „ : *Toddalia lanceolata* Lam.; *Rutac.;* Kap., O.-Af.
 „ , westafrikanisches (d): *Lophira procera* Chev.; *Ochnac.;* W.-Af.
 „ , westindisches (d): *Colubrina ferruginosa* Brongn.; *Rhamnac.;* W.-Ind.
Eisenrindenbaum (d): *Eucalyptus leucoxylon* F. v. M.; *Myrtac.;* ö. Au.
 „ „ : *Eucalyptus siderophloia* Bth.; *Myrtac.;* N.-S.-W., Queensl.
Ejem: *Albizzia fastigiata* Oliv.; *Mimosac.;* tr. Af.
Ejen: *Distemonanthus benthamianus* Baill.; *Caesalpiniac.;* Kam.
Ej(u)ong: *Sterculia oblonga* Mast.; *Sterculiac.;* Kam.
Ejuong: *Triplochiton scleroxylon* K. Schum.; *Sterculiac.;* Kam.
Ek: *Quercus sessiliflora* Salisb.; *Fagac.;* Schwed.
Ekak: *Cynometra sp.; Caesalpiniac.;* Kam.
Ekaling: *Taxus baccata* L.; *Conifer.;* nw. O.-Ind. (Punjab).
Ekango: *Pachypodanthium confine* Engl. et Diels; *Anonac.;* Gab.
Ekbié: *Khaya ivorensis* Chev.; *Meliac.;* Elf.
Ekadania: *Bridelia retusa* Spreng.; *Euphorbiac.;* n. O.-Ind.
Eké: *Sarcocephalus trillesi* Pierre; *Rubiac.;* Kam.
Ekelle: *Klainedoxa longifolia* Pierre; *Simarubac.;* B.-K. (Lukolela).
Ekeng: *Sarcocephalus trillesi* Pierre, S. *pobeguini* Pob.; *Rubiac.;* Kam.
Ekep: *Heisteria trillesiana* Pierre; *Olacac.;* Gab.
Ekévazingo: *Didelotia africana* Pierre; *Caesalpiniac.;* Gab.
Ekhuié: *Khaya ivorensis* Chev.; *Meliac.;* Elf.
Ekimi: *Millettia sp.; Papilionac.;* Elf.
Ekjarepore: *Paypayrola guianensis* Aubl.; *Violac.;* Sur.
Ekó: *Ricinodendron africanum* Muell. Arg.; *Euphorbiac.;* Kam.
 „ : *Lannea welwitschi* Engl.; *Anacardiac.;* Kam.
Ekoako: *Conopharyngia brachyantha* Stapf; *Apocynac.;* Kam.
Ekob: *Cynometra afzelii* Harms; *Caesalpiniac.;* Kam.
Ekoe: *Lannea welwitschi* Engl.; *Anacardiac.;* Kam.
Ekok: *Ricinodendron africanum* Muell. Arg.; *Euphorbiac.;* sp. Guin.
Ekoka: *Lophira procera* Chev.; *Ochnac.;* Kam., Gab.
Ekokom: *Hymenostegia aff. afzelii* Harms; *Caesalpiniac.;* Kam.
Ekol: *Ficus vogeliana* Miq.; *Morac.;* Kam.
Ekombilé: *Piptadenia africana* Hook. f.; *Mimosac.;* Kam.
Ekombolo: ? *Parkia klainei* Pierre, P. *spp.; Mimosac.;* Kam.
Ekon: *Trichoscypha sp.; Anacardiac.;* Kam.
Ekona: *Alstonia congensis* Engl.; *Apocynac.;* Gab.
Ekondjo: *Piptadenia sp.; Mimosac.;* Gab.
Ekonge: *Sterculia oblonga* Mart.; *Sterculiac.;* tr. Af.

Ekop: *Berlinia sp.; Caesalpiniac.;* Kam.
„ : *Staudtia gabonensis* Warb.; *Myristicac.;* Kam.
Ekopa: *Lonchocarpus sericeus* H. B. K.; *Papilionac.;* Elf.
Ekosuba: *Deinbollia indeniensis* Chev.; *Sapindac.;* Elf.
Ekoua: *Alstonia congensis* Engl.; *Apocynac.;* Gab.
Ekouané: *Bridelia speciosa* Muell. Arg.; *Euphorbiac.;* Elf.
Ekouk: *Alstonia macrophylla* Wall.; *Apocynac.;* Kam.
„ : *Alstonia congensis* Engl.; *Apocynac.;* Kam.
Eku: *Bombax buonopozense* Beauv.; *Bombacac.;* Go.
Ekua: *Macaranga heudeloti* Baill.; *Euphorbiac.;* Elf.
Ekualokpoe: *Lannea acida* A. Rich.; *Anacardiac.;* Togo.
Ekuama: *Pentaclethra macrophylla* Bth.; *Mimosac.;* Go.
Ekubé: *Borassus flabellifer* L. *var. aethiopum* Warb.; *Palmac.;* Elf.
Ekuie: *Khaya ivorensis* Chev.; *Meliac.;* Elf.
Ekur: *Bombax buonopozense* Beauv.; *Bombacac.;* Go.
Ekuro: *Baphia nitida* Lodd.; *Papilionac.;* Elf.
Ekusamba: *Sarcocephalus trillesi* Pierre, *S. pobeguini* Pob.; *Rubiac.;* Elf.
Ekusawa: *Sarcocephalus esculentus* Afzel.; *Rubiac.;* Go.
Ela palol: *Sterospermum suaveolens* DC.; *Bignoniac.;* O.-Ind.
Elaka: *Feronia elephantum* Correa; *Rutac.;* w. O.-Ind. (s. Deccan).
Elakoumi: *Carapa sp.; Meliac.;* Elf.
Elamtäk: ? *Randia sp.; Rubiac.;* Kam.
Elandap-pazham: *Zizyphus jujuba* Lamk.; *Rhamnac.;* s. Ind.
Elandei: *Zizyphus jujuba* Lamk.; *Rhamnac.;* s. Ind.
Elands Boontjes: *Elephantorrhiza burchelli* Bth.; *Mimosac.;* S.-Af.
Elang: *Mimusops djave* Engl. *M. congolensis* Russell et Hédin; *Sapotac.;* Kam.
Elavangam: *Cinnamomum zeylanicum* Breyn; *Laurac.;* O.-Ind. (Mal.).
Elder: *Piper aduncum* L.; *Piperac.;* br. Hond.
„ : *Sambucus racemosa* L. *var. sieboldiana* Miq.; *Caprifoliac.;* Jap.
„ , Box-: *Acer negundo* L.; *Acerac.;* N.-Am.
„ , Box-, californian: *Acer negundo-californicum* Sarg.; *Acerac.;* USA.
„ , wild: *Nuxia floribunda* Bth.; *Loganiac.;* Kap.
Elderberry, Native: *Sambucus xanthocarpus* F. v. M.; *Caprifoliac.;* Queensl.
Flégué-mouani: *Erythrophloeum guineense* G. Don; *Caesalpiniac.;* Elf.
Elekene: *Erythrina sp.; Papilionac.;* C.-R.
Elékhua: *Uapaca bingervillensis* Beille; *Euphorbiac.;* Elf.
Elel: *Nerium oleander* L.; *Apocynac.;* Sahara.
Elelom: *Mitragyne macrophylla* Hiern; *Rubiac.;* Kam., Gab.
Eielom'zame: *Mitragyne macrophylla* Hiern; *Rubiac.;* Gab.
Elelommzame: *Mitragyne macrophylla* Hiern; *Rubiac.;* Gab.
Flem: ? *Diospyros sp., Maba sp.; Ebenac.;* Kam.
Elemi: *Protium heptaphyllum* March.; *Burserac.;* Bras. (Pará).
„ , african: *Canarium schweinfurthi* Engl.; *Burserac.;* Lib., Go.
„ , Manila-: *Canarium luzonicum* A. Gray; *Burserac.;* Phil.
„ de Mexico: *Bursera sp.; Burserac.;* Mex.
„ -tree: *Protium elemigera* ? March., *P. aracouchini* March.; *Burserac.;* fr. Gu.
Elémier d'Afrique (f): *Canarium schweinfurthi* Engl.; *Burserac.;* W.-Af.
Elengi: *Mimusops elengi* Roxb.; *Sapotac.;* O.-Ind. (Mal.).
Elephant's wood: *Bolusanthus speciosus* Harms; *Papilionac.;* ö. Transv.
Elequene: *Erythrina sp.; Papilionac.;* Nic.
Eleten: ? *Caesalpinia granadillo* Pittier; *Caesalpiniac.;* Nic.
Eleutharay: *Chukrasia tabularis* Adr. Juss.; *Meliac.;* w. O.-Ind. (s. Deccan).
Fli-Bengan: *Distemonanthus benthamianus* Baill.; *Caesalpiniac.;* Gab.
Elilom: *Mitragyne macrophylla* Hiern; *Rubiac.;* Kam.
Elim Nkak: *Dracaena fragrans* Ker-Gawl.; *Liliac.;* Kam.
Elimitcham-maron: *Citrus limonum* Risso; *Rutac.;* s. Ind.
Elkwood: *Magnolia tripetala* L.; *Magnoliac.;* USA.
Ella: *Piper aduncum* L.; *Piperac.;* br. Hond.
Ellen grypho: *Peltogyne sp.; Caesalpiniac.;* Bras.
Eller: *Alnus glutinosa* Gaertn.; *Betulac.;* Eu.
Ellupi: *Bassia longifolia* L.; *Sapotac.;* w. O.-Ind. (Mal.).
Elm: *Ulmus campestris* L.; *Ulmac.;* E.

Elm: *Ulmus americana* L.; *Ulmac.*; USA.
„ : *Guazuma ulmifolia* Lamk.; *Sterculiac.*; Trin.
„ : *Ulmus parvifolia* Jacq.; *Ulmac.*; Jap.
Elm, american: *Ulmus americana* L.; *Ulmac.*; ö. N.-Am.
„ , Bastard-: *Celtis occidentalis* L.; *Ulmac.*; ö. N.-Am.
„ , Cork-: *Ulmus racemosa* Thomas; *Ulmac.*; sö. Kan., ö. USA.
„ , „ -, northern: *Ulmus racemosa* Thomas; *Ulmac.*; ö. N.-Am.
„ , Cork-barked: *Ulmus racemosa* Thomas; *Ulmac.*; sö. Kan.
„ , Crow's foot-: *Tarrietia argyrodendron* Bth.; *Sterculiac.*; Au.
„ , Dutch (e): *Ulmus montana* Withering; *Ulmac.*; Holl.
„ , english: *Ulmus campestris* L.; *Ulmac.*; nö. USA.
„ , european: *Ulmus campestris* L.; *Ulmac.*; USA.
„ , false: *Celtis occidentalis* L.; *Ulmac.*; ö. N.-Am.
„ , Florida-: *Ulmus floridiana* Chapm.; *Ulmac.*; N.-Am.
„ , Hickory-: *Ulmus racemosa* Thomas; *Ulmac.*; ö. N.-Am.
„ , indian (H): *Holoptelea integrifolia* Planch.; *Ulmac.*; Ind.
„ . mexican: *Ulmus mexicana* Planch.; *Ulmac.*; Mex.
„ , native: *Aphananthe philippinensis* Planch.; *Ulmac.*; Au.
„ , red: *Ulmus alata* Michx., *U. fulva* Michx.; *Ulmac.*; N.-Am.
„ , Rock-: *Ulmus racemosa* Thomas, *U. americana* L.; *Ulmac.*; ö. N.-Am.
„ , Sand-: *Ulmus major* L.; *Ulmac.*; Holl.
„ , silky: *Tarrietia argyrodendron* Bth.; *Sterculiac.*; Au.
„ , slippery: *Ulmus fulva* Michx.; *Ulmac.*; ö. N.-Am.
„ , slippery-barked: *Ulmus fulva* Michx.; *Ulmac.*; sö. Kan.
„ , soft: *Ulmus fulva* Michx.; *Ulmac.*; sö. Kan.
„ , spnanish: *Cordia gerascanthus* L.; *Borraginac.*; Jam., br. W.-Ind.
„ , „ : *Colubrina rufa* Reisseck; *Rhamnac.*; Pan.
„ , Swamp-: *Ulmus americana* L., *U. racemosa* Thomas; *Ulmac.*; ö. USA.
„ , Trinidad-: *Guazuma ulmifolia* Lam.; *Sterculiac.*; Trin.
„ , Water-: *Ulmus americana* L.; *Ulmac.*; ö. N.-Am.
„ , „ : *Planera aquatica* Gmel.; *Ulmac.*; sö. N.-Am.
„ , Westindian: *Guazuma ulmifolia* Lam.; *Sterculiac.*; br. W.-Ind.
„ , white: *Ulmus racemosa* Thomas; *Ulmac.*; sö. Kan., nö. USA.
„ , „ : *Ulmus americana* L.; *Ulmac.*; N.-Am.
„ , winged: *Ulmus alata* Michx.; *Ulmac.*; ö. USA.
„ , Wych-: *Ulmus scabra* Mill.; *Ulmac.*; E.
Elon: *Fillaeopsis discophora* Harms; *Mimosac.*; tr. W.-Af.
Elondo: *Piptadenia unijuga* Pierre; *Mimosac.*; Gab.
„ : *Erythrophloeum guineense* G. Don; *Caesalpiniac.*; W.-Af.
Elongankouma: *Pithecolobium sp.*; *Mimosac.*; Gab.
Elongo: *Fagara macrophylla* Engl.; *Rutac.*; Kam.
„ : *Oxystigma manni* Harms; *Caesalpiniac.*; Kam.
Elongolongo: *Panda oleosa* Pierre; *Pandac.*; Gab.
Eloué: *Saccoglottis gabonensis* Urban; *Humiriac.*; Kam.
Eloum: *Erythrophloeum guineense* G. Don; *Caesalpiniac.*; Kam.
Eloun: *Daniella spp.*; *Caesalpiniac.*; Kam.
„ : *Chlorophora excelsa* Bth. et Hook., *C. regia* Chev.; *Morac.*; Gab.
„ : *Erythrophloeum guineense* G. Don; *Caesalpiniac.*; Kam., Gab.
Elozi-zegue: *Ximenia americana* L.; *Olacac.*; Gab.
Els, Kaboo-: *Mystroxylon kubu* Eckl. et Zeyh.; *Celastrac.*; Kap.
„ , Klip-: *Rhus thunbergi* Hook.; *Anacardiac.*; Kap.
„ , rood: *Cunonia capensis* L.; *Cunoniac.*; Kap.
„ , white: *Platylophus trifoliatus* D. Don; *Cunoniac.*; Kap.
Elsbeerbaum (d): *Sorbus torminalis* Cranţ; *Rosac.*; Eu.
Elui: *Chlorophora excelsa* Bth. et Hook.; *Morac.*; Elf.
Elumpurukki: *Litsea sebifera* Pers.; *Laurac.*; s. O.-Ind.
Elun: *Erythrophloeum guineense* G. Don; *Caesalpiniac.*; Kam.
Elunli: *Chlorophora excelsa* Bth. et Hook.; *Morac.*; Go.
Eluwi: *Chlorophora excelsa* Bth. et Hook.; *Morac.*; Go.
Elwi: *Chlorophora excelsa* Bth. et Hook.; *Morac.*; Elf.
Elzebeere: *Sorbus torminalis* Cranţ; *Rosac.*; Mitt.- & s. Eu.
Emajagua: *Hibiscus elatus* Swarţ; *Malvac.*; P.-R.

Emajagua excelsa: *Hibiscus elatus* Swartz; *Malvac.*; P.-R.
Emajaguilla: *Thespesia populnea* Correa; *Malvac.*; P.-R.
Emajazua: *Hibiscus tiliaceus* L.; *Malvac.*; P.-R.
Emang: *Cylicodiscus gabunensis* Harms; *Mimosac.*; Kam.
Emanga: *Ceiba pentandra* Gaertn.; *Bombacac.*; Elf.
Emátabi: *Bixa orellana* L.; *Bixac.*; W.-Ind. (Carib).
Embauba: *Cecropia spp.*; *Morac.*; Bras. (Amaz.)
Embi-Siembi: *Rauwolfia vomitoria* Afzel.; *Apocynac.*; Elf.
Embira branca: *Daphnopsis brasiliensis* Mart.; *Thymelaeac.*; Bras.
  „   „   : *Funifera utilis* Leandr.; *Thymelaeac.*; Bras.
Embirussú: *Bombax spp.*; *Bombacac.*; Bras.
Embrum: *Cordia myxa* L.; *Borraginac.*; O.-Ind.
Embuia (d, u): *Phoebe porosa* Mez; *Laurac.*; s. Bras.
  „ amarello: *Phoebe porosa* Mez; *Laurac.*; s. Bras.
  „ vermelha: *Phoebe porosa* Mez; *Laurac.*; s. Bras.
Emburana: *Bursera leptophloeos* Mart.; *Burserac.*; Bras.
Embu-râna: *Bursera leptophloeos* Mart.; *Burserac.*; Bras.
Embuya: *Phoebe porosa* Mez; *Laurac.*; s. Bras.
Embyú branco: ? *Oxandra lanceolata* Baill.; *Anonac.*; Bras.
Emeri: *Vochysia hondurensis* Sprague; *Vochysiac.*; br. Hond.
  „   : *Terminalia ivorensis* Chev.; *Combretac.*; Go.
Emery: *Vochysia hondurensis* Sprague; *Vochysiac.*; br. Hond.
Emien: *Alstonia congensis* Engl.; *Apocynac.*; W.-Af.
Emiengré: *Pachypodanthium staudti* Engl. et Diels; *Anonac.*; Elf.
Emigbegeri: *Pseudocedrela kotschyi* Harms; *Meliac.*; tr. W.-Af.
Emil: *Terminalia ivorensis* Chev.; *Combretac.*; Go.
Emion: *Alstonia congensis* Engl.; *Apocynac.*; Kam.
Empadu: *Vatica spp., Cotylelobium spp.*, ? *Shorea spp.*; *Dipterocarpac.*; tr. As.
Emri: *Terminalia ivorensis* Chev.; *Combretac.*; Go.
Emroi: *Ulmus wallichiana* Planch.; *Ulmac.*; n. O.-Ind. (Jaunsar).
Emuinquim: *Maesobotrya stapfiana* Beille; *Euphorbiac.*; Elf.
Ena da: *Mallotus philippinensis* Muell. Arg.; *Euphorbiac.*; Ind.-Ch.
Enak: *Macrolobium palisoti* Bth.; *Caesalpiniac.*; Kam.
Enbotta-koenatjepie: *Tecoma sp., Tabebuia sp.*; *Bignoniac.*; Sur.
Encens: *Hedwigia balsamifera* Swartz; *Burserac.*; Guad.
  „   : *Protium spp.*; *Burserac.*; fr. Gu.
  „   : *Protium guianense* March.; *Burserac.*; fr. Gu.
  „ blanc: *Protium guianense* March.; *Burserac.*; fr. Gu.
  „ grand bois: *Protium guianense* March.; *Burserac.*; Gu.
  „ gris: *Protium sagotianum* March.; *Burserac.*; Sur.
  „ „ : *Protium heptaphyllum* March. et Pulle; *Burserac.*; Sur.
  „ „ : *Protium puberulum* Engl.; *Burserac.*; Sur.
  „ „ : *Protium aracouchini* Mart., *P. altissimum* March.; *Burserac.*; Sur.
  „ rose: ? *Protium sp.*; *Burserac.*; fr. Gu.
  „ rosé: *Tetragastris sp.*; *Burserac.*; Sur.
  „ rouge: *Hedwigia balsamifera* Swartz, *Protium spp.*; *Burserac.*; Sur.
  „   „ : *Paypayrola guianensis* Aubl.; *Violac.*; fr. Gu.
Encina: *Quercus ilex* L.; *Fagac.*; Sp.
  „   : *Quercus sp.*; *Fagac.*; sp. Am.
  „   : *Quercus virens* Ait.; *Fagac.*; Cuba.
  „   : *Licania sp.*; *Rosac.*; Guat., Hond.
  „   : *Weinmannia sp.*; *Cunoniac.*; Kol.
Encinillo: *Weinmannia sp.*; *Cunoniac.*; Kol.
Encinillos: *Weinmannia sp.*; *Cunoniac.*; Kol.
Encino: *Quercus ilex* L.; *Fagac.*; Sp.
  „   : *Quercus spp., Q. brachystachys* Bth.; *Fagac.*; Guat.
  „   : *Quercus spp.*; *Fagac.*; br. Hond., Hond.
  „   : *Quercus virens* Ait.; *Fagac.*; Hond.
  „   : *Weinmannia sp.*; *Cunoniac.*; Kol.
  „ negro: *Quercus oleoides* Cham. et Schlecht.; *Fagac.*; Guat.
  „   „ : *Quercus callosa* Bth.; *Fagac.*; Hond.
End: *Terminalia superba* Engl. et Diels; *Combretac.*; Kam.

Endé: *Rhizophora racemosa* G. F. Meyer; *Rhizophorac.;* Elf.
Endielé: *Calpocalyx dinklagei* Harms; *Mimosac.;* Kam.
Endjaè: *Elaeis guineensis* Jacq. *var. virescens* Chev.; *Palmac.;* Gab.
Endoa: *Calpocalyx klainei* Pierre; *Mimosac.;* Kam.
Endrino: *Prunus spinosa* L.; *Rosac.;* Sp.
„ blanco: *Prunus insititia* L.; *Rosac.;* Sp.
Endwi: *Lophira alata* Banks; *Ochnac.;* Lib.
Enébien: *Uapaca bingervillensis* Beille; *Euphorbiac.;* Elf.
Enebro: *Juniperus communis* L., *J. oxycedrus* L.; *Conifer.;* Sp.
„ : *Juniperus spp.; Conifer.;* Mex.
„ criollo: *Juniperus lucayana* Britton; *Conifer.;* Cuba.
„ de la miera: *Juniperus oxycedrus* L.; *Conifer.;* Sp.
Enem: *Symphonia gabonensis* Pierre; *Guttifer.;* Kam.
Enésisć: *? Copaifera sp.; Caesalpiniac.;* Gab.
Eng (d, e): *Dipterocarpus tuberculatus* Roxb.; *Dipterocarpac.;* tr. As.
„ : *Dipterocarpus turbinatus* Gaertn. f., *D. pilosus* Roxb.; *Dipterocarpac.;* Burma.
Engakom: *Myrianthus arboreus* Beauv.; *Morac.;* Kam.
? Engan: *Turraeanthus aff. vignei* Hutch.; *Meliac.;* Kam.
Engan: *Sorindeia ochracea* Engl.; *Anacardiac.;* Kam.
? Engelboom: *Dicorynia paraënsis* Bth.; *Caesalpiniac.;* fr. Gu.
Engessan: *Ricinodendron africanum* Muell. Arg.; *Euphorbiac.;* Gab.
Engó: *Casearia sp.; Flacourtiac.;* Kam.
Engom: *Coula edulis* Baill.; *Olacac.;* Kam.
Engong: *Eriocoelum kerstingi* Gilg; *Sapindac.;* Kam.
„ : *Phialodiscus unijugatus* Radlk.; *Sapindac.;* Kam.
Engongoui: *Antrocaryon klaineanum* Pierre; *Anacardiac.;* Kam.
Engoom: *Barteria fistulosa* Hook. f.; *Flacourtiac.;* Kam.
Engosso: *Eschweilera corrugata* Miers; *Lecythidac.;* Sur.
Engraver's wood: *Sideroxylon pohlmanianum* Bth. et Hook.; *Sapotac.;* Au.
Engyin: *Hopea suavis* Wall.; *Dipterocarpac.;* And.
Eniamianga: *Berlinia bracteosa* Bth.; *Caesalpiniac.;* Gab.
Enianga: *Ceiba pentandra* Gaertn.; *Bombacac.;* Elf.
Eniangua: *Ceiba pentandra* Gaertn.; *Bombacac.;* Elf.
Enieme: *Ceiba pentandra* Gaertn.; *Bombacac.;* Elf.
Enju: *Sophora japonica* L.; *Papilionac.;* Jap.
Enkagouma: *Tetrapleura thonningi* Bth.; *Mimosac.;* Gab.
Ennei: *Dipterocarpus indicus* Bedd.; *Dipterocarpac.;* w. O.-Ind. (s. Deccan).
Enok: *Croton sp.; Euphorbiac.;* Kam.
Enoki: *Celtis sinensis* Pers. *var. japonica* Nakai; *Ulmac.;* Jap.
Enonee: *Lovoa klaineana* Pierre; *Meliac.;* W.-Af.
Enoumnouma: *Pithecolobium sp.; Mimosac.;* Gab.
Ensale: *Newtonia sp., Piptadenia africana* Bth.; *Mimosac.;* Gab.
Ensésang: *Ricinodendron africanum* Muell. Arg.; *Euphorbiac.;* Gab.
En-shu: *Sophora japonica* L.; *Papilionac.;* Jap.
Entangor: *Calophyllum spp.; Guttifer.;* Tr.
Entédua: *Copaifera salikounda* Heckel; *Caesalpiniac.;* Go.
Entié: *Cleistopholis patens* Engl. et Diels; *Anonac.;* Elf.
Entranel: *Lagerstroemia spp.; Lythrac.;* Kamb.
„ : *Lagerstroemia flos-reginae* Reţ.; *Lythrac.;* Ind.-Ch.
Entravel: *Lagerstroemia loudoni* Teijsm. et Binn.; *Lythrac.;* Ind.-Ch.
„ : *Lagerstroemia flos-reginae* Reţ.; *Lythrac.;* Co.-Ch.
Entrenel: *Lagerstroemia sp.; Lythrac.;* Ind.-Ch.
Envé: *Pterocarpus soyauxi* Taub.; *Papilionac.;* sp. Guin.
Envira: *? Oxandra lanceolata* Baill.; *Anonac.;* Bras.
„ preta: *Guatteria poeppigiana* Mart.; *Anonac.;* Bras.
„ „ do igapó: *Guatteria inundata* Mart.; *Anonac.;* Bras.
Envireira: *Sterculia pruriens* K. Schum.; *Sterculiac.;* Bras.
„ : *Anona sp.; Anonac.;* Bras.
„ : *Guatteria poeppigiana* Mart.; *Anonac.;* Bras.
En wiwan: *Sterculia rhinopetala* K. Schum.; *Sterculiac.;* Nig., s. Kam.
Enya: *Ceiba pentandra* Gaertn.; *Bombacac.;* Go.
Enyiña: *Ceiba pentandra* Gaertn.; *Bombacac.;* Go.

Enzésang: *Ricinodendron africanum* Muell. Arg.; *Euphorbiac.;* Gab.
Eon: *Fillaeopsis discophora* Harms; *Mimosac.;* tr. W.-Af.
Eoumé: *Coula edulis* Baill.; *Olacac.;* Kam.
Eouoleveu: *Bridelia speciosa* Muell. Arg.; *Euphorbiac.;* Gab.
Epakotroubo: ? *Bridelia speciosa* Muell. Arg.; *Euphorbiac.;* Elf.
Epel: *Intsia bijuga* O. Ktze.; *Caesalpiniac.;* Mad., tr. As., Au., Südsee.
Epéru: *Eperua falcata* Aubl.; *Caesalpiniac.;* fr. Gu.
Epfog: *Pterygota macrocarpa* K. Schum.; *Sterculiac.;* Kam.
Epi de Blé: *Vouacapoua americana* Aubl.; *Caesalpiniac.;* fr. Gu.
Epicea: *Picea morinda* Link; *Conifer.;* nw. O.-Ind. (Punjab).
Epindé pindé: *Diospyros spp., D. aggregata* Gürke; *Ebenac.;* Kam.
  „   „   emassangué: *Diospyros aggregata* Gürke; *Ebenac.;* Kam.
Epine vinette: *Berberis vulgaris* L.; *Berberidac.;* Fr.
Epinette bâtarde: *Picea mariana* B. S. P.; *Conifer.;* w. Kan.
  „   blanche: *Picea canadensis* B. S. P.; *Conifer.;* Kan.
  „   grise: *Picea canadensis* B. S. P.; *Conifer.;* Kan.
  „   jaune: *Picea mariana* B. S. P.; *Conifer.;* w. Kan.
  „   noire: *Picea mariana* B. S. P.; *Conifer.;* w. Kan.
  „   rouge: *Larix laricina* C. Koch; *Conifer.;* nö. Kan.
  „   à bière: *Picea canadensis* B. S. P.; *Conifer.;* Kan.
  „   d'Engelmann: *Picea engelmanni* Engelm.; *Conifer.;* w. Kan.
  „   de Sitka: *Picea sitchensis* Carr.; *Conifer.;* w. Kan.
Epineux blanc: *Zanthoxylum emarginatum* Sw.; *Rutac.;* Guad.
  „   jaune: *Zanthoxylum clava-herculis* L.; *Rutac.;* Kam.
Epion: *Pterocarpus soyauxi* Taub.; *Papilionac.;* Kam.
Epoi: *Pycnanthus kombo* Warb.; *Myristicac.;* Elf.
Epok: *Cola heterophylla* Schott et Endl.; *Sterculiac.;* Kam.
Epopa: *Pachylobus balsamifera* Guill., *P. sp.; Burserac.;* Gab.
Epro: *Cistanthera papaverifolia* Chev.; *Tiliac.;* Go.
Epuwi: *Ricinodendron africanum* Muell. Arg.; *Euphorbiac.;* Go.
Equi: *Funtumia africana* Stapf; *Apocynac.;* Kam.
Equipal: *Trema micrantha* Blume; *Ulmac.;* Mex.
Erable (f): *Acer sp.; Acerac.;* n. Hem.
  „   bâtard: *Acer spicatum* Lam.; *Acerac.;* sö. Kan.
  „   blanc· *Acer saccharum* Marsh.; *Acerac.;* ö. Kan.
  „   „   de Montagne: *Acer pseudoplatanus* L.; *Acerac.;* Eu.
  „   circiné: *Acer circinatum* Pursh; *Acerac.;* sw. Kan.
  „   dur: *Acer saccharum* Marsh.; *Acerac.;* ö. Kan.
  „   grand: *Acer pseudoplatanus* L.; *Acerac.;* Eu.
  „   moiré: *Acer saccharum* Marsh.; *Acerac.;* ö. Kan.
  „   moucheté d'Amérique: *Acer saccharum* Marsh.; *Acerac.;* N.-Am.
  „   nain: *Acer douglasi* Hook.; *Acerac.;* w. Kan.
  „   négundo: *Acer negundo* L.; *Acerac.;* C.-Kan.
  „   ondé: *Acer saccharum* Marsh.; *Acerac.;* ö. Kan.
  „   piqué: *Acer saccharum* Marsh.; *Acerac.;* ö. Kan.
  „   rouge: *Acer rubrum* L.; *Acerac.;* sö. Kan.
  „   tendre: *Acer rubrum* L.; *Acerac.;* sö. Kan.
  „   à épis: *Acer spicatum* Lam.; *Acerac.;* sö. Kan.
  „   à feuilles de frêne: *Acer negundo* L.; *Acerac.;* C.-Kan.
  „   à fruits cotonneux: *Acer saccharum* L.; *Acerac.;* ö. Kan.
  „   à grandes feuilles: *Acer macrophyllum* Pursh; *Acerac.;* w. Kan.
  „   à sucre: *Acer saccharum* Marsh.; *Acerac.;* ö. Kan.
  „   de Pennsylvanie: *Acer pennsylvanicum* L.; *Acerac.;* sö. Kan.
Erafi: *Acacia seyal* Del.; *Mimosac.;* n. Nig.
Eravadi: *Dalbergia latifolia* Roxb.; *Papilionac.;* w. O.-Ind. (s. Deccan).
Erba cornetta: *Coronilla emerus* L.; *Papilionac.;* Arg.
Erbse, angolische (d): *Cajanus indicus* Spreng.; *Papilionac.;* Tr.
Erbsenstrauch (d): *Caragana arborescens* L.; *Papilionac.;* Mitt.- & s. Eu.
Erdbeerbaum (d): *Arbutus unedo* L.; *Ericac.;* Eu.
  „   , westamerikanischer (d): *Arbutus menziesi* Pursh.; *Ericac.;* w. N.-Am.
Ereh: *Funtumia africana* Stapf; *Apocynac.;* W.-Af. (Lagos).
Erejoeloe: *Hymenolobium flavum* Kleinh., *H. sp.; Mimosac.;* Sur.

Erhoné: *Erythrophloeum guineense* G. Don.; *Caesalpiniac.;* Elf.
Erigei: *Dalbergia lanceolaria* L.; *Papilionac.;* w. O.-Ind. (s. Deccan).
Erikissé: *Diospyros sanza-minika* Chev.; *Ebenac.;* Elf.
Eringolam: *Cinnamomum zeylanicum* Breyn; *Laurac.;* sw. O.-Ind. (Mal.)
Erizo: *Erinacea pungens* Boiss.; *Papilionac.;* Sp.
  ,, : *Apeiba tibourbou* Aubl.; *Tiliac.;* Ven.
Erle (d): *Alnus spp.; Betulac.;* n. Hem.
  ,, , afrikanische (d): *Turraeanthus africana* Pellegr.; *Meliac.;* W.-Af.
  ,, ,   ,,   ,, : *Mitragyne macrophylla* Hiern; *Rubiac.;* W.-Af.
  ,, , Alpen-: *Alnus viridis* DC.; *Betulac.;* Mitt.-Eu.
  ,, , Berg-: *Alnus viridis* DC.; *Betulac.;* Mitt.-Eu.
  ,, , Grau-: *Alnus incana* Willd.; *Betulac.;* Eu.
  ,, , Grün-: *Alnus viridis* DC.; *Betulac.;* Mitt.-Eu.
  ,, , Oregon-: *Alnus rubra* Bongard; *Betulac.;* N.-Am.
  ,, , Rot-: *Alnus glutinosa* Gaertn.; *Betulac.;* Eu.
  ,, , Schwarz-: *Alnus glutinosa* Gaertn.; *Betulac.;* Eu.
  ,, , Weiss-: *Alnus incana* Willd.; *Betulac.;* Eu.
Erma: *Albizzia odoratissima* Bth.; *Mimosac.;* C.-O.-Ind.
Erramaddi: *Terminalia arjuna* Bedd.; *Combretac.;* ö. O.-Ind. (Madras).
Erüi: *Erythrophloeum guineense* G. Don; *Caesalpiniac.;* Elf.
Erui: *Chlorophora excelsa* Bth. et Hook.; *Morac.;* Go.
Eruvalu: *Xylia xylocarpa* Taub.; *Mimosac.;* ö. O.-Ind. (n. Madras).
Esa: *Celtis soyauxi* Engl.; *Ulmac.;* Go.
Esafra (f): *Eugenia rowlandi* Sprague; *Myrtac.;* Elf.
Esang: *Piptadenia sp.; Mimosac.;* Gab.
Esangué: *Haematostaphis barteri* Hook. f.; *Anacardiac.;* Elf.
Esanhé: *Haematostaphis barteri* Hook. f.; *Anacardiac.;* Elf.
Esaonsow: *Pseudospondias microcarpa* Engl.; *Anacardiac.;* B.-K. (Likimi).
Esarghébem: *Chytranthus sp.; Sapindac.;* Gab.
Esasanghi: *Pseudospondias microcarpa* Engl.; *Anacardiac.;* B.-K.
Esasia: *Pachylobus edulis* G. Don; *Burserac.;* Gab.
Escambrón: *Pisonia aculeata* L.; *Nyctaginac.;* P.-R.
  ,, : *Adelia triloba* Hemsl.; *Euphorbiac.;* Hond.
  ,, : *Casearia aculeata* Jacq.; *Flacourtiac.;* Hond.
Esche (d): *Fraxinus spp.; Oleac.;* n. Hem.
  ,, , Blau-: *Fraxinus quadrangulata* Michx.; *Oleac.;* N.-Am.
  ,, , Blumen-: *Fraxinus ornus* L.; *Oleac.;* s. Eu.
  ,, , gemeine: *Fraxinus excelsior* L.; *Oleac.;* Eu.
  ,, , Gift-: *Chionanthus virginica* L.; *Oleac.;* N.-Am.
  ,, , Grau-: *Fraxinus alba* Marsh.; *Oleac.;* ö. N.-Am.
  ,, , mandschurische: *Fraxinus mandshurica* Rupr.; *Oleac.;* O.-As.
  ,, , Weiss-: *Fraxinus alba* Marsh.; *Oleac.;* ö. N.-Am.
Escoba: *Cytisus eriocarpus* Boiss.; *Papilionac.;* Sp.
  ,, : *Cytisus linifolius* Lam.; *Papilionac.;* Sp.
  ,, : *Spartium junceum* L.; *Papilionac.;* Arg.
  ,, blanca: *Cytisus albus* Link; *Papilionac* Sp.
  ,, negra (s): *Cytisus scoparius* Link; *Papilionac.;* Eu.
Escobetilla: ? *Ceiba sp.,* ? *Pachira sp.,* ? *Bombax sp.; Bombacac.;* Mex.
Escobilla: *Spiraea flabellata* Bertol.; *Rosac.;* Sp.
Escobón: *Cytisus affinis* Presl, *C. baeticus* Steud.; *Papilionac.;* Sp.
  ,, : *Cytisus eriocarpus* Boiss.; *Papilionac.;* Sp.
  ,, : *Cytisus triflorus* L'Hér.; *Papilionac.;* Sp.
  ,, : *Dorycnium suffruticosum* Vill.; *Papilionac.;* Sp.
  ,, morisco: *Sarothamnus welwitchi* Boiss. et Reut.; *Papilionac.;* Sp.
Esem: *Tetrapleura tetraptera* Taub.; *Mimosac.;* Go.
Esèmé: *Baphia nitida* Lodd.; *Papilionac.;* Elf.
Esensau: *Pseudospondias microcarpa* Engl.; *Anacardiac.;* B.-K. (Lukolela).
Eséé: *Afzelia africana* Smith; *Caesalpiniac.;* Kam.
Esiä: *Petersia africana* Welw.; *Lecythidac.;* Go.
  ,, -kokobin: *Petersia africana* Welw.; *Lecythidac.;* Go.
Esiape: *Petersia africana* Welw.; *Lecythidac.;* Go.
Esivé: *Petersia viridiflora* Chev.; *Lecythidac.;* Elf.

Eso: *Lophira procera* Chev.; *Ochnac.;* Elf.
Esoang: *Ricinodendron africanum* Muell. Arg.; *Euphorbiac.;* Gab.
Esoma: *Rauwolfia macrophylla* Stapf; *Apocynac.;* Gab.
Eso-matsu: *Pinus glehni* Masters; *Conifer.;* n. Jap.
Esomatzu: *Picea alcockiana* Carr.; *Conifer.;* Jap.
Eson: *Macaranga heudeloti* Baill.; *Euphorbiac.;* Elf.
Esong: ? *Irvingia klainei* Pierre; *Simarubac.;* Gab.
Esongosongo: *Pseudospondias microcarpa* Engl.; *Anacardiac.;* Gab.
Esonousoua: *Lavalleopsis densivenia* Engl.; *Olacac.;* Gab.
Esoré: *Lophira procera* Chev.; *Ochnac.;* Elf.
Esosanga: *Pseudospondias microcarpa* Engl.; *Anacardiac.;* B.-K.
Esoua: *Diospyros aggregata* Gürke; *Ebenac.;* Kam.
 „ : *Saccoglottis gabonensis* Urban; *Humiriac.;* Gab.
Esoula: *Placodiscus pseudostipularis* Radlk.; *Sapindac.;* Gab.
Esouna: *Hymenostegia sp.;* *Caesalpiniac.;* Gab.
Espadeira: *Eperua falcata* Aubl.; *Caesalpiniac.;* Bras.
Espantalobos (s): *Colutea arborescens* L.; *Papilionac.;* M.-M.
Espavá: *Anacardium rhinocarpus* DC.; *Anacardiac.;* Pan.
Espavé: *Anacardium rhinocarpus* DC.; *Anacardiac.;* Pan.
Espavel: *Anacardium rhinocarpus* DC.; *Anacardiac.;* C.-R.
Espe (d): *Populus tremula* L.; *Salicac.;* n. Eu., n. As.
Espejuelo: *Sarcomphalus reticulatus* Urban; *Rhamnac.;* P.-R.
 „ : *Krugiodendron ferreum* Urban; *Rhamnac.;* P.-R.
Espigol: *Lavandula vera* DC.; *Labiat.;* Sp.
Espin de bobo: *Zanthoxylum sp.;* *Rutac.;* fr. W.-Ind.
Espina: ? *Gleditschia amorphoides* Taub.; *Caesalpiniac.;* Cuba.
 „ Christi: *Gleditschia amorphoides* Taub.; *Caesalpiniac.;* Arg.
 „ corona: *Gleditschia amorphoides* Taub.; *Caesalpiniac.;* Arg.
 „ de arroyo: *Cormonema spinosum* Reisseck; *Rhamnac.;* Arg.
 „ „ bañado: *Citharexylum barbinerve* Cham.; *Verbenac.;* n. Arg.
 „ „ bobo: *Zanthoxylum clava-herculis* L.; *Rutac.;* Trin.
 „ „ corona: *Gleditschia amorphoides* Taub.; *Caesalpiniac.;* Arg.
 „ „ „ Christi: *Gleditschia amorphoides* Taub.; *Caesalpiniac.;* Arg.
Espineux blanc: *Zanthoxylum sp.;* *Rutac.;* fr. W.-Ind.
 „ jaune: *Zanthoxylum sp.;* *Rutac.;* fr. W.-Ind.
Espinha de meicha: *Ximenia americana* L.; *Olacac.;* Bras.
Espinheiro branco: *Chlorophora tinctoria* Gaudich.; *Morac.;* Bras.
Espinho roxo: *Piptadenia sp.;* *Mimosac.;* Bras.
 „ de Cristo: *Gleditschia amorphoides* Taub.; *Caesalpiniac.;* Bras.
 „ do vintem: *Zanthoxylum sp.;* *Rutac.;* Bras.
Espinilho: *Gleditschia amorphoides* Taub.; *Caesalpiniac.;* Bras.
Espinillenhout (h): *Prosopis ñandubay* Lorentz; *Mimosac.;* n. Arg.
 „ „ : ? *Pithecolobium vinhatico* Record; *Mimosac.;* Bras., Arg.
Espinillo: *Zanthoxylum flavum* Vahl; *Rutac.;* St.-D.
 „ : *Prosopis ñandubay* Lorentz; *Mimosac.;* n. Arg.
 „ : *Acacia farnesiana* Willd., *A. lutea* N. L. Britton; *Mimosac.;* Arg.
 „ : *Pithecolobium scalare* Griseb.; *Mimosac.;* Arg.
 „ : *Pithecolobium vinhatico* Record; *Mimosac.;* Arg.
 „ : *Gleditschia amorphoides* Taub.; *Caesalpiniac.;* Arg.
 „ amarillo: *Gleditschia amorphoides* Taub.; *Caesalpiniac.;* Arg.
 „ aromita: *Acacia aroma* Gill.; *Mimosac.;* Arg.
 „ ñandubay: *Prosopis sp.;* *Mimosac.;* Arg.
 „ rubial: *Zanthoxylum caribaeum* Lam.; *Rutac.;* P.-R.
Espino: *Crataegus monogyna* Jacq.; *Rosac.;* Sp.
 „ : *Zanthoxylum sp.;* *Rutac.;* P.-R., Cuba.
 „ : *Acacia vera* Willd.; *Mimosac.;* Hond.
 „ : *Pithecolobium dulce* Bth.; *Mimosac.;* Salv.
 „ : *Guettarda foliacea* Standl.; *Rubiac.;* Pan.
 „ : *Piptadenia communis* Bth.; *Mimosac.;* Kol.
 „ : *Bumelia sp.;* *Sapotac.;* Kol.
 „ albar: *Crataegus monogyna* Jacq.; *Rosac.;* Sp.
 „ amarillo: *Hippophaë rhamnoides* L.; *Elaeagnac.;* Sp.

Espino blanco: *Crataegus oxyacantha* L.; *Rosac.;* Sp.
„   „   : *Acacia glomerosa* Bth.; *Mimosac.;* Hond.
„   „   : *Bumelia lankesteri* Standl.; *Sapotac.;* C.-R.
„ cambrón: *Berberis hispanica* Boiss. et Reut.; *Berberidac.;* Sp.
„ cerval: *Rhamnus cathartica* L.; *Rhamnac.;* Sp.
„ madroño: *Calycophyllum candidissimum* DC.; *Rubiac.;* Nic.
„ negro: *Rhamnus oleoides* L., *R. lycioides* L.; *Rhamnuc.;* Sp.
„ prieto: *Rhamnus lycioides* L.; *Rhamnac.;* Sp.
„ rubial: *Fagara caribaea* Kr. et Urb., *F. martinicensis* Lam.; *Rutac.;* P.-R.
„   „   : *Fagara monophylla* Lam., *F. pterota* L.; *Rutac.;* P.-R.
„   „   : *Zanthoxylum ochroxylum* DC.; *Rutac.;* P.-R.
„ sur: *Ormosia panamensis* Bth.; *Papilionac.;* Pan.
„ de playa: *Acacia macracantha* Humb. et Bonpl.; *Mimosac.;* Hond.
„   „   „   : *Pithecolobium dulce* Bth., *P. unguis-cati* Bth.; *Mimosac.;* Nic.
Espinuelo: *Pithecolobium unguis-cati* Bth.; *Mimosac.;* Ven.
Espliego: *Lavandula vera* DC., *L. latifolia* Vill.; *Labiat.;* Sp.
Espuela de caballero: *Belairia speciosa* A. Rich.; *Papilionac.;* Cuba.
„ del diablo: *Pisonia sp.; Nyctaginac.;* Salv.
Espuelita: *Inga microphylla* Salzm.; *Mimosac.;* Ven.
Essabem (f): *Macrolobium aff. limba* Sc. Ell., *Berlinia sp.; Caesalpiniac.;* Kam.
Essabème: *Chytranthus sp.; Sapindac.;* Gab.
Essac: *Albizzia fastigiata* Oliv., *A. welwitschi* Oliv.; *Mimosac.,* Kam.
„   : *Albizzia brownei* Oliv., *A. angolensis* Welw.; *Mimosac.;* Kam.
Essafra: *Eugenia rowlandi* Sprague; *Myrtac.;* Elf.
·? Essai: *Mansonia altissima* Chev.; *Sterculiac.;* Kam.
Essamé: *Haematostaphis barteri* Hook. f.; *Anacardiac.;* Elf.
Essané: *Anthostema aubryanum* Baill.; *Euphorbiac.;* Kam.
Essang: *Ricinodendron africanum* Muell. Arg.; *Euphorbiac.;* Kam., Gab.
„   : *Piptadenia sp.; Mimosac.;* Gab.
Essanga sanga: *Tetrapleura thonningi* Bth.; *Mimosac.;* Kam.
Essarghebem: ? *Chytranthus sp.; Sapindac.;* Gab.
Essasanga: *Ricinodendron africanum* Muell. Arg.; *Euphorbiac.;* Gab.
Essat: *Bridelia micrantha* Baill.; *Euphorbiac.;* Kam.
Esschenhout (h): *Ekebergia capensis* Sparrm.; *Meliac.;* S.-Af. (Natal).
Essè-Hessé: *Tetrapleura thonningi* Bth.; *Mimosac.;* Elf.
Essenboom (h): *Ekebergia capensis* Sparrm.; *Meliac.;* S.-Af.
Essenhout (h): *Ekebergia capensis* Sparrm.; *Meliac.;* S.-Af.
Essenlipen: *Diospyros aggregata* Gürke; *Ebenac.;* Kam.
Essenwood (e): *Ekebergia capensis* Sparrm.; *Meliac.;* S.-Af.
Essessang: *Ricinodendron africanum* Muell. Arg.; *Euphorbiac.;* Kam., Gab.
Essessé: *Tetrapleura thonningi* Bth.; *Mimosac.;* Kam.
Essigbaum (d): *Rhus typhina* L.; *Anacardiac.;* N.-Am.
Essin: *Baphia nitida* Lodd.; *Papilionac.;* Elf.
Essingang: *Didelotia africana* Pierre; *Caesalpiniac.;* Kam.
Essivé: *Petersia viridiflora* Chev.; *Lecythidac.;* Elf.
Esso: *Lophira procera* Chev.; *Ochnac.;* Elf.
„ belia: *Stereospermum kunthianum* Cham.; *Bignoniac.;* Togo.
Essok: *Garcinia punctata* Oliv.; *Guttifer.;* Kam.
Essomba: *Rauwolfia macrophylla* Stapf; *Apocynac.;* Kam.
Essombe: *Rauwolfia macrophylla* Stapf; *Apocynac.;* Kam.
Essombo: *Rauwolfia macrophylla* Stapf; *Apocynac.;* Kam.
Esson: *Oldfieldia africana* Bth. et Hook. f.; *Euphorbiac.;* Elf.
„ angouaran: *Oldfieldia africana* Bth. et Hook. f.; *Euphorbiac.;* Elf.
Essona: *Saccoglottis gabonensis* Urban; *Humiriac.;* Gab.
Essoua: *Saccoglottis gabonensis* Urban; *Humiriac.;* Kam., Gab.
Essoubé: *Sarcocephalus esculentus* Afzel.; *Rubiac.;* Elf.
Essoubo: *Sarcocephalus esculentus* Afzel.; *Rubiac.;* Elf.
Essoula: *Placodiscus pseudostipularis* Radlk.; *Sapindac.;* W.-Af.
Essoulé: *Berlinia sp.; Caesalpiniac.;* Kam.
Essoule: *Tylostemon sp.; Laurac.;* Kam.
Essoun: *Scorodophloeus zenkeri* Harms; *Caesalpiniac.;* Kam., Gab.
„   : *Hua gaboni* Pierre; *Sterculiac.;* Gab.

Essussouk: *Spathodea campanulata* Beauv.; *Bignoniac.;* Kam.
Estepa: *Cistus albidus* L., *C. laurifolius* L.; *Cistac.;* Sp.
Estoragne: *Styrax tomentosum* H. B. K.; *Styracac.;* Ven.
Estoraque: *Liquidambar styraciflua.; L.; Hamamelidac.;* Mex.
Estrella: *Crataeva tapia* L.; *Capparidac.;* Pan.
Estribeira: *Luehea divaricata* Mart.; *Tiliac.;* Bras.
Estribillo: *Trichilia sp.; Meliac.;* Mex.
Estriveira: *Luehea divaricata* Mart.; *Tiliac.;* Bras.
Esui: *Oldfieldia africana* Bth. et Hook. f.; *Euphorbiac.;* Elf.
Esukatia: *Pterocarpus santalinoides* L'Hérit.; *Papilionac.;* Go.
Esuro: *Enantia chlorantha* Oliv.; *Anonac.;* Elf.
Eta-balli: *Vochysia melinoni* Beckmann; *Vochysiac.;* Sur.
   „   „ : *Vochysia curvata* Klotzsch, *V. guianensis* Aubl.; *Vochysiac.;* br. Gu.
Etagenbaum (d): *Terminalia catappa* L.; *Combretac.;* O.-Af., Mad.
Etama: *Pycnanthus kombo* Warb.; *Myristicac.;* Elf.
Etamba: *Mangifera zeylanica* Hook. f.; *Anacardiac.;* Cey.
Etan: *Pycnanthus kombo* Warb.; *Myristicac.;* Gab.
Etang: *Pycnanthus kombo* Warb.; *Myristicac.;* Gab.
Etchoa: *Macaranga heudeloti* Baill.; *Euphorbiac.;* Elf.
Etchui: *Ceiba pentandra* Gaertn.; *Bombacac.;* Elf.
Eteng: *Pycnanthus kombo* Warb.; *Myristicac.;* Kam., Gab.
Ethel: *Tamarix articulata* Vahl; *Tamaricac.;* Arab.
Etheraliya: *Kurrimia zeylanica* Arn.; *Celastrac.;* Cey.
Etié: *Funtumia elastica* Stapf; *Apocynac.;* Kam.
Eting: *Pycnanthus kombo* Warb.; *Myristicac.;* Gab.
Etjannaka: *Diospyros monbuttensis* Gürke; *Ebenac.;* W.-Af. (? Togo).
Etonn: *Conopharyngia brachyantha* Stapf; *Apocynac.;* Kam.
Etotom: *Fillaeopsis discophora* Harms; *Mimosac.;* tr. W.-Af.
Etotum: *Filaeopsis discophora* Harms; *Mimosac.;* tr. W.-Af.
Etou: *Treculia sp.; Morac.;* Gab.
Etoup: *Treculia africana* Dcne.; *Morac.;* Kam.
Etsa: *Erythrophloeum micranthum* Harms; *Caesalpiniac.;* Go.
Etsien: *Alstonia congensis* Engl.; *Apocynac.;* Elf.
? Etsu: *Pycnanthus kombo* Warb.; *Myristicac.;* Go.
Etu: *Irvingia barteri* Hook. f.; *Simarubac.;* Kam.
Etub: *Pterygota kamerunensis* K. Schum.; *Sterculiac.;* Kam.
Etue: *Irvingia barteri* Hook. f.; *Simarubac.;* Kam.
Etui: *Oldfieldia africana* Bth. et Hook. f.; *Euphorbiac.;* Elf.
Ety: *Funtumia africana* Stapf; *Apocynac.;* Gab.
Eucalypt, apple-scented: *Eucalyptus stuartiana* F. v. M.; *Myrtac.;* Vict.
   „   , Giant: *Eucalyptus amygdalina* Labill.; *Myrtac.;* Au.
Eucalyptus vermelho: *Eucalyptus resinifera* Sm.; *Myrtac.;* s. Bras.
Eurabbie: *Eucalyptus globulus* Labill.; *Myrtac.;* N.-S.-W.
Eutié: *Cleistopholis patens* Engl. et Diels; *Anonac.;* Elf.
Evamba: *Placodiscus pseudostipularis* Radlk.; *Sapindac.;* Gab.
Evel: *Klainedoxa trillesi* Pierre; *Simarubac.;* Gab.
Evès: *Klainedoxa gabonensis* Pierre; *Simarubac.;* Gab.
Evess: *Klainedoxa latifolia* Pierre; *Simarubac.;* Gab.
Eveuss: *Klainedoxa latifolia* Pierre; *Simarubac.;* Kam., Gab.
   „ : *Klainedoxa sphaerocarpa* van Tiegh.; *Simarubac.;* Kam., Gab.
Evila: *Diospyros evila* Pierre, *D. flavescens* Gürke; *Ebenac.;* Gab.
Evino: *Vitex pachyphylla* Baker; *Verbenac.;* Kam., Gab.
Evoma: *Hexalobus crispiflorus* A. Rich.; *Anonac.;* Gab.
Evong Evong: *Spathodea campanulata* Beauv.; *Bignoniac.;* Gab.
Evonghelé-Vonghelé: *Spathodea campanulata* Beauv.; *Bignoniac.;* Gab.
Evonlvonlé: *Spathodea campanulata* Beauv.; *Bignoniac.;* Gab.
Evonoué: *Hibiscus tiliaceus* L.; *Malvac.;* Gab.
Evonymo: *Evonymus europaea* L.; *Celastrac.;* Port.
Evou: *Vitex pachyphylla* Baker; *Verbenac.;* Kam.
Evoui: *Rauwolfia sp.; Apocynac.;* Kam.
   „ : *Vitex grandifolia* Gürke; *Verbenac.;* Kam.

Evoula: *Vitex grandifolia* Gürke; *Verbenac.;* Kam.
Evoumanga: *Berlinia sp.; Caesalpiniac.;* Gab.
Evous: *Vitex grandifolia* Gürke; *Verbenac.;* Gab.
Evovonne: *Spathodea campanulata* Beauv.; *Bignoniac.;* Kam.
Ewai: *Ricinodendron africanum* Muell. Arg.; *Euphorbiac.;* Go.
Ewaletombo: *Drypetes paxi* Hutch.; *Euphorbiac.;* Kam.
Ewan: *Ricinodendron africanum* Muell. Arg.; *Euphorbiac.;* Go.
Ewäti: *Mimusops multinervis* Baker; *Sapotac.;* Togo.
Ewawa: *Panda oleosa* Pierre; *Pandac.;* Gab.
Ewé: *Guarea gomma* Pulle; *Meliac.;* Sur.
Eweli somu: *Mimusops kerstingi* Engl., *M. multinervis* Baker; *Sapotac.;* Togo.
Ewoe: *Microdesmis zenkeri* Pax; *Euphorbiac.;* Kam.
Ewoei: *Microdesmis puberula* Hook. f.; *Euphorbiac.;* Kam.
Ewomaé: *Coula edulis* Baill.; *Olacac.;* Kam.
Ewomc: *Coula edulis* Baill.; *Olacac.;* Kam.
Ewomoe: *Coula edulis* Baill.; *Olacac.;* Gab.
Ewotobé: *Allanblackia parviflora* Chev.; *Guttifer.;* Elf.
Ewoui: *Rauwolfia sp.; Apocynac.;* Kam.
Ewoulet: *Bridelia micrantha* Baill.; *Euphorbiac.;* Kam.
   „ m'lon: ? *Oncocalamus sp.; Palmac.;* Kam.
Ewoumé: *Coula edulis* Baill.; *Olacac.;* Gab.
Ewu: *Vitex grandifolia* Gürke, *V. rivularis* Gürke; *Verbenac.;* Kam.
Exoyamagi: *Salix rorida* Lackschewitz; *Salicac.;* Jap.
Eyaban: *Chrysophyllum lacourtianum* de Wild.; *Sapotac.;* Kam.
Eyabe: *Cola ballayi* Cornu; *Sterculiac.;* Kam.
Eyec: *Pachyelasma tesmanni* Harms; *Mimosac.;* Kam.
Eyen: *Distemonanthus benthamianus* Baill.; *Caesalpiniac.;* Kam., Gab.
Eyêne: *Distemonanthus benthamianus* Baill.; *Caesalpiniac.;* Kam., Gab.
Eye-opening tree: *Duboisia myoporoides* R. Br.; *Solanac.;* Au.
Eyeyongo: *Fagara macrophylla* Engl.; *Rutac.;* Kam.
Eyo: *Erythrophloeum guineense* G. Don; *Caesalpiniac.;* Gab.
Eyom: *Dialium aff. connaroides* Harms; *Caesalpiniac.;* Gab.
Eyôme: *Dialium aff. connaroides* Harms; *Caesalpiniac.;* Gab.
Eyon: *Sterculia oblonga* Mast.; *Sterculiac.;* Kam.
Eyonguinaninego: *Psychotria gabonica* Hiern; *Rubiac.;* Gab.
Eyou: *Trichilia gilleti* de Wild.; *Meliac.;* Gab.
Eyoum: *Dialium aff. connaroides* Harms; *Caesalpiniac.;* Gab.
Ezigo: *Pterocarpus soyauxi* Taub.; *Papilionac.;* Gab.
   „ : *Pachylobus büttneri* Engl.; *Burserac.;* Gab.
Ezo-matsu: *Picea ajanensis* Fischer, *P. jezoënsis* Carr.; *Conifer.;* Jap.
Ezoyanagi: *Salix rorida* Lackschewitz; *Salicac.;* Jap.

F

Facheiro: *Xylopia ligustrifolia* Dunal; *Anonac.;* Bras.
Fackelbaum (d): *Pinus silvestris* L.; *Conifer.;* Eu., As., Sib.
Fafaraha: *Malacantha warneckeana* Engl.; *Sapotac.;* Go.
Fafoo laut: *Pygeum maingayi* Hook. f.; *Rosac.;* Malak.
Faggio: *Fagus silvatica* L.; *Fagac.;* It.
Fago: *Fagus silvatica* L.; *Fagac.;* Sp.
   „ : *Hibiscus tiliaceus* L.; *Malvac.;* Südsee (Guam).
Fah: *Placodiscus pseudostipularis* Radlk.; *Sapindac.;* Lib.
Fai: *Pentaclethra macrophylla* Bth.; *Mimosac.;* Lib.
Faisán: *Sideroxylon amygdalinum* Standl.; *Sapotac.;* Guat.
   „ : *Dipholis minutiflora* Pittier; *Sapotac.;* C.-R.
   „ , red: *Calocarpum viride* Pittier; *Sapotac.;* br. Hond.
   „ , white: *Calocarpum viride* Pittier; *Sapotac.;* br. Hond.
   „ , zapote: *Dipholis stevensoni* Standl.; *Sapotac.;* br. Hond.
Faix: *Fagus silvatica* L.; *Fagac.;* Sp.
Fakpo: *Albizzia sp.; Mimosac.;* W.-Af.
Fanam: *Ochna afzelii* R. Br.; *Ochnac.;* Togo.

M e y e r , Buch der Holznamen

Fandamane: *Aphloia theaeformis* Benn.; *Flacourtiac.;* Réu.
Fandaqe-hindi: *Sapindus emarginatus* Vahl; *Sapindac.;* Arab.
Fandrianakanga: *Albizzia boinensis* R. Vig.; *Mimosac.;* Mad.
Fang: *Poinciana pulcherrima* L.; *Caesalpiniac.;* Ind.-Ch.
Fangalitra: *Stereospermum sp.; Bignoniac.;* Mad.
Faoulé: *Irvingia sp.; Simarubac.;* Elf.
„ -Blé: *Irvingia sp.; Simarubac.;* Elf.
„ -kokolé: *Parinarium tenuifolium* Chev.; *Rosac.;* Elf.
Fara doka: *Isoberlinia dalzieli* Craib et Stapf; *Caesalpiniac.;* n. Nig.
„ geza: *Combretum micranthum* G. Don; *Combretac.;* n. Nig.
„ kaya: *Acacia sieberiana* DC.; *Mimosac.;* n. Nig., Go.
Farafata: *Sonneratia alba* Smith; *Sonneratiac.;* Mad.
Faràs: *Tamarix articulata* Vahl; *Tamaricac.;* Ind.
Farash: *Tamarix articulata* Vahl; *Tamaricac.;* nw. O.-Ind. (Punjab).
Farayemile: *Terminalia ivorensis* Chev.; *Combretac.;* Go.
Farayen: *Terminalia superba* Engl. et Diels; *Combretac.;* Go.
Farber-maclura: *Chlorophora tinctoria* Gaudich.; *Morac.;* tr. Am.
Farichin shafo: *Acacia campylacantha* Hochst.; *Mimosac.;* n. Nig.
Farin baure: *Ficus capensis* Thunb.; *Morac.;* n. Nig.
„ rura: *Monotes kerstingi* Gilg; *Dipterocarpac.;* n. Nig.
„ sansame: *Lonchocarpus laxiflorus* Guill. et Perr.; *Papilionac.;* n. Nig.
„ taramniya: *Combretum verticillatum* Engl. et Diels; *Combretac.;* n. Nig.
Farinha seca: *Ouratea castaneaefolia* Engl.; *Ochnac.;* Arg.
Farkleberry: *Vaccinium arboreum* Marsh.; *Ericac.;* tr. Am.
Farnetto: *Quercus conferta* Kit.; *Fagac.;* It.
Farnia: *Quercus pedunculata* Ehrh.; *Fagac.;* It.
Faro: ? *Daniella sp.; Caesalpiniac.;* Elf.
Farri: *Grewia elastica* Royle; *Tiliac.;* nw. O.-Ind. (Punjab).
? Faru: *Odina acida* Walp., *O. barteri* Oliv.; *Anacardiac.;* n. Nig.
Farun doya: *Odina barteri* Oliv.; *Anacardiac.;* n. Nig.
„ mutane: *Odina acida* Walp.; *Anacardiac.;* n. Nig.
Fasazacuru: *Euptelea polyandra* S. et Z.; *Trochodendrac.;* Jap.
Fäsero: *Pavetta crassipes* K. Schum.; *Rubiac.;* Togo.
Faskara giwa: *Ormocarpum bibracteatum* Baker; *Papilionac.;* n. Nig.
Fasteque (f): *Chlorophora tinctoria* Gaudich.; *Morac.;* tr. Am.
Fatajuba: *Bagassa guianensis* Aubl.; *Morac.;* Sur.
Fat pork: *Chrysobalanus icaco* L.; *Rosac.;* tr. Am., W.-Af.
Fau: *Hibiscus tiliaceus* L.; *Malvac.;* Südsee (Samoa, Tahiti).
Fauce: *Fagus silvatica* L.; *Fagac.;* Eu.
Fauh: *Lophira alata* Banks; *Ochnac.;* Lib.
Faulbaum (d): *Rhamnus frangula* L.; *Rhamnac.;* Eu.
Faurestina: *Albizzia lebbek* Bth.; *Mimosac.;* Cuba.
Fava de bezouro: *Cassia xinguensis* Ducke; *Caesalpiniac.;* Bras.
Faveira do matto: *Pithecolobium sp.; Mimosac.;* Bras.
Faveiro: *Pterodon pubescens* Bth.; *Mimosac.;* Bras. (Sao P).
„ amarello: *Pterodon sp.; Mimosac.;* Bras. (Sao P).
„ vermelho: *Pterodon sp.; Mimosac.;* Bras. (Sao P).
„ do Campo: *Pterodon sp.; Mimosac.;* Bras. (Sao P).
„ „ Matto: *Pterodon sp.; Mimosac.;* Bras. (Sao P).
Fay: *Duabanga sonneratioides* Buch.-Ham.; *Sonneratiac.;* Ind.-Ch.
„ chân: *Wrightia annamensis* Eberh. et Dubard; *Apocynac.;* Ind.-Ch.
Faya: *Fagus silvatica* L.; *Fagac.;* Sp.
„ : *Myrica faya* Ait.; *Myricac.;* Madeira.
Fé: *Trichilia candollei* Chev.; *Meliac.;* Elf.
„ : *Dialium guineense* Willd.; *Caesalpiniac.;* Elf.
Fe de gozo: *Trichilia sp.; Meliac.;* Arg.
Feasshur: *Buxus sempervirens* L.; *Buxac.;* Kl.-As.
Feather-bush: *Cercocarpus ledifolius* Nutt.; *Rosac.;* USA.
Feather wood: *Polyosma cunninghami* J. J. Benn.; *Saxifragac.;* ö. Au.
Fegblo: *Cussonia barteri* Seem.; *Araliac.;* Togo.
Feigenbaum, echter (d): *Ficus carica* L.; *Morac.;* s. USA.
„ , wilder (d): *Ficus aurea* Nutt.; *Morac.;* Fl., Cuba, Bah.

Feijãosinho rasteiro: *Canavalia albiflora* Dudke; *Papilionac.;* Bras.
„ da matta: *Calopogonium coeruleum* Bth.; *Papilionac.;* Bras.
Fei-tsao-tou: *Gymnocladus chinensis* Baill.; *Caesalpiniac.;* Mitt.-Ch.
Fekigo: *Sterculia platanifolia* L. f.; *Sterculiac.;* Jap.
Féla: *Terminalia ivorensis* Chev.; *Combretac.;* Elf.
Felsenbirne: *Amelanchier vulgaris* Mönch; *Rosac.;* Ind.
Fenchelholz (d): *Sassafras officinalis* Nees; *Laurac.;* ö. N.-Am.
Fenian: *Strombosia pustulata* Oliv.; *Olacac.;* Elf.
Fenoll de rabosa: *Bupleurum fruticescens* L.; *Umbellifer.;* Sp.
Fer: *Stadmannia sideroxylon* DC.; *Sapindac.;* Maur.
Féranga: *Strombosia pustulata* Oliv.; *Olacac.;* Elf.
Fern tree: *Jagera pseudorhus* Radlk.; *Sapindac.;* Au.
Fernambucco (i): *Caesalpinia echinata* Lam.; *Caesalpiniac.;* ö. Bras.
Fernambucowood (e, u): *Caesalpinia echinata* Lam.; *Caesalpiniac.;* ö. Bras.
Fernambukholz (d): *Caesalpinia echinata* Lam.; *Caesalpiniac.;* ö. Bras.
Fernansánchez: *Triplaris guaiaquilensis* Wedd.; *Polygonac.;* Ek.
Feronier (f): *Feronia elephantum* Roxb.; *Rutac.;* tr. As.
Ferréol: *Swartzia tomentosa* DC.; *Papilionac.;* fr. Gu.
Fersig: *Tamarix gallica* L.; *Tamaricac.;* Sahara.
Fessouba: ? *Phyllanthus diswideus* Muell. Arg.; *Euphorbiac.;* Elf.
Fetid: *Aesculus glabra* Willd.; *Hippocastanac.;* USA. (Ohio).
Fèttèj-ie: *Jacaranda copaïa* D. Don; *Bignoniac.;* Sur.
Feuille rude (f): *Curatella americana* L.; *Dilleniac.;* fr. Gu.
„ à polir (f): *Curatella americana* L.; *Dilleniac.;* fr. Gu.
Fever-bark (e): *Alstonia constricta* F. v. M.; *Apocynac.;* Queensl., N.-S.-W.
Fever-tree (e): *Acacia xanthophloea* Bth.; *Mimosac.;* S.-Af.
Fez: *Thespesia populnea* Correa; *Malvac.;* Go.
Fi: *Torreya nucifera* S. et Z.; *Conifer.;* Jap.
Fibrewood (e): *Laportea photiniphylla* Wedd., *L. gigas* Wedd.; *Urticac.;* ö. Au.
Ficha: *Hippomane mancinella* L.; *Euphorbiac.;* Peru.
Fichte (d): *Picea* spp.; *Conifer.;* n. Hem.
„ : *Picea excelsa* Link; *Conifer.;* Eu.
„ , Blau-: *Picea pungens* Engelm.; *Conifer.;* w. N.-Am.
„ , Douglas-: *Pseudotsuga douglasi* Carr.; *Conifer.;* USA. (Oregon).
„ , Engelmann-: *Picea engelmanni* Engelm.; *Conifer.;* w. USA.
„ , Hasel-: *Picea excelsa* Link; *Conifer.;* Eu.
„ , Hudson-: *Picea rubra* Link; *Conifer.;* ö. Kan.
„ , Kauri-: *Agathis australis* Steud.; *Conifer.;* N.-Seel.
„ , Rot-: *Picea rubra* Link; *Conifer.;* ö. Kan.
„ , Schwarz-: *Picea mariana* B. S. P.; *Conifer.;* Kan.
„ , Sitka-: *Picea sitchensis* Carr.; *Conifer.;* w. N.-Am.
„ , Stech-: *Picea pungens* Engelm.; *Conifer.;* w. N.-Am.
Fiddlewood: *Citharexylum fruticosum* L.; *Verbenac.;* USA., s. Fl., Ant.
„ : *Petitia domingensis* Jacq.; *Verbenac.;* W.-Ind.
„ : *Citharexylum villosum* Jacq.; *Verbenac.;* Fl.
„ : *Citharexylum lucidum* Cham. et Schl.; *Verbenac.;* Cuba.
„ : *Citharexylum quadrangulare* Jacq.; *Verbenac.;* Berm., Trin.
„ : *Vitex divaricata* Sw.; *Verbenac.;* Jam., Trin., Tob.
„ : *Citharexylum melanocardium* Sw.; *Verbenac.;* Tob.
„ : *Citharexylum cinereum* L.; *Verbenac.;* W.-Ind. (St. Lucia).
„ : *Citharexylum surrectum* Griseb.; *Verbenac.;* W.-Ind. (St. Lucia).
„ : *Chimarrhis parviflora* Standl.; *Rubiac.;* Pan.
„ : *Tabebuia pentaphylla* Hemsl.; *Bignoniac.;* br. Hond.
? „ : *Dracaena americana* Donn. Smith; *Liliac.;* br. Hond.
„ : *Vitex longeracemosa* Pittier; *Verbenac.;* br. Hond.
„ , black: *Vitex divaricata* Sw., *V. capitata* Vahl; *Verbenac.;* Trin., Tob.
Fieberbaum (d): *Eucalyptus globulus* Lab., *E. amygdalina* Lab.; *Myrtac.;* Au.
Fieroberoe sokoné: *Helicostylis poeppigiana* Tréc.; *Morac.;* Gu.
Fierrillo: *Eugenia guatemalensis* Donn. Smith; *Myrtac.;* Hond.
Fifei: *Gardenia thunbergi* L. f.; *Rubiac.;* tr. Af.
Fig (e): *Ficus carica* L.; *Morac.;* Tr.
„ : *Ficus* spp., *F. oerstediana* Miq.; *Morac.;* br. Hond.

Fig: *Ficus radula* Willd.; *Morac.;* br. Hond., Pan.
„ : *Ficus colubrinae* Standl., *F. tonduzii* Standl.; *Morac.;* Pan.
„ : *Clusia minor* L.; *Guttifer.;* Pan.
„ , Balsam-: *Clusia rosea* Jacq.; *Guttifer.;* br. W.-Ind.
„ , black: *Ficus macrophylla* Desf.; *Morac.;* ö. Au.
„ , blue: *Elaeocarpus grandis* F. v. M.; *Elaeocarpac.;* ö. Au.
„ , Cherry-: *Ficus pedunculata* Willd.; *Morac.;* USA.
„ , common: *Ficus carica* L.; *Morac.;* s. USA.
„ , Country-: *Sarcocephalus esculentus* Afz.; *Rubiac.;* Lib.
„ , Creek-: *Ficus stephanocarpa* Warb.; *Morac.;* ö. Au.
„ , golden: *Ficus aurea* Nutt.; *Morac.;* Fl., Cuba, Bah.
„ , Illawarra-: *Ficus rubiginosa* Desf.; *Morac.;* Au.
„ , Johnstone River-, ribbed: *Ficus pleurocarpa* F. v. M.; *Morac.;* Queensl.
„ , Moreton Bay-: *Ficus macrophylla* Desf.; *Morac.;* Au., N.-S.-W.
„ , „ „ -, green-leaved: *Ficus watkinsiana* F. M. Bailey; *Morac.;* ö. Au.
„ , „ „ -, small-leaved: *Ficus platypoda* A. Cunn.; *Morac.;* ö. Au.
„ , Port Jackson-: *Ficus rubiginosa* Desf.; *Morac.;* sö. Au.
„ , Purple-: *Ficus aspera* Forst.; *Morac.;* Queensl., N.-S.-W.
„ , rough: *Ficus aspera* Forst.; *Morac.;* Queensl., N.-S.-W.
„ , rusty: *Ficus rubiginosa* Desf.; *Morac.;* sö. Au.
„ , Sandpaper-: *Ficus stenocarpa* F. v. M., *F. stephanocarpa* Warb.; *Morac.;* sö. Au.
„ , small-leaved: *Ficus eugenioides* F. v. M.; *Morac.;* sö. Au.
„ , Strangler-: *Ficus involuta* Miq.; *Morac.;* Pan.
„ , strangling: *Ficus sp.; Morac.;* br. Hond.
„ , Sycamore-: *Ficus sycomorus* L.; *Morac.;* B.-K.
„ , white: *Ficus infectoria* Roxb.; *Morac.;* sö. Au.
„ , wild: *Ficus brevifolia* Nutt.; *Morac.;* tr. Am.
„ , „ : *Ficus glabrata* H. B. K.; *Morac.;* br. Hond., Hond.
„ , „ : *Ficus oerstediana* Miq.; *Morac.;* br. Hond.
„ , „ : *Ficus radula* Willd., *Ficus spp.; Morac.;* br. Hond.
„ , „ : *Sapium jamaicense* Sw.; *Euphorbiac.;* Pan.
„ , „ : *Ficus colubrina* Standl., *F. glabrata* H. B. K.; *Morac.;* Pan.
„ , „ : *Sycomorus capensis* Miq.; *Morac.;* Kap.
„ , „ : *Ficus spp.; Morac.;* Phil.
„ tree: ? *Ficus vogeliana* Miq.; *Morac.;* Lib.
„ „ , small: *Ficus thonningi* Blume; *Morac.;* Lib.
„ „ , wild: *Ficus natalensis* Hochst.; *Morac.;* Kap.
Figeira: *Ficus doliaria* Mart.; *Morac.;* Bras.
Figue du pays (f): *Sarcocephalus esculentus* Afzel.; *Rubiac.;* Elf.
Figueira brava: *Ficus anthelminthica* Mart.; *Morac.;* Bras.
Figuer: *Ficus sp.; Morac.;* fr. W.-Ind.
Figuera borda: *Rhamnus alpina* L.; *Rhamnac.;* Sp.
Figuier (f): *Ficus carica* L.; *Morac.;* s. USA.
„ : *Ficus sp.; Morac.;* Guad.
„ : *Ficus teloukat* Battand. et Trab.; *Morac.;* Sahara.
„ doré: *Ficus aurea* Nutt.; *Morac.;* Fl., Cuba, Bah.
„ faux: *Guettarda speciosa* L.; *Rubiac.;* N.-Kal.
„ gros: *Ficus sp.; Morac.;* fr. W.-Ind.
„ marron: *Clusia rosea* Jacq.; *Guttifer.;* Ant.
„ maudit: *Clusia rosea* Jacq.; *Guttifer.;* Ant.
„ petites feuilles: *Ficus sp.; Morac.;* fr. W.-Ind.
Figuiera brava: *Apodytes dimidiata* E. Meyer; *Olacac.;* port. Af.
Figured tree: *Pinus palustris* Miller; *Conifer.;* s. & sö. N.-Am.
Figwood: *Ficus watkinsiana* F. M.Bailey; *Morac.;* sö. Au.
Filao: *Casuarina equisetifolia* L.; *Casuarinac.;* Senl., Mad.
„ : *Casuarina spp.; Casuarinac.;* Réu.
„ : *Casuarina equisetifolia* Forst.; *Casuarinac.;* Ind.-Ch.
„ du Sénégal: *Casuarina equisetifolia* L.; *Casuarinac.;* Senl.
Filaos: *Casuarina sp.; Casuarinac.;* Senl., Ind.-Ch.
Fimbo-ya-mtume: *Asparagus sp.; Liliac.;* D.-O.-Af. (Tabora).
Findi: *Lavalleopsis densivenia* Engl.; *Olacac.;* Kam.
Fingerkaut, strauchiges (d): *Potentilla fruticosa* L.; *Rosac.;* Eu. (Pyr.)

Finzan: *Blighia sapida* Koenig; *Sapindac.;* Elf.
Fir (e): *Abies spp.; Conifer.;* n. Hem.
„ : *Abies balsamea* Miller; *Conifer.;* n. N.-Am.
„ , Alpine-: *Abies lasiocarpa* Nutt.; *Conifer.;* w. Kan.
„ , „ : *Abies amabilis* Forbes; *Conifer.;* w. Kan.
„ , Amabilis-: *Abies amabilis* Forbes; *Conifer.;* w. Kan.
„ , Balm of Gilead-: *Abies balsamea* Miller; *Conifer.;* USA.
„ , Balsam-: *Abies balsamea* Miller; *Conifer.;* ö. N.-Am.
„ , „ -, western: *Abies lasiocarpa* Nutt.; *Conifer.;* w. Kan.
„ , canadian: *Abies balsamea* Miller; *Conifer.;* Kan.
„ , Caribou-: *Abies lasiocarpa* Nutt.; *Conifer.;* w. Kan.
„ , Douglas-: *Pseudotsuga taxifolia* Britton; *Conifer.;* w. USA.
„ , -: *Pseudotsuga macrocarpa* Mayr; *Conifer.;* w. USA. (Kalif.)
„ , „ -, British Columbia: *Pseudotsuga douglasi* Carr.; *Conifer.;* Kan.
„ , „ -, canadian: *Pseudotsuga douglasi* Carr.; *Conifer.;* Kan.
„ , himalayan: *Abies pindrow* Spach; *Conifer.;* n. Ind.
„ , indian: *Polyalthia longifolia* Bth. et Hook. f.; *Anonac.;* Ind., Cey.
„ , Lowland-: *Abies grandis* Lindl.; *Conifer.;* sw. Kan.
„ , Mountain-: *Abies lasiocarpa* Nutt.; *Conifer.;* w. Kan.
„ , red: *Abies amabilis* Forbes; *Conifer.;* w. Kan.
„ , „ : *Pseudotsuga taxifolia* Britton, *P. macrocarpa* Mayr; *Conifer.;* w. USA.
„ , scotch: *Pinus silvestris* L.; *Conifer.;* Eu.
„ , Silver-: *Abies pectinata* DC.; *Conifer.;* Eu.
„ , „ -: *Pinus halepensis* Miller; *Conifer.;* M.-M.
„ , „ -: *Abies grandis* Lindl.; *Conifer.;* w. N.-Am.
„ , „ -: *Abies pindrow* Spach, *A. webbiana* Lindl.; *Conifer.;* Ind.
„ , „ -, american: *Abies balsamifera* Michx. f.; *Conifer.;* ö. N.-Am.
„ , „ -, himalayan (e): *Abies pindrow* Spach, *A. webbiana* Lindl.; *Conifer.;* Ind.
„ , white: *Abies grandis* Lindl., *A. lasiocarpa* Nutt.; *Conifer.;* w. Kan.
„ , „ : *Abies concolor* Lindl.; *Conifer.;* w. USA., n. Mex.
Fire tree: *Stenocarpus sinuatus* Endl.; *Dilleniac.;* N.-S.-W.
Firnisbaum (d): *Ailanthus glandulosa* Desf.; *Simarubac.;* ö. As.
„ „ : *Rhus vernicifera* DC.; *Anacardiac.;* Ch.
Fiselholz (d): *Chlorophora tinctoria* Gaudich.; *Morac.;* tr. Am.
Fisetholz (d): *Chlorophora tinctoria* Gaudich.; *Morac.;* tr. Am.
Fisettholz (d): *Rhus cotinus* L.; *Anacardiac.;* s. Eu.
Fish-poison (e): *Ichthyomethia piscipula* Hitch.; *Papilionac.;* Bah.
Fitsiriky: *Neodypsis gracilis* H. Jum.; *Palmac.;* Mad.
Fiú: *Calocarpum mammosum* Pierre; *Sapotac.;* C.-R.
Fiume: *Lomatia ferruginea* R. Br.; *Proteac.;* S.-Am. (Anden).
Flambeau rouge: *Guenetia macrosperma* Sagot; *Tiliac.;* fr. Gu.
Flamboyan: *Poinciana regia* Bojer; *Caesalpiniac.;* P.-R.
Flamboyant (f): *Poinciana regia* Bojer; *Caesalpiniac.;* Tr.
„ : *Caesalpinia pulcherrima* Sw.; *Caesalpiniac.;* fr. Gu.
„ blanco: *Bauhinia kappleri* Sagot; *Caesalpiniac.;* P.-R.
Flame tree (e): *Sterculia acerifolia* A. Cunn.; *Sterculiac.;* sö. Au.
Flan-chu: *Caloncoba echinata* Gilg; *Flacourtiac.;* Lib.
Flaschenbaum (d): *Anona glabra* L.; *Anonac.;* Fl., Bah.
Flatcrown: *Albizzia gummifera* C. A. Smith; *Mimosac.;* s. C.-Af.
„ : *Albizzia fastigiata* Oliv.; *Mimosac.;* S.-Af.
Flat-pod tree: *Platypodium maxonianum* Pittier; *Papilionac.;* Pan., Kol.
Fleur jaune (f): *Hypericum lanceolatum* Lamk.; *Guttifer.;* Réu.
„ de Paradis (f): *Caesalpinia pulcherrima* Sw.; *Caesalpiniac.;* fr. Gu.
Flieder (d): *Syringa vulgaris* L.; *Oleac.;* Eu., Pers.
Fliegenholz (d): *Quassia amara* L.; *Simarubac.;* S.-Am.
Flindersia: *Flindersia australis* R. Br.; *Rutac.;* sö. Au.
Flindosa: *Flindersia australis* R. Br.; *Rutac.;* Queensl.
Flindosy: *Flindersia australis* R. Br.; *Rutac.;* sö. Au.
Flint bark: ? *Diospyros sp.; Ebenac.;* Go.
Flor amarillo: ? *Tecoma spectabilis* Pl. et Linden; *Bignoniac.;* Ven.
Flor azul: *Vitex kuyleni* Standl.; *Verbenac.;* Hond.
„ de corazón: *Talauma sp.; Magnoliac.;* Mex.

Flor de muerto: *Gustavia yaracuyensis* Pittier; *Lecythidac.;* Ven.
„ „ Palomita: *Bourreria huanita* Hemsl.; *Borraginac.;* Mex.
„ „ papagallo: *Ichthyomethia piscipula* Hitch.; *Papilionac.;* Mex.
„ „ rayo: *Parkinsonia aculeata* L.; *Caesalpiniac.;* s. USA., W.-Ind., Bah.
„ „ seda: *Crataeva tapia* L.; *Capparidac.;* Arg.
„ „ „ : *Calliandra parvifolia* Speg.; *Mimosac.;* Arg.
„ „ venadillo: *Swietenia sp.; Meliac.;* Mex.
„ del Angel: *Caesalpinia pulcherrima* Sw.; *Caesalpiniac.;* Peru.
„ „ indio: *Caesalpinia gilliesi* Wall.; *Caesalpiniac.;* Arg.
Floral: *Sapium aucuparium* Jacq.; *Euphorbiac.;* Kol.
Flußholz (d): *Populus nigra* L.; *Salicac.;* Sib.
Foa: *Amyris elemifera* L.; *Rutac.;* P.-R.
„ : *Erithalis fruticosa* L.; *Rubiac.;* P.-R.
Foambark: *Jagera pseudorhus* Radlk.; *Sapindac.;* sö. Au.
Fobo: *Carapa obovata* Blume; *Meliac.;* Mad.
Foby: *Carapa obovata* Blume; *Meliac.;* Mad.
Foengoe: *Parinarium campestre* Aubl., *P. spp.; Rosac.;* Sur.
„ : *Licania pachystachya* Kleinh., *L. micrantha* Miq.; *Rosac.;* Sur.
„ hoedoe: *Hernandia guianensis* Aubl.; *Hernandiac.;* Gu.
„ „ : *Hirtella caudata* Kleinh., *H. americana* Aubl.; *Rosac.;* Sur.
„ „ : *Hirtella hirsuta* Lam.; *Rosac.;* Sur.
„ „ : *Parinarium campestre* Aubl., *? P. brachystachyum* Bth.; *Rosac.;* Sur.
Foengoepau: *Parinarium campestre* Aubl., *? P. brachystachyum* Bth.; *Rosac.;* Sur.
„ : *Hirtella caudata* Kleinh., *H. americana* Aubl.; *Rosac.;* Sur.
„ : *Hirtella hirsuta* Lam.; *Rosac.;* Sur.
Foengoe-pau: *Licania micrantha* Miq., *L. pachystachya* Kleinh.; *Rosac.;* Sur.
Foete-ie: *Jacaranda copaia* D. Don; *Bignoniac.;* Sur.
Fofoi: *Myrianthus serratus* Bth. et Hook. f.; *Morac.;* Lib.
Fogel kop: *Sideroxylon sp.; Sapotac.;* br. Gu.
Fognian: *Strombosia pustulata* Oliv.; *Olacac.;* Elf.
Föhre (d): *Pinus silvestris* L.; *Conifer.;* Eu., w. As., Sib.
„ , Schwarz-: *Pinus nigra* Arnold; *Conifer.;* Eu. (Alpen, Karp.)
„ , Weiss-: *Pinus silvestris* L.; *Conifer.;* Eu.
Fohre (d): *Pinus silvestris* L.; *Conifer.;* Eu., w. As., Sib.
Foji: *Securidaca longipedunculata* Fresen.; *Polygalac.;* Togo.
Fok: *Afzelia africana* Smith; *Caesalpiniac.;* Sudan.
Fola: *Ficus exasperata* Vahl; *Morac.;* Togo.
Folha larga: *Phoebe porosa* Mez; *Laurac.;* Bras. (Sao P).
„ de Bolo: *Platycyamus regnelli* Bth.; *Papilionac.;* Bras.
„ „ comminâo: *Duroia saccifera* Hook. f.; *Rubiac.;* Bras.
Folhas róxas: *Tecoma sp.; Bignoniac.;* Bras. (Sao P).
Fondé: *Xylopia aethiopica* A. Rich.; *Anonac.;* Elf.
Fonian: *Strombosia pustulata* Oliv.; *Olacac.;* Elf.
Fô noki: *Magnolia hypoleuca* S. et Z.; *Magnoliac.;* Jap.
Fontolo: *Protium sessiliflorum* Standl.; *Burserac.;* Hond.
Fooraha: *Calophyllum inophyllum* L.; *Guttifer.;* Mad.
Forche (d): *Pinus silvestris* L.; *Conifer.;* Eu., w. As.
Forestina: *Albizzia lebbek* Bth.; *Mimosac.;* Cuba.
Forked leaf: *Quercus catesbaei* Michx.; *Fagac.;* ö. USA.
Forle (d): *Pinus silvestris* L.; *Conifer.;* Eu., w. As.
Formigueira: *Triplaris sp.; Polygonac.;* Bras.
Formosa: *Prunus triflora* Roxb. var. ?; *Rosac.;* w. USA (Kalif.).
Fosforito: *Protium copal* Engl.; *Burserac.;* Nic.
Fosi: *Securidaca longipedunculata* Fresen.; *Polygalac.;* Togo.
Fo-ti: *Vitex cienkowskyi* Kotschy et Peyr.; *Verbenac.;* Togo.
Fotie: *Cleistopholis patens* Bth.; *Anonac.;* Go.
Foto: *Bombax buonopozense* Beauv.; *Bombacac.;* Togo.
Fotui: *Jacaranda copaia* D. Don; *Bignoniac.;* br. Gu.
Fou: *Oldfieldia africana* Bth. et Hook. f.; *Euphorbiac.;* Elf.
Fougères arborescentes: *Cyathea spp.; Cyatheac.;* Réu.
Foum: *Chrysophyllum autranianum* Chev.; *Sapotac.;* Kam.
Foundi: *Scorodophloeus zenkeri* Harms; *Caesalpiniac.;* Kam.

Fouon: *Acacia arabica* Willd.; *Mimosac.;* Sudan.
Fourha: *Calophyllum inophyllum* L.; *Guttifer.;* Mad.
Foxwood, red: ? *Dalbergia retusa* Hemsl.; *Papilionac.;* C.-Am.
Foyard: *Fagus silvatica* L.; *Fagac.;* Eu.
Frafrah: *Millettia stapfiana* Dunn; *Papilionac.;* Go.
Fragna: *Quercus macedonica* A. DC.; *Fagac.;* It.
Frahé: *Lophira procera* Chev.; *Ochnac.;* Elf.
Frailecillo: *Licania sp.;* *Rosac.;* Mex.
Fraké: *Terminalia altissima* Chev.; *Combretac.;* Elf., Kam.
„ : *Terminalia superba* Engl. et Diels; *Combretac.;* Elf., Kam.
Frakuan: *Daniella oblonga* Oliv.; *Caesalpiniac.;* Elf.
Fram: *Terminalia altissima* Chev.; *Combretac.;* Elf.
„ : *Terminalia superba* Engl. et Diels; *Combretac.;* Go.
Frambueso: *Rubus idaeus* L.; *Rosac.;* Sp.
Frameri: *Terminalia ivorensis* Chev.; *Combretac.;* Go.
„ : *Terminalia superba* Engl. et Diels; *Combretac.;* Go.
Framiré (f): *Terminalia ivorensis* Chev.; *Combretac.;* Elf.
Framo: *Terminalia superba* Engl. et Diels; *Combretac.;* Go.
Franchipanier: *Plumiera sp.;* *Apocynac.;* fr. W.-Ind.
„ rose: *Plumiera sp.;* *Apocynac.;* fr. W.-Ind.
Francisco Alvarez: *Luehea divaricata* Mart.; *Tiliac.;* Arg.
Frane: *Terminalia superba* Engl. et Diels; *Combretac.;* Go.
Frangipani, native: *Hymenosporum flavum* F. v. M.; *Pittosporac.;* Au.
Frangipanier: *Plumiera sp.;* *Apocynac.;* Cuba.
Frangipanni: *Plumiera sp.;* *Apocynac.;* br. W.-Ind.
Fransi mopé: *Spondias sp.;* *Anacardiac.;* Sur.
Franzosenholz (d): *Guaiacum officinale* L.; *Zygophyllac.;* W.-Ind., C.-Am., n. S.-Am.
Frash: *Tamarix articulata* Vahl; *Tamaricac.;* nw. O.-Ind. (Punjab).
Frassine: *Fraxinus excelsior* L.; *Oleac.;* It.
Fref: *Thespesia populnea* Correa; *Malvac.;* Go.
Frefi: *Thespesia populnea* Correa; *Malvac.;* Go.
Freijo: *Cordia goeldiana* Huber; *Borraginac.;* Bras. (Amaz.).
Frei-jo: *Cordia goeldiana* Huber; *Borraginac.;* Bras. (Amaz.).
Frei-jorge: *Cordia goeldiana* Huber; *Borraginac.;* Bras. (Amaz.).
Frêne (f): *Fraxinus spp.;* *Oleac.;* n. Hem.
„ : *Storckiella pancheri* Baill.; *Caesalpiniac.;* N.-Kal.
„ anguleux: *Fraxinus quadrangulata* Michx.; *Oleac.;* sö. Kan.
„ blanc: *Fraxinus americana* L.; *Oleac.;* ö. Kan.
„ épineux: *Zanthoxylum fraxineum* Willd.; *Rutac.;* USA., Mex.
„ franc: *Fraxinus americana* L.; *Oleac.;* ö. Kan.
„ noir: *Fraxinus nigra* Marsh.; *Oleac.;* sö. Kan.
„ oxyphylle: *Fraxinus oxyphylla* Bieb.; *Oleac.;* Marok.
„ piquant (u): *Zanthoxylum sp.;* *Rutac.;* W.-Ind.
„ puant: *Ailanthus glandulosa* Desf.; *Simarubac.;* Moluk., Jap.
„ pubescent: *Fraxinus pennsylvanica* Marsh.; *Oleac.;* C.-Kan.
„ rouge: *Fraxinus pennsylvanica* Marsh.; *Oleac.;* C.-Kan.
„ vert: *Fraxinus pennsylvanica* Marsh.; *Oleac.;* ö. Kan.
„ à feuilles de sureau: *Fraxinus nigra* Marsh.; *Oleac.;* sö. Kan.
„ de savanne: *Fraxinus pennsylvanica* Marsh.; *Oleac.;* C.-Kan.
Fresno: *Rhus juglandifolia* H. B. K.; *Anacardiac.;* Kol.
„ amargo: ? *Picraena excelsa* Lindl.; *Simarubac.;* Kol.
„ comun: *Fraxinus excelsior* L.; *Oleac.;* Sp.
„ de America: *Chlorophora tinctoria* Gaudich.; *Morac.;* Cuba.
Friega-plato: *Solanum umbellatum* Mill.; *Solanac.;* Hond.
Frijol de mico: *Pithecolobium sophorocarpum* Bth.; *Mimosac.;* Guat.
„ „ palo: *Cajanus indicus* Spreng.; *Papilionac.;* Tr.
Frijolillo: *Gliricidia platycarpa* Griseb.; *Papilionac.;* Cuba.
„ : *Pithecolobium arboreum* Urban; *Mimosac.;* Mex.
„ : *Poeppigia procera* Presl; *Caesalpiniac.;* Salv.
„ : ? *Andira excelsa* H. B. K.; *Papilionac.;* Guat.
„ : *Astronium sp.;* *Anacardiac.;* Pan.
„ amarillo: *Lonchocarpus latifolius* H. B. K.; *Papilionac.;* Cuba.

Frimu: *Khaya senegalensis* A. Juss.; *Meliac.;* Togo.
„ -abalu: *Ekebergia senegalensis* A. Juss.; *Meliac.;* Togo.
Fringe-tree: *Chionanthus virginica* L.; *Oleac.;* USA.
Fromage de Hollande: *Bombax malabaricum* DC.; *Bombacac.;* tr. As., n. Au.
Fromager (f): *Ceiba pentandra* Gaertn.; *Bombacac.;* Tr.
„ à kapok: *Ceiba pentandra* Gaertn.; *Bombacac.;* tr. As., W.-Af.
„ du Soudan: *Ceiba pentandra* Gaertn.; *Bombacac.;* W.-Af.
Fromagier: *Ceiba pentandra* Gaertn.; *Bombacac.;* Trin.
Frotón: *Protium sessiliflorum* Standl.; *Burserac.;* Hond.
Fructa de Arara: *Joannesia princeps* Vell.; *Euphorbiac.;* Bras. (Sao P).
„ „ pomba: *Erythroxylon sp.; Erythroxylac.;* Bras.
Fructeira de burro: *Capparis sp.; Capparidac.;* Bras.
Fruit défendu: *Citrus decumana* L.; *Rutac.;* Guad.
Fruta dorada: *Virola panamensis* Warb.; *Myristicac.;* Pan.
„ de chacha: *Callicarpa acuminata* H. B. K.; *Verbenac.;* Guat.
„ „ conde: *Anona sp.; Anonac.;* Bras.
„ „ danto: *Cynometra retusa* Britt. et Rose; *Caesalpiniac.;* Hond.
„ „ mono: *Posoqueria latifolia* Roem. et Schult.; *Rubiac.;* Pan.
„ „ paloma: *Rudgea hostmanniana* Bth.; *Rubiac.;* Ven.
„ „ „ : *Trema micrantha* Blume; *Ulmac.;* Arg.
„ „ tucano: *Vochysia tucanorum* Mart.; *Vochysiac.;* Bras.
? Fruto del diablo: *Rauwolfia macrocarpa* Standl.; *Apocynac.;* Pan.
Fu: *Oldfieldia africana* Bth. et Hook. f.; *Euphorbiac.;* Elf.
Fuchsbaum (d): *Ulmus fulva* Michx.; *Ulmac.;* ö. N.-Am.
Fuchsrebe (d): *Vitis labrusca* Michx.; *Vitac.;* N.-Am.
Fugáyong: *Cassia javanica* L.; *Caesalpiniac.;* Phil. (Cag.).
Fuji: *Millettia floribunda* Matsum.; *Papilionac.;* Jap.
Fukadi: *Terminalia buceras* Hook.; *Combretac.;* br. Gu.
Fuka-noki: *Schefflera octophylla* Harms; *Araliac.;* Jap.
Fullài: *Acacia seyal* Delile var. *fistula* Oliv.; *Mimosac.;* O.-Af. (it. Somali).
Funera: *Dalbergia lineata* Pittier; *Papilionac.;* C.-Am.
Funga nyumba: *Dichrostachys nutans* Bth.; *Mimosac.;* D.-O.-Af. (Kilossa).
Funtum: *Funtumia elastica* Stapf; *Apocynac.;* Go.
Funtumia rubber tree: *Funtumia elastica* Stapf; *Apocynac.;* Go.
Furu bukoku: ? *Ficus vallis-choudae* Delile; *Morac.;* Togo.
„ kissem: *Ficus capensis* Thunb.; *Morac.;* Togo.
Furze (e): *Ulex europaeus* L.; *Papilionac.;* Eu.
Fusaggine: *Evonymus europaea* L.; *Celastrac.;* It.
Fusain d'Europe (f): *Evonymus europaea* L.; *Celastrac.;* Fr.
Fusazakura: *Euptelea polyandra* S. et Z.; *Trochodendrac.;* Jap.
Fusi-noki: *Chamaecyparis obtusa* S. et Z.; *Conifer.;* Eu., w. As.
Fusteque (f): *Chlorophora tinctoria* Gaudich.; *Morac.;* tr. Am.
Fustet: *Chlorophora tinctoria* Gaudich.; *Morac.;* Elf.
Fustete: *Chlorophora tinctoria* Gaudich.; *Morac.;* tr. Am.
Fustic (e): *Chlorophora tinctoria* Gaudich.; *Morac.;* tr. Am.
„ , old: *Chlorophora tinctoria* Gaudich.; *Morac.;* tr. Am.
„ , young: *Rhus cotinus* L.; *Anacardiac.;* M.-M.
Fustick: *Chlorophora tinctoria* Gaudich.; *Morac.;* tr. Am.
Fustik (d): *Chlorophora tinctoria* Gaudich.; *Morac.;* tr. Am.
Fustik, alter (d): *Chlorophora tinctoria* Gaudich.; *Morac.;* tr. Am.
Fustteholz: *Chlorophora tinctoria* Gaudich.; *Morac.;* tr. Am.
Fustuk: *Pistacia vera* Mill.; *Anacardiac.;* s. Eu.
Futeiba: *Chlorophora tinctoria* Gaudich.; *Morac.;* tr. Am.
Futi: *Jacaranda copaia* D. Don; *Bignoniac.;* Gu.
Futu: *Barringtonia speciosa* Forst.; *Lecythidac.;* Südsee (Tonga).
Fuzibaside: *Engelhardtia formosana* Hayata; *Juglandac.;* Form.

G

Gab: *Diospyros peregrina* Gürke; *Ebenac.;* Ind.
Gaba chara: *Acacia dalzieli* Craib; *Mimosac.;* n. Nig.

Gabaruwa: *Acacia arabica* Willd.; *Mimosac.;* n. Nig.
Gab-boh: *Macrolobium macrophyllum* Macbr.; *Caesalpiniac.;* Lib.
Gabellón: *Phyllanthus sp.; Euphorbiac.;* Kol.
Gabilán: *Schizolobium parahybum* Blake; *Caesalpiniac.;* Nic.
Gaboon (e): *Aucoumea klaineana* Pierre; *Burserac.;* Gab.
Gabur: *Acacia arabica* Willd.; *Mimosac.;* ö. Beng.
Gada lopong: *Trewia nudiflora* L.; *Euphorbiac.;* Beng.
  „ sigric: *Salix tetrasperma* Roxb.; *Salicac.;* Beng.
Gadapad: *Erythrina suberosa* Roxb.; *Papilionac.;* O.-Ind.
Gadava: *Careya arborea* Roxb.; *Lecythidac.;* ö. O.-Ind. (Madras).
Gadda: *Cordia macleodi* Hook. f. et Thoms.; *Borraginac.;* O.-Ind. (Jeypore).
Gadha palas: *Erythrina suberosa* Roxb.; *Papilionac.;* O.-Ind. (Nimar).
Gadichora: *Erythrina suberosa* Roxb.; *Papilionac.;* C.-O.-Ind. (Merwara).
Gading: *Hunteria corymbosa* Roxb.; *Apocynac.;* br. Malay.
Gadkinu: *Acer pictum* Thunb.; *Acerac.;* n. O.-Ind. (Garhwal).
Gadoeng: *Premna tomentosa* Willd.; *Verbenac.;* Ind., Cey., Java.
Gadoengan: *Premna tomentosa* Willd.; *Verbenac.;* Ind., Cey., Java.
Gadog: *Bischofia javanica* Blume; *Euphorbiac.;* Java.
Gadok: *Bischofia javanica* Blume; *Euphorbiac.;* Sum.
Gadru: *Cordia macleodi* Hook. f. et Thoms.; *Borraginac.;* O.-Ind. (Ajmere).
Gadzawuwu: *Acacia suma* Kurz; *Mimosac.;* Togo.
Gafalholz (d): *Commiphora erythraea* Engl.; *Burserac.;* Dahlak.
Gagelstrauch (d): *Myrica gale* L.; *Myricac.;* n. Hem.
Gaggaia: *Robinia pseudacacia* L.; *Papilionac.;* It.
Gagil: *Koordersiodendron pinnatum* Merr.; *Anacardiac.;* Phil. (Zamb.)
  „ : *Parashorea plicata* Brandis; *Dipterocarpac.;* br. N.-Bor.
  „ : *Pentacme contorta* Merr. et Rolfe; *Dipterocarpac.;* br. N.-Bor.
  „ : *Hopea acuminata* Merr., *H. spp.; Dipterocarpac.;* br. N.-Bor.
Gahta: *Sterculia villosa* Roxb.; *Sterculiac.;* Burma.
Gai bóm: *Scolopia chinensis* Clos; *Flacourtiac.;* Ind.-Ch.
Gaïac: *Guaiacum spp.; Zygophyllac.;* tr. Am., W.-Ind.
  „ : *Dipteryx odorata* Willd.; *Papilionac.;* Sur., fr. Gu.
  „ : *Intsia bijuga* O. Ktze.; *Caesalpiniac.;* Mad.
  „ blanc: *Guaiacum officinale* L.; *Zygophyllac.;* W.-Ind., n. S.-Am.
  „ brun: *Dialium cochinchinense* Pierre; *Caesalpiniac.;* Co.-Ch.
  „ faux: *Acacia spirorbis* Labill.; *Mimosac.;* N.-Kal.
  „ franc: *Dipteryx odorata* Willd.; *Papilionac.;* Sur.
  „ noir: *Guaiacum officinale* L.; *Zygophyllac.;* Haiti.
  „ de Cayenne: *Dipteryx odorata* Willd.; *Papilionac.;* fr. Gu.
  „ du Chile: *Porliera hygrometrica* Ruiz et Pav.; *Zygophyllac.;* Chile.
Gaiacholz (d): *Dipteryx odorata* Willd.; *Papilionac.;* n. S.-Am.
Gaiachout (h): *Dipteryx odorata* Willd.; *Papilionac.;* tr. Am.
Gaik: *Pterospermum acerifolium* Willd.; *Sterculiac.;* Burma.
Gainier du Canada (f): *Cercis canadensis* L.; *Caesalpiniac.;* Kan.
Gair: *Olea glandulifera* Wall.; *Oleac.;* O.-Ind.
Gaita: *Exothea paniculata* Radlk.; *Sapindac.;* s. Fl., Bah., Ant.
  „ : *Hypelate paniculata* Cambess.; *Sapindac.;* P.-R.
? „ : *Trichilia sp.; Meliac.;* P.-R.
Gakakan: *Cyclostemon spp.; Euphorbiac.;* Phil. (Cag.).
Gal mendora: *Cynometra ramiflora* L.; *Caesalpiniac.;* Ind.
Gal-siyambala: *Dialium ovoideum* Thw.; *Caesalpiniac.;* Cey.
Gala-gala: *Buxus macowani* Oliv., *Notobuxus natalensis* Oliv.; *Buxac.;* S.-Af.
Galarigal: *Maba buxifolia* Pers.; *Ebenac.;* tr. Af., Au.
Galba: *Calophyllum calaba* Jacq., *C. antillanum* Britton; *Guttifer.;* Trin., Guad., Tob.
Galeh: *Nauclea grandifolia* Bl.; *Rubiac.;* Java.
Gali: *Randia dumetorum* Lamk.; *Rubiac.;* Beng.
Galimetaholz (d): *Dipholis salicifolia* A. DC.; *Sapotac.;* W.-Ind.
Galing libor: *Amoora wallichi* King; *Meliac.;* Assam.
Galingasing: *Amoora rohituka* Wight et Arn.; *Meliac.;* Assam.
Galipápa: *Vitex aherniana* Merr.; *Verbenac.;* Phil. (Cag.).
Galiya: *Acer oblongum* Wall.; *Acerac.;* nw. O.-Ind.
Galla: *Schrebera swietenioides* Roxb.; *Oleac.;* ö. O.-Ind. (Madras).

Gallinazo: *Cassia sp.; Caesalpiniac.;* Nic.
Gallito: *Celtis sp.; Ulmac.;* St.-D.
„ : *Triplaris sp.; Polygonac.;* Salv.
Galopa: *Calophyllum calaba* Jacq.; *Guttifer.;* Trin.
Galpa: *Calophyllum calaba* Jacq.; *Guttifer.;* Trin.
Gamari: *Trewia nudiflora* L.; *Euphorbiac.;* nw. O.-Ind. (Nepal).
Gamatúlai: *Eugenia claviflora* Roxb.; *Myrtac.;* Phil. (Cag.).
Gamazumi: *Viburnum dilatatum* Thunb.; *Caprifoliac.;* Jap.
Gambari: *Gmelina arborea* L.; *Verbenac.;* nö. O.-Ind. (Orissa).
„ : *Callicarpa arborea* Roxb.; *Verbenac.;* nö. O.-Ind. (Bihal, Orissa).
Gamboge, westindian: *Clusia rosea* Jacq.; *Guttifer.;* br. W.-Ind.
Gambogewood: *Garcinia spp.; Guttifer.;* Ind.
Gambir: *Calophyllum inophyllum* L.; *Guttifer.;* Bor.
Gamelleira: *Ficus sp.; Morac.;* Bras.
„ branca: *Ficus sp.; Morac.;* Bras.
„ preta: ? *Piratinera guianensis* Aubl.; *Morac.;* Bras.
„ de lombrigueira: *Ficus sp.; Morac.;* Bras.
Gamji: *Ficus platyphylla* Delile; *Morac.;* n. Nig.
Gamma fada: *Swartzia madagascariensis* Desv.; *Papilionac.;* n. Nig.
Gammala: *Pterocarpus marsupium* Roxb.; *Papilionac.;* Ind., Cey.
Gammasagon: *Couepia surinamensis* Kleinh.; *Rosac.;* Sur.
„ : ? *Couepia leptostachys* Bth., ? *C. glandulosa* Miq.; *Rosac.;* Sur.
Gamolomie: *Bursera gummifera* L.; *Burserac.;* Bah.
Gamri: *Gmelina arborea* L.; *Verbenac.;* Beng.
Gan: *Randia tomentosa* Hook.; *Rubiac.;* Ind.-Ch.
Gan'án: *Dipterocarpus vernicifluus* Blco.; *Dipterocarpac.;* Phil. (Cam.).
Gandal: *Sorindeia ochracea* Engl.; *Anacardiac.;* Kam.
Gandala: *Santalum album* L.; *Santalac.;* sw. O.-Ind. (Kantara).
Gandoe: *Swartzia tomentosa* DC.; *Papilionac.;* Sur.
Gang: *Randia dumetorum* Lamk.; *Rubiac.;* Annam.
„ com: *Canthium parvifolium* Roxb.; *Rubiac.;* Ind.-Ch.
„ cuôm: *Canthium parvifolium* Roxb.; *Rubiac.;* Ind.-Ch.
„ trang: *Randia tomentosa* Hook.; *Rubiac.;* Ind.-Ch.
„ vàng: *Canthium tomentosum* D. Dietr.; *Rubiac.;* Ind.-Ch.
Gañga: *Buchanania spp.; Anacardiac.;* Phil. (Cag.).
Ganga: *Boswellia serrata* Roxb.; *Burserac.;* C.-O.-Ind.
„ -iesè: *Protium sagotianum* March., *P. puberulum* Engl.; *Burserac.;* Sur.
„ „ : *Protium heptaphyllum* March. et Pulle; *Burserac.;* Sur.
„ „ : *Protium aracouchini* Mart., *P. altissimum* March.; *Burserac.;* Sur.
„ -pisie: *Protium sagotianum* March., *P. puberulum* Engl.; *Burserac.;* Sur.
„ „ : *Protium heptaphyllum* March. et Pulle; *Burserac.;* Sur.
„ „ : *Protium aracouchini* Mart., *P. altissimum* March.; *Burserac.;* Sur.
Gangai: *Mallotus philippinensis* Muell. Arg.; *Euphorbiac.;* Assam.
Gangaji: *Pongamia glabra* Vent.; *Papilionac* C.-O.-Ind.
Gangala: *Pachylobus fraxinifolius* Engl., *P. le-testuï* Pellegr.; *Burserac.;* Gab. [(Mayombe).
Ganganeruchettu: *Thespesia populnea* Correa; *Malvac.;* Ind.
Gañgáuan: *Palaquium spp.; Sapotac.;* Phil. (Mind., Mas.).
Gangaw: *Mesua ferrea* L.; *Guttifer.;* Burma, And.
Ganger: *Grewia populifolia* Vahl; *Tiliac.;* tr. Af., O.-Ind.
Gangerum: *Grewia populifolia* Vahl; *Tiliac.;* tr. Af., O.-Ind.
Ganggo: *Mesua ferrea* L.; *Guttifer.;* Burma, And.
Gangi: *Ximenia americana* L.; *Olacac.;* Af. (Kongo).
Gangow: *Mesua ferrea* L.; *Guttifer.;* And.
Gangwa: *Excoecaria agallocha* L.; *Euphorbiac.;* Beng.
Ganitri: *Elaeocarpus angustifolius* Blume; *Elaeocarpac.;* Nied.-Ind.
Ganjher: *Sterculia urens* Roxb., *S. villosa* Roxb.; *Sterculiac.;* w. Beng.
Gansilai: *Cratoxylon spp.; Guttifer.;* Phil. (Bas.).
Gante melle: *Chukrasia tabularis* A. Juss.; *Meliac.;* sö. Ind. (Madras).
Ganzepruim: *Chrysobalanus icaco* L.; *Rosac.;* fr. Gu.
Gao: *Adina sessilifolia* Hook. f.; *Rubiac.;* Ind.-Ch.
„ : *Bombax malabaricum* DC.; *Bombacac.;* Annam.
„ : *Adina cordifolia* Hook. f.; *Rubiac.;* Annam.

Gao: *Sarcocephalus cordatus* Miq.; *Rubiac.;* Co.-Ch.
„ : *Anthocephalus indica* A. Rich. *var. macrophylla* Pierre; *Rubiac.;* Ind.-Ch.
„ rùng: *Adina cordifolia* Hook. f.; *Rubiac.;* Ind.-Ch.
„ vang: *Adina sessilifolia* Hook. f.; *Rubiac.;* Ind.-Ch.
„ „ : *Canthium parvifolium* Roxb.; *Rubiac.;* Ind.-Ch.
„ „ : *Adina cordifolia* Hook. f.; *Rubiac.;* Annam.
„ vay: *Adina cordifolia* Hook. f.; *Rubiac.;* Ind.-Ch.
Gaos de gallo: *Celtis aculeata* Sw.; *Ulmac.;* tr. Am.
Gaoudi: *Acacia arabica* Willd.; *Mimosac.;* Sudan.
Gapas-gápas: *Cumingia philippinensis* Vid.; *Bombacac.;* Phil. (Zamb., Neg.).
Gaque: *Clusia sp.; Guttifer.;* Kol.
Gara: *Lagerstroemia flos-reginae* Retz.; *Lythrac.;* C.-O.-Ind. (Nagpur).
„ : *Mallotus philippinensis* Muell. Arg.; *Euphorbiac.;* O.-Ind.
„ bursu: *Grewia laevigata* Vahl; *Tiliac.;* ö. Beng.
„ -hesel: *Anogeissus acuminata* Wall.; *Combretac.;* n. O.-Ind.
„ loa: *Trewia nudiflora* L.; *Euphorbiac.;* O.-Ind.
„ patana: *Terminalia arjuna* Bedd.; *Combretac.;* n. O.-Ind.
Garabato: *Celtis sp.; Ulmac.;* Mex.
„ : *Acacia furcata* Gill.; *Mimosac.;* Arg.
„ blanco: *Celtis sp.; Ulmac.;* Mex.
„ „ : *Acacia bonariensis* Gill.; *Mimosac.;* Arg.
Garais: *Evonymus europaea* L.; *Celastrac.;* Fr.
Garapa: *Apuleia praecox* Mart.; *Caesalpiniac.;* Bras. (R.-J.).
„ amarella: *Apuleia praecox* Mart.; *Caesalpiniac.;* Bras. (Sao P).
Carapacunta: *Conomorpha peruviana* C. DC.; *Myrsinac.;* Bras.
Garapriapunha: *Apuleia praecox* Mart.; *Caesalpiniac.;* Bras. (Sao P).
Garauna: *Melanoxylon brauna* Schott; *Caesalpiniac.;* Bras. (Sao P).
Garbancillera borde: *Ononis fruticosa* L.; *Papilionac.;* Sp.
Garbancillo: *Trichilia sp.; Meliac.;* Mex.
Garbha: *Canthium didymum* Roxb.; *Rubiac.;* Beng.
Garcero: *Licania arborea* Seem.; *Rosac.;* Kol.
Garésu: *Mimusops heckeli* H. Lec.; *Sapotac.;* Elf.
Garfoe: *Macaranga sp.; Euphorbiac.;* Lib.
Garga: *Garuga pinnata* Roxb.; *Burserac.;* w. O.-Ind. (n. Bombay).
Garjanka-tel: *Dipterocarpus alatus* Roxb.; *Dipterocarpac.;* Annam.
Garlic tree: *Crataeva tapia* L.; *Capparidac.;* Jam.
„ wood: *Ilex panamensis* Standl.; *Aquifoliac.;* Pan.
„ „ : *Gallesia scorododendrum* Casaretto; *Phytolaccac.;* Bras. (Bahia), Peru.
Garo tsjampaca: *Michelia velutina* Blume, *M. sericea* Pers.; *Magnoliac.;* Nied.-Ind.
Garonton tangah: *Chisocheton divergens* Bl.; *Meliac.;* Malay.
Garpipal: *Populus ciliata* Wall.; *Salicac.;* nw. O.-Ind. (Kumaon).
Garrapata de playa: *Caesalpinia bonducella* L.; *Caesalpiniac.;* Ven.
Garapatillo: *Trichilia sp.; Meliac.;* Mex.
Garrar: *Lebidieropsis orbicularis* Muell. Arg.; *Euphorbiac.;* Ind.
Garroncha: *Erica ciliaris* L., *E. tetralix* L.; *Ericac.;* Sp.
Garso: *Albizzia procera* Bth.; *Mimosac.;* sw. Beng. (Orissa).
Garuga (H): *Garuga pinnata* Roxb.; *Burserac.;* w. O.-Ind. (n. Bombay).
Garugo: *Garuga pinnata* Roxb.; *Burserac.;* O.-Ind.
Garum: *Trewia nudiflora* L.; *Euphorbiac.;* O.-Ind., Sunda.
Garunián: *Neonauclea spp.; Rubiac.;* Phil. (I.-S.).
Gasa kura: *Monotes kerstingi* Gilg; *Dipterocarpac.;* n. Nig.
Gasátan: *Palaquium spp.; Sapotac.;* Phil. (Ilocos).
Gasparil: *Esenbeckia febrifuga* Mart.; *Rutac.;* Trin.
Gasparillo: *Esenbeckia febrifuga* Mart.; *Rutac.;* Trin.
„ : *Licania sp.; Rosac.;* Trin.
Gas-penela: *Sapindus emarginatus* Vahl; *Sapindac.;* Ind.-Ch.
Gât nai: *Gatnaia annamica* Gagnep.; *Euphorbiac.;* Ind.-Ch.
Gatásan: *Garcinia venulosa* Choisy, *G. spp.; Guttifer.;* Phil.
„ : *Santiria nitida* Merr.; *Burserac.;* Phil. (w. Neg.).
„ : *Palaquium spp.; Sapotac.;* Phil. (Cag., I.-S., Pang.).
Gatasgátas: *Bassia ramiflora* Merr.; *Sapotac.;* Phil. (Ley.).
Gatátan: *Dysoxylum turczaninowi* C. DC.; *Meliac.;* Phil. (Cag.).

Gateado: *Coccoloba laurifolia* Jacq.; *Polygonac.;* Fl., W.-Ind., n. S.-Am.
„ : *Coccoloba sp.; Polygonac.;* P.-R.
„ : *Swietenia sp.; Meliac.;* Mex.
„ : *Astronium graveolens* Jacq.; *Anacardiac.;* C.-R., Ven.
„ : ? *Piratinera guianensis* Aubl.; *Morac.;* Bras.
Gatia: ? *Piratinera guianensis* Aubl.; *Morac.;* Bras.
Gatillo: *Ochroma limonensis* Rowlee; *Bombacac.;* Nic.
„ : *Capparis linearis* Jacq.; *Capparidac.;* Ven.
Gatip pahit: *Samandura indica* L.; *Simarubac.;* Malay.
Gatottie: *Matayba surinamensis* Warb., *M. fallax* Radlk.; *Sapotac.;* Sur.
„ : *Matayba arborescens* Radlk., *M. guianensis* Aubl.; *Sapotac.;* Sur.
„ : *Cupania scrobiculata* L. C. Rich.; *Sapindac.;* Sur.
Gattátan: *Palaquium spp.; Sapotac.;* Phil. (Nueva Ecija).
Gatuna: *Ulex europaeus* L.; *Papilionac.;* Arg.
Gaude: *Gardenia erubescens* Stapf et Hutch.; *Rubiac.;* n. Nig.
Gauden kura: *Gardenia ternifolia* Thunb.; *Rubiac.;* n. Nig.
Gaulette bâtard: *Cupania alternifolia* Pers.; *Sapindac.;* Réu.
„ blanc: *Cupania alternifolia* Pers.; *Sapindac.;* Réu.
„ marron: *Cupania laevis* Pers.; *Sapindac.;* Réu.
„ rouge: *Licania sp.; Rosac.;* Guad.
„ „ : *Melicocca diversifolia* Juss.; *Sapindac.;* Réu., Maur.
Gauli: *Bridelia retusa* Spreng.; *Euphorbiac.;* n. O.-Ind. (Garhwal).
Gausal: *Miliusa velutina* Hook. f. et Thoms.; *Anonac.;* n. O.-Ind. (Garhwal).
Gausam: *Schleichera trijuga* Willd.; *Sapindac.;* n. O.-Ind.
Gauwetic: *Matayba surinamensis* Warb., *M. fallax* Radlk.; *Sapotac.;* Sur.
„ : *Matayba arborescens* Radlk., *M. guianensis* Aubl.; *Sapotac.;* Sur.
„ : *Cupania scrobiculata* L. C. Rich.; *Sapindac.;* Sur.
Gauwitie: *Matayba surinamensis* Warb., *M. fallax* Radlk.; *Sapotac.;* Sur.
„ : *Matayba arborescens* Radlk., *M. guianensis* Aubl.; *Sapotac.;* Sur.
„ : *Cupania scrobiculata* L. C. Rich.; *Sapindac.;* Sur.
Gauze: *Lagetta lintearia* Lam.; *Thymelaeac.;* W.-Ind.
Gavarrera: *Rosa arvensis* Hudson; *Rosac.;* Sp.
Gavilán: *Engelhardtia pterocarpa* Standl.; *Juglandac.;* ö. C.-R.
„ : *Pentaclethra filamentosa* Bth.; *Mimosac.;* Pan.
Gavó: *Ononis aragonensis* Asso; *Papilionac.;* Sp.
Gavuldu: *Careya arborea* Roxb.; *Lecythidac.;* sw. O.-Ind. (Mysore).
Gaw: *Piptadenia africana* Hook. f.; *Mimosac.;* Lib.
Gawasa: *Parinarium macrophyllum* Sabine; *Rosac.;* n. Nig.
Gawkngu: *Cassia fistula* L.; *Caesalpiniac.;* O.-Ind.
Gawknguchyamang: *Lagerstroemia flos-reginae* Retz.; *Lythrac.;* O.-Ind.
Gawo: *Acacia albida* Delile; *Mimosac.;* n. Nig.
Gaya: *Bridelia retusa* Spreng.; *Euphorbiac.;* n. O.-Ind. (Garhwal).
Gayac (f): *Guaiacum officinale* L.; *Zygophyllac.;* W.-Ind., n. S.-Am.
„ : *Dipteryx odorata* Willd.; *Papilionac.;* fr. Gu.
„ : *Intsia bijuga* O. Ktze.; *Caesalpiniac.;* Sey.
„ faux: *Dipteryx odorata* Willd.; *Papilionac.;* fr. Gu.
„ jaune vert: *Guaiacum officinale* L.; *Zygophyllac.;* Mart.
„ male: *Dipteryx odorata* Willd.; *Papilionac.;* fr. Gu.
„ de Cayenne: *Dipteryx odorata* Willd.; *Papilionac.;* Sur., Guad.
Gayacan: *Bulnesia arborea* Engl.; *Zygophyllac.;* Ven.
„ : *Porliera hygrometrica* Ruiz et Pav.; *Zygophyllac.;* Chile.
Gayumahin: *Terminalia edulis* Blco.; *Combretac.;* Phil. (Zam.).
Gayumba: *Spartium junceum* L.; *Papilionac.;* Sp.
Gazumaru: *Ficus retusa* L.; *Morac.;* Jap.
Gbah-chu: *Androsiphonia adenostegia* Stapf; *Flacourtiac.;* Lib.
„ „ : *Bertiera racemosa* K. Sch. *var. glabrata* Hutch. et Dalz.; *Rubiac.;* Lib.
Gban-gbah: *Microglossa volubilis* DC.; *Composit.;* Lib.
Gbar-bee-mleh: *Uvaria afzelii* Scott Ell.; *Anonac.;* Lib.
„ chu: *Allanblackia parviflora* Chev.; *Guttifer.;* Lib.
Gbay: *Xylopia quintasii* Engl. et Diels; *Anonac.;* Lib.
„ -dee: *Xylopia quintasii* Engl. et Diels; *Anonac.;* Lib.
„ -vlehn: *Macrolobium macrophyllum* Macbr.; *Caesalpiniac.;* Lib.

Gbé: *Ceiba pentandra* Gaertn.; *Bombacac.;* Elf.
Gbéé: *Ceiba pentandra* Gaertn. *var. dehiscens* Ulb.; *Bombacac.;* Dah.
Gbeh: *Trichoscypha ferruginea* Engl.; *Anacardiac.;* Lib.
„ -sch: *Paullinia pinnata* L.; *Sapindac.;* Lib.
Gboe-dah: ? *Angylocalyx obligophyllus* Baker f.; *Papilionac.;* Lib.
„ -kpar: *Desmostachys vogeli* Stapf; *Icacinac.;* Lib.
„ -kpay: *Diospyros thomasi* Hutch. et Dalz.; *Ebenac.;* Lib.
Gboite: *Monodora myristica* Dunal; *Anonac.;* Lib.
Gbolei: *Ricinodendron africanum* Muell. Arg.; *Euphorbiac.;* Lib.
G'Bon: *Piptadenia africana* Hook. f.; *Mimosac.;* Elf.
Gbor-du-orh: *Berlinia spp.; Caesalpiniac.;* Lib.
Gbu-aye: ? *Fagara angolensis* Engl.; *Rutac.;* Lib.
Ge-ahn: *Vismia leonensis* Hook. f. *var. macrophylla* Hutch. et Dalz.; *Guttifer.;* Lib.
Ge-ahn-de-pay: *Vismia leonensis* Hook. f.; *Guttifer.;* Lib.
Ge-ay: *Chlorophora excelsa* Bth. et Hook.; *Morac.;* Lib.
Gebarú: *Eperua sp.; Caesalpiniac.;* Bras.
Gedar kurumi: *Pterocarpus santalinoides* L'Hérit.; *Papilionac.;* Go.
Geebong: *Persoonia media* R. Br.; *Proteac.;* Queensl., N.-S.-W.
Geelhart (h): *Platonia insignis* Mart.; *Guttifer.;* Sur.
„ , surinaamsch: *Platonia insignis* Mart.; *Guttifer.;* Sur.
Geelhout: *Chlorophora tinctoria* Gaudich.; *Morac.;* tr. Am.
Gêgamba: *Heisteria trillesiana* Pierre; *Olacac.;* Gab.
Geigenholz (d): *Casearia parviflora* Willd.; *Flacourtiac.;* Guad.
Geiger-tree: *Cordia sebestena* L.; *Borraginac.;* Fl., tr. Am.
Geio: *Bridelia retusa* Spreng.; *Euphorbiac.;* Beng.
Geissblatt (d): *Lonicera caprifolium* L.; *Caprifoliac.;* s. Eu.
„ , wildes (d): *Lonicera periclymenum* L.; *Caprifoliac.;* Eu.
Gékadi: *Elaeis guineensis* Jacq. *var. nigrescens* Chev.; *Palmac.;* Gab.
„ sa mavasa: *Elaeis guineensis* Jacq. *var.* ?; *Palmac.;* Gab.
„ sa-Montangaé: *Cocos nucifera* L.; *Palmac.;* Gab.
Gelam: *Eugenia spp.; Myrtac.;* Malay.
Gelbholz (d): *Chlorophora tinctoria* Gaudich.; *Morac.;* tr. S.-Am.
„ , Amur-: *Cladrastis amurensis* Bth.; *Papilionac.;* Ch., Jap.
„ , asiatisches: *Cladrastis amurensis* Bth.; *Papilionac.;* Ch., Jap.
„ , brasilianisches: *Chlorophora tinctoria* Gaudich.; *Morac.;* tr. Am.
„ , Carthagena-: *Chlorophora tinctoria* Gaudich.; *Morac.;* Kol.
Gelbes Holz (d): *Ceanothus chloroxylon* Nees; *Rhamnac.;* Jam.
Gele Pisi: *Nectandra sp., Ocotea sp.; Laurac.;* Sur.
Gellu: *Cupressus torulosa* D. Don; *Conifer.;* nw. O.-Ind. (Punjab).
Genasing: *Stereospermum xylocarpum* Bth. et Hook. f.; *Bignoniac.;* w. O.-Ind.
Gendeli de poma: *Garuga pinnata* Roxb.; *Burserac.;* Assam.
Genêt d'Espagne (f): *Spartium junceum* L.; *Papilionac.;* M.-M., Kanar.
„ des teinturies (f): *Genista tinctoria* L.; *Papilionac.;* Eu., As., Sib.
Genévrier (f): *Juniperus spp.; Conifer.;* n. Hem., Tr.
„ méridional: *Juniperus barbadensis* L.; *Conifer.;* s. USA., W.-Ind.
„ rouge: *Juniperus virginiana* L.; *Conifer.;* ö. Kan.
Gengri: *Dalbergia lanceolaria* L.; *Papilionac.;* O.-Ind.
Geni: *Loesenera kalantha* Harms; *Caesalpiniac.;* Lib.
Genip: *Melicocca bijuga* L.; *Sapindac.;* Jam., Curaçao.
„ tree: *Melicocca paniculata* Juss.; *Sapindac.;* USA.
Genipa: *Genipa americana* L.; *Rubiac.;* tr. S.-Am.
„ : *Genipa caruto* H. B. K.; *Rubiac.;* Trin.
Genipapa: *Genipa americana* L.; *Rubiac.;* Bras.
Genipapeiro: *Genipa americana* L.; *Rubiac.;* Bras.
Genipapo: *Genipa americana* L.; *Rubiac.;* Bras. (Sao P).
„ rosa: *Palicourea corymbifera* Standl.; *Rubiac.;* Bras.
„ do campo: *Tocoyena formosa* K. Schum.; *Rubiac.;* Bras.
„ „ matto: *Genipa americana* L.; *Rubiac.;* Bras.
Geniparana: *Gustavia sp.; Lecythidac.;* Bras.
„ : *Eschweilera carri* Standl.; *Lecythidac.;* n. Bras. (Pará).
„ da matta: *Gustavia sp.; Lecythidac.;* Bras.
Genipat: *Genipa americana* L.; *Rubiac.;* fr. W.-Ind.

Genipayer: *Genipa americana* L.; *Rubiac.;* fr. W.-Ind.
Genipo: *Genipa americana* L.; *Rubiac.;* fr. W.-Ind.
Genisero: *Enterolobium cyclocarpum* Griseb.; *Mimosac.;* Mex., Nic.
Genista de Portugal: *Cytisus multiflorus* Sweet-Bailey; *Papilionac.;* Arg.
Genitri: *Elaeocarpus angustifolius* Blume; *Elaeocarpac.;* Nied.-Ind.
Genizero (u): *Enterolobium cyclocarpum* Griseb.; *Mimosac.;* tr. Am.
Genjan: *Odina wodier* Roxb.; *Anacardiac.;* C.-O.-Ind. (Nagpur).
Geno Geno: *Lonchocarpus violaceus* H. B. K.; *Papilionac.;* P.-R.
Genthi: *Boehmeria rugulosa* Wedd.; *Urticac.;* n. O.-Ind. (Garhwal).
Genti: *Boehmeria rugulosa* Wedd.; *Urticac.;* n. O.-Ind. (Garhwal).
Geo: *Pisonia aculeata* L.; *Nyctaginac.;* P.-R.
Géomba: *Pycnanthus kombo* Warb.; *Myristicac.;* Gab.
Geor: *Excoecaria agallocha* L.; *Euphorbiac.;* Beng.
Georgia bark: *Pinckneya pubescens* Michx.; *Rubiac.;* sö. USA.
Geria: *Excoecaria agallocha* L.; *Euphorbiac.;* Beng.
Geritta: *Turpinia pomifera* DC.; *Staphyleac.;* tr. As.
Geroenggang: *Cratoxylon sp.;* *Guttifer.;* Java.
Geronggang: *Cratoxylon arborescens* Blume; *Guttifer.;* Malay.
Gerunggang: *Cratoxylon glaucum* Korth., *Cratoxylon spp.;* *Guttifer.;* Bor.
Gésanga: *Ricinodendron africanum* Muell. Arg.; *Euphorbiac.;* Gab.
Gésantgala: *Ricinodendron africanum* Muell. Arg.; *Euphorbiac.;* Gab.
Getah-seundik: *Payena leeri* Kurz; *Sapotac.;* Malay, Bor., Sum., Banka.
Geti: *Boehmeria rugulosa* Wedd.; *Urticac.;* Ind.
Geva: *Excoecaria agallocha* L.; *Euphorbiac.;* sw. O.-Ind.
Gévèi: *Elaeis guineensis* Jacq. var. ?; *Palmac.;* Gab.
Geweihbaum (d): *Gymnocladus canadensis* Lam.; *Caesalpiniac.;* N.-Am.
Geza: *Combretum micranthum* G. Don; *Combretac.;* n. Nig.
Ghaja: *Bridelia retusa* Spreng.; *Euphorbiac.;* nw. O.-Ind. (Punjab).
Ghangru: *Randia dumetorum* Lamk.; *Rubiac.;* C.-O.-Ind.
Ghant: *Schrebera swietenioides* Roxb.; *Oleac.;* n. O.-Ind.
Ghanti: *Cordia macleodi* Hook. f. et Thoms.; *Borraginac.;* sw. Beng.
Ghanto: *Schrebera swietenioides* Roxb.; *Oleac.;* C.-O.-Ind.
Ghari am: *Mangifera indica* L.; *Anacardiac.;* Assam.
Gharri: *Garuga pinnata* Roxb.; *Burserac.;* O.-Ind. (Baiga).
Ghatbor: *Zizyphus xylopyrus* Willd.; *Rhamnac.;* C.-O.-Ind.
Ghato: *Schrebera swietenioides* Roxb.; *Oleac.;* C.-O.-Ind.
Ghattol: *Zizyphus xylopyrus* Willd.; *Rhamnac.;* C.-O.-Ind.
Ghé: *Glochidion obliquum* Dcne.; *Euphorbiac.;* Ind.-Ch.
Ghela: *Randia dumetorum* Lamk.; *Rubiac.;* sw. O.-Ind.
Ghesi: *Quercus semecarpifolia* Smith; *Fagac.;* Beng.
Ghian: *Litsea polyantha* Juss.; *Laurac.;* nw. O.-Ind. (Punjab).
Ghiriya: *Chloroxylon swietenia* DC.; *Rutac.;* C.-O.-Ind.
Ghittoe of North Queensland: *Halfordia scleroxyla* F. v. M.; *Rutac.;* Au.
  „ , southern: *Halfordia drupifera* F. v. M.; *Rutac.;* Au.
Ghiwala: *Callicarpa arborea* Roxb.; *Verbenac.;* n. O.-Ind.
Ghoent: *Zizyphus xylopyrus* Willd.; *Rhamnac.;* O.-Ind. (Baiga).
Ghogar: *Gardenia latifolia* Ait.; *Rubiac.;* C.-O.-Ind.
  „ : *Garuga pinnata* Roxb.; *Burserac.;* C.-O.-Ind.
Ghogari: *Gardenia latifolia* Ait.; *Rubiac.;* sw. O.-Ind.
Ghont: *Zizyphus xylopyrus* Willd.; *Rhamnac.;* n. O.-Ind.
Ghorkaranj: *Ailanthus excelsa* Roxb.; *Simarubac.;* O.-Ind. (n. Deccan).
Ghot-ber: *Zizyphus xylopyrus* Willd.; *Rhamnac.;* nw. O.-Ind. (Bombay).
Ghota: *Zizyphus xylopyrus* Willd.; *Rhamnac.;* C.-O.-Ind.
Ghoti: *Zizyphus xylopyrus* Willd.; *Rhamnac.;* C.-O.-Ind.
Ghoto: *Prunus capuli* Cav.; *Rosac.;* Mex.
Ghout: *Zizyphus xylopyrus* Willd.; *Rhamnac.;* C.-O.-Ind.
Ghunza: *Crataegus oxyacantha* L.; *Rosac.;* Eu., As., n. Af.
Gia: *Excoecaria agallocha* L.; *Euphorbiac.;* Ind.-Ch.
  „ ca: *Wendlandia paniculata* DC.; *Rubiac.;* Tonk.
  „ da trang: *Taractogenos serrata* Pierre; *Flacourtiac.;* Ind.-Ch.
  „   „   „ : *Walsura villosa* Wall.; *Meliac.;* Ind.-Ch.
  „   „   „ : *Hydnocarpus heterophylla* Blume; *Flacourtiac.;* Co.-Ch.

Gia thi: *Berrya mollis* Wall., *B. ammonilla* Roxb.; *Tiliac.*; Ind.-Ch.
„ „ : *Tectona grandis* L.; *Verbenac.*; Ind.-Ch.
„ trang: *Dracontomelum duperreanum* Pierre; *Anacardiac.*; Ind.-Ch.
„ „ : *Walsura villosa* Wall.; *Meliac.*; Ind.-Ch.
Giáe: *Pithecolobium balansae* Oliv.; *Mimosac.*; Ind.-Ch.
Giai ma: *Chisocheton thoreli* Pierre *var. tonkinensis* Pierre; *Meliac.*; Annam.
Giâm: *Chukrasia tabularis* A. Juss.; *Meliac.*; Ind.-Ch.
„ : *Grewia eriocarpa* Jussieu; *Tiliac.*; Ind.-Ch.
„ : *Stephegyne parvifolia* Korth.; *Rubiac.*; Annam.
„ : *Hopea lowi* Dyer; *Dipterocarpac.*; br. Malay.
„ : *Hopea nutans* Ridl.; *Dipterocarpac.*; Malay.
Giân: *Ixonanthes cochinchinensis* Pierre; *Linac.*; Annam.
Giang cua: *Viburnum colebrookeanum* Wall.; *Caprifoliac.*; Ind.-Ch.
„ giàng: *Spatholobus sp.*; *Papilionac.*; Ind.-Ch.
„ huong: *Pterocarpus pedatus* Pierre; *Papilionac.*; Annam.
„ „ : *Pterocarpus macrocarpus* King; *Papilionac.*; Ind.-Ch.
„ „ : *Pterocarpus cambodianus* Pierre; *Papilionac.*; Ind.-Ch.
„ núi: *Ternstroemia japonica* Thunb.; *Theac.*; Annam.
„ nùong: *Ternstroemia penangiana* Choisy; *Theac.*; Annam.
Giant of the forest: *Sequoia sempervirens* Endl.; *Conifer.*; w. USA. (Kalif.).
Giau: *Morinda citrifolia* L.; *Rubiac.*; Ind.-Ch.
Giâu: *Morus indica* L.; *Morac.*; Ind.-Ch.
„ da dât: *Baccaurea sapida* Muell. Arg.; *Euphorbiac.*; Ind.-Ch.
„ dât: *Aporosa microcalyx* Hassk.; *Euphorbiac.*; Ind.-Ch.
„ „ : *Baccaurea cauliflora* Lour.; *Euphorbiac.*; Ind.-Ch.
„ den: *Morus indica* L.; *Morac.*; Ind.-Ch.
„ dô: *Carpinus pubescens* Burkill; *Betulac.*; Ind.-Ch.
„ gia: *Baccaurea cauliflora* Lour.; *Euphorbiac.*; Ind.-Ch.
„ „ dât: *Baccaurea sapida* Muell. Arg.; *Euphorbiac.*; Ind.-Ch.
„ „ soan: *Spondias lakonensis* Pierre; *Anacardiac.*; Ind.-Ch.
„ giat: *Baccaurea sapida* Muell. Arg.; *Euphorbiac.*; Ind.-Ch.
„ sat: *Baccaurea sapida* Muell. Arg.; *Euphorbiac.*; Ind.-Ch.
Giây: *Pterospermum pierrei* Hance; *Sterculiac.*; Annam.
„ : *Pterospermum jackianum* Wall.; *Sterculiac.*; Ind.-Ch.
„ : *Mallotus cochinchinensis* Lour.; *Euphorbiac.*; Ind.-Ch.
„ rung gia: *Mallotus philippinensis* Muell. Arg.; *Euphorbiac.*; Ind.-Ch.
Gibatão: ? *Astronium commune* Jacq., *A. concinnum* Schott; *Anacardiac.*; Bras.
„ : *Astronium gracile* Engl.; *Anacardiac.*; Bras. (Sao P).
Gibitan: *Astronium commune* Jacq., *A. gracile* Engl.; *Anacardiac.*; Bras.
„ : *Astronium concinnum* Schott; *Anacardiac.*; Bras.
Gibofoyoi: *Carpolobia spp.*; *Polygalac.*; Lib.
Gibowali: *Pausinystalia lane-poolei* Hutch.; *Rubiac.*; Lib.
Gibuaya: *Litsea spp.*; *Laurac.*; Phil. (Lag.).
Gichtbeere (d): *Ribes nigrum* L.; *Saxifragac.*; N.-Am.
Gidda: *Wrightia tomentosa* Roem. et Schult.; *Apocynac.*; w. O.-Ind. (Bombay).
Gidgea: *Acacia cambagei* R. T. Baker; *Mimosac.*; N.-S.-W.
Gidia: *Acacia homalophylla* A. Cunn.; *Mimosac.*; sö. Au.
Gidjikó: *Vitex cienkowskyi* Kotschy et Peyr.; *Verbenac.*; W.-Af.
Giduri: *Cordia myxa* L.; *Borraginac.*; nw. O.-Ind. (Bombay).
Gidya: *Acacia homalophylla* A. Cunn.; *Mimosac.*; Au.
Gié: *Quercus spp., Pasania spp.*; *Fagac.*; Ind.-Ch.
„ : *Mallotus hookerianus* Muell. Arg.; *Euphorbiac.*; Ind.-Ch.
„ bôp: *Quercus poilanei* Hickel et A. Camus; *Fagac.*; Ind.-Ch.
„ cau: *Pasania areca* Hickel et A. Camus; *Fagac.*; Ind.-Ch.
„ den: *Quercus glauca* Thunb.; *Fagac.*; Ind.-Ch.
„ do: *Quercus wallichiana* Lindl.; *Fagac.*; Ind.-Ch.
„ goi: *Castanopsis lecomtei* Hickel et A. Camus; *Fagac.*; Ind.-Ch.
„ mö gà: *Pasania tubulosa* Hickel et A. Camus; *Fagac.*; Ind.-Ch.
„ núi: *Mitrephora thoreli* Pierre; *Anonac.*; Ind.-Ch.
„ quông: *Quercus pseudocornea* Chev.; *Fagac.*; Ind.-Ch.
„ sôi: *Pasania tubulosa* Hickel et A. Camus; *Fagac.*; Ind.-Ch.
„ trang: *Quercus poilanei* Hickel et A. Camus; *Fagac.*; Ind.-Ch.

Gié xanh: *Pasania pseudo-sundaica* Hickel et A. Camus; *Fagac.;* Ind.-Ch.
Giembé: *Uapaca bingervillensis* Beille; *Euphorbiac.;* Elf.
Giembi: *Uapaca bingervillensis* Beille; *Euphorbiac.;* Elf.
Gièn: *Pentace siamensis* Kurz; *Tiliac.;* Ind.-Ch.
   „   do: *Xylopia vielana* Pierre; *Anonac.;* Annam.
   „   doc: *Bridelia minutiflora* Hook. f.; *Euphorbiac.;* Ind.-Ch.
   „   trang: *Xylopia pierrei* Hance; *Anonac.;* Annam.
Giêng giêng: *Butea frondosa* Roxb.; *Papilionac.;* Ind.-Ch.
Gienhatti: *Tecoma leucoxylon* Mart. *var. pentaphylla* Juss.; *Bignoniac.;* Sur.
   „   : *Tecoma araliacea* P. DC.; *Bignoniac.;* Sur.
Giginya: *Borassus flabellifer* L. *var. aethiopum* Warb.; *Palmac.;* n.Nig.
Giho: *Shorea guiso* Blume; *Dipterocarpac.;* Phil. (Zam., Bat.).
Giji: *Ocotea sp.; Laurac.;* S.-L.
Gik: *Acacia seyal* Delile *var. fistula* Oliv.; *Mimosac.;* O.-Af. (it. Somali).
Gill, Turkey-: *Pithecolobium arboreum* Urban; *Mimosac.;* br. Hond.
Gimaimaí: *Parinarium spp.; Rosac.;* Phil. (Lanac).
Gimák: *Schizolobium diffusum* Merr., *S. acutiflorum* Munro; *Gramin.;* Phil. (Bat.).
Gimlet: *Eucalyptus salubris* F. v. M.; *Myrtac.;* w. Au.
   „   , silver-topped: *Eucalyptus campaspe* Spencer Moore; *Myrtac.;* w. Au.
   „   , Swamp-: *Eucalyptus spathulata* Hook.; *Myrtac.;* w. Au.
Gin: *Ginkgo biloba* L.; *Ginkgoac.;* nö. Ind.
? Gina: *Haronga madagascariensis* Choisy; *Guttifer.;* Gab.
Ginabang: *Endospermum peltatum* Merr.; *Euphorbiac.;* Phil. (Ben.).
Gináiang: *Parinarium spp.; Rosac.;* Phil. (Riz., Batg., Tay.).
Ginesta (s): *Spartium junceum* L.; *Papilionac.;* M.-M., Kanar.
Ginestra (i): *Spartium junceum* L.; *Papilionac.;* M.-M., Kanar.
   „   odorosa: *Spartium junceum* L.; *Papilionac.;* M.-M., Kanar.
   „   de'carbonai (i): *Cytisus scoparius* Link; *Papilionac.;* Eu.
Ginestrella (i): *Genista tinctoria* L.; *Papilionac.;* Eu., w. As., Sib.
Ginestrilla borde: *Cytisus fontanei* Spach; *Papilionac.;* Sp.
Ginestrone (i): *Ulex europaeus* L.; *Papilionac.;* Eu.
Gingepau: *Maprounea guianensis* Aubl.; *Euphorbiac.;* Sur.
Ginja Gong: *Ixonanthes reticulata* Jack; *Linac.;* Malak.
Gin-nemu: *Leucaena glauca* Bth.; *Mimosac.;* Jap.
Ginsa: *Castanopsis indica* A. DC., *Castanea indica* Roxb.; *Fagac.;* O.-Ind.
Ginster, Färber- (d): *Genista tinctoria* L.; *Papilionac.;* Eu., w. As., Sib.
   „   , spanischer (d): *Spartium junceum* L.; *Papilionac.;* Sp.
Ginsterbaum, stachliger (d): *Parkinsonia aculeata* L.; *Caesalpiniac.;* s. USA., W.-Ind.,
Ginugal: *Myrtus beckleri* F. v. M.; *Myrtac.;* N.-S.-W.   [Bah.
Gio: *Mallotus furetianus* Muell. Arg.; *Euphorbiac.;* Ind.-Ch.
Gió: *Rhamnoneuron balansae* Gilg; *Thymelaeac.;* Ind.-Ch.
Gio tom: *Unona thoreli* Pierre; *Anonac.;* Annam.
Gioc: *Garcinia tonkinensis* Vesque; *Guttifer.;* Annam.
   „   khê: *Cipadessa fruticosa* Blume; *Meliac.;* Ind.-Ch.
   „   son: *Dillenia indica* L.; *Dilleniac.;* Ind.-Ch.
Giôi: *Talauma gioi* Chev.; *Magnoliac.;* Ind.-Ch.
   „   : *Michelia tonkinensis* Chev.; *Magnoliac.;* Ind.-Ch.
   „   : *Michelia baviensis* Finet et Gagnep.; *Magnoliac.;* Annam.
   „   : *Clausena wampi* Oliv.; *Rutac.;* Ind.-Ch.
   „   : *Clausena excavata* Burm.; *Rutac.;* Ind.-Ch.
   „   : *Eugenia ternifolia* Roxb.; *Myrtac.;* Ind.-Ch.
   „   dât: *Clausena excavata* Burm.; *Rutac.;* Ind.-Ch.
   „   gang: *Manglietia gioi* ?; *Magnoliac.;* Ind.-Ch.
   „   lua: *Manglietia gioi* ?; *Magnoliac.;* Ind.-Ch.
   „   mo: *Michelia baviensis* Finet et Gagnep.; *Magnoliac.;* Ind.-Ch.
   „   mo gà: *Manglietia gioi* ?; *Magnoliac.;* Ind.-Ch.
Giombi: *Uapaca bingervillensis* Beille; *Euphorbiac.;* Elf.
Giramong: *Pittosporum ferrugineum* Ait.; *Pittosporac.;* Malay, Au.
Girangi: *Pongamia glabra* Vent.; *Papilionac.;* C.-O.-Ind.
Girchi: *Holarrhena antidysenterica* Wall.; *Apocynac.;* C.-O.-Ind.
Girét: *Canarium villosum* F. Vill.; *Burserac.;* Phil. (Cag.).
Girigitík: *Cyclostemon spp.; Euphorbiac.;* Phil. (Sor.).

Gisang: *Tarrietia javanica* Blume; *Sterculiac.;* Phil.
Gísek: *Shorea guiso* Blume; *Dipterocarpac.;* Phil. (Zamb., Bas.).
Gishishiya: *Acacia seyal* Delile; *Mimosac.;* n. Nig.
Gisían: *Pentacme contorta* Merr. et Rolfe; *Dipterocarpac.;* Phil. (Cam.).
Gisihan: *Aglaia laevigata* Merr.; *Meliac.;* Phil.
Gísik: *Shorea guiso* Blume, *S. balangeran* Dyer; *Dipterocarpac.;* Phil.
Gísit: *Terminalia edulis* Blco.; *Combretac.;* Phil. (Nu.-V.).
Gíso': *Shorea guiso* Blume; *Dipterocarpac.;* Phil.
Gísok: *Shorea balangeran* Dyer, *S. guiso* Blume; *Dipterocarpac.;* Phil.
„ : *Shorea polysperma* Merr.; *Dipterocarpac.;* Phil. (Ley.).
„ : *Hopea philippinensis* Dyer; *Dipterocarpac.;* Phil. (Neg., Ag.).
„ -gísok: *Hopea philippinensis* Dyer; *Dipterocarpac.;* Phil. (Ley., Zamb.)
„ „ : *Hopea plagata* Vid.; *Dipterocarpac.;* Phil. (Sor.).
„ : *Shorea balangeran* Dyer; *Dipterocarpac.;* Phil.
„ -madláu: *Vatica mangachapoi* Blco.; *Dipterocarpac.;* Phil. (Sam., Ley.).
„ -purá': *Shorea polysperma* Merr.; *Dipterocarpac.;* Phil. (Cam.).
„ -takpáng: *Isoptera borneensis* Scheff.; *Dipterocarpac.;* Phil. (Zamb.)
Gissara: *Euterpe oleracea* Mart.; *Palmac.;* Bras. (Sao P).
Gitahy: *Hymenaea stignocarpa* Mart.; *Caesalpiniac.;* Bras. (Sao P).
Gitakí: *Kayea paniculata* Merr.; *Guttifer.;* Phil. (Cam.).
Gitarón: *Suriana maritima* L.; *Simarubac.;* P.-R.
Gitó: *Guarea trichilioides* L.; *Meliac.;* Bras.
Giuâ ngua: *Aglaia cucullata* Pellegr.; *Meliac.;* Ind.-Ch.
Gium hin: *Chukrasia tabularis* A. Juss.; *Meliac.;* Ind.-Ch.
Giung: *Symplocos laurina* Wall.; *Symplocac.;* Annam.
Giuôc: *Abacca sp.;* *Scitamin.;* Ind.-Ch.
Giuong: *Broussonetia papyrifera* Vent.; *Morac.;* Tonk.
Giusho: *Cinnamomum kanehirai* Hayata; *Laurac.;* Form.
Giyaiya: *Mitragyne africana* Korth.; *Rubiac.;* n.Nig.
Giyeya: *Mitragyne africana* Korth.; *Rubiac.;* n. Nig.
Glabe: *Vernonia conferta* Bth.; *Composit.;* Lib.
Glamberry: *Byrsonima lucida* DC.; *Malpighiac.;* s. Fl., W.-Ind.
Glantori: *Hannoa undulata* Planch.; *Simarubac.;* Go.
Glasstree, Looking-: *Heritiera littoralis* Dryander; *Sterculiac.;* tr. As.
Glass-wood: *Myrica serrata* Lam.; *Myricac.;* Kap.
Glassy wood: *Astronium graveolens* Jacq.; *Anacardiac.;* br. Hond.
„ „ : *Guettarda seleriana* Standl.; *Rubiac.;* br. Hond.
Glateado: *Coccoloba sp.;* *Polygonac.;* P.-R.
Glégléti (f): *Lannea acidissima* Chev.; *Anacardiac.;* Elf.
Glicina: *Wistaria sinensis* Sweet; *Papilionac.;* Arg.
„ blanca: *Wistaria sinensis* Swect var. *alba* Lindl.; *Papilionac.;* Arg.
„ de la China: *Wistaria sinensis* Sweet; *Papilionac.;* Arg.
Glicine (i): *Wistaria sinensis* Sweet; *Papilionac.;* Ch.
Gluta (H): *Gluta travancorica* Bedd.; *Anacardiac.;* Ind.
„ , black (H): *Gluta tavoyana* Hook. f.; *Anacardiac.;* Ind.
„ , Burma- (H): *Gluta tavoyana* Hook. f.; *Anacardiac.;* Burma.
Gnang: *Dipterocarpus alatus* Roxb.; *Dipterocarpac.;* Ind.-Ch.
Gniagnia: *Virola surinamensis* Warb.; *Myristicac.;* Sur.
Gniangon: *Cola proteiformis* Chev.; *Sterculiac.;* Elf.
„ : *Tarrietia utilis* Sprague; *Sterculiac.;* Lib., Elf., Go.
Gnibi: *Cola mirabilis* Chev.; *Sterculiac.;* Elf.
Gnili: *Cola mirabilis* Chev.; *Sterculiac.;* Elf.
Gnom: *Tarrietia cochinchinensis* Pierre; *Sterculiac.;* Ind.-Ch.
Gnong: *Dipterocarpus alatus* Roxb.; *Dipterocarpac.;* Ind.-Ch.
Gnorpin: *Eucalyptus robusta* Sm.; *Myrtac.;* sö. Au.
Gnoué: *Staudtia gabunensis* Warb.; *Myristicac.;* Gab.
Gnoum: *Pygeum arboreum* Engl.; *Rosac.;* Ind.-Ch.
Go: *Ceiba pentandra* Gaertn. var. *dehiscens* Ulb.; *Bombacac.;* Elf.
„ : *Sindora cochinchinensis* Baill.; *Caesalpiniac.;* Ind.-Ch.
„ : *Sterculia dongnaiensis* Pierre; *Sterculiac.;* Ind.-Ch.
„ ba mia: *Sindora cochinchinensis* Baill.; *Caesalpiniac.;* Ind.-Ch.
„ bông lau: *Sindora cochinchinensis* Baill.; *Caesalpiniac.;* Ind.-Ch.

Meyer, Buch der Holznamen 13

Go ca tac: *Sindora cochinchinensis* Baill.; *Caesalpiniac.;* Ind.-Ch.
„ „ té: *Pahudia cochinchinensis* Pierre; *Caesalpiniac.;* Annam.
„ „ „: *Intsia bijuga* O. Ktze.; *Caesalpiniac.;* Ind.-Ch.
„ chou: *Sterculia dongnaiensis* Pierre; *Sterculiac.;* Annam.
„ do: *Pahudia cochinchinensis* Pierre; *Caesalpiniac.;* Annam.
„ huong: *Cinnamomum ilicioides* Chev.; *Laurac.;* Ind.-Ch.
„ mât: *Sindora cochinchinensis* Baill.; *Caesalpiniac.;* Ind.-Ch.
„ mioc: *Intsia bijuga* O. Ktze.; *Caesalpiniac.;* Annam.
„ nuoc: *Intsia bijuga* O. Ktze.; *Caesalpiniac.;* Ind.-Ch.
„ rut: *Ilex godajam* Colebr. var. *capitellata* Andr.; *Aquifoliac.;* Annam.
„ ta-hi: *Sindora cochinchinensis* Baill.; *Caesalpiniac.;* Co.-Ch.
„ „ ki: *Sindora maritima* Pierre; *Caesalpiniac.;* Ind.-Ch.
„ „ „: *Sindora cochinchinensis* Baill.; *Caesalpiniac.;* Ind.-Ch.
„ to te: *Intsia bijuga* O. Ktze.; *Caesalpiniac.;* Ind.-Ch.
„ „ „: *Pahudia cochinchinensis* Pierre; *Caesalpiniac.;* Annam.
„ vang: *Sindora cochinchinensis* Baill.; *Caesalpiniac.;* Ind.-Ch.
„ wood: *Lindera cochinchinensis* Pierre; *Laurac.;* Ind.-Ch.
„ xiém: *Pahudia cochinchinensis* Pierre; *Caesalpiniac.;* Ind.-Ch.
Goa: *Psidium guajava* L.; *Myrtac.;* Go.
Goatwood: *Cassipourea podantha* Standl.; *Rhizophorac.;* Pan.
Gob: *Odina wodier* Roxb.; *Anacardiac.;* C.-O.-Ind. (Ajmere).
Gobaïa-hout (h): *Jacaranda copaia* D. Don; *Bignoniac.;* Sur.
Gobalu: *Acacia arabica* Willd.; *Mimosac.;* w. O.-Ind. (Bombay).
Gob-boh: *Macrolobium heudeloti* Planch.; *Caesalpiniac.;* Lib.
Gobli: *Acacia arabica* Willd.; *Mimosac.;* w. O.-Ind. (Bombay).
Gobra nairul: *Bischofia javanica* Blume; *Euphorbiac.;* ö. O.-Ind. (Madras).
Gobre: *Echinocarpus dasycarpus* Bth.; *Elaeocarpac.;* nö. O.-Ind. (Nepal).
Gobria: *Echinocarpus dasycarpus* Bth.; *Elaeocarpac.;* nö. O.-Ind. (Nepal).
Goda: *Vitex peduncularis* Wall.; *Verbenac.;* Beng.
Godachi: *Zizyphus xylopyrus* Willd.; *Rhamnac.;* w. O.-Ind. (Bombay).
Godakaduru: *Strychnos nux-vomica* Roxb.; *Loganiac.;* Cey.
Godapara: *Dillenia retusa* Thunb.; *Dilleniac.;* Cey.
Godda: *Odina wodier* Roxb.; *Anacardiac.;* w. O.-Ind (Bombay).
Godé: *Myrianthus serratus* Bth. et Hook.; *Morac.;* Elf.
Godgudala: *Sterculia villosa* Roxb.; *Sterculiac.;* n. O.-Ind.
Godhunchi: *Albizzia stipulata* Boivin; *Mimosac.;* w. O.-Ind. (Bombay).
Godi Babul: *Acacia scorpioides* W. F. Wight; *Mimosac.;* Af.
Godo: *Rhizophora racemosa* E. Meyer; *Rhizophorac.;* Go.
Goe: *Klainedoxa gabonensis* Pierre; *Simarubac.;* Lib.
Goe-doo: *Dioncophyllum peltatum* Hutch. et Dalz.; *Flacourtiac.;* Lib.
„ -quehn: *Canarium schweinfurthi* Engl.; *Burserac.;* Lib.
Goebai: *Jacaranda brasiliana* Pers.; *Bignoniac.;* fr. Gu.
Goehlo: *Callicarpa arborea* Roxb.; *Verbenac.;* nw. O.-Ind. (Nepal).
Goei: *Koordersiodendron pinnatum* Merr.; *Anacardiac.;* Nied.-Ind.
Goela: *Aphanomyxis grandifolia* Bl.; *Meliac.;* Java.
Goenfoloe: *Qualea albiflora* Warm., *Q. rosea* Aubl.; *Vochysiac.;* Sur.
Goerana: *Solanum auriculatum* Ait.; *Solanac.;* Bras. (Sao P).
Goewanna kwarie: *Vochysia melinoni* Beckmann; *Vochysiac.;* Sur.
Gofa: *Mitragyne macrophylla* Hiern; *Rubiac.;* Elf.
Gogarni: *Gardenia latifolia* Ait.; *Rubiac.;* C.-O.-Ind.
Gogo: *Carapa gogo* Chev.; *Meliac.;* W.-Af. (St. Thomé).
Gógo: *Albizzia saponaria* Blume; *Mimosac.;* Phil. (Tay.).
„ -kásai: *Albizzia saponaria* Blume; *Mimosac.;* Phil. (Tay.).
Gogong casay: *Albizzia saponaria* Blume; *Mimosac.;* Phil.
„ malatokó: *Albizzia saponaria* Blume; *Mimosac.;* Phil. (Riz.).
„ -tokó: *Albizzia saponaria* Blume; *Mimosac.;* Phil.
Goguldhup: *Canarium sikkimense* King; *Burserac.;* Beng.
Gogwi: *Erythrophloeum guineense* G. Don; *Caesalpiniac.;* Lib.
Gohira: *Acacia leucophloea* Willd.; *Mimosac.;* w. Beng. (Orissa).
Goi: *Bucklandia tonkinensis* H. Lec.; *Hamamelidac.;* Ind.-Ch.
„ : *Aglaia gigantea* Pellegr.; *Meliac.;* Annam.
„ : *Chisocheton cochinchinensis* Pierre; *Meliac.;* Annam.

Goi: *Amoora montana* Bth.; *Meliac.;* Co.-Ch.
„ an: *Aglaia gigantea* Pellegr.; *Meliac.;* Ind.-Ch.
„ báng súng: *Aglaia euphorioides* Pierre; *Meliac.;* Ind.-Ch.
„ duong: *Chisocheton cochinchinensis* Pierre; *Meliac.;* Ind.-Ch.
„ hang: *Aglaia polystachya* Wall.; *Meliac.;* Ind.-Ch.
„ mât: *Aglaia sp.; Meliac.;* Ind.-Ch.
„ „ : *Chisocheton cochinchinensis* Pierre; *Meliac.;* Annam.
„ mit: *Aglaia gigantea* Pellegr.; *Meliac.;* O.-As.
„ mu: *Amoora rohituka* Wight et Arn.; *Meliac.;* Annam.
„ muôc: *Aglaia euphorioides* Pierre; *Meliac.;* Ind.-Ch.
„ nêp: *Aglaia aquatica* Pierre; *Meliac.;* Ind.-Ch.
„ núi: *Dysoxylum juglans* Finet et Pellegr.; *Meliac.;* Ind.-Ch.
„ „ : *Aglaia gigantea* Pellegr.; *Meliac.;* Co.-Ch.
„ nuoc: *Aglaia quocensis* Pierre; *Meliac.;* Ind.-Ch.
„ nuôi: *Chisocheton cochinchinensis* Pierre; *Meliac.;* Annam.
„ oi: *Aglaia pisifera* Hance; *Meliac.;* Ind.-Ch.
„ oy: *Aglaia pisifera* Hance; *Meliac.;* Co.-Ch.
„ té: *Aglaia sp.; Meliac.;* Ind.-Ch.
„ tia: *Aglaia gigantea* Pellegr.; *Meliac.;* Annam.
„ tôm: *Chisocheton coriaceus* Pierre; *Meliac.;* Ind.-Ch.
Goiaba do matto: *Psidium sp.; Myrtac.;* Bras. (Amaz.).
Goiabeira: *Psidium sp.; Myrtac.;* Bras. (Amaz.).
„ do matto: ? *Psidium sp.; Myrtac.;* Bras. (Sao P).
Goiabeiras: *Psidium guaiava* L.; *Myrtac.;* Bras.
Goindu: *Diospyros montana* Roxb.; *Ebenac.;* Ind.
Goit: *Zizyphus xylopyrus* Willd.; *Rhamnac.;* C.-O.-Ind. (Nagphur).
Goita: *Trichilia sp.; Meliac.;* P.-R.
Goitituruba: *Lucuma sp.; Sapotac.;* Bras.
Gojha: *Canthium didymum* Roxb.; *Rubiac.;* ö. Beng.
Gok: *Michelia excelsa* Blume; *Magnoliac.;* nö. O.-Ind. (Sikkim).
Goki: *Schrebera swietenioides* Roxb.; *Oleac.;* n. O.-Ind.
Gokiru: *Albizzia lebbek* Bth.; *Mimosac.;* nw. O.-Ind. (Kumaon).
Gokotoo: *Garcinia morella* Desr.; *Guttifer.;* Cey.
Gokuldhup: *Canarium sikkimense* King; *Burserac.;* Beng.
Gol: *Odina wodier* Roxb.; *Anacardiac.;* C.-O.-Ind. (Merwara).
Golden Deal: *Doryphora sassafras* Endl.; *Monimiac.;* Au.
Goldia: *Anogeissus latifolia* Wall.; *Combretac.;* O.-Ind. (Raip.).
Goldregen: *Cytisus laburnum* L.; *Papilionac.;* Eu.
Goldwood: *Pithecolobium vinhatico* Record; *Mimosac.:* Bras. (Bahia).
Golia: *Anogeissus latifolia* Wall.; *Combretac.;* O.-Ind. (Raip).
Golli-nyum-buy-ambei: *Sarcocephalus esculentus* Afz.; *Rubiac.;* Lib.
Golol: *Acacia bussei* Harms; *Mimosac.;* O.-Af. (it. Somali).
Golra: *Anogeissus latifolia* Wall.; *Combretac.;* O.-Ind. (Raip.).
Gom anime boom: ? *Hymenaea courbaril* L.; *Caesalpiniac.;* W.-Ind.
Goma colorado: *Liquidambar styraciflua* L.; *Hamamelidac.;* Arg.
Goma-dake: *Phyllostachys puberula* Munro *forma nigropunctata* Makino; *Gramin.;*
„ de limón: *Bursera sp.; Burserac.;* Mex. [Jap.
Gomale: *Sideroxylon tomentosum* Roxb.; *Sapotac.;* w. O.-Ind. (Bombay).
Gomard: *Bursera gummifera* L.; *Burserac.;* fr. W.-Ind.
Gomari: *Gmelina arborea* L.; *Verbenac.;* Assam.
Gomart d'Amérique: *Bursera gummifera* L.; *Burserac.;* W.-Ind., C.-Am., n. S.-Am.
Gombolimbo: *Bursera gummifera* L.; *Burserac.;* br. Hond.
Gomita: *Cordia sp.; Borraginac.;* Arg.
Gomma: *Guarea gomma* Pulle; *Meliac.;* Sur.
Gommart: *Protium obtusifolium* March.; *Burserac.;* Mask.
„ balsamifère: *Hedwigia balsamifera* Swartz; *Burserac.;* tr. Am., Ant.
„ d'Amérique: *Bursera gummifera* L.; *Burserac.;* Ant.
„ des Antilles: *Bursera gummifera* L.; *Burserac.;* Ant.
Gommeiro azul: *Eucalyptus globulus* Labill.; *Myrtac.;* s. Bras.
Gommier: *Bursera gummifera* L.; *Burserac.;* Ant.
„ : *Cordia myxa* L.; *Borraginac.;* N.-Kal.
„ blanc: *Bombax malabaricum* DC.; *Bombacac.;* Guad.

Gommier blanc: *Dacryodes hexandra* Griseb.; *Burserac.;* Guad.
„ bleu de Tasmanie: *Eucalyptus globulus* Labill.; *Myrtac.;* Tasm.
„ gris: *Bursera gummifera* L.; *Burserac.;* Guad.
„ jaune des carrières: *Protium sp.; Burserac.;* Guad.
„ rouge: *Hedwigia balsamifera* Swartʒ; *Burserac.;* Sur.
„ „ : *Protium sagotianum* March., *P. polybotryum* Engl.; *Burserac.;* Sur
„ de montagne: *Hedwigia balsamifera* Swartʒ; *Burserac.;* Sur., Ant.
„ „ la Guadeloupe: *Dacryodes hexandra* Griseb.; *Burserac.;* Guad.
Gommo limo: *Bursera gummifera* L.; *Burserac.;* Bah.
Gomorrow: *Dipteryx odorata* Willd.; *Papilionac.;* br. Gu.
Gòn: *Ceiba pentandra* Gaertn.; *Bombacac.;* Ind.-Ch.
„ rung: *Bombax malabaricum* DC.; *Bombacac.;* Annam.
Gonakié: *Acacia arabica* Willd.; *Mimosac.;* Sudan.
„ : *Acacia decurrens* Willd.; *Mimosac.;* Senl.
Gonçalo Alves: *Astronium fraxinifolium* Schott; *Anacardiac.;* Bras.
„ „ rajado branco: *Astronium fraxinifolium* Schott; *Anacardiac.;* Bras.
„ „ „ preto: *Astronium fraxinifolium* Schott; *Anacardiac.;* Bras.
Gondan: *Cordia myxa* L.; *Borraginac.;* C.-O.-Ind.
Gondhori: *Cinnamomum cecidodaphne* Meissn.; *Laurac.;* Assam.
Gondri: *Cinnamomum cecidodaphne* Meissn.; *Laurac.;* Assam.
Gondserai: *Cinnamomum cecidodaphne* Meissn.; *Laurac.;* Assam.
Gondui: *Bridelia retusa* Spreng.; *Euphorbiac.;* n. O.-Ind.
Gondurú: *Brosimum angustifolium* Ducke; *Morac.;* Bras. (Pará).
„ : *Brosimum conduru* Allem.; *Morac.;* Bras. (Pará).
Gong: *Helicia petiolaris* Benn.; *Proteac.;* Malay.
Gonggang: *Cratoxylon arborescens* Blume; *Guttifer.;* Malay.
Gonguonkiur: *Rauwolfia vomitoria* Afzel.; *Apocynac.;* Elf.
Goni: *Pterocarpus erinaceus* Lam.; *Papilionac.;* Sudan.
Gontatsoe: *Xylocarpus sp.; Meliac.;* Nied.-Ind.
Gonwi: ? *Ficus vogeliana* Miq.; *Morac.;* Lib.
Gonyer: *Grewia elastica* Royle; *Tiliac.;* w. Beng. (Bihar, Orissa).
Gonyo: *Antrocaryon nannani* de Wild.; *Anacardiac.;* B.-K.
Goo mao mah: *Laportea gigas* Wedd.; *Urticac.;* N.-S.-W.
Go-onje: *Careya australis* F. v. M.; *Lecythidac.;* nö. Au.
Googgilapa-karra: *Shorea tumbuggaria* Roxb.; *Dipterocarpac.;* w. O.-Ind. (n. Bombay).
Googoolapoo-chittoo: *Boswellia serrata* Roxb.; *Burserac.;* w. O.-Ind.
Goojoo: *Bridelia retusa* Spreng.; *Euphorbiac.;* w. O.-Ind. (Bombay).
Gooler: *Ficus glomerata* Roxb.; *Morac.;* O.-Ind.
Gopher wood: *Cladrastis lutea* Koch; *Papilionac.;* ö. USA.
Gorchi: *Zizyphus xylopyrus* Willd.; *Rhamnac.;* w. O.-Ind. (Bombay).
Goré: *Ongokea klaineana* Pierre; *Olacac.;* Gab.
Gore: *Stephegyne parvifolia* Korth.; *Rubiac.;* ö. Beng.
Gorgoran: *Didymopanax morototoni* Dcne. et Planch.; *Araliac.;* Pan.
Goria nim: *Toona ciliata* M. Roem.; *Meliac.;* O.-Ind.
Goriba: *Hyphaene thebaica* Mart.; *Palmac.;* n. Nig.
Gormi-kawat: *Ailanthus excelsa* Roxb.; *Simarubac.;* sw. Beng.
Goro: *Garcinia kola* Heckel; *Guttifer.;* Kam.
Goroçahy: *Moldenhauera floribunda* Schrader; *Caesalpiniac.;* Bras. (Sao P).
Gorongo: *Rhizophora mucronata* Lamk.; *Rhizophorac.;* O.-Af. (Moz., Quel.).
Gorse (e): *Ulex europaeus* L.; *Papilionac.;* Eu.
Gorukh-chenʒ: *Adansonia digitata* L.; *Bombacac.;* Ind.
Gor-vah: *Sabicea lasiocalyx* Stapf; *Rubiac.;* Lib.
Gorwi: *Zizyphus xylopyrus* Willd.; *Rhamnac.;* w. O.-Ind. (Bombay).
Gosum: *Schleichera trijuga* Willd.; *Sapindac.;* n. O.-Ind.
Got: *Zizyphus xylopyrus* Willd.; *Rhamnac.;* w. O.-Ind. (Bombay).
Goterero: *Psychotria carthaginensis* Jacq.; *Rubiac.;* Kol.
Goti: *Zizyphus xylopyrus* Willd.; *Rhamnac.;* w. O.-Ind. (n. Bombay).
Goting: *Terminalia belerica* Roxb.; *Combretac.;* sw. O.-Ind.
Goto: *Zizyphus xylopyrus* Willd.; *Rhamnac.;* sw. Beng.
Gotoboro: *Zizyphus xylopyrus* Willd.; *Rhamnac.;* sw. Beng.
Gotogiri: *Sterculia platanifolia* L. f.; *Sterculiac.;* Jap.
Götterbaum (d): *Ailanthus glandulosa* Desf.; *Simarubac.;* Ch.

Götterbaum, drüsiger: *Ailanthus glandulosa* Desf.; *Simarubac.;* Ch.
Gotti: *Zizyphus xylopyrus* Willd.; *Rhamnac.;* Ind.
Gouahimbondou: *Spondianthus preussi* Engl.; *Anacardiac.;* Gab.
Gouaré: *Guarea trichilioides* L.; *Meliac.;* tr. Am.
Gouenesso: *Ceiba pentandra* Gaertn. *var. clausa* Ulb.; *Bombacac.;* Dah.
Gouit: *Adansonia digitata* L.; *Bombacac.;* Senl.
Goulougou-albani: *Sloanea sp.; Elaeocarpac.;* Gu.
Gouma: *Ceiba pentandra* Gaertn. *var. dehiscens* Ulb.; *Bombacac.;* Dah.
Gounou gounou: *Macrolobium palisoti* Bth.; *Caesalpiniac.;* Gab.
Goupi: *Goupia tomentosa* Aubl., *G. glabra* Aubl.; *Celastrac.;* fr. Gu.
„   franc: *Goupia glabra* Aubl.; *Celastrac.;* Sur.
„   glabre: *Goupia glabra* Aubl.; *Celastrac.;* Sur.
„   jaune: *Goupia tomentosa* Aubl.; *Celastrac.;* fr. Gu.
„   red: *Goupia glabra* Aubl.; *Celastrac.;* fr. Gu.
Goura: *Cola acuminata* R. Br.; *Sterculiac.;* Senl.
Goure: *Stephegyne parvifolia* Korth.; *Rubiac.;* ö. Beng.
Gourou: *Cola acuminata* R. Br.; *Sterculiac.;* Senl.
Gouty stem tree: *Adansonia gregori* F. v. M.; *Bombacac.;* n. Au.
Gouty tree: *Adansonia gregori* F. v. M.; *Bombacac.;* n. Au.
Govwi: *Musanga smithi* R. Br.; *Morac.;* Lib.
Goyabarana: *Mouriria sp.; Melastomatac.;* Bras.
Goyavier (f): *Psidium spp., P. guajava* L.; *Myrtac.;* tr. Am.
„   commun: *Psidium guajava* Raddi; *Myrtac.;* tr. Am.
„   fraise: *Psidium cattleyanum* Sabine; *Myrtac.;* Guat.
„   marron: *Ludia sessiliflora* Lam.; *Flacourtiac.;* Réu.
„   „ blanc: *Ludia heterophylla* Lamk.; *Flacourtiac.;* Réu.
„   montagne: *Myrcia deflexa* DC.; *Myrtac.;* Guad.
„   sauvage: *Ludia heterophylla* Bory; *Flacourtiac.;* Réu.
„   „   : *Prockia theaeformis* Willd.; *Flacourtiac.;* Réu.
„   à fruits: *Psidium guajava* Raddi; *Myrtac.;* W.-Ind.
Goyomatsu: *Pinus parviflora* S. et Z.; *Conifer.;* Jap.
Goyomatzu: *Pinus parviflora* S. et Z.; *Conifer.;* Jap.
G'Pon: *Fagara macrophylla* Engl.; *Rutac.;* Elf.
Gracuhy: *Ferreirea spectabilis* Allem.; *Papilionac.;* Bras. (Sao P).
Graines bleues: *Symplocos martinicensis* Jacq.; *Symplocac.;* Guad.
Grainier: *Cercis siliquastrum* L.; *Caesalpiniac.;* s. Eu., Kl.-As.
Gran: *Picea excelsa* Link; *Conifer.;* Schwed.
Granada: *Punica granatum* L.; *Punicac.;* P.-R.
Granadilla: *Gymnanthes lucida* Sw.; *Euphorbiac.;* St.-D.
„   : *Caesalpinia sclerocarpa* Standl.; *Caesalpiniac.;* Mex.
„   : *Couroupita odoratissima* Seem.; *Lecythidac.;* Ven.
Granadillo: *Brya ebenus* DC.; *Papilionac.;* Cuba, Jam.
„   : *Buchenavia capitata* Eichler; *Combretac.;* P.-R.
„   : *Dalbergia granadillo* Pittier; *Papilionac.;* Mex.
? „   : *Dalbergia retusa* Hemsl.; *Papilionac.;* Mex., Hond.
„   : *Dalbergia cubilquitzensis* Pittier; *Papilionac.;* Guat.
„   : *Dalbergia aff. lineata* Pittier, *D. sp.; Papilionac.;* Hond.
„   : *Platymiscium polystachyum* Donn. Smith; *Papilionac.;* Hond., Salv.
„   : ? *Platymiscium sp.; Papilionac.;* br. Hond.
„   : *Couroupita odoratissima* Seem.; *Lecythidac.;* Pan.
„   : *Caesalpinia paucijuga* Bth.; *Caesalpiniac.;* Kol.
„   : ? *Dialium divaricatum* Vahl; *Caesalpiniac.;* Kol.
„   : *Caesalpinia granadillo* Pittier; *Caesalpiniac.;* Ven., Mex.
„   : *Podocarpus coriaceus* Rich.; *Conifer.;* Ven.
„   : *Rhamnus sp.; Rhamnac.;* Par.
Granado: *Punica granatum* L.; *Punicac.;* Sp.
Grand Bassam: *Khaya ivorensis* Chev.; *Meliac.;* Elf.
Grandul: *Cajanus indicus* Spreng.; *Papilionac.;* Tr.
Granévano: *Astragalus clusi* Boiss.; *Papilionac.;* Sp.
Grangeno: *Celtis sp.; Ulmac.;* Mex.
Granjeno: *Celtis sp.; Ulmac.:* Mex.
„   huasteco: *Celtis sp.; Ulmac.;* Mex.

Granodoro: *Punica granatum* L.; *Punicac.;* s. Eu., w. As.
Grão de gallo: *Cordia calocephala* Cham., *C. insignis* Cham.; *Borraginac.;* Bras.
„ „ „ : *Lucuma sp.; Sapotac.;* Bras.
Graos grandes de gallo: *Celtis glycycarpa* Mart.; *Ulmac.;* Bras.
Grape, Mangrove-: *Coccoloba uvifera* L.; *Polygonac.;* Fl., br. W.-Ind.
„ , Sea-: *Coccoloba uvifera* L.; *Polygonac.;* Fl., tr. Am., W.-Ind.
„ , „ : *Coccoloba barbadensis* Jacq.; *Polygonac.;* br. Hond.
„ , Sea-side-: *Coccoloba uvifera* L.; *Polygonac.;* tr. Am.
„ , wild: *Coccoloba barbadensis* Jacq.; *Polygonac.;* br. Hond. ·
„ , „ : *Sloanea sp.; Elaeocarpac.;* br. Hond.
„ , „ : *Pentagonia macrophylla* Bth.; *Rubiac.;* Pan.
Grapeapunha: *Apuleia praecox* Mart.; *Caesalpiniac.;* Bras. (Sao P).
Grapefruit: *Citrus grandis* Osbeck; *Rutac.;* Hond.
Grapiapunha: *Apuleia praecox* Mart.; *Caesalpiniac.;* Bras.
„ branca: *Apuleia praecox* Mart.; *Caesalpiniac.;* Bras.
Grapiapuña: *Apuleia praecox* Mart.; *Caesalpiniac.;* Arg.
Grass tree: *Xanthorrhoea arborea* R. Br.; *Liliac.;* Queensl., N.-S.-W.
„ „ : *Panax crassifolium* Dcne. et Planch.; *Araliac.;* N.-Seel.
Grataúba: *Vitex sp.; Verbenac.;* Bras. (Sao P).
Graúna: *Melanoxylon brauna* Schott; *Caesalpiniac.;* Bras. (Sao P).
„ parda: *Melanoxylon brauna* Schott; *Caesalpiniac.;* Bras. (Sao P).
„ preta: *Melanoxylon brauna* Schott; *Caesalpiniac.;* Bras. (Sao P).
Gravelin: *Quercus pedunculata* Ehrh.; *Fagac.;* Eu., Kl.-As.
Gravilea: *Grevillea robusta* Cunn.; *Proteac.;* Hond.
Grawang: *Palaquium javense* Burck.; *Sapotac.;* Java.
Graw-grawp: *Dillenia pentagyna* Roxb.; *Dilleniac.;* O.-Ind.
Grayume: *Didymopanax morototoni* Dcne. et Planch.; *Araliac.;* P.-R.
„ macho: *Didymopanax morototoni* Dcne. et Planch.; *Araliac.;* P.-R.
Grayumo: *Didymopanax morototoni* Dcne. et Planch.; *Araliac.;* P.-R.
„ macho: *Didymopanax morototoni* Dcne. et Planch.; *Araliac.;* P.-R.
Grease nut: *Hernandia bivalvis* Bth.; *Hernandiac.;* Queensl.
Greenheart: *Tecoma sp.; Bignoniac.;* Tob.
„ : *Laurus chloroxylon* L.; *Laurac.;* Jam.
„ (e): *Ocotea rodioei* Mez; *Laurac.;* br. Gu.
„ „ : *Tecoma leucoxylon* Mart.; *Bignoniac.;* Sur.
„ : *Geijera salicifolia* Schott; *Rutac.;* Au.
„ , african (e): *Piptadenia africana* Hook. f.; *Mimosac.;* W.-Af.
„ , „ : *Cylicodiscus gabunensis* Harms; *Papilionac.;* Nig., Go.
„ , „ , light (e): *Piptadenia africana* Hook. f.; *Mimosac.;* W.-Af.
„ , black: *Nectandra sp.; Laurac.;* br. Gu.
„ , „ : *Tecoma sp.; Bignoniac.;* Sur.
„ , brown: *Nectandra sp.; Laurac.;* br. Gu.
„ , Demerara-: *Ocotea rodioei* Mez; *Laurac.;* Sur.
„ , Jamaica-: *Ceanothus chloroxylon* Nees; *Rhamnac.;* Jam.
„ , Queensland-: *Endiandra compressa* C. T. Wight; *Laurac.;* Au.
„ , Sloane's-: *Sloanea sp.; Elaeocarpac.:* Jam.
„ , Surinam-: *Tecoma leucoxylon* Mart.; *Bignoniac.;* Sur.
„ , Westindian-: *Colubrina ferruginea* Brongn.; *Rhamnac.;* Ant.
„ , „ : *Colubrina reclinata* Brongn.; *Rhamnac.;* Fl., W.-Ind.
„ , white: *Nectandra sp.; Laurac.;* br. Gu.
„ , yellow: *Nectandra sp.; Laurac.;* br. Gu.
„ de la Guyane anglaise: *Ocotea rodioei* Mez; *Laurac.;* fr. Gu.
Greenweed, Dyers-: *Genista tinctoria* L.; *Papilionac.;* Eu., w. As., Sib.
Grenada: *Hirtella americana* L.; *Rosac.;* br. Hond.
„ : *Coccoloba barbadensis* Jacq.; *Polygonac.;* br. Hond.
„ cimarrón: *Eugenia axillaris* Willd.; *Myrtac.;* br. Hond.
Grenadier (f): *Punica granatum* L.; *Punicac.;* Tr.
Grenadille: *Brya ebenus* DC.; *Papilionac.;* W.-Ind.
„ d'Afrique (f): *Dalbergia melanoxylon* Guill. et Perr.; *Papilionac.;* tr. Af.
Grenadillholz (d): *Brya ebenus* DC.; *Papilionac.;* W.-Ind.
„ , afrikanisch (d): *Dalbergia melanoxylon* Guill. et Perr.; *Papilionac.;* tr. Af.
Gréou (f): *Ilex aquifolium* L.; *Aquifoliac.;* w. Eu.

Gretado amarillo: *Cupressus sp.; Conifer.;* Mex.
„ galán: *Cupressus sp.; Conifer.;* Mex.
Gretski Aryekh: *Juglans regia* L.; *Juglandac.;* R.
Greyanger: *Panax elegans* F. v. M.; *Araliac.;* Queensl., N.-S.-W.
Greywood (e): *Acer pseudoplatanus* L.; *Acerac.;* Eu.
„ , Silver-: *Terminalia bialata* Wall.; *Combretac.;* Ind., Cey.
„ , „ , indian: *Terminalia bialata* Wall.; *Combretac.;* Ind.
Griegrie: *Dryobalanops aromatica* Gaertn. f.; *Dipterocarpac.;* Sunda.
Grienharti: *Tecoma leucoxylon* Mart. *var. pentaphylla* Juss.; *Bignoniac.;* Sur.
„ : *Tecoma araliacea* P. DC.; *Bignoniac.;* Sur.
Griesholz (d): *Ligustrum vulgare* L.; *Oleac.;* s. Eu.
Grignon: *Conocarpus erecta* L.; *Combretac.;* fr. Gu.
„ : *Bucida angustifolia* DC.; *Combretac.;* fr. Gu.
„ : *Ocotea rubra* Mez; *Laurac.;* fr. Gu.
„ fou: *Parkia pendula* Bth.; *Mimosac.;* fr. Gu.
„ „ : *Qualea coerulea* Aubl.; *Vochysiac.;* fr. Gu.
„ „ rouge: *Qualea coerulea* Aubl., *Q. rosea* Aubl.; *Vochysiac.;* Sur.
„ „ „ : *Qualea altiflora* Warm.; *Vochysiac.;* Sur.
„ franc: *Ocotea rubra* Mez; *Laurac.;* fr. Gu.
„ rouge: ? *Nectandra sp., Ocotea rubra* Mez; *Laurac.;* Sur.
„ „ : ? *Terminalia buceras* Hook.; *Combretac.;* Sur.
Gri-gri: *Parinarium campestre* Aubl.; *Rosac.;* Sur.
Grijano: *Phillyrea media* L.; *Oleac.;* Sp.
Grijsappel: *Parinarium mobola* Oliv.; *Rosac.;* S.-Af.
Gris gris: *Licania galibica* R. Ben.; *Rosac.;* fr. Gu.
„ - „ : *Licania heteromorpha* Bth.; *Rosac.;* Sur.
„ „ : *Licania divaricata* Bth., *L. crassifolia* Bth.; *Rosac.;* Sur.
„ „ : *Licania macrophylla* Bth.; *Rosac.;* Sur.
„ „ coumaté: *Licania macrophylla* Bth.; *Rosac.;* Sur.
„ „ gaulette: *Licania galibica* R. Ben.; *Rosac.;* fr. Gu.
„ „ grande feuille: *Licania macrophylla* Bth.; *Rosac.;* fr. Gu.
„ „ rouge: *Licania galibica* R. Ben.; *Rosac.;* fr. Gu.
Griting: *Lumnitzera littorea* Voigt; *Combretac.;* tr. As., tr. Au., Südsee.
Groçahy: *Moldenhauera floribunda* Schrader; *Caesalpiniac.;* Bras. (Sao P).
Groenheart (h): ? *Tecoma leucoxylon* Mart.; *Bignoniac.;* Sur.
„ , Demerara-: *Ocotea rodioei* Mez; *Laurac.;* Sur.
„ , surinaamsch: *Tecoma leucoxylon* Mart. *var. pentaphylla* Juss.; *Bignoniac.;* Sur.
„ , „ : *Tecoma araliacea* P. DC.; *Bignoniac.;* Sur.
„ , Water-: *Sweetia nitens* Bth.; *Papilionac.;* Sur.
Groenhartboom (h): *Ocotea rodioei* Mez; *Laurac.;* Sur.
Groenhati: *Tecoma sp.; Bignoniac.;* Sur.
Groenhout (h): *Ceanothus chloroxylon* Nees; *Rhamnac.;* Jam.
Groenhoutboom (h): *Ceanothus chloroxylon* Nees; *Rhamnac.;* Jam.
Grofoloe: *Qualea albiflora* Warm.; *Vochysiac.;* Sur.
Gronfoeloe: *Qualea albiflora* Warm.; *Vochysiac.;* Sur.
Gronfoloe: *Qualea albiflora* Warm.; *Vochysiac.;* Sur.
Grosellero: *Ribes grossularia* L.; *Saxifragac.;* Sp.
Grossahy: *Moldenhauera floribunda* Schrader; *Caesalpiniac.;* Bras. (Sao P).
„ azeite: *Moldenhauera sp.; Caesalpiniac.;* Bras. (Sao P).
Groundnut, Forest-: *Pterocarpus santalinoides* L'Hérit.; *Papilionac.;* Go.
Gru-gru: *Acrocomia sclerocarpa* Mart.; *Palmac.;* br. Hond., Trin.
Grumichaba: *Eugenia brasiliensis* L.; *Myrtac.;* Bras. (Sao P.)
„ branca: *Eugenia sp.; Myrtac.;* Bras. (Sao P).
„ vermelha: *Eugenia sp.; Myrtac.;* Bras. (Sao P).
Grünherz (d): *Ocotea rodioei* Mez; *Laurac.;* Sur.
„ : : *Tecoma leucoxylon* Mart.; *Bignoniac.;* Sur.
Grünholz (d): *Ocotea rodioei* Mez; *Laurac.;* br. Gu.
Grünes Holz (d): *Ceanothus chloroxylon* Nees; *Rhamnac.;* Jam.
Gtoul: *Holoptelea integrifolia* Planch.; *Ulmac.;* Ind.-Ch.
Gu: *Aesculus indica* Colebr.; *Hippocastanac.;* n. O.-Ind. (Kengra).
„ *Sindora spp., S. cochinchinensis* Baill.; *Caesalpiniac.;* Ind.-Ch.
„ cau: *Sterculia dongnaiensis* Pierre; *Sterculiac.;* Ind.-Ch.

Gu huong: *Cinnamomum ilicioides* Chev.; *Laurac.;* Ind.-Ch.
„ lau: *Sindora sp.; Caesalpiniac.;* Ind.-Ch.
„ mât: *Sindora tonkinensis* Chev.; *Caesalpiniac.;* Ind.-Ch.
„ toul: *Adina cordifolia* Hook.; *Rubiac.;* Ind.-Ch.
„ tui: *Adina cordifolia* Hook.; *Rubiac.;* Ind.-Ch.
Gua: *Psidium guajava* L.; *Myristicac.;* Go.
Guaba: *Inga vera* H. B. K.; *Mimosac.;* P.-R.
Guabán: *Trichilia sp.; Meliac.;* Cuba.
Guabillo: *Casearia sp.; Flacourtiac.;* Guat.
Guabira: *Campomanesia crenata* Berg; *Myrtac.;* Arg.
Guabiraba: *Campomanesia crenata* Berg; *Myrtac.;* Arg.
Guabiróba: *Campomanesia sp.; Myrtac.;* Bras. (Sao P).
Guabito peludo: *Inga sp.; Mimosac.;* Pan.
Guabiyú: *Eugenia guabiju* Berg, *E. pungens* Berg; *Myrtac.;* Arg.
Guabo: *Inga spp.; Mimosac.;* Ek.
Guacá: *Lucuma sp.; Sapotac.;* Bras. (Sao P).
„ branco: *Lucuma sp.; Sapotac.;* Bras. (Sao P).
„ vermelho: *Lucuma sp.; Sapotac.;* Bras. (Sao P).
Guacacoa: *Daphnopsis cubensis* Meissn.; *Thymelaeac.;* Cuba.
Guacal: *Crescentia cujete* L.; *Bignoniac.;* Salv.
Guacalote prieto: *Caesalpinia crista* L.; *Caesalpiniac.;* Cuba.
Guacamarí común: *Wallenia laurifolia* Sw.; *Myrsinac.;* Cuba.
Guacamaya: *Delonix regia* Rafin.; *Caesalpiniac.;* Hond.
Guacamayo: ? *Andira vermifuga* Mart.; *Papilionac.;* Guat., Hond.
„ : *Caesalpinia pulcherrima* Sw.; *Caesalpiniac.;* Hond.
„ : *Acacia sp.; Mimosac.;* Kol.
„ : *Protium sp.; Burserac.;* Ven., Kol.
„ -caspi: *Coutarea hexandra* K. Schum.; *Rubiac.;* Peru.
Guacarán: *Exothea paniculata* Radlk.; *Sapindac.;* s. Fl., Bah., Ant., Guat.
Guacayamo: ? *Andira excelsa* H. B. K.; *Papilionac.;* Guat.
Guachapelí: *Lysiloma guachapele* Bth.; *Mimosac.;* Ek.
Guacharaco: *Protium sp.; Burserac.;* Kol.
„ : *Anisomeria polyantha* Rusby; *Rubiac.;* Ven.
„ : *Tabernaemontana sp.; Apocynac.;* Bras.
Guache jobonillo: *Bulnesia bonariensis* Griseb.; *Zygophyllac.;* Arg.
Guachimole: *Pithecolobium dulce* Bth.; *Mimosac.;* Salv.
Guachipelí: *Diphysa carthaginensis* Jacq.; *Papilionac.;* Salv., C.-R.
Guachipilín: *Diphysa carthaginensis* Jacq.; *Papilionac.;* Salv., C.-R.
„ : *Diphysa robinioides* Bth.; *Papilionac.;* C.-Am.
Guacima: *Guazuma ulmifolia* Lam.; *Sterculiac.;* Mex.
„ : *Luehea seemanni* Triana et Planch.; *Tiliac.;* Pan.
„ amarilla: *Luehea sp.; Tiliac.;* Cuba.
„ Baria: *Luehea platypetala* Rich.; *Tiliac.;* Cuba.
„ blanca: *Goethalsia isthmica* Pittier; *Tiliac.;* Pan.
„ boba: *Guazuma ulmifolia* Lam.; *Sterculiac.;* Cuba.
„ del Norte: *Guazuma guazuma* Cockerell; *Sterculiac.;* P.-R.
„ „ „ : *Guazuma ulmifolia* Lam.; *Sterculiac.;* P.-R.
Guacimilla: *Trema micrantha* Blume; *Ulmac.;* P.-R.
„ de costa: *Prockia crucis* L.; *Flacourtiac.;* Ven.
Guacimillo: *Guazuma ulmifolia* Lam.; *Sterculiac.;* Nic.
Guácimo: *Luehea sp.; Tiliac.;* C.-R.
„ : *Guazuma ulmifolia* Lam.; *Sterculiac.;* tr. Am.
„ : *Luehea seemanni* Triana et Planch.; *Tiliac.;* Pan.
„ : *Bursera gummifera* L.; *Burserac.;* Kol.
? „ : *Laetia americana* L.; *Flacourtiac.;* Kol.
„ blanco: *Luehea sp.; Tiliac.;* Ven.
„ „ : *Goethalsia meiantha* Burret; *Tiliac.;* Pan.
„ cimarrón: *Guazuma tomentosa* H. B. K.; *Sterculiac.;* Ven.
„ colorado: *Luehea seemanni* Triana et Planch.; *Tiliac.;* Hond.
? „ dulce: *Guazuma ulmifolia* Lam.; *Sterculiac.;* Ven.
„ macho: *Luehea sp.; Tiliac.;* C.-R.
„ „ : *Guazuma ulmifolia* Lam.; *Sterculiac.;* Ven.

Guácimo molenillo: *Luehea sp.; Tiliac.;* C.-R.
    „   „ : *Luehea seemanni* Triana et Planch.; *Tiliac.;* Pan.
  „ molinero: *Luehea seemanni* Triana et Planch.; *Tiliac.;* Nic.
  „ de ternero: *Guazuma ulmifolia* Lam.; *Sterculiac.;* Pan.
Guacle: *Bulnesia bonariensis* Griseb.; *Zygophyllac.;* Arg.
Guaco: *Pisonia sp.; Nyctaginac.;* Salv.
Guacoco: *Capparis sp.; Capparidac.;* Salv.
Guacolote: *Caesalpinia crista* L.; *Caesalpiniac.;* Mex.
Guacuco: ? *Casearia aculeata* Jacq.; *Flacourtiac.;* Guat.
Guadua: *Guadua latifolia* Kunth, *G. angustifolia* Kunth; *Gramin.;* Ek.
Guaiacan: *Guaiacum officinale* L.; *Zygophyllac.;* tr. Am.
Guaiacholz (d): *Guaiacum officinale* L.; *Zygophyllac.;* tr. Am.
Guaiaco (p): *Guaiacum officinale* L.; *Zygophyllac.;* tr. Am.
Guaiaco banco (i): *Guaiacum sanctum* L.; *Zygophyllac.;* W.-Ind., Mex.
  „ nero (i): *Guaiacum officinale* L.; *Zygophyllac.;* tr. Am.
Guaiacum: *Guaiacum officinale* L.; *Zygophyllac.;* Ven.
Guaiacumwood (e): *Guaiacum officinale* L.; *Zygophyllac.;* tr. Am.
Guaiavirai: *Rhamnus sp.; Rhamnac.;* Par.
Guaicá: *Ocotea sp.; Laurac.;* ? Arg.
Guaímaro: ? *Piratinera guianensis* Aubl.; *Morac.;* Pan.
    „   : *Brosimum columbianum* Blake; *Morac.;* Kol.
  „ macho: *Couma sapida* Pittier; *Apocynac.;* Ven.
Guaimí-piré: *Guarea sp.; Meliac.;* Arg.
    „   „ : *Lonchocarpus costatus* Bth.; *Papilionac.;* Arg.
Guaita: *Trichilia sp.; Meliac.;* P.-R.
Guajaba: *Psidium guajava* L.; *Myrtac.;* tr. Am.
Guajabara: *Coccoloba uvifera* L.; *Polygonac.;* Bras.
Guajacan: *Guaiacum officinale* L.; *Zygophyllac.;* tr. Am.
    „   : *Caesalpinia melanocarpa* Griseb.; *Caesalpiniac.;* Arg.
  „ blanco: *Guaiacum officinale* L.; *Zygophyllac.;* W.-Ind.
  „ negro: *Guaiacum officinale* L.; *Zygophyllac.;* W.-Ind.
Guajakholz (d): *Guaiacum spp.; Zygophyllac.;* tr. Am.
    „   „ : *Guaiacum officinale* L.; *Zyhophyllac.;* W.-Ind.
Guajak, Bastard-: *Tecoma leucoxylon* Mart.; *Bignoniac.;* S.-Am., Ant.
Guajará: *Lucuma spp., Chrysophyllum sp.; Sapotac.;* Bras.
Guajava: *Psidium guajava* L.; *Myrtac.;* s. USA.
Guaje: *Leucaena brachycarpa* Urban; *Mimosac.;* Salv.
Guaje cirial: *Crescentia alata* H. B. K.; *Bignoniac.;* Mex.
Guajilla: *Pithecolobium brevifolium* Bth.; *Mimosac.;* Mex.
Guajillo: *Leucaena lanceolata* Watson; *Mimosac.;* Mex. (Sinaloa).
Guajilote: *Bombax palmeri* S. Watson; *Bombacac.;* Mex.
Guajiniquil: *Inga edulis* Mart.; *Mimosac.;* Nic., Pan.
Guajurú: *Chrysobalanus icaco* L.; *Rosac.;* Bras.
Guajuvira: *Patagonula americana* L.; *Borraginac.;* Bras.
  „ branca: *Patagonula americana* L.; *Borragniac.;* Bras.
Gualanday: *Jacaranda sp.; Bignoniac.;* Kol.
Gualeguay: *Schinus molle* L.; *Anacardiac.;* Arg.
Guamá: *Inga sp.; Mimosac.;* Cuba.
  „ : *Lonchocarpus sericeus* H. B. K.; *Papilionac.;* Cuba.
  „ : *Inga laurina* Willd.; *Mimosac.;* P.-R.
  „ : *Inga bonplandiana* H. B. K., *I. vera* Willd.; *Mimosac.;* Ven.
  „ común: *Lonchocarpus sericeus* H. B. K.; *Papilionac.;* Cuba.
  „ hediondo: *Ichthyomethia piscipula* Hitch.; *Papilionac.;* Cuba.
  „ tapaculo: *Inga sp.; Mimosac.;* Kol.
  „ de Costa: *Lonchocarpus latifolius* H. B. K.; *Papilionac.;* Cuba.
  „ „ machete: *Inga sp.; Mimosac.;* Kol.
Guamachi: *Pithecolobium dulce* Bth.; *Mimosac.;* Mex.
Guamachile: *Pithecolobium dulce* Bth.; *Mimosac.;* Mex.
Guamacho: *Pereskia guamacho* Weber; *Cactac.;* Ven.
    „   : *Xylosma prunifolium* Griseb.; *Flacourtiac.;* Kol.
    „   : *Pereskia colombiana* Britt. et Rose; *Cactac.;* Kol.
Guamare: *Dipteryx odorata* Willd.; *Papilionac.;* Sur.

Guambo: *Phoebe ambigens* Blake; *Laurac.;* Guat., Hond.
Guamichava: *Trichilia sp.; Meliac.;* Bras. (Sao P).
Guamirí: *Trichilia sp.; Meliac.;* Arg.
Guamito: *Inga sp.; Mimosac.;* Kol.
  „ macho: *Caesalpinia platyloba* Watson; *Caesalpiniac.;* Kol.
Guamo: *Inga spp.; Mimosac.;* tr. Am.
  „ : *Inga rodrigueziana* Pittier; *Mimosac.;* Guat.
  „ : *Inga insignis* Kunth; *Mimosac.;* Ven.
  „ bejuco: *Inga sp.; Mimosac.;* Ven.
  „ blanco: *Guarea sp.; Meliac.;* Kol.
  „ caraota: *Inga sp.; Mimosac.;* Ven.
  „ cimarrón: *Guarea sp.; Meliac.;* Kol.
  „ machete: *Inga spectabilis* Willd.; *Mimosac.;* Kol.
  „ macho: *Inga recordi* Britt. et Rose; *Mimosac.;* Guat.
? „   „ : *Pithecolobium pilulosum* Pittier; *Mimosac.;* Kol.
  „   „ : *Inga sp.; Mimosac.;* Kol.
  „ mico: *Inga sp.; Mimosac.;* Kol.
  „ peludo: *Inga sp.; Mimosac.;* Ven.
  „ prieto: *Pithecolobium longifolium* Standl.; *Mimosac.;* Kol.
  „ de hierro: *Inga sp.; Mimosac.;* Ven.
Guamuche: *Pithecolobium dulce* Bth.; *Mimosac.;* Mex.
Guamúchil: *Pithecolobium dulce* Bth.; *Mimosac.;* Mex.
Guamúchitl: *Pithecolobium dulce* Bth.; *Mimosac.;* Mex.
Guanaba: *Anona sp.; Anonac.;* Salv.
Guanábana: *Anona muricata* L.; *Anonac.;* tr. Am.
  „ cimarrón: *Anona sp.; Anonac.;* P.-R.
Guanábancilla cimarrona: *Vincetoxycum sp.; Asclepiadac.;* P.-R.
Guanabanillo: ? *Oxandra lanceolata* Baill.; *Anonac.;* Ven.
Guanábano: *Anona sp.; Anonac.;* Bras.
  „ cimarrón: *Anona glabra* L.; *Anonac.;* P.-R.
  „ de corcho: *Anona glabra* L.; *Anonac.;* St.-D.
Guana-berry: *Byrsonima sp.; Malpighiac.;* br. W.-Ind.
Guanacaste (u): *Enterolobium cyclocarpum* Griseb.; *Mimosac.;* C.-Am.
  „ blanco: *Cassia sp.; Caesalpiniac.;* Nic.
Guanahé: *Guarea cedrata* Pellegr.; *Meliac.;* Elf.
Guanaké: *Guarea cedrata* Pellegr.; *Meliac.;* Elf.
Guanandi: *Calophyllum brasiliense* Camb.; *Guttifer.;* Bras. (Sao P).
  „ amarello: *Calophyllum sp.; Guttifer.;* Bras. (Sao P).
  „ cedro: *Calophyllum sp.; Guttifer.;* Bras. (Sao P).
  „ piolho: *Calophyllum sp.; Guttifer.;* Bras. (Sao P).
  „ vermelho: *Calophyllum sp.; Guttifer.;* Bras. (Sao P).
Guanandy: *Calophyllum brasiliense* Camb.; *Guttifer.;* Bras. (Sta. Catharina).
Guanco blanco: *Guarea trichilioides* L.: *Meliac.;* Kol.
Guandú: *Cajanus indicus* Spreng.; *Papilionac.;* Arg.
Guango: *Pithecolobium saman* Bth.; *Mimosac.;* Pan., Jam.
Guano: *Ochroma lagopus* Swartz; *Bombacac.;* P.-R.
  „ : *Ochroma limonensis* Rowlee; *Bombacac.;* Hond.
  „ espinosa: *Copernicia pauciflora* Burret; *Palmac.;* Cuba.
  „ hediondo: *Copernicia clarensis* León; *Palmac.;* Cuba.
  „ jata: *Copernicia macroglossa* H. Wendl.; *Palmac.;* Cuba.
  „ prieto: *Copernicia pauciflora* Burret; *Palmac.;* Cuba.
Guao: *Comocladia platyphylla* Rich.; *Anacardiac.;* Cuba.
  „ de costa: *Rhus metopium* L.; *Anacardiac.;* Cuba.
  „ „ „ : *Metopium toxiferum* Krug et Urb., *M. brownei* Urban; *Anacardiac.;*
Guapaque: *Ostrya guatemalensis* Winkler; *Betulac.;* Mex.   [Cuba.
Guapariba: *Rhizophora mangle* L.; *Rhizophorac.;* Bras.
Guapéba: *Lucuma sp.; Sapotac.;* Bras.
Guaperúvú: *Schizolobium excelsum* Vog.; *Caesalpiniac.;* Bras. (Sao P).
Guapéva: *Lucuma lancifolia* A. DC.; *Sapotac.;* Bras. (Sao P).
Guapinol: *Hymenaea courbaril* L.; *Caesalpiniac.;* Guat., Mex.
Guapinole: *Pithecolobium confine* Standl.; *Mimosac.;* Mex. (Sinaloa).
  „   : *Hymenaea courbaril* L.; *Caesalpiniac.;* Mex.

Guapuruvú: *Schizolobium excelsum* Vog.; *Caesalpiniac.*; Bras.
Guara: *Cupania glabra* Swartz; *Sapindac.*; Cuba.
„ : *Cupania americana* L.; *Sapindac.*; P.-R.
„ blanca: *Cupania glabra* Swartz; *Sapindac.*; W.-Ind.
„ común: *Cupania americana* L.; *Sapindac.*; Cuba.
„ de costa: *Cupania glabra* Swartz; *Sapindac.*; W.-Ind.
Guarabú: *Astronium fraxinifolium* Schott; *Anacardiac.*; Bras.
„ : *Peltogyne confertiflora* Bth.; *Caesalpiniac.*; Bras. (Sao P).
„ batata: *Astronium fraxinifolium* Schott; *Anacardiac.*; Bras.
„ branco: *Peltogyne sp.*; *Caesalpiniac.*; Bras. (Sao P).
„ encirado: *Astronium fraxinifolium* Schott; *Anacardiac.*; Bras.
„ preto: *Peltogyne sp.*; *Caesalpiniac.*; Bras. (Sao P).
„ „ : *Astronium concinnum* Schott; *Anacardiac.*; Bras. (Matto Grosso).
„ rajado: *Astronium fraxinifolium* Schott; *Anacardiac.*; Bras. (Sao P).
„ „ : *Peltogyne sp.*; *Caesalpiniac.*; Bras. (Sao P).
„ roxo: *Peltogyne sp.*; *Caesalpiniac.*; Bras.
„ vermelho: *Peltogyne sp.*; *Caesalpiniac.*; Bras.
Guarabussu: *Peltogyne sp.*; *Caesalpiniac.*; Bras.
Guaracahy: *Moldenhauera floribunda* Schrader; *Caesalpiniac.*; Bras. (Sao P).
Guaraguao: ? *Guarea trichilioides* L.; *Meliac.*; P.-R.
„ : *Trichilia spondioides* Jacq.; *Meliac.*; P.-R.
„ : *Guarea sp.*; *Meliac.*; P.-R., Mex.
Guaraja: *Sideroxylon sp.*; *Sapotac.*; Bras.
Guarajuba: *Terminalia sp.*; *Combretac.*; Bras.
Guaranga: *Caesalpinia spinosa* Kuntze; *Caesalpiniac.*; Ek.
Guarango: *Caesalpinia spinosa* Kuntze; *Caesalpiniac.*; Kol.
Guaran-guaran: *Stenolobium stans* Seem.; *Bignoniac.*; Arg.
Guaranguay amarillo: *Stenolobium stans* Seem.; *Bignoniac.*; Arg.
Guaranhé: *Pradosia latescens* Radlk.; *Sapotac.*; Bras.
Guaranhen: *Pradosia latescens* Radlk.; *Sapotac.*; Bras.
Guarantá: *Esenbeckia leiocarpa* Engl.; *Rutac.*; Bras. (Sao P).
Guarantán: *Esenbeckia leiocarpa* Engl.; *Rutac.*; Bras.
Guararema: *Gallesia scorododendron* Casaretto; *Phytolaccac.*; Bras.
Guarataro: *Couratari guianensis* Aubl.; *Lecythidac.*; Ven.
Guarayta: *Astronium sp.*; *Anacardiac.*; Bras. (Sao P).
Guardalobo: *Osyris alba* L.; *Santalac.*; Sp.
Guariare: *Capparis tenuisiliqua* Lam.; *Capparidac.*; Ven.
Guaribu: *Olmedia erythrorrhiza* Huber; *Morac.*; Bras.
Guarichamaca: *Tocoyena foetida* Poepp. et Endl.; *Rubiac.*; Ven.
Guariuba: *Clarisia racemosa* Ruiz et Pav.; *Morac.*; Bras.
Guarri, Bosch-: *Euclea lanceolata* E. Meyer; *Ebenac.*; Kap.
Guarucaia: *Pithecolobium sp.*; *Mimosac.*; Bras. (Sao P).
Guarumbo: *Cecropia sp.*; *Morac.*; Mex.
Guarumo: *Cecropia mexicana* Hemsl.; *Morac.*; C.-Am.
„ : *Cecropia spp.*; *Morac.*; Guat.
„ : *Cecropia asperrima* Pittier; *Morac.*; Hond.
„ : *Cecropia arachnoidea* Pittier; *Morac.*; Pan.
„ : *Cecropia peltata* L., *C. adenopus* Mart.; *Morac.*; Kol.
„ : *Cecropia sp.*; *Morac.*; Ek.
„ macho: *Pourouma aspera* Trecul; *Morac.*; Nic.
„ de montaña: *Pourouma aspera* Trécul; *Morac.*; Hond.
Guasabara: *Mouriria domingensis* Spach; *Melastomatac.*; P.-R.
„ : *Eugenia aeruginea* DC., *E. eggersi* Kiaersk.; *Myrtac.*; P.-R.
„ : *Eugenia tetrasperma* Bello; *Myrtac.*; P.-R.
Guasavará: *Eugenia eggersi* Kiaersk., *E. aeruginea* DC.; *Myrtac.*; P.-R.
Guascanal: *Acacia cooki* Safford; *Mimosac.*; br. Hond., Guat.
Guásima: *Guazuma guazuma* Cockerell; *Sterculiac.*; P.-R.
Guasimo nogal: *Cordia sp.*; *Borraginac.*; Kol.
Guassacu: *Hura crepitans* L.; *Euphorbiac.*; Bras.
Guassalongo amarillo: ? *Guarea pendula* ?; *Meliac.*; S.-Am.
Guastapana: *Caesalpinia coriaria* Willd.; *Caesalpiniac.*; St.-D.
Guatacare: *Lecythis laevifolia* Griseb.; *Lecythidac.*; Trin., Tob.

Guatacare: *Chytroma idatimon* Miers; *Lecythidac.*; Ven.
Guatamblu amarillo: *Aspidosperma olivaceum* Muell. Arg.; *Apocynac.*; Arg.
Guatambú: *Aspidosperma sessiliflorum* Muell. Arg.; *Apocynac.*; Bras.
„   : *Aspidosperma tomentosum* Mart.; *Apocynac.*; Bras.
„   : *Aspidosperma macrocarpum* Mart.; *Apocynac.*; Bras. (Sao P).
„ amarello: *Aspidosperma tomentosum* Mart.; *Apocynac.*; Bras. (Sao P).
„ moroti: *Balfourodendron riedelianum* Engl.; *Rutac.*; Arg.
„ vermelho: *Aspidosperma sp.*; *Apocynac.*; Bras. (Sao P).
Guatapan: *Caesalpinia coriaria* Willd.; *Caesalpiniac.*; Ven.
Guatapana: *Roupala sp.*; *Proteac.*; Trin.
Guatapanare: *Caesalpinia coriaria* Willd.; *Caesalpiniac.*; Ven.
Guatayapóca: *Raputia magnifica* Engl.; *Rutac.*; Bras. (Sao P).
Guatecare: *Chytroma idatimon* Miers; *Lecythidac.*; Trin.
„   : *Lecythis laevifolia* Griseb.; *Lecythidac.*; Trin., Tob.
Guatecaro: *Chytroma idatimon* Miers; *Lecythidac.*; Trin.
Guatope: *Inga sp.*; *Mimosac.*; Mex.
Guatuso: *Hasseltia mexicana* Standl.; *Flacourtiac.*; Hond.
Guautecomate: *Crescentia cucurbitina* L.; *Bignoniac.*; C.-Am., Ven., W.-Ind.
Guava: *Inga sp.*; *Mimosac.*; P.-R.
„   : *Psidium guajava* Raddi; *Myrtac.*; Hond.
„   : *Inga spuria* Willd.; *Mimosac.*; Pan.
„   : *Psidium pomiferum* L.; *Myrtac.*; fr. Gu.
„ blossom: *Lonchocarpus sp.*; *Papilionac.*; br. Hond.
„ , false: *Tristania decorticata* Merr.; *Myrtac.*; Phil.
„ , Florida-: *Psidium buxifolium* Nutt.; *Myrtac.*; USA.
„ , wild: *Calycolpus glaber* Berg; *Myrtac.*; br. Gu.
„ , „ : *Eugenia origanoides* Berg; *Myrtac.*; br. Hond.
„ , „ : *Psidium guajava* Raddi; *Myrtac.*; br. Hond.
„ del mono: *Inga punctata* Willd.; *Mimosac.*; Pan.
„ , Mountain-: *Psidium montanum* Sw.; *Myrtac.*; Jam.
Guavito: *Quassia amara* L.; *Simarubac.*; Pan.
Guavo: *Inga edulis* Mart., *I. punctata* Willd.; *Mimosac.*; Nic., Pan.
„   : *Inga spectabilis* Willd.; *Mimosac.*; Nic., Pan.
„ machete: *Inga edulis* Mart.; *Mimosac.*; Pan.
„ , wild: *Psidium friedrichsthalianum* Bth. et Hook.; *Myrtac.*; Pan.
Guaximbé: *Machaerium sp.*; *Papilionac.*; Bras. (Sao P).
Guayaba: *Psidium guajava* Raddi; *Myrtac.*; tr. Am.
„   : *Psidium guajava* Raddi var. *piriferum* Raddi; *Myrtac.*; S.-Am.
„ agria: *Psidium friedrichsthalianum* Niedenzu; *Myrtac.*; Hond.
„ león: *Terminalia sp.*; *Combretac.*; Kol.
Guayabacoa: *Rheedia portoricensis* Urban; *Guttifer.*; P.-R.
Guayabacón: *Eugenia aeruginea* DC.; *Myrtac.*; P.-R.
„   : *Myrcia divaricata* DC., *M. leptoclada* DC.; *Myrtac.*; P.-R.
Guayabí: *Patagonula americana* L.; *Borraginac.*; Bras. (Sao P).
„ amarillo: *Patagonula americana* L.; *Borraginac.*; Arg.
„ blanco: *Patagonula americana* L.; *Borraginac.*; Arg.
„ moroti: *Patagonula americana* L.; *Borraginac.*; Arg.
„ negro: *Patagonula americana* L.; *Borraginac.*; Arg.
„ sayyu: *Terminalia sp.*; *Combretac.*; Arg.
„ amarillo: *Terminalia sp.*; *Combretac.*; Arg.
Guayabilla: *Psidium molle* Bertol.; *Myrtac.*; Hond.
Guayabillo: *Pterocarpus orbiculatum* DC.; *Papilionac.*; Mex.
„   : *Terminalia obovata* Eichler; *Combretac.*; Guat.
„   : *Chrysophyllum mexicanum* Brandeg.; *Sapotac.*; Salv.
„   : *Quararibea asterolepis* Pittier; *Bombacac.*; Pan.
Guayabira: *Patagonula americana* L.; *Borraginac.*; Arg.
„ amarillo: *Patagonula americana* L.; *Borraginac.*; Arg.
„ blanco: *Patagonula americana* L.; *Borraginac.*; Arg.
„ crespo: *Patagonula americana* L.; *Borraginac.*; Arg.
„ negro: *Patagonula americana* L.; *Borraginac.*; Arg.
Guayabo: *Psidium guajava* Raddi; *Myrtac.*; tr. Am.
„   : *Terminalia obovata* Eichler; *Combretac.*; Guat., Hond.

Guayabo: *Terminalia hayesi* Pittier; *Combretac.*; Hond.
„ : ? *Quararibea asterolepsis* Pittier; *Bombacac.*; Pan.
„ : *Guettarda foliacea* Standl.; *Rubiac.*; Pan.
„ : *Ruprechtia ramiflora* Mez; *Polygonac.*; Kol.
„ : *Eugenia spp., Psidium spp.*; *Myrtac.*; Ven.
„ alazano: *Calycophyllum candidissimum* DC.; *Rubiac.*; Pan.
„ bolador: *Terminalia obovata* Eichler; *Combretac.*; Guat.
„ cascudo: *Eugenia sp.*; *Myrtac.*; Kol.
„ encarnado: *Psidium sp.*; *Myrtac.*; Ven.
„ hormiguero: *Triplaris americana* L.; *Polygonac.*; Pan.
„ león: *Terminalia sp.*; *Combretac.*; Kol.
„ negro: *Hamelia axillaris* Sw.; *Rubiac.*; Pan.
„ pauji: ?*Bumelia buxifolia* Willd.; *Sapotac.*; Ven.
„ sabanero: *Psidium molle* Bertol.; *Myrtac.*; Kol.
„ de agua: *Psidium friedrichsthalianum* Niedenzu; *Myrtac.*; Pan.
„ „ montaña: *Chloroleucum guatemalense* Britt. et Rose; *Mimosac.*; Hond.
„ „ „ : *Terminalia hayesi* Pittier; *Combretac.*; Pan.
Guayabota: *Eugenia stahli* Krug et Urban; *Myrtac.*; P.-R.
„ : *Diospyros sp.*; *Ebenac.*; P.-R.
„ nispero: *Maba sp.*; *Ebenac.*; P.-R.
Guayacan: *Porliera angustifolia* A. Gray; *Zygophyllac.*; N.-Am.
„ : *Guaiacum officinale* L.; *Zygophyllac.*; tr. Am.
„ : *Guaiacum coulteri* A. Gray, *G. palmeri* Vail; *Zygophyllac.*; w. Mex.
„ : *Guaiacum guatemalense* Planch.; *Zygophyllac.*; Guat.
„ : ? *Myrospermum sp.*; *Papilionac.*; Salv.
„ : *Tecoma guayacan* Seem.; *Bignoniac.*; Pan., Kol., Ven.
„ : *Bulnesia arborea* Engl.; *Zygophyllac.*; Kol., Ven.
„ : *Porliera lorentzi* Engl.; *Zygophyllac.*; Arg. (Cord.)
„ : *Porliera hygrometrica* Ruiz et Pav.; *Zygophyllac.*; Arg.
„ : *Caesalpinia melanocarpa* Griseb.; *Caesalpiniac.*; Arg.
„ blanco: *Calliandra portoricensis* Bth.; *Mimosac.*; P.-R.
„ chaparro: *Pithecolobium sp.*; *Mimosac.*; Kol.
„ ciénaga: *Pithecolobium sp.*; *Mimosac.*; Kol.
„ coco: *Chytroma sp.*; *Lecythidac.*; Kol.
„ , Colombia-: *Bulnesia arborea* Engl.; *Zygophyllac.*; Kol.
„ jobo: *Centrolobium sp.*; *Papilionac.*; Kol.
„ negro: *Caesalpinia melanocarpa* Griseb.; *Caesalpiniac.*; Arg.
„ polvillo: *Tecoma spectabilis* Planch. et Linden; *Bignoniac.*; Kol.
„ prieto: *Guaiacum officinale* L., *G. sanctum* L.; *Zygophyllac.*; Cuba.
„ , yellow: *Tecoma leucoxylon* Mart. *var. pentaphylla* Juss.; *Bignoniac.*; Sur.
„ , „ „ : *Tecoma araliacea* P. DC.; *Bignoniac.*; Sur.
„ , yellow-flowering: *Tecoma guayacan* Seem.; *Bignoniac.*; Pan.
Guayacancillo: *Behaimia cubensis* Griseb.; *Papilionac.*; Cuba.
„ : *Guaiacum sanctum* L.; *Zygophyllac.*; Cuba.
Guayacillo: *Rinorea squamata* Blake; *Violac.*; Pan.
Guayaibi blanco: *Patagonula americana* L.; *Borraginac.*; Arg. (Formosa, Chaco).
„ sayyú: ? *Terminalia balansae* Hassler; *Combretac.*; Arg.
Guayaibira: *Patagonula americana* L.; *Borraginac.*; Arg.
Guayamero: *Brosimum columbianum* Blake; *Morac.*; Kol.
Guayaniquil: *Inga sp.*; *Mimosac.*; C.-R.
Guayarote: *Meliosma obtusifolia* Krug et Urban; *Sabiac.*; P.-R.
Guayarrote: *Elaeodendron sp.*; *Celastrac.*; P.-R.
Guayatil: *Sickingia maxoni* Standl.; *Rubiac.*; Pan.
„ blanco: *Genipa americana* L.; *Rubiac.*; Pan.
„ colorado: *Sickingia maxoni* Standl.; *Rubiac.*; Pan.
Guayavacon: *Trichilia sp.*; *Meliac.*; P.-R.
Guayavi: *Patagonula americana* L.; *Borraginac.*; tr. S.-Am.
Guayavira: *Patagonula americana* L.; *Borraginac.*; Arg.
Guayavo: *Psidium pyriferum* L.; *Myrtac.*; Ven.
Guayaybi: *Patagonula americana* L.; *Borraginac.*; Arg.
„ -ra: *Terminalia sp.*; *Combretac.*; Arg.
Guayo colorado: *Acacia peregrina* Kunth; *Mimosac.*; Ven.

Guayti: *Moquilea tomentosa* Bth.; *Rosac.;* Bras.
Guayubira: *Patagonula americana* L.; *Borraginac.;* Arg.
„ amarillo: *Patagonula sp.; Borraginac.;* tr. S.-Am.
Guayul: *Colubrina ferruginosa* Brongn.; *Rhamnac.;* Mex.
Guayun: *Rhaphithamnus cyanocarpus* Miers; *Verbenac.;* S.-Am. (Anden).
Guayuvira: *Ruprechtia sp.; Polygonac.;* Bras. (? Sao P).
Guay-ve-ney: *Diospyros thomasi* Hutch. et Dalz.; *Ebenac.;* Lib.
Guazatumba: *Casearia sp.; Flacourtiac.;* Arg.
Guazima blanca: *Goethalsia isthmica* Pittier; *Tiliac.;* Pan.
Guazuma: *Guazuma ulmifolia* Lamk.; *Sterculiac.;* Trin.
Guazymillo: *Trema micrantha* Blume; *Ulmac.;* P.-R.
Gubas: *Endospermum peltatum* Merr.; *Euphorbiac.;* Phil. (Lag.).
Gudayong: *Parkia roxburghi* G. Don; *Mimosac.;* Malay.
Gude: *Crataeva adansoni* DC.; *Capparidac.;* n. Nig.
Gueba del gato: *Sloanea sp.; Elaeocarpac.;* Kol.
Guedebounsou: *Ceiba pentandra* Gaertn. *var. dehiscens* Ulb.; *Bombacac.;* Dah.
Guédj: *Anogeissus leiocarpus* Guill. et Perr.; *Combretac.;* Sudan.
? Güégüiro Wa Baka: *Berlinia acuminata* Soland.; *Caesalpiniac.;* Elf.
Guélé: *Prosopis oblonga* Bth.; *Mimosac.;* fr. Guin., Sudan.
Guenepa: *Melicocca bijuga* L.; *Sapindac.;* C.-Am.
Guenepe: *Melicocca bijuga* L.; *Sapindac.;* Trin.
Guenlé: *Chlorophora excelsa* Bth. et Hook.; *Morac.;* Elf.
Guenlo: *Chlorophora excelsa* Bth. et Hook.; *Morac.;* Elf.
Guénou: *Pterocarpus erinaceus* Lam.; *Papilionac.;* Sudan.
Guento: *Chlorophora excelsa* Bth. et Hook.; *Morac.;* Elf.
Gueppois: *Eugenia octopleura* Krug et Urban; *Myrtac.;* Guad.
Guerbe hono: *Balanites aegyptiaca* Delile; *Simarubac.;* Sudan.
Guericina: *Cordia sp.; Borraginac.;* s. Mex.
Guerrero: *Eupatorium dalea* L., *E. portoricense* Urb.; *Composit.;* P.-R.
„ : *Eupatorium resinifluum* Urb.; *Composit.;* P.-R.
Gugal (f): *Boswellia serrata* Roxb.; *Burserac.;* Ind.
Gugal: *Shorea robusta* Gaertn. f.; *Dipterocarpac.;* w. O.-Ind. (n. Bombay).
„ : *Balsamodendron mukul* Hook.; *Burserac.;* Ind.
Gugera: *Schima wallichi* Choisy; *Theac.;* n. Assam.
Guggar: *Boswellia serrata* Roxb.; *Burserac.;* nw. O.-Ind.
Gugilam: *Shorea robusta* Gaertn.; *Dipterocarpac.;* Ind.
Gugle: *Vateria indica* L.; *Dipterocarpac.;* sw. O.-Ind.
Gúgo: *Albizzia saponaria* Blume; *Mimosac.;* Phil. (Tayabas).
Gugul: *Odina wodier* Roxb.; *Anacardiac.;* w. O.-Ind.
Gugulukundrikam: *Boswellia serrata* Roxb.; *Burserac.;* w. O.-Ind. (s. Deccan).
Gugumkún: *Shorea palosapis* Merr.; *Dipterocarpac.;* Phil. (Isabella).
Guiacan: *Bulnesia arborea* Engl.; *Zygophyllac.;* tr. Am.
Guiaguia: ? *Virola surinamensis* Warb.; *Myristicac.;* fr. Gu.
„ : ? *Virola sebifera* Aubl.; *Myristicac.;* fr. Gu.
Guibi: *Cola mirabilis* Chev.; *Sterculiac.;* Elf.
Güicume: *Lucuma palmeri* Fernald; *Sapotac.;* Salv.
Guié: *Erythrophloeum guineense* G. Don; *Caesalpiniac.;* Elf.
Guie-Biche: *Apoplanesia paniculata* Presl; *Papilionac.;* s. Mex.
Guielachi: *Talauma sp.; Magnoliac.;* Mex.
Guiendé: *Desbordesia sp.; Simarubac.;* Kam.
Guiexoba: *Bourreria huanita* Hemsl.; *Borraginac.;* Mex.
Guigaganga: *Copaifera sp.; Caesalpiniac.;* Gab.
Guijarro: *Tabernaemontana sp.; Apocynac.;* C.-R.
Guijo: *Shorea guiso* Blume; *Dipterocarpac.;* Phil. (Zam., Bat.).
Guíjo blanco: *Parashorea malaanonan* Merr.; *Dipterocarpac.;* Phil. (Zamb.)
Guiliqueme: *Erythrina glauca* Willd.; *Papilionac.;* Hond.
Guillomo: *Amelanchier vulgaris* Moench; *Rosac.;* Sp.
„ : *Cotoneaster granatensis* Boiss.; *Rosac.;* Sp.
Guima: *Musanga smithi* R. Br.; *Morac.;* Elf.
Guimanim: *Anona paludosa* Aubl.; *Anonac.;* fr. Gu.
Guimatsu: *Larix dahurica* Turcz. *var. japonica* Maxim.; *Conifer.;* Jap.
Guindo: *Nothofagus sp.; Fagac.;* Chile.

Guingnamadoa: *Virola surinamensis* Warb.; *Myristicac.;* Sur.
Guinguamadou: *Virola sp.; Myristicac.;* Gu.
„ del monte: *Virola sp.; Myristicac.;* Gu.
Guiozoc: *Mimusops congolensis* Russel et Hédin; *Sapotac.;* Kam.
Guira: *Crescentia cujete* L.; *Bignoniac.;* Cuba.
„ cimarrona: *Crescentia cujete* L.; *Bignoniac.;* Cuba.
Guiral: *Bauhinia purpurea* L.; *Caesalpiniac.;* n. O.-Ind. (Garhwal).
Guiri-biche: *Pinus sp.; Conifer.;* Mex.
Guiro: *Crescentia cujete* L.; *Bignoniac.;* Mex., C.-Am.
Guisache: *Prosopis juliflora* DC.; *Mimosac.;* Mex.
Guiscoyol: *Bactris horrida* Oerst.; *Palmac.;* br. Hond.
„ : *Bactris sp.; Palmac.;* Guat.
Guisoc: *Shorea balangeran* Burck.; *Dipterocarpac.;* Bor., Phil.
Guitaran: *Colubrina ferruginosa* Brongn.; *Rhamnac.;* P.-R.
Gui zoc: *Mimusops congolensis* Russell et Hédin; *Sapotac.;* Kam.
Gujerkota: *Randia longiflora* Lam.; *Rubiac.;* Assam.
Gujo: *Shorea guiso* Blume; *Dipterocarpac.;* Phil.
Gul-kandar: *Sterculia villosa* Roxb.; *Sterculiac.;* nw. O.-Ind. (Punjab).
Gula: *Pterocarpus cabrae* de Wild.; *Papilionac.;* Gab.
„ : *Pinus longifolia* Roxb.; *Conifer.;* n. O.-Ind.
Gulal: *Diospyros ramiflora* Roxb.; *Ebenac.;* Ind.
Gular: *Ficus glomerata* Roxb.; *Morac.;* C.-O.-Ind.
„ : *Sterculia urens* Roxb.; *Sterculiac.;* n. O.-Ind.
Gulbodla: *Sterculia villosa* Roxb.; *Sterculiac.;* nw. O.-Ind. (Punjab).
Gulili: *Olea glandulifera* Wall.; *Oleac.;* n. O.-Ind.
Gulipápa: *Cyclostemon spp.; Euphorbiac.;* Phil. (Cag.).
Gulmavu: *Machilus macrantha* Nees; *Laurac.;* w. O.-Ind.
Gulmaw: *Machilus macrantha* Nees; *Laurac.;* w. O.-Ind.
Gulnashtar: *Erythrina suberosa* Roxb.; *Papilionac.;* nw. O.-Ind. (Punjab).
Gulrai: *Cupressus torulosa* D. Don; *Conifer.;* nw. O.-Ind. (Punjab).
Gulu: *Sarcocephalus trillesi* Pierre; *Rubiac.;* Gab. (Mayombe).
„ : *Sterculia urens* Roxb.; *Sterculiac.;* n. O.-Ind.
Gulum: *Machlus macrantha* Nees; *Laurac.;* w. O.-Ind. (Bombay).
Gúlus: *Taxotrophis ilicifolia* Vid.; *Morac.;* Phil. (Mind.).
Gum (u): *Liquidambar straciflua* L.; *Hamamelidac.;* sö. USA., C.-Am.
„ : *Barringtonia angulata* Gaertn.; *Lecythidac.;* Ind.
„ , American-: *Bursera gummifera* L.; *Burserac.;* fr. Gu.
„ animi: *Hymenaea courbaril* L.; *Caesalpiniac.;* br. Gu.
„ arabic: *Acacia arabica* Willd.; *Mimosac.;* Go., Ind.
„ , Bastard-: *Eucalyptus goniocalyx* F. v. M.; *Myrtac.;* s. Au.
„ , black: *Nyssa silvatica* Marsh.; *Cornac.;* ö. N.-Am.
„ „ : *Haplormosia monophylla* Harms; *Papilionac.;* Lib.
„ , blue: *Eucalyptus globulus* Labill.; *Myrtac.;* ö. Au., Tasm.
„ , „ , south-australian: *Eucalyptus leucoxylon* F. v. M.; *Myrtac.;* Au.
„ , „ , spotted: *Eucalyptus maideni* F. v. M.; *Myrtac.;* Au.
„ , „ , Sydney: *Eucalyptus saligna* Sm.; *Myrtac.;* N.-S.-W.
„ , „ , Tasmanien: *Eucalyptus globulus* Labill.; *Myrtac.;* Tasm.
„ , Bolly-: *Litsea reticulata* Bth.; *Laurac.;* Au.
„ , „ -, hard: *Beilschmiedia obtusifolia* Bth.; *Laurac.;* Au.
„ , „ -, brown: *Litsea ferruginea* Bth. et Hook.; *Laurac.;* Au.
„ , Botany Bay-: *Eucalyptus resinifera* Smith; *Myrtac.;* Queensl.
„ , Box-: *Eucalyptus stellulata* Sieb.; *Myrtac.;* Au. (Vict.).
„ , „ -: *Eucalyptus hemiphloia* F. v. M.; *Myrtac.;* s. Au.
„ , brown: *Eucalyptus robusta* Sm.; *Myrtac.;* sö. Au.
„ , Cabbage-: *Eucalyptus papuana* F. v. M.; *Myrtac.;* w. Au.
„ , „ -: *Eucalyptus risdoni* Hook. var. *elata* Dehnh.; *Myrtac.;* Tasm.
„ , Carana-: *Protium altissimum* March.; *Burserac.;* fr. Gu.
„ , Cider-: *Eucalyptus gunni* Hook.; *Myrtac.;* Tasm.
„ , Coastal-, white: *Eucalyptus decipiens* Endl.; *Myrtac.;* w. Au.
„ , Copal-: *Daniella thurifera* Bennett; *Caesalpiniac.;* Lib.
„ , coral-flowered: *Eucalyptus torquata* Luehmann; *Myrtac.;* w. Au.
„ , Cotton-: *Nyssa uniflora* Wangenh.; *Cornac.;* N.-Am.

Gum, Desert-: *Eucalyptus cliftoniana* W. V Fitg.; *Myrtac.*; w. Au.
„ „ -: *Eucalyptus papuana* F. v. M.; *Myrtac.*; w. Au.
„ , Doctor-: *Symphonia globulifera* L.; *Guttifer.*; br. Ant.
„ , Drooping: *Eucalyptus viminalis* Labill.; *Myrtac.*; Au. (Vict.).
„ , „ : *Eucalyptus risdoni* Hook.; *Myrtac.*; Tasm.
„ , Dwarf-: *Eucalyptus vernicosa* Hook.; *Myrtac.*; Tasm.
„ , Elemi-: *Bursera sp.*; *Burserac.*; tr. Am.
„ , figured: *Liquidambar styraciflua* L.; *Hamamelidac.*; USA. (Tennessee).
„ . flooded: *Eucalyptus tereticornis* Sm.; *Myrtac.*; Au.
„ , „ : *Eucalyptus rostrata* Schlecht.; *Myrtac.*; N.-S.-W.
„ , „ : *Eucalyptus rudis* Endl.; *Myrtac.*; w. Au.
„ , „ : *Eucalyptus saligna* Smith; *Myrtac.*; N.-Seel.
„ , Forest-, red: *Eucalyptus tereticornis* Sm.; *Myrtac.*; ö. Au.
„ , Giant-: *Eucalyptus amygdalina* Labill.; *Myrtac.*; Au. (Vict.).
„ , Goldfields-, red-flowered: *Eucalyptus torquata* Luehm.; *Myrtac.*; w. Au.
„ , „ , yellow-flowered: *Eucalyptus stricklandi* Maiden; *Myrtac.*; w. Au.
„ , Gray: *Eucalyptus propinqua* D. et M.; *Myrtac.*; Au.
„ , „ : *Eucalyptus punctata* DC.; *Myrtac.*; Au.
„ , green: *Eucalyptus stellulata* Sieb.; *Myrtac.*; Au. (Vict.).
„ , grey: *Eucalyptus punctata* DC.; *E. propinqua* D. et M.; *Myrtac.*; ö. Au.
„ , „ : *Eucalyptus griffithi* Maiden; *Myrtac.*; w. Au.
„ , „ : *Eucalyptus saligna* Smith; *Myrtac.*; Queensl., N.-S.-W.
„ , heartleaved: *Eucalyptus cordata* Labill.; *Myrtac.*; Tasm.
„ , Hog-: *Metopium toxiferum* Krug et Urban; *Anacardiac.*; Fl., W.-Ind.
„ , „ -: *Symphonia globulifera* L.; *Guttifer.*; br. W.-Ind.
„ , Iron-: *Eucalyptus raveretiana* F. v. M.; *Myrtac.*; Queensl.
„ , Karri-: *Eucalyptus diversicolor* F. v. M.; *Myrtac.*; w. Au.
„ , large: *Nyssa grandidentata* Michx. f.; *Cornac.*; USA.
„ , Lead-: *Eucalyptus stellulata* Sieb.; *Myrtac.*; Au. (Vict.).
„ , lemon scented: *Eucalyptus citriodora* Hook.; *Myrtac.*; Au.
„ , Locust-: *Hymenaea courbaril* L.; *Caesalpiniac.*; tr. Am.
„ , Manna-: *Eucalyptus viminalis* Labill.; *Myrtac.*; s. Au., Tasm.
„ , marbled: *Eucalyptus maculata* Hook.; *Myrtac.*; s. Au.
„ , Mountain-, grey: *Eucalyptus goniocalyx* F. v. M.; *Myrtac.*; sö. Au.
„ , „ , red: *Eucalyptus muelleri* Moore; *Myrtac.*; Tasm.
„ , „ , white: *Eucalyptus pauciflora* Sieb.; *Myrtac.*; Au. (Vict.).
„ , Peppermint-: *Eucalyptus odorata* Behr; *Myrtac.*; s. Au.
„ , „ : *Eucalyptus amygdalina* Labill.; *Myrtac.*; Tasm.
„ , pink: *Eucalyptus fasciculosa* F. v. M.; *Myrtac.*; s. Au.
„ , Poplar-: *Eucalyptus alba* Reinw.; *Myrtac.*; Queensl. [Neu-Guin.
„ , , broad-leaved: *Eucalyptus platyphylla* F. v. M.; *Myrtac.*; n. Au.,
„ , poplar-leaved: *Eucalyptus polyanthemos* Schauer; *Myrtac.*; Au. (Vict.).
„ , red (u, e): *Liquidambar styraciflua* L.; *Hamamelidac.*; sö. USA., C.-Am.
„ , red: *Eucalyptus tereticornis* Sm.; *Myrtac.*; Au.
„ , „ : *Eucalyptus calophylla* R. Br.; *Myrtac.*; w. Au.
„ , „ : *Eucalyptus resinifera* Sm.; *Myrtac.*; Queensl., N.-S.-W.
„ , „ : *Eucalyptus robusta* Sm.; *Myrtac.*; w. Au.
„ , „ : *Eucalyptus rostrata* Schlecht.; *Myrtac.*; sö. & s. Au.
„ , „ : *Eucalyptus acervula* Hook.; *Myrtac.*; Tasm.
„ , River-, red: *Eucalyptus rostrata* Schlecht.; *Myrtac.*; Au. (Vict.).
„ , Ribbony-: *Eucalyptus viminalis* Labill.; *Myrtac.*; Au.
„ , Ridge-: *Eucalyptus alba* Reinw.; *Myrtac.*; nw. Au.
„ , rusty: *Angophora lanceolata* Cav.; *Myrtac.*; Queensl., N.-S.-W.
„ , Salmon-: *Eucalyptus salmonophloia* F. v. M.; *Myrtac.*; w. Au.
„ , Sambo-: *Symphonia globulifera* L. f.; *Guttifer.*; Pan.
„ , scarlet-flowering: *Eucalyptus ficifolia* F. v. M.; *Myrtac.*; w. Au.
„ , scribbly: *Eucalyptus haemastoma* Sm.; *Myrtac.*; sö. Au., Tasm.
„ , Semla-: *Bauhinia retusa* Ham.; *Caesalpiniac.*; Ind.
„ , Slaty-: *Eucalyptus tereticornis* Sm.; *Myrtac.*; Au.
„ . „ : *Eucalyptus dawsoni* R. T. B.; *Myrtac.*; N,-S.-W.
„ , sour: *Nyssa multiflora* Wangenh.; *Cornac.*; N.-Am.
„ , „ : *Nyssa capitata* Walt.; *Cornac.*; USA.

Gum spotted: *Eucalyptus haemastoma* Sm.; *Myrtac.;* N.-S.-W.
„ , „ : *Eucalyptus maculata* Hook.; *Myrtac.;* Queensl., N.-S.-W.
„ , star-leaved: *Liquidambar· styraciflua* L.; *Hamamelidac.;* sö. USA.
„ , Sugar-: *Eucalyptus cladocalyx* F. v. M.; *Myrtac.;* s. Au.
„ , Swamp-: *Eucalyptus paniculata* Sm.; *Myrtac.;* s. Au.
„ , „ : *Eucalyptus viminalis* Labill.; *Myrtac.;* sö. Au., Tasm.
„ , „ : *Eucalyptus regnans* F. v. M.; *Myrtac.;* Tasm.
„ , „ , black: *Nyssa biflora* Michx.; *Cornac.;* sö. USA.
„ , „ , mahogany: *Eucalyptus robusta* Sm.; *Myrtac.;* Queensl., N.-S.-W.
„ , sweet: *Liquidambar styraciflua* L.; *Hamamelidac.;* sö. USA., C.-Am.
„ , Tar-: *Clusia minor* L.; *Guttifer.;* Pan.
„ , Top-: *Eucalyptus sieberiana* F. v. M.; *Myrtac.;* Au. (Vict.).
„ , Tupelo-: *Nyssa aquatica* Marsh.; *Cornac.;* USA.
„ , urn-bearing: *Eucalyptus urnigera* Hook. f.; *Myrtac.;* Tasm.
„ , Water-: *Tristania laurina* R. Br.; *Myrtac.;* N.-S.-W.
„ , „ : *Tristania francisi* F. M. B.; *Myrtac.;* N.-S.-W.
„ , „ : *Tristania neriifolia* R. Br.; *Myrtac.;* N.-S.-W.
„ , „ : *Eugenia ventenati* Bth.; *Myrtac.;* Queensl., N.-S.-W.
„ , „ : *Callistemon lanceolatus* DC.; *Myrtac.;* Queensl.
„ , „ , giant: *Eugenia francisi* F. M. B.; *Myrtac.;* Au.
„ , „ , small-leaved: *Eugenia luehmanni* F. v. M.; *Myrtac.;* Au.
„ , water-rooted: *Eucalyptus oleosa* F. v. M.; *Myrtac.;* Au. (Vict.).
„ , weeping: *Eucalyptus coriacea* A. Cunn.; *Myrtac.;* Tasm.
„ , white: *Eucalyptus maculosa* R. T. Baker; *Myrtac.;* Au.
„ , „ : *Eucalyptus viminalis* Labill.; *Myrtac.;* s. Au., Tasm.
„ , „ : *Eucalyptus redunca* Schauer; *Myrtac.;* w. Au.
„ , „ : *Eucalyptus stellulata* Sieb.; *Myrtac.;* Au. (Vict.).
„ , „ : *Eucalyptus amygdalina* Labill.; *Myrtac.;* Au. (Vict.).
„ , „ : *Eucalyptus goniocalyx* F. v. M.; *Myrtac.;* Au. (Vict.).
„ , white: *Eucalyptus haemastoma* Sm.; *Myrtac.;* ö. Au., Tasm.
„ , „ : *Eucalyptus risdoni* Hook. var. elata Dehnh.; *Myrtac.;* Tasm.
„ , yellow: *Eucalyptus leucoxylon* F. v. M.; *Myrtac.;* sö. & s. Au.
„ , York-: *Eucalyptus loxophleba* Bth.; *Myrtac.;* w. Au.
Gúma: *Cordia myxa* L.; *Borraginac.;* Phil. (Bal.).
Gumadi: *Gmelina arborea* Roxb.; *Verbenac.;* w. O.-Ind.(s. Deccan).
Gumaldi: *Gmelina arborea* Roxb.; *Verbenac.;* Ind., Burma, Cey.
Gumamela: *Hibiscus rosa-sinensis* L.; *Malvac.;* Phil.
Gumar tek: *Gmelina arborea* Roxb.; *Verbenac.;* w. O.-Ind. (n. Bombav).
Gumbar: *Gmelina arborea* Roxb.; *Verbenac.;* Beng.
Gumbi: *Mimosa asperata* L.; *Mimosac.;* n. Nig.
Gumbijava: *Sideroxylon gardnerianum* A. DC.; *Sapotac.;* Bras.
Gumbixama: *Sideroxylon sp.; Sapotac.;* Bras.
Gumbixava: *Sideroxylon mastichodendron* Jacq.; *Sapotac.;* Bras.
Gumbo-limbo: *Bursera gummifera* L.; *Burserac.;* br. Hond.
„ „ , white: *Gilibertia concinna* Standl.; *Araliac.;* br. Hond.
Gumhar (H): *Gmelina arborea* L.; *Verbenac.;* n. O.-Ind.
„ : *Trewia nudiflora* L.; *Euphorbiac.;* n. O.-Ind. (Punjab).
Gummadi: *Gmelina arborea* Roxb.; *Verbenac.;* Ind. (? Burma).
Gummar: *Careya arborea* Roxb.; *Lecythidac.;* C.-O.-Ind.
Gummier, large: *Bursera gummifera* L.; *Burserac.;* br. W.-Ind.
Gummilera: *Coussapoa currani* Blake; *Morac.;* Bras. (Bahia).
Gumni: *Parkia bicolor* Chev.; *Mimosac.;* Lib.
Gumpini: *Odina wodier* Roxb.; *Anacardiac.;* w. O.-Ind. (n. Bombay).
Gumudu: *Gmelina arborea* Roxb.; *Verbenac.;* sw. O.-Ind.
Gun: *Aesculus indica* Colebr.; *Hippocastanac.;* n. O.-Ind. (Kangra).
Gunda: *Cordia myxa* L.; *Borraginac.;* O.-Ind. (Jeypore).
Gundada: *Santalum album* L.; *Santalac.;* w. O.-Ind. (Bombay).
Gundroi: *Cinnamomum cecidodaphne* Meissn.; *Laurac.;* Assam (Cachar).
Gung: *Ceiba pentandra* Gaertn.; *Bombacac.;* Go.
Gunserai: *Cinnamomum cecidodaphne* Meissn.; *Laurac.;* n. O.-Ind., Assam.
Gunsi: *Podocarpus neriifolia* D. Don; *Conifer.;* nö. O.-Ind. (Nepal), Beng.
Guntapái: *Alangium salviifolium* Merr.; *Cornac.;* Phil.

Guntapái: *Alangium longiflorum* Merr.; *Cornac.;* Phil.
Guntha-Marrah: *Careya australis* F. v. M.; *Lecythidac.;* n. Au.
Gúog: *Petersianthus quadrialatus* Merr.; *Lecythidac.;* Phil. (Mas.).
Gupariba: *Tecoma spp.; Bignoniac.;* fr. Gu.
Gupil: *Pygeum glandulosum* Merr., *P. presli* Merr.; *Rosac.;* Phil.
Gupit: *Pygeum glandulosum* Merr., *P. presli* Merr.; *Rosac.;* Phil.
Gupri marra: *Odina wodier* Roxb.; *Anacardiac.;* C.-O.-Ind.
Guranhem: *Pradosia latescens* Radlk.; *Sapotac.;* Bras.
Guranhen: *Pradosia latescens* Radlk.; *Sapotac.;* Bras.
Gurar: *Albizzia procera* Bth.; *Mimosac.;* n. & C.-O.-Ind.
Guras: *Rhododendron arboreum* Sm.; *Ericac.;* nw. O.-Ind. (Nepal), Beng.
Gurenham: *Pradosia latescens* Radlk.; *Sapotac.;* Bras.
Gurgu: *Garuga pinnata* Roxb.; *Burserac.;* C.-O.-Ind.
Gurgura: *Reptonia buxifolia* A. DC.; *Sapotac.;* nö. O.-Ind. (Punjab).
Gurjan: *Dipterocarpus turbinatus* Gaertn. f.; *Dipterocarpac.;* And.
Gurjiya: *Bombax buonopozense* Beauv.; *Bombacac.;* n. Nig.
Gurjum-oil-tree: *Dipterocarpus turbinatus* Gaertn. f.; *Dipterocarpac.;* Ind.
Gurjun (H): *Dipterocarpus costatus* Gaertn. f.; *Dipterocarpac.;* Burma.
    „   „ : *Dipterocarpus alatus* Roxb.; *Dipterocarpac.;* Burma.
    „   „ : *Dipterocarpus turbinatus* Gaertn. f.; *Dipterocarpac.;* nö. O.-Ind., And.
    „   „ : *Dipterocarpus griffithi* Miq.; *Dipterocarpac.;* Burma, And.
    „   : *Dipterocarpus pilosus* Roxb.; *Dipterocarpac.;* Beng. (Chitt.).
    „   , long-leaf: *Dipterocarpus griffithi* Miq.; *Dipterocarpac.;* And.
Gurmala: *Cassia fistula* L.; *Caesalpiniac.;* O.-Ind.
Gurn-kina: *Calophyllum tomentosum* Wight; *Guttifer.;* Cey.
Guru: *Sterculia urens* Roxb.; *Sterculiac.;* C.-O.-Ind (Berar).
Gurudu: *Gardenia gummifera* L.; *Rubiac.;* w. Beng. (Orissa).
Gurunduba: *Elaeis guineensis* Jacq. *var. virescens* Chev.; *Palmac.;* Gab.
Gurupia: *Celtis tala* Gill.; *Ulmac.;* Bras.
Gusanero: *Astronium fraxinifolium* Schott; *Anacardiac.;* Kol.
Gushiocho: *Balanites aegyptiana* Delile; *Zygophyllac.;* Togo.
Gustanapare: *Caesalpinia coriaria* Willd.; *Caesalpiniac.;* tr. Am.
Gutah-sundik: *Payena leeri* Kurz; *Sapotac.;* Malay, Bor., Sum., Banka.
Gutta-percha: *Palaquium spp.; Sapotac.;* tr. O.-As.
    „   „ : *Sapium sp.; Euphorbiac.;* Peru.
    „   „ : *Isonandra gutta* Hook.; *Sapotac.;* Malay.
    „   „ : *Excoecaria parviflora* Muell. Arg.; *Euphorbiac.;* n. Au., Queensl.
Gutti: *Polyalthia cerasoides* Bth. et Hook. f.; *Anonac.;* w. O.-Ind. (n. Bombay).
Guttier du Gabon: *Haronga madagascariensis* Choisy; *Guttifer.;* Mad., Kam.
Gutul: *Zizyphus xylocarpus* Willd.; *Rhamnac.;* O.-Ind.
Guva: *Cassine aquifolium* Fiori, *C. schweinfurthi* Loes.; *Celastrac.;* O.-Af. (it. Somali).
Guvetri: *Lasianthera austrocaledonica* Baill.; *Icacinac.;* N.-Kal.
Guya: *Dipterocarpus indicus* Bedd.; *Dipterocarpac.;* sw. O.-Ind.
Guyacan: *Guaiacum officinale* L.; *Zygophyllac.;* sp. Am.
Gúyong: *Anisoptera thurifera* Blume; *Dipterocarpac.;* Phil. (n. Ilocos).
Guyongguyong: *Polyscias nodosa* Seem.; *Araliac.;* Phil. (Bat.).
Guyung-guyung: *Cratoxylon spp., C. blancoi* Blume; *Guttifer.;* Phil.
Gwa: *Litsea polyantha* Juss.; *Laurac.;* n. O.-Ind.
Gwabale: *Sideroxylon tomentosum* Roxb.; *Sapotac.;* w. O.-Ind. (Bombay).
Gwabsa: *Cussonia nigerica* Hutch.; *Araliac.;* n. Nig.
Gwandar daji: *Anona senegalensis* Pers.; *Anonac.;* n. Nig.
Gwanja kusa: *Trichilia emetica* Vahl; *Meliac.;* n. Nig.
Gwanna: *Vochysia melinoni* Beckmann; *Vochysiac.;* Sur.
Gwanno: *Acacia dalzieli* Craib; *Mimosac.;* n. Nig.
Gwaria: *Acacia leucophloea* Willd.; *Mimosac.;* sw. Beng.
Gwaska: *Andira inermis* H. B. K.; *Papilionac.;* n. Nig.
Gwayral: *Bauhinia retusa* Ham.; *Caesalpiniac.;* n. O.-Ind.
Gwazkiya: *Swartzia madagascariensis* Desv.; *Papilionac.;* n. Nig.
Gwè: *Spondias mangifera* Willd.; *Anacardiac.;* Burma, And.
Gyagyen: *Macrolobium limba* Scott Ell.; *Caesalpiniac.;* Go.
Gyamantoa: *Pterocarpus santalinoides* L'Hérit; *Papilionac.;* Go.
Gyan: ? *Hopea basilanica* Foxw.; *Dipterocarpac.;* Phil.

Gyaung-byu-obein: *Ehretia laevis* Roxb.; *Borraginac.;* Burma.
Gympie: *Laportea photiniphylla* Wedd.; *Urticac.;* Au.
Gyo: *Schleichera trijuga* Willd.; *Sapindac.;* Burma.
Gyoban: *Polyalthia cerasoides* Bth. et Hook. f.; *Anonac.;* Burma.
Gyok: *Quercus semiserrata* Roxb.; *Fagac.;* Burma.

H

Ha ngam: *Aporosa sphaerosperma* Gagnep.; *Euphorbiac.;* Ind.-Ch.
„ nu: *Ixonanthes cochinchinensis* Pierre; *Linac.;* Annam.
„ tan: *Millingtonia hortensis* L. f.; *Bignoniac.;* Ind.-Ch.
„ tau: *Cinnamomum iners* Wall.; *Laurac.;* Ind.-Ch.
Haabi: *Ichthyomethia piscipula* Hitch.; *Papilionac.;* Mex.
Haabin: *Ichthyomethia piscipula* Hitch.; *Papilionac.;* Mex.
Haakdoorn: *Acacia detinens* Burch.; *Mimosac.;* S.-Af.
Haaz: *Calocarpum sp.; Sapotac.;* Mex.
Haba: *Hura polyandra* Baill.; *Euphorbiac.;* Mex. (s. Sinaloa).
„ de Guatemala: *Hura sp.; Euphorbiac.;* Mex.
„ „ indio: *Hura sp.; Euphorbiac.;* Mex.
„ „ San Antonio: *Caesalpinia crista* L.; *Caesalpiniac.;* Mex.
„ „ San Ignacio: *Hura sp.; Euphorbiac.;* Mex.
Habás: *Dracontomelum dao* Merr. et Rolfe; *Anacardiac.;* Phil. (Ag.).
Habeem: *Cordia gerascanthus* L.; *Borraginac.;* Mex.
Habi: *Ichthyomethia piscipula* Hitch.; *Papilionac.;* Mex.
Habilla: *Hura polyandra* Baill.; *Euphorbiac.;* Mex. (s. Sinaloa).
Habillo: *Hura crepitans* L.; *Euphorbiac.;* Ven.
Habim: *Piscidia communis* Standl.; *Papilionac.;* br. Hond.
Habizacué: *Baccaurea bonneti* Beille; *Euphorbiac.;* Elf.
Hac ho: *Randia longiflora* Lamk.; *Rubiac.;* Ind.-Ch.
Hacana: *Lucuma sp.; Sapotac.;* P.-R.
Hachia: *Tecoma sp., Tabebuia sp.; Bignoniac.;* Sur.
Hachiku: *Phyllostachys puberula* Munro; *Gramin.;* Jap.
Hackberry: *Celtis occidentalis* L.; *Ulmac.;* ö. N.-Am.
Hackia: *Tecoma sp.; Bignoniac.;* br. Gu.
„ : *Ixora triflorum* Bth. et Hook. f.; *Rubiac.;* br. Gu.
Hackmatac: *Larix americana* Michx.; *Conifer.;* ö. N.-Am.
Hackmatack: *Larix americana* Michx.; *Conifer.;* ö. N.-Am.
Hadang (H): *Cordia macleodi* Hook. f. et Thoms.; *Borraginac.;* w. O.-Ind.
Haddoka: *Chaetocarpus castanicarpus* Thw.; *Euphorbiac.;* O.-Ind., Malay.
Hadri: *Terminalia tomentosa* Wight et Arn.; *Combretac.;* O.-Ind.
Hafa: *Triplochiton scleroxylon* K. Schum.; *Sterculiac.;* Elf.
Hagachác: *Dipterocarpus pilosus* Roxb.; *Dipterocarpac.;* Phil.
„ : *Dipterocarpus speciosus* Brandis; *Dipterocarpac.;* Phil.
„ : *Dipterocarpus hasselti* Blume; *Dipterocarpac.;* Phil.
„ : *Dipterocarpus affinis* Brandis; *Dipterocarpac.;* Phil.
Hagad: *Pterocarpus spp.; Papilionac.;* Phil. (Cag.).
Hagakhák: *Dipterocarpus affinis* Brandis; *Dipterocarpac.;* Phil. (Tay.).
„ : *Dipterocarpus grandiflorus* Blco.; *Dipterocarpac.;* Phil. (Cam., Sibu.).
„ : *Dipterocarpus pilosus* Roxb.; *Dipterocarpac.;* Phil.
Hagdán-anák: *Polyscias nodosa* Seem.; *Araliac.;* Phil. (Cebu).
Hagís: *Eugenia spp.; Myrtac.;* Phil. (Luzon).
Hagua: *Genipa americana* L.; *Rubiac.;* tr. Am .
Hagué: *Turraeanthus africana* Pellegr.; *Meliac.;* Elf.
„ : *Guarea cedrata* Pellegr.; *Meliac.;* Elf.
Haguey: *Calatola costaricensis* Standl.; *Icacinac.;* Pan.
Haguguwa: *Ficus capensis* Thunb.; *Morac.;* n. Nig.
„ : *Grias fendleri* Seem.; *Lecythidac.;* Pan.
Hagumuná: *Garcinia spp.; Guttifer.;* Phil. (Cam.).
Hagy: *Lespedeza bicolor* Turcz.; *Papilionac.;* Jap., Ch.
Ha-ha: *Coelostegia griffithi* Bth. et Hook. f.; *Bombacac.;* Malay.
Hahuma: *Ostryoderris impressa* Dunn; *Papilionac.;* Go.

Hai: *Ficus sp.; Morac.;* Ind.-Ch.
„ sanh: *Cassia garrettiana* Craib; *Caesalpiniac.;* Ind.-Ch.
Haiariballi: *Diplotropis sp.; Papilionac.;* br. Gu.
Haiawá: *Protium sp.; Burserac.;* br. Gu.
Haiawaballi: *Connarus guianensis* Lamb.; *Connarac.;* fr. Gu.
Haiefai: *Hannoa klaineana* Pierre; *Simarubac.;* Elf.
Haiga: *Hopea wightiana* Wall.; *Dipterocarpac.;* sw. O.-Ind.
Haimatsu: *Pinus pumila* Parl.; *Conifer.;* Jap.
Haimaʒu: *Pinus pumila* Parl.; *Conifer.;* Jap., ö. Sib.
Haindé: *Pynaertia occidentalis* Chev.; *Meliac.;* Elf.
„ : *Anopyxis occidentalis* Chev.; *Rhizophorac.;* Gab.
Haindi: *Pynaertia occidentalis* Chev.; *Meliac.;* Elf.
Hainfain: *Picralima ellioti* Stapf; *Apocynac.;* Elf.
Haiowá: *Protium sp.; Burserac.;* br. Gu.
Haipi: *Ricinodendron africanum* Muell. Arg.; *Euphorbiac.;* Elf.
Haité: *Gymnanthes lucida* Sw.; *Euphorbiac.;* Cuba.
Hajawa: *Protium spp.; Burserac.;* Sur.
Hajawaballi: *Protium spp.; Burserac.;* Sur.
„ -hororodikoro: *Protium spp.; Burserac.;* Sur.
Hak: *Melanorrhoea usitata* Wall.; *Anacardiac.;* Burma.
Háket: *Terminalia nitens* Presl; *Combretac.;* Phil. (Zam.).
Hákit: *Terminalia oocarpa* Merr., *T. pellucida* Presl; *Combretac.;* Phil. (Zam.).
Hakué: *Turraeanthus africana* Pellegr.; *Meliac.;* Elf.
Hakunuboku: *Styrax obassia* S. et Z.; *Styracac.;* Jap.
Hakuunboku: *Styrax obassia* S. et Z.; *Styracac.;* Jap.
Hakuyo: *Populus tremula* L. var. *villosa* Wesm., *P. alba* L.; *Salicac.;* Jap.
Hal: *Vateria acuminata* Hayne; *Dipterocarpac.;* Cey.
„ -tumbri: *Bassia latifolia* Roxb.; *Sapotac.;* w. O.-Ind. (Bombay).
Halabalagi: *Garuga pinnata* Roxb.; *Burserac.;* w. O.-Ind. (Bombay).
Haladbera: *Chloroxylon swietenia* DC.; *Rutac.;* C.-O.-Ind.
Halauíhau: *Dracontomelum lamijo* Merr.; *Anacardiac.;* Phil.
„ : *Dracontomelum edule* Skeels; *Anacardiac.;* Phil.
Halban: *Vitex pubescens* Vahl; *Verbenac.;* tr. As.
Halda: *Chloroxylon swietenia* DC.; *Rutac.;* sw. O.-Ind.
Haldu (H): *Adina cordifolia* Hook. f.; *Rubiac.;* n. & C.-O.-Ind.
Hale: *Alstonia scholaris* R. Br.; *Apocynac.;* w. O.-Ind. (Bombay).
Half crown: *Mouriria spp.; Melastomatac.;* br. Hond.
Hal-gass: *Vateria indica* L.; *Dipterocarpac.;* s. Ind.
Halippajan: *Allophylus sundanus* Miq.; *Sapindac.;* Malay.
Hallab: *Periploca laevigata* Ait.; *Asclepiadac.;* Sahara.
Hallarin: *Pseudotsuga sp.; Conifer.;* Mex.
Halmalille: *Berrya ammonilla* Roxb.; *Tiliac.;* tr. As.
Halmendora: *Stemonoporus wighti* Thw.; *Dipterocarpac.;* Cey.
Halmililla: *Berrya ammonilla* Roxb.; *Tiliac.;* Ind., Cey.
Halmilla: *Berrya ammonilla* Roxb.; *Tiliac.;* Cey.
Halra: *Terminalia chebula* Reʒ.; *Combretac.;* sw. O.-Ind.
Halshen Sa: *Lonchocarpus laxiflorus* Guill. et Perr.; *Papilionac.;* n. Nig.
Halu: *Cola vera* K. Schum.; *Sterculiac.;* Elf.
Halupág: *Euphoria didyma* Blco.; *Sapindac.;* Phil.
Halupák: *Euphoria didyma* Blco.; *Sapindac.;* Phil.
Ham: *Aglaia gigantea* Pellegr.; *Meliac.;* Ind.-Ch.
„ : *Talauma gioi* Chev.; *Magnoliac.;* Ind.-Ch.
„ hom: *Dalbergia kurzi* Prain; *Papilionac.;* Ind.-Ch.
„ khom: *Manglietia fordiana* Oliv.; *Magnoliac.;* Ind.-Ch.
Hamaga-Kashi: *Quercus hondai* Makino; *Fagac.;* Jap.
Hamamélis de Virginie: *Hamamelis virginiana* L.; *Hamamelidac.;* ö. & sö. Kan.
Hamanasu: *Rosa rugosa* Thunb.; *Rosac.;* Jap.
Hama-nezu: *Juniperus formosana* Hayata var. *concolor* Hayata; *Conifer.;* Form.
Hamanthein: *Cinnamomum obtusifolium* Nees; *Laurac.;* Burma.
Hamárak: *Dracontomelum dao* Merr. et Rolfe; *Anacardiac.;* Phil. (Luzon).
Hàmba: *Ceiba pentandra* Gaertn.; *Bombacac.;* Mad.
Hambabáiud: *Neonauclea spp.; Rubiac.;* Phil. (Surigas).

Hambabálud: *Neonauclea reticulata* Merr., *N. spp.; Rubiac.;* Phil.
Hamerang: *Sterculia colorata* Roxb.; *Sterculiac.;* Sunda.
Hami: *Pangium edule* Reinw.; *Flacourtiac.;* Moluk. (Buru).
Hamigí: *Artocarpus lamellosa* Blco.; *Morac.;* Phil. (Cam.).
  „ : *Artocarpus rubrovenia* Warb.; *Morac.;* Phil. (Mind.).
Hamislág: *Securinega flexuosa* Muell. Arg.; *Euphorbiac.;* Phil.
Hammatti: *Cavanillesia platanifolia* H. B. K.; *Bombacac.;* Pan.
Hammock, Monkey-: *Peltogyne paradoxa* Ducke; *Caesalpiniac.;* Bras. (Pará).
Hamuráuon: *Vitex parviflora* Juss.; *Verbenac.;* Phil. (Cam., Alb.).
  „ -ásu: *Vitex turczaninowi* Merr.; *Verbenac.;* Phil. (Alb.).
Hamuyáuon: *Vitex parviflora* Juss.; *Verbenac.;* Phil. (Surigao).
Han: *Aesculus indica* Colebr.; *Hippocastanac.;* nw. O.-Ind. (Kasch.).
Hanagás: *Semecarpus sp.; Anacardiac.;* Phil. (Mind., Neg., Guim.).
Hanchiwakaede: *Acer japonicum* Thunb.; *Acerac.;* Jap.
Handang: *Cordia macleodi* Hook. f. et Thoms.; *Borraginac.;* Ind.
Handeong: *Commersonia echinata* Forst.; *Sterculiac.;* Nied.-Ind.
Handle-wood, grey: *Aphananthe philippinensis* Planch.; *Ulmac.;* Au.
  „ „ „ : *Pseudomorus brunoniana* Bur.; *Morac.;* Au.
Hand-tree: *Chiranthodendron sp.; Sterculiac.;* Mex.
Hanè: *Monodora myristica* Dunal; *Anonac.;* Elf.
Hane: *Aesculus indica* Colebr.; *Hippocastanac.;* nw. O.-Ind. (Kasch.).
Hanfuru: *Chrysobalanus ellipticus* Soland.; *Rosac.;* Elf.
Hang: *Shorea vulgaris* Pierre; *Dipterocarpac.;* Ind.-Ch.
  „ : *Terminalia tomentosa* Bedd.; *Combretac.;* Ind.-Ch.
  „ hên: *Pterospermum diversifolium* Blume; *Sterculiac.;* Ind.-Ch.
  „ quay: *Polyalthia jucunda* Finet et Gagnep.; *Anonac.;* Ind.-Ch.
Hañgálai: *Bruguiera parviflora* Wight et Arn.; *Rhizophorac.;* Phil.
Hañgárai: *Bruguiera parviflora* Wight et Arn.; *Rhizophorac.;* Phil.
Hangilang: *Cananga odorata* Hook. f.; *Anonac.;* tr. O.-Au.
Hani: *Lagerstroemia flos-reginae* Retz,; *Lythrac.;* O.-Ind.
Hanigigari: *Canthium didymum* Roxb.; *Rubiac.;* w. O.-Ind.
Hannoki: *Alnus japonica* S. et Z.; *Betulac.;* Jap.
Hano: *Boswellia dalzieli* Hutch.; *Burserac.;* n. Nig.
Hantap batoe: *Sterculia blumei* G. Don; *Sterculiac.;* Sunda.
  „ hoelan: *Sterculia colorata* Roxb.; *Sterculiac.;* Sunda.
  „ hulang: *Sterculia colorata* Roxb.; *Sterculiac.;* Sunda.
  „ passang: *Sterculia nobilis* Smith; *Sterculiac.;* Nied.-Ind.
  „ passoeng: *Sterculia nobilis* Smith; *Sterculiac.;* Nied.-Ind.
Hantep batoe: *Sterculia blumèi* G. Don; *Sterculiac.;* Sunda.
Hantige: *Acrocarpus fraxinifolius* Wight et Arn.; *Caesalpiniac.;* sw. O.-Ind.
Hanudun: *Aesculus indica* Colebr.; *Hippocastanac.;* nw. O.-Ind. (Kasch.).
Hanwgo: *Fagara macrophylla* Engl.; *Rutac.;* Elf.
Hanwogo: *Fagara macrophylla* Engl.; *Rutac.;* Elf.
Hao: *Pinus khasya* Royle; *Conifer.;* Ind.-Ch., w. Annam.
  „ : *Mytilaria laosensis* H. Lec.; *Hamamelidac.;* Ind.-Ch.
  „ : *Parashorea stellata* Kurz; *Dipterocarpac.;* Ind.-Ch.
  „ : *Artocarpus tonkinensis* Chev.; *Morac.;* Ind.-Ch.
Haoul: *Betula alnoides* Ham.; *Betulac.;* nw. O.-Ind. (Kumaon).
Hapènié: *Spondias lutea* L.; *Anacardiac.;* Elf.
Haperrié: *Spondias lutea* L.; *Anacardiac.;* Elf.
Hapi: *Ricinodendron africanum* Muell. Arg.; *Euphorbiac.;* Elf.
Hapnít: *Shorea polysperma* Merr.; *Dipterocarpac.;* Phil. (Cam.).
  „ : *Shorea teysmanniana* Dyer; *Dipterocarpac.;* Phil. (Cam.).
Hapo: *Cola vera* K. Schum; *Sterculiac.;* Elf.
Haqué-Haqué: *Dodonaea viscosa* Jacq.; *Sapindac.;* N.-Seel.
Har: *Terminalia chebula* Retz,; *Combretac.;* n. O.-Ind.
Hara: *Terminalia chebula* Retz,; *Combretac.;* C.-O.-Ind.
  „ : *Excoecaria agallocha* L.; *Euphorbiac.;* w. O.-Ind. (Bombay).
Haragiri Sen: *Kalopanax ricinifolius* Miq.; *Araliac.;* Jap.
Harahara: *Neobaronia xylophylloides* Taub.; *Papilionac.;* Mad.
  „ : *Neobaronia phyllanthoides* Baker; *Papilionac.;* Mad.
Harar: *Acacia albida* Delile; *Mimosac.;* br. Sudan.

Hàrar: *Terminalia polycarpa* Engl. et Diels; *Combretac.;* O.-Af. (it. Somali).
Harar: *Terminalia chebula* Retz.; *Combretac.;* nw. O.-Ind. (Punjab).
Hararh: *Terminalia chebula* Retz.; *Combretac.;* n. O.-Ind.
Háras: *Garcinia spp.; Guttifer.;* Phil. (Cap.).
„   : *Hopea plagata* Vid.; *Dipterocarpac.;* Phil. (Tab.).
Hardi: *Morinda tinctoria* Roxb.; *Rubiac.;* Beng.
Hardu: *Adina cordifolia* Hook. f.; *Rubiac.;* C.-O.-Ind.
Hardwood, Langdon's: *Backhousia bancrofti* F. M. Bailey et F. v. M.; *Myrtac.;* Queensl.
„   , Luya's: *Xanthostemon oppositifolius* F. M. B.; *Myrtac.;* Au.
Harepang: *Canarium altissimum* Blume; *Burserac.;* Sunda.
Harewood (e, u): *Acer pseudoplatanus* L.; *Acerac.;* Eu.
„   „   „   : *Acer campestre* L.; *Acerac.;* Eu.
„   , artificial (e, u): *Acer pseudoplatanus* L.; *Acerac.;* Eu.
„   , english (e, u): *Acer pseudoplatanus* L.; *Acerac.;* Eu.
„   , San Domingo-: *Zanthoxylum sp.; Rutac.;* St.-D.
„   , white: *Acer pseudoplatanus* L.; *Acerac.;* Eu.
Hargesa: *Dillenia indica* L.; *Dilleniac.;* Beng.
Harh: *Terminalia chebula* Retz.; *Combretac.;* n. O.-Ind.
Hari: *Cassia fistula* L.; *Caesalpiniac.;* w. Beng. (Orissa).
„   -taki: *Terminalia chebula* Retz.; *Combretac.;* Beng.
Hariekoekoen: *Pterospermum suberifolium* Willd.; *Sterculiac.;* Sunda.
Harigiri: *Kalopanax ricinifolius* Miq.; *Araliac.;* Jap.
Harikakoen: *Pterospermum suberifolium* Willd.; *Sterculiac.;* Sunda.
Harikoekem: *Schoutenia ovata* Korth.; *Tiliac.;* Nied.-Ind.
Harina: *Enterolobium cyclocarpum* Griseb.; *Mimosac.;* Pan.
Harinharra: *Amoora rohituka* Wight et Arn.; *Meliac.;* n. O.-Ind.
Harinkhana: *Amoora rohituka* Wight et Arn.; *Meliac.;* n. O.-Ind.
Harino: *Dipterodendron costaricense* Radlk.; *Sapindac.;* C.-R.
Hariraro-Jaroeroe: *Aspidosperma nitidum* Bth.; *Apocynac.;* Sur.
„   „   : ? *Aspidosperma oblongum* A. DC.; *Apocynac.;* Sur.
„   „   : *Aspidosperma excelsum* Bth.; *Apocynac.;* Sur.
„   -maballi: *Conceveiba guianensis* Aubl., *C. hostmanni* Bth.; *Euphorbiac.;* Sur.
„   walaba: *Eperua schomburgkiana* Bth.; *Caesalpiniac.;* Sur.
„   „   : ? *Eperua jenmanni* Oliv.; *Caesalpiniac.;* Sur.
Hari-taka: *Terminalia chebula* Retz.; *Combretac.;* O.-Ind.
Harizii-no-ki: *Castanopsis brachycantha* Hayata; *Fagac.;* Form.
Harmon: *Tarrietia utilis* Sprague; *Sterculiac.;* Lib.
Haro: *Pterocarpus spp.; Papilionac.;* Ven.
Haron: *Sterculia alata* Roxb.; *Sterculiac.;* O.-Ind.
Harpolli: *Harpullia cupanioides* Roxb.; *Sapindac.;* Beng.
Harpulli: *Harpullia cupanioides* Roxb.; *Sapindac.;* Ind.
Harra: *Terminalia chebula* Retz.; *Combretac.;* tr. As.
Harrani: *Dalbergia lanceolaria* L.; *Papilionac.;* w. O.-Ind. (s. Bombay).
Harré: *Talauma hodgsoni* Hook. f. et Thoms.; *Magnoliac.;* nö. O.-Ind. (Nepal).
Harreri: *Albizzia lebbek* Bth.; *Mimosac.;* sw. O.-Ind.
Harro: *Terminalia chebula* Retz.; *Combretac.;* C.-O.-Ind.
Harsinghar: *Nyctanthes arbor-tristis* L.; *Oleac.;* Ind.
Hart naar buiten: *Diplotropis guianensis* Bth.; *Papilionac.;* Sur.
„   „   „   : *Diplotropis leptophylla* Kleinh.; *Papilionac.;* Sur.
Hartriegel (d): *Cornus spp. Cornac.;* n. Hem.
„   , Blumen-: *Cornus florida* L.; *Cornac.;* ö. USA.
„   , gemeiner: *Cornus mas* L., *C. sanguinea* L.; *Cornac.;* Eu.
Haru-Nire: *Ulmus campestris* Sm. *var. laevis* Planch.; *Ulmac.;* Jap.
Haruwa: *Erythrina suberosa* Roxb.; *Papilionac.;* C.-O.-Ind.
Harwar: *Acacia leucophloea* Willd.; *Mimosac.;* w. O.-Ind. (n. Bombay).
Hasa-dhamin: *Grewia elastica* Royle; *Tiliac.;* C.-O.-Ind. (Merwara).
Hasakura: *Prunus sp.; Rosac.;* Jap.
Hasel (d): *Corylus avellana* L.; *Betulac.;* Eu.
„   , Baum-: *Corylus colurna* L.; *Betulac.;* sö. Eu., Kl.-As., As.
„   , gemeine: *Corylus avellana* L.; *Betulac.;* Eu.
„   , türkische: *Corylus colurna* L.; *Betulac.;* sö. Eu., s. As.
Haselnuss (d): *Corylus avellana* L.; *Betulac.;* Eu.

Hashidoi: *Syringa japonica* Nakai; *Oleac.*; Jap.
Hasi: *Aquilaria agallocha* Roxb.; *Thymelaeac.*; Assam.
Hasy: *Faguetia falcata* March.; *Anacardiac.*; Mad.
Hat: *Holarrhena antidysenterica* Wall.; *Apocynac.*; ö. Beng.
„ : *Celtis australis* L.; *Ulmac.*; Ind.-Ch.
„ : *Millingtonia hortensis* L. f.; *Bignoniac.*; Ind.-Ch.
Hatana: *Terminalia tomentosa* Wight et Arn.; *Combretac.*; n. O.-Ind.
Hatchanda: *Sterculia urens* Roxb.; *Sterculiac.*; Assam.
Haté: *Mannia africana* Hook. f.; *Simarubac.*; Elf.
Hathi-Khatyan: *Adansonia digitata* L.; *Bombacac.*; Ind.
Hattier: *Anona sp.*; *Anonac.*; fr. Gu.
Hau: *Hibiscus tiliaceus* L.; *Malvac.*; Südsee (Marquesas, Sandwich).
„ phât: *Ficus religiosa* L.; *Morac.*; Ind.-Ch.
„ „ : *Cinnamomum iners* Reinw.; *Laurac.*; Annam.
Hauchiwakayede: *Acer japonicum* Thunb. var. *typicum* v. Schwerin; *Acerac.*; Jap.
Haudan: *Gustavia angusta* L., *Eschweilera corrugata* Miers; *Lecythidac.*; Sur.
Hauoean: *Elaeocarpus floribundus* Blume; *Elaeocarpac.*; Nied.-Ind.
Haupea: *Allophylus cobbe* Swartz; *Sapindac.*; Südsee (Tahiti).
Haur: *Betula alnoides* Ham.; *Betulac.*; n. O.-Ind.
Hautuiquas Essen: *Ekebergia capensis* Sparrm.; *Meliac.*; Kap.
Havarilla: *Hura crepitans* L.; *Euphorbiac.*; P.-R.
Havibaval: *Acacia leucophloea* Willd.; *Mimosac.*; O.-Ind.
Havillo: *Hura crepitans* L.; *Euphorbiac.*; P.-R.
Havu-gandha: *Bridelia retusa* Spreng.; *Euphorbiac.*; w. O.-Ind. (Bombay).
Haw, black: *Crataegus douglasi* Lindl.; *Rosac.*; USA.
„ , yellow: *Crataegus flava* Ait.; *Rosac.*; N.-Am.
Hawoean: *Elaeocarpus floribundus* Blume; *Elaeocarpac.*; Nied.-Ind.
„ : *Elaeocarpus macrophyllus* Blume; *Elaeocarpac.*; Nied.-Ind.
Haxomena: *Khaya madagascariensis* Jum. et Perr.; *Meliac.*; Mad.
Hay: *Cassia garrettiana* Craib; *Caesalpiniac.*; Ind.-Ch.
„ san: *Melia azedarach* L.; *Meliac.*; Ind.-Ch.
Haya: *Fagus silvatica* L.; *Fagac.*; Sp.
„ : *Oxandra laurifolia* A. Rich.; *Anonac.*; P.-R.
„ blanca: ? *Oxandra lanceolata* Baill.; *Anonac.*; P.-R.
„ „ : *Oxandra laurifolia* A. Rich.; *Anonac.*; P.-R.
„ criolla: ? *Roupala polystachya* H. B. K.; *Proteac.*; Ven.
„ prieta: ? *Oxandra lanceolata* Baill.; *Anonac.*; P.-R.
Hayawa: *Protium guianense* March.; *Burserac.*; fr. Gu.
Hayo: *Acacia peregrina* Kunth; *Mimosac.*; Ven.
Haze: *Rhus succedanea* L.; *Anacardiac.*; O.-As.
Hazel (e): *Corylus avellana* L.; *Betulac.*; Eu.
„ : *Pomaderris apetala* Labill.; *Rhamnac.*; Au.
„ , indian: *Corylus spp.*; *Betulac.*; Ind.
„ , snapping: *Hamamelis virginiana* L.; *Hamamelidac.*; ö. & sö. Kan.
„ , Witch-: *Hamamelis virginiana* L.; *Hamamelidac.*; ö. N.-Am.
„ , Wych-: *Betula lenta* L.; *Betulac.*; Kan. (N.-Fundl.).
Hazelnut, common (e): *Corylus avellana* L.; *Betulac.*; Eu.
Hazelwood (e): *Liquidambar styraciflua* L.; *Hamamelidac.*; USA.
Haze-no-ki: *Rhus succedanea* L.; *Anacardiac.*; Jap.
Hazodrano: *Ilex monticola* Tul.; *Aquifoliac.*; Mad.
Hazomainty: *Diospyros spp.*; *Ebenac.*; Mad.
Hazo-mainty: ? *Cinnamosma fragrans* Baill.; *Canellac.*; Mad.
Hazomalanga: *Hernandia voyroni* H. Jum.; *Hernandiac.*; Mad.
Hazomaly: *Hernandia voyroni* H. Jum.; *Hernandiac.*; Mad.
Hazomena: *Khaya madagascariensis* Jum. et Perr.; *Meliac.*; Mad.
„ : *Weinmannia rutenbergi* Engl.; *Cunoniac.*; Mad.
Hazonema: *Khaya madagascariensis* Jum. et Perr.; *Meliac.*; Mad.
Hazotokana: *Veronia sp.*, *Brachylaena ramiflora* Humb.; *Composit.*; Mad.
Hazozoby: *Diospyros perrieri* H. Jum.; *Ebenac.*; w. Mad.
Heartwood: *Notelaea ligustrina* Vent.; *Oleac.*; Tasm.
Hebalasu: *Artocarpus hirsuta* Lamk.; *Morac.*; w. O.-Ind.
Heb-bevu: *Melia azedarach* L.; *Meliac.*; nw. O.-Ind. (Punjab).

Hebhalasu: *Artocarpus hirsuta* Lamk.; *Morac.;* O.-Ind.
Hebhulsina: *Artocarpus hirsuta* Lamk.; *Morac.;* w. O.-Ind.
Hecksame (d): *Ulex europaeus* L.; *Papilionac.;* w. & s. Eu.
Heckensame, gemeine: *Ulex europaeus* L.; *Papilionac.;* Eu.
Hedaggal: *Myristica attenuata* Wall.; *Myristicac.;* w. O.-Ind. (Bombay).
Hedde: *Adina cordifolia* Hook. f.; *Rubiac.;* w. O.-Ind. (Bombay).
Heddi: *Adina cordifolia* Hook. f.; *Rubiac.;* O.-Ind.
Hedge plant: *Maclura aurantiaca* Nutt.; *Morac.;* N.-Am.
Hediondo: *Copernicia excelsa* León; *Palmac.;* Cuba.
Hedoka: *Chaetocarpus castanocarpus* Thw.; *Euphorbiac.;* Cey.
Hedra: *Hedera helix* L.; *Araliac.;* Sp.
Hedu: *Stephegyne parvifolia* Korth.; *Rubiac.;* w. O.-Ind. (n. Bombay).
Heela: *Garcinia cambogia* Desr.; *Guttifer.;* Nied.-Ind.
Heglig: *Balanites aegyptiaca* Delile; *Simarubac.;* Aeg.
Hegronbébé: *Pterocarpus rhori* Vahl; *Papilionac.;* Sur.
Hégron-Taproepa: *Gustavia pterocarpa* Poit.; *Lecythidac.;* Sur.
Hehe: *Anogeissus schimperi* Hochst., Hutch. et Dalz.; *Combretac.;* Go.
Héhé: *Canarium schweinfurthi* Engl.; *Burserac.;* Kam.
Heide, Baum- (d): *Erica arborea* L.; *Ericac.;* De.
  „ , fleischrote: *Erica carnea* L.; *Ericac.;* De. •
  „ , gemeine: *Calluna vulgaris* Salisb.; *Ericac.;* De.
  „ , Hasen-: *Cytisus scoparius* Link; *Papilionac.;* De.
Heidekraut: *Calluna vulgaris* Salisb.; *Ericac.;* De.
Heidelbeere, gemeine: *Vaccinium myrtillus* L.; *Ericac.;* De.
Heinan: *Meliosma wallichi* Hook. f.; *Sabiac.;* ö. Himal.
Heistère rouge: *Swartzia tomentosa* DC.; *Caesalpiniac.;* fr. Gu.
Heiwortel (h): *Erica arborea* L.; *Ericac.;* M.-M.
Hela: *Terminalia belerica* Roxb.; *Combretac.;* w. O.-Ind.
Helbeva: *Ailanthus excelsa* Roxb.; *Simarubac.;* w. O.-Ind.
Helengomaash: *Spirostachys africana* Sond.; *Euphorbiac.;* O.-Af., S.-Af.
Hemelboom (h): *Ailanthus glandulosa* Desf.; *Simarubac.;* Nied.-Ind.
Hemlock: *Tsuga spp.;* *Conifer.;* N.-Am.
  „ : *Tsuga heterophylla* Sarg.; *Conifer.;* w. Kan.
  „ : *Tsuga canadensis* Carr.; *Conifer.;* ö. N.-Am.
  „ , black: *Tsuga canadensis* Carr.; *Conifer.;* ö. Kan.
  „ , „ : *Tsuga mertensiana* Carr.; *Conifer.;* w. Kan.
  „ , British Columbia-: *Tsuga heterophylla* Sarg.; *Conifer.;* w. Kan.
  „ , canadian: *Tsuga canadensis* Carr.; *Conifer.;* ö. Kan.
  „ , eastern: *Tsuga canadensis* Carr.; *Conifer.;* ö. Kan.
  „ , Mountain-: *Tsuga mertensiana* Carr.; *Conifer.;* w. N.-Am.
  „ , red: *Tsuga canadensis* Carr.; *Conifer.;* ö. Kan.
  „ , western: *Tsuga heterophylla* Sarg.; *Conifer.;* w. N.-Am.
  „ , „ : *Tsuga mertensiana* Carr.; *Conifer.;* w. Kan.
  „ , white: *Tsuga canadensis* Carr.; *Conifer.;* ö. Kan.
Hemp, Blackfellow's: *Commersonia fraseri* J. Gay; *Sterculiac.;* w. Au.
Hendjé: *Zanthoxylum parvifolium* Chev.; *Rutac.;* Elf.
Hendol: *Barringtonia acutangula* Gaertn.; *Lecythidac.;* Assam.
Henduri poma: *Toona ciliata* M. Roem.; *Meliac.;* Assam.
Hengüé: *Zanthoxylum parvifolium* Chev.; *Rutac.;* Elf.
Hengun: *Balanites roxburghi* Planch.; *Zygophyllac.;* Ind.
Henhja: *Anthocephalus cadamba* Miq.; *Rubiac.;* sö. As., Neu-Guin.
Henna, egyptian: *Lawsonia inermis* L.; *Lythrac.;* tr. Am.
Hérbedo: *Arbutus unedo* L.; *Ericac.;* Sp.
Hercules, yellow: *Zanthoxylum sp.;* *Rutac.;* Jam.
Hercules club: *Aralia chinensis* L. var. *glabrescens* Fr. et Sav.; *Araliac.;* Jap.
Heréri: *Terminalia somalilensis* Engl. et Diels; *Combretac.;* O.-Af. (Somali).
Heriso: *Apeiba sp.;* *Tiliac.;* Ven.
Heritsika: *Weinmannia rutenbergi* Engl.; *Cunoniac.;* Mad.
Herka: *Buchanania latifolia* Roxb.; *Anacardiac.;* C.-O.-Ind.
Hermang: *Cratoxylon hornschuchi* Blume; *Guttifer.;* Sunda.
Hernandier de la Guyane (f): *Hernandia guianensis* Aubl.; *Hernandiac.;* Gu.
Herran: *Pongamia glabra* Vent.; *Papilionac.;* Beng.

Herva cidreira: *Lippia alba* N. E. Brown; *Verbenac.;* Bras.
Hesa: *Ficus religiosa* L.; *Morac.;* n. O.-Ind.
Hesak: *Ficus religiosa* L.; *Morac.;* ö. Beng.
Hesel: *Anogeissus latifolia* Wall.; *Combretac.;* ö. Beng.
Hesomomi: *Abies umbellata* Mayr; *Conifer.;* Jap.
Hetatra: *Podocarpus thunbergi* Hook.; *Conifer.;* Mad.
Hete-Baké: *Hannoa klaineana* Pierre; *Simarubac.;* Elf.
Heteré: *Pycnanthus kombo* Warb.; *Myristicac.;* Elf.
Hétéré: *Aucoumea klaineana* Pierre; *Burserac.;* Gab.
Hêtre (f): *Fagus silvatica* L.; *Fagac.;* Eu.
  „ : *Fagus grandifolia* Ehrh.; *Fagac.;* ö. Kan.
  „ blanc: *Grevillea gillivrayi* Hook. f.; *Proteac.;* N.-Kal.
  „ gris: *Grevillea gillivrayi* Hook. f.; *Proteac.;* N.-Kal.
  „ „ : *Guazuma ulmifolia* Lamk.; *Sterculiac.;* Ant.
  „ noir: *Stenocarpus trinervis* Guill.; *Proteac.;* N.-Kal.
  „ rouge: *Fagus grandifolia* Ehrh.; *Fagac.;* ö. Kan.
  „ „ : *Grevillea rubiginosa* Brongn. et Griseb.; *Proteac.;* N.-Kal.
Hetsuka-Nigaki: *Adina racemosa* Miq.; *Rubiac.;* Jap.
Heuschreckenbaum (d): *Hymenaea courbaril* L.; *Caesalpiniac.;* tr. Am.
Hevea: *Hevea guianensis* Aubl.; *Euphorbiac.;* Sur.
Heyderie: *Libocedrus decurrens* Torr.; *Conifer.;* w. USA.
Heymascoli: *Ximenia americana* L.; *Olacac.;* Gu.
Hia: *Homalium foetidum* Roxb.; *Flacourtiac.;* Nied.-Ind.
Hiang: *Ailanthus glandulosa* Desf.; *Simarubac.;* Ch. (Shanghai).
  „ lân mou: *Toona sinensis* M. Roem.; *Meliac.;* Ch.
Hiarakoro kakeralli: *Eschweilera corrugata* Miers; *Lecythidac.;* Sur.
Hiaroe kakeralli: *Eschweilera corrugata* Miers; *Lecythidac.;* Sur.
Hiava: *Protium guianense* March.; *Burserac.;* fr. Gu.
Hiawa: *Protium guianense* March., *P. heptaphyllum* March.; *Burserac.;* Gu.
  „ -Balli: *Omphalobium lamberti* DC.; *Connarac.;* br. Gu.
Hiba: *Thujopsis dolabrata* S. et Z.; *Conifer.;* Jap.
Hibiscus, tall: *Hibiscus elatus* Swarʒ; *Malvac.;* Barb.
Hiburu: *Pterocarpus spp.;* *Papilionac.;* br. Gu.
Hicaco: *Chrysobalanus sp.;* *Rosac.;* Mex., C.-Am., W.-Ind.
  „ : *Chrysobalanus icaco* L.; *Rosac.;* P.-R.
  „ prieto: *Hirtella sp.;* *Rosac.;* Cuba.
Hickory (e, u): *Carya spp.;* *Juglandac.;* N.-Am.
  „ : *Carya alba* Nutt.; *Juglandac.;* USA.
  „ : *Pterocarya rhoifolia* S. et Z.; *Juglandac.;* Jap.
  „ : *Acacia falcata* Willd.; *Mimosac.;* Au.
  „ : *Tarrietia argyrodendron* Bth.; *Sterculiac.;* Au.
  „ , Big Bud-: *Carya alba* Nutt.; *Juglandac.;* ö. N.-Am.
  „ , Bird's-eye:- *Carya alba* Nutt.; *Juglandac.;* sö. USA.
  „ , Bitternut-: *Carya cordiformis* C. Koch; *Juglandac.;* N.-Am.
  „ , black: *Carya porcina* Nutt.; *Juglandac.;* sö. Kan.
  „ , brown: *Carya porcina* Nutt.; *Juglandac.;* ö. USA.
  „ , Butternut-: *Carya cordiformis* C. Koch; *Juglandac.;* N.-Am.
  „ , curly: *Carya alba* Nutt.; *Juglandac.;* sö. USA.
  „ , grossfrüchtige (d): *Carya sulcata* Nutt.; *Juglandac.;* USA.
  „ , kleinfrüchtige (d): *Carya microcarpa* Nutt.; *Juglandac.;* USA.
  „ , Mockernut-: *Carya alba* Nutt.; *Juglandac.;* ö. USA.
  „ , Native-: *Acacia leprosa* Sieber; *Mimosac.;* Au. (Vict.).
  „ , Nutmeg-: *Carya myristicaeformis* Michx.; *Juglandac.;* USA.
  „ of Chile: ? *Myrtus luma* Barn.; *Myrtac.;* Chile.
  „ , Pignut-: *Carya porcina* Nutt.; *Juglandac.;* ö. USA.
  „ , red: *Carya tomentosa* Nutt.; *Juglandac.;* N.-Am.
  „ , Shagbark-: *Carya alba* Nutt.; *Juglandac.;* ö. N.-Am.
  „ , Shellbark-: *Carya alba* Nutt.; *Juglandac.;* ö. N.-Am.
  „ , smallfruited: *Carya microcarpa* Nutt.; *Juglandac.;* N.-Am.
  „ , spiʒfrüchtige (d): *Carya sulcata* Nutt.; *Juglandac.;* USA.
  „ , Swamp-: *Carya cordiformis* C. Koch; *Juglandac.;* ö. N.-Am.
  „ ,   „ -: *Carya aquatica* Nutt.; *Juglandac.;* N.Am.

Hickory, Thick-bark-: *Carya laciniosa* Loud.; *Juglandac.;* USA.
„ , Water-: *Carya aquatica* Nutt.; *Juglandac.;* USA.
„ , weisse (d): *Carya alba* Nutt.; *Juglandac.;* USA.
„ , White heart-: *Carya tomentosa* Nutt., *C. alba* Nutt.; *Juglandac.;* ö. USA.
„ , Lignum vitae-: *Acacia falcata* Willd.; *Mimosac.;* N.-S.-W.
Hid: *Pterocarpus marsupium* Roxb.; *Papilionac.;* n. O.-Ind.
Hiên: *Pygeum arboreum* Engl.; *Rosac.;* Ind.-Ch.
Hierba de la conchuda: *Laurelia aromatica* Juss.; *Monimiac.;* Mex.
„ „ las mataduras: *Talauna sp.; Magnoliac.;* Mex.
„ del talaje: *Laurelia aromatica* Juss.; *Monimiac.;* Mex.
Hietsjieté: *Gustavia pterocarpa* Poit.; *Lecythidac.;* Sur.
Higo: *Ficus carica* L.; *Morac.;* Mex.
„ : *Ficus colubrinae* Standl.; *Morac.;* Pan.
„ : *Ficus spp.; Morac.;* Hond.
Higuera: *Ficus carica* L.; *Morac.;* s. USA., Mex.
„ : *Crescentia cujete* L.; *Bignoniac.;* P.-R.
„ blanca: *Ficus sp.; Morac.;* Arg.
„ loca: *Ficus carica* L.; *Morac.;* Sp.
„ morada: *Ficus sp.; Morac.;* Arg.
„ del agua: *Ficus cestrifolia* Schott; *Morac.;* Arg.
„ „ Chaco: *Ficus sp.; Morac.;* Arg.
Higuerillo: *Vitex divaricata* Sw.; *Verbenac.;* P.-R.
„ : *Citharexylum fruticosum* L.; *Verbenac.;* P.-R.
Higuerita: *Ruprechtia sp.; Polygonac.;* Arg.
Higuero: *Crescentia cucurbitina* L.; *Bignoniac.;* Fl., W.-Ind., C.-Am., Ven.
„ : *Ficus spp.; Morac.;* P.-R., Hond.
„ : *Crescentia cujete* L.: *Bignoniac.;* P.-R.
Higuerón: *Ficus tonduzi* Standl., *F. glabrata* H. B. K.; *Morac.;* Pan.
„ : *Ficus sp.; Morac.;* Kol., Ek.
„ colorado: *Ficus sp.; Morac.;* P.-R.
Higuerton: *Didymopanax morototoni* Dcne. et Planch.; *Araliac.;* Trin.
Higuillo: *Ficus spp.; Morac.;* Hond.
„ preto: *Ficus sp.; Morac.;* P.-R.
Hiiragi: *Osmanthus aquifolium* Sieb.; *Oleac.;* Jap.
„ -gasi: *Quercus spinosa* A. David *var. miyabei* Hayata; *Fagac.;* Form.
Hiiran-kurikasi: *Castanopsis subacuminata* Hayata; *Fagac.;* Form.
Hijal: *Barringtonia acutangula* Gaertn.; *Lecythidac.;* Beng.
Hik: *Lannea grandis* Engl.; *Anacardiac.;* Ind., Burma, Cey.
Hikob: *Staudtia gabunensis* Warb., *S. kamerunensis* Warb.; *Myristicac.;* Kam.
Hila anwal: *Vitex peduncularis* Wall.; *Verbenac.;* Assam (Cachar).
Hilda: *Terminalia chebula* Reʒ.; *Combretac.;* C.-O.-Ind.
Hilem: *Drypetes gabonensis* Pierre; *Euphorbiac.;* Gab.
Hilikha: *Terminalia chebula* Reʒ.; *Combretac.;* Assam.
Hill toon (H): *Toona serrata* M. Roem.; *Meliac.;* Ind.
Himala: *Bombax malabaricum* DC.; *Bombacac.;* Assam.
Himbabálud: *Neonauclea spp.; Rubiac.;* Phil.
„ : *Barringtonia spp.; Lecythidac.;* Phil. (Cap.).
Himbaba-ó: *Allaeanthus luzonicus* Vid.; *Morac.;* Phil.
„ : *Radermachera sp.; Bignoniac.;* Phil (Mind.).
Himbeere (d): *Rubus idaeus* L.; *Rosac.;* De.
Himeko-matsu: *Pinus parviflora* S. et Z.; *Conifer.;* Jap.
Himekomaʒu: *Pinus parviflora* S. et Z.; *Conifer.;* Jap.
Himpagkatán: *Dipterocarpus grandiflorus* Blco.; *Dipterocarpac.;* Phil. (Sam.).
Himu: *Morus serrata* Roxb.; *Morac.;* n. C.-O.-Ind. (Dehra Dun).
Himuláuon: *Vitex parviflora* Juss.; *Verbenac.;* Phil. (Sulu).
Hinabuád: *Terminalia comintana* Merr.; *Combretac.;* Phil. (Mind.).
Hinabusí: *Terminalia comintana* Merr.; *Combretac.;* Phil. (Mind.).
Hinahina: *Melicytus ramiflorus* Forst.; *Violac.;* N.-Seel.
Hinau: *Elaeagnus hinau* ?; *Elaeagnac.;* N.-Seel.
„ : *Elaeocarpus dentatus* Vahl; *Elaeocarpac.;* N.-Seel.
Hincha huevos: *Hippomane mancinella* L.; *Euphorbiac.;* Mex.
Hinchador: *Rhus juglandifolia* Willd.; *Anacardiac.;* C.-Am., nw. S.-Am.

Hindáng: *Litsea spp.; Laurac.;* Phil. (Cag., Ley., Surigao).
Hindi: *Schizostachyum diffusum* Merr.; *Gramin.;* Phil.
Hindurugú: *Myristica spp.; Myristicac.;* Phil. (Bat.).
Hiñgálai: *Bruguiera parviflora* Wight et Arn.; *Rhizophorac.;* Phil. (? Pal.).
Hiñgáli: *Bruguiera cylindrica* Blume; *Rhizophorac.;* Phil. (Neg.).
Hingla: *Litsea polyantha* Juss.; *Laurac.;* Assam.
Hingoota: *Balanites roxburghi* Planch.; *Zygophyllac.;* Ind.
Hingori: *Castanopsis hystrix* A. DC.; *Fagac.;* Assam.
Hinh: *Keteleeria davidiana* Beissn.; *Conifer.;* Ind.-Ch.
Hiniesta: *Genista cinerea* DC.; *Papilionac.;* Sp.
  „ de escoba: *Cytisus scoparius* Link; *Papilionac.;* Sp.
  „ „ tintes: *Genista tinctoria* L.; *Papilionac.;* Sp.
Hinjolo: *Barringtonia acutangula* Gaertn.; *Lecythidac.;* nö. O.-Ind. (Orissa).
Hinjoo: *Barringtonia acutangula* Gaertn.; *Lecythidac.;* ö. Beng.
Hinlagasí': *Shorea negrosensis* Foxw.; *Dipterocarpac.;* Phil. (Sibu.).
  „ : *Shorea polysperma* Merr.; *Dipterocarpac.;* Phil. (Sibu., Cap.).
  „ : *Shorea teysmanniana* Dyer; *Dipterocarpac.;* Phil. (Sibu.).
Hinlap-bagió: *Cyclostemon spp.; Euphorbiac.;* Phil. (Sam.).
Hinoki: *Chamaecyparis obtusa* S. et Z.; *Conifer.;* Jap.
Hintsy: *Intsia bijuga* O. Ktze.; *Caesalpiniac.;* Mad.
Hintзy: *Intsia bijuga* O. Ktze.; *Caesalpiniac.;* Mad.
Hir: *Terminalia chebula* Retз.; *Combretac.;* C.-O.-Ind.
Hiral bogi: *Hopea wightiana* Wall.; *Dipterocarpac.;* sw. O.-Ind.
Hirda: *Terminalia chebula* Retз.; *Combretac.;* sw. O.-Ind.
Hirek: *Diospyros montana* Roxb.; *Ebenac.;* O.-Ind.
Hirih: *Albizzia lebbek* Bth.; *Mimosac.;* Assam.
Hi-sakaki: *Eurya japonica* Thunb.; *Theac.;* Jap.
Hisasage: *Catalpa kaempferi* S. et Z.; *Bignoniac.;* Jap.
Hisigata-kurikasi: *Castanopsis stellato-spina* Hayata; *Fagac.;* Form.
Hisopillo: *Satureia montana* L.; *Labiat.;* Sp.
Hisopo: *Hyssopus officinalis* L.; *Labiat.;* Sp.
Hitchia: *Byrsonima sp.; Malpighiac.;* Sur.
  „ -balli: *Archytaea multiflora* Bth.; *Theac.;* br. Gu.
Hitriribouraballi: *Machaerium schomburgki* Bth.; *Papilionac.;* br. Gu.
Hitum: *Sterculia urens* Roxb.; *Sterculiac.;* C.-O.-Ind.
Hiwar: *Acacia leucophloea* Willd.; *Mimosac.;* C.-O.-Ind.
Hka-mari: *Salix tetrasperma* Roxb.; *Salicac.;* O.-Ind.
  „ -shatawi: *Bischofia javanica* Blume; *Euphorbiac.;* O.-Ind.
Hkala-shawng: *Anthocephalus cadamba* Miq.; *Rubiac.;* O.-Ind.
Hkri: *Melanorrhoea usitata* Wall.; *Anacardiac.;* O.-Ind.
Hlafuta: *Afzelia quanzensis* Welw.; *Caesalpiniac.;* O.-Af. (Moz.).
Hlekanan: *Bridelia retusa* Spreng.; *Euphorbiac.;* Burma (Ten.).
Hlerm: *Melia azedarach* L.; *Meliac.;* O.-Ind.
Hlim: *Melia azedarach* L.; *Meliac.;* O.-Ind.
Hlosunli: *Betula alnoides* Ham.; *Betulac.;* Beng.
Hmanbyu: *Gardenia turgida* Roxb.; *Rubiac.;* Burma.
Hmanthein: *Cinnamomum zeylanicum* Breyn; *Laurac.;* Burma.
Hmantheinpo: *Cinnamomum tavoyanum* Meissn.; *Laurac.;* Burma (s. Ten.).
Hmetkaung: *Zanthoxylum budrunga* Wall.; *Rutac.;* Burma.
Hminkya: *Elaeocarpus robustus* Roxb.; *Elaeocarpac.;* Burma (Ten.).
Hnaw: *Adina cordifolia* Hook. f.; *Rubiac.;* Burma.
Hnawbinga: *Stephegyne diversifolia* Hook. f.; *Rubiac.;* Burma.
Hnawthein: *Stephegyne diversifolia* Hook. f.; *Rubiac.;* Burma.
Ho: *Magnolia hypoleuca* S. et Z.; *Magnoliac.;* Jap.
  „ : *Aleurites cordata* R. Br.; *Euphorbiac.;* Ind.-Ch.
  „ -bi: *Pahudia cochinchinensis* Pierre; *Caesalpiniac.;* Annam.
  „ -médiriya: *Diospyros thwaitesi* Bedd.; *Ebenac.;* Cey.
Hoa dông: *Eugenia sp.; Myrtac.;* Ind.-Ch.
  „ su nam: *Michelia champaca* L.; *Magnoliac.;* Annam.
Hoac quang: *Wendlandia paniculata* DC.; *Rubiac.;* Ind.-Ch.
Hoan-Dan: *Dacrydium elatum* Wall.; *Conifer.;* Ind.-Ch.
  „ linh: *Peltophorum dasyrachis* Kurz; *Caesalpiniac.;* Annam.

Hoâng bá: *Pterocarpus flavus* Lour.; *Papilionac.*; Ind.-Ch.
„ dán: *Dacrydium elatum* Wall.; *Conifer.*; Ind.-Ch.
„ -linh: *Peltophorum dasyrachis* Kurz, P. *tonkinense* Pierre; *Caesalpiniac.*; Ind.-Ch.
„ „ : *Peltophorum ferrugineum* Bth.; *Caesalpiniac.*; Ind.-Ch.
„ mang: *Cryptocarya impressa* Miq.; *Laurac.*; Annam.
Hoanh: *Linociera macrophylla* Wall.; *Oleac.*; w. Annam.
Hoary: *Campnosperma panamensis* Standl.; *Anacardiac.*; Pan.
Hoayacan: *Guaiacum officinale* L.; *Zygophyllac.*; Mex.
Hoaxacan: *Guaiacum officinale* L.; *Zygophyllac.*; P.-R.
Hobbo: *Spondias lutea* L., S. *dulcis* Forst.; *Anacardiac.*; Sur.
Hobi: *Pahudia cochinchinensis* Pierre; *Caesalpiniac.*; Ind.-Ch.
Hobo: *Spondias sp.*; *Anacardiac.*; Kol.
„ : *Ricinodendron africanum* Muell. Arg.; *Euphorbiac.*; Elf.
Hoboballi: *Loxopterygium sagoti* Hook. f.; *Anacardiac.*; Sur.
Hoc: *Averrhoa carambola* L.; *Oxalidac.*; Tonk.
Hockmatack: *Larix americana* Michx.; *Conifer.*; nö. Kan.
Hoé: *Irvingia sp.*; *Simarubac.*; Elf.
„ : *Xerospermum noronhianum* Blume; *Sapindac.*; Nied.-Ind.
„ : *Averrhoa carambola* L.; *Oxalidac.*; Ind.-Ch.
„ : *Illicium verum* Hook.; *Magnoliac.*; Ind.-Ch.
„ : *Sophora japonica* L.; *Papilionac.*; Ind.-Ch.
Hoeboe: *Spondias lutea* L., S. *dulcis* Forst.; *Anacardiac.*; Sur.
Hoeboeballi: *Loxopterygium sagoti* Hook. f.; *Anacardiac.*; Sur.
Hoelia: *Byrsonima sp.*; *Malpighiac.*; Sur.
Hoenderspor: *Phoberos zeyheri* Presl; *Flacourtiac.*; Kap.
Hoepel: *Copaifera guianensis* Desf.; *Caesalpiniac.*; Sur.
Hoepelboom (h): *Copaifera guianensis* Desf.; *Caesalpiniac.*; Sur.
Hoepelhout (h): *Copaifera guianensis* Desf.; *Caesalpiniac.*; Sur.
Hoepfroe-hoedoe: *Copaifera guianensis* Desf.; *Caesalpiniac.*; Sur.
Hoeproe: *Copaifera guianensis* Desf.; *Caesalpiniac.*; Sur.
Hoeroe batoe: *Schima wallichi* Choisy; *Theac.*; Sunda.
„ katjang: *Phoebe opaca* Blume; *Laurac.*; Java.
Hoeroewassa: *Pithecolobium pedicellare* Bth.; *Mimosac.*; Sur.
„ : *Sweetia nitens* Bth.; *Papilionac.*; Sur.
Hofa: *Triplochiton scleroxylon* K. Schum.; *Sterculiac.*; Elf.
Hohere: *Hoheria populnea* A. Cunn.; *Malvac.*; N.-Seel.
Hoho: *Ricinodendron africanum* Muell. Arg.; *Euphorbiac.*; Elf.
Hô Hônoki: *Magnolia hypoleuca* S. et Z.; *Magnoliac.*; Jap.
Hôi: *Illicium verum* Hook.; *Magnoliac.*; Ind.-Ch.
Hoja ancha: ? *Nectandra polyphylla* Nees; *Laurac.*; Ven.
„ blanca: *Solanum verbascifolium* L.; *Solanac.*; Hond.
„ -chigüe: *Curatella americana* L.; *Dilleniac.*; C.-R.
„ menuda: *Calytranthes sintenisi* Kiaersk.; *Myrtac.*; P.-R.
„ „ : *Eugenia spp.*, *Myrcia spp.*; *Myrtac.*; P.-R.
„ ramón: *Brosimum alicastrum* Swartj; *Morac.*; Mex. (Yucatán).
„ tamal: *Hernandia guianensis* Aubl.; *Hernandiac.*; Hond.
Hojaman: *Curatella americana* L.; *Dilleniac.*; Mex.
Ho-kai: *Camellia japonica* L.; *Theac.*; Ch.
Hol: *Afzelia africana* Smith; *Caesalpiniac.*; Senl.
Holda: *Terminalia tomentosa* Wight et Arn.; *Combretac.*; O.-Ind. (Haid.).
Holder, schwarzer (d): *Sambucus nigra* L.; *Caprifoliac.*; Eu.
Hole-dasal: *Lagerstroemia flos-reginae* Retj.; *Lythrac.*; sw. O.-Ind. (Bombay).
„ -kauva: *Barringtonia acutangula* Gaertn.; *Lecythidac.*; w. O.-Ind. (Bombay).
„ -lakki: *Vitex leucoxylon* L.; *Verbenac.*; w. O.-Ind. (Bombay).
Holematti: *Terminalia arjuna* Bedd.; *Combretac.*; w. O.-Ind. (Bombay).
Holia: *Byrsonima sp.*; *Malpighiac.*; Sur.
Holigar: *Holigarna arnottiana* Hook. f.; *Anacardiac.*; sw. O.-Ind.
Hollock (H): *Spondias mangifera* Willd.; *Anacardiac.*; Ind.
„ : *Terminalia myriocarpa* Heurck et Muell. Arg.; *Combretac.*; Assam.
Hollong (H): *Dipterocarpus pilosus* Roxb.; *Dipterocarpac.*; Assam, Burma, Beng.
Holloray: *Odina wodier* Roxb.; *Anacardiac.*; n. O.-Ind. (Nepal).
Hollunder (d): *Syringa vulgaris* L.; *Oleac.*; Pers.

Hollunder, gemeiner: *Sambucus nigra* L.; *Caprifoliac.;* Eu.
„ , Berg-: *Sambucus racemosa* L.; *Caprifoliac.;* Eu.
„ , schwarzer: *Sambucus nigra* L.; *Caprifoliac.;* Eu.
„ , Trauben-: *Sambucus racemosa* L.; *Caprifoliac.;* Eu.
Holly (e): *Ilex aquifolium* L.; *Aquifoliac.;* Eu.
„ : *Ilex opaca* Ait.; *Aquifoliac.;* ö. USA.
„ : *Ilex pedunculosa* Miq.; *Aquifoliac.;* Jap.
„ , american: *Ilex opaca* Ait.; *Aquifoliac.;* USA.
„ , Cassena-: *Ilex cassine* L.; *Aquifoliac.;* sö. USA.
„ , deciduous: *Ilex decidua* Walt.; *Aquifoliac.;* s. USA.
„ , european: *Ilex aquifolium* L.; *Aquifoliac.;* Eu., N.-Am.
„ , Mountain-: *Nemopanthus mucronata* Trel.; *Aquifoliac.;* nö. USA.
„ , prickly: *Zanthoxylum panamense* P. Wilson; *Rutac.;* Pan.
„ , Swamp-: *Ilex decidua* Walt.; *Aquifoliac.;* s. USA.
„ , white: *Ilex opaca* Ait.; *Aquifoliac.;* ö. USA.
Hollywood: *Ilex spp.; Aquifoliac.;* N.-Am.
„ , Satin-: *Vitex lignum-vitae* A. Cunn.; *Verbenac.;* Queensl.
„ , yellow: *Vitex lignum-vitae* A. Cunn.; *Verbenac.;* Au.
Holo: *Hibiscus sp.; Malvac.;* Mex.
Holonda: *Adina cordifolia* Hook. f.; *Rubiac.;* nö. O.-Ind. (Orissa).
Holowasa: *Pithecolobium gonggrijpi* Kleinh.; *Mimosac.;* Sur.
„ : *Pithecolobium trapezifolium* Bth., *P. kegeli* Meissn.; *Mimosac.;* Sur.
? Holwesca: *Aralia crassifolia* Soland.; *Araliac.;* N.-Seel.
Holy pole: *Triplaris sp.; Polygonac.;* tr. Am.
Holy wood: *Guaiacum officinale* L.; *Zygophyllac.;* W.-Ind., n. S.-Am., Pan.
Hom: *Stephegyne parvifolia* Korth.; *Rubiac.;* Ind.-Ch.
„ dang: *Dysoxylum loureiri* Pierre; *Meliac.;* Ind.-Ch.
„ thom: *Ailanthus fauveliana* Pierre; *Simarubac.;* Ind.-Ch.
Homaïde: *Bussea occidentalis* Hutch.; *Caesalpiniac.;* Elf.
Homakutupon: *Ventilago africana* Exell.; *Rhamnac.;* Go.
Hombre viejo: *Cereus sp.; Cactac.;* Mex.
Homé: *Coula edulis* Baill.; *Olacac.;* Kam.
Homiry (f): *Humiria floribunda* Mart., *H. balsamifera* Aubl.; *Humiriac.;* Sur.
Hon: *Sapindus mukorossi* Gaertn.; *Sapindac.;* Ind.-Ch.
Hona: *Calophyllum inophyllum* L.; *Guttifer.;* w. O.-Ind.
Honal: *Terminalia paniculata* Roth; *Combretac.;* w. O.-Ind.
Hond: *Pterocarpus marsupium* Roxb.; *Papilionac.;* w. O.-Ind.
Honey-berry: *Melicocca paniculata* Juss.; *Sapindac.;* USA.
„ „ of Guiana: *Melicocca bijuga* L.; *Sapindac.;* Gu.
„ pod: *Prosopis juliflora* DC.; *Mimosac.;* USA.
„ shucks: *Gleditschia triacanthus* L.; *Caesalpiniac.;* ö. USA.
Honeysuckle: *Banksia integrifolia* L. f.; *Proteac.;* Queensl.;
„ : *Banksia marginata* Cav.; *Proteac.;* s. Au., Tasm.
„ : *Knightia excelsa* R. Br.; *Proteac.;* N.-Seel.
„ , Coast-: *Banksia integrifolia* L. f.; *Proteac.;* Vict., N.-S.-W.
„ , New Zealand-: *Knightia excelsa* R. Br.; *Proteac.;* N.-Seel.
Hông: *Diospyros kaki* L.; *Ebenac.;* Ind.-Ch.
„ : *Shorea vulgaris* Pierre; *Dipterocarpac.;* Ind.-Ch.
„ bi: *Clausena wampi* Oliv.; *Rutac.;* Ind.-Ch.
„ phât: *Garcinia cambodgiensis* Vesque; *Guttifer.;* Ind.-Ch.
Hongal: *Terminalia paniculata* Roth; *Combretac.;* w. O.-Ind.
Honge: *Galedupa pinnata* Taub.; *Papilionac.;* As., Au.
„ : *Pongamia glabra* Vent.; *Papilionac.;* w. O.-Ind.
Hongué: *Alstonia congensis* Engl.; *Apocynac.;* Elf.
Honguié: *Alstonia congensis* Engl.; *Apocynac.;* Elf.
Honi: *Plagianthus betulinus* A. Cunn.; *Malvac.;* N.-Seel.
Honka: *Rhizophora mucronata* Lam.; *Rhizophorac.;* Mad.
? Honkolafy: *Rhizophora mucronata* Lam.; *Rhizophorac.;* Mad.
Honkolahy: *Rhizophora mucronata* Lam.; *Rhizophorac.;* Mad.
Honkolavy: *Ceriops candolleana* Arn.; *Rhizophorac.;* Mad.
Honkon-gasi: *Quercus championi* Bth.; *Fagac.;* Form.
Honkovavy: *Ceriops candolleana* Arn., *C. boiviniana* Tul.; *Rhizophorac.;* Mad.

Honkovavy: *Rhizophora mucronata* Lam.; *Rhizophorac.;* Mad.
Honmaki: *Podocarpus macrophylla* D. Don; *Conifer.;* Jap., Ch.
Honné: *Pterocarpus marsupium* Roxb.; *Papilionac.;* w. O.-Ind.
Honoki: *Magnolia hypoleuca* S. et Z.; *Magnoliac.;* Jap.
„ : *Alnus japonica* S. et Z.; *Betulac.;* Jap.
Hono-ki: *Magnolia obovata* Thunb.; *Magnoliac.;* Jap.
Honsugi: *Cryptomeria japonica* D. Don; *Conifer.;* Jap.
Honton: *Antiaris africana* Engl.; *Morac.;* Go.
Hooboo: *Spondias lutea* L., *S. dulcis* Forst.; *Anacardiac.;* Sur.
Hoobooballi: *Loxopterygium sagoti* Hook. f.; *Anacardiac.;* Sur.
Hooboodia: *Anacardium giganteum* Hancock; *Anacardiac.;* br. Gu.
Hooday: *Stereospermum suaveolens* DC.; *Bignoniac.;* w. O.-Ind. (Bombay).
Hoogia: *Tetrameles nudiflora* R. Br.; *Datiscac.;* nö. O.-Ind. (Nepal).
Hoolanghik-gass: *Chukrasia tabularis* A. Juss.; *Meliac.;* s. Ind., Cey.
Hoolgeri: *Holigarna arnottiana* Hook. f.; *Anacardiac.;* w. O.-Ind.
Hoo-loop: *Bravaisia tubiflora* Hemsl.; *Acanthac.;* br. Hond.
Hoom: *Polyalthia cerasoides* Bth. et Hook. f.; *Anonac.;* sw. O.-Ind.
Hoonoki: *Magnolia hypoleuca* S. et Z.; *Magnoliac.;* Jap.
Hoorihea: *Humiria spp.; Humiriac.;* br. Gu.
Hoorihee: *Humiria spp.; Humiriac.;* br. Gu.
Hooroowassa: *Pithecolobium trapezifolium* Bth.; *Mimosac.;* br. Gu.
Hopbeam: *Ostrya carpinifolia* Scopoli; *Betulac.;* s. Eu., Kl.-As.
Hop bush: *Dodonaea viscosa* Jacq.; *Sapindac.;* Queensl., N.-Seel.
„ „ : *Dodonaea triquetra* Wendl.; *Sapindac.;* ö. Au.
Hopea (H): *Hopea parviflora* Bedd.; *Dipterocarpac.;* sw. O.-Ind.
Hopwood: *Coccoloba uvifera* L.; *Polygonac.;* Fl., br. W.-Ind.
Hora: *Dipterocarpus zeylanicus* Thw.; *Dipterocarpac.;* Cey.
Horco cebil: *Piptadenia macrocarpa* Bth.; *Mimosac.;* Arg.
„ „ : *Piptadenia excelsa* Lillo, *P. communis* Bth.; *Mimosac.;* Arg.
„ molle: *Bumelia obtusifolia* Roem. et Schult.; *Sapotac.;* Arg.
„ sauce: *Escallonia montana* Phil.; *Saxifragac.;* Arg.
Horie: *Byrsonima sp.; Malpighiac.;* Sur.
Horizontal: *Anodopetalum biglandulosum* A. Cunn.; *Cunoniac.;* Tasm.
Hormigo: *Platymiscium polystachyum* Bth.; *Papilionac.;* Guat.
„ : *Platymiscium spp.; Papilionac.;* Hond.
Hornbaum (d): *Carpinus betulus* L.; *Betulac.;* Eu.
Hornbeam (e): *Carpinus betulus* L.; *Betulac.;* Eu.
„ : *Ostrya virginiana* C. Koch; *Betulac.;* ö. Kan.
„ : *Carpinus americana* Michx.; *Betulac.;* ö. Kan.
„ : *Carpinus laxiflora* Blume; *Betulac.;* Jap.
„ , american: *Carpinus americana* Michx.; *Betulac.;* USA.
„ , dyed: *Carpinus betulus* L.; *Betulac.;* Mitt.-Eu.
„ , Hop-: *Carpinus caroliniana* Walt.; *Betulac.;* ö. Kan.
„ , „ : *Ostrya virginiana* C. Koch; *Betulac.;* ö. N.-Am.
„ , „ : *Ostrya japonica* Sarg.; *Betulac.;* Jap.
„ , silky: *Lucuma amorphosperma* F. M. Bailey; *Sapotac.;* Au.
Hornstrauch, blutroter (d): *Cornus sanguinea* L.; *Cornac.;* Eu.
Horoeka: *Panax crassifolium* Dcne. et Planch.; *Araliac.;* N.-Seel.
Horopito: *Drimys axillaris* Forst.; *Magnoliac.;* N.-Seel.
Horroebatoe: *Antidesma ghaesembilla* Gaertn.; *Euphorbiac.;* Java.
Horse bean: *Parkinsonia aculeata* L.; *Caesalpiniac.;* s. USA., W.-Ind.
Horseflesh: *Lysiloma sabicu* Bth.; *Mimosac.;* Bah.
Horsefleshwood: *Mimusops spp.; Sapotac.;* tr. Am.
Horse-raddish-tree: *Moringa oleifera* Lam.; *Moringac.;* Tr.
Horse-sugar: *Symplocos myrtacea* S. et Z.; *Symplocac.;* Jap.
Horsewood: *Hippobromus alatus* Eckl. et Zeyh.; *Sapindac.;* Kap.
Hortensie: *Hydrangea spp.; Saxifragac.;* As., Am.
Hosoba-shira-kashi: *Quercus pseudo-myrsinaefolia* Hayata; *Fagac.;* Form.
„ -sirakasi: *Quercus longinux* Hayata; *Fagac.;* Form.
Hou liên: *Melia azedarach* L.; *Meliac.;* Annam.
Houbooballi: *Loxopterygium sagoti* Hook. f.; *Anacardiac.;* Gu.
Houétin: *Triplochiton scleroxylon* K. Schum.; *Sterculiac.;* Dah.

Houhere: *Hoheria populnea* A. Cunn.; *Malvac.;* N.-Seel.
Houi: *Plagianthus betulinus* A. Cunn.; *Malvac.;* N.-Seel.
Houmiri: *Humiria balsamifera* Aubl.; *Humiriac.;* Gu.
„ boumier: *Humiria spp.; Humiriac.;* fr. Gu.
Houngo: *Khaya ivorensis* Chev.; *Meliac.;* Kam.
Hounti: *Ceiba pentandra* Gaertn. var. *dehiscens* Ulb.; *Bombacac.;* Dah.
Houp: *Montrouziera cauliflora* Planch. et Triana; *Guttifer.;* N.-Kal.
„ : *Montrouziera sphaeraeflora* Panch., *M. sphaeroidea* Panch.; *Guttifer.;* N.-Kal.
„ , faux: *Garcinia collina* Vieill.; *Guttifer.;* N.-Kal.
Hou-Po: *Magnolia hypoleuca* S. et Z.; *Magnoliac.;* Jap., Ch.
Housson (f): *Ilex aquifolium* L.; *Aquifoliac.;* Eu.
Houx (f): *Ilex aquifolium* L.; *Aquifoliac.;* Eu.
„ commun (f): *Ilex aquifolium* L.; *Aquifoliac.;* Eu.
Howadanni: *Oxandra lanceolata* Baill.; *Anonac.;* br. Gu.
Hoza tsiketra: *Louvelia albicans* H. Jum.; *Palmac.;* Mad.
Hpadawng: *Mallotus philippinensis* Muell. Arg.; *Euphorbiac.;* O.-Ind.
Hpak-lü: *Ficus glomerata* Roxb.; *Morac.;* Burma.
„ -mong: *Cordia myxa* L.; *Borraginac.;* Burma.
Hpang: *Stephegyne parvifolia* Korth.; *Rubiac.;* Burma.
„ : *Calophyllum inophyllum* L.; *Guttifer.;* Burma.
Hpawng-awn: *Mallotus philippinensis* Muell. Arg.; *Euphorbiac.;* Burma.
Hpunam-makawk: *Spondias mangifera* Willd.; *Anacardiac.;* O.-Ind.
Hpunja: *Aegle marmelos* Correa; *Rutac.;* O.-Ind.
Hpunmang: *Diospyros burmanica* Kurz; *Ebenac.;* O.-Ind.
Hpu-no: *Careya arborea* Roxb.; *Lecythidac.;* Burma.
Hpunsha: *Anogeissus acuminata* Wall.; *Combretac.;* O.-Ind.
Hseik-khyae: *Erioglossum edule* Blume; *Sapindac.;* And.
Htakyi: *Phyllanthus emblica* L.; *Euphorbiac.;* O.-Ind.
Hu: *Commersonia echinata* Forst. var. *platyphylla* Andr.; *Sterculiac.;* Ind.-Ch.
„ mong: *Commersonia echinata* Forst.; *Sterculiac.;* Ind.-Ch.
„ -tiao: *Juglans sieboldiana* Maxim.; *Juglandac.;* Ch.
Huachipilín: *Diphysa carthaginensis* Jacq.; *Papilionac.;* Salv., C.-R.
Huacux: *Sideroxylon capiri* Pittier; *Sapotac.;* Mex.
Huahuan: *Laurelia aromatica* Juss.; *Monimiac.;* Chile.
Huajillo: *Leucaena lanceolata* Watson; *Mimosac.;* Mex. (Sinaloa).
Hualhua: *Talauma sp.; Magnoliac.;* Mex.
Hualle: *Nothofagus sp.; Fagac.;* Chile.
Huamaga: *Hibiscus tiliaceus* L.; *Malvac.;* Ek.
Huamuche: *Pithecolobium dulce* Bth.; *Mimosac.;* Mex.
Huamuchil costeno: *Pithecolobium dulce* Bth.; *Mimosac.;* Mex.
Huanaxtle: *Enterolobium cyclocarpum* Griseb.; *Mimosac.;* Mex.
Huan-bai: *Odina wodier* Roxb.; *Anacardiac.;* Burma.
„ huan: *Laurelia aromatica* Juss.; *Monimiac.;* Chile.
Huang-yang: *Buxus microphylla* S. et Z. var. *sinica* R. et W.; *Buxac.;* Ch.
Huanita: *Beureria huanita* Hemsl.; *Borraginac.;* Mex.
Huara: *Litsea polyantha* Juss.; *Laurac.;* Assam (Cachar).
Huarango: *Caesalpinia spinosa* Kuntze; *Caesalpiniac.;* Ek.
„ : *Acacia macracantha* Humb. et Bonpl.; *Mimosac.;* Peru.
Huaw: *Adina cordifolia* Hook. f.; *Rubiac.;* Ind.
Huayu: *Kageneckia oblonga* Ruiz et Pav.; *Rosac.;* S.-Am. (Anden).
Hubaballi: *Loxopterygium sagoti* Hook. f.; *Anacardiac.;* holl. Gu.
Hublás: *Tristania spp.; Myrtac.;* Phil. (Neg.).
Hubu: *Spondias lutea* L., *S. dulcis* Forst.; *Anacardiac.;* Sur.
Hububalli: *Loxopterygium sagoti* Hook. f.; *Anacardiac.;* Gu.
Húcar blanco: *Bucida buceras* L.; *Combretac.;* P.-R.
Huck: *Celtis crassifolia* Lam.; *Ulmac.;* USA.
Hucuya: *Swartzia tomentosa* DC.; *Caesalpiniac.;* fr. Gu.
Hudoke (u): *Bowdichia nitida* Spruce, *B. virgilioides* H. B. K.; *Papilionac.;* Bras.
Huê môc: *Dalbergia sp.; Papilionac.;* Ind.-Ch.
„ „ : *Osmanthus fragrans* Lour.; *Oleac.;* Ind.-Ch.
„ -Môe: *Pterocarpus pedatus* Pierre var. ?; *Papilionac.;* Ind.-Ch.
Huejocote: *Salix humboldtiana* Willd.; *Salicac.;* Mex.

Huele noche: *Solanum nudum* H. B. K.; *Solanac.;* Nic.
„ de noche: *Bumelia retusa* Swartz; *Sapotac.;* Guat.
„ „ „ : *Cestrum spp.; Solanac.;* Hond.
Huesillo: *Sweetia panamensis* Bth.; *Papilionac.;* Mex.
„ : *Allophylus occidentalis* Radlk.; *Sapindac.;* br. Hond.
Huesito: *Sweetia panamensis* Bth.; *Papilionac.;* Mex.
„ : *Faramea occidentalis* A. Rich.; *Rubiac.;* s. Mex.
„ : *Trichilia hirta* L.; *Meliac.;* Pan.
„ : ? *Prockia flava* Karst.; *Flacourtiac.;* w. Pan.
„ : *Prockia crucis* L.; *Flacourtiac.;* Ven.
Hueso: *Drypetes glauca* Poit.; *Euphorbiac.;* Cuba.
„ : *Drypetes keyensis* Urban; *Euphorbiac.;* sp. W.-Ind.
„ : *Picramnia pentandra* Sw.; *Simarubac.;* P.-R.
„ blanco: *Mayepea domingensis* Krug et Urban; *Oleac.;* P.-R.
„ prieto: *Ilex nitida* Maxim.; *Aquifoliac.;* P.-R.
„ de anta: *Aphelandra deppeana* Cham. et Schl.; *Acanthac.;* Kol.
„ „ finado: *Psychotria pinularis* Sessé et Moç.; *Rubiac.;* br. Hond.
Huevo de gato: *Psidium molle* Bertol.; *Myrtac.;* Hond.
Huevos de gato: ? *Pterocarpus belizensis* Standl.; *Papilionac.;* Pan.
Huexotl: *Salix humboldtiana* Willd.; *Salicac.;* Mex.
Huicon: *Lucuma palmeri* Fernald; *Sapotac.;* Mex.
Huilca: *Piptadenia sp.; Mimosac.;* Peru.
„ romana: *Piptadenia sp.; Mimosac.;* Peru.
Huiliguiste: *Karwinskia calderoni* Standl.; *Rhamnac.;* Guat.
Huiloche: *Diphysa occidentalis* Rose; *Papilionac.;* Mex.
Huiñag: *Tabebuia nodosa* Griseb.; *Bignoniac.;* Arg.
Huinecaztle: *Enterolobium cyclocarpum* Griseb.; *Mimosac.;* Mex.
Huinh: *Tarrietia javanica* Blume; *Sterculiac.;* Annam.
„ -dân: *Dysoxylum loureiri* Pierre; *Meliac.;* Annam.
„ duong: *Dysoxylum loureiri* Pierre; *Meliac.;* Annam.
„ nùong: *Ternstroemia penangiana* Choisy; *Theac.;* Annam.
Huinke: *Eucryphia cordifolia* Cav.; *Eucryphiac.;* Arg.
Huinque: *Lomatia ferruginea* R. Br.; *Proteac.;* Chile.
Huisache: *Pithecolobium albicans* Bth.; *Mimosac.;* Mex. (Yucatán).
„ bola: *Caesalpinia sclerocarpa* Standl.; *Caesalpiniac.;* Mex. (Sinaloa).
Huitoc: *Genipa sp.; Rubiac.;* Peru.
Hujed: *Adansonia digitata* L.; *Bombacac.;* Arab.
Hukup: *Bursera gummifera* L.; *Burserac.;* br. Hond.
Hula: *Pterocarpus tinctorius* Welw.; *Papilionac.;* Gab.
Hule: *Castilla elastica* Cerv.; *Morac.;* Hond.
„ macho: *Castilla fallax* Cook; *Morac.;* C.-Am.
Huligano: *Pterocymbium tinctorium* Merr.; *Sterculiac.;* Phil. (Nueva Ecija).
Hulluch: *Terminalia belerica* Roxb.; *Combretac.;* Assam.
Hullung: *Dipterocarpus pilosus* Roxb.; *Dipterocarpac.;* nö. Assam.
Hülse (d): *Ilex aquifolium* L.; *Aquifoliac.;* Eu.
Hulstboom (h): *Ilex aquifolium* L.; *Aquifoliac.;* Eu.
Hulu-bai: *Bravaisia tubiflora* Hemsl.; *Acanthac.;* br. Hond.
Hulzenboom (h): *Ilex aquifolium* L.; *Aquifoliac.;* Eu.
Hum: *Fraxinus excelsior* L.; *Oleac.;* nw. O.-Ind. (Kasch.).
Humbug: *Schizomeria ovata* D. Don; *Cunoniac.;* ö. Au.
Humiri: *Humiria balsamifera* Aubl.; *Humiriac.;* Gu.
Humo: *Pithecolobium dulce* Bth.; *Mimosac.;* Mex.
„ de sabana: *Pithecolobium obovale* Sauv.; *Mimosac.;* Cuba.
Humpé: *Khaya ivorensis* Chev.; *Meliac.;* Elf.
Hunab: *Terminalia paniculata* Roth; *Combretac.;* sw. O.-Ind.
Hunalu: *Litsea polyantha* Juss.; *Laurac.;* Assam.
Hundsholz, Jamaika- (d): *Ichthyomethia piscipula* Hitch.; *Papilionac.;* W.-Ind., s. Mex.
Hung-Chai: *Ormosia sp.; Papilionac.;* Ch.
„ chi mou: *Quercus chevalieri* Hicker et A. Camus; *Fagac.;* Ind.-Ch.
Hunnagere: *Canthium didymum* Roxb.; *Rubiac.;* sw. O.-Ind.
Hunti: *Ceiba pentandra* Gaertn. *var. dehiscens* Ulb.; *Bombacac.;* Dah.
Hunúg: *Pygeum glandulosum* Merr., *P. presli* Merr.; *Rosac.;* Phil.

Hupong-húpong: *Buchanania spp.; Anacardiac.;* Phil.
Huragalu: *Chloroxylon swietenia* DC.; *Rutac.;* sw. O.-Ind. (Mysore).
Hura-wood (u): *Hura crepitans* L., *H. polyandra* Baill.; *Euphorbiac.;* tr. Am.
Huria: *Byrsonima coccolobaefolia* H. B. K.; *Malpighiac.;* br. Gu.
Hurihi: *Albizzia odoratissima* Bth.; *Mimosac.;* Cey.
Hurinhura: *Amoora rohituka* Wight et Arn.; *Meliac.;* Hind.
Huriri: *Albizzia odoratissima* Bth.; *Mimosac.;* O.-Ind.
Hurkli: *Pistacia integerrima* Stewart; *Anacardiac.;* nw. O.-Ind. (Punjab).
Hurowassa: *Pithecolobium gonggrijpi* Kleinh., *P. kegeli* Meissn.; *Mimosac.;* Gu.
    „   : *Pithecolobium trapezifolium* Bth.; *Mimosac.;* Gu.
Huruk: *Toona ciliata* M. Roem.; *Meliac.;* w. O.-Ind.
Hutu tawhai: *Nothofagus fusca* Oerst.; *Fagac.;* N.-Seel.
Huyet muong: *Knema conferta* Warb.; *Myristicac.;* Ind.-Ch.
Huynduong: *Dysoxylum loureiri* Pierre; *Meliac.;* Co.-Ch.
Huynh: *Tarrietia cochinchinensis* Pierre; *Sterculiac.;* Annam.
  „ ba: *Sarcocephalus officinalis* Pierre; *Rubiac.;* Kamb.
  „ dàn: *Dysoxylum loureiri* Pierre; *Meliac.;* Ind.-Ch.
  „ duong: *Dysoxylum loureiri* Pierre; *Meliac.;* Ind.-Ch.
  „ , faux: ? *Peltophorum sp.; Caesalpiniac.;* Kamb.
  „ mai: *Ochna harmandi* H. Lec.; *Ochnac.;* Ind.-Ch.
  „ nuong: *Ternstroemia penangiana* Choisy; *Theac.;* Ind.-Ch.
Hwenchwenti: *Sarcocephalus russegeri* Kotschy; *Rubiac.;* Go.
Hya-hya: *Tabernaemontana utilis* Wight et Arn.; *Apocynac.;* fr. Gu.
Hyawá: *Protium sp.; Burserac.;* br. Gu.
Hyawana: *Protium sp.; Burserac.;* br. Gu.
Hymarikushi: *Mimusops spp.; Sapotac.;* fr. Gu.
Hypernic (u): *Haematoxylon brasiletto* Karst.; *Caesalpiniac.;* C.-Am., n. S.-Am., W.-Ind.

I

Iba: *Pachylobus balsamifera* Guill.; *Burserac.;* Gab.
 „ : *Irvingia gabonensis* Baill.; *Simarubac.;* Gab.
 „ -Eé: *Acanthosyris spinescens* Griseb.; *Santalac.;* Arg.
 „ -hé-hé: *Acanthosyris spinescens* Griseb.; *Santalac.;* Arg., Bras.
 „ -jai: *Eugenia edulis* Vell.; *Myrtac.;* Arg.
 „ rá: *Pterocarpus micheli* Britt.; *Papilionac.;* Arg.
 „ hay guazú: *Eugenia uvalha* Cambess.; *Myrtac.;* Arg.
Ibahai: *Eugenia edulis* Vell.; *Myrtac.;* tr. Am.
Ibajay mi: *Eugenia euscidifolia* Engl.; *Myrtac.;* Arg.
Ibamba: *Cocos nucifera* L.; *Palmac.;* Gab.
Iban: *Achras sp.; Sapotac.;* Nic.
Ibanda: *Didelotia africana* Pierre; *Caesalpiniac.;* Gab.
I bapohy: *Ficus i bapohy* ?; *Morac.;* Arg.
Ibapohy caa guih: *Ficus sp.; Morac.;* Arg.
Ibapoi: *Ficus sp.; Morac.;* Arg.
  „ -moroti: *Ficus sp.; Morac.;* Arg.
Ibapoy say: *Ficus cestrifolia* Schott; *Morac.;* Arg.
Ibéca: *Cystanthe sp.; Epacridac.;* Gab.
  „ : *Mammea sp., Ochrocarpus sp.; Guttifer.;* Gab.
Ibéka: *Mammea klaineana* Pierre; *Guttifer.;* Gab.
Ibel: *Cola acuminata* Schott et Endl.; *Sterculiac.;* Kam.
Ibera-obi: *Helietta cuspidata* Chod. et Hassl.; *Rutac.;* Arg.
Iberraobi: *Helietta cuspidata* Chod. et Hassl.; *Rutac.;* Arg.
Ibirá-berá: *Caesalpinia melanocarpa* Griseb.; *Caesalpiniac.;* Arg.
  „ -catú: *Phyllostylon brasiliensis* Capanema; *Ulmac.;* Arg.
  „ „ : *Phyllostylon rhamnoides* Taub.; *Ulmac.;* Arg. (Formosa).
  „ -katú: *Phyllostylon rhamnoides* Taub.; *Ulmac.;* Arg. (Chaco).
  „ hembrá: *Ruprechtia polystachya* Griseb.; *Polygonac.;* Arg.
  „ -hú: *Acanthosyris spinescens* Griseb.; *Santalac.;* Arg. (Mis., Corr., S.-Fé).
  „ - „ : *Achatocarpus spinulosus* Griseb.; *Phytolaccac.;* Arg. (S.-Fé).
  „ itá: *Muellera glaziovi* Chod. et Hassl.; *Papilionac.;* Arg.
  „ moroti: *Calycophyllum candidissimum* DC.; *Rubiac.;* Arg.

Ibirá-nirá: *Bumelia obtusifolia* Roem. et Schult.; *Sapotac.;* Arg.
„ pepé: *Holocalyx balansae* Micheli; *Caesalpiniac.;* Arg.
„ peré: *Apuleia praecox* Mart.; *Caesalpiniac.;* Arg.
„ piapuna: *Apuleia pogomana* Allem.; *Caesalpiniac.;* tr. S.-Am.
„ -pi-hú: *Diatenopteryx sorbifolia* Radlk.; *Sapindac.;* Arg. (Mis.).
„ -Pita: *Peltophorum vogelianum* Walp.; *Caesalpiniac.;* Arg.
„ -pita-bi: *Ruprechtia sp.; Polygonac.;* Arg.
„ -puitá: *Peltophorum vogelianum* Walp.; *Caesalpiniac.;* Arg. (Mis.).
„ „ : *Ruprechtia polystachya* Griseb.; *Polygonac.;* Arg. (Chaco).
„ „ guassú: *Peltophorum vogelianum* Walp.; *Caesalpiniac.;* Arg. (S.-Fé).
„ „ y: *Ruprechtia polystachya* Griseb.: *Polygonac.;* Arg.
„ „ „ rá: *Ruprechtia polystachya* Griseb.; *Polygonac.;* Arg.
„ -pyita: *Ruprechtia virarú* Griseb.; *Polygonac.;* Arg.
„ rá: *Pterogyne nitens* Tul.; *Caesalpiniac.;* Arg.
„ -ré: *Piptadenia sp.; Mimosac.;* Arg.
„ -ró: *Pterogyne nitens* Tul.; *Caesalpiniac.;* Arg.
„ saiyú: *Bergeronia sericea* Micheli; *Papilionac.;* Arg.
„ yú: *Pithecolobium haesleri* Chod.; *Mimosac.;* Arg.
Ibirarema: *Gallesia scorododendrum* Casar.; *Phytolaccac.;* Bras.
Ibiráro: *Pterogyne nitens* Tul.; *Caesalpiniac.;* Arg.
„ -mi: *Ruprechtia sp.; Polygonac.;* Arg.
Ibiratai: *Pilocarpus pennatifolius* Lem.; *Rutac.;* Bras.
Ibirapitanga: *Caesalpinia echinata* Lamk.; *Caesalpiniac.;* Bras. (Sao P).
Ibiriribá-rána: ? *Eschweilera sp.; Lecythidac.;* Bras.
Ibiripatanga: *Caesalpinia echinata* Lamk.; *Caesalpiniac.;* Bras.
Ibiri pitanga: *Caesalpinia echinata* Lamk.; *Caesalpiniac.;* Bras.
Ibol: *Trichadenia philippinensis* Merr.; *Flacourtiac.;* Phil. (Pang.).
Ibopé-hu: *Prosopis sp.; Mimosac.;* Arg.
„ -moroti: *Prosopis sp.; Mimosac.;* Arg.
„ -pyitá: *Prosopis sp.; Mimosac.;* Arg.
„ -saiyú: *Prosopis sp.; Mimosac.;* Arg.
Iboré: *Mammea klaineana* Pierre; *Guttifer.;* Gab.
Ibota: *Ligustrum ibota* Sieb.; *Oleac.;* Jap.
Ibotou: *Guarea cedrata* Pellegr.; *Meliac.;* Elf.
Ibu: *Pometia pinnata* Forst.; *Sapindac.;* Phil. (Neg., Cebú).
Ibuki: *Juniperus chinensis* L.; *Conifer.;* Ch., Jap.
Ibyrá hú: *Achatocarpus obovatus* Schinz et Autran.; *Phytolaccac.;* Arg.
„ ovi: *Helietta cuspidata* Chod. et Hassl.; *Rutac.;* Arg.
Icacillo: *Hirtella sp.; Rosac.;* P.-R.
Icaco: *Chrysobalanus icaco* L.; *Rosac.;* Hond., br. Hond., P.-R.
„ prieto: *Hirtella sp.; Rosac.;* Cuba.
„ de aura: *Hirtella sp.; Rosac.;* Cuba.
Icaque: *Hirtella triandra* Swartz; *Rosac.;* Guad.
„ montagne: *Licania sp.; Rosac.;* Guad.
„ des bois: *Chrysobalanus sp.; Rosac.;* Guad.
Icaquier: *Chrysobalanus icaco* L.; *Rosac.;* s. Fl., C.-Am., W.-Ind.
Icaquillo: *Hirtella sp.; Rosac.;* Mex.
Icaquito: *Licania sp.; Rosac.;* Ven.
Icaroré catinga: *Ardisia sp.; Myrsinac.;* Bras.
Iccaaya: *Jagera pseudorhus* Radlk.; *Sapindac.;* N.-S.-W.
Icé ts'baki: *Camellia japonica* L.; *Theac.;* Jap.
Ich-bahatsch: *Trichilia cuneta* Radlk.; *Meliac.;* br. Hond.
Icha: *Ficus nervosa* Heyne; *Morac.;* O.-Ind.
Ichii: *Taxus baccata* L. *subsp. cuspidata* Pilger; *Conifer.;* Jap.
„ -gashi: *Quercus gilva* Blume; *Fagac.;* Form.
Icho: *Ginkgo biloba* L.; *Ginkgoac.;* Jap.
Ici: *Myrocarpus fastigiatus* Allem.; *Papilionac.;* Bras.
Iciboicica: *Adansonia digitata* L.; *Bombacac.;* Bras.
Icica: *Protium icicariba* March.; *Burserac.;* Bras.
„ -riba: *Protium icicariba* March.; *Buserac.;* Bras.
Icinillo: *Weinmannia sp.; Cunoniac.;* S.-Am.
Icipo guaica: *Guazuma ulmifolia* Lam.; *Sterculiac.;* S.-Am.

Icipo guaica: *Myrtus mucronata* Cambess.; *Myrtac.;* S.-Am.
Iciquier (f): *Protium altissimum* March.; *Burserac.;* W.-Ind., Gu., Bras.
Icounindou: *Coula edulis* Baill.; *Olacac.;* Gab.
Idatimon: *Chytroma idatimon* Miers; *Lecythidac.;* fr. Gu.
Idauk: *Rhizophora mucronata* Lamk.; *Rhizophorac.;* Burma.
Iddáng: *Litsea spp.; Laurac.;* Phil. (Cag.).
Idé: *Klainedoxa latifolia* Pierre; *Simarubac.;* Kam.
Idewa: ? *Petersia viridiflora* Chev.; *Lecythidac.;* Gab.
Idou: *Saccoglottis gabonensis* Urban; *Humiriac.;* Kam.
Idoumbeni: *Cola ballayi* Cornu; *Sterculiac.;* Gab.
Idulanhu: *Daemia angolensis* Dcne.; *Asclepiadac.;* D.-O.-Af. (Tabora).
If (f): *Taxus baccata* L.; *Conifer.;* Eu.
Iffe (d): *Ulmus effusa* Willd.; *Ulmac.;* Mitt.-Eu.
Ifi-lele: *Intsia bijuga* O. Ktʒe.; *Caesalpiniac.;* Südsee (Samoa).
Ifil: *Intsia bijuga* O. Ktʒe.; *Caesalpiniac.;* Südsee (Guam).
Ifondo: *Strombosia grandifolia* Hook. f.; *Olacac.;* Af.
Igamba: *Heisteria trillesiana* Pierre; *Olacac.;* Gab.
Igáng: *Vitex aherniana* Merr.; *Verbenac.;* Phil. (Tay.).
Igedudu: *Diospyros atropurpurea* Gürke; *Ebenac.;* tr. W.-Af.
Iggwanxe: *Olea laurifolia* Lam.; *Oleac.;* Kap.
Igíu: *Dysoxylum decandrum* Merr.; *Meliac.;* Phil. (Batg.).
Igogozo: *Ficus vogeliana* Miq.; *Morac.;* Gab.
Igopé: *Prosopis alba* Griseb.; *Mimosac.;* Arg.
„ -guazú: *Prosopis nigra* Hieron.; *Mimosac.;* Arg. (Corr.).
„ -pará: *Prosopis alba* Griseb.; *Mimosac.;* Arg.
Igót: *Eugenia spp.; Myrtac.;* Phil. (Luzon).
Igoumou: ˏ*Coula edulis* Baill.; *Olacac.;* Gab.
Igoungou: *Pterocarpus soyauxi* Taub.; *Papilionac.;* Gab.
Iguajari dulce: *Eugenia edulis* Bth.; *Myrtac.;* S.-Am.
Iguanero: *Citharexylum macrochlamys* Pittier; *Verbenac.;* Pan.
Iguano: *Caesalpinia eriostachya* Bth.; *Caesalpiniac.;* Mex.
„ : ? *Leucaena sp.; Mimosac.;* Pan.
„ blanco: *Goldmania constricta* Micheli et Rose; *Mimosac.;* Mex.
Iguersel: *Ilex aquifolium* L.; *Aquifoliac.;* n. Af. (Algier).
Igüi: *Bumelia sp.; Sapotac.;* Ven.
Igusi (e): *Baikiaea plurijuga* Harms; *Caesalpiniac.;* Rhod.
Ihand: *Prosopis spicigera* L.; *Mimosac.;* Ind.
Iharberi: *Zizyphus jujuba* Lamk.; *Rhamnac.;* Hind.
Ijal: *Barringtonia acutangula* Gaertn.; *Lecythidac.;* n. O.-Ind.
Ijar: *Careya arborea* Roxb.; *Lecythidac.;* n. Beng. (Mongyr).
Iju: *Schima noronhae* Reinw.; *Theac.;* Jap.
Ijwgi: *Adenanthera pavonina* L.; *Mimosac.;* And.
Ijzerhart (h): *Swartʒia spp.; Caesalpiniac.;* Sur.
„ : *Tecoma sp., Tabebuia sp.; Bignoniac.;* Sur.
„ : *Chrysophyllum glabrum* Jacq.; *Sapotac.;* Sur.
„ , Bastard-: *Swartʒia tomentosa* DC.; *Caesalpiniac.;* Sur.
„ ; Gandoe-: *Swartʒia tomentosa* DC.; *Caesalpiniac.;* Sur.
„ , Gewone: *Swartʒia-spp.; Caesalpiniac.;* Sur.
Ijzerhout (h): *Sideroxylon spp.; Sapotac.;* fr. Gu.
„ : *Mouriria sp.; Melastomatac.;* fr. Gu.
„ : *Swartʒia minutiflora* Kleinh.; *Caesalpiniac.;* Sur.
„ : *Aglaia eusideroxylon* Koord. et Valet.; *Meliac.;* Java.
„ , Man-: *Siderodendrum triflorum* Vahl; *Rubiac.;* Sur.
Ikandika: *Desplatʒia trillesiana* Pierre; *Tiliac.;* Gab.
Ikèle: *Klainedoxa longifolia* Pierre; *Simarubac.;* B.-K. (Lukolela).
Ikoko: *Chlorophora regia* Chev.; *Morac.;* Gab.
Ikola hindi: *Xylopia aethiopica* A. Rich.; *Anonac.;* Kam.
Ikomba: *Musanga smithi* R. Br.; *Morac.;* tr. W.-Af.
Ikop: *Staudtia gabunensis* Warb.; *Myristicac.;* Kam.
Ikoum: *Pycnanthus kombo* Warb.; *Myristicac.;* Gab.
Ikoumbi: *Dialium guineense* Willd.; *Caesalpiniac.;* Gab.
Ikouninou: *Coula edulis* Baill.; *Olacac.;* Gab.

Ikusi: *Baikiaea plurijuga* Harms; *Caesalpiniac.*; s. Rhod.
Ikwahobo:. *Lovoa klaineana* Pierre; *Meliac.*; Nig.
Ilama: *Anona sp.*; *Anonac.*; Mex.
Ilana de Tehuantepec: *Anona sp.*; *Anonac.*; Mex.
Ilandampajam: *Zizyphus jujuba* Lamk.; *Rhamnac.*; s. Ind.
Ilang-ilang: *Canangium odoratum* Baill.; *Anonac.*; Phil.
„ „ gúbat: *Litsea spp.*; *Laurac.*; Phil. (Lag.)
„ „ laláki: *Litsea spp.*; *Laurac.*; Phil. (Riz.)
Ilangi: *Zizyphus jujuba* Lamk.; *Rhamnac.*; w. O.-Ind.
Ilantai: *Zizyphus jujuba* Lamk.; *Rhamnac.*; w. O.-Ind.
Ilantha: *Zizyphus jujuba* Lamk.; *Rhamnac.*; w. O.-Ind. (s. Deccan).
Ilapongu: *Hopea wightiana* Wall.; *Dipterocarpac.*; w. O.-Ind. (s. Deccan).
Ilar: *Miliusa velutina* Hook. f. et Thoms.; *Anonac.*; ö. O.-Ind. (Naini-Tal).
Ilavam: *Bombax malabaricum* DC.; *Bombacac.*; w. O.-Ind. (s. Deccan).
Ilavu: *Bombax malabaricum* DC.; *Bombacac.*; w. O.-Ind. (s. Deccan).
Ile: *Uapaca heudeloti* Baill.; *Euphorbiac.*; tr. W.-Af.
Ilenden-marom: *Zizyphus jujuba* Lamk.; *Rhamnac.*; s. Ind.
Iliahi: *Santalum yasi* Seem.; *Santalac.*; Fid.
„ aloe: *Santalum littorale* Rock.; *Santalac.*; Fid.
Ilimba: *Pycnanthus kombo* Warb.; *Myristicac.*; Kam.
Iliya: *Kydia calycina* Roxb.; *Malvac.*; sw. O.-Ind.
Il-lagasí: *Shorea spp.*; *Dipterocarpac.*; Phil. (Sibu., Cap.)
Illaponga: *Hopea glabra* Wight et Arn.; *Dipterocarpac.*; sw. O.-Ind.
Illarega: *Angophora subvelutina* F. v. M.; *Myrtac.*; N.-S.-W.
Illomba: *Pycnanthus kombo* Warb.; *Myristicac.*; Gab.
Illupai: *Bassia longifolia* Willd.; *Sapotac.*; Cey.
Illupathla: *Vateria indica* L.; *Dipterocarpac.*; sw. O.-Ind.
Illupei: *Bassia latifolia* Roxb., *B. longifolia* L.; *Sapotac.*; w. O.-Ind. (s. Deccan).
Illupi wood: *Bassia longifolia* Willd.; *Sapotac.*; Ind.
Illyarrie: *Eucalyptus erythrocorys* F. v. M.; *Myrtac.*; w. Au.
Ilolo: *Mammea klaineana* Pierre, *Ochrocarpus africana* Oliv.; *Guttifer.*; Gab
Ilomba (f): *Pycnanthus kombo* Warb.; *Myristicac.*; Kam., Gab.
Ilombo-Bolo: *Berlinia bracteosa* Bth.; *Caesalpiniac.*; Gab.
Ilouyé: *Saccoglottis gabonensis* Urban; *Humiriac.*; Kam.
Ilukabbán: *Sonneratia caseolaris* Engl.; *Sonneratiac.*; Phil. (Cag.)
Iluku: *Mitragyne macrophylla* Hiern; *Rubiac.*; B.-K.
Ilztamatl: *Bombax ellipticum* H. B. K.; *Bombacac.*; Mex.
Im: *Hopea parviflora* Bedd.; *Dipterocarpac.*; sw. O.-Ind.
Imbali: *Macrolobium dewevrei* de Wild.; *Caesalpiniac.*; B.-K.
Imbauba: *Cecropia sp.*; *Morac.*; Bras.
Imbaya: *Xylopia striata* Engl. *Anonac.*; Kam.
Imbilo: *Pterocarpus angolensis* DC.; *Papilionac.*; port. O.-Af. (Moz.)
Imbira: *Bombax sp.*; *Bombacac.*; Bras. (Sao P).
„ : *Rollinia exalbida* Mart.; *Anonac.*; s. Bras.
„ -guassú: *Bombax spp.*; *Bombacac.*; Bras.
„ quiaba: *Sterculia sp.*; *Sterculiac.*; Bras.
„ de folha: *Bombax spp.*; *Bombacac.*; Bras.
Imbirassu: *Bombax spp.*; *Bombacac.*; Bras.
Imbirinha: *Bombax sp.*; *Bombacac.*; Bras. (Sao P).
Imbirussú: *Bombax cyathophorum* K. Schum.; *Bombacac.*; Bras. (Sao P).
Imbondeiro: *Adansonia digitata* L.; *Bombacac.*; port. Af.
Imborana: *Bursera sp.*; *Burserac.*; Bras.
„ de espinho: *Bursera sp.*; *Burserac.*; Bras.
Imbuia (eu, u): *Phoebe porosa* Mez; *Laurac.*; s. Bras.
Imburana: *Bursera leptophloeos* Mart.; *Burserac.*; Bras.
Imbuya (u): *Phoebe porosa* Mez; *Laurac.*; s. Bras.
Imbuzeiro: *Spondias sp.*; *Anacardiac.*; Bras.
Immortal: *Erythrina glauca* Willd.; *Papilionac.*; Pan.
Immortel (f): *Erythrina corallodendron* L.; *Papilionac.*; fr. Gu.
„ du pays: *Erythrina crista-galli* L.; *Papillionac.*; Guad.
Imonoki. *Acanthopanax innovans* S. et Z.; *Araliac.*; Jap.
Imputeiro: *Adansonia digitata* L.; *Bombacac.*; port. O.-Af. (s. Moz.)

Imroi: *Ulmus wallichiana* Planch.; *Ulmac.;* n. O.-Ind. (Jaunṣar).
Imyracem: *Lucuma sp., Pradosia latescens* Radlk.; *Sapotac.;* Bras.
In: *Dipterocarpus spp.; Dipterocarpac.;* Burma.
Inai: *Quercus incana* Roxb.; *Fagac.;* n. O.-Ind. (Jaunsar).
Inajárana envira: *Matisia lasiocalyx* Schum.; *Bombacac.;* Bras.
Inambuquissáua: *Caraipa insidiosa* B. Rodr.; *Guttifer.;* Bras. (Amaz.)
Inang: *Chrysophyllum sp.; Sapotac.;* Gab.
Inanue: *Bursera gummifera* L.; *Burserac.;* Mex.
Inbo: *Dipterocarpus obtusifolius* Teijsm., *D. pilosus* Roxb.; *Dipterocarpac.;* Burma.
Incati: *Rhus juglandifolia* H.B.K.; *Anacardiac.;* Peru.
Incense, Frank- (e): *Boswellia serrata* Roxb.; *Burserac.;* Ind.
  „ tree (e): *Protium heptaphyllum* March.; *Burserac.;* Trin.
  „ wood (e): *Protium guianense* March.; *Burserac.;* fr. Gu.
Inchènepolo: *Chrysophyllum sp.; Sapotac.;* Gab.
Inchenu: *Afzelia quanzensis* Welw.; *Caesalpiniac.;* port. O.-Af. (Moz.)
Incienso: *Myrocarpus frondosus* Allem., *M. fastigiatus* Allem.; *Papilionac.;* n. Arg.
  „ cabrioba: *Myrocarpus frondosus* Allem.; *Papilionac.;* Arg.
  „ de costa: *Amyris sp.; Rutac.;* Cuba.
  „ del pais: *Bursera sp.; Burserac.;* Mex.
Incorruptible (f): *Minquartia guianensis* Aubl.; *Olacac.;* Sur.
Inda-guiassú: *Joannesia princeps* Vell.; *Euphorbiac.;* Bras. (Sao P).
Indaco bastardo (i): *Amorpha fruticosa* L.; *Papilionac.;* N.-Am.
Indak: *Cordia vestita* Hook. f. et Thoms.; *Borraginac.;* n. O.-Ind.
Indáng: *Artocarpus cumingiana* Tréc.; *Morac.;* Phil. (Lag., Riz.)
  „ : *Endospermum peltatum* Merr.; *Euphorbiac.;* Phil. (Lag.)
  „ : *Litsea sp.; Laurac.;* Phil. (Cag.)
Indano: *Byrsonima sp.; Malpighiac.;* Peru.
Indi: *Schizostachyum diffusum* Merr.; *Gramin.;* Phil. (Alb., Sor.)
Indian wood: *Guaiacum officinale* L.; *Zygophyllac.;* s. C.-Am., n. S.-Am., W.-Ind.
Indigo, Bastard- (e): *Amorpha fruticosa* L.; *Papilionac.;* N.-Am.
  „ , gemeiner (d): *Amorpha fruticosa* L.; *Papilionac.;* N.-Am.
  „ , bastardo (s): *Amorpha fruticosa* L.; *Papilionac.;* N.-Am.
  „ , Big-leaf-: *Lonchocarpus cyanescens* Bth.; *Papilionac.;* Lib.
  „ , false (e): *Amorpha fruticosa* L.; *Papilionac.;* N.-Am.
  „ , falso: *Amorpha fruticosa* L.; *Papilionac.;* Arg.
  „ , faux (f): *Amorpha fruticosa* L.; *Papilionac.;* N.-Am.
  „ , wild (e): *Amorpha fruticosa* L.; *Papilionac.;* N.-Am.
Indio desnudo: *Arbutus xalapensis* H.B.K.; *Ericac.;* C.-Am., Ven.
  „ „ : *Bursera gummifera* L.; *Burserac.;* C.-Am., Ven.
Indjoe: *Prosopis juliflora* DC.; *Mimosac.;* Curacao.
Indrajan: *Holarrhena antidysenterica* Wall.; *Apocynac.;* C.-O.-Ind.
? Inessa: *Maprounea membranacea* Pax et Hoffm.; *Euphorbiac.;* Gab.
Inetoma: *Sarcocephalus sp.; Rubiac.;* Gab.
Infinze: *Bruguiera gymnorrhiza* Lamk.; *Rhizophorac.;* port. O.-Af. (Moz., Quel.)
Infize: *Rhizophora mucronata* Lamk.; *Rhizophorac.;* port. O.-Af. (Moz., Quel.)
  „ : *Bruguiera gymnorrhiza* Lamk.; *Rhizophorac.;* port. O.-Af. (Moz., Quel.)
Ingá: *Inga spp.; Mimosac.;* Bras. (Sao P).
  „ : *Inga affinis* DC.; *Mimosac.;* Arg.
  „ amargo: *Inga uruguensis* Hook. et Arn.; *Mimosac.;* tr. S.-Am.
  „ assú: *Inga sp.; Mimosac.;* Bras.
  „ blanca: *Trema micrantha* Blume; *Ulmac.;* Arg.
  „ chichi: *Inga sp.; Mimosac.;* Bras.
  „ chichica: *Inga scabriuscula* Bth.; *Mimosac.;* Bras.
  „ cipo: *Inga sp.; Mimosac.;* Bras.
  „ colorado: *Inga uruguensis* Hook. et. Arn.; *Mimosac.;* Arg.
  „ doce: *Inga sp.; Mimosac.;* Bras.
  „ dulce: *Inga fagifolia* Willd.; *Mimosac.;* S.-Am.
  „ ferradura: *Inga sp.; Mimosac.;* Bras.
  „ guazú: *Inga sp.; Mimosac.;* Arg.
  „ hú: *Inga marginata* Willd.; *Mimosac.;* Arg.
  „ morotí: *Trema micrantha* Blume; *Ulmac.;* Arg.
  „ negro: *Guazuma ulmifolia* Lam.; *Sterculiac.;* Arg.

Ingá puitá: *Inga uruguensis* Hook. et Arn.; *Mimosac.;* Arg.
„ pyitá: *Inga sp.; Mimosac.;* Arg.
„ râna: *Bursera leptophloeos* Mart.; *Burserac.;* Bras.
„ -y: *Inga sp.; Mimosac.;* Arg.
„ de comer: *Inga sp.; Mimosac.;* Arg.
„ del Cerro: *Inga marginata* Willd.; *Mimosac.;* Arg.
Ingaina: *Roupala sp.; Proteac.;* Peru.
Ingajero: *Inga sp.; Mimosac.;* Bras.
Ingarana: *Inga sp.; Mimosac.;* Bras.
„ de beira: *Pithecolobium panurense* Spruce; *Mimosac.;* Bras.
Iñgás: *Semecarpus sp.; Anacardiac.;* Phil. (Cam., Zamb.)
Ingazylocarva: *Xylia dolabriformis* Bth.; *Mimosac.;* Burma.
Ingerto: *Calocarpum viride* Pittier; *Sapotac.;* Guat., Salv.
„ de montaña: *Lucuma laeteviridis* Pittier; *Sapotac.;* Guat.
Ingibarki: *Licania heteromorpha* Bth.; *Rosac.;* Sur.
Ingi pipa: *Couratari guianensis* Aubl.; *Lecythidac.;* Gu.
„   „ : *Lecythis ollaria* L.; *Lecythidac.;* Sur.
Ingie pipa: *Couratari guianensis* Aubl.; *C. sp.; Lecythidac.;* Sur.
Ingikandra: *Protium guianense* March.; *Burserac.;* fr. Gu.
Inglez: *Calophyllum brasiliense* Camb.; *Guttifer.;* Bras. (Sao P).
Ingli: *Barringtonia acutangula* Gaertn.; *Lecythidac.;* sw. O.-Ind.
Ingoe ingoe: *Saurauja sp.; Theac.;* Sum.
Ingou: *Firmiana spireana* Pierre; *Sterculiac.;* Gab.
Ingyin: *Pentacme suavis* A. DC.; *Dipterocarpac.;* Burma.
Ingyn: *Pentacme suavis* A. DC.; *Dipterocarpac.;* Burma.
Inhahyba: *Lecythis ollaria* L.; *Lecythidac.;* Bras. (Bahia).
Inhé branco: *Xylopia africana* Oliv.; *Anonac.;* W.-Af. (St. Thomé).
Injar: *Barringtonia acutangula* Gaertn.; *Lecythidac.;* C.-O.-Ind. (Oudh).
Injo: *Ongokea kamerunensis* Engl.; *Olacac.;* Kam.
Inkala: *Xylopia aethiopica* A. Rich.; *Anonac.;* B.-K. (Mayombe).
Ink-wood: *Hypelate paniculata* Cambess.; *Sapindac.;* tr. Am.
„   „ : *Exothea paniculata* Radlk.; *Sapindac.;* s. Fl., W.-Ind.
Inoutsougné: *Ilex crenata* Thunb.; *Aquifoliac.;* Jap.
Inri: *Schizostachyum diffusum* Merr.; *Gramin.;* Phil. (Alb., Sor.)
Insena: *Afzelia quanzensis* Welw.; *Caesalpiniac.;* port. O.-Af. (Moz.)
Intolwana: *Elephantorrhiza burchelli* Bth.; *Mimosac.;* S.-Af. (Natal).
Intyú: *Mitragyne inermis* K. Schum.; *Rubiac.;* tr .W.-Af.
Inubuna: *Fagus japonica* Maxim.; *Fagac.;* Jap.
Inuenju: *Cladrastis amurensis* Bth.; *Papilionac.;* Jap.
Inugaya: *Cephalotaxus drupacea* S. et Z.; *Conifer.;* Jap.
Inu tsuge: *Ilex crenata* Thunb., *I. integra* Thunb.; *Aquifoliac.;* Jap.
Inuyenju: *Maackia amurensis* Rupr. et Maxim.; *Papilionac.;* Jap.
Inuzansho: *Fagara schinifolia* Engl.; *Rutac.;* Jap.
Inyám: *Beilschmiedia aff. cairocan* Vidal; *Laurac.;* Phil. (w. Neg.)
Ipadú: *Erythroxylon sp.; Erythroxylac.;* Bras.
Ipana: *Parkia pendula* Bth.; *Mimosac.;* Sur.
Ipapane: *Cynometra sp.; Caesalpiniac.;* Kam.
Ipé: *Tecoma odontodiscus* Bur. et K. Sch., *T. spp.; Bignoniac.;* s. Bras.
„ amarelo: *Tecoma araliacea* P. DC.; *Bignoniac.;* Bras.
„ amarello: *Tecoma ochracea* Cham.; *Bignoniac.;* Bras. (Sao P).
„ branco: *Patagonula americana* L.; *Borraginac.;* Bras.
„ cascudo: *Tecoma sp.; Bignoniac.;* Bras. (Sao P).
„ folhas rôxas: *Tecoma sp.; Bignoniac.;* Bras.
„ jabotiá: *Tecoma sp.; Bignoniac.;* Bras. (Sao P).
„ peroba: *Paratecoma peroba* Kuhlmann; *Bignoniac.;* Bras.
„ preto: *Tecoma impetiginosa* Mart.; *Bignoniac.;* Bras. (Sao P).
„ roxo: *Tecoma heptaphylla* Mart.; *Bignoniac.;* Bras.
„   „ : *Tecoma impetiginosa* Mart.; *Bignoniac.;* Bras. (Sao P).
„ tabaco: *Tecoma chrysotricha* Mart.; *Bignoniac.;* Bras. (R.-J.)
„   „ : *Tecoma pedicellata* Bureau et K. Schum.; *Bignoniac.;* Bras. (Sao P).
„ tobaco: *Tecoma pedicellata* Bureau et K. Schum.; *Bignoniac.;* Bras.
„ una: *Tecoma sp.; Bignoniac.;* Bras. (Sao P.).

Ipé da varzea: *Tecoma spp.; Bignoniac.;* Bras.
„ do campo: *Tecoma araliacea* P. DC.; *Bignoniac.;* Bras. (Sao P).
Ipil: *Intsia bijuga* O. Ktҙe., *I. acuminata* Merr.; *Caesalpiniac.;* Phil.
„ : *Pahudia rhomboidea* Prain; *Caesalpiniac.;* Phil. (I.-N., Cag.)
„ : *Adenanthera intermedia* Merr.; *Mimosac.;* Phil. (Cag., Zam., Bat.)
„ -tanglín: *Adenanthera intermedia* Merr.; *Mimosac.;* Phil. (Cag., Zam., Bat.)
Ipovo: *Sclerosperma walkeri* Chev.; *Palmac.;* Gab.
Ippa: *Bassia latifolia* Roxb., *B. longifolia* I..; *Sapotac.;* ö. O.-Ind. (Madras).
Ippi: *Bassia latifolia* Roxb., *B. longifolia* L.; *Sapotac.;* w. O.-Ind. (Bombay).
Ipus-ípus: *Pygeum glandulosum* Merr., *P. presli* Merr.; *Rosac.;* Phil. (Cebu).
Iquimite: *Erythrina crista-galli* L.; *Papilionac.;* Mex.
Iqumza elinameva: *Scolopia zeyheri* Szyszylowicz; *Flacourtiac.;* Kap.
Ir: *Prosopis oblonga* Bth.; *Mimosac.;* Sudan.
„ : *Prosopis dubia* H. B. K.; *Mimosac.;* Senl.
Ira: *Khaya sp.; Meliac.;* Elf.
Irai: *Calophyllum wightianum* Wall.; *Guttifer.;* sw. O.-Ind.
Irak: *Salvadora persica* L.; *Salvadorac.;* Sahara.
Irambaratthan: *Canthium didymum* Roxb.; *Rubiac.;* w. O.-Ind. (Deccan).
Iramomi: *Picea bicolor* Mayr, *P. alcockiana* Carr.; *Conifer.;* Jap.
Irayol: *Genipa spp., G. caruto* H. B. K.; *Rubiac.;* C.-Am.
„ : *Grias fendleri* Seem., *Gustavia integrifolia* Standl.; *Lecythidac.;* Hond.
„ de loma: *Genipa americana* L. var. *caruto* H. B. K.; *Rubiac.;* Guat.
„ „ montaña: *Coccoloba tuerckheimi* Donn. Smith; *Polygonac.;* Guat.
Ircuma: *Bourreria havanensis* Miers; *Borraginac.;* Fl., Ant., Bah.
Iré: *Funtumia africana* Stapf; *Apocynac.;* W.-Af. (Lagos).
Ireh: *Funtumia africana* Stapf; *Apocynac.;* B.-K. (Mongalla).
Iri: *Quercus incana* Roxb.; *Fagac.;* nw. O.-Ind. (Kasch.)
Iria-kopie: *Qualea coerulea* Aubl., *Qu. albiflora* Warm., *Qu. rosea* Aubl.;
Iriki: *Cordia myxa* L.; *Borraginac.;* sw. O.-Ind. (Madras).   [*Vochysiac.;* Sur.
Irikirina: *Diospyros spp.; Ebenac.;* Mad.
Iril: *Coccoloba spp.; Polygonac.;* br. Hond.
Irire: *Coccoloba sp.; Polygonac.;* Salv.
Iririsi ja mou: *Illicium anisatum* L.; *Magnoliac.;* Jap.
Iroko: *Chlorophora excelsa* Bth. et Hook., ? *Ch. regia* Chev.; *Morac.;* W.-Af.
Ironbark: *Eucalyptus spp.; Myrtac.;* Au., Tasm.
„ , broad-leaved: *Eucalyptus siderophloia* Bth.; *Myrtac.;* ö. Au.
„ , grey: *Eucalyptus paniculata* Smith; *Myrtac.;* ö. Au.
„ , Isdele River-: *Eucalyptus melanophloia* F. v. M.; *Myrtac.;* w. Au.
„ , lemon-scented: *Eucalyptus staigeriana* F. v. M.; *Myrtac.;* Au. (Queensl.)
„ , narrow-leaved: *Eucalyptus crebra* F. v. M.; *Myrtac.;* ö. Au.
„ , „ „ , red: *Eucalyptus crebra* F. v. M.; *Myrtac.;* ö. Au.
„ , red: *Eucalyptus siderophloia* Bth.; *Myrtac.;* ö. Au.
„ , red-flowering: *Eucalyptus sideroxylon* A. Cunn.; *Myrtac.;* Au.
„ , scrub-: *Bridelia exaltata* F. v. M.; *Euphorbiac.;* ö. Au.
„ , silver-leaved: *Eucalyptus melanophloia* F. v. M.; *Myrtac.;* Queensl., N.-S.-W.
„ , white: *Eucalyptus paniculata* Smith; *Myrtac.;* ö. Au.
„ , „ : *Eucalyptus leucoxylon* F. v. M.; *Myrtac.;* Au.
Iron tree: *Tecoma spp.; Bignoniac.;* br. Gu.
„ wood: *Acacia farnesiana* Willd.; *Mimosac.;* Tr.
„ „ : *Carpinus ostrya* L., *C. caroliniana* Walt.; *Betulac.;* ö. N.-Am.
„ „ : *Bumelia lycioides* Gaertn. f.; *Sapotac.;* tr. Am.
„ „ : *Sloanea jamaicensis* Hook. f.; *Elaeocarpac.;* Bah.
„ „ : *Laplacea haematoxylon* G. Don; *Theac.;* Jam.
„ „ : *Dialium divaricatum* Vahl; *Caesalpiniac.;* br. Hond.
„ „ : *Krugiodendron ferreum* Urban; *Rhamnac.;* Hond.
„ „ : *Swartҙia spp.; Caesalpiniac.;* Sur.
„ „ : *Tecoma sp., Tabebuia sp.; Bignoniac.;* Sur.
„ „ : *Swartҙia tomentosa* DC.; *Caesalpiniac.;* fr. Gu.
„ „ : *Caesalpinia ferrea* Mart.; *Caesalpiniac.;* Bras.
„ „ : *Lophira procera* Chev., *L. alata* Banks; *Ochnac.;* W.-Af.
„ „ : *Olea capensis* L., *O. foveolata* E. Meyer; *Oleac.;* Kap.
„ „ : *Cassia siamea* Lamk.; *Caesalpiniac.;* D.-O.-A.
„ „ : *Vateria seychellarum* Dyer; *Dipterocarpac.;* Sey.

Iron wood: *Maba buxifolia* Pers.; *Ebenac.;* tr. As., tr. Af., tr. Au.
„ „ : *Memecylon edule* Roxb.; *Melastomatac.;* O.-Ind., Cey., Malay.
„ „ : *Casuarina equisetifolia* Forst., *C. sumatrana* Miq.; *Casuarinac.;*
„ „ : *Acacia ferruginea* DC.; *Mimosac.;* Ind.   [tr. As., tr. Au.
„ „ : *Mesua ferrea* L.; *Guttifer.;* Ind.
„ „ : *Xylia xylocarpa* Taub.; *Mimosac.;* Ind., Burma, Co.-Ch.
„ „ : *Mimusops hexandra* Roxb.; *Sapotac.;* Ind.
„ „ : *Mimusops kauki* L.; *Sapotac.;* Malay, Java, Cel.
„ „ : *Afzelia bakeri* Prain; *Caesalpiniac.;* Malay, Sunda.
„ „ : *Eusideroxylon zwageri* Teijsm. et Binn.; *Laurac.;* Borneo.
„ „ : *Myrus hilli* Bth.; *Myrtac.;* ö. Au.
„ „ : *Acacia excelsa* Bth.; *Mimosac.;* ö. Au.
„ „ : *Geijera salicifolia* Schott; *Rutac.;* Au. (Queensl.)
„ „_ : *Tarrietia argyrodendron* Bth.; *Sterculiac.;* Queensl., N.-S.-W.
„ „ : *Metrosideros lucida* A. Rich.; *Myrtac.;* N.-Seel.
„ „ : *Notelae ligustrina* Vent.; *Oleac.;* Tasm.
„ „ : ? *Casuarina sp.;* *Casuarinac.;* Südsee (N.-Kal., Tonga).
„ „ , Bastard-: *Zanthoxylum pterota* H. B. K.; *Rutac.;* USA.
„ „ , black: *Condalia terrea* Griseb.; *Rhamnac.;* tr. Am.
„ „ , „ : *Krugiodendron ferreum* Urban; *Rhamnac.;* W.-Ind., C.-Am.
„ „ , „ : *Olea latifolia* Salisb., *O. laurifolia* Lam.; *Oleac.;* S.-Af.
„ „ , Borneo-: *Eusideroxylon zwageri* Teijsm. et Binn.; *Laurac.;* Phil., Sunda.
„ „ , Bourbon-: *Cupania sideroxylon* Cambess.; *Sapindac.;* Maur.
„ „ , „ : *Stadmannia sideroxylon* DC.; *Sapindac.;* Maur.
„ „ , brazilian: *Caesalpinia ferrea* Mart.; *Caesalpiniac.;* Bras.
„ „ , Ceylon-: *Mesua ferrea* L.; *Guttifer.;* O.-Ind.
„ „ , East Indian: *Mesua ferrea* L.; *Guttifer.;* O.-Ind.
„ „ , giant: *Syncarpia subargentea* C. T. White; *Myrtac.;* ö. Au.
„ „ , Highland-: *Casuarina stricta* Ait.; *Casuarinac.;* Hawaii.
„ „ , New Zealand-: *Metrosideros lucida* A. Rich.; *Myrtac.;* N.-Seel.
„ „ , Philippine-: *Xanthostemon verdugonianus* Naves; *Myrtac.;* Phil.
„ „ , red (e): *Lophira procera* Chev., *L. alata* Banks; *Ochnac.;* W.-Af.
„ „ , rough-barked: *Ostrya virginiana* C. Koch; *Betulac.;* ö. Kan.
„ „ , Scrub-: ? *Myrtus hilli* Bth.; *Myrtac.;* ö. Au.
„ „ , smooth-barked: *Carpinus caroliniana* Walt.; *Betulac.;* ö. Kan.
„ „ , white: *Hypelate trifoliata* Swartz; *Sapindac.;* Fl., W.-Ind.
„ „ , „ : *Toddalia lanceolata* Lam.; *Rutac.;* S.-Af.
Irool-marum: *Mesua ferrea* L.; *Guttifer.;* s. O.-Ind.
Iroul-marom: *Mesua ferrea* L.; *Guttifer.;* s. O.-Ind.
? Irriariadan: *Cassia sp.;* *Caesalpiniac.;* br. Gu.
Irrud: *Bassia latifolia* Roxb.; *Sapotac.;* C.-O.-Ind.
Irsel: *Ilex aquifolium* L.; *Aquifoliac.;* Algier, Tunis.
Irtalie: *Castanospermum australe* A. Cunn.; *Papilionac.;* N.-S.-W.
Iru: *Bassia latifolia* Roxb.; *Sapotac.;* C.-O.-Ind.
Irubogam: *Hopea parviflora* Bedd.; *Dipterocarpac.;* sw. O.-Ind. (Mal.)
Irul (H): *Xylia xylocarpa* Taub.; *Mimosac.;* w. O.-Ind. (n. Bombay).
Irum: *Bischofia javanica* Blume; *Euphorbiac.;* C.-O.-Ind. (Oudh).
Irumpala: *Wrightia tinctoria* R. Br.; *Apocynac.;* O.-Ind. (Trav. Hills).
Irup: *Bassia latifolia* Roxb.; *Sapotac.;* C.-O.-Ind.
Iruppa: *Hopea parviflora* Bedd.; *Dipterocarpac.;* sw. O.-Ind. (Coorg).
Isa: *Celtis zenkeri* Engl.; *Ulmac.;* Go.
„ , black: *Celtis adolphi-frederici* Engl.; *Ulmac.;* Go.
„ kukubin: *Celtis adolphi-frederici* Engl.; *Ulmac.;* Go.
„ nkesua: *Celtis adolphi-frederici* Engl.; *Ulmac.;* Go.
„ pui: *Dalbergia glaucescens* Bth.; *Papilionac.;* Arg.
Isák: *Hopea pierrei* Hance; *Dipterocarpac.;* Phil. (Tay.)
Isaka: *Albizzia welwitschi* Oliv.; *Mimosac.;* Kam.
Isamere: *Celtis adolphi-frederici* Engl.; *Ulmac.;* Go.
Isanhiamne: *Diospyros sp.;* *Ebenac.;* Nig.
Isano: *Ongokea klaineana* Pierre; *Oleac.;* Gab., B.-K. (Mayombe).
Ischii-gashi: *Quercus gilva* Blume; *Fagac.;* s. Jap.
Ischo: *Ginkgo biloba* L.; *Conifer.;* Ch., Jap.
Ise bubaki: *Camellia japonica* L.; *Theac.;* Jap.

Ishan: *Lophira alata* Banks; *Ochnac.;* Nig. (Benin).
Ishim-ché: *Andira inermis* H. B. K.; *Papilionac.;* br. Hond.
„ „ : *Casearia nitida* Jacq.; *Flacourtiac.;* br. Hond.
Ishtatén: *Avicennia nitida* Jacq.; *Verbenac.;* Salv.
Ising-p'i-hsiang: *Ilex sp.; Aquifoliac.;* Ch.
Is-ís: *Artocarpus cumingiana* Trécul; *Morac.;* Phil. (Neg.)
Isi-tomboti: *Acalypha glabrata* Thunb.; *Euphorbiac.;* S.-Af. (East London).
Isombé: *Scyphocephalium ochocoa* Warb.; *Myristicac.;* Gab.
„ : *Testulea gabonensis* Pellegr.; *Ochnac.;* Gab. (Bas-Ogooué)
Isombo: *Scyphocephalium ochocoa* Warb.; *Myristicac.;* Gab.
Isonguin: *Mimusops lacera* Baker; *Sapotac.;* Elf.
Isote: *Dracaena americanan* Donn. Smith, *Yucca elephantipes* Regel; *Liliac.;* Hond.
Ispundio: *Bumelia leiogyna* Donn. Smith; *Sapotac.;* Salv.
Isrihati: *Swartzia sp.; Caesalpiniac.;* Sur.
Issanguila: *Ricinodendron africanum* Muell. Arg.; *Euphorbiac.;* Gab.
Issingani: *Piptadenia sp.; Mimosac.;* Gab.
Issógbe: *Bosquiea welwitschi* Engl.; *Morac.;* Kam.
Issombe: *Rauwolfia macrophylla* Stapf; *Apocynac.;* Kam.
Issombo: *Scyphocephalium ochocoa* Warb.; *Myristicac.;* Gab.
Issote: *Cyrtogonone argentea* Prain; *Euphorbiac.;* Kam.
Issoua: *Saccoglottis gabonensis* Urban; *Humiriac.;* Gab.
Issouk: *Dialium macranthum* Chev.; *Caesalpiniac.;* Kam.
Issoula: *Sarcocephalus trillesi* Pierre; *Rubiac.;* Gab.
Istatén: *Avicennia nitida* Jacq.; *Verbenac.;* Salv.
Isu: *Distylium racemosum* S. et Z.; *Hamamelidac.;* Jap.
Isunoki: *Distylium racemosum* S. et Z.; *Hamamelidac.;* Jap.
Isu-nuki: *Symplocos myrtacea* S. et Z.; *Symplocac.;* Jap.
Itaballi: *Vochysia guianensis* Aubl.; *Vochysiac.;* Gu.
Itaiba: *Hymenaea courbaril* L.; *Caesalpiniac.;* Gu.
Itaka: *Machaerium schomburgki* Bth.; *Papilionac.;* br. Gu.
Itam: *Diospyros ebenum* Koenig; *Ebenac.;* tr. As.
Itanda: *Rhizophora mangle* L.; *Rhizophorac.;* Kam., Gab.
Itang: *Pycnanthus sphaerocarpa* Stapf; *Myristicac.;* Kam., Gab.
„ -ítang: *Alstonia macrophylla* Wall.; *Apocynac.;* Phil. (Guim.)
Itapiúna: *Callisthene major* Mart.; *Vochysiac.;* Bras. (Minas Geraes)
Itare: *Piptadenia africana* Hook. f.; *Mimosac.;* Nig.
Itaúba: *Ocotea rodioei* Mez; *Laurac.;* Bras.
„ amarella: *Silvia ita-uba* Pax; *Laurac.;* Bras. (Amaz.)
„ preta: *? Oreodaphne sp.; Laurac.;* Bras.
„ rana· *Sweetia nitens* Bth.; *Papilionac.;* Bras. (Amaz.)
„ verdadeira: *Silvia ita-uba* Pax; *Laurac.;* Bras. (Amaz.)
Itaya-kaede: *Acer pictum* Thunb.; *Acerac.;* Korea.
Itaya-kayede: *Acer pictum* Thunb.; *Acerac.;* Jap.
Itchia: *Byrsonima sp.; Malpighiac.;* Sur.
Itchiki-boura: *Pterocarpus spp.; Papilionac.;* br. Gu.
Itchou-y-n'jogou: *Mimusops africana* H. Lec.; *Sapotac.;* Kam., Gab.
„ „ „ : *Diospyros aggregata* Gürke; *Ebenac.;* Gab.
Ite: *Mauritia flexuosa* L. f.; *Palmac.;* br. Gu.
„ -balli: *Vochysia melinoni* Beckmann; *Vochysiac.;* Sur.
Iteng: *Pycnanthus sphaerocarpa* Stapf; *Myristicac.;* Kam., Gab.
Iteruku: *Piptadenia africana* Hook. f.; *Mimosac.;* Nig.
Iti: *Dalbergia latifolia* Roxb.; *Papilionac.;* w. O.-Ind. (Deccan)
Itii-gasi: *Quercus gilva* Blume; *Fagac.;* Form.
Itik-boura: *Pterocarpus spp.; Papilionac.;* br. Gu.
Itiki: *Machaerium schomburgki* Bth.; *Papilionac.;* br. Gu.
„ boeraballi: *Machaerium schomburgki* Bth.; *Papilionac.;* Sur.
„ boeroe: *Pterocarpus draco* L., *P. rhori* Vahl; *Papilionac.;* Sur.
„ boro: *Pterocarpus draco* L., *P. rhori* Vahl; *Papilionac.;* Sur.
„ „ -hororadikoro: *Pterocarpus rhori* Vahl; *Papilionac.;* Sur.
„ boura: *Pterocarpus draco* L., *P. rhori* Vahl; *Papilionac.;* Sur.
„ bouraballi: *? Machaerium schomburgki* Bth.; *Papilionac.;* br. Gu.
„ bourballi: *Machaerium schomburgki* Bth.; *Papilionac.;* br. Gu.

Itikie boeroeballi: *Swartzia tomentosa* DC.; *Caesalpiniac.*; Sur.
Itílan: *Vatica mangachapoi* Blco.; *Dipterocarpac.*; Phil. (Rizal)
Itin: *Prosopis kuntzi* Harms; *Mimosac.*; Arg.
Itjoeranano-anakoko: *Ormosia fastigiata* Tul.; *Papilionac.*; Sur.
Itjoeroe-anakoko: *Ormosia coccinea* Jacks.; *Papilionac.*; Sur.
   „ -tanomoetoesi: *Pterocarpus rhori* Vahl; *Papilionac.*; Sur.
Itjoetano perapisie: *Maprounea guianensis* Aubl.; *Euphorbiac.*; Sur.
Ito-balli: *Vochysia tetraphylla* DC.; *Vochysiac.*; br. Gu.
Itoerie ballaba: *Eperua sp.*; *Caesalpiniac.*; Sur.
Itom-itóm: *Diospyros mindanaënsis* Merr.; *Ebenac.*; Phil. (Lanao)
Itoundoulou: *Ficus vogeliana* Miq.; *Morac.*; Gab.
Itschille: *Phoenix spinosa* Schum. et Thonn.; *Palmac.*; Togo.
It slips: *Ulmus fulva* Michx.; *Ulmac.*; ö. N.-Am.
Itú: *Dialium divaricata* Vahl; *Caesalpiniac.*; Bras.
Itué: *Conopharyngia durissima* Stapf; *Apocynac.*; Gab.
Ituri: *Eperua jenmani* Oliv.; *Caesalpiniac.*; br. Gu.
   „ -walaba: *Eperua schomburgkiana* Bth., *E. jenmani* Oliv.; *Caesalpiniac.*; Gu.
Ivahehé: *Acanthosyris spinescens* Griseb.; *Santalac.*; Arg.
Ivaro: *Ruprechtia sp.*; *Polygonac.*; Arg.
Ivatingi: *Luehea divaricata* Mart.; *Tiliac.*; Arg.
Ivatingy: *Luehea paniculata* Mart.; *Tiliac.*; Bras. (Sao P).
Ivira pitá: *Peltophorum vogelianum* Bth.; *Caesalpiniac.*; Arg.
   „ „ -guazú: *Peltophorum vogelianum* Bth.; *Caesalpiniac.*; Arg.
Ivira-ro: *Ruprechtia vivaru* Griseb.; *Polygonac.*; W.-Ind.
Ivitinga: *Luehea divaricata* Mart., *L. paniculata* Mart.; *Tiliac.*; Bras. (Sao P).
Ivivaroa: *Ruprechtia sp.*; *Polygonac.*; Par.
Ivoire végétal: *Phytelephas macrocarpa* R. et Pav., *P. microcarpa* R. et Pav.;
Ivondé: *Daniella soyauxi* Harms; *Caesalpiniac.*; Gab.   *[Palmac.*; Ven.
Ivorywood: *Siphonodon australe* Bth.; *Celastrac.*; ö. Au.
   „ , pink: *Rhamnus zeyheri* Sond.; *Rhamnac.*; S.-Af.
   „ , red: *Rhamnus zeyheri* Sond.; *Rhamnac.*; S.-Af.
Ivouyra: *Luehea divaricata* Mart.; *Tiliac.*; Bras.
Ivy, english (e): *Hedera helix* L.; *Araliac.*; Eu.
   „ , Poison-: *Rhus toxicodendron* L.; *Anacardiac.*; ö. USA.
Iwungowungo: *Landolphia parvifolia* K. Schum.; *Apocynac.*; D.-O.-Af. (Tabora)
Ixcanal: *Acacia sp., Acacia hindsi* Bth.; *Mimosac.*; Guat., Hond.
Ixpepe: *Trema sp.*; *Ulmac.*; Mex.
Ixtepeque: *Poeppigia procera* Presl; *Caesalpiniac.*; Salv.
Iya: *Tabernaemontana sp.*; *Apocynac.*; fr. Gu.
Izaquente: *Treculia africana* L.; *Morac.*; Elf.
Izé: *Klainedoxa latifolia* Pierre; *Simarubac.*; Kam.
Izgi: *Quercus suber* Kotschy; *Fagac.*; N.-Af.
Izingana: *Cynometra aff. leplaei* de Wild.; *Caesalpiniac.*; Kam., Gab.
Izogozo: *Ficus vogeliana* Miq.; *Morac.*; Gab.
Izombé (f): *? Scyphocephalium ochocoa* Warb.; *Myristicac.*; Gab.
   „ : *Testulea gabonensis* Pellegr.; *Ochnac.*; Gab.
Izote de montaña: *Dracaena americana* Donn. Smith; *Liliac.*; Guat.

J

Jabani kestané: *Aesculus hippocastanum* L.; *Hippocastanac.*; Türkei.
Jabilla: *Caesalpinia crista* L.; *Caesalpiniac.*; Mex.
Jabillo: *Hura crepitans* L.; *Euphorbiac.*; Guat., Kol., Ven.
Jabon: *Citrus decumana* L.; *Rutac.*; Jap.
Jabón-ché: *Sapindus saponaria* L.; *Sapindac.*; br. Hond.
Jaboncelle: *Sapindus saponaria* L.; *Sapindac.*; Fl., C.-Am., n. S.-Am.
Jaboncillo: *Sapindus saponaria* L.; *Sapindac.*; P.-R., Cuba, Kol., Ek.
   „ : *Bursera sp.*; *Burserac.*; Mex.
Jabopfouey: *Caloncoba glauca* Gilg; *Flacourtiac.*; Kam.
Jaboticabeira: *Eugenia edulis* Bth. et Hook.; *Myrtac.*; Bras. (Sao P)
Jaboty: *Erisma calcaratum* Warm.; *Vochysiac.*; Bras.
   „ da Terra Firma: *Erisma uncinatum* Warm.; *Vochysiac.*; n. Bras. (Amaz.)

Jácana: *Lucuma multiflora* A. DC.; *Sapotac.*; P.-R.
Jacapucaya: *Lecythis sp.; Lecythidac.;* Bras.
Jacaranda (d): *Dalbergia nigra* Allem.; *Papilionac.;* Bras.
„ : *Jacaranda cuspidifolia* Mart.; *Bignoniac.;* Bras. (R.-J., Sao P)
„ : *Jacaranda acutifolia* Humb. et Bonpl.; *Bignoniac.;* Arg.
„ : *Jacaranda brasiliana* Pers.; *Bignoniac.;* fr. Gu.
„ : *Jacaranda mimosaefolia* D. Don; *Bignoniac.;* S.-Af.
„ : *Machaerium scleroxylon* Tul.; *Papilionac.;* Sur.
„ : *Machaerium villosum* Vogel, *M. acutifolium* Vogel; *Papilionac.;* Bras. (Sao P)
„ : *Caesalpinia echinata* Lam.; *Caesalpiniac.;* nö. Bras.
„ : *Platypodium elegans* Vogel; *Papilionac.;* Bras. (Sao P)
„ : *Prosopis kuntzei* Harms; *Mimosac.;* Arg.
„ bico de pato: *Machaerium sp.; Papilionac.;* Bras.
„ branca: *Dalbergia nigra* Allem.; *Papilionac.;* ö. Bras.
„ „ : *Platypodium elegans* Vogel; *Papilionac.;* Bras. (Sao P)
„ cabiuna: *Dalbergia nigra* Allem., *D. miscolobium* Bth.; *Papilionac.;* Bras.
.. capitao: *Terminalia januarensis* DC.; *Combretac.;* Bras. (Sao P)
„ , mimosa-leaved: *Jacaranda mimosaefolia* D. Don; *Bignoniac.;* O.-Af. (Nyassa)
„ paulista: *Machaerium villosum* Vogel; *Papilionac.;* Bras. (Sao P)
„ preto: *Machaerium legale* Bth., *Dalbergia nigra* Allem.; *Papilionac.;* s. Bras.
„ „ : *Jacaranda brasiliana* Pers.; *Bignoniac.;* s. Bras.
„ rosa: *Swartzia sp.; Caesalpiniac.;* Bras. (Sao P)
„ „ : *Machaerium sp., Dalbergia nigra* Allem.; *Papilionac.;* ö. Bras.
„ , rotes: *Machaerium firmum* Bth.; *Papilionac.;* Bras.
„ roxa: *Dalbergia nigra* Allem.; *Papilionac.;* ö. Bras.
„ roxo: *Machaerium firmum* Bth.; *Papilionac.;* Bras. (Sao P)
„ tã: *Machaerium allemani* Bth.; *Papilionac.;* Bras. (Sao P)
„ „ amarello: *Machaerium sp.; Papilionac.;* Bras. (Sao P)
„ tan: *Dalbergia nigra* Allem.; *Papilionac.;* ö. Bras.
„ vermelho: *Dalbergia sp.; Papilionac.;* Bras. (Sao P)
„ violeta: *Dalbergia nigra* Allem., *Machaerium violaceum* Vogel; *Papilionac.;* Bras.
„ de Bahia: *Dalbergia nigra* Allem.; *Papilionac.;* Bras. (Sao P)
„ „ campo: *Platypodium elegans* Vogel; *Papilionac.;* Bras. (Sao P)
„ do matto: *Machaerium villosum* Vogel; *Papilionac.;* Bras. (Sao P)
Jacaré: *Piptadenia communis* Bth.; *Mimosac.;* Bras. (Sao P)
„ do matto: *Cybianthus detergens* Mart.; *Myrsinac.;* Bras.
Jacaréuba: *Calophyllum brasiliense* Camb.; *Guttifer.;* Bras. (Sao P)
Jacataúba: *Vitex sellowiana* Cham.; *Verbenac.;* Bras. (Sao P)
Jacatirão: *Miconia brasiliensis* Triana; *Melastomatac.;* Bras.
Jacatirôes: *Miconia trianaei* Cogn., *M. brasiliensis* Triana, *M. sp.; Melastomatac.;* Bras.
Jacifate: *Protium altissimum* March.; *Burserac.;* Ven.
Jacitará: *Desmoncus sp.; Palmac.;* Bras.
Jack, black: *Tarrietia actinophylla* F. M. Bailey; *Sterculiac.;* ö. Au.
„ , long: *Flindersia oxleyana* F. v. M.; *Rutac.;* Au. (Queensl.)
„ tree: *Artocarpus integrifolia* L.; *Morac.;* Tr.
„ „ , wild: *Artocarpus hirsuta* Lam.; *Morac.;* Ind.-Ch.
„ fruit tree: *Artocarpus integrifolia* L., *A. communis* Forst.; *Morac.;* Malay.
„ wood: *Cordia alba* Roem. et Schultes; *Borraginac.;* br. Hond.
Jacquier: *Artocarpus integrifolia* L.; *Morac.;* Tr.
„ : *Artocarpus hirsuta* Lam.; *Morac.;* Ind.-Ch.
Jadali: *Sarcocephalus esculentus* Afzel.; *Rubiac.;* W.-Af. (? Elf.)
Jadhirdah: *Cocos nucifera* L.; *Palmac.;* Arab.
Jagaruwa: *Cassia fistula* L.; *Caesalpiniac.;* C.-O.-Ind.
Jagidambar: *Ficus glomerata* Roxb.; *Morac.;* Beng.
Jagrikat: *Machilus gammieana* King; *Laurac.;* nö. O.-Ind. (Nepal)
Jagua: *Genipa americana* L.; *Rubiac.;* Cuba, P.-R., Hond., Kol.
„ : *Genipa caruto* H. B. K.; *Rubiac.;* Hond., Sur.
„ : *Genipa spp.; Rubiac.;* Ven., Ek.
„ amarilla: *Chimarrhis latifolia* Standl., *? Genipa sp.; Rubiac.;* Pan.
„ azul: *Genipa americana* L.; *Rubiac.;* Mex.
„ blanca: *Genipa americana* L.; *Rubiac.;* Pan.
„ dulce: *Genipa sp.; Rubiac.;* Ek.
„ negra: *Genipa americana* L.; *Rubiac.;* Pan.

Jagua de lagarto: *Crataeva tapia* L.; *Capparidac.*; Ek.
„ „ montaña: *Genipa americana* L., *Sickingia maxoni* Standl.; *Rubiac.*; Pan.
Jaguarzo: *Cistus monspeliensis* L.; *Cistac.*; Sp.
Jaguatirão: *Tibouchina mutabilis* Cogn.; *Melastomatac.*; Bras. (Sao P).
Jaguay: *Pithecolobium littorale* Britt. et Rose; *Mimosac.*; Guat.
„ cimarrón: *Pithecolobium microstachyum* Standl.; *Mimosac.*; Guat.
„ de llano: *Pithecolobium microstachyum* Standl.; *Mimosac.*; Guat.
Jaguey: *Ficus sp.*; *Morac.*; Cuba.
„ hembra: *Ficus sp.*; *Morac.*; Cuba.
„ macho: *Ficus sp.*; *Morac.*; Cuba.
Jaguillo: *Gustavia integrifolia* Standl.; *Lecythidac.*; Hond.
Jagya: *Ficus glomerata* Roxb.; *Morac.*; Beng.
Jahar: *Cassia siamea* Lamk.; *Caesalpiniac.*; Malay.
Jaimangal: *Stereospermum suaveolens* DC., *S. xylocarpum* Wight; *Bignoniac.*; C.-O.-Ind.
Jaimiqui: *Byrsonima lucida* DC.; *Malpighiac.*; Cuba.
Jaint: *Sesbania aegyptiaca* Pers.; *Papilionac.*; Tr.
Jaiphal: *Myristica laurifolia* Hook. f.; *Myristicac.*; Cey.
Jaïvie: *Jacaranda copaia* D. Don; *Bignoniac.*; Sur.
Jak: *Artocarpus integrifolia* L. f.; *Morac.*; Ind., Cey., Malay.
Jakaranda, ostindisches (d): *Dalbergia latifolia* Roxb.; *Papilionac.*; O.-Ind.
Jakkelsbessie: *Diospyros mespiliformis* Hochst.; *Ebenac.*; S.-Af.
Jakopi: *Qualea coerulea* Aubl., *Q. albiflora* Warm., *Q. rosea* Aubl.; *Vochysiac.*; Sur.
Jakopie: *Qualea albiflora* Warm.; *Vochysiac.*; Sur.
Jakoppi-tapirin: *Qualea albiflora* Warm.; *Vochysiac.*; Sur.
Jala: *Shorea talura* Roxb.; *Dipterocarpac.*; sw. O.-Ind. (Coorg)
Jalari: *Shorea tumbuggaia* Roxb., *S. talura* Roxb.; *Dipterocarpac.*; nw. O.-Ind.
Jali: *Acacia arabica* Willd.; *Mimosac.*; nw. O.-Ind. (Bombay)
„ Salei: *Acacia planifrons* Wight et Arn.; *Mimosac.*; w. O.-Ind.
Jalicote: *Pinus sp.*; *Conifer.*; Mex.
Jalmala: *Salix tetrasperma* Roxb.; *Salicac.*; n. O.-Ind.
Jalna: *Terminalia myriocarpa* Heurck et Muell. Arg.; *Combretac.*; Hawai.
Jalpai: *Elaeocarpus robustus* Roxb.; *Elaeocarpac.*; Assam.
Jam: *Schima wallichi* Choisy; *Theac.*; Assam (Cachar).
„ : *Eugenia jambolana* Lamk.; *Myrtac.*; O.-Ind.
„ wood: *Acacia acuminata* Bth.; *Mimosac.*; Au.
„ „ , Raspberry-: *Acacia acuminata* Bth.; *Mimosac.*; w. Au.
Jama: *Alchornea cordifolia* Muell. Arg.; *Euphorbiac.*; Go.
Jaman: *Eugenia jambolana* Lamk.; *Myrtac.*; O.-Ind.
Jamaquey: *Belairia mucronata* Griseb.; *Papilionac.*; Cuba.
Jamba: *Xylia xylocarpa* Taub.; *Mimosac.*; sw. O.-Ind.
Jambe: *Xylia xylocarpa* Taub.; *Mimosac.*; sw. O.-Ind.
Jambhul: *Eugenia jambolana* Lamk.; *Myrtac.*; C.-O.-Ind.
Jamblonnier: *Syzygium multipetalum* Panch.; *Myrtac.*; N.-Kal.
Jambobohnem: *Acronychia pedunculata* Forst.; *Rutac.*; O.-Ind.
Jambolanem: *Acronychia pedunculata* Forst.; *Rutac.*; O.-Ind.
Jambosier: *Eugenia jambos* L.; *Myrtac.*; tr. As.
Jambu: *Eugenia spp.*; *Myrtac.*; Malay.
„ Ayer: *Eugenia aquea* Burm.; *Myrtac.*; Malay.
„ Biji: *Psidium guajava* L.; *Myrtac.*; Malay.
„ Bol: *Eugenia malaccensis* L.; *Myrtac.*; Malay.
Jambul: *Eugenia jambolana* Lamk.; *Myrtac.*; sw. O.-Ind.
Jamelonnier: *Syzygium multipetalum* Panch.; *Myrtac.*; N.-Kal.
Jamfuta: *Afzelia quanzensis* Welw.; *Caesalpiniac.*; port. O.-Af. (Lor. Marqu.)
Jamo: *Eugenia jambolana* Lamk.; *Myrtac.*; nö. O.-Ind. (Orissa)
Jamoi: *Prunus padus* L.; *Rosac.*; n. O.-Ind. (Jaunsar)
Jamplond: *Calophyllum inophyllum* L.; *Guttifer.*; Java.
Jamroi: *Prunus padus* L.; *Rosac.*; n. O.-Ind. (Jaunsar)
Jamrosa: *Jambosa vulgaris* DC.; *Myrtac.*; Réunion.
Jamu: *Eugenia jambolana* Lamk.; *Myrtac.*; Assam.
Jamun: *Eugenia jambolana* Lamk.; *Myrtac.*; n. & C.-O.-Ind., And.
„ : *Prunus padus* L.; *Rosac.*; nw. O.-Ind. (Punjab)
Jan: *Didelotia africana* Pierre; *Caesalpiniac.*; Kam.
„ Ganyi: *Combretum hypopilinum* Diels; *Combretac.*; n. Nig.

Jan iche: *Hymenocardia acida* Tul.; *Euphorbiac.*; n. Nig.
„ saye: *Trichilia emetica* Vahl; *Meliac.*; n. Nig.
„ Taramniya: *Combretum hypopilinum* Diels; *Combretac.*; n. Nig.
„ Yaro: *Hymenocardia acida* Tul.; *Euphorbiac.*; n. Nig.
Jana: *Grewia tiliaefolia* Vahl; *Tiliac.*; w. O.-Ind. (n. Bombay)
Janaledan: *Scerolobium sp.*; *Caesalpiniac.*; Sur.
Janauba: *Plumiera sp.*; *Apocynac.*; Bras.
Janboka: *Pouteria sp.*; *Sapotac.*; Sur.
Jand: *Prosopis spicigera* L.; *Mimosac.*; O.-Ind. (Punjab)
Jandakhai: *Callicarpa arborea* Roxb.; *Verbenac.*; w. Beng.
Jangada: *Apeiba sp.*; *Tiliac.*; Bras.
„ brava: *Heliocarpus americanus* L.; *Tiliac.*; Bras.
Jangkang: *Hopea spp.*; *Dipterocarpac.*; Malay.
Jangli: *Populus ciliata* Wall.; *Salicac.*; nw. O.-Ind. (Kaschmir)
„ -Kali-mirchi: *Toddalia asiatica* Baill.; *Rutac.*; O.-Ind.
Jangsiris: *Albizzia odoratissima* Bth.; *Mimosac.*; ö. Beng.
Janipha: *Genipa americana* L.; *Rubiac.*; tr. Am.
Jankang: *Hopea intermedia* King; *Dipterocarpac.*; Malay.
Jan Snijder: *Pouteria guianensis* Aubl.; *Sapotac.*; Sur.
Jantan: *Gordonia excelsa* Blume; *Theac.*; H.-Ind., Sunda.
Jantia: *Gardenia latifolia* Aiton; *Rubiac.*; nö. O.-Ind.
Jaowa: *Icica sp.*; *Burserac.*; fr. Gu.
Jaoz: *Juglans regia* L.; *Juglandac.*; Pers.
Japopalli: *Licania heteromorpha* Bth., *L. divaricata* Bth., *L. sp.*; *Rosac.*; Sur.
„ : *Couepia surinamensis* Kleinh., *? C. spp.*; *Rosac.*; Sur.
Jappoparé: *Licania heteromorpha* Bth., *L. divaricata* Bth., *L. sp.*; *Rosac.*; Sur.
„ : *Couepia surinamensis* Kleinh., *? C. spp.*; *Rosac.*; Sur.
Japud: *Albizzia stipulata* Boivin; *Mimosac.*; n. O.-Ind.
Jaqueira: *Artocarpus integrifolia* L.; *Morac.*; Bras.
Jara blanca: *Cistus albidus* L.; *Cistac.*; Sp.
„ comun: *Cistus ladaniferus* L.; *Cistac.*; Sp.
„ del diablo: *Halimium sp.*; *Cistac.*; Sp.
Jarana: *Holopyxidium jarana* Ducke; *Amarantac.*; Bras. (Amaz.)
Jarandeuva: *Pithecolobium sp.*; *Mimosac.*; Bras.
Jari: *Ficus religiosa* L.; *Morac.*; O.-Ind.
Jarilla: *Halimium umbellatum* Spach; *Cistac.*; Sp.
Jarillo: *Escallonia floribunda* H. B. K.; *Saxifragac.*; Ven.
Jarina: *Enterolobium cyclocarpum* Griseb.; *Mimosac.*; Pan.
Jarino: *Dipterodendron costaricense* Radlk.; *Sapindac.*; w. Pan., w. C.-R.
Jaris: *Bombacopsis jaris* Pitt.; *Bombacac.*; Ven.
Jarjo: *Schrebera swietenioides* Roxb.; *Oleac.*; n. O.-Ind.
Jaroeroe: *Aspidosperma nitidum* Bth., *A. exelsum* Bth.; *Apocynac.*; Sur.
Jarón: *Cistus populifolius* L.; *Cistac.*; Sp.
Jarora-harararo: *Aspidosperma nitidum* Bth., *A. excelsum* Bth.; *Apocynac.*; Sur.
Jaroroballi: *Swartzia acuminata* Willd., *S. spp.*; *Caesalpiniac.*; Sur.
Jarrah: *Eucalyptus marginata* J. Smith; *Myrtac.*; w. Au.
„ , Bastard-: *Eucalyptus botryoides* Smith; *Myrtac.*; ö. Au.
Jarre-ewę: *Guarea gomma* Pulle; *Meliac.*; Sur.
Jarul: *Lagerstroemia flos-reginae* Retz.; *Lythrac.*; Beng.
Jasmin, unechter (d): *Philadelphus coronarius* L.; *Saxifragac.*; N.-Am.
Jasmine: *Asimina triloba* Dunal; *Anonac.*; ö. Kan.
„ , Mountain-: *Stemmadenia macrantha* Standl.; *Apocynac.*; Pan.
„ , spanish: *Plumiera sp.*; *Apocynac.*; br. W.-Ind.
„ , wild: *Aegiphila martinicensis* L.; *Verbenac.*; Pan.
Jasminier: *Asimina triloba* Dunal; *Anonac.*; ö. Kan.
Jaspeada: *Piptadenia cebil* Griseb.; *Mimosac.*; Bras.
Jassie-hoedoe: *Jacaranda copaia* D. Don; *Bignoniac.*; Sur.
Jata: *Copernicia sp.*; *Palmac.*; Cuba.
Jatahy: *Hymenaea courbaril* L., *H. spp.*; *Caesalpiniac.*; Bras. (Sao P)
„ amarello: *Apuleia praecox* Mart.; *Caesalpiniac.*; Bras.
„ péba: *Hymenaea ? courbaril* L.; *Caesalpiniac.*; Bras. (Sao P)
„ vermelho: *Hymenaea sp.*; *Caesalpiniac.*; Bras. (Sao P)
Jataiba: *Hymenaea courbaril* L.; *Caesalpiniac.*; Bras.
Jati: *Tectona grandis* L. f.; *Verbenac.*; O.-Ind., Siam, Java.

Jati-korai: *Albizzia odoratissima* Bth.; *Mimosac.*; Assam.
Jatia: *Phyllostylon brasiliensis* Capanema; *Ulmac.*; Cuba.
Jatoba: *Hymenaea courbaril* L., *H. stilbocarpa* Hayne; *Caesalpiniac.*; Bras.
  „ : *Anisosperma passiflora* Manso; *Cucurbitac.*; Bras. (Amaz.)
  „ rôxo: *Hymenaea sp.*; *Caesalpiniac.*; Bras. (Sao P)
Jatuáuba: *Guarea paraënsis* C. DC.; *Meliac.*; Bras.
  „ preta: *Guarea spp.*; *Meliac.*; Bras.
Jaty: *Hymenaea stilbocarpa* Hayne; *Caesalpiniac.*; Bras. (Sao P)
Jaul: *Alnus sp.*; *Betulac.*; C.-R.
Jaune d'oeuf: *Mimusops bidentata* DC.; *Sapotac.*; fr. Gu.
  „   „ : *Lucuma mammosa* Gaertn. f., *L. rivicoa* Gaertn. f.; *Sapotac.*; fr. Gu.
  „   „ : *Pouteria guianensis* Aubl.; *Sapotac.*; Gu.
Javillo: *Hura crepitans* L.; *Euphorbiac.*; C.-Am.
  „ amarillo: *Hura crepitans* L.; *Euphorbiac.*; Trin.
  „ blanco: *Hura crepitans* L.; *Euphorbiac.*; Trin.
Javin: *Ichthyomethia piscipula* Hitch.; *Papilionac.*; Mex.
Jawalidan: *Sclerolobium paniculatum* Vogel; *Caesalpiniac.*; Sur.
Jawalidanie: *Minquartia guianensis* Aubl.; *Olacac.*; Gu.
Jawané-lolotin-kwatéré: *Eschweilera sp.*, *? Lecythis sp.*; *Lecythidac.*; Sur.
Jawaredan: *Sclerolobium paniculatum* Vogel; *Caesalpiniac.*; Sur.
Jawie: *Jacaranda copaia* D. Don; *Bignoniac.*; Sur.
Jawora: *Garcinia cornea* L.; *Guttifer.*; Sunda.
Jay-wree· *Memecylon sp.*; *Melastomatac.*; Lib.
Jazmín: *Clerodendron ligustrinum* R. Br.; *Verbenac.*; Nic.
  „ cimarrón: *Randia armata* DC.; *Rubiac.*; Hond.
  „ de estrella: *Faramea occidentalis* A. Rich.; *Rubiac.*; Ven.
  „ „ Malabar: *Gardenia jasminoides* Ellis; *Rubiac.*; Ven.
  ., „ montaña: *Tabernaemontana sp.*; *Apocynac.*; Cuba.
Jazminorro: *Jasminum fruticans* L.; *Oleac.*; Sp.
  „ silvestre: *Jasminum fruticans* L.; *Oleac.*; Sp.
Jea: *Virola sp.*; *Myristicac.*; Gu.
Je-ah-chu: *Stenanthera yalensis* Hutch. et Dalz.; *Anonac.*; Lib.
Jeamadou: *Virola sp.*; *Myristicac.*; Gu.
Jebarú: *Eperua purpurea* Bth.; *Caesalpiniac.*; n. Bras.
  „ -rana: *Eperua falcata* Aubl.; *Caesalpiniac.*; Bras.
Jee-jeray-flay: *Mussaenda afzelii* G. Don; *Rubiac.*; Lib.
Jejerecou: *Oxandra lanceolata* Baill.; *Anonac.*; fr. Gu.
Jejubier: *Ziziphus lotus* L.; *Rhamnac.*; Sahara.
Jejuira: *Astronium fraxinifolium* Schott; *Euphorbiac.*; Bras.
Jekoena: *Triplaris surinamensis* Cham.; *Polygonac.*; Sur.
Jekoeroe: *Nectandra sp.*, *Ocotea sp.*; *Laurac.*; Sur.
Jekona: *Triplaris surinamensis* Cham.; *Polygonac.*; Sur.
Jelängerjelieber: *Lonicera caprifolium* L.; *Caprifoliac.*; Eu.
Jeli: *Conocarpus erecta* L.; *Combretac.*; Ek.
Jelutong: *Dyera spp.*; *Apocynac.*; Malay, Sum.
Jemerelang: *Peltophorum ferrugineum* Bth.; *Caesalpiniac.*; Malay.
Jenequite: *Bursera gummifera* L.; *Burserac.*; Guat.
Jengi-barki: *Licania heteromorpha* Bth., *L. spp.*; *Rosac.*; Sur.
  „ kanda: *Hymenaea courbaril* L.; *Caesalpiniac.*; Sur.
  „ kendra: *Hymenaea courbaril* L.; *Caesalpiniac.*; Sur.
  „ -siri: *Nectandra sp.*, *Ocotea sp.*; *Laurac.*; Sur.
Jengie sopo: *Pithecolobium gonggrijpii* Kleinh., *P. spp.*; *Mimosac.*; Sur.
Jenisero (u): *Enterolobium cyclocarpum* Griseb.; *Mimosac.*; tr. Am.
Jen-nee: *Loesenera kalantha* Harms; *Caesalpiniac.*; Lib.
Jenny wood (u): *Cordia goeldiana* Huber; *Borraginac.*; Bras. (Amaz.)
Jequitiba: *Cariniana excelsa* Casar., *C. legalis* Kuntze; *Lecythidac.*; Bras.
  „ amarella: *Cariniana excelsa* Casar., *C. legalis* Kuntze; *Lecythidac.*; Bras.
  „ branca: *Cariniana excelsa* Casar., *C. legalis* Kuntze; *Lecythidac.*; Bras.
  „ rosa: *Cariniana excelsa* Casar., *C. legalis* Kuntze; *Lecythidac.*; Bras.
  „ vermelha: *Cariniana excelsa* Casar., *C. legalis* Kuntze; *Lecythidac.*; Bras.
Je-rah-kpar: *Bersama paullinioides* Baker; *Melianthac.*; Lib.
Je-ray-krehn: *Voacanga obtusa* K. Schum.; *Apocynac.*; Lib.
Je-ray-war-be-deh: *Urophyllum linderi* Hutch. et Dalz.; *Rubiac.*; Lib.
Jérérécou: *Xylopia frutescens* Aubl.; *Anonac.*; Gu.

Jereton: *Didymopanax morototoni* Dcne. et Planch.; *Araliac.*; Trin., Tob.
Jerinu: *Acer pictum* Thunb.; *Acerac.*; nw. O.-Ind. (Punjab)
Jermala: *Tetrameles nudiflora* R. Br.; *Datiscac.*; sw. O.-Ind.
Jesriharti: *Swartʒia minutiflora* Kleinh., *S. spp.*; *Caesalpiniac.*; Sur.
Jessamine: *Plumiera sp.*; *Apocynac.*; br. W.-Ind.
Jetahy preta: *Dialium divaricatum* Vahl; *Caesalpiniac.*; Bras.
Jetay: *Hymenaea courbaril* L.; *Caesalpiniac.*; Bras.
Jetè-balli-bèlèro: *Erisma uncinatum* Warm.; *Vochysiac.*; Sur.
Jetiballi: *Vochysia tomentosa* DC.; *Vochysiac.*; Sur.
Jetikiboralli: *Swartʒia minutiflora* Kleinh.; *Caesalpiniac.*; Sur.
Jetiki boroballi hororadikoro: *Swartʒia spp.*; *Caesalpiniac.*; Sur.
Jetjoenban karaäpa: *Carapa procera* DC., *C. surinamensis* Miq.; *Meliac.*; Sur.
Jetoeri walaba: *Eperua schomburgkiana* Bth., *? E. jenmani* Oliv.; *Caesalpiniac.*; Sur.
Jeve: *Hevea sp.*; *Euphorbiac.*; Peru.
Jew: *Taxus baccata* L.; *Conifer.*; Eu.
Jewalidanni: *Minquartia guianensis* Aubl.; *Olacac.*; Sur.
Je-ye-neh-doo: *Oligostemon pictus* Bth.; *Caesalpiniac.*; Lib.
Jhal: *Salvadora oleoides* Dcne.; *Salvadorac.*; O.-Ind. (Punjab)
Jhall mara: *Shorea talura* Roxb.; *Dipterocarpac.*; sw. O.-Ind.
Jhallanda: *Shorea talura* Roxb.; *Dipterocarpac.*; sw. O.-Ind. (Mysore)
Jhan: *Schrebera swietenioides* Roxb.; *Oleac.*; O.-Ind.
Jhar-beri: *Zizyphus jujuba* Lamk.; *Rhamnac.*; n. O.-Ind. (United-Prov.)
Jhingan: *Lannea grandis* Engl.; *Anacardiac.*; C.-O.-Ind. (Chota-Nagpur)
Jhinghan: *Lannea grandis* Engl.; *Anacardiac.*; n. O.-Ind. (United-Prov.)
Jhinjit: *Bauhinia malabarica* Roxb., *B. retusa* Ham.; *Caesalpiniac.*; ö. Beng.
Jhinkri: *Altingia excelsa* Noronha; *Hamamelidac.*; Assam.
Jia blanca: *Casearia sp.*; *Flacourtiac.*; Cuba.
„ brava: *Casearia sp.*; *Flacourtiac.*; Cuba.
„ manzanilla: *Ximenia americana* L.; *Olacac.*; Cuba.
Jibá de costa: *Erythroxylon sp.*; *Erythroxylac.*; Cuba.
Jicaco: *Chrysobalanus sp.*; *Rosac.*; W.-Ind., C.-Am., Mex.
Jicara: *Crescentia cujete* L.; *Bignoniac.*; Mex., C.-Am.
Jicarillo: *Cochlospermum vitifolium* Spreng.; *Cochlospermac.*; Hond.
„ : *Clethra lanata* Mart. et Gal.; *Clethrac.*; Mex.
Jícaro: *Crescentia cujete* L.; *Bignoniac.*; Hond., Pan.
Jicote: *Bursera gummifera* L.; *Burserac.*; Hond.
Jiga: *Maerua crassifolia* Forsk.; *Capparidac.*; n. Nig.
Jiggerwood: *Bravaisia floribunda* A. DC.; *Acanthac.*; Trin.
Jigiri: *Idesia polycarpa* Maxim.; *Flacourtiac.*; Jap.
Jigüe: *Lysiloma latisiliqua* Bth.; *Mimosac.*; Cuba.
„ blanco: *Lysiloma latisiliqua* Bth.; *Mimosac.*; Cuba.
Jiguerillo: *Ficus sp.*; *Morac.*; P.-R.
Jilinsuche: *Bombax spp.*; *Bombacac.*; Salv.
Jimshi: *Acacia seyal* Delile; *Mimosac.*; n. Nig.
Jinari: *Podocarpus neriifolia* D. Don; *Conifer.*; Assam (Cachar).
Jindai-Sugi: *Cryptomeria japonica* D. Don; *Conifer.*; Jap.
Ji-ndimu: *Xymalos usambarensis* Engl.; *Monimiac.*; D.-O.-A. (w. Usambara)
Jingbawng: *Zanthoxylum budrunga* Wall.; *Rutac.*; O.-Ind.
Jinicuile: *Inga sp.*; *Mimosac.*; Mex.
Jinicuite: *Bursera simaruba* Sarg.; *Burserac.*; Hond.
Jinjajong: *Ixonanthes reticulata* Jack; *Linac.*; Malak.
Jiñocuabo: *Bursera simaruba* Sarg.; *Burserac.*; P.-R.
Jiñocuave: *Bursera simaruba* Sarg.; *Burserac.*; C.-R.
Jiñocuavo: *Bursera simaruba* Sarg.; *Burserac.*; Kol.
Jiote: *Bursera simaruba* Sarg.; *Burserac.*; Hond.
Jipigapa: *Carludovica palmata* R. et P.; *Cyclanthac.*; Kol.
Jique: *Pera sp.*; *Euphorbiac.*; Cuba.
Jiqui espinoso: *Bumelia sp.*; *Sapotac.*; Cuba.
„ de costa: *Malpighia obovata* H. B. K.; *Malpighiac.*; Cuba.
Jiquimite: *Erythrina christa-galli* L.; *Papilionac.*; Mex.
Jiquitibá: *Cariniana brasiliensis* Casar., *C. domestica* Miers; *Lecythidac.*; Bras.
Jirga: *Bauhinia rufescens* Lam.; *Caesalpiniac.*; n. Nig.
Jiri: *Stereospermum kunthianum* Cham.; *Bignoniac.*; n. Nig.
Jissára: *Euterpe oleracea* Mart.; *Palmac.*; Bras. (Sao P)

Jitahy: *Hymenaea stigonocarpa* Mart.; *Caesalpiniac.*; Bras. (Sao P)
„ amarello: *Apuleia praecox* Mart.; *Caesalpiniac.*; Bras.
Jitangi: *Dalbergia latifolia* Roxb.; *Papilionac.*; w. O.-Ind. (n. Bombay)
Jitegi: *Dalbergia latifolia* Roxb.; *Papilionac.*; w. O.-Ind. (n. Bombay)
Jito: *Guarea trichilioides* L.; *Meliac.*; Bras.
Jiyal: *Lannea grandis* Engl.; *Anacardiac.*; Beng.
Jizerhart, Gandoe-: *Swartzia tomentosa* DC.; *Caesalpiniac.*; Sur.
„ , gewone: *Swartzia spp.*; *Caesalpiniac.*; Sur.
Jizerhout: *Tecoma sp.*, *Tabebuia sp.*; *Bignoniac.*; tr. S.-Am.
„ : *Aglaia minahassae* Koord.; *Meliac.*; Cel., Ceram.
Jizo-kanba: *Betula globispica* Shirai; *Betulac.*; Jap.
Jo: *Butyrospermum parki* Kotschy; *Sapotac.*; Togo.
Joa minda: *Celtis aculeata* Sw.; *Ulmac.*; tr. Am.
Joao dormido: *Pisonia sp.*; *Nyctaginac.*; Bras.
„ molle: *Pisonia sp.*; *Nyctaginac.*; Bras.
Jobillo: *Tapirira guianensis* Aubl.; *Anacardiac.*; Ven.
Jobitillo: *Phyllanthus conami* Swartz; *Euphorbiac.*; Pan.
Jobito: *Spondias lutea* L.; *Anacardiac.*; Pan.
Jobo: *Spondias lutea* L.; *Anacardiac.*; P.-R., C.-Am., Kol., Ven.
„ espino: *Spondias sp.*; *Anacardiac.*; Mex.
„ francés: *Spondias sp.*; *Anacardiac.*; P.-R.
„ hembra: *Spondias sp.*; *Anacardiac.*; Cuba.
„ lagarto: *Sciadodendron excelsum* Griseb.; *Araliac.*; Pan.
„ negro: *Spondias sp.*; *Anacardiac.*; Cuba.
„ roñoso: *Spondias sp.*; *Anacardiac.*; Mex.
Jocomico: *Protium sessiliflorum* Standl.; *Burserac.*; Hond.
„ : *Rheedia edulis* Triana et Planch.; *Guttifer.*; Nic.
Jocote: *Spondias lutea* L., *S. spp.*, *Anacardium occidentale* L.; *Anacardiac.*; Guat.
„ : *Phytolacca rivinoides* Kth. et Bouché; *Phytolaccac.*; br. Hond.
„ jobo: *Spondias lutea* L.; *Anacardiac.*; Guat.
„ marañón: *Anacardium occidentale* L.; *Anacardiac.*; Hond.
„ montero: *Spondias lutea* L.; *Anacardiac.*; Nic.
„ tronador: *Spondias sp.*; *Anacardiac.*; C.-R.
„ de jobo: *Spondias sp.*; *Anacardiac.*; Nic.
„ „ mico: *Vitex longeracemosa* Pittier; *Verbenac.*; Guat., Hond.
„ „ „ : *Simaruba glauca* DC.; *Simarubac.*; Guat.
Jocotillo: *Trichilia sp.*; *Meliac.*; Salv.
Jocuma amarilla: *Sideroxylon foetidissimum* Jacq.; *Sapotac.*; Cuba.
„ prieta: *Sideroxylon foetidissimum* Jacq.; *Sapotac.*; Cuba.
Joebush: *Jacquinia keyensis* Mez; *Theophrastac.*; Fl.
Joekoejapoi: *Nectandra sp.*, *Ocotea sp.*; *Laurac.*; Sur.
Joekoetoena: *Guarea gomma* Pulle; *Meliac.*; Sur.
Joekoetoena: ? *Swartzia triphylla* Willd.; *Caesalpiniac.*; Sur.
Joeliballi fierberoebana: *Hedwigia balsamifera* Swartz, *Protium spp.*; *Burserac.*; Sur.
Joelieballi: *Hedwigia balsamifera* Swartz, *Protium spp.*; *Burserac.*; Sur.
Joenoepé: *Couepia surinamensis* Kleinh., ? *C. spp.*; *Rosac.*; Sur.
Joeriballi tataro: *Hedwigia balsamifera* Swartz, *Protium spp.*; *Burserac.*; Sur.
Joe wood: *Jacquinia keyensis* Mez; *Myrsinac.*; tr. Am.
Joge: *Albizzia sp.*; *Mimosac.*; br. O.-Af. (Uganda/Mabira)
Johannisbeere: *Ribes rubrum* L.; *Saxifragac.*; Eu.
„ , Alpen-: *Ribes alpinum* L.; *Saxifragac.*; Eu.
„ , rote (d): *Ribes rubrum* L.; *Saxifragac.*; Eu.
„ , schwarze (d): *Ribes nigrum* L.; *Saxifragac.*; Eu.
Johannisbrotbaum: *Ceratonia siliqua* L.; *Caesalpiniac.*; M.-M.
John crow: *Pithecolobium arboreum* Urban; *Mimosac.*; br. Hond.
„ „ bead: *Pithecolobium donnell-smithi* Standl.; *Mimosac.*; br. Hond.
Joki: *Bischofia javanica* Blume; *Euphorbiac.*; Assam.
Joli coeur: *Celastrus undulatus* Lamk.; *Celastrac.*; Kap, Mask.
Jolocin: *Heliocarpus sp.*; *Tiliac.*; Mex.
„ blanco: *Heliocarpus sp.*; *Tiliac.*; Mex.
Joloßín: *Heliocarpus sp.*; *Tiliac.*; Mex.
Jom-janum: *Zizyphus jujuba* Lamk.; *Rhamnac.*; C.-O.-Ind. (Chota-Nagpur)
Jomirim: *Rapanea laetevirens* Mez; *Myrsinac.*; Bras.
Jondo: *Piptadenia africana* Hook. f.; *Mimosac.*; Kam.

Jono: *Dryobalanops aromatica* Gaertn. f.; *Dipterocarpac.*; Sum.
Jonote: *Heliocarpus sp.; Tiliac.;* Mex.
„ blanco: *Hampea sp.; Bombacac.;* Mex.
Jonze-hindie: *Cocos nucifera* L.; *Palmac.;* Arab.
Jorco: *Rheedia sp.; Guttifer.;* C.-R.
Joro-joro-pisi: *Nectandra sp., Ocotea sp.; Laurac.;* Sur.
Jorokan-pomoire: *Nectandra sp., Ocotea sp.; Laurac.;* Sur.
Jor-wee: *Canthium acutiflorum* Hiern; *Rubiac.;* Lib.
Jovo: *Spondias sp.; Anacardiac.;* Mex.
Jowalidovi: *Minquartia guianensis* Aubl.; *Olacac.;* fr. Gu
Jru: *Erythrophloeum guineense* G. Don; *Caesalpiniac.;* Lib.
Juá: *Solanum toxicarium* Lam.; *Solanac.;* Bras.
Juan blanco: *Capparis sp.; Capparidac.;* Ven.
„ garote: *Coccoloba leptostachya* Bth.; *Polygonac.;* Kol.
Juanco: *Koeberlinia spinosa* Zuccarini; *Koeberliniac.;* s. USA., Mex.
Juargarzo blanco: *Halimium halimifolium* Willk.; *Cistac.;* Sp.
„ prieto: *Cistus crispus* L.; *Cistac.;* Sp.
Jubabán: *Trichilia sp.; Meliac.;* Cuba.
Jubatan: *Astronium concinnum* Schott, *A. gracile* Engl.; *Anacardiac.;* Bras.
Jubatao: *Astronium fraxinifolium* Schott; *Anacardiac.;* ö. Bras.
Juca: *? Caesalpinia ferrea* Mart.; *Caesalpiniac.;* nö. Bras.
Jucaro: *Bucida buceras* L.; *Combretac.;* Cuba.
„ negro: *Terminalia buceras* Wright; *Combretac.;* Cuba.
„ prieto: *Buchenavia capitata* Eichl.; *Combretac.;* Cuba.
„ de playa: *Terminalia buceras* Wright; *Combretac.;* Cuba.
Juche: *Plumiera sp.; Apocynac.;* C.-R.
Jucumico: *Simaruba sp.; Simarubac.;* Salv.
Judas: *Cossignia borbonica* DC.; *Sapindac.;* Réu.
Judasbaum: *Cercis siliquastrum* L.; *Caesalpiniac.;* M.-M.
Judas tree: *Cercis canadensis* L.; *Caesalpiniac.;* ö. N.-Am.
Ju-eh-ye-neh-chu: *Desmostachys vogeli* Stapf; *Icacinac.;* Lib.
Ju-ehn-jrah: *Cassipourea afzelii* Alston; *Rhizophorac.;* Lib.
Ju- „ „ : *Millettia spp.; Papilionac.;* Lib.
Ju-ihn: *Bombax brevicuspe* Sprague; *Bombacac.;* Lib.
Ju-wrah: *Smeathmannia pubescens* Soland.; *Flacourtiac.;* Lib.
Jug: *Mouriria sp.; Melastomatac.;* br. Hond.
Jug tree: *Cordia diversifolia* Pav.; *Borraginac.;* Pan.
Jugia: *Cordia macleodi* Hook. f. et Thoms.; *Borraginac.;* ö. Beng.
Jugini: *Strychnos potatorum* L. f.; *Loganiac.;* O.-Ind.
Jugli: *Barringtonia acutangula* Gaertn.; *Lecythidac.;* C.-O.-Ind.
Jugon: *Firmiana klaineana* Pierre; *Sterculiac.;* Gab.
Jujuba: *Zizyphus spp.; Rhamnac.;* Tr.
Jujube, indian: *Zizyphus jujuba* Lam.; *Rhamnac.;* tr. Af., tr. As., tr. Au.
Jujubier: *Zizyphus lotus* Lamk.; *Rhamnac.;* n. Af., (s. Tunis)
„ sauvage: *Zizyphus lotus* Lamk.; *Rhamnac.;* n. Af.
„ de l'Inde: *Zizyphus jujuba* Lamk.; *Rhamnac.;* Ch., Ind., tr. Af.
Julchihout: *Hymenaea courbaril* L.; *Caesalpiniac.;* Sur.
Jum: *Garuga pinnata* Roxb.; *Burserac.;* Beng.
Jumbeba: *Cerus sp.; Cactac.;* Bras.
Jumero: *Dalbergia sp.; Ebenac.;* Guat.
Jummina: *Zanthoxylum rhetsa* DC.; *Rutac.;* w. O.-Ind.
Junco: *Koeberlinia spinosa* Zuccarini; *Koeberliniac.;* Mex.
June, Japanese Red: *Prunus triflora* Roxb.; *Rosac.;* nö. USA.
Juneberry: *Amelanchier canadensis* Medic.; *Rosac.;* ö. N.-Am.
Junero: *Dalbergia cubiliquitzensis* Pittier; *Papilionac.;* Guat.
Junghardt (norweg.): *Dipterocarpus grandiflorus* Blco.; *Dipterocarpac.;* O.-As.
Jungli Mohwa: *Bassia butyracea* Roxb.; *Sapotac.;* And.
Junipa: *Genipa americana* L.; *Rubiac.;* tr. Am.
Junipapo: *Genipa americana* L.; *Rubiac.;* Bras.
Junipappeeywa: *Genipa americana* L.; *Rubiac.;* ö. Bras.
Juniper: *Juniperus communis* L.; *Conifer.;* Eu.
„ : *Larix americana* Michx.; *Conifer.;* ö. N.-Am.
„ : *Pinus banksiana* Lamb.; *Conifer.;* ö. N.-Am.
„ , Alligator-: *Juniperus pachyphloea* Torr.; *Conifer.;* USA.

Juniper, Common: *Juniperus communis* L.; *Conifer.; n. USA.
,, , Dwarf-: *Juniperus communis* L.; *Conifer.; ö. USA.
,, , East African-: *Juniperus procera* Hochst.; *Conifer.; O.-Af.
,, , Indian: *Juniperus macropoda* Boiss.; *Conifer.; O.-Ind.
,, , Mountain-: *Juniperus sabinoides* Kunth; *Conifer.; Texas, Mex.
,, , Pencil-: *Juniperus virginiana* L.; *Conifer.; ö. Kan.
,, , red: *Juniperus virginiana* L.; *Conifer.; ö. N.-Am.
,, , Rocky Mountain-: *Juniperus scopulorum* Sarg.; *Conifer.; w. N.-Am.
,, , Uganda-: *Juniperus procera* Hochst.; *Conifer.; O.-Af. (Nyassa)
Jur: *Canthium didymum* Roxb.; *Rubiac.; w. Beng.
Jurema: *Pithecolobium reticulata* Bth.; *Mimosac.; Bras.
Jurighas: *Filicium decipiens* Thw.; *Sapindac.; Cey.
Juril: *Coccoloba sp.; Polygonac.; Salv.
Jurubeba do campo: *Solanum toxicarium* Lam.; *Solanac.; Bras.
Jussara: *Euterpe oleracea* Mart.; *Palmac.; Bras. (Sao P)
Jutahy: *Hymenaea spp.; Caesalpiniac.; Bras.
,, amarello: *Apuleia praecox* Mart.; *Caesalpiniac.; Bras.
,, assú: *Hymenaea courbaril* L.; *Caesalpiniac.; Bras.
,, café: *Hymenaea courbaril* L.; *Caesalpiniac.; Bras.
,, catinga: *Hymenaea courbaril* L.; *Caesalpiniac.; Bras.
,, miry: *Hymenaea courbaril* L.; *Caesalpiniac.; Bras.
,, peba: *Hymenaea courbaril* L., *Dialium divaricatum* Vahl; *Caesalpiniac.; Bras.
,, pororoca: *Hymenaea courbaril* L.; *Caesalpiniac.; Bras.
,, roxo: *Hymenaea courbaril* L.; *Caesalpiniac.; Bras.
,, do campo: *Hymenaea parvifolia* Huber; *Caesalpiniac.; Bras.
Juti: *Murraya exotica* L.; *Rutac.; n. O.-Ind. (United-Prov.)
,, mersolo: *Murraya exotica* L.; *Rutac.; n. O.-Ind. (Garhwal)
Jutili: *Altingia excelsa* Noronha; *Hamamelidac.; O.-Ind.
Jutraj: *Amoora rohituka* Wight et Arn.; *Meliac.; Beng.
Jutry: *Taxus baccata* L.; *Conifer.; Eu.
Jutuli: *Altingia excelsa* Noronha; *Hamamelidac.; Assam.
Juvia: *Bertholletia excelsa* H. B. K., *B. nobilis* Miers; *Lecythidac.; Gu., n. Bras.
Juwaredan: *Sclerolobium sp.; Caesalpiniac.; Sur.

K

Ka: *Diplanthemum viridiflorum* Hutch.; *Tiliac.; Kam.
,, : *Cornus capitata* Wall.; *Cornac.; Ind.-Ch.
,, gnuom: *Shorea cochinchinensis* Pierre; *Dipterocarpac.; Ind.-Ch.
,, kao: *Aleurites montana* Pierre; *Euphorbiac.; Ind.-Ch.
,, kas: *Sindora cochinchinensis* Baill.; *Caesalpiniac.; Ind.-Ch.
,, nhung: *Dalbergia cochinchinensis* Pierre; *Papilionac.; Ind.-Ch.
,, sanh: *Feronia lucida* Teijsm. et Binn.; *Rutac.; Ind.-Ch.
,, thang: *Mesua ferrea* L.; *Guttifer.; Ind.-Ch.
,, vao: *Eugenia chanlos* Gagnep.; *Myrtac.; Ind.-Ch.
,, zo: *Cassia fistula* L.; *Caesalpiniac.; O.-Ind.
Kaa pororó: *Rapanea sp.; Myrsinac.; Arg.
Kaala: *Bridelia retusa* Spreng.; *Euphorbiac.; O.-Ind., Cey.
Kaäpa: *Carapa procera* DC., *C. surinamensis* Miq.; *Meliac.; Sur.
Kaäpaä: *Carapa procera* DC., *C. surinamensis* Miq.; *Meliac.; Sur.
Kaáporoka: *Rapanea laetevirens* Mez; *Myrsinac.; Arg.
Kaá pororó: *Rapanea laetevirens* Mez; *Myrsinac.; Arg.
? Kaba: *Conocarpus erecta* L.; *Combretac.; Go.
Kaba: *Betula sp.; Betulac.; Jap.
Kabag: *Diospyros aherni* Merr.; *Ebenac.; Phil. (Isa.)
Kabak: *Nauclea sp.; Rubiac.; Phil. (Samar, Agusan)
Kabakabát: *Pometia pinnata* Forst.; *Sapindac.; Phil. (Ilocos-Norte)
Kabakally: *Dicorynia paraënsis* Bth.; *Caesalpiniac.; Sur.
Kabamba Katoka: *Brachystegia sp.; Caesalpiniac.; B.-K. (Katanga)
,, lungwaluole: *Brachystegia sp.; Caesalpiniac.; B.-K. (Katanga)
Kabana: *Jacaranda filicifolia* D. Don; *Bignoniac.; Sur.
,, : *Pithecolobium gonggrijpi* Kleinh., *P. spp.; Mimosac.; Sur.
Kabaránga: *Celtis soyauxi* Engl.; *Ulmac.; Togo.
Kabasale: *Hydnocarpus wightiana* Blume; *Flacourtiac.; sw. O.-Ind. (Bombay)

Kabashi: *Acer thomsoni* Miq., *A. campbelli* Hook. f. et Thoms.; *Acerac.;* Beng.
Kabasi: *Acer pictum* Thunb.; *Acerac.;* n. O.-Ind. (Jaunsar)
Kabaski: *Acer campbelli* Hook. f. et Thoms.; *Acerac.;* ö. Himal.
Kabba: *Parashorea stellata* Kurz; *Dipterocarpac.;* s. Burma.
Kabbes: *? Andira inermis* H. B. K.; *Papilionac.;* Sur.
„ . levarte: *Diplotropis sp.; Papilionac.;* Sur.
„ , roode: *Andira coriacea* Pulle; *Papilionac.;* Sur.
„ , witte: *? Andira retusa* H. B. K.; *Papilionac.;* Sur.
„ , zwarte: *Diplotropis guianensis* Bth., *D. leptophylla* Kleinh.; *Papilionac.;* Sur.
„ hout: *Andira spp.; Papilionac.;* Sur.
Kabbi: *Vouacapoua americana* Aubl.; *Papilionac.;* Sur.
Kabeljauw-hout: *? Vitex sp.; Verbenac.;* fr. Gu.
Kabi: *Swartzia madagascariensis* Desv.; *Caesalpiniac.;* B.-K.
Kabiki: *Mimusops parvifolia* R. Br.; *Sapotac.;* Phil. (Manila, Camarines)
Kabinethout: *Philippia chamissonis* Klotzsch; *Ericac.;* Kap.
Kabisi: *? Andira retusa* H. B. K.; *Papilionac.;* Sur.
Kabit: *Feronia elephantum* Correa; *Rutac.;* C.-O.-Ind.
Kaboekallie: *Goupia glabra* Aubl.; *Celastrac.;* Sur.
Kabokali: *Goupia glabra* Aubl.; *Celastrac.;* Sur.
Kabolóan: *Bambusa vulgaris* Schrad., *Dendrocalamus latiflorus* Munro; *Gramin.;* Phil.
Kabong: *Arenga pinnata* Merr.; *Palmac.;* Malay.
Kabugáuan: *Bambusa spinosa* Roxb.; *Gramin.;* Phil.
Kabu-kabu: *Ceiba pentandra* Gaertn.; *Malvac.;* Malay.
Kabukalli: *Goupia glabra* Aubl., *G. tomentosa* Aubl.; *Celastrac.;* br. Gu.
Kabuli: *Acacia sp.; Mimosac.;* nw. O.-Ind.
Kabulo: *Nothophoebe malabonga* Merr.; *Laurac.;* Phil. (Mindoro)
Kaburo: *Dehaasia triandra* Merr., *Phoebe 'sterculioides* Merr.; *Laurac.;* Phil. (Mindoro)
Kabyi: *Artocarpus lakoocha* Roxb.; *Morac.;* O.-Ind. (? Burma)
Kachal: *Picea morinda* Link; *Conifer.;* nw. O.-Ind. (Punjab, Kaschmir)
Kachein: *Melia azedarach* L.; *Meliac.;* nw. O.-Ind. (Punjab)
Kachila: *Strychnos nux-vomica* L.; *Loganiac.;* Beng.
Kachirachirai: *Michelia kachirachirai* K. et Y.; *Magnoliac.;* Form.
Kachnar: *Bauhinia variegata* L.; *Caesalpiniac.;* O.-Ind.
Kachúchis: *Avicennia alba* Blume; *Verbenac.;* Phil. (Surigao)
Kad-kanagala: *Dillenia indica* L.; *Dilleniac.;* w. O.-Ind. (s. Bombay)
Kadaburichi: *Oxandra lanceolata* Baill.; *Anonac.;* br. Gu.
Kada-Kongi-Cheddi: *Murraya exotica* L.; *Rutac.;* w. O.-Ind. (Deccan)
Kada-kukkn: *Crataeva roxburghi* R. Br.; *Capparidac.;* O.-Ind.
Kadada: *Dichrostachys nutans* Bth.; *Mimosac.;* s. Sudan.
Kadai: *Butyrospermum parki* Kotschy; *Sapotac.;* n. Nig.
Kadakai: *Terminalia chebula* Retz.; *Combretac.;* w. O.-Ind. (Deccan)
Kadali: *Lagerstroemia flos-reginae* Retz.; *Lythrac.;* O.-Ind. (Madras)
Kadam: *Anthocephalus cadamba* Miq.; *Rubiac.;* Beng.
Kadambe: *Anthocephalus cadamba* Miq.; *Rubiac.;* O.-Ind. (Madras)
Kadambo: *Anthocephalus cadamba* Miq.; *Rubiac.;* sw. Beng. (Orissa)
Kadambu· *Adina cordifolia* Hook. f.; *Rubiac.;* sw. O.-Ind. (Malabar)
Kadang-ísol: *Kayea paniculata* Merr.; *Guttifer.;* Phil. (Camarines)
Kadani: *Stephegyne parvifolia* Korth.; *Rubiac.;* O.-Ind. (Madras)
Kadanya: *Butyrospermum parki* Kotschy; *Sapotac.;* n. Nig.
Kadanyar rafi: *Adina microcephala* Hiern; *Rubiac.;* n. Nig.
Kadara bobo: *Sterculia tragacantha* Lindl.; *Sterculiac.;* Togo.
Kadashing: *Stereospermum xylocarpum* Bth. et Hook. f.; *Bignoniac.;* sw. O.-Ind.
Kadatnyan: *Canangium odoratum* Baill.; *Anonac.;* Burma.
Kadaura: *Paradaniella oliveri* Rolfe; *Caesalpiniac.;* n. Nig.
Kadawar: *Adina cordifolia* Hook. f., *Stephegyne parvifolia* Korth.; *Rubiac.;* w. O.-Ind.
Kaddam: *Anthocephalus cadamba* Miq., *Adina cordifolia* Hook. f.; *Rubiac.;* sw. O.-Ind.
„ : *Stephegyne parvifolia* Korth.; *Rubiac.;* sw. O.-Ind.
Kaddu: *Sterculia urens* Roxb.; *Sterculiac.;* C.-O.-Ind. (Jeypore)
Kadé: *Butyrospermum parki* Kotschy; *Sapotac.;* Sudan.
Kade besji: *Capparis sp.; Capparidac.;* Curacao.
Kaderi: *Acacia catechu* Willd.; *Mimosac.;* w. O.-Ind. (Bombay)
Kadiala: *Stephegyne parvifolia* Korth.; *Rubiac.;* sw. O.-Ind. (Coorg)
Kadíl: *Duabanga moluccana* Blume; *Sonneratiac.;* Phil. (Abra)
Kadír: *Duabanga moluccana* Blume; *Sonneratiac.;* Phil. (Ilocos-N., Cagayan)

Kadír: *Erythrophloeum densiflorum* Merr.; *Caesalpiniac.;* Phil. (Ilocos.-N.)
Kadjand: *Uvaria grandiflora* Rich.; *Anonac.;* Java.
Kadjoe mattoe: *Dimorphandra latifolia* Tul.; *Caesalpiniac.;* Sur.
Kadjoesie-awha: *Stryphnodendron flammatum* Kleinh.; *Mimosac.;* Sur.
„   „   : *Pithecolobium racemiflorum* Ducke; *Mimosac.;* Sur.
„   „   : *Hymenolobium flavum* Kleinh., *H. sp.; Mimosac.;* Sur.
Kadmero: *Litsea polyantha* Juss.; *Laurac.;* nö. O.-Ind. (Nepal)
Kadoe: *Durio zibethinus* Murr.; *Bombacac.;* Sunda.
Kadoeng: *Parkia roxburghi* G. Don; *Mimosac.;* Malay.
Kadok: *Pterospermum javanicum* Jungh.; *Sterculiac.;* Java.
Kadol: *Careya sphaerica* Roxb.; *Lecythidac.;* Ind.-Ch.
Kadondong mata hari: *Trigonochlamys griffithi* Hook. f.; *Burserac.;* Malay.
„   outan: *Canarium kadondon* A. W. Benn.; *Burserac.;* Malak.
Kadongan: *Bouea macrophylla* Griff.; *Anacardiac.;* Malay.
Kadrupala: *Bridelia retusa* Spreng.; *Euphorbiac.;* w. Beng. (Chota-Nagpur)
Kadsige: *Albizzia amara* Boivin; *Mimosac.;* sw. O.-Ind. (Coorg)
Kadsoura: *Cercidiphyllum japonicum* S. et Z.; *Trochodendrac.;* Jap.
Kadsura: *Cercidiphyllum japonicum* S. et Z.; *Trochodendrac.;* Jap.
Kadu: *Gynocardia odorata* R. Br.; *Flacourtiac.;* nö. O.-Ind. (Nepal)
„   -kajar: *Melia composita* Willd.; *Meliac.;* O.-Ind.
Kadumberiya: *Diospyros gardneri* Thw.; *Ebenac.;* Cey.
Kadung: *Bombax insigne* Wall. *var. andamanica* Prain; *Bombacac.;* O.-Ind.
„   : *Bombax insigne* Wall. *var. wighti* Prain; *Bombacac.;* O.-Ind.
Kadusale: *Alstonia scholaris* R. Brown; *Apocynac.;* w. O.-Ind. (Deccan)
Kadut: *Parashorea stellata* Kurz; *Dipterocarpac.;* s. Burma (Tenasserim)
Kadutni: *Parashorea stellata* Kurz; *Dipterocarpac.;* s. Burma (Tenasserim)
Kadutpyu: *Parashorea stellata* Kurz; *Dipterocarpac.;* s. Burma (Tenasserim)
Kadwal: *Anthocephalus cadamba* Miq.; *Rubiac.;* w. O.-Ind. (Bombay)
Kadwar: *Stephegyne parvifolia* Korth.; *Rubiac.;* O.-Ind. (Madras)
Kaealéka: *Anogeissus sp.; Combretac.;* Elf.
Kaeda: *Planchonia timorensis* Blume; *Lecythidac.;* Neu-Guinea.
Kaede: *Acer palmatum* Thunb., *A. sp.; Acerac.;* Jap.
Kaem: *Stephegyne parvifolia* Korth.; *Rubiac.;* n. O.-Ind. (United-Prov.)
Kaen: *Bischofia javanica* Blume; *Euphorbiac.;* O.-Ind. (Garhwal)
Kafa: *Triplochiton scleroxylon* K. Schum.; *Sterculiac.;* Elf.
Kafafogo: *Uapaca guineensis* Muell. Arg.; *Euphorbiac.;* n. Nig.
Kafferboom: *Erythrina caffra* Thunb.; *Papilionac.;* sö. S.-Af.
Kaffi: *Guarea thompsoni* Sprague et Hutch.; *Meliac.;* Lib.
Kaffirboom: *Erythrina caffra* Thunb.; *Papilionac.;* S.-Af.
Kafungu: *Dialium angolense* Welw.; *Caesalpiniac.;* B.-K. (Lukafu)
Kagasákas: *Litchi philippinensis* Radlk.; *Sapindac.;* Phil. (Albay)
Kagatúngan: *Pygeum sp.; Rosac.;* Phil. (Rizal)
Kagbhalai: *Aporosa lindleyana* Baill.; *Euphorbiac.;* O.-Ind.
Kagemkém: *Parinarium sp.; Rosac.;* Phil. (Ilocos-Sur)
Kagi: *Drimycarpus racemosa* Hook. f.; *Anacardiac.;* ö. Himal.
Kagli: *Acacia catechu* Willd.; *Mimosac.;* w. O.-Ind. (Bombay)
Kagokó: *Eugenia mananquil* Blco.; *Myrtac.;* Phil. (Leyte, Lanao, Cotabato)
Kagonoki: *Actinodaphne lancifolia* Meissn.; *Laurac.;* Jap.
Kagsákan: *Litchi philippinensis* Radlk.; *Sapindac.;* Phil. (Albay)
Kagunga nischwa: *Acacia holsti* Taub.; *Mimosac.;* O.-Af.
Kah: *Haplormosia monophylla* Harms; *Papilionac.;* Lib.
Kakaralli, witte: *Eschweilera corrugata* Miers; *Lecythidac.;* br. Gu.
Kahi: *Strychnos potatorum* L. f.; *Loganiac.;* C.-O.-Ind.
Kahikatea: *Podocarpus dacrydioides* A. Rich., *Pinus strobus* L.; *Conifer.;* N.-Seel.
Kahikatoa: *Leptospermum scoparium* Forst.; *Myrtac.;* N.-Seel.
Kahimbi: *Erythrophloeum africanum* Harms; *Caesalpiniac.;* B.-K. (Katanga)
Kahn: *Oxystigma stapfianum* Chev.; *Caesalpiniac.;* Lib.
Kahoedjan: *Shorea selanica* Blume; *Dipterocarpac.;* Sunda-Ins.
Kahrn: *Diospyros gabunensis* Gürke; *Ebenac.;* Lib.
Kahta: *Sterculia villosa* Roxb.; *Sterculiac.;* Burma.
Kahu: *Olea cuspidata* Wall.; *Oleac.;* n. O.-Ind. (United-Prov.)
Kahua: *Terminalia arjuna* Bedd.; *Combretac.;* ö. Beng.
Kahud: *Garuga pinnata* Roxb.; *Burserac.;* C.-O.-Ind.
Kahur: *Strychnos potatorum* L. f.; *Loganiac.;* C.-O.-Ind.

Kahur: *Dipterocarpus tuberculatus* Roxb.; *Dipterocarpac.;* n. Burma.
Kai: *Lophira procera* Chev.; *Ochnac.;* Kam.
„ ainsi: *Cullenia excelsa* Wight; *Bombacac.;* sw. O.-Ind.
„ bevu: *Azadirachta indica* A. Juss.; *Meliac.;* w. O.-Ind. (s. Bombay)
„ mara: *Myristica attenuata* Wall.; *Myristicac.;* sw. O.-Ind. (Madras)
Kaiarima: *Ternstroemia sp.; Theac.;* br. Gu.
Kaidaji: *Mimosa asperata* L.; *Mimosac.;* n. Nig.
Kaido kwa: *Pittosporum tobira* Ait.; *Pittosporac.;* Jap.
Kaieriballi: *Licania heteromorpha* Bth., *L. spp.; Rosac.;* Sur.
Kaierieballi: *Couepia surinamensis* Kleinh., *? C. spp.; Rosac.;* Sur.
Kaiger: *Acacia ferruginea* DC.; *Mimosac.;* nw. O.-Ind. (Gujarat)
Kaigüigo: *Entandrophragma septentrionalis* Chev., *A. rufa* Chev.; *Meliac.;* Elf.
Kaiisa: *Deinbollia indeniensis* Chev.; *Sapindac.;* Elf.
Kaikar: *Garuga pinnata* Roxb.; *Burserac.;* n. O.-Ind. (United-Prov.)
Kaikumba: *Ochrocarpus africanus* Oliv.; *Guttifer.;* Lib.
Kaïl: *Khaya senegalensis* A. Juss.; *Meliac.;* Sudan.
„ : *Pinus excelsa* Wall.; *Conifer.;* nw. O.-Ind. (Punjab, Kaschmir)
Kaim: *Adina cordifolia* Hook. f., *Stephegyne parvifolia* Korth.; *Rubiac.;* C.-O.-Ind.
Kaimoni: *Eugenia operculata* Roxb.; *Myrtac.;* Beng.
Kain: *Ulmus wallichiana* Planch.; *Ulmac.;* nw. O.-Ind. (Punjab)
Kainchli: *Acer pictum* Thunb.; *Acerac.;* n. O.-Ind. (Jaunsar)
Kaing-sha: *Acacia sp.; Mimosac.;* Burma.
Kainjal: *Bischofia javanica* Blume; *Euphorbiac.;* Beng.
Kainjli: *Acer pictum* Thunb.; *Acerac.;* n. O.-Ind. (Jaunsar)
Kainju: *Acer caesium* Wall.; *Acerac.;* n. O.-Ind. (Jaunsar)
Kainya: *Diospyros mespiliformis* Hochst.; *Ebenac.;* n. Nig.
Kairiballi: *Licania heteromorpha* Bth., *L. spp.; Rosac.;* Sur., br. Gu.
Kairókan: *Vatica mangachapoi* Blco.; *Dipterocarpac.;* Phil. (Bataan)
Kairu: *Sterculia urens* Roxb.; *Sterculiac.;* w. O.-Ind. (Deccan)
„ : *Pinus excelsa* Wall.; *Conifer.;* nw. O.-Ind. (Kaschmir)
Kairúkan: *Beilschmiedia cairocan* Vid.; *Laurac.;* Phil. (Zambales)
Kait: *Feronia elephantum* Correa; *Rutac.;* nö. O.-Ind. (United-Prov.)
Kaitatanág: *Eugenia claviflora* Roxb.; *Myrtac.;* Phil. (Laguna)
Kaitha: *Albizzia amara* Boivin; *Mimosac.;* C.-O.-Ind.
Kaiwa: *Diospyros mespiliformis* Hochst.; *Ebenac.;* n. Nig.
Kajana: *Hampea trilobata* Standl.; *Bombacac.;* br. Hond.
Kajat: *Pterocarpus angolensis* DC.; *Papilionac.;* s. C.-Af.
Kajatenhout: *Pterocarpus angolensis* DC.; *P. erinaceus* Lam.; *Papilionac.;* S.-Af.
„ , Transvaal-: *Pterocarpus angolensis* DC.; *Papilionac.;* S.-Af.
Kajinoki: *Broussonetia papyrifera* Vent.; *Morac.;* Jap.
Kajoe bapa: *Shorea selanica* Blume; *Dipterocarpac.;* Borneo, Moluk.
„ besi: *Metrosideros vera* Lindl.; *Myrtac.;* Malay, Sunda.
„ „ : *Eusideroxylon zwageri* Teijsm. et Binn.; *Laurac.;* Malay.
„ gaboes: *Alstonia scholaris* R. Br.; *Apocynac.;* Nied.-Ind.
„ garoe: *Gonystylus bancanus* Baill.; *Gonystylac.;* Java.
„ kaleh: *Harpullia cupanioides* Roxb.; *Sapindac.;* Malay.
„ koening: *Stephegyne parvifolia* Korth.; *Rubiac.;* Malay.
„ „ : *Anthocephalus cadamba* Miq.; *Rubiac.;* Malay.
„ laka (h): *Dalbergia parviflora* Roxb.; *Papilionac.;* O.-Ind.
„ limoeta: *Styrax benzoin* Dryand.; *Styracac.;* H.-Ind., Sum., Java.
„ mas: *Stephegyne parvifolia* Korth.; *Rubiac.;* O.-Ind., Cey., Burma.
„ „ : *Sarcocephalus cordatus* Miq.; *Rubiac.;* Malay.
„ oelar: *Strychnos nux-vomica* L.; *Loganiac.;* Java.
„ palaka: *Octomeles sumatrana* Miq.; *Datiscac.;* Malay.
„ pelleth: *Kleinhovia hospita* L.; *Sterculiac.;* Nied.-Ind.
„ poetih: *Melaleuca leucadendron* L.; *Myrtac.;* Nied.-Ind.
„ pokken: *Heritiera littoralis* Dryand.; *Sterculiac.;* Annam.
„ radjah: *Fagraea fragrans* Roxb.; *Loganiac.;* Malay.
„ ramo: *Crypteronia paniculata* Blume; *Crypteroniac.;* Malay.
„ reboeng: *Meliosma nitida* Blume; *Sapindac.;* Sum.
„ sabel: *Oroxylum indicum* Bth.; *Bignoniac.;* Malay.
Kajra: *Strychnos nux-vomica* L.; *Loganiac.;* n. O.-Ind. (U.-Prov.)
Kajramta: *Miliusa velutina* Hook. f. et Thoms.; *Anonac.;* O.-Ind.
Kajrauta: *Miliusa velutina* Hook. f. et Thoms.; *Anonac.;* n. O.-Ind. (Oudh)

Kak: *Diplanthemum viridiflorum* Hutch.; *Tiliac.*; Kam.
Kaka: *Pachystela cinerea* Pierre; *Sapotac.*; Elf.
„ : *Blighia sapida* Koenig; *Sapindac.*; Elf.
„ suroli: *Diospyros silvatica* Roxb.; *Ebenac.*; s. O.-Ind., Cey.
„ tati: *Diospyros ebenum* Koenig; *Ebenac.*; sw. O.-Ind. (Madras)
Kakaboekoe: *Swartzia tomentosa* DC.; *Caesalpiniac.*; Sur.
Kakad: *Garuga pinnata* Roxb.; *Burserac.*; w. O.-Ind. (Bombay)
Kakadikro: *Trichilia prieuriana* A. Juss.; *Meliac.*; Go.
Kakamut: *Acacia suma* Kurz; *Mimosac.*; sö. Sudan, O.-Af. (Uganda)
Kakana: *Picralima ellioti* Stapf; *Apocynac.*; Elf.
Kakandan: *Bruguiera caryophylloides* Blume; *Rhizophorac.*; w. O.-Ind. (Deccan)
Kakani: *Pleiocarpa mutica* Bth.; *Apocynac.*; Go.
Kakanjan boesoe: *Hymenaea courbaril* L.; *Caesalpiniac.*; Sur.
Kakánla: *Anogeissus leiocarpus* Guill. et Perr.; *Combretac.*; Togo.
Kakar: *Garuga pinnata* Roxb.; *Burserac.*; C.-O.-Ind.
„ : *Pistacia integerrima* Stewart; *Anacardiac.*; nw. O.-Ind. (Punjab)
„ singi: *Pistacia integerrima* Stewart; *Anacardiac.*; nw. O.-Ind. (Kumaon)
Kakaralli: *Lecythis ollaria* L.; *Lecythidac.*; fr. Gu.
„ : *Eschweilera laevifolia* Miers, *? Couratari sp.*; *Lecythidac.*; br. Gu.
„ , black: *? Eschweilera corrugata* Miers; *Lecythidac.*; br. Gu.
„ balli: *Eschweilera sp.*, *Lecythis sp.*; *Lecythidac.*; Sur.
„ wadilie-koro: *Eschweilera longipes* Miers, *E. subglandulosa* Miers; *Lecythidac.*;
Kakatara: *Ilex martiniana* D. Don; *Aquifoliac.*; br. Gu.   [Sur.
Kakataraballi: *Ilex sp.*; *Aquifoliac.*; br. Gu.
Kakatoddah: *Toddalia aculeata* Pers.; *Rutac.*; Mal.
Kakauati: *Gliricidia sepium* Steud.; *Papilionac.*; Phil.
Kakauballi: *Paypayrola guianensis* Aubl.; *Violac.*; Sur.
Kakera: *Melia azedarach* L.; *Meliac.*; Nied.-Ind.
Kakeralli: *Eschweilera spp.*, *Lecythis spp.*; *Lecythidac.*; Sur.
„ , black: *? Lecythis sp.*, *? Eschweilera sp.*, *? Jugastrum sp.*; *Lecythidac.*; br. Gu
Kakhtar: *Cedrus deodara* Loudon; *Conifer.*; Afg.
Kaki: *Prosopis oblonga* Bth.; *Mimosac.*; Togo.
„ : *Cassia fistula* L.; *Caesalpiniac.*; w. O.-Ind. (Deccan)
„ pala: *Zizyphus rugosa* Lamk.; *Rhamnac.*; w. O.-Ind. (n. Bombay)
Kakira: *Ximenia americana* L.; *Olacac.*; Tr.
Kakkai: *Cassia fistula* L.; *Caesalpiniac.*; w. O.-Ind. (Deccan)
Kakke: *Cassia fistula* L.; *Caesalpiniac.*; w. O.-Ind. (Deccan)
Kako: *Mimusops clitandrifolia* Chev.; *Sapotac.*; Elf.
„ : *Lophira alata* Banks; *Ochnac.*; Go.
Kakoclita: *Maximiliana regia* Mart.; *Palmac.*; fr. Gu.
Kakoma: *Dialium angolense* Welw.; *Caesalpiniac.*; B.-K. (Lukafu)
Kakoro: *Cordia tetrandra* Aubl.; *Borraginac.*; Sur.
„ : *Lannea acidissima* Chev.; *Anacardiac.*; Elf.
Kakra: *Bruguiera gymnorrhiza* Lamk.; *Rhizophorac.*; s. Beng.
Kakria: *Butea frondosa* Roxb.; *Papilionac.*; n. & w. O.-Ind.
„ : *Lagerstroemia parviflora* Roxb.; *Lythrac.*; w. O.-Ind. (n. Bombay)
Kakrian: *Pistacia integerrima* Stewart; *Anacardiac.*; nw. O.-Ind. (Punjab)
Kakroi: *Pistacia integerrima* Stewart; *Anacardiac.*; n. O.-Ind. (Jaunsar)
Kakru: *Acer pictum* Thunb.; *Acerac.*; nw. O.-Ind. (Punjab)
Kaku: *Lophira alata* Banks, *L. procera* Chev.; *Ochnac.*; Go.
Kakuabiui: *Talauma sambuensis* Pittier; *Magnoliac.*; Pan.
Kakulu: *? Pseudocedrela sp.*; *Meliac.*; B.-K. (Katanga)
Kakunguni: *Cleome hirta* Oliv.; *Capparidac.*; D.-O.-Af. (Tabora)
Kakur siris: *Albizzia odoratissima* Bth.; *Mimosac.*; Beng.
„ chita: *Litsea polyantha* Juss.; *Laurac.*; Beng.
Kakure-mino: *Gilibertia trifida* Makino; *Araliac.*; Jap.
Kakuri: *Litsea polyantha* Juss.; *Laurac.*; n. O.-Ind. (United-Prov.)
Kakurkat: *Hymenodictyon excelsum* Wall.; *Rubiac.*; n. O.-Ind. (United-Prov.)
Kal baghi: *Albizzia lebbek* Bth.; *Mimosac.*; O.-Ind.
Kala: *Lannea acida* A. Rich.; *Anacardiac.*; Togo.
„ chuglam: *Terminalia manni* King; *Combretac.*; And.
„ dhaura: *Anogeissus pendula* Edgew.; *Combretac.*; nw. O.-Ind. (Raiputana)
„ inderjow: *Wrightia tomentosa* Roem. et Schultes; *Apocynac.*; sw. O.-Ind.
„ lakri: *Diospyros oocarpa* Thwaites; *Ebenac.*; And.

Kala lakuch: *Artocarpus gomeziana* Wall.; *Morac.;* And.
„ pendra: *Randia dumetorum* Lamk.; *Rubiac.;* C.-O.-Ind.
„ phulas: *Ougeinia dalbergioides* Bth.; *Papilionac.;* sw. O.-Ind.
„ rukh: *Dalbergia latifolia* Roxb.; *Papilionac.;* w. O.-Ind. (Deccan-Berar)
„ sarasis: *Albizzia odoratissima* Bth.; *Mimosac.;* w. O.-Ind. (Gujarat)
„ siris: *Albizzia odoratissima* Bth., *A. stipulata* Boivin; *Mimosac.;* C.- & n. O.-Ind.
„ takri: *Diospyros marmorata* Parker; *Ebenac.;* And.
„ tendu: *Diospyros peregrina* Gürke; *Ebenac.;* O.-Ind.
Kalabau: *Intsia bijuga* O. Ktze., *I. retusa* O. Ktze., *I. bakeri* Prain; *Caesalpiniac.;* Malay.
Kalabose: *Crescentia cujete* L.; *Bignoniac.;* Curacao.
Kalagoru: *Stereospermum suaveolens* DC.; *Bignoniac.;* O.-Ind. (Madras)
Kalái: *Albizzia procera* Bth.; *Mimosac.;* Phil. (Abra)
Kaláitik: *Litsea sp.; Laurac.;* Phil. (Mindoro)
Kalakat: *Prunus padus* L.; *Rosac.;* nw. O.-Ind. (Punjab)
Kalakdoum: *Diospyros spp.; Ebenac.;* Kam.
Kalakh: *Stephegyne parvifolia* Korth.; *Rubiac.;* O.-Ind. (Punjab)
Kalakudi: *Wrightia tinctoria* R. Brown; *Apocynac.;* sw. O.-Ind. (Bombay)
Kalakura: *Wrightia tinctoria* R. Brown; *Apocynac.;* sw. O.-Ind. (Bombay)
Kalaloe: *Amaranthus oleraceus* L.; *Amarantac.;* Sur.
Kalam: *Duabanga sonneratioides* Ham.; *Sonneratiac.;* Burma.
Kalama: *Anogeissus leiocarpa* Guill. et Perr.; *Combretac.;* Sudan.
Kalamági: *Tamarindus indica* L.; *Caesalpiniac.;* Phil. (n. Luzon)
Kalamansákat: *Terminalia blancoi* Merr.; *Combretac.;* Phil.
Kalamansáli: *Terminalia calamansanai* Rolfe; *Combretac.;* Phil. (Zambales, N.-Ec.)
Kalamansanai: *Neonauclea spp.; Rubiac.;* Phil.
„ : *Mangifera monandra* Merr.; *Anacardiac.;* Phil. (Rizal, Laguna)
„ : *Terminalia calamansanai* Rolfe, *T. edulis* Blco.; *Combretac.;* Phil.
„ : *Bassia ramiflora* Merr.; *Sapotac.;* Phil. (Laguna)
Kalamátau: *Erythrophloeum densiflorum* Merr.; *Caesalpiniac.;* Phil. (Tayabas)
Kalamb: *Adina cordifolia* Hook. f.; *Rubiac.;* sw. O.-Ind. (Bombay)
„ : *Stephegyne parvifolia* Korth.; *Rubiac.;* sw. O.-Ind., C.-O.-Ind.
Kalambac: *Aquilaria agallocha* Roxb.; *Thymelaeac.;* Burma.
Kalamel: *Cordia fragrantissima* Kurz; *Borraginac.;* O.-Ind. (Burma)
Kalamet: *Euphorbia antiquorum* L.; *Euphorbiac.;* ö. As.
„ : *Mansonia gagei* J. R. Drumm.; *Sterculiac.;* s. Burma (Tenasserim)
„ : *Cordia fragrantissima* Kurz; *Borraginac.;* Burma.
Kalamismís: *Aglaia luzoniensis* Merr. et Rolfe; *Meliac.;* Phil. (Dinaga)
Kalampak: *Euphorbia antiquorum* L.; *Euphorbiac.;* Siam (Bangkok)
Kalamungús: *Sideroxylon ferrugineum* Hook. et Arn.; *Sapotac.;* Phil. (Tayabas)
Kalang-giáuan: *Sterculia sp.; Sterculiac.;* Phil. (Ilocos-Sur)
Kalangigíng: *Beilschmiedia cairocan* Vid.; *Laurac.;* Phil. (Ilocos-Sur)
Kalanigen: *Vatica mangachapoi* Blco.; *Dipterocarpac.;* Phil. (Ilocos-Sur)
Kalantas: *Toona calantas* Merr. et Rolfe; *Meliac.;* Phil.
Kalapa: *Cocos nucifera* L.; *Palmac.;* Malay.
Kalapía: *Palaquium sp.; Sapotac.;* Phil. (Zamboanga)
Kalapíni: *Avicennia officinalis* L.; *Verbenac.;* Phil.
„ : *Lumnitzera littorea* Voigt; *Combretac.;* Phil. (Zambales)
Kalaruk: *Dalbergia latifolia* Roxb.; *Papilionac.;* w. O.-Ind. (Konkan)
Kalasan: *Lannea grandis* Engl.; *Anacardiac.;* sw. O.-Ind. (Malabar)
Kalatúche: *Alstonia macrophylla* Wall.; *Apocynac.;* Phil. (Pangasinan)
Kalauahan: *Artocarpus cumingiana* Tréc.; *Morac.;* Phil. (Bontok)
Kalaum: *Eugenia glaucicalyx* Merr.; *Myrtac.;* Phil. (Culion)
Kalaupí: *Terminalia nitens* Presl; *Combretac.;* Phil. (Cagayan)
Kalauri: *Sterculia urens* Roxb.; *Sterculiac.;* w. O.-Ind. (Gujarat)
Kalautít: *Terminalia spp.; Combretac.;* Phil.
Kalay: *Nyssa sessiliflora* Hook. f.; *Cornac.;* Beng.
Kalbow: *Hopea wightiana* Wall.; *Dipterocarpac.;* sw. O.-Ind.
Kalei: *Albizzia odoratissima* Bth.; *Mimosac.;* w. O.-Ind. (Ajmer)
Kalempajan: *Allophylus sundanus* Miq.; *Sapindac.;* Malak.
Kalgari: *Stereospermum suaveolens* DC.; *Bignoniac.;* sw. O.-Ind. (Bombay)
Kalgo: *Bauhinia reticulata* DC.; *Caesalpiniac.;* n. Nig.
Kalhain: *Pinus longifolia* Roxb.; *Conifer.;* n. O.-Ind. (Jaunsar)
Kalhoni: *Hopea wightiana* Wall.; *Dipterocarpac.;* sw. O.-Ind. (s. Bombay)
Kali: *Holarrhena antidysenterica* Wall.; *Apocynac.;* C.-O.-Ind.

Kali dudhi: *Wrightia tinctoria* R. Brown; *Apocynac.;* C.-O.-Ind.
„ saras: *Albizzia odoratissima* Bth.; *Mimosac.;* w. O.-Ind. (Gujarat)
„ siris: *Albizzia odoratissima* Bth., *A. stipulata* Boivin; *Mimosac.;* n. & w. O.-Ind.
Kaliáan: *Shorea palosapis* Merr.; *Dipterocarpac.;* Phil.
Kalikikar: *Acacia arabica* Willd.; *Mimosac.;* nw. O.-Ind. (Sindh)
Kalimantáu: *Uitex turczaninowi* Merr.; *Uerbenac.;* Phil. (Tayabas)
Kalíngag: *Cinnamomum mercadoi* Vid.; *Laurac.;* Phil.
Kaliót: *Hopea spp., Uatica mangachapoi* Blco.; *Dipterocarpac.;* Phil.
„ : *Neonauclea sp.; Rubiac.;* Phil. (Ilocos-Sur)
Kalipápa: *Uitex parviflora* Juss.; *Uerbenac.;* Phil.
„ -ásu: *Uitex celebica* Koord., *U. pentaphylla* Merr.; *Uerbenac.;* Phil.
„ -mádam: *Uitex celebica* Koord.; *Uerbenac.;* Phil. (Agusan, Cotabato)
Kalipapri: *Holoptelea integrifolia* Planch.; *Ulmac.;* n. O.-Ind. (United-Prov.)
Kalipáya: *Palaquium sp.; Sapotac.;* Phil. (Zamboanga)
Kalíwas: *Kayea paniculata* Merr.; *Guttifer.;* Phil. (Bataan)
Kaliyen: *Bridelia retusa* Spreng.; *Euphorbiac.;* C.-O.-Ind.
Kallu: *Diospyros gardneri* Thw.; *Ebenac.;* Cey.
Kalm: *Stephegyne parvifolia* Korth.; *Rubiac.;* C.-O.-Ind.
Kalmi: *Stephegyne parvifolia* Korth.; *Rubiac.;* C.-O.-Ind.
Kalo-sarasis: *Albizzia stipulata* Boivin; *Mimosac.;* w. O.-Ind. (Gujarat)
Kalobkób: *Eugenia sp.; Myrtac.;* Phil. (Luzon, Samar, Leyte)
Kalokatingan: *Pterospermum sp.; Sterculiac.;* Phil.
Kalokatmón: *Dillenia sp.; Dilleniac.;* Phil.
Kalot: *Dillenia pentagyna* Roxb.; *Dilleniac.;* w. O.-Ind.
Kalpayin: *Dipterocarpus indicus* Bedd.; *Dipterocarpac.;* sw. O.-Ind. (Malabar)
Kalsis: *Albizzia lebbek* Bth.; *Mimosac.;* O.-Ind.
Kalubkúb: *Eugenia calubcob* C. B. Rob., *E. benthami* A. Gray; *Myrtac.;* Phil.
Kalukalumpángan: *Sterculia sp.; Sterculiac.;* Phil. (Rizal)
Kalúkoi: *Endospermum pellatum* Merr.; *Euphorbiac.;* Phil. (Bataan)
Kalumágon: *Terminalia edulis* Blco.; *Combretac.;* Phil.
Kalúmai: *Diospyros multiflora* Blco.; *Ebenac.;* Phil.
Kalumángog: *Terminalia edulis* Blco.; *Combretac.;* Phil.
Kalumánog: *.Koordersiodendron pinnatum* Merr.; *Anacardiac.;* Phil. (Biscaya)
„ : *Neonauclea sp.; Rubiac.;* Phil. (Masbate)
„ : *Terminalia edulis* Blco.; *Combretac.;* Phil.
Kalumback: *Aquilaria agallocha* Roxb.; *Thymelaeac.;* Burma.
Kalumpág: *Sterculia sp.; Sterculiac.;* Phil. (Rizal)
Kalumpáng: *Sterculia sp., Sterculia foetida* L.; *Sterculiac.;* Phil.
Kalumpít: *Terminalia spp.; Combretac.;* Phil.
„ -babáe: *Terminalia calamansanai* Rolfe; *Combretac.;* Phil. (Bataan)
Kalungi: *Entandrophragma pierrei* Chev., *E. leplaei* Verm.; *Meliac.;* B.-K.
Kabulungu: *Autranella congolensis* Chev.; *Sapotac.;* B.-K.
Kalunguti: *Terminalia sericea* Burch.; *Combretac.;* n. Rhod.
Kalúnti: *Shorea palosapis* Merr., *S. mindanaensis* Foxw.; *Dipterocarpac.;* Phil. (Zambo.)
Kalupáng: *Sterculia sp.; Sterculiac.;* Phil. (Negrosinsel)
Kalupí: *Terminalia edulis* Blco.; *Combretac.;* Phil. (Cagayan)
Kalupít: *Terminalia calamansanai* Rolfe; *Combretac.;* Phil. (Cagayan)
Kaluríg: *Terminalia edulis* Blco.; *Combretac.;* Phil. (Cagayan)
Kalusi: *Terminalia edulis* Blco.; *Combretac.;* Phil. (Cagayan, Ilocos-Sur)
Kalusít: *Terminalia edulis* Blco.; *Combretac.;* Phil. (Cagayan, Ilocos-Sur)
Kalusúban: *Dipterocarpus vernicifluus* Blco.; *Dipterocarpac.;* Phil. (Ilocos)
Kaluwara: *Diospyros spp.; Ebenac.;* O.-Ind.
Kam: *Parinarium annamense* Hance; *Rosac.;* Annam.
„ Khol: *Juglans regia* L.; *Juglandac.;* s. Eu., w. As.
„ lam: *Phyllanthus emblica* L.; *Euphorbiac.;* Ind.-Ch.
„ pak: *Belotia campbelli* Sprague; *Tiliac.;* br. Hond.
„ tel: *Citrus decumana* L.; *Rutac.;* Ind.-Ch.
Kamachile: *Pithecolobium dulce* Bth.; *Mimosac.;* Phil.
Kamada: *Adina cordifolia* Hook. f.; *Rubiac.;* O.-Ind. (Madras)
Kamadani: *Palicourea guianensis* Aubl.; *Rubiac.;* br. Gu.
Kamadanni: *Glandonia sp.; Malpighiac.;* br. Gu.
Kamagáhai: *Homalium luzoniense* F. Vill.; *Flacourtiac.;* Phil. (Camarines)
Kamagáhi: *Homalium luzoniense* F. Vill.; *Flacourtiac.;* Phil. (Camarines)
Kamagon: *Diospyros spp.; Ebenac.;* Phil.

Kamagong: *Diospyros discolor* Willd., *D. spp.; Ebenac.;* Phil.
Kamahi: *Weinmannia racemosa* L. f.; *Cunoniac.;* N.-Seel.
Kamai: *Weinmannia racemosa* L. f.; *Cunoniac.;* N.-Seel.
Kamaing: *Duabanga sonneratioides* Ham.; *Sonneratiac.;* Burma.
Kamakutshi: *Bombax sp.; Bombacac.;* br. Gu.
Kamálan: *Vitex turczaninowi* Merr.; *Verbenac.;* Phil. (Rizal)
Kamalia: *Coccoloba sp.; Polygonac.;* W.-Ind.
Kamanak-manák: *Cordia subcordata* Lam.; *Borraginac.;* Phil. (Cotabato)
Kamandág: *Artocarpus cumingiana* Tréc.; *Morac.;* Phil. (Cagayan)
Kamandíis: *Garcinia sp.; Guttifer.;* Phil. (Mindoro)
Kamani: *Terminalia catappa* L.; *Combretac.;* Hawai.
  „ : *Calophyllum inophyllum* L.; *Guttifer.;* Hawai.
Kamanji: *Bridelia retusa* Spreng.; *Euphorbiac.;* O.-Ind. (Madras)
Kamanou: *Calophyllum inophyllum* L.; *Guttifer.;* Hawai.
Kamansí: *Artocarpus communis* Forst.; *Morac.;* Phil. (Leyte)
Kamantíis: *Garcinia sp.; Guttifer.;* Phil. (Sorsogon)
Kamap: *Strombosia rotundifolia* King; *Olacac.;* Malay.
Kamárag: *Pterocarpus sp.; Papilionac.;* Phil. (Abra)
Kamaraga: *Averrhoa carambola* L.; *Oxalidac.;* Bras.
Kamarak: *Averrhoa carambola* L.; *Oxalidac.;* O.-Ind.
  „ : *Dracontomelum dao* Merr. et Rolfe; *Anacardiac.;* Phil.
Kamaranga: *Averrhoa carambola* L.; *Oxalidac.;* O.-Ind.
Kamare: *Gmelina arborea* L.; *Verbenac.;* nö. O.-Ind. (Nepal)
Kamaris: *Terminalia edulis* Blco.; *Combretac.;* Phil. (Palawan)
Kamarunga: *Averrhoa carambola* L.; *Oxalidac.;* O.-Ind.
Kamani: *Gonioma kamassi* E. Meyer; *Apocynac.;* Hawaii
Kamassie: *Celastrus ellipticus* Thunb.; *Celastrac.;* S.-Af.
Kamassihout: *Gonioma kamassi* E. Meyer; *Apocynac.;* S.-Af.
Kamassi wood: *Gonioma kamassi* E. Meyer; *Apocynac.;* Kap.
Kamatog: *Erythrophloeum densiflorum* Merr.; *Caesalpiniac.;* Phil. (Tayabas)
Kamatroe: *Schima wallichi* Choisy; *Theac.;* Malay.
Kamatti: *Beilschmiedia roxburghiana* Nees; *Laurac.;* O.-Ind.
Kamaung-pyu: *Lagerstroemia floribunda* Jack; *Lythrac.;* s. Burma (s. Tenass.)
Kamaungni: *Lagerstroemia flos-reginae* Retz.; *Lythrac.;* s. Burma (s. Tenass.)
Kamaungthwe: *Lagerstroemia tomentosa* Presl; *Lythrac.;* s. Burma (s. Tenass.)
Kamayúan: *Strombosia philippinensis* Rolfe; *Olacac.;* Phil.
  „ : *Coccoloba philippinensis* Rolfe; *Polygonac.;* Phil.
Kamba: *Heisteria trillesiana* Pierre; *Olacac.;* Kam., Gab.
  „ : *Lavalleopsis densivenia* Engl.; *Olacac.;* Kam., Gab.
  „ : *Chlorophora excelsa* Bth. et Hook.; *Morac.;* B.-K. (Mayombe)
  „ -Kamba: *Chlorophora excelsa* Bth. et Hook.; *Morac.;* B.-K. (Mayombe)
Kambagam: *Hopea parviflora* Bedd.; *Dipterocarpac.;* sw. O.-Ind. (Malabar)
Kambai Hutan: *Dillenia spp., Wormia spp.; Dilleniac.;* Malay.
Kambál: ? *Pygeum glandulosum* Merr., ? *P. presli* Merr.; *Rosac.;* Phil.
Kambala: *Chlorophora excelsa* Bth. et Hook., *C. regia* Chev.; *Morac.;* Gab.
  „ (H): *Chlorophora excelsa* Bth. et Hook.; *Morac.;* Lib.
Kambala: *Sonneratia apetala* Ham.; *Sonneratiac.;* Burma.
Kambel: *Mallotus philippinensis* Muell. Arg.; *Euphorbiac.;* n. O.-Ind. (Jaunsar)
Kambli: *Stephegyne parvifolia* Korth.; *Rubiac.;* s. O.-Ind. (Travancore)
Kambola-mbola: *Fadogia sp.; Rubiac.;* D.-O.-Af. (Tabora)
Kamdob: *Calophyllum polyanthum* Wall.; *Guttifer.;* n. Beng.
Kamé: *Sonneratia apetala* Ham.; *Sonneratiac.;* Burma.
Kameeldoorn: *Acacia giraffae* Willd.; *Mimosac.;* S.-Af.
Kameldorn: *Acacia giraffae* Willd.; *Mimosac.;* S.-Af.
Kamellappel: *Anona sp.; Anonac.;* Sur.
Kamenee: *Murraya exotica* L.; *Rutac.;* Beng.
Kamerara: *Diospyros ebenum* Koenig; *Ebenac.;* O.-Ind.
Kamferboom, japanische (h): *Cinnamomum camphora* Nees et Eberm.; *Laurac.;* Jap.
Kamian: *Styrax benzoin* Dryand.; *Styracac.;* Siam, Malak., Sum., Java.
Kamíing: *Semecarpus sp.; Anacardiac.;* Phil. (Pampanga, Zambales, Bataan)
Kamila: *Mallotus philippinensis* Muell. Arg.; *Euphorbiac.;* nw. O.-Ind. (Punjab)
Kamíling: *Semecarpus sp.; Anacardiac.;* Phil. (Abra)
Kamíngi: *Santiria nitida* Merr.; *Burserac.;* Phil. (Bataan)
Kamini: *Murraya exotica* L.; *Rutac.;* Beng.

Kamíring: *Semecarpus sp.; Anacardiac.;* Phil. (Cagayan, Union)
Kamkampílan: *Oroxylum indicum* Vent.; *Bignoniac.;* Phil. (Ilocos-Sur, Isa.)
Kamlai: *Lannea grandis* Engl.; *Anacardiac.;* nw. O.-Ind. (Punjab)
Kamma-regu: *Artocarpus lakoocha* Roxb.; *Morac.;* sw. O.-Ind. (Madras)
Kamna: *Anogeissus leiocarpa* Guill. et Perr.; *Combretac.;* N.-Af.
Kamo: *Rhizophora mucronata* Lamk.; *Rhizophorac.;* nw. O.-Ind. (Sindh)
Kamoening: *Murraya exotica* L., *M. sumatrana* Roxb.; *Rutac.;* Malay, Sum.
Kampferholz, Borneo-: *Dryobalanops aromatica* Gaertn. f.; *Dipterocarpac.;* Malay.
Kampferhout (h): *Dryobalanops aromatica* Gaertn. f.; *Dipterocarpac.;* Malay, Sunda.
Kampílan: *Oroxylum indicum* Vent.; *Bignoniac.;* Phil. (Ilocos-Sur)
Kampong: *Nephelium lappaceum* L.; *Sapindac.;* Java, Sum.
Kamra: *Hardwickia binata* Roxb.; *Caesalpiniac.;* w. O.-Ind. (Deccan)
Kamsaw: *Bassia longifolia* L.; *Sapotac.;* Burma.
Kamulitingan: *Parinarium sp.; Rosac.;* Phil. (Pampanga)
Kamúng: *Buchanania sp.; Anacardiac.;* Phil. (Cagayan, Zambales)
Kamunie: *Murraya sumatrana* Roxb.; *Rutac.;* Cel. (Menado)
Kamúning: *Murraya paniculata* Jack, *M. exotica* L.; *Rutac.;* Malay, Phil.
Kamunog: *? Pygeum glandulosum* Merr., *? P. presli* Merr.; *Rosac.;* Phil. (Camar.)
Kamúyau: *Dipterocarpus spp.; Dipterocarpac.;* Phil.
  „   : *Homalium bracteatum* Bth.; *Flacourtiac.;* Phil. (Abra)
Kan: *Petersia viridiflora* Chev.; *Lecythidac.;* Elf.
  „ Dol: *Careya sphaerica* Roxb., *C. arborea* Roxb.; *Lecythidac.;* Ind.-Ch.
  „ lúong: *Adina cordifolia* Hook. f.; *Rubiac.;* Ind.-Ch.
  „ yaung: *Dipterocarpus pilosus* Roxb.; *Dipterocarpac.;* s. Burma (Bassein).
Kana-goraka-gass: *Garcinia morella* Desrouss.; *Guttifer.;* s. Cey.
Kanaga: *Pongamia glabra* Vent.; *Papilionac.;* w. O.-Ind. (Deccan)
Kanagola: *Dillenia pentagyna* Roxb.; *Dilleniac.;* w. O.-Ind. (Deccan)
Kanak-champa: *Pterospermum acerifolium* Willd.; *Sterculiac.;* w. O.-Ind., Beng.
Kanakpa: *Evodia fraxinifolia* Hook. f.; *Rutac.;* n. Beng.
Kanakugi: *Lindera erythrocarpa* Makino; *Laurac.;* Jap.
Kanalla: *Bauhinia retusa* Ham.; *Caesalpiniac.;* nw. O.-Ind. (Kumaon)
Kanálum: *Diospyros ahernii* Merr., *D. nitida* Merr.; *Ebenac.;* Phil.
Kaname: *Poupartia fordi* Hemsl.; *Anacardiac.;* Jap.
Kanamemochi: *Photinia glabra* Thunb.; *Rosac.;* Jap.
Kananga: *Canangium odoratum* Baill.; *Anonac.;* Malay.
Kanapa: *Barringtonia acutangula* Gaertn.; *Lecythidac.;* O.-Ind.
Kanapu: *Adina cordifolia* Hook. f.; *Rubiac.;* O.-Ind. (Madras)
Kanar: *Acer caesium* Wall.; *Acerac.;* nw. O.-Ind. (Kaschmir)
Kanarém: *Bischofia javanica* Blume; *Euphorbiac.;* Phil. (Union-Prov.)
Kanarie: *Canarium commune* L.; *Burserac.;* Malay.
Kanazo: *Heritiera fomes* Buch.-Ham.; *Sterculiac.;* O.-Ind., Burma.
Kanchana: *Bauhinia purpurea* L.; *Caesalpiniac.;* O.-Ind. (? s. Beng.)
Kanchanam: *Bauhinia purpurea* L.; *Caesalpiniac.;* O.-Ind. (Madras)
Kanchia: *Diospyros silvatica* Roxb.; *Ebenac.;* s. O.-Ind., Cey.
Kanchivala: *Bauhinia purpurea* L.; *Caesalpiniac.;* w. O.-Ind. (Deccan)
Kanchurai: *Strychnos nux-vomica* L.; *Loganiac.;* sw. O.-Ind. (Madras)
Kandaba-esseri: *Cola chlamydantha* K. Schum.; *Sterculiac.;* Go.
Kandal· *Sonneratia apetala* Ham.; *Sonneratiac.;* w. O.-Ind. (Bombay)
  „   : *Rhizophora mucronata* Lamk.; *Rhizophorac.;* w. O.-Ind. (Deccan)
Kandar: *Aesculus indica* Colebr.; *Hippocastanac.;* n. O.-Ind. (Jaunsar)
Kandari: *Terminalia macroptera* Guill. et Perr.; *Combretac.;* n. Nig.
Kandeb: *Calophyllum polyanthum* Wall.; *Guttifer.;* Beng.
Kandelaka: *Melia azedarach* L.; *Meliac.;* Mad.
Kandep: *Calophyllum polyanthemum* Wall.; *Guttifer.;* Beng.
Kandeys: *Garcinia nigro-lineata* Planch.; *Guttifer.;* Malak.
Kandiawa: *Bauhinia retusa* Ham.; *Caesalpiniac.;* n. O.-Ind. (Garhwal)
Kandiis: *Garcinia sp.; Guttifer.;* Phil. (Zamboanga, Palawan)
Kandika: *Pentadesma leptonema* Pierre; *Guttifer.;* Kam.
Kandiyar: *Bursera serrata* Colebr.; *Burserac.;* Beng., Assam, n. Burma.
Kandla: *Bauhinia retusa* Ham.; *Caesalpiniac.;* nw. O.-Ind. (Kumaon)
Kandol: *Sterculia urens* Roxb.; *Sterculiac.;* sw. O.-Ind. (Bombay)
  „   : *Careya sphaerica* Roxb.; *Lecythidac.;* Kamb.
Kandol-kandól: *Sterculia sp.; Sterculiac.;* Phil. (Mindoro)
Kandong-ísol: *Euphoria didyma* Blco.; *Sapindac.;* Phil. (Masbate)

Kandra-hoedoe: *Jacaranda filicifolia* D. Don; *Bignoniac.;* Sur.
Kandur: *Aesculus indica* Colebr.; *Hippocastanac.;* n. O.-Ind. (Jaunsar)
Kandwer: *Garuga pinnata* Roxb.; *Burserac.;* O.-Ind. (? Burma)
Kane: *Anogeissus schimperi* Hochst.; *Combretac.;* Go.
Kaneelappel: *Anona sp.; Anonac.;* Sur.
Kaneelhart: *Acrodiclidium canella* Mez, *A. chrysophyllum* Meissn.; *Laurac.;* Sur.
Kaneelpisi: *Nectandra sp., Ocotea sp.; Laurac.;* Sur.
   „ , zware: *Nectandra sp., Ocotea sp.; Laurac.;* Sur.
Kaneerjoe: *Acrodiclidium canella* Mez, *A. chrysophyllum* Meissn.; *Laurac.;* Sur.
Kanelo: *Drimys winteri* Forst.; *Magnoliac.,* Arg.
Kancre: *Vitex littoralis* A. Cunn.; *Verbenac.;* N.-Seel.
Kanerie hoedoe: *Acrodiclidium canella* Mez, *A. chrysophyllum* Meissn.; *Laurac.;* Sur.
Kanfu: *Fagara xanthoxyloides* Lam.; *Rutac.;* Go.
Kang: *Albizzia lebbekoides* Bth.; *Mimosac.;* Ind.-Ch.
   „ : *Sterculia lanceolata* Ham.; *Sterculiac.;* Ind.-Ch.
   „ dor: *Swintonia pierrei* Hance; *Anacardiac.;* Ind.-Ch.
Kangali: *Bauhinia retusa* Ham.; *Caesalpiniac.;* C.-O.-Ind. (Gondwana)
Kangar: *Pistacia integerrima* Stewart; *Anacardiac.;* nw. O.-Ind. (Punjab)
Kangi: *Wendlandia exserta* DC.; *Rubiac.;* tr. Himal.
Kangiliam: *Canarium strictum* Roxb.; *Burserac.;* sw. O.-Ind. (Madras)
Kangji: *Ficus bengalensis* L.; *Morac.;* nö. O.-Ind.
Kangkángan: *Parinarium sp.; Rosac.;* Phil. (Davao)
Kangok meas: *Caesalpinia pulcherrima* Sw.; *Caesalpiniac.;* Ind.-Ch.
Kangra: *Grewia laevigata* Vahl; *Tiliac.;* nw. O.-Ind. (Punjab)
Kanier: *Cassia fistula* L.; *Caesalpiniac.;* tr. As.
Kanifing: *Xylopia aethiopica* A. Rich.; *Anonac.;* Gab.
Kaníla: *Cinnamomum mercadoi* Vid.; *Laurac.;* Phil.
Kaniongan: *Hopea acuminata* Merr., *H. spp.; Dipterocarpac.;* holl. O.-Borneo.
Kaníwi: *Aglaia diffusa* Merr.; *Meliac.;* Phil. (Rizal)
Kaniwi-putí: *Aglaia harmsiana* Perk.; *Meliac.;* Phil. (Laguna)
Kanjar: *Acer pictum* Thunb.; *Acerac.;* nw. O.-Ind. (Punjab)
Kanjeram: *Strychnos nux-vomica* L.; *Loganiac.;* sw. O.-Ind. (Malabar)
Kanji: *Pongamia glabra* Vent.; *Papilionac.;* C.-O.-Ind.
Kanju: *Holoptelea integrifolia* Planch.; *Ulmac.;* n. O.-Ind. (United-Prov.)
Kankala: *Anogeissus leiocarpus* Guill. et Perr.; *Combretac.;* Togo.
Kankan: *Ceiba pentandra* Gaertn.; *Bombacac.;* holl. Gu.
Kankantrie: *Ceiba pentandra* Gaertn.; *Bombacac.;* Sur.
Kankono: *Diospyros mespiliformis* Hochst.; *Ebenac.;* B.-K.
Kankor: *Zizyphus xylopyrus* Willd.; *Rhamnac.;* w. Beng. (Chota-Nagpur)
Kankra: *Bruguiera gymnorrhiza* Lamk.; *Rhizophorac.;* s. Beng.
Kankrei: *Butea frondosa* Roxb.; *Papilionac.;* n. O.-Ind. (United-Prov.)
Kanloo: *Bauhinia retusa* Ham.; *Caesalpiniac.;* n. O.-Ind.
Kanman: *Garuga pinnata* Roxb.; *Burserac.;* nw. O.-Ind. (Kumaon)
Kannä: *Anogeissus leiocarpus* Guill. et Perr.; *Combretac.;* Togo.
Kanoewaballi: *Nectandra sp., Ocotea sp.; Laurac.;* Sur.
Kanómoi: *Diospyros multiflora* Blco.; *Ebenac.;* Phil. (n. Luzon, Camar., Zamboanga)
Kanooka: *Tristania laurina* R. Br.; *Myrtac.;* ö. Au. (Vict.)
Kanor: *Aesculus indica* Colebr.; *Hippocastanac.;* n. O.-Ind.
Kansak: *Acacia stefanini* Chiov.; *Mimosac.;* O.-Af. (it. Somali)
Kansel: *Hydnocarpus wightiana* Blume; *Flacourtiac.;* sw. O.-Ind. (Bombay)
Kanshin: *Acer caesium* Wall.; *Acerac.;* n. O.-Ind. (Tibet)
Kansilai: *Cratoxylon sp.; Guttifer.;* Phil. (Mindoro, Negrosinsel)
Kansúlud: *Aglaia multifoliola* Merr., *A. clarki* Merr.; *Meliac.;* Phil.
Kansúyod: *Aglaia clarki* Merr.; *Meliac.;* Phil.
   „ : *Aglaia clarki* Merr.; *Meliac.;* Phil. (Tablas)
Kanta bohul: *Zizyphus xylopyrus* Willd.; *Rhamnac.;* O.-Ind. (Uriya)
Kanta-gotti: *Zizyphus xylopyrus* Willd.; *Rhamnac.;* sw. O.-Ind. (Bombay)
   „ harina: *Zanthoxylum budrunga* Wall.; *Rutac.;* ö. Beng. (Chittagong)
   „ kauchi: *Bridelia retusa* Spreng.; *Euphorbiac.;* sw. O.-Ind. (Bombay)
   „ kumla: *Sideroxylon tomentosum* Roxb.; *Sapotac.;* sw. O.-Ind. (Bombay)
Kantabahul: *Sideroxylon tomentosum* Roxb.; *Sapotac.;* w. Beng. (Orissa)
Kantabora: *Sideroxylon tomentosum* Roxb.; *Sapotac.;* w. Beng. (Orissa)
Kanteng: *Buchanania sp.; Anacardiac.;* Phil. (Abra)
Kanthekera: *Carallia lucida* Roxb.; *Rhizophorac.;* Assam.

Kanti: *Acacia ferruginea* DC.; *Mimosac.;* O.-Ind.
Kantingan: *Pterospermum sp.; Sterculiac.;* Phil. (Mindoro)
Kantíngen: *Albizzia marginata* Merr.; *Mimosac.;* Phil. (Ilocos-Sur)
„ : *Koordersiodendron pinnatum* Merr.; *Anacardiac.;* Phil. (Union-Prov.)
„ : *Toona calantas* Merr. et Rolfe; *Meliac.;* Phil. (Ilocos-N., Zambales)
Kanton: *Zanthoxylum parvifolium* Chev.; *Rutac.;* Elf.
Kantonka-sema: *Bombax malabaricum* DC.; *Bombacac.;* O.-Ind.
Kantrock: *Clausena excavata* Burm.; *Rutac.;* Ind.-Ch.
Kanu: *Adina cordifolia* Hook. f., *Stephegyne parvifolia* Korth.; *Rubiac.;* w. O.-Ind.
„ : *Evodia fraxinifolia* Hook. f.; *Rutac.;* nö. O.-Ind.
Kanublíng: *Artocarpus cumingiana* Tréc.; *Morac.;* Phil.
Kanudun: *Aesculus indica* Colebr.; *Hippocastanac.;* nw. Himal.
Kanuga: *Pongamia glabra* Vent.; *Papilionac.;* sw. O.-Ind. (Madras)
Kanujerla: *Albizzia stipulata* Boivin; *Mimosac.;* n. O.-Ind. (United-Prov.)
Kanuka: *Leptospermum ericoides* A. Rich.; *Myrtac.;* N.-Seel.
Kanumai: *Diospyros spp.; Ebenac.;* Phil.
Kanumi: *Diospyros multiflora* Blco.; *Ebenac.;* Phil.
Kanumung: *Alstonia scholaris* R. Brown; *Apocynac.;* O.-Ind. (? w. Beng.)
Kanun palle: *Mimusops hexandra* Roxb.; *Sapotac.;* O.-Ind. (Madras)
Kanvel: *Litsea zeylanica* C. & Fr. Nees; *Laurac.;* sw. O.-Ind., Cey.
Kanvini: *Pleiocarpa mutica* Bth.; *Apocynac.;* Go.
Kanwala: *Litsea umbrosa* Nees; *Laurac.;* n. O.-Ind. (United-Prov.)
Kanwi: *Pleiocarpa mutica* Bth.; *Apocynac.;* Go.
Kanwo: *Afzelia africana* Smith; *Caesalpiniac.;* Nig.
Kanyâma: *Kigellaria serrata* Warb.; *Flacourtiac.;* D.-O.-Af.
Kanyin: *Dipterocarpus spp.; Dipterocarpac.;* Burma.
Kayinbyan: *Dipterocarpus griffithi* Miq.; *Dipterocarpac.;* s. Burma (s. Tenass.)
Kanyinbyu: *Dipterocarpus alatus* Roxb.; *Dipterocarpac.;* Burma.
Kanyingok: *Dipterocarpus obtusifolius* Teijsm.; *Dipterocarpac.;* s. Burma.
Kanyin-ni: *Dipterocarpus turbinatus* Gaertn. f.; *Dipterocarpac.;* Burma.
„ „ : *Dipterocarpus costatus* Gaertn. f.; *Dipterocarpac.;* Burma.
Kanzal: *Acer caesium* Wall.; *Acerac.;* nw. O.-Ind. (Punjab)
Kao: *Olea cuspidata* Wall.; *Oleac.;* nw. O.-Ind. (Punjab)
„ : *Randia longiflora* Lamk.; *Rubiac.;* Ind.-Ch.
Káong: *Arenga pinnata* Merr.; *Palmac.;* Phil.
Kaori: *Agathis lanceolata* Warb.; *Conifer.;* N.-Kal.
Kaorika: *Terminalia catappa* L.; *Combretac.;* Südsee (Rarotonga)
Kaoris: *Agathis ovata* Warb.; *Conifer.;* N.-Kal.
Kaowui: *Mussaenda afzelii* G. Don; *Rubiac.;* W.-Af. (S.-L.)
Kapa: *Clitandra orientalis* K. Schum.; *Apocynac.;* br. O.-Af. (Uganda/Mabira)
Kapanga: *Amblygonocarpus obtusangulus* Harms; *Mimosac.;* n. Rhod.
Kapansulo: *Piptadenia kerstingi* Harms; *Mimosac.;* Togo.
Kapar naga: *Calophyllum inophyllum* L.; *Guttifer.;* Borneo.
Kápas-sanglái: *Ceiba pentandra* Gaertn.; *Bombacac.;* Phil. (Bontok, Pangasinan)
Kapasi: *Populus ciliata* Wall.; *Salicac.;* n. O.-Ind. (United-Prov.)
Kapatupatu: *Pachylobus pubescens* Verm.; *Burserac.;* B.-K.
? Kapaussulo: *Piptadenia kerstingi* Harms; *Mimosac.;* Togo.
Kapaussuto: *Piptadenia kerstingi* Harms; *Mimosac.;* W.-Af.
Kapayang: *Pangium edule* Reinw.; *Flacourtiac.;* Borneo.
Kapetanzovu: *Albizzia sassa* McBride; *Mimosac.;* B.-K.
Kapgángan: *Parinarium sp.; Rosac.;* Phil. (Davao)
Kapila: *Mallotus philippinensis* Muell. Arg.; *Euphorbiac.;* sw. O.-Ind.
Kapinango: *Dysoxylum densiflorum* Miq.; *Meliac.;* Java.
Kapiníg: *Eugenia xanthophylla* C. B. Rob.; *Myrtac.;* Phil. (Sorsogon)
Kapiteinhout: *Perebea sp., Piratinera spp.; Morac.;* Sur.
Kapitin hoedoe: *Helicostylis poeppigiana* Tréc.; *Morac.;* Sur.
Kapittka: *Feronia elephantum* Correa; *Rutac.;* O.-Ind., Cey.
Kapli: *Mallotus philippinensis* Muell. Arg.; *Euphorbiac.;* sw. O.-Ind. (Madras)
Kapoekoeroe japoekoetjare: *Swartzia spp.; Caesalpiniac.;* Sur.
Kapoe-ratja: *Calophyllum inophyllum* L.; *Guttifer.;* Malay.
Kapok: *Ceiba pentandra* Gaertn.; *Bombacac.;* Gab., Malay, Phil.
Kapokboom (h): *Ceiba pentandra* Gaertn.; *Bombacac.;* Tr.
Kapokier: *Ceiba pentandra* Gaertn.; *Bombacac.;* Elf.
„ à fleurs rouges: *Bombax buonopozense* Beauv.; *Bombacac.;* Dah.

Kapoktree: *Ceiba pentandra* Gaertn.; *Bombacac.;* Sur.
Kapoor rantjang: *Calophyllum inophyllum* L.; *Guttifer.;* Malay.
Kapor: *Dryobalanops aromatica* Gaertn. f.; *Dipterocarpac.;* Sunda.
Kapor bennar: *Dryobalanops kayanensis* Becc.; *Dipterocarpac.;* br. Borneo.
„ gunong: *Dryobalanops beccari* Dyer; *Dipterocarpac.;* Malay.
„ paya: *Dryobalanops beccari* Dyer; *Dipterocarpac.;* Malay, br. Borneo.
Kappeweri letri: *Piratinera spp.; Morac.;* Sur.
Kappla: *? Cinnamomum sp.; Laurac.;* br. Borneo (Sarawak)
Kapsin: *Tetrameles nudiflora* R. Br.; *Datiscac.;* sw. O.-Ind. (Bombay)
Kapúlau: *Petersianthus quadriolatus* Merr.; *Lecythidac.;* Phil. (Cebu)
„ : *Terminalia quadrialata* Merr.; *Combretac.;* Phil. (Cebu)
Kapur: *Dryobalanops aromatica* Gaertn. f.; *Dipterocarpac.;* Malay, Borneo.
„ barus: *Dryobalanops aromatica* Gaertn. f.; *Dipterocarpac.;* Malay.
„ tohori: *Cinnamomum camphora* Nees et Eberm.; *Laurac.;* Malay.
Kaputu: *Brachystegia hocki* de Wild.; *Caesalpiniac.;* B.-K.
Kar: *Garcinia manni* Oliv.; *Guttifer.;* Lib.
„ : *Strychnos nux-vomica* L.; *Loganiac.;* sw. O.-Ind. (Bombay)
„ ayani: *Cullenia excelsa* Wight; *Bombacac.;* sw. O.-Ind. (Malabar)
Kara: *Eugenia claviflora* Roxb.; *Myrtac.;* Phil. (Bataan)
Karaál: *Albizzia procera* Bth.; *Mimosac.;* Phil. (Cagayan, Pangasinan)
Karaba: *Carapa spp.; Meliac.;* Gu.
Karabababalli: *Guarea gomma* Pulle; *Meliac.;* Sur.
Karababalli: *Guarea sp.; Meliac.;* br. Gu.
„ ewéda koro-aboe: *Guarea gomma* Pulle; *Meliac.;* Sur.
Karachi: *Hardwickia binata* Roxb.; *Caesalpiniac.;* w. O.-Ind. (Deccan)
Karagil: *Dysoxylum binectariferum* Hook. f.; *Meliac.;* sw. O.-Ind. (Madras)
Karai: *Miliusa velutina* Hook. f. et Thoms.; *Anonac.;* C.-O.-Ind.
„ : *Sterculia urens* Roxb.; *Sterculiac.;* C.-O.-Ind. (Berar).
Karail: *Dendrocalamus strictus* Nees; *Gramin.;* Beng.
„ : *Albizzia procera* Bth.; *Mimosac.;* Phil. (Zambales)
Karaka: *Terminalia chebula* Retz.; *Combretac.;* O.-Ind. (Madras)
Karakil: *Dysoxylum binectariferum* Hook. f.; *Meliac.;* sw. O.-Ind. (Madras)
Karakong: *Balanocarpus utilis* Bedd.; *Dipterocarpac.;* s. O.-Ind. (Madras)
Karalli: *Bauhinia purpurea* L.; *Caesalpiniac.;* nw. O.-Ind. (Punjab)
„ : *Carallia lucida* Roxb.; *Rhizophorac.;* C.-O.-Ind.
„ , black: *Eschweilera longipes* Miers, *E. subglandulosa* Miers; *Lecythidac.;* Sur.
Karam: *Anthocephalus cadamba* Miq.; *Rubiac.;* Beng.
„ : *Adina cordifolia* Hook. f.; *Rubiac.;* n. & C.-O.-Ind.
Karamatsu: *Larix leptolepis* Gord.; *Conifer.;* Jap.
Karambel: *Dillenia pentagyna* Roxb.; *Dilleniac.;* sw. O.-Ind. (Bombay)
Karambola: *Carapa obovata* Blume; *Meliac.;* s. Beng.
Karamu: *Coprosma linariifolia* Hook. f.; *Rubiac.;* N.-Seel.
Karamunga: *Averrhoa carambola* L.; *Oxalidac.;* Beng., Hind.
Karamútan: *Casuarina equisetifolia* Forst.; *Casuarinac.;* Phil. (Moro)
Karan-galli: *Acacia catechu* Willd.; *Mimosac.;* w. O.-Ind. (Deccan)
Karang: *Canarium strictum* Roxb.; *Burserac.;* sw. O.-Ind. (Madras)
„ : *Pongamia glabra* Vent.; *Papilionac.;* sw. O.-Ind. (Bombay)
Karanga: *Pongamia glabra* Vent.; *Papilionac.;* w. Beng.
Karangi: *Pongamia glabra* Vent.; *Papilionac.;* w. O.-Ind. (Gujarat)
Karani: *Cullenia excelsa* White; *Bombacac.;* O.-Ind.
Karanj: *Pongamia glabra* Vent.; *Papilionac.;* C.-O.-Ind.
Karanja: *Pongamia glabra* Vent.; *Papilionac.;* Beng.
Karanjalam: *Holoptelea integrifolia* Planch.; *Ulmac.;* C.-O.-Ind.
Karanji: *Holoptelea integrifolia* Planch.; *Ulmac.;* C.-O.-Ind.
„ : *Pongamia glabra* Vent.; *Papilionac.;* C.-O.-Ind.
Karantáng: *Buchanania sp.; Anacardiac.;* Phil. (Palawan)
Karánung: *Diospyros nitida* Merr.; *Ebenac.;* Phil. (Mindoro)
Karapa: *Carapa spp.; Meliac.;* Sur.
Karapu: *Canarium strictum* Roxb.; *Burserac.;* sw. O.-Ind. (Madras)
Karar: *Bauhinia purpurea* L.; *Caesalpiniac.;* nw. O.-Ind. (Punjab)
Karaseri: *Oxandra lanceolata* Baill.; *Anonac.;* br. Gu.
Karasuzansho: *Zanthoxylum ailanthoides* S. et Z.; *Rutac.;* Jap.
Karatákat: *Parinarium sp.; Rosac.;* Phil. (Ilocos, Cagayan)
Karatta: *Corynocarpus laevigata* Forst.; *Corynocarpac.;* N.-Seel.

Karawara: *Cordia sp.; Borraginac.;* Curacao.
Karawé: *Cinnamomum inunctum* Meissn.; *Laurac.;* Burma.
Karay: *Randia dumetorum* Lam.; *Rubiac.;* ö. N.-Af., tr. As.
Karayan: *Myristica attenuata* Wall.; *Myristicac.;* sw. O.-Ind.
Kardahi: *Anogeissus pendula* Edgew.; *Combretac.;* nw. O.-Ind. (Gwalior)
Kardaji: *Mimosa asperata* L.; *Mimosac.;* n. Nig.
Karedha: *Terminalia chebula* Reţ.; *Combretac.;* O.-Ind. (Uriya)
Kareeboom: *Rhus viminalis* Ait.; *Anacardiac.;* S.-Af.
Karei: *Randia dumetorum* Lamk.; *Rubiac.;* O.-Ind. (Madras)
Karemara: *Diospyros ebenum* Koenig; *Ebenac.;* s. O.-Ind. (Madras)
Karenga: *Gardenia lucida* Roxb.; *Rubiac.;* O.-Ind. (Madras)
Karen wood· *Heterophragma adenophyllum* Seem.; *Sterculiac.;* O.-Ind.
Karfu bili: *Ficus sp.; Morac.;* Elf.
Kargo: *Bauhinia reticulata* DC.; *Caesalpiniac.;* n. Nig., Kam.
Karha: *Albizzia procera* Bth.; *Mimosac.;* n. O.-Ind. (United-Prov.)
Karhai: *Albizzia procera* Bth.; *Mimosac.;* n. O.-Ind. (United-Prov.)
Karhar: *Gardenia turgida* Roxb.; *Rubiac.;* C.-O.-Ind.
    „   : *Albizzia procera* Bth.; *Mimosac.;* O.-Ind.
Karhotta: *Dillenia pentagyna* Roxb.; *Dilleniac.;* w. Beng.
Kari: *Miliusa velutina* Hook. f. et Thoms.; *Anonac.;* C.-O.-Ind.
    „   : *Limonia acidissima* L.; *Rutac.;* C.-O.-Ind. (Ajmer)
    „   : *Litsea polyantha* Juss.; *Laurac.;* n. & C.-O.-Ind.
    „ gadda: *Randia dumetorum* Lamk.; *Rubiac.;* w. O.-Ind. (Deccan)
    „ matti: *Terminalia tomentosa* Wight et Arn.; *Combretac.;* w. O.-Ind. (Deccan)
    „ nyaral: *Eugenia gardneri* Thw.; *Myrtac.;* sw. O.-Ind. (Malabar)
Karia: *Sterculia urens* Roxb.; *Sterculiac.;* sw. O.-Ind. (Bombay)
    „ seja: *Lagerstroemia parviflora* Roxb.; *Lythrac.;* C.-O.-Ind.
Kariar: *Cassia fistula* L.; *Caesalpiniac.;* nw. O.-Ind. (Punjab)
Karibevan: *Melia composita* Willd.; *Meliac.;* sw. O.-Ind. (Bombay)
Karifurog: *Lumniţera coccinea* Wight et Arn.; *Combretac.;* Phil. (Cagayan)
Kárig: *Duabanga moluccana* Blume; *Sonneratiac.;* Phil. (Cagayan)
    „   : *Mangifera monandra* Merr.; *Anacardiac.;* Phil. (Bataan)
    „   : *Vatica mangachapoi* Blco.; *Dipterocarpac.;* Phil. (Bataan)
Karije gatari: *Burkea africana* Hook. f.; *Caesalpiniac.;* n. Nig.
Karikawdi: *Grewia laevigata* Vahl; *Tiliac.;* w. O.-Ind. (Bombay)
Karil: *Capparis aphylla* Roth; *Capparidac.;* nw. O.-Ind. (Punjab)
Karima: *Diospyros mespiliformis* Hochst.; *Ebenac.;* Sudan.
Karimanni: *Symphonia globulifera* L.; *Guttifer.;* br. Gu.
Karindi: *Schrebera swietenioides* Roxb.; *Oleac.;* w. Beng. (Chota Nagpur)
Karingi: *Wrightia tomentosa* Roem. et Schultes; *Apocynac.;* nö. O.-Ind. (Nepal)
Karingkura: *Stereospermum chelonioides* DC.; *Bignoniac.;* sw. O.-Ind. (Malabar)
Kariókan: *Vatica mangachapoi* Blco.; *Dipterocarpac.;* Phil. (Bataan)
Kariraro-koeraharoe: *Calophyllum sp.; Guttifer.;* Sur.
Kariskís: *Albizzia lebbekoides* Bth.; *Mimosac.;* Phil.
Karité: *Butyrospermum parki* Kotschy; *Sapotac.;* Dah., fr. Guin.
Karíwas: *Kayea paniculata* Merr.; *Guttifer.;* Phil. (Bataan)
Kariya: *Combretum abbreviatum* Engl.; *Combretac.;* n. Nig.
Karjara: *Bridelia retusa* Spreng.; *Euphorbiac.;* nw. O.-Ind. (Jeypore)
Karkapilly: *Pithecolobium dulce* Bth.; *Mimosac.;* w. O.-Ind.
Karkar: *Pistacia integerrima* Stewart; *Anacardiac.;* nw. O.-Ind. (Kaschmir)
Karkata: *Zizyphus xylopyrus* Willd.; *Rhamnac.;* w. Beng. (Chota-Nagpur)
Karkawa: *Litsea polyantha* Juss.; *Laurac.;* nw. O.-Ind. (Agra, Dehra Dun)
Karkaya: *Terminalia tomentosa* Wight et Arn.; *Combretac.;* O.-Ind. (Haiderabad)
Karke-anum: *Bridelia retusa* Spreng.; *Euphorbiac.;* w. Beng. (Chota Nagpur)
Karkho: *Bridelia retusa* Spreng.; *Euphorbiac.;* C.-O.-Ind.
Karki: *Acacia campylacantha* Hochst.; *Mimosac.;* n. Nig.
Karku: *Litsea polyantha* Juss.; *Laurac.;* nw. O.-Ind. (Agra-Dehra Dun)
Karlu: *Sterculia urens* Roxb.; *Sterculiac.;* C.-O.-Ind.
Karmai: *Bauhinia malabarica* Roxb.; *Caesalpiniac.;* w. Beng.
Karmal: *Dillenia pentagyna* Roxb.; *Dilleniac.;* sw. O.-Ind. (Bombay)
Karmari: *Gardenia gummifera* L.; *Rubiac.;* w. O.-Ind. (n. Bombay)
Karmaru: *Albizzia odoratissima* Bth.; *Mimosac.;* O.-Ind.
Karmi: *Stephegyne parvifolia* Korth.; *Rubiac.;* O.-Ind. (Madras)
Karmkara: *Pterospermum acerifolium* Willd.; *Sterculiac.;* w. O.-Ind. (Bombay)

Karnikar: *Pterospermum acerifolium* Willd.; *Sterculiac.*; w. O.-Ind. (Bombay)
Karo: *Acacia campylacantha* Hochst.; *Mimosac.*; n. Nig.
Karobkób: *Eugenia sp.*; *Myrtac.*; Phil. (s. Luzon, Samar, Leyte)
Karoekoro: *Swartӡia tomentosa* DC.; *Caesalpiniac.*; Sur.
Karogkóg: *Eugenia calubcob* C. B. Rob., *E. sp.*; *Myrtac.*; Phil.
  „ : *Koordersiodendron pinnatum* Merr.; *Anacardiac.*; Phil.
Karoi: *Albizzia procera* Bth.; *Mimosac.*; Beng.
Karoto: *Inga spp.*; *Mimosac.*; Sur.
  „ Waikey: *Inga spp.*; *Mimosac.*; Sur.
Karpamara: *Cinnamomum zeylanicum* Breyn; *Laurac.*; sw. O.-Ind. (Coorg)
Karr: *Sterculia urens* Roxb.; *Sterculiac.*; O.-Ind. (Jeypore)
Karra: *Holarrhena antidysenterica* Wall.; *Apocynac.*; n. O.-Ind. (Un.-Prov.)
  „ marda: *Terminalia tomentosa* Wight et Arn.; *Combretac.*; w. O.-Ind. (Deccan)
Karrai: *Sterculia urens* Roxb.; *Sterculiac.*; n. O.-Ind. (Un.-Prov.)
Karre nembu: *Garuga pinnata* Roxb.; *Bursrac.*; w. O.-Ind. (Deccan)
  „ vemba: *Garuga pinnata* Roxb.; *Burserac.*; w. O.-Ind. (Deccan)
Karrei: *Butyrospermum parki* Kotschy; *Sapotac.*; Sudan.
Karri: *Grewia laevigata* Vahl; *Tiliac.*; n. O.-Ind. (United-Prov.)
  „ : *Eucalyptus divaricata* F. v. M.; *Myrtac.*; w. Au.
  „ -vembou-marom: *Garuga pinnata* Roxb.; *Burserac.*; w. O.-Ind. (Deccan)
Karshu: *Quercus semecarpifolia* Smith; *Fagac.*; n. O.-Ind. (United-Prov.)
Karsu: *Quercus semecarpifolia* Smith; *Fagac.*; n. O.-Ind. (Bushahr)
Kartivo: *Prioria copaifera* Griseb.; *Caesalpiniac.*; Pan.
Karu: *Diospyros ebenum* Koenig; *Ebenac.*; sw. O.-Ind. (Malabar)
  „ maruthu: *Terminalia tomentosa* Wight et Arn.; *Combretac.*; w. O.-Ind. (Deccan)
  „ vagei: *Albizzia odoratissima* Bth.; *Mimosac.*; w. O.-Ind. (Deccan)
Karud: *Allaeanthus glaber* Warb.; *Morac.*; Phil. (Misamis)
Karukuva: *Zizyphus rugosa* Lamk.; *Rhamnac.*; w. O.-Ind. (Deccan)
Karum: *Morus serrata* Roxb.; *Morac.*; nw. O.-Ind. (Punjab)
  „ : *Canarium strictum* Roxb.; *Burserac.*; w. O.-Ind. (Madras)
  „ -charei: *Holigarna arnottiana* Hook. f.; *Anacardiac.*; sw. O.-Ind. (Madras)
  „ -kali: *Diospyros ebenum* Koenig; *Ebenac.*; sw. O.-Ind. (Madras)
Karun: *Morus serrata* Roxb.; *Morac.*; nw. Himal.
  „ -phul: *Protium heptaphyllum* March.; *Burserac.*; Hind.
  „ thagara: *Albizzia procera* Bth.; *Mimosac.*; sw. O.-Ind. (Malabar)
Karunda: *Carissa spinarum* A. DC.; *Apocynac.*; O.-Ind.
Karunián: *Pometia pinnata* Forst.; *Sapindac.*; Phil. (Mindoro)
Karunko: *Terminalia superba* Engl. et Diels; *Combretac.*; W.-Af. (S.-L.)
Karunthali: *Diospyros ebenum* Koenig; *Ebenac.*; s. O.-Ind. (Madras)
Karunyáu: *Dipterocarpus vernicifluus* Blco.; *Dipterocarpac.*; Phil. (Cagayan)
Karuvelam: *Acacia arabica* Willd.; *Mimosac.*; w. O.-Ind. (Deccan)
Karvanari: *Hymenaea courbaril* L.; *Caesalpiniac.*; Sur.
Kar-we-eh: *Cola buntingi* Baker f.; *Sterculiac.*; Lib.
Karwiroe: *Bignonia chica* Humb. et Bonpl.; *Bignoniac.*; fr. Gu.
Kasa: *Erythrophloeum guineense* G. Don; *Caesalpiniac.*; tr. Af.
Kasai: *Bridelia retusa* Spreng.; *Euphorbiac.*; C.-O.-Ind.
  „ : *Albizzia procera* Bth., *A. retusa* Bth.; *Mimosac.*; Phil.
Kasambee: *Canarium secundum* A. W. Benn.; *Burserac.*; Malak.
Kasambi: *Schleichera trijuga* Willd.; *Sapindac.*; O.-Ind., Malay.
Kasanda: *Swartӡia madagascariensis* Desv.; *Caesalpiniac.*; D.-O.-Af. (Tabora)
Kasarkana-mara: *Strychnos nux-vomica* L.; *Loganiac.*; w. O.-Ind. (Bombay)
Kasase: *Pentaclethra eetveldeana* de Wild. et Th. Dur.; *Mimosac.*; B.-K. (Bukaka)
Kaschu-baum: *Anacardium occidentale* L.; *Anacardiac.*; tr. Am.
Kasenga: *Flueggea bailloniana* Pax; *Euphorbiac.*; D.-O.-Af. (Tabora)
Kasfiya: *Crossopteryx kotschyana* Fenzl; *Rubiac.*; n. Nig.
Kash: *Alnus nitida* Endl.; *Betulac.*; nw. O.-Ind. (ö. Punjab/Kangra)
Kashekaji: *Psorospermum senegalense* Spach; *Guttifer.*; n. Nig.
Kasheshi: *Rhus insignis* Vell.; *Anacardiac.*; n. Nig.
Kashi: *Bridelia retusa* Spreng.; *Euphorbiac.*; Assam.
  „ : *Quercus sp.*; *Fagac.*; Jap.
Kashin awaki: *Crossopteryx kotschyana* Fenzl; *Rubiac.*; n. Nig.
Kashioron: *Castanopsis indica* A. DC.; *Fagac.*; nö. O.-Ind.
Kashit: *Pentace burmanica* Kurz; *Tiliac.*; Burma.
Kashiwa: *Quercus dentata* Thunb.; *Fagac.*; Jap.

Kasisa: *Celtis kraussiana* Bernh.; *Ulmac.;* br. O.-Af. (Uganda/Mabira)
Kasiwa: *Quercus dentata* Thunb.; *Fagac.;* Form.
Kaskawami: *Psorospermum senegalense* Spach; *Guttifer.;* n. Nig.
Kasondeh: *Cordia myxa* L.; *Borraginac.;* O.-Ind. (? Burma)
Kasru: *Quercus semecarpifolia* Smith; *Fagac.;* nö. O.-Ind. (Nepal, n. Beng.)
Kassa: *Erythrophloeum guineense* G. Don; *Caesalpiniac.;* B.-K.
Kassamar: *Cordia macleodi* Hook. f. et Thoms.; *Borraginac.;* C.-O.-Ind.
Kasséhui: *Pynaertia occidentalis* Chev.; *Meliac.;* Elf.
Kasserüi: *Pynaertia occidentalis* Chev.; *Meliac.;* Elf.
Kassi: *Bridelia retusa* Spreng.; *Euphorbiac.;* C.-O.-Ind.
Kassikui: *Pynaertia occidentalis* Chev.; *Meliac.;* Elf.
  „ : *Anopyxis occidentalis* Chev.; *Rhizophorac.;* Gab.
Kastanie, australische: *Castanospermum australe* A. Cunn.; *Papilionac.;* s. Au.
  „ , Edel- (d): *Castanea sativa* Mill.; *Fagac.;* Eu., n. Af.
  „ , Haest- (sch): *Aesculus hippocastanum* L.; *Hippocastanac.;* Schwed.
  „ , Heste- (dä): *Aesculus hippocastanum* L.; *Hippocastanac.;* Dänemark.
  „ , japanische (d): *Castanea crenata* S. et Z.; *Fagac.;* Jap.
  „ , Ross- (d): *Aesculus hippocastanum* L.; *Hippocastanac.;* Eu.
  „ , „ , Ohio-: *Aesculus glabra* Willd.; *Hippocastanac.;* N.-Am.
  „ , wilde: *Calodendron capense* Thunb.; *Rutac.;* Kap, Natal.
  „ , Zwerg-: *Castanea pumila* Mill.; *Fagac.;* N.-Am.
Kastanje, Brazilië-: *Bertholletia exelsa* Humb. et Bonpl.; *Lecythidac.;* fr. Gu.
Kastanjeboom, indiaansche (h): *Aesculus hippocastanum* L.; *Hippocastanac.;* n. O.-Ind.,
Kastel: *Hydnocarpus wightiana* Blume; *Flacourtiac.;* sw. O.-Ind.   [Pers.
Kasu-besi-prempuan: *Wallaceodendron celebicum* Koord.; *Mimosac.;* Malay.
Kasuku: *? Pachylobus sp.;* *Burserac.;* B.-K. (Katanga)
Kasul: *Grewia tiliaefolia* Vahl; *Tiliac.;* C.-O.-Ind. (Gondwana)
Kasumba: *Antidesma ghaesembilla* Gaertn.; *Euphorbiac.;* Malay.
  „ : *Canarium secundum* A. W. Benn.; *Burserac.;* Malak.
Kat-bel: *Feronia elephantum* Correa; *Rutac.;* Burma.
  „ illupei: *Palaquium ellipticum* Bth.; *Sapotac.;* sw. O.-Ind. (Madras)
  „ kumbla: *Trewia nudiflora* L.; *Euphorbiac.;* w. O.-Ind. (Bombay)
  „ maä: *Buchanania latifolia* Roxb.; *Anacardiac.;* O.-Ind.
  „ maa: *Spondias mangifera* Willd.; *Anacardiac.;* w. O.-Ind. (Deccan)
  „ mangri: *Randia dumetorum* Lamk.; *Rubiac.;* w. O.-Ind. (Bombay)
  „ pindi: *Myristica attenuata* Wall.; *Myristicac.;* sw. O.-Ind. (Madras)
Kata bilek: *Castanopsis sp.;* *Fagac.;* Malay.
  „ narunga: *Atalantia monophylla* Correa; *Rutac.;* O.-Ind. (Uriya)
  „ tangga: *Castanopsis sp.;* *Fagac.;* Malay.
Katábang: *Quercus sp.;* *Fagac.;* Phil. (Zambales, Bataan, Laguna)
Katakyidua: *Microdesmis puberula* Hook. f.; *Euphorbiac.;* Go.
Katamba: *Spondias mangifera* Willd.; *Anacardiac.;* C.-O.-Ind.
Katambiri: *Gardenia ternifolia* Thunb.; *Rubiac.;* n. Nig.
Katan: *Bischofia javanica* Blume; *Euphorbiac.;* Form.
Katangai: *Toona ciliata* M. Roem.; *Meliac.;* O.-Ind.
Katank: *Lophira alata* Banks; *Ochnac.;* W.-Af. (S.-L.)
Katapang: *Terminalia catappa* L.; *Combretac.;* Malay.
  „ : *Garcinia sp.;* *Guttifer.;* Phil. (Agusan)
Katarang: *Gardenia latifolia* Aiton; *Rubiac.;* w. Beng. (Orissa)
Katarin: *Capparis grandis* L. f.; *Capparidac.;* O.-Ind.
Katbhilawa: *Buchanania latifolia* Roxb.; *Anacardiac.;* n. O.-Ind. (Garhwal)
Katdoorn: *Scutia capensis* Eckl. et Zeyh.; *Rhamnac.;* Kap.
Kateran nujus: *Cedrus libani* Barrel.; *Conifer.;* Kl.-As.
Katerika: *Drypetes sp.;* *Euphorbiac.;* Go.
Katgeru: *Holigarna arnottiana* Hook. f.; *Anacardiac.;* w. O.-Ind. (s. Bombay)
Kath bhewal: *Grewia laevigata* Vahl; *Tiliac.;* n. O.-Ind. (United-Prov.)
Kathit: *Erythrina suberosa* Roxb.; *Papilionac.;* Burma.
Kathitka: *Pentace burmanica* Kurz; *Tiliac.;* Burma.
Kathitsu: *Zanthoxylum budrunga* Wall.; *Rutac.;* Burma.
Kati: *Petersia viridiflora* Chev.; *Lecythidac.;* Elf.
Katiain: *Bridelia retusa* Spreng.; *Euphorbiac.;* C.-O.-Ind.
Katilmá: *Diospyros nitida* Merr.; *Ebenac.;* Phil. (Rizal)
Katilúk: *Quercus sp., Q. jordanae* Lag.; *Fagac.;* Phil.
Katjang: *Adinandra sp.;* *Theac.;* Sum.

Katjapie: *Sandoricum indicum* Cav.; *Meliac.;* Sunda.
Katjé: *Clathrotropis brachypetalum* Kleinh.; *Papilionac.;* Sur.
Katjeputbaum: *Melaleuca leucadendron* L.; *Myrtac.;* sö. As., Au.
Katkaula: *Phoebe lanceolata* Nees; *Laurac.;* O.-Ind. (Kumaon)
Katma: *Buchanania latifolia* Roxb.; *Anacardiac.;* O.-Ind. (Madras)
Katmarra: *Litsea polyantha* Juss.; *Laurac.;* O.-Ind.
Katmon: *Dillenia sp.; Dilleniac.;* Phil.
Káto: *Amoora aherniana* Merr.; *Meliac.;* Phil. (Bataan)
Katoelampa: *Elaeocarpus macrophyllus* Blume; *Elaeocarpac.;* Nied.-Ind.
Katoelimia: *Dipteryx odorata* Willd.; *Papilionac.;* Sur.
Katoenboom: *Ceiba pentandra* Gaertn.; *Bombacac.;* Sur.
Katoenpruim: *Chrysobalanus icaco* L.; *Rosac.;* fr. Gu.
Katoka: *Brachystegia sp.; Caesalpiniac.;* B.-K. (Katanga)
Katondo tondo: *Pterocarpus sp.; Papilionac.;* B.-K.
Katong-bakulau: *Kayea paniculata* Merr.; *Guttifer.;* Phil. (Bataan)
Katongzu: *Dysoxylum binectariferum* Hook. f.; *Meliac.;* nö. O.-Ind.
Katori: *Feronia elephantum* Correa; *Rutac.;* w. O.-Ind. (n. Bombay)
„ : *Stereospermum xylocarpum* Bth. et Hook. f.; *Bignoniac.;* C.-O.-Ind. (Gondwana)
Katos: *Protium javanicum* Burm.; *Burserac.;* Nied.-Ind.
Katpali: *Mimusops littoralis* Kurz; *Sapotac.;* Burma.
Katsari: *Albizzia chevalieri* Harms; *Mimosac.;* n. Nig.
Katschikautscha: *Cynometra megalophylla* Harms; *Caesalpiniac.;* Togo.
Katsura: *Cercidiphyllum japonicum* S. et Z.; *Trochodendrac.;* Jap.
Katta: *Zanthoxylum rhetsa* DC.; *Rutac.;* sw. O.-Ind. (Malabar)
„ : *Bridelia retusa* Spreng.; *Euphorbiac.;* Cey.
„ -naragam: *Atalantia monophylla* Correa; *Rutac.;* O.-Ind. (Madras)
Kattenagels: *Bignonia unguis-cati* L.; *Bignoniac.;* fr. Gu.
Kattiber: *Zizyphus xylopyrus* Willd.; *Rhamnac.;* n. O.-Ind. (United-Prov.)
Kattito: *Wendlandia exserta* DC.; *Rubiac.;* tr. Himal.
Kattou-ilenden-marom: *Zizyphus xylopyrus* Willd.; *Rhamnac.;* w. O.-Ind. (Deccan)
„ -maga-marom: *Cupania canescens* Pers.; *Sapindac.;* w. O.-Ind. (Deccan)
Kattra: *Bauhinia malabarica* Roxb.; *Caesalpiniac.;* Assam.
Kattu puvarasu: *Rhododendron arboreum* Sm.; *Ericac.;* s. O.-Ind. (Madras)
Katupuveras: *Filicium decipiens* Thw.; *Sapindac.;* sw. O.-Ind. (Malabar)
Katur: *Mangifera silvatica* Roxb.; *Anacardiac.;* nö. O.-Ind.
Katúrai: *Radermachera sp.; Bignoniac.;* Phil. (Zambales)
Katúri: *Garcinia sp.; Guttifer.;* Phil. (Cagayan)
Katus: *Castanopsis hystrix* A. DC.; *Fagac.;* nö. O.-Ind. (Nepal)
Katyilenga: *Morinda citrifolia* L.; *Rubiac.;* Togo.
Kau: *Gardenia ternifolia* Schum. et Thonn.; *Rubiac.;* Togo.
„ : *Olea cuspidata* Wall.; *Oleac.;* n. O.-Ind. (Punjab, Jaunsar)
„ solo: *Podocarpus cupressina* R. Br.; *Conifer.;* Fid.
„ tambua: *Podocarpus cupressina* R. Br.; *Conifer.;* Fid.
Kauál: *Cyclostemon sp.; Euphorbiac.;* Phil. (Cagayan)
Kauarika: *Terminalia catappa* L.; *Combretac.;* Südsee (Rarotonga)
Kauchia: *Holoptelea integrifolia* Planch.; *Ulmac.;* O.-Ind. (Madras)
Kauha: *Terminalia arjuna* Bedd.; *Combretac.;* Beng.
Kauila: *Alphitonia ponderosa* Hillebr.; *Rhamnac.;* Hawai.
Kauílan: *Lagerstroemia speciosa* Pers.; *Lythrac.;* Phil. (Guimaras)
Kauklaw-taro: *Dipterocarpus tuberculatus* Roxb.; *Dipterocarpac.;* Burma.
Kaukutoku: *Gardenia thunbergi* L. f.; *Rubiac.;* Togo.
Kaula: *Machilus odoratissima* Nees; *Laurac.;* O.-Ind., Himal.
Kaulia: *Acacia scorpioides* W. F. Wight; *Mimosac.;* O.-Ind. (Berar Poona)
Kaulu: *Machilus odoratissima* Nees; *Laurac.;* n. O.-Ind. (United-Prov.)
Kaumakka: *Acrocomia sclerocarpa* Mart.; *Palmac.;* Sur.
Kaunghmu: *Parashorea stellata* Kurz; *Dipterocarpac.;* s. Burma.
Kaunglaung: *Duabanga sonneratioides* Ham.; *Sonneratiac.;* s. Burma (s. Tenass.)
Kaurchi: *Dalbergia lanceolaria* L.; *Papilionac.;* sw. O.-Ind. (Bombay)
Kauri: *Acer caesium* Wall.; *Acerac.;* nw. O.-Ind. (Punjab)
„ : *Agathis australis* Steud., *A. spp.; Conifer.;* ö. Au.
„ : *Agathis vitiensis* Bth. et Hook. f.; *Conifer.;* Fid.
„ , mottled: *Agathis australis* Steud.; *Conifer.;* N.-Seel.
„ , Noumean- (Sydney): *Agathis lanceolata* Pancher; *Conifer.;* Au.
„ , New Zealand- (Sydney): *Agathis australis* Steud.; *Conifer.;* N.-Seel.

Kauri Queensland-: *Agathis spp.; Conifer.;* Au. (Queensl.)
„ , southern (Sydney): *Agathis robusta* F. v. M.; *Conifer.;* Au. (Queensl.)
„ , Vanikoro- (Sydney): *Agathis macrophylla* Masters; *Conifer.;* Au.
Kauria: *Acacia sp.; Mimosac.;* nw. O.-Ind.
Kaurobandikoro: *Hirtella caudata* Kleinh., *H. spp.; Rosac.;* Sur.
Kausa: *Deinbollia indeniensis* Chev.; *Sapindac.;* Elf.
Kauston: *Licania macrophylla* Bth.; *Rosac.;* Sur.
Kauta: *Licania apetala* Fritsch, *Hirtella sp.; Rosac.;* Sur., br. Gu.
Kautaballi: *Licania venosa* Rusby, *L. mollis* Bth.; *Rosac.;* br. Gu.
Kauwi: *Alstonia congensis* Engl.; *Apocynac.;* tr. W.-Af.
Kava: *Erythroxylon sp.; Erythroxylac.;* Südsee.
Kaval: *Careya arborea* Roxb.; *Lecythidac.;* w. O.-Ind. (Bombay)
Kavalam: *Sterculia urens* Roxb.; *Sterculiac.;* O.-Ind. (Madras)
Kavali: *Sterculia urens* Roxb.; *Sterculiac.;* O.-Ind. (Madras)
Kavat: *Feronia elephantum* Correa; *Rutac.;* sw. O.-Ind. (Bombay)
Kavsi: *Hopea wightiana* Wall.; *Dipterocarpac.;* sw. O.-Ind. (Bombay)
Kaw: *Randia dumetorum* Lamk.; *Rubiac.;* Burma.
Kawajang: *Pygeum parviflorum* Teijsm. et Binn.; *Rosac.;* Java.
Kawaka: *Libocedrus doniana* Endl.; *Conifer.;* N.-Seel.
Kawakami-gasi: *Lithocarpus kawakamii* Hay.; *Fagac.;* Form.
Kawakanalli: *Hymenaea courbaril* L.; *Caesalpiniac.;* Sur.
Kawala: *Machilus spp.; Laurac.;* n. O.-Ind., Beng.
Kawanari: *Hymenaea courbaril* L.; *Caesalpiniac.;* Sur.
Kawang: *Hopea fagifolia* Miq.; *Dipterocarpac.;* Java, Banks.
„ : *Palaquium javense* Burck; *Sapotac.;* Java.
Kawarahannoki: *Alnus glutinosa* Willd.; *Betulac.;* Jap.
Kawarri: *Ficus sp.; Morac.;* n. Nig. (Kontagora)
Kawat: *Limonia acidissima* L.; *Rutac.;* sw. O.-Ind. (Bombay)
Kawatha: *Feronia elephantum* Correa; *Rutac.;* nw. O.-Ind. (Bombay)
Kaway: *Pterocarpus belizensis* Standl., *P. officinalis* Jacq.; *Papilionac.;* C.-Am.
„ , Swamp-: *Pterocarpus belizensis* Standl.; *Papilionac.;* Guat., br. Hond.
„ , „ : *Platymiscium polystachium* Bth.; *Papilionac.;* Pan.
Kawáyan: *Bambusa spinosa* Roxb.; *Gramin.;* Phil.
„ -bayúgin: *Bambusa vulgaris* Schrad.; *Gramin.;* Phil. (Nueva Ecija)
„ -bobéro: *Bambusa vulgaris* Schrad.; *Gramin.;* Phil. (Laguna)
„ -géd: *Bambusa spinosa* Roxb.; *Gramin.;* Phil. (Bisaya)
„ -hobéro: *Bambusa vulgaris* Schrad.; *Gramin.;* Phil. (Laguna)
„ -kilíng: *Bambusa vulgaris* Schrad.; *Gramin.;* Phil.
„ -matiník: *Bambusa spinosa* Roxb.; *Gramin.;* Phil.
„ -nga-bulilau: *Bambusa spinosa* Roxb.; *Gramin.;* Phil. (w. Negros Prov.)
„ - „ -dalusa: *Bambusa vulgaris* Schrad.; *Gramin.;* Phil. (w. Negros Prov.)
„ -sa-china: *Bambusa vulgaris* Schrad.; *Gramin.;* Phil. (Cebu)
„ -siítan: *Bambusa spinosa* Roxb.; *Gramin.;* Phil. (Ilocos-Sur, Union Prov.)
„ -sína: *Dendrocalamus latiflorus* Munro; *Gramin.;* Phil.
„ „ : *Gigantochloa levis* Merr.; *Gramin.;* Phil. (Bulacan)
„ -sungsúng: *Schizostachyum lumampao* Merr.; *Gramin.;* Phil. (Laguna)
„ -tiník: *Bambusa spinosa* Roxb.; *Gramin.;* Phil.
„ totóo: *Bambusa spinosa* Roxb.; *Gramin.;* Phil.
Kawayanagi: *Salix purpurea* L. *subsp. gymnolepis* Koidz.; *Salicac.;* Jap.
Kawkaka: *Libocedrus bidwilli* Hook. f.; *Conifer.;* N.-Seel.
Kawies watoe: *Feronia elephantum* Roxb.; *Rutac.;* Java.
Kawiesta: *Feronia elephantum* Roxb.; *Rutac.;* Java.
Kawo: *Afzelia africana* Smith, *A. quanzensis* Welw.; *Caesalpiniac.;* n. Nig.
Kawogei: *Rauwolfia vomitoria* Afz.; *Apocynac.;* Lib.
Kawraw: *Diospyros mespiliformis* Hochst.; *Ebenac.;* tr. Af.
Kawri: *Grewia laevigata* Vahl; *Tiliac.;* w. O.-Ind. (Deccan)
„ : *Agathis australis* Steud.; *Conifer.;* N.-Seel.
Kawtanok: *Mesua ferrea* L.; *Guttifer.;* s. Burma.
Kawthu: *Parashorea stellata* Kurz; *Dipterocarpac.;* s. Burma.
Kawti: *Hydnocarpus wightiana* Blume; *Flacourtiac.;* sw. O.-Ind. (Bombay)
Kawuri: *Ficus kawuri* Hutch.; *Morac.;* n. Nig.
Kawwa: *Parashorea stellata* Kurz; *Dipterocarpac.;* s. Burma.
Kaya: *Mimusops elengi* Roxb.; *Sapotac.;* Burma.
„ : *Torreya nucifera* S. et Z.; *Conifer.;* Jap.

Kayahmwe: *Elaeocarpus robustus* Roxb.; *Elaeocarpac.;* s. Burma.
Kayaing-bataing: *Dipterocarpus alatus* Roxb.; *Dipterocarpac.;* s. Burma.
Kayanoki: *Torreya nucifera* S. et Z.; *Conifer.;* Jap.
Kayátau: *Dysoxylum turczaninowi* C. DC.; *Meliac.;* Phil. (Palawan, Busuanga)
Kayaw: *Excoecaria agallocha* L.; *Euphorbiac.;* Burma.
Kay-ne-doo: *Canthium venosum* Hiern; *Rubiac.;* Lib.
Kayo: *Uapaca bingervillensis* Beille; *Euphorbiac.;* Elf.
    „   : *Ceiba pentandra* Gaertn.; *Bombacac.;* Phil. (Camarines, Albay, Sorsogon)
Kayogkóg: *Eugenia sp.; Myrtac.;* Phil. (Luzon, Samar, Leyte)
Kayokó: *Eugenia xanthophylla* C. B. Rob.; *Myrtac.;* Phil. (Tayabas)
Kayongkóng: *Celtis sp.; Ulmac.;* Phil. (Cebu)
Kayu arang: *Diospyros spp.; Ebenac.;* Malay, Borneo.
    „   baru: *Hibiscus macrophyllus* Roxb., *H. tiliaceus* L.; *Malvac.;* Jap.
    „   „   : *Thespesia populnea* Correa; *Malvac.;* Jap.
    „   gading: *Canthium spp., Petunga venulosa* Hook. f.; *Rubiac.;* Malay.
    „   „   : *Urophyllum glabrum* Wall.; *Rubiac.;* Malay.
    „   „   : *Hunteria corymbosa* Roxb.; *Apocynac.;* Malay.
    „   gálu: *Sindora inermis* Merr.; *Caesalpiniac.;* Phil. (Cotabato, Davao)
    „   garu: *Aquilaria agallocha* Roxb.; *Thymelaeac.;* Malay.
    „   „   : *Gonystylus bancanus* Gilg; *Gonystylac.;* Malay, Java.
    „   hitam: *Diospyros spp.; Ebenac.;* Malay.
    „   klat: *Eugenia sp.; Myrtac.;* Malay.
    „   loh: *Anisoptera sp., A. thurifera* Blume; *Dipterocarpac.;* Malay (Bentong)
    „   malam: *Diospyros spp.; Ebenac.;* Borneo.
    „   manis: *Cinnamomum sp.; Laurac.;* Malay.
    „   -rajah: *Koompassia excelsa* Taub.; *Caesalpiniac.;* Borneo.
Kayun: *Albizzia procera* Bth., *A. stipulata* Boivin; *Mimosac.;* O.-Ind.
Kazcat: *Luehea speciosa* Willd.; *Tiliac.;* br. Hond.
Kda cong hen: *Mitrephora thoreli* Pierre; *Anonac.;* Ind.-Ch.
Kdol: *Adina cordifolia* Hook. f.; *Rubiac.;* Ind.-Ch.
    „   : *? Wrightia sp.; Apocynac.;* Kamb.
Ké: *Stereospermum annamense* Chev.; *Bignoniac.;* Ind.-Ch.
Keaki: *Zelkowa acuminata* Planch.; *Ulmac.;* Jap.
Kebarre: *Afzelia africana* Smith; *Caesalpiniac.;* Togo.
Kechi-lebet: *Mitragyne macrophylla* Hiern; *Rubiac.;* Nig.
Keda: *Ficus bembicicarpa* Warb., *F. dicranostyla* Mildbr. et Burret; *Morac.;* Togo.
Kedanja: *Butyrospermum parki* Kotschy; *Sapotac.;* n. Nig.
Kedawoeng: *Parkia roxburghi* G. Don; *Mimosac.;* Java.
Kedemponasi: *Pseudocedrela kotschyi* Harms; *Meliac.:* Togo.
Kedia: *Ficus rokko* Warb. et Schweinf., *F. spp.; Morac.;* Togo.
Keditia: *Tamarindus indica* L.; *Caesalpiniac.;* Togo.
Kedjimas: *Duabanga moluccana* Blume; *Sonneratiac.;* Phil., Sunda.
Kedondong: *Canarium kakondon* A. W. Benn.; *Burserac.;* Malay.
Kedra: *Pinus sp.; Conifer.;* ö. Sib.
Kedyetyelo: *Sarcocephalus sambucinus* K. Schum.; *Rubiac.;* Togo.
Keelu: *Cedrus deodara* Loudon; *Conifer.;* mw. O.-Ind. (Punjab)
Keena-tel: *Calophyllum tomentosum* Wight; *Guttifer.;* Cey.
Kees: *Bactris mimax* Miq.; *Palmac.;* Sur.
Keesiemakka: *Bactris mimax* Miq.; *Palmac.;* Sur.
Keeskeesi makka: *Bactris gasipaes* Kunth; *Palmac.;* Sur.
Keg: *Diospyros kaki* L. f.; *Ebenac.;* Ch., Jap., O.-Ind.
Kegblale: *Terminalia superba* Engl. et Diels; *Combretac.;* Go.
Kéguigo: *Khaya ivorensis* Chev.; *Meliac.;* Elf.
Kei pala: *Palaquium ellipticum* Baill.; *Sapotac.:* sw. O.-Ind. (Madras)
Keiri: *Limonia acidissima* L.; *Rutac.;* w. O.-Ind. (Ajmer, Merwara)
Kejai: *Santiria griffithi* Engl.; *Burserac.;* Malak.
Keka: *Blighia sapida* Koenig; *Sapindac.;* Togo.
Kekala: *Cynthocalyx zeylanicus* Champ.; *Anonac.;* O.-Ind., Cey.
Kekange: *Terminalia ivorensis* Chev.; *Combretac.;* Nig.
Kekda: *Garuga pinnata* Roxb.; *Burserac.;* C.-O.-Ind.
    „   : *Spondias mangifera* Willd.; *Anacardiac.;* C.-O.-Ind.
Kekëu: *Erythrophloeum guineense* G. Don; *Caesalpiniac.;* Togo.
Kekosi: *Cola proteiformis* Chev., *Tarrietia utilis* Sprague; *Sterculiac.;* Elf.
Kékoʒi: *Tarrietia utilis* Sprague; *Sterculiac.;* Elf.

Kekpili: *Ormosia laxiflora* Bth.; *Papilionac.;* Togo.
Kekrefinde: *Lophira alata* Banks; *Ochnac.;* Togo.
Kelaba: *Carapa procera* DC., *C. surinamensis* Miq.; *Meliac.;* Sur.
Keladan: *Dryobalanops oblongifolia* Dyer; *Dipterocarpac.;* Malay.
Kelambir: *Cocos nucifera* L.; *Palmac.;* Malay.
Kelantoro: *Hannoa undulata* Planch.; *Simarubac.;* Togo.
Kelapa: *Cocos nucifera* L.; *Palmac.;* Malay.
Kelappa-bosie: *Dipteryx odorata* Willd.; *Papilionac.;* Sur.
Kelat: *Eugenia spp.; Myrtac.;* Malay.
   „ layu: *Erioglossum edule* Blume; *Sapindac.;* Malay.
Kelatrang: *Aromadendron elegans* Blume; *Magnoliac.;* Nied.-Ind.
Keledang: *Artocarpus lancaefolia* Roxb.; *Morac.;* Malay.
Kelempening: *Pasania sp., Quercus sp.; Fagac.;* Malay.
Kelengan: *Harpullia sp.; Sapindac.;* Celebes.
Kelengmau: *Fagara xanthoxyloides* Lam.; *Rutac.;* Togo.
Kelengoi: *Parinarium glabrum* Oliv.; *Rosac.;* B.-K. (Azande)
Kelinkeki: *Dialium yambataense* Verm.; *Caesalpiniac.;* B.-K. (Kisantu)
Kelinting: *Campnosperma auriculata* Hook. f.; *C. spp.; Anacardiac.;* Malay.
Kélipé: *Ochrocarpus africanus* Oliv.; *Guttifer.;* Elf.
Kelipoto: *Sterculia* Guill. et Perr.; *Sterculiac.;* Togo.
Kelle: *Adansonia digitata* L.; *Bombacac.;* Togo.
   „ : *Sapindus senegalensis* Poir.; *Sapindac.;* Senl.
Kellerhals (d): *Daphne mezereun* L.; *Thymelaeac.;* Eu., Kauk.
Kello: *Parinarium annamense* Hance; *Rosac.;* Kamb.
Kelmung: *Cedrus deodara* Loudon; *Conifer.;* nw. O.-Ind. (Kunawar)
Keloekoep: *Vatica sublacunosa* Miq.; *Dipterocarpac.;* Sunda (Bangka)
Keloetoem: *Adinandra sp.; Theac.;* Sum.
Kelon: *Cedrus deodara* Loudon; *Conifer.;* n. O.-Ind. (Jaunsar)
Kelor: *Artocarpus integrifolia* L.; *Morac.;* Malay.
   „ -pante: *Albizzia littoralis* Teijsm. et Binn.; *Mimosac.;* Malay.
Kelsey: *Prunus triflora* Roxb.; *Rosac.;* w. USA. (Kalif.)
Keluang: *Tarrietia spp.; Sterculiac.;* Malay.
Kem: *Adina cordifolia* Hook. f.; *Rubiac.;* C.-O.-Ind. (Gondwana)
   „ càng: *Elaeocarpus dubius* A. DC.; *Elaeocarpac.;* Ind.-Ch.
Kemajan: *Styrax benzoin* Dryand.; *Styracac.;* Malay, Sum., Java.
Kemarek: *Tamarix articulata* Vahl; *Tamaricac.;* n. Af.
Kembal: *Lannea grandis* Engl.; *Anacardiac.;* nw. O.-Ind. (Punjab)
Kembang-Kantjil: *Michelia champaca* L.; *Magnoliac.;* Nied.-Ind.
Kemlandingan: *Albizzia montana* Bth.; *Mimosac.;* Java.
Kemouning: *Murraya exotica* L.; *Rutac.;* Malay.
Kempas: *Koompassia malaccensis* Maing.; *Caesalpiniac.;* Malay.
Kemponashi: *Hovenia dulcis* Thunb.; *Rhamnac.;* Jap.
Kemuning: *Merrillia caloxylon* Swingle, *Murraya exotica* L.; *Rutac.;* Malay.
   „ Kampong: *Murraya exotica* L.; *Rutac.;* Malay.
   „ lada: *Murraya exotica* L.; *Rutac.;* Malay.
Kemunting: *Rhodomyrtus tomentosa* Wight; *Myrtac.;* Malay.
Ken: *Aesculus chinensis* Bunge; *Hippocastanac.;* Ind.-Ch.
   „ : *Hopea odorata* Roxb.; *Dipterocarpac.;* Ind.-Ch.
   „ nang: *Cyanodaphne cuneata* Blume; *Laurac.;* Ind.-Ch.
   „ tây: *Miliusa bailloni* Pierre; *Anonac.;* Ind.-Ch.
Kenang: *Cananga odorata* Hook. f. et Thoms.; *Anonac.;* Java.
Kenanga hutan: *Cananga odorata* Hook. f. et Thoms.; *Anonac.;* Malay.
Kenari: *Canarium commune* L.; *Burserac.;* Malay.
Kenchna: *Bauhinia purpurea* L.; *Caesalpiniac.;* O.-Ind. (? s. Beng.)
Kend: *Diospyros tomentosa* Roxb.; *Ebenac.;* w. Beng.
Kendal: *Ehretia javanica* Blume; *Borraginac.;* Java.
Kendhu: *Diospyros ebenum* Koenig; *Ebenac.;* w. Beng. (Orissa)
Kendu: *Diospyros tomentosa* Roxb.; *Ebenac.;* n. O.-Ind. (United Prov.)
Kenep: *Melicocca bijuga* L.; *Sapindac.;* Guad.
Kenghe vumba: *Conopharyngia smithi* Stapf; *Apocynac.;* B.-K. (Mayombe)
Kengüe: *Fagara macrophylla* Engl.; *Rutac.;* Elf.
Kenkar: *Garuga pinnata* Roxb.; *Burserac.;* C.-O.-Ind.
Kenponashi: *Hovenia dulcis* Thunb.; *Rhamnac.;* Jap.
Kensagui: *Erismadelphus baudoni* Chev. et Buss.; *Vochysiac.;* fr. Kongo (Haut-Ogoué)

261

Keo: *Acacia farnesiana* Willd., *Leucaena glauca* Bth.; *Mimosac.;* Ind.-Ch.
„ : *Murraya exotica* L.; *Rutac.;* Ind.-Ch.
„ ta: *Acacia farnesiana* Willd.; *Mimosac.;* Ind.-Ch.
„ tay: *Pithecolobium dulce* Bth.; *Mimosac.;* Ind.-Ch.
Keolar: *Bauhinia purpurea* L.; *Caesalpiniac.;* C.-O.-Ind.
Keolari: *Bauhinia purpurea* L.; *Caesalpiniac.;* C.-O.-Ind.
Keor: *Holarrhena antidysenterica* Wall.; *Apocynac.;* nw. O.-Ind. (Punjab-Kangra)
Kéowazingo: *Didelotia africana* Pierre; *Caesalpiniac.;* Gab.
Keowra: *Sonneratia apetala* Ham.; *Sonneratiac.;* s. Beng.
Kepoh: *Sterculia foetida* L.; *Sterculiac.;* Sunda.
Kepong: *Shorea macroptera* Dyer, *S. sericea* Dyer; *Dipterocarpac.;* Malay.
„ hantu: *Shorea macroptera* Dyer; *Dipterocarpac.;* Malay.
„ labu: *Shorea macroptera* Dyer, *S. sericea* Dyer; *Dipterocarpac.;* Malay.
„ segar: *Shorea sericea* Dyer; *Dipterocarpac.;* Malay.
Keraba: *Carapa procera* DC., *C. surinamensis* Miq.; *Meliac.;* Sur.
Kerandang: *Carissa carandas* L.; *Apocynac.;* Malay.
Keranji: *Dialium sp.; Caesalpiniac.;* Malay.
„ papan: *Dialium laurinum* Baker; *Caesalpiniac.;* Malay.
Kerapa: *Carapa procera* DC., *C. surinamensis* Miq.; *Meliac.;* Sur.
Keraserom: *Albizzia stipulata* Boivin; *Mimosac.;* O.-Ind. (? Burma)
Kérnegué: *Pachystela cinerea* Pierre; *Sapotac.;* Elf.
Kerewowo: *Pterocarpus esculentus* Schum. et Thonn.; *Papilionac.;* Togo.
Kerian: *Eugenia spp.; Myrtac.;* Malay.
Kers: *Pterocelastrus variabilis* Sond.; *Celastrac.;* S.-Af. (Knysna)
„ , inlandsche: *Eugenia uniflora* L.; *Myrtac.;* fr. Gu.
„ , zoete: *Isertia coccinea* Vahl; *Rubiac.;* Sur.
„ , „ : *Eugenia pumila* Gardn.; *Myrtac.;* Sur.
Kersehout: *Pterocelastrus variabilis* Sond.; *Celastrac.;* S.-Af.
Kershout: *Pterocelastrus variabilis* Sond.; *Celastrac.;* S.-Af.
Kertak tangga: *Castanopsis spp.; Fagac.;* Malay.
Keru: *Quercus semecarpifolia* Smith; *Fagac.;* nw. O.-Ind. (Kaschmir)
Kerubing: *Dipterocarpus spp.; Dipterocarpac.;* Malay.
Keruing: *Dipterocarpus cornutus* Dyer, *D. spp.; Dipterocarpac.;* Malay.
„ dadi: *Dipterocarpus cornutus* Dyer; *Dipterocarpac.;* Malay.
Kerukup siam: *Muntingia calabura* L.; *Tiliac.;* Malay.
Kerwara: *Cassia fistula* L.; *Caesalpiniac.;* C.-O.-Ind. (Gondwana)
Kés: *Xylia kerri* Craib et Hutchinson; *Mimosac.;* Ind.-Ch.
Kesambi: *Schleichera trijuga* Willd., *Cupania sideroxylon* Camb.; *Sapindac.;* Malay,
Kesang: *Monotes kerstingi* Gilg; *Dipterocarpac.;* Togo.   [Nied.-Ind.
Kesau: *Monotes kerstingi* Gilg; *Dipterocarpac.;* Togo.
Kesumba: *Bixa orellana* L.; *Bixac.;* Malay.
Két: *Gleditschia sinensis* Lam.; *Caesalpiniac.;* Ind.-Ch.
Ketapang: *Terminalia catappa* L.; *Combretac.;* Malay.
„ -ketek: *Guettarda speciosa* L.; *Rubiac.;* Malay.
Ketemaha: *Kleinhovia hospita* L.; *Sterculiac.;* Java.
Ketenggah: *Merrillia caloxylon* Swingle; *Rutac.;* Malay.
Ketey: *Sterculia trichosiphon* Bth.; *Sterculiac.;* Au. (Queensl.)
Ketiau: *Ganua motleyana* Pierre; *Sapotac.;* Malay.
„ : *Madhuca spp., Palaquium spp., Payena spp.; Sapotac.;* Malay.
Kétibu: *Omphalocarpum ahia* Chev.; *Sapotac.;* Elf.
Ketimoho: *Kleinhovia hospita* L.; *Sterculiac.;* Java.
Ketjapi: *Sandoricum indicum* Cav.; *Meliac.;* Malay.
Ketschikantschá: *Entada abyssinica* Steud.; *Mimosac.;* Togo.
Keu: *Sterculia pierrei* Gagnep.; *Sterculiac.;* Ind.-Ch.
Keua han: *Albizzia milleti* Bth.; *Mimosac.;* Ind.-Ch.
Keulenbaum: *Casuarina equisetifolia* Forst.; *Casuarinac.;* Au.
Keurboom: *Virgilia capensis* Lam.; *Papilionac.;* Kap.
Këure: *Gardenia ternifolia* Schum. et Thonn.; *Rubiac.;* Togo.
Kevazingo (f): *Copaifera demeusi* Harms; *Caesalpiniac.;* Kam., Gab.
„ „ : *Copaifera tessmanni* Harms; *Caesalpiniac.;* Kam., Gab.
Kewar: *Holarrhena antidysenterica* Wall.; *Apocynac.;* nw. O.-Ind. (Punjab-Kangra)
Kewer: *Sapindus senegalensis* Poir.; *Sapindac.;* Senl.
Keyaki: *Zelkova formosana* Hayata; *Ulmac.;* Form.
„ : *Zelkova acuminata* Planch., *Z. serrata* Makino; *Ulmac.;* Jap.

262

Keyi: *Diospyros mespiliformis* Hochst.; *Ebenac.;* Go.
„ -keyi: *Diospyros mespiliformis* Hochst.; *Ebenac.;* Go.
Kha giom: *Shorea cochinchinensis* Pierre; *Dipterocarpac.;* Ind.-Ch.
„ jung: *Dalbergia cochinchinensis* Pierre; *Papilionac.;* Ind.-Ch.
„ nhom: *Shorea cochinchinensis* Pierre; *Dipterocarpac.;* Ind.-Ch.
„ nhung: *Dalbergia cochinchinensis* Pierre; *Papilionac.;* Ind.-Ch.
Khadka: *Bridelia retusa* Spreng.; *Euphorbiac.;* C.-O.-Ind.
Khai: *Miliusa bailloni* Pierre; *Anonac.;* Ind.-Ch.
„ deng: *Croton joufra* Roxb.; *Euphorbiac.;* Ind.-Ch.
Khaia: *Strychnos potatorum* L. f.; *Loganiac.;* C.-O.-Ind. (Gondwana)
Khaiger: *Acacia ferruginea* DC.; *Mimosac.;* O.-Ind.
Khair: *Acacia catechu* Willd.; *Mimosac.;* sw. O.-Ind. (Bombay)
„ bora: *Acacia ferruginea* DC.; *Mimosac.;* C.-O.-Ind.
Khairwal: *Bauhinia purpurea* L.; *Caesalpiniac.;* nw. O.-Ind. (Punjab-Simla)
„ papri: *Bauhinia purpurea* L.; *Caesalpiniac.;* n. O.-Ind. (United Prov.)
Khaja: *Bridelia retusa* Spreng.; *Euphorbiac.;* n. O.-Ind. (United Prov.)
Khakda: *Butea frondosa* Roxb.; *Papilionac.;* O.-Ind.
Khakra: *Butea frondosa* Roxb.; *Papilionac.;* w. O.-Ind. (Ajmer-Merwara)
Khalawa: *Wrightia tomentosa* Roem. et Schultes; *Apocynac.;* nw. O.-Ind. (Punjab)
Khalemeroe: *Protium spp.; Burserac.; Qualea albiflora* Warm.; *Vochysiac.;* Sur.
„ -koelisirie: *Matayba spp., Cupania scrobiculata* L. C. Rich.; *Sapindac.;* Sur.
Kham: *Phyllanthus emblica* L.; *Euphorbiac.;* Ind.-Ch.
„ : *Tamarindus indica* L.; *Caesalpiniac.;* Ind.-Ch.
„ : *Gironniera sinensis* Bth.; *Ulmac.;* Ind.-Ch.
Khamara: *Trewia nudiflora* L.; *Euphorbiac.;* nw. O.-Ind. (Kumaon)
Khamer: *Gmelina arborea* L.; *Verbenac.;* C.-O.-Ind.
Khamnar: *Gmelina arborea* L.; *Verbenac.;* n. O.-Ind. (United Prov.)
Khanak: *Celtis australis* L.; *Ulmac.;* n. O.-Ind. (Dehra Dun)
Khandergai: *Garuga pinnata* Roxb.; *Burserac.;* w. O.-Ind. (Deccan)
Khao: *Cinnamomum sp.; Laurac.;* Ind.-Ch.
„ : *Symplocos ferruginea* Roxb.; *Symplocac.;* Ind.-Ch.
„ : *Bignonia longissima* Jacq.; *Bignoniac.;* Ind.-Ch.
„ : *Randia longiflora* Lamk.; *Rubiac.;* Ind.-Ch.
„ boc: *Vatica tonkinensis* Chev.; *Dipterocarpac.;* Ind.-Ch.
„ cai: *Vatica tonkinensis* Chev.; *Dipterocarpac.;* Ind.-Ch.
„ chuông: *Laurus camphorata* Buch.-Ham.; *Laurac.;* Ind.-Ch.
„ khinh: *Litsea citrata* Blume; *Laurac.;* Ind.-Ch.
„ lech: *Vatica tonkinensis* Chev.; *Dipterocarpac.;* Ind.-Ch.
„ mât: *Cinnamomum sp.; Laurac.;* Ind.-Ch.
„ thôi: *Symplocos ferruginea* Roxb.; *Symplocac.;* Ind.-Ch.
„ tia: *Machilus odoratissima* Nees; *Laurac.;* Ind.-Ch.
„ vàng: *Elaeocarpus madopetalus* Pierre; *Elaeocarpac.;* Ind.-Ch.
Khar: *Tamarix articulata* Vahl; *Tamaricac.;* O.-Ind.
„ phulsa: *Grewia laevigata* Vahl; *Tiliac.;* sw. O.-Ind. (Bombay)
Kharak: *Celtis australis* L.; *Ulmac.;* n. O.-Ind. (Jaunsar, Garhwal)
Kharamb: *Machilus gamblei* King; *Laurac.;* nw. O.-Ind. (Punjab)
Kharani: *Symplocos theaefolia* Ham.; *Symplocac.;* O.-Ind.
Kharanj: *Quercus incana* Roxb.; *Fagac.;* nw. O.-Ind.
Kharé-chotor: *Alhagi persarum* Boiss. et Buhse; *Papilionac.;* Pers.
Kharemero-jaroro: *Swartzia acuminata* Willd., *S. spp.; Caesalpiniac.;* Sur.
Kharemeroe-tataboe: *Diplotropis guianensis* Bth., *D. leptophylla* Kleinh.; *Papilionac.;*
Kharik: *Celtis australis* L.; *Ulmac.;* nw. O.-Ind. (Jaunsar, Garhwal)   [Sur.
Kharot: *Juglans regia* L.; *Juglandac.;* nw. O.-Ind. (Kumaon)
Kharpat: *Garuga pinnata* Roxb.; *Burserac.;* w. Beng.
Kharréméroe: *Qualea albiflora* Warm.; *Vochysiac.;* Sur.
Kharshu: *Quercus semecarpifolia* Smith; *Fagac.;* n. O.-Ind. (United Prov.)
Kharsing: *Stereospermum xylocarpum* Bth. et Hook. f.; *Bignoniac.;* sw. O.-Ind.
Kharsu: *Quercus semecarpifolia* Smith; *Fagac.;* nw. O.-Ind.
Kharub: *Bauhinia reticulata* DC.; *Caesalpiniac.;* s. Sudan.
Kharya: *Quercus semecarpifolia* Smith; *Fagac.;* nw. O.-Ind. (Kaschmir)
Khas: *Borassus aethiopum* Warb. *var. khas* ?; *Palmac.;* Senl.
Khassu: *Quercus semecarpifolia* Smith; *Fagac.;* nw. O.-Ind. (Kaschmir)
Khat: *Catha edulis* Forsk.; *Celastrac.;* Arab.
„ -papri: *Bauhinia malabarica* Roxb.; *Caesalpiniac.;* n. O.-Ind. (Unit.-Prov.)

Khatsawar: *Bombax malabaricum* DC.; *Bombacac.;* C.-O.-Ind. (sö. Berar)
Khatsaweri: *Bombax malabaricum* DC.; *Bombacac.;* C.-O.-Ind. (sö. Berar)
Khattajhinjhora: *Bauhinia malabarica* Roxb.; *Caesalpiniac.;* n. O.-Ind. (Unit.-Prov.)
Khatua: *Bauhinia malabarica* Roxb.; *Caesalpiniac.;* C.-O.-Ind.
Khatyan-Ka-Kalli: *Ceiba pentandra* Gaertn.; *Bombacac.;* O.-Ind.
Khau: *Olea cuspidata* Wall.; *Oleac.;* O.-Ind.
Khaya: *Mimusops elengi* Roxb.; *Sapotac.;* And.
„ mulsari: *Mimusops elengi* Roxb.; *Sapotac.;* And.
Khé: *Markhamia stipulata* Seem., *Stereospermum annamense* Chev.; *Bignoniac.;* Ind.-Ch.
„ : *Averrhoa carambola* L.; *Oxalidac.;* Annam.
„ chua: *Averrhoa carambola* L.; *Oxalidac.;* Ind.-Ch.
Khen: *Hopea odorata* Roxb.; *Dipterocarpac.;* Ind.-Ch. (Laos)
Kherwa: *Holarrhena antidysenterica* Wall.; *Apocynac.;* w. Beng. (Orissa)
Khet: *Pithecolobium clypearia* Bth. *var. acuminatum* Gagnep.; *Mimosac.;* Annam.
Kheu: *Melanorrhoea usitata* Wall.; *Anacardiac.;* Assam (Manipur)
Kheunia: *Ficus glomerata* Roxb.; *Morac.;* nw. O.-Ind. (Kumaon)
Khi cay: *Cipadessa fruticosa* Blume; *Meliac.;* Ind.-Ch.
„ lech: *Cassia siamea* Lamk.; *Caesalpiniac.;* Ind.-Ch.
„ si may hong: *Shorea vulgaris* Pierre; *Dipterocarpac.;* Ind.-Ch.
Khina: *Sapium insigne* Bth.; *Euphorbiac.;* O.-Ind.
Khiri: *Mimusops elengi* L., *M. hexandra* Roxb.; *Sapotac.;* O.-Ind., w. Beng.
Khirk: *Celtis australis* L.; *Ulmac.;* nw. O.-Ind. (Punjab-Hazara)
Khirkuli: *Mimusops hexandra* Roxb.; *Sapotac.;* w. Beng. (Bihar, Orissa)
Khirni: *Mimusops hexandra* Roxb.; *Sapotac.;* C.-O.-Ind.
Khleum borout: *Eugenia jambos* L. *var. silvatica* Lam.; *Myrtac.;* Ind.-Ch.
Khli: *Randia dumetorum* Lamk.; *Rubiac.;* Kamb.
Khlok: *Parinarium annamense* Hance; *Rosac.;* Ind.-Ch.
Khlong: *Dipterocarpus tuberculatus* Roxb.; *Dipterocarpac.;* Kamb.
Khmau· *Diospyros mun* H. Lec.; *Ebenac.;* Ind.-Ch.
Khnor: *Artocarpus integrifolia* L.; *Morac.;* Ind.-Ch.
Khoai: *Artocarpus tonkinensis* Chev.; *Morac.;* Ind.-Ch.
Khofo: *Dalbergia melanoxylon* Guill. et Perr.; *Papilionac.;* Sudan.
Khogra: *Acacia ferruginea* DC.; *Mimosac.;* O.-Ind. (Mandavi)
Khoira: *Acacia catechu* Willd.; *Mimosac.;* Assam.
Khoiru: *Acacia catechu* Willd.; *Mimosac.;* O.-Ind. (Uriya)
Khoja: *Callicarpa arborea* Roxb.; *Uerbenac.;* Assam.
Khor: *Juglans regia* L.; *Juglandac.;* nw. O.-Ind.
Khosla: *Grewia tiliaefolia* Vahl; *Tiliac.;* C.-O.-Ind. (Gondwana)
Khraikata: *Nauclea sessilifolia* Roxb.; *Rubiac.;* Burma.
Khtiaou: *Shorea thoreli* Pierre; *Dipterocarpac.;* Ind.-Ch.
Khting: *Calophyllum dryobalanoides* Pierre; *Guttifer.;* Ind.-Ch.
Khtung: *Podocarpus pedatus* ?; *Conifer.;* Ind.-Ch.
Khuat suôn: *Citrus sp.; Rutac.;* Ind.-Ch.
Khurpendra: *Gardenia turgida* Roxb.; *Rubiac.;* C.-O.-Ind. (Gondwana)
Khurrur: *Gardenia turgida* Roxb.; *Rubiac.;* n. O.-Ind. (United Prov.)
Khursi: *Gmelina arborea* L.; *Uerbenac.;* C.-O.-Ind. (Gondwana)
Khuyon vi: *Markhamia stipulata* Seem.; *Bignoniac.;* Tonkin.
Khuyu: *Elaeocarpus dubius* A. DC.; *Elaeocarpac.;* Ind.-Ch.
Khvao: *Mangifera foetida* Lour.; *Anacardiac.;* Ind.-Ch.
„ : *Bignonia longissima* Jacq.; *Bignoniac.;* Ind.-Ch.
Khwan: *Olea cuspidata* Wall.; *Oleac.;* nw. O.-Ind. (Punjab)
Khwaw: *Dillenia pentagyna* Roxb.; *Dilleniac.;* Burma.
Ki-bako: *Ehretia javanica* Blume; *Borraginac.;* Java.
„ -djoelang: *Pahudia javanica* Miq.; *Caesalpiniac.;* Java.
„ -hidoeng: *Pancovia edulis* Willd.; *Sapindac.;* Sunda.
„ -ich-ché: *Guettarda elliptica* Swartz; *Rubiac.;* br. Hond.
„ -koetjing: *Sophora tomentosa* L.; *Papilionac.;* Malay.
„ -kou: *Hovenia dulcis* Thunb.; *Rhamnac.;* Jap.
„ -lep pa: *Cassia timoriensis* DC.; *Caesalpiniac.;* Ind.-Ch.
„ -manoek: *Tarrietia javanica* Blume; *Sterculiac.;* Nied.-Ind.
„ -mbidi: *Canarium schweinfurthi* Engl.; *Burserac.;* B.-K.
„ -pantjar: *Pterospermum sp.; Sterculiac.;* Java.
„ -sapie: *Gordonia excelsa* Blume; *Theac.;* Sunda.
„ -sasage: *Catalpa kaempferi* S. et Z.; *Bignoniac.;* Jap.

Ki sereh: *Cinnamomum parthenoxylon* Meissn.; *Laurac.;* Java.
„ sona louang: *Casearia kerri* Craib, *C. greviaefolia* Vent.; *Flacourtiac.;* Ind.-Ch.
„ -tandjoeng: *Mimusops elengi* L.; *Sapotac.;* Malay.
„ -tarrietie: *Tarrietia javanica* Blume; *Sterculiac.;* Nied.-Ind.
„ -thmar: *Dipterocarpus sp.; Dipterocarpac.;* Ind.-Ch.
„ -tokè: *Albizzia lebbek* Bth.; *Mimosac.;* Java.
„ -yen: *Homalium dictyoneurum* Pierre; *Flacourtiac.;* Ind.-Ch.
Kia-kía: *Dracontomelum dao* Merr. et Rolfe; *Anacardiac.;* Phil. (Leyte)
„ „ : *Pometia pinnata* Forst.; *Sapindac.;* Phil. (Samar, Leyte, Cebu)
Kiabanj: *Carallia lucida* Roxb.; *Rhizophorac.;* Beng.
Kiachalom: *Dalbergia lanceolaria* L.; *Papilionac.;* n. O.-Ind.
Kiaga: *Omphalocarpum anocentrum* Pierre; *Sapotac.;* Elf.
Kiagua: *Omphalocarpum anocentrum* Pierre; *Sapotac.;* Elf.
Kiahong: *Melanorrhoea usitata* Wall.; *Anacardiac.;* O.-Ind.
Kiakia: *Dracontomelum dao* Merr. et Rolfe; *Anacardiac.;* Phil.
Kiamil: *Lannea grandis* Engl.; *Anacardiac.;* O.-Ind. (Hind.)
Kiang: *Pycnanthus kombo* Warb.; *Myristicac.;* Kam.
Kibalebale: *Afzelia quanzensis* Welw.; *Caesalpiniac.;* B.-K. (Katanga)
Kibawang-bodas: *Dysoxylum mollissimum* Miq.; *Meliac.;* Java.
Kibewog peutjang: *Aglaia elaeagnoidea* Bth.; *Meliac.;* Nied.-Ind.
Kibewok: *Aglaia elaeagnoidea* Bth.; *Meliac.;* Nied.-Ind.
Kibiroe: *Canarium hispidum* Blume; *Burserac.;* Sunda.
Kiboenaga: *Calophyllum inophyllum* L.; *Guttifer.;* Nied.-Ind.
Kidgelin: *Uapaca togoensis* Pax; *Euphorbiac.;* Togo.
Kidgelung: *Uapaca togoensis* Pax; *Euphorbiac.;* Togo.
Kidjoedjoel: *Amoora grandifolia* C. DC.; *Meliac.;* Nied.-Ind.
Kièerèoe: *Dicorynia paraënsis* Bth.; *Caesalpiniac.;* Sur.
Kiefer (d): *Pinus silvestris* L.; *Conifer.;* Eu.
„ , Aleppo-: *Pinus halepensis* Mill., *P. brutia* Tenore; *Conifer.;* s. Eu.
„ , Berg-: *Pinus montana* Mill.; *Conifer.;* Eu.
„ , Emodi-: *Pinus longifolia* Roxb.; *Conifer.;* w. Himal.
„ , Gelb-: *Pinus ponderosa* Dougl.; *Conifer.;* w. USA.
„ , gemeine: *Pinus silvestris* L.; *Conifer.;* Eu.
„ , Hacken-: *Pinus montana* Mill. *var. uncinata* Willk.; *Conifer.;* Mitt.-Eu.
„ , Hänge-: *Pinus flexilis* James; *Conifer.;* N.-Am.
„ , Igel-: *Pinus pinaster* Sol.; *Conifer.;* M.-M.
„ , „ : *Pinus echinata* Mill.; *Conifer.;* N.-Am.
„ , kleinblütige: *Pinus parviflora* S. et Z.; *Conifer.;* Jap.
„ , Monterey-: *Pinus insignis* Dougl.; *Conifer.;* N.-Am.
„ , Murray-: *Pinus murrayana* Balf.; *Conifer.;* W.-Af.
„ , Panzer-: *Pinus heldreichi* Christ *var. typica* Mgf.; *Conifer.;* s. Eu.
„ , „ : *Pinus heldreichi* Christ *var. leucodermis* Mgf.; *Conifer.;* s. Eu.
„ , Pech-: *Pinus rigida* Mill.; *Conifer.;* N.-Am.
„ , Schlangenhaut-: *Pinus heldreichi* Christ *var. leucodermis* Mgf.; *Conifer.;* s. Eu.
„ . Schwarz-: *Pinus nigra* Arn.; *Conifer.;* s. Eu. (It.)
„ , „ , österreichische: *Pinus nigra* Arn. *var. austriaca* Höss.; *Conifer.;* Eu.
„ , Seestrand-: *Pinus pinaster* Sol.; *Conifer.;* It.
„ , Sumpf-: *Pinus palustris* Mill.; *Conifer.;* N.-Am.
„ , Weymouths-: *Pinus strobus* L.; *Conifer.;* N.-Am.
„ , „ , Himalaya-: *Pinus excelsa* Wall.; *Conifer.;* Himal.
„ , Wiesen-: *Pinus rigida* Mill., *P. serotina* Michx.; *Conifer.;* N.-Am.
„ , Zirbel-: *Pinus cembra* L.; *Conifer.;* Eu.
„ , Zucker-, kalifornische: *Pinus lambertiana* Dougl.; *Conifer.;* w. N.-Am.
„ , Zwerg-: *Pinus banksiana* Lamb.; *Conifer.;* N.-Am.
Kièjère-oe: *Dicorynia paraënsis* Bth.; *Caesalpiniac.;* Sur.
Kiendoh: *Xanthopyllum vitellinum* Blume; *Polygonac.;* Malay.
Kiên-kiên: *Hopea pierrei* Hance; *Dipterocarpac.;* Annam.
„ - „ : *Adenanthera microsperma* Teijsm. et Binn.; *Mimosac.;* Annam.
Kienbaum: *Pinus taeda* L.; *Conifer.;* N.-Am.
Kiêng: *Pentace tonkinensis* Chev.; *Tiliac.;* Ind.-Ch.
„ Kiêng: *Pentace tonkinensis* Chev.; *Tiliac.;* Ind.-Ch.
Kierpa: *Carallia lucida* Roxb.; *Rhizophorac.;* Beng.
Kieselholz, Antillen-: *Pithecolobium unguis-cati* Bth.; *Mimosac.;* W.-Ind.
„ , mexikanisches: *Calliandra tetragona* Bth.; *Mimosac.;* Mex., C.-Am.

Kiesie-ma: *Humiria floribunda* Mart., *H. balsamifera* Aubl.; *Humiriac.;* Sur.
Kifabakasi: *Spathodea campanulata* Beauv.; *Bignoniac.;* br. O.-Af. (Uganda)
Kifoo-umbië: *Bauhinia thonningi* K. Schum.; *Caesalpiniac.;* B.-K.
Kifungwela: *Kigelia aethiopica* Dcne.; *Bignoniac.;* B.-K.
Kigogwèka: *Elaeodendron stuhlmanni* Loes.; *Celastrac.;* D.-O.-Af. (w. Usambara)
Kigwándi: *Dasylepis leptophylla* Gilg; *Flacourtiac.;* D.-O.-Af. (w. Usambara)
Kihada: *Phellodendron amurense* Rupr.; *Rutac.;* Jap.
Ki-hadji: *Dysoxylum macrocarpum* Blume; *Meliac.;* Java.
Kihagílo: *Maba buxifolia* Pers.; *Ebenac.;* D.-O.-Af. (w. Usambara)
Kiharpan: *Canarium altissimum* Blume; *Burserac.;* Nied.-Ind.
Kihoë: *Xerospermum noronhianum* Blume; *Sapindac.;* Nied.-Ind.
Kihoeh: *Xerospermum noronhianum* Blume; *Sapindac.;* Nied.-Ind.
Kijai: *Trigonochlamys griffithi* Hook. f.; *Burserac.;* Malay.
Kijiy: *Symphonia fasciculata* Baill.; *Guttifer.;* Mad.
Kijy: *Symphonia sp.; Guttifer.;* Mad.
Kikar: *Acacia arabica* Willd.; *Mimosac.;* nw. O.-Ind.
Kikatjang: *Strombosia javanica* Blume; *Olacac.;* Sunda.
„ laut: *Strombosia javanica* Blume; *Olacac.;* Sunda.
Kikatjapie: *Sandoricum indicum* Cav.; *Meliac.;* Java.
Kikera: *Melia azedarach* L.; *Meliac.;* Nied.-Ind.
Kikeujup: *Strombosia javanica* Blume; *Olacac.;* Sunda.
Kikobwa: *? Stereospermum katangensis* de Wild.; *Bignoniac.;* B.-K. (Katanga)
Kikuakua: *Caloncoba welwitschi* Gilg; *Flacourtiac.;* Kam.
„ -bukala: *Caloncoba glauca* Gilg; *Flacourtiac.;* B.-K.
Kikuwa-kuwa: *Datura sp.; Solanac.;* D.-O.-Af. (Tabora)
Kikuyu m'tarakma: *Juniperus procera* Hochst.; *Conifer.;* br. O.-Af.
Kila: *Shorea negrosensis* Foxw.; *Dipterocarpac.;* Phil. (Sorsogon)
Kilabe: *Fagara usambarensis* Engl.; *Rutac.;* D.-O.-Af. (w. Usambara)
Kilai: *Albizzia procera* Bth.; *Mimosac.;* O.-Ind. (Bombay)
Kilalayoe: *Pancovia edulis* Willd.; *Sapindac.;* Java.
„ hiedung: *Pancovia edulis* Willd.; *Sapindac.;* Java.
Kilaoet: *Strombosia javanica* Blume; *Olacac.;* Sunda.
Kileloh: *Saurauia sp.; Dilleniac.;* Java.
Kili: *Albizzia procera* Bth.; *Mimosac.;* Assam;
Kilindira-mgunda: *Gardenia thunbergi* L. f.; *Rubiac.;* D.-O.-Af. (Tabora)
Kilkíl: *Cassia javanica* L.; *Caesalpiniac.;* Phil. (Zamboanga)
Kiloeng-loeng: *Aromadendron elegans* Blume; *Magnoliac.;* Nied.-Ind.
Kilóg: *Quercus sp., Q. luzoniensis* Merr.; *Fagac.;* Phil.
Kilonj: *Quercus dilatata* Lindley; *Fagac.;* n. O.-Ind. (Kumaon, Garhwal)
Kilpattar: *Acer pictum* Thunb.; *Acerac.;* nw. O.-Ind. (Punjab)
Kilu: *Acer caesium* Wall.; *Acerac.;* nw. O.-Ind. (Kumaon)
Kilung-lung: *Aromadendron elegans* Blume; *Magnoliac.;* Nied.-Ind.
Kim giao: *Podocarpus latifolia* Wall.; *Conifer.;* Annam.
„ phuong: *Caesalpinia pulcherrima* Sw.; *Caesalpiniac.;* Ind.-Ch.
„ cang: *Elaeocarpus lacunosus* Wall.; *Elaeocarpac.;* Ind.-Ch.
Kimanga: *Erythrophloeum coumingo* Baill.; *Caesalpiniac.;* Mad.
Kimangal: *Schima wallichi* Choisy; *Theac.;* Sunda.
Kimanjal: *Gordonia excelsa* Blume; *Theac.;* Sunda.
Kimba: *Xylopia aethiopica* A. Rich.; *Anonac.;* Kam.
„ : *Plectronia lamprophylla* K. Schum.; *Rubiac.;* D.-O.-Af. (w. Usambara)
Kimbu: *Morus laevigata* Wall.; *Morac.;* nö. O.-Ind. (Nepal)
Kimeong: *Lasianthera macrophylla* Miq.; *Icacinac.;* Sunda.
Kimiseh: *Baphia laurenti* de Wild.; *Papilionac.;* B.-K. (Eala)
Kimpampa: *Monotes katangensis* de Wild., *M. glabra* Sprague; *Dipterocarpac.;* B.-K.
Kimu: *Morus serrata* Roxb.; *Morac.;* n. O.-Ind. (Punjab, Jaunsar)   [(Katanga)
Kimul: *Lannea grandis* Engl.; *Anacardiac.;* n. O.-Ind. (United Prov.)
Kin jung: *Sciadopitys verticillata* S. et Z.: *Conifer.;* s. Jap.
„ kori: *Hovenia dulcis* Thunb.; *Rhamnac.;* Jap.
„ nemu: *Acacia farnesiana* Willd.; *Mimosac.;* Jap.
„ sung: *Sciadopitys verticillata* S. et Z.; *Conifer.;* Jap.
Kina: *Myroxylon toluiferum* H. B. K.; *Papilionac.;* Arg.
„ : *Calophyllum walkeri* Wight; *Guttifer.;* Cey.
„ morado: *Myroxylon toluiferum* H. B. K.; *Papilionac.;* Arg.
„ du Rio-Nunez: *Sarcocephalus esculentus* Afzel.; *Rubiac.;* Elf.

Kina kina: *Myroxylon toluiferum* H. B. K.; *Papilionac.;* Arg.
Kinasaikásai: *Adenanthera intermedia* Merr.; *Mimosac.;* Phil.
Kindal: *Terminalia paniculata* Roth; *Combretac.;* sw. O.-Ind. (Bombay)
Kindi: *Uapaca guineensis* Muell. Arg.; *Euphorbiac.;* Lib.
Kindzou: *Dialium aff. connaroides* Harms; *Caesalpiniac.;* Gab.
King nut: *Carya sulcata* Nutt.; *Juglandac.;* USA.
King of the timber: *Khaya anthotheca* C. DC.; *Meliac.;* W.-Af. (Angola)
Kingele: *Ricinodendron africanum* Muell. Arg.; *Euphorbiac.;* W.-Af. (Kisantu)
Kingelu: *Spondias lutea* L.; *Anacardiac.;* Togo.
Kingué: *Fagara macrophylla* Engl.; *Rutac.;* Elf.
  „   : *Bombax brevicuspe* Sprague; *Bombacac.;* Lib.
Kingwood (H): *Astronium fraxinifolium* Schott; *Anacardiac.;* ö. Bras.
  „   (u): *Dalbergia cearensis* Ducke; *Papilionac.;* nö. Bras. (Ceara)
  „   : *Fagraea fragrans* Roxb.; *Loganiac.;* tr. As.
  „   , brazilian: *Dalbergia cearensis* Ducke; *Papilionac.;* nö. Bras. (Ceara)
Kinhai: *Albizzia procera* Bth.; *Mimosac.;* sw. O.-Ind. (Bombay)
Kinhi: *Albizzia lebbek* Bth., *A. procera* Bth.; *Mimosac.;* C.-O.-Ind.
Kinhiphata: *Albizzia procera* Bth.; *Mimosac.;* C.-O.-Ind.
Kinjal: *Terminalia paniculata* Roth; *Combretac.;* sw. O.-Ind. (Bombay)
Kinkiri: *Burkea africana* Hook.; *Caesalpiniac.;* Togo.
Kinkono-won: *Coelocaryon oxycarpum* Stapf; *Myristicac.;* Elf.
Kinkuésin: *? Cynometra sp., ? Cryptosepalum sp.; Caesalpiniac.;* Elf.
Kinnu: *Diospyros tomentosa* Roxb.; *Ebenac.;* nw. O.-Ind. (Punjab-Kangra)
Kino, african: *Pterocarpus erinaceus* Poir.; *Papilionac.;* Go.
  „   , american: *Coccoloba uvifera* Jacq.; *Polygonac.;* Fl., W.-Ind., ö. S.-Am.
  „   gum: *Coccoloba uvifera* Jacq.; *Polygonac.;* Fl., W.-Ind., ö. S.-Am.
  „   tree, West Indian-: *Coccoloba uvifera* Jacq.; *Polygonac.;* Fl., br. W.-Ind.
  „   d'Amérique: *Coccoloba uvifera* Jacq.; *Polygonac.;* Ant.
Kinsa: *Castanopsis indica* A. DC.; *Fagac.;* Burma.
Kinsi: *Rhizophora racemosa* G. F. W. Meyer; *Rhizophorac.;* Elf.
Kinsonghia: *Symphonia gabonensis* Pierre; *Guttifer.;* B.-K. (Kisantu)
Kinthat-putgyi: *Chukrasia tabularis* Adr. Juss.; *Meliac.;* s. Burma (s. Tenass.)
Kinuyanagi: *Salix viminalis* L.; *Salicac.;* Jap.
Kion: *Pterygota kamerunensis* K. Schum.; *Sterculiac.;* Kam.
Kioray: *Commersonia echinata* Forst.; *Sterculiac.;* Nied.-Ind.
Kipada: *Phellodendron amurense* Rupr.; *Rutac.;* Jap.
Kipampa: *Monotes spp.; Dipterocarpac.;* n. Rhod.
Kipapa: *Afzelia quanzensis* Welw.; *Caesalpiniac.;* B.-K. (Katanga)
Kiparai: *Lepisanthes montana* Blume; *Sapindac.;* Java.
Kirakuli: *Mimusops hexandra* Roxb.; *Sapotac.;* w. Beng. (Orissa)
Kirala: *Cassia fistula* L.; *Caesalpiniac.;* n. O.-Ind. (United Prov.)
Kiralboghi: *Hopea parviflora* Bedd.; *Dipterocarpac.;* sw. O.-Ind. (Bombay)
Kiranji: *Albizzia procera* Bth.; *Mimosac.;* C.-O.-Ind.
Kirballi: *Poeciloneuron indicum* Bedd.; *Guttifer.;* sw. O.-Ind. (Bombay)
Kiri: *Paulownia imperialis* S. et Z.; *Scrophulariac.;* Jap.
  „   -mawaa: *Holarrhena mitis* R. Br.; *Apocynac.;* O.-Ind.
Kiria: *Cynometra megalophylla* Harms; *Caesalpiniac.;* Togo.
  „   : *Entada abyssinica* Steud.; *Mimosac.;* Togo.
Kiricowa: *Virola sebifera* Aubl.; *Myristicac.;* br. Gu.
Kirikawa: *Virola sebifera* Aubl.; *Myristicac.;* br. Gu.
  „   : *Iryanthera sagotiana* Warb., *I. hostmani* Warb.; *Myristicac.;* Sur.
  „   -wade-likoro: *Iryanthera sagotiana* Warb., *I. hostmani* Warb.; *Myristicac.;* Sur.
Kiriwalla: *Holarrhena mitis* R. Br.; *Apocynac.;* O.-Ind.
Kiriya: *Prosopis oblonga* Bth.; *Mimosac.;* n. Nig.
  „   ta mata: *Amblygonocarpus schweinfurthi* Harms; *Mimosac.;* n. Nig.
Kirmalia: *Cassia fistula* L.; *Caesalpiniac.;* w. O.-Ind. (Ajmer-Merwara)
Kirmola: *Acer oblongum* Wall.; *Acerac.;* n. O.-Ind. (Garhwal)
Mirmoti: *Acer oblongum* Wall.; *Acerac.;* n. O.-Ind.
Kirnee: *Mimusops hexandra* Roxb.; *Sapotac.;* O.-Ind.
Kirni: *Bridelia ferruginea* Bth.; *Euphorbiac.;* n. Nig., Kam.
Kironli: *Calophyllum polyanthum* Wall.; *Guttifer.;* nö. O.-Ind. (Nepal)
Kirra: *Wrightia tomentosa* Roem. et Schultes; *Apocynac.;* nö. O.-Ind. (Nepal)
Kirrne: *Bridelia ferruginea* Bth.; *Euphorbiac.;* n. Nig. (Kontagora)
Kirroah: *Ehretia abyssinica* R. Br.; *Borraginac.;* nö. Af. (n. Abessinien)

Kirsche: *Prunus spp.; Rosac.;* Eu.
„ , Ahl-: *Prunus padus* L.; *Rosac.;* Eu.
„ , Felsen-: *Prunus mahaleb* L.; *Rosac.;* w. & s. Eu.
„ , Hecken-, gemeine: *Lonicera xylosteum* L.; *Caprifoliac.;* Eu.
„ ,   „ , schwarze: *Lonicera nigra* L.; *Caprifoliac.;* Mitt.-Eu.
„ , Kornel-: *Cornus mas* L.; *Cornac.;* Mitt.- & S.-Eu.
„ , Sauer-: *Prunus cerasus* L.; *Rosac.;* Kl.-As.
„ , Trauben-: *Prunus padus* L.; *Rosac.;* Eu.
„ ,   „ , spätblühende: *Prunus serotina* Ehrh.; *Rosac.;* w. N.-Am.
„ , Vogel-: *Prunus avium* L.; *Rosac.;* Eu.
„ , Weichsel-: *Prunus mahaleb* L.; *Rosac.;* w. & s. Eu.
„ , Wild-: *Prunus avium* L.; *Rosac.;* Eu.
Kirschhout: *Pterocelastrus variabilis* Sond.; *Celastrac.;* Kap.
Kirsel: *Stereospermum chelonoides* DC.; *Bignoniac.;* sw. O.-Ind. (Bombay)
Kirton wood: *Elaeocarpus dentatus* Vahl; *Elaeocarpac.;* ö. Au.
Kisafou: *Pachylobus edulis* G. Don; *Burserac.;* Gab.
„ : *Pachylobus edulis* G. Don *var. sativa* Ficalho; *Burserac.;* Aeq. Af.
Kisafoukala: *Pachylobus edulis* G. Don; *Burserac.;* Gab.
„ : *Pachylobus edulis* G. Don *var. sativa* Ficalho; *Burserac.;* Aeq. Af.
Kisamfi: *Uapaca guineensis* Muell. Arg.; *Euphorbiac.;* B.-K.
Kisampan: *Evodia aromatica* Blume; *Rutac.;* Sunda.
Ki-semir: *Hura crepitans* L.; *Euphorbiac.;* Java.
Kiserogol: *Tarrietia javanica* Blume; *Sterculiac.;* Nied.-Ind.
Kishia: *Sarcocephalus esculentus* Afz.; *Rubiac.;* Lib.
Kishing: *Aesculus indica* Colebr.; *Hippocastanac.;* nw. O.-Ind. (Kaschmir)
Kishongo: *Ricinodendron africanum* Muell. Arg.; *Euphorbiac.;* br. O.-Af. (Uganda)
Kisi-hinau: *Elaeocarpus dentatus* Vahl; *Elaeocarpac.;* N.-Seel.
Kisia: *Sarcocephalus esculentus* Afz.; *Rubiac.;* Go.
Kisigoeng: *Echinocarpus sigun* Blume; *Elaeocarpac.;* Sunda.
Kiskidee: *Vismia cayennensis* Pers.; *Guttifer.;* Trin., Tob.
Kiskissi kamkam: *? Apeiba aspera* Aubl.; *Tiliac.;* Sur.
Kisni: *Bridelia scleroneura* Muell. Arg., *B. ferruginea* Bth.; *Euphorbiac.;* n. Nig.
Kisoko: *Trichilia kisoko* de Wild.; *Meliac.;* Mitt.-Kongo.
Kisonghi: *Allanblackia kisonghi* ?; *Guttifer.;* B.-K. (Pese)
Kisonghia: *Allanblackia kisonghi* ?; *Guttifer.;* B.-K. (Pese)
Kispur: *Mallotus philippinensis* Muell. Arg.; *Euphorbiac.;* Beng.
Kistenbaum: *Allophylus occidentalis* Radlk.; *Sapindac.;* tr. Am.
Kitakíta: *Albizzia acle* Merr.; *Mimosac.;* Phil.
Kitek: *? Sterculia* sp.; *Sterculiac.;* Java.
Kitelohr: *Xanthophyllum vitellinum* Blume; *Polygalac.;* Malay, Java.
Kiti: *Hydnocarpus wightiana* Blume; *Flacourtiac.;* sw. O.-Ind. (Bombay)
Kitililundu: *Ricinodendron africanum* Muell. Arg.; *Euphorbiac.;* B.-K. (Katanga)
Kiting-kíting: *Kayea paniculata* Merr.; *Guttifer.;* Phil. (Bataan)
Kitiwoe: *Meliosma nitida* Blume; *Sapindac.;* Sunda.
Kitjapi: *Sandoricum indicum* Cav.; *Meliac.;* Malay.
Kitoewak: *Canarium commune* L.; *Burserac.;* Sunda.
Kitola: *Cassia fistula* L.; *Caesalpiniac.;* nw. O.-Ind. (Kumeon)
Kittir: *Acacia mellifera* Bth.; *Mimosac.;* Aegypten (Sudan)
Kiukuésin: *Cynometra* sp.; *Caesalpiniac.;* Elf.
Kiukuésin: *? Cynometra* sp., *? Cryptosepalum* sp.; *Caesalpiniac.;* Elf.
Kiukuma-uon: *Coelocaryon oxycarpum* Stapf; *Myristicac.;* Elf.
Kiuma: *Pseudocinchona africana* Chev.; *Rubiac.;* Elf.
Kiumbra: *Pseudocinchona africana* Chev.; *Rubiac.;* Elf.
Kivindi: *Parinarium glabrum* Oliv.; *Rosac.;* B.-K. (Ganda Sundi)
Kiwada: *Evodia glauca* Miq., *Phellodendron amurense* Rupr.; *Rutac.;* Jap.
Kizalahy: *Symphonia laevis* Jumelle; *Guttifer.;* Mad.
Kizavavy: *Symphonia louveli* Jumelle; *Guttifer.;* Mad.
Kjair: *Acacia catechu* Willd.; *Mimosac.;* O.-Ind.
Kjèkè-awha: *Stryphnodendron flammatum* Kleinh., *S. guianensis* Bth.; *Mimosac.;* Sur.
Klalar: *Dipterocarpus trinervis* Blume; *Dipterocarpac.;* Java, Sum.
Klampis: *Acacia tomentosa* Willd.; *Mimosac.;* Malay.
Klapak: *Shorea polysperma* Merr.; *Dipterocarpac.;* ö. holl. Borneo.
Klaw: *Aquilaria agallocha* Roxb.; *Thymelaeac.;* Burma.
Klayoe: *Pancovia edulis* Willd.; *Sapindac.;* Sunda, Java.

Klê: *Drypetes floribunda* Hutch.; *Euphorbiac.;* Go.
Klehn: *Caloncoba brevipes* Gilg; *Flacourtiac.;* Lib.
Kleipeo: *Pithecolobium gonggrijpi* Kleinh., *P. spp.; Mimosac.;* Sur.
Klepe-oë: *Pithecolobium gonggrijpi* Kleinh., *P. spp.; Mimosac.;* Sur.
Klêti: *Drypetes floribunda* Hutch.; *Euphorbiac.;* Go.
Kliajur: *Phoenix dactylifera* L.; *Palmac.;* sw. As.
Klike: *Dichrostachys glomerata* Hutch. et Dalz.; *Mimosac.;* Go.
Klipdoorn: *Scolopia mundti* Warb.; *Flacourtiac.;* Kap.
Kliphout: *Anaphrenium argenteum* E. Meyer; *Anacardiac.;* S.-Af.
Klo: *Bauhinia reticulata* DC.; *Caesalpiniac.;* Togo.
Kloetoom bassah: *Gordonia sp.; Theac.;* Sum.
Kloh: *Bombax insigne* Wall. *var. andamanica* Prain; *Bombacac.;* Burma.
   „ : *Bombax insigne* Wall. *var. wighti* Prain; *Bombacac.;* Burma.
Klongbau: *Fagara xanthoxyloides* Lam.; *Rutac.;* Togo.
Kloum tuk: *Stephegyne parvifolia* Korth.; *Rubiac.;* Kamb.
Klut Lyoo: *Pancovia edulis* Willd.; *Sapindac.;* Malak.
Knépier: *Melicocca bijuga* L.; *Sapindac.;* Ant., C.-Am., Gu., Ven., Par.
   „ : *Melicocca diversifolia* Juss.; *Sapindac.;* Réu., Maur.
Knet: *Bruguiera caryophylloides* Blume; *Rhizophorac.;* s. Burma (s. Tenasserim)
   „ -ywethe: *Bruguiera parviflora* Wight et Arn.; *Rhizophorac.;* s. Burma (s. Ten.)
Knipnee: *Melicocca bijuga* L.; *Sapindac.;* C.-Am.
Knippa: *Melicocca bijuga* L.; *Sapindac.;* Curaçao.
Knobhout: *Xanthoxylum capense* Harv.; *Rutac.;* O.-Af.
Knobthorn: *Xanthoxylum capense* Harv.; *Rutac.;* O.-Af.
Knobwood: *Xanthoxylum capense* Harv.; *Rutac.;* S.-Af.
Knock-me-back: *Hyperbaena winzerlingi* Standl.; *Menispermac.;* br. Hond.
Knong: *Pterocarpus pedatus* Pierre, *P. cambodianus* Pierre; *Papilionac.;* Ind.-Ch.
Knor prey: *Artocarpus polyphema* Pers.; *Morac.;* Ind.-Ch.
Knoto-epu: *Diplotropis guianensis* Bth., *D. leptophylla* Kleinh.; *Papilionac.;* Sur.
Ko: *Olea cuspidata* Wall.; *Oleac.;* nw. O.-Ind. (Punjab)
   „ : *Ceiba pentandra* Gaertn.; *Bombacac.;* Kamb.
   „ ka: *Cornus capitata* Wall.; *Cornac.;* Ind.-Ch.
   „ yé: *Canarium commune* L.; *Burserac.;* Annam.
Koa: *Lophira procera* Chev.; *Ochnac.;* Kam.
   „ : *Acacia koa* A. Gray; *Mimosac.;* Hawai.
   „ , curly: *Acacia koa* A. Gray; *Mimosac.;* Hawai.
   „ , figured: *Acacia koa* A. Gray; *Mimosac.;* Hawai.
Koan: *Tamarix articulata* Vahl; *Tamaricac.;* O.-Ind.
Kobadi: *Sarcocephalus pobeguini* Pob.; *Rubiac.;* fr. Guin.
Kobak: *Sideroxylon sp.; Sapotac.;* C.-Am.
Koberu: *Mimusops surinamensis* Miq., *M. spp.; Sapotac.;* Sur.
Kobi: *Carapa microcarpa* Chev.; *Meliac.;* Elf.
   „ : *Carapa touloucouna* Guill. et Perr.; *Meliac.;* fr. Guin.
   „ : *Carapa velutina* C. DC.; *Meliac.;* Sudan.
Kobo tree: *Copaifera copallifera* O. Ktze.; *Caesalpiniac.;* fr. Guin.
Kobunire: *Ulmus campestris* Sm. *var. vulgaris* Planch.; *Ulmac.;* Jap.
Kobushi: *Magnolia kobus* DC.; *Magnoliac.;* Jap.
Kocha molle: *Calyptranthes oreophila* Speg.; *Myrtac.;* Arg.
Kochbel: *Feronia elephantum* Correa; *Rutac.;* w. Beng. (Chota Nagpur)
Koda: *Ehretia acuminata* R. Br.; *Borraginac.;* Au.
Kodaga: *Holarrhena antidysenterica* Wall.; *Apocynac.;* sw. O.-Ind. (Madras)
Kôdaiho-gasi: *Lithocarpus kodaihoensis* Hay.; *Fagac.;* Form.
Kodale: *Alstonia scholaris* R. Brown; *Apocynac.;* w. O.-Ind. (Deccan)
Kodari: *Bridelia retusa* Spreng.; *Euphorbiac.;* O.-Ind. (Madras)
Kodelea: *Ormosia laxiflora* Bth.; *Papilionac.;* Togo.
Kodelia: *Anogeissus leiocarpus* Guill. et Perr.; *Combretac.;* Togo.
Kodi nirvitti: *Hydnocarpus wightiana* Blume; *Flacourtiac.;* sw. O.-Ind.
Kodjo oedoe: *Guarea gomma* Pulle; *Meliac.;* Sur.
Kodmurki: *Wrightia tinctoria* R. Brown; *Apocynac.;* w. O.-Ind. (Deccan)
Kodohure: *Sterculia barteri* Mast., *S. tragacantha* Lindl.; *Sterculiac.;* Go.
Kodzo: *Antiaris africana* Engl.; *Morac.;* Go.
Koé: *Phyllanthus discoideus* Muell. Arg.; *Euphorbiac.;* Elf.
Koe: *Alnus nepalensis* D. Don; *Betulac.;* nw. O.-Ind. (Punjab)
Koebésini: *Parinarium campestre* Aubl., ? *P. brachystachyum* Bth.; *Rosac.;* Sur.

Koeda Sondak: *Flacourtia rukam* Zoll. et Moritʒi; *Flacourtiac.;* Sunda.
Koeipjarie: *Loxopterygium sagoti* Hook. f.; *Anacardiac.;* Sur.
Koejakè fehoeta: *Guarea gomma* Pulle; *Meliac.;* Sur.
Koejari itabatje: *Parkia pendula* Bth.; *Mimosac.;* Sur.
„ -topaitji: *Parkia microcephala* Kleinh.; *Mimosac.;* Sur.
Koejoeng: *Shorea eximia* Scheff., *Vatica eximia* Miq.; *Dipterocarpac.;* Sum., Bangka.
Koekoeran: *Gynotroches axillaris* Blume; *Rhizophorac.;* Malay.
Koekoon: *Schoutenia ovata* Korth.; *Tiliac.;* Nied.-Ind.
Koele: *Schleichera trijuga* Willd.; *Sapindac.;* Nied.-Ind. (Solor)
Koelekoe: *Clathrotropis brachypetalum* Kleinh.; *Papilionac.;* Sur.
Koelero: *Piratinera guianensis* Aubl.; *Morac.;* Sur.
Koeliki-kokoja: *Xylopia frutescens* Aubl., *X. spp.; Anonac.;* Sur.
Koelisiri-wokoeloe: *Inga heterophylla* Willd., *I. lateriflora* Miq.; *Mimosac.;* Sur.
Koelisirie khaléméroe: *Matayba spp., Cupania scrobiculata* L. C. Rich.; *Sapindac.;* Sur.
Koelit-bawang: *Dysoxylum mollissimum* Miq.; *Meliac.;* Java.
Koeloeboerelli: *Peltogyne spp., Copaifera bracteata* Bth.; *Caesalpiniac.;* Sur.
Koeloetoem bajah: *Gordonia sp.; Theac.;* Sum.
Koemaka: *Ceiba pentandra* Gaertn., *Bombax sp., Pachira sp.; Bombacac.;* Sur.
Koemantie kwatie-ie: *Eschweilera sp., ? Lecythis sp.; Lecythidac.;* Sur.
Koemarie: *Dipteryx odorata* Willd.; *Papilionac.;* n. S.-Am.
Koemaroe: *Dipteryx odorata* Willd.; *Papilionac.;* Sur.
Koematalie: *Mouriria anomala* Pulle, *M. plaschaerti* Pulle; *Melastomatac.;* Sur.
Koemboe: *Oenocarpus bacaba* Mart.; *Palmac.;* Sur.
Koembotasje-poto: *? Swartʒia triphylla* Willd.; *Caesalpiniac.,* Sur.
Koembotassi: *Hirtella caudata* Kleinh., *H. spp.; Rosac.;* Sur.
Koenapetie: *Platymiscium polystachyum* Bth.; *Papilionac.;* Sur.
Koenatepi: *Cedrela sp.; Meliac.;* Sur.
Koengè: *Xylopia frutescens* Aubl., *X. spp.; Anonac.;* Sur.
Koenore-èreparè: *Qualea albiflora* Ducke, *Q. sp.; Vochysiac.;* Sur.
Koenotto-epoei: *Diplotropis guianensis* Bth., *D. leptophylla* Kleinh.; *Papilionac.;* Sur.
Koenter gauwa: *Harpullia cupanioides* Roxb.; *Sapindac.;* Südsee (Neu-Guinea)
Koepa: *Eugenia polycephala* Miq.; *Myrtac.;* Java.
Koepaia: *Tabebuia sp.; Bignoniac.;* Sur.
Koepaja: *Jacaranda copaia* D. Don; *Bignoniac.;* Sur.
Koepajoewa: *Copaifera guianensis* Desf.; *Caesalpiniac.;* Sur.
Koeparé-èrèpalé: *Paypayrola guianensis* Aubl.; *Violac.;* Sur.
Koepawa: *Copaifera guianensis* Desf.; *Caesalpiniac.;* Sur.
Koepésini: *Licania micrantha* Miq.; *Rosac.;* Sur.
Koepesjini: *Parinarium campestre* Aubl., *? P. brachystachyum* Bth.; *Rosac.;* Sur.
Koepi: *Couepia guianensis* Aubl.; *Rosac.;* Sur.
Koepie: *Goupia glabra* Aubl.; *Celastrac.;* Gu.
Koerahara: *Calophyllum sp.; Guttifer.;* Sur.
Koerali: *Calophyllum sp.; Guttifer.;* Sur.
Koerallaballi: *Nectandra sp., Ocotea sp.; Laurac.;* Sur.
Koerara-hara: *Calophyllum sp.; Guttifer.;* Sur.
Koeraroe: *Andira coriacea* Pulle; *Papilionac.;* Sur.
Koerekoe: *Clathrotropis brachypetalum* Kleinh.; *Papilionac.;* Sur.
Koerero-ceteballie: *Vochysia melinoni* Beckmann; *Vochysiac.;* Sur.
Koereroe: *Piratinera guianensis* Aubl.; *Morac.;* Sur.
„ -janaledan: *Sclerolobium sp.; Caesalpiniac.;* Sur.
„ , Sokoné-: *Helicostylis poeppigiana* Tréc.; *Morac.;* Gu.
„ -talaboe: *Andira coriacea* Pulle; *Papilionac.;* Sur.
Koeroebahara: *Pentaclethra filamentosa* Bth.; *Mimosac.;* Sur.
Koeroeballi: *Pentaclethra filamentosa* Bth.; *Mimosac.;* Sur.
Koeroekai: *Protium spp.; Burserac.;* Sur.
Koerroe-arara: *Xylopia frutescens* Aubl., *X. spp.; Anonac.;* Sur.
Koesali-jepo: *Erisma uncinatum* Warm.; *Vochysiac.;* Sur.
Koesaljeppo: *Chaetocarpus sp., ? C. schomburgkiana* Pax et Hoffm.; *Euphorbiac.;* Sur.
Koesalli-epo: *Chaetocarpus sp., ? C. schomburgkiana* Pax et Hoffm.; *Euphorbiac.;* Sur.
Koesari jepo: *Pouteria sp.; Sapotac.;* Sur.
Koesarie jeppo: *Pouteria sp.; Sapotac.;* Sur.
Koesirie-parataré: *Sideroxylon guianensis* A. DC.; *Sapotac.;* Sur.
Koewalli: *Vochysia tomentosa* DC.; *Vochysiac.;* Sur.
Koewotto-epoe: *Diplotropis guianensis* Bth., *D. leptophylla* Kleinh.; *Papilionac.;* Sur.

Kofa: *Triplochiton scleroxylon* K. Schum.; *Sterculiac.;* Elf.
Kofé: *Garcinia kola* Heckel; *Guttifer.;* Lib.
Koffie mama: *Erythrina corallodendron* L.; *Papilionac.;* Sur.
Ko-fina: *Dialium guineense* Willd.; *Caesalpiniac.;* fr. Guin., Sudan.
Kofo: *Mitragyne macrophylla* Hiern; *Rubiac.;* B.-K. (Azande, Uelé)
Koga: *Actinodaphne lancifolia* Meissn.; *Laurac.;* Jap.
? Kogelboom: *Guarea aubleti* A. Juss.; *Meliac.;* fr. Gu.
Koghia béra: *Rhizophora racemosa* G. F. W. Meyer; *Rhizophorac.;* Elf.
Koha: *Terminalia arjuna* Bedd.; *Combretac.;* C.-O.-Ind.
Kohan: *Schleichera trijuga* Willd.; *Sapindac.;* sw. O.-Ind. (Bombay)
Kohe kohe: *Dysoxylum spectabile* Hook. f.; *Meliac.;* N.-Seel.
Kohekoe: *Dysoxylum spectabile* Hook. f.; *Meliac.;* N.-Seel.
Kohi: *Alnus nepalensis* D. Don; *Betulac.;* nw. O.-Ind. (Punjab)
Kohigüs: *Nothofagus betuloides* Oerst.; *Fagac.;* s. Arg.
Kohikohe: *Dysoxylum spectabile* Hook. f.; *Meliac.;* N.-Seel.
Kohlbaum (d): ? *Andira inermis* H. B. K.; *Papilionac.;* tr. Am.
Kohomba: *Azadirachta indica* A. Juss.; *Meliac.;* Cey.
 „ -gass: *Azadirachta indica* A. Juss.; *Meliac.;* Cey.
Kohope: *Odyendea gabunensis* Engl.; *Simarubac.;* Kam.
Kohr: *Diospyros sanza-minika* Chev.; *Ebenac.;* Lib.
Kohu: *Terminalia arjuna* Bedd.; *Combretac.;* C.-O.-Ind.
 „ : *Olea cuspidata* Wall.; *Oleac.;* O.-Ind.
Kohutuhtu: *Fuchsia excorticata* L. f.; *Onagrac.;* N.-Seel.
Kohutukuta: ? *Fuchsia excorticata* L. f.; *Onagrac.;* N.-Seel.
Koiguibé: *Guarea thompsoni* Sprague et Hutch.; *Meliac.;* Elf.
Koilar: *Bauhinia retusa* Ham., *B. purpurea* L.; *Caesalpiniac.;* C.-O.-Ind.
Koïli: *Dracontomelum dao* Merr. et Rolfe; *Anacardiac.;* Cel.
Koir: *Acacia catechu* Willd.; *Mimosac.;* Assam.
Koit: *Petersia viridiflora* Chev.; *Lecythidac.;* Elf.
Koito: *Feronia elephantum* Correa; *Rutac.;* O.-Ind. (Uriya)
Koïval: *Bauhinia purpurea* L.; *Caesalpiniac.;* Beng.
Kojageï: *Terminalia superba* Engl. et Diels; *Combretac.;* Lib., S.-L.
Koka: *Bombax albidum* Gagnep.; *Bombacac.;* Ind.-Ch.
Kokan: *Duabanga sonneratioides* Buch.-Ham.; *Sonneratiac.;* Assam.
Kokank: *Lophira procera* Chev.; *Ochnac.;* S.-L.
Kokatmón: *Buchanania sp.;* *Anacardiac.;* Phil. (Surigao)
Kokeleko: *Bertholletia excelsa* H. B. K.; *Lecythidac.;* Sur.
Kokeriki: *Ormosia coccinea* Jacks., *Ormosia spp.;* *Papilionac.;* Sur.
Kokerit: *Maximiliana regia* Mart.; *Palmac.;* br. Gu.
Kokerite: *Cecropia sp.;* *Morac.;* br. Gu.
 „ -balli: *Lucuma sp., Sideroxylon sp.;* *Sapotac.;* br. Gu.
Kokhé: *Dipterocarpus pilosus* Roxb.; *Dipterocarpac.;* Burma (Bassein)
Koki: *Hopea odorata* Roxb.; *Dipterocarpac.;* Kamb.
 „ dêk: *Hopea dealbata* Hance; *Dipterocarpac.;* Kamb.
 „ phong: *Hopea odorata* Roxb.; *Dipterocarpac.;* Kamb.
Kokiya: *Strychnos spinosa* Lam., *S. triclisioides* Baker; *Loganiac.;* n. Nig.
Kokko (H): *Albizzia lebbek* Bth.; *Mimosac.;* O.-Ind., Burma, And.
Koko (u): *Albizzia lebbek* Bth.; *Mimosac.;* O.-Ind., And.
Kokoa: *Celtis integrifolia* Lam.; *Ulmac.;* Togo.
Kokoh: *Albizzia lebbek* Bth.; *Mimosac.;* O.-Ind.
Kokoho: *Licania pachystachya* Kleinh., *L. micrantha* Miq.; *Rosac.;* Sur.
Kokojá: *Celtis integrifolia* Lam.; *Ulmac.;* Togo.
Kokomatt: *Alstonia congensis* Engl.; *Apocynac.;* Kam.
Kokomayur: *Spathodea campanulata* Beauv.; *Bignoniac.;* Elf.
Kokoni-breta: *Gustavia angusta* L.; *Lecythidac.;* Sur.
Kokoona: *Kokoona sp.;* *Celastrac.;* Java.
Kokop: *Ailanthus peekeli* H. Melchior; *Simarubac.;* Südsee (Neu-Mecklenburg)
Kokorietjie-balli: *Pouteria guianensis* Aubl.; *Sapotac.;* Sur.
Kokosan: *Lansium domesticum* Jack; *Meliac.;* Sunda.
Kokosi: *Cola proteiformis* Chev.; *Sterculiac.;* Elf.
Kokossan: *Lansium domesticum* Jack; *Meliac.;* Sunda.
Kokote: *Anopyxis ealaensis* Sprague; *Rhizophorac.;* Go.
Kokoti: *Anopyxis ealaensis* Sprague; *Rhizophorac.;* Go.
Kokotsi: *Cola proteiformis* Chev.; *Sterculiac.;* Elf.

Kokoʒi: *Tarrietia utilis* Sprague; *Sterculiac.*; Lib., Elf., Go.
Kokou: *Parinarium curatellifolium* Planch.; *Rosac.*; W.-Af. (Boumouali)
Kokpara: *Albizzia browni* Oliv.; *Mimosac.*; Togo.
Kokra wood: *Aporosa dioica* Muell. Arg.; *Euphorbiac.*; O.-Ind.
Kokriki: *Ormosia coccinea* Jacks., *O. spp.*; *Papilionac.*; Sur.
Kokué: *Alstonia congensis* Engl.; *Apocynac.*; Elf.
Koky: *Hopea odorata* Roxb.; *Dipterocarpac.*; Ind.-Ch.
  „ ma san: *Hopea recopei* Pierre; *Dipterocarpac.*; Ind.-Ch.
  „ tsat: *Hopea pierrei* Hance; *Dipterocarpac.*; Ind.-Ch.
Kôl: *Juglans regia* L.; *Juglandac.*; nö. O.-Ind.
Kola: *Cola acuminata* Schott et Endl.; *Sterculiac.*; fr. Guin., Lib.
  „ : *Acacia suma* Kurz; *Mimosac.*; Togo.
  „ : bitter: *Cola nitida* Chev.; *Sterculiac.*; Lib.
  „ , Bush-: *Cola buntingi* Baker f.; *Sterculiac.*; Lib.
  „ -nut: *Cola acuminata* Schott et Endl.; *Sterculiac.*; W.-Af., C.-Af.
  „ sampige: *Michelia champaca* L.; *Magnoliac.*; sw. O.-Ind. (Bombay)
Kolahi: *Fagraea fragrans* Roxb.; *Loganiac.*; Nied.-Ind.
Kolain: *Pinus longifolia* Roxb.; *Conifer.*; n. O.-Ind. (Garhwal)
Kolakka: *Parinarium sp.*; *Rosac.*; Nied.-Ind.
Kolamava: *Buchanania angustifolia* Roxb.; *Anacardiac.*; O.-Ind. (Madras)
Kolar: *Bauhinia purpurea* L.; *Caesalpiniac.*; O.-Ind., Cey., Java.
Kolatier: *Cola vera* K. Schum.; *Sterculiac.*; Elf.
  „ : *Cola ballayi* Cornu; *Sterculiac.*; Gab.
Kolavu: *Hardwickia pinnata* Roxb.; *Caesalpiniac.*; sw. O.-Ind. (Coorg)
Kole-kole: *Cleistopholis verschuereni* de Wild.; *Anonac.*; B.-K. (Malela)
Kolei: *Detarium senegalense* Gmel.; *Caesalpiniac.*; Lib.
Kolero: *Piratinera guianensis* Aubl.; *Morac.*; Sur.
Koli-al: *Ficus callosa* Willd.; *Morac.*; w. O.-Ind. (Deccan)
  „ kurkhunda: *Zizyphus jujuba* Lamk.; *Rhamnac.*; O.-Ind.
  „ payar: *Olea cuspidata* Wall.; *Oleac.*; O.-Ind.
Kólis: *Eugenia brevistylis* C. B. Rob.; *Myrtac.*; Phil. (Tayabas)
Koliúng: *Artocarpus cumingiana* Tréc.; *Morac.*; Phil. (Abra)
Kolkol: *Acacia senegal* Willd.; *Mimosac.*; Nig.
Kolla mavu: *Machilus macrantha* Nees; *Laurac.*; sw. O.-Ind. (Madras)
Kolland: *Heritiera littoralis* Dryander; *Sterculiac.*; sw. O.-Ind. (Bombay)
Kollopello: *Garcinia manni* Oliv.; *Guttifer.*; W.-Af.
Kolo: *Amblygonocarpus schweinfurthi* Harms; *Mimosac.*; n. Nig.
  „ : *Burkea africana* Hook.; *Caesalpiniac.*; n. Nig.
  „ : *Mimusops congolensis* de Wild.; *Sapotac.*; Gab.
  „ -kolo: *Afrormosia sp.*, *Ormosia laxiflora* Bth.; *Papilionac.*; Sudan.
Kolokotingan: *Pterospermum sp.*; *Sterculiac.*; Phil.
Kololo: *Cynometra sp.*; *Caesalpiniac.*; Sudan.
Kolu: *? Bridelia ferruginea* Bth.; *Euphorbiac.*; Togo.
Kolumbolu: *Ceiba pentandra* Gaertn. var. *clausa* Ulb.; *Bombacac.*; Togo.
Kom: *Terminalia superba* Engl. et Diels; *Combretac.*; Kam.
  „ : *Nauclea sessilifolia* Roxb.; *Rubiac.*; Beng.
  „ -kia: *Calocarpum mammosum* Pierre; *Guttifer.*; C.-R.
Komami-gasi: *Lithocarpus rhombocarpa* Hay.; *Fagac.*; Form.
Komaro: *Dipteryx odorata* Willd.; *Papilionac.*; Sur.
Komba: *Adina cordifolia* Hook. f.; *Rubiac.*; O.-Ind. (? Beng.)
Kombo: *Pycnanthus kombo* Warb.; *Myristicac.*; Gab.
Kombo-kombo: *Musanga smithi* R. Br.; *Morac.*; B.-K. (Kurubu)
  „ man: *Strombosiopsis tetrandra* Engl.; *Olacac.*; Kam.
Komboe: *Oenocarpus bataua* Mart.; *Palmac.*; Sur.
Kombolo: *Pentaclethra macrophylla* Bth.; *Mimosac.*; Kam.
Kombulu: *Macrolobium dewevrei* de Wild.; *Caesalpiniac.*; B.-K. (Eulé)
Kombuluka: *Sterculia tragacantha* Lindl. var. *cruciata* Verm.; *Sterculiac.*; B.-K.
Kometsuga: *Tsuga diversifolia* Maxim.; *Conifer.*; Jap.   [(Mayombe)
Kominoholz: *Aniba perutilis* Hemsl.; *Laurac.*; Kol.
Komla-neboo: *Citrus aurantium* L.; *Rutac.*; Beng.
Kommol: *Coula edulis* Baill.; *Olacac.*; Kam.
Kommott: *Pentaclethra macrophylla* Bth.; *Mimosac.*; Kam.
Komo: *Elaeis guineensis* Jacq. var. *virescens* Chev.; *Palmac.*; Gab.
Komoi: *Diospyros spp.*, *Maba spp.*; *Ebenac.*; Malay.

Komon: *Pygeum sp.; Rosac.;* Phil. (Cagayan)
Komondip: *Lavalleopsis densivenia* Engl.; *Olacac.;* Kam.
Komotolie: *Mouriria anomala* Pulle, *M. plaschaerti* Pulle; *Melastomatac.;* Sur.
Komotori-balli: *Mouriria acutiflora* Naud.; *Melastomatac.;* Sur.
Komotorie: *Mouriria anomala* Pulle, *M. plaschaerti* Pulle; *Melastomatac.;* Sur.
Komou aguiré: *Cola mirabilis* Chev.; *Sterculiac.;* Elf.
Kompeng reack: *Sandoricum indicum* Cav.; *Meliac.;* Ind.-Ch.
Komu: *Ceiba pentandra* Gaertn.; *Bombacac.;* Togo.
Kon-sha: *Acacia sp.; Mimosac.;* Burma.
Konaheri: *Pachira aquatica* Aubl.; *Bombacac.;* br. Gu.
Konangbri: *Trichilia sp.; Meliac.;* Elf.
Konara: *Quercus glandulifera* Blume; *Fagac.;* Jap.
Konatepie: *Platymiscium polystachyum* Bth.; *Papilionac.;* Sur.
Konda chiragu: *Albizzia stipulata* Boivin; *Mimosac.;* sw. O.-Ind. (Madras)
   „   kasinda: *Toddalia asiatica* Baill.; *Rutac.;* w. O.-Ind. (n. Bombay)
   „   tangedu: *Xylia dolabriformis* Bth.; *Mimosac.;* sw. O.-Ind. (Madras)
   „   vaghe: *Albizzia procera* Bth.; *Mimosac.;* sw. O.-Ind. (Madras)
Kondé: *Ceiba pentandra* Gaertn. *var. dehiscens* Ulb.; *Bombacac.;* Guin.
Kondjo: *Sclerosperma walkeri* Chev.; *Palmac.;* Gab.
   „   : *Klainedoxa gabonensis* Pierre; *Simarubac.;* Gab.
Kondricam: *Vateria indica* L.; *Dipterocarpac.;* sw. O.-Ind. (Madras)
Kondricum: *Vateria indica* L.; *Dipterocarpac.;* w. O.-Ind. (Deccan)
Kong: *Hopea glabra* Wight et Arn., *Balanocarpus utilis* Bedd.; *Dipterocarpac.;*
   „   kang: *Rhizophora conjugata* L.; *Rhizophorac.;* Ind.-Ch.   [s. O.-Ind.
Konghélé: *Pseudospondias microcarpa* Engler; *Anacardiac.;* Gab.
Kongki: *Prunus puddum* Roxb.; *Rosac.;* nö. O.-Ind.
Kongkong: *Voacanga africana* Stapf; *Apocynac.;* Togo.
Kongo: *Copaifera demeusei* Harms; *Caesalpiniac.;* B.-K. (Batetela)
Kongofura: *Strychnos buettneri* Gilg; *Loganiac.;* Togo.
Kongoli: *Vernonia conferta* Bth.; *Composit.;* Lib.
Kongolu: *Cussonia barteri* Seem.; *Araliac.;* Togo.
Kongora: *Xylia dolabriformis* Bth.; *Mimosac.;* w. Beng. (Orissa)
Konhlé: *Pseudospondias microcarpa* Engler; *Anacardiac.;* Gab.
Koniaphul: *Bauhinia malabarica* Roxb.; *Caesalpiniac.;* C.-O.-Ind. (Gondwana)
Königsholz (d): *Dalbergia greveana* Baill.; *Papilionac.;* Mad. (Majunga)
   „   „   : *Dalbergia sp.; Papilionac.;* Mad. (Tamatave)
Konina: *Protium heptaphyllum* March.; *Burserac.;* Gu.
Koningshout: *Clusia rosea* Jacq.; *Guttifer.;* fr. Gu.
Konishigusu: *Cryptocarya konishii* Hayata; *Laurac.;* Form.
Konisi-gasi: *Lithocarpus konishii* Hayata; *Fagac.;* Form.
Konju: *Hopea parviflora* Bedd.; *Dipterocarpac.;* sw. O.-Ind. (Madras)
Konkona: *Phyllanthus discoideus* Muell. Arg.; *Euphorbiac.;* Togo.
Konkruma: ? *Morinda citrifolia* L.; *Rubiac.;* Go.
Konné: *Acrocarpus fraxinifolius* Wight; *Caesalpiniac.;* sw. O.-Ind. (Madras)
Konnei: *Cassia fistula* L.; *Caesalpiniac.;* O.-Ind. (Madras)
Konnie-konnie-hoedoe: *Pouteria sp.; Sapotac.;* Sur.
Konoko: *Licania sp.; Rosac.;* br. Gu.
   „   : *Sideroxylon guianensis* A. DC.; *Sapotac.;* Sur.
Konora: *Quercus glandulifera* Blume; *Fagac.;* Jap.
Konote-kaskiwa: *Thuja orientalis* L.; *Conifer.;* Jap.
Konsa: *Hymenodictyon excelsum* Wall.; *Rubiac.;* w. Beng. (Orissa)
Konsakka wiwiri: *Peperomia pellucida* H. et K.; *Piperac.;* Sur.
Konskoi kaztan: *Aesculus hippocastanum* L.; *Hippocastanac.;* n. O.-Ind., Pers.
Konta: *Afzelia africana* Smith; *Caesalpiniac.;* W.-Af. (S.-L.)
Kontah wood: *Afzelia africana* Smith; *Caesalpiniac.;* W.-Af.
Kontebora: *Cussonia barteri* Seem.; *Araliac.;* Togo.
Konthout: *Minquartia guianensis* Aubl.; *Olacac.;* Sur.
Kontoehoedoe: *Minquartia guianensis* Aubl.; *Olacac.;* Sur.
Kontol kambieng: *Heritiera littoralis* Dryander; *Sterculiac.;* Sunda.
Konvavola: *Sterculia quadrifida* R. Br.; *Sterculiac.;* Au. (Queensl.)
Koo-gbeh: *Isolona cooperi* Hutch. et Dalz.; *Anonac.;* Lib.
Koodsaloo: *Holarrhena antidysenterica* Wall.; *Apocynac.;* w. O.-Ind. (Deccan)
Kooël pialli: *Loxopterygium sagoti* Hook. f.; *Anacardiac.;* Gu.
Kookerite-balli: *Ilex sp.; Aquifoliac.;* br. Gu.

Kookeriti-balli, Bastard-: *Chrysophyllum sp.; Sapotac.;* br. Gu.
Kool: *Zizyphus jujuba* Lamk.; *Rhamnac.;* Beng.
Kooleshiri: *Matayba spp., Cupania scrobiculata* L. C. Rich.; *Sapindac.;* Sur.
Koomuttie-ghenzaloo: *Sapindus emarginatus* Vahl; *Sapindac.;* w. O.-Ind.(n. Bombay)
Koon: *Schleichera trijuga* Willd.; *Sapindac.;* sw. O.-Ind. (Bombay)
Koondon belookar: *Garcinia andersoni* Hook. f.; *Guttifer.;* Malay.
Koonookoo: *Licania sp.; Rosac.;* br. Gu.
Koor: *Ricinodendron africanum* Muell. Arg.; *Euphorbiac.;* Lib.
Koorak: *Garuga pinnata* Roxb.; *Burserac.;* Ind.-Ch.
Koorangal: *Pongamia glabra* Vent.; *Papilionac.;* w. O.-Ind. (Deccan)
Koorka-poolee: *Garcinia cornea* L.; *Guttifer.;* w. O.-Ind. (Deccan)
Koorooboo-elli: *Peltogyne spp.; Caesalpiniac.;* Sur.
Kooroobooilli: *Peltogyne spp., Copaifera bracteata* Bth.; *Caesalpiniac.;* Sur.
Kooroobovelli: *Peltogyne spp.; Caesalpiniac.;* Sur.
Koortskruid: *Lantana camara* L.; *Verbenac.;* Sur.
Kooruk: *Toona ciliata* M. Roem.; *Meliac.;* w. O.-Ind. (Bombay)
Koosga: *Stereospermum chelonioides* DC.; *Bignoniac.;* sw. O.-Ind.
Kooudoo: *Chartocalyx accrescens* Mast.; *Tiliac.;* Malak.
Koouru: *Garcinia manni* Oliv.; *Guttifer.;* W.-Af. (Sudan)
Kooveruka: *Toona ciliata* M. Roem.; *Meliac.;* O.-Ind.
Kooying doodoo: *Dipterocarpus turbinatus* Gaertn.; *Dipterocarpac.;* Malak.
Kopaia: *Jacaranda copaia* D. Don; *Bignoniac.;* Sur.
Kopakópa: *Eugenia mananquil* Blco.; *Myrtac.;* Phil. (Ilocos-Sur)
Kopasia: *Kydia calycina* Roxb.; *Malvac.;* w. Beng. (Orissa)
Kopi: *Coffea abeocuta* Cramer; *Rubiac.;* Malay.
Kopie: *Goupia glabra* Aubl.; *Celastrac.;* Sur.
Kopó: *Ficus sp.; Morac.;* Mex.
Kopu: *Tamarindus indica* L.; *Caesalpiniac.;* Togo.
Kor-dru-boe: *Palisota thyrsifolia* Bth.; *Commelinac.;* Lib.
„ -wah: *Psychotria sp.; Rubiac.;* Lib.
Koraalhout: *Adenanthera pavonina* L.; *Mimosac.;* As.
Korallenbaum (d): *Erythrina crista-galli* L.; *Papilionac.;* Arg.
Korallenholz: *Adenanthera pavonina* L.; *Mimosac.;* tr. As.
„ : *Pterocarpus dalbergioides* Roxb.; *Papilionac.;* And.
Korallenstrauch: *Erythrina crista-galli* L.; *Papilionac.;* Arg.
Koramaru: *Bridelia retusa* Spreng.; *Euphorbiac.;* s. O.-Ind. (Madras)
Koranju: *Pongamia glabra* Vent.; *Papilionac.;* w. Beng. (Orissa)
Korchu: *Holarrhena antidysenterica* Wall.; *Apocynac.;* w. O.-Ind. (Deccan)
Kordi: *Wrightia tinctoria* R. Brown; *Apocynac.;* O.-Ind. (Bombay)
Korhoa: *Berlinia heudelotiana* Baill.; *Caesalpiniac.;* Togo.
Koriba: *Croton amabilis* Muell. Arg.; *Euphorbiac.;* n. Nig.
Kórig: *Santiria nitida* Merr.; *Burserac.;* Phil. (Bataan)
Koriyanagi: *Salix rubra* L.; *Salicac.;* Jap.
Korkrinde: *Melaleuca leucadendron* L.; *Myrtac.;* Au., sö. As., Malay.
Korku: *Mallotus philippinensis* Muell. Arg.; *Euphorbiac.;* C.-O.-Ind. (Gondwana)
Korna-neboo: *Citrus limonum* Risso; *Rutac.;* Beng.
Kornelkirsche, gemeine: *Cornus mas* L.; *Cornac.;* Mitt.- & O.-Eu., As.
Korób: *Achras sapota* L.; *Sapotac.;* C.-R.
Koroballi: *Jacaranda filicifolia* D. Don; *Bignoniac.;* Sur.
„ : *Pentaclethra filamentosa* Bth.; *Mimosac.;* Sur.
Koroborelli: *Peltogyne spp., Copaifera bracteata* Bth.; *Caesalpiniac.;* Sur.
Koroh: *Shorea robusta* Gaertn. f.; *Dipterocarpac.;* n. O.-Ind. (Oudh)
Koroi: *Albizzia odoratissima* Bth., *A. procera* Bth.; *Mimosac.;* Beng., Assam.
Korokororo: *? Diplotropis sp.; Papilionac.;* br. Gu.
Koromiko: *Veronica salicifolia* Forst.; *Scrophulariac.;* N.-Seel.
Koroti waikey: *Inga sp.; Mimosac.;* br. Gu.
Korounka: *Calotropis procera* Ait.; *Asclepiadac.;* Sahara.
Korra: *Strychnos nux-vomica* L.; *Loganiac.;* w. Beng. (Orissa)
Korsa: *Bischofia javanica* Blume; *Euphorbiac.;* O.-Ind.
Korsoe wiwiri: *Lantana camara* L.; *Verbenac.;* Sur.
Koruburelli: *Peltogyne spp., Copaifera bracteata* Bth.; *Caesalpiniac.;* Sur.
Kos: *Artocarpus integrifolia* L.; *Morac.;* Cey.
Kosai: *Diosbyros chloroxylon* Roxb.; *Ebenac.;* C.-O.-Ind., s. O.-Ind.
Kosambie: *Schleichera trijuga* Willd.; *Sapindac.;* Java, Malay, Sunda.

Koshiabura: *Kalopanax sciadophylloides* Harms; *Araliac.;* Jap.
Kôshun-nagi: *Podocarpus nagi* Zoll. *var. koshunensis* Kanehira; *Conifer.;* Form.
Kosipo: *Entandrophragma candollei* Harms; *Meliac.;* Elf.
Kossé-kuli: *Diospyros montana* Roxb.; *Ebenac.;* O.-Ind., Cey.
Kosum: *Schleichera trijuga* Wiild.; *Sapindac.;* sw. O.-Ind. (Bombay)
Kot semla: *Bischofia javanica* Blume; *Euphorbiac.;* n. O.-Ind. (Garhwal)
Kota ranga: *Gardenia latifolia* Aiton; *Rubiac.;* w. Beng. (Orissa)
Kotaki: *Strychnos potatorum* L. f.; *Loganiac.;* w. Beng. (Orissa)
Kotekyi: *Sterculia tragacantha* Lindl.; *Sterculiac.;* Go.
Kotésima: *Parinarium tenuifolium* Chev.; *Rosac.;* Elf.
Koti: *Cistanthera papaverifera* Chev.; *Tiliac.;* Elf.
  „ : *Petersia viridiflora* Chev.; *Lecythidac.;* Elf.
Kotia: *Allophylus africanus* Beauv.; *Sapindac.;* D.-O.-Af. (Usambara), Sans.
Ko-tiao: *Juglans regia* L.; *Juglandac.;* Ch.
Kotibé: *Cistanthera sp.; Tiliac.;* Elf.
Kotie: *Cistanthera papaverifera* Chev.; *Tiliac.;* Elf.
Kotmó: *Dillenia sp.; Dilleniac.;* Phil. (Camarines, Albay, Sorsogon)
Kotmók: *Dillenia sp.; Dilleniac.;* Phil. (Camarines, Albay, Sorsogon)
  „ : *Terminalia edulis* Blco.; *Combretac.;* Phil. (Camarines)
Koto: *Pterygota sp.; Sterculiac.;* Elf.
Kotokié: *Sterculia tragacantha* Lindl.; *Sterculiac.;* Elf.
Kotokorowa: *Monodora myristica* Dunal; *Anonac.;* Go.
Kotopapa: *Macrolobium limba* Scott Elliott; *Caesalpiniac.;* Go.
Kotoprepre: *Berlinia heudelotiana* Baill., *Bussea occidentalis* Hutch.; *Caesalpiniac.;* Go.
  „ : *Calpocalyx sp., Xylia sp.; Mimosac.;* Go.
Kotopuan: *Cleistopholis patens* Engl. et Diels; *Anonac.;* Elf.
Kotta: *Zizyphus xylopyrus* Willd.; *Rhamnac.;* sw. O.-Ind. (Malabar)
Kottamba: *Terminalia catappa* L.; *Combretac.;* Cey.
Kottei: *Zizyphus xylopyrus* Willd.; *Rhamnac.;* sw. O.-Ind. (Madras)
  „ : *Strychnos potatorum* L. f.; *Loganiac.;* sw. O.-Ind. (Madras)
Kotublassu: *Lophira alata* Banks; *Ochnac.;* Togo.
Kou ao: *Adina cordifolia* Hook. f.; *Rubiac.;* Ind.-Ch.
  „ kan: *Elaeocarpus tomentosus* Blume; *Elaeocarpac.;* Ind.-Ch.
  „ kong: *Acacia farnesiana* Willd.; *Mimosac.;* Ind.-Ch.
Kouala: *Cola vera* K. Schum.; *Sterculiac.;* Elf.
Kouan: *Copaifera aff. arnoldiana* de Wild. et Th. Dur.; *Caesalpiniac.;* Kam.
Kouanda: *Cola proteiformis* Chev., *Tarrietia utilis* Sprague; *Sterculiac.;* Elf.
Kouc móc: *Sterculia alata* Roxb.; *Sterculiac.;* Annam.
Koudou: *Vitex sp.; Verbenac.;* fr. Guin.
Koué: *Lophira procera* Chev.; *Ochnac.;* Kam., Gab.
Kouilli: *Entandrophragma congoense* Chev.; *Meliac.;* Kongo.
Koukinjok: *Entandrophragma utile* Sprague; *Meliac.;* Kam.
Koum: *Crataeva religiosa* R. Br.; *Capparidac.;* Ind.-Ch.
Koumé: *Coula edulis* Baill.; *Olacac.;* Kam.
Koumi: *Aucoumea klaineana* Pierre; *Burserac.;* Gab.
Koumoun: *Coula edulis* Baill.; *Olacac.;* Kam.
Koumounou: *Coula edulis* Baill.; *Olacac.;* Aeq. Af. (Loango)
Koung: *Dipterocarpus tuberculatus* Roxb.; *Dipterocarpac.;* Ind.-Ch.
Kouni Kong: *Entandrophragma sp.; Meliac.;* Kam.
Kounikol: *Borassus aethiopum* Warb.; *Palmac.;* Senl.
Kounitaye: *Borassus aethiopum* Warb.; *Palmac.;* Senl.
Kouoc môc: *Sterculia alata* Roxb.; *Sterculiac.;* Ind.-Ch.
Koura: *Parinarium excelsum* Sabine; *Rosac.;* fr. Guin.
Kouratary: *Couratari guianensis* Aubl.; *Lecythidac.;* fr. Guiana.
Kouri babul: *Acacia scorpioides* W. F. Wight; *Mimosac.;* O.-Ind. (Berar, Poona)
Kourka-pouli: *Garcinia cornea* L.; *Guttifer.;* w. O.-Ind. (Deccan)
Koussé: *Palaquium sp.; Sapotac.;* fr. Guin.
Kovalam: *Aegle marmelos* Correa; *Rutac.;* sw. O.-Ind. (Malabar)
Kovit: *Feronia elephantum* Correa; *Rutac.;* sw. O.-Ind. (Bombay)
Kow: *Olea cuspidata* Wall.; *Oleac.;* O.-Ind.
Kowa: *Terminalia arjuna* Bedd.; *Combretac.;* C.-O.-Ind.
Kowal: *Alnus nepalensis* D. Don; *Betulac.;* nö. O.-Indien.
Kowaré: *Vochysia tomentosa* DC.; *Vochysiac.;* Sur.
Kowhai: *Sophora tetraptera* J. Mill.; *Papilionac.;* N.-Seel.

Kowi: *Carapa procera* DC.; *Meliac.;* Lib.
Koyamaki: *Sciadopitys verticillata* S. et Z.; *Conifer.;* Jap.
Koyar: *Strychnos potatorum* L. f.; *Loganiac.;* w. Beng. (Orissa)
Koyetchi: *Rollinia multiflora* Splitg.; *Anonac.;* br. Gu.
Koyop: *Strombosia javanica* Blume; *Olacac.;* Sunda.
Koyosan: *Cunninghamia sinensis* R. Br.; *Conifer.;* Jap.
Kozo: *Broussonetia kaempferi* Sieb.; *Morac.;* Jap.
Kpa: *Triplochiton scleroxylon* K. Schum.; *Sterculiac.;* Nig.
Kpahn-wee: *Drypetes ivorensis* Hutch. et Dalz.; *Euphorbiac.;* Lib.
Kpaini: *Enantia chlorantha* Oliv.; *Anonac.;* Lib.
Kpakpa: *Afzelia africana* Smith; *Caesalpiniac.;* Togo.
Kpakpei: *Albizzia browni* Walp.; *Mimosac.;* W.-Af. (S.-L.)
Kpánenã: *Prosopis oblonga* Bth.; *Mimosac.;* Togo.
Kpar: *Parinarium excelsum* Sabine; *Rosac.;* Lib.
Kpar-seh: *Vitex spp.; Verbenac.;* Lib.
   „ -wee: *Vangueriopsis discolor* Robyns; *Rubiac.;* Lib.
Kpay: *Detarium senegalense* Gmel.; *Caesalpiniac.;* Lib.
Kpedi: *Ricinodendron africanum* Muell. Arg.; *Euphorbiac.;* Go.
Kpendei: *Berlinia acuminata* Soland., *B. spp.; Caesalpiniac.;* S.-L., Lib.
Kpendi-deli: *Afzelia africana* Smith; *Caesalpiniac.;* W.-Af. (S.-L.)
   „ -guli: *Afzelia africana* Smith; *Caesalpiniac.;* W.-Af. (S.-L.)
   „ -kpoli: *Afzelia africana* Smith; *Caesalpiniac.;* W.-Af. (S.-L.)
Kpetekplebi: *Gardenia tenuifolia* Schum. et Thonn.; *Rubiac.;* Go.
Kploe: *Schefflera sp.; Araliac.;* Lib.
Kpoe: *Anopyxis ealaensis* Sprague; *Rhizophorac.;* Lib.
   „ : *Oncoba spinosa* Forsk.; *Flacourtiac.;* Togo.
   „ -kohn: *Oligostemon pictus* Bth.; *Caesalpiniac.;* Lib.
Kpome: *Funtumia africana* Stapf; *Apocynac.;* Go.
Kpomli: *Funtumia africana* Stapf; *Apocynac.;* Go.
Kpon-doe: *Ixora divaricata* Hutch. et Dalz.; *Rubiac.;* Lib.
Kpu-ah: *Calpocalyx brevibracteatus* Harms; *Mimosac.;* Lib.
Kpwa: *Tetrapleura thonningi* Bth.; *Mimosac.;* Kam.
Krå: *Dichrostachys glomerata* Hutch. et Dalz.; *Mimosac.;* Go.
   „ ka sbek: *Sindora maritima* Pierre; *Caesalpiniac.;* Ind.-Ch.
   „ kirck prach: *Casearia glomerata* Roxb.; *Flacourtiac.;* Ind.-Ch.
   „ -kpar-doo: *Psychotria sp.; Rubiac.;* Lib.
   „ lanh: *Dialium cochinchinense* Pierre; *Caesalpiniac.;* Ind.-Ch.
   „ nhung: *Dalbergia cochinchinensis* Pierre; *Papilionac.;* Ind.-Ch.
   „ san: *Feronia lucida* Teijsm. et Binn.; *Rutac.;* Kamb.
   „ sang si phlé: *Feronia lucida* Teijsm. et Binn.; *Rutac.;* Ind.-Ch.
   „ sanh: *Feronia lucida* Teijsm. et Binn.; *Rutac.;* Ind.-Ch.
Krabas soua: *Taractogenos microcarpa* Pierre; *Flacourtiac.;* Ind.-Ch.
Krabau: *Hydnocarpus anthelmintica* Pierre; *Flacourtiac.;* Ind.-Ch.
Krabise: *Carapa procera* DC.; *Meliac.;* Go.
Kraboo: *Xanthophyllum rufum* A. W. Benn.; *Polygonac.;* Malay.
Krah-dro: *? Angylocalyx obligophyllus* Baker f.; *Papilionac.;* Lib.
Kraï: *Myrcia integrifolia* Roxb.; *Myricac.;* w. Annam.
   „ sar: *Elaeocarpus tomentosus* Blume; *Elaeocarpac.;* Ind.-Ch.
Kraipèoe: *Pithecolobium gonggrijpi* Kleinh., *P. spp.; Mimosac.;* Sur.
Krâka: *Sindora cochinchinensis* Baill.; *Caesalpiniac.;* Ind.-Ch.
   „ sbek: *Sindora cochinchinensis* Baill.; *Caesalpiniac.;* Ind.-Ch.
Krakas: *Crypteronia paniculata* Blume; *Crypteroniac.;* Malay.
   „ : *Sindora cochinchinensis* Baill., *S. siamensis* Teijsm.; *Caesalpiniac.;* Ind.-Ch.
   „ meng: *Sindora cochinchinensis* Baill.; *Caesalpiniac.;* Kamb.
   „ sbec: *Sindora cochinchinensis* Baill.; *Caesalpiniac.;* Kamb.
Krakun: *Panda oleosa* Pierre; *Pandac.;* Go.
Krala: *Khaya anthotheca* C. DC., *Entandrophragma sp.; Meliac.;* Elf.
Kralanh: *Dialium cochinchinense* Pierre; *Caesalpiniac.;* Kamb.
Krám rémia: *Garcinia harmandi* Pierre; *Guttifer.;* Kamb.
   „ rémir: *Garcinia harmandi* Pierre; *Guttifer.;* Kamb.
Kramas: *Vatica astrotricha* Hance; *Dipterocarpac.;* Ind.-Ch.
Kramol: *Acronychia lauriflora* Blume; *Rutac.;* Ind.-Ch.
Kranch chhmar: *Citrus limonum* Risso; *Rutac.;* Kamb.
   „ inôn: *Citrus limonum* Risso; *Rutac.;* Kamb.

276

Kranch pôursat: *Citrus aurantium* L.; *Rutac.*; Kamb.
Krandja-haïgué: *Canarium occidentale* Chev.; *Burserac.*; Elf.
Krandjan-haïgué: *Canarium occidentale* Chev.; *Burserac.*; Elf.
Kranioung: *Dalbergia cochinchinensis* Pierre; *Papilionac.*; Kamb.
Kranji: *Dialium indum* L.; *Caesalpiniac.*; Java.
Krank thlông: *Citrus decumana* L.; *Rutac.*; Kamb.
Kranku: *Butyrospermum parki* Kotschy; *Sapotac.*; Elf.
Krantie: *Santiria apiculata* A. W. Benn.; *Burserac.*; Malak.
Krapa: *Carapa procera* DC., *C. guianensis* Aubl., *C. surinamensis* Miq.; *Meliac.*; Sur.
Krapaboom: *Carapa guianensis* Aubl.; *Meliac.*; Sur.
Krapabosie: *Dipteryx odorata* Willd.; *Papilionac.*; Sur.
Krappa: *Carapa guianensis* Aubl., *C. procera* DC., *C. spp.*; *Meliac.*; Sur.
„ , roode: *Carapa sp.*; *Meliac.*; Sur.
„ hout: *Carapa procera* DC., *C. surinamensis* Miq.; *Meliac.*; Sur.
Kras: *Kayea eugeniaefolia* Pierre; *Guttifer.*; Ind.-Ch.
Krasang: *Feronia lucida* Teijsm. et Binn.; *Rutac.*; Kamb.
Krasna: *Melaleuca leucadendron* L.; *Myrtac.*; Ind.-Ch.
Krassa: *Calophyllum calaba* Jacq.; *Guttifer.*; Nic.
Krassé: *Guarea cedrata* Pellegr.; *Meliac.*; Elf.
Krassi-pisi: *Nectandra sp.*, *Ocotea sp.*; *Laurac.*; Sur.
Krat: *Xylopia vielana* Pierre; *Anonac.*; Ind.-Ch.
Kraul: *Acacia farnesiana* Willd.; *Mimosac.*; Ind.-Ch.
„ : *Melanorrhoea laccifera* Pierre; *Anacardiac.*; Ind.-Ch.
Kray-bu: *Monodora spp.*; *Anonac.*; Lib.
„ doo: *Cissus producta* Afz.; *Ampelidac.*; Lib.
„ sar: *Elaeocarpus tomentosus* Blume; *Elaeocarpac.*; Ind.-Ch.
Krayie: *Pterocarpus erinaceus* Poir.; *Papilionac.*; Go.
Kré-kré: *Anogeissus leiocarpa* Guill. et Perr.; *Combretac.*; fr. Guin.
Kred-kred: *Crataeva religiosa* Forst.; *Capparidac.*; Senl.
Krééte: *Cassia fistula* L.; *Caesalpiniac.*; Ind.-Ch.
Krékété: *Anogeissus sp.*, *A. leiocarpa* Guill. et Perr.; *Combretac.*; Elf., Sudan
Kreland: *Mesua ferrea* L.; *Guttifer.*; Kamb.
Kremuon chăm bàch: *Irvingia malayana* Oliv.; *Simarubac.*; Kamb.
Krendja: *Canarium occidentale* Chev.; *Burserac.*; Elf.
Krere-krere: *Caesalpinia pulcherrima* Sw.; *Caesalpiniac.*; Sur.
Krerk: *Mangifera indica* L.; *Anacardiac.*; O.-Ind. (? Burma)
Krès: *Pentacme suavis* A. DC.; *Dipterocarpac.*; Ind.-Ch.
Kretty: *Nectandra sp.*, *Ocotea sp.*, *Aniba sp.*; *Laurac.*; br. Gu.
Kreu: *Quercus semecarpifolia* Smith; *Fagac.*; nw. O.-Ind. (Punjab-Chamba)
Kreuzdorn: *Rhamnus cathartica* L.; *Rhamnac.*; Eu.
Kripa: *Lumnitzera racemosa* Willd.; *Combretac.*; Beng.
Kroebara: *Pentaclethra filamentosa* Bth.; *Mimosac.*; Sur.
Kroesa-berry: *Vaccinium vitis-idaea* L.; *Ericac.*; Schweden, Dänemark.
Kroeul: *Melanorrhoea laccifera* Pierre; *Anacardiac.*; Kamb.
Kroewing: *Dipterocarpus eurhynchus* Miq.; *Dipterocarpac.*; Malay.
Krok: *Abies pindrow* Spach; *Conifer.*; nw. O.-Ind. (Punjab-Kumaon)
Krokoué: *Funtumia africana* Stapf; *Apocynac.*; Elf.
Kroma (f): *Klainedoxa gabonensis* Pierre, *K. sp.*; *Simarubac.*; Elf., Kam.
Kromi: *Holarrhenå wulfsbergi* Stapf; *Apocynac.*; Go.
Krong: *Aporosa ficifolia* Baill.; *Euphorbiac.*; Ind.-Ch.
Kronsbeere (d): *Vaccinium vitis-idaea* L.; *Ericac.*; Mitt. & n. Eu.
Krot: *Juglans regia* L.; *Juglandac.*; nw. O.-Ind. (Kaschmir)
Kru: *Quercus semecarpifolia* Smith; *Fagac.*; nw. O.-Ind. (Kaschmir)
„ -ada-wrah: *Leptonychia urophylla* Welw.; *Sterculiac.*; Lib.
Krubeta: *Pseudocedrela kotschyi* Harms; *Meliac.*; Go.
Krubete: *Pseudocedrela kotschyi* Harms; *Meliac.*; Togo.
Krubna: *Khaya grandifolia* C. DC.; *Meliac.*; Go.
Kruen: *Dipterocarpus grandiflorus* Blco., *D. spp.*; *Dipterocarpac.*; Malay, n. & w. Bor.
Kruk: *Mangifera indica* L.; *Anacardiac.*; O.-Ind. (? Burma)
Krun: *Prunus padus* L.; *Rosac.*; nw. O.-Ind. (Punjab)
K'runtum: *Helicia sp.*; *Proteac.*; Malay, br. Borneo (Sarawak)
Krupi: *Carapa procera* DC.; *Meliac.*; Go.
Krupia: *Carapa procera* DC.; *Meliac.*; Go.
Krutu: *Oncoba spinosa* Forsk.; *Flacourtiac.*; Togo.

Kruyn: *Dipterocarpus lowi* Hook. f.; *Dipterocarpac.;* br. Borneo (Sarawak)
Kthúng: *Calophyllum inophyllum* L.; *Guttifer.;* Kamb.
K'tiaou: *Shorea thoreli* Pierre; *Dipterocarpac.;* Kamb.
K'tim: *Shorea thoreli* Pierre; *Dipterocarpac.;* Kamb.
Ku Deuk: *Phyllostachys nigra* Munro; *Gramin.;* Ch. (Fukien)
„ len: *Litchi sinensis* J. F. Gmel.; *Sapindac.;* Ind.-Ch.
Kuakua: *Caloncoba welwitschi* Gilg; *Flacourtiac.;* Kam.
Kuanda: *Cola proteiformis* Chev.; *Sterculiac.;* Elf.
Kuanga: *Daniella oblonga* Oliv.; *Caesalpiniac.;* Elf.
„ -iniama: *? Piptadenia africana* Hook. f.; *Mimosac.;* Elf.
Kuangan: *Albizzia fastigiata* Oliv.; *Mimosac.;* Elf.
Kuanguariniama: *Piptadenia africana* Hook. f.; *Mimosac.;* Elf.
Kuanguli: *Afzelia africana* Smith; *Caesalpiniac.;* Go.
Kuapa: *Cistanthera papaverifera* Chev.; *Tiliac.;* Elf.
Kuatieigualé: *Baccaurea bonneti* Beille; *Euphorbiac.;* Elf.
Kuayral: *Bauhinia retusa* Ham.; *Caesalpiniac.;* n. O.-Ind. (United Prov.)
Kuba: *Careya arborea* Roxb.; *Lecythidac.;* sw. O.-Ind. (Bombay)
Kube: *Borassus flabellifer* L. var. *aethiopum* Warb.; *Palmac.;* Go.
Kubi: *Artocarpus spp.; Morac.;* Phil.
„ : *Beilschmiedia cairocan* Vid.; *Laurac.;* Phil. (Zamboanga, Basilan)
„ : *Phoebe sterculioides* Merr.; *Laurac.;* Phil. (Negrosinsel)
„ : *Pometia pinnata* Forst.; *Sapindac.;* Phil. (Iloilo)
Kubílan: *Litsea sp.; Laurac.;* Phil. (Benguet)
Kubindé: *Kydia calycina* Roxb.; *Malvac.;* O.-Ind.
Kuchale: *Strychnos nux-vomica* L.; *Loganiac.;* C.-O.-Ind.
Kuché: *Cedrela spp.; Meliac.;* Mex.
Kuchila: *Strychnos potatorum* L. f.; *Loganiac.;* w. Beng. (Chota Nagpur)
Kuchla: *Strychnos nux-vomica* L.; *Loganiac.;* n. O.-Ind. (United Prov.)
Kud: *Sterculia urens* Roxb.; *Sterculiac.;* w. O.-Ind. (Deccan)
„ champa: *Michelia champaca* L.; *Magnoliac.;* sw. O.-Ind. (Bombay)
Kuda: *Eugenia jambolana* Lamk.; *Myrtac.;* Beng.
„ : *Holarrhena antidysenterica* Wall.; *Apocynac.;* sw. & C.-O.-Ind.
„ : *Wrightia tinctoria* R. Brown; *Apocynac.;* w. O.-Ind. (Deccan)
Kudak: *Garuga pinnata* Roxb.; *Burserac.;* sw. O.-Ind. (Bombay)
Kudal: *Sterculia villosa* Roxb.; *Sterculiac.;* C.-O.-Ind.
Kudibutshi: *Sideroxylon guianensis* A. DC.; *Sapotac.;* Sur.
Kudkúd: *Neonauclea sp.; Rubiac.;* Phil. (Cagayan)
Kudupóu: *Lannea egregia* Engl., *L. buettneri* Engl.; *Anacardiac.;* Togo.
Kudwal: *Adina cordifolia* Hook. f., *Stephegyne parvifolia* Korth.; *Rubiac.;* w. O.-Ind.
Kudwari: *Bauhinia purpurea* L.; *Caesalpiniac.;* C.-O.-Ind.
Küé: *Albizzia rhombifolia* Bth.; *Mimosac.;* Elf.
„ : *Irvingia sp.; Simarubac.;* Elf.
„ hoé: *Irvingia sp.; Simarubac.;* Elf.
Kuentrillo: *Fagara hiemalis* Engl.; *Rutac.;* Arg.
Kufa: *Clusia colorans* Engl.; *Guttifer.;* br. Gu.
Kugelbaum: *Mimusops balata* Gaertn.; *Sapotac.;* Gu., Ant.
Kugík: *Pometia pinnata* Forst.; *Sapindac.;* Phil. (Catanduane)
Kugoni: *Limonia warneckei* Engl.; *Rutac.;* Togo.
Kugonu: *Limonia warneckei* Engl.; *Rutac.;* Togo.
Kuhumb: *Schleichera trijuga* Willd.; *Sapindac.;* O.-Ind. (? w. Beng.)
Kuin: *Dipterocarpus spp.; Dipterocarpac.;* Malay.
Kujarra: *Strychnos nux-vomica* L.; *Loganiac.;* O.-Ind.
Kuka: *Adansonia digitata* L.; *Bombacac.;* n. Nig., Togo.
„ : *Alstonia gilleti* de Wild.; *Apocynac.;* B.-K. (Bukaka)
Kukate-kayi: *Sapindus laurifolius* Vahl; *Sapindac.;* w. O.-Ind. (s. Bombay)
Kuku: *Mallotus philippinensis* Muell. Arg.; *Euphorbiac.;* C.-O.-Ind.
„ -bacha: *Mallotus philippinensis* Muell. Arg.; *Euphorbiac.;* C.-O.-Ind.
Kukudu-kayalu: *Sapindus laurifolius* Vahl; *Sapindac.;* w. O.-Ind. (n. Bombay)
Kukuinoki: *Aleurites moluccana* Willd.; *Euphorbiac.;* Jap.
Kukuki: *Sterculia tomentosa* Guill. et Perr.; *Sterculiac.;* n. Nig.
Kukunaria: *Abies cephalonica* Link; *Conifer.;* Griech.
Kukur chita: *Litsea sebifera* Pers.; *Laurac.;* Beng.
Kukuru-man: *Randia dumetorum* Lam.; *Rubiac.;* tr. As., Sunda.
Kul: *Zizyphus jujuba* Lamk.; *Rhamnac.;* Beng.

Kula: *Caryocar glabrum* Pers.; *Caryocarac.;* br. Gu.
„ telu: *Stereospermum xylocarpum* Bth. et Hook. f.; *Bignoniac.;* C.-O.-Ind.
„ maruthu: *Terminalia arjuna* Bedd.; *Combretac.;* O.-Ind. (Madras)
Kúlak: *Ceiba pentandra* Gaertn.; *Bombacac.;* Phil. (Abra)
Kulambak: *Aquilaria agallocha* Roxb.; *Thymelaeac.;* Burma.
Kulanim: *Millingtonia hortensis* L. f.; *Bignoniac.;* Burma (Awa, Tenasserim)
Kulási: *Lumnitzera littorea* Voigt, *L. racemosa* Willd.; *Combretac.;* Phil.
„ : *Avicennia officinalis* L.; *Verbenac.;* Phil. (Cotabato)
Kulatingan: *Pterospermum obliquum* Blco., *P. sp.; Sterculiac.;* Phil.
„ : *Parinarium sp.; Rosac.;* Phil. (Nueva Ecija, Tarlac)
Kulau: *Taractogenos kurzi* King; *Flacourtiac.;* Malay.
Kulché: *Cedrela sp.; Meliac.;* Mex.
Kulen: *Litchi chinensis* Radlk.; *Sapindac.;* Ind.-Ch.
Kuliáan: *Shorea negrosensis* Foxw.; *Dipterocarpac.;* Phil. (Sorsogon)
Kulilem: *Garcinia sp.; Guttifer.;* Phil. (Cagayan)
Kulim: *Scorodocarpus borneensis* Becc.; *Olacac.;* Malay.
Kulimbáning: *Xylocarpus granatum* Koenig; *Meliac.;* Phil. (Culion)
Kulimpápa: *Vitex parviflora* Juss.; *Verbenac.;* Phil.
Kuling-bábui: *Dysoxylum euphlebium* Merr.; *Meliac.;* Phil. (Laguna)
„ -manúk: *Aglaia luzoniensis* Merr. et Rolfe; *Meliac.;* Phil. (Laguna)
„ „ : *Dysoxylum turczaninowi* C. DC.; *Meliac.;* Phil. (Laguna)
Kulipápa: *Vitex aherniana* Merr., *V. parviflora* Juss.; *Verbenac.;* Phil.
Kulipia: *Carapa microcarpa* Chev.; *Meliac.;* Elf.
Kulis: *Cratoxylon sp.; Guttifer.;* Phil. (Nueva Ecija)
Kuliskúlis: *Taxotrophis ilicifolia* Vid.; *Morac.;* Phil. (Palawan)
Kulitingan: *Parinarium sp.; Rosac.;* Phil. (Nueva Ecija, Tarlac)
Kulitóm: *Diospyros foveo-reticulata* Merr.; *Ebenac.;* Phil. (Zamboana)
Kulíuan: *Cinnamomum mercadoi* Vid.; *Laurac.;* Phil. (Cagayan)
Kulkul: *Bauhinia rufescens* Lam.; *Caesalpiniac.;* s. Sudan.
Kulm: *Stephegyne parvifolia* Korth.; *Rubiac.;* O.-Ind.
Kulmuri: *Gardenia gummifera* L.; *Rubiac.;* C.-O.-Ind.
Kulo: *Albizzia zygia* MacBride; *Mimosac.;* Go.
Kulong-kulóng: *Pterocymbium tinctorium* Merr.; *Sterculiac.;* Phil. (Marinduque)
Kulu: *Sterculia urens* Roxb.; *Sterculiac.;* C.-O.-Ind. (Berar)
„ layo hitam: *Arytera littoralis* Blume; *Sapindac.;* Malay.
Kulukutingan: *Pterospermum sp.; Sterculiac.;* Phil.
Kum: *Adina sessilifolia* Hook. f.; *Rubiac.;* Beng.
„ barola: *Cordia vestita* Hook. f. et Thoms.; *Borraginac.;* n. O.-Ind. (Un.-Pr.)
„ paiman: *Cordia vestita* Hook. f. et Thoms.; *Borraginac.;* n. O.-Ind. (Un.-Pr.)
„ tai: *Acacia farnesiana* Willd.; *Mimosac.;* Ind.-Ch.
Kumaka: *Ceiba pentandra* Gaertn.; *Bombacac.;* br. Gu.
„ -balli: *Ficus gleasoni* Standley; *Morac.;* br. Gu.
Kumakuafo: *Gymnosporia senegalensis* Loes.; *Celastrac.;* Go.
Kuma-kuma: *Klainedoxa grandifolia* Engl.; *Simarubac.;* B.-K. (Ganda Sungi)
Kumala: *Gmelina arborea* L.; *Verbenac.;* sw. O.-Ind. (Travancore Hills)
„ : *Mallotus philippinensis* Muell. Arg.; *Euphorbiac.;* w. Beng. (Orissa)
Kumalamaram: *Gmelina arborea* L.; *Verbenac.;* O.-Ind. (Madras)
Kumara: *Dipteryx odorata* Willd.; *Papilionac.;* Sur., n. Bras.
„ -mara: *Amioua guianensis* Aubl.; *Rubiac.;* br. Gu.
Kumashide: *Carpinus japonica* Blume, *C. distegocarpus* Koidzumi; *Betulac.;* Jap.
Kumbalam: *Gmelina arborea* L.; *Verbenac.;* O.-Ind. (Madras)
Kumbar shafo: *Acacia campylacantha* Hochst.; *Mimosac.;* n. Nig.
Kumbay: *Gardenia latifolia* Aiton; *Rubiac.;* O.-Ind. (Madras)
Kumbhi: *Careya arborea* Roxb.; *Lecythidac.;* n. O.-Ind. (United Prov.)
Kumbi: *Lannea welwitschi* Engl.; *Anacardiac.;* B.-K. (Mayombe)
„ : *Careya arborea* Roxb.; *Lecythidac.;* C.-O.-Ind.
„ : *Gardenia lucida* Roxb.; *Rubiac.;* O.-Ind. (Madras)
Kumbia: *Careya arborea* Roxb.; *Lecythidac.;* sw. O.-Ind. (Bombay)
Kumbil: *Gmelina arborea* L.; *Verbenac.;* sw. O.-Ind. (Malabar)
„ : *Trewia nudiflora* L.; *Euphorbiac.;* sw. O.-Ind. (Malabar)
Kumbir: *Careya arborea* Roxb.; *Lecythidac.;* w. Beng. (Chota Nagpur)
Kumboi: *Adina cordifolia* Hook. f.; *Rubiac.;* O.-Ind.
Kumbul: *Sideroxylon tomentosum* Roxb.; *Sapotac.;* sw. O.-Ind. (Bombay)
Kumenini: *Lannea aff. acidissima* Chev.; *Anacardiac.;* Go.

Kumhar: *Gmelina arborea* L.; *Verbenac.;* n. O.-Ind. (Punjab, United-Prov.)
„ : *Trewia nudiflora* L.; *Euphorbiac.;* nw. O.-Ind. (Punjab)
Kumher: *Gmelina arborea* L.; *Verbenac.;* C.-O.-Ind.
Kumhi: *Careya arborea* Roxb.; *Lecythidac.;* C.-O.-Ind.
Kumili: *Gmelina arborea* L.; *Verbenac.;* O.-Ind. (Madras)
Kumiska: *Gmelina arborea* L.; *Verbenac.;* sw. O.-Ind. (Malabar)
Kumkuma: *Mallotus philippinensis* Muell. Arg.; *Euphorbiac.;* O.-Ind. (Madras)
Kumal: *Citrus aurantium* L.; *Rutac.;* Beng.
Kummar: *Careya arborea* Roxb.; *Lecythidac.;* C.-O.-Ind.
Kumpas: *Koompassia malaccensis* Maingay; *Caesalpiniac.;* Malay.
„ : *Koompassia beccariana* Taub.; *Caesalpiniac.;* Borneo.
„ -ruman: *Trigonochlamys griffithi* Hook. f.; *Burserac.;* Malay.
Kumpoli: *Sideroxylon tomentosum* Roxb.; *Sapotac.;* w. O.-Ind. (Bombay)
Kumulu: *Gmelina arborea* L.; *Verbenac.;* O.-Ind. (Madras)
Kumunu: *Coula edulis* Baill.; *Olacac.;* B.-K. (Mayombe)
Kumus: *Shorea ciliata* King, *S. costata* King; *Dipterocarpac.;* Malay.
„ hitam: *Shorea ciliata* King, *S. costata* King; *Dipterocarpac.;* Malay.
„ merah: *Shorea ciliata* King, *S. costata* King; *Dipterocarpac.;* Malay.
Kun: *Areca catechu* L.; *Palmac.;* Sunda.
Kunbhi: *Cordia vestita* Hook. f. et Thoms.; *Borraginac.;* nw. O.-Ind. (Punjab)
Kunch: *Alnus nepalensis* D. Don; *Betulac.;* nw. O.-Ind.
Kundangan: *Bouea macrophylla* Griff.; *Anacardiac.;* Malay.
Kunding: *Parkia sherfeseei* Merrill; *Mimosac.;* Phil. (Agusan)
Kundu baenge: *Millettia laurenti* de Wild.; *Papilionac.;* Aeq. Af.
Kundur: *Boswellia serrata* Roxb.; *Burserac.;* Pers.
Kùng: *Calophyllum inophyllum* L.; *Guttifer.;* Kamb.
Kunghas: *Schleichera trijuga* Willd.; *Sapindac.;* s. Cey.
Kungkur: *Pithecolobium confertum* Bth.; *Mimosac.;* Malay.
Kungli: *Boswellia serrata* Roxb.; *Burserac.;* O.-Ind. (Madras)
Kunis: *Alnus nepalensis* D. Don, *Betula alnoides* Ham.; *Betulac.;* n. O.-Ind. (Garhwal)
„ : *Alnus nitida* Endl.; *Betulac.;* n. O.-Ind. (United Prov.)
Kùnji: *Kydia calycina* Roxb.; *Malvac.;* O.-Ind. (Madras)
Kunkoodoo: *Sapindus emarginatus* Vahl; *Sapindac.;* w. O.-Ind. (n. Bombay)
Kunkuma: *Mallotus philippinensis* Muell. Arg.; *Euphorbiac.;* w. & C.-O.-Ind.
Kunkushewa: *Gymnosporia senegalensis* Loes.; *Celastrac.;* n. Nig.
Kunlai: *Dichrostachys cinerea* Wight et Arn.; *Mimosac.;* O.-Ind. (Merwara)
Kunli: *Murraya exotica* L.; *Rutac.;* w. O.-Ind. (Bombay)
Kunsung: *Grewia vestita* Wall.; *Tiliac.;* nö. O.-Ind.
Kuntan: *Uapaca heudeloti* Baill.; *Euphorbiac.;* Go.
Kunthay: *Olea glandulifera* Wall.; *Oleac.;* O.-Ind.
Kunthirikka-payin: *Canarium strictum* Roxb.; *Burserac.;* sw. O.-Ind. (Malabar)
Kunthrikam: *Canarium strictum* Roxb.; *Burserac.;* sw. O.-Ind. (Madras)
Kunti: *Murraya exotica* L.; *Rutac.;* w. O.-Ind. (Bombay)
Kunt-ich: *Acacia farnesiana* Willd.; *Mimosac.;* br. Hond.
Kuntunkum: *Bombax buonopozense* Beauv.; *Bombacac.;* Go.
? Kuntunkuri: *Khaya senegalensis* A. Juss.; *Meliac.;* Go.
Kuntz: *Alnus nepalensis* D. Don; *Betulac.;* nw. O.-Ind. (Punjab)
Kunugi: *Quercus acutissima* Carruth., *Q. serrata* Thunb.; *Fagac.;* Jap.
Kúpang: *Albizzia marginata* Merr., *Parkia timoriana* Merr.; *Mimosac.;* Phil.
„ -babáe: *Albizzia marginata* Merr.; *Mimosac.;* Phil. (Bataan)
„ bundúk: *Albizzia marginata* Merr.; *Mimosac.;* Phil. (Bataan)
„ : *Wallaceodendron celebicum* Koord.; *Mimosac.;* Phil. (n. Tayabas)
„ -kúpang: *Parishia malabog* Merr.; *Anacardiac.;* Phil. (Sibuyan)
Kupanssulo: *Piptadenia kerstingi* Harms; *Mimosac.;* W.-Af.
Kupaussulo: *Albizzia angolensis* Welw.; *Mimosac.;* Togo.
Kupé: *Clusia sp.; Guttifer.;* br. Gu.
Kuperf: *Piptadenia africana* Hook. f; *Mimosac.;* W.-Af. (S.-L.)
Kupferholz: *Jambosa aromatica* Miq.; *Myrtac.;* Java.
Kura: *Holarrhena antidysenterica* Wall., *Wrightia tinctoria* R. Brown; *Apocynac.;*
Kuradia: *Polyalthia longifolia* Bth. et Hook. f.; *Anonac.;* O.-Ind. [O.-Ind.
Kurahara: *Calophyllum lucidum* Bth., *C. sp.; Guttifer.;* br. Gu., Sur.
Kurahura: *Calophyllum sp.; Guttifer.;* Gu.
Kuraiya: *Holarrhena antidysenterica* Wall.; *Apocynac.;* C.-O.-Ind.
Kuraka: *Protium guianense* March.; *Burserac.;* br. Gu.

Kurakat: *Holarrhena antidysenterica* Wall.; *Apocynac.;* C.-O.-Ind.
Kural: *Bauhinia retusa* Ham.; *Caesalpiniac.;* nw. O.-Ind. (Punjab)
Kuran: *Dialium sp.; Caesalpiniac.;* Malay.
Kurana: *Cedrela odorata* L., *C. sp.; Meliac.;* Sur.
Kurangadi: *Acrocarpus fraxinifolius* Wight et Arn.; *Caesalpiniac.;* sw. O.-Ind.
Kurangan: *Acrocarpus fraxinifolius* Wight et Arn.; *Caesalpiniac.;* sw. O.-Ind.
Kuraru: *Andira coriacea* Pulle; *Papilionac.;* Sur.
Kuras: *Dryobalanops oblongifolia* Dyer; *Dipterocarpac.;* Malay.
Kurasám: *Eugenia claviflora* Roxb.; *Myrtac.;* Phil. (Cagayan)
Kurayong: *Parkia roxburghi* G. Don; *Mimosac.;* Malay.
Kurbu: *Gardenia gummifera* L.; *Rubiac.;* w. Beng. (Bihar, Orissa)
Kurchi: *Holarrhena antidysenterica* Wall.; *Apocynac.;* Beng., Burma.
Kurdi: *Burkea africana* Hook.; *Caesalpiniac.;* n. Nig.
Kurdu: *Gardenia gummifera* L.; *Rubiac.;* w. Beng. (Bihar, Orissa)
 „ : *Holarrhena antidysenterica* Wall.; *Apocynac.;* w. Beng. (Chota Nagpur)
Kuré: *Albizzia rhombifolia* Bth.; *Mimosac.;* Elf.
 „ : *Irvingia sp.; Simarubac.;* Elf.
Kuri: *Holarrhena antidysenterica* Wall.; *Apocynac.;* n. O.-Ind. (United Prov.)
 „ : *Castanea vulgaris* Lam. *var. japonica* DC., *C. pubinervis* Schneid.; *Fagac.;* Jap.
 „ -kasi: *Castanopsis taiwaniana* Hayata; *Fagac.;* Form.
 „ -mutal: *Ougeinia dalbergioides* Bth.; *Papilionac.;* w. O.-Ind. (Deccan)
Kariaput: *Melia composita* Willd.; *Meliac.;* sw. O.-Ind. (Bombay)
Kuríat: *Shorea guiso* Blume; *Dipterocarpac.;* Phil. (Nueva Ecija)
Kuribu: *Shorea guiso* Blume; *Dipterocarpac.;* Phil. (Isa.)
Kurig: *Mangifera monandra* Merr.; *Anacardiac.;* Phil. (Zambales)
Kurihouga: *Hopea wightiana* Wall.; *Dipterocarpac.;* sw. O.-Ind. (Bombay)
Kurimau: *Dipterocarpus vernicifluus* Blco.; *Dipterocarpac.;* Phil. (Cagayan)
Kuringhura: *Stereospermum chelonioides* DC.; *Bignoniac.;* O.-Ind., Cey.
Kuripal: *Durio sp.; Bombacac.;* Malay.
Kuriya: *Bombax buonopozense* Beauv.; *Bombacac.;* n. Nig.
Kurk-moutouchi: *Pterocarpus draco* L.; *Papilionac.;* fr. Gu.
Kurku: *Schleichera trijuga* Willd.; *Sapindac.;* C.-O.-Ind.
Kurma: *Machilus macrantha* Nees; *Laurac.;* sw. O.-Ind. (Madras)
Kurmi: *Adina cordifolia* Hook. f.; *Rubiac.;* C.-O.-Ind. (Gondwana)
Kurmuri: *Gardenia gummifera* L.; *Rubiac.;* C.-O.-Ind.
Kurmurunga: *Averrhoa carambola* L.; *Oxalidac.;* O.-Ind.
Kurna: *Zizyphus jujuba* Lam., *Z. spina-christi* Willd.; *Rhamnac.;* n. Nig. (Kontagora)
 „ , foreign: *Melia azedarach* L.; *Meliac.;* n. Nig. (Kontagora)
 „ -na-sara: *Melia azedarach* L.; *Meliac.;* n. Nig. (Kontagora)
Kurogaki: *Diospyros discolor* Willd.; *Ebenac.;* Form.
Kuroganemochi: *Ilex rotunda* Thunb.; *Aquifoliac.;* Jap.
Kurók: *Calocarpum mammosum* Pierre; *Sapotac.;* C.-R.
Kurokaki: *Diospyros lotus* L.; *Ebenac.;* Jap.
Kuromatsu: *Pinus thunbergi* Parl.; *Conifer.;* Jap.
Kuromoji: *Lindera sericea* Blume; *Laurac.;* Jap.
Kurong: *Trewia nudiflora* L.; *Euphorbiac.;* nö. O.-Ind. (Nepal)
Kurosugi: *Cryptomeria japonica* D. Don; *Conifer.;* Jap., Ch.
Kurpa: *Barringtonia acutangula* Gaertn.; *Lecythidac.;* O.-Ind. (Madras)
Kurpendra: *Gardenia turgida* Roxb.; *Rubiac.;* sw. O.-Ind. (Bombay)
Kurpodol: *Toona ciliata* M. Roem.; *Meliac.;* C.-O.-Ind. (Gondwana)
Kurrajong: *Sterculia gregori* ?; *Sterculiac.;* w. Au.
 „ , brown: *Commersonia echinata* Forst.; *Sterculiac.;* ö. Au.
 „ , flame-: *Brachychiton acerifolius* F. v. M.; *Sterculiac.;* ö. Au.
 „ , white: *Brachychiton discolor* F. v. M.; *Sterculiac.;* ö. Au.
Kuru: *Lophira procera* Chev.; *Ochnac.;* Nig.
 „ : *Gardenia gummifera* L.; *Rubiac.;* C.-O.-Ind. (Gondwana)
Kurudu: *Gardenia latifolia* Aiton; *Rubiac.;* w. Beng. (Orissa)
Kuruinj: *Pongamia glabra* Vent.; *Papilionac.;* w. Beng. (Chota Nagpur)
Kurukubi: *Grewia mollis* Juss.; *Tiliac.;* n. Nig.
Kurum: *Stephegyne parvifolia* Korth.; *Rubiac.;* C.-O.-Ind.
Kurumba: *Nauclea cordifolia* Roxb.; *Rubiac.;* O.-Ind. (? Beng.)
Kurumi: *Juglans sieboldiana* Maxim.; *Juglandac.;* Jap.
Kurupikai: *Sapium haematospermum* Muell. Arg.; *Euphorbiac.;* Arg.
Kuruskúrus: *Taxotrophis ilicifolia* Vid.; *Morac.;* Phil. (Camarines)

Kurutwe: *Canarium schweinfurthi* Engl.; *Burserac.;* Go.
Kurweil: *Dillenia pentagyna* Roxb.; *Dilleniac.;* sw. O.-Ind. (Bombay)
Kusa atte: *Thujopsis dolabrata* S. et Z.; *Conifer.;* Jap.
„ maki: *Podocarpus macrophylla* D. Don; *Conifer.;* Jap.
Kusaba: *Chlorophora excelsa* Bth. et Hook.; *Morac.;* Go.
Kusagi: *Clerodendron trichotomum* Thunb.; *Verbenac.;* Jap.
Kusam: *Schleichera trijuga* Willd.; *Sapindac.;* C.-O.-Ind., n. O.-Ind. (Un. Prov.)
Kusambi: *Schleichera trijuga* Willd.; *Sapindac.;* O.-Ind.
Kusamo: *Schleichera trijuga* Willd.; *Sapindac.;* w. Beng. (Orissa)
Kusan: *Hymenodictyon excelsum* Wall.; *Rubiac.;* Burma.
Kusano-kurikasi: *Castanopsis kusanoi* Hayata; *Fagac.;* Form.
Kush-uh-ché: *Croton sp.; Euphorbiac.;* br. Hond.
Kushub: *Bixa orellana* L.; *Bixac.;* br. Hond.
Kusia: *Sarcocephalus esculentus* Afz.; *Rubiac.;* Go.
Kusibiru: *Diospyros sanza-minika* Chev.; *Ebenac.;* Elf.
Kusiepoetoe-joele-malédie: *Couratari guianensis* Aubl., *C. sp.; Lecythidac.;* Sur.
Kusipapa: *Macrolobium limba* Scott Elliott; *Caesalpiniac.;* Go.
Kusmar: *Gmelina arborea* L.; *Verbenac.;* O.-Ind. (? Beng.)
Kusmus: *Shorea ciliata* King; *Dipterocarpac.;* Malay.
Kussam: *Schleichera trijuga* Willd.; *Sapindac.;* nw. O.-Ind. (Punjab)
Kussamar: *Gmelina arborea* L.; *Verbenac.;* C.-O.-Ind.
Kussi: *Stereospermum suaveolens* DC.; *Bignoniac.;* O.-Ind. (? Beng.)
Kussumar: *Schleichera trijuga* Willd.; *Sapindac.;* w. O.-Ind. (Panch-Mahals)
Kussumb: *Schleichera trijuga* Willd.; *Sapindac.;* O.-Ind.
Kusu: *Cinnamomum camphora* Nees et Eberm.; *Laurac.;* Jap.
Kusua: *Haronga paniculata* Sprague; *Guttifer.;* Go.
Kusum: *Schleichera trijuga* Willd.; *Sapindac.;* O.-Ind.
Kusumb: *Schleichera trijuga* Willd.; *Sapindac.;* C.-O.-Ind., sw. O.-Ind. (Bomb.)
Kusunoki: *Cinnamomum camphora* Nees et Eberm.; *Laurac.;* Jap.
Kuthada: *Sterculia villosa* Roxb.; *Sterculiac.;* C.-O.-Ind.
Kuthan: *Hymenodictyon excelsum* Wall.; *Rubiac.;* Burma.
Kutki: *Miliusa velutina* Hook. f. et Thoms.; *Anonac.;* C.-O.-Ind. (Gondwana)
„ : *Bridelia retusa* Spreng.; *Euphorbiac.;* sw. O.-Ind. (Bombay)
Kuttú: *Cratoxylon sp.; Guttifer.;* Phil. (Cagayan)
Kutul: *Randia dumetorum* Lamk.; *Rubiac.;* C.-O.-Ind.
Kut-wahn: *Cassipourea firestoneana* Hutch. et Dalz.; *Rhizophorac.;* Lib.
Kuviy: *Araucaria brasiliana* Lamb.; *Conifer.;* Arg.
Kuwa: *Xylopia aethiopica* A. Rich.; *Anonac.;* B.-K. (Malela)
Kuwondön: *Anthocleista kerstingi* Gilg; *Loganiac.;* Togo.
Kuyau-kuyáu: *Alstonia macrophylla* Wall.; *Apocynac.;* Phil. (Camarines)
Kuyché: *? Ceiba sp., ? Pachira sp., ? Bombax sp.; Bombacac.;* Mex.
Kuyus kuyus: *Taxotrophis ilicifolia* Vid.; *Morac.;* Phil.
Kuzoe-arak: *Vitex pubescens* Vahl; *Verbenac.;* tr. As.
Kvao: *Adina cordifolia* Hook. f.; *Rubiac.;* Kamb.
Kwa-dwuma: *Guarea sp.; Meliac.;* w. Go.
„ -ie: *Qualea dinizii* Ducke, *Vochysia melinoni* Beckmann; *Vochysiac.;* Sur.
Kwabohora: *Entandrophragma sp.; Meliac.;* Go.
Kwabohori: *Khaya anthotheca* C. DC.; *Meliac.;* W.-Af. (Angola)
Kwabohoro: *Entandrophragma spp.; Meliac.;* Go.
Kwadwo-amanin: *Lonchocarpus sericeus* H. B. K.; *Papilionac.;* Go.
Kwadwuma: *Guarea thompsoni* Sprague et Hutch.; *Meliac.;* Go.
„ : *Turraeanthus aff. africana* Pellegr.; *Meliac.;* Go.
K'wahak: *Chamaecyparis pisifera* S. et Z.; *Conifer.;* Jap.
Kwai: *Vochysia tomentosa* DC.; *Vochysiac.;* Sur.
Kwajemu akenka: *Limonia sp.; Rutac.;* Go.
Kwaka: *Calophyllum sp.; Guttifer.;* Sur.
Kwaku bise: *Carapa procera* DC.; *Meliac.;* Go.
Kwali: *Vochysia guianensis* Aubl., *Qualea coerulea* Aubl.; *Vochysiac.;* Sur.
Kwalidan: *Erisma uncinatum* Warm.; *Vochysiac.;* Sur.
Kwan-kwan: *Oncoba welwitschi* Oliv.; *Flacourtiac.;* Kam.
Kwanari: *Hymenaea courbaril* L.; *Caesalpiniac.;* br. Gu.
Kwanedua: *Parinarium tenuifolium* Chev.; *Rosac.;* Go.
Kwang booloo: *Dipterocarpus crinitus* Dyer; *Dipterocarpac.;* Malak.
Kwangtung-aburagiri: *Aleurites montana* E. H. Wilson; *Euphorbiac.;* Jap.

Kwantanura: *Lovoa klaineana* Pierre; *Meliac.;* Go.
Kwantanuro: *Lovoa klaineana* Pierre; *Meliac.;* Go.
Kwantenura: *Lovoa klaineana* Pierre; *Meliac.;* Go.
Kwaré: *Vochysia tomentosa* DC.; *Vochysiac.;* Sur.
Kwarerian: *Erisma uncinatum* Warm.; *Vochysiac.;* Sur.
Kwari: *Vochysia guianensis* Aubl., *Erisma uncinatum* Warm.; *Vochysiac.;* Sur.
„ , Waddie-wassie-: *Qualea dinizii* Ducke, *Q. sp.; Vochysiac.;* Sur.
Kwarie: *Vochysia tomentosa* DC., *V. melinoni* Beckmann; *Vochysiac.;* Sur.
„ , echte: *Vochysia tomentosa* DC.; *Vochysiac.;* Sur.
„ , Gwanna-: *Vochysia melinoni* Beckmann; *Vochysiac.;* Sur.
„ , Sabana-: *Byrsonima sp.; Malpighiac.;* Sur.
„ , Savanna-: *Byrsonima sp.; Malpighiac.;* Sur.
„ , Singrie-: *Erisma uncinatum* Warm.; *Vochysiac.;* Sur.
„ , Wassie-wassie-: *Qualea dinizii* Ducke, *Q. sp.; Vochysiac.;* Sur.
„ , Water-: *Qualea coerulea* Aubl., *Q. albiflora* Warm., *Q. rosea* Aubl.; *Vochysiac.;*
„ , Watra: *Qualea spp.; Vochysiac.;* Sur.   [Sur.
„ , witte: *Vochysia tomentosa* DC.; *Vochysiac.;* Sur.
Kwarrie: *Erisma uncinatum* Warm.; *Vochysiac.;* Sur.
„ , Gwana-: *Vochysia sp.; Vochysiac.;* Sur.
„ , Sengrie-: *Erisma uncinatum* Warm.; *Vochysiac.;* Sur.
„ , Wane-: *Vochysia tomentosa* DC.; *Vochysiac.;* Sur.
„ , Wassie-wassie-: *Pithecolobium pedicellare* Bth.; *Mimosac.;* Sur.
„ , Watra: *Qualea coerulea* Aubl.; *Vochysiac.;* Sur.
„ , witte: *Vochysia tomentosa* DC.; *Vochysiac.;* Sur.
Kwassi: *Quassia amara* L.; *Simarubac.;* Sur.
„ bitterhout: *Quassia sp.; Simarubac.;* Sur.
„ , grammam-: *Picrasma sp.; Simarubac.;* Sur.
Kwassie: *Quassia amara* L.; *Simarubac.;* Sur.
„ hout: *Picrasma sp.; Simarubac.;* Sur.
„ -kaan-asoebi: *? Andira retusa* H. B. K.; *Papilionac.;* Sur.
Kwata: *Couepia spp., Licania apetala* Fritsch; *Rosac.;* Sur.
„ kama: *Parkia pendula* Bth.; *Mimosac.;* Sur.
„ -werie: *Stryphnodendron flammatum* Kleinh.; *Mimosac.;* Sur.
Kwatendro: *Entandrophragma sp.; Meliac.;* Go.
Kwatéré: *Eschweilera corrugata* Miers; *Lecythidac.;* Sur.
Kwati: *Teclea unifoliolata* Baill.; *Rutac.;* D.-O.-Af. (w. Usambara)
Kwatie-ie: *Eschweilera corrugata* Miers; *Lecythidac.;* Sur.
Kwatta: *Parkia pendula* Bth.; *Mimosac.;* Sur.
„ kama: *Parkia pendula* Bth.; *Mimosac.;* Sur.
„ patoe: *Lecythis grandiflora* Aubl.; *Lecythidac.;* Sur.
Kwayemu akenka: *Limonia sp.; Rutac.;* Go.
Kwè: *Dolichandrone stipulata* Bth.; *Bignoniac.;* Burma.
Kwebie: *Licania apetala* Fritsch; *Rosac.;* Sur.
Kweig yabra: *Rhizophora racemosa* G. F. W. Meyer; *Rhizophorac.;* Go.
Kwepi: *Couepia guianensis* Aubl., *Licania apetala* Fritsch; *Rosac.;* Sur.
Kwepie: *Couepia spp., Licania apetala* Fritsch; *Rosac.;* Sur.
Kweranyani: *Piptadenia buchanani* Baker; *Mimosac.;* br. O.-Af. (Nyassa)
Kwet: *Feronia elephantum* Correa; *Rutac.;* O.-Ind. (? Burma)
Kwetayaw: *Grewia laevigata* Vahl; *Tiliac.;* Burma (Tenasserim)
Kweti: *Spondias mangifera* Willd.; *Anacardiac.;* Burma.
Kwiengé: *Xylopia frutescens* Aubl., *X. spp.; Anonac.;* Sur.
Kwillar: *Bauhinia purpurea* L.; *Caesalpiniac.;* n. O.-Ind. (United Prov.)
Kwo: *Lovoa klaineana* Pierre; *Meliac.;* Nig.
„ -kwo: *Mitragyne macrophylla* Hiern; *Rubiac.;* Elf.
Ky tchaò: *Hovenia dulcis* Thunb.; *Rhamnac.;* Ch.
Ky yên: *Homalium dictyoneurum* Pierre; *Flacourtiac.;* w. Assam.
Kyalanki: *Sonneratia apetala* Ham.; *Sonneratiac.;* ö. O.-Ind. (n. Madras)
Kyama: *Carapa moluccensis* Lamk.; *Meliac.;* Burma.
Kyana: *Carapa moluccensis* Lamk.; *Meliac.;* Burma.
Kyathnan: *Carapa moluccensis* Lamk.; *Meliac.;* Burma (s. Tenasserim)
Kyauk-tinyu: *Taxus baccata* L.; *Conifer.;* n. Burma.
Kyaw: *Nauclea sessilifolia* Roxb.; *Rubiac.;* n. Burma.
Kye: *Gluta tavoyana* Hook. f.; *Anacardiac.;* Burma.
Kyeneboa: *Cordia irvingi* Baker; *Borraginac.;* Go.

Kyenedua: *Cordia irvingi* Baker; *Borraginac.*; Go.
Kyeng-lwai: *Elaeocarpus robustus* Roxb.; *Elaeocarpac.*; Burma.
Kyenedua: *Cordia irvingi* Baker; *Borraginac.*; Go.
Kyeng-lwai: *Elaeocarpus robustus* Roxb.; *Elaeocarpac.*; Burma.
Kyeni: *Barringtonia acutangula* Gaertn.; *Lecythidac.*; Burma.
Kyenkyen: *Antiaris africana* Engl.; *Morac.*; Go.
Kyet-mouk: *Nephelium longana* Camb.; *Sapindac.*; Burma.
Kyettayaw: *Grewia laevigata* Vahl; *Tiliac.*; Burma.
Kyetyo: *Vitex peduncularis* Wall., *V. heterophylla* Roxb.; *Verbenac.*; Burma.
Kyi: *Barringtonia acutangula* Gaertn.; *Lecythidac.*; Burma.
Kyirafunti: *Lophira alata* Banks; *Ochnac.*; Go.
Kyirebente: *Lophira alata* Banks; *Ochnac.*; Go.
Kyon: *Diospyros tomentosa* Roxb.; *Ebenac.*; w. Beng.
Kyonix (u): *Liquidambar styraciflua* L.; *Hamamelidac.*; s. USA.
Kyumba: *Diospyros sp.*; *Ebenac.*; D.-O.-Af. (w. Usambara)
Kyun: *Tectona grandis* L. f.; *Verbenac.*; Burma.
Kyunalin: *Callicarpa arborea* Roxb.; *Verbenac.*; Burma.
Kywèdanyin: *Pithecolobium dulce* Bth.; *Mimosac.*; O.-Ind.
Kywèmagyolein: *Stereospermum suaveolens* DC.; *Bignoniac.*; Burma.
Kywetho: *Bischofia javanica* Blume; *Euphorbiac.*; Burma.

L

La: *Parkia agboensis* Chev.; *Mimosac.*; Elf.
„ : *Hibiscus tiliaceus* L.; *Malvac.*; Ind.-Ch.
„ linh: *Ailanthus malabarica* DC.; *Simarubac.*; Ind.-Ch.
„ crai: *Vismia cayennensis* Pers.; *Guttifer.*; Trin., Tob.
„ loa: *Kurrimia pulcherrima* Wall.; *Celastrac.*; Ind.-Ch.
„ mâ: *Saraca dives* Pierre; *Caesalpiniac.*; Ind.-Ch. (Laos)
„ ngoa: *Mallotus barbatus* Muell. Arg.; *Euphorbiac.*; Ind.-Ch.
Laanga: *Platysepalum chevalieri* Harms; *Papilionac.*; B.-K.
Laau ala: *Santalum yasi* Seemann; *Santalac.*; Fid.
Laba: *Sonneratia griffithi* Kurz; *Sonneratiac.*; Burma.
Labakh: *Albizzia lebbek* Bth.; *Mimosac.*; Sudan.
Laba-laba: *Qualea sp.*; *Vochysiac.*; Bras.
Labalaba: *Zizyphus sp.*; *Rhamnac.*; Phil. (Pangasinan)
Laban-hout: *Vitex pubescens* Vahl; *Verbenac.*; O.-Ind.
Labáng: *Vatica mangachapoi* Blco.; *Dipterocarpac.*; Phil. (Ilocos Sur)
Labau: *Stephegyne diversifolia* Hook. f.; *Rubiac.*; Burma.
Labé: *Canarium occidentale* Chev.; *Burserac.*; Elf.
„ : *Sonneratia apetala* Ham.; *Sonneratiac.*; Burma.
L'abeille: *Mimusops sp.*; *Sapotac.*; fr. Gu.
Labíd: *Livistona rotundifolia* Mart.; *Palmac.*; Phil. (Ilocos, Pampanga)
Labiernago: *Phillyrea angustifolia* L.; *Oleac.*; Sp.
Labíg: *Livistona rotundifolia* Mart.; *Palmac.*; Phil. (Ilocos, Pampanga)
„ : *Livistona rotundifolia* Mart. var. *luzonensis* Becc.; *Palmac.*; Phil.
Labík: *Livistona rotundifolia* Mart.; *Palmac.*; Phil. (Cagayan)
Labinkong: *Dracaena fragrans* Ker-Gawl.; *Liliac.*; Kam.
Labladzi: *Lonchocarpus sericeus* H. B. K.; *Papilionac.*; Go.
Labri: *Morus alba* L., *M. laevigata* Wall.; *Morac.*; Burma.
Laburno (it): *Cytisus laburnum* L.; *Papilionac.*; Eu.
Laburnum (e): *Cytisus laburnum* L.; *Papilionac.*; Eu.
„ : *Goodia lotifolia* Salisb.; *Papilionac.*; Tasm.
„ , indian: *Cassia fistula* L.; *Caesalpiniac.*; tr. As.
„ , Scotch: *Laburnum alpinum* Griseb.; *Papilionac.*; Eu.
Lace bark: *Plagianthus betulinus* A. Cunn., *Hoheria populnea* A. Cunn.; *Malvac.*;
Lace-flower tree: *Pithecolobium grandiflorum* Bth.; *Mimosac.*; ö. Au. [N.-Seel.
„ , indian: *Lagetta linearia* Lam.; *Thymelaeac.*; W.-Ind.
Lacewood (u): *Platanus occidentalis* L.; *Platanac.*; sö. USA.
Lacre: *Vismia guianensis* Pers.; *Guttifer.*; Ven.
„ blanco: *Vismia sp.*; *Guttifer.*; Ven.
„ branco: *Vismia guianensis* Pers.; *Guttifer.*; Bras.
Lac tree: *Schleichera trijuga* Willd.; *Sapindac.*; tr. As.

Ladder wood: *Machilus spp.; Laurac.;* O.-Ind.
Ladiángau: *Agathis alba* Foxw.; *Conifer.;* Phil. (Tayabas, Camarines, Albay)
Ladierna: *Rhamnus alaternus* L.; *Rhamnac.;* Sp.
Lagan: *Dipterocarpus trinervis* Blume, *D. sp.; Dipterocarpac.;* Java, Sum.
Lagaña de perro: *Caesalpinia gilliesi* Wall.; *Caesalpiniac.;* Arg.
Lagárai: *Bruguiera cylindrica* Blume; *Rhizophorac.;* Phil. (Cotabato)
Lagartillo: *Zanthoxylum sp.; Rutac.;* C.-R.
Lagarto: *Zanthoxylum kellermani* P. Wilson; *Rutac.;* Guat.
„   : *Zanthoxylum microcarpum* Griseb.; *Rutac.;* Nic.
„   : *Zanthoxylum panamense* P. Wilson; *Rutac.;* Pan.
„ amarillo: *Zanthoxylum kellermani* P. Wilson; *Rutac.;* Hond.
Lagásá: *Bruguiera sexangula* Poir.; *Rhizophorac.;* Phil. (Cagayan)
Lagásak: *Bruguiera sexangula* Poir.; *Rhizophorac.;* Phil. (Palaui)
Lagat: *Ficus religiosa* L.; *Morac.;* Burma.
Lagetto: *Lagetta lintearia* Lam.; *Thymelaeac.;* W.-Ind.
Lagi-lagí: *Eugenia brevistylis* C. B. Rob.; *Myrtac.;* Phil. (Surigao)
Lágo: *Myristica sp.; Myristicac.;* Phil. (Basilan, Palawan)
„   : *Pygeum presli* Merr.; *Rosac.;* Phil.
Lagopus: *Ochroma lagopus* Sw.; *Bombacac.;* W.-Ind.
Lagundi: *Erythroxylon cuneatum* Kurz; *Erythroxylac.;* Malay, Phil.
Lagunero: *Pterocarpus sp.; Papilionac.;* Ven.
Lagut-út: *Xylocarpus moluccensis* M. Roem.; *Meliac.;* Phil. (Guimaras)
Lahan kudu: *Gardenia gummifera* L.; *Rubiac.;* C.-O.-Ind. (Gondwana)
Lahi-láhi: *Santiria nitida* Merr.; *Burserac.;* Phil. (Surigao)
Lahkyik: *Callicarpa arborea* Roxb.; *Verbenac.;* Burma.
Láho: *Myristica sp.; Myristicac.;* Phil. (Culion)
Lai: *Aleurites moluccana* Willd.; *Euphorbiac.;* Ind.-Ch.
Láian: *Quercus sp.; Fagac.;* Phil. (Agusan)
Laidré: *Stemmadenia grandiflora* Miers; *Apocynac.;* Kol.
Laila: *Salix tetrasperma* Roxb.; *Salicac.;* n. O.-Ind. (United Prov.)
Laini mienzam: *Dracaena fragrans* Ker-Gawl.; *Liliac.;* Kam.
Laintlaintain: *Lophira alata* Banks; *Ochnac.;* W.-Af. (S.-L.)
Laiúsin: *Parinarium sp.; Rosac.;* Phil. (Pampanga, Camarines, Laguna)
Laka-hout: *Dalbergia parviflora* Roxb.; *Papilionac.;* Nied.-Ind.
Lakkey: *Albizzia julibrissin* Durazzini; *Mimosac.;* O.-Ind. (Nepal)
Laknáb: *Terminalia comintana* Merr.; *Combretac.;* Phil. (Nueva Ecija)
Lako-láko: *Palaquium sp.; Sapotac.;* Phil. (Guimaras)
Lakolako: *Koordersiodendron pinnatum* Merr.; *Anacardiac.;* Phil. (Bisaya)
Lakooch: *Artocarpus lakoocha* Roxb.; *Morac.;* O.-Ind.
Lakori: *Wightia gigantea* Wall.; *Scrophulariac.;* ö. Himal.
Lakoshe: *Monodora myristica* Dunal; *Anonac.;* tr. Af.
Laksana: *Acacia farnesiana* Willd.; *Mimosac.;* Malay.
Lakshmi am: *Mangifera silvatica* Roxb.; *Anacardiac.;* Assam (Sylhet)
Laksiri: *Protium guianense* March.; *Burserac.;* Sur.
Laku-chamma: *Artocarpus lakoocha* Roxb.; *Morac.;* sw. O.-Ind. (Madras)
Lakuch: *Artocarpus lakoocha* Roxb.; *Morac.;* And.
Lal chandan: *Pterocarpus santalinus* L. f.; *Papilionac.;* O.-Ind. (Madras)
„ devdari: *Chukrasia tabularis* Adr. Juss.; *Meliac.;* sw. O.-Ind. (Bombay)
„ -khair: *Acacia catechu* Willd.; *Mimosac.;* sw. O.-Ind. (Bombay)
„ lakri: *Rhizophora mucronata* Lamk.; *Rhizophorac.;* And.
Lalanjo: *Allanblackia floribunda* Oliv.; *Guttifer.;* W.-Af. (nw. Angola)
Lalar: *Dipterocarpus littoralis* Blume; *Dipterocarpac.;* Java.
Lalchini: *Calophyllum spectabile* Willd.; *Guttifer.;* And.
„   : *Amoora wallichi* King; *Meliac.;* Assam.
Lali: *Canarium sikkimense* King; *Burserac.;* nw. O.-Ind. (Sikkim)
„ : *Machilus gammieana* King; *Laurac.;* nö. O.-Ind. (Nepal)
Lali-kawala: *Machilus gammieana* King; *Laurac.;* Beng.
Laljhau: *Tamarix articulata* Vahl; *Tamaricac.;* n. O.-Ind. (Punjab/Assam)
Lallei: *Albizzia amara* Boivin; *Mimosac.;* w. O.-Ind. (Bombay)
Lalli: *Phyllanthus emblica* L.; *Euphorbiac.;* C.-O.-Ind. (Gondwana)
Lalona: *Weinmannia bojeriana* Tul.; *Cunoniac.;* Mad.
Lamao (u): *Pentacme contorta* Merr. et Rolfe; *Dipterocarpac.;* Phil.
Lamau: *Artocarpus lakoocha* Roxb.; *Morac.;* Burma.
Lambapatti: *Sideroxylon longepetiolatum* King et Prain; *Sapotac.;* O.-Ind., And.

Lambda patti: *Xylopia parviflora* L.; *Anonac.*; And.
Lamboben: *Buchanania latifolia* Roxb.; *Anacardiac.*; O.-Ind.
Lambrán: *Alnus sp.; Betulac.*; Peru.
Lambu: *Sonneratia griffithi* Kurz; *Sonneratiac.*; s. Burma (s. Tenasserim)
„ : *Evodia fraxinifolia* Hook. f.; *Rutac.*; Assam (Khasis)
Lamé: *Sonneratia griffithi* Kurz; *Sonneratiac.*; Burma.
Lamer: *Prosopis sp.; Mimosac.*; Arg.
Lami: *Pentadesma butyracea* Sabine; *Guttifer.*; fr. Guin.
Lamílan: *Dipterocarpus vernicifluus* Blco.; *Dipterocarpac.*; Phil. (Isa.)
Lamió: *Dracontomelum cumingianum* Baill., *D. edule* Skeels; *Anacardiac.*; Phil.
Lamisi: *Wrightia lanti* Merr.; *Loganiac.*; Phil.
Lamkana: *Bridelia retusa* Spreng.; *Euphorbiac.*; O.-Ind. (Ajmere)
Lamóg: *Planchonia spectabilis* Merr.; *Lecythidac.*; Phil. (Bataan, Laguna, Tayabas)
Lampati: *Duabanga sonneratioides* Ham.; *Sonneratiac.*; n. Beng.
Lampatia: *Duabanga sonneratioides* Ham.; *Sonneratiac.*; O.-Ind. (Nepal)
Lampatti: *Duabanga sonneratioides* Ham.; *Sonneratiac.*; O.-Ind.
Lampo: *Polyosma cambodiana* Gagnep.; *Saxifragac.*; Ind.-Ch.
Lamshing: *Pinus excelsa* Wall.; *Conifer.*; nw. O.-Ind. (Kumaon)
Lamung: *Mangifera indica* L.; *Anacardiac.*; Burma.
Lan: *Lagerstroemia siamica* Gagnep.; *Lythrac.*; Ind.-Ch.
„ han: *Cratoxylon formosum* Bth. et Hook. f.; *Guttifer.*; Ind.-Ch.
„ sao: *Lagerstroemia spireana* Gagnep.; *Lythrac.*; Ind.-Ch.
Lana: *Ochroma sp.; Bombacac.*; Guat.
„ : *Genipa americana* L., *? G. caruto* H. B. K.; *Rubiac.*; Sur., fr. W.-Ind.
Lanaballi: *Gustavia angusta* L.; *Lecythidac.*; Sur.
„ -diamaroe: *Gustavia pterocarpa* Poit.; *Lecythidac.*; Sur.
Lanat: *Litsea sp.; Laurac.*; Phil. (Masbate)
„ : *Neolitsea vidali* Merr.; *Laurac.*; Phil. (Guimaras)
Lanca: *Artocarpus integrifolia* L.; *Morac.*; O.-Ind., Malay.
Lancetilla: *Euterpe edulis* Mart.; *Palmac.*; br. Hond.
„ : *Astrocaryum cohune* Standl.; *Palmac.*; Hond.
Lancewood: *Amelanchier canadensis* Medic.; *Rosac.*; sö. Kan.
„ : *Oxandra lanceolata* Baill.; *Anonac.*; W.-Ind., Hond., tr. S.-Am.
„ : *Rollinia multiflora* St. Hil., *R. longifolia* St. Hil.; *Anonac.*; tr. S.-Am.
„ : *Dissiliaria baloghioides* F. v. M.; *Euphorbiac.*; ö. Au.
„ : *Acacia doratoxylon* A. Cunn.; *Mimosac.*; Au.
„ : *Eriostemon squameus* Lab.; *Rutac.*; Tasm.
„ , black: *Oxandra lanceolata* Baill.; *Anonac.*; Jam.
„ , Burma-: *Homalium tomentosum* Bth.; *Flacourtiac.*; O.-Ind.
„ , Moulmein-: *Homalium tomentosum* Bth.; *Flacourtiac.*; Burma.
„ , red: *Mimusops spp.; Sapotac.*; br. Gu.
„ , Swamp-: *Oxandra sp.; Anonac.*; br. Gu.
„ , Tulip-: *Harpullia pendula* Planch.; *Sapindac.*; ö. Au.
„ , white: *Oxandra sp.; Anonac.*; Jam.
Lancitillo: *Chamaedora sp.; Palmac.*; Guat.
Lancitollo: *Vismia hamani* Blake; *Guttifer.*; Ven.
Landi: *Calophyllum brasiliense* Camb.; *Guttifer.*; Bras.
„ : *Terminalia superba* Engl. et Diels; *Combretac.*; Kam.
Landim: *Calophyllum brasiliense* Camb.; *Guttifer.*; Bras. (Sao P)
Lándo: *Peucedanum araliaceum* Bth. et Hk. var. *fraxinifolium* Hiern; *Umbellifer.*;
Lanéte: *Wrightia sp., W. laniti* Merr., *Kickxia sp.; Apocynac.*; Phil.   [Togo.
„ : *Allaeanthus luzonicus* Vid.; *Morac.*; Phil. (Laguna, Tayabas)
Lang: *Dillenia heterosepala* Finet et Gagnep.; *Dilleniac.*; Ind.-Ch.
„ : *Dialium cochinchinense* Pierre; *Caesalpiniac.*; Ind.-Ch.
„ : *Erythrina indica* Lam.; *Papilionac.*; Ind.-Ch.
„ : *Randia tomentosa* Hook.; *Rubiac.*; Ind.-Ch.
„ cay: *Toddalia aculeata* Pers.; *Rutac.*; Ind.-Ch.
„ dam: *Diospyros siamensis* Hochr.; *Ebenac.*; Ind.-Ch.
„ ktiai: *Buchanania pallida* Pierre; *Anacardiac.*; Ind.-Ch.
„ mang: *Pterospermum diversifolium* Blume; *Sterculiac.*; Ind.-Ch.
Langa, wilde: *Aglaia minahassae* Koord.; *Meliac.*; Nied.-Ind.
Lánganā: *Gardenia thunbergi* L. f.; *Rubiac.*; Togo.
Langárai: *Bruguiera parviflora* Wight et Arn.; *Rhizophorac.*; Phil.
„ : *Bruguiera cylindrica* Blume; *Rhizophorac.*; Phil. (Cotabato)

Langári: *Bruguiera parviflora* Wight et Arn.; *Rhizophorac.;* Phil.
Langil: *Albizzia lebbek* Bth., *A. retusa* Bth., *A. saponaria* Blume; *Mimosac.;* Phil.
Langin: *Albizzia acle* Merr.; *Mimosac.;* Phil. (Masbate)
Langíp: *Albizzia acle* Merr.; *Mimosac.;* Phil. (Bisaya)
Langká: *Artocarpus integra* Merr.; *Morac.;* Phil.
Langkángan: *Parinarium sp.; Rosac.;* Phil. (Lanao)
Langóg: *Parinarium sp.; Rosac.;* Phil. (Agusan)
Langóog: *Parinarium sp.; Rosac.;* Phil. (Agusan)
Langoussi: *Terminalia tanibouca* Rich.; *Combretac.;* fr. Gu.
Langsa oetan: *Aglaia minahassae* Koord.; *Meliac.;* Nied.-Ind.
Langsat: *Lansium domesticum* Jack; *Meliac.;* Malay.
„ -loetoeng: *Aglaia eusideroxylon* Koord. et Valet.; *Meliac.;* Java.
„ oetan: *Lansium sp.; Meliac.;* Bangka.
Langsep: *Lansium domesticum* Jack; *Meliac.;* Java.
„ biedjietan: *Lansium domesticum* Jack; *Meliac.;* Java.
„ biesietan: *Lansium domesticum* Jack; *Meliac.;* Java.
Lanh: *Ailanthus malabarica* DC.; *Simarubac.;* Ind.-Ch.
„ nganh: *Cratoxylon formosum* Bth.; *Guttifer.;* Ind.-Ch.
Lanigda: *Toona calantas* Merr. et Rolfe; *Meliac.;* Phil.
Lanillo: *Ochroma sp.; Bombacac.;* Guat.
Lanipgá: *Toona calantas* Merr. et Rolfe; *Meliac.;* Phil.
Laníti: *Wrightia sp.; Apocynac.;* Phil.
Lanna: *Genipa americana* L.; *? G. caruto* H. B. K.; *Rubiac.;* Sur.
Lano: *Ochroma lagopus* Swartz; *Bombacac.;* Ven.
Lanseh: *Lansium domesticum* Jack; *Meliac.;* Java.
Lanshout: *Duguetia guianensis* DC.; *Anonac.;* Sur.
Lansjee: *Lansium domesticum* Jack; *Meliac.;* Sum.
Lansones-gúbat: *Aglaia luzoniensis* Merr. et Rolfe; *Meliac.;* Phil. (Tayabas)
Lant: *Elaeocarpus floribundus* Blume, *E. lacunosus* Wall.; *Elaeocarpac.;* Ind.-Ch.
Lantana: *Lantana camara* L.; *Verbenac.;* N.-Kal.
Lanúk: *Radermachera sp.; Bignoniac.;* Phil. (Cagayan)
Lanusi: *Artocarpus lamellosa* Blco.; *Morac.;* Phil. (Cagayan)
Lanutan: *Bombycidendron vidalianum* Merr. et Rolfe; *Malvac.;* Phil.
„   : *Bombycidendron campylosiphon* Warb.; *Malvac.;* Phil.
„   : *Cyathocalyx globosus* Merr.; *Anonac.;* Phil. (Bataan, Tayabas)
„ -putí: *Neolitsea vidali* Merr.; *Laurac.;* Phil. (Mindoro)
„ , white: *Alphonsea sp., Goniothalamus sp., Orophea sp.; Anonac.;* Phil.
„ , „ : *Mitrephora sp., Phaeanthus sp.; Polyalthia sp.; Anonac.;* Phil.
„ , „ : *Popowia sp., Saccopetalum sp.; Anonac.;* Phil.
„ , „ : *Sageraea sp., Xylopia sp.; Anonac.;* Phil.
Lanutí: *Wrightia sp.; Apocynac.;* Phil.
Lanza amarilla: *Chuncoa triflora* Griseb.; *Combretac.;* tr. Am.
„ „ : *Terminalia sp.; Combretac.;* Arg.
„ blanca: *Chrysophyllum sp.; Sapotac.;* Arg.
„ „ : *Myrsine melanophleos* R. Br.; *Myrsinac.;* Arg.
„ colorado: *Bumelia obtusifolia* Roem. et Schultes; *Sapotac.;* Arg.
Lanzenholz (d): *Duguetia quitarensis* Bth.; *Anonac.;* Gu.
Laot: *Acacia nubica* Bth.; *Mimosac.;* Aegypten (Sudan/Kordofan)
Lapachillo: *Blepharocalyx cisplatensis* Griseb.; *Myrtac.;* tr. S.-Am.
„ : *Tecoma araliacea* P. DC.; *Bignoniac.;* Arg. (Chaco, Santa Fe)
„ : *Terminalia sp.; Combretac.;* Arg.
„ : *Poecilanthe parviflora* Bth., *Sweetia elegans* Bth.; *Papilionac.;* Arg.
„ : *Heliocarpus americanus* L.; *Tiliac.;* Arg.
„ morado: *Poecilanthe parviflora* Bth.; *Papilionac.;* Arg.
Lapacho: *Tecoma sp., T. lapacho* K. Schum.; *Bignoniac.;* Bras., Arg.
„ amarillo: *Tecoma ochracea* Cham.; *Bignoniac.;* Arg.
„ blanco: *Tecoma ochracea* Cham.; *Bignoniac.;* Arg.
„ colorado: *Tabebuia avellaneda* Lorentz; *Bignoniac.;* tr. S.-Am.
„ „ : *Tecoma ochracea* Cham.; *Bignoniac.;* Arg.
„ crespo: *Tecoma ipé* Liais; *Bignoniac.;* s. Bras., Par., Urug., Arg.
„ morado: *Tabebuia avellaneda* Lorentz; *Bignoniac.;* tr. S.-Am.
„ negro: *Tecoma ipé* Liais; *Bignoniac.;* Arg.
„ rosa: *Tecoma sp.; Bignoniac.;* Arg.
Lapachoholz (d): *Tabebuia flavescens* Bth. et Hook. f.; *Bignoniac.;* Arg.

Lapák: *Knema glomerata* Merr.; *Myristicac.;* Phil. (Laguna)
Lapanga: *Autranella congolensis* Chev.; *Sapotac.;* B.-K.
Lapat: *Palaquium sp.; Sapotac.;* Phil. (Mindoro)
Lapi: *Ulmus lancifolia* Roxb.; *Ulmac.;* nö. O.-Ind. (Nepal)
Lapong: *Terminalia belerica* Roxb.; *Combretac.;* w. Beng. (Chota Nagpur)
Lara: *Aerua javanica* Juss.; *Amarantac.;* Aegypten (Sudan)
„ : *Metrosideros vera* Rumph.; *Myrtac.;* Nied.-Ind.
Laraija: *Lagerstroemia parviflora* Roxb.; *Lythrac.;* C.-O.-Ind.
Larak: *Strombosia philippinensis* Rolfe; *Olacac.;* Phil. (Isa.)
Larakusana: *Swartzia acuminata* Willd., *S. spp.; Caesalpiniac.;* Sur.
Laran: *Michelia champaca* L.; *Magnoliac.;* Burma.
Larana: *Triplochiton scleroxylon* K. Schum.; *Sterculiac.;* Nig.
Larangeira: *Citrus aurantium* L.; *Rutac.;* Bras.
„ do matto· *Esenbeckia febrifuga* A. Juss.; *Rutac.;* Bras. (Sao P)
Laranjeira do Mato: *Esenbeckia febrifuga* A. Juss.; *Rutac.;* Bras.
Larat api: *Isoptera borneensis* Scheff.; *Dipterocarpac.;* Malay.
Larch (e): *Larix europaea* DC.; *Conifer.;* Eu.
„ : *Larix americana* Michx.; *Conifer.;* nö. N.-Am.
„ : *Abies grandis* Lindl., *A. amabilis* Forbes; *Conifer.;* w. Kan.
„ : *Larix occidentalis* Nutt.; *Conifer.;* w. Kan.
„ : *Larix leptolepis* Gordon; *Conifer.;* Jap.
„ , Alaska-: *Larix alaskensis* W. F. Wight; *Conifer.;* nw. USA (Alaska)
„ , Alpine-: *Larix lyalli* Parl.; *Conifer.;* w. Kan.
„ , american: *Larix americana* Michx.; *Conifer.;* nö. N.-Am.
„ , black: *Larix americana* Michx.; *Conifer.;* nö. Kan.
„ , european: *Larix europaea* DC.; *Conifer.;* N.-Seel.
„ , Lyall's-: *Larix lyalli* Parl.; *Conifer.;* w. Kan.
„ , Mountain-: *Larix lyalli* Parl.; *Conifer.;* w. Kan.
„ , western: *Larix occidentalis* Nutt.; *Conifer.;* w. N.-Am.
Lärche (d): *Larix europaea* DC.; *Conifer.;* Eu.
„ , dünnschuppige: *Larix leptolepis* Gordon; *Conifer.;* Jap.
„ , gemeine: *Larix europaea* DC.; *Conifer.;* Eu.
„ , Gold-, chinesische: *Pseudolarix kaempferi* Gord.; *Conifer.;* nö. Ch.
„ , Hondo-: *Larix leptolepis* Gordon; *Conifer.;* Jap.
„ , japanische: *Larix leptolepis* Gordon; *Conifer.;* Jap.
„ , ostamerikanische: *Larix americana* Michx.; *Conifer.;* USA.
„ , sibirische: *Larix sibirica* Ledeb.; *Conifer.;* Sib.
„ , westamerikanische: *Larix occidentalis* Nutt.; *Conifer.;* w. N.-Am.
Large-leaved tree: *Eucalyptus siderophloia* Bth.; *Myrtac.;* ö. Au.
Larice: *Larix europaea* DC.; *Conifer.;* It.
Lärk: *Larix europaea* DC., *L. sibirica* Ledeb.; *Conifer.;* Schwed.
Laro: *Shorea leprosula* Miq.; *Dipterocarpac.;* Malay.
Lashawng: *Anthocephalus cadamba* Miq.; *Rubiac.;* Burma.
Lasie: *Adina fagifolia* Valeton; *Rubiac.;* Nied.-Ind.
Lasíla: *Lagerstroemia piriformis* Koehne; *Lythrac.;* Phil. (Cagayan, Iloko)
„ : *Terminalia comintana* Merr.; *Combretac.;* Phil. (Ilocos, Cagayan)
Lasílak: *Lagerstroemia piriformis* Koehne; *Lythrac.;* Phil. (Cagayan, Ilocos)
„ : *Terminalia comintana* Merr.; *Combretac.;* Phil. (Ilocos, Cagayan)
Lasilásan: *Terminalia comintana* Merr.; *Combretac.;* Phil. (Ilocos, Cagayan)
Lasóna: *Dysoxylum turczaninowi* C. DC.; *Meliac.;* Phil. (Leyte)
Lasora: *Cordia myxa* L.; *Borraginac.;* tr. As.
Lassi-hout: *Adina fagifolia* Valeton; *Rubiac.;* Nied.-Ind.
Lassora: *Cordia myxa* L.; *Borraginac.;* n. O.-Ind., C.-O.-Ind.
Lassuni: *Dysoxylum binectariferum* Hook. f.; *Meliac.;* Beng.
Lassura: *Cordia myxa* L.; *Borraginac.;* n. O.-Ind. (United Prov.)
Lasura: *Cordia myxa* L.; *Borraginac.;* nw. O.-Ind. (Punjab)
Lát: *Chukrasia tabularis* A. Juss.; *Meliac.;* Ind.-Ch.
„ gia dông: *Chukrasia tabularis* A. Juss.; *Meliac.;* Ind.-Ch.
„ hoa: *Chukrasia tabularis* A. Juss.; *Meliac.;* Ind.-Ch.
„ „ tía: *Chukrasia tabularis* A. Juss.; *Meliac.;* Ind.-Ch.
„ hoat: *Chukrasia tabularis* A. Juss.; *Meliac.;* Ind.-Ch.
„ khet: *Toona febrifuga* Roem.; *Meliac.;* Ind.-Ch. (Tonkin)
„ xoan: *Spondias mangifera* Willd., *S. tonkinensis* Willd.; *Anacardiac.;* Ind.-Ch.
Lata: *Ruprechtia sp.; Polygonac.;* Arg.

Latang: *Lannea grandis* Engl.; *Anacardiac.*; Burma.
Latanier: *Borassus sp.; Palmac.;* W.-Af.
   „ : *Corypha laevis* Chev.; *Palmac.*; Annam.
Latauán: *Cyathocalyx globosus* Merr.; *Anonac.;* Phil. (Bataan)
Lating'è: *Pycnanthus kombo* Warb.; *Myristicac.;* Gab.
Lati: *Amphimas pterocarpoides* Harms; *Papilionac.;* Elf.
Lationj: *Alstonia scholaris* R. Brown; *Apocynac.;* nw. O.-Ind. (Kumaon)
Latoena: *Canarium sp.; Burserac.;* ? Nied.-Ind.
Latonero: *Celtis australis* L.; *Ulmac.;* Sp.
Latora: *Cordia vestita* Hook. f. et Thoms.; *Borraginac.;* n. O.-Ind. (Oudh)
? Latsa: *Lophira alata* Banks; *Ochnac.;* Go.
Latsai: *Toona ciliata* M. Roem.; *Meliac.;* Burma.
Latsche: *Pinus montana* Miller; *Conifer.;* M.-M.
Latuán: *Cyathocalyx globosus* Merr.; *Anonac.;* Phil. (Bataan)
Latubá: *Barringtonia spp.; Lecythidac.;* Phil. (Cagayan)
Lâu: *Bassia pasquieri* H. Lec.; *Sapotac.;* Ind.-Ch.
   „ : *Sonneratia acida* L.; *Sonneratiac.;* Ind.-Ch.
   „ táu: *Vatica astrotricha* Hance, *V. spp.; Dipterocarpac.;* Ind.-Ch.
   „ „ giac: *Vatica sp.; Dipterocarpac.;* Ind.-Ch.
   „ „ gíam: *Vatica astrotricha* Hance; *Dipterocarpac.;* Annam.
   „ „ nui: *Vatica dyeri* Pierre; *Dipterocarpac.;* Annam.
   „ „ núi: *Vatica thoreli* Pierre; *Dipterocarpac.;* Ind.-Ch.
   „ „ núoc.; *Vatica philastreana* Pierre; *Dipterocarpac.;* Annam.
   „ „ trang: *Vatica astrotricha* Hance; *Dipterocarpac.;* Ind.-Ch.
   „ „ xanh: *Vatica dyeri* King, *V. astrotricha* Hance; *Dipterocarpac.;* Ind.-Ch.
   „ „ „ : *Sideroxylon eburneum* Chev.; *Sapotac.;* Ind.-Ch.
Lauáan: *Dipterocarpus vernicifluus* Blco.; *Dipterocarpac.;* Phil. (Rizal, Nueva-Ecija)
   „ : *Parashorea malaanonan* Merr.; *Dipterocarpac.;* Phil.
   „ : *Pentacme contorta* Merr. et Rolfe; *Dipterocarpac.;* Phil.
   „ : *Shorea eximia* Scheff.; *Dipterocarpac.;* Phil. (Sorsogon)
   „ : *Shorea polita* Vid.; *Dipterocarpac.;* Phil. (Zambales, Tayabas, Nueva Ecija)
   „ : *Shorea polysperma* Merr.; *Dipterocarpac.;* Phil. (Mindoro)
   „ -na-puti: *Pentacme contorta* Merr. et Rolfe; *Dipterocarpac.;* Phil.
Lauán: *Anisoptera thurifera* Blume; *Dipterocarpac.;* Phil. (Rizal)
   „ : *Parashorea malaanonan* Merr.; *Dipterocarpac.;* Phil.
   „ : *Shorea spp., Pentacme contorta* Merr. et Rolfe; *Dipterocarpac.;* Phil.
   „ blanc: *Pentacme contorta* Merr. et Rolfe; *Dipterocarpac.;* Phil.
   „ -na-puti: *Shorea polita* Vid.; *Dipterocarpac.;* Phil.
   „ , red: *Shorea negrosensis* Foxw.; *Dipterocarpac.;* Phil. (Negrosinsel)
   „ , „ : *Shorea spp.,* ? *Hopea spp.; Dipterocarpac.;* Phil.
   „ rouge: *Shorea sp.; Dipterocarpac.;* Phil.
   „ , white: *Parashorea plicata* Brandis; *Dipterocarpac.;* Phil.
   „ , „ : *Pentacme contorta* Merr. et Rolfe; *Dipterocarpac.;* Phil.
Laukya: *Schima wallichi* Choisy; *Theac.;* Burma.
   „ : *Litsea polyantha* Juss.; *Laurac.;* Burma.
Laupé: *Lannea grandis* Engl.; *Anacardiac.;* Burma.
Laur: *Acer pictum* Thunb.; *Acerac.;* nw. O.-Ind. (Punjab)
Laurac (u): *Calycophyllum candidissimum* DC.; *Rubiac.;* Cuba.
Laurat api: *Isoptera borneensis* Scheff.; *Dipterocarpac.;* Malay (Pahang)
Laurel: *Anona bullata* A. Rich.; *Anonac.;* Cuba.
   „ : *Cordia alliodora* Cham.; *Borraginac.;* C.-Am., S.-Am.
   „ : *Nectandra membranacea* Griseb., *N. sanguinea* Rottb.; *Laurac.;* br. Hond.
   „ : *Phoebe helicterifolia* Mez; *Laurac.;* br. Hond.
   „ : *Ocotea veraguensis* Mez, *Nectandra concinna* Nees; *Laurac.;* Kol.
   „ : *Laurelia aromatica* Juss.; *Monimiac.;* Chile.
   „ : *Schleichera trijuga* Willd.; *Sapindac.;* C.-O.-Ind., (Bihar)
   „ : *Terminalia tomentosa* Wight et Arn.; *Combretac.;* C.-O.-Ind.., Burma.
   „ , alexandrian: *Calophyllum inophyllum* L.; *Guttifer.;* tr. Af., tr. As., tr. Au.
   „ amarillo: *Ocotea suaveolens* Bth. et Hook. f.; *Laurac.;* Arg.
   „ angelino: *Nectandra discolor* Nees; *Laurac.;* Ven.
   „ , australian (u): *Endiandra palmerstoni* C. T. White; *Laurac.;* Au.
   „ avispillo: *Ocotea sp., Nectandra sp.; Laurac.;* P.-R.
   „ , Big- (u): *Magnolia grandiflora* L.; *Magnoliac.;* sö. N.-Am.
   „ blanco: *Cordia alliodora* Cham.; *Borraginac.;* C.-Am.

Laurel blanco: *Nectandra glabrescens* Bth.; *Laurac.;* br. Hond.
„ „ : *Ocotea suaveolens* Bth. et Hook. f.; *Laurac.;* Arg.
„ , Bobo-: *Phoebe sp.; Laurac.;* P.-R.
„ , californian: *Umbellularia californica* Nutt.; *Laurac.;* w. USA.
„ canelon: *Ocotea sp., Nectandra sp.; Laurac.;* P.-R.
„ crespo: *Phoebe sp.; Laurac.;* Arg.
„ espada: *Casearia sp.; Flacourtiac.;* P.-R.
„ , european: *Laurus nobilis* L.; *Laurac.;* Eu.
„ , geo-: *Phoebe sp.; Laurac.;* W.-Ind.
„ , Hedge-: *Pittosporum eugenioides* A. Cunn.; *Pittosporac.;* N.-Seel.
„ hediondo: *Ocotea sp.; Laurac.;* Arg.
„ hú: *Ocotea sp.; Laurac.;* Arg.
„ , indian: *Ficus nitida* Thunb.; *Morac.;* Kol.
„ , „ : *Terminalia tomentosa* Wight et Arn.; *Combretac.;* O.-Ind., Cey.
„ , Jamaica-: *Ceanothus chloroxylon* Nees; *Rhamnac.;* Jam.
„ macho: *? Cordia sp.; Borraginac.;* Salv.
„ maestro: *Nectandra sp.; Laurac.;* Ven.
„ mestizo: *Ocotea sp.; Laurac.;* Arg.
„ , Native-: *Anopterus glandulosus* Lab.; *Saxifragac.;* Tasm.
„ negro: *Cordia alliodora* Cham.; *Borraginac.;* C.-Am.
„ „ : *Nectandra sp.; Laurac.;* Kol.
„ „ : *Ocotea glomerata* Mez; *Laurac.;* Ven.
„ „ : *Ocotea sp., O. spectabilis* Mez, *Nectandra porphyra* Griseb.; *Laurac.;* Arg.
„ overo: *Ocotea sp.; Laurac.;* Arg.
„ prieto: *? Cordia sp.; Borraginac.;* Salv.
„ , Queensland-: *Endiandra palmerstoni* C. T. White; *Laurac.;* ö. Au.
„ quina: *Nectandra moritziana* Klotzsch; *Laurac.;* Ven.
„ rosada: *Nectandra coriacea* Griseb.; *Laurac.;* Ven.
„ , Rose-: *Cryptocarya erythroxylon* Maiden et Betche; *Laurac.;* ö. Au.
„ roseta: *Ocotea sp., Nectandra sp.; Laurac.;* P.-R.
„ sabino: *Magnolia splendens* Urban; *Magnoliac.;* Mex., C.-Am., W.-Ind.
„ sassafras: *Ocotea sp., Nectandra sp.; Laurac.;* P.-R.
„ sayiú: *Ocotea sp.; Laurac.;* Arg.
„ , Swamp- (u): *Magnolia glauca* L.; *Magnoliac.;* ö. N.-Am.
„ tulipán: *Talauma sp.; Magnoliac.;* Mex.
„ , Victorian-: *Pittosporum undulatum* Vent.; *Pittosporac.;* Au.
„ , white (u): *Magnolia glauca* L.; *Magnoliac.;* ö. N.-Am.
„ de cuabal: *Anona sp.; Anonac.;* Cuba.
„ „ la falda: *Phoebe porphyria* Mez; *Laurac.;* Arg.
„ „ „ sierra: *Persea sp.; Laurac.;* Mex. (Sonora)
Laurela: *Laurelia aromatica* Juss.; *Monimiac.;* Chile.
Laurelón: *Sassifridium sp.; Laurac.;* Mex.
Lauri: *Cordia macleodi* Hook. f. et Thoms.; *Borraginac.;* O.-Ind.
Laurier: *Nectandra spp., Ocotea spp., Aniba spp.; Laurac.;* Trin., Tob.
„ blanc: *Ocotea sp., Nectandra sp.; Laurac.;* Guad., Mart.
„ bois jaune: *Ceanothus chloroxylon* Nees; *Rhamnac.;* Jam.
„ bord de mer: *Ocotea sp., Nectandra sp.; Laurac.;* Guad., Mart.
„ canelle: *Oreodaphne strumosa* Griseb.; *Laurac.;* Trin., Tob.
„ grec (f): *Melia azedarach* L.; *Meliac.;* Tr.
„ marbré: *Nectandra concinna* Nees; *Laurac.;* Mart.
„ mattack: *Aniba megacarpa* Hemsl.; *Laurac.;* Trin., Tob.
„ montagne: *Ocotea sp., Nectandra sp.; Laurac.;* Guad., Mart.
„ rose: *Podocarpus coriaceus* L. C. Rich.; *Conifer.;* Guad.
„ „ : *Nerium oleander* L.; *Apocynac.;* Sahara.
Lauro amarello: *Persea splendens* Meissn. *var. chrysophylla* ?; *Laurac.;* Bras.
„ congade porco: *Persea alba* Nees; *Laurac.;* Bras.
„ pardo: *Cordia gerascanthoides* H. B. K.; *Borraginac.;* Bras.
„ preto: *Cordia gerascanthoides* H. B. K.; *Borraginac.;* Bras.
Laut: *Anneslea fragrans* Wall.; *Theac.;* w. Annam.
Lautea: *Lautea serrata* F. Brown, *L. stokesiana* F. Brown; *Cornac.;* Südsee (Marquesas)
„ : *Lautea stokesiana* F. Brown *var. integrifolia* F. Brown; *Cornac.;* Marqu.
„ : *Lautea stokesiana* F. Brown *var. denticulata* F. Brown; *Cornac.;* Marqu.
„ : *Lautea stokesiana* F. Brown *var. primaeva* F. Brown; *Cornac.;* Marquesas.
Lavuan: *Neonauclea sp.; Rubiac.;* Phil. (Nueva Viscaya)

Lawa: *Croton sp.; Euphorbiac.;* Sans.
„ : *Ficus glomerata* Roxb.; *Morac.;* C.-O.-Ind.
Lawan: *Anisoptera thurifera* Blume; *Dipterocarpac.;* Phil.
Lawang: *Cinnamomum spp.; Laurac.;* Malay.
Lawari: *Pithecolobium dulce* Bth.; *Mimosac.;* Curacao.
Lay: *Aesculus chinensis* Bunge; *Hippocastanac.;* Ind.-Ch.
„ : *Sideroxylon eburneum* Chev.; *Sapotac.;* Ind.-Ch.
„ : *Garcinia fagraeoides* Chev.; *Guttifer.;* Ind.-Ch.
„ : *Carallia lucida* Roxb.; *Rhizophorac.;* Ind.-Ch.
„ lay: *Cordia locharti* O. Ktze., *C. sulcata* P. DC.; *Borraginac.;* Trin., Tob.
Layá: *Alstonia scholaris* R. Br.; *Apocynac.;* Phil. (Ilocos Norte)
Layang: *Betula alnoides* Ham.; *Betulac.;* Burma.
Lê hê: *Semecarpus tonkinensis* H. Lec.; *Anacardiac.;* Ind.-Ch.
Leadwood: *Combretum porphyrolepis* Engl. et Diels; *Combretac.;* S.-Af.
„ : *Combretum imberbe* Wawra *var. petersi* Engl.; *Combretac.;* ö. Transval.
Leal craham: *Dipterocarpus artocarpifolius* Pierre; *Dipterocarpac.;* Kamb.
„ „ : *Dipterocarpus insularis* Hance; *Dipterocarpac.;* Kamb.
Leang: *Dalbergia bariensis* Pierre, *D. spp.; Papilionac.;* Ind.-Ch.
Leather Jacket: *Geissois benthami* F. v. M.; *Cunoniac.;* ö. Au.
„ „ : *Ceratopetalum apetalum* D. Don; *Cunoniac.;* ö. Au.
„ „ : *Callicoma serratifolia* Andr.; *Cunoniac.;* ö. Au.
Leatherwood: *Flindersia collina* F. M. Bailey; *Rutac.;* ö. Au.
„ : *Eucryphia billardieri* Spach; *Rosac.;* Tasm.
Leauri: *Cupressus torulosa* D. Don; *Conifer.;* n. O.-Ind. (United Prov.)
Leban: *Vitex pubescens* Vahl, *V. spp.; Verbenac.;* Malay.
Lebbek: *Albizzia lebbek* Bth.; *Mimosac.;* Go.
Lebensbaum, abendländischer: *Thuja occidentalis* L.; *Conifer.;* ö. N.-Am.
„ , beilblättriger: *Thujopsis dolabrata* S. et Z.; *Conifer.;* C.-Jap.
„ , chinesischer: *Thuja orientalis* L.; *Conifer.;* Ch.
„ , gemeiner: *Thuja occidentalis* L.; *Conifer.;* ö. N.-Am.
„ , japanischer: *Thuja japonica* Maxim.; *Conifer.;* Jap.
„ , orientalischer: *Thuja orientalis* L.; *Conifer.;* Ch.
„ , Riesen-: *Thuja gigantea* Nutt.; *Conifer.;* w. N.-Am.
Leberwurstbaum: *Kigelia africana* Bth.; *Bignoniac.;* Togo.
Lebingué: *Homalium africanum* Bth.; *Flacourtiac.;* Kam.
Lebisa: *Acrodiclidium sp.; Laurac.;* Cuba.
Lebrero: *Mouriria sp.; Melastomatac.;* Cuba.
Lebrisa: *Acrodiclidium sp.; Laurac.;* Cuba.
Lebvola: *Odyendea gabunensis* Engl.; *Simarubac.;* Gab.
Lebwe: *Ulmus lancifolia* Roxb.; *Ulmac.;* Burma.
Leccio: *Quercus ilex* L.; *Fagac.;* It.
Leche amarilla: *Symphonia globulifera* L. f.; *Guttifer.;* C.-Am.
„ caspi: *Couma sp.; Apocynac.;* Peru.
„ de maría: *Calophyllum chiapense* Standley; *Guttifer.;* Mex. (Chiapas)
„ „ „ : *Sapium lateriflorum* Hemsl.; *Euphorbiac.;* br. Hond.
Lechecillo: *Chrysophyllum argenteum* Jacq.; *Sapotac.;* P.-R.
Lechee: *Nephelium litchi* Camb.; *Sapindac.;* Beng.
Leche-leche: *Ficus sp.; Morac.;* Peru.
Lechero: *Euphorbia spp., Sapium spp.; Euphorbiac.;* Ven.
Lecherón: *Sapium spp.; Euphorbiac.;* Arg.
„ curupi: *Sapium aucuparium* Jacq. *var. stenophyllum* Huber; *Euphorbiac.;* tr. Am.
Lechesillo: *Ficus sp.; Morac.;* P.-R.
„ : *Chrysophyllum sp.* (? *cainito* L.); *Sapotac.;* P.-R.
Lecheso: *Trophis macrostachya* Donn. Smith; *Morac.;* Pan.
Lechillo: *Carpinus caroliniana* Walt.; *Betulac.;* Mex.
Lechoso: *Calotropis gigantea* R. Br.; *Asclepiadac.;* Kol.
Leeboo: *Citrus limonum* Risso; *Rutac.;* Hind.
Leemoo: *Citrus limonum* Risso; *Rutac.;* Hind.
Legno amaranto: *Peltogyne spp.; Caesalpiniac.;* Bras.
„ azzurro: *Haematoxylon campechianum* L.; *Caesalpiniac.;* tr. Am.
„ benedetto: *Guaiacum officinale* L.; *Zygophyllac.;* W.-Ind., s. Mex.
„ carne di cavallo: *Swartzia tomentosa* DC.; *Caesalpiniac.;* Gu.
„ ferro: *Swartzia tomentosa* DC.; *Caesalpiniac.;* Gu.
„ giallo: *Chlorophora tinctoria* Gaudich.; *Morac.;* C.-Am.

Legno granadillo: *Brya ebenus* DC.; *Papilionac.;* tr. Am.
„ locusta: *Hymenaea courbaril* L.; *Caesalpiniac.;* W.-Ind.
„ nero: *Haematoxylon campechianum* L.; *Caesalpiniac.;* tr. Am.
„ porpora: *Peltogyne sp.; Caesalpiniac.;* n. S.-Am.
„ rosso: *Caesalpinia echinata* Lam.; *Caesalpiniac.;* Bras.
„ santo: *Guaiacum officinale* L.; *Zygophyllac.;* W.-Ind., s. Mex.
„ serpente: *Piratinera guianensis* Aubl.; *Caesalpiniac.;* Gu.
„ tigre: *Brosimum aubleti* Poepp. et Endl.; *Morac.;* Gu.
„ violetto: *Peltogyne sp.; Caesalpiniac.;* Gu.
„ del Brasile: *Caesalpinia echinata* Lam.; *Caesalpiniac.;* ö. Bras.
„ di campeggio: *Haematoxylon campechianum* L.; *Caesalpiniac.;* tr. Am.
„ di Cuba: *Cedrela sp.; Meliac.;* tr. Am.
„ di guaiaco: *Guaiacum officinale* L.; *Zygophyllac.;* W.-Ind., s. Mex.
„ di jacaranda: *Dalbergia nigra* Allem.; *Papilionac.;* Bras.
„ di Noce Indiano (i): *Albizzia lebbek* Bth.; *Mimosac.;* As.
„ „ „ satinato: *Liquidambar styraciflua* L.; *Hamamelidac.;* sö. USA.
„ „ Santa Marta: *Caesalpinia echinata* Lam.; *Caesalpiniac.;* Bras.
„ „ Sirsa (i): *Albizzia lebbek* Bth.; *Mimosac.;* As.
„ „ Spagna: *Cedrela sp.; Meliac.;* tr. Am.
Leh-bohn: *Discoglypremna caloneura* Prain; *Euphorbiac.;* Lib.
Lehua: *Metrosideros polymorpha* Forst.; *Myrtac.;* ö. Au., N.-Seel.
Lei: *? Plumiera sp.; Apocynac.;* S.-Am.
„ : *Tamarix articulata* Vahl; *Tamaricac.;* n. O.-Ind. (Punjab/Assam)
Lein: *Terminalia pyrifolia* Kurz, *T. bialata* Steudel; *Combretac.;* Burma.
Leiteira: *Brosimopsis latescens* Sp. Moore, *B. diandre* Blake; *Morac.;* Bras.
Leiteiro: *Sapium biglandulosum* Muell. Arg.; *Euphorbiac.;* Bras. (Sao P)
Leja: *Litsea polyantha* Juss.; *Laurac.;* C.-O.-Ind. (Gondwana)
Lek: *Litsea vang* H. Lec.; *Laurac.;* Ind.-Ch.
„ deuk: *Bambusa pallida* Munro; *Gramin.;* Ch. (Fukien)
Leko: *Lannea welwitschi* Engl.; *Anacardiac.;* Kam.
Lekoko: *Ficus vogeliana* Miq.; *Morac.;* Kam.
Lékutuku: *Sclerosperma walkeri* Chev.; *Palmac.;* Gab.
Leli: *Cedrela sp.; Meliac.;* Curacao.
Lem: *Peltophorum tonkinensis* Pierre; *Caesalpiniac.;* Ind.-Ch.
Lémbèdi: *Elaeis guineensis* Jacq. *var. nigrescens* Chev.; *Palmac.;* Gab.
„ la bitsima byolé: *Elaeis guineensis* Jacq. *var. ?; Palmac.;* Gab.
Lémbida: *Elaeis guineensis* Jacq. *var. nigrescens* Chev.; *Palmac.;* Gab.
Lemboe: *Gordonia sp.; Theac.;* Java.
Lemesu: *Fagraea gigantea* Ridl., *F. fragrans* Roxb.; *Loganiac.;* Malay.
Lemo-epo: *Sideroxylon guianensis* A. DC.; *Sapotac.;* Sur.
Lemon: *Citrus limonia* Osbeck; *Rutac.;* Hond.
„ : *Citrus limonum* Risso; *Rutac.;* O.-Ind.
Lemonwood (u): *Calycophyllum candidissimum* DC.; *Rubiac.;* W.-Ind.
„ : *Xymalos monospora* Baill.; *Monimiac.;* S.-Af.
Lempa: *Hernandia guianensis* Aubl.; *Hernandiac.;* Pan.
Lemtan: *Gynocardia odorata* R. Br.; *Flacourtiac.;* Assam.
Lemusu: *Shorea sp.; Dipterocarpac.;* Malay (Pahang)
Len: *Ceiba pentandra* Gaertn. *var. dehiscens* Ulb.; *Bombacac.;* Senl.
Leña dura: *Maytenus boaria* Mol.; *Celastrac.;* Arg.
„ nefritica: *Eysenhardtia polystachia* Sargent; *Caesalpiniac.;* Mex. (Otoni)
Lendé: *Calpocalyx klainei* Pierre; *Mimosac.;* Gab.
Lende: *Lagerstroemia parviflora* Roxb.; *Lythrac.;* sw. O.-Ind. (Bombay)
Lendi: *Lagerstroemia parviflora* Roxb.; *Lythrac.;* O.-Ind.
„ : *Eugenia jambolana* Lamk.; *Myrtac.;* C.-O.-Ind. (Gondwana)
Lendia: *Lagerstroemia parviflora* Roxb.; *Lythrac.;* C.-O.-Ind.
Lendo: *Irvingia grandifolia* Engl.; *Simarubac.;* W.-Af. (Boumouali)
Lendya: *Lagerstroemia parviflora* Roxb.; *Lythrac.;* n. O.-Ind. (United Prov.)
Lenga: *Nothofagus pumilio* Reiche; *Fagac.;* Chile.
Lenggadai: *Bruguiera parviflora* Wight et Arn.; *Rhizophorac.;* Malay.
Lenggadi: *Bruguiera parviflora* Wight et Arn.; *Rhizophorac.;* Phil.
Lenggapus: *Mesua ferrea* L.; *Guttifer.;* Malay.
Leng-keng: *Nephelium longana* Camb.; *Sapindac.;* Java.
„ „ : *Nephelium litchi* Camb.; *Sapindac.;* Malay.
Lengsar: *Pometia pinnata* Forst.; *Sapindac.;* Phil., Java.

19*

Lengsarhout: *Pometia sp.; Sapindac.;* Nied.-Ind.
Lengua de buey: *Cordia heterophylla* Roem. et Schultes; *Borraginac.;* Pan.
   „   „ vaca: *Curatella americana* L.; *Dilleniac.;* Salv.
Lengue: *Nothofagus sp.; Fagac.;* Arg.
   „ de vaca: *Piptadenia sp.; Mimosac.;* Salv.
Léngyè: *Pycnanthus kombo* Warb.; *Myristicac.;* Gab.
Lenja: *Litsea sebifera* Pers., *L. polyantha* Juss.; *Laurac.;* C.-O.-Ind.
Lenjo: *Litsea sebifera* Pers.; *Laurac.;* C.-O.-Ind.
Lentille: *Didymopanax morototoni* Dcne. et Planch.; *Araliac.;* Trin.
Lentisco: *Pistacia lentiscus* L.; *Anacardiac.;* Sp., It.
   „   :   *Phillyrea media* L.; *Oleac.;* Sp.
Leopard tree, broad-leaved: *Flindersia collina* F. M. Bailey; *Rutac.;* ö. Au.
Leopardwood: *Flindersia maculosa* Lindl.; *Rutac.;* N.-S.-W.
Leoué: *Hexalobus crispiflorus* A. Rich.; *Anonac.;* Kam.
Lepacho del campo: *Sweetia elegans* Bth.; *Papilionac.;* Arg.
Lepelhout: *Cassine capensis* L.; *Celastrac.;* Kap.
L'Épinet: *Zanthoxylum microcarpum* Griseb.; *Rutac.;* Trin., Tob.
L'épineux: *Zanthoxylum sp.; Rutac.;* Trin.
Lerek: *Dittelasma rarak* Hook. f.; *Sapindac.;* Java.
Leroï: *Alstonia congensis* Engl.; *Apocynac.;* Elf.
Lerué: *Alstonia congensis* Engl.; *Apocynac.;* Elf.
Lésombé: *Elaeis guineensis* Jacq. *var. ?; Palmac.;* Gab.
Lessuri: *Cordia myxa* L.; *Borraginac.;* w. O.-Ind. (Sindh)
Letchi: *Nephelium litchi* Camb.; *Sapindac.;* s. Ch.
Lethôk: *Sterculia alata* Roxb.; *Sterculiac.;* Burma, And.
Létin: *Funtumia sp.; Apocynac.;* Dah.
Letis: *Anisoptera thurifera* Blume, *Shorea polysperma* Merr.; *Dipterocarpac.;* Phil.
Letkok: *Sterculia alata* Roxb.; *Sterculiac.;* And.
Letpan: *Bombax malabaricum* DC.; *Bombacac.;* Burma.
Letri: *Brosimum aubleti* Poepp. et Endl.; *Morac.;* Sur.
   „ , Basra-:   *Brosimum aubleti* Poepp. et Endl.; *Morac.;* Sur.
   „ , „ -:   *Helicostylis poeppigiana* Trec.; *Morac.;* Gu., n. Bras., C.-Am.
   „ , Kapeweri-:   *Brosimum aubleti* Poepp. et Endl.; *Morac.;* Sur.
   „ , Man-:   *Helicostylis poeppigiana* Trec.; *Morac.;* Gu., n. Bras., C.-Am.
   „ sabana: *Helicostylis poeppigiana* Trec.; *Morac.;* Gu., n. Bras., C.-Am.
Letterhout: *Brosimum aubleti* Poepp. et Endl.; *Morac.;* Sur.
   „ , gespikkeld: *Brosimum aubleti* Poepp. et Endl.; *Morac.;* Sur.
   „ , Man-:   *Brosimum aubleti* Poepp. et Endl.; *Morac.;* Sur.
   „ , roode: *Brosimum aubleti* Poepp. et Endl.; *Morac.;* Sur.
Letternholz (d):   *Brosimum aubleti* Poepp. et Endl.; *Morac.;* Gu.
Letterwood: *Brosimum aubleti* Poepp. et Endl.; *Morac.;* Gu.
Lettôk: *Alstonia scholaris* R. Br., *Holarrhena antidysenterica* Wall.; *Apocynac.;* Burma.
   „ -thein: *Wrightia tomentosa* Roem. et Schultes; *Apocynac.;* Burma.
Lettôkgyi: *Holarrhena antidysenterica* Wall.; *Apocynac.;* Burma.
Lettre marbré: *Machaerium schomburgki* Bth.; *Papilionac.;* fr. Gu.
   „ moucheté: *Brosimum aubleti* Poepp. et Endl.; *Morac.;* fr. Gu.
   „ rubané: *Amanoa guianensis* Aubl.; *Euphorbiac.;* fr. Gu.
Leuri: *Cupressus torulosa* D. Don; *Conifer.;* n. O.-Ind. (Jaunsar)
Levisa: *Acrodiclidium sp.; Laurac.;* Cuba.
Leza: *Lagerstroemia tomentosa* Presl; *Lythrac.;* Burma.
   „ -byu: *Lagerstroemia calyculata* Kurz; *Lythrac.;* s. Burma.
Lezard, black: *Vitex divaricata* Sw.; *Verbenac.;* Trin., Tob.
Lia: *Irvingia oblonga* Chev.; *Simarubac.;* Kam.
Liä-nuwasaure: *Diospyros monbuttensis* Gürke; *Ebenac.;* Togo.
Liaan appel: *Passiflora laurifolia* L.; *Passiflorac.;* fr. Gu.
Lialiadan koelocra: *Hymenolobium flavum* Kleinh.; *Mimosac.;* Sur.
   „ koleroe: *Hymenolobium flavum* Kleinh., *H. sp.; Mimosac.;* Sur.
   „ tataroe: *Parkia pendula* Bth.; *Mimosac.;* Sur.
Lián: *Croton glabellus* L.; *Euphorbiac.;* Hond.
   „ :   *Irvingia oblonga* Chev.; *Simarubac.;* Kam.
Liane-acrochet: *Uncaria guianensis* Gmel.; *Rubiac.;* Ven.
   „ jaune: *Danais fragrans* Gaertn. f.; *Rubiac.;* Maur.
   „ meurtrière: *Clusia rosea* Jacq.; *Guttifer.;* Ant.
   „ papaye: *Omphalea diandra* L.; *Euphorbiac.;* Mart.

Liane à caoutchouc: *Landolphia owariensis* Beauv.; *Apocynac.*; Elf.
„ à l'anse: *Omphalea diandra* L.; *Euphorbiac.*; Mart.
Liard: *Populus deltoides* Marsh.; *Salicac.*; sö. Kan.
Liasá: *Uvaria chamae* Beauv.; *Anonac.*; Togo.
Liba: *Elaeis guineensis* Jacq. var. *nigrescens* Chev.; *Palmac.*; Gab.
„ : *Allaeanthus glaber* Warb.; *Morac.*; Phil. (Davao)
„ la bikindi byolè: *Elaeis guineensis* Jacq. var. *?*; *Palmac.*; Gab.
„ -la-etangani: *Cocos nucifera* L.; *Palmac.*; Gab.
Libamba: *Albizzia fastigiata* Oliv.; *Mimosac.*; B.-K.
Libás: *Garuga sp.*; *Burserac.*; Phil. (Tayabas)
„ : *Octomeles sumatrana* Miq.; *Datiscac:*; Phil. (Laguna, Tayabas)
„ -na-puti: *Octomeles sumatrana* Miq.; *Datiscac.*; Phil. (Laguna, Tayabas)
Libáto: *Lumnitzera littorea* Voigt; *Combretac.*; Phil. (Tayabas, Palawan)
„ -puti: *Cumingia philippinensis* Vid.; *Bombacac.*; Phil. (Tayabas)
Libelokote: *Fagara altissima* Engl.; *Rutac.*; B.-K.
Liben: *Azara spp.*; *Bixac.*; Chile.
Liberokuta: *Fagara altissima* Engler; *Rutac.*; B.-K.
Libi-dibi: *Caesalpinia coriaria* Willd.; *Caesalpiniac.*; Kol.
Libtúk: *Pterocymbium tinctorium* Merr.; *Sterculiac.*; Phil. (Cagayan)
Libuyu: *Entandrophragma sp.*; *Meliac.*; B.-K. (Katanga)
Licari kanali: *Licaria guianensis* Aubl.; *Laurac.*; fr. Gu.
Licca tree: *Zanthoxylum sp.*; *Rutac.*; Jam.
Lichopwa: *Pouzolzia hypoleuca* Wedd.; *Urticac.*; br. O.-Af. (Nyassa)
Licondo: *Klainedoxa dybowskii* van Tiegh.; *Simarubac.*; Gab.
Licupiraholz: *Bowdichia nitida* Spruce; *Papilionac.*; Bras.
Lidah kerbau putih: *Marlea ebenacea* C. B. Clarke; *Cornac.*; Malay.
Liège des Antilles: *Hibiscus tiliaceus* L.; *Malvac.*; Guad.
„ du pays: *Pterocarpus sp.*; *Papilionac.*; fr. Gu.
Liembán: *Alangium meyeri* Merr.; *Cornac.*; Phil. (Ilocos Sur)
Liên: *Xylia kerri* Craib et Hutch.; *Mimosac.*; Ind.-Ch.
„ : *Eusideroxylon sp.*; *Laurac.*; Ind.-Ch.
„ : *Melia azedarach* L.; *Meliac.*; Ind.-Ch.
„ tu: *Antidesma bunius* Spreng.; *Euphorbiac.*; Ind.-Ch.
Lienkapatoe: *Macoubea guianensis* Aubl.; *Apocynac.*; Sur.
Lifake: *Entandrophragma congoense* Chev.; *Meliac.*; B.-K.
Lifaki: *Entandrophragma congoense* Chev., *E. roburoides* Verm.; *Meliac.*; B.-K.
? Lifui: *Newbouldia laevis* Seem.; *Bignoniac.*; Go.
Ligaá: *Zizyphus sp.*, *Z. inermis* Merr.; *Rhamnac.*; Phil.
Ligaán: *Palaquium sp.*; *Sapotac.*; Phil. (Cotabato)
Ligáian: *Mimusops parvifolia* R. Br.; *Sapotac.*; Phil. (Zamboanga, Sulu)
Ligámen: *Zizyphus sp.*; *Rhamnac.*; Phil. (Ilocos Norte, I. Sur)
Ligás: *Semecarpus cuneiformis* Blco.; *Anacardiac.*; Phil.
„ : *Buchanania sp.*; *Anacardiac.*; Phil. (Camarines)
Ligáson: *Garuga sp.*; *Burserac.*; Phil. (Batangas, Tayabas)
Lightwood: *Ceratopetalum apetalum* D. Don, *Callicoma serratifolia* Andr.; *Cunoniac.*;
Ligkeng: *Euphoria longana* Lamk.; *Sapindac.*; Java.   [ö. Au.
Lign aloes: *Aquilaria agallocha* Roxb.; *Thymelaeac.*; O.-Ind. (Sylhet)
Lignaloë-wood (e): *Bursera delpechiana* J. Poiss.; *Burserac.*; Mex.
Lignoaloe: *Bursera delpechiana* J. Poiss.; *Burserac.*; Mex.
Lignum agallochi: *Aquilaria agallocha* Roxb.; *Thymelaeac.*; Burma.
„ aloes: *Aquilaria agallocha* Roxb.; *Thymelaeac.*; Burma.
„ aquila: *Aquilaria agallocha* Roxb.; *Thymelaeac.*; Burma.
„ aquilariae: *Aquilaria agallocha* Roxb.; *Thymelaeac.*; Burma.
„ aspalathi: *Aquilaria agallocha* Roxb.; *Thymelaeac.*; Burma.
„ dorum: *Ocotea sp.*, *Nectandra sp.*; *Laurac.*; Jam.
„ guaiaci: *Guaiacum sanctum* L.; *Zygophyllac.*; W.-Ind., s. Mex.
„ literatum: *Brosimum aubleti* Poepp. et Endl.; *Morac.*; Gu.
„ muricum: *Albizzia montana* Bth.; *Mimosac.*; Java, Sum., N.-Kal., Au.
„ quassiae (H): *Quassia amara* L.; *Simarubac.*; S.-Am.
„ Rhodium: *Rhodorrhiza scoparia* Webb.; *Convolvulac.*; Kanar.
„ rorum: *Zanthoxylum sp.*; *Rutac.*; Jam.
„ sanctum: *Guaiacum officinale* L.; *Zygophyllac.*; P.-R.
„ vitae (u): *Guaiacum officinale* L.; *Zygophyllac.*; W.-Ind., Kol.
„   „ : *Guaiacum sp.*; *Zygophyllac.*; Mex., Hond.

Lignum vitae: *Gymnanthes lucida* Swartz; *Euphorbiac.;* C.-Am.
„ „ : *Bulnesia arborea* Engl.; *Zygophyllac.;* Kol.
„ „ : *Dialium cochinchinense* Pierre; *Caesalpiniac.;* Ind.-Ch.
„ „ : *Acacia falcata* Willd.; *Mimosac.;* Au.
„ „ : *Vitex lignum-vitae* A. Cunn.; *Verbenac.;* ö. Au.
„ „ : *Dodonaea viscosa* L.; *Sapindac.;* Au.
„ „ , Bastard-: *Tecoma spp.; Bignoniac.;* S.-Am.
„ „ , „ (u): *Guaiacum sanctum* L.; *Zygophyllac.;* W.-Ind.
„ „ , Makaraiba-: *Bulnesia arborea* Engl.; *Zygophyllac.;* tr. Am.
„ „ , Philippine-: *Xanthostemon verdugonianus* Naves; *Myrtac.;* Phil.
„ „ , red: *Schinopsis sp.; Anacardiac.;* Mex.
„ „ , Victorian: *Dodonaea viscosa* L.; *Sapindac.;* Au.
Liguster: *Ligustrum vulgare* L.; *Oleac.;* Eu., N.-As., N.-Am.
Ligustro do Japão: *Ligustrum japonicum* Thunb.; *Oleac.;* Bras.
Lihnihi: *Ceiba pentandra* Gaertn. var. *dehiscens* Ulb.; *Bombacac.;* Dah.
Lihos: *Lophira procera* Chev.; *Ochnac.;* Kam., Gab.
Lihung: *Terminalia belerica* Roxb.; *Combretac.;* O.-Ind.
Likari: *? Aniba panurensis* Mez; *Laurac.;* fr. Gu.
Lila: *Melia azedarach* L.; *Meliac.;* Mex.
„ da Percia: *Melia azedarach* L.; *Meliac.;* Bras. (Sao P)
Lilac, persian: *Melia azedarach* L.; *Meliac.;* Tr.
„ des Indes: *Melia azedarach* L.; *Meliac.;* Tr.
„ wood: *Syringa vulgaris* L.; *Oleac.;* sö. Eu.
Lilas: *Melia azedarach* L.; *Meliac.;* Réunion.
„ de la Chine: *Melia azedarach* L.; *Meliac.;* s. Fr.
„ „ „ Perse: *Melia azedarach* L.; *Meliac.;* s. Fr.
„ des Indes (f): *Melia azedarach* L.; *Meliac.;* Tr.
„ du Japon: *Melia azedarach* L.; *Meliac.;* Ind.-Ch.
„ du pays: *Melia azedarach* L.; *Meliac.;* Guad.
Lilen: *Azara spp.; Bixac.;* Chile.
Lilít: *Schizostachyum diffusum* Merr.; *Gramin.;* Phil. (Pampanga)
Lillipilli: *Eugenia smithi* Poir., *E. australis* Wendl.; *Myrtac.;* Au.
Lilombo: *Pycnanthus kombo* Warb.; *Myristicac.;* Gab.
Lilywood, swedish (u): *Betula sp.; Betulac.;* Eu.
Lim: *Erythrophloeum fordi* Oliv.; *Caesalpiniac.;* Annam.
„ : *Toona ciliata* M. Roem.; *Meliac.;* n. O.-Ind. (United Prov.)
„ : *Pinus excelsa* Wall.; *Conifer.;* nw. O.-Ind. (Kunawar, Chamba)
„ set: *Peltophorum tonkinense* Pierre; *Caesalpiniac.;* Ind.-Ch.
„ vang: *Peltophorum dasyrachis* Kurz, *P. ferrugineum* Bth.; *Caesalpiniac.;* Ind.-Ch.
„ vangh: *Peltophorum ferrugineum* Bth.; *Caesalpiniac.;* Co.-Ch.
„ vert: *Erythrophloeum fordi* Oliv.; *Caesalpiniac.;* Ind.-Ch.
„ xanh: *Erythrophloeum fordi* Oliv.; *Caesalpiniac.;* Ind.-Ch.
„ xet: *Peltophorum ferrugineum* Bth., *P. tonkinense* Pierre; *Caesalpiniac.;* Ind.-Ch.
„ „ : *Peltophorum dasyrachis* Kurz; *Caesalpiniac.;* Annam.
„ de l'Indo-Chine: *Baryxylon sp.; Caesalpiniac.;* Elf.
? Lima: *Casearia praecox* Griseb.; *Flacourtiac.;* Ven.
Lima wood: *Haematoxylon brasiletto* Karst.; *Caesalpiniac.;* C.- & S.-Am.
Limah broh: *Xanthophyllum griffithi* Hook. f.; *Polygalac..;* Malay.
Limao: *Chlorophora tinctoria* Gaudich.; *Morac.;* Bras.
Limau asam: *Citrus medica* L. var. *acida* Pers.; *Rutac.;* Malay.
„ besar: *Citrus decumana* L.; *Rutac.;* Malay.
„ chumbol: *Citrus limonum* Risso; *Rutac.;* Malay.
„ hijau: *Citrus medica* L. var. *acida* Pers.; *Rutac.;* Malay.
„ manis: *Citrus aurantium* L., *C. aurantium* L. var. *bergamia* Risso; *Rutac.;* Malay.
Limba (f): *Terminalia superba* Engl. et Diels; *Combretac.;* Kam., B.-K. (Mayombe)
Limbali: *Macrolobium dewevrei* de Wild.; *Caesalpiniac.;* B.-K. (Katanga)
Limbarra: *Melia composita* Willd.; *Meliac.;* sw. O.-Ind. (Bombay)
Limbo (f): *Terminalia superba* Engl. et Diels; *Combretac.;* Kam., Gab., B.-K.
„ blanc (f): *Terminalia superba* Engl. et Diels; *Combretac.;* Kam., B.-K. (Mayombe)
„ noir: *Terminalia superba* Engl. et Diels; *Combretac.;* Gab.
Limboro: *Macrolobium dewevrei* de Wild.; *Caesalpiniac.;* B.-K.
Limbosi: *Azadirachta indica* A. Juss.; *Meliac.;* C.-O.-Ind.
Limbu: *Citrus limonum* Risso; *Rutac.;* Hind.

Lime: *Citrus medica* L. *var. acida* Pers.; *Rutac.;* Malay.
„ tree (u): *Tilia americana* L.; *Tiliac.;* N.-Am.
„ „ (e): *Tilia cordata* Mill.; *Tiliac.;* Eu.
„ „ : *Tilia glabra* Vent.; *Tiliac.;* Kan.
„ : *Citrus aurantifolia* Swingle; *Rutac.;* br. Hond., Hond.
„ : *Trichilia minutiflora* Standl.; *Meliac.;* br. Hond.
„ : *Tilia cordata* Mill. *var. japonica* Miq.; *Tiliac.;* Jap.
„ , wild: *Trichilia minutiflora* Standl.; *Meliac.;* br. Hond.
„ , „ : *Citharexylum cooperi* Standl.; *Verbenac.;* Pan.
„ , „ : *Zanthoxylum sp.;* *Rutac.;* Bras. (Bahia)
Liméné: *Schumanniophytum klaineanum* Pierre; *Rubiac.;* Gab.
Limene: *Citrus limonum* Risso; *Rutac.;* Südsee (Tahiti)
Limes: *Sideroxylon ferrugineum* Hook. et Arn.; *Sapotac.;* Phil. (Cagayan)
Limiggién: *Zizyphus sp.;* *Rhamnac.;* Phil. (Cagayan)
Limo: *Citrus aurantifolia* Swingle; *Rutac.;* Hond.
Limoeira brava: *Laurelia aromatica* Juss.; *Monimiac.;* Bras.
Limo-limó: *Vitex turczaninowi* Merr.; *Verbenac.;* Phil. (Ilocos Sur)
Limón: *Citrus limonum* Risso; *Rutac.;* P.-R., Arg.
„ : *Citrus aurantifolia* Swingle; *Rutac.;* Peru.
„ café: *Zanthoxylum* sp.; *Rutac.;* Cuba.
„ cimarrón: *Laurelia aromatica* Juss.; *Monimiac.;* Kol.
„ real: *Citrus limonia* Osbeck; *Rutac.;* Hond.
„ sacha: *Zanthoxylum naranjillo* Griseb.; *Rutac.;* Arg.
„ de piedra: *Borreria pygmaea* Spruce; *Rubiac.;* Ven.
Limoncillo: *Amomis caryophyllata* Krug et Urb.; *Myrtac.;* P.-R.
„ : *Calyptranthes sintenisi* Kiaersk.; *Myrtac.;* P.-R.
„ : *Eugenia stahli* Krug et Urb.; *Myrtac.;* P.-R.
„ : *Bumelia leiogyna* Donn. Smith; *Sapotac.;* Salv.
„ : *Bumelia panamensis* Standl.; *Sapotac.;* Pan.
„ : *Trichilia havanensis* Jacq.; *Meliac.;* Guat., Hond.
„ : *Rheedia edulis* Triana et Planch.; *Guttifer.;* br. Hond.
„ : *Aspidosperma vargasi* A. DC.; *Apocynac.;* Ven.
? „ : *Casearia praecox* Griseb.; *Flacourtiac.;* Ven.
„ de cordóba: *Calophyllum calaba* Jacq.; *Guttifer.;* Mex.
Limone: *Citrus limonum* Risso; *Rutac.;* It.
Limonero: *Citrus limonum* Risso; *Rutac.;* Sp., Mex.
Limonier (f): *Citrus limonum* Risso; *Rutac.;* M.-M., O.-Ind.
Limoun: *Citrus limonum* Risso; *Rutac.;* Arab.
Limpiadientes: *Exandra rhodoclada* Standley; *Rubiac.;* Salv.;
Limu: *Citrus limonum* Risso; *Rutac.;* Hind.
Lin, american (u): *Tilia americana* L.; *Tiliac.;* N.-Am.
Lináb: *Trichadenia philippinensis* Merr.; *Flacourtiac.;* Phil. (Capiz, Negros)
Linaloe (f): *Bursera delpechiana* J. Poiss.; *Burserac.;* Mex.
Linaloeholz (d): *Aquilaria agallocha* Roxb.; *Thymelaeac.;* Burma.
Linaloé de Cayenne: *? Aniba panurensis* Mez; *Laurac.;* fr. Gu.
Linaloe wood: *? Aniba panurensis* Mez; *Laurac.;* Gu.
Linalué: *Bursera sp.;* *Burserac.;* Mex.
Linanué: *Bursera delpechiana* J. Poiss.; *Burserac.;* Mex.
Línau: *Lagerstroemia piriformis* Koehne; *Lythrac.;* Phil. (Sorsogon)
Lind: *Tilia cordata* Mill.; *Tiliac.;* Schwed., Dänemark.
Lindaki: *Klainedoxa gabonensis* Pierre; *Simarubac.;* Gab.
Linde (d): *Tilia cordata* Mill.; *Tiliac.;* Eu.
„ , amerikanische: *Tilia americana* L.; *Tiliac.;* N.-Am.
„ , großblättrige: *Tilia grandifolia* Ehrh.; *Tiliac.;* Eu.
„ , kleinblättrige: *Tilia cordata* Mill.; *Tiliac.;* Eu.
„ , Sommer-: *Tilia grandifolia* Ehrh.; *Tiliac.;* Eu.
„ , Stein-: *Phillyrea latifolia* L.; *Oleac.;* M.-M.
„ , Winter-: *Tilia cordata* Mill.; *Tiliac.;* Eu.
Linden (e): *Tilia cordata* Mill.; *Tiliac.;* Eu.
„ : *Tilia glabra* Vent.; *Tiliac.;* N.-Am.
„ , american (u): *Tilia americana* L.; *Tiliac.;* N.-Am.
Lingabunu: *Buchanania sp.;* *Anacardiac.;* Phil. (Basilan)
Lingo-língo: *Vitex turczaninowi* Merr.; *Verbenac.;* Phil.
Lingóg: *Avicennia officinalis* L.; *Verbenac.;* Phil. (Cagayan)

Lingon: *Vaccinium vitis-idaea* L.; *Ericac.*; Schwed., Dänemark.
Lingué: *Afzelia africana* Smith; *Caesalpiniac.*; fr. Guin.
Lingue: *Persea lingue* Nees; *Laurac.*; Chile.
„ : *Mussaenda sp.*; *Rubiac.*; Réunion.
Lingyaw: *Dillenia parviflora* Griff.; *Dilleniac.*; s. Burma.
Linké: *Afzelia africana* Smith; *Caesalpiniac.*; Sudan.
Lino blanco: *Linum suffruticosum* L.; *Linac.*; Sp.
Linyaw: *Dillenia parviflora* Griff.; *Dilleniac.*; s. Burma.
Linzin: *Duabanga sonneratioides* Buch.-Ham.; *Sonneratiac.*; Burma.
Lioczo: *Scorodophloeus zenkeri* Harms; *Caesalpiniac.*; Gab.
„ : *Hua gaboni* Pierre; *Sterculiac.*; Gab.
Lipa: *Tilia cordata* Mill.; *Tiliac.*; Russland.
Lipáuen: *Alstonia scholaris* R. Br.; *Apocynac.*; Phil.
Lipiti: *Daniella thurifera* J. J. Benn.; *Caesalpiniac.*; Togo.
Lipóte: *Eugenia sp.*; *Myrtac.*; Phil. (C.-Luzon)
Lipús: *Dipterocarpus affinis* Brandis, *D. vernicifluus* Blco.; *Dipterocarpac.*; Phil.
Liput: *Dipterocarpus affinis* Brandis; *Dipterocarpac.*; Phil. (Sur.)   [(Surigao)
Lipúut: *Dipterocarpus affinis* Brandis; *Dipterocarpac.*; Phil. (Surigao)
Liquidambar (u): *Liquidambar styraciflua* L.; *Hamamelidac.*; Mex., C.-Am.
„ : *Liquidambar formosana* Hance, *L. tonkinensis* Chev.; *Hamamelidac.*;
Lirio: *Alibertia edulis* A. Rich.; *Rubiac.*; Hond.   [Ind.-Ch.
„ amarillo: *Plumiera sp.*; *Apocynac.*; Cuba.
„ morado: *Plumiera sp.*; *Apocynac.*; Cuba.
„ de Costa: *Plumiera alba* L.; *Apocynac.*; Cuba.
„ „ dulce: *Plumiera sp.*; *Apocynac.*; Cuba.
„ „ parecloa: *Plumiera sp.*; *Apocynac.*; Cuba.
„ „ playa: *Plumiera sp.*; *Apocynac.*; Cuba.
Lísak: *Neonauclea sp.*; *Rubiac.*; Phil. (Tayabas)
Lisíkan: *Vatica mangachapoi* Blco.; *Dipterocarpac.*; Phil. (Rizal)
Lissamba: *Uapaca staudti* Pax; *Euphorbiac.*; Kam.
Litchi: *Nephelium litchi* Camb.; *Sapindac.*; s. Ch.
„ ponceau: *Nephelium litchi* Camb.; *Sapindac.*; Ch., O.-Ind.
Lítis: *Dipterocarpus grandiflorus* Blco.; *Dipterocarpac.*; Phil. (Capiz)
Litjeh: *Nephelium litchi* Camb.; *Sapindac.*; Java.
Litjik: *Nephelium litchi* Camb.; *Sapindac.*; Java.
Lito: *Hopea pierrei* Hance; *Dipterocarpac.*; Phil. (Cagayan)
Litok: *Shorea polita* Vid.; *Dipterocarpac.*; Phil. (Cagayan)
Litre: *Schinus sp.*; *Anacardiac.*; Arg.
„ : *Lithraea venenosa* Miers; *Anacardiac.*; Chile.
Li-tschi: *Nephelium litchi* Camb.; *Sapindac.*; Ch.
Litschi ponceau: *Nephelium litchi* Camb.; *Sapindac.*; Ch., O.-Ind.
Liur: *Juniperus macropoda* Boiss.; *Conifer.*; nw. O.-Ind. (Punjab)
Liusin: *Parinarium griffithianum* Bth., *P. corymbosum* Miq.; *Rosac.*; Phil.
„ -gúbat: *Pygeum sp.*; *Rosac.*; Phil. (Bataan)
„ -pulá: *Kayea paniculata* Merr.; *Guttifer.*; Phil. (Bataan)
Liviza: *Hua gaboni* Pierre; *Sterculiac.*; Gab.
„ : *Scorodophloeus zenkeri* Harms; *Caesalpiniac.*; Gab.
Liwamba: *Afzelia africana* Smith; *Caesalpiniac.*; B.-K.
Lizard wood: *Vitex longeracemosa* Pittier; *Verbenac.*; br. W.-Ind.
Llagrume macho: *Didymopanax morototoni* Dcne. et Planch.; *Araliac.*; P.-R.
Llagrumo: *Cecropia sp.*; *Morac.*; P.-R.
Llallo: *Luehea seemanni* Triana et Planch.; *Tiliac.*; Guat.
Llalo: *Luehea divaricata* Mart.; *Tiliac.*; Guat., Hond.
Llcuque: *Prumnopitys elegans* Phil.; *Conifer.*; Chile.
Llema de huevo: *Morinda panamensis* Seem.; *Rubiac.*; Guat.
„ „ „ : *Chimarrhis parviflora* Standl.; *Rubiac.*; Pan.
Llengue: *Podocarpus andina* Poepp. et Endl.; *Conifer.*; Chile.
Llora-sangre: *Pterocarpus orbiculatus* DC.; *Papilionac.*; Mex.
Lluchán: *Chorisia speciosa* St. Hil.; *Bombacac.*; Arg.
Lluicho vainilla: *Guazuma ulmifolia* Lam.; *Sterculiac.*; Peru.
Lluvia de oro: *Cytisus laburnum* L.; *Papilionac.*; Arg.
Lmanza: *Cedrus deodara* Loudon; *Conifer.*; Afg.
Lo: *Cola vera* K. Schum., *C. nitida* Schott et Endl.; *Sterculiac.*; Elf.
„ : *Parkia agboensis* Chev.; *Mimosac.*; Elf.

Lo: *Erythrophloeum guineense* G. Don; *Caesalpiniac.;* Elf.
„ : *Lonchocarpus sericeus* H. B. K.; *Papilionac.;* Togo.
„ bo: *Brownlowia tabularis* Pierre; *Tiliac.;* Ind.-Ch.
„ „ la long: *Brownlowia denysiana* Pierre; *Tiliac.;* Ind.-Ch.
„ ka vi: *Glochidion obliquum* Dcne.; *Euphorbiac.;* Ind.-Ch.
„ le: *Chaetocarpus castanocarpus* Thw.; *Euphorbiac.;* Ind.-Ch.
„ mann: *Excoecaria oppositifolia* Muell. Arg.; *Euphorbiac.;* Ind.-Ch.
Loa: *Ficus glomerata* Roxb.; *Morac.;* w. Beng. (Chota Nagpur)
Loang: *Carallia lucida* Roxb.; *Rhizophorac.;* Ind.-Ch.
Loc buc: *Mallotus barbatus* Muell. Arg.; *Euphorbiac.;* Ind.-Ch.
„ chui: *Castanopsis tribuloides* A. DC.; *Fagac.;* Ind.-Ch.
„ ma: *Claoxylon indicum* Hassk.; *Euphorbiac.;* Ind.-Ch.
Locma: *Lucuma sp.; Sapotac.;* Peru.
Loctob: *Duabanga moluccana* Blume; *Sonneratiac.;* Phil.
Locus: *Hymenaea courbaril* L.; *Caesalpiniac.;* Sur.
„ , Basra-: *Dicorynia paraensis* Bth.; *Caesalpiniac.;* Sur.
„ , Bastard-: *Dicorynia paraensis* Bth.; *Caesalpiniac.;* Sur.
„ , groot: *Dipteryx odorata* Willd.; *Papilionac.;* Sur.
„ , Zand-: *Dicorynia paraensis* Bth.; *Caesalpiniac.;* Sur.
„ , zwarte: *Hymenaea sp.; Caesalpiniac.;* Sur.
Locusboom: *Hymenaea courbaril* L.; *Caesalpiniac.;* W.-Ind., S.-Am.
Locust: *Hymenaea courbaril* L.; *Caesalpiniac.;* br. Hond., Ant., Gu., Bras.
„ : *Gleditschia horrida* Mak.; *Caesalpiniac.;* Jap.
„ berry: *Byrsonima sp.; Malpighiac.;* br. W.-Ind.
„ , black: *Robinia pseudacacia* L.; *Papilionac.;* br. W.-Ind.
„ , clammy- (e): *Robinia viscosa* Vent.; *Papilionac.;* N.-Am.
„ , Honey-: *Gleditschia triacanthos* L.; *Caesalpiniac.;* ö. N.-Am.
„ , leather-leaved: *Hymenaea courbaril* L.; *Caesalpiniac.;* W.-Ind., S.-Am.
„ , rose-flowering (e): *Robinia viscosa* Vent.; *Papilionac.;* USA.
„ , south-american: *Hymenaea courbaril* L.; *Caesalpiniac.;* W.-Ind., S.-Am.
„ , Water-: *Gleditschia monosperma* Walt.; *Caesalpiniac.;* N.-Am.
„ , west-indian: *Hymenaea courbaril* L.; *Caesalpiniac.;* W.-Ind., S.-Am.
„ , wild: *Pentaclethra macrophylla* Bth.; *Mimosac.;* Lib.
„ , yellow: *Robinia pseudacacia* L.; *Papilionac.;* N.-Am.
Locustbaum (d): *Hymenaea courbaril* L.; *Caesalpiniac.;* W.-Ind. S.-Am.
„ bean: *Parkia bicolor* Chev.; *Mimosac.;* Lib.
Locustrier: *Hymenaea courbaril* L.; *Caesalpiniac.;* fr. Gu.
Lod: *Symplocos beddomei* C. B. Clarke; *Symplocac.;* w. O.-Ind. (Bombay)
Loeloe-oë: *Parkia nitida* Miq., *P. microcephala* Kleinh.; *Mimosac.;* Sur.
Loewis-madang: *Peronema canescens* Jack; *Uerbenac.;* Borneo.
Logo: ? *Ficus barbidens* Warb., *Antiaris africana* Engl.; *Morac.;* Togo.
„ : *Terminalia catappa* L.; *Combretac.;* Phil. (Cagayan, Ilocos Sur, Union)
„ asagu: *Chlorophora excelsa* Bth. et Hook.; *Morac.;* Togo.
Logotsi: *Antiaris africana* Engl.; *Morac.;* Go.
Logwood, Bastard-: ? *Pithecolobium sp.; Mimosac.;* br. Hond.
„ brush: *Mimosa hemiendyta* Rose et Robinson; *Mimosac.;* br. Hond.
„ , Catseem-: *Mimosa hemiendyta* Rose et Robinson; *Mimosac.;* br. Hond.
Lohagasi: *Aegle marmelos* Correa; *Rutac.;* O.-Ind.
Lohero: *Tecoma undulata* G. Don; *Bignoniac.;* nw. O.-Ind., Arab.
Loho: *Musanga smithi* R. Br.; *Morac.;* Elf.
Lohoedoe: *Sideroxylon guianense* A. DC.; *Sapotac.;* Sur.
Lôi: *Crypteronia paniculata* Blume; *Crypteroniac.;* Annam.
„ : *Bischofia javanica* Blume; *Euphorbiac.;* Ind.-Ch.
„ tho: *Gmelina arborea* Roxb.; *Uerbenac.;* Annam.
„ tia: *Bischofia javanica* Blume; *Euphorbiac.;* Ind.-Ch.
Loiad: *Dalbergia lanceolaria* L.; *Papilionac.;* O.-Ind.
Lokar bhadi: *Lannea grandis* Engl.; *Anacardiac.;* Beng.
Lokisi kaka: *Hymenaea courbaril* L.; *Caesalpiniac.;* Sur.
Lokisie: *Hymenaea courbaril* L.; *Caesalpiniac.;* Sur.
„ wakie: *Inga spp.; Mimosac.;* Sur.
Loko: *Chlorophora excelsa* Bth. et Hook.; *Morac.;* Nig.
„ : *Khaya senegalensis* A. Juss.; *Meliac.;* Go., Togo.
Lokoba: *Entandrophragma ferruginea* Chev.; *Meliac.;* Elf.
Lokobo: *Entandrophragma ferruginea* Chev.; *Meliac.;* Elf.

Lokobua: *Khaya ivorensis* Chev.; *Meliac.;* Elf.
Lokolo: *Entandrophragma ferruginea* Chev.; *Meliac.;* tr. Af.
Loksi: *Hymenaea sp.; Caesalpiniac.;* Sur.
Loktób: *Duabanga moluccana* Blume; *Sonneratiac.;* Phil.
Lokus: *Hymenaea courbaril* L.; *Caesalpiniac.;* Gu.
„ , Basra-: *Dicorynia paraënsis* Bth.; *Caesalpiniac.;* Sur.
Lokustbaum: *Hymenaea courbaril* L.; *Caesalpiniac.;* tr. Am.
Lole lutoke: *Xylopia sp.; Anonac.;* B.-K. (Bukaka)
„ mutoke: *Xylopia sp.; Anonac.;* B.-K. (Bukaka)
Lolihoedoe: *Nectandra sp., Ocotea sp.; Laurac.;* Sur
Lolo: *Klainedoxa longifolia* Pierre; *Simarubac.;* B.-K.
Loloe: *Sterculia tragacantha* Lindl.; *Sterculiac.;* Go., Togo.
Loloti: *Lannea acidissima* Chev.; *Anacardiac.;* Elf.
Lolyondo: *Linociera welwitschi* Knobl.; *Oleac.;* D.-O.-Af.
Lom: *Sterculia oblonga* Mast.; *Sterculiac.;* Kam.
„ côm: *Elaeocarpus dubius* A. DC.; *Elaeocarpac.;* Ind.-Ch.
„ vang: *Ailanthus malabarica* DC.; *Simarubac.;* Annam.
Loma de caiman: *Platypodium maxonianum* Pittier; *Papilionac.;* Kol.
Lomangené: *Triplochiton scleroxylon* K. Schum.; *Sterculiac.;* Kam.
Lomba· *Pycnanthus kombo* Warb.; *Myristicac.;* Gab., B.-K.
Lombelómbe: *Adina rubrostipulata* K. Schum.; *Rubiac.;* D.-O.-Af. (w. Usambara)
Lombi: *Saccoglottis gabonensis* Urban; *Humiriac.;* Gab.
Lombo: *Pseudocedrela kotschyi* Harms; *Meliac.;* W.-Af.
„ : *Pycnanthus kombo* Warb.; *Myristicac.;* Gab.
Lombricero: *? Andira sp.; Papilionac.;* Trin.
Lombrigueira: *Andira sp.; Papilionac.;* Bras.
„ : *Ficus anthelmintica* Mart.; *Morac.;* Bras. (Amaz.)
Lomburu: *Sterculia tragacantha* Lindl.; *Sterculiac.;* Elf.
Lomvogo: *Fagara sp.; Rutac.;* Gab.
Lon bon: *Aglaia domestica* Pellegr.; *Meliac.;* Ind.-Ch.
„ gieng: *Cratoxylon formosum* Bth. et Hook. f.; *Guttifer.;* Kamb.
„ hirn: *Cratoxylon formosum* Bth. et Hook. f.; *Guttifer.;* Kamb.
„ putte: *Hopea grisea* Brandis; *Dipterocarpac.;* br. Borneo (Sarawak)
Long: *Dillenia heterosepala* Finet et Gagnep.; *Dilleniac.;* Ind.-Ch.
„ : *Wrightia annamensis* Eberh. et Dubard; *Apocynac.;* Ind.-Ch.
„ côm: *Elaeocarpus dubius* A. DC.; *Elaeocarpac.;* Ind.-Ch.
„ deuk: *Phyllostachys nigra* Munro; *Gramin.;* Ch. (Fukien)
„ giêng: *Cratoxylon polyanthum* Korth.; *Guttifer.;* Ind.-Ch.
„ hien: *Cratoxylon prunifolium* Dyer; *Guttifer.;* Kamb.
„ jack: *Flindersia oxleyana* F. v. M.; *Rutac.;* ö. Au.
„ John: *Triplaris surinamensis* Cham.; *Polygonac.;* br. Gu.
„ mân: *Pterospermum grewiaefolium* Pierre; *Sterculiac.;* Annam.
„ mang: *Pterospermum spp.; Sterculiac.;* Ind.-Ch.
„ „ : *Macaranga triloba* Muell. Arg.; *Euphorbiac.;* Ind.-Ch.
„ mûc: *Wrightia annamensis* Eberh. et Dubard; *Apocynac.;* Annam.
„ nâo: *Cinnamomum camphora* Nees et Eberm.; *Laurac.;* Annam.
„ nhân: *Nephelium longana* Cambess.; *Sapindac.;* Annam.
„ yan: *Nephelium longana* Cambess.; *Sapindac.;* Ch.
„ yèn: *Nephelium longana* Cambess.; *Sapindac.;* Annam.
„ yeng: *Cratoxylon formosum* Bth.; *Guttifer.;* Ind.-Ch.
Longan: *Nephelium longana* Cambess.; *Sapindac.;* O.-Ind., Cey.
Longanier (f): *Nephelium longana* Cambess.; *Sapindac.;* s. Ch.
Longbói: *Eugenia sp., E. cumini* Merr.; *Myrtac.;* Phil. (n. & s. Luzon, Biscaya)
Longieng: *Cratoxylon prunifolium* Dyer; *Guttifer.;* Kamb.
Longotra: *Cryptocarya louveli* Danguy, *C. perrieri* Danguy; *Laurac.;* Mad.
„ Fotsy: *Anthocleista sp.; Loganiac.;* Mad.
„ „ : *Cryptocarya perrieri* Danguy; *Laurac.;* Mad.
„ mena: *Cryptocarya louveli* Danguy; *Laurac.;* Mad.
Longué: *Fagara macrophylla* Engl.; *Rutac.;* Kam.
Lonlaviol: *Daniella sp.; Caesalpiniac.;* Gab.
Lontong: *Sterculia sp.; Sterculiac.;* Phil. (Zamboanga)
Loob: *Schizostachyum dielsianum* Merr.; *Gramin.;* Phil. (Bohol)
Loon putih: *Hopea grisea* Brandis; *Dipterocarpac.;* br. Borneo (Sarawak)
Loong nhàn: *Nephelium longana* Cambess.; *Sapindac.;* Annam.

Lopinga: *Diospyros perrieri* Jum.; *Ebenac.;* Mad.
Lopingo: *Diospyros perrieri* Jum.; *Ebenac.;* Mad.
Loppo: *Acer italum* Lauth; *Acerac.;* It.
Loquat: *Eriobotrya japonica* Lindl.; *Rosac.;* Hond.
Lorbeer, kalifornischer: *Umbellularia californica* Nutt.; *Laurac.;* w. USA.
„ , Madeira-: *Persea indica* Spreng.; *Laurac.;* Kanar.
Lorito: *Weinmannia trichosperma* Cav.; *Cunoniac.;* C.-R.
Lorito: *Pithecolobium arboreum* Urban; *Mimosac.;* C.-R.
Loro: *Pithecolobium arboreum* Urban; *Mimosac.;* C.-R.
„ : *Butyrospermum parki* Kotschy; *Sapotac.;* Sudan.
„ amarillo: *Cordia hypoleuca* A. DC.; *Borraginac.;* Arg.
„ blanco: *Sida densiflora* Hook. et Arn.; *Malvac.;* Arg. (Misiones)
„ negro: *Cordia alliodora* Cham.; *Borraginac.;* Arg.
Losóban: *Bombycidendron campylosiphon* Warb.; *Malvac.;* Phil. (Abra)
„ : *Bombycidendron vidalianum* Merr. et Rolfe; *Malvac.;* Phil. (Abra)
Lota amara: *Amoora rohituka* Wight et Arn.; *Meliac.;* Assam.
Lotier (f): *Zizyphus lotus* Lamk.; *Rhamnac.;* n. Af.
Loto: *Randia dumetorum* Lamk.; *Rubiac.;* w. Beng. (Chota Nagpur)
Lotofa: *Cola sp.; Sterculiac.;* Elf.
Lotong: *Aglaia eusideroxylon* Koord. et Valet.; *Meliac.;* Java.
Lotos (f): *Zizyphus lotus* Lamk.; *Rhamnac.;* n. Af.
„ des anciens: *Zizyphus lotus* Lamk.; *Rhamnac.;* n. Af.
Louambo: *Mitragyne macrophylla* Hiern; *Rubiac.;* Kam.
Louan-tan: *Vantanea guianensis* Aubl.; *Humiriac.;* fr. Gu.
Loundi: *Erythrophloeum guineense* G. Don; *Caesalpiniac.;* Kam.
Louro: *Ocotea opifera* Mart.; *Laurac.;* Bras. (Para)
„ : *Ocotea teleiandra* Mez; *Laurac.;* Bras. (Bahia)
„ : *Cordia spp.; Borraginac.;* Bras. (Rio)
„ amarello: *Nectandra lanceolata* Nees, *N. nitida* Nees; *Laurac.;* Bras.
„ „ : *Cordia sp.; Borraginac.;* Bras.
„ branco: *? Ocotea opifera* Mart., *O. guianensis* Aubl.; *Laurac.;* Bras. (Amaz.)
„ cachoeira: *Nectandra sp., Ocotea sp.; Laurac.;* Bras. (Amaz.)
„ camphora: *Ocotea costulata* Mez; *Laurac.;* Bras. (Para)
„ canella: *Nectandra sp, Ocotea sp.; Laurac.;* Bras. (Amaz.)
„ cheiroso: *Dicypellium caryophyllatum* Nees; *Laurac.;* Bras. (Amaz.)
„ cheiroso: *Ocotea pretiosa* Mez; *Laurac.;* s. Bras.
„ chumbo: *Nectandra sp., Ocotea sp.; Laurac.;* Bras. (Amaz.)
„ congonha: *Nectandra sp., Ocotea sp.; Laurac.;* Bras. (Amaz.)
„ cravo: *Nectandra sp., Ocotea sp.; Laurac.;* Bras. (Amaz.)
„ cunuaru: *Nectandra sp., Ocotea sp.; Laurac.;* Bras. (Amaz.)
„ faia: *Roupala sp.; Proteac.;* Bras. (Para)
„ inamuy: *Nectandra elaiophora* Barb. Rodr.; *Laurac.;* Bras. (Manaos)
„ pederneira: *Nectandra sp., Ocotea sp.; Laurac.;* Bras. (Amaz.)
„ pimenta: *Ocotea canaliculata* Mez; *Laurac.;* Bras. (Amaz.)
„ preto: *Nectandra puberula* Nees, *N. mollis* Nees; *Laurac.;* s. Bras.
„ „ : *Ocotea spectabilis* Mez; *Laurac.;* s. Bras.
„ rosa: *Aniba terminalis* Ducke; *Laurac.;* ö. Bras. (Amaz.)
„ tachy: *Nectandra sp., Ocotea sp.; Laurac.;* Bras. (Amaz.)
„ tamancão: *Ocotea guianensis* Aubl.; *Laurac.;* Bras. (Amaz.)
„ tamanco: *? Nectandra cuspidata* Nees; *Laurac.;* n. Bras. (Para)
„ vermelho: *Nectandra globosa* Mez, *Ocotea rubra* Mez; *Laurac.;* nö. Bras.
„ da Varzea: *Nectandra amazonum* Nees; *Laurac.;* nö. Bras. (Para)
„ de cheiro: *Nectandra sp., Ocotea sp.; Laurac.;* Bras. (Amaz.)
„ do igapó: *Nectandra amazonum* Nees; *Laurac.;* Bras. (Amaz.)
„ „ Parana: *Cordia sp.; Borraginac.;* s. Bras. (Sao P)
Loustau: *Antirrhoea verticillata* DC.; *Rubiac.;* Maur.
Louveira: *Cyclolobium sp.; Papilionac.;* s. Bras. (Sao P)
Love tree: *Cercis siliquastrum* L.; *Caesalpiniac.;* s. Eu.
Lovieng: *Randia tomentosa* Hook.; *Rubiac.;* Kamb.
Lovoa wood (u): *Lovoa klaineana* Pierre; *Meliac.;* W.-Af.
Lowi: *Artocarpus lakoocha* Roxb.; *Morac.;* sw. O.-Ind. (Bombay)
Loxa bark: *Cinchona officinalis* L.; *Rubiac.;* tr. S.-Am., Jam.
Lu: *Garcinia cambodiensis* Vesque; *Guttifer.;* Ind.-Ch.
„ uoi: *Sterculia lychnophora* Hance; *Sterculiac.;* Ind.-Ch.

Luang xuòng: *Anneslea fragrans* Wall.; *Theac.;* Annam.
Lubagán: *Eugenia clausa* C. B. Rob.; *Myrtac.;* Phil. (Isa.)
Luban: *Boswellia serrata* Roxb.; *Burserac.;* Beng.
Lubanáyong: *Xylocarpus granatum* Koenig; *Meliac.;* Phil. (Cagayan)
Lubég: *Eugenia saligna* C. B. Rob.; *Myrtac.;* Phil. (Cagayan)
Lubigniati: *Irvingia gabonensis* Baill.; *Simarubac.;* Elf.
Lubota: *Platysepalum chevalieri* Harms; *Papilionac.;* B.-K.
Lucée: *Malpighia punicifolia* L.; *Malpighiac.;* fr. Gu.
Lucet (f): *Vaccinium myrtillus* L.; *Ericac.;* Eu.
Luch: *Crescentia cujete* L.; *Bignoniac.;* br. Hond.
Lucmo: *Lucuma sp.; Sapotac.;* Peru.
Lucula: *Pterocarpus tinctoria* Welw.; *Papilionac.;* Gab.
Lucuma: *Lucuma obovata* H. B. K.; *Sapotac.;* Peru.
Lucumo: *Lucuma sp.; Sapotac.;* Peru.
Lud: *Toona ciliata* M. Roem.; *Meliac.;* Beng.
Lufefe: *Apodytes dimidiata* E. Meyer; *Icacinac.;* s. C.-Af.
Luftholz (d): *Copaifera bracteata* Bth.; *Caesalpiniac.;* n. S.-Am., Gu.
Lukabbán: *Sonneratia caseolaris* Engl.; *Sonneratiac.;* Phil. (Cagayan)
Luki: *Vitex leucoxylon* L. f.; *Verbenac.;* O.-Ind. (Madras)
Lukóan: *Artocarpus subrotundifolia* Elmer; *Morac.;* Phil. (Sorsogon)
Luktúb: *Duabanga moluccana* Blume; *Sonneratiac.;* Phil.
Lukumanju: *Cassia bacillaris* L. f.; *Caesalpiniac.;* br. Gu.
Lulu: *Butyrospermum parki* Kotschy; *Sapotac.;* Sudan, br. O.-Af. (Uganda)
Luma: *Myrtus luma* Barn.; *Myrtac.;* Chile.
„ : *Myrceugenia luma* Berg, *M. apiculata* Ndz.; *Myrtac.;* Chile.
Lumakáu: *Cyclostemon sp.; Euphorbiac.;* Phil. (Camarines)
Lumampáu: *Schizostachyum lumampao* Merr.; *Gramin.;* Phil. (Tagalog)
Lumángog: *Terminalia calamansanai* Rolfe; *Combretac.;* Phil. (Surigao)
„ : *Petersianthus quadrialatus* Merr.; *Lecythidac.;* Phil.
Lumánog: *Terminalia calamansanai* Rolfe; *Combretac.;* Phil. (Surigao)
Lumáti: *Lagerstroemia piriformis* Koehne; *Lythrac.;* Phil. (Ilocos-Sur)
Lumbanau: *Aglaia everetti* Merr.; *Meliac.;* Phil. (Surigao)
Lumbáng: *Aleurites moluccana* Willd.; *Euphorbiac.;* Phil. (Rizal, Laguna, Batg.)
„ -banukalád: *Aleurites trisperma* Blco.; *Euphorbiac.;* Phil. (Cav., Laguna)
„ -batú: *Aleurites moluccana* Willd.; *Euphorbiac.;* Phil. (Cav.)
„ -gúbat: *Aleurites trisperma* Blco.; *Euphorbiac.;* Phil. (Cav., Batg.)
Lumbayao: *Tarrietia javanica* Blume; *Sterculiac.;* Phil. (Zamboanga, Basilán)
„ -batu: *? Tarrietia sp.; Sterculiac.;* Phil. (Zamboanga, Basilán)
Lumbiá: *Metroxylon rumphi* Mart.; *Palmac.;* Phil. (Mindanao, Biscaya)
Lumbói: *Eugenia sp., E. cumini* Merr.; *Myrtac.;* Phil.
Lumbor: *Shorea sp.; Dipterocarpac.;* Ind.-Ch.
Lumpían: *Lagerstroemia piriformis* Koehne; *Lythrac.;* Phil. (Samar)
Lumpundu: *Alstonia gilleti* de Wild.; *Apocynac.;* B.-K.
Lumuluás: *Parinarium sp.; Rosac.;* Phil. (Cotabato)
Lúnas: *Bambusa vulgaris* Schrad.; *Gramin.;* Phil. (Leyte, Bohol)
„ : *Cyclostemon sp.; Euphorbiac.;* Phil. (Laguna)
„ : *Neonauclea sp.; Rubiac.;* Phil. (Bataan)
„ -batú: *Celtis sp.; Ulmac.;* Phil. (Rizal)
Lunbo: *Buchanania latifolia* Roxb.; *Anacardiac.;* Burma.
Lung: *Ficus religiosa* L.; *Morac.;* Ind.-Ch.
„ yen: *Nephelium longana* Cambess.; *Sapindac.;* s. Ch.
Luni: *Cotoneaster bacillaris* Wall.; *Rosac.;* nw. O.-Ind. (Hazara)
Luong: *Berrya mollis* Wall.; *Tiliac.;* Ind.-Ch.
„ khôm: *Manglietia fordiana* Oliv.; *Magnoliac.;* Ind.-Ch.
Luot: *Knema conferta* Warb.; *Myristicac.;* Ind.-Ch.
Lupák: *Euphoria didyma* Blco.; *Sapindac.;* Phil. (Camarines)
Lupál: *Euphoria didyma* Blco.; *Sapindac.;* Phil. (Cotabato)
Lupangá: *Pometia pinnata* Forst.; *Sapindac.;* Phil. (Zamboanga, Sulu, Manukmanka)
Lupangán: *Pometia pinnata* Forst.; *Sapindac.;* Phil. (Zamboanga, Sulu, Manukm.)
Lupigi: *Dracontomelum dao* Merr. et Rolfe; *Anacardiac.;* Phil. (Cagayan)
„ : *Wallaceodendron celebicum* Koord.; *Mimosac.;* Phil. (Babuyan, Cagayan)
Lupuna: *Chorisia speciosa* St. Hil.; *Bombacac.;* Peru.
Lupung: *Terminalia belerica* Roxb.; *Combretac.;* O.-Ind.

301

Luruche: *Jacquinia aurantiaca* Ait.; *Theophrastac.;* Guat.
„ : *Gliricidia sepium* Steud.; *Papilionac.;* Guat.
Lusambya: *Dolichandrone platycalyx* Baker; *Bignoniac.;* br. O.-Af. (Uganda)
Lusenga: *Staudtia congensis* ?; *Myristicac.;* B.-K. (Bukaka)
Lusengasenga: *Pycnanthus kombo* Warb.; *Myristicac.;* B.-K. (Bukaka)
Lut-ter: *Artocarpus chaplasha* Roxb.; *Morac.;* nö. O.-Ind. (Nepal)
Lutub: *Vatica mangachapoi* Blco.; *Dipterocarpac.;* Phil. (Zamboanga)
Luxwezo: *Notobuxus natalensis* Oliver; *Buxac.;* S.-Af.
Luyong: *Livistona rotundifolia* Mart. var. *luzonensis* Becc.; *Palmac.;* Phil.
„ : *Livistona rotundifolia* Mart.; *Palmac.;* Phil. (Zambales, Bataan)
Luyur: *Casearia glomerata* Roxb.; *Flacourtiac.;* O.-Ind., Sunda.
Luzerne en arbre: *Medicago arborea* L.; *Papilionac.;* Syr., Pers.
Ly: *Carallia lucida* Roxb.; *Rhizophorac.;* Ind.-Ch.
„ : *Aleurites moluccana* Willd.; *Euphorbiac.;* Ind.-Ch.
„ : *Garcinia fragraeoides* Chev.; *Guttifer.;* Tonkin.
„ -tchê: *Nephelium litchi* Cambess.; *Sapindac.;* Annam, Ch.
Lychee: *Nephelium litchi* Cambess.; *Sapindac.;* s. Ch.

M

Ma: *Vitex glabrata* R. Br.; *Verbenac.;* Ind.-Ch.
„ : *Saraca dives* Pierre; *Caesalpiniac.;* Ind.-Ch.
„ -atte: *Thujopsis dolabrata* S. et Z.; *Conifer.;* Jap.
„ deuk: *Phyllostachys sp.;* *Gramin.;* Ch. (Fukien)
„ monton: *Mangifera indica* L.; *Anacardiac.;* Burma.
„ nam: *Maba buxifolia* Pers.; *Ebenac.;* Annam.
„ tien: *Strychnos nux-vomica* L.; *Loganiac.;* Ind.-Ch.
„ trinh: *Unona jucunda* Pierre; *Anonac.;* Annam.
„ -u: *Duabanga sonneratioides* Ham.; *Sonneratiac.;* Burma.
„ -u: *Sarcocephalus cordatus* Miq.; *Rubiac.;* Burma.
„ -u: *Anthocephalus cadamba* Miq.; *Rubiac.;* Burma, And.
„ -u-gyi: *Sarcocephalus cordatus* Miq.; *Rubiac.;* Burma.
„ -u-kadon: *Sarcocephalus cordatus* Miq.; *Rubiac.;* Burma.
„ -u-lelan: *Sarcocephalus cordatus* Miq.; *Rubiac.;* Burma.
„ -u-lettan: *Duabanga sonneratioides* Ham.; *Sonneratiac.;* n. Burma.
„ -u-lettanshe: *Sarcocephalus cordatus* Miq.; *Rubiac.;* Burma.
Maa: *Mangifera indica* L.; *Anacardiac.;* O.-Ind. (Madras)
Maas: ? *Tarrietia sp.;* *Sterculiac.;* Moluk.
Maba: *Maba klaineana* Pierre; *Ebenac.;* Gab.
Maballi-harriraro: *Conceveiba guianensis* Aubl., *C. hostmanni* Bth.; *Euphorbiac.;* Sur.
Mabalógo: *Kingiodendron alternifolium* Merr.; *Papilionac.;* Phil. (Samar)
Mabálud: *Neonauclea sp.;* *Rubiac.;* Phil. (Guimaras)
Maban mvoulé: *Rauwolfia sp.;* *Apocynac.;* Kam.
Mabanga: *Barteria nigritiana* Hook. f.; *Flacourtiac.;* Kam.
Mabantút: *Terminalia calamansanai* Rolfe; *Combretac.;* Phil. (Bataan)
Mabayáon: *Eugenia saligna* C. B. Rob.; *Myrtac.;* Phil. (Tarlac)
Mabéré: *Cyrtogonome argentea* Prain; *Euphorbiac.;* Gab.
Mabi: *Colubrina reclinata* Brongn.; *Rhamnac.;* Cuba, P.-R.
„ -hout: *Conceveiba guianensis* Aubl., *C. hostmanni* Bth.; *Euphorbiac.;* Sur.
Mabolo: *Diospyros discolor* Willd.; *Ebenac.;* Phil.
Maboro: *Mammea ebboro* Pierre; *Guttifer.;* Gab.
Mabouron: *Ochrocarpus africanus* Oliv.; *Guttifer.;* Gab.
Mabouya: *Capparis ferruginea* L.; *Capparidac.;* Ant.
„ peau: *Morisonia americana* L.; *Capparidac.;* Guad.
Mabula: *Pentaclethra macrophylla* Bth.; *Mimosac.;* B.-K. (Maniema)
Mabúlo: *Diospyros discolor* Willd.; *Ebenac.;* Phil.
„ : *Myristica sp.;* *Myristicac.;* Phil. (Pangasinan)
Mabúlong-gúbat: *Myristica sp.;* *Myristicac.;* Phil. (Zambales)
Mâc: *Wrightia annamensis* Eberh. et Dub.; *Apocynac.;* Ind.-Ch.
„ chác: *Illicium verum* Hook. f.; *Magnoliac.;* Ind.-Ch.
„ , faux: *Geissois pruinosa* Brongn. et Gris.; *Saxifragac.;* N.-Kal.
„ kua: *Diospyros ebenum* Koenig; *Ebenac.;* Ind.-Ch.
„ meo: *Caesalpinia bonducella* Fleming; *Caesalpiniac.;* Ind.-Ch.

302

Maca do cobra: *Anona sp.; Anonac.;* Bras.
Macaásim: *Eugenia sp., E. benthami* A. Gray; *Myrtac.;* Phil.
Macaca poranga: *Dicypellium caryophyllatum* Nees; *Laurac.;* nö. Bras.
„ vermelha: *Lucuma procera* Mart.; *Sapotac.;* Bras.
Macacahuba: *Platymiscium ulei* Harms; *Papilionac.;* Bras.
Macacarecuya: *Chytroma turbinata* Miers, *? Eschweilera sp.; Lecythidac.;* Bras.
Macacariba: *Platymiscium polystachyum* Bth.; *Papilionac.;* Sur.
Macacauba: *Platymiscium spp.; Papilionac.;* Bras.
„ : *Acrocomia sclerocarpa* Mart.; *Palmac.;* Bras.
„ da varaea: *Platymiscium paraënse* Huber; *Papilionac.;* Bras.
„ „ terra firma: *Platymiscium polystachyum* Bth.; *Papilionac.;* Bras.
Macagua: *Aveledoa nucifera* Pittier; *Olacac.;* Ven.
Macallo: *Andira sp.; Papilionac.;* Mex.
Macambrará: *Craterispermum montanum* Hiern; *Rubiac.;* St. Thomé (W.-Af.)
Macana: *Diphysa carthaginensis* Jacq.; *Papilionac.;* Pan.
Macano: *Diphysa carthaginensis* Jacq.; *Papilionac.;* Pan.
Macaqueiro: *Guarea spp.; Meliac.;* Bras.
Macasuba: *Acrocomia sclerocarpa* Mart.; *Palmac.;* Sur.
Macata: *Caesalpinia pulcherrima* Sw.; *Caesalpiniac.;* fr. Gu.
Macawood: *Platymiscium sp., ? P. ulei* Harms; *Papilionac.;* Bras.
Macaxocotl: *Spondias sp.; Anacardiac.;* Mex.
Macayo: *Andira sp.; Papilionac.;* Mex.
Macayxtli: *Erythrina sp.; Papilionac.;* Mex.
Macceira brava: *Ziziphus jujuba* Lamk.; *Rhamnac.;* port. Af.
Maçcranduba: *Mimusops spp.; Sapotac.;* Bras.
Mach: *Hopea ferrea* Pierre; *Dipterocarpac.;* Ind.-Ch.
Machete: *Erythrina rubrinervia* H. B. K.; *Papilionac.;* Pan.
Machie: *Dillenia pentagyna* Roxb.; *Dilleniac.;* O.-Ind.
Machkund: *Pterospermum acerifolium* Willd.; *Sterculiac.;* C.-O.-Ind.
Macholder (d): *Juniperus communis* L.; *Conifer.;* n. Hem.
Mackheng: *Dialium cochinchinensis* Pierre; *Caesalpiniac.;* Ind.-Ch.
Macla: *Vitex altissima* L. f.; *Verbenac.;* O.-Ind.
Maclura tinctorial: *Chlorophora tinctoria* Gaudich.; *Morac.;* tr. Am.
Maco: *Melicocca bijuga* L.; *Sapindac.;* Trin.
Macondo: *Cavanillesia platanifolia* H. B. K.; *Bombacac.;* Kol.
Macoucou: *Chrysophyllum macoucou* Aubl.; *Sapotac.;* fr. Gu.
Macoupi: *Attalea sp.; Palmac.;* fr. Gu.
Macreleaf: *? Conostegia puberula* Cogn.; *Melastomatac.;* Pan.
Macrocarpa: *Cupressus macrocarpa* Hartw.; *Conifer.;* N.-Seel.
Macucú: *Ilex macoucoua* Pers.; *Aquifoliac.;* Bras.
„ do Rio Negro: *Licania sp.; Rosac.;* Bras.
Macuelizo: *Tabebuia pentaphylla* Hemsl.; *Bignoniac.;* Hond.
Macuil: *Tabebuia pentaphylla* Hemsl.; *Bignoniac.;* Mex.
Macuiliz: *Tabebuia pentaphylla* Hemsl.; *Bignoniac.;* Salv.
Maculigua: *Tabebuia pentaphylla* Hemsl.; *Bignoniac.;* Salv.
Maculiz: *Tabebuia pentaphylla* Hemsl.; *Bignoniac.;* Mex.
„ prieto: *Tabebuia pentaphylla* Hemsl.; *Bignoniac.;* Mex.
Macurige: *Matayba apetala* Radlk.; *Sapotac.;* Cuba.
Maçutahyba: *Zollernia ilicifolia* Vogel; *Caesalpiniac.;* Bras.
Madaburi: *Clusia cuneata* Bth.; *Guttifer.;* br. Gu.
„ : *Ternstroemia aff. delicatula* Choisy; *Theac.;* br. Gu.
Madachi: *Khaya senegalensis* A. Juss.; *Meliac.;* n. Nig.
Madachin dutsi: *Ekebergia senegalensis* A. Juss.; *Meliac.;* n. Nig.
Madagari vembu: *Chukrasia tabularis* Adr. Juss.; *Meliac.;* O.-Ind. (Madras)
Madake: *Phyllostachys bambusoides* S. et Z.; *Gramin.;* Jap.
Madale: *Alstonia scholaris* R. Brown; *Apocynac.;* w. O.-Ind. (Deccan)
Madaló: *Pometia pinnata* Forst.; *Sapindac.;* Phil. (Cagayan)
Madang boenghar: *Schima crenata* Korth.; *Theac.;* Sum.
Madar: *Erythrina suberosa* Roxb.; *Papilionac.;* n. O.-Ind. (Un.-Prov.)
Madati: *Terminalia tomentosa* Wight et Arn.; *Combretac.;* w. O.-Ind. (Deccan)
Madayan samprani: *Hardwickia pinnata* Roxb.; *Caesalpiniac.;* sw. O.-Ind.
Maddad: *Cedrus libani* Barrel. *var. atlantica* Manetti; *Conifer.;* Marok., Algier.
Maddi: *Terminalia tomentosa* Wight et Arn.; *Combretac.;* O.-Ind. (Madras)
Madeira: *Swietenia spp.; Meliac.;* Fl., Bah.

Mädelbaum (d): *Pinus silvestris* L.; *Conifer.;* Eu., As., Sib.
Madera manchada: *Apuleia praecox* Mart.; *Caesalpiniac.;* Arg.
„ negra: *Gliricidia sepium* Steud.; *Papilionac.;* Hond., C.-R.
„ „ : *Tabebuia rufescens* J. R. Johnston; *Bignoniac.;* Ek.
„ pino: *Pinus palustris* Mill.; *Conifer.;* sö. N.-Am.
„ purpurea: *Peltogyne spp.; Caesalpiniac.;* tr. Am.
„ de boj (s): *Buxus sempervirens* L.; *Buxac.;* Eu.
„ „ Nogal: *Juglans spp.; Juglandac.;* tr. Am.
„ „ palisandro: *Dalbergia nigra* Allem.; *Papilionac.;* Bras.
„ do buxo (p): *Buxus sempervirens* L.; *Buxac.;* s. Eu., w. As., n. Af.
Madi: *Bursera serrata* Colebr.; *Burserac.;* Burma.
Madiabug: *Aglaia luzoniensis* Merr. et Rolfe; *Meliac.;* Phil. (Tayabas)
Madja: *Aegle marmelos* Correa, *Feronia elephantum* Correa; *Rutac.;* Java, Sunda.
Madja: *Aromadendron elegans* Blume; *Magnoliac.;* Nied.-Ind.
Madkom: *Bassia latifolia* Roxb.; *Sapotac.;* w. Beng. (Chota Nagpur)
Madmandi: *Holarrhena antidysenterica* Wall.; *Apocynac.;* Assam (Garo)
Mado chulia: *Vitex peduncularis* Wall.; *Verbenac.;* w. Beng. (Orissa)
Madobia: *Andira inermis* H. B. K., *Pterocarpus erinaceus* Poir.; *Papilionac.;* n. Nig.
Madobiar rafi: *Albizzia brownei* Oliv.; *Mimosac.;* n. Nig.
Madras thorn: *Pithecolobium dulce* Bth.; *Mimosac.;* Malay.
Madre brava: *Erythrina sp.; Papilionac.;* Mex.
„ cacao: *Gliricidia sepium* Steud.; *Papilionac.;* br. Hond., Phil.
„ de cacáo: *Erythrina sp.; Papilionac.;* Trin.
Madreselva: *Lonicera etrusca* Santi, *L. periclymenum* L.; *Caprifoliac.;* Sp.
Madriado: *Gliricidia sepium* Steud.; *Papilionac.;* Hond., Nic.
Madrial: *Gliricidia sepium* Steud.; *Papilionac.;* Hond.
Madroña: *Calycophyllum candidissimum* DC.; *Rubiac.;* Pan.
„ : *Arbutus menziesi* Purch; *Ericac.;* USA. (Kalif.)
Madrone tree: *Arbutus menziesi* Purch; *Ericac.;* w. N.-Am.
Madroño: *Arbutus unedo* L.; *Ericac.;* Sp.
„ : *Calycophyllum candidissimum* DC.; *Rubiac.;* C.-Am.
„ : *Weinmannia sp.; Cunoniac.;* C.-R.
Madu: *Dialium guineense* Willd.; *Caesalpiniac.;* Togo.
„ : *Randia dumetorum* Lamk.; *Rubiac.;* O.-Ind. (Madras)
Maduga: *Erythrina suberosa* Roxb.; *Papilionac.;* O.-Ind. (Madras)
Mafoureire: *Trichilia emetica* Vahl; *Meliac.;* O.-Af.
Mafumeira: *Ceiba pentandra* Gaertn.; *Bombacac.;* port. Af.
„ encarnada: *Bombax buonopozense* Beauv.; *Bombacac.;* port. Af.
Mag: *Anona sp.; Anonac.;* Mex.
Maga: *Thespesia grandiflora* DC.; *Malvac.;* P.-R.
Magabolíng: *Shorea negrosensis* Foxw.; *Dipterocarpac.;* Phil. (Agusan)
Magabuluán: *Neonauclea sp.; Rubiac.;* Phil. (Cotabato)
Magabúyo: *Celtis sp., C. luzonica* Warb.; *Ulmac.;* Phil.
Magadam: *Mimusops elengi* Roxb.; *Sapotac.;* O.-Ind. (Madras)
Magaklúd: *Lagerstroemia piriformis* Koehne; *Lythrac.;* Phil. (Cotabato)
Magakombó: *Eugenia similis* Merr.; *Myrtac.;* Phil. (Davao)
Magálapaláli: *Dillenia sp.; Dilleniac.;* Phil. (n. & C.-Luzon)
Magalayáu: *Pahudia rhomboidea* Prain; *Caesalpiniac.;* Phil. (Cagayan)
Magale: *Hopea sp.; Dipterocarpac.;* Burma.
Magaleto: *Xylopia frutescens* Aubl.; *Anonac.;* Guat.
Magalibás: *Koordersiodendron pinnatum* Merr.; *Anacardiac.;* Phil. (Zamboanga)
Magalolo: *Lumnitzera littorea* Voigt; *Combretac.;* Phil. (Palawan)
Maganhop-sa-bukid: *Albizzia lebbekoides* Bth.; *Mimosac.;* Phil. (Sibuyan)
Magapisa: *Cyclostemon sp.; Euphorbiac.;* Phil. (Cotabato)
Magar: *Thespesia grandiflora* DC.; *Malvac.;* P.-R.
Magariya: *Ziziphus jujuba* Lam.; *Rhamnac.;* n. Nig.
Magariyar kura: *Ziziphus mucronata* Willd.; *Rhamnac.;* n. Nig.
Magárua: *Acacia arabica* Willd.; *Mimosac.;* P.-O.
Magas: *Thespesia grandiflora* DC.; *Malvac.;* P.-R.
Magasinóyo: *Shorea palosapis* Merr.; *Dipterocarpac.;* Phil. (Agusan)
Magatalulóng: *Lagerstroemia piriformis* Koehne; *Lythrac.;* Phil. (Ilocos Sur)
Magatáru: *Cyclostemon sp.; Euphorbiac.;* Phil. (Misamis)
Magatlí: *Dillenia sp.; Dilleniac.;* Phil. (Cagayan)
Magáyau: *Heritiera littoralis* Dry.; *Sterculiac.;* Phil. (Cagayan)

Magbuábang: *Phoebe sterculioides* Merr.; *Laurac.*; Phil. (Surigao)
Magdau: *Wallaceodendron celebicum* Koord.; *Mimosac.*; Phil. (Negrosinsel)
Maggiociondolo (i):   *Cytisus laburnum* L.; *Papilionac.*; Eu.
Maghabúyo: *Celtis sp.*; *Ulmac.*; Phil. (Masbate, Ticao)
Maghúbo: *Terminalia comintana* Merr.; *Combretac.*; Phil. (Rizal)
Magidam: *Mimusops elengi* Roxb.; *Sapotac.*; O.-Ind. (Madras)
Magilík: *Litsea sp.*; *Laurac.*; Phil. (Bataan)
Magitlumbói: *Aglaia turczaninowi* C. DC.; *Meliac.*; Phil. (Guimaras)
Magkonó: *Xanthostemon verdugonianus* Naves; *Myrtac.*; Phil.
Maglalopoi: *Terminalia comintana* Merr.; *Combretac.*; Phil. (Pangasinam)
Maglangká: *Zizyphus sp.*; *Rhamnac.*; Phil. (Palawan)
Magnolia: *Magnolia grandiflora* L.; *Magnoliac.*; N.-Am.
   „   :   *Magnolia spp.*; *Magnoliac.*; O.-Ind., Jap.
   „   :   *Talauma sambuensis* Pittier, *T. sp.*; *Magnoliac.*; Pan., Cuba.
   „   :   *? Talauma elegans* Miq.; *Magnoliac.*; Nied.-Ind.
   „   , large flowered evergreen: *Magnolia grandiflora* L.; *Magnoliac.*; USA.
   „   , Mountain-:   *Magnolia acuminata* L.; *Magnoliac.*; N.-Am.
   „   , pointed-leaved: *Magnolia acuminata* L.; *Magnoliac.*; sö. Kan.
   „   , sweet: *Magnolia glauca* L.; *Magnoliac.*; USA.
   „   do brejo: *Talauma sp.*; *Magnoliac.*; Bras.
   „   „   matto: *Talauma sp.*; *Magnoliac.*; Bras.
Magnolie, immergrüne, großblütige: *Magnolia grandiflora* L.; *Magnoliac.*; USA.
   „   , spitzblättrige: *Magnolia acuminata* L.; *Magnoliac.*; N.-Am.
   „   , weissrückige: *Magnolia hypoleuca* S. et Z.; *Magnoliac.*; Jap., Ch.
Magnolier (f):   *Magnolia grandiflora* L.; *Magnoliac.*; USA.
   „   bleu: *Magnolia glauca* L.; *Magnoliac.*; USA.
   „   à feuilles acuminées: *Magnolia acuminata* L.; *Magnoliac.*; sö. Kan.
   „   des marais: *Magnolia glauca* L.; *Magnoliac.*; USA.
Mago: *Hernandia sonora* L.; *Hernandiac.*; P -R.
Magona: *Lannea welwitschi* Engl.; *Anacardiac.*; Kam.
Magongo: *Irvingia gabonensis* Baill.; *Simarubac.*; Gab.
Magsantól: *Sandoricum vidali* Merr.; *Meliac.*; Phil. (Zambales)
Magsinólo: *Shorea eximia* Scheff.; *Dipterocarpac.*; Phil. (Misamis)
Magsinóyo: *Aglaia luzoniensis* Merr. et Rolfe; *Meliac.*; Phil. (Surigao)
Magtabígi: *Aglaia diffusa* Merr.; *Meliac.*; Phil. (Leyte)
Magtalísai: *Terminalia spp.*; *Combretac.*; Phil.
   „   :   *Petersianthus quadrialatus* Merr.; *Lecythidac.*; Phil. (Masbate)
Magtalísi: *Terminalia nitens* Presl; *Combretac.*; Phil.
Magtongód: *Ceriops tagal* C. B. Rob.; *Rhizophorac.*; Phil. (Mindoro)
Magtongóg: *Bruguiera cylindrica* Blume; *Rhizophorac.*; Phil. (Masbate)
Magua: *Carallia lucida* Roxb.; *Rhizophorac.*; Phil. (Mindoro)
Magugáhum: *Lagerstroemia piriformis* Koehne; *Lythrac.*; Phil. (Agusan)
Maguliók: *Buchanania sp.*; *Anacardiac.*; Phil. (Tayabas)
Magulipák: *Sterculia sp.*; *Sterculiac.*; Phil. (Masbate, Ticao)
Magundo: *Chlorophora excelsa* Bth. et Hook.; *Morac.*; port. O.-Af.
Maguripák: *Sterculia sp.*; *Sterculiac.*; Phil. (Masbate, Ticao)
Maguyabud: *Koordersiodendron pinnatum* Merr.; *Anacardiac.*; Phil. (Samar)
Magwi-napa: *Pterospermum acerifolium* Willd.; *Sterculiac.*; Burma.
Magyi: *Tamarindus indica* L.; *Caesalpiniac.*; tr. Af.
Maha-limbo: *Melia azedarach* L.; *Meliac.*; C.-O.-Ind.
   „   limbu: *Toona ciliata* M. Roem.; *Meliac.*; w. Beng. (Orissa)
   „   nim: *Toona ciliata* M. Roem.; *Meliac.*; w. O.-Ind. (Bombay)
   „   nimb: *Melia azedarach* L.; *Meliac.*; O.-Ind.
Mahagoñi (d):   *Swietenia spp.*; *Meliac.*; tr. Am.
   „   :   *Swietenia mahagoni* Jacq.; *Meliac.*; W.-Ind.
   „   :   *Swietenia macrophylla* King; *Meliac.*; C.-Am., nö. Peru.
   „   :   *Swietenia humilis* Zucc.; *Meliac.*; sw. Mex., Guat., Salv.
   „   :   *Swietenia cirrhata* Blake; *Meliac.*; nw. Mex.
   „   :   *Swietenia candollei* Pittier; *Meliac.*; Ven.
   „   :   *Khaya spp., Entandrophragma spp.*; *Meliac.*; W.-Af.
   „   :   *Khaya senegalensis* A. Juss.; *Meliac.*; D.-O.-Af.
   „   :   *Tarrietia cochinchinensis* Pierre; *Sterculiac.*; Kamb.
   „   :   *Melanorrhoea laccifera* Pierre; *Anacardiac.*; Ind.-Ch.
   „   , afrikanisch (d):   *Khaya ivorensis* Chev., *K. spp.*; *Meliac.*; W.-Af.

Mahagoni, Annam-: *Aglaia gigantea* Pellegr.; *Meliac.;* Annam.
„ , Antillen-: *Swietenia mahagoni* Jacq.; *Meliac.;* W.-Ind.
„ , australisch: *Dysoxylum fraseranum* Bth.; *Meliac.;* ö. Au. (N.-S.-W.)
„ , Benin-: *Khaya sp., Khaya grandifolia* C. DC.; *Meliac.;* Nig.
„ , Berg-: *Cercocarpus parvifolius* Nutt., *C. ledifolius* Nutt.; *Rosac.;* w. N.-Am.
„ , Cap Lopez-: *Khaya sp.; Meliac.;* Gab.
„ , echtes (d): *Swietenia mahagoni* Jacq.; *Meliac.;* W.-Ind.
„ , falsches: *Eucalyptus botryoides* Smith; *Myrtac.;* Au.
„ , Gabun-: *Khaya ivorensis* Chev.; *Meliac.;* Gab.
„ , Gambia-: *Khaya senegalensis* A. Juss., *? K. ivorensis* Chev.; *Meliac.;* Togo.
„ , geflecktes: *Entandrophragma candollei* Harms; *Meliac.;* Kongo.
„ , gelbes: *Terminalia superba* Engl. et Diels; *Combretac.;* W.-Af.
„ , Grand Bassam-: *Khaya ivorensis* Chev.; *Meliac.;* Elf.
„ , Honduras-: *Swietenia macrophylla* King; *Meliac.;* Hond.
„ , indisches: *Cedrela odorata* L.; *Meliac.;* C.-Am.
„ , Insel-: *Swietenia mahagoni* Jacq.; *Meliac.;* Cuba, St.-D., Haiti, Jam.
„ , Kamerun-: *Khaya sp., Entandrophragma rederi* Harms; *Meliac.;* Kam.
„ , kapensisches: *Ptaeroxylon obliquum* Radlk.; *Meliac.;* S.-Af.
„ , Kilimandscharo-: *Lovoa spp.; Meliac.;* D.-O.-Af.
„ , Kongo-: *Mimusops djave* Engl.; *Sapotac.;* tr. W.-Af.
„ , Konta-: *Afzelia bracteata* Vog.; *Caesalpiniac.;* Af.
„ , Kuba-, echtes: *Swietenia mahagoni* Jacq.; *Meliac.;* Cuba.
„ , Lagos-: *Khaya spp.; Meliac.;* W.-Af.
„ , Laguna de Terminos-: *Swietenia macrophylla* King; *Meliac.;* s. Mex., br. Hond.
„ , Locobo-: *Entandrophragma candollei* Harms; *Meliac.;* Elf.
„ , Madagaskar-: *Khaya madagascariensis* Jum. et Perrier; *Meliac.;* Mad.
„ , Makua-: *Entandrophragma macrophyllum* Chev.; *Meliac.;* Elf.
„ , Mangona-: *Khaya anthotheca* C. DC.; *Meliac.;* Kam.
„ , Mbolo-: *Entandrophragma sp.; Meliac.;* Kam.
„ , Mbossé-: *Guarea thompsoni* Sprague et Hutch.; *Meliac.;* tr. W.-Af.
„ , Mouondue-: *Entandrophragma sp.; Meliac.;* W.-Af.
„ , Muandara-: *Ptaeroxylon obliquum* Radlk.; *Meliac.;* S.-Af.
„ , Natal-: *Kigellaria drageana* Turcz.; *Flacourtiac.;* S.-Af.
„ , Nicaragua-: *Swietenia sp.; Meliac.;* Nic.
„ , Nigeria-: *Khaya sp.; Meliac.;* W.-Af.
„ , Nsamgilla-: *Khaya ivorensis* Chev.; *Meliac.;* W.-Af.
„ , Nussangue-: *Khaya anthotheca* C. DC.; *Meliac.;* W.-Af.
„ , Obega-: *Canarium schweinfurthi* Engl.; *Burserac.;* W.-Af. (? Gab.)
„ , Oganwe-: *Khaya senegalensis* A. Juss., *K. grandifolia* C. DC.; *Meliac.;* W.-Af.
„ , Oganwo-: *Khaya ivorensis* Chev.; *Meliac.;* W.-Af.
„ , Ogipogo-: *Entandrophragma utile* Sprague; *Meliac.;* W.-Af.
„ , Ogwangu-: *Khaya grandifolia* C. DC.; *Meliac.;* W.-Af.
„ , Okoume-: *Aucoumea klaineana* Pierre; *Burserac.;* Gab.
„ , Opapao-: *Afzelia africana* Smith; *Caesalpiniac.;* W.-Af.
„ , ostindisches: *Toona serrata* Roemer; *Meliac.;* O.-Ind.
„ , Panama-: *Anacardium rhinocarpus* DC.; *Anacardiac.;* C.-Am.
„ , Penkwa-: *Entandrophragma cylindricum* Sprague; *Meliac.;* W.-Af.
„ , Philippinen-: *Tarrietia javanica* Blume; *Sterculiac.;* Phil.
„ , Port Gentil- (d): *Khaya sp.; Meliac.;* W.-Af.
„ , Punkwa-: *Entandrophragma cylindricum* Sprague; *Meliac.;* W.-Af.
„ , Quibaba da Queta-: *Entandrophragma angolensis* C. DC.; *Meliac.;* Ang.
„ , „ de Mussangue-: *Khaya anthotheca* C. DC.; *Meliac.;* Ang.
„ , Rolo-: *Dysoxylum fraseranum* Bth.; *Meliac.;* Au.
„ , rotes: *Eucalyptus gomphocephala* A. DC.; *Myrtac.;* N.-S.-W.
„ , „ : *Eucalyptus marginata* Smith; *Myrtac.;* w. Au.
„ , weisses: *Entandrophragma macrophylla* Chev.; *Meliac.;* W.-Af.
„ , „ : *Eucalyptus acmenoides* Schau.; *Myrtac.;* Au.
„ , westindisches: *Swietenia mahagoni* Jacq.; *Meliac.;* Cuba, St.-D., Haiti, Jam.
„ , Zitronen-: *Alstonia scholaris* R. Br.; *Apocynac.;* O.-Ind.
Mahagua: *Ficus sp.; Morac.;* Ek.
„ : *Hibiscus elatus* Swartz; *Malvac.;* P.-R.
„ de Indio: *? Eschweilera sp.; Lecythidac.;* Pan.
„ del mar: *Hibiscus tiliaceus* L.; *Malvac.;* Trin.
Mahahlegabyu: *Bauhinia acuminata* L.; *Caesalpiniac.;* Burma.

Mahajo: *Casearia silvestris* Sw.; *Flacourtiac.;* Kol.
Mahajua: *Bombax sp.; Bombacac.;* Kol.
Mahalansot: *Aglaia dulcis* Teijsm. et Binn.; *Meliac.;* Nied.-Ind.
Mahalim: *Toona ciliata* M. Roem.; *Meliac.;* C.-O.-Ind.
Mahalum: *Toona ciliata* M. Roem.; *Meliac.;* C.-O.-Ind.
Mahan besi: *Hopea beccariana* Burck; *Dipterocarpac.;* br. Borneo (Sarawak)
Mahang puteh: *Mallotus paniculata* Muell. Arg.; *Euphorbiac.;* Malay.
  „   serindit: *Macaranga triloba* Muell. Arg.; *Euphorbiac.;* Malay.
Mahanim: *Ailanthus excelsa* Roxb.; *Simarubac.;* w. Beng. (Orissa)
  „   : *Toona ciliata* M. Roem.; *Meliac.;* n. O.-Ind. (United Prov.)
Maharuk: *Ailanthus excelsa* Roxb.; *Simarubac.;* w. O.-Ind. (Bombay)
Mahass: *Quararibea funebris* Standl.; *Bombacac.;* br. Hond.
Mahau: *Heliocarpus donnell-smithi* Rose; *Tiliac.;* Guat., br. Hond.
Mahaujo: *Muntingia calabura* L.; *Tiliac.;* Ven.
Mahault chardon: *Apeiba sp.; Tiliac.;* Trin.
  „   de londres: *Thespesia populnea* Correa; *Malvac.;* Trin.
  „   du bord de mer: *Hibiscus tiliaceus* L.; *Malvac.;* Trin.
Mahaut, Sea side-: *Hibiscus tiliaceus* L.; *Malvac.;* Trin.
Mahban: *Virola merendonis* Pittier; *Myristicac.;* br. Hond.
  „   (u): *Myristica aff. panamensis* Hemsl.; *Myristicac.;* br. Hond.
Mahila: *Mimusops elengi* Roxb.; *Sapotac.;* O.-Ind. (Madras)
Mahkaw: *Zizyphus jujuba* Lamk.; *Rhamnac.;* Burma.
Mahlwa: *Dolichandrone stipulata* Bth.; *Bignoniac.;* Burma.
Maho: *Sterculia sp.; Sterculiac.;* br. Gu.
  „   : *Couratari guianensis* Aubl., *C. sp.; Lecythidac.;* Sur.
  „   : *Lecythis spp.; Lecythidac.;* fr. Gu.
  „   : *Hibiscus tiliaceus* L.; *Malvac.;* Sur.
  „   couratari: *Couratari guianensis* Aubl.; *Lecythidac.;* fr. Gu.
  „   fer: *Couratari guianensis* Aubl.; *Lecythidac.;* fr. Gu.
  „   jaune: *Couratari guianensis* Aubl., *C. sp.; Lecythidac.;* Sur.
? „   „   : *Eschweilera longipes* Mart., *E. subglandulosa* Miers; *Lecythidac.;* Sur.
  „   sigare: *Couratari guianensis* Aubl., *C. sp.; Lecythidac.;* Sur.
Mahobo-hobo: *Uapaca kirkiana* Muell. Arg.; *Euphorbiac.;* s. C.-Af.
Mahoe: *Sterculia caribaea* R. Br.; *Sterculiac.;* Trin.
  „   : *Hibiscus elatus* Swartz; *Malvac.;* Jam.
  „   : *Melicytus ramiflorus* Forst.; *Violac.;* N.-Seel.
  „   , black: *Rollinia multiflora* Splitg.; *Anonac.;* br. Gu.
  „   , blue: *Hibiscus elatus* Swartz; *Malvac.;* Jam.
  „   , Mountain-: *Hibiscus elatus* Swartz; *Malvac.;* Jam.
  „   , Seaside-: *Hibiscus tiliaceus* L.; *Malvac.;* Trin.
Mahogany (e): *Swietenia spp.; Meliac.;* tr. Am.
  „   : *Swietenia mahagoni* Jacq.; *Meliac.;* W.-Ind.
  „   : *Swietenia macrophylla* King; *Meliac.;* C.-Am., nö. Peru.
  „   : *Swietenia humilis* Zucc.; *Meliac.;* sw. Mex., Guat., Salv.
  „   : *Swietenia cirrhata* Blake; *Meliac.;* nw. Mex.
  „   : *Swietenia candollei* Pittier; *Meliac.;* Ven.
  „   : *Khaya senegalensis* A. Juss.; *Meliac.;* n. Nig.
  „   : *Khaya nyasica* Stapf; *Meliac.;* n. Rhod.
  „   : *Saccoglottis gabonensis* Urban; *Humiriac.;* Lib.
  „   : *Afzelia quanzensis* Welw.; *Caesalpiniac.;* O.-Af.
  „   : *Eucalyptus hemilampra* F. v. M.; *Myrtac.;* N.-S.-W.
  „   : *Dysoxylum fraseranum* Bth.; *Meliac.;* ö. Au. (Queensl.)
  „   , african (e): *Khaya ivorensis* Chev.; *Meliac.;* W.-Af.
  „   , „   : *Khaya senegalensis* A. Juss.; *Meliac.;* n. Nig.
  „   , „   : *Khaya caudata* Stapf; *Meliac.;* br. W.-Af.
  „   , „   : *Khaya grandifolia* C. DC.; *Meliac.;* Go.
  „   , „   : *Afzelia africana* Smith; *Caesalpiniac.;* W.-Af.
  „   , „   : *Tarrietia utilis* Sprague; *Sterculiac.;* Go.
  „   , „   : *Xylocarpus sp.; Meliac.;* O.-Af.
  „   , „   , figured: ? *Entandrophragma sp.; Meliac.;* Nig.
  „   , australian: *Dysoxylum fraseranum* Bth.; *Meliac.;* ö. Au. (N.-S.-W.)
  „   , Bastard-: *Ochrocarpus africanus* Oliv.; *Guttifer.;* Lib.
  „   , „   : *Eucalyptus marginata* Sm.; *Myrtac.;* w. Au.
  „   , „   : *Eucalyptus botryoides* Smith; *Myrtac..;* ö. Au.

Mahogany, Bataan-: *Shorea polysperma* Merr.; *Dipterocarpac.;* Phil.
„ , Bay-: *Swietenia macrophylla* King; *Meliac.;* W.-Ind.
„ , Benin: *Khaya grandifolia* Stapf; *Meliac.;* s. Nig. (Benin)
„ , „ : *Guarea thompsoni* Sprague et Hutch.; *Meliac.;* s. Nig. (Benin)
„ , black: *Lovoa klaineana* Pierre; *Meliac.;* W.-Af.
„ , „ : *Endiandra palmerstoni* C. T. White; *Laurac.;* Au.
„ , Borneo-: *Shorea spp.; Dipterocarpac.;* n. Borneo.
„ , Bossé: *Guarea cedrata* Pellegr.; *Meliac.;* Elf.
„ , brazilian: *Carapa procera* DC. *C. surinamensis* Miq.; *Meliac.;* Sur.
„ , British Guiana-: *Carapa procera* DC., *C. surinamensis* Miq.; *Meliac.;* Sur.
„ , brown: *Entandrophragma septentrionale* Chev.; *Meliac.;* W.-Af.
„ , „ : *Lovoa swynnertoni* E. G. Baker; *Meliac.;* s. C.-Af.
„ , Burma (u): *Pentace burmanica* Kurz; *Tiliac.;* Burma.
„ , Cambodge-: *Tarrietia cochinchinensis* Pierre; *Sterculiac.;* Ind.-Ch.
„ , Cape-: *Trichilia emetica* Vahl; *Meliac.;* S.-Af.
„ , Cedar-: *Guarea thompsoni* Sprague et Hutch.; *Meliac.;* Nig.
„ , „ , hard: *Pseudocedrela kotschyi* Harms; *Meliac.;* W.-Af.
„ , Cherry-: *Khaya sp.; Meliac.;* Nig.
„ , close-grained: *Guarea thompsoni* Sprague et Hutch.; *Meliac.;* s. Nig. (Benin)
„ , Coast-: *Khaya ivorensis* Chev.; *Meliac.;* W.-Af.
„ , Colombian: *Cariniana pyriformis* Miers; *Lecythidac.;* Kol.
„ , Curlet-: *Swietenia mahagoni* Jacq.; *Meliac.;* Jam.
„ , Demerara-: *Carapa procera* DC., *C. surinamensis* Miq.; *Meliac.;* Sur.
„ , Derham- (u): *Wallaceodendron celebicum* Koord.; *Mimosac.;* Phil.
„ , Dry zone-: *Khaya senegalensis* A. Juss.; *Meliac.;* Go.
„ , East Indian: *Chukrasia tabularis* A. Juss.; *Meliac.;* tr. As.
„ , „ „ : *Soymida febrifuga* A. Juss.; *Meliac.;* O.-Ind., Cey.
„ , Eucalypt-, red: *Eucalyptus resinifera* Smith; *Myrtac.;* ö. Au.
„ , false: *Eucalyptus botryoides* Smith; *Myrtac.;* Au.
„ , Fiji-: *Calophyllum burmanni* Wight, *C. spectabile* Willd.; *Guttifer.;* Fid.
„ , Forest-: *Eucalyptus resinifera* Smith; *Myrtac.;* Au.
„ , Gaboon (e): *Aucoumea klaineana* Pierre; *Burserac.;* Gab.
„ , Gambia-: *Khaya senegalensis* A. Juss.; *Meliac.;* W.-Af. (Gambia)
„ , Gippsland-: *Eucalyptus botryoides* Smith; *Myrtac.;* ö. Au.
„ , Grand Bassam-: *Khaya ivorensis* Chev.; *Meliac.;* Kam.
„ , Honduras-: *Swietenia macrophylla* King; *Meliac.;* C.-Am.
„ , Horseflesh-: *Lysiloma sabicu* Bth.; *Mimosac.;* Bah.
„ , indian: *Toona ciliata* M. Roem., *T. serrata* M. Roem.; *Meliac.;* O.-Ind.
„ , „ , white: *Canarium euphyllum* Kurz; *Burserac.;* O.-Ind.
„ , Iron-: *Erythrophloeum sp.; Caesalpiniac.;* Au.
„ , Juana Costa- (u): *Enterolobium cyclocarpum* Griseb.; *Mimosac.;* tr. Am.
„ , Lagos-: *Khaya ivorensis* Chev.; *Meliac.;* Nig.
„ , large-leaved: *Swietenia macrophylla* King; *Meliac.;* C.-Am.
„ , Red lauan-: *Shorea negrosensis* Foxw.; *Dipterocarpac.;* Phil.
„ , Mountain-: *Cercocarpus ledifolius* Nutt., *C. parvifolius* Nutt.; *Rosac.;* w. N.-Am.
„ , „ „ : *Betula lutea* Michx., *B. lenta* L.; *Betulac.;* w. N.-Am.
„ , Red Narra-: *Pterocarpus indicus* Willd.; *Papilionac.;* Phil.
„ , Natal-: *Kigellaria drageana* Turcz.; *Flacourtiac.;* S.-Af.
„ , New Zealand-: *Dysoxylum spectabile* Hook. f.; *Meliac.;* N.-Seel.
„ , Nyasaland-: *Khaya nyasica* Stapf; *Meliac.;* s. C.-Af.
„ , peruvian: *Swietenia macrophylla* King; *Meliac.;* Peru.
„ , Philippine- (u): *Shorea negrosensis* Foxw.; *Dipterocarpac.;* Phil.
„ , „ : *Shorea polysperma* Merr.; *Dipterocarpac.;* Phil.
„ , Picus- (u): *Aucoumea klaineana* Pierre; *Burserac.;* W.-Af.
„ , pink: *Guarea cedrata* Pellegr.; *Meliac.;* Elf.
„ , „ : *Gossweilerodendron balsamiferum* Harms; *Flacourtiac.;* Nig.
„ , Pod- (e): *Afzelia quanzensis* Welw.; *Caesalpiniac.;* O.-Af., C.-Af.
„ , red (e): *Khaya ivorensis* Chev.; *Meliac.;* Kam.
„ , „ : *Eucalyptus resinifera* Smith, *E. spp.; Myrtac.;* Au.
„ , rhodesian: *Khaya nyasica* Stapf; *Meliac.;* Rhod.
„ , „ : *Copaifera coleosperma* Bth.; *Caesalpiniac.;* s. Rhod.
„ , „ : *Afzelia quanzensis* Welw.; *Caesalpiniac.;* C.-Af., O.-Af.
„ , Roba- (u): *Carapa guianensis* Aubl.; *Meliac.;* Gu.
„ , Rose-: *Dysoxylum fraseranum* Bth.; *Meliac.;* ö. Au.

Mahogany, Sapeli-: *Entandrophragma septentrionale* Chev., *E. spp.; Meliac.;* Nig.
„   , Satin-: *Guarea africana* Welw.; *Meliac.;* Lib.
„   , scented: *Entandrophragma utile* Sprague; *Meliac.;* Go., Nig., Kam.
„   ,   „   : *Guarea thompsoni* Sprague et Hutch.; *Meliac.;* s. Nig. (Benin)
„   ,   „   : *Guarea glomerulata* Harms; *Meliac.;* Kam.
„   , south american: *Mimusops spp.; Sapotac.;* S.-Am.
„   , South Pacific- (u): *Shorea negrosensis* Foxw.; *Dipterocarpac.;* Phil.
„   , spanish: *Swietenia mahagoni* Jacq.; *Meliac.;* Trin., Tob.
„   , stinking: *Entandrophragma candollei* Harms; *Meliac.;* Nig.
„   , Straits-: *Melanorrhoea mangayi* Hook. f.; *Anacardiac.;* Malay, Sunda.
„   , Swamp-: *Eucalyptus robusta* Sm., *E. botryoides* Smith; *Myrtac.;* Au.
„   , Tenasserim-: *Pterocarpus macrocarpa* Kurz; *Papilionac.;* s. Burma.
„   , true: *Swietenia mahagoni* Jacq.; *Meliac.;* W.-Ind.
„   , unscented: *Entandrophragma candollei* Harms; *Meliac.;* Go., Nig.
„   , west african: *Khaya ivorensis* Chev., *K. senegalensis* A. Juss.; *Meliac.;* W.-Af.
„   , west indian: *Swietenia mahagoni* Jacq.; *Meliac.;* Trin., Tob.   [(Gambia)
„   , white: *Tabebuia donnell-smithi* Rose; *Bignoniac.;* C.-Am.
„   ,   „   : *Uochysia hondurensis* Sprague; *Uochysiac.;* br. Hond.
„   ,   „   (e): *Aucoumea klaineana* Pierre; *Burserac.;* Gab.
„   ,   „   : *Khaya anthotheca* C. DC.; *Meliac.;* Go.
„   ,   „   : *Antrocaryon klaineanum* Pierre; *Anacardiac.;* Kam.
„   ,   „   : *Canarium sp.; Burserac.;* And.
„   ,   „   : *Eucalyptus spp.; Myrtac.;* Au.
„   , wild: *Mosquitoxylum jamaicense* Krug et Urb.; *Anacardiac.;* br. Hond.
„   , yellow: *Plathymenia reticulata* Bth.; *Mimosac.;* Bras.
„   bean tree: *Afzelia quanzensis* Welw.; *Caesalpiniac.;* C.-Af., O.-Af.
Mahok: *Swietenia spp.; Meliac.;* tr. Am.
Mahoka: *Terminalia chebula* Retz.; *Combretac.;* C.-O.-Ind. (Gondwana)
Mahonie: *Swietenia spp.; Meliac.;* tr. Am.
„   , Corantijnsch-: *Hedwigia balsamifera* Swartz, *Protium spp.; Burserac.;* Sur.
„   , Grootblad-: *Swietenia macrophylla* King; *Meliac.;* C.-Am.
„   , surinaamsch: *Carapa procera* DC., *C. surinamensis* Miq.; *Meliac.;* Sur.
„   hout: *Swietenia spp.; Meliac.;* Sur.
„   „   , echt: *Swietenia spp.; Meliac.;* Sur.
Mahonijboom (h): *Swietenia mahagoni* Jacq., *S. spp.; Meliac.;* tr. Am.
Mahot blanc: *Hibiscus tiliaceus* L.; *Malvac.;* fr. Gu.
„   chardon: *Apeiba aspera* Aubl.; *Tiliac.;* Gu.
„   cochin: *Sterculia pruriens* Schum.; *Sterculiac.;* fr. Gu.
„   couatari: *Couratari guianensis* Aubl.; *Lecythidac.;* fr. Gu.
„   fleur: *Kleinhovia hospita* L.; *Sterculiac.;* Réunion.
„   forestier: *Hibiscus tiliaceus* L.; *Malvac.;* fr. Gu.
„   franc: *Hibiscus tiliaceus* L.; *Malvac.;* Guad.
„   grand: *Hibiscus tiliaceus* L.; *Malvac.;* N.-Kal.
„   mahotière: *Hibiscus tiliaceus* L.; *Malvac.;* Guad.
„   noir: *Guetteria ouregou* Dunal; *Anonac.;* Guad.
?   „   rouge: *Dimorphandra excelsa* Baill.; *Caesalpiniac.;* Sur.
„   rouge à petites feuilles: *Dombeya punctata* Cav.; *Sterculiac.;* Réunion.
„   à feuilles rondes: *Dombeya umbellata* Cav.; *Sterculiac.;* Réunion.
„   de Londres: *Thespesia populnea* Correa; *Malvac.;* Trin.
„   du bord de la mer: *Hibiscus tiliaceus* L.; *Malvac.;* Trin.
Mahou couratarie: *Couratari guianensis* Aubl., *C. sp.; Lecythidac.;* Sur.
Mahr-chu: *Mimusops clitandrifolium* Chev.; *Sapotac.;* Lib.
Mahua: *Bassia latifolia* Roxb.; *Sapotac.;* n. O.-Ind. (United Prov.)
„   tree: *Illipe malabrorum* Koenig; *Sapotac.;* sw. O.-Ind., Cey.
Mahula: *Bassia latifolia* Roxb.; *Sapotac.;* Beng.
Mahura: *Aegle marmelos* Correa; *Rutac.;* O.-Ind.
Mahuwa karar: *Hymenodictyon excelsum* Wall.; *Rubiac.;* C.-O.-Ind.
Mahúyan: *Diospyros whitfordi* Merr.; *Ebenac.;* Phil. (Surigao)
Mahwa: *Mimusops littoralis* Kurz; *Sapotac.;* O.-Ind.
Mai: *Dillenia pentagyna* Roxb.; *Dilleniac.;* Burma.
„   : *Podocarpus spicata* R. Br.; *Conifer.;* N.-Seel.
„   chan: *Mansonia gagei* J. R. Drumm.; *Sterculiac.;* Burma.
„   chi-tawk: *Pterocarpus macrocarpus* Kurz; *Papilionac.;* Burma.
„   dasak: *Litsea sebifera* Pers.; *Laurac.;* nw. O.-Ind. (Punjab)

Mai deng: *Xylia dolabriformis* Bth.; *Mimosac.;* Siam.
„ hak: *Melanorrhoea usitata* Wall.; *Anacardiac.;* Burma.
„ hao: *Dipterocarpus turbinatus* Gaertn. f., *D. alatus* Roxb.; *Dipterocarpac.;* Burma.
„ hawktik: *Schrebera swietenioides* Roxb.; *Oleac.;* Burma.
„ hen: *Terminalia belerica* Roxb.; *Combretac.;* Burma.
„ hio: *Anogeissus acuminata* Wall.; *Combretac.;* Burma.
„ hkai: *Eugenia operculata* Roxb.; *Myrtac.;* Burma.
„ „ : *Salix tetrasperma* Roxb.; *Salicac.;* Burma.
„ hkam: *Lannea grandis* Engl.; *Anacardiac.;* Burma.
„ „ : *Phyllanthus emblica* L.; *Euphorbiac.;* Burma.
„ hkao-long: *Holarrhena antidysenterica* Wall.; *Apocynac.;* Burma.
„ ho-gwan: *Dalbergia oliveri* Gamble; *Papilionac.;* Burma.
„ hok-pha: *Shorea obtusa* Wall.; *Dipterocarpac.;* Burma.
„ hpawng-tun: *Mallotus philippinensis* Muell. Arg.; *Euphorbiac.;* Burma.
„ ka-yong: *Duabanga sonneratioides* Buch.-Ham.; *Sonneratiac.;* Burma.
„ kam-kaw: *Mesua ferrea* L.; *Guttifer.;* Burma.
„ kan-ang: *Homalium tomentosum* Bth.; *Flacourtiac.;* Burma.
„ kao: *Butea frondosa* Roxb.; *Papilionac.;* Burma.
„ keik: *Salix tetrasperma* Roxb.; *Salicac.;* Burma.
„ khaine: *Hopea odorata* Roxb.; *Dipterocarpac.;* Ind.-Ch. (Laos)
„ kham: *Garuga pinnata* Roxb.; *Burserac.;* Burma.
„ khao lam: *Cephalostachyum pergracile* Munro; *Gramin.;* Siam.
„ khaw: *Schleichera trijuga* Willd.; *Sapindac.;* Burma.
„ kokko: *Albizzia lebbek* Bth.; *Mimosac.;* Burma.
„ kyang: *Schleichera trijuga* Willd.; *Sapindac.;* Burma.
„ kye: *Dolichandrone stipulata* Bth.; *Bignoniac.;* Burma.
„ kying-lwai: *Albizzia odoratissima* Bth.; *Mimosac.;* Burma.
„ laï: *Sideroxylon eburneum* Chev.; *Sapotac.;* Ind.-Ch.
„ lang: *Wrightia tomentosa* Roem. et Schultes; *Apocynac.;* Burma.
„ lim: *Elaeocarpus robustus* Roxb.; *Elaeocarpac.;* Burma.
„ löng: *Berrya ammonilla* Roxb.; *Tiliac.;* Burma.
„ lum: *Cassia fistula* L.; *Caesalpiniac.;* Burma.
„ mahen: *Terminalia belerica* Roxb.; *Combretac.;* Burma.
„ mak-na: *Terminalia chebula* Retz.; *Combretac.;* Burma.
„ makho-ling: *Diospyros burmanica* Kurz; *Ebenac.;* Burma.
„ makkawk: *Spondias mangifera* Willd.; *Anacardiac.;* Burma.
„ mannah: *Terminalia chebula* Retz.; *Combretac.;* Burma.
„ mawt-sao: *Lagerstroemia parviflora* Roxb.; *Lythrac.;* Burma.
„ mi-myen: *Litsea sebifera* Pers.; *Laurac.;* Burma.
„ nangsang: *Miliusa velutina* Hook. f. et Thoms.; *Anonac.;* Burma.
„ nao-nau: *Tetrameles nudiflora* R. Br.; *Datiscac.;* Burma.
„ naw: *Stephegyne diversifolia* Hook. f.; *Rubiac.;* Burma.
„ ngye: *Shorea obtusa* Wall.; *Dipterocarpac.;* Burma.
„ nio: *Bombax malabaricum* DC.; *Bombacac.;* Burma.
„ nyawng: *Ficus religiosa* L.; *Morac.;* Burma.
„ nye: *Shorea obtusa* Wall.; *Dipterocarpac.;* Burma.
„ ong-tong: *Litsea sebifera* Pers.; *Laurac.;* Burma.
„ padu (d): *Pterocarpus pedatus* Pierre; *Papilionac.;* H.-Ind.
„ pai: *Bambusa arundinacea* Willd.; *Gramin.;* Siam.
„ pao: *Pentacme suavis* A. DC.; *Dipterocarpac.;* Burma.
„ pheu: *Bursera serrata* Colebr.; *Burserac.;* Ind.-Ch. (Laos)
„ phoot: *Gardenia sp., G. obtusifolia* Roxb.; *Rubiac.;* Siam.
„ „ : *Ervatamia coronaria* Stapf, *Holarrhena similis* Craib; *Apocynac.;* Siam.
„ „ : *Zizyphus jujuba* Lam.; *Rhamnac.;* Siam.
„ „ : *Limonia acidissima* L.; *Rutac.;* Siam.
„ „ : *Hibiscus mutabilis* L.; *Malvac.;* Siam.
„ „ : *Jasminum auriculatum* Vahl; *Oleac.;* Siam.
„ pi: *Anogeissus acuminata* Wall.; *Combretac.;* Burma.
„ tawk: *Pterocarpus macrocarpus* Kurz; *Papilionac.;* Burma.
„ pinngo: *Careya arborea* Roxb.; *Myrtac.;* Burma.
„ pradoo: *Pterocarpus indicus* Willd.; *Papilionac.;* Siam.
„ put: *Gardenia sp.; Rubiac.;* Siam.
„ pyele: *Stephegyne parvifolia* Korth.; *Rubiac.;* Burma.
„ rak: *Melanorrhoea usitata* Wall.; *Anacardiac.;* Siam.

Mai rang: *Pentacme suavis* A. DC.; *Dipterocarpac.;* Siam.
„ sa-hpong: *Lagerstroemia flos-reginae* Retz.; *Lythrac.;* Burma.
„ sa-lan: *Xylia dolabriformis* Bth.; *Mimosac.;* Burma.
„ „ lao-long: *Duabanga sonneratioides* Buch.-Ham.; *Sonneratiac.;* Burma.
„ sak: *Tectona grandis* L. f.; *Verbenac.;* Burma.
„ salao: *Lagerstroemia tomentosa* Presl; *Lythrac.;* Siam.
„ saw: *Gmelina arborea* L.; *Verbenac.;* Burma.
„ sein: *Prunus puddum* Roxb.; *Rosac.;* Burma.
„ sonpu: *Hymenodictyon excelsum* Wall.; *Rubiac.;* Burma.
„ taung: *Tetrameles nudiflora* R. Br.; *Datiscac.;* Burma.
„ tawn: *Albizzia odoratissima* Bth., *A. procera* Bth.; *Mimosac.;* Burma.
„ teen: *Dipterocarpus tuberculatus* Roxb.; *Dipterocarpac.;* Burma.
„ teng: *Shorea obtusa* Wall.; *Dipterocarpac.;* Siam.
„ ti: *Bursera serrata* Colebr.; *Burserac.;* Burma.
„ ting: *Mesua ferrea* L.; *Guttifer.;* Burma.
„ tong: *Dipterocarpus tuberculatus* Roxb.; *Dipterocarpac.;* Burma.
„ tonghu: *Berrya ammonilla* Roxb.; *Tiliac.;* Burma.
„ viet: *Dalbergia cultrata* Grah.; *Papilionac.;* Burma.
„ wan: *Sideroxylon tomentosum* Roxb.; *Sapotac.;* Burma.
„ yang: *Holarrhena antidysenterica* Wall.; *Apocynac.;* Burma.
„ „ hkao-awn: *Wrightia tomentosa* Roem. et Schultes; *Apocynac.;* Burma.
„ yum: *Toona ciliata* M. Roem.; *Meliac.;* Burma.
„ yum-myen: *Tetrameles nudiflora* R. Br.; *Datiscac.;* Burma.
Maiang: *Ganua motleyana* Pierre, *Madhuca spp., Palaquium spp.; Sapotac.;* Malay.
„ : *Payena lucida* A. DC., *P. maingayi* Clarke; *Sapotac.;* Malay.
Maibau: *Alnus nepalensis* D. Don; *Betulac.;* Burma.
Maida: *Litsea sebifera* Pers.; *Laurac.;* O.-Ind.
„ lakadi: *Litsea sebifera* Pers.; *Laurac.;* O.-Ind. (Bombay)
„ lakri: *Litsea sebifera* Pers.; *Laurac.;* C.-O.-Ind., n. O.-Ind. (Un. Prov.)
Maidbaid: *Oroxylum indicum* Vent.; *Bignoniac.;* Phil. (Camarines)
Maidenhair tree: *Ginkgo biloba* L.; *Ginkgoac.;* Ch.
Maiden's Blush: *Echinocarpus australis* Bth.; *Elaeocarpac.;* ö. Au.
„ „ : *Euroschinus falcatus* Hook. f.; *Anacardiac.;* N.-S.-W.
Maidi: *Boswellia frereana* Birdw.; *Burserac.;* O.-Af. (it. Somali)
„ : *Bursera serrata* Colebr.; *Burserac.;* Burma.
Maidjin kadaii: *Lophira alata* Banks; *Ochnac.;* Kam.
Maidou: *Pterocarpus pedatus* Pierre; *Papilionac.;* Ind.-Ch.
Maihakhpu: *Buchanania latifolia* Roxb.; *Anacardiac.;* Burma.
Maihkaw: *Machilus odoratissima* Nees; *Laurac.;* Burma.
Maika: *Aegle marmelos* Correa; *Rutac.;* C.-O.-Ind. (Gondwana)
Maikang: *Albizzia stipulata* Boivin; *Mimosac.;* Burma.
Maikye: *Stereospermum suaveolens* DC.; *Bignoniac.;* Burma.
Maila: *Vitex altissima* L. f.; *Verbenac.;* O.-Ind. (Madras)
Mailam pala: *Wrightia tomentosa* Roem. et Sch.; *Apocynac.;* sw. O.-Ind. (Mal.)
Maimakhat: *Artocarpus lakoocha* Roxb.; *Morac.;* Burma.
Main: *Dracaena perrotteti* Baker; *Liliac.;* Elf.
„ : *Chukrasia tabularis* Adr. Juss.; *Meliac.;* O.-Ind. (Haiderabad)
Mainakat: *Tetrameles nudiflora* R. Br.; *Datiscac.;* nö. O.-Ind. (Nepal)
Maindal: *Randia dumetorum* Lamk.; *Rubiac.;* n. O.-Ind. (United Prov.)
Maini: *Randia dumetorum* Lamk.; *Rubiac.;* n. O.-Ind. (United Prov.)
Mainpal: *Randia dumetorum* Lamk.; *Rubiac.;* C.-O.-Ind.
Mainphal: *Randia dumetorum* Lamk.; *Rubiac.;* n. O.-Ind. (United Prov.)
Maipiorie kerapori: *Paypayrola guianensis* Aubl.; *Violac.;* Sur.
Mairac: *Melanorrhoea laccifera* Pierre; *Anacardiac.;* Kamb.
Maire: *Olea cunninghami* Hook. f., *O. sp. prob. lanceolata* Hook. f.; *Oleac.;* N.-Seel.
„ : *Santalum cunninghami* Hook. f.; *Santalac.;* N.-Seel.
„ , black: *Olea cunninghami* Hook. f.; *Oleac.;* N.-Seel.
„ -rau-nui: *Olea cunninghami* Hook. f.; *Oleac.;* N.-Seel.
„ , white: *Olea lanceolata* Hook. f.; *Oleac.;* N.-Seel.
Mais cocido: *Coccoloba acuminata* H. B. K.; *Polygonac.;* Kol.
Maisjoeré-ekjarepore: *Paypayrola guianensis* Aubl.; *Violac.;* Sur.
Maïswe: *? Androstachys johnstoni* Prain; *Euphorbiac.;* port. O.-Af.
Maiten: *Maytenus boaria* Molina; *Celastrac.;* Chile.
„ grande: *Maytenus boaria* Molina; *Celastrac.;* Arg., Chile.

Maiten de Magellanes: *Maytenus magellanica* Hook. f.; *Celastrac.*; Chile.
Maiyang: *Dipterocarpus tuberculatus* Roxb.; *Dipterocarpac.*; O.-Ind.
Maiyanghkao: *Alstonia scholaris* R. Brown; *Apocynac.*; Burma.
Maiz tostado: *Comocladia sp.*; *Anacardiac.*; P.-R.
Majaäriran: *Jacaranda filicifolia* D. Don; *Bignoniac.*; Sur.
Majagua: *Hibiscus tiliaceus* L.; *Malvac.*; Ant., Cuba, Jam., Hond.
„ : *Bombax barrigon* Dcne.; *Bombacac.*; Kol.
„ azul: *Hibiscus elatus* Swartz; *Malvac.*; Cuba.
„ blanca: *Hibiscus elatus* Swartz; *Malvac.*; Cuba.
„ común: *Hibiscus elatus* Swartz; *Malvac.*; Cuba.
„ de Florida: *Thespesia populnea* Correa; *Malvac.*; Cuba.
„ de la maestra: *Hibiscus elatus* Swartz; *Malvac.*; Cuba.
„ „ playa: *Hibiscus tiliaceus* L.; *Malvac.*; Pan.
Majagüillo: *Muntingia calabura* L.; *Elaeocarpac.*; Pan., Ven.
Majagüito de playa: *Hibiscus tiliaceus* L.; *Malvac.*; Kol.
Majahás: *Quararibea fieldi* Millsp.; *Bombacac.*; br. Hond.
Majahua: *Hibiscus tiliaceus* L.; *Malvac.*; W.-Ind., Mex., C.-Am.
„ de toro: *Guazuma ulmifolia* Lam.; *Sterculiac.*; Mex.
Majaine: *Swietenia macrophylla* King; *Meliac.*; Hond.
Majana: *Hampea trilobata* Standl.; *Bombacac.*; br. Hond.
Majao: *Hampea stipitata* S. Watson; *Bombacac.*; Hond.
„ : *Hibiscus tiliaceus* L.; *Malvac.*; Hond.
„ : *Heliocarpus spp.*; *Tiliac.*; Hond.
„ blanco: *Heliocarpus glanduliferus* Robinson; *Tiliac.*; Hond.
„ colorado: *Heliocarpus spp.*; *Tiliac.*; Hond.
„ „ : *Hibiscus tiliaceus* L.; *Malvac.*; Hond.
„ „ : *Hampea stipitata* S. Watson; *Bombacac.*; Hond.
„ de indio: *Naucleopsis naga* Pittier; *Morac.*; Hond.
Maje: *Daniella oliveri* Hutch. et Dalz.; *Caesalpiniac.*; Go., n. Nig.
Majiriya: *Erythrina senegalensis* DC.; *Papilionac.*; n. Nig.
Majomo: *Lonchocarpus fendleri* Bth.; *Papilionac.*; Ven.
Majugera: *Hibiscus tiliaceus* L.; *Malvac.*; ö. Cuba.
Majugua: *Belotia greviaefolia* A. Rich.; *Tiliac.*; Cuba.
„ : *Hibiscus tiliaceus* L.; *Malvac.*; C.-Am., Ant.
Majume: *Evonymus sieboldiana* Blume; *Celastrac.*; Jap.
Mak-hkaw: *Zizyphus jujuba* Lamk.; *Rhamnac.*; Burma.
„ kawng-tawp: *Bridelia retusa* Spreng.; *Euphorbiac.*; Burma.
„ mong: *Mangifera indica* L.; *Anacardiac.*; Burma.
„ pyin: *Aegle marmelos* Correa; *Rutac.*; Burma.
„ san: *Dillenia pentagyna* Roxb.; *Dilleniac.*; Burma.
Makaásim: *Eugenia benthami* A. Gray; *Myrtac.*; Phil.
Makabálo: *Lagerstroemia speciosa* Pers.; *Lythrac.*; Phil. (Pangasinan)
Makabongló: *Dysoxylum turczaninowi* C. DC.; *Meliac.*; Phil. (Camarines)
Makad-bhirand: *Carallia lucida* Roxb.; *Rhizophorac.*; w. O.-Ind. (Bombay)
Makadáag: *Dracontomelum dao* Merr. et Rolfe; *Anacardiac.*; Phil. (Ilocos Norte)
Maka-grien: *Tecoma sp., Tabebuia sp.*; *Bignoniac.*; Sur.
Makai: *Shorea assamica* Dyer; *Dipterocarpac.*; Assam.
Makak: *Imbricaria petiolaris* A. DC.; *Sapotac.*; Au.
„ , white: *Imbricaria petiolaris* A. DC.; *Sapotac.*; Maur.
Makakoku: *Ternstroemia japonica* Thunb.; *Theac.*; tr. As.
Makala: *Uapaca guineensis* Muell. Arg.; *Euphorbiac.*; B.-K.
Makalae: *Terminalia belerica* Roxb.; *Combretac.*; Burma.
Makan nim: *Melia azedarach* L.; *Meliac.*; O.-Ind. (Madras)
Makanchi: *Callicarpa arborea* Roxb.; *Verbenac.*; Assam (Garo)
Makarfo: *Afrormosia laxiflora* Harms; *Papilionac.*; n. Nig.
Makaru: *Mimusops heckeli* Lecomte; *Sapotac.*; Elf.
Makarufa: *Dalbergia sp.*; *Papilionac.*; n. Nig. (Kontagora)
Makarukh: *Ailanthus excelsa* Roxb.; *Simarubac.*; n. O.-Ind., C.-O.-Ind.
Makaton: *Gardenia turgida* Roxb.; *Rubiac.*; Burma.
Makatu: *Beilschmiedia cairocan* Vid.; *Laurac.*; Phil. (Palaui)
Makau: *Agathis alba* Foxw.; *Conifer.*; Phil. (Misamis)
„ : *Dracontomelum dao* Merr. et Rolfe; *Anacardiac.*; Phil. (Agusan, Colabato)
Makawk: *Spondias mangifera* Willd.; *Anacardiac.*; Burma.
Maki: *Podocarpus chinensis* Wall., *P. macrophylla* D. Don; *Conifer.*; Jap.

Makil: *Mimusops elengi* Roxb.; *Sapotac.;* s. O.-Ind. (Madras)
Makinduri: *Croton elliottanus* Engl. et Pax; *Euphorbiac.;* br. O.-Af.
Makitarim: *Hopea philippinensis* Dyer; *Dipterocarpac.;* Phil. (Pangasinan, Tayabas)
Makiti: *Parinarium laurinum* Λ. Gray; *Rosac.;* Fid.
Makka: *Minquartia guianensis* Aubl.; *Olacac.;* Sur.
„ groenhart: *Tecoma sp., Tabebuia sp.; Bignoniac.;* Sur.
Makkam: *Schrebera swietenioides* Roxb.; *Oleac.;* O.-Ind. (Madras)
Makle: *Dalbergia latifolia* Roxb.; *Papilionac.;* n. C.-O.-Ind.
Makmong-sang-yip: *Mangifera silvatica* Roxb.; *Anacardiac.;* Burma.
Mako: *Aristotelia racemosa* Hook. f.; *Elaeocarpac.;* N.-Seel.
Makob: *Schrebera swietenioides* Roxb.; *Oleac.;* O.-Ind. (Madras)
Makoom-makooba: *? Uapaca kirkiana* Muell. Arg.; *Euphorbiac.;* B.-K.
Makópa: *Aglaia clarki* Merr.; *Meliac.;* Phil. (Batangas, Cav.)
„ : *Eugenia calubcob* C. B. Rob.; *Myrtac.;* Phil. (Laguna)
Makópang-gúbat: *Aglaia clarki* Merr.; *Meliac.;* Phil. (Batangas, Cav.)
Makoré (f): *Mimusops heckeli* Lecomte; *Sapotac.;* Elf.
Makpyensum: *Feronia elephantum* Correa; *Rutac.;* Burma.
Makra: *Butea frondosa* Roxb.; *Papilionac.;* C.-O.-Ind.
Makriah: *Schima wallichi* Choisy; *Theac.;* Assam.
Makua: *Entandrophragma rufa* Chev.; *Meliac.;* Elf.
Makube: *Borassus flabellifer* L. var. *aethiopum* Warb.; *Palmac.;* Elf., Go.
Makuláb: *Quercus sp.; Fagac.;* Phil. (Cagayan)
Makurjal: *Schima wallichi* Choisy; *Theac.;* Hind.
Makurlimbu: *Atalantia monophylla* Correa; *Rutac.;* sw. O.-Ind. (Bombay)
Makusal: *Schima wallichi* Choisy; *Theac.;* Assam.
Makuwa: *Xylopia aethiopica* A. Rich.; *Anonac.;* B.-K.
Makwe: *Mimusops prob. djave* Engl.; *Sapotac.;* Go.
Mal de ojos: *Caesalpinia gilliesi* Wall.; *Caesalpiniac.;* Arg.
Malaadúas: *Dysoxylum decandrum* Merr.; *Meliac.;* Phil. (w. Negros)
Malaanónang: *Shorea polita* Vid.; *Dipterocarpac.;* Phil. (Rizal, Tayabas)
„ : *Parashorea malaanonan* Merr.; *Dipterocarpac.;* Phil. (Rizal, Laguna)
„ : *Pentacme contorta* Merr. et Rolfe; *Dipterocarpac.;* Phil.
Malabágo: *Adenanthera intermedia* Merr.; *Mimosac.;* Phil. (Masbate)
Malabalabáng: *Shorea palosapis* Merr.; *Dipterocarpac.;* Phil. Cagayan)
Malabalúnu: *Buchanania sp.; Anacardiac.;* Phil. (Zamboanga)
Malabangkál: *Neonauclea sp.; Rubiac.;* Phil. (Bataan)
Malabapaya: *Polyscias nodosa* Seem.; *Araliac.;* Phil.
Malabar: *Gardenia jasminoides* Ellis; *Rubiac.;* Ven.
Malabató: *Hopea philippinensis* Dyer; *Dipterocarpac.;* Phil. (Agusan)
Malabatíno: *Paralstonia clusiacea* Baill.; *Apocynac.;* Phil. (Tayabas)
Malabayabas: *Tristania decorticata* Merr.; *Myrtac.;* Phil., Borneo (Sarawak)
Malabera: *Fagraea crenulata* Maingay, *F. fastigiata* Blume; *Loganiac.;* Malay.
Malabignay: *Anacolosa luzoniensis* Merr.; *Olacac.;* Phil.
Malabingan: *Allaeanthus luzonicus* Vid.; *Morac.;* Phil.
Malabira: *Fagraea crenulata* Maingay; *Loganiac.;* Malay.
Malabobónau: *Sandoricum vidali* Merr.; *Meliac.;* Phil. (Samar)
Malábog: *Parishia malabog* Merr.; *Anacardiac.;* Phil.
Malabóho: *Sterculia sp., S. oblongata* R. Br.; *Sterculiac.;* Phil.
Malabóhok: *Casuarina equisetifolia* Forst.; *Casuarinac.;* Phil. (Bisaya)
Malabokbók: *Gymnartocarpus woodi* Merr.; *Morac.;* Phil. (Zambales)
Malabonot: *Sterculia sp.; Sterculiac.;* Phil. (Bataan, Rizal, Manila)
Malabúlak: *Bombax malabaricum* DC.; *Bombacac.;* Phil. (Bataan, Rizal, Laguna)
Malabúnga: *Nothophoebe malabonga* Merr.; *Laurac.;* Phil. (Laguna)
„ : *Sterculia sp.; Sterculiac.;* Phil. (Tayabas)
Malabúnog: *Garcinia sp.; Guttifer.;* Phil. (Cagayan, Bontoc)
Malacacáo: *Sterculia sp.; Sterculiac.;* Phil. (Bataan, Laguna, Mindoro)
Malacacheta: *Tabebuia leucoxyla* DC.; *Bignoniac.;* Bras. (Sao P)
Malacadiós: *Beilschmiedia cairocan* Vidal; *Laurac.;* Phil. (Tayabas)
„ : *Derris cumingi* Bth.; *Papilionac.;* Phil.
Malacatmón: *Dillenia sp., D. luzoniensis* Merr.; *Dilleniac.;* Phil.
Malacmálac: *Palaquium sp., P. philippinense* C. B. Rob.; *Sapotac.;* Phil.
Maladísai: *Bischofia javanica* Blume; *Euphorbiac.;* Phil. (Agusan)
Maladitá: *Paralstonia clusiacea* Baill.; *Apocynac.;* Phil. (Batangas, Tayabas)
Malady: *Aspidosperma megalocarpon* Muell. Arg.; *Apocynac.;* br. Hond.

313

Malady, white: *Aspidosperma sp.; Apocynac.;* br. Hond.
Malagábi: *Terminalia nitens* Presl, *T. oocarpa* Merr.; *Combretac.;* Phil. (Mindoro)
Malagahánip: *Albizzia lebbekoides* Bth.; *Mimosac.;* Phil. (Bataan, Rizal, Lag., N. Ec.)
„ : *Albizzia marginata* Merr.; *Mimosac.;* Phil. (Tayabas)
Malagaitmán: *Diospyros mindanaensis* Merr., *D. plicata* Merr.; *Ebenac.;* Phil. (Tay.)
„ : *Diospyros currani* Merr., *Ebenac.;* Phil. (Tay., Camar., Marinduque)
Malagaitmon: *Diospyros currani* Merr.; *Ebenac.;* Phil. (Tay., Camar., Marinduque)
Malagángau: *Anisoptera curtisi* Dyer; *Dipterocarpac.;* Phil. (Camarines)
„ : *Shorea balangeran* Dyer; *Dipterocarpac.;* Phil. (Tayabas)
„ : *Shorea negrosensis* Foxw.; *Dipterocarpac.;* Phil. (Albay)
Malagánit: *Albizzia lebbekoides* Bth.; *Mimosac.;* Phil.
Malagano: *Luehea speciosa* Willd.; *Tiliac.;* Kol.
Malagasáha: *Sterculia sp.; Sterculiac.;* Phil. (Tayabas)
Malagibúyo: *Celtis sp.; Ulmac.;* Phil. (Mindoro, Capiz)
Malagíso: *Shorea polysperma* Merr., *S. teysmanniana* Dyer; *Dipterocarpac.;* Phil.
Malagmát: *Pygeum sp.; Rosac.;* Phil. (Pampanga) [(Bulácan)
„ : *Shorea polysperma* Merr.; *Dipterocarpac.;* Phil. (Laguna)
Malagueta: *Amomis caryophylla* Krug et Urb.; *Myrtac.;* P.-R.
„ : *Guatteria aeruginosa* Standl.; *Anonac.;* Pan.
„ prieto: *Guatteria slateri* Standl.; *Anonac.;* Pan.
Malahagís: *Eugenia sp.; Myrtac.;* Phil. (s. Luzon)
Malaígang: *Dillenia sp.; Dilleniac.;* Phil. (Zamboanga)
„ : *Parinarium sp.; Rosac.;* Phil. (Samar)
„ : *Vitex aherniana* Merr.; *Verbenac.;* Phil. (Lubang)
Malaigmó: *Celtis sp.; Ulmac.;* Phil. (Nueva Ecija, Camarines)
Malaigót: *Eugenia sp.; Myrtac.;* Phil. (s. Luzon)
Malaikmó: *Celtis sp., C. philippinensis* Blco.; *Ulmac.;* Phil.
Malaing: *Morus laevigata* Wall.; *Morac.;* Burma.
Malaióhot: *Sideroxylon duclitan* Blco.; *Sapotac.;* Phil. (Mindoro)
Malaíyau: *Dracontomelum dao* Merr. et Rolfe; *Anacardiac.;* Phil. (Tayabas)
Malakadiós: *Allaeanthus glaber* Warb.; *Morac.;* Phil. (Masbate)
„ : *Beilschmiedia cairocan* Vidal; *Laurac.;* Phil. (Tayabas)
Malakalumpáng: *Sterculia sp.; Sterculiac.;* Phil. (Camarines)
Malakalumpít: *Terminalia calamansanai* Rolfe; *Combretac.;* Phil. (Bat., Lag., Camar.)
Malakamánga: *Homalium bracteatum* Bth.; *Flacourtiac.;* Phil. (Bataan)
Malakapái: *Alangium longiflorum* Merr.; *Cornac.;* Phil. (Laguna, Tayabas, Camar.)
„ : *Alangium meyeri* Merr.; *Cornac.;* Phil. (Tayabas)
Malakayán: *Parashorea malaanonan* Merr.; *Dipterocarpac.;* Phil. (Zamboanga)
„ : *Pentacme contorta* Merr. et Rolfe; *Dipterocarpac.;* Phil. (Zamboanga)
„ : *Shorea spp.; Dipterocarpac.;* Phil. (Zamboanga, Basilan)
Malakná: *Eugenia costulata* C. B. Rob.; *Myrtac.;* Phil. (Rizal)
Malako: *Peltogyne spp., Copaifera bracteata* Bth.; *Caesalpiniac.;* Sur.
Malakópa: *Eugenia calubcob* C. B. Rob.; *Myrtac.;* Phil. (Negrosinsel)
Malalamba: *Hopea philippinensis* Dyer; *Dipterocarpac.;* Phil. (Albay)
Malaligás: *Buchanania sp.; Anacardiac.;* Phil. (Tayabas)
Malalimón: *Taxotrophis ilicifolia* Vidal; *Morac.;* Phil. (Zamboanga)
Malaliring: *Dillenia sp.; Dilleniac.;* Phil. (Bataan)
Malam: *Acrocarpus fraxinifolius* Wight et Arn.; *Caesalpiniac.;* sw. O.-Ind.
Malamabúlo: *Myristica sp.; Myristicac.;* Phil. (Pangasinan)
Malamanggá: *Semecarpus sp.; Anacardiac.;* Phil. (Pampanga, Zambales, Tay., Min.)
„ : *Sideroxylon duclitan* Blco.; *Sapotac.;* Phil. (Bataan)
Malambíngan: *Allaeanthus glaber* Warb.; *Morac.;* Phil. (Basilan)
Malamuláuin: *Vitex turczaninowi* Merr.; *Verbenac.;* Phil. (Rizal)
Malanangká: *Gymnartocarpus woodi* Merr.; *Morac.;* Phil. (Bataan, Laguna)
Malanga: *Markhamia lutea* K. Schum.; *Bignoniac.;* Kam.
Malángau: *Securinega flexuosa* Muell. Arg.; *Euphorbiac.;* Phil. (Agusan)
Mala-oko: *Peltogyne spp., Copaifera bracteata* Bth.; *Caesalpiniac.;* Sur.
Malapáho: *Anisoptera curtisi* Dyer; *Dipterocarpac.;* Phil. (Camarines)
„ : *Dipterocarpus vernicifluus* Blco.; *Dipterocarpac.;* Phil. (Pol.)
„ : *Mangifera monandra* Merr., *M. altissima* Blco.; *Anacardiac.;* Phil.
Malapalikpík: *Duabanga moluccana* Blume; *Sonneratiac.;* Phil. (Rizal)
Malapánau: *Shorea balangeran* Dyer; *Dipterocarpac.;* Phil. (Camarines)
Malapángi: *Trichadenia philippinensis* Merr.; *Flacourtiac.;* Phil. (Negrosinsel)
Malapano: *Dipterocarpus grandiflorus* Blco.; *Dipterocarpac.;* Phil.

Malapapáya: *Polyscias nodosa* Seem.; *Araliac.;* Phil.
„ : *Sterculia sp.; Sterculiac.;* Phil. (Bataan)
Malapíga: *Parinarium sp.; Rosac.;* Phil. (Tayabas)
„ : *Tristania sp.; Myrtac.;* Phil. (Tayabas)
„ : *Xanthostemon verdugonianus* Naves; *Myrtac.;* Phil. (Culion)
Malapinggán: *Trichadenia philippinensis* Merr.; *Flacourtiac.;* Phil. (Tay., Camar.)
Malapútat: *Terminalia oocarpa* Merr.; *Combretac.;* Phil. (Rizal)
Malapuyau: *Diospyros pilosanthera* Blco.; *Ebenac.;* Phil. (Tayabas)
„ : *Parinarium sp.; Rosac.;* Phil. (Tayabas)
Malarambó: *Sandoricum vidali* Merr.; *Meliac.;* Phil. (Rizal)
Malaresa: *Euphoria cinerea* Radlk.; *Sapindac.;* Phil.
Malarmo: *Bumelia sp.; Sapotac.;* Ven.
Malarubat-na-puti: *Eugenia claviflora* Roxb.; *Myrtac.;* Phil. (Bataan)
Malarúhat: *Eugenia spp.; Myrtac.;* Phil.
Malarúngon: *Heritiera littoralis* Dry.; *Sterculiac.;* Phil. (Tayabas)
Malasabón: *Endospermum peltatum* Merr.; *Euphorbiac.;* Phil. (Laguna)
Malasagád: *Adenanthera intermedia* Merr.; *Mimosac.;* Phil. (Rizal)
Malaságing: *Aglaia diffusa* Merr., *A. harmsiana* Perk.; *Meliac.;* Phil. (Tay., Lag.)
Malasamát: *Celtis sp.; Ulmac.;* Phil. (Pampanga, Bataan)
Malasambóng-batú: *Sideroxylon luzoniense* Merr.; *Sapotac.;* Phil. (Rizal)
Malasampálok: *Albizzia lebbekoides* Bth., *A. marginata* Merr.; *Mimosac.;* Phil. (Bat.,
Malasantól: *Aglaia everetti* Merr.; *Meliac.;* Phil. (Cebu)   [N. E.
„ : *Sandoricum vidali* Merr.; *Meliac.;* Phil.
Malasapsáp: *Polyscias nodosa* Seem.; *Araliac.;* Phil. (Pampanga)
„ : *Pterocymbium tinctorium* Merr.; *Sterculiac.;* Phil. (Pampanga)
Malasíko: *Litsea sp.; Laurac.;* Phil. (Bataan)
Malasinóro: *Pentacme contorta* Merr. et Rolfe; *Dipterocarpac.;* Phil. (Albay)
„ : *Shorea spp.; Dipterocarpac.;* Phil.
Malatagúm: *Hopea philippinensis* Dyer; *Dipterocarpac.;* Phil. (Albay)
„ : *Hopea pierrei* Hance; *Dipterocarpac.;* Phil. (Albay, Sorsogon)
„ : *Planchonia spectabilis* Merr.; *Lecythidac.;* Phil. (Camarines)
„ : *Terminalia comintana* Merr.; *Combretac.;* Phil. (Zamboanga)
„ : *Wallaceodendron celebicum* Koord.; *Mimosac.;* Phil. (Nueva Ecija)
Malatálang: *Diospyros pilosanthera* Blco.; *Ebenac.;* Phil. (Pampanga)
Malatampúi: *Eugenia sp.; Myrtac.;* Phil. (s. Luzon, Samar, Leyte)
Malatapái: *Alangium longiflorum* Merr.; *Cornac.;* Phil. (Laguna, Tayabas, Camar.)
„ : *Alangium brachyanthum* Merr.; *Cornac.;* Phil. (Tayabas)
„ : *Cyathocalyx globosus* Merr.; *Anonac.;* Phil. (Negrosinsel)
Malatbáng: *Shorea negrosensis* Foxw.; *Dipterocarpac.;* Phil. (Tayabas)
Malatiáong: *Pentacme contorta* Merr. et Rolfe; *Dipterocarpac.;* Phil. (Pol.)
„ : *Shorea teysmanniana* Dyer; *Dipterocarpac.;* Phil. (Pol.)
Malatibíg: *Neonauclea sp.; Rubiac.;* Phil. (Tayabas)
Malatigí: *Albizzia marginata* Merr.; *Mimosac.;* Phil. (Rizal, Laguna)
Malatínta: *Diospyros maritima* Blume, *D. nitida* Merr.; *Ebenac.;* Phil. (Pampanga)
Malatokó: *Albizzia saponaria* Blume; *Mimosac.;* Phil. (Bataan, Pamp., Rizal, Lag.)
Malatubíg: *Neonauclea sp.; Rubiac.;* Phil. (Tayabas)
Malatugon: *Litsea sp.; Laurac.;* Phil. (Isa.)
Malatumbága: *Aglaia spp.; Meliac.;* Phil. (Bataan, Laguna)
„ : *Homalium luzoniense* F. Vill.; *Flacourtiac.;* Phil. (Cotabato)
„ : *Neonauclea sp.; Rubiac.;* Phil. (Zamboanga, Samal)
„ : *Chisocheton philippinus* Harms; *Meliac.;* Phil.
„ -babáe: *Aglaia turczaninowi* C. DC.; *Meliac.;* Phil. (Bataan)
Malausá: *Vitex turczaninowi* Merr.; *Verbenac.;* Phil. (Bataan)
Malaut: *Hopea anomala* Foxw.; *Dipterocarpac.;* Malay (Kedah)
Malay-á: *Nothophoebe malabonga* Merr.; *Laurac.;* Phil. (Camarines)
Malayambo: *Eugenia xanthophylla* C. B. Rob.; *Myrtac.;* Phil. (Tayabas)
Malayapoo-toon-marum: *Toona ciliata* M. Roem.; *Meliac.;* w. O.-Ind. (Deccan)
Malay-apple: *Eugenia malaccensis* L.; *Myrtac.;* O.-As.
Malbira: *Fagraea fastigiata* Blume; *Loganiac.;* Malay, Java.
Malbóg: *Nauclea sp.; Rubiac.;* Phil. (Samar)
Malcajaco: *Curatella americana* L.; *Dilleniac.;* Salv.
Male of the Ire: *Holarrhena wulfsbergi* Stapf; *Apocynac.;* W.-Af.
Malebera: *Fagraea fastigiata* Blume; *Loganiac.;* Malay, Java.
Malei veppu: *Chukrasia tabularis* Adr. Juss.; *Meliac.;* s. O.-Ind. (Madras)

315

Mâlé-vimbou: *Melia azedarach* L.; *Meliac.;* w. O.-Ind. (Deccan)
Malga: *Cassia arereh* Delile, *C. sieberiana* DC.; *Caesalpiniac.;* n. Nig.
Mali: *Gardenia lucida* Roxb.; *Rubiac.;* C.-O.-Ind.
Malibató: *Shorea balangeran* Dyer; *Dipterocarpac.;* Phil. (Agusan)
Malíbog: *Parishia malabog* Merr.; *Anacardiac.;* Phil.
Malicia: *Acacia piparia* Bth.; *Mimosac.;* Bras.
Maligáng: *Aglaia diffusa* Merr.; *Meliac.;* Phil. (Zamboanga)
„ : *Eugenia sp.; Myrtac.;* Phil. (s. Luzon)
Malikmík: *Palaquium sp., P. cuneatum* Vidal; *Sapotac.;* Phil.
Malináb: *Albizzia retusa* Bth.; *Mimosac.;* Phil. (Cagayan)
Malingaggók: *Celtis sp.; Ulmac.;* Phil. (Cagayan)
Malinké: *Lophira alata* Banks; *Ochnac.;* Sudan.
Malinsot: *Aglaia minahassae* Koord.; *Meliac.;* Cel., Ceram.
Malipaga: *Zizyphus sp.; Rhamnac.;* Phil. (Tayabas)
Malitambíng: *Cyclostemon sp.; Euphorbiac.;* Phil. (Sorsogon)
Maliúm: *Hopea plagata* Vidal; *Dipterocarpac.;* Phil. (Mindoro)
Malla nim: *Melia azedarach* L.; *Meliac.;* C.-O.-Ind.
Mallay nangal: *Mesua ferrea* L.; *Guttifer.;* sw. O.-Ind. (Madras)
„ vembu: *Melia azedarach* L., *M. composita* Willd.; *Meliac.;* O.-Ind. (Madras)
Mallet-bark: *Eucalyptus occidentalis* Endl.; *Myrtac.;* Au.
Malletwood, brown: *Rhodamnia trinervia* Blume; *R. argentea* Bth.; *Myrtac.;* ö. Au.
Malligiri: *Cinnamomum cecicodaphne* Meissn.; *Laurac.;* nö. O.-Ind. (Nepal)
„ : *Cinnamomum glanduliferum* Meissn.; *Laurac.;* Assam.
Malmo: *Eugenia guineensis* DC.; *Myrtac.;* n. Nig.
Malobo: *Sterculia sp.; Sterculiac.;* Phil. (Tayabas)
Maloucou: *Mitragyne macrophylla* Hiern; *Rubiac.;* B.-K. (Bangala)
Malsari: *Mimusops elengi* Roxb.; *Sapotac.;* w. Beng. (Orissa)
Maltesewood: *Hibiscus elatus* Swartz; *Malvac.;* W.-Ind.
Malubágo: *Hibiscus tiliaceus* L.; *Malvac.;* Phil.
Maluco: *Genipa americana* L.; *Rubiac.;* Mex.
Malucu: *Mitragyne macrophylla* Hiern; *Rubiac.;* B.-K. (Bangala)
Maludbálud: *Neonauclea sp.; Rubiac.;* Phil. (Albay)
Malugai: *Pometia pinnata* Forst.; *Sapindac.;* Phil.
Malugay: *Pometia pinnata* Forst.; *Sapindac.;* Phil.
Malukka: *Garcinia cornea* L.; *Guttifer.;* Malay.
Maluklík: *Parinarium sp.; Rosac.;* Phil. (Zamboanga)
Maluku: *Mitragyne macrophylla* Hiern; *Rubiac.;* B.-K.
Malukut jantan: *Eurya acuminata* DC.; *Theac.;* tr. As.
Malumbayábas: *Tristania sp.; Myrtac.;* Phil. (Zamboanga, Basilan)
Malut: *Vatica spp., Shorea spp., Hopea spp., Isoptera spp.; Dipterocarpac.;* Malav.
„ : *Hopea anomala* Foxw.; *Dipterocarpac.;* Malay (Kedah)
Malva branca santarem: *Waltheria americana* L.; *Sterculiac.;* Bras.
„ velluda: *Waltheria americana* L.; *Sterculiac.;* Bras.
„ do campo: *Killmeyera coriacea* Mart.; *Guttifer.;* s. & ö. Bras.
Malvecino: *Lonchocarpus lucidus* Pittier; *Papilionac.;* Pan.
Malwini: *Diospyros cf. gracilescens* Gürke; *Ebenac.;* Kam.
Mâm: *Avicennia officinalis* L.; *Verbenac.;* Ind.-Ch.
„ dá: *Sideroxylon eburneum* Chev.; *Sapotac.;* Ind.-Ch.
„ pulicchi: *Spondias mangifera* Willd.; *Anacardiac.;* O.-Ind. (Madras)
Mama malhuaztili: *Clethra lanata* Mart. et Gal.; *Clethrac.;* Mex.
Mamadi: *Mangifera indica* L.; *Anacardiac.;* O.-Ind. (Madras)
Mamag: *Pistacia chinensis* Bunge; *Anacardiac.;* Phil.
Mamakau: *Dracontomelum dao* Merr. et Rolfe; *Anacardiac.;* Phil. (Davao)
Mamangkás: *Sideroxylon ferrugineum* Hook. et Arn.; *Sapotac.;* Phil. (Burias)
Mamani: *Sophora chrysophylla* Seem.; *Papilionac.;* Hawai.
Mamaorana: *Jacaratia spinosa* C. DC.; *Caricac.;* Bras.
Mamatá: *Euphoria didyma* Blco.; *Sapindac.;* Phil. (Olutanga)
Mamayahooka: *Isertia hypoleuca* Bth.; *Rubiac.;* br. Gu.
Mambali: *Mimusops africana* Lecomte; *Sapotac.;* W.-Af. (? Gab.)
Mambarklak: *Eschweilera longipes* Miers; *Lecythidac.;* holl. & fr. Gu.
Mambóg: *Nauclea junghuhni* Merr., *Neonauclea sp.; Rubiac.;* Phil.
Mamé de mono: *Pachira villosula* Pittier; *Bombacac.;* Pan.
Mame-gaki: *Diospyros lotus* L.; *Ebenac.;* Jap.

Mamecillo: *Lucuma calistophylla* Standl.; *Sapotac.;* Pan.
,, blanco: *Guarea chiricana* Standl.; *Meliac.;* Pan.
Mamee-apple: *Calocarpum mammosum* Pierre; *Sapotac.;* br. Hond.
,, ciruela: *Lucuma heyderi* Standl.; *Sapotac.;* br. Hond.
,, -sapote: *Calocarpum mammosum* Pierre; *Sapotac.;* br. W.-Ind.
,, , wild: *Alseis yucatana* Standl.; *Rubiac.;* br. Hond.
Mamegaki: *Diospyros lotus* L.; *Ebenac.;* Jap.
Mameicillo: *Lucuma calistophylla* Standl.; *Sapotac.;* Pan.
,, colorado: *Sloanea sp.; Rhamnac.;* Pan.
Mameira: *Uitex sp.; Uerbenac.;* Bras.
Mamey: *Mammea americana* L.; *Guttifer.;* Ant., P.-R., Guat.
,, : *Calocarpum mammosum* Pierre; *Sapotac.;* Pan.
,, cerera: *Lucuma campechiana* H. B. K.; *Sapotac.;* br. Hond.
,, colorado: *Calocarpum mammosum* Pierre; *Sapotac.;* Cuba.
? ,, montero: *Rheedia edulis* Triana et Planch.; *Guttifer.;* Nic.
,, rojo: *Mammea americana* L.; *Guttifer.;* Cuba.
,, sapote: *Calocarpum mammosum* Pierre; *Sapotac.;* P.-R.
,, serilla: *Lucuma campechiana* H. B. K.; *Sapotac.;* br. Hond.
,, zapote: *Calocarpum mammosum* Pierre; *Sapotac.;* Cuba.
,, de Cartagena: *Mammea americana* L.; *Guttifer.;* Pan.
,, ,, Santo Domingo: *Mammea americana* L.; *Guttifer.;* Cuba.
Mameyito: *Saurauia sp.; Dilleniac.;* Mex.
,, blanco: *Saurauia sp.; Dilleniac.;* Mex.
,, negro: *Clethra lanata* Mart. et Gal.; *Clethrac.;* Mex.
Mameyuelo: *Ardisia coriacea* Ledru, *A. spp.; Myrsinac.;* P.-R.
Mamica de cadela negro: ? *Zanthoxylum sp.; Rutac.;* Arg.
,, ,, cadella: *Fagara hiemalis* Engl.; *Rutac.;* Arg. (Mis.)
,, ,, candela negra: *Zanthoxylum sp.; Rutac.;* tr. Am.
,, ,, porca: *Zanthoxylum rhoifolium* Lam.; *Rutac.;* Bras. (Sao P)
Mamié kini: *Garcinia polyantha* Oliv.; *Guttifer.;* Elf.
Mamilier: *Anona reticulata* L.; *Anonac.;* Ant.
Mamme sapote: *Mammea americana* L.; *Guttifer.;* Trin.
Mammee apple, wild: *Platonia insignis* Mart.; *Guttifer.;* br. Gu.
Mammeetree: *Mammea americana* L.; *Guttifer.;* W.-Ind.
Mammica de cadella: *Zanthoxylum sp.; Rutac.;* Bras.
Mammoeira: *Carica papaya* L.; *Caricac.;* Bras.
Mamoeri-balli: *Mouriria acutiflora* Naud.; *Melastomatac.;* Sur.
Mamón: *Melicocca bijuga* L.; *Sapindac.;* Ven., Kol.
,, de leche: *Sideroxylon colombianum* Standl.; *Sapotac.;* Kol.
Mamonák: *Aglaia harmsiana* Perk.; *Meliac.;* Phil. (Zamboanga)
Mamoncillo: *Melicocca bijuga* L.; *Sapindac.;* Cuba, Kol.
Mamonhinho: *Esenbeckia febrifuga* Mart.; *Rutac.;* Bras.
Mamorana: *Pachira aquatica* Aubl.; *Bombacac.;* Bras. (Amaz.)
Mampat: *Cratoxylon formosum* Dyer; *Guttifer.;* Java.
Mamungkalión: *Palaquium sp.; Sapotac.;* Phil. (Laguna)
Mamusaru: *Rinorea sp.; Uiolac.;* br. Gu.
Mamushi: *Mimusops sp.; Sapotac.;* fr. Gu.
Man: *Irvingia oblonga* Chev.; *Simarubac.;* Kam.
,, : *Prunus triflora* Roxb.; *Rosac.;* Ind.-Ch.
,, au: *Aporosa microcalyx* Hassk.; *Euphorbiac.;* Ind.-Ch.
,, bau: *Macaranga henricorum* Hemsl.; *Euphorbiac.;* Ind.-Ch.
,, cau tia: *Atalantia disticha* Merr.; *Rutac.;* Ind.-Ch.
,, cho: *Macaranga poilanei* Gagnep.; *Euphorbiac.;* Ind.-Ch.
,, côi: *Acacia farnesiana* Willd.; *Mimosac.;* Ind.-Ch.
,, -dalli: *Uirola sebifera* Aubl.; *Myristicac.;* br. Gu.
,, deuk: *Phyllostachys nigra* Munro; *Gramin.;* Ch. (Fukien)
,, foengoe: *Licania pachystachya* Kleinh.; *Rosac.;* Sur.
,, gorro-gorro: *Licania micrantha* Miq.; *Rosac.;* Sur.
,, idalei: *Olea dioica* Roxb.; *Oleac.;* O.-Ind.
,, kêt: *Gleditschia australis* Hemsl.; *Caesalpiniac.;* Ind.-Ch.
,, mahoe: *Sterculia sp.; Sterculiac.;* br. Gu.
,, mây: *Gelonium multiflorum* A. Juss.; *Euphorbiac.;* Ind.-Ch.
,, nanoi: *Artocarpus lakoocha* Roxb.; *Morac.;* Burma.
,, pa: *Fagraea fragrans* Roxb.; *Loganiac.;* Ind.-Ch.

Man tang: *Litsea citrata* Blume; *Laurac.;* Ind.-Ch.
„   trey: *Adenanthera pavonina* L., *A. microsperma* T. et B.; *Mimosac.;* Ind.-Ch.
Mana: *Lophira alata* Banks; *Ochnac.;* fr. Guin.
„   : *Terminalia chebula* Retz.; *Combretac.;* Burma.
Manaballi: *Amyris sp.; Rutac.;* Gu.
Manaca: *Attalea cohune* Morris; *Palmac.;* C.-Am.
Manaliballi: *Pithecolobium spp.; Mimosac.;* Sur.
Manalliballi: *Stryphnodendron flammatum* Kleinh., *S. sp.; Mimosac.;* Sur.
„   -hororodikoro: *Pithecolobium pedicellare* Bth.; *Mimosac.;* Sur.
Manalu: *Semecarpus sp.; Anacardiac.;* Phil. (Zamboanga)
Manamu: *Vitex longifolia* Merr.; *Verbenac.;* Phil. (Agusan)
Manan: *Funtumia africana* Stapf; *Apocynac.;* Elf.
„   -wala: *Funtumia africana* Stapf; *Apocynac.;* Elf.
Manang: *Shorea polysperma* Merr.; *Dipterocarpac.;* Phil. (Cebu)
Manangkil: *Eugenia mananquil* Blco.; *Myrtac.;* Phil.
Manangolo: *Randia spinosa* Karst.; *Rubiac.;* Kol.
Manáong: *Terminalia pellucida* Presl; *Combretac.;* Phil. (Pangasinan)
Manápo: *Anisoptera curtisi* Dyer; *Dipterocarpac.;* Phil. (Pol.)
„   : *Sindora supa* Merr.; *Caesalpiniac.;* Phil. (n. Tayabas)
Manapuyog: *Shorea polysperma* Merr.; *Dipterocarpac.;* Phil. (Capiz)
Manariballi: *Pithecolobium spp.; Mimosac.;* Sur.
„   -koeréroe: *Pithecolobium pedicellare* Bth.; *Mimosac.;* Sur.
Manáring: *Quercus sp.; Fagac.;* Phil. (Isa., Nueva Viscaya)
Manarizoby: *Dalbergia obtusa* Lecomte, *D. retusa* Baill.; *Papilionac.;* Mad.
Manary: *Dalbergia greveana* Baill.; *Papilionac.;* w. Mad.
Manash: *Pseudolmedia sp.; Morac.;* br. Hond.
Manati: *Coccoloba sp.; Polygonac.;* Cuba.
Manatu: *Plagianthus betulinus* A. Cunn.; *Malvac.;* N.-Seel.
Manay-poongungkai: *Sapindus laurifolius* Vahl; *Sapindac.;* O.-Ind.
Manbalúnu: *Buchanania sp.; Anacardiac.;* Phil. (Zamboanga)
Manbarklak: *Eschweilera spp., Lecythis ollaria* L.; *Lecythidac.;* Sur.
Manbarkraki: *Eschweilera longipes* Miers, *E. sp.; Lecythidac.;* Sur.
Manca montero: *Pithecolobium unguis-cati* Bth.; *Mimosac.;* Cuba.
„   caballo: *Prosopis juliflora* DC.; *Mimosac.;* Pan.
Mançanila: *Hippomane mancinella* L.; *Euphorbiac.;* Ven., Kol.
Mancenilheira: *Hippomane mancinella* L.; *Euphorbiac.;* Bras.
Mancenillier: *Hippomane mancinella* L.; *Euphorbiac.;* tr. Am.
Manchador: *Parathesis macranthera* J. D. Smith; *Myrsinac.;* Guat.
Manchi-bikki: *Gardenia gummifera* L.; *Rubiac.;* O.-Ind. (Madras)
„   moyadi: *Eugenia alternifolia* Wight; *Myrtac.;* s. O.-Ind. (Madras)
„   pala: *Mimusops hexandra* Roxb.; *Sapotac.;* O.-Ind. (Madras)
Manchineel (u): *Hippomane mancinella* L.; *Euphorbiac.;* C.-Am., Kol., Ant.
„   , Mountain-: *Metopium toxiferum* Krug et Urb.; *Anacardiac.;* Fl., br. W.-Ind.
Mancinille: *Hippomane mancinella* L.; *Euphorbiac.;* tr. Am.
Manconilier: *Hippomane mancinella* L.; *Euphorbiac.;* tr. Am.
Manconó: *Xanthostemon verdugonianus* Naves; *Myrtac.;* Phil.
Mancuernillo: *Tabernaemontana sp.; Apocynac.;* Guat.
Manda: *Ulmus wallichiana* Planch.; *Ulmac.;* nw. O.-Ind. (Kaschmir)
Mandal: *Rhododendron arboreum* Sm.; *Ericac.;* nw. O.-Ind. (Punjab)
Mandania: *Acrocarpus fraxinifolius* Wight et Arn.; *Caesalpiniac.;* Beng.
Mandar: *Acer caesium* Wall.; *Acerac.;* nw. O.-Ind. (Punjab)
Mandaracu: *Cereus sp.; Cactac.;* Bras.
Mandarauá: *Shorea palosapis* Merr.; *Dipterocarpac.;* Phil. (Cagayan)
Mandari: *Bauhinia purpurea* L.; *Caesalpiniac.;* O.-Ind. (Madras)
Mandat: *Erythrina suberosa* Roxb.; *Papilionac.;* Assam (Garo)
Mandiyu-ra: *Ceiba sp.; Bombacac.;* Arg.
Mandioca: *Didymopanax longipetiolatum* March.; *Araliac.;* Bras. (Sao P)
Mandioqueira: *Didymopanax sp.; Araliac.;* Bras. (Sao P).
Mandji: *Chlorophora excelsa* Bth. et Hook.; *Morac.;* Gab., Mitt.-Kongo.
Mandjoc.: *Strombosiopsis tetrandra* Engl.; *Olacac.;* Kam.
Mandling: *Acrocarpus fraxinifolius* Wight et Arn.; *Caesalpiniac.;* nö. O.-Ind.
Mandos: *Bischofia javanica* Blume; *Euphorbiac.;* Phil. (Agusan)
Mandukam: *Bassia latifolia* Roxb.; *Sapotac.;* w. Beng. (Chota Nagpur)
Mandur: *Clusia sp.; Guttifer.;* Ek.

Manduvina: *? Zanthoxylum sp.; Rutac.;* Arg.
Manduvira guazu: *? Sterculia sp.; Sterculiac.;* Par.
Manduyirá rá: *? Talisia sp.; Sapindac.;* Arg.
Mang: *Pterospermum truncatolobatum* Gagnep.; *Sterculiac.;* Ind.-Ch.
„ cut: *Garcinia mangostana* L.; *Guttifer.;* Annam.
„ dam: *Alangium chinense* Rehder; *Cornac.;* Ind.-Ch.
„ kieng: *Pterospermum truncatolobatum* Gagnep.; *Sterculiac.;* Ind.-Ch.
„ mang: *Pterospermum greviaefolium* Pierre; *Sterculiac.;* Annam.
Manga: *Macrolobium coeruleoides* de Wild.; *Caesalpiniac.;* B.-K.
„ : *Cocos nucifera* L., *Sclerosperma walkeri* Chev.; *Palmac.;* Gab.
„ : *Randia dumetorum* Lamk.; *Rubiac.;* O.-Ind. (Madras)
„ da praia: *Clusia sp.; Guttifer.;* Bras.
„ -dja-motangani: *Cocos nucifera* L.; *Palmac.;* Gab.
Mangabé: *Didymopanax morototoni* Dcne. et Planch.; *Araliac.:* Pan.
Mangabeira: *Hancornia speciosa* Gomez; *Apocynac.;* Bras.
Mangachapuy: *Hopea acuminata* Merr., *H. sp.; Dipterocarpac.;* Phil.
Mangacitara: *Trichilia sp.; Meliac.;* Arg.
Mangaladi: *Gymnosporia senegalensis* Loes.; *Celastrac.;* n. Nig.
Mangalargo: *Xylopia xylopioides* Standl.; *Anonac.;* Nic.
Mangalito: *Trichilia currani* Blake; *Meliac.;* Kol.
Mangalô: *Dolichos lablab* L.; *Papilionac.;* Bras. (Bahia, Minas Geraes)
„ : *Uouacapoua sp.; Papilionac.;* Bras. (Sao P.)
„ : *Platycyamus regnelli* Bth.; *Papilionac.;* Bras.
„ : *Peraltea erythrinaefolia* Sald.; *Papilionac.;* Bras. (R.-J.)
Mangaluás: *Zizyphus sp.; Rhamnac.;* Phil. (Cotabato)
Mangarahara: *Stereospermum euphorioides* Engl.; *Bignoniac.;* Mad.
Mangari: *Randia dumetorum* Lamk.; *Rubiac.;* w. O.-Ind. (Deccan)
Mangaris: *Koompassia excelsa* Taub.; *Caesalpiniac.;* Borneo.
Mangasinóro: *Shorea philippinensis* Brandis, *S. spp.; Dipterocarpac.;* Phil. (Tay.)
Mangatschapuy: *Uatica mangachapoi* Blco.; *Dipterocarpac.;* Phil.
Mangbesi: *Hopea beccariana* Burck; *Dipterocarpac.;* br. Borneo (Sarawak)
Mangeao: *Litsea calicaris* Kirk; *Laurac.;* N.-Seel.
Mangel: *Sarcocephalus junghuhni* Miq.; *Rubiac.;* Malay.
Mángga: *Mangifera indica* L.; *Anacardiac.;* Phil.
„ dodol: *Mangifera pentandra* Hook. f.; *Anacardiac.;* Malay.
Manggachapúi: *Hopea spp., Shorea spp.; Dipterocarpac.;* Phil.
Manggala: *Garcinia sp.; Guttifer.;* Phil. (Zamboanga)
Manggapóle: *Mangifera altissima* Blco.; *Anacardiac.;* Phil. (Olutanga, Zamboanga)
Manggasinóro: *Shorea spp., Hopea acuminata* Merr.; *Dipterocarpac.;* Phil.
„ : *Parashorea malaanonan* Merr.; *Dipterocarpac.;* Phil. (Masbate)
Manggasiríki: *Quercus sp., Q. ovalis* Blco.; *Fagac.;* Phil.
Manggatarím: *Hopea philippinensis* Dyer; *Dipterocarpac.;* Phil. (Pang., Tay.)
Manggatsapúi: *Mangifera monandra* Merr.; *Anacardiac.;* Phil. (Rizal)
Manggis: *Garcinia mangostana* L.; *Guttifer.;* Malay.
„ : *Koompassia excelsa* Taub.; *Caesalpiniac.;* Phil. (Palawan)
„ : *Pometia pinnata* Forst.; *Sapindac.;* Phil. (Palawan)
„ hutan: *Garcinia hombroniana* Pierre; *Guttifer.;* Malay.
„ outan: *Garcinia malaccensis* Hook. f.; *Guttifer.;* Malak.
Manggoórang: *Shorea polysperma* Merr.; *Dipterocarpac.;* Phil. (Camar., Albay)
Manghás: *Cyclostemon sp.; Euphorbiac.;* Phil. (Tawitawi)
Mangi: *Terminalia arjuna* Bedd.; *Combretac.;* C.-O.-Ind. (Gondwana)
Mangián: *Buchanania sp.; Anacardiac.;* Phil. (Lanao)
Mangienhatti: *Tecoma sp., Tabebuia sp.; Bignoniac.;* Sur.
Mangies boom: *Garcinia mangostana* L.; *Guttifer.;* Nied.-Ind.
„ oetan: *Garcinia javanica* Blume, *G. oxyphylla* Miq.; *Guttifer.;* Nied.-Ind.
Mangir: *Ganophyllum falcatum* Blume; *Sapindac.;* Java.
Mangka: *Artocarpus integrifolia* L. f.; *Morac.;* Nied.-Ind. (n. Cel.)
Mangkabang assu: *Shorea brachyptera* Heim; *Dipterocarpac.;* Borneo (Sarawak)
Mangkás: *Sideroxylon ferrugineum* Hook. et Arn.; *Sapotac.;* Phil. (Sibutu)
Mangkonó: *Xanthostemon verdugonianus* Naves; *Myrtac.;* Phil.
Mangláti: *Lagerstroemia piriformis* Koehne; *Lythrac.;* Phil. (Bisaya)
Mangle: *Rhizophora mangle* L.; *Rhizophorac.;* tr. Am.
„ : *Maytenus phyllanthoides* Bth.; *Celastrac.;* Mex.
„ : *Conocarpus erecta* L.; *Combretac.;* P.-R., St.-D., Mex.

Mangle: *Symmeria paniculata* Bth.; *Polygonac.;* Kol.
  „ aguabola: *Maytenus phyllanthoides* Bth.; *Celastrac.;* Mex.
  „ amarillo: *Laguncularia racemosa* Gaertn. f.; *Combretac.;* Cuba.
  „ blanc: *Laguncularia racemosa* Gaertn. f.; *Combretac.;* Guad.
  „ blanco: *Avicennia nitida* Jacq.; *Verbenac.;* P.-R.
  ··   „ : *Laguncularia racemosa* Gaertn. f.; *Combretac.;* P.-R., Hond., Ven.
  „   „ : *Bravaisia integerrima* Standl.; *Acanthac.;* Nic.
  „   „ : *Trichilia sp.; Meliac.;* Kol.
  „ bobo: *Laguncularia racemosa* Gaertn. f.; *Combretac.;* P.-R.
  „ „ : *Avicennia nitida* Jacq.; *Verbenac.;* P.-R.
  „ boton: *Conocarpus erecta* L.; *Combretac.;* tr. Am.
  „ botoncillo: *Conocarpus erecta* L.; *Combretac.;* P.-R.
  „ , China-: *Avicennia tomentosa* Jacq.; *Verbenac.;* Trin.
  „ colorado: *Rhizophora mangle* L.; *Rhizophorac.;* Cuba, Hond., Ven., Ek.
  „   „ : *Tovomitopsis multiflora* Standl.; *Guttifer.;* Pan.
  „ dulce: *Rhizophora mangle* L.; *Rhizophorac.;* Mex.
  „ „ : *Maytenus phyllanthoides* Bth.; *Celastrac.;* Mex.
  „ gateador: *Rhizophora mangle* L.; *Rhizophorac.;* C.-R.
  „ negro: *Avicennia nitida* Jacq.; *Verbenac.;* br. Hond.
  „ oseille: *Dodonaea viscosa* Jacq. *var. angustifolia* Bth.; *Sapindac.;* Jam.
  „ piñuelo: *Conocarpus erecta* L.; *Combretac.;* Pan.
  „ prieto: *Avicennia officinalis* L.; *Verbenac.;* Ven.
  „ roche: *Conocarpus erecta* L.; *Combretac.;* Trin.
  „ rouge: *Rhizophora mangle* L.; *Rhizophorac.;* Guad.
  „ salado: *Avicennia nitida* Jacq.; *Verbenac.;* Ek.
  „ súre: *Dodonaea viscosa* Jacq. *var. angustifolia* Bth.; *Sapindac.;* Jam.
  „ tinto: *Rhizophora mangle* L.; *Rhizophorac.;* Mex.
  „ de agua: *Bravaisia integerrima* Standl.; *Acanthac.;* Pan.
Manglesito: *Trichilia sp.; Meliac.;* Kol.
Manglid: *Michelia montana* Blume; *Magnoliac.;* Java.
Manglier: *Sideroxylon bojeranum* A. DC.; *Sapotac.;* Maur.
  „ blanc: *Avicennia officinalis* L.; *Verbenac.;* Sey.
  „ chandelle: *Rhizophora mangle* L.; *Rhizophorac.;* fr. Gu.
Mangliet: *Manglietia glauca* Blume; *Magnoliac.;* Nied.-Ind.
Manglillo: *Heisteria macrophylla* Oerst.; *Olacac.;* C.-R.
Mango: *Mangifera indica* L.; *Anacardiac.;* Tr.
  „   : *Casuarina equisetifolia* Forst.; *Casuarinac.;* Au.
  „ , african: *Irvingia gabonensis* Baill.; *Simarubac.;* W.-Af.
  „ bark: *Protium australasicum* T. A. Sprague; *Burserac.;* ö. Au.
  „ oetan: *Garcinia javanica* Blume; *Guttifer.;* Nied.-Ind.
  „ , Sabana-: *Byrsonima sp.; Malpighiac.;* Sur.
  „ , wild: *Irvingia gabonensis* Baill.; *Simarubac.;* Gab.
Mangoe: *Garcinia mangostana* L.; *Guttifer.;* Nied.-Ind.
  „ lowong: *Garcinia javanica* Blume; *Guttifer.;* Nied.-Ind.
Mangollano: *Pithecolobium dulce* Bth.; *Mimosac.;* Salv.
Mangoma: *Klaineana latifolia* Pierre; *Simarubac.;* Gab.
Mangona: *Khaya anthotheca* C. DC.; *Meliac.;* Kam.
  „ , white: *? Khaya anthotheca* C. DC.; *Meliac.;* Kam.
Mangongkóng: *Celtis sp.; Ulmac.;* Phil. (Negrosinsel)
Mangostan: *Garcinia mangostana* L.; *Guttifer.;* Nied.-Ind.
Mangosteen: *Garcinia mangostana* L.; *Guttifer.;* Malay.
  „ , wild: *Sandoricum indicum* Cav.; *Meliac.;* Malay.
Mangot: *Mangifera indica* L.; *Guttifer.;* Guad.
Mangottier: *Mangifera indica* L.; *Guttifer.;* Guad.
Mangourra: *Hymenostegia sp.; Caesalpiniac.;* Gab.
Mangoustan: *Garcinia mangostana* L.; *Guttifer.;* Ind.-Ch.
  „   : *Garcinia malabarica* Lamk.; *Guttifer.;* sw. O.-Ind. (Mal.)
  „ corné: *Garcinia cornea* L.; *Guttifer.;* Mart.
  „ sauvage: *Sandoricum indicum* Cav.; *Meliac.;* tr. As.
  „ du Malabar: *Garcinia malabarica* Lamk.; *Guttifer.;* sw. O.-Ind. (Mal.)
Mangrai: *Salix tetrasperma* Roxb.; *Salicac.;* Burma.
Mangro: *Rhizophora mangle* L.; *Rhizophorac.;* Sur.
  „ , witte: *Conocarpus erecta* L.; *Combretac.;* Sur.
  „ , Zwamp-: *Rhizophora mangle* L.; *Rhizophorac.;* Sur.

Mangrove: *Rhizophora spp., Bruguiera spp., Ceriops spp.; Rhizophorac.;* Tr.
„   : *Xylocarpus sp.; Meliac.;* O.-Af. (it. Somali)
„ , amerikanische: *Rhizophora mangle* L.; *Rhizophorac.;* tr. Am.
„ , Berg-: *Tabebuia dentata* Miers; *Bignoniac.;* fr. Gu.
„ , black: *Avicennia nitida* Jacq.; *Verbenac.;* Trin., Tob., Hond., br. Hond.
„ , Button-: *Conocarpus erecta* L.; *Combretac.;* Hond.
„ , red: *Rhizophora mangle* L.; *Rhizophorac.;* tr. Am.
„ , „ : *Rhizophora racemosa* G. F. W. Meyer; *Rhizophorac.;* Lib., Go.
„ , red-flowered: *Lumniţera coccinea* Wight et Arn.; *Combretac.;* tr. As., tr. Au.
„ , roode: *Rhizophora mangle* L.; *Rhizophorac.;* fr. Gu.
„ , white: *Avicennia nitida* Jacq.; *Verbenac.;* Pan., Go.
„ , witte: *Avicennia nitida* Jacq.; *Verbenac.;* fr. Gu.
„ , Zaragoza-: *Conocarpus erecta* L.; *Combretac.;* br. Hond.
Mangsteen: *Garcinia mangostana* L.; *Guttifer.;* Nied.-Ind.
Mangu: *Zizyphus mucronata* Willd.; *Rhamnac.;* Togo.
„ -mangu: *Symphonia gabonensis* Pierre; *Guttifer.;* B.-K.
Mangue: *Avicennia africana* Beauv.; *Verbenac.;* port. W.-Af.
„ apareiba: *Rhizophora mangle* L.; *Rhizophorac.;* fr. Gu.
„ blanc: *Moronobea sp.; Guttifer.;* fr. Gu.
„ branco: *Laguncularia racemosa* Gaertn. f.; *Combretac.;* Bras.
„ encarnado: *Rhizophora mucronata* Lamk.; *Rhizophorac.;* port. O.-Af.
„   „ : *Bruguiera gymnorrhiza* Lamk.; *Rhizophorac.;* port. O.-Af.
„   „ : *Ceriops candolleana* Arn.; *Rhizophorac.;* port. O.-Af.
„ rôxo: *? Rhizophora mangle* L.; *Rhizophorac.;* Angola (Zaire)
„ vermelho: *Rhizophora mangle* L.; *Rhizophorac.;* fr. Gu., Bras.
„   „ : *Rhizophora sp., Bruguiera sp., Ceriops sp.; Rhizophorac.;* p. O.-Af.
„ , yellow: *Moronobea coccinea* Aubl.; *Guttifer.;* Trin.
„ da praia: *? Rhizophora mangle* L.; *Rhizophorac.;* Angola (Zaire)
Mangueira: *Mangifera indica* L.; *Anacardiac.;* tr. Am.
Manguier: *Mangifera indica* L.; *Anacardiac.;* Guad., Guat.
„ fétide: *Mangifera foetida* Lour.; *Anacardiac.;* Ind.-Ch.
„ marron: *Celastrus undulatus* Lamk.; *Celastrac.;* Kap, Mask.
„ sauvage: *Irvingia gabonensis* Baill.; *Simarubac.;* Kam., Gab.
„   „ : *Mangifera indica* L.; *Anacardiac.;* Ind.-Ch.
Manh pha: *Fagraea fragrans* Roxb.; *Loganiac.;* Ind.-Ch.
Mani: *Moronobea coccinea* Aubl.: *Guttifer.;* Gu.
„ maruthu: *Lagerstroemia flos-reginae* Reţ.; *Lythrac.;* sw. O.-Ind. (Mal.)
Maniapotano wéwé: *Nectandra sp., Ocotea sp.; Laurac.;* Sur.
Maniawga: *Carallia lucida* Roxb.; *Rhizophorac.;* Burma.
Maniawya: *Carallia lucida* Roxb.; *Rhizophorac.;* And.
Maniballi: *Platonia insignis* Mart., *Moronobea coccinea* Aubl.; *Guttifer.;* br. Gu.
„   : *Amyris balsamifera* L.; *Rutac.;* fr. Gu.
Mani kwaho: *Calophyllum sp.; Guttifer.;* Sur.
„ kwaka: *Calophyllum sp.; Guttifer.;* Sur.
Manicníc: *Palaquium sp.; Sapotac.;* Phil.
Maniçoba: *Manihot glaziovi* Muell. Arg.; *Euphorbiac.;* Bras.
Manicole: *Euterpe edulis* L.; *Palmac.;* br. Gu.
Manie: *Moronobea coccinea* Aubl.; *Guttifer.;* Sur.
Maniglia: *Filicium decipiens* Thw.; *Sapindac.;* sw. O.-Ind., Cey.
Maníhai: *Dehaasia triandra* Merr.; *Laurac.;* Phil. (Mindoro)
Manikbub'uyo: *Celtis sp.; Ulmac.;* Phil. (Iloilo)
Manikník: *Palaquium sp.; Sapotac.;* Phil.
Manil: *Moronobea coccinea* Aubl., *Platonia insignis* Mart.; *Guttifer.;* Sur., fr. Gu.
„ kara: *Mimusops balata* L.; *Sapotac.;* Gu.
„ pacouri: *Moronobea coccinea* Aubl., *Platonia insignis* Mart.; *Guttifer.;* fr. Gu.
Maniíí: *Shorea polysperma* Merr.; *Dipterocarpac.;* Phil. (Camarines, Albay)
Manilig: *Bassia betis* Merr.; *Sapotac.;* Phil. (Cotabato)
Mañin: *Saxegothaea sp.; Conifer.;* Chile.
Maniníla: *Garcinia sp.; Guttifer.;* Phil. (Camarines)
Manipau: *Platonia insignis* Mart., *Moronobea coccinea* Aubl.; *Guttifer.;* Sur.
Manipika: *Dalbergia trichocarpa* Baker, *D. greveana* Baill.; *Papilionac.;* Mad.
Manipníp: *Palaquium sp.; Sapotac.;* Phil. (Bataan)
Manipounga-marom: *Sapindus laurifolius* Vahl; *Sapindac.;* w. O.-Ind. (Deccan)
Manirito: *Anona sp.; Anonac.;* Ven.

Manirote: *Anona sp.; Anonac.;* Ven.
Mañiu: *Podocarpus chilina* Rich., *P. andina* Poepp.; *Conifer.;* Chile.
Manjakadambe: *Adina cordifolia* Hook. f.; *Rubiac.;* O.-Ind. (Madras)
Mank-barklak: *Lecythis ollaria* L., *Eschweilera longipes* Miers; *Lecythidac.;* Sur.
Mankrappa: *Tapirira aff. guianensis* Aubl.; *Anacardiac.;* Sur.
Manletri: *Piratinera guianensis* Aubl.; *Morac.;* Sur.
Manletterhout: *Piratinera spp.; Morac.;* Sur., fr. Gu.
Manloáb: *Quercus sp.; Fagac.;* Phil. (Rizal)
Manni: *Moronobea coccinea* Aubl.; *Guttifer.;* Sur., br. Gu.
Manniballi: *Moronobea coccinea* Aubl.; *Guttifer.;* br. Gu.
Mannie: *Moronobea coccinea* Aubl.; *Guttifer.;* Sur.
Mano de león: *Hernandia guianensis* Aubl.; *Hernandiac.;* Hond.
  „ „ „ : *Oreopanax salvini* Hemsl.; *Araliac.;* Hond.
  „ „ „ : *Tabebuia pentaphylla* Hemsl.; *Bignoniac.;* C.-Am.
  „ „ „ : *Elaeodendron sp.; Celastrac.;* Mex.
  „ „ pilón: *Mouriria sp.; Melastomatac.;* Cuba.
Manoa papos: *Anona squamosa* L.; *Anonac.;* W.-Ind.
Manoc: *Parinarium griffithianum* Bth.; *Rosac.;* Java., Cel.
Manocarpa: *Casearia sp.; Flacourtiac.;* Guat.
Manogaróm: *Diospyros nitida* Merr.; *Ebenac.;* Phil. (Guimaras)
Manogtalísai: *Palaquium sp.; Sapotac.;* Phil. (Capiz)
Manoka mavo: *Asteropeia rhopaloides* Baker; *Flacourtiac.;* Mad.
  „ mena: *Asteropeia micraster* Haller; *Flacourtiac.;* Mad.
Manoko: *Ficus melleri* Baker; *Morac.;* Mad.
Manománo: *Polyscias nodosa* Seem.; *Araliac.;* Phil. (Basilan)
Manoon: *Ulmus wallichiana* Planch.; *Ulmac.;* nw. O.-Ind. (Kaschmir)
Manoué: *Flindersia fournieri* Panch. et Sebert; *Rutac.;* N.-Kal.
Manschinellapfel: *Hippomane mancinella* L.; *Euphorbiac.;* tr. Am.
Manschinellenbaum: *Hippomane mancinella* L.; *Euphorbiac.;* C.-Am.
Manshû-gurumi (jap): *Juglans mandschurica* Maxim.; *Juglandac.;* O.-As.
  „ kurumi: *Juglans mandschurica* Maxim.; *Juglandac.;* Jap., Korea.
Mantabíg: *Terminalia nitens* Presl; *Combretac.;* Phil. (Zamboanga)
Mantapoepa: *Gustavia angusta* L.; *Lecythidac.;* Sur.
Manteca: *Celtis hottlei* Standl.; *Ulmac.;* Hond.
  „ : *Tetrorchidium rotundatum* Standl.; *Euphorbiac.;* Hond.
Manteco: *Croton fragrans* H. B. K.; *Euphorbiac.;* Kol.
Manú: *Minquartia guianensis* Aubl.; *Olacac.;* C.-R.
Manuka: *Leptospermum scoparium* Forst., *L. ericoides* A. Rich.; *Myrtac.;* N.-Seel.
  „ , black: *Leptospermum scoparium* Forst.; *Myrtac.;* N.-Seel.
Manumbága: *Myristica sp.; Myristicac.;* Phil. (Zamboanga)
Manuno: *Cordia diversifolia* Pavón; *Borraginac.;* Guat.
Manuton: *Leptogonum ericoides* A. Rich.; *Polygonac.;* N.-Seel.
Manutra: *? Leptospermum ericoides* A. Rich.; *Myrtac.;* N.-Seel.
Manvonkhout: *Licania pachystachya* Kleinh.; *Rosac.;* Sur.
Manwood: *Nectandra sp.; Laurac.;* Pan.
  „ : *Minquartia guianensis* Aubl.; *Olacac.;* Pan.
  „ , black: *Minquartia guianensis* Aubl.; *Olacac.;* Pan.
  „ , red: *Calocarpum mammosum* Pierre; *Sapotac.;* Pan.
  „ , yellow: *Vitex cooperi* Standl.; *Verbenac.;* Pan.
Manyul: *Randia dumetorum* Lam.; *Rubiac.;* O.-Ind.
Manzana: *Eugenia jambos* L.; *Myrtac.;* Hond.
  „ : *Malus silvestris* Mill.; *Rosac.;* Hond.
  „ rosa: *Eugenia jambos* L.; *Myrtac.;* Nic.
Manzanilla: *Crataegus stipulosa* Steud.; *Rosac.;* Guat.
  „ : *Ximenia americana* L.; *Olacac.;* Hond.
  „ de playa: *Hippomane mancinella* L.; *Euphorbiac.;* C.-Am.
Manzanillo: *Hippomane mancinella* L.; *Euphorbiac.;* W.-Ind., C.-Am., Ek.
  „ de la costa: *Hippomane mancinella* L.; *Euphorbiac.;* Cuba.
  „ „ „ playa: *Hippomane mancinella* L.; *Euphorbiac.;* C.-R.
Manzanita: *Arctostaphylos glauca* Lindl., *A. bicolor* A. Gray; *Ericac.;* w. USA.
  „ : *Arbutus menziesi* Pursh; *Ericac.;* w. N.-Am.
Manzano silvestre: *Pirus malus* L.; *Rosac.;* Sp.
  „ del campo: *Ruprechtia corylifolia* Gris., *R. triflora* Gris.; *Polygonac.;* Arg.
Manzanote: *Pereskia autumnalis* Rose; *Cactac.;* Guat.

Mao dá: *Mallotus philippinensis* Muell. Arg.; *Euphorbiac.*; Ind.-Ch.
„ de picao: *Licania turiuva* Cham.; *Rosac.*; Bras.
Maoranaballi: *Qualea dinizi* Ducke, *Q. sp.*; *Vochysiac.*; Sur.
Maou: *Couratari guianensis* Aubl.; *Lecythidac.*; fr. Gu.
Mapa: *Macoubea guianensis* Aubl., *Couma guianensis* Aubl.; *Apocynac.*; Sur.
Mapalapa: *Hevea guianensis* Aubl.; *Euphorbiac.*; Sur.
Maparajuba: *Mimusops sp.*; *Sapotac.*; Bras.
Maparakoni: *Inga sp.*; *Mimosac.*; br. Gu.
Maparapa: *Hevea guianensis* Aubl.; *Euphorbiac.*; Sur.
Mapiche: *Myrcia lanceolata* Camb. *var. angustifolia* Ndz.; *Myrtac.*; Peru.
Mapingo: *Diospyros perrieri* H. Jum.; *Ebenac.*; w. Mad.
Mapipí: *Litsea sp.*; *Laurac.*; Phil. (Ticao)
Maple (e): *Acer pseudoplatanus* L.; *Acerac.*; Eu.
„ : *Acer saccharum* Marsh.; *Acerac.*; ö. N.-Am.
„ : *Acer spp.*; *Acerac.*; Jap.
„ : *Flindersia chatawaiana* F. M. Bailey; *Rutac.*; nö. Au.
„ : *Villaresia moorei* F. v. M.; *Icacinac.*; N.-S.-W.
„ , african: *Triplochiton scleroxylon* K. Schum.; *Sterculiac.*; Nig.
„ , ash-leaved: *Acer negundo* L.; *Sapindac.*; N.-Am.
„ , australian: *Flindersia brayleyana* F. v. M.; *Rutac.*; Au.
„ , big-leaf-: *Acer macrophyllum* Pursh; *Acerac.*; nw. Südsee.
„ , Bird's-eye-: *Acer saccharum* Marsh.; *Acerac.*; N.-Am.
„ , black: *Acer saccharum* Marsh., *A. nigrum* Michx.; *Acerac.*; N.-Am.
„ , British-Columbia-: *Acer macrophyllum* Pursh; *Acerac.*; sw. Kan.
„ , broad-leaf-: *Acer macrophyllum* Pursh; *Acerac.*; w. USA.
„ , Bush-: *Triplochiton scleroxylon* K. Schum.; *Sterculiac.*; Kam., Nig.
„ , californian: *Acer macrophyllum* Pursh; *Acerac.*; USA.
„ , canadian: *Acer saccharum* Marsh.; *Acerac.*; N.-Am.
„ , Curly-: *Acer saccharum* Marsh.; *Acerac.*; USA.
„ , Dwarf-: *Acer glabrum* Torr.; *Acerac.*; nw. N.-Am.
„ , „ : *Acer spicatum* Lam.; *Acerac.*; sö. Kan.
„ , dyed: *Acer campestre* L.; *Acerac.*; Eu.
„ , himalayan: *Acer spp.*; *Acerac.*; O.-Ind.
„ , Manitoba-: *Acer negundo* L.; *Acerac.*; Kan.
„ , Moose-: *Acer pennsylvanicum* L.; *Acerac.*; ö. Kan.
„ , Mountain-: *Acer spicatum* Lam.; *Acerac.*; sö. Kan.
„ , „ : *Acer circinatum* Pursh; *Acerac.*; w. N.-Am.
„ , Norway-: *Acer platanoides* L.; *Acerac.*; Eu.
„ , Oregon-: *Acer macrophyllum* Pursh; *Acerac.*; sw. Kan.
„ , Queensland-: *Flindersia spp.*; *Rutac.*; ö. Au.
„ , red: *Acer rubrum* L., *A. glabrum* Torr.; *Acerac.*; N.-Am.
„ , River-: *Acer saccharum* Marsh.; *Acerac.*; sö. Kan.
„ , Rock-: *Acer saccharum* Marsh.; *Acerac.*; sö. Kan.
„ , „ -, canadian: *Acer saccharum* Marsh.; *Acerac.*; Kan.
„ , Scarlet-: *Acer rubrum* L.; *Acerac.*; sö. Kan.
„ , Silver-: *Acer dasycarpum* Ehrh.; *Acerac.*; N.-Am.
„ , soft: *Acer dasycarpum* Ehrh.; *Acerac.*; N.-Am.
„ , southern: *Cryptocarya erythroxylon* Maid. et Betche; *Laurac.*; ö. Au.
„ , striped: *Acer pennsylvanicum* L.; *Acerac.*; ö. N.-Am.
„ , Sugar-: *Acer saccharum* Marsh.; *Acerac.*; USA.
„ , Swamp-: *Acer dasycarpum* Ehrh.; *Acerac.*; sö. Kan.
„ , Sycamore-: *Acer pseudoplatanus* L.; *Acerac.*; Eu.
„ , Water-: *Acer dasycarpum* Ehrh.; *Acerac.*; sö. Kan.
Mapoa gris: *Pisonia obtusata* Jacq.; *Nyctaginac.*; Guad.
Mapola: *Luehea seemanni* Triana et Planch.; *Tiliac.*; br. Hond.
? „ : *Bernoullia flammea* Oliver; *Bombacac.*; br. Hond.
„ : *Malvaviscus grandiflorus* H. B. K.; *Malvac.*; Nic.
„ : *Malvaviscus arboreus* Cav.; *Malvac.*; Pan.
Mapoöpo: ? *Sterculia sp.*; *Sterculiac.*; Celebes.
Mapoortza: *Afzelia quanzensis* Welw.; *Caesalpiniac.*; S.-Af. (Transwal)
Maporokon: *Inga alba* Willd., *I. sertulifera* DC.; *Mimosac.*; Sur.
Mapou: *Monimia sp.*; *Monimiac.*; Réunion.
„ blanc: *Cordia colococca* L.; *Borraginac.*; Guad.
„ à petites feuilles: *Monimia ovalifolia* Pet.-Th.; *Monimiac.*; Réunion.

Mapourita: *Zanthoxylum sp.; Rutac.;* Trin.
Mappa: *Macoubea guianensis* Aubl.; *Apocynac.;* Sur.
Maprokon: *Inga alba* Willd., *I. sertulifera* DC.; *Mimosac.;* Sur.
Maprokonie: *Inga alba* Willd., *I. sertulifera* DC.; *Mimosac.;* Sur.
Mapurito: *Fagara monophylla* Lam.; *Rutac.;* P.-R.
   „   : *Zanthoxylum clava-herculis* L.; *Rutac.;* Trin.
Maqueliz: *Tabebuia pentaphylla* Hemsl.; *Bignoniac.;* br. Hond.
Maquelizo: *Tabebuia pentaphylla* Hemsl.; *Bignoniac.;* Guat.
Maqui: *Aristotelia macqui* L'Hérit.; *Elaeocarpac.;* Chile.
   „ sapa ñac-cha: *Apeiba sp.; Tiliac.;* Peru.
Maquiligua: *Tabebuia pentaphylla* Hemsl.; *Bignoniac.;* Salv.
Maquilihuatl: *Quercus sp.; Fagac.;* Mex.
Maquiliz: *Tabebuia pentaphylla* Hemsl.; *Bignoniac.;* W.-Ind., C.-Am., n. S.-Am.
Mar akka: *Hopea treubi* Heim; *Dipterocarpac.;* br. Borneo (Sarawak)
   „ pinang: *Diospyros sp.; Ebenac.;* Borneo.
Mara blanca: *Bursera karsteniana* Engl.; *Burserac.;* Ven.
   „ colorada: *Bursera heterophylla* Engl.; *Burserac.;* Ven.
   „ kata: *Uitex peduncularis* Wall.; *Uerbenac.;* w. Beng. (Chota Nagpur)
Marabikál: *Diospyros pilosanthera* Blco.; *Ebenac.;* Phil. (Camarines)
Marabitáog: *Calophyllum cumingi* Triana et Planch.; *Guttifer.;* Phil. (Union)
Marabóhok: *Casuarina equisetifolia* Forst.; *Casuarinac.;* Phil. (Biscaya)
Maradu: *Terminalia tomentosa* Wight et Arn.; *Combretac.;* O.-Ind., Cey.
   „   : *Aegle marmelos* Correa; *Rutac.;* O.-Ind. (Madras)
Maragabúto: *Alangium meyeri* Merr.; *Cornac.;* Phil. (Tayabas)
Maragatáu: *Neonauclea sp.; Rubiac.;* Phil. (Cagayan)
Maragauéd: *Celtis sp.; Ulmac.;* Phil.
Maragutau: *Aglaia luzoniensis* Merr. et Rolfe; *Meliac.;* Phil. (Cagayan)
Marakata: *Uitex peduncularis* Wall.; *Uerbenac.;* O.-Ind.
Marakh: *Leptadenia spartium* Wight; *Asclepiadac.;* O.-Af. (n. Sudan)
Marako: *Peltogyne spp., Copaifera bracteata* Bth.; *Caesalpiniac.;* Sur.
Marama: *Sonneratia apetala* Ham.; *Lythrac.;* O.-Ind. (Madras)
Maramatám: *Eugenia claviflora* Roxb.; *Myrtac.;* Phil. (Cagayan)
Maran: *Pinus khasya* Royle; *Conifer.;* Burma.
Márang: *Litsea sp., L. perrotteti* F. Vill.; *Laurac.;* Phil.
   „   : *Neolitsea vidali* Merr.; *Laurac.;* Phil. (Laguna)
Marang-inang: *Dysoxylum densiflorum* Miq.; *Meliac.;* Nied.-Ind.
   „ jowar: *Grewia laevigata* Vahl; *Tiliac.;* Burma.
Maranga calalú: *Moringa oleifera* Lam.; *Moringac.;* br. Hond.
Maránggo: *Azadirachta integrifoliola* Merr.; *Meliac.;* Phil. (Palawan)
Marañón: *Anacardium occidentale* L.; *Anacardiac.;* Hond., Nic., Peru.
Maranti: *Hopea mengarawan* Miq., *H. spp.; Dipterocarpac.;* Nied.-Ind.
   „ -betoel: *Hopea maranti* Miq.; *Dipterocarpac.;* Nied.-Ind.
   „ bras: *Shorea nitens* Miq.; *Dipterocarpac.;* Sum.
   „ chingal: *Shorea bracteolata* Dyer; *Dipterocarpac.;* Malak.
   „ kerrap: *Shorea parvifolia* Dyer; *Dipterocarpac.;* Malak.
   „ payah: *Shorea acuminata* Dyer; *Dipterocarpac.;* Malay.
Marapiáo: *Zanthoxylum macrophyllum* Oliv.; *Rutac.;* port. Af.
Marara: *Weinmannia lachnocarpa* F. v. M.; *Cunoniac.;* ö. Au.
   „ , rose: *Weinmannia lachnocarpa* F. v. M.; *Cunoniac.;* ö. Au.
   „ , rose-leaf-: *Ackama paniculata* Engl.; *Cunoniac.;* ö. Au.
   „ , yellow: *Sideroxylon australe* Bth. et Hook. f.; *Sapotac.;* Queensl.
Mararie: *Weinmannia lachnocarpa* F. v. M.; *Cunoniac.;* ö. Au.
Marasantog: *Koordersiodendron pinnatum* Merr.; *Anacardiac.;* Phil. (n. Luzon)
Marat: *Aegle marmelos* Correa; *Rutac.;* O.-Ind. (Madras)
Maratakka: *Ocotea rodioei* Mez; *Laurac.;* Sur.
Maratayúm: *Adenanthera intermedia* Merr.; *Mimosac.;* Phil. (Nueva Ecija)
Maratekka: *Albizzia saponaria* Blume; *Mimosac.;* Phil. (Ilocos Sur)
Marati-moggu: *Ceiba pentandra* Gaertn.; *Bombacac.;* w. O.-Ind. (Deccan)
Maratigá: *Albizzia saponaria* Blume; *Mimosac.;* Phil. (Ilocos Sur)
Maraútong: *Euphoria didyma* Blco.; *Sapindac.;* Phil. (Cagayan)
   „   : *Zizyphus sp.; Rhamnac.;* Phil. (Ilocos Norte)
Maravatti: *Hydnocarpus wightiana* Blume; *Flacourtiac.;* sw. O.-Ind. (Madras)
Maraw: *Schima wallichi* Choisy; *Theac.;* Burma.
Marawinazoo: *Peltogyne spp., Copaifera bracteata* Bth.; *Caesalpiniac.;* Sur.

Marawineroo: *Peltogyne sp.; Caesalpiniac.;* br. Gu.
Marban tayan: *Dryobalanops sp.; Dipterocarpac.;* Nied.-Ind. (? Borneo)
Marblewood: *Diospyros kurzi* Hiern, *D. oocarpa* Thw.; *Ebenac.;* And.
  „   : *Acacia bakeri* J. H. Maid.; *Mimosac.;* ö. Au.
  „ , andaman: *Diospyros kurzi* Hiern, *D. oocarpa* Thw.; *Ebenac.;* O.-Ind.
Marbre: *Hymenaea courbaril* L.; *Caesalpiniac.;* Guad.
Marbri: *Richeria grandis* Vahl; *Euphorbiac.;* Guad.
Marchi: *Murraya exotica* L.; *Rutac.;* nw. O.-Ind. (Kumaon)
Marchula: *Murraya exotica* L.; *Rutac.;* n. O.-Ind. (United Prov.)
Marchulajuti: *Murraya exotica* L.; *Rutac.;* w. O.-Ind. (Bombay)
Marda: *Terminalia tomentosa* Wight et Arn.; *Combretac.;* C.-O.-Ind. (Gondwana)
Mareen: *Ulmus wallichiana* Planch.; *Ulmac.;* nw. O.-Ind. (Punjab)
Marenballi wadilikoro: *Licania pachystachya* Kleinh.; *Rosac.;* Sur.
Marfil: *Homalium pittieri* Bth.; *Flacourtiac.;* Ven.
  „ vegetal: *Astronium graveolens* Jacq.; *Anacardiac.;* Kol.
Marfim: *Balfourodendron riedelianum* Engl.; *Rutac.;* Bras. (Sao P)
Marfinzeiro: *Agonandra brasiliensis* Bth.; *Olacac.;* nö. Bras. (Amaz.)
Margabólo: *Cyclostemon sp.; Euphorbiac.;* Phil. (Laguna)
  „   : *Knema sp.; Myristicac.;* Phil. (Rizal, Laguna)
Margadiláu: *Neonauclea sp.; Rubiac.;* Phil. (Laguna)
  „   : *Phoebe sterculioides* Merr.; *Laurac.;* Phil. (Laguna)
Margapáli: *Dehaasia triandra* Merr., *Nothophoebe malabonga* Merr.; *Laurac.;* Phil.
  „ -kutilisiáu: *Machilus philippinensis* Merr.; *Laurac.;* Phil. (Lag.)   [(Lag.)
Margarita: *Karwinskia humboldtiana* Zucc.; *Rhamnac.;* Mex. (Sinaloa)
Marghang: *Quercus dilatata* Lindley; *Fagac.;* nw. O.-Ind. (Punjab)
Margosa tree: *Azadirachta indica* A. Juss.; *Meliac.;* O.-Ind., Malay.
Margosier: *Azadirachta indica* A. Juss.; *Meliac.;* tr. As.
Mari: *Caryota urens* L.; *Palmac.;* O.-Ind., Beng., Assam.
  „ : *Ficus bengalensis* L.; *Morac.;* O.-Ind. (Madras)
Maria: *Calophyllum spp.; Guttifer.;* tr. Am.
  „ blanca: *Tabebuia pentaphylla* Hemsl.; *Bignoniac.;* Mex.
  „ preta: *Melanoxylon brauna* Schott; *Caesalpiniac.;* Bras. (Sao P)
  „   „ : *Thouinia ornifolia* Griseb.; *Sapindac.;* Arg.
  „   „ : *Diatenopterix sorbifolia* Radlk.; *Sapindac.;* Arg.
Mariamol: *Pisonia inermis* Jacq.; *Nyctaginac.;* Bras.
Maricao: *Byrsonima spicata* L. C. Rich.; *Malpighiac.;* P.-R.
Marie crabe: *Lantana camara* L.; *Verbenac.;* fr. Gu.
Marignia: *Protium obtusifolium* March.; *Burserac.;* Mask.
Mariíg: *Eugenia glaucicalyx* Merr.; *Myrtac.;* Phil. (Bataan)
Marika: *Anogeissus leiocarpus* Guill. et Perr.; *Combretac.;* n. Nig.
Marike: *Anogeissus schimperi* ?; *Combretac.;* Nig.
Mari-mari: *Jacaranda filicifolia* D. Don; *Bignoniac.;* Sur.
Marimáu: *Cyclostemon sp.; Euphorbiac.;* Phil. (Agusan)
Marinheiro: *Guarea trichilioides* L.; *Meliac.;* Bras.
Mario: *Calophyllum rekoi* Standl.; *Guttifer.;* Guat.
Marion: *Bucida buceras* L.; *Combretac.;* Pan.
Maripa: *Attalea excelsa* Mart.; *Palmac.;* fr. Gu.
  „ : *Maximiliana maripa* Drude, *M. regia* Mart.; *Palmac.;* Bras.
Maripenda: *Liquidambar styraciflua* L.; *Hamamelidac.;* Mex.
Marishiballi: *Licania sp.; Rosac.;* br. Gu.
Marisiballi: *Licania pachystachya* Kleinh., *L. micrantha* Miq.; *Rosac.;* Sur.
  „ kibelekoberoe: *Licania pachystachya* Kleinh.; *Rosac.;* Sur.
  „ tataroe: *Licania micrantha* Miq.; *Rosac.;* Sur.
Mark: *Acer oblongum* Wall.; *Acerac.;* nw. O.-Ind. (Punjab)
Marka: *Mangifera indica* L.; *Anacardiac.;* O.-Ind., Cey.
  „ marra: *Mangifera indica* L.; *Anacardiac.;* C.-O.-Ind. (Gondwana)
Marlberry: *Ardisia sp.; Myrsinac.;* Fl.
Marmalada naturelle: *Calocarpum mammosum* Pierre; *Sapotac.;* fr. W.-Ind.
Marmelada tree: *Calocarpum mammosum* Pierre; *Sapotac.;* br. W.-Ind.
Marmelade doosjes-boom: *Duroia eriopila* L. f.; *Rubiac.;* Sur.
Marmelero: *Ruprechtia sp.; Polygonac.;* Arg.
  „   : *Guazuma ulmifolia* Lam.; *Sterculiac.;* Arg.
  „ colorado: *Guidonia ulmifolia* Baill.; *Flacourtiac.;* tr. S.-Am.
Marmelos: *Aegle marmelos* Correa; *Rutac.;* O.-Ind.

Marmite de singe: *Lecythis grandiflora* Aubl.; *Lecythidac.;* fr. Gu.
„   „   „ : *Lecythis pisonis* Camb.; *Lecythidac.;* Bras., Kol.
Marmolero: *Ruprechtia sp.; Polygonac.;* Arg.
Marofranc: *Hibiscus tiliaceus* L.; *Malvac.;* Guad.
Marong: *Cratoxylon hornschuchi* Blume; *Guttifer.;* Sunda.
Maronnier d'Inde: *Aesculus hippocastanum* L.; *Hippocastanac.;* n. O.-Ind., Pers.
Marrochie nut: *Macadamia lowi* F. v. M.; *Proteac.;* ö. Au.
Maroom-meira: *Commersonia echinata* Forst.; *Sterculiac.;* Nied.-Ind.
Maroserano: *Moringa hildebrandti* Engl., *M. drouhardi* Jum.; *Moringac.;* sw. Mad.
Marouba: *Simaruba amara* Aubl.; *Simarubac.;* n. Bras., Gu.
Marri: *Eucalyptus calophylla* R. Br.; *Myrtac.;* Au.
Marrouba-masaki: *Evonymus japonica* Thunb.; *Celastrac.;* Jap.
Marsan: *Murraya exotica* L.; *Rutac.;* O.-Ind.
Marsiballi: *Tecoma leucoxylon* Mart.; *Bignoniac.;* fr. Gu.
Maru: *Quercus dilatata* Lindley; *Fagac.;* nw. O.-Ind.
„   : *Leptospermum ericoides* A. Rich.; *Myrtac.;* N.-Seel.
„ kinchiram: *Anogeissus latifolia* Wall.; *Combretac.;* sw. O.-Ind. (Malabar)
Maruba: *Simaruba amara* Aubl., *S. versicolor* St. Hil.; *Simarubac.;* Sur.
Maribakaede: *Acer distylum* S. et Z.; *Acerac.;* Jap.
Marukh: *Ailanthus excelsa* Roxb.; *Simarubac.;* C.-O.-Ind.
Marumangkás-na-laláki: *Sideroxylon ferrugineum* Hook. et Arn.; *Sapotac.;* Phil.
Marup: *Ailanthus excelsa* Roxb.; *Simarubac.;* C.-O.-Ind. [(n. Tay.)
Marupá: *Simaruba amara* Aubl.; *Simarubac.;* Bras. (Para, Sao P)
„ falso: *Jacaranda copaia* D. Don; *Bignoniac.;* n. Bras. (Amaz.)
„ -rana: *Simaruba sp.; Simarubac.;* Bras.
Marupaúba: *Simaruba amara* Aubl.; *Simarubac.;* Bras.
Marurúgi: *Bambusa spinosa* Roxb.; *Gramin.;* Phil. (Bikol)
Marwa: *Terminalia paniculata* Roth; *Combretac.;* sw. O.-Ind. (Mangalore)
Mas-moca: *Adenanthera pavonina* L.; *Mimosac.;* Cey.
„ de ? pilão: *Licania turiuva* Cham.; *Rosac.;* Bras.
Masa: *Tetragastris balsamifera* O. Ktze.; *Burserac.;* P.-R.
Masa: *Shorea obtusa* Wall.; *Dipterocarpac.;* Burma.
„ colorado: *Tetragastris balsamifera* O. Ktze.; *Burserac.;* P.-R.
„ -magu: *Litsea polyantha* Juss.; *Laurac.;* Burma.
Masábalo: *Carapa guianensis* Aubl.; *Meliac.;* Kol.
Masahua: *Hibiscus tiliaceus* L.; *Malvac.;* Mex.
Masaki: *Evonymus japonica* Thunb.; *Celastrac.;* Jap.
Masamuko: *Strombosiopsis tetrandra* Engl.; *Olacac.;* Kam.
Masaquila: *Trema micrantha* Blume· *Ulmac.;* Ven.
Masarandú: *Mimusops sp.; Sapotac.;* Ven.
Masasamba: *Anona sp.; Anonac.;* Peru.
Masavalo: *Centrolobium sp.; Papilionac.;* tr. S.-Am.
Mascal wood: *Vatica lanceaefolia* Blume, *V. scaphula* Dyer; *Dipterocarpac.;* Assam,
Másche: *Daniella thurifera* J. J. Benn.; *Caesalpiniac.;* Togo. [Burma.
Maseco: *Brosimum terrabanum* Pittier; *Morac.;* Guat., Hond.
Mashwal: *Chloroxylon swietenia* DC.; *Rutac.;* w. O.-Ind. (Deccan)
Masica: *Brosimum terrabanum* Pittier; *Morac.;* Hond.
Masicarán: *Astronium graveolens* Jacq.; *Anacardiac.;* Hond.
„   : *Tabebuia chrysantha* Nicholson; *Bignoniac.;* Hond.
„   : *Brosimum costaricanum* Liebm., *B. terrabanum* Pitt.; *Morac.;* Hond.,
Masicarón: *Brosimum costaricanum* Liebm.; *Morac.;* Hond. [br. Hond.
Masico: *Brosimum terrabanum* Pittier; *Morac.;* C.-Am.
„   : *Brosimum alicastrum* Swartz; *Morac.;* br. Hond.
Massaranduba: *Mimusops bidentata* DC., *M. surinamensis* Miq.; *Sapotac.;* Sur.
„   : *Mimusops balata* Gaertn., *M. globosa* Gaertn.; *Sapotac.;* Sur.
„   : *Mimusops amazonica* Huber, *M. huberi* Ducke; *Sapotac.;* Bras. (Para)
„   : *Mimusops aff. elata* Allem.; *Sapotac.;* Bras. (Para)
„ amarella: *Mimusops sp.; Sapotac.;* Bras. (Sao P)
„ branca: *Lucuma procera* Mart.; *Sapotac.;* Bras.
„ vermelha: *Mimusops sp.; Sapotac.;* Bras. (Sao P)
„ de leite: *Mimusops sp.; Sapotac.;* Bras. (Sao P)
Massellerholz (d): *Acer campestre* L.; *Acerac.;* Mitt.-Eu.
Massholder (d): *Acer campestre* L.; *Acerac.;* Eu.

Massikou: *Pachylobus buettneri* Engler; *Burserac.;* Gab.
Masson: *Ziziphus jujuba* Lamk.; *Rhamnac.;* Maur.
Massu: *Sterculia villosa* Roxb.; *Sterculiac.;* nw. O.-Ind. (Punjab)
Mast tree: *Polyalthia longifolia* Bth. et Hook. f.; *Anonac.;* O.-Ind., Cey.
Mastate: *Inophloeum armatum* Pittier; *Morac.;* Pan.
Mastic bully: *Sideroxylon mastichodendron* Jacq.; *Sapotac.;* Bah.
Mastixstrauch (d): *Pistacia lentiscus* L.; *Anacardiac.;* s. Eu., n. Af.
Masuka: *Warburgia ugandensis* Sprague; *Canellac.;* O.-Af.
Mat: *Clausena wampi* Oliv.; *Rutac.;* Ind.-Ch.
„ vai: *Spondias lakonensis* Pierre; *Anacardiac.;* Ind.-Ch.
Mata buaya: *Bruguiera eriopetala* Wight et Arn.; *Rhizophorac.;* Malay.
„ cansada: *Eschweilera calyculata* Pittier; *Lecythidac.;* Pan.
„ fome: *Cordia sp.;* *Borraginac.;* Bras.
„ keli: *Gynotroches axillaris* Blume; *Rhizophorac.;* Malay.
„ koetjieng: *Hopea mengarawana* Miq.; *Dipterocarpac.;* Nied.-Ind.
„ kuching: *Nephelium malaiense* Griff.; *Sapindac.;* Malak.
„ negra: *Ruprechtia sp.;* *Polygonac.;* Arg.
„ ojos: *Pouteria salicifolia* Radlk.; *Sapotac.;* Arg.
„ paja: *Vernonia patens* H. B. K.; *Composit.;* Kol.
„ passeh: *Trigoniastrum hypoleucum* Miq.; *Polygonac.;* Malay.
„ piojo: *Machaerium latifolium* Pittier; *Papilionac.;* Hond.
„ piskao: *Jacquinia sp.;* *Sapotac.;* Curacao.
„ rancha: *Vismia dealbata* H. B. K.; *Guttifer.;* Nic.
„ ratón: *Gliricidia sepium* Steud.; *Papilionac.;* Kol.
Matagallos: *Phlomis purpurea* L.; *Labiat.;* Sp.
Matagrie: *Moronobea coccinea* Aubl.; *Guttifer.;* Sur.
Matai: *Podocarpus spicata* R. Br.; *Conifer.;* N.-Seel.
Matakkie: *Moronobea coccinea* Aubl.; *Guttifer.;* Sur.
Matallo: *Sorbus aria* Cranß; *Rosac.;* It.
Matamatá: *Aglaia spp.;* *Meliac.;* Phil.
„ : *Euphoria didyma* Blco.; *Sapindac.;* Phil. (Zamboanga)
Matambokál: *Homalium bracteatum* Bth.; *Flacourtiac.;* Phil. (Ilocos Sur)
Matandua: *Bertiera montana* Hiern; *Rubiac.;* Go.
Matáng-hípon: *Ziziphus sp.;* *Rhamnac.;* Phil. (Rizal)
„ -uláng: *Adenanthera intermedia* Merr.; *Mimosac.;* Phil. (Laguna, Tayabas)
„ „ : *Aglaia harmsiana* Perk.; *Meliac.;* Phil. (Bataan)
„ -usá: *Litsea sp.;* *Laurac.;* Phil. (Laguna)
Matangál: *Ceriops roxburghiana* Arn.; *Rhizophorac.;* Phil. (Bataan)
Matángan: *Shorea polysperma* Merr.; *Dipterocarpac.;* Phil. (Cagayan)
Matangíd: *Shorea polysperma* Merr.: *Dipterocarpac.;* Phil (Cagayan)
Matapal: *Ficus benjamina* L.; *Morac.;* Peru.
Matapale: *Clusia rosea* Jacq.; *Guttifer.;* Trin.
Matapalo: *Coussapoa panamensis* Pittier; *Morac.;* Hond.
„ : *Ficus padifolia* H. B. K.; *Morac.;* Guat.
„ : *Ficus spp.;* *Morac.;* Hond., br. Hond.
„ : *Clusia sp.;* *Guttifer.;* br. Hond.
„ blanco: *Ficus sp.;* *Morac.;* Ek.
„ colorado: *Coussapoa sp.;* *Morac.;* Ek.
Mata-páo: *Ficus sp.;* *Morac.;* Bras.
„ „ , cipo-: *Ficus sp.;* *Morac.;* Bras.
Matapez: *Ichthyomethia sp.;* *Papilionac.;* Mex.
Matapiojo: *Trichilia sp.;* *Meliac.;* Nic.
Mataratú: *Capparis verrucosa* Jacq.; *Capparidac.;* Kol.
Matarro: *Piper auritum* H. B. K.; *Piperac.;* Hond.
Matasano: *Casimiroa tetramera* Millsp.; *Rutac.;* Hond.
„ de mico: *Diospyros ebenaster* Reß.; *Ebenac.;* Guat.
Matazama: *Platonia insignis* Mart.; *Guttifer.;* Ek.
Matazano: *Casimiroa edulis* Llave et Lexarza; *Rutac.;* Salv.
Matchwood: *Didymopanax morototoni* Dcne. et Planch.; *Araliac.;* Trin., Tob.
Mate arbol: *Zanthoxylum sp.;* *Rutac.;* Cuba.
Mateba-shii: *Pasania edulis* Makino; *Fagac.;* Jap.
Mathagiri vembu: *Toona ciliata* M. Roem.; *Meliac.;* sw. O.-Ind. (Malabar)
Maticki: ? *Morinda citrifolia* L.; *Rubiac.;* Togo.
Matiliscuate: *Tabebuia pentaphylla* Hemsl.; *Bignoniac.;* Guat.

Matilisquate: *Tabebuia pentaphylla* Hemsl.; *Bignoniac.;* Guat.
Matipo: *Myrsine australis* Colenso; *Myrsinac.;* N.-Seel.
Matkom: *Bassia latifolia* Roxb.; *Sapotac.;* w. Beng. (Chota-Nagpur)
Matlalquahuitl: *Guaiacum officinale* L.; *Zygophyllac.;* Mex.
Mato: *Eugenia mato* Griseb., *E. pungens* Beng; *Myrtac.;* Arg.
„ azul: *Caesalpinia echinata* Lam.; *Caesalpiniac.;* P.-R.
„ palo: *Ficus hemsleyana* Standl.; *Morac.;* Guat., Hond.
„ piojo: *Machaerium latifolium* Pittier; *Papilionac.;* Hond.
„ de playa: *Caesalpinia echinata* Lam.; *Caesalpiniac.;* P.-R.
Matobató: *Cyclostemon sp.;* *Euphorbiac.;* Phil. (Masbate)
„ : *Homalium villarianum* Vid.; *Flacourtiac.;* Phil. (Samar)
Matopus: *Mesua ferrea* L.; *Guttifer.;* Malay.
Matosirian: *Pterocarpus draco* L., *P. rhori* Vahl; *Papilionac.;* Sur.
Matowo: *Dombeya spectabilis* Boj.; *Sterculiac.;* O.-Af. (Nyasa)
Matsagi: *Bauhinia rufescens* Lam.; *Caesalpiniac.;* n. Nig.
Matsawi: *Castanopsis indica* A. DC.; *Fagac.;* Burma.
Matsi noki: *Trochodendron aralioides* S. et Z.; *Magnoliac.;* Jap.
Matsu: *Pinus sp.;* *Conifer.;* Jap.
Mattaquahuitl: *Guaiacum sp.;* *Zygophyllac.;* Mex.
Matti: *Terminalia tomentosa* Wight et Arn.; *Combretac.;* s. O.-Ind. (Coorg)
Mattoe: *Tabebuia sp.;* *Bignoniac.;* Sur.
„ gwégwé: *Pterocarpus rhori* Vahl; *Papilionac.;* Sur.
Matumi: *Adina galpini* Oliv.; *Rubiac.;* S.-Af.
Matzuhada: *Picea alcockiana* Carr.; *Conifer.;* Jap.
Mau: *Zanthoxylum rhetsa* DC.; *Rutac.;* O.-Ind. (Madras)
„ : *Glycosmis montana* Pierre; *Rutac.;* Ind.-Ch.
„ : *Antidesma eberhardti* Gagnep.; *Euphorbiac.;* Ind.-Ch.
„ : *Commersonia echinata* Forst.; *Sterculiac.;* Südsee (Tahiti)
„ cao: *Glycosmis montana* Pierre; *Rutac.;* Ind.-Ch.
„ cau: *Glycosmis montana* Pierre; *Rutac.;* Ind.-Ch.
„ chó: *Knema conferta* Warb., *K. corticosa* Lour.; *Myristicac.;* Ind.-Ch.
„ día: *Pithecolobium acuminatum* Bth.; *Mimosac.;* Ind.-Ch.
„ kadon: *Sarcocephalus cordatus* Miq.; *Rubiac.;* Burma.
„ lettan-she: *Anthocephalus cadamba* Miq.; *Rubiac.;* Burma.
„ tan: *Zanthoxylum avicennae* DC.; *Rutac.;* Ind.-Ch.
Maubin: *Spondias lutea* L., *S. dulcis* Forst.; *Anacardiac.;* Sur.
Maul: *Bassia latifolia* Roxb.; *Sapotac.;* Beng.
Maula: *Parinarium nobola* Oliv.; *Rosac.;* O.-Af. (Nyassa), n. Rhod.
Maulbeerbaum (d): *Morus alba* L.; *Morac.;* O.-As.
„ , Färber- (d): *Chlorophora tinctoria* Gaudich.; *Morac.;* tr. Am.
„ , indischer: *Morus indica* L.; *Morac.;* Himal., H.-Ind.
„ , roter: *Morus rubra* L.; *Morac.;* N.-Am.
Maupat: *Terminalia comintana* Merr.; *Combretac.;* Phil. (Palawan)
Maurecie: *Byrsonima altissima* DC.; *Malpighiac.;* Guad.
Mauret: *Vaccinium myrtillus* L.; *Ericac.;* Fr.
Mauricie: *Byrsonima sp.;* *Malpighiac.;* Guad.
Mauricif: *Byrsonima spicata* L. C. Rich.; *Malpighiac.;* Guad.
„ : *Byrsonima sp.;* *Malpighiac.;* Sur.
Mauricio: *Magnolia portoricensis* Bello; *Magnoliac.;* P.-R.
Maurier sauvage: *Morus rubra* L.; *Morac.;* N.-Am.
Mauto: *Lysiloma divaricata* Macbr.; *Mimosac.;* w. C.-Mex.
„ blanco: *Acacia standleyana* Rose; *Mimosac.;* Mex.
Mauwa: *Bassia latifolia* Roxb.; *Sapotac.;* n. O.-Ind. (United Prov.)
Mavavé: *Homalium sp.;* *Flacourtiac.;* fr. Gu.
Mavévé: *Potalia amara* Aubl.; *Loganiac.;* fr. Gu.
„ : *Homalium racoubea* Sw.; *Flacourtiac.;* fr. Gu.
Mavin-kubia: *Barringtonia acutangula* Gaertn.; *Lecythidac.;* w. O.-Ind. (Deccan)
Mavina-mara: *Mangifera indica* L.; *Anacardiac.;* w. O.-Ind. (Deccan)
Mavini: *Diospyros sp.;* *Ebenac.;* Kam.
Mavu: *Mangifera indica* L.; *Anacardiac.;* w. O.-Ind. (Deccan)
Mawar ayer: *Eugenia aquea* Burm.; *Myrtac.;* Malay.
Mawaranaballi: *Qualea dinizi* Ducke, *Q. sp.;* *Vochysiac.;* Sur.
? Mawassi: *Terminalia tanibouca* Rich.; *Combretac.;* Sur.
Mawbee stick: *Colubrina ferruginosa* Brongn.; *Rhamnac.;* Trin.

Mawksam-lung: *Michelia champaca* L.; *Magnoliac.;* Burma.
Mawna tree: *Moronobea coccinea* Aubl.; *Guttifer.;* Jam.
Mawon: *Morus alba* L.; *Morac.;* Burma.
Maxunasti: *Olmedia erythrorrhiza* Huber; *Morac.;* ö. Peru.
May: *Schleichera trijuga* Willd.; *Sapindac.;* O.-Ind. (Madras)
„ : *Ochrocarpus siamensis* T. Anders.; *Guttifer.;* Ind.-Ch.
„ chau: *Carya tonkinensis* H. Lec.; *Juglandac.;* Tonkin.
„ cho chi: *Parashorea stellata* Kurz; *Dipterocarpac.;* Ind.-Ch.
„ cô den: *Quercus chevalieri* Hick. et A. Camus; *Fagac.;* Ind.-Ch.
„ lay: *Sideroxylon eburneum* Chev.; *Sapotac.;* Ind.-Ch.
„ yang: *Dipterocarpus spp.; Dipterocarpac.;* Ind.-Ch. (Laos)
Maya: *Miconia impetiolaris* D. Don, *M. leucocephala* Naud.; *Melastomatac.;* br. Hond.
„ , red: *Miconia spp.; Melastomatac.;* br. Hond.
„ , white: *Miconia aff. argentea* DC.; *Melastomatac.;* br. Hond.
Mayahnin: *Acrocarpus fraxinifolius* Wight et Arn.; *Caesalpiniac.;* Burma.
Mayamaya: *Pithecolobium spp.; Mimosac.;* Hond.
Mayan-itara: *Trichilia hieronymi* Griseb.; *Meliac.;* Arg. (Misiones)
Mayanin: *Zanthoxylum budrunga* Wall.; *Rutac.;* And.
„ -kyetsu: *Zanthoxylum budrunga* Wall.; *Rutac.;* Burma.
Mayápis: *Anisoptera thurifera* Blume; *Dipterocarpac.;* Phil.
„ : *Parashorea malaanonan* Merr.; *Dipterocarpac.;* Phil. (Tay., Zamboanga)
„ : *Pentacme contorta* Merr. et Rolfe; *Dipterocarpac.;* Phil. (Pol.)
„ : *Shorea spp.; Dipterocarpac.;* Phil.
Mayaúban: *Eugenia similis* Merr.; *Myrtac.;* Phil. (Surigao)
Mayen: *Ficus glomerata* Roxb.; *Morac.;* Burma.
Mayeng: *Pterospermum acerifolium* Willd.; *Sterculiac.;* n. O.-Ind. (Jaunsar)
Mayflower: *Tabebuia pentaphylla* Hemsl.; *Bignoniac.;* tr. Am.
Mayiladi: *Vitex altissima* L. f.; *Verbenac.;* O.-Ind. (Madras)
Mayilai: *Vitex altissima* L. f.; *Verbenac.;* sw. O.-Ind. (Malabar)
Maylay: *Sideroxylon eburneum* Chev.; *Sapotac.;* Ind.-Ch.
Mayo: *Zizyphus melastomoides* Pittier; *Rhamnac.;* Ven.
Mayoumi: *Evonymus sieboldiana* Blume; *Celastrac.;* Jap.
Mayten: *Maytenus boaria* Molina; *Celastrac.;* s. Chile, s. Arg.
Mayu-de: *Dolichandrone stipulata* Bth.; *Bignoniac.;* Burma.
Mayumi: *Evonymus oxyphylla* Miq.; *Celastrac.;* Jap.
„ : *Evonymus europaea* L. var. *hamiltoniana* Maxim.; *Celastrac.;* Jap.
Mayusip: *Palaquium sp.; Sapotac.;* Phil. (Mindoro)
Maza: *Mitragyne macrophylla* Hiern; *Rubiac.;* B.-K.
Mazahua: *Hibiscus tiliaceus* L.; *Malvac.;* Mex.
Mazapán: *Artocarpus communis* Forst.; *Morac.;* Hond.
Mazaxochotli: *Spondias sp.; Anacardiac.;* Mex.
Mazi-mazi: *Funtumia latifolia* Stapf; *Apocynac.;* B.-K.
Mba: *Pentaclethra macrophylla* Bth.; *Mimosac.;* Kam.
„ : *Cola acuminata* Schott et Endl.; *Sterculiac.;* Kam.
„ : *Entada scandens* Bth.; *Mimosac.;* Ind.-Ch.
Mbabi: *Casearia sp., C. gossypiosperma* Briq.; *Flacourtiac.;* Arg.
M'babou: *Mimusops heckeli* H. Lec.; *Sapotac.;* Elf.
M'babu: *Mimusops heckeli* H. Lec.; *Sapotac.;* Elf.
Mbafoughélé: *Homalium africanum* Bth.; *Flacourtiac.;* Gab.
Mbag-bold-ede: *Leptonychia urophylla* Welw.; *Sterculiac.;* Lib.
Mbaga: *Mammea klaineana* Pierre; *Guttifer.;* Gab.
Mbagha: *Khaya sp.; Meliac.;* Gab.
Mbaghesa: *Tetrapleura thonningi* Bth.; *Mimosac.;* Gab.
Mbaka: *Copaifera demeusei* Harms; *Caesalpiniac.;* B.-K.
Mbala: *Pentaclethra macrophylla* Bth.; *Mimosac.;* Gab.
Mbalu: *Macrolobium dewevrei* de Wild.; *Caesalpiniac.;* B.-K.
Mbama: *Albizzia fastigiata* Oliv.; *Mimosac.;* B.-K.
Mbamb: *Croton oligandrum* Pierre; *Euphorbiac.;* Gab. (Loango)
Mbamba: *Croton oligandrum* Pierre; *Euphorbiac.;* Gab. (Loango)
Mbambakosi: *Afzelia quanzensis* Welw.; *Caesalpiniac.;* br. O.-Af.
M'bambana: *Xylopia rubescens* Oliv.; *Anonac.;* Kam.
Mban: *Cola acuminata* Schott et Endl.; *Sterculiac.;* Kam.
„ : *Guarea sp.; Meliac.;* Gab.
Mbana: *Brachystegia robusta* Pierre; *Caesalpiniac.;* Gab.

Mbanda: *Hylodendron gabunense* Taub.; *Caesalpiniac.;* Kam.
M'bandé: *? aff. Brachystegia sp.; Caesalpiniac.;* Kam.
Mbang: *Chlorophora excelsa* Bth. et Hook.; *Morac.;* W.-Af., O.-Af.
„ bazoa: *Strombosiopsis tetrandra* Engl.; *Olacac.;* Kam.
Mbanga: *Mammea klaineana* Pierre; *Guttifer.;* Gab.
„ : *Cocos nucifera* L.; *Palmac.;* Gab.
„ -ntanga: *Cocos nucifera* L.; *Palmac.;* Gab.
„ -putu: *Cocos nucifera* L.; *Palmac.;* Gab.
M'bango: *Afzelia bipendensis* Harms; *Caesalpiniac.;* Kam.
M'bangué: *? aff. Brachystegia sp.; Caesalpiniac.;* Kam.
Mbangwe: *Allophylus griseotomentosus* Gilg; *Sapindac.;* D.-O.-Af. (w. Usambara)
Mbapa: *Markhamia lanata* K. Schum.; *Bignoniac.;* D.-O.-Af. (Tabora)
Mbapfolo: *Homaliùm africanum* Bth.; *Flacourtiac.;* Kam.
Mbara: *Chlorophora excelsa* Bth. et Hook.; *Morac.;* B.-K.
Mbari: *Elaeis guineensis* Jacq. *var. nigrescens* Chev.; *Palmac.;* Gab.
„ -divasa: *Élaeis guineensis* Jacq. *var. ?; Palmac.;* Gab.
„ -ibamba: *Cocos nucifera* L.; *Palmac.;* Gab.
„ -mutangeni: *Cocos nucifera* L.; *Palmac.;* Gab.
„ -putu: *Cocos nucifera* L.; *Palmac.;* Gab.
Mbarra: *Khaya sp.; Meliac.;* Gab.
Mbau: *Macrolobium dewevrei* de Wild.; *Caesalpiniac.;* B.-K.
Mbavy: *Casearia sp.; Flacourtiac.;* Arg.
Mbawa: *Khaya senegalensis* A. Juss.; *Meliac.;* br. O.-Af. (Nyassa)
„ : *Khaya nyasica* Stapf; *Meliac.;* n. Rhod., s. C.-Af.
Mbawé: *Enantia chlorantha* Oliv.; *Anonac.;* Elf.
M'bazoge: *Strombosiopsis tetrandra* Engl.; *Olacac.;* Kam.
Mbe: *Pterocarpus soyauxi* Taub.; *Papilionac.;* Kam.
„ miki: *Pterocarpus soyauxi* Taub.; *Papilionac.;* Kam.
M'béa: *Pterocarpus soyauxi* Taub.; *Papilionac.;* Kam.
Mbébame: *Gambeya africana* Pierre; *Sapotac.;* Kam., Mitt.-Kongo.
„ : *Chrysophyllum belemba* de Wild.; *Sapotac.;* s. Kam., n. Gab.
„ : *Anthocleista nobilis* G. Don; *Loganiac.;* Kam.
Mbéga: *Entandrophragma sp., Khaya ivorensis* Chev.; *Meliac.;* Gab., Mitt.-Kongo.
Mbègo: *Sclerosperma walkeri* Chev.; *Palmac.;* Gab.
Mbel: *Pterocarpus soyauxi* Taub.; *Papilionac.;* Kam., Gab.
„ : *Canarium velutinum* Guill.; *Burserac.;* Gab.
„ ebon ngon: *Omphalocarpum procerum* Beauv.; *Sapotac.;* Kam.
Mbeli: *Haronga madagascariensis* Choisy; *Guttifer.;* W.-Af. (S.-L.)
„ -gulin: *Piptadenia africana* Hook. f.; *Mimosac.;* W.-Af. (S.-L.)
Mbémame: *Chrysophyllum sp.; Sapotac.;* W.-Af.
Mbemba kofi: *Afzelia quanzensis* Welw.; *Caesalpiniac.;* O.-Af. (Mozamb.)
Mben: *Sapium ellipticum* Pax; *Euphorbiac.;* Gab.
Mbende: *Monodora myristica* Dunal; *Anonac.;* B.-K.
Mbengué: *Ficus punctata* Lamk.; *Morac.;* Gab.
Mbeta: *Symphonia gabonensis* Pierre; *Guttifer.;* B.-K.
Mbidi: *Canarium schweinfurthi* Engl.; *Burserac.;* B.-K. (Mayombe)
„ nkala: *Canarium schweinfurthi* Engl.; *Burserac.;* B.-K.
Mbié: *Pterocarpus soyauxi* Taub.; *Papilionac.;* Kam.
Mbikam: *Newbouldia laevis* Bth.; *Bignoniac.;* Kam.
Mbil: *Pachylobus edulis* G. Don; *Burserac.;* Kam.
„ : *Pterocarpus soyauxi* Taub.; *Papilionac.;* Kam.
Mbila: *Ficus fischeri* Warb.; *Morac.;* D.-O.-Af. (Tabora)
Mbili: *Pterocarpus soyauxi* Taub.; *Papilionac.;* Kam.
„ : *Canarium schweinfurthi* Engl.; *Burserac.;* Gab., Kam.
„ : *Copaifera sp.; Caesalpiniac.;* Gab.
Mbimo: *Pachystela cinerea* Pierre, *Mimusops lacera* Baker; *Sapotac.;* Gab.
„ : *Petersia viridiflora* Chev.; *Lecythidac.;* Gab.
Mbimou: *Mimusops lacera* Baker; *Sapotac.;* Gab.
Mbims: *Mimusops lacera* Baker; *Sapotac.;* Gab.
Mbiya: *Elaeis guineensis* Jacq. *var. nigrescens* Chev.; *Palmac.;* Gab.
Mboasoo: *Strombosia sp.; Olacac.;* Kam.
Mbob: *Ricinodendron africanum* Muell. Arg.; *Euphorbiac.;* Elf.
Mboba: *Ceiba pentandra* Gaertn., *Bombax buonopozense* Beauv.; *Bombacac.;* Elf.
„ : *Myrianthus arboreus* Beauv.; *Morac.;* Gab.

Mbobotso: *Ochrocarpus africanus* Oliv.; *Guttifer.*; Gab.
Mbocaya: *Acrocomia totai* Mart.; *Palmac.*; Arg.
Mboha: *Bombax buonopozense* Beauv.; *Bombacac.*; Elf.
Mbokora: *Afzelia africana* Smith; *Caesalpiniac.*; Nig.
Mbol: *Paivaeusa ? gabonensis* Chev.; *Euphorbiac.*; Gab.
Mbon: *Zanthoxylum parvifolium* Chev.; *Rutac.*; Elf.
Mbona: *Staudtia gabonensis* Warb.; *Myristicac.*; Gab.
Mbonda: *Staudtia stipitata* Warb.; *Myristicac.*; Kam.
Mbong: *Monodora myristica* Dunal; *Anonac.*; Elf.
Mbonn: *Staudtia gabonensis* Warb.; *Myristicac.*; Gab.
Mbonoi-cauri: *Terminalia ivorensis* Chev.; *Combretac.*; Elf.
Mbopou: *Antiaris toxicaria* Lesch., *A. africana* Engl.; *Morac.*; Elf.
Mborebi caá guazu: *Roupala gardneri* Meissn.; *Proteac.*; Arg.
Mboro: *Oubanguia denticulata* van Tiegh.; *Tiliac.*; Gab.
Mbossa: *Guarea cedrata* Pellegr.; *Meliac.*; Elf.
Mbossé (f): *Guarea cedrata* Pellegr.; *Meliac.*; Elf.
Mbota: *Vitex pachyphylla* Baker; *Verbenac.*; Gab.
Mboti: *Terminalia ivorensis* Chev.; *Combretac.*; Elf.
Mboto: *Vitex pachyphylla* Baker; *Verbenac.*; Gab.
Mbotu: *Millettia laurenti* de Wild., *Platysepalum chevalieri* Harms; *Papilionac.*; B.-K.
Mbou: *Irvingia excelsa* Mildbr.; *Simarubac.*; W.-Af.
Mbouambo mvoul: *Piptadenia sp.*; *Mimosac.*; Kam.
Mbouangha: *Cola ballayi* Cornu; *Sterculiac.*; Gab.
Mboull: *Sapindus saponaria* L.; *Sapindac.*; Senl.
Mboun: *Staudtia gabonensis* Warb.; *Myristicac.*; Gab.
Mboundé: *Staudtia gabonensis* Warb., *S. kamerunensis* Warb.; *Myristicac.*; Kam.
Mboune: *Staudtia gabonensis* Warb.; *Myristicac.*; Kam.
Mbourou: *Oubanguia denticulata* van Tiegh.; *Tiliac.*; Gab.
Mbourwé: *Oubanguia denticulata* van Tiegh.; *Tiliac.*; Gab.
Mbout: *Diospyros aggregata* Gürke; *Ebenac.*; Kam.
Mbowwa: *? Cedrela sp.*; *Meliac.*; br. C.-Af. (Mangoche)
Mboy: *Mitragyne stipulosa* O. Kџe.; *Rubiac.*; Lib.
Mbrahu: *Pseudocinchona africana* Chev.; *Rubiac.*; Elf.
Mbraoua: *Cola proteiformis* Chev.; *Sterculiac.*; Elf.
Mbuambua: *Guettarda speciosa* L.; *Rubiac.*; Fid.
Mbudika: *Schrebera koiloneura* Gilg; *Oleac.*; D.-O.-Af. (Tabora)
Mbuessé: *Cola vera* K. Schum.; *Sterculiac.*; Elf.
Mbula: *Sorindeia maxima* Verm.; *Anacardiac.*; B.-K.
„ : *Parinarium curatellifolium* Planch.; *Rosac.*; D.-O.-Af. (Tabora).
Mbumbu: *Lannea barteri* Engl.; *Anacardiac.*; D.-O.-Af. (Tabora)
Mbumbuli: *Diospyros holtzi* Gürke; *Ebenac.*; D.-O.-Af. (Tabora)
Mbunze: *Allanblackia sacleuxi* Hua; *Guttifer.*; nw. Angola.
Mbussenge: *Musanga smithi* R. Br.; *Morac.*; tr. W.-Af.
Mbuzungo: *Pittosporum abyssinicum* Del.; *Pittosporac.*; D.-O.-Af. (w. Usambara)
Mbwangha: *Cola ballayi* Cornu; *Sterculiac.*; Gab.
Mchenga: *Brachystegia spicaeformis* Bth.; *Caesalpiniac.*; O.-Af. (Nyassa)
Mchere: *Ficus sp.*; *Morac.*; O.-Af. (Nyassa)
Mchibi: *Copaifera coleosperma* Bth.; *Caesalpiniac.*; Rhod.
Mdaga: *Trinervia tropica* Burgk.; *Flacourtiac.*; D.-O.-Af.
Mdanánda: *Neumannia theiformis* A. Rich.; *Flacourtiac.*; D.-O.-Af. (w. Usambara)
Mdara: *Euclea kellau* Hochst.; *Ebenac.*; D.-O.-A. (w. Usambara)
Mdati: *Grewia cf. conocarpa* K. Schum.; *Tiliac.*; D.-O.-Af. (Tabora)
Mdayen: *Pentadesma butyraceum* Don; *Guttifer.*; Lib.
Mdonho: *Commiphora fischeri* Engl.; *Burserac.*; D.-O.-Af. (Tabora)
Mdoni: *Eugenia gerrardi ?*; *Myrtac.*; S.-Af. (Zulu)
„ , Coast-: *Syzygium cordatum* Hochst.; *Myrtac.*; S.-Af.
Mduku: *Uapaca kirkiana* Muell. Arg.; *Euphorbiac.*; s. C.-Af.
Me: *Dillenia pentagyna* Roxb.; *Dilleniac.*; Ind.-Ch.
„ : *Tamarindus indica* L.; *Caesalpiniac.*; Ind.-Ch.
„ công: *Canthium didymum* Roxb.; *Rubiac.*; Kamb.
„ diêt: *Anisoptera cochinchinensis* Pierre; *Dipterocarpac.*; Tonkin.
„ kang: *Canthium didymum* Roxb.; *Rubiac.*; Ind.-Ch.
„ roi: *Dillenia pentagyna* Roxb.; *Dilleniac.*; w. Annam.
„ rùng: *Phyllanthus emblica* L.; *Euphorbiac.*; Ind.-Ch.

Meaperros: *Santolina rosmarinifolia* L.; *Composit.;* Sp.
Meba: *Irvingia gabonensis* Baill.; *Simarubac.;* B.-K.
Mebe men gon: *Omphalocarpum procerum* Beauv.; *Sapotac.;* Kam.
Mebrou: *Entandrophragma utile* Sprague; *Meliac.;* Elf.
Mecate blanco: *Heliocarpus donnell-smithi* Rose; *Tiliac.;* Guat.
„   colorado: *Belotia campbelli* Sprague; *Tiliac.;* Guat.
„   de agua: *Heliocarpus appendiculatus* Turcz.; *Tiliac.;* Hond.
Mechi: *Lagerstroemia parviflora* Roxb.; *Lythrac.;* Assam.
Mécou: *Piscidia erythrina* L.; *Papilionac.;* fr. Gu.
Meda: *Litsea polyantha* Juss.; *Laurac.;* n. O.-Ind. (United Prov.)
Medad: *Cedrus atlantica* Manetti; *Conifer.;* Marokko.
Medang: *Erythroxylon cuneatum* Kurz; *Erythroxylac.;* Malay, Phil.
„   : *Litsea perrotteti* F. Vill.; *Laurac.;* Phil.
„   : *Actinodaphne sp., Alseodaphne sp., Beilschmiedia sp.; Laurac.;* Malay.
„   : *Cinnamomum sp., Cryptocarya sp., Dehaasia sp.; Laurac.;* Malay.
„   : *Endiandra sp., Lindera sp., Litsea sp., Machilus sp.; Laurac.;* Malay.
„   : *Micropora sp., Neolitsea sp., Nothophoebe sp.; Laurac.;* Malay.
„   : *Phoebe sp., Stemmatodaphne sp.; Laurac.;* Malay.
„   bekwi: *Schima noronhae* Reinw.; *Theac.;* tr. As.
„   buaya: *Pithecolobium confertum* Bth.; *Mimosac.;* Malay (Kedah)
„   kemangi: *Cinnamomum sp.; Laurac.;* Malay.
„   kok: *Pithecolobium confertum* Bth.; *Mimosac.;* Malay (Kedah)
„   lada: *Cinnamomum sp.; Laurac.;* br. Borneo (Sarawak).
„   rimoeëng: *Campnosperma auriculata* Hook. f.; *Anacardiac.;* Sum.
„   sawa: *Anisoptera spp., A. thurifera* Blume; *Dipterocarpac.;* Malay.
Medeloa: *Albizzia procera* Bth.; *Mimosac.;* Beng.
Medi: *Ficus glomerata* Roxb.; *Morac.;* O.-Ind. (Madras)
Medong: *Elaeocarpus spp.; Elaeocarpac.;* Malay.
Medzimkogo: *Psychotria gabonensis* ?; *Rubiac.;* Gab.
Medzimkouro: *Psychotria gabonensis* ?; *Rubiac.;* Gab.
Meetha-kamarunga: *Averrhoa carambola* L.; *Oxalidac.;* Beng.
Mefinze: *Bruguiera gymnorrhiza* Lamk.; *Rhizophorac.;* port. O.-Af. (Moz., Quel.)
Megi: *Berberis thunbergi* DC.; *Berberidac.;* Jap.
Mehame-megoma: *Caloncoba glauca* Gilg; *Flacourtiac.;* Kam.
Méhamé memoga: *Caloncoba glauca* Gilg; *Flacourtiac.;* Gab.
Mehibi: *Copaifera coleosperma* Bth.; *Caesalpiniac.;* Rhod.
Mehlbeerbaum (d): *Sorbus aria* Cranß; *Rosac.;* s. Eu.
„   , breitblättriger: *Sorbus latifolia* Pers.; *Rosac.;* Eu.
„   , gemeiner: *Sorbus aria* Cranß; *Rosac.;* Mitt.-Eu.
Mehlbeere, nordische: *Pirus scandica* Fries; *Rosac.;* n. Eu.
„   , schwedische: *Pirus scandica* Fries; *Rosac.;* n. Eu.
Mehr: *Randia malleifera* Bth. et Hook. f.; *Rubiac.;* Lib.
„   -chu: *Scottellia coriacea* Chev.; *Flacourtiac.;* Lib.
Meijú: *Duguetia sp.; Anonac.;* Bras.
Meikkyè: *Albizzia odoratissima* Bth.; *Mimosac.;* Burma.
Mekeluang: *Tarrietia spp., T. simplicifolia* Mast.; *Sterculiac.;* Malay.
Mekhi: *Ficus sp.; Morac.;* Elf.
Mekoa: *Symphonia gabonensis* Pierre; *Guttifer.;* Kam.
Mekoe-koewaré: *Eschweilera sp.; Lecythidac.;* Sur.
„   -kwarie: *Eschweilera sp.; Lecythidac.;* Sur.
Melambo: *Drimys winteri* Forst.; *Magnoliac.;* Bras.
Melankan katong: *Cynometra inaequifolia* A. Gray; *Caesalpiniac.;* Malay.
Melantai: *Shorea macroptera* Dyer; *Dipterocarpac.;* Malay.
Melarancio: *Citrus aurantium* L.; *Rubiac.;* It.
Melebekan: *Shorea palembanica* Miq.; *Dipterocarpac.;* Sum.
Melegra: *Albizzia sp.; Mimosac.;* Elf.
Mélèze (f): *Larix spp.; Conifer.;* n. Hem.
„   occidental: *Larix occidentalis* Nutt.; *Conifer.;* w. Kan.
„   d'Amérique: *Larix americana* Michx.; *Conifer.;* nö. Kan.
„   d'Europe: *Larix europaea* DC.; *Conifer.;* Mitt.-Eu.
„   de Lyall: *Larix lyalli* Parl.; *Conifer.;* w. Kan.
Meli: *Myrtus meli* Phil.; *Myrtac.;* Chile.
Meliko: *Calophyllum sp.; Guttifer.;* Nied.-Ind.
Melilin: *Dillenia spp., Wormia spp.; Dilleniac.;* Malay.

Melima: *Tarrietia spp., T. simplicifolia* Mast.; *Sterculiac.;* Malay.
Melkhout: *Tabernaemontana utilis* Wight et Arn.; *Apocynac.;* fr. Gu.
Melmél: *Wallaceodendron celebicum* Koord.; *Mimosac.;* Phil. (Cagayan)
Melmér: *Wallaceodendron celebicum* Koord.; *Mimosac.;* Phil. (Cagayan)
Melón: *Amyris elemifera* L.; *Rutac.;* Salv.
Melon tree: *Carica papaya* L.; *Caricac.;* Bras.
Meloncillo: *Agonandra excelsa* Griseb.; *Olacac.;* Arg.
Melumut: *Campnosperma auriculata* Hook. f., *C. wallichi* Griff.; *Anacardiac.;* Malay.
Membaloe: *Dysoxylum acutangulum* Miq.; *Meliac.;* Nied.-Ind.
Membatu laiang: *Helicia excelsa* Blume; *Proteac.;* Malay.
Memble: *Poeppigia procera* Presl; *Caesalpiniac.;* Salv.
Memboki: *Dialium yambataense* Verm.; *Caesalpiniac.;* B.-K.
Membomélé: *Byrsanthus sp.; Flacourtiac.;* Gab.
Membrillo: *Cydonia vulgaris* Pers.; *Rosac.;* Sp.
    „   : *Cydonia oblonga* Mill.; *Rosac.;* Hond.
    „   : *Terminalia hayesi* Pittier; *Combretac.;* Hond.
    „   : *Phyllostylon brasiliensis* Capanema; *Ulmac.;* Ven.
    „   : *Grias fendleri* Seem.; *Lecythidac.;* Pan.
    „ macho: *Grias fendleri* Seem.; *Lecythidac.;* Pan.
    „ de monte: *Gustavia pubescens* Ruiz et Pavon; *Lecythidac.;* Ek.
Memiso: *Trema micrantha* Blume; *Ulmac.;* St.-Domingo.
Mempanjor: *Dialium spp.; Caesalpiniac.;* Malay.
Mempayan: *Rhodamnia trinervia* Blume; *Myrtac.;* Malay.
Mempelam: *Mangifera indica* L.; *Anacardiac.;* Malay.
Mempelu: *Dillenia spp., Wormia spp.; Dilleniac.;* Malay.
Mempening: *Pasania sp., Quercus sp.; Fagac.;* Malay.
    „ merah: *Quercus sp.; Fagac.;* Malay.
Memrentim: *Treculia aff. africana* Dcne.; *Morac.;* Go.
Menamosina: *Chrysalidocarpus ruber* H. Jum.; *Palmac.;* Mad. (Masoala)
Menangi: *Chlorophora excelsa* Bth. et Hook.; *Morac.;* Kam.
Mencoar: *Minquartia guianensis* Aubl.; *Olacac.;* fr. Gu.
Menda: *Litsea sebifera* Pers.; *Laurac.;* Beng.
    „ : *Litsea polyantha* Juss.; *Laurac.;* C.-O.-Ind. (Gondwana)
Mendjantong: *Amoora grandifolia* C. DC.; *Meliac.;* Java.
Mendong: *Elaeocarpus spp.; Elaeocarpac.;* Malay.
Méné: *Lophira alata* Banks; *Ochnac.;* fr. Guin.
Ménéhané: *Fagara macrophylla* Engl.; *Rutac.;* Elf.
Mengdeng: *Liquidambar formosana* Hance; *Hamamelidac.;* Ind.-Ch.
Mengarawan: *Hopea mengarawan* Miq., *H. intermedia* King; *Dipterocarpac.;* Malay.
    „   : *Hopea myrtifolia* Miq.; *Dipterocarpac.;* Sum.
Menggeris: *Koompassia malaccensis* Maing.; *Caesalpiniac.;* Java.
Mengkabang pinang: *Shorea ghysbertsiana* Burck; *Dipterocarpac.;* Borneo.
    „   „   : *Shorea falcifera* Dyer; *Dipterocarpac.;* br. Borneo (Sarawak)
Mengkeluang: *Tarrietia spp., T. simplicifolia* Mast.; *Sterculiac.;* Malay.
Mengkudu: *Morinda tinctoria* Roxb.; *Rubiac.;* tr. As.
Mengkulang: *Tarrietia spp.; Sterculiac.;* Malay.
Mengrawan: *Hopea mengarawan* Miq.; *Dipterocarpac.;* Sum., Bangka.
    „ batoe: *Hopea sp.; Dipterocarpac.;* Nied.-Ind.
    „ boenga: *Hopea sp.; Dipterocarpac.;* Nied.-Ind.
    „ djangoet: *Hopea sp.; Dipterocarpac.;* Nied.-Ind.
    „ telor: *Hopea sp.; Dipterocarpac.;* Nied.-Ind.
    „ tjingkang: *Hopea mengarawan* Miq.; *Dipterocarpac.;* Nied.-Ind.
Meni: *Lophira alata* Banks; *Ochnac.;* Nig.
Meni oil tree: *Lophira alata* Banks; *Ochnac.;* Nig.
Meniridan: *Qualea coerulea* Aubl., *Q. spp.; Vochysiac.;* Sur.
    „ -hororadi-korro: *Qualea albiflora* Warm.; *Vochysiac.;* Sur.
    „ -oewe-bera-bandikoro: *Qualea coerulea* Aubl., *Q. spp.; Vochysiac.;* Sur.
Meniriolan: *Qualea coerulea* Aubl., *Q. spp.; Vochysiac.;* Sur.
Menjan: *Styrax benzoin* Dryand.; *Styracac.;* Siam, Malak., Sum., Java.
Mentangor: *Calophyllum sp.; Guttifer.;* Malay.
Mentaos: *Wrightia javanica* A. DC.; *Apocynac.;* Java.
Mentasawah: *Anisoptera spp., A. thurifera* Blume; *Dipterocarpac.;* Malay.
Mentatai: *Ochanostachys amentacea* Mast.; *Olacac.;* Malay (Kedah)
Mentawa: *Artocarpus superba* Becc.; *Morac.;* Borneo.

Mentelor: *Parinarium rubiginosum* Ridl., *Angelisia splendens* Korth.; *Rosac.;* Malay.
Menzomu: *Dialium yambataense* Verm.; *Caesalpiniac.;* B.-K.
Mepareyba: *Rhizophora mangle* L.; *Rhizophorac.;* Bras.
Meraga: *Adina rubescens* Hemsl.; *Rubiac.;* Malay.
„ bukit: *? Canthium didymum* Gaertn. f.; *Rubiac.;* Malay.
Merambong: *Vernonia arborea* Ham.; *Composit.;* Malay.
Merana: *Vernonia merana* Baker, *Brachylaena merana* Humbert; *Composit.;* Ma⸫
„ -hazotokana: *Brachylaena sp.; Composit.;* Mad.
Meranti: *Shorea spp., ? Hopea spp.; Dipterocarpac.;* Malay.
„ bakau: *Shorea sp.; Dipterocarpac.;* Malay.
? „ boaya: *Shorea platycarpa* Heim; *Dipterocarpac.;* br. Borneo (Sarawak)
„ bunga: *Shorea leprosula* Miq.; *Dipterocarpac.;* Malay.
„ daun kechil: *Shorea parvifolia* Dyer; *Dipterocarpac.;* Malay.
„ hijau: *Shorea acuminata* Dyer; *Dipterocarpac.;* Malay.
„ kepong: *Shorea sericea* Dyer; *Dipterocarpac.;* Malay.
„ kerap: *Shorea parvifolia* Dyer; *Dipterocarpac.;* Malay.
„ ketapak: *Shorea bracteolata* Dyer; *Dipterocarpac.;* Malay.
„ kunyit: *Shorea sp.; Dipterocarpac.;* Malay (Johore)
„ labu: *Shorea macroptera* Dyer, *S. bracteolata* Dyer; *Dipterocarpac.;* Malay
„ lebah: *Shorea globifera* Ridl.; *Dipterocarpac.;* Malay (w. Pahang)
„ pa'ang: *Shorea bracteolata* Dyer; *Dipterocarpac.;* Malay.
„ pahang: *Shorea scutulata* King, *S. spp.; Dipterocarpac.;* Malay.
„ pasir: *Shorea globifera* Ridl.; *Dipterocarpac.;* Malay (w. Pahang)
„ paya: *Shorea palustris* Ridl.; *Dipterocarpac.;* Malak.
„ pipit: *Shorea globifera* Ridl.; *Dipterocarpac.;* Malay (w. Pahang)
„ rambei daun: *Shorea acuminata* Dyer; *Dipterocarpac.;* Malay.
„ red: *Shorea leprosula* Miq.; *Dipterocarpac.;* Malay.
„ sabut: *Shorea parvifolia* Dyer; *Dipterocarpac.;* Malay.
„ samak: *Shorea pauciflora* King, *S. parvifolia* Dyer; *Dipterocarpac.;* Malay.
„ sarang punai: *Shorea parvifolia* Dyer; *Dipterocarpac.;* Malay.
„ sepang: *Hopea maranti* Miq.; *Dipterocarpac.;* Sum., Borneo, Bangka.
„ seraya: *Shorea curtisi* Dyer; *Dipterocarpac.;* Malay.
„ temak: *Shorea bracteolata* Dyer; *Dipterocarpac.;* Malay.
„ tembaga: *Shorea leprosula* Miq.; *Dipterocarpac.;* Malay.
Meraut: *Musanga smithi* R. Br.; *Morac.;* Elf.
Merawan: *Hopea spp.; Dipterocarpac.;* Malay.
„   : *Hopea mengarawan* Miq.; *Dipterocarpac.;* Sum.
„ jangkang: *Hopea intermedia* King; *Dipterocarpac.;* Malay.
„ -hout: *Hopea mengarawan* Miq.; *Dipterocarpac.;* ö. As.
„ kunyit: *Hopea intermedia* King; *Dipterocarpac.;* Malay.
„ mera: *Shorea elliptica* Burck; *Dipterocarpac.;* br. Borneo (Sarawak)
Merbaju: *Tarrietia spp., T. simplicifolia* Mast.; *Sterculiac.;* Malay.
Merbatu: *Parinarium rubiginosum* Ridl., *Angelisia splendens* Korth.; *Rosac.;* Malay.
„ layang: *Parinarium griffithianum* Bth.; *Rosac.;* Malay.
Merbau: *Intsia spp.; Caesalpiniac.;* Malay.
„   : *Intsia bakeri* Prain; *Caesalpiniac.;* Malay, Sunda.
„   : *Intsia retusa* O. Ktze., *I. bakeri* Prain; *Caesalpiniac.;* br. Borneo (Sarawak)
„   : *Intsia bijuga* O. Ktze.; *Caesalpiniac.;* Java.
„ apil: *Intsia bijuga* O. Ktze.; *Caesalpiniac.;* Malay.
Merbo: *Intsia amboinensis* Thouars; *Caesalpiniac.;* Nied.-Ind.
Mercurio do campo: *Erythroxylon sp.; Erythroxylac.;* Bras.
Merendyba: *Terminalia sp.; Combretac.;* Bras.
„ -bagre: *Terminalia aff. januarensis* DC.; *Combretac.;* Bras.
Merey: *Anacardium occidentalis* L.; *Anacardiac.;* Kol.
Meriang: *Fagraea gigantea* Ridl., *F. fragrans* Roxb.; *Loganiac.;* Malay.
Meribut: *Diospyros spp., Maba spp.; Ebenac.;* Malay.
Mérie: *Humiria floribunda* Mart., *H. balsamifera* Aubl.; *Humiriac.;* Sur.
Merisier: *Prunus avium* L.; *Rosac.;* Eu.
„   : *Prunus serotina* Ehrh.; *Rosac.;* sö. Kan.
„   : *Betula lutea* Michx.; *Betulac.;* sö. Kan.
„   : *Malpighia glabra* L.; *Malpighiac.;* Guad.
„ doré: *Byrsonima spicata* L. C. Rich.; *Malpighiac.;* Mart.
„ jaune: *Betula lutea* Michx.; *Betulac.;* sö. Kan.
„   „   : *Eugenia sp.; Myrtac.;* Guad.

Merisier odorant: *Betula lenta* L.; *Betulac.*; sö. Kan.
„ ondé: *Betula lutea* Michx.; *Betulac.*; sö. Kan.
„ , petit: *Prunus pennsylvanica* L. f.; *Rosac.*; s. Kan.
„ rouge: *Betula lenta* L.; *Betulac.*; sö. Kan.
Merombong: *Adina rubescens* Hemsl.; *Rubiac.*; Malay.
Meroola: *Sclerocarya caffra* Sond.; *Anacardiac.*; S.-Af.
Mersawa: *Anisoptera spp.*, *A. thurifera* Blume; *Dipterocarpac.*; Malay (Bentong)
Mertajam: *Erioglossum edule* Blume; *Sapindac.*; Malay.
Mescal: *Eugenia sp.*; *Myrtac.*; br. Hond.
Mescrew: *Prosopis sp.*; *Mimosac.*; USA.
Mesegar: *Dipterocarpus trinervis* Blume; *Dipterocarpac.*; Java, Sum.
Meshirié: *Necepsia afzeli* Prain; *Euphorbiac.*; Elf.
Mesica: *Brosimum terrabanum* Pittier; *Morac.*; nö. Nic.
Meslier épineux: *Ilex aquifolium* L.; *Aquifoliac.*; Eu., n. Af.
Mesquite: *Prosopis juliflora* DC.; *Mimosac.*; tr. Am.
„ : *Caesalpinia sclerocarpa* Standl.; *Caesalpiniac.*; Mex. (Sinaloa)
„ , screwbean: *Prosopis sp.*; *Mimosac.*; Mex.
„ , screw-pod-: *Prosopis sp.*; *Mimosac.*; Mex.
Mesquito: *Prosopis juliflora* DC.; *Mimosac.*; C.-Am.
Messère gèvis: *Terminalia parvula* Pampan.; *Combretac.*; O.-Af. (it. Somali)
Mess-guch: *Albizzia lucida* Bth.; *Mimosac.*; Assam.
Messmate: *Eucalyptus spp.*; *Myrtac.*; ö. Au.
Mestizo: *Guarea trichilioides* L.; *Meliac.*; Kol.
Mesu: *Shorea pauciflora* King, *S. spp.*; *Dipterocarpac.*; Malay (Pahang)
Mesua: *Mesua ferrea* L.; *Guttifer.*; O.-Ind.
Metchi: *Pontya excelsa* Chev.; *Urticac.*; Elf.
Metchibanaïe: *Guarea thompsoni* Sprague et Hutch.; *Meliac.*; Elf.
Metis: *Payena utilis* Ridl., *P. spp.*, *Madhuca spp.*; *Sapotac.*; Malay.
„ : *Palaquium stellatum* King et Gamble; *Sapotac.*; Malay.
Metlein: *Quercus serrata* Thunb.; *Fagac.*; Burma.
Metrobeantina: *Cinnamomum fragrans* Baill.; *Laurac.*; Mad.
Mêtʒa: *Quercus sp.*; *Fagac.*; Mex.
Meu lac: *Barringtonia longipes* Gagnep.; *Lecythidac.*; Ind.-Ch.
Meurapiranga: *Brosimum paraense* Huber; *Morac.*; Bras.
Meurinoki: *Acer crataegifolium* S. et Z.; *Acerac.*; Jap.
Mézali: *Cassia siamea* Lamk.; *Caesalpiniac.*; O.-Ind., Malay.
Mezcal: *Ulmus mexicana* Planch.; *Ulmac.*; Salv.
Mèzè: *Bassia longifolia* L.; *Sapotac.*; Burma.
Mezquite: *Prosopis juliflora* DC.; *Mimosac.*; Mex.
„ amarillo: *Prosopis juliflora* DC.; *Mimosac.*; Mex.
„ blanco: *Prosopis juliflora* DC.; *Mimosac.*; Mex.
„ chino: *Prosopis juliflora* DC.; *Mimosac.*; Mex.
„ colorado: *Prosopis juliflora* DC.; *Mimosac.*; Mex.
M'fan: *Distemonanthus benthamianus* Baill.; *Caesalpiniac.*; Kam.
Mfan: *Dialium guineense* Willd.; *Caesalpiniac.*; Kam.
Mfantu: *Pachystela cinerea* Pierre; *Sapotac.*; Elf.
Mfass: *Ouratea calophylla* Engl.; *Ochnac.*; Kam.
M'fé: *Alstonia congensis* Engl.; *Apocynac.*; Kam.
Mfère: *Pterygopodium balsamiferum* Verm.; *Papilionac.*; B.-K.
Mfila: *Anona senegalensis* Pers.; *Anonac.*; D.-O.-Af. (Tabora)
M'findja: *Rhizophora mucronata* Lamk.; *Rhizophorac.*; port. O.-Af. (Moz., Quel.)
Mfintima: *Ostryoderris impressa* Dunn; *Papilionac.*; Go.
Mfo: *Poga oleosa* Pierre; *Rhizophorac.*; Gab.
M'fo: *Enantia chlorantha* Oliv.; *Anonac.*; Kam.
Mfoenga: *Xylopia quintasi* Pierre; *Anonac.*; Kam.
Mfomba: *Alstonia congensis* Engl.; *Apocynac.*; Gab.
M'foum: *Ceiba pentandra* Gaertn.; *Bombacac.*; Gab.
Mfoum: *Spondianthus preussi* Engler; *Euphorbiac.*; Gab.
Mfouma: *Ceiba pentandra* Gaertn.; *Bombacac.*; Gab.
Mfoumé: *Ceiba pentandra* Gaertn.; *Bombacac.*; Gab.
Mfrikiu: *Albizzia pospischili* Harms; *Mimosac.*; D.-O.-Af. (Kilossa)
Mfubu: *Ficus sp.*; *Morac.*; S.-Af. (Natal)
Mfunguru: *Pappea capensis* Eckl. et Zeyh.; *Sapindac.*; D.-O.-Af. (Usambara)
Mfuranji: *Albizzia sp.*; *Mimosac.*; D.-O.-Af.

Mgambo: *Mimusops eicki* Engler; *Sapotac.;* D.-O.-Af. (w. Usambara)
M'ganaké: *Guarea cedrata* Pellegr.; *Meliac.;* Elf.
Mgando-mkalatè: *Pterocarpus sp.; Papilionac.;* D.-O.-Af. (Tabora)
Mgembe: *Dalbergia melanoxylon* Guill. et Perr.; *Papilionac.;* D.-O.-Af. (Tabora)
Mgenda: *Acacia catechu* Willd.; *Mimosac.;* D.-O.-Af. (Usambara)
Mgera: *? Berlinia emini* Taub.; *Caesalpiniac.;* D.-O.-Af. (Tabora)
Mghogho: *Malacantha sp.; Sapotac.;* D.-O.-Af. (w. Usambara)
Mgogondi: *Phyllanthus reticulatus* Poir.; *Euphorbiac.;* D.-O.-Af. (Tabora)
M'gongo: *Ricinodendron rautaneni* Schinz; *Euphorbiac.;* Rhod.
M'goronda: *Bruguiera gymnorrhiza* Lamk.; *Rhizophorac.;* port. O.-Af. (Moz.)
Mgouanha: *Cola ballayi* Cornu; *Sterculiac.;* Gab.
M'gouin: *Pterocarpus erinaceus* Poir.; *Papilionac.;* fr. Guin.
Mgua: *Celtis integrifolia* Lamk.; *Ulmac.;* Elf.
  „ : *Sterculia sp.; Sterculiac.;* D.-O.-Af. (Tabora)
Mgualo: *Uangueria infausta* Burch.; *Rubiac.;* D.-O.-Af. (Tabora)
M'guangua: *Octoknema affinis* Pierre; *Olacac.;* Elf.
Mgukulama: *? Combretum taborense* Engl.; *Combretac.;* D.-O.-Af.
Mgukuti: *Moñotes elegans* Gilg; *Dipterocarpac.;* D.-O.-Af. (Tabora)
Mgulugunga: *? Acacia suma* Kurz; *Mimosac.;* D.-O.-Af. (Tabora)
Mgumo: *Ficus stuhlmanni* Warb.; *Morac.;* D.-O.-Af. (Tabora)
Mgunde: *Chlorophora excelsa* Bth. et Hook.; *Morac.;* port. O.-Af.
Mgunga: *Acacia catechu* Willd.; *Mimosac.;* D.-O.-Af. (Usambara)
Mgunumbwe: *Combretum petersi* Engl.; *Combretac.;* D.-O.-Af. (Kilossa)
Mgurangurane: *Randia pierrei* Chev.; *Rubiac.;* Gab.
Mguruka: *Maerua sp.; Capparidac.;* D.-O.-Af. (Tabora)
Mgurunguja: *Strychnos engleri* Gilg; *Loganiac.;* D.-O.-Af. (Kilossa)
Mgwägwe: *Strychnos stuhlmanni* Gilg; *Loganiac.;* D.-O.-Af. (Tabora)
Mhafa: *Millettia oblata* Dunn; *Papilionac.;* D.-O.-Af. (w. Usambara)
Mhande: *Lonchocarpus aff. brevicaudatus* Vatke; *Papilionac.;* D.-O.-Af. (w. Us.)
Mhangáti: *Juniperus procera* Hochst.; *Conifer.;* D.-O.-Af. (w. Usambara)
Mhombo: *Calodendron eicki* Engler; *Rutac.;* D.-O.-Af. (w. Usambara)
Mhosia: *Sterculia sp.; Sterculiac.;* D.-O.-Af. (Tabora)
Mhowra: *Bassia latifolia* Roxb.; *Sapotac.;* O.-Ind. (Bombay)
Mhullula-mbuli: *Daemia cordifolia* K. Schum.; *Asclepiadac.;* D.-O.-Af. (Tabora)
Mhundu: *Strychnos unguacha* A. Rich. *var. polyantha* Gilg; *Loganiac.;* D.-O.-Af. (Tab.)
Mi: *Artocarpus integrifolia* L. f.; *Morac.;* Ind.-Ch.
Mi-itu: *Treculia aff. africana* Dcne.; *Morac.;* Go.
Mia-maté-si: *Diospyros burmanica* Kurz; *Ebenac.;* Burma.
Miama: *Calpocalyx klainei* Pierre; *Mimosac.;* Kam., sp. Guin., Gab.
Miamen gomo: *Caloncoba glauca* Gilg; *Flacourtiac.;* Kam.
  „ gomo: *Caloncoba welwitschi* Gilg; *Flacourtiac.;* s. Kam., n. Gab.
Miami-ngoma: *Caloncoba glauca* Gilg; *Flacourtiac.;* Kam., Gab.
Miáo: *Dysoxylum euphlebium* Merr.; *Meliac.;* Phil. (Negrosinsel)
Miápi: *Avicennia officinalis* L.; *Uerbenac.;* Phil. (Leyte)
Mibella: *Uitex altissima* L. f.; *Uerbenac.;* O.-Ind., Cey.
Micaquahuitl: *Morisonia americana* L.; *Capparidac.;* Mex.
Michai: *Berberis chilensis* Gill.; *Berberidac.;* Chile.
Micocoulier: *Celtis australis* L.; *Ulmac.;* s. Eu.
  „ occidental: *Celtis occidentalis* L.; *Ulmac.;* ö. Kan.
Miedieng: *Calophyllum sp.; Guttifer.;* Nied.-Ind.
Miedzo: *Pachypodanthium staudti* Engl. et Diels; *Anonac.;* Elf.
Mielga real: *Medicago arborea* L.; *Papilionac.;* Arg.
Mién: *Litchi chinensis* Sonn.; *Sapindac.;* Kamb.
  „ kêt: *Gleditschia australis* Hemsl.; *Caesalpiniac.;* Ind.-Ch.
  „ mou: *Ceiba pentandra* Gaertn.; *Bombacac.;* Annam.
  „ prey: *Aglaia gigantea* Pellegr.; *Meliac.;* Ind.-Ch.
  „ „ : *Euphoria cambodiana* H. Lec.; *Sapindac.;* Ind.-Ch.
Miendjoe: *Hirtella spp.; Rosac.;* Sur.
Miendjou: *Licania spp.; Rosac.;* Sur.
Miengumo: *Caloncoba glauca* Gilg; *Flacourtiac.;* W.-Af. (sp. Muni)
Mierenhout: *Triplaris surinamensis* Cham.; *Polygonac.;* Sur.
Mihiriya: *Gordonia zeylanica* Wight; *Theac.;* Ceylon.
Mihuasungu: *Randia dumetorum* Lamk.; *Rubiac.;* D.-O.-Af. (Tabora)
Mija: *Anacardium rhinocarpus* DC.; *Anacardiac.;* Ven.

Mijaguo: *Anacardium rhinocarpus* DC.; *Anacardiac.;* Ven.
Mijao: *Anacardium rhinocarpus* DC.; *Anacardiac.;* Ven.
Mijico: *Dipholis salicifolia* A. DC.; *Sapotac.;* br. Hond.
Mijin kadai: *Lophira alata* Banks; *Ochnac.;* n. Nig.
„ kade: *Lophira alata* Banks; *Ochnac.;* Nig.
Mikosa: *Rhizophora racemosa* G. F. W. Meyer; *Rhizophorac.;* Gab. (Loango)
Milachithayam: *Bischofia javanica* Blume; *Euphorbiac.;* O.-Ind. (Madras)
Milakaranai: *Toddalia aculeata* Pers.; *Rutac.;* O.-Ind.
Mileucaraney-cheddi: *Toddalia aculeata* Pers.; *Rutac.;* O.-Ind.
Milho cosido: *Licania incana* Aubl.; *Rosac.;* Bras. (Sao P)
Milili: *Polyalthia cerasoides* Bth. et Hook. f.; *Anonac.;* O.-Ind. (Madras)
Milipíli: *Canarium villosum* F. Vill.; *Burserac.;* Phil. (Cebu)
Milki hoedoe: *Tabernaemontana utilis* Wight et Arn.; *Apocynac.;* Sur.
„ „ : *Clerodendron spp.; Verbenac.;* Sur.
Milk-tree: *Mimusops sp.; Sapotac.;* n. S.-Am.
Milkwood: *Tabernaemontana citrifolia* L.; *Apocynac.;* Pan.
„ : *Sideroxylon inerme* L.; *Sapotac.;* Kap.
„ , white: *Sideroxylon inerme* L.; *Sapotac.;* S.-Af.
Milla: *Vitex altissima* L. f.; *Verbenac.;* Ceylon.
Millai: *Lophira alata* Banks; *Ochnac.;* W.-Af. (S.-L.)
Mille: *Lophira alata* Banks; *Ochnac.;* W.-Af. (S.-L.)
„ branches: *Phyllanthus nobilis* Muell. Arg.; *Euphorbiac.;* Guad.
„ -pieds: *Clusia rosea* Jacq.; *Guttifer.;* Ant.
Milola: *Hibiscus tiliaceus* L.; *Malvac.;* port. Af.
Miloló: *Anona reticulata* L.; *Anonac.;* Bras.
Mil-pesos: *Hura crepitans* L.; *Euphorbiac.;* Kol.
Milua: *Strychnos sp.; Loganiac.;* D.-O.-Af. (Tabora)
Mimbre: *Salix chilensis* Molina; *Salicac.;* Kol.
Mimbrera: *Salix viminalis* L.; *Salicac.;* Sp.
Mimosa: *Acacia spp., Mimosa spegazzinii* Pirotta; *Mimosac.;* Arg.
„ : *Acacia horrida* Willd.; *Mimosac.;* S.-Af.
Mimpanda: *Hexalobus crispiflorus* A. Rich.; *Anonac.;* Kam.
Minanúkai: *Pometia pinnata* Forst.; *Sapindac.;* Phil. (Negrosinsel)
Mincouart: *Minquartia guianensis* Aubl.; *Olacac.;* fr. Gu., Sur.
Mincouin: *Minquartia guianensis* Aubl.; *Olacac.;* Gu.
Mindhal: *Randia dumetorum* Lamk.; *Rubiac.;* O.-Ind. (Bombay)
Mindi: *Melia azedarach* L.; *Meliac.;* Nied.-Ind.
Mindo: *Polyalthia simiarum* Bth. et Hook. f.; *Anonac.;* Annam.
Ming deuk: *Phyllostachys nigra* Munro; *Gramin.;* Ch. (Fukien)
Mingris: *Koompassia beccariana* Taub.; *Caesalpinac.;* Borneo.
Minguar: *Minquartia guianensis* Aubl.; *Olacac.;* Gu.
Minguella: *Ricinodendron africanum* Muell. Arg.; *Euphorbiac.;* Angola.
Mininga: *Pterocarpus bussei* Harms, *P. erinaceus* Poir.; *Papilionac.;* D.-O.-Af.
Minjak tengkawang: *Shorea aptera* Burck; *Dipterocarpac.;* Borneo.
Minjinjiem: *Calophyllum sp.; Guttifer.;* Nied.-Ind.
Minjiriya: *Erythrina senegalensis* DC.; *Papilionac.;* n. Nig.
Minquart: *Minquartia guianensis* Aubl.; *Olacac.;* Gu.
Minsi: *Calpocalyx dinklagei* Harms; *Mimosac.;* Kam.
Mint tree: *Prosthanthera lasianthos* Labill.; *Labiat.;* sö. Au., Tasm.
Minuang: *Octomeles sumatrana* Miq.; *Datiscac.;* Malay.
Minyak berok: *Xanthophyllum obscurum* A. W. Benn.; *Polygalac.;* Malay.
„ keruing: *Dipterocarpus spp.; Dipterocarpac.;* Malay.
Minzu: *Petersia africana* Welw.; *Lecythidac.;* B.-K.
Mio: *Thespesia populnea* Correa; *Malvac.;* Südsee (Marquesas)
Miovu: *Entandrophragma utilis* Sprague; *Meliac.;* br. O.-Af. (Uganda)
Mira amae: *Thespesia populnea* Correa; *Malvac.;* Südsee (Tahiti)
„ -pishuna: *Swartzia sp.; Caesalpiniac.;* Bras.
Mirabau: *Intsia sp.; Caesalpiniac.;* br. N.-Borneo.
Miraboo: *Intsia bakeri* Prain; *Caesalpiniac.;* Malay, Sunda.
„ laut: *Intsia bijuga* O. Ktze.; *Caesalpiniac.;* Malay.
Mirabow: *Intsia bakeri* Prain; *Caesalpiniac.;* Malay, Sunda.
Mirahoedoe: *Triplaris surinamensis* Cham.; *Polygonac.;* Sur.
Mirahoudou: *Triplaris sp.; Polygonac.;* Sur.
Miral: *Phyllanthus emblica* L.; *Euphorbiac.;* w. Beng. (Chota Nagpur)

Mirauan: *Anisoptera spp.; Dipterocarpac.;* br. Borneo (Sarawak)
Mirim: *Bowdichia virgilioides* H. B. K.; *Papilionac.;* Ven., Bras.
Mirindiba: *Terminalia sp.; Combretac.;* Bras.
Miriti: *Mauritia sp.; Palmac.;* Ek.
Mirlang: *Irvingia malayana* Oliv.; *Simarubac.;* Malak.
Miro: *Podocarpus ferruginea* G. Benn.; *Conifer.;* N.-Seel,
„ : *Thespesia populnea* Correa; *Malvac.;* Südsee (Tahiti)
„ toromiro: *Podocarpus ferruginea* G. Benn.; *Conifer.;* N.-Seel.
Mirobalanenbaum (d): *Phyllanthus emblica* L.; *Euphorbiac.;* tr. As.
Mirobolan: *Hernandia guianensis* Aub.; *Hernandiac.;* Gu.
Mirto: *Myrtus communis* L.; *Myrtac.;* Sp.
Mispel, Berg-, filzige (d): *Cotoneaster tomentosa* Lindl.; *Rosac.;* s. Eu.
„ , „ -, gemeine: *Cotoneaster vulgaris* Lindl.; *Rosac.;* Eu.
Mispelbaum, gemeiner: *Mespilus germanica* L.; *Rosac.;* s. Eu., Or.
Mispelboom: *Achras sapota* L.; *Sapotac.;* Ant.
Misquitl: *Prosopis juliflora* DC.; *Mimosac.;* Mex.
Missa amandan: *Millettia sp.; Papilionac.;* fr. Guin.
Mistel, nordische: *Viscum album* L.; *Loranthac.;* Eu.
Mistol: *Zizyphus mistol* Griseb.; *Rhamnac.;* Arg.
Mit: *Artocarpus integrifolia* L., *A. hirsuta* Lamk.; *Morac.;* Ind.-Ch.
„ giai: *Artocarpus sp.; Morac.;* Ind.-Ch.
„ mât: *Artocarpus sp.; Morac.;* Ind.-Ch.
„ này: *Artocarpus hirsuta* Lamk.; *Morac.;* Annam.
Mitla: *Lagerstroemia speciosa* Pers.; *Lythrac.;* Phil. (Pampanga)
Mitsudekaede: *Acer cissifolium* C. Koch; *Acerac.;* Jap.
Miula-swagallo: *Allophylus africanus* Beauv.; *Sapindac.;* D.-O.-Af. (Tabora)
Miwelie: *Strophanthus emini* Aschers. et Pax; *Apocynac.;* D.-O.-Af. (Tabora)
Miyamahannoki: *Alnus viridis* DC. var. *sibirica* Rgl.; *Betulac.;* Jap.
Miyama-inuzakura: *Prunus ssiori* Fr. Schmidt; *Rosac.;* Jap.
Miyamazakura: *Prunus maximowiczi* Rupr.; *Rosac.;* Jap.
Miyo: *Thespesia populnea* Correa; *Malvac.;* Südsee (Marquesas)
Mizuki: *Cornus controversa* Hemsl.; *Cornac.;* Jap.
Mizunara: *Quercus grosseserrata* Blume; *Fagac.;* n. Jap.
Mjense: *Brachystegia sp.; Caesalpiniac.;* D.-O.-Af. (Tabora)
Mjimo: *Sarcocephalus esculentus* Afzel.; *Rubiac.;* Elf.
Mjombo: *Brachystegia* sp.; *Caesalpiniac.;* O.-Af. (Nyassa)
Mkaka: *Rhizophora mucronata* Lam.; *Rhizophorac.;* D.-O.-Af.
Mkala: *Rhizophora mucronata* Lam.; *Rhizophorac.;* Sans.
Mkalati: *Burkea sp.; Caesalpiniac.;* O.-Af. (Nyassa)
Mkam: *Vangueria infausta* Burch.; *Rubiac.;* D.-O.-Af. (Tabora)
Mkambala: *Acacia brosigi* Harms; *Mimosac.;* D.-O.-Af. (Kilossa)
Mkame: *Piptadenia hildebrandti* Vatke; *Mimosac.;* O.-Af. (Usambara)
Mkandala: *Ceriops candolleana* Arn.; *Rhizophorac.;* tr. Af.
Mkande: *Stereospermum dentatum* A. Rich.; *Bignoniac.;* D.-O.-Af. (Usambara)
Mkangi: *Allanblackia stuhlmanni* Engl.; *Guttifer.;* D.-O.-Af. (Usambara)
Mkani: *Allanblackia stuhlmanni* Engl.; *Guttifer.;* D.-O.-Af. (Usambara)
Mkeli: *Piptadenia africana* Hook. f.; *Mimosac.;* Lib.
Mkemini: *Cassia fistula* L.; *Caesalpiniac.;* D.-O.-Af. (Kilossa)
Mkerembeke: *Heeria mucronifolia* Bernh.; *Anacardiac.;* O.-Af.
Mkéyu: *Dombeya malacoxylon* K. Schum.; *Sterculiac ;* D.-O.-Af. (w. Usambara)
Mkoba: *Baikiaea emini* Taub.; *Caesalpiniac.;* D.-O.-Af.
Mkoko: *Conopharyngia holsti* Stapf; *Apocynac.;* D.-O.-Af. (w. Usambara)
„ : *Faurea speciosa* Welw.; *Proteac.;* D.-O.-Af. (Kilossa)
Mkola: *Afzelia quanzensis* Welw.; *Caesalpiniac.;* D.-O.-Af. (Tabora)
Mkole: *Grewia microcarpa* K. Schum.; *Tiliac.;* D.-O.-Af. (Kilossa)
Mkom wi tale: *Ficus glumosa* Delile; *Morac.;* D.-O.-Af. (Tabora)
Mkoma: *Grewia bicolor* Juss.; *Tiliac.;* D.-O.-Af. (Tabora)
„ -mkandja: *Bridelia sp.; Euphorbiac.;* D.-O.-Af. (Tabora)
Mkomahoya: *Pygeum africanum* Hook. f.; *Rosac.;* D.-O.-Af. (w. Usambara)
Mkome: *Strychnos pungens* Solered.; *Loganiac.;* D.-O.-Af. (Tabora)
Mkomohoyo: *Pygeum africanum* Hook. f.; *Rosac.;* D.-O.-Af.
Mkonga: *Balanites aegyptiaca* Delile; *Zygophyllac.;* D.-O.-Af. (Kilossa)
Mkongolo: *Combretum schumanni* Engl.; *Combretac.;* D.-O.-Af. (Usambara)
Mkongomwa: *Afzelia quanzensis* Welw.; *Caesalpiniac.;* s. C.-Af.

Mkonko: *Rhizophora mucronata* Lam.; *Rhizophorac.;* Sans.
Mkonse: *Mimusops densiflora* Engl.; *Sapotac.;* D.-O.-Af. (Tabora)
Mkora: *Afzelia quanzensis* Welw.; *Caesalpiniac.;* D.-O.-Af. (Tabora)
Mkowe: *Garcinia kilossana* Engl.; *Guttifer.;* D.-O.-Af. (Kilossa)
Mkufi: *Piptadenia schweinfurthi* Vatke; *Mimosac.;* D.-O.-Af.
Mkuguza: *Widdringtonia whytei* Rendle; *Conifer.;* br. C.-Af.
Mkuju: *Ficus glumosa* Delile; *Morac.;* D.-O.-Af. (Tabora)
Mkuka: *Ficalhoa laurifolia* Hiern; *Ericac.;* D.-O.-Af. (w. Usambara)
    „   : *Royena nyassae* Gürke, *R. sp.; Ebenac.;* D.-O.-Af.
Mkulo: *Ocotea usambarensis* Engl.; *Laurac.;* D.-O.-Af. (w. Usambara)
Mkumba: *Macaranga usambarica* Pax; *Euphorbiac.;* D.-O.-Af. (w. Usambara)
Mkunazi: *Zizyphus jujuba* Lamk.; *Rhamnac.;* tr. Af., tr. As., Au.
Mkungu: *Terminalia catappa* L.; *Combretac.;* Malay.
Mkunguna: *Sorindeia usambarensis* Engl.; *Anacardiac.;* O.-Af.
Mkunguni: *Fagaropsis oppositifolia* Mildbr.; *Rutac.;* D-.O.-Af. (w. Usambara)
    „   : *Commiphora zimmermanni* Engl.; *Burserac.;* D.-O.-Af.
Mkungusa: *Widdringtonia whytei* Rendle; *Conifer.;* s. C.-Af.
Mkuru: *Cardiogyne africana* Bureau; *Morac.;* O.-Af.
Mkuruka: *Maerua angolensis* DC.; *Capparidac.;* D.-O.-Af. (Kilossa)
Mkurungu: *Pterocarpus chrysothrix* Taub.; *Papilionac.;* D.-O.-Af. (Tabora)
Mkushi: *Baikiaea plurijuga* Harms; *Caesalpiniac.;* n. Rhod.
Mkusi: *Baikiaea plurijuga* Harms; *Caesalpiniac.;* n. Rhod.
Mkusu: *Ricinodendron rautaneni* Schinz; *Euphorbiac.;* n. Rhod.
Mkute: *Podocarpus milanjianus* Rendle; *Conifer.;* O.-Af. (Nyassa)
Mkuyu: *Ficus sycomorus* L.; *Morac.;* O.-Af. (Nyassa)
Mkwadju: *Tamarindus indica* L.; *Caesalpiniac.;* D.-O.-Af. (Tabora)
Mkwambéti: *Trichocladus ellipticus* Sond.; *Hamamelidac.;* D.-O.-Af. (w. Usamb.)
Mkweo: *Tylostemon kweo* Mildbr.; *Laurac.;* D.-O.-Af. (Usambara)
Mkwerenyani: *Piptadenia buchanani* Baker; *Mimosac.;* O.-Af. (Nyassa)
Mkwisimkwi: *Cassia fistula* L.; *Caesalpiniac.;* D.-O.-Af. (Kilossa)
Mlago: *Heeria pulcherrima* O. Kt̃ze.; *Anacardiac.;* D.-O.-Af. (Tabora)
Mlama: *Combretum oblongum* F. Hoffm.; *Combretac.;* D.-O.-Af. (Tabora)
    „ ekundu: *Combretum sp.; Combretac.;* D.-O.-Af. (Kilossa)
    „ meupe: *Combretum spp.; Combretac.;* D.-O.-Af. (Kilossa)
Mlazi: *Schrebera holsti* Gilg; *Oleac.;* D.-O.-Af. (w. Usambara)
Mlefu: *Ehretia silvatica* Gürke; *Borraginac.;* D.-O.-Af. (w. Usambara)
Mlifu: *Warburgia stuhlmanni* Engler; *Canellac.;* D.-O.-Af. (w. Usambara)
Mlihua-hulu: *Tarchonanthus camphoratus* L.; *Composit.;* D.-O.-Af. (Tabora)
Mlombwa: *Pterocarpus angolensis* DC.; *Papilionac.;* O.-Af. (Nyassa)
Mlugalla: *Weihea insignis* Engl.; *Rhizophorac.;* D.-O.-Af. (Tabora)
Mlungulungu: *Fagara chalybaea* Engl.; *Rutac.;* D.-O.-Af. (Tabora)
Mlunguschigiti: *Vepris glomerata* Engl.; *Rutac.;* D.-O.-Af. (Tabora)
M'man: *Irvingia oblonga* Chev.; *Simarubac.;* Kam.
M'muaga: *Ochna holsti* Engl.; *Ochnac.;* D.-O.-Af. (Tabora)
Mnaze: *Cocos nucifera* L.; *Palmac.;* D.-Af.
Mnembu: *Cordia dioica* Boj.; *Borraginac.;* D.-O.-Af. (Tabora)
Mnemwua: *Ximenia americana* L.; *Olacac.;* D.-O.-Af. (Tabora)
Mninga: *Pterocarpus erinaceus* Poir., *P. bussei* Harms; *Papilionac.;* D.-O.-Af.
Mnyassa: *Piptadenia buchanani* Taub.; *Mimosac.;* D.-O.-Af. (w. Usambara)
Mnyowe: *Eugenia cordata* Laws.; *Myrtac.;* O.-Af. (Nyassa)
Mo: *Litsea polyantha* Juss.; *Laurac.;* Ind.-Ch.
    „ : *Deutzianthus tonkinensis* Gagnep.; *Euphorbiac.;* Ind.-Ch.
    „ : *Manglietia ? fordiana* Oliv., *M. glauca* Blume; *Magnoliac.;* Ind.-Ch.
    „ cao: *Glycosmis montana* Pierre; *Rutac.;* Ind.-Ch.
    „ choul: *Aglaia pirifera* Hance; *Meliac.;* Ind.-Ch.
    „ cua: *Terminalia catappa* L.; *Combretac.;* Ind.-Ch.
    „   „ : *? Alstonia scholaris* R. Br.; *Apocynac.;* Ind.-Ch.
    „   „ : *Bassia edulis* Choisy; *Sapotac.;* Ind.-Ch.
    „   „ : *Albizzia procera* Bth.; *Mimosac.;* Ind.-Ch.
    „ -giôi: *Michelia baviensis* Finet et Gagnep.; *Magnoliac.;* Tonkin.
    „ hú: *Calophyllum inophyllum* L.; *Guttifer.;* Annam.
? „ kak: *Spondias mangifera* Willd.; *Anacardiac.;* Ind.-Ch.
    „ trang: *Claoxylon indicum* Hassk.; *Euphorbiac.;* Ind.-Ch.
    „ vang tâm: *Manglietia fordiana* Oliv., *M. glauca* Blume; *Magnoliac.;* Ind.-Ch.

Moa: *Flindersia australis* R. Br.; *Rutac.*; tr. O.-Au.
Moabi (f): *Mimusops djave* Engl., *M. pierreana* Engl.; *Sapotac.*; Kam., Gab.
„ : *Mimusops congolensis* Russel et Hédin; *Sapotac.*; Kam., Gab.
„ : *Sterculia quinqueloba* K. Schum.; *Sterculiac.*; B.-K. (Katanga)
Moabiéré: *Mimusops heckeli* Lecomte; *Sapotac.*; Gab.
Moahholz: *Flindersia australis* R. Br.; *Rutac.*; tr. O.-Au.
Moamba: *Polyalthia suaveolens* Engl. et Diels; *Anonac.*; B.-K.
Moambe jaune (f): *Enantia chlorantha* Oliv.; *Anonac.*; Kam.
Moana: *Hylodendron gabunense* Taub.; *Caesalpiniac.*; Gab.
Mobaka: *Copaifera demeusei* Harms; *Caesalpiniac.*; B.-K.
Moc: *Eugenia bullochi* Hance; *Myrtac.*; Ind.-Ch.
„ : *Osmanthus fragrans* Lour.; *Oleac.*; Ind.-Ch.
„ tông: *Irvingia oliveri* Pierre; *Simarubac.*; Annam.
Moca: *Andira jamaicensis* Urban; *Papilionac.*; P.-R.
„ amarilla: *? Andira inermis* H. B. K.; *Papilionac.*; P.-R.
„ blanca: *? Andira inermis* H. B. K.; *Papilionac.*; P.-R.
„ colorada: *? Andira inermis* H. B. K.; *Papilionac.*; Mex.
Mocaya: *Acrocomia sclerocarpa* Mart.; *Palmac.*; fr. Gu.
Mochi: *Ilex integra* Thunb.; *Aquifoliac.*; Jap.
Mochigüiste: *Pithecolobium dulce* Bth.; *Mimosac.*; C.-R.
Mochinoki: *Ilex integra* Thunb., *I. othera* Spreng.; *Aquifoliac.*; Jap.
Moçitahyba: *Zollernia falcata* Nees, *Z. ilicifolia* Vog.; *Caesalpiniac.*; Bras.
Mock orange: *Pittosporum undulatum* Vent.; *Pittosporac.*; ö. Au.
Moco: *Afzelia sp., prob. quanzensis* Welw.; *Caesalpiniac.*; port. O.-Af. (L.-M.)
Mocoso: *Ziziphus guatemalensis* Hemsl.; *Rhamnac.*; Guat.
Modugu: *Butea frondosa* Roxb.; *Papilionac.*; O.-Ind. (Madras)
Moee: *Lannea grandis* Engl.; *Anacardiac.*; sw. O.-Ind. (Bombay)
Moejipan latoe: *Piratinera spp.*; *Morac.*; Sur.
Moejepaulétoe: *Piratinera guianensis* Aubl.; *Morac.*; Sur.
Moejepau lettoé: *Helicostylis poeppigiana* Trec.; *Morac.*; Gu., n. Bras.
Moeleidan: *Byrsonima sp.*; *Malpighiac.*; Sur.
Moelè-ie: *Byrsonima sp.*; *Malpighiac.*; Sur.
Moeloe: *Evodia sp.*; *Rutac.*; Sum.
Moendoe: *Garcinia dulcis* Kurz; *Guttifer.*; Sunda.
Moerbei: *Morus nigra* L.; *Morac.*; s. Eu., w. As.
Moeré-ie: *Byrsonima sp.*; *Malpighiac.*; Sur.
Moereiran: *Byrsonima sp.*; *Malpighiac.*; Sur.
Moetoesi: *Pterocarpus draco* L., *P. rhori* Vahl; *Papilionac.*; Sur.
Moetoesiran: *Pterocarpus draco* L., *P. rhori* Vahl; *Papilionac.*; Sur.
Mofoumou: *Aucoumea klaineana* Pierre; *Burserac.*; Gab.
Mofu: *Entandrophragma sp.*; *Meliac.*; n. Rhod.
Mofwe: *Entandrophragma delevoyi* de Wild.; *Meliac.*; B.-K.
Mofwe: *? Pseudocedrela sp.*; *Meliac.*; B.-K. (Katanga)
Mogadigila: *Pentadesma butyraceum* Don; *Guttifer.*; Gab.
Mogalinga: *Schrebera swietenioides* Roxb.; *Oleac.*; O.-Ind. (Madras)
Mogameba: *Lavalleopsis densivenia* Engl., *Strombosia grandifolia* Hook. f.; *Olacac.*;
Moganga: *Copaifera sp.*; *Caesalpiniac.*; Gab.   [Gab.
Mogano (i): *Swietenia spp.*; *Meliac.*; tr. Am.
„ Benin: *Khaya grandifolia* Stapf; *Meliac.*; s. Nig.
Mogariza: *Erica umbellata* L.; *Ericac.*; Sp.
Mogi: *Eugenia alternifolia* Wight; *Myrtac.*; O.-Ind. (Madras)
Mogno (p): *Swietenia spp.*; *Meliac.*; tr. Am.
„ branco: *Eucalyptus robusta* Sm.; *Myrtac.*; Bras. (Sao P)
Mogomogo: *Miconia incarum* Macbride; *Melastomatac.*; Peru.
Mogongoongo: *Sideroxylon inerme* L.; *Sapotac.*; O.-Af. (Muoa)
Mogoubi: *Staudtia gabonensis* Warb.; *Myristicac.*; Gab.
Moguinzi: *Barteria dewevrei* de Wild. et Dur.; *Flacourtiac.*; Gab.
Moh: *Afzelia africana* Smith; *Caesalpiniac.*; Kam.
„ : *Bassia latifolia* Roxb.; *Sapotac.*; C.-O.-Ind.
Moha: *Bassia latifolia* Roxb.; *Sapotac.*; w. Beng. (Orissa)
Mohamma: *Borassus flabellifer* Murr. *var. aethiopica* ?; *Palmac.*; B.-K. (Katanga)
Mohan: *Cassia garrettiana* Craib; *Caesalpiniac.*; Ind.-Ch.
Mohi: *Lannea grandis* Engl.; *Anacardiac.*; C.-O.-Ind.

Mohingue: *Pterocarpus soyauxi* Taub.; *Papilionac.*; Kam.
„ mossoumbé: *Pterocarpus soyauxi* Taub.; *Papilionac.*; Kam.
Mohkwa: *Carica papaya* L.; *Caricac.*; Jap.
Mohni: *Lannea grandis* Engl.; *Anacardiac.*; C.-O.-Ind. (Berar)
Moho: *Hampea trilobata* Standl.; *Bombacac.*; br. Hond.
„ : *Belotia sp.*, *Heliocarpus sp.*; *Tiliac.*; br. Hond.
„ : *Bassia latifolia* Roxb.; *Sapotac.*; C.-O.-Ind.
„ , blue: *Hibiscus tiliaceus* L.; *Malvac.*; br. Hond.
„ , broad-leaved: *Heliocarpus donnell-smithi* Rose; *Tiliac.*; br. Hond.
„ , narrow-leaved: *Belotia campbelli* Sprague; *Tiliac.*; br. Hond.
„ , white: *Belotia campbelli* Sprague; *Tiliac.*; br. Hond.
„ , „ : *Heliocarpus donnell-smithi* Rose; *Tiliac.*; br. Hond.
„ , „ : *Hibiscus sp.*; *Malvac.*; br. Hond.
„ , yellow: *Heliocarpus donnell-smithi* Rose; *Tiliac.*; br. Hond.
Mohobohobo: *Uapaca kirkiana* Muell. Arg.; *Euphorbiac.*; n. Rhod.
Mohola: *Bassia latifolia* Roxb.; *Sapotac.*; w. Beng. (Orissa)
Mohr-chu: *Gaertnera salicifolia* Hutch. et Gillett; *Loganiac.*; Lib.
Mohrenpfeffer: *Xylopia aethiopica* A. Rich.; *Anonac.*; W.-Af.
Mohroo: *Quercus dilatata* Lindley; *Fagac.*; nw. O.-Ind. (Kaschmir)
Mohru: *Quercus dilatata* Lindley; *Fagac.*; nw. O.-Ind. (Kaschmir)
Mohu: *Calophyllum inophyllum* L.; *Guttifer.*; Co.-Ch.
Mohwa: *Bassia latifolia* Roxb., *B. longifolia* L.; *Sapotac.*; O.-Ind.
„ : *Mimusops littoralis* Kurz; *Sapotac.*; And.
Mohwai: *Lannea grandis* Engl.; *Anacardiac.*; C.-O.-Ind.
Moi: *Lannea grandis* Engl.; *Anacardiac.*; w. Beng. (Orissa)
Moi-moua: *Canarium schweinfurthi* Engl.; *Burserac.*; Elf.
„ xùong: *Podocarpus cupressina* R. Brown; *Conifer.*; Ind.-Ch.
Moichi: *Ficus glomerata* Roxb.; *Morac.*; O.-Ind. (Madras)
Moigbwamy: *Cleistopholis patens* Bth.; *Anonac.*; Lib.
Moina: *Nectandra sp.*; *Laurac.*; Peru.
„ -caoba: *Platymiscium polystachyum* Bth.; *Papilionac.*; Bras.
„ pinima: *Piratinera guianensis* Aubl.; *Morac.*; Bras.
„ -piranga: *Brosimum paraense* Huber; *Morac.*; Bras.
Moj: *Albizzia lucida* Bth.; *Mimosac.*; Assam.
Mojabi: *Mimusops pierreana* Engl.; *Sapotac.*; fr. Kongo (Haut-Ogooué)
Mojo: *Hibiscus tiliaceus* L.; *Malvac.*; C.-Am.
Moka: *Schrebera swietenioides* Roxb.; *Oleac.*; O.-Ind.
Mokalapu: *Schrebera swietenioides* Roxb.; *Oleac.*; O.-Ind. (Madras)
Mokari: *Schrebera swietenioides* Roxb.; *Oleac.*; O.-Ind. (Bombay)
Mokha: *Schrebera swietenioides* Roxb.; *Oleac.*; C.-O.-Ind.
Mokkokou: *Ternstroemia japonica* Thunb.; *Theac.*; Jap.
Mokkoku: *Ternstroemia japonica* Thunb.; *Theac.*; Form.
Mokoanhehe: *Combretum bruchhausenianum* Engl. et Diels; *Combretac.*; D.-O.-Af.
Mokonghe: *Millettia laurenti* de Wild.; *Papilionac.*; B.-K. [(Kil.)
Mokonko: *Copaifera demeusei* Harms; *Caesalpiniac.*; B.-K.
Mokoumi: *Aucoumea klaineana* Pierre; *Burserac.*; Gab.
Moksongayok: *Murraya exotica* L.; *Rutac.*; Burma.
Mokukenju: *Koelreuteria paniculata* Lam.; *Sapindac.*; Jap.
Mokukoku: *Ternstroemia japonica* Thunb.; *Theac.*; Jap.
Mokuma: *Ceiba pentandra* Gaertn.; *Bombacac.*; B.-K. (Kurubu)
Mokwa: *Xylopia aethiopica* A. Rich.; *Anonac.*; B.-K.
Molakarunnay: *Toddalia aculeata* Pers.; *Rutac.*; O.-Ind.
Molambeira: *Adansonia digitata* L.; *Bombacac.*; port. O.-Af. (Mozamb.)
? Molan: *Baccaurea annamensis* Gagnep.; *Euphorbiac.*; Ind.-Ch.
Molave: *Vitex spp.*; *Verbenac.*; Phil.
„ , Dog-: *Premna nauseosa* Blco.; *Verbenac.*; Phil.
„ , Female-: *Vitex spp.*; *Verbenac.*; Phil.
„ , Hairyleaf-: *Vitex pubescens* Vahl; *Verbenac.*; Phil.
„ , Longleaf-: *Vitex longifolia* Merr.; *Verbenac.*; Phil. (Agusan)
Moleinole: *Parinarium curatellifolium* Planch.; *Rosac.*; C.-Af.
Molemole: *Parinarium curatellifolium* Planch.; *Rosac.*; Togo.
Molengó: *Uapaca guineensis* Muell. Arg.; *Euphorbiac.*; B.-K.
Molenillo: *Luehea sp.*; *Tiliac.*; C.-R.
Molie: *Byrsonima sp.*; *Malpighiac.*; Sur.

Molinillo: *Geonoma simplicifrons* Willd.; *Palmac.;* Ven.
Molle: *Schinus molle* L.; *Anacardiac.;* Ek.
„ : *Schinus dependens* Ortega; *Anacardiac.;* Arg.
„ blanco: *Moya spinosa* Griseb.; *Celastrac.;* Arg.
„ crespo: *Bumelia obtusifolia* Roem. et Sch.; *Sapotac.;* Arg.
„ dulce: *Lithraea aroerinha* March.; *Anacardiac.;* S.-Am. (? Arg.)
„ guazú: *Schinus dependens* Ortega; *Anacardiac.;* Arg.
„ morado: *Schinus latifolius* Engl.; *Anacardiac.;* Arg.
„ negro: *Castela coccinea* Griseb.; *Simarubac.;* Arg.
„   „ : *Moya spinosa* Griseb.; *Celastrac.;* Arg.
„ de Bolivia: *Schinus molle* L.; *Anacardiac.;* Arg.
„ „ castilla: *Schinus molle* L.; *Anacardiac.;* Arg.
„ „ curtir: *Schinus sp.; Anacardiac.;* Arg.
„. del monte: *Bumelia obtusifolia* Roem. et Sch.; *Sapotac.;* Arg.
„ del Peru: *Schinus molle* L.; *Anacardiac.;* Arg.
Molongo: *Chlorophora excelsa* Bth. et Hook.; *Morac.;* B.-K.
Molongotoh: *? Murraya sp.; Rutac.;* n. Celebes (Gorontalo)
Molundu: *Chlorophora excelsa* Bth. et Hook.; *Morac.;* B.-K.
Momaka: *Salix tetrasperma* Roxb.; *Salicac.;* Burma.
Momangi: *Chlorophora excelsa* Bth. et Hook.; *Morac.;* Kam.
Momba periwede: *Afzelia quanzensis* Welw.; *Caesalpiniac.;* port. O.-Af. (Moz.)
Mombanda: *Lannea welwitschi* Engler; *Anacardiac.;* B.-K.
Mombé: *Spondias lutea* L., *S. dulcis* Forst.; *Anacardiac.;* Sur.
Mombele: *Canarium schweinfurthi* Engler; *Burserac.;* B.-K.
Mombende: *Monodora myristica* Dunal; *Anonac.;* B.-K.
Mombin: *Spondias lutea* L.; *Anacardiac.;* Guat.
„ bâtard: *Trichilia hirta* L.; *Meliac.;* Guad.
Mombolu: *Irvingia wombolu* Verm.; *Simarubac.;* B.-K.
Mombulu: *Irvingia gabonensis* Baill.; *Simarubac.;* fr. Kongo.
Momi: *Abies firma* S. et Z.; *Conifer.;* Jap.
Momo: *Carolinea princeps* L. f.; *Bombacac.;* Sur.
Momoen: *Jacaranda brasiliana* Pers.; *Bignoniac.;* Sur.
Momoi: *Jacaranda filicifolia* D. Don; *Bignoniac.;* Sur.
Mompin: *Spondias sp.; Anacardiac.;* Mex.
Mompón: *Dehaasia triandra* Merr.; *Laurac.;* Phil. (Camarines)
Mon: *Buchholzia macrophylla* Engler; *Capparidac.;* Elf.
„ : *Morus indica* L.; *Morac.;* Ind.-Ch.
„ tia: *Alangium chinense* Rehder; *Cornac.;* Ind.-Ch.
„ trong trang: *Erismanthus indochinensis* Gagnep.; *Euphorbiac.;* Ind.-Ch.
Monangi: *Chlorophora excelsa* Bth. et Hook.; *Morac.;* Elf.
Monbet: *Ricinodendron africanum* Muell. Arg.; *Euphorbiac.;* Senb.
Monbin: *Spondias lutea* L., *S. dulcis* Forst.; *Anacardiac.;* Sur., fr. Gu.
„ : *Spondias lutea* L.; *Anacardiac.;* Gab.
„ jamaique: *Spondias lutea* L.; *Anacardiac.;* fr. Gu.
Monca prieto: *Cyphomandra heterophylla* D. Sm.; *Solanac.;* Pan.
Monda: *Trewia nudiflora* L.; *Euphorbiac.;* w. Beng. (Orissa)
„ dhup: *Canarium strictum* Roxb.; *Burserac.;* w. O.-Ind. (Bombay)
Mondelu: *Sterculia tomentosa* Guill. et Perr.; *Sterculiac.;* Togo.
Mondo: *Garcinia dulcis* Kurz; *Guttifer.;* Java.
„ : *Capparis mitchelli* Lindl.; *Capparidac.;* ö. Au. (Queensl.)
Monesia: *Pradosia latescens* Radlk.; *Sapotac.;* Bras. (Sao P)
Mong khut: *Garcinia mangostana* L.; *Guttifer.;* Lamb.
„ quân: *Flacourtia cataphracta* Roxb.; *Flacourtiac.;* Annam.
„ thau dàu: *Acer tonkinense* H. Lec.; *Acerac.;* Ind.-Ch.
„ -tog: *Ochna wallichi* Planch.; *Ochnac.;* Annam.
Mongègè: *Ongokea klaineana* Pierre; *Olacac.;* Gab.
Mongheo: *Garuga pinnata* Roxb.; *Burserac.;* Co.-Ch.
Mongk'ho: *Ekebergia rüppelhana* A. Rich.; *Meliac.;* D.-O.-Af. (w. Usambara)
Mongo: *Randia pierrei* Chev.; *Rubiac.;* Gab.
Mongollano: *Pithecolobium dulce* Bth.; *Mimosac.;* Salv.
Mongomo: *Caloncoba glauca* Gilg; *Flacourtiac.;* Kam.
Mongongome: *Ricinodendron africanum* Muell. Arg.; *Euphorbiac.;* B.-K.
Mongoumougouma: *Klainedoxa gabonensis* Pierre; *Simarubac.;* Gab.
Mongubeira: *Pachira aquatica* Aubl.; *Bombacac.;* Bras.

Mongue sapeteiro: *Rhizophora mangle* L.; *Rhizophorac.*; Bras.
Monia: *Lannea grandis* Engl.; *Anacardiac.*; w. O.-Ind. (n. Bombay)
Monigeli: *Randia dumetorum* Lamk.; *Rubiac.*; O.-Ind. (Bombay)
Monjoleiro preto: *Pithecolobium sp.*; *Mimosac.*; Bras. (Sao P)
Monjôlo: *Piptadenia spp.*, *Enterolobium sp.*; *Mimosac.*; Bras.
  „ cambuy: *Enterolobium sp.*; *Mimosac.*; Bras.
  „ lizo: *Enterolobium sp.*; *Mimosac.*; Bras.
  „ vermelho: *Enterolobium sp.*; *Mimosac.*; Bras.
Monkeka: *Conopharyngia smithi* Stapf; *Apocynac.*; B.-K.
Monkey bread: *Adansonia digitata* L.; *Bombacac.*; Lib.
  „ comb: *Apeiba sp.*; *Tiliac.*; Bras. (Amaz.)
Monkey's dinner bell: *Hura crepitans* L.; *Euphorbiac.*; br. W.-Ind.
Monkey fiddle: *Uitex spp.*; *Uerbenac.*; br. Hond.
  „ fruit: *? Acioa barteri* Engl.; *Rosac.*; Lib.
  „ „ : *Urophyllum linderi* Hutch. et Dalz.; *Rubiac.*; Lib.
  „ guave: *Diospyros mespiliformis* Hochst.; *Ebenac.*; tr. Af.
  „ jack: *Artocarpus rigida* Blume; *Morac.*; Malay.
  „ -pot: *Lecythis ollaria* L.; *Lecythidac.*; Bras. (Bahia)
  „ „ , small: *Eschweilera sp.*; *Lecythidac.*; br. Gu.
  „ puzzle: *Araucaria imbricata* Pav.; *Conifer.*; s. Chile.
  „ rattle: *Phyllanthus glauce.cens* H. B. K.; *Euphorbiac.*; br. Hond.
  ., tail: *Chamaedorea sp.*; *Palmac.*; br. Hond.
Monpa-gasi: *Lithocarpus hypophaea* Hay.; *Fagac.*; Form.
Monpeque: *Ximenia americana* L.; *Olacac.*; port. O.-Af. (Moz.)
Monsafou: *Pachylobus edulis* G. Don; *Burserac.*; Gab.
Monsan-houé: *Phyllanthus discoideus* Muell. Arg.; *Euphorbiac.*; Elf.
Mont hé: *Memecylon edule* Roxb.; *Melastomatac.*; Ind.-Ch.
Montano: *Tecoma sp.*; *Bignoniac.*; Mex.
Monterillo: *Brownea ariza* Bth.; *Caesalpiniac.*; Ek.
Montjoly: *Cordia martinicensis* Roem. et Schultes; *Borraginac.*; fr. Gu.
  „ : *Cordia graveolens* H. B. K.; *Borraginac.*; Sur.
Montol: *Sonneratia caseolaris* Engl.; *Sonneratiac.*; Phil., Malay.
Montole: *Cleistopholis grandiflora* de Wild.; *Anonac.*; B.-K.
Montouchy: *Swartzia benthamiana* Miq.; *Caesalpiniac.*; fr. Gu.
Monty: *Carissa densiflora* Baker var. *microphylla* P. Dang.; *Apocynac.*; Mad.
Moo-aye: *Securidaca longepedunculata* Fresen.; *Polygalac.*; B.-K.
Moobanga: *Afrormosia angolensis* Harms; *Papilionac.*; B.-K.
Moochukoonda: *Pterospermum suberifolium* Lam.; *Sterculiac.*; Beng.
Moodooda: *Chloroxylon swietenia* DC.; *Meliac.*; O.-Ind.
Moo-enge: *Diplorrhynchus mossambicensis* Bth.; *Apocynac.*; B.-K.
Moofungo: *Anisophyllea laurina* R. Br.; *Rhizophorac.*; B.-K.
Moolamma: *Combretum odontopetalum* Engl. et Diels; *Combretac.*; B.-K.
Moolumbwa: *Pterocarpus erinaceus* Poir.; *Papilionac.*; B.-K.
Moonba: *Uirola surinamensis* Warb.; *Myristicac.*; Sur.
Moonbé: *Spondias lutea* L., *S. dulcis* Forst.; *Anacardiac.*; Sur.
Moontangoo: *Calophyllum canum* Hook. f.; *Guttifer.*; Malak.
Moon trefoil: *Medicago arborea* L.; *Papilionac.*; Pers., Syr.
Moopaala: *Baphia bequaerti* de Wild.; *Papilionac.*; B.-K.
Moorbeere (d): *Uaccinium uliginosum* L.; *Ericac.*; Eu.
Moosa: *Kigelia moosa* Sprague; *Bignoniac.*; br. O.-Af. (Uganda)
Moosaalie: *Pseudolachnostylis glauca* Hutch.; *Euphorbiac.*; B.-K.
Moosaasi: *Erythrophloeum africanum* Harms; *Caesalpiniac.*; B.-K.
Moosbeere (d): *Uaccinium oxycoccus* L.; *Ericac.*; Eu.
Moosesjie: *Marquesia macroura* Gilg; *Flacourtiac.*; B.-K.
Mooum: *Staudtia gabonensis* Warb.; *Myristicac.*; Gab.
Móp: *Aporosa microcalyx* Hassk.; *Euphorbiac.*; Ind.-Ch.
Mopaani: *Copaifera mopane* J. Kirk; *Caesalpiniac.*; n. Rhod.
Mopan: *Myrsine urvillei* A. DC.; *Myrsinac.*; N.-Seel.
Mopani: *Copaifera mopane* J. Kirk; *Caesalpiniac.*; s. C.-Af.
Mopé: *Spondias lutea* L., *S. dulcis* Forst.; *Anacardiac.*; Sur.
Mopie: *Mouriria anomala* Pulle, *M. plaschaerti* Pulle; *Melastomatac.*; Sur.
Mopou à grandes feuilles: *Monimia rotundifolia* Pet. Thou.; *Monimiac.*; Réunion.
Moppie: *Mouriria anomala* Pulle, *M. plaschaerti* Pulle; *Melastomatac.*; Sur.
Moquillo: *Saurauia sp.*; *Dilleniac.*; Ven.

Moquombire-bire: *Pterocarpus angolensis* DC.; *Papilionac.;* port. O.-Af. (Moz.)
Mora: *Chlorophora tinctoria* Gaudich.; *Morac.;* tr. Am.
„ : *Chaetoptelea mexicana* Liebm.; *Ulmac.;* Hond.
„ : *Dimorphandra mora* Bth. et Hook.; *Caesalpiniac.;* Trin., Tob., Hond., Gu.
„ : *Mora oleifera* Ducke; *Caesalpiniac.;* Pan.
„ : *Maclura aurantiaca* Nutt.; *Morac.;* Arg.
„ amarilla: *Chlorophora tinctoria* Gaudich.; *Morac.;* tr. Am.
„ „ : *Maclura aurantiaca* Nutt.; *Morac.;* Arg.
„ blanca: *Chlorophora tinctoria* Gaudich.; *Morac.;* Cuba.
„ , British Guiana-: *Dimorphandra mora* Bth. et Hook.; *Caesalpiniac.;* W.-Ind.
„ colorada: *Chlorophora mora* Lillo; *Morac.;* Arg.
„ hedionda: *Cassia emarginata* L.; *Caesalpiniac.;* Mex.
„ , roode: *Dimorphandra mora* Bth. et Hook.; *Caesalpiniac.;* Sur.
„ , witte: *Dimorphandra mora* Bth. et Hook.; *Caesalpiniac.;* Sur.
„ de Guayana: *Dimorphandra mora* Bth. et Hook.; *Caesalpiniac.;* Ven.
„ „ loma: *Chlorophora tinctoria* Gaudich.; *Morac.;* Cuba.
„ „ pais: *Chlorophora tinctoria* Gaudich.; *Morac.;* Cuba.
„ „ piedra: *Chlorophora tinctoria* Gaudich.; *Morac.;* Cuba.
„ of Peto: *Dimorphandra mora* Bth. et Hook.; *Caesalpiniac.;* Sur.
„ yellow wood: *Chlorophora tinctoria* Gaudich.; *Morac.;* tr. Am.
Moraballi: *? Peltogyne sp.; Caesalpiniac.;* br. Gu.
„ : *Pouteria sp.; Sapotac.;* Sur.
„ -diamaro: *Pouteria sp.; Sapotac.;* Sur.
„ hariraroe: *Chaetocarpus sp., ? C. schomburgkiana* Pax et Hoff.; *Euphorbiac.;* Sur.
Morabucquia: *Dimorphandra gonggrijpi* Kleinh.; *Caesalpiniac.;* Sur.
Morabukea: *Dimorphandra gonggrijpi* Kleinh.; *Caesalpiniac.;* Sur., br. Hond.
Morada: *Boswellia serrata* Roxb.; *Burserac.;* O.-Ind. (Madras)
Moradilla: *Chlorophora tinctoria* Gaudich.; *Morac.;* Mex.
Morado: *Peltogyne sp.; Caesalpiniac.;* Bol.
Moraeda: *Boswellia serrata* Roxb.; *Burserac.;* C.-O.-Ind.
Moraingy: *Ceiba pentandra* Gaertn.; *Bombacac.;* Mad.
Morakokuru: *Mimusops sp.; Sapotac.;* fr. Gu.
Moral: *Cordia sulcata* DC.; *Borraginac.;* P.-R.
„ : *Buchanania latifolia* Roxb.; *Anacardiac.;* O.-Ind. (Madras)
„ : *Ulmus wallichiana* Planch.; *Ulmac.;* nw. O.-Ind. (Punjab)
„ amarillo: *Chlorophora tinctoria* Gaudich.; *Morac.;* Mex.
„ bobo: *Chlorophora tinctoria* Gaudich.; *Morac.;* Ek.
„ liso: *Chlorophora tinctoria* Gaudich.; *Morac.;* Mex.
„ de clavo: *Chlorophora tinctoria* Gaudich.; *Morac.;* Mex.
„ „ paz: *Cordia sp.; Borraginac.;* P.-R.
Morala: *Buchanania latifolia* Roxb.; *Anacardiac.;* O.-Ind. (Madras)
Moralón: *Coccoloba grandifolia* Jacq.; *Polygonac.;* P.-R.
Moras prou phnom: *Dysoxylum loureiro* Pierre; *Meliac.;* Ind.-Ch.
Morcegueira: *? Andira inermis* H. B. K.; *Papilionac.;* Bras.
Morcielago: *Cornutia grandifolia* Schauer; *Uerbenac.;* Pan.
Morelon: *Coccoloba sp.; Polygonac.;* P.-R.
Morera: *Morus multicaulis* Perr.; *Morac.;* Hond.
Mori-gasi: *Quercus morii* Hayata; *Fagac.;* Form.
„ -yanagi: *Salix morii* Hayata; *Salicac.;* Form.
Morillo: *Trophis sp.; Morac.;* Pan.
Morinda: *Picea morinda* Link; *Conifer.;* n. O.-Ind. (Garhwal)
„ : *Abies pindrow* Spach; *Conifer.;* n. O.-Ind. (Jaunsar)
Morli: *Buchanania latifolia* Roxb.; *Anacardiac.;* O.-Ind.
„ sara: *Buchanania angustifolia* Roxb.; *Anacardiac.;* O.-Ind. (Madras)
Moro: *Quararibea asterolepis* Pittier; *Bombacac.;* Guat.
Moroballi: *Cupania sp.; Sapindac.;* br. Gu.
Morocolo: *? Lonchocarpus violaceus* H. B. K.; *Papilionac.;* Kol.
Moro-cy: *Byrsonima sp.; Malpighiac.;* Bras.
Moroi: *Albizzia odoratissima* Bth.; *Mimosac.;* Assam.
Moroló: *Anona sp.; Anonac.;* Bras.
Moropié: *Rhaptopetalum tieghemi* Chev.; *Olacac.;* Elf.
Mororó: *Bauhinia macrostachya* Bth.; *Caesalpiniac.;* Bras.
Morroroto: *Didymopanax morototoni* Dcne. et Planch.; *Araliac.;* Bras.
Moroto: *Pterocarpus angolensis* DC.; *Papilionac.;* S.-Af. (Transval)

Morotobi: *Calycophyllum sp.; Rubiac.;* Arg.
Morototó: *Didymopanax morototoni* Aubl.; *Araliac.;* Bras. (Amaz.)
Morrell: *Eucalyptus longicornis* F. v. M.; *Myrtac.;* w. Au.
Morrionera: *Viburnum lantana* L.; *Caprifoliac.;* Sp.
Morrito: *Crescentia cujete* L.; *Bignoniac.;* Salv.
Morro: *Crescentia cujete* L.; *Bignoniac.;* Guat., Hond.
Mortiño: *Vaccinium floribundum* H. B. K.; *Ericac.;* w. S.-Am. (Anden)
Moru: *Quercus dilatata* Lindley; *Fagac.;* nw. & n. O.-Ind.
Morubi: *Artocarpus gomeziana* Wall.; *Morac.;* Malay.
Morung-sal: *Shorea robusta* Gaertn. f.; *Dipterocarpac.;* O.-Ind.
Moruro: *Pithecolobium arboreum* Urban; *Mimosac.;* Cuba.
  „   prieto: *Pithecolobium arboreum* Urban; *Mimosac.;* Cuba.
  „   roja: *Peltophorum adnatum* Griseb.; *Caesalpiniac.;* Cuba.
  „   , Sabicú-: *Peltophorum adnatum* Griseb.; *Caesalpiniac.;* Cuba.
  „   de Sabana: *Peltophorum adnatum* Griseb.; *Caesalpiniac.;* Cuba.
Mor-way-dah: *Eugenia whytei* Sprague; *Myrtac.;* Lib.
Mosa-gass: *Nephelium longana* Camb.; *Sapindac.;* Ceylon.
Mosal: *Araucaria brasiliana* Lamb.; *Conifer.;* s. Bras.
Moscon: *Acer campestre* L.; *Acerac.;* Sp.
Mosmote: *Ceiba sp.; Bombacac.;* Mex.
Moso: *Phyllostachys edulis* A. et C. Rivière; *Gramin.;* Jap.
Mosonea: *Litsea polyantha* Juss.; *Laurac.;* w. Beng. (Orissa)
Mosongo: *Ricinodendron africanum* Muell. Arg.; *Euphorbiac.;* Gab.
Mosote: *Desmodium canum* Gmel.; *Papilionac.;* Nic.
Mosoukouga: *Saccoglottis gabonensis* Urban; *Humiriac.;* Gab.
Mosqueta: *Rosa sempervirens* L.; *Rosac.;* Sp.
Mosquito wood: *Mosquitoxylum jamaicense* Krug et Urb.; *Anacardiac.;* W.-Ind.
  „   „   : *Posoqueria latifolia* Roem. et Sch.; *Rubiac.;* Pan.
Mossanghe: *Xylopia aethiopica* A. Rich.; *Anonac.;* B.-K.
Mossanki: *Xylopia aethiopica* A. Rich.; *Anonac.;* B.-K.
Mossinga: *Piptadenia sp.; Mimosac.;* Gab.
Mossomba: *Baphia acuminata* de Wild.; *Papilionac.;* B.-K.
Mossotahyba: *Zollernia ilicifolia* Vogel; *Caesalpiniac.;* Bras. (Sao P)
  „   parda: *Zollernia sp.; Caesalpiniac.;* Bras. (Sao P)
  „   preta: *Zollernia sp.; Caesalpiniac.;* Bras. (Sao P)
  „   vermelha: *Zollernia sp.; Caesalpiniac.;* Bras. (Sao P)
Mostajo: *Sorbus aria* Crantz, *S. torminalis* Crantz; *Rosac.;* Sp.
Mót: *Aporosa microcalyx* Hassk.; *Euphorbiac.;* Ind.-Ch.
  „   mo: *Taractogenos serrata* Pierre; *Flacourtiac.;* Ind.-Ch.
  „   trang: *Antidesma ghaesembilla* Gaertn.; *Euphorbiac.;* Ind.-Ch.
Mota kermal: *Dillenia indica* L.; *Dilleniac.;* sw. O.-Ind.
Motamayu: *Olea chrysophylla* Lam.; *Oleac.;* br. O.-Af.
Motillo: *Sloanea berteriana* Choisy; *Elaeocarpac.;* P.-R.
Motin: *Gustavia fustis-mortui* Pittier; *Lecythidac.;* Ven.
Moto-aduso: *Ailanthus excelsa* Roxb.; *Simarubac.;* w. O.-Ind. (n. Bombay)
Motoko: *Platysepalum chevalieri* Harms; *Papilionac.;* B.-K.
Motong-bótong: *Planchonia spectabilis* Merr.; *Lecythidac.;* Phil. (Camar., Alb.)
Motoudon: *Calpocalyx klainei* Pierre; *Mimosac.;* Kam.
Motségui: *Pachylobus edulis* G. Don; *Burserac.;* Gab.
Motukrodua: *Monodora myristica* Dunal, *Uvaria sp.; Anonac.;* Go.
Motukundu: *Dialium lacourtianum* de Wild.; *Caesalpiniac.;* B.-K.
Mou: *Garcinia collina* Vieill., *Clusia pedicellata* Forst.; *Guttifer.;* N.-Kal.
Mouabi: *Mimusops djave* Engl., *M. pierreana* Engl.; *Sapotac.;* Kam., Gab.
Moualékoumbi: *Cola ballayi* Cornu; *Sterculiac.;* Gab.
Moualékoumei: *Cola ballayi* Cornu; *Sterculiac.;* Gab.
Moubili: *Canarium schweinfurthi* Engl.; *Burserac.;* Gab.
Moubouba: *Myrianthus arboreus* Beauv.; *Morac.;* Gab.
Moucaya: *Acrocomia sclerocarpa* Mart.; *Palmac.;* fr. Gu.
Mouchigo: *Virola sp.; Myristicac.;* Gu.
  „   rouge: *Virola sp.; Myristicac.;* Gu.
Moudouka: *Mimusops africana* H. Lec.; *Sapotac.;* Kam., Gab.
Moudzandji: *Anthostema aubryanum* Baill.; *Euphorbiac.;* Gab.
Mouengue bossico: *Pterocarpus soyauxi* Taub.; *Papilionac.;* Kam.
  „   mossoumbé: *Pterocarpus soyauxi* Taub.; *Papilionac.;* Kam.

Moufilou: *Vitex pachyphylla* Baker; *Verbenac.*; Gab.
Moufinzi: *Diospyros aggregata* Gürke; *Ebenac.*; Gab.
Moufouma: *Ceiba pentandra* Gaertn.; *Bombacac.*; s. Kam., n. Gab.
Mouganga: *Pachylobus le-testui* Pell., *P. fraxinifolius* Engl.; *Burserac.*; Gab.
   „   : *Spondias lutea* L.; *Anacardiac.*; Kam.
Mougouma: *Coula edulis* Baill.; *Olacac.*; Gab.
Mouiba: *Irvingia gabonensis* Baill.; *Simarubac.*; Gab.
Mouin: *Musanga smithi* R. Br.; *Morac.*; Elf.
Moukeba: *Macrolobium sp.*; *Caesalpiniac.*; Gab.
Moukiba: *Macrolobium sp.*; *Caesalpiniac.*; Gab.
Moukonia malamba: *Mitragyne macrophylla* Hiern; *Rubiac.*; Kam.
   „ mamoundi: *Sarcocephalus trillesi* Pierre; *Rubiac.*; Kam.
   „ mamundi: *Sarcocephalus trillesi* Pierre; *Rubiac.*; Kam.
   „ weiss: *Terminalia superba* Engl. et Diels; *Combretac.*; Kam.
Moukoumi: *Aucoumea klaineana* Pierre; *Burserac.*; Gab.
Moukourodji: *Sapindus mukorosi* Gaertn.; *Sapindac.*; Jap.
Moulenda: *Distemonanthus benthamianus* Baill.; *Caesalpiniac.*; Gab.
Moulimba: *Terminalia superba* Engl. et Diels; *Combretac.*; Kam.
Moulinda: *Desbordesia insignis* Pierre; *Simarubac.*; Gab.
Moulindé: *Desbordesia insignis* Pierre; *Simarubac.*; Gab.
Moulla-panza: *Pentaclethra macrophylla* Bth.; *Mimosac.*; Gab.
Moullou-iléven marom: *Bombax malabaricum* DC.; *Bombacac.*; w. O.-Ind. (n. Bombay)
Moulomba: *Pycnanthus kombo* Warb.; *Myristicac.*; Gab.
Moulombo: *Pachypodanthium confine* Engl. et Diels; *Anonac.*; Gab.
Moum: *Elaeocarpus floribundus* Blume; *Elaeocarpac.*; Ind.-Ch.
Moumba: *Terminalia brosigiana* Engl. et Diels; *Combretac.*; D.-O.-Af. (Kilossa)
Moumfilou: *Vitex pachyphylla* Baker; *Verbenac.*; Gab.
Moumou: *Erythrophloeum guineense* G. Don; *Caesalpiniac.*; Kam.
Mounia nianga: *Symphonia gabonensis* Pierre; *Guttifer.*; Gab.
Mounorou: *Bombax buonopozense* Beauv.; *Bombacac.*; Dah.
Mountain cabbage: *Oreodoxa oleracea* Morris; *Palmac.*; br. Hond.
   „ cow: *Symphonia globulifera* L. f.; *Guttifer.*; br. Hond.
Mouondue: *Entandrophragma bussei* Harms; *Meliac.*; O.-Af.
Mouong: *Ceiba pentandra* Gaertn.; *Bombacac.*; Elf.
Mouou: *Calophyllum inophyllum* L.; *Guttifer.*; Co.-Ch.
Moureil: *Byrsonima altissima* DC.; *Malpighiac.*; Gu.
Moureiller: *Byrsonima altissima* DC., *B. coriacea* DC.; *Malpighiac.*; Guad.
   „ : *Byrsonima sp.*; *Malpighiac.*; Sur.
   „ piquant: *Malpighia martinicensis* Jacq.; *Malpighiac.*; Guad.
   „ de montagne: *Byrsonima crassifolia* DC.; *Malpighiac.*; Guad.
Mourotoki: *Balanites aegyptiaca* Delile; *Zygophyllac.*; Sudan.
Mouroumou: *Astrocaryum acaule* Mart.; *Palmac.*; fr. Gu.
Mourresif: *Byrsonima sp.*; *Malpighiac.*; fr. Gu.
Mousan koé: *Phyllanthus discoideus* Muell. Arg.; *Euphorbiac.*; Elf.
Mousewood: *Acer sp.*; *Acerac.*; USA.
Mousighou: *Pachylobus buettneri* Engl.; *Burserac.*; Gab.
   „ bisiya: *Pachylobus buettneri* Engl.; *Burserac.*; Gab.
Moussan hoé: *? Phyllanthus discoideus* Muell. Arg.; *Euphorbiac.*; Elf.
   „ houé: *? Phyllantus discoideus* Muell. Arg.; *Euphorbiac.*; Elf.
   „ koué: *Phyllanthus discoideus* Muell. Arg.; *Euphorbiac.*; Elf.
Moussanfi: *Uapaca sp.*; *Euphorbiac.*; Gab.
   „ -benga: *Uapaca sp.*; *Euphorbiac.*; Gab.
Moussara: *Brosimum alicastrum* Swartz; *Morac.*; Trin.
Moussé: *Parinarium tenuifolium* Chev.; *Rosac.*; Elf.
Moussigo: *Virola sp.*; *Myristicac.*; Gu.
Moussiguiri: *Odyendea gabunensis* Engl.; *Simarubac.*; Gab.
Moussikou: *Pachylobus buettneri* Engl.; *Burserac.*; Gab.
Moussinga: *Musanga smithi* R. Br.; *Morac.*; Gab.
   „ : *Piptadenia sp.*; *Mimosac.*; Gab.
Moutangani: *Daniella sp.*; *Caesalpiniac.*; Gab.
Moutchégou: *Pachylobus edulis* G. Don; *Burserac.*; Gab.
Moutocochi: *Inga spp.*; *Mimosac.*; Sur.
Moutouchi: *Pterocarpus draco* L., *P. rhori* Vahl; *Papilionac.*; Sur.
   „ : *Inga alba* Willd., *I. sertulifera* DC.; *Mimosac.*; Sur.

Moutouchi bouchon: *Pterocarpus draco* L.; *Papilionac.;* fr. Gu.
„ grand bois: *Machaerium schomburgki* Bth.; *Papilionac.;* Sur.
„ de savane: *Pterocarpus sp.; Papilionac.;* fr. Gu.
Moutsetsendi: *Bridelia speciosa* Muell. Arg.; *Euphorbiac.;* Gab.
Mouvindou: *Vitex pachyphylla* Baker; *Verbenac.;* Gab.
Mouvougou: *Anthocleista nobilis* G. Don; *Loganiac.;* Gab.
Mouvounda: *Hexalobus crispiflorus* A. Rich.; *Anonac.;* Gab.
Mova: *Engelhardtia spicata* Blume; *Juglandac.;* n. O.-Ind., Burma.
Movaro: *Bassia longifolia* L.; *Sapotac.;* sw. O.-Ind. (Bombay)
Movingui (f): *Distemonanthus benthamianus* Baill.; *Caesalpiniac.;* Kam., Gab.
Movounda: *Hexalobus crispiflorus* A. Rich.; *Anonac.;* Gab.
Mowa tree: *Illipe latifolia* Engl., *I. malabrorum* Koenig; *Sapotac.;* sw. O.-Ind., Cey.
Mowana: *Adansonia digitata* L.; *Bombacac.;* port. Af.
Mowbulan whitewood: *Panax elegans* C. Moore et F. v. M.; *Araliac.;* nö. Au.
Mowha: *Mimusops littoralis* Kurz; *Sapotac.;* And.
Mowingui: *Distemonanthus benthamianus* Baill.; *Caesalpiniac.;* Kam., Gab.
Mowongda: *Didelotia duparquetiana* ?; *Caesalpiniac.;* Gab.
Moyabi: *Mimusops pierreana* Engl.; *Sapotac.;* Gab.
Moyal: *Lannea grandis* Engl.; *Anacardiac.;* C.-O.-Ind.
Moyan: *Lannea grandis* Engl.; *Anacardiac.;* n. O.-Ind. (United Prov.)
Moyen: *Lannea grandis* Engl.; *Anacardiac.;* C.-O.-Ind. (Berar)
Mozote: *Heliocarpus glanduliferus* Robinson; *Tiliac.;* Guat.
Mozotillo: *Heliocarpus sp.; Tiliac.;* Salv.
M'pahé: *Fagara macrophylla* Engl.; *Rutac.;* Elf.
Mpai: *? Ficus sp.; Morac.;* S.-Af. (Natal)
Mpaka: *Copaifera demeusei* Harms; *Caesalpiniac.;* B.-K.
Mpampro: *Oxytenanthera ? abyssinica* Munro; *Gramin.;* Go.
M'panda: *Panda oleosa* Pierre; *Pandac.;* Gab.
Mpandja: *Didelotia duparquetiana* ?; *Caesalpiniac.;* Gab.
M'pandya: *Hylodendron gabunense* Taub.; *Caesalpiniac.;* Gab.
Mpango: *Hylodendron gabunense* Taub.; *Caesalpiniac.;* Gab.
Mpanza: *Pentaclethra macrophylla* Bth.; *Mimosac.;* Gab.
M'paqua: *Cynometra sp.; Caesalpiniac.;* Kam.
Mpaya: *Guarea thompsoni* Sprague et Hutch.; *Meliac.;* Go.
Mpebi: *Chrysophyllum sp.; Sapotac.;* Gab.
Mpele: *Pterygopodium balsamiferum* Verm.; *Papilionac.;* B.-K.
Mpelemese: *Grewia platyclada* K. Schum.; *Tiliac.;* D.-O.-Af. (Tabora)
Mpengwa: *Lovoa klaineana* Pierre; *Meliac.;* Go.
Mpenwa: *Lovoa klaineana* Pierre; *Meliac.;* Go.
Mpére: *Pterygopodium balsamiferum* Verm.; *Papilionac.;* B.-K.
Mpevi: *Chrysophyllum sp.; Sapotac.;* Gab.
Mpevu: *Trema bracteolata* Blume; *Ulmac.;* br. O.-Af. (Nyassa)
Mpewere: *Piptadenia africana* Hook. f.; *Mimosac.;* br. O.-Af. (Uganda)
„ of Chagwe: *? Celtis soyauxi* Engl.; *Ulmac.;* br. O.-Af. (Uganda)
Mpfo: *Poga oleosa* Pierre; *Rhizophorac.;* Kam.
Mpimbia: *Maba aff. abyssinica* Hiern; *Ebenac.;* br. O.-Af. (Uganda)
Mpimbyi: *Maba abyssinica* Hiern; *Ebenac.;* br. O.-Af. (Uganda)
Mpindimbi: *Vitex shirensis* Baker; *Verbenac.;* s. C.-Af.
Mpingo: *Dalbergia melanoxylon* Guill. et Perr.; *Papilionac.;* D.-O.-Af. (Kilossa)
M'pingu: *Diospyros sp.; Ebenac.;* br. O.-Af. (Nyassa)
? Mpingue: *? Dalbergia melanoxylon* Guill. et Perr.; *Papilionac.;* port. O.-Af.
Mpini: *Terminalia sericea* Burch.; *Combretac.;* br. O.-Af. (Nyassa)
Mpiwere: *Piptadenia africana* Hook. f.; *Mimosac.;* br. O.-Af. (w. Uganda)
Mpoasa: *Leplaea coalescens* Verm.; *Meliac.;* B.-K.
Mpoba: *Acrosepalum poba* Pierre; *Tiliac.;* Gab.
Mpoga: *Poga oleosa* Pierre; *Rhizophorac.;* Gab.
Mpohe: *Strombosia pustulata* Oliv.; *Olacac.;* Elf.
Mpoi: *Poga oleosa* Pierre; *Rhizophorac.;* Kam.
M'poie: *Strombosia pustulata* Oliv.; *Olacac.;* Elf.
M'polai: *Enantia chlorantha* Oliv.; *Anonac.;* Kam.
M'polé: *Khaya sp.; Meliac.;* Elf.
Mponda: *Commiphora fischeri* Engl., *C. krausei* Engl.; *Burserac.;* D.-O.-Af. (Tabora)
Mpopo: *Areca catechu* L.; *Palmac.;* O.-Af., O.-Ind., Sunda.
Mpopwe: *Fagara homblei* de Wild.; *Rutac.;* B.-K. (Katanga)

Mpossa: *Berlinia acuminata* Soland.; *Caesalpiniac.;* Gab.
„ : *Macrolobium sp.; Caesalpiniac.;* B.-K.
Mpouley: *Enantia chlorantha* Oliv.; *Anonac.;* Kam.
M'poundé: *Omphalocarpum procerum* Beauv.; *Sapotac.;* Kam.
Mpoussa: *Monodora myristica* Dunal; *Anonac.;* Gab.
M'poussan: *Monodora myristica* Dunal; *Anonac.;* tr. W.-Af.
? Mpowuo: *Xylia evansi* Hutch.; *Mimosac.;* Go.
Mpulu: *Vitex sp.; Verbenac.;* D.-O.-Af. (Tabora)
„ -genge: *Vitex mombassae* Vatke; *Verbenac.;* D.-O.-Af. (Tabora)
„ -legea: *Vitex sp.; Verbenac.;* D.-O.-Af. (Tabora)
Mpumbe: *Ficus sp.; Morac.;* br. O.-Af. (Nyassa)
? Mranti boaya: *Shorea platycarpa* Heim; *Dipterocarpac.;* br. Borneo (Sarawak)
M'rihi: *Albizzia sp.; Mimosac.;* br. O.-Af.
Mringi: *Cordia holsti* Gürke; *Borraginac.;* D.-O.-Af.
M'roi: *Dillenia pentagyna* Roxb.; *Dilleniac.;* w. Annam.
Mroma: *Cordyla africana* Lour.; *Caesalpiniac.;* D.-O.-Af.
Mruziruzi: *Hagenia abyssinica* Gmel.; *Rosac.;* D.-O.-Af. (w. Usambara)
Msagasi: *Commiphora pilosa* Engl.; *Burserac.;* D.-O.-Af. (Tabora)
Msahua: *Kigelia aethiopica* Bth.; *Bignoniac.;* D.-O.-Af. (Tabora)
Msala: *Oxyanthus speciosus* DC.; *Rubiac.;* D.-O.-Af. (Usambara)
Msambo: *Allanblackia stuhlmanni* Engl.; *Guttifer.;* D.-O.-Af. (Tanganjika)
Msaur: *Lovoa spp.; Meliac.;* D.-O.-Af.
Mschem: *Cordia dioica* Boj.; *Borraginac.;* D.-O.-Af. (Tabora)
Mschinde: *Diospyros holtzi* Gürke; *Ebenac.;* D.-O.-Af. (Tabora)
Mse: *Podocarpus milanjanus* Radlk.; *Conifer.;* D.-O.-Af. (w. Usambara)
„ : *Podocarpus usambariensis* Pilger; *Conifer.;* D.-O.-Af. (w. Usambara)
M'senha: *Strombosia scheffleri* Engl.; *Olacac.;* nw. Angola.
Msewa: *Lonchocarpus stuhlmanni* Auct.; *Papilionac.;* s. C.-Af.
Mshafusho: *Dombeya gilgiana* K. Schum.; *Sterculiac.;* D.-O.-Af. (w. Usambara)
Msharagi: *Olea laurifolia* Lam.; *Oleac.;* br. O.-Af.
Mshegéshe: *Myrica kilimanjarica* Engl.; *Myriac.;* D.-O.-Af. (w. Usambara)
Mshihui: *Syzygium guineense* DC.; *Myrtac.;* D.-O.-Af. (w. Usambara)
Mshinda'ngurúwe: *Bersama holsti* Gürke; *Melianthac.;* D.-O.-Af. (w. Usambara)
Mshinga: *Trema guineensis* Engler; *Ulmac.;* D.-O.-Af. (w. Usambara)
Mshinzi: *Bruguiera gymnorrhiza* Lam.; *Rhizophorac.;* D.-O.-Af.
Mshunduzi: *Croton macrostachys* A. Rich.; *Euphorbiac.;* D.-O.-Af. (w. Usambara)
Mshungúti: *Acocanthera venenata* Don; *Apocynac.;* D.-O.-Af. (w. Usambara)
Mshwizo: *Rapanea usambarensis* Gilg; *Myrsinac.;* D.-O.-Af. (w. Usambara)
Msiagembe: *Olea chrysophylla* Lam.; *Oleac.;* D.-O.-Af. (Usambara)
Msiga: *Dobera glabra* Juss. *var. subcoriacea* Engl. et Gilg; *Salvadorac.;* D.-O.-Af.
Msikiri: *Trichilia emetica* Vahl; *Meliac.;* n. Rhod. [(Kilos.)
Msima: *Terminalia sericea* Burch.; *Combretac.;* D.-O.-Af. (Tabora)
Msindui: *Anisophyllea boehmi* Engl.; *Rhizophorac.;* D.-O.-Af. (Tabora)
Msirangembe: *Olea chrysophylla* Lam.; *Oleac.;* D.-O.-Af. (w. Usambara)
Msolo: *Chrysophyllum msolo* Engl.; *Sapotac.;* D.-O.-Af. (Usambara)
Msonga: *Diplorrhynchus mossambicensis* Klotzsch; *Apocynac.;* D.-O.-Af. (Tabora)
Msopa: *Bridelia micrantha* Baill.; *Euphorbiac.;* s. C.-Af.
Msoro: *Maprounea africana* Stapf; *Euphorbiac.;* D.-O.-Af. (Tabora)
Mssumbiye: *Combretum petersi* Engl.; *Combretac.;* D.-O.-Af. (Kilossa)
Msuku: *Uapaca kirkiana* Muell. Arg.; *Euphorbiac.;* br. O.-Af. (Nyassa)
Msungui: *Vitex mombassae* Vatke; *Verbenac.;* D.-O.-Af. (Tabora)
Msungururu: *Strophanthus emini* Aschers. et Pax; *Apocynac.;* D.-O.-Af.
Msurura: *Maerua sp.; Capparidac.;* D.-O.-Af. (Tabora)
Msussu: *Celtis kraussiana* Bernh.; *Ulmac.;* D.-O.-Af. (w. Usambara)
Msutue: *Dodonaea viscosa* Jacq.; *Sapindac.;* Tr. (? O.-Af.)
Mtakale: *Terminalia spinosa* Engl.; *Combretac.;* D.-O.-Af. (Kilossa)
Mtakula: *Ochna holsti* Engl.; *Ochnac.;* D.-O.-Af. (Usambara)
Mtalali: *Vitex mombassae* Vatke; *Verbenac.;* D.-O.-Af. (Kikonongo)
M'tamyu: *Olea chrysophylla* Lam.; *Oleac.;* br. O.-Af.
M'tarakma: *Juniperus procera* Hochst.; *Conifer.;* br. O.-Af.
Mtejo-ya-hasi: *Phyllanthus sp.; Euphorbiac.;* D.-O.-Af. (Tabora)
Mtere: *Dobera glabra* Juss. *var. subcoriacea* Engl. et Diels; *Salvadorac.;* D.-O.-Af.
Mtini: *Ficus carica* L.; *Morac.;* O.-Ind. [(Kilos.)
Mtinje: *Commiphora rugosa* Engl.; *Burserac.;* D.-O.-Af.

Mtoa mada: *Ocotea usambarensis* Engl.; *Laurac.;* O.-Af.
Mtonga: *Strychnos engleri* Gilg; *Loganiac.;* D.-O.-Af. (Kilossa)
Mtónge: *Drypetes subdentata* Mildbr.; *Euphorbiac.;* D.-O.-Af. (w. Usambara)
Mtowo: *Dombeya sp.; Sterculiac.;* D.-O.-Af. (Tabora)
Mtshigizi: *Trichilia emetica* Vahl; *Meliac.;* Arab., tr. Af.
Mtulla-mpálla: *Scolopia stuhlmanni* Warb.; *Flacourtiac.;* D.-O.-Af. (w. Usamb.)
M'tulu: *Rhizophora mucronata* Lam.; *Rhizophorac.;* port. O.-Af. (Mozamb.)
M'tumassi: *Bruguiera gymnorrhiza* Lamk.; *Rhizophorac.;* port. O.-Af. (Moz.)
Mtumball: *Pterocarpus erinaceus* Poir.; *Papilionac.;* D.-O.-Af. (Kilossa)
Mtunda: *Sideroxylon inerme* L.; *Sapotac.;* D.-O.-Af. (Tanga)
Mtundu: *? Brachystegia appendiculata* Bth.; *Caesalpiniac.;* D.-O.-Af.
Mtunguru: *Maprounea africana* Pax; *Euphorbiac.;* D.-O.-Af. (Tabora)
Mtunungu: *Tephrosia vogeli* Hook. f.; *Papilionac.;* D.-O.-Af. (Tabora)
Mtuzi: *Acacia usambarensis* Taub.; *Mimosac.;* D.-O.-Af. (Usambara)
Mu: *Bassia latifolia* Roxb.; *Sapotac.;* C.-O.-Ind.
„ cua: *Albizzia procera* Bth.; *Mimosac.;* Ind.-Ch.
„ khoa deng: *Styrax tonkinensis* Pierre; *Styracac.;* Ind.-Ch.
„ roi: *Dillenia pentagyna* Roxb.; *Dilleniac.;* Ind.-Ch.
„ u: *Calophyllum inophyllum* L.; *Guttifer.;* Ind.-Ch.
Mua: *Dodonaea viscosa* Jacq.; *Sapindac.;* Tr. (? O.-Af.)
„ : *Melastoma sp.; Melastomatac.;* Ind.-Ch.
„ eua: *Alstonia scholaris* R. Br.; *Apocynac.;* Ind.-Ch.
Muabi: *Mimusops ebolowensis* Engl. et K. Krause; *Sapotac.;* nw. Angola.
Muage: *Strychnos sp.; Loganiac.;* D.-O.-Af.
Muaka: *Xylopia sp.; Anonac.;* O.-Af.
Muanamal: *Mimusops kauki* L.; *Sapotac.;* Burma, Malay.
Muandwe: *Theodora fischeri* Taub.; *Caesalpiniac.;* D.-O.-Af. (Usamb.), Sans.
Muanga: *Combretum petersi* Engl.; *Combretac.;* D.-O.-Af. (Kilossa)
Muangati: *Juniperus procera* Hochst.; *Conifer.;* D.-O.-Af.
Muata-pana: *Caesalpinia coriaria* Willd.; *Caesalpiniac.;* S.-Am. (? Ven.)
Muavi: *Erythrophloeum guineense* G. Don; *Caesalpiniac.;* tr. Af. (? Kongo)
Mubajangabo: *Zanthoxylum sp.; Rutac.;* br. O.-Af. (Uganda)
Mubajangalabi: *Rauwolfia sp.; Apocynac.;* br. O.-Af. (Uganda)
Mubaka: *Panda oleosa* Pierre; *Pandac.;* Gab.
Mubako: *Erythrophloeum africanum* Harms; *Caesalpiniac.;* n. Rhod.
Mubanga: *Ormosia brasseuriana* de Wild.; *Papilionac.;* B.-K. (Katanga)
„ : *Afrormosia angolensis* Harms; *Papilionac.;* s. C.-Af., n. Rhod.
Mubata: *Omphalocarpum laurenti* de Wild.; *Sapotac.;* B.-K. (Bukaka)
Mubaya: *Pachylobus pubescens* Verm.; *Burserac.;* B.-K. (Bukaka)
Mubôna: *Ochocoa gaboni* Pierre; *Myristicac.;* Gab.
Mubula: *Parinarium mobola* Oliv.; *Rosac.;* n. Rhod.
Mubulwebulwe: *Erythrophloeum africanum* Harms; *Caesalpiniac.;* n. Rhod.
Mubwabwa: *Petersia africana* Welw.; *Lecythidac.;* B.-K. (Bukaka)
Mubyimbyi: *Amblygonocarpus obtusangulus* Harms; *Mimosac.;* n. Rhod.
Muc: *Wrightia annamensis* Eberh. et Dub.; *Apocynac.;* Ind.-Ch.
„ bát: *Wrightia ovata* A. DC.; *Apocynac.;* Ind.-Ch.
Mucaça-mumbi: *Carapa procera* DC.; *Meliac.;* Seng.
„ n'cumbi: *Carapa procera* DC.; *Meliac.;* port. Af.
Mucandara: *Ceriops candolleana* Arn.; *Rhizophorac.;* port. O.-Af. (Moz., Quel.)
Mucandra: *Ceriops candolleana* Arn ; *Rhizophorac.;* port. O.-Af. (Moz., Quel.)
Muchenja: *Diospyros mespiliformis* Hochst.; *Ebenac.;* n. Rhod.
Muchenje: *Diospyros mespiliformis* Hochst.; *Ebenac.;* n. Rhod.
Muchite: *Pithecolobium dulce* Bth.; *Mimosac.;* Mex.
Muchokunda: *Pterospermum acerifolium* Willd.; *Sterculiac.;* Beng.
Muco: *Couroupita guianensis* Aubl.; *Lecythidac.;* br. Gu.
Muconite: *Excoecaria africana* Muell. Arg.; *Euphorbiac.;* port. O.-Af.
Muconti: *Excoecaria africana* Muell. Arg.; *Euphorbiac.;* O.-Af.
Mucuri: *Lucuma procera* Mart.; *Sapotac.;* Bras. (Bahia)
Mudamah· *Buchanania angustifolia* Roxb.; *Anacardiac.;* O.-Ind. (Madras)
Mudang asam: *Elaeocarpus stipularis* Blume; *Elaeocarpac.;* Malak.
Mudayat: *Bassia latifolia* Roxb.; *Sapotac.;* Burma.
Mudhol: *Alstonia scholaris* R. Br.; *Apocynac.;* O.-Ind. (Bombay)
Mudoka: *Mimusops africana* H. Lec.; *Sapotac.;* Gab.
Mududad: *Scleroxylon swietenia* DC.; *Rutac.;* O.-Ind. (Madras)

Muduka: *Mimusops africana* H. Lec.; *Sapotac.;* Gab.
Mudus: *Parkia filicoidea* Welw.; *Mimosac.;* Sudan.
Mueba: *Irvingia gabonensis* Baill.; *Simarubac.;* B.-K.
Muelle: *Schinus molle* L.; *Anacardiac.;* Kol.
Muelo: *Drimys winteri* Forst.; *Magnoliac.;* C.-R.
Muema: *Rhizophora mangle* L.; *Rhizophorac.;* B.-K., Angola (Zaire)
Muembe: *Mangifera indica* L.; *Anacardiac.;* O.-Ind., Cey.
Muenge: *Pterocarpus soyauxi* Taub.; *Papilionac.;* Kam.
Muerdago: *Viscum album* L., *V. laxum* Boiss. et Reut.; *Loranthac.;* Sp.
Muermo: *Eucryphia cordifolia* Cav.; *Eucryphiac.;* Chile.
Mufinsa des bois: *? Syzygium oweriensis* Bth.; *Myrtac.;* B.-K.
Mufula: *Clorophora excelsa* Bth. et Hook.; *Morac.;* B.-K. (Katanga)
Mufuma: *Ceiba pentandra* Gaertn.; *Bombacac.;* port. Af.
Mufutu: *Vitex cienkowskyi* Kotschy et Peyr.; *Verbenac.;* B.-K. (Katanga)
Mugaita: *Rapanea rhododendroides* Mez; *Myrsinac.;* br. O.-Af. (Kenya)
Mugamba: *Heisteria trillesiana* Pierre; *Olacac.;* Gab.
Mugavu: *Albizzia zygia* Macbride, *A. coriaria* Welw.; *Mimosac.;* br. O.-Af. (Uganda)
Mugèla: *Ricinodendron africanum* Muell. Arg.; *Euphorbiac.;* Gab.
Mugga: *Eucalyptus sideroxylon* A. Cunn.; *Myrtac.;* Au.
Mugila: *Acer oblongum* Wall.; *Acerac.;* n. O.-Ind. (Nepal)
Múgis: *Koordersiodendron pinnatum* Merr.; *Anacardiac.;* Phil.
Mugoberere: *Afzelia quanzensis* Welw.; *Caesalpiniac.;* port. O.-Af. (Mozamb.)
Mugoma: *Klainedoxa gabonensis* Pierre; *Simarubac.;* Gab.
Muguba: *Panda oleosa* Pierre; *Pandac.;* Gab.
Mugubi: *Staudtia gabonensis* Warb.; *Myristicac.;* Gab.
Mugumini: *Coula edulis* Baill.; *Olacac.;* Gab.
Muguminu: *Coula edulis* Baill.; *Olacac.;* Gab.
Mugumunu: *Coula edulis* Baill.; *Olacac.;* Gab.
Muguvi: *Staudtia gabonensis* Warb.; *Myristicac.;* Gab.
Muhama: *Acacia senegal* Willd.; *Mimosac.;* D.-O.-Af. (Tabora)
Muhengere: *Acacia brosigi* Harms; *Mimosac.;* D.-O.-Af. (Kilossa)
Muhi: *Mauritia carana* Wallace; *Palmac.;* Bras.
Muhinge: *Ximenia americana* L.; *Olacac.;* Angola.
Muhli: *Melia azedarach* L.; *Meliac.;* C.-O.-Ind.
Muhonono: *Terminalia sericea* Burch.; *Combretac.;* n. Rhod.
Muhumula: *Maesopsis emini* Engl.; *Rhamnac.;* D.-O.-Af.
Mui: *Chloraphora excelsa* Bth. et Hook.; *Morac.;* Elf.
   „ -kurékaá: *Capparis sp.; Capparidac.;* Arg.
Muiracoatiára: *Centrolobium paraënse* Tul.; *Papilionac.;* Bras. (Amaz.)
Muiracutáca: *Swartzia sp.; Caesalpiniac.;* Bras.
Muiragyboia: *? Pterocarpus sp.; Papilionac.;* Bras.
Muirajussára: *Aspidosperma duckei* Huber; *Apocynac.;* Bras. (Amaz.)
Muirapenima: *Brosimum sp.; Morac.;* Bras.
Muirapinima: *Piratinera guianensis* Aubl.; *Morac.;* Bras.
   „ preta: *Zollernia paraënse* Huber; *Caesalpiniac.;* Bras.
Muirapiranga: *Brosimum paraënse* Huber, *? B. angustifolium* Ducke; *Morac.;*Bras.(Amaz.)
   „   : *Mimusops sp.; Sapotac.;* Bras.
   „.   : *Ferolia guianensis* Aubl.; *Rosac.;* Sur.
Muiratinga: *Olmedia sp.; Morac.;* Bras.
   „ verdadeira: *Olmedia caloneura* Huber; *Morac.;* Bras. (Amaz.)
Muiraúba da terra firma: *Qualea sp.; Vochysiac.;* Bras.
   „ da varzea: *Qualea sp.; Vochysiac.;* Bras.
Mujua: *Alstonia congensis* Engl.; *Apocynac.;* br. O.-Af. (Uganda)
Mukala: *Malacantha superba* Verm.; *Sapotac.;* B.-K.
   „ tanda: *Rhizophora mangle* L. var. *?; Rhizophorac.;* Kam.
Mukama: *Funtumia africana* Stapf; *Apocynac.;* Kam.
Mukamba: *Afzelia quanzensis* Welw.; *Caesalpiniac.;* n. Rhod.
Mukampalei: *Alstonia scholaris* R. Brown; *Apocynac.;* O.-Ind. (Madras)
Mukangala: *Albizzia spp.; Mimosac.;* n. Rhod.
Mukarti: *Stereospermum chelonioides* DC.; *Bignoniac.;* sw. O.-Ind. (Bombay)
Mukebu: *? Aleurites sp.; Euphorbiac.;* br. O.-Af. (Uganda)
Mukeke: *Podocarpus milanjiana* Rendle; *Conifer.;* br. O.-Af. (Uganda)
Mukha: *Schrebera swietenioides* Roxb.; *Oleac.;* w. Beng. (Orissa)
Mukoge: *Tamarindus indica* L.; *Caesalpiniac.;* br. O.-Af. (Uganda)

Mukoi: *Piptadenia buchanani* Baker; *Mimosac.;* br. O.-Af. (Kenya)
Mukokowe: *Ficus sp.; Morac.;* br. O.-Af. (Uganda)
Mukole: *Dombeya mukole* Sprague; *Sterculiac.;* br. Af. (Uganda)
Mukongoru: *Lovoa klaineana* Pierre; *Meliac.;* br. O.-Af. (Kenya)
Mukonia-malamba: *Mitragyne macrophylla* Hiern; *Rubiac.;* Kam.
„ mamoundi: *Sarcocephalus trillesi* Pierre; *S. pobeguini* Pob.; *Rubiac.;* Kam.
„ weiss: *Terminalia superba* Engl. et Diels; *Combretac.;* Kam.
Mukonjaholz: *Terminalia superba* Engl. et Diels; *Combretac.;* Go., Kam.
Mukonja weiss: *Terminalia superba* Engl. et Diels; *Combretac.;* Go., Kam.
Mukorosi: *Sapindus mukorosi* Gaertn.; *Sapindac.;* Jap.
Mukose n'kose: *Malacantha sp.; Sapotac.;* B.-K. (Maniema)
Mukoso: *Erythrophloeum africanum* Harms; *Caesalpiniac.;* n. Rhod.
Mukotokoto: *Acacia pallens* Rolfe; *Mimosac.;* n. Rhod.
Mukuchumswa: *Diospyros mespiliformis* Hochst.; *Ebenac.;* n. Rhod.
Mukuenoki: *Trema orientalis* Blume; *Ulmac.;* Form.
Mukuhakuha: *Macaranga kilimandscharica* Pax; *Euphorbiac.;* br. O.-Af. (Kenya)
Mukui: *Piptadenia buchanani* Baker; *Mimosac.;* br. O.-Af. (Kenya)
Mukula: *Pterocarpus spp.; Papilionac.;* B.-K. (Katanga)
Mukungu: *Ceiba pentandra* Gaertn.; *Bombacac.;* B.-K. (Maniema)
Mukunoki· *Aphanante aspera* Planch.; *Ulmac.;* Jap.
Mukunti: *Excoecaria africana* Muell. Arg.; *Euphorbiac.;* O.-Af.
Mukunyu: *Ficus aff. corylifolia* Warb.; *Morac.;* br. O.-Af. (Uganda)
Mukunzanume: *Warburgia ugandensis* Sprague; *Canellac.;* br. O.-Af. (Uganda)
Mukura: *Coula edulis* Baill.; *Olacac.;* Gab.
Mukurodji: *Sapindus mukorosi* Gaertn.; *Sapindac.;* Jap.
Mukuroji: *Sapindus mukorosi* Gaertn.; *Sapindac.;* Jap.
Mukurozi: *Sapindus mukorosi* Gaertn.; *Sapindac.;* Jap.
Mukusi: *Baikiaea plurijuga* Harms; *Caesalpiniac.;* n. Rhod.
Mukusu: *Ricinodendron rautaneni* Schinz; *Euphorbiac.;* B.-K. (Katanga)
Mukuu: *Juniperus procera* Hochst.; *Conifer.;* br. O.-Af. (Kenya)
Mukuwe: *? Parinarium polyandrum* Bth.; *Rosac.;* B.-K. (Katanga)
„ : *Parinarium bequaerti* de Wild.; *Rosac.;* n. Rhod.
Mukuyenoki: *Aphanante aspera* Planch.; *Ulmac.;* Jap.
Mukwa: *Xylopia aethiopica* A. Rich.; *Anonac.;* B.-K.
Mukwa: *Pterocarpus angolensis* DC.; *Papilionac.;* Rhod.
Mukwe: *Berlinia aff. angolensis* Welw.; *Caesalpiniac.;* O.-Af.
Mula: *Parinarium holsti* Engl.; *Rosac.;* D.-O.-Af. (w. Usambara)
Mulábog: *Parishia malabog* Merr.; *Anacardiac.;* Phil.
Mulah: *Erythroxylon cuneatum* Kurz; *Erythroxylac.;* Malay, Phil.
Mulambi: *Pterocarpus angolensis* DC.; *Papilionac.;* n. Rhod.
Mulanga: *Afrormosia angolensis* Harms; *Papilionac.;* B.-K.
Mulanga: *Staudtia gabonensis* Warb.; *Myristicac.;* Gab.
Mulatinho: *Rudgea dahlgreni* Standl.; *Rubiac.;* Bras.
Mulato: *Triplaris americana* L.; *Polygonac.;* Guat.
„ : *Bursera simaruba* Sarg.; *Burserac.;* Hond.
„ : *Pentaclethra filamentosa* Bth.; *Mimosac.;* Ven.
Mulatto: *Platymiscium polystachyum* Bth.; *Papilionac.;* Guat., Hond.
„ tree: *Calycophyllum sp.; Rubiac.;* tr. Am.
Muláuin: *Uitex parviflora* Juss.; *Uerbenac.;* Phil.
„ -ásu: *Uitex turczaninowi* Merr.; *Uerbenac.;* Phil. (Laguna, Mindoro)
Mulawin-asu: *Premna nauseosa* Blco., *Uitex spp.; Uerbenac.;* Phil.
„ -babáe: *Uitex sp.; Uerbenac.;* Phil.
Mulberry: *Morus spp.; Morac.;* O.-Ind.
„ : *Liquidambar styraciflua* L.; *Hamamelidac.;* N.-Am.
„ : *Chlorophora excelsa* Bth. et Hook.; *Morac.;* Lib.
„ , black: *Morus rubra* L.; *Morac.;* ö. Kan.
„ , Dyer's-: *Chlorophora tinctoria* Gaudich.; *Morac.;* tr. Am.
„ , Fustic-: *Chlorophora tinctoria* Gaudich.; *Morac.;* tr. Am.
„ , indian: *Morus indica* L.; *Morac.;* Malay.
„ , Paper-: *Broussonetia kaempferi* Sieb.; *Morac.;* Jap.
„ , red: *Morus rubra* L.; *Morac.;* ö. N.-Am.
„ , yellow: *Morus laevigata* Wall.; *Morac.;* Burma.
Mulèko: *Poga oleosa* Pierre; *Rhizophorac.;* Gab.
Muleku: *Ongokea klaineana* Pierre; *Olacac.;* B.-K. (Bukaka)

Mulela: *Ricinodendron africanum* Muell. Arg.; *Euphorbiac.*; B.-K. (Katanga)
Mulemu: *Landolphia florida* Bth.; *Apocynac.*; br. O.-Af. (Uganda)
Mulga: *Acacia aneura* F. v. M.; *Mimosac.*; C.-Af., S.-Af.
Mulgal: *Morinda tinctoria* Roxb.; *Rubiac.*; w. O.-Ind. (Bombay)
Mul-ilava: *Bombax malabaricum* DC.; *Bombacac.*; w. O.-Ind. (n. Bombay)
Mulillam: *Zanthoxylum rhetsa* DC.; *Rutac.*; O.-Ind. (Madras)
Muljane: *Bridelia retusa* Spreng.; *Euphorbiac.*; w. O.-Ind. (Bombay)
Mullan-gayum: *Bridelia retusa* Spreng.; *Euphorbiac.*; sw. O.-Ind. (Malabar)
Mulle-honne: *Bridelia retusa* Spreng.; *Euphorbiac.*; w. O.-Ind. (Bombay)
„ -kare: *Zizyphus xylopyrus* Willd.; *Rhamnac.*; w. O.-Ind. (Deccan)
Mulli: *Schinus molle* L.; *Anacardiac.*; Peru.
Mullu buraga: *Bombax malabaricum* DC.; *Bombacac.*; sw. O.-Ind. (Kanara)
Mulomba: *Pycnanthus kombo* Warb.; *Myristicac.*; Gab.
Mulombe: *Pterocarpus angolensis* DC.; *Papilionac.*; n. Rhod.
Mulombo: *Pterocarpus angolensis* DC.; *Papilionac.*; B.-K.
Mulombva: *Pterocarpus angolensis* DC.; *Papilionac.*; B.-K.
Mulombwa: *Pterocarpus angolensis* DC.; *Papilionac.*; B.-K., n. Rhod.
„   : *Pterocarpus erinaceus* Poir.; *Papilionac.*; B.-K. (Katanga)
Mulu modugu: *Erythrina suberosa* Roxb.; *Papilionac.*; O.-Ind. (Madras)
„ -vengai: *Bridelia retusa* Spreng.; *Euphorbiac.*; O.-Ind. (Madras)
Mulule: *Macrolobium dewevrei* de Wild.; *Caesalpiniac.*; B.-K. (Maniema)
Mululu: *Khaya nyasica* Stapf; *Meliac.*; n. Rhod.
Mulumba: *Pterocarpus angolensis* DC.; *Papilionac.*; B.-K.
Mulumut: *Campnosperma oxyrachis* Engl.; *Anacardiac.*; Malay.
Mulungu: *Erythrina crista-galli* L., *E. sp.*; *Papilionac.*; Bras.
Mulut: *Hopea anomala* Foxw.; *Dipterocarpac.*; Malay.
Mûm: *Barringtonia eberhardti* Gagnep.; *Lecythidac.*; Ind.-Ch.
Mum pesand: *Polyalthia jenkinsi* Bth.; *Anonac.*; Malak.
Mumbanga: *Cocos nucifera* L.; *Palmac.*; Gab.
Mumpala: *Platysepalum chevalieri* Harms; *Papilionac.*; B.-K.
Mumpampi: *Chrysophyllum lacourtianum* de Wild., *C. sp.*; *Sapotac.*; B.-K.
Mumpilai klat: *Aphania paucijuga* Radlk.; *Sapindac.*; Malay.
Mun: *Diospyros mun* H. Lec., *D. tonkinensis* Chev., *D. ebenum* Koenig; *Ebenac.*;
„   : *Calophyllum inophyllum* L.; *Guttifer.*; Co.-Ch.   [Ind.-Ch.
„   : *Barringtonia sp.*; *Lecythidac.*; Ind.-Ch.
„ haneca: *Pterocarpus angolensis* DC.; *Papilionac.*; port. W.-Af. (Angola)
„ kokoski: *Hovenia dulcis* Thunb.; *Rhamnac.*; nö. O.-Ind. (Nepal)
„ si: *Peltophorum dasyrachis* Kurz; *Caesalpiniac.*; Ind.-Ch.
„ -thi: *Diospyros decandra* Loureiro; *Ebenac.*; Ind.-Ch.
Munaabenaabe: *Pterocarpus angolensis* DC.; *Papilionac.*; S.-Af. (Transval)
Munamal: *Mimusops elengi* L.; *Sapotac.*; w. O.-Ind., Cey.
Munangára: *Combretum petersi* Engl.; *Combretac.*; D.-O.-Af. (Kilossa)
Mundambi: *Millettia laurenti* de Wild.; *Papilionac.*; B.-K. (Bukaka)
Mundani (H): *Acrocarpus fraxinifolius* Wight; *Caesalpiniac.*; O.-Ind.
Munderendu: *Toddalia nobilis* Hook. f.; *Rutac.*; br. O.-Af. (Kenya)
Mundi: *Stephegyne parvifolia* Korth.; *Rubiac.*; O.-Ind. (Gondwana, w. Beng.)
„ -mundi: *Stephegyne parvifolia* Korth.; *Rubiac.*; O.-Ind., Malay.
Mundillos: *Viburnum opulus* L.; *Caprifoliac.*; Sp.
Mundzangala: *Ricinodendron africanum* Muell. Arg.; *Euphorbiac.*; Gab.
Muñeca: *Cordia borinquensis* Urban; *Borraginac.*; P.-R.
„   : *Hasseltia floribunda* H. B. K.; *Flacourtiac.*; Nic.
„ amarilla: *Cordia collococca* L.; *Borraginac.*; Pan.
Muñeco: *Bombax ellipticum* H. B. K.; *Bombacac.*; Guat.
Mung khut: *Garcinia mangostana* L.; *Guttifer.*; Kamb.
„ quân: *Flacourtia cataphracta* Roxb.; *Flacourtiac.*; Ind.-Ch.
„ sau: *Barringtonia annamica* Gaertn.; *Lecythidac.*; Ind.-Ch.
„ -tu: *Zanthoxylum usitatum* Pierre; *Rutac.*; Co.-Ch.
Munga: *Mitragyne macrophylla* Hiern; *Rubiac.*; Angola (Golungo Alto)
„   : *Acacia albida* Delile; *Mimosac.*; n. Rhod.
„ pera: *Buchanania latifolia* Roxb.; *Anacardiac.*; sw. O.-Ind. (Madras)
Mungagnya: *Bosquiea welwitschi* Engl.; *Morac.*; B.-K.
Mungèka: *Ongokea klaineana* Pierre; *Olacac.*; Gab.
Mungèkè: *Ongokea klaineana* Pierre; *Olacac.*; Gab.

Mungèmbè: *Ricinodendron africanum* Muell. Arg.; *Euphorbiac.;* Gab.
Munghegnye: *Bosquiea welwitschi* Engl.; *Morac.;* B.-K.
Mungilkíl: *Eugenia mananquil* Blco.; *Myrtac.;* Phil. (Mindoro)
Mungongo: *Sclerocarya birrea* Hochst.; *Anacardiac.;* D.-O.-Af. (Tabora)
„ : *Ricinodendron rautaneni* Schinz; *Euphorbiac.;* n. Rhod.
Munguba: *Bombax munguba* Mart.; *Bombacac.;* s. Bras.
Mungubeira: *Bombax munguba* Mart.; *Bombacac.;* s. Bras.
Munguella: *Ricinodendron africanum* Muell. Arg.; *Euphorbiac.;* Angola.
Mungundu: *Parinarium glabrum* Oliv.; *Rosac.;* B.-K.
Mungurra: *Eucalyptus tereticornis* Smith; *Myrtac.;* ö. Au.
Muni: *Erythrina suberosa* Roxb.; *Papilionac.;* O.-Ind. (Madras)
Muniah: *Anogeissus latifolia* Wall.; *Combretac.;* C.-O.-Ind. (Gondwana)
Muniengili: *Scorodophloeus zenkeri* Harms; *Caesalpiniac.;* B.-K. (Maniema)
Munkombo: *Lannea welwitschi* Engl.; *Anacardiac.;* B.-K.
Munseera: *Ilex cymosa* Blume; *Aquifoliac.;* Malak., Java.
Muntahwun: *Roucheria griffithiana* Planch.; *Linac.;* Malak.
Muntangoo booga: *Calophyllum canum* Hook. f.; *Guttifer.;* Malak.
Munyama: *Khaya anthotheca* C. DC.; *Meliac.;* br. O.-Af. (Uganda)
Munyenye: *Tetrapleura sp.; Mimosac.;* B.-K. (Katanga)
„ : *Albizzia sp.; Mimosac.;* D.-O.-Af.
Munzap: *Couepia dodecandra* Hemsl.; *Rosac.;* Hond.
Muôi: *Rhus semialata* Murr.; *Anacardiac.;* Ind.-Ch.
Muôm: *Mangifera foetida* Lour.; *Anacardiac.;* Annam.
Muông: *Glycosmis cochinchinensis* Pierre; *Rutac.;* Ind.-Ch.
„ : *Aporosa microcalyx* Hassk.; *Euphorbiac.;* Ind.-Ch.
„ : *Cassia timorensis* DC., *C. siamea* Lamk.; *Caesalpiniac.;* Ind.-Ch.
„ chek: *Cassia garettiana* Craib; *Caesalpiniac.;* Ind.-Ch.
„ do: *Cassia timorensis* DC.; *Caesalpiniac.;* Ind.-Ch.
„ gai: *Cassia arabica* ?; *Caesalpiniac.;* Ind.-Ch.
„ núi: *Cassia siamea* Lamk.; *Caesalpiniac.;* Ind.-Ch.
„ rút: *Cassia tonkinensis* Chev., *C. timorensis* DC.; *Caesalpiniac.;* Ind.-Ch.
„ ta: *Zanthoxylum avicennae* DC.; *Rutac.;* Ind.-Ch.
„ „ : *Cassia arabica* ?; *Caesalpiniac.;* Ind.-Ch.
„ tráng: *Cassia tonkinensis* Chev., *C. timorensis* DC.; *Caesalpiniac.;* Ind.-Ch.
„ „ : *Lysidice sp.; Caesalpiniac.;* Ind.-Ch.
„ troung: *Zanthoxylum rhetsa* DC.; *Rutac.;* Ind.-Ch.
„ xoan: *Cassia tonkinensis* Chev., *C. timorensis* DC.; *Caesalpiniac.;* Ind.-Ch.
Mupaka: *Pterygopodium balsamiferum* Verm.; *Papilionac.;* B.-K. (Bukaka)
Mupandji: *Pentaclethra macrophylla* Bth.; *Mimosac.;* Gab.
Mupane: *Copaifera mopane* J. Kirk; *Caesalpiniac.;* n. Rhod.
Mupapa: *Afzelia quanzensis* Welw.; *Caesalpiniac.;* n. Rhod., B.-K.
Mupatu: *Canarium schweinfurthi* Engl.; *Burserac.;* B.-K.
Mupayu: *? Irvingia klainei* Pierre; *Simarubac.;* Gab.
Mupumena: *Entandrophragma sp.; Meliac.;* n. Rhod.
Mupundu: *Parinarium mobola* Oliv.; *Rosac.;* B.-K., n. Rhod.
„ blanc: *Parinarium sp.; Rosac.;* B.-K. (Katanga)
„ gris: *Parinarium sp.; Rosac.;* B.-K. (Katanga)
Muquesu: *Cola acuminata* Schott et Endl.; *Sterculiac.;* port. W.-Af.
Mur: *Butea frondosa* Roxb.; *Papilionac.;* C.-O.-Ind. (Gondwana)
Murara: *Juniperus procera* Hochst.; *Conifer.;* br. O.-Af. (Kenya)
Muratthan: *Sterculia villosa* Roxb.; *Sterculiac.;* O.-Ind. (Madras)
Muráuin: *Vitex parviflora* Juss.; *Verbenac.;* Phil. (Masbate, Ticao)
Muráuon: *Vitex parviflora* Juss.; *Verbenac.;* Phil. (Guimaras)
Murcielago: *Benthamantha glabrescens* Alef.; *Papilionac.;* Kol.
Mureche: *Byrsonima sp.; Malpighiac.;* Trin.
Mureci: *Byrsonima* sp.; *Malpighiac.;* Bras.
Murecy: *Byrsonima* sp.; *Malpighiac.;* Bras.
„ vermelho: *Byrsonima sp.; Malpighiac.;* Bras.
Murei: *Byrsonima* sp.; *Malpighiac.;* Bras.
Mureila: *Byrsonima sp.; Malphighiac.;* Bras.
Murènda: *Garcinia klaineana* Pierre; *Guttifer.;* Gab.
Mureré: *Piratinera guianensis* Aubl., *Brosimum acutifolium* Huber; *Morac.;* Bras.
„ vermelho: *Brosimum sp.; Morac.;* Bras. (Amaz.) [(Amaz.)

Murga: *Pterocarpus marsupium* Roxb.; *Papilionac.;* w. Beng.
Muri: *Humiria sp.; Humiriac.;* br. Gu.
„ kupé: *Clusia nemorosa* G. F. W. Meyer; *Guttifer.;* br. Gu.
Muria: *Buchanania latifolia* Roxb.; *Anacardiac.;* n. O.-Ind. (Garhwal)
Muriangombe du Benguella: *Maerua angolensis* DC.; *Capparidac.;* port. W.-Af. (Ang.)
Murici: *Byrsonima verbascifolia* Rich.; *Malpighiac.;* Bras.
„ do campo: *Byrsonima crassifolia* H. B K.; *Malpighiac.;* Bras.
Muricy: *Byrsonima coriacea* DC.; *Malpighiac.;* n. Bras. (Pará)
„ : *Vochysia sp.; Vochysiac.;* Bras.
Murier: *Morus indica* L.; *Morac.;* Ind.-Ch.
„ blanc: *Morus alba* L.; *Morac.;* ö. Eu., As.
„ rouge: *Morus rubra* L.; *Morac.;* ö. Kan.
„ sauvage: *Morus rubra* L.; *Morac.;* ö. Kan.
„ à papier: *Broussonetia papyrifera* Vent.; *Morac.;* Ind.-Ch.
„ „ rameaux épineux: *Chlorophora tinctoria* Gaudich.; *Morac.;* tr. Am.
„ de Java: *Zanthoxylum montanum* Blume; *Rutac.;* Java.
„ des teinturiers: *Chlorophora tinctoria* Gaudich.; *Morac.;* tr. Am.
Muriri: *Mouriria sp.; Melastomatac.;* Bras.
Muro: *Dimorphandra mora* Bth. et Hook.; *Caesalpiniac.;* Trin.
Murogoro: *Spathodea nilotica* Seem.; *Bignoniac.;* br. O.-Af. (Uganda)
Murongo: *Albizzia zygia* MacBride; *Mimosac.;* br. O.-Af. (Uganda)
Murrakku: *Zanthoxylum rhetsa* DC.; *Rutac.;* sw. O.-Ind. (Malabar)
Murrogun: *Cryptocarya microneura* Meissn.; *Laurac.;* ö. Au.
Murr-rung: *Alphitonia excelsa* Reisseck; *Rhamnac.;* ö. Au.
Murta: *Coccoloba ramosissima* Wedd.; *Polygonac.;* n. Kol.
„ : *Eugenia sp.; Myrtac.;* Bras.
Murtenga: *Bursera serrata* Colebr.; *Burserac.;* Assam.
Murucuru: ? *Lonchocarpus violaceus* H. B. K.; *Papilionac.;* Kol.
Muruku: *Sapium mannianum* Bth.; *Euphorbiac.;* br. O.-Af. (Uganda)
Murungurungu: *Baphia wollastoni* Bak. f.; *Papilionac.;* br. O.-Af. (Uganda)
Murup: *Butea frondosa* Roxb.; *Papilionac.;* Burma.
Murupa: *Simaruba amara* Aubl.; *Simarubac.;* Bras.
Murupita: *Sapium lanceolatum* Huber; *Euphorbiac.;* Bras.
Mururé: *Brosimopsis diandra* Blake, *B. acutifolia* Ducke; *Morac.;* Bras. (Bahia)
Mus: *Pterospermum acerifolium* Willd.; *Sterculiac.;* Beng.
Musa: *Kigelia moosa* Sprague; *Bignoniac.;* br. O.-Af. (Uganda)
Musabera: *Albizzia zygia* MacBride; *Mimosac.;* br. O.-Af. (Uganda)
„ : *Albizzia grandibracteata* Taub.; *Mimosac.;* br. O.-Af. (Uganda)
Musacosse: *Afzelia quanzensis* Welw.; *Caesalpiniac.;* port. O.-Af. (Mozamb.)
Musale: *Garcinia buchanani* Baker; *Guttifer.;* br. O.-Af. (Uganda)
Musali: *Garcinia buchanani* Baker; *Guttifer.;* br. O.-Af. (Uganda)
Musamba: *Brachystegia bequaeti* de Wild.; *Caesalpiniac.;* B.-K. (Katanga)
Musamwu: *Monodora myristica* Dunal; *Anonac.;* br. O.-Af. (Uganda)
Musanda: *Funtumia elastica* Stapf, *F. latifolia* Stapf; *Apocynac.;* br. O.-Af. (Uganda)
Musangu: *Acacia albida* Delile; *Mimosac.;* n. Rhod.
Musaniki: *Clausena anisata* Oliv.; *Rutac.;* br. O.-Af. (Uganda)
Musanke: *Canarium schweinfurthi* Engl.; *Burserac.;* br. O.-Af. (Uganda)
Musa-noué: *Phyllanthus discoideus* Muell. Arg.; *Euphorbiac.;* Elf.
Musanu: *Ongokea klaineana* Pierre; *Olacac.;* Gab.
Musanvuma: *Sapium mannianum* Bth.; *Euphorbiac.;* br. O.-Af. (Uganda)
Musanya: *Marquesia macroura* Gilg; *Dipterocarpac.;* n. Rhod.
Musasa: *Sapium mannianum* Bth.; *Euphorbiac.;* br. O.-Af. (Uganda)
Musase: *Albizzia katangensis* de Wild., *A. brownei* Walp.; *Mimosac.;* B.-K. (Katanga)
„ : *Albizzia antunesiana* Harms; *Mimosac.;* n. Rhod.
Muscadier faux: *Pycnanthus kombo* Warb.; *Myristicac.;* Gab.
„ sauvage: *Rheedia lateriflora* L.; *Guttifer.;* Trin.
„ à suif: *Virola sp.; Myristicac.;* Gu.
Musekankusu: *Barteria fistulosa* Mast.; *Flacourtiac.;* br. O.-Af. (Uganda)
Museke: *Arundinaria alpina* K. Schum.; *Gramin.;* br. O.-Af. (Uganda)
Musekeh: *Pterocarpus angolensis* DC.; *Papilionac.;* B.-K.
Musekera: *Rhus villosa* L. f.; *Anacardiac.;* br. O.-Af. (Uganda)

Musembo: *Haronga paniculata* Lodd.; *Guttifer.;* B.-K. (Mohanga)
Museneka: *Podocarpus gracilior* Pilger; *Conifer.;* br. O.-Af. (Kenya)
Musenene: *Podocarpus gracilior* Pilger; *Conifer.;* br. O.-Af. (Uganda)
   „   : *Podocarpus milanjianus* Rendle; *Conifer.;* br. O.-Af. (Uganda)
Musengera: *Podocarpus gracilior* Pilgeı; *Conifer.;* br. O.-Af. (Kenya)
   „   : *Podocarpus milanjianus* Rendle; *Conifer.;* br. O.-Af. (Kenya)
Musese: *Rhus glaucescens* A. Rich.; *Anacardiac.;* br. O.-Af. (Uganda)
Museshe: *Marquesia macroura* Gilg; *Dipterocarpac.;* n. Rhod.
Musesji: *Parinarium sp.; Rosac.;* B.-K. (Katanga)
   „   : *Monotes sapini* de Wild.; *Dipterocarpac.;* B.-K.
Musesjie: *Marquesia macroura* Gilg; *Dipterocarpac.;* n. Rhod. (Katanga)
Museta: *Terminalia cycloptera* R. Br.; *Combretac.;* br. O.-Af. (Uganda)
Mushakashela: *Swartzia madagascariensis* Desv.; *Caesalpiniac.;* n. Rhod.
Mushalaka: *Excoecaria africana* Muell. Arg.; *Euphorbiac.;* D.-O.-Af. (Tanganjika)
Mushasha: *Sapium mannianum* Bth.; *Euphorbiac.;* br. O.-Af. (Uganda)
   „   : *Macaranga neomildbraediana* Lebrun; *Euphorbiac.;* B.-K. (Mohanga)
Mushavu: *Markhamia lutea* K. Schum.; *Bignoniac.;* B.-K. (Beni, Semliki)
Mushebere: *Albizzia gummifera* C. A. Smith, *A. grandistipulata* Taun.; *Mimosac.;*
Musherage: *Olea hochstetteri* Baker; *Oleac.;* br. O.-Af. (Kenya)   [B.-K.
Mushese: *Burkea africana* Hook.; *Caesalpiniac.;* n. Rhod.
Mushikishi: *Trichilia emetica* Vahl; *Meliac.;* n. Rhod.
Mush-timbi: *Diospyros ebenum* Koenig; *Ebenac.;* sw. O.-Ind. (Malabar)
Mushyi: *Croton megalobotryoides* Muell. Arg.; *Euphorbiac.;* B.-K. (Mohanga)
Musibe: *Copaifera coleosperma* Bth.; *Caesalpiniac.;* n. Rhod.
Musibi: *Copaifera coleosperma* Bth.; *Caesalpiniac.;* n. Rhod.
Musigiri: *Odyendea gabunensis* Engl.; *Simarubac.;* Gab.
Musigiti: *Odyendea gabunensis* Engl.; *Simarubac.;* Gab.
Musika: *Tamarindus indica* L.; *Caesalpıniac.;* n. Rhod.
Musikili: *Trichilia emetica* Vahl; *Meliac.;* n. Rhod.
Musimba: *Trichilia volkensi* Harms; *Meliac.;* B.-K.
Musinda: *Artocarpus sp.; Morac.;* br. O.-Af. (Uganda)
Musira: *Ficus eryobotryoides* Kunth et Bouché; *Morac.;* br. O.-Af. (Uganda)
Musisa: *Albizzia coriaria* Welw.; *Mimosac.;* br. O.-Af. (Uganda)
   „   : *Trema guineensis* Engl.; *Ulmac.;* br. O.-Af. (Uganda)
Musisi: *Maesopsis emini* Engl.; *Rhamnac.;* br. O.-Af. (Uganda)
Musita: *Albizzia coriaria* Welw.; *Mimosac.;* br. O.-Af. (Uganda)
Musizi: *Maesopsis berchemioides* Nob., *M. emini* Engl.; *Rhamnac.;* br. O.-Af. (Uganda)
Musk: *Olearia argophylla* F. v. M.; *Composit.;* sö. Au., Tasm.
Muskatholz: *Piratinera guianensis* Aubl.; *Moıac.;* n. S.-Am.
Muskatnussholz, wildes (d): *Pycnanthus kombo* Warb.; *Myristicac.;* tr. W.-Af.
Muskheart, black: *Marlea vitiensis* Bth.; *Cornac.;* ö. Au.
Muskwood: *Guarea trichilioides* L.; *Meliac.;* Jam., S.-D., Trin.
   „   : *Guarea gomma* Pulle; *Meliac.;* Sur.
   „   : *Olearia argophylla* F. v. M.; *Composit.;* Au.
   „   : *Marlea vitiensis* Bth.; *Cornac.;* ö. Au.
Muslini: *Sterculia alata* Roxb.; *Sterculiac.;* nö. O.-Ind. (Nepal)
Musodo: *Ricinodendron africanum* Muell. Arg.; *Euphorbiac.;* br. O.-Af. (Uganda)
Musoga: *Croton macrostachys* A. Rich.; *Euphorbiac.;* br. O.-Af. (Uganda)
   „   : *Alstonia congensis* Engl.; *Apocynac.;* br. O.-Af. (Uganda)
Musogasoga: *Croton macrostachys* A. Rich.; *Euphorbiac.;* br. O.-Af. (Uganda)
Musokolindu: *Clausena anisata* Oliv.; *Rutac.;* br. O.-Af. (Uganda)
Musokoto: *Faurea discolor* Welw.; *Proteac.;* n. Rhod.
Musombè: *Elaeis guineensis* Jacq. var. ?; *Palmac.;* Gab.
Musomo: *Elaeis guineensis* Jacq. var. ?; *Palmac.;* Gab.
Musonga: *Ricinodendron heudeloti* Pierre; *Euphorbiac.;* B.-K. (Beni, Semliki)
Musorongo: *Albizzia ferruginea* Bth.; *Mimosac.;* br. O.-Af. (Uganda)
Musow: *Petersia africana* Welw.; *Lecythidac.;* B.-K. (Maniema)
Mussutahyba: *Zollernia ilicifolia* Vogel; *Caesalpiniac.;* Bras. (Sao P.)
Musti: *Strychnos nux-vomica* L.; *Loganiac.;* sw. O.-Ind. (Madras)
Musuku: *Ochocoa gaboni* Pierre; *Myristicac.;* Gab.
   „   : *Canarium schweinfurthi* Engl.; *Burserac.;* B.-K. (Beni, Semliki)

Musuku: *Uapaca kirkiana* Muell. Arg.; *Euphorbiac.;* n.-Rhod.
Musung: *Dialium yambataënse* Verm.; *Caesalpiniac.;* B.-K.
Mususu: *Dialium lacourtianum* de Wild.; *Caesalpiniac.;* B.-K.
Mut sanya: *Cistanthera leplaei* Verm.; *Tiliac.;* B.-K.
Mutala: *Uitex fischeri* Gürke, *U. cienkowskii* Kotschy et Peyr.; *Uerbenac.;* br. O.-Af.
Mutamaiyu: *Olea chrysophylla* Lam.; *Oleac.;* br. O.-Af. (Kenya)   [(Uganda)
Mutambá: *Guazuma ulmifolia* Lamk.; *Sterculiac.;* Bras.
Mutampindi: *Albizzia coriaria* Welw.; *Mimosac.;* br. O.-Af. (Uganda)
Mutangeni: *Cocos nucifera* L.; *Palmac.;* Gab.
Mutango: *Bersama holsti* Gürke; *Melianthac.;* br. O.-Af. (Uganda)
Mutarai: *Podocarpus milanjianus* Rendle; *Conifer.;* br. O.-Af. (Kenya)
Mutarakwa: *Juniperus procera* Hochst.; *Conifer.;* br. O.-Af. (Kenya)
Mutatabamkubebe: *Hymenocardia acida* Tul.; *Euphorbiac.;* br. O.-Af. (Uganda)
Mutchibanaye: *Lovoa klaineana* Pierre; *Meliac.;* Elf.
Mutèli: *Copaifera ? salikounda* Heckel; *Caesalpiniac.;* Gab.
Mutembo: *Monotes spp.; Dipterocarpac.;* n. Rhod.
Mutendi: *Conopharyngia usambarensis* Stapf.; *Apocynac.;* B.-K.
Mutete: *Pterocarpus angolensis* DC.; *Papilionac.;* port. W.-Af. (Angola)
    „   : *Allophylus subcoriaceus* Baker f.; *Sapindac.;* br. O.-Af. (Uganda)
Muti: *Acacia stenocarpa* Hochst.; *Mimosac.;* br. O.-Af. (Uganda)
Mutie: *Antiaris africana* Engl., *A. toxicaria* Lesch.; *Morac.;* Elf.
Mutigbanaïe: *Guarea thompsoni* Sprague et Hutch.; *Meliac.;* Elf.
Mutigbanaye: *Guarea thompsoni* Sprague et Hutch.; *Meliac.;* Elf.
    „   : *Guarea sp. aff. bipindeana* C. DC.; *Meliac.;* Elf.
    „   duveteux: *Trichilia sp. aff. vestita* C. DC.; *Meliac.;* Elf.
    „   rasso: *Trichilia sp.; Meliac.;* Elf.
Mutirai: *Chloroxylon swietenia* DC.; *Rutac.;* O.-Ind. (Madras)
Mutobo: *Isoberlinia tomentosa* Hutch.; *Caesalpiniac.;* n. Rhod.
Mutobotobo: *Solanum aculeatissimum* Jacq.; *Solanac.;* br. O.-Af. (Uganda)
    „   : *Solanum campylacanthum* Hochst.; *Solanac.;* br. O.-Af. (Uganda)
Mutokelebwe: *Parinarium bequaerti* de Wild.; *Rosac.;* n. Rhod.
Mutoko: *Platysepalum chevalieri* Harms; *Papilionac.;* B.-K.
Mutoma: *Ficus eryobotryoides* Kunth et Bouché; *Morac.;* br. O.-Af. (Uganda)
Mutombé: *Copaifera ? salikounda* Heckel; *Caesalpiniac.;* Gab.
Mutombi: *Copaifera ? salikounda* Heckel; *Caesalpiniac.;* Gab.
Mutombo: *Guazuma ulmifolia* Lamk.; *Sterculiac.;* Bras.
Mutondi: *Conopharyngia usambarensis* Stapf; *Apocynac.;* B.-K. (Mohanga)
Mutondo: *Berlinia globifera* Hutch. et B. Davy; *Caesalpiniac.;* B.-K.
    „   : *Isoberlinia paniculata* Hutch.; *Caesalpiniac.;* n. Rhod.
    „   : *Pterocarpus angolensis* DC., *P. erinaceus* Poir.; *Papilionac.;* B.-K. (Katanga)
    „   : *Alstonia gilletti* de Wild.; *Apocynac.;* B.-K. (Kurubu)
Mutongana: *Carapa grandiflora* Sprague; *Meliac.;* br. O.-Af. (Uganda)
Mutongole: *Acacia hebecladioides* Harms; *Mimosac.;* br. O.-Af. (Uganda)
Mutono: *Elaeis guineensis* Jacq. *var. virescens* Chev.; *Palmac.;* Gab.
Mutontwe: *Parinarium sp.; Rosac.;* B.-K. (Katanga)
Mutonwa: *Clausena anisata* Oliv.; *Rutac.;* br. O.-Af. (Uganda)
Mutsiku: *Mitragyne macrophylla* Hiern; *Rubiac.;* B.-K. (Mohanga)
Muttala: *Butea frondosa* Roxb.; *Papilionac.;* w. O.-Ind. (sw. Deccan)
Muttee-pal: *Ailanthus malabarica* DC.; *Simarubac.;* O.-Ind.
Mutti: *Terminalia tomentosa* Wight et Arn.; *Combretac.;* O.-Ind.
Muttonwood: *Rapanea variabilis* Mez; *Myrsinac.;* ö. Au.
Muttugal: *Butea frondosa* Roxb.; *Papilionac.;* w. O.-Ind. (sw. Deccan)
Mutty: *Ailanthus malabarica* DC.; *Simarubac.;* O.-Ind.
Mutuba: *Ficus sycomorus* L.; *Moiac.;* br. O.-Af. (Uganda)
Mutugunda: *Vangueria apiculata* K. Schum.; *Rubiac.;* br. O.-Af. (Uganda)
Mutumba: *Cordia unyorensis* Stapf; *Borraginac.;* br. O.-Af. (Uganda)
    „   : *Chlorophora excelsa* Bth. et Hook.; *Morac.;* br. O.-Af. (Uganda)
Mutumbwi: *Sterculia sp.; Sterculiac.;* br. O.-Af. (Uganda)
Mutunata: *Chytroma idatimon* Miers; *Lecythidac.;* Bras.
? Mutune: *Haronga madagascariensis* Choisy; *Guttifer.;* O.-Af. (Zambesi)
Mutunguru: *Lovoa klaineana* Pierre; *Meliac.;* br. O.-Af. (Kenya)

Mutunku: *Aberia macrocalyx* Oliv.; *Flacourtiac.;* br. O.-Af. (Uganda)
Mutushi: *Pterocarpus spp.; Papilionac.;* br. Gu.
Mututy: *Pterocarpus sp.; Papilionac.;* Bras.
„ grand bois: *Pterocarpus draco* L.; *Papilionac.;* n. Bras. (Amaz.)
„ da terra firma: *Pterocarpus sp.; Papilionac.;* Bras.
„ „ varzea: *Pterocarpus sp.; Papilionac.;* Bras.
Mutwinda: *Myristica irya* Gaertn.; *Myristicac.;* Burma.
Mu-u: *Calophyllum inophyllum* L.; *Guttifer.;* Annam.
Muui: *Chlorophora excelsa* Bth. et Hook.; *Morac.;* Elf.
Muvadju: *? Irvingia klainei* Pierre; *Simarubac.;* Gab.
Muvaga: *Panda oleosa* Pierre; *Pandac.;* Gab.
Muvamba: *Panda oleosa* Pierre; *Pandac.;* Gab.
Muvandji: *Pentaclethra macrophylla* Bth.; *Mimosac.;* Gab.
Muvanga: *Afrormosia angolensis* Harms; *Papilionac.;* D.-O.-Af.
Muvayu: *? Irvingia klainei* Pierre; *Simarubac.;* Gab.
Muvëi: *Elaeis guineensis* Jacq. *var.* ?; *Palmac.;* Gab.
Muvèndji: *Pentaclethra macrophylla* Bth.; *Mimosac.;* Gab.
Muvengea: *Podocarpus gracilior* Pilger; *Conifer.;* br. O.-Af. (Kenya)
Muvoga: *Poga oleosa* Pierre; *Rhizophorac.;* Gab.
Muvule: *Chlorophora excelsa* Bth. et Hook.; *Morac.;* br. O.-Af. (Uganda)
„ : *Sterculia sp.; Sterculiac.;* br. O.-Af. (Uganda)
Muvumu: *Erythrophloeum guineense* G. Don; *Caesalpiniac.;* br. O.-Af. (Uganda)
Muwafu: *Canarium schweinfurthi* Engl.; *Burserac.;* br. O.-Af. (Uganda)
Muwala: *Garcinia klaineana* Pierre; *Guttifer.;* Gab.
Muwanika: *Dichrostachys glomerata* Hutch. et Dalz.; *Mimosac.;* br. O.-Af. (Uganda)
Muwawa: *Acacia sieberiana* DC.; *Mimosac.;* br. O.-Af. (Uganda)
Muwogi: *Arduina edulis* K. Schum.; *Apocynac.;* br. O.-Af. (Uganda)
Muwondji: *Grewia coriacea* Mast.; *Tiliac.;* Gab.
Muwumbo: *Odina barteri* Oliv.; *Anacardiac.;* br. O.-Af. (Uganda)
Muy: *Achras sapota* L.; *Sapotac.;* Guat.
„ : *Aglaia odorata* Loureiro *var. repanensis* ?; *Meliac.;* Ind.-Ch.
Muyavi: *Mimusops pierreana* Engl.; *Sapotac.;* Gab.
Muyèbi: *Mimusops pierreana* Engl.; *Sapotac.;* Gab.
Muyonza: *Arduina edulis* K. Schum.; *Apocynac.;* br. O.-Af. (Uganda)
Muyovu: *Entandrophragma spp.; Meliac.;* br. O.-Af. (Uganda)
Muyozapot: *Achras sapota* L.; *Sapotac.;* Salv.
Muyunyui: *Podocarpus milanjianus* Rendle; *Conifer.;* br. O.-Af. (Uganda)
Muzaiti: *Ocotea usambarensis* Engl.; *Laurac.;* br. O.-Af. (Kenya)
Muzanvuma: *Sapium mannianum* Bth.; *Euphorbiac.;* br. O.-Af. (Uganda)
Muzaule: *Copaifera coleosperma* Bth.; *Caesalpiniac.;* n. Rhod.
Muziga: *Warburgia ugandensis* Sprague; *Canellac.;* br. O.-Af. (Kenya)
Muzinda: *Treculia africana* Dcne.; *Morac.;* br. O.-Af. (Uganda)
Muziru: *Pseudospondias microcarpa* Engl.; *Anacardiac.;* br. O.-Af. (Uganda)
„ : *Sorindeia sp.; Anacardiac.;* br. O.-Af. (Uganda)
Muzu: *Toddalia nobilis* Hook. f.; *Rutac.;* br. O.-Af. (Uganda)
Muzwamaloa: *Pterocarpus angolensis* DC.; *Papilionac.;* n. Rhod.
Mvagasi: *Pterocarpus angolensis* DC.; *Papilionac.;* s. Rhod.
Mvakfina: *Diospyros aggregata* Gürke; *Ebenac.;* Gab.
Mvala: *Pentaclethra macrophylla* Bth.; *Mimosac.;* Gab.
Mvambe: *Olinia usambarensis* Gilg; *Oliniac.;* D.-O.-Af. (w. Usambara)
Mvan: *Sterculia oblonga* Mast.; *Sterculiac.;* Kam.
Mvana: *Hylodendron gabunense* Taub.; *Caesalpiniac.;* Gab.
Mvane: *Hylodendron gabunense* Taub.; *Caesalpiniac.;* Gab.
M'vans: *Pentaclethra macrophylla* Bth.; *Mimosac.;* Gab. (Loango)
M'vanza: *Pentaclethra macrophylla* Bth.; *Mimosac.;* B.-K. (Mayombé)
Mvarfina: *Diospyros aggregata* Gürke; *Ebenac.;* Gab.
M'vé: *Haronga madagascariensis* Choisy; *Guttifer.;* Kam.
Mvénézo: *Homalium macropterum* Gilg; *Flacourtiac.;* Gab.
Mvenzi: *Scorodophloeus sp.; Caesalpiniac.;* B.-K.
Mviè: *Sclerosperma walkeri* Chev.; *Palmac.;* Gab.
Mvindo: *Vitex pachyphylla* Baker; *Verbenac.;* Gab. (Loango)

M'vindou: *Vitex pachyphylla* Baker; *Verbenac.;* Gab.
Mvindya: *Casuarina equisetifolia* Forst.; *Casuarinac.;* Tr. (? O.-Af.)
Mvinje: *Casuarina equisetifolia* Forst.; *Casuarinac.;* D.-O.-Af.
Mviru: *Vangueria edulis* Vahl; *Rubiac.;* D.-O.-Af. (w. Usambara)
Mvirumwitu: *Plectronia siebenlisti* Krause; *Rubiac.;* D.-O.-Af. (w. Usambara)
Mvoka: *Poga oleosa* Pierre; *Rhizophorac.;* Gab.
Mvoma: *Xylopia striata* Engl.; *Anonac.;* Kam.
Mvomba: *Xylopia acutiflora* A. Rich.; *Anonac.;* Kam.
M'voulé: *Vitex grandifolia* Gürke; *Verbenac.;* Kam.
M'vouley: *Spathodea campanulata* Beauv.; *Bignoniac.;* Kam.
M'voum: *Ceiba pentandra* Gaertn.; *Bombacac.;* Kam.
M'vouma: *Ceiba pentandra* Gaertn.; *Bombacac.;* Kam.
Mvout: *Trichoscypha sp.; Anacardiac.;* Gab.
Mvouta: *Trichoscypha sp.; Anacardiac.;* Gab.
Mvuku: *Mitragyne macrophylla* Hiern; *Rubiac.;* B.-K.
Mvule: *Chlorophora excelsa* Bth. et Hook.; *Morac.;* O.-Af. (Tanganjika, Uganda)
Mvuli: *Chlorophora excelsa* Bth. et Hook.; *Morac.;* s. C.-Af.
Mvunguti: *Kigelia pinnata* DC.; *Bignoniac.;* br. O.-Af. (Niassa)
Mvutiruánda: *Nuxia siebenlisti* Gilg; *Loganiac.;* D.-O.-Af. (w. Usambara)
Mwabi: *Sterculia quinqueloba* K. Schum.; *Sterculiac.;* B.-K. (Katanga)
Mwabvi: *Erythrophloeum guineense* G. Don; *Caesalpiniac.;* br. O.-Af. (Niassa)
Mwafi: *Erythrophloeum guineense* G. Don; *Caesalpiniac.;* B.-K. (Katanga), n. Rhod.
Mwafu: *Canarium schweinfurthi* Engl.; *Burserac.;* br. O.-Af. (Uganda)
Mwaganza: *Neoboutonia macrocalyx* Pax, *N. micelli* ?; *Euphorbiac.;* br. O.-Af.
Mwaifi: *Erythrophloeum guineense* G. Don; *Caesalpiniac.;* B.-K.   [(Uganda)
Mwamba: *Myrianthus holsti* Engl.; *Morac.;* B.-K. (Mohanga)
    „   : *Myrianthus arborea* Beauv.; *Morac.;* B.-K. (Beni, Semliki)
Mwandâma: *Catha edulis* Forsk.; *Celastrac.;* D.-O.-Af. (w. Usambara)
    „   : *Salicia sp.; Celastrac.;* D.-O.-Af.
Mwandara: *Ptaeroxylon obliquum* Radlk.; *Sapindac.;* D.-O.-Af. (w. Usambara)
Mwande: *Afzelia quanzensis* Welw.; *Caesalpiniac.;* n. Rhod.
Mwanga: *Afrormosia angolensis* Harms; *Papilionac.;* s. C.-Af.
Mwangula: *Pterocarpus stevensoni* Burtt Davy; *Papilionac.;* n. Rhod.
Mwangura: *Pterocarpus stevensoni* Burtt Davy; *Papilionac.;* n. Rhod., s. C.-Af.
Mwani: *Copaifera mopane* J. Kirk; *Caesalpiniac.;* n. Rhod.
Mwansabuso: *Burkea africana* Hook.; *Caesalpiniac.;* n. Rhod.
Mwasa: *Parinarium excelsum* Sabine; *Rosac.;* br. O.-Af. (Uganda)
Mwasanyiko: *Apodytes dimidiata* E. Meyer; *Icacinac.;* D.-O.-Af. (w. Usambara)
Mwavi: *Erythroploeum guineense* G. Don; *Caesalpiniac.;* s. C.-Af.
Mwehera: *Trema guineensis* Engl.; *Morac.;* B.-K. (Beni, Semliki)
Mwei: *Celtis sp.; Ulmac.;* Kam.
Mwenya: *Adina microcephala* Hiern; *Rubiac.;* br. O.-Af. (Niassa), s. C.-Af.
Mwepo: *Podocarpus milanjianus* Rendle; *Conifer.;* br. O.-Af. (Uganda)
Mwera: *Flueggea microcarpa* Blume; *Euphorbiac.;* br. O.-Af. (Uganda)
Mweramanyo: *Acacia rehmanniana* Schinz; *Mimosac.;* br. O.-Af. (Uganda)
Mwerene: *Combretum racemosum* Beauv.; *Combretac.;* br. O.-Af. (Uganda)
Mwêve: *Mystroxylon aethiopicum* Loes.; *Celastrac.;* D.-O.-Af. (w. Usambara)
Mwié: *Celtis sp.; Ulmac.;* Kam.
Mwindo: *Vitex pachyphylla* Baker; *Verbenac.;* Gab.
Mwindou: *Vitex pachyphylla* Baker; *Verbenac.;* Gab.
Mwinja: *Casuarina equisetifolia* Forst.; *Casuarinac.;* D.-O.-Af.
Mwipe: *Podocarpus milanjianus* Rendle; *Conifer.;* B.-K.
Mwitankoko: *Psorospermum sp.; Hypericac.;* br. O.-Af. (Uganda)
Mwizi: *Pygeum africanum* Hook. f.; *Rosac.;* br. O.-Af.
Mwolola: *Entada abyssinica* Steud.; *Mimosac.;* br. O.-Af. (Uganda)
Mwule (d): *Chlorophora excelsa* Bth. et Hook.; *Morac.;* D.-O.-Af.
My: *Lysidice rhodostegia* Hance; *Caesalpiniac.;* Ind.-Ch.
    „ ap: *Columbia floribunda* Wall.; *Tiliac.;* Ind.-Ch.
Myall: *Acacia pendula* A. Cunn., *A. homalophylla* A. Cunn.; *Mimosac.;* ö. Au.
    „   : *Acacia aneura* F. v. M.; *Mimosac.;* ö. Au.
    „ , Bastard-: *Acacia falcata* Willd., *A. cunninghami* Hook.; *Mimosac.;* ö. Au.

Myaukchaw: *Homalium tomentosum* Bth.; *Flacourtiac.;* Burma.
Myauk-laung: *Artocarpus lakoocha* Roxb.; *Morac.;* Burma.
Myauklaung: *Canthium didymum* Roxb.; *Rubiac.;* Burma.
Myauklôk: *Artocarpus lakoocha* Roxb.; *Morac.;* s. Burma (s. Tenass.), And.
Myauklôkni: *Artocarpus lakoocha* Roxb.; *Morac.;* Burma (s. Tenasserim)
Myaukngo: *Homalium tomentosum* Bth.; *Flacourtiac.;* Burma.
„ : *Duabanga sonneratioides* Buch.-Ham.; *Sonneratiac.;* Burma.
Myaukseik: *Holoptelea integrifolia* Planch.; *Ulmac.;* Burma.
Myethlwa: *Kydia calycina* Roxb.; *Malvac.;* Burma.
Myinkabin: *Cynometra ramiflora* L.; *Caesalpiniac.;* Burma.
Myinva: *Dendrocalamus strictus* Nees; *Gramin.;* Burma.
My lady: *Aspidosperma megalocarpon* Muell. Arg.; *Apocynac.;* br. Hond., Guat.
„ „ , white: *Aspidosperma sp.; Apocynac.;* br. Hond.
Mylai: *Vitex altissima* L. f.; *Verbenac.;* O.-Ind. (Madras)
Myrabolan wood: *Terminalia chebula* Retz.; *Combretac.;* O.-Ind.
Myrobalan: *Prunus myrobalana* L.; *Rosac.;* As., N.-Af.
Myrole: *Vitex altissima* L. f.; *Verbenac.;* O.-Ind. (Madras)
Myrrh: *Commiphora myrrha* Engl.; *Burserac.;* n. O.-Af., Arab.
Myrsine: *Rapanea sp.; Myrsinac.;* Bah.
Myrte (d): *Myrtus nivelli* Battand. et Trab.; *Myrtac.;* n. Af. (Sahara)
„ , Gerber-: *Myrica gale* L.; *Myricac.;* n. Hem.
Myrtille (f): *Vaccinium myrtillus* L.; *Ericac.;* w. Eu.
Myrtle (e): *Myrtus communis* L.; *Myrtac.;* M.-M.
„ : *Umbellularia californica* Nutt.; *Laurac.;* w. USA.
„ : *Nothofagus cunninghami* Oerst.; *Fagac.;* ö. Au.
„ : *Diospyros pentamera* F. v. M.; *Ebenac.;* ö. Au.
„ , black: *Diospyros pentamera* F. v. M.; *Ebenac.;* ö. Au.
„ , Ceylon-: *Rhodomyrtus tomentosa* Wight; *Myrtac.;* Malay.
„ commun: *Myrtus communis* L.; *Myrtac.;* M.-M.
„ , grey: *Backhousia myrtifolia* Hook. f. et Harv.; *Myrtac.;* ö. Au.
„ , Oregon- (u): *Umbellularia californica* Nutt.; *Laurac.;* w. USA.
„ , Sea-: *Jacquinia keyensis* Mez; *Myrtac.;* tr. Am.
„ , Tasmanian: *Nothofagus cunninghami* Oerst.; *Fagac.;* Tasm.
„ , Wax-: *Myrica cerifera* L.; *Myricac.;* N.-Am.
„ , Weeping: *Eugenia ventenati* Bth.; *Myrtac.;* ö. Au.
„ , White: *Myrtus acmenioides* F. v. M., *Rhodamnia argentea* Bth.; *Myrtac.;* ö. Au.
Mzensi: *Mimusops sulcata* Engl.; *Sapotac.;* D.-O.-Af. (Usambara), Sans.
Mzeza: *Tarchonanthus camphoratus* L.; *Composit.;* D.-O.-Af. (Usambara)
? Mzimbiti: *? Androstachys johnsoni* Prain; *Euphorbiac.;* port. O.-Af.
Mzulu: *Osyris tenuifolia* Engl.; *Santalac.;* D.-O.-Af. (w. Usambara)
Mzuzu: *Acacia usambarensis* Taub.; *Mimosac.;* D.-O.-Af. (Usambara)

N

Na: *Cola vera* K. Schum.; *Sterculiac.;* Elf.
„ : *Uapaca bingervillensis* Beille; *Euphorbiac.;* Elf.
„ : *Alangium costatum* Valeton; *Cornac.;* Ind.-Ch.
„ kua: *Diospyros ebenum* Koenig; *Ebenac.;* Ind.-Ch.
Naba: *Liquidambar styraciflua* L.; *Hamamelidac.;* Mex.
Nabanche: *Bursera sp.; Burserac.;* Mex.
Nabè: *Lannea grandis* Engl.; *Anacardiac.;* Burma.
„ boumba: *Ficus sp.; Morac.;* Kam.
Nabhay: *Lannea grandis* Engl.; *Anacardiac.;* Burma.
Nabig: *Ziziphus jujuba* Lamk.; *Rhamnac.;* O.-Ind., Ch., Au.
Nabk: *Ziziphus spina-christi* Willd.; *Rhamnac.;* N.-Af. (Tunis, Tripolis)
Nabu: *Macrolobium dewevrei* de Wild.; *Caesalpiniac.;* B.-K.
Nabukwi: *Cordia abyssinica* R. Br.; *Borraginac.;* br. O.-Af. (Niassa)
Nabuli: *Gardenia ternifolia* Schum. et Thonn.; *Rubiac.;* Togo.
Nabulong: *Lagerstroemia speciosa* Pers.; *Lythrac.;* Phil. (Cagayan)
Nacapuli: *Ficus sp.; Morac.;* Mex.
Nacascol: *Caesalpinia coriaria* Willd.; *Caesalpiniac.;* Salv.

Nacascol: *Prosopis chilensis* Stuntz; *Mimosac.;* Guat.
Nacascolo: *Caesalpinia coriaria* Willd.; *Caesalpiniac.;* Salv.
Nacascolote: *Caesalpinia coriaria* Willd.; *Caesalpiniac.;* Guat.
„ : *Prosopis chilensis* Stuntz; *Mimosac.;* Guat.
Nacascolotl: *Caesalpinia coriaria* Willd.; *Caesalpiniac.;* Mex.
Nacascul: *Caesalpinia coriaria* Willd.; *Caesalpiniac.;* Mex.
Nacaspilo: *Inga sp.; Mimosac.;* Salv.
Nacaxtle: *Enterolobium cyclocarpum* Griseb.; *Mimosac.;* Mex.
Nacazle: *Enterolobium cyclocarpum* Griseb.; *Mimosac.;* Mex.
Nacescul: *Caesalpinia coriaria* Willd.; *Caesalpiniac.;* Mex.
Nachal: *Salix tetrasperma* Roxb.; *Salicac.;* w. Beng. (Bihar)
Naga: *Eugenia jambolana* Lamk.; *Myrtac.;* sw. O.-Ind. (Malabar)
„ : *Pterocarpus blancoi* Merr., *P. echinatus* Pers.; *Papilionac.;* Phil.
„ : *Pterocarpus indicus* Willd.; *Papilionac.;* Phil.
„ golunga: *Murraya exotica* L.; *Rutac.;* O.-Ind. (Madras)
„ kesara: *Mesua ferrea* L.; *Guttifer.;* sw. O.-Ind. (Madras)
„ -sampighe: *Mesua ferrea* L.; *Guttifer.;* w. O.-Ind. (Deccan)
Nagaba-inubiwa (jap.): *Ficus caudato-longifolia* Sata; *Morac.;* Form.
„ -sii-no-ki (jap.): *Shiia longicaudata* Kudo et Masamuna; *Fagac.;* Form.
Nagae-gasi (jap.): *Quercus pachyloma* O. Seem.; *Fagac.;* Form.
Nagaed wood (u): *Dalbergia sp.; Papilionac.;* br. Hond.
Nagal: *Premna tomentosa* Willd.; *Verbenac.;* Ceylon.
Nagari: *Calophyllum tomentosum* Wight; *Guttifer.;* sw. O.-Ind. (Bombay)
Nagarunga: *Citrus aurantium* L.; *Rubiac.;* br. O.-Ind.
Nagasampige: *Mesua ferrea* L.; *Guttifer.;* w. O.-Ind. (Bombay)
Nagasari: *Mesua ferrea* L.; *Guttifer.;* O.-Ind. (Hind.)
Nagasholz (d): *Mesua ferrea* L.; *Guttifer.;* O.-Ind.
Nagchampa: *Mesua ferrea* L.; *Guttifer.;* sw. O.-Ind. (Bombay)
Nagchumpa: *Mesua ferrea* L.; *Guttifer.;* O.-Ind. (Hind.)
Nagesar: *Mesua ferrea* L.; *Guttifer.;* Beng.
Nageshvaro: *Mesua ferrea* L.; *Guttifer.;* w. Beng. (Orissa)
Naggara: *Elaeocarpus tuberculatus* Roxb.; *Elaeocarpac.;* sw. O.-Ind. (Malabar)
Naghas: *Mesua ferrea* L.; *Guttifer.;* Co.-Ch.
Naghubó: *Lagerstroemia piriformis* Koehne; *Lecythidac.;* Phil. (Laguna)
„ : *Terminalia comintana* Merr.; *Combretac.;* Phil. (Rizal)
Nagi (jap.): *Podocarpus nagi* Pilger; *Conifer.;* Jap.
Nagkaksur: *Mesua ferrea* L.; *Guttifer.;* O.-Ind. (Hind.)
Nagkesar: *Mesua ferrea* L.; *Guttifer.;* Beng.
Nagosse: *Terminalia sp.; Combretac.;* fr. Gu.
Nagossi: *Terminalia tanibouca* Rich.; *Combretac.;* fr. Gu.
Nagre hoedoe: *Acrodiclidium canella* Mez, *A. chrysophyllum* Meissn.; *Laurac.;* Sur.
„ -pisi: *Nectandra sp., Ocotea sp.; Laurac.;* Sur.
Nagrepisie: *Nectandra sp., Ocotea sp.; Laurac.;* Sur.
Nagrunga: *Citrus aurantium* L.; *Rubiac.;* br. O.-Ind.
Nagthada: *Erythrina suberosa* Roxb.; *Papilionac.;* C.-O.-Ind.
Nagudi: *Uapaca togoënsis* Pax; *Euphorbiac.;* Togo.
Nagusip: *Beilschmiedia cairocan* Vidal; *Laurac.;* Phil. (Antigue Prov.)
Naharu: *Terminalia sp.; Combretac.;* br. Gu.
Nahor: *Mesua ferrea* L.; *Guttifer.;* Assam.
Nai: *Dialium cochinchinense* Pierre; *Caesalpiniac.;* Ind.-Ch.
„ : *Viburnum inopinatum* Craib, *V. colebrookeanum* Wall.; *Caprifoliac.;* Ind.-Ch.
„ -tek: *Dillenia pentagyna* Roxb.; *Dilleniac.;* w. O.-Ind.
Naibela: *Limonia acidissima* L.; *Rutac.;* w. O.-Ind. (sw. Deccan)
Naibullal: *Limonia acidissima* L.; *Rutac.;* w. O.-Ind. (sw. Deccan)
Naida: *Sloanea dentata* L.; *Elaeocarpac.;* Gu.
Naindi: *Eugenia jambolana* Lamk.; *Myrtac.;* C.-O.-Ind. (Gondwana)
Nainvi: *? Piptadenia africana* Hook. f.; *Mimosac.;* Elf.
Naio: *Myoporum sandwicense* A. Gray; *Myoporac.;* Südsee (Hawaii)
Nakai-gasi (jap.): *Lithocarpus nakaii* Hayata; *Fagac.;* Form.
Nakasanga: *Terminalia sericea* Burch.; *Combretac.;* n. Rhod.
Naked wood: *Colubrina ferruginosa* Brongn., *C. reclinata* Brongn.; *Rhamnac.;* s. Fl.,

Naked wood: *Eugenia dicrana* Berg; *Myrtac.;* s. Fl., Bah., Ant.   [Bah., W.-Ind.
Nakka-renu: *Artocarpus lakoocha* Roxb.; *Morac.;* sw. O.-Ind. (Madras)
Nakkeri: *Cordia myxa* L.; *Borraginac.;* O.-Ind. (Madras)
Nakkini: *Canthium didymum* Roxb.; *Rubiac.;* O.-Ind. (Madras)
Nakla: *Phoenix dactylifera* L.; *Palmac.;* n. Sudan.
Naku: *Mesua ferrea* L.; *Guttifer.;* sw. O.-Ind.
Nakulsi: *Polyalthia cerasoides* Bth. et Hook. f.; *Anonac.;* O.-Ind. (Madras)
Nal-valanga: *Dalbergia lanceolaria* L.; *Papilionac.;* O.-Ind. (Madras)
Nála: *Pterocarpus blancoi* Merr., *P. echinatus* Pers.; *Papilionac.;* Phil. (Mindanao)
„ : *Pterocarpus indicus* Willd.; *Papilionac.;* Phil. (Mindanao)
Nali: *Cocos nucifera* L.; *Palmac.;* O.-Ind.
„ : *Holoptelea integrifolia* Planch.; *Ulmac.;* O.-Ind. (Madras)
„ : *Landolphia leonensis* Stapf; *Apocynac.;* Lib.
Nalla balasu: *Canthium didymum* Roxb.; *Rubiac.;* O.-Ind. (Madras)
„ duduga: *Miliusa velutina* Hook. f. et Thoms.; *Anonac.;* O.-Ind. (n. Madras)
„ maddi: *Terminalia tomentosa* Wight et Arn.; *Combretac.;* O.-Ind. (Madras)
„ renga: *Albizzia amara* Boivin; *Mimosac.;* O.-Ind. (Madras)
„ tuma: *Acacia arabica* Willd.; *Mimosac.;* O.-Ind. (Madras)
Nalli: *Phyllanthus emblica* L.; *Euphorbiac.;* C.-O.-Ind. (Gondwana)
Nalluti: *Diospyros ebenum* Koenig; *Ebenac.;* s. O.-Ind. (Madras)
Nam kiêng: *Melanorrhoea laccifera* Pierre; *Anacardiac.;* Ind.-Ch.
Nam-nu: *Randia dumetorum* Lamk.; *Rubiac.;* Burma.
Nambar: *Dalbergia retusa* Hemsl., *D. hypoleuca* Pitt., *D. granadillo* Pitt.; *Papilionac.;*
„ bastardo: *Platymiscium sp.; Papilionac.;* Nic.   [C.-R.
„ legitimo: *Dalbergia retusa* Hemsl., *D. spp.; Papilionac.;* Nic.
„ de agui: *Dalbergia retusa* Hemsl.; *Papilionac.;* C.-R.
Nambyong: *Morus laevigata* Wall.; *Morac.;* nö. O.-Ind. (Nepal, Sikkim, Bhutan)
Namijin kadai: *Lophira alata* Banks; *Ochnac.;* n. Nig.
„ kade: *Lophira alata* Banks; *Ochnac.;* Nig.
„ kiriya: *Burkea africana* Hook.; *Caesalpiniac.;* n. Nig.
„ tsada: *Gymnosporia senegalensis* Loes.; *Celastrac.;* n. Nig.
Namli: *Holoptelea integrifolia* Planch.; *Ulmac.;* O.-Ind. (Madras)
Namme: *Anogeissus latifolia* Wall.; *Combretac.;* O.-Ind. (Madras)
Nam-nam: *Cynometra cauliflora* L.; *Caesalpiniac.;* Malay.
Namoek: *Hemicyclia serrata* J. J. Smith; *Euphorbiac.;* Java (Tjemara)
Nampelas: *Ficus pisifera* Wall.; *Morac.;* Java (? Pantjoer)
Namprane: *Gardenia ternifolia* Schum. et Thonn.; *Rubiac.;* Go.
Namta-yok: *Altingia excelsa* Noronha; *Hamamelidac.;* Java.
Namú: *Tobagoa maleolens* Urban; *Rubiac.;* Ven.
Namukago: *Funtumia latifolia* Stapf; *Apocynac.;* br. O.-Af. (Uganda)
Namukagu: *Funtumia elastica* Stapf, *F. latifolia* Stapf; *Apocynac.;* br. O.-Af. (Uganda)
„ : *Conopharyngia holsti* Stapf; *Apocynac.;* br. O.-Af. (Uganda)
„ , broad-leaved: *Funtumia latifolia* Stapf; *Apocynac.;* br. O.-Af. (Uganda)
Nan: *Uapaca bingervillensis* Beille; *Euphorbiac.;* Elf.
Nan-mu: *Persea nan-muh* Oliv.; *Laurac.;* Ch.
Nan-mu-hua: *Phoebe sp.; Laurac.;* Ch.
Nan-ta-yoh: *Altingia excelsa* Noronha; *Hamamelidac.;* Burma.
Ñaña: *Lumnitzera littorea* Voigt; *Combretac.;* Südsee (Guam)
Nana (H): *Lagerstroemia lanceolata* Wall.; *Lythrac.;* sw. O.-Ind. (Bombay)
Nanakamado: *Pirus aucuparia* Gaertn. *var. japonica* Thunb.; *Rosac.;* Jap.
Nanamenoki: *Ilex oldhami* Miq.; *Aquifoliac.;* Jap.
Nananche: *Byrsonima crassifolia* H. B. K.; *Malphighiac.;* Mex.
Nanari: *Canarium commune* L.; *Burserac.;* Nied.-Ind. (Banda)
Nanary: *Canarium silvestre* Gaertn.; *Burserac.;* Nied.-Ind.
Nanban-gasi (jap.): *Lithocarpus ternaticupula* Hay.; *Fagac.;* Form.
„ -inumaki (jap.): *Podocarpus philippinensis* Foxw.; *Conifer.;* Form.
Nanca: *Artocarpus integrifolia* L. f.; *Morac.;* O.-Ind., Malay.
Nance: *Byrsonima crassifolia* H. B. K.; *Malpighiac.;* Guat., Hond.
„ : *Malpighia punicifolia* L.; *Malpighiac.;* Salv.
„ : *Clethra lanata* Mart. et Gal.; *Clethrac.;* C.-R.
„ blanco: *Byrsonima crassifolia* H. B. K.; *Malpighiac.;* Pan.

Nance colorado: *Byrsonima crassifolia* H. B. K.; *Malpighiac.;* Pan.
„ verde: *Byrsonima crassifolia* H. B. K.; *Malpighiac.;* Salv.
Nancenes: *Malpighia glabra* L.; *Malpighiac.;* Mex.
Nanche: *Byrsonima crassifolia* H. B. K.; *Malpighiac.;* s. Mex.
„ agrio: *Byrsonima crassifolia* H. B. K.; *Malpighiac.;* Mex.
„ de perro: *Byrsonima crassifolia* H. B. K.; *Malpighiac.;* Mex.
Nanchi: *Byrsonima crassifolia* H. B. K.; *Malpighiac.;* Mex.
Nanci: *Byrsonima crassifolia* H. B. K.; *Malpighiac.;* Kol.
Nancite: *Byrsonima crassifolia* H. B. K.; *Malpighiac.;* Salv.
Nancito: *Byrsonima crassifolia* H. B. K.; *Malpighiac.;* Hond.
„ : *Hieronyma alchorneoides* Allem.; *Euphorbiac.;* Nic.
Nandi: *Lagerstroemia parviflora* Roxb., *L. lanceolata* Wall.; *Lythrac.;* s. O.-Ind.
Nandipá: *Genipa americana* L.; *Rubiac.;* Arg.
Nandiroba: *Carapa guianensis* Aubl.; *Meliac.;* Bras.
Nandirova: *Carapa guianensis* Aubl.; *Meliac.;* Bras.
Ñandubay: *Prosopis ñandubay* Lorentz; *Mimosac.;* Arg.
? Nanelipa: *Genipa brasiliensis* Baill.; *Rubiac.;* Bras.
Nang: *Sapium baccatum* Roxb.; *Euphorbiac.;* Ind.-Ch.
„ gia: *Aglaia cochinchinensis* Pellegrin; *Meliac.;* Ind.-Ch.
„ kiêng: *Melanorrhoea laccifera* Pierre; *Anacardiac.;* Ind.-Ch.
„ mon: *Dalbergia robusta* Bth.; *Papilionac.;* Ind.-Ch.
Nangal: *Mesua ferrea* L.; *Guttifer.;* sw. O.-Ind. (Madras)
Ñangapire: *Eugenia uniflora* L.; *Melastomatac.;* Arg.
Nangha: *Artocarpus integrifolia* L. f.; *Morac.;* Malak., Phil.
Nangka: *Artocarpus integrifolia* L. f., *A. communis* Forst.; *Morac.;* Malay, Phil.,
„ -an: *Artocarpus gomeziana* Wall.; *Morac.;* Java.   [Nied.-Ind.
„ -blanda: *Anona muricata* L.; *Anonac.;* Java.
„ -landa: *Anona muricata* L.; *Anonac.;* Java.
„ -landi: *Anona muricata* L.; *Anonac.;* Java.
„ -manila: *Anona muricata* L.; *Anonac.;* Java.
„ -nangká: *Sideroxylon ferrugineum* Hook. et Arn.; *Sapotac.;* Phil. (Basilan)
„ -sabrang: *Anona muricata* L.; *Anonac.;* Java.
„ -walanda: *Anona muricata* L.; *Anonac.;* Java.
Nangocy: *Terminalia sp.;* *Combretac.;* fr. Gu.
Nangsi: *Boehmeria pilosiuscula* Hassk., *Villebrunea scabra* Wedd.; *Urticac.;* Java.
„ areuj: *Pipturus repandus* Wedd.; *Urticac.;* Java (Bobodjong)
Nangu: *Mesua ferrea* L.; *Guttifer.;* sw. O.-Ind. (Madras)
Nangwesu: *Brachystegia longifolia* Bth.; *Caesalpiniac.;* br. O.-Af. (Niassa)
Nanhaolat: *Grewia vestita* Wall.; *Tiliac.;* w. Beng.
Nani: *Metrosideros vera* Lindl.; *Myrtac.;* Malay.
„ ayer: *Gordonia sp.;* *Theac.;* Amboina.
Nanka-nanká: *Sideroxylon ferrugineum* Hook. ct Arn.; *Sapotac.;* Phil. (Basilan)
Nannal: *Bischofia javanica* Blume; *Euphorbiac.;* sw. O.-Ind. (Trav. Hills)
Nanny berries: *Mallotus japonicus* Muell. Arg.; *Euphorbiac.;* Japan.
Nansali: *Landolphia dawei* Stapf; *Apocynac.;* br. O.-Af. (Uganda)
Nantayok: *Altingia excelsa* Noronha; *Hamamelidac.;* Burma (s. Tenasserim)
Nantô-gasi (jap.): *Lithocarpus nantoënsis* Hayata; *Fagac.;* Form.
Nantzinxocotl: *Byrsonima crassifolia* H. B. K.; *Malpighiac.;* Mex.
Nanzinquahuitl: *Byrsonima crassifolia* H. B. K.; *Malpighiac.;* Mex.
Nao: *Chukrasia sp.;* *Meliac.;* Ind.-Ch.
Náp: *Unona debilis* Pierre; *Anonac.;* Annam.
„ nui: *Unona jucunda* Pierre; *Anonac.;* Annam.
Naparli: *Detarium microcarpum* Guill. et Perr.; *Caesalpiniac.;* Togo.
Ñapindá: *Acacia bonariensis* Gill.; *Mimosac.;* Arg.
„ hú: *Acacia paniculata* Willd.; *Mimosac.;* Arg.
„ negro: *Acacia paniculata* Willd.; *Mimosac.;* Arg.
Napini: *Terminalia sericea* Burch.; *Combretac.;* n. Rhod.
Naprámpogo: *Strychnos laxa* Solereder; *Loganiac.;* Togo.
Nar-yepi: *Hardwickia binata* Roxb.; *Caesalpiniac.;* O.-Ind. (Madras)
Nára: *Pterocarpus blancoi* Merr., *P. spp.;* *Papilionac.;* Phil. (n. & C.-Luzon, Mindoro,
Nara: *Quercus glandulifera* Blume, *Q. crispula* Blume; *Fagac.;* Jap.   [Masbate]

Nara noki: *Quercus glandulifera* Blume; *Fagac.;* Jap.
Narala: *Eugenia jambolana* Lamk.; *Myrtac.;* w. O.-Ind. (sw. Deccan)
Narang-dauél: *Cassia javanica* L.; *Caesalpiniac.;* Phil. (Ilocos Sur)
  „ „ : *Wallaceodendron celebicum* Koord.; *Mimosac.;* Phil. (Cagayan)
Naranja: *Tapura cubensis* Griseb.; *Dichapetalac.;* Cuba.
  „ : *Citrus sinensis* Osbeck; *Rutac.;* Hond.
  „ ácida: *Citrus aurantium* L.; *Rutac.;* Hond.
  „ agria: *Citrus bigaradia* Duham.; *Rutac.;* Salv.
  „ amarga: *Citrus aurantium* L.; *Rutac.;* Arg.
  „ dulce: *Citrus sinensis* Osbeck; *Rutac.;* Hond.
  „ silvestre: *Citrus bigaradia* Duham.; *Rutac.;* Sp.
  „ vigueta: *Tapura cubensis* Griseb.; *Dichapetalac.;* Cuba.
? Naranjillo: *Caesearia praecox* Griseb.; *Flacourtiac.;* W.-Ind.
Naranjillo: *Swartzia darienensis* Pittier; *Caesalpiniac.;* Hond., Pan.
  „ : *Trichanthera gigantea* H. B. K.; *Acanthac.;* Ven.
  „ : *Crataeva tapia* L.; *Capparidac.;* Kol.
  „ : *Aspidosperma vargasi* A. DC.; *Apocynac.;* Ven.
  „ : *Eschweilera moritziana* Miers; *Lecythidac.;* Ven.
  „ : *Eugenia grandiflora* Berg; *Myrtac.;* Ven.
  „ : *Zanthoxylum naranjillo* Griseb.; *Rutac.;* Arg., Par.
  „ bobo: *Bravaisia floribunda* DC.; *Acanthac.;* Ven.
  „ bravo: *Esenbeckia febrifuga* A. Juss.; *Rutac.;* Bras., Arg.
  „ colorado: *Heisteria macrophylla* Oerst.; *Olacac.;* Pan.
  „ del campo: *? Zanthoxylum sp.; Rutac.;* Arg.
Naranjito: *Swartzia sp.; Caesalpiniac.;* Mex.
  „ : *Crataeva tapia* L.; *Capparidac.;* n. Kol.
  „ : *Amyris sp.; Rutac.;* Kol.
  „ : *Maytenus boaria* Molina; *Celastrac.;* Arg.
Naranjo: *Citrus aurantium* L., *C. bigaradia* Duham.; *Rutac.;* P.-R.
  „ : *Terminalia obovata* Eichl.; *Combretac.;* Guat.
  „ : *Bumelia nicaraguensis* Loes.; *Sapotac.;* Nic.
  „ agrio: *Citrus bigaradia* Duham.; *Rutac.;* Arg.
  „ amargo: *Citrus bigaradia* Duham.; *Rutac.;* Arg.
  „ dulce: *Citrus aurantium* L.; *Rutac.;* tr. Am.
Naranjuelo: *Ilex sp.; Aquifoliac.;* Kol.
Narave: *Crataeva religiosa* Forst.; *Capparidac.;* Südsee (Gesellsch. Ins.)
Narek: *Balanocarpus cagayanensis* Foxw.; *Dipterocarpac.;* Phil.
Narel: *Cocos nucifera* L.; *Palmac.;* O.-Ind. (Hind.)
Narendj: *Citrus aurantium* L.; *Rutac.;* Pers.
Narenga: *Vitex cienkowskyi* Kotschy et Peyr.; *Verbenac.;* Togo.
Nargil: *Cocos nucifera* L.; *Palmac.;* Pers.
Narguni: *Atalantia monophylla* Correa; *Rutac.;* w. Beng. (Orissa)
Nargusta: *Terminalia obovata* Eichl., *T. hayesi* Pittier; *Combretac.;* br. Hond.
Nari: *Cocos nucifera* L.; *Palmac.;* O.-Ind.
Nariel: *Cocos nucifera* L.; *Palmac.;* Beng.
Narig: *Vatica spp., Cotylelobium spp.; Dipterocarpac.;* tr. As.
  „ : *Vatica mangachapoi* Blco.; *Dipterocarpac.;* Phil. (Zamb., Cotabato)
Nárik: *Vatica mangachapoi* Blco.; *Dipterocarpac.;* Phil. (Ilocos Sur, Cagayan)
Narikeleekool: *Zizyphus jujuba* Lamk.; *Rhamnac.;* Beng.
Narikera: *Cocos nucifera* L.; *Palmac.;* ? O.-Ind.
Nariyal: *Cocos nucifera* L.; *Palmac.;* O.-Ind. (Hind.)
Narlei: *Tamarix articulata* Vahl; *Tamaricac.;* O.-Ind.
Narock-pa: *Canarium sikkimense* King; *Burserac.;* nö. O.-Ind. (Nepal, Sikkim, Bhotan)
Narra: *Pterocarpus indicus* Willd., *P. echinatus* Pers.; *Papilionac.;* Phil.
  „ : *Dipterocarpus sp.; Dipterocarpac.;* Phil.
  „ Alagi: *Litsea sebifera* Pers.; *Laurac.;* O.-Ind. (Madras)
  „ , Blanco's: *Pterocarpus blancoi* Merr.; *Papilionac.;* Phil.
  „ , Prickly: *Pterocarpus echinatus* Pers.; *Papilionac.;* Phil.
  „ , Red: *Pterocarpus indicus* Willd.; *Papilionac.;* Phil.
Nartem-marom: *Citrus bigaradia* Duham.; *Rutac.;* w. O.-Ind. (Deccan)
Narungee: *Citrus aurantium* L.; *Rutac.;* O.-Ind. (Hind.)

Narunj: *Citrus aurantium* L.; *Rutac.;* Arab.
Nasamba: *Pycnanthus kombo* Warb.; *Myristicac.;* Kam.
Naseberry: *Achras sapota* L.; *Sapotac.;* W.-Ind., C.-Am., n. S.-Am.
Nashi: *Pirus sinensis* Link; *Rosac.;* s. Jap.
Nassarli: *Gardenia ternifolia* Schum. et Thonn.; *Rubiac.;* Togo.
Nasut: *Erythrina suberosa* Roxb.; *Papilionac.;* n. O.-Ind. (United Prov.)
Nat vadom: *Terminalia catappa* L.; *Combretac.;* O.-Ind. (Madras)
Nathubó: *Lagerstroemia piriformis* Koehne; *Lythrac.;* Phil. (Laguna)
Natilég: *Quercus sp.; Fagac.;* Phil. (Benguet)
Natki-mokha: *Schrebera swietenioides* Roxb.; *Oleac.;* w. O.-Ind. (Vindhja)
Nato: *Dimorphandra oleifera* Triana; *Caesalpiniac.;* Kol.
   „ : *Mimusops commersoni* Engl.; *Sapotac.;* nw. Mad.
   „ : *Imbricaria coriacea* A. DC.; *Sapotac.;* Mad.
   „ : *Faucherea hexandra* H. Lec.; *Sapotac.;* Mad.
   „ : *Sideroxylon rubrocostatum* Jum. et Perr.; *Sapotac.;* Mad. (Bassindu)
   „ : *Palaquium barnesi* Merr., *P. luzoniense* Vidal; *Sapotac.;* Phil.
   „ : *Sideroxylon duclitan* Blco.; *Sapotac.;* Phil. (Tayabas)
   „ : *Myoporum sandwicense* A. Gray; *Myoporac.;* Südsee (Hawaii)
   „ lahy: *Calophyllum sp.; Guttifer.;* Mad.
   „   „ : *Faucherea sp.; Sapotac.;* Mad.
   „ putí: *Sideroxylon macranthum* Merr.; *Sapotac.;* Phil. (Tayabas)
   „ , white: *Sideroxylon macranthum* Merr.; *Sapotac.;* Phil. (Tayabas)
Natsoumé: *Zizyphus vulgaris* Lamk.; *Rhamnac.;* Jap.
Natsume: *Zizyphus vulgaris* Lamk.; *Rhamnac.;* Jap.
Natsutsubaki: *Stewartia pseudocamellia* Maxim.; *Theac.;* Jap.
Natte: *Imbricaria maxima* Poir.; *Sapotac.;* Maur.
   „ , grand: *Imbricaria maxima* Poir.; *Sapotac.;* Réunion.
   „ legentil: *Labourdonnaisia calophylloides* Bojer; *Sapotac.;* Maur.
   „ , petite: *Imbricaria petiolaris* A. DC.; *Sapotac.;* Réunion.
Nattier: *Imbricaria maxima* Poir.; *Sapotac.;* Maur.
Nau: *Alangium costatum* Wang.; *Cornac.;* Ind.-Ch.
   „ : *Aegle marmelos* Correa; *Rutac.;* Annam.
? Nauelipa: *Genipa brasiliensis* Baill.; *Rubiac.;* Bras.
Nauladimara: *Vitex altissima* L. f.; *Verbenac.;* w. O.-Ind. (sw. Deccan)
Naulmitik: *Vitex altissima* L. f.; *Verbenac.;* w. O.-Ind. (sw. Deccan)
Navadi: *Elaeocarpus tuberculatus* Roxb.; *Elaeocarpac.;* sw. O.-Ind. (Travanc. Hills)
Naval: *Eugenia jambolana* Lamk.; *Myrtac.;* O.-Ind. (Madras)
Naviladi: *Vitex altissima* L. f.; *Verbenac.;* sw. O.-Ind. (Coorg)
Navili: *Holoptelea integrifolia* Planch.; *Ulmac.;* O.-Ind. (Madras)
Nawashirogumi: *Elaeagnus pungens* Thunb.; *Elaeagnac.;* Jap.
Nawi: *Myrica salicifolia* Hochst.; *Myricac.;* n. O.-Af. (Abessinien)
Nayíle: *Spondias lutea* L.; *Anacardiac.;* Togo.
Nazareno: *Brosimum sp.; Morac.;* Mex.
   „ : *Peltogyne purpurea* Pittier; *Caesalpiniac.;* Pan.
   „ : *Hymenaea floribunda* H. B. K.; *Caesalpiniac.;* Ven.
   „ : *Hymenaea courbaril* L., *Peltogyne sp.; Caesalpiniac.;* Kol.
   „ morado: *Maytenus lineata* Wright; *Celastrac.;* Cuba.
Nazuc: *Zizyphus jujuba* Lamk.; *Rhamnac.;* O.-Ind. (Hind.)
Nbalawi: *Randia dumetorum* Lamk.; *Rubiac.;* Burma.
N'bilinga: *Sarcocephalus trillesi* Pierre; *Rubiac.;* s. Kongo.
N'bondo: *Adansonia digitata* L.; *Bombacac.;* port. Af.
Nbor: *Ricinodendron africanum* Muell. Arg.; *Euphorbiac.;* Elf.
Nbounde: *Staudtia gabonensis* Warb.; *Myristicac.;* Gab.
N'cannang: *Sterculia rhinopetala* K. Schum.; *Sterculiac.;* Kam.
Nchioumbou: *Piptadenia africana* Hook. f.; *Mimosac.;* Gab.
Nchiumba: *Pentaclethra griffoniana* Baill.; *Mimosac.;* Gab.
N'coula: *Coula edulis* Baill.; *Olacac.;* Gab.
Noubé: *Hannoa klaineana* Pierre et Engl.; *Simarubac.;* Elf.
N'daale: *Swartzia madagascariensis* Desv.; *Caesalpiniac.;* B.-K.
Ndakua makandre: *Agathis vitiensis* Bth. et Hook. f.; *Conifer.;* Fid.
   „ salusalu: *Podocarpus cupressina* R. Br., *P. vitiensis* Seem.; *Conifer.;* Fid.

Ndale: *Swartzia madagascariensis* Desv.; *Caesalpiniac.*; B.-K. (Katanga) n. Rhod.
Ndamaru: *Calophyllum burmanni* Wight, *C. spectabile* Willd.; *Guttifer.*; Fid.
Ndamba: *Holarrhena wulfsbergi* Stapf; *Apocynac.*; Kam.
„ : *Blighia sapida* Koenig; *Sapindac.*; Kam.
Ndambe: *Xylocarpus sp.*; *Meliac.*; Fid.
Ndanwi-badgi: *Parinarium excelsum* Sabine; *Rosac.*; Lib.
N'dar: *Borassus aethiopum* Warb. *var. n'dar* ?; *P.almac.*; Senl.
Ndarakwa: *Juniperus procera* Hochst.; *Conifer.*; br. O.-Af. (Kenya)
Ndebera: *Stereospermum kunthianum* Cham.; *Bignoniac.*; br. O.-Af. (Uganda)
Ndébéré: *Sarcocephalus pobeguini* Pob.; *Rubiac.*; Elf.
Ndgokom: *Myrianthus arboreus* Beauv.; *Morac.*; Gab.
N'di: *Chlorophora excelsa* Bth. et Hook.; *Morac.*; Elf.
N'diago bobfouo: *Caloncoba glauca* Gilg; *Flacourtiac.*; Kam.
Ndiai: *Pachyelasma tessmanni* Harms; *Mimosac.*; Kam.
N'diar: *Uvaria parviflora* Rich.; *Anonac.*; Senl.
N'dimb: *Cola cordifolia* R. Br.; *Sterculiac.*; Senl.
N'dimba: *Cordyla africana* Lour.; *Caesalpiniac.*; Sudan.
„ : *Terminalia superba* Engl. et Diels; *Combretac.*; Kam., B.-K. (Mayombe)
Ndimbo: *Pterocarpus soyauxi* Taub.; *Papilionac.*; W.-Af.
Ndimbu-ndimbu: *Funtumia latifolia* Stapf; *Apocynac.*; B.-K.
N-dina: *Dialum sp.*; *Caesalpiniac.*; Gab.
Ndisok: *Myrianthus arboreus* Beauv.; *Morac.*; Nig.
„ : *Irvingia gabonensis* Baill.; *Simarubac.*; Nig. (Old Calabar)
N'djavé: *Mimusops djave* Engl., *M. pierreana* Engl.; *Sapotac.*; Kam., Gab.
N'djengue: *Quassia gabonensis* Pierre; *Simarubac.*; fr. Kongo (Haut Ogoué)
Ndjole: *Staudtia gabonensis* Warb.; *Myristicac.*; Gab.
N'djondjokon: *Antrocaryon klaineanum* Pierre; *Anacardiac.*; Gab.
Ndoa: *Irvingia gabonensis* Baill.; *Simarubac.*; Kam.
Ndock: *Irvingia gabonensis* Baill.; *Simarubac.*; Gab.
N'dogo: *Irvingia gabonensis* Baill.; *Simarubac.*; Nig.
N'dok: *Irvingia gabonensis* Baill.; *Simarubac.*; Gab.
Ndoka: *Irvingia gabonensis* Baill.; *Simarubac.*; Kam.
N'dola: *Khaya ivorensis* Chev.; *Meliac.*; Gab., Mitt.-Kongo.
Ndombe: *Pterocarpus angolensis* DC.; *Papilionac.*; n. Rhod.
Ndongi: *Polyscias fulva* Harms; *Araliac.*; B.-K. (Mohanga)
Ndongo: *Bruguiera gymnorrhiza* Lam.; *Rhizophorac.*; Fid.
Ndongosan: *Morus mesozygia* Stapf; *Morac.*; Elf.
Ndongué: *Rauwolfia macrophylla* Stapf; *Apocynac.*; Kam.
Ndonka: *Mimusops africana* H. Lec.; *Sapotac.*; Gab.
N'donon: *Croton oligandrum* Pierre; *Euphorbiac.*; Kam.
Ndooson: *Antrocaryon klaineanum* Pierre; *Anacardiac.*; Kam.
N'doua: *Irvingia oblonga* Chev.; *Simarubac.*; Kam.
N'douka: *Mimusops africana* H. Lec.; *Sapotac.*; Kam., Gab.
N'doum: *Ceiba pentandra* Gaertn.; *Bombacac.*; Kam.
Ndouon: *Croton sp.*; *Euphorbiac.*; Kam.
N'duka: *Mimusops africana* H. Lec.; *Sapotac.*; Kam., Gab.
Ndukwon: *Macrolobium limba* Scott Elliott; *Caesalpiniac.*; Go.
Ndulu: *Parinarium glabrum* Oliv.; *Rosac.*; B.-K.
Ndunga: *Xylopia aethiopica* A. Rich., *X. sp.*; *Anonac.*; B.-K.
Ndzara: *Uapaca guineensis* Muell. Arg.; *Euphorbiac.*; B.-K.
Neang leang: *Dalbergia lanceolaria* L.; *Papilionac.*; Ind.-Ch.
„ nuon: *Dalbergia dongnaiensis* Pierre; *Papilionac.*; Ind.-Ch.
„ nuong: *Dalbergia fusca* Pierre, *D. ? bariensis* Pierre; *Papilionac.*; Kamb.
„ paek: *Terminalia tomentosa* Wight et Arn.; *Combretac.*; Kamb.
„ phach: *Cyanodaphne cuneata* Blume; *Laurac.*; Ind.-Ch.
„ phaëc: *Terminalia tomentosa* Wight et Arn.; *Combretac.*; Ind.-Ch.
Nebga: *Ziziphus spina-christi* Willd.; *Rhamnac.*; n. Af. (Tunis, Tripolis)
Neb-neb: *Acacia arabica* Willd.; *Mimosac.*; Sudan.
Ne-bor-vah: ? *Poponia ferruginea* Engl. et Diels; *Anonac.*; Lib.
Ne-mor-ba-day: *Scottellia coriacea* Chev.; *Flacourtiac.*; Lib.
Nedun: *Pericopsis mooniana* Thw.; *Papilionac.*; Cey.

Neeboo: *Citrus limonum* Risso; *Rutac.;* Beng.
Neece: *Dialium guineense* Willd.; *Caesalpiniac.;* Nig.
Needle wood: *Schima wallichi* Choisy; *Theac.;* Assam.
„ „ : *Xylosma panamensis* Turcz.; *Flacourtiac.;* Pan.
„ „ : *Hakea leucoptera* R. Br.; *Proteac.;* ö. Au. (N.-S.-W.)
Neem: *Azadirachta indica* A. Juss.; *Meliac.;* C.-O.-Ind., Go.
Ncemeeri: *Terminalia paniculata* Roth; *Combretac.;* sw. O.-Ind. (Madras)
Neemoo: *Citrus limonum* Risso; *Rutac.;* O.-Ind. (Hind.)
Neem wood, Malabar-: *Melia composita* Willd.; *Meliac.;* O.-Ind.
Neeriya: *Elaeodendron glaucum* Pers.; *Celastrac.;* w. O.-Ind. (Deccan)
Neesberry: *Achras sapota* L.; *Sapotac.;* br. W.-Ind.
Nee-wahn-johr: *Cleistopholis patens* Bth.; *Anonac.;* Lib.
Nefle: *Parinarium campestre* Aubl.; *Rosac.;* fr. Gu.
Néflier: *Mespilus germanica* L.; *Rosac.;* ö. Eu.
„ du Japon: *Eriobotrya japonica* Lindl.; *Rosac.;* Guat.
Negra lora: *Guatteria blaini* Urban; *Ebenac.;* P.-R.
„ de Cuba: *Maga grandiflora* Urban; *Malvac.;* Cuba.
Negrillo: *Machaerium whitfordi* Macbride; *Papilionac.;* Kol.
Negrito: *Simaruba glauca* DC.; *Simarubac.;* br. Hond., Hond.
„ : *Diphysa carthaginensis* Jacq.; *Papilionac.;* Kol.
Negundo (u): *Acer negundo* L.; *Acerac.;* N.-Am.
Neh-mle-chu: *Randia acuminata* Bth.; *Rubiac.;* Lib.
Nehoumbou: *Piptadenia africana* Hook. f.; *Mimosac.;* Gab.
Nein: *Eurya japonica* Thunb.; *Theac.;* Annam.
Neinei: *Dracophyllum latifolium* A. Cunn.; *Epacridac.;* N.-Seel.
Nejoer: *Cocos nucifera* L.; *Palmac.;* Java (Batavia)
Nekazito: *Weihea eicki* Engl., *W. africana* Bth.; *Rhizophorac.;* D.-O.-Af. (w. Usamb.)
Neko-yanagi: *Salix gracilistyla* Miq.; *Salicac.;* Jap.
Nékoe: *Piscidia erythrina* L., *Lonchocarpus sp.;* *Papilionac.;* Sur.
? „ -hoedoe: *Clathrotropis brachypetala* Kleinh.; *Papilionac.;* Sur.
Nekoehoedoe: *Conceveiba guianensis* Aubl., *C. hostmanni* Bth.; *Euphorbiac.;* Sur.
Nelbé: *Diospyros mespiliformis* Hochst.; *Ebenac.;* Sudan.
Nellekai: *Phyllanthus emblica* L.; *Euphorbiac.;* ö. As.
Nelli: *Phyllanthus emblica* L.; *Euphorbiac.;* O.-Ind. (Madras)
Nellimara: *Phyllanthus emblica* L.; *Euphorbiac.;* O.-Ind. (Bombay)
Nel mal: *Strychnos potatorum* L. f.; *Loganiac.;* O.-Ind.
Nembar mohi: *Bursera serrata* Colebr.; *Burserac.;* w. Beng. (Orissa)
Nembura moi: *Bursera serrata* Colebr.; *Burserac.;* w. Beng. (Orissa)
Nemesu: *Shorea pauciflora* King, *S. sp.;* *Dipterocarpac.;* Malay.
Nemiliadage: *Vitex altissima* L. f.; *Verbenac.;* sw. O.-Ind. (Madras)
Nemoca: *Ocotea sp., Nectandra sp.;* *Laurac.;* P.-R.
Nemunoki: *Albizzia julibrissin* Durazzini; *Mimosac.;* Jap.
Neng ôi: *Eurya japonica* Thunb.; *Theac.;* Annam.
Nengre notto: *Caryocar glabrum* Pers.; *Caryocarac.;* Sur.
Neoza: *Pinus gerardiana* Wall.; *Conifer.;* O.-Ind. (nw. Himalaya)
Nerale: *Eugenia jambolana* Lamk.; *Myrtac.;* sw. O.-Ind. (Mysore)
Néré: *Parkia biglobosa* Bth.; *Mimosac.;* Elf., fr. Guin.
Nerebi: *Canarium bengalense* Roxb.; *Burserac.;* w. Assam.
Neredu: *Eugenia jambolana* Lamk.; *Myrtac.;* O.-Ind. (Madras)
Nerek: *Artocarpus cumingiana* Trécul; *Morac.;* Phil.
Nerie: *Nothofagus antarctica* Oerst.; *Fagac.;* Chile.
Nerlu: *Eugenia jambolana* Lamk.; *Myrtac.;* w. O.-Ind. (sw. Deccan)
Nermali: *Strychnos potatorum* L. f.; *Loganiac.;* w. O.-Ind. (Bombay)
Nerprun alaterne (f): *Rhamnus alaternus* L.; *Rhamnac.;* Mitt.-Eu., n. Af.
Nerpum ferrugineaux (f): *Colubrina ferruginosa* Brongn.; *Rhamnac.;* Mart.
Nerunga: *Citrus aurantium* L.; *Rutac.;* O.-Ind. (Hind.)
? Nesipolela: *Excoecaria africana* Muell. Arg.; *Euphorbiac.;* port. O.-Af. (Quel.)
Nesko: *Willardia mexicana* Rose; *Papilionac.;* w. Mex.
Nespera: *Sterculia tragacantha* Lindl.; *Sterculiac.;* port. W.-Af.
Nespero: *Sterculia tragacantha* Lindl.; *Sterculiac.;* port. W.-Af. (Angola)
Net: *Ceriops candolleana* Arn.; *Rhizophorac.;* Ind.-Ch.

Nete: *? Piptadenia sp.; Mimosac.;* Elf.
Neté: *Parkia biglobosa* Bth.; *Mimosac.;* Dah., Elf.
Nété: *Piptadenia africana* Hook. f.; *Mimosac.;* Elf.
Netra: *Pterocarpus marsupium* Roxb.; *Papilionac.;* w. O.-Ind. (sw. Deccan)
Nettle tree: *Celtis occidentalis* L.; *Ulmac.;* ö. N.-Am.
„   „   : *Celtis sinensis* Pers. *var. japonica* Nakai; *Ulmac.;* Jap.
„   „   , Jamaican: *Trema micrantha* Blume; *Ulmac.;* Jam.
„   wood: *Celtis australis* L.; *Ulmac.;* O.-Ind.
Neuguineaholz (d): *Intsia bijuga* O. Ktze.; *Caesalpiniac.;* Neuguinea, n. Südsee
Newn: *Alnus nepalensis* D. Don; *Betulac.;* nw. O.-Ind. (Punjab-Sutlej)
New Zealand wood (e): *Dracontomelum dao* Merr. et Rolfe; *Anacardiac.;* ö. As.
Ney kee (ch): *Derris elliptica* Bth.; *Papilionac.;* Singapore.
Nezuko: *Thuja japonica* Maxim.; *Conifer.;* Jap.
Nezumimochi: *Ligustrum japonicum* Thunb.; *Oleac.;* Jap.
Nezumisashi: *Juniperus rigida* S. et Z.; *Conifer.;* Jap.
N'fomba: *Alstonia congensis* Engl.; *Apocynac.;* W.-Af. (Loango, Angola)
Nga'chié: *Ceiba pentandra* Gaertn.; *Bombacac.;* Elf.
N'gaha: *Ricinodendron africanum* Muell. Arg.; *Euphorbiac.;* Kam.
Ngai sahn: *Erythrophloeum sp.; Caesalpiniac.;* Kamb.
Ngakombo: *Staudtia gabonensis* Warb.; *Myristicac.;* Gab.
Ngalati: *Brachystegia mimosaefolia* Hutch. et Burtt Davy; *Caesalpiniac.;* B.-K. (Katanga)
Ngale: *Poga oleosa* Pierre; *Rhizophorac.;* Kam.
Ngali: *Poga oleosa* Pierre; *Rhizophorac.;* Kam.
Ngamè: *Dracontomelum dao* Merr. et Rolfe; *Anacardiac.;* Cel.
Ngan: *Carapa procera* DC.; *Meliac.;* Kam., Gab.
Ngân: *Nephelium longana* Cambess.; *Sapindac.;* Ind.-Ch.
„   ngan: *Cratoxylon formosum* Bth. et Hook.; *Guttifer.;* Ind.-Ch.
„   „   : *Cratoxylon prunifolium* Dyer; *Guttifer.;* Annam.
„   ngank: *Cratoxylon polyanthemum* Korth.; *Guttifer.;* Annam.
N'ganaké: *Guarea cedrata* Pellegr.; *Meliac.;* Elf.
Ngandja: *Schumanniophytum klaineanum* Pierre; *Rubiac.;* Gab.
Ngando: *Spondias lutea* L.; *Anacardiac.;* Kam.
Ngâng-ngang-do: *Cratoxylon formosum* Bth. et Hook. f.; *Guttifer.;* Annam.
Nganga: *Antrocaryon klaineanum* Pierre; *Anacardiac.;* Kam.
„   : *Pachylobus le-testui* Pellegr., *P. fraxinifolius* Engl.; *Burserac.;* Gab.
Ngành nganh: *Cratoxylon formosum* Bth. et Hook.; *Guttifer.;* Ind.-Ch.
Ngao: *Morinda citrifolia* L.; *Rubiac.;* Ind.-Ch.
Ngap-set-ting: *Morus alba* L.; *Morac.;* n. Burma (Chin)
Ngarawan: *Hopea mengarawan* Miq.; *Dipterocarpac.;* Sum., Bangka.
Ngarít: *Eugenia saligna* C. B. Rob.; *Myrtac.;* Phil. (Palaui)
Ngarosángis: *Cyclostemon sp.; Euphorbiac.;* Phil. (Ilocos Sur)
Ngát: *Gironniera chinensis* Bth.; *Ulmac.;* Ind.-Ch.
„   tráng: *Gironniera chinensis* Bth.; *Ulmac.;* Ind.-Ch.
„   vàng: *Gironniera chinensis* Bth.; *Ulmac.;* Ind.-Ch.
„   xanh: *Gironniera chinensis* Bth.; *Ulmac.;* Ind.-Ch.
Ngau: *Aglaia duperreana* Pierre; *Meliac.;* Ind.-Ch.
Ngau dai: *Aglaia odorata* Loureiro *var. repanensis* ?; *Meliac.;* Ind.-Ch.
„   rùng: *Aglaia pleuropteris* Pierre; *Meliac.;* Ind.-Ch.
Ngay xanh: *Cassia garrettiana* Craib; *Caesalpiniac.;* Ind.-Ch.
Ngaysheek: *Rhododendron arboreum* Sm.; *Ericac.;* n. Burma (Chin)
Ngbora: *Entandrophragma sp.; Meliac.;* Nig.
Ngedi: *Bursera sp.; Burserac.;* Mex.
Ngega: *Parkia filicoidea* Welw.; *Mimosac.;* br. O.-Af. (Uganda)
Ngele: *Pterocarpus soyauxi* Taub.; *Papilionac.;* B.-K.
Ngèlè: *Ricinodendron africanum* Muell. Arg.; *Euphorbiac.;* Gab.
Ngé-ngé: *Garcinia klaineana* Pierre; *Guttifer.;* Gab.
Ngeul: *Croton oligandrum* Pierre; *Euphorbiac.;* Gab.
Nghang-nghanh: *Cratoxylon formosum* Bth. et Hook. f.; *Guttifer.;* Annam.
Nghât sanh: *Gironniera chinensis* Bth.; *Ulmac.;* Annam.
„   vang: *Gironniera chinensis* Bth.; *Ulmac.;* Annam.
N'ghiarr: *Xylopia aethiopica* A. Rich.; *Anonac.;* Senl.

Nghiên: *Pentace tonkinensis* Chev., *P. burmanica* Kurz; *Tiliac.;* Ind.-Ch.
„ mât: *Pentace tonkinensis* Chev.; *Tiliac.;* Ind.-Ch.
„ trang: *Pentace tonkinensis* Chev.; *Tiliac.;* Tonkin.
„ trung: *Pentace tonkinensis* Chev.; *Tiliac.;* Tonkin.
Nghieu bân: *Nephelium lappaceum* L.; *Sapindac.;* Ind.-Ch.
Nghiu: *Bombax malabaricum* DC.; *Bombacac.;* Ind.-Ch.
Ngidi: *Bursera sp.; Burserac.;* Mex.
Ngihié: *Ceiba pentandra* Gaertn.; *Bombacac.;* Elf.
N'gilica ia muchito: *Ekebergia senegalensis* A. Juss.; *Meliac.;* port. Af.
Ngillasondo: *Pterocarpus angolensis* DC.; *Papilionac.;* S.-Af. (Transval)
Ngion: *Klainedoxa gabonensis* Pierre; *Simarubac.;* Kam.
Ngíran: *Cananga odorata* Hook. f. et Thoms.; *Anonac.;* Phil. (Agusan)
Ngirik-ngírik: *Erythrophloeum densiflorum* Merr.; *Caesalpiniac.;* Phil. (Cagayan)
Ngirikngík: *Erythrophloeum densiflorum* Merr.; *Caesalpiniac.;* Phil. (Cagayan)
Ngné-ngné: *Garcinia klaineana* Pierre; *Guttifer.;* Gab.
N'gneké: *Ceiba pentandra* Gaertn.; *Bombacac.;* Elf.
Ngo: *Celtis sp.; Ulmac.;* W.-Af. (? Gab.)
„ : *Pinus khasya* Royle, *P. insularis* Endl.; *Conifer.;* Annam.
„ : *Randia exaltata* Griff. var. *griffithiana* Pierre; *Rubiac.;* Ind.-Ch.
„ dông: *Careya arborea* Roxb.; *Lecythidac.;* Ind.-Ch.
„ -haila: *Pinus khasya* Royle; *Conifer.;* Annam.
„ ne nkyene: *Cleistopholis patens* Bth.; *Anonac.;* Go.
Ngoa: *Salix tetrasperma* Roxb.; *Salicac.;* Ind.-Ch.
Ngoát: *Mallotus eberhardti* Gagnep.; *Euphorbiac.;* Ind.-Ch.
Ngobyè: *Staudtia gabonensis* Warb.; *Myristicac.;* Gab.
Ngoc am: *Cupressus funebris* Endl.; *Conifer.;* Ind.-Ch.
Ngodua: *Haronga paniculata* Sprague; *Guttifer.;* Go.
Ngoiongói: *Wrightia sp.; Apocynac.;* Phil. (Nueva Viscaya)
N'gokoum: *Barteria dewevrei* de Wild. et Dur.; *Flacourtiac.;* Gab.
Ngole: *Albizzia fastigiata* Oliv.; *Mimosac.;* B.-K.
Ngollo: *Khaya ivorensis* Chev.; *Meliac.;* Kam.
N'gollon: *Khaya ivorensis* Chev.; *Meliac.;* Kam.
N'golo: *Khaya ivorensis* Chev.; *Meliac.;* Kam.
Ngolo ngoloti: *Lannea acidissima* Chev.; *Anacardiac.;* Elf.
N'gom: *Klainedoxa latifolia* Pierre; *Simarubac.;* Kam.
Ngom: *Aporosa microcalyx* Hassk.; *Euphorbiac.;* Ind.-Ch.
Ngoma: *Barteria dewevrei* de Wild. et Dur.; *Flacourtiac.;* Gab.
N'gomaban: *Funtumia africana* Stapf; *Apocynac.;* Gab.
N'gome: *Sindora klaineana* Pierre; *Caesalpiniac.;* Gab.
N'gomi: *Copaifera sp.; Caesalpiniac.;* Gab.
Ngon: *Klainedoxa gabonensis* Pierre, *K. gabonensis* Pierre var. *oblongifolia* Engl.;
„ : *Chytranthus macrophyllus* Gilg; *Sapindac.;* Gab.   [*Simarubac.;* Kam.
N'gon: *Sorindeia ochracea* Engl.; *Anacardiac.;* Kam.
Ngondi: *Klainedoxa gabonensis* Pierre; *Simarubac.;* Kam.
Ngondo: *Klainedoxa gabonensis* Pierre; *Simarubac.;* Kam.
Ngong: *Gelonium multiflorum* A. Juss.; *Euphorbiac.;* Ind.-Ch.
Ngonga: *Funtumia elastica* Stapf; *Apocynac.;* Kam.
N'gonga: *Antrocaryon klaineanum* Pierre; *Anacardiac.;* Kam.
Ngonhugoué: *Corynanthe gabonensis* Chev.; *Rubiac.;* Gab.
Ngonokung: *Fillaeopsis discophora* Harms; *Mimosac.;* Kam.
N'goré: *Ongokea klaineana* Pierre; *Olacac.;* Kongo.
Ngôs: *Heisteria trillesiana* Pierre; *Olacac.;* Gab.
Ngôsou vyas: *Heisteria trillesiana* Pierre; *Olacac.;* Gab.
Ngotto: *Terminalia superba* Engl. et Diels; *Combretac.;* B.-K.
N'goué: *Ceiba pentandra* Gaertn. var. *dehiscens* Ulb.; *Bombacac.;* Elf.
„ -yonaye: *Funtumia africana* Stapf; *Apocynac.;* Gab.
Ngouen: *Diplanthemum viridiflorum* Hutch.; *Tiliac.;* Kam.
Ngouma: *Coula edulis* Baill.; *Olacac.;* Kam.
N'goumi: *Aucoumea klaineana* Pierre; *Burserac.;* Gab.
Ngouonn: *Trichoscypha sp.; Anacardiac.;* Kam.
Ngourou: *Cola acuminata* Schott et Endl.; *Sterculiac.;* Senl.

Ngozé: *Schumanniophytum klaineanum* Pierre; *Rubiac.;* Gab.
Ngraem: *Albizzia lucida* Bth.; *Mimosac.;* nö. O.-Ind. (Nepal, Sikkim, Bhotan)
Ngrawan-tjinkang: *Hopea mengarawan* Miq.; *Dipterocarpac.;* Nied.-Ind.
N'gron: *Testulea gabonensis* Pellegr.; *Ochnac.;* Kam. (Kribi)
Ngu: *Albizzia zygia* Macbride; *Mimosac.;* Nig.
Ngua: *Spondias lutea* L.; *Anacardiac.;* Elf.
„ abo: *Deinbollia indeniensis* Chev.; *Sapindac.;* Elf.
Nguanahé: *Guarea cedrata* Pellegr.; *Meliac.;* Elf.
Nguangua: *Mareya spicata* Baill.; *Euphorbiac.;* Lib.
Ngubô: *Staudtia gabonensis* Warb.; *Myristicac.;* Gab.
N'gué: *Erythrophloeum guineense* G. Don; *Caesalpiniac.;* Elf.
„ -ngué: *Plagiostylis africana* Prain; *Euphorbiac.;* Gab.
Nguéhié: *Ceiba pentandra* Gaertn.; *Bombacac.;* Elf.
Nguembé: *Mangifera africana* Oliv.; *Anacardiac.;* Gab.
„ : *Lecomtedoxa klaineana* Pierre; *Sapotac.;* Gab.
Nguépé: *Hasskarlia didymostemon* Baill.; *Euphorbiac.;* Elf.
N'gueul: *Croton oligandrus* Pierre; *Euphorbiac.;* Gab.
N'guié: *Erythrophloeum guineense* G. Don; *Caesalpiniac.;* Elf.
N'guindi: *Coelocaryum klainei* Pierre; *Myristicac.;* Kam., Kongo.
N'guissou: *Entandrophragma aff. rufa* Chev., *E. aff. utile* Sprague; *Meliac.;* Elf.
Ngula: *Pterocarpus cabrae* de Wild., *P. soyauxi* Taub.; *Papilionac.;* Gab., B.-K.
N'hulu: *Sarcocephalus trillesi* Pierre, *S. pobeguini* Pob.; *Rubiac.;* Kongo.
„ : *Corynanthe paniculata* Welw.; *Rubiac.;* B.-K.
„ maza: *Sarcocephalus trillesi* Pierre; *Rubiac.;* Gab., B.-K.
Nguni: *Limonia warneckei* Engl.; *Rutac.;* Togo.
Nguobi: *Diospyros sansa-minika* Chev.; *Ebenac.;* Elf.
Ngusat: *Cassia fistula* L.; *Caesalpiniac.;* Burma.
Ngusch: *Manniophyton africcnum* Muell. Arg.; *Euphorbiac.;* sp. Guin.
Ngut: *Clausena wampi* Oliv.; *Rutac.;* Ind.-Ch.
„ : *Mallotus eberhardti* Gagnep.; *Euphorbiac.;* Ind.-Ch.
„ : *Cordia bantamensis* Blume; *Borraginac.;* Tonkin.
Nguyet qui: *Murraya exotica* L., *Chalcas paniculata* L.; *Rutac.;* Ind.-Ch.
„ -qui-tau: *Murraya exotica* L.; *Rutac.;* Co.-Ch.
Ngwagiri: *Grewia coriacea* Mast.; *Tiliac.;* Gab.
Ngwaki: *Testulea gabonensis* Pellegr.; *Ochnac.;* Kongo (Mayombe)
Ngwama: *Ricinodendron africanum* Muell. Arg.; *Euphorbiac.;* Go.
Ngwe: *Olea hochstetteri* Baker; *Oleac.;* D.-O.-Af. (w. Usambara)
Nhâ: *Bertholletia excelsa* Bonpl., *B. nobilis* Miers; *Lecythidac.;* n. Bras.
Nhân: *Styrax benzoin* Dryand.; *Styracac.;* Ind.-Ch.
„ : *Nephelium longana* Cambess., *N. litchi* Cambess.; *Sapindac.;* Annam.
„ rùng: *Nephelium sp.;* *Sapindac.;* Ind.-Ch.
„ xa: *Nephelium longana* Cambess.; *Sapindac.;* Ind.-Ch.
Nhang: *Dipterocarpus alatus* Roxb.; *Dipterocarpac.;* Ind.-Ch. (Laos)
Nháp: *Columbia thoreli* Gagnep.; *Tiliac.;* Ind.-Ch.
Nhare: *Taxus baccata* L.; *Conifer.;* O.-Ind. (Tibet)
Nhatanzira: *Rhizophora mucronata* Lamk.; *Rhizophorac.;* port. O.-Af. (Moz., Quel.)
Nhau: *Morinda tinctoria* Roxb.; *Rubiac.;* Ind.-Ch.
„ núi: *Morinda citrifolia* L.; *Rubiac.;* Ind.-Ch.
„ rang: *Tarenna citrina* Pitard; *Rubiac.;* Annam.
Nhè xanh: *Cinnamomum tonkinense* Pitard; *Laurac.;* Ind.-Ch.
Nhioc: *Terminalia papilio* Hance; *Combretac.;* Ind.-Ch.
Nhoc: *Terminalia papilio* Hance; *Combretac.;* Ind.-Ch.
„ : *Polyalthia corticosa* Finet et Gagnep.; *Anonac.;* Ind.-Ch.
„ quich: *Polyalthia corticosa* Finet et Gagnep.; *Anonac.;* Ind.-Ch.
Nhôi: *Bischofia javanica* Blume; *Euphorbiac.;* Ind.-Ch.
„ : *Elaeocarpus quercifolius* Gagnep.; *Elaeocarpac.;* Ind.-Ch.
Nhom: *Tarrietia cochinchinensis* Pierre; *Sterculiac.;* Ind.-Ch.
„ nhom: *Elaeocarpus dubius* A. DC.; *Elaeocarpac.;* Ind.-Ch.
Nhoncutdec: *Euphoria informis* Steud.; *Sapindac.;* Annam.
Nhong: *Dipterocarpus alatus* Roxb.; *Dipterocarpac.;* Ind.-Ch.
Nhót vàng: *Mallotus hookerianus* Muell. Arg.; *Euphorbiac.;* Ind.-Ch.

Nhoum: *Pygeum arboreum* Engl.; *Rosac.;* Ind.-Ch.
Nhu: *Bombax malabaricum* DC.; *Bombacac.;* Ind.-Ch.
„ xoan: *Spondias mangifera* Willd.; *Anacardiac.;* Ind.-Ch.
Nhum: *Mallotus barbatus* Muell. Arg.; *Euphorbiac.;* Ind.-Ch.
Nhyu: *Anthocephalus cadamba* Miq.; *Rubiac.;* sw. O.-Ind. (Bombay)
Ni: *Alnus nepalensis* D. Don; *Betulac.;* nw. O.-Ind. (Punjab-Sutlej)
Nia: *Canarium commune* L.; *Burserac.;* Moluk. (Insel Ternate)
Niam: *Lophira alata* Banks; *Ochnac.;* Nig.
Niamprani: *Gardenia jovis-tonantis* Hiern; *Rubiac.;* Go.
Nianga: *Myrianthus serratus* Bth. et Hook. f.; *Morac.;* Elf.
Niangamagni: *Myrianthus serratus* Bth. et Hook. f.; *Morac.;* Elf.
Niango: *Pachypodanthium staudti* Engl. et Diels; *Anonac.;* Elf.
Niangon (f): *Tarrietia utilis* Sprague, *Cola proteiformis* Chev.; *Sterculiac.;* Elf.
Niankuma: *Tarrietia utilis* Sprague; *Sterculiac.;* Go.
„ : *Myrianthus arboreus* Beauv.; *Morac.;* Go.
Niaouli: *Melaleuca viridiflora* Soland.; *Myrtac.;* N.-Kal.
„ blanc de plaine: *Melaleuca leucadendron* L.; *Myrtac.;* N.-Kal.
„ rouge de montagne: *Melaleuca leucadendron* L.; *Myrtac.;* N.-Kal.
Niáp: *Unona debilis* Pierre; *Anonac.;* Annam.
Niassa: *Piptadenia aff. buchanani* Baker; *Mimosac.;* D.-O.-Af.
Niato balam: *Payena lucida* A. DC.; *Sapotac.;* br. O.-Ind., Malay.
Nibasé: *Morinda tinctoria* Roxb.; *Rubiac.;* Burma.
Nibong: *Oncosperma sp.; Palmac.;* Borneo.
Nicaragua wood: *Haematoxylon brasiletto* Karst.; *Caesalpiniac.;* Bras.
Nichikigni: *Evonyma alatus* S. et Z.; *Celastrac.;* Jap.
Nickers: *Caesalpinia echinata* Lam.; *Caesalpiniac.;* W.-Ind. (St. Croix)
„ , grey: *Caesalpinia echinata* Lam.; *Caesalpiniac.;* W.-Ind. (St. Thomas)
Nicker tree: *Gymnocladus dioica* C. Koch; *Caesalpiniac.;* ö. USA.
Nictaa: *Mosquitoxylum jamaicense* Krug et Urb.; *Anacardiac.;* br. Hond.
Niéné: *Lophira sp.; Ochnac.;* Elf.
Niéri: *Humiria floribunda* Mart.; *Humiriac.;* fr. Gu.
Nies hout: *Ptaeroxylon utile* Eck. et Zeyh.; *Sapindac.;* S.-Af. (Kap, Natal)
Nieuxbladboom (h): *Swietenia mahagoni* Jacq.; *Meliac.;* tr. Am.
Nigaki: *Picrasma quassioides* Benn.; *Simarubac.;* Jap.
Niganire: *Ulmus campestris* Sm. *var. major* Planch.; *Ulmac.;* Jap.
Night bloom: *Melia azedarach* L.; *Meliac.;* br. Hond.
Nígi: *Carapa moluccensis* Lamk., *C. obovata* Blume; *Meliac.;* Phil. (Tayabas, Palawan)
„ -putí: *Cumingia philippinensis* Vidal; *Bombacac.;* Phil. (Tayabas)
Nignaki: *Picrasma quassioides* Benn.; *Simarubac.;* Jap.
Nigue: *Carapa obovata* Blume; *Meliac.;* Phil.
Nigüito: *Cordia heterophylla* Roem. et Schult.; *Borraginac.;* Pan.
Niitaka-byakusin (jap.): *Juniperus squamata* Lambert; *Conifer.;* Form.
„ -sirodamo (jap.): *Neolitsea acuto-trinervia* Kanehira et Sasaki; *Laurac.;* Form.
„ -todomatsu (jap.): *Abies kawakamii* Ito; *Conifer.;* Form.
„ -tôki (jap.): *Picea morrisonicola* Hayata; *Conifer.;* Form.
Nikari: *Castanopsis indica* A. DC.; *Fagac.;* w. Assam (Sylhet)
Níket: *Beilschmiedia cairocan* Vidal; *Laurac.;* Phil. (Ilocos Norte)
Nikli: *Microdesmis puberula* Hook. f.; *Euphorbiac.;* Lib.
Nil-bhandi: *Garuga pinnata* Roxb.; *Burserac.;* w. Beng.
Nila palai: *Wrightia tinctoria* R. Brown; *Apocynac.;* O.-Ind. (Madras)
Nilan: *Guioa pubescens* Radlk.; *Sapindac.;* Malay.
Nili-mara: *Bischofia javanica* Blume; *Euphorbiac.;* w. O.-Ind. (Bombay-Deccan)
Nilika: *Phyllanthus emblica* L.; *Euphorbiac.;* O.-Ind. (Madras)
Nim: *Azadirachta indica* A. Juss.; *Meliac.;* Go., O.-Ind.
„ : *Rhodomyrtus tomentosa* Wight; *Myrtac.;* Ind.-Ch.
Nimat: *Cordia myxa* L.; *Borraginac.;* nö. O.-Ind. (Nepal, Sikkim, Bhotan)
Nimba: *Azadirachta indica* A. Juss.; *Meliac.;* br. O.-Ind.
Nimbar: *Acacia leucophloea* Willd.; *Mimosac.;* n. O.-Ind. (United Prov.)
Nimbarra: *Melia composita* Willd.; *Meliac.;* sw. O.-Ind. (Bombay)
Nimbo: *Azadirachta indica* A. Juss.; *Meliac.;* Beng., Hindustan.
Nimbooka: *Citrus limonum* Risso; *Rutac.;* br. O.-Ind.

Nimbu: *Citrus limonum* Risso; *Rutac.*; O.-Ind. (Hindustan)
Nimbuka: *Citrus limonum* Risso; *Rutac.*; br. O.-Ind.
Nimuri: *Azadirachta indica* A. Juss.; *Meliac.*; w. O.-Ind. (Bombay-Sindh)
Nimwau-peso: *Eriocoelum kerstingi* Gilg; *Sapindac.*; Togo.
Ninégoné: *Staudtia gabonensis* Warb.; *Myristicac.*; Gab.
Ningo: *Spondias lutea* L.; *Anacardiac.*; Elf.
Ningre notto: *Caryocar tomentosum* Willd., *C. glabrum* Pers., *C. sp.*; *Caryocarac.*; Sur.
Ninh: *Crudia chrysantha* Pierre; *Caesalpiniac.*; Ind.-Ch.
„ : *Pahudia cochinchinensis* Pierre; *Caesalpiniac.*; Ind.-Ch.
Niñoarzoté: *Calliandra tetragona* Bth.; *Mimosac.*; Arg.
Niobé: *Staudtia gabonensis* Warb.; *Myristicac.*; Gab.
Niogo: *Staudtia gabonensis* Warb.; *Myristicac.*; Gab.
Niohé: *Staudtia gabonensis* Warb.; *Myristicac.*; Gab.
Niohue: *Staudtia gabonensis* Warb.; *Myristicac.*; Gab.
Nioikobushi: *Magnolia salicifolia* Maxim.; *Magnoliac.*; Jap.
Nioi-tabu (jap.): *Machilus zuihoensis* Hayata; *Laurac.*; Form.
Niondobi: *Uapaca benguelensis* Muell. Arg.; *Euphorbiac.*; Elf.
Niori: *Humiria floribunda* Mart.; *Humiriac.*; Gu.
Niotoko: *Platysepalum chevalieri* Harms; *Papilionac.*; B.-K.
Niové (f): *Staudtia gabonensis* Warb., *S. kamerunensis* Warb.; *Myristicac.*; Kam., Gab
Nipis kulit: *Memecylon edule* Roxb.; *Melastomatac.*; O.-Ind., Cey., Malay.
Nir kadambe: *Stephegyne parvifolia* Korth.; *Rubiac.*; O.-Ind. (Madras)
„ naval: *Eugenia gardneri* Thw.; *Myrtac.*; sw. O.-Ind. (Madras)
„ nocchi: *Vitex leucoxylon* L.; *Verbenac.*; sw. O.-Ind. (Madras)
„ perzha: *Barringtonia acutangula* Gaertn.; *Lecythidac.*; sw. O.-Ind. (Malabar)
Nira: *Bischofia javanica* Blume; *Euphorbiac.*; sw. O.-Ind. (Malabar)
Nirambali: *Podocarpus latifolia* Wall.; *Conifer.*; O.-Ind.
Niranji: *Salix tetrasperma* Roxb.; *Salicac.*; O.-Ind. (Bombay, Madras)
Niratti niralam: *Hydnocarpus wightiana* Blume; *Flacourtiac.*; sw. O.-Ind. (Malabar)
Nire: *Ulmus campestris* Sm. *var. laevis* Planch.; *Ulmac.*; Jap.
Nireh: *Carapa obovata* Blume; *Meliac.*; Malay.
Niretimara: *Ficus callosa* Willd.; *Morac.*; O.-Ind. (? Bombay)
Niri: *Carapa borneensis* Becc.; *Meliac.*; Borneo, Phil.
„ : *Carapa obovata* Blume, *C. moluccensis* Lamk.; *Meliac.*; Malay.
Nirkongu: *Hopea parviflora* Bedd.; *Dipterocarpac.*; s. O.-Ind. (Madras-Tinnevelly)
Nirmali: *Strychnos potatorum* L. f.; *Loganiac.*; sw. O.-Ind., C.-O.-Ind.
Nirre: *Nothofagus antarctica* Oerst.; *Fagac.*; Chile.
Nirvali: *Filicium decipiens* Thw.; *Sapindac.*; sw. O.-Ind. (Travancore)
Nisani: *Cinnamomum zeylanicum* Breyn; *Laurac.*; sw. O.-Ind. (Bombay)
Nisberry: *Achras sapota* L.; *Sapotac.*; tr. Am.
Nishikigi: *Evonymus alata* Maxim.; *Celastrac.*; Jap.
Nisne: *Cinnamomum zeylanicum* Breyn; *Laurac.*; sw. O.-Ind. (Bombay)
Nispero: *Mespilus germanica* L.; *Rosac.*; Sp.
„ : *Achras sapota* L.; *Sapotac.*; tr. Am.
„ : *Achras chicle* Pittier; *Sapotac.*; Guat.
„ : *Eriobotrya japonica* Lindl.; *Rosac.*; Hond.
„ : *Mimusops spectabilis* Pittier; *Sapotac.*; C.-R.
„ : *Achras calcicola* Pittier, *Mimusops darienensis* Pittier; *Sapotac.*; Pan.
„ : *Lucuma chiricana* Standl.; *Sapotac.*; Pan.
„ : ? *Achras sapota* L.; *Sapotac.*; Ven.
„ amarillo: *Dipholis minutiflora* Pittier; *Sapotac.*; C.-R.
„ blanco: ? *Ficus sp.*; *Morac.*; Pan.
„ colorado: *Lucuma chiricana* Standl.; *Sapotac.*; Pan.
„ negro: *Minquartia guianensis* Aubl.; *Olacac.*; Pan.
„ quitense: *Achras sapota* L.; *Sapotac.*; Ek.
„ de montaña: *Achras chicle* Pittier; *Sapotac.*; Salv.
„ „ monte: ? *Achras calcicola* Pittier; *Sapotac.*; n. Kol.
„ „ saíno: *Morisonia americana* L.; *Capparidac.*; n. Kol.
„ del Japón: *Eriobotrya japonica* Lindl.; *Rosac.*; Jap.
Nitaka-akamatsu (jap.): *Pinus taiwanensis* Hayata; *Conifer.*; Form.
Nivali: *Strychnos potatorum* L. f.; *Loganiac.*; sw. O.-Ind. (Bombay)

Nivar: *Barringtonia racemosa* Blume; *Lecythidac.;* sw. O.-Ind.
Nivoli: *Filicium decipiens* Thw.; *Sapindac.;* sw. O.-Ind. (Travancore)
Niwatoko: *Sambucus racemosa* L. var. *sieboldiana* Miq.; *Caprifoliac.;* Jap.
Niyor: *Cocos nucifera* L.; *Palmac.;* Malay.
Nja: *Laguncularia racemosa* Gaertn. f.; *Combretac.;* Elf.
Njabi (d, f): *Mimusops djave* Engl., *M. pierreana* Engl.; *Sapotac.;* Kam., Gab.
N'jabi: *Mimusops heckeli* Lecomte; *Sapotac.;* Kam.
Njãkpẽ: *Kigelia africana* Bth.; *Bignoniac.;* Togo.
Njakpokpo: *Kigelia africana* Bth.; *Bignoniac.;* Togo.
Njale: *Artocarpus sp.; Morac.;* br. O.-Af. (Niassa)
Njambokka: *Pouteria guianensis* Aubl.; *Sapotac.;* Sur.
Njamploeng: *Calophyllum inophyllum* L.; *Guttifer.;* Java.
Njangsang: *Ricinodendron africanum* Muell. Arg.; *Euphorbiac.;* Kam.
Njansang: *Ricinodendron africanum* Muell. Arg.; *Euphorbiac.;* Kam.
Njanting-mahambong: *Shorea balangeran* Burck; *Dipterocarpac.;* Malay.
N'jap: *Mimusops djave* Engl., *M. pierreana* Engl.; *Sapotac.;* Kam., Gab.
Njating: *Vatica rassak* Korth.; *Dipterocarpac.;* nied. Borneo.
Njatoe: *Palaquium javense* Burck; *Sapotac.;* Java.
   „ -balam-bringin: *Payena leeri* Kurz; *Sapotac.;* Malay.
   „ -doerian: *Palaquium borneense* Burck; *Sapotac.;* Borneo.
Njatoh: *Palaquium javense* Burck; *Sapotac.;* Java.
Nje: *Enantia chlorantha* Oliv.; *Anonac.;* Elf., Nig., Kam.
N'jeck: *Ongokea gore* Engl.; *Olacac.;* Kam.
Njeli: *Entandrophragma utile* Sprague; *Meliac.;* W.-Af. (S.-L.)
Njié: *Ongokea klaineana* Pierre; *Olacac.;* Kam.
   „ : *Enantia chlorantha* Oliv., *Xylopia quintasi* Engl. et Diels; *Anonac.;* Kam.
N'jion: *Scyphocephalium ochocoa* Warb.; *Myristicac.;* Kam.
Njion: *Klainedoxa gabonensis* Pierre; *Simarubac.;* Kam.
Njoc: *Afzelia bipindensis* Harms; *Caesalpiniac.;* Kam.
   „ élé: *Afzelia bipindensis* Harms; *Caesalpiniac.;* Kam.
Njokuboré: *Entandrophragma rederi* Harms; *Meliac.;* Kam.
Njokubwale: *Entandrophragma rederi* Harms; *Meliac.;* Kam.
Njokubwele: *Entandrophragma rederi* Harms; *Meliac.;* Kam.
Njore-njolé: *Poga oleosa* Pierre; *Rhizophorac.;* Kam.
Njouli: *Ricinodendron africanum* Muell. Arg.; *Euphorbiac.;* Kam.
Nkachi: *Podocarpus milanjianus* Rendle; *Conifer.;* br. O.-Af. (Niassa)
Nkala: *Pachylobus pubescens* Verm.; *Burserac.;* B.-K.
   „ : *Bridelia speciosa* Muell. Arg.; *Euphorbiac.;* Gab. (Loango)
Nkalate: *Sideroxylon brevipes* Baker; *Sapotac.;* br. O.-Af. (Uganda)
Nkamba: *Chlorophora excelsa* Bth. et Hook.; *Morac.;* B.-K.
   „ : *Heisteria trillesiana* Pierre; *Olacac.;* Gab.
Nkaméba: *Lavalleopsis densivenia* Engl.; *Olacac.;* Gab.
Nkampon: *Oxytenanthera ? abyssinica* Munro; *Gramin.;* Go.
Nkampro: *Oxytenanthera ? abyssinica* Munro; *Gramin.;* Go.
N'kandella: *Ceriops candolleana* Arn.; *Rhizophorac.;* port. O.-Af. (Moz., Quel.)
Nkanguelé: *Pseudospondias microcarpa* Engl.; *Anacardiac.;* Kam.
Nkanna: *Sterculia rhinopetala* K. Schum.; *Sterculiac.;* Kam.
Nkännang: *Sterculia rhinopetala* K. Schum.; *Sterculiac.;* Kam.
Nkarl: *Maesopsis berchemioides* Nob.; *Rhamnac.;* Gab.
N'kassa: *Erythrophloeum guineense* G. Don; *Caesalpiniac.;* Gab.
N'keng: *Morinda sp.; Rubiac.;* Kam.
N'kéva: *Copaifera sp.; Caesalpiniac.;* Gab.
N'kévazingo: *Didelotia africana* Pierre; *Caesalpiniac.;* Gab.
N'kéwa: *Copaifera sp.; Caesalpiniac.;* Gab.
Nkewazingo: *Didelotia africana* Pierre; *Caesalpiniac.;* Gab.
Nkiébé: *Dracaena perrotteti* Baker; *Liliac.;* Elf.
Nkiembelé: *Strombosiopsis tetrandra* Engl.; *Olacac.;* Kam.
Nkikimbo: *Myrica kandtiana* Engl.; *Myricac.;* br. O.-Af. (Uganda)
Nkoba: *Lovoa browni* Sprague; *Meliac.;* br. O.-Af. (Uganda)
   „ : *Baikiaea emini* Taub.; *Caesalpiniac.;* br. O.-Af. (Uganda)
N'kolla: *Mimusops africana* Lecomte; *Sapotac.;* s. Kam., n. Gab.

Nkom: *Triplochiton scleroxylon* K. Schum.; *Sterculiac.;* Kam.
  „ : *Terminalia superba* Engl. et Diels; *Combretac.;* Go., S.-L., Kam.
N'koma: *Pycnanthus kombo* Warb.; *Myristicac.;* Gab.
Nkoma: *Elaeis guineensis* Jacq. *var. virescens* Chev.; *Palmac.;* Gab.
Nkombe: *Terminalia ivorensis* Chev.; *Combretac.;* Nig.
Nkombo: *Pycnanthus kombo* Warb.; *Myristicac.;* Gab.
  „ : *Lannea welwitschi* Engl.; *Anacardiac.;* B.-K.
N-komo: *Pycnanthus kombo* Warb.; *Myristicac.;* Gab.
Nkondi: *Terminalia ivorensis* Chev.; *Combretac.;* Nig.
Nkondjo: *Klainedoxa gabonensis* Pierre; *Simarubac.;* Gab.
Nkop: *Berlinia sp.; Caesalpiniac.;* Kam.
N'koubi: *Staudtia gabonensis* Warb.; *Myristicac.;* Gab.
Nkoul: *Mansonia altissima* Chev.; *Sterculiac.;* Kam.
N'koumi: *Aucoumea klaineana* Pierre; *Burserac.;* Gab. (Loango)
N'kouninou: *Coula edulis* Baill.; *Olacac.;* Gab.
Nkowe-mbani: *Elaeis guineensis* Jacq. *var. ?; Palmac.;* Gab.
Nkpi: *Carapa procera* DC.; *Meliac.;* Gab.
Nkrunya: *Sterculia barteri* Mast.; *Sterculiac.;* Go.
Nkuba: *Panda oleosa* Pierre; *Pandac.;* Gab.
Nkubi: *Staudtia gabonensis* Warb.; *Myristicac.;* Gab.
Nkula: *Coula edulis* Baill.; *Olacac.;* Kam.
  „ : *Pterocarpus soyauxi* Taub., *P. cabrae* de Wild.; *Papilionac.;* B.-K.
Nkulajo: *Trema guineense* Engl.; *Euphorbiac.;* br. O.-Af. (Uganda)
Nkumbi: *Lannea welwitschi* Engl.; *Anacardiac.;* B.-K.
Nkumunu: *Coula edulis* Baill.; *Olacac.;* Gab.
Nkundi: *Parkia filicoidea* Welw.; *Mimosac.;* br. O.-Af. (Niassa)
Nkungusa: *Widdringtonia whytei* Rendle; *Conifer.;* br. O.-Af. (Niassa)
Nkunya: *Mimusops cuneifolia* Baker; *Sapotac.;* br. O.-Af. (Uganda)
Nkurunyan: *Hymenostegia afzeli* Oliv.; *Caesalpiniac.;* Go.
Nkwasi: *Albizzia grandibracteata* Taub., *A. zygia* MacBride; *Mimosac.;* br. O.-Af.
Nkweranyani: *Piptadenia buchanani* Baker; *Mimosac.;* s. C.-Af.   [(Uganda)
Nkwerenyani: *Piptadenia buchanani* Baker; *Mimosac.;* br. O.-Af. (Niassa)
Nkyene ne ngo: *Cleistopholis patens* Bth.; *Anonac.;* Go.
N'limba: *Terminalia superba* Engl. et Diels; *Combretac.;* Kam., Gab. (Mayombe)
Nlombo: *Pycnanthus kombo* Warb.; *Myristicac.;* Gab.
N'looue: *Pausinystalia yohimbe* Pierre; *Rubiac.;* Kam.
Nloué: *Pausinystalia yohimbe* Pierre, *P. macroceras* Pierre; *Rubiac.;* Kam.
  „ : *Pausinystalia gabonensis* Chev., *Pseudocinchona pachyceras* K. Schum.; *Rubiac.;*
Nnambar: *Dalbergia retusa* Hemsl.; *Papilionac.;* C.-R.   [Kam.
N'nou: *Lannea acidissima* Chev.; *Anacardiac.;* Elf.
Nobunoki: *Platycarya strobilacea* S. et Z.; *Juglandac.;* Jap.
Noce (i): *Juglans spp.; Juglandac.;* Eu.
  „ nero (i): *Juglans nigra* L.; *Juglandac.;* N.-Am.
  „ satin (i): *Liquidambar styraciflua* L.; *Hamamelidac.;* s. N.-Am., C.-Am.
  „ satinato, legno di (i): *Liquidambar styraciflua* L.; *Hamamelidac.;* s. N.-Am.,
No-choc-ché: *Guarea chichon* C. DC.; *Meliac.;* br. Hond.   [C.-Am.
Noemi: *Orophea hexandra* Blume; *Anonac.;* Java (Pantjoer)
Nogabe: *Schima wallichi* Choisy; *Theac.;* Assam.
Nogakat: *Schima wallichi* Choisy; *Theac.;* Assam.
Nogal (s): *Juglans spp.; Juglandac.;* sp. Am.
  „ : *Juglans jamaicensis* C. DC.; *Juglandac.;* S.-D.
  „ : *Juglans sp.; Juglandac.;* Hond.
  „ : *Juglans colombiensis* Dode; *Juglandac.;* Ven.
  „ : *Juglans sp.; Juglandac.;* Ek.
  „ : *Juglans peruviana* Dode; *Juglandac.;* Peru.
  „ : *Juglans australis* Griseb.; *Juglandac.;* Arg.
  „ : *Zanthoxylum sp.; Rutac.;* Trin.
  „ blanco: *Juglans sp., ? Juglans boliviana* Dode; *Juglandac.;* w. Peru.
  „ cayure: *Juglans australis* Griseb.; *Juglandac.;* Arg.
  „ cayuri: *Juglans australis* Griseb.; *Juglandac.;* s. S.-Am.
  „ cimarron: *Cedrela sp.; Meliac.;* Mex.

Nogal cimarron: *Juglans australis* Griseb.; *Juglandac.;* Arg.
„ corriente: *Cedrela sp.; Meliac.;* Mex.
„ criollo: *Juglans australis* Griseb.; *Juglandac.;* Arg.
„ negro: *Juglans sp., ? J. boliviana* Dode; *Juglandac.;* w. Peru (Ucayali)
„ silvestre: *Juglans australis* Griseb.; *Juglandac.;* tr. Am.
„ del país: *Juglans australis* Griseb.; *Juglandac.;* Cuba.
Nogé: *Toona ciliata* M. Roem.; *Meliac.;* sw. O.-Ind. (Coorg)
Nogueira (p): *Juglans spp.; Juglandac.;* n. Hem.
„ de iguape: *Aleurites moluccana* Willd.; *Euphorbiac.;* Bras. (Sao P)
Noi: *Oligoceras eberhardti* Gagnep.; *Euphorbiac.;* Ind.-Ch.
Noibwood (u): *Tecoma leucoxylon* Mart.; *Bignoniac.;* holl. Gu.
Noisetier: *Corylus avellana* L.; *Betulac.;* Eu.
Noisette des bois: *Heisteria trillesiana* Pierre; *Olacac.;* Gab.
Noix de Bancoul: *Aleurites triloba* Forst.; *Euphorbiac.;* Gu.
„ „ beurre: *Caryocar tomentosum* Willd.; *Caryocarac.;* Gu.
„ „ N'gore: *Ongokea klaineana* Pierre; *Olacac.;* Gab.
Noka: *Odyendea gabonensis* Engl.; *Simarubac.;* Gab.
Noki: *Distylium racemosum* S. et Z.; *Hamamelidac.;* Jap.
Nokoba-no-naragasiwa (jap.): *Quercus aliena* Blume *var. acuteserrata* Maxim.; *Fagac.;*
Nokué: *Lophira procera* Chev.; *Ochnac.;* Elf. [Form.
„ : *Alstonia congensis* Engl.; *Apocynac.;* Elf.
Nom eten: *Coelocaryon sp.; Myristicac.;* W.-Af. (? Kam.)
Nombo: *Chlorophora regia* Chev., *C. excelsa* Bth. et Hook.; *Morac.;* Gab.
Nomélé: *Didelotia africana* Pierre; *Caesalpiniac.;* Kam.
Nometen: *Scyphocephalium ochocoa* Warb.; *Myristicac.;* Kam.
Nomotcho: *Bussea occidentalis* Hutch.; *Caesalpiniac.;* Elf.
Nona: *Anona reticulata* L.; *Anonac.;* w. Java.
„ : *Lophira procera* Chev.; *Ochnac.;* Elf.
„ kapri: *Anona reticulata* L.; *Anonac.;* Malay.
Nonblé: *Sarcocephalus esculentus* Afzel.; *Rubiac.;* Elf.
Nông: *Saurauia tristyla* Wall.; *Dilleniac.;* Ind.-Ch.
„ lua: *Alchornea tiliaefolia* Muell. Arg.; *Euphorbiac.;* Ind.-Ch.
Nongko: *Artocarpus integrifolia* L. f.; *Morac.;* Java.
„ poenjoe: *Artocarpus integrifolia* L. f.; *Morac.;* Java.
Nongo: *Fagara macrophylla* Engl.; *Rutac.;* Gab.
„ : *Albizzia grandibracteata* Taub.; *Mimosac.;* br. O.-Af. (Uganda)
„ : *Albizzia zygia* MacBride, *A. coriaria* Welw.; *Mimosac.;* br. O.-Af. (Uganda)
„ : *? Cynometra sp.; Caesalpiniac.;* br. O.-Af. (Uganda-Mabira)
Nono: *Gyranthera caribensis* Pittier; *Bombacac.;* Ven.
Nonon giwa: *Kigelia aethiopica* Dcne. *var. bornuensis* Sprague; *Bignoniac.;* n. Nig.
Noogoora burr: *Xanthium pungens* Wallr.; *Composit.;* ö. Au. (N.-S.-W.)
Noot (h): *Juglans regia* L.; *Juglandac.;* Eu.
Norh thom: *Morinda tinctoria* Roxb.; *Rubiac.;* Ind.-Ch.
Nori-no-ki: *Hydrangea paniculata* Sieb.; *Saxifragac.;* n. Jap.
Noten, satijn (h): *Liquidambar styraciflua* L.; *Hamamelidac.;* s. N.-Am., C.-Am.
Notenhout (h): *Juglans spp.; Juglandac.;* Eu.
„ , zwart, amerikaansch: *Juglans nigra* L.; *Juglandac.;* USA.
Notra-ciruelillo: *Embothrium coccineum* Forst.; *Proteac.;* s. Chile.
Notru: *Embothrium coccineum* Forst.; *Proteac.;* s. Chile.
Noum-etan: *Scyphocephalium manni* Warb.; *Myristicac.;* Kam.
„ eten: *Coelocaryon sp.; Myristicac.;* W.-Af. (? Kam.)
Noumbaouan: *Entandrophragma sp., Khaya sp.; Meliac.;* Kam.
Noumeten: *Scyphocephalium ochocoa* Warb.; *Myristicac.;* Kam.
Noumgou: *Mimusops djave* Engl., *M. africana* Lecomte; *Sapotac.;* Gab.
Noungou: *Mimusops africana* Lecomte; *Sapotac.;* Kam., Gab.
Nouogui: *Croton sp.; Euphorbiac.;* Kam.
Nourrouc: *Elaeodendron orientale* Jacq.; *Celastrac.;* Maur.
Nouvé: *Pinus cembra* L.; *Conifer.;* Mitt.-Eu.
Novire: *Balanites aegyptiaca* Delile; *Zygophyllac.;* D.-O.-Af. (Kilossa)
Noyer (f): *Juglans spp.; Juglandac.;* Eu.
„ : *Picrasma excelsa* Planch.; *Simarubac.;* Guad.

Noyer bassam: *Lovoa klaineana* Pierre; *Meliac.;* W.-Af.
„ blanc: *Carya alba* Nutt.; *Juglandac.;* sö. Kan.
„ „ d'Amérique: *Carya ovata* C. Koch; *Juglandac.;* Kan.
„ cendré: *Juglans cinerea* L.; *Juglandac.;* sö. Kan.
„ commun: *Juglans regia* L.; *Juglandac.;* sö. Eu., As., Jap.
„ dur: *Carya cordiformis* C. Koch, *C. alba* Nutt.; *Juglandac.;* Kan.
„ écailleux: *Carya ovata* C. Koch; *Juglandac.;* sö. Kan.
„ gris: *Juglans cinerea* L.; *Juglandac.;* sö. Kan.
„ noir: *Juglans nigra* L.; *Juglandac.;* sö. Kan.
„ satiné (f): *Liquidambar styraciflua* L.; *Hamamelidac.;* s. N.-Am., C.-Am.
„ tendre: *Juglans cinerea* L.; *Juglandac.;* sö. Kan.
„ à noix de cochon: *Carya porcina* Nutt.; *Juglandac.;* Kan.
„ „ „ douces: *Carya alba* Nutt.; *Juglandac.;* sö. Kan.
„ des Antilles: *Fagara tragodes* Jacq.; *Rutac.;* Guad., Mart.
„ du Gabon (f): *Lovoa klaineana* Pierre; *Meliac.;* W.-Af.
„ „ „ : *Terminalia superba* Engl. et Diels; *Combretac.;* W.-Af.
„ „ „ , vrai (f): *Lovoa klaineana* Pierre; *Meliac.;* Kam.
„ „ „ Mayombe (f): *Terminalia superba* Engl. et Diels; *Combretac.;* Kam., Kongo.
„ d'Afrique (f): *Lovoa klaineana* Pierre; *Meliac.;* W.-Af.
„ „ : *Coula edulis* Baill.; *Olacac.;* W.-Af.
Noz moscada do Brasil: *Cryptocarya moschata* Nees et Mart.; *Laurac.;* Bras.
„ de galha: *Caryocar villosum* Pers.; *Caryocarac.;* n. Bras. (Pará)
N'roné: *Testulea gabonensis* Pellegr.; *Ochnac.;* Kam. (Kribi)
N'sa evila: *Diospyros aggregata* Gürke; *Ebenac.;* Kam.
Nsa-sana: ? *Ricinodendron africanum* Muell. Arg.; *Euphorbiac.;* s. Nig. (Calabar)
N'safo: *Pachylobus edulis* G. Don *var. sativa* Ficalho; *Burserac.;* Gab. (Mayombe)
N'safou: *Pachylobus edulis* G. Don *var. sativa* Ficalho; *Burserac.;* Gab. (Mayombe)
Nsagalane: *Xylopia emini* Engl.; *Anonac.;* br. O.-Af. (Uganda)
Nsagalanyi: *Xylopia emini* Engl.; *Anonac.;* br. O.-Af. (Uganda)
N'sahou: *Ongokea klaineana* Pierre; *Olacac.;* Gab. (Loango)
Nsaki: *Pterocarpus santalinoides* L'Hérit.; *Papilionac.;* Go.
Nsamanguila: *Khaya klainei* Pierre; *Meliac.;* sp. Guin.
Nsambya: *Markhamia platycalyx* Sprague; *Bignoniac.;* br. O.-Af. (Uganda)
Nsamgilla: *Khaya klainei* Pierre; *Meliac.;* W.-Af.
N'sangha: *Bakeriella longistyla* Dubard; *Sapotac.;* s. Kam., n. Gab.
Nsangha: *Mimusops lacera* Baker; *Sapotac.;* Gab.
Nsangôme: *Pentadesma butyraceum* Don; *Guttifer.;* Gab.
Nsangome: *Pentadesma leptonema* Pierre; *Guttifer.;* Kam.
Nsangomo: *Allanblackia klainei* Pierre; *Guttifer.;* Kam.
Nsano: *Ongokea klaineana* Pierre; *Olacac.;* Gab.
Nsasanga: *Ricinodendron africanum* Muell. Arg.; *Euphorbiac.;* Gab.
Nsass: *Pentaclethra eetveldeana* de Wild.; *Mimosac.;* Kam.
Nsate: *Ficus sp.;* *Morac.;* br. O.-Af. (Uganda)
Nsempe: *Symphonia gabonensis* Pierre; *Guttifer.;* B.-K.
Nshene: *Afzelia quanzensis* Welw.; *Caesalpiniac.;* S.-Af. (Transval)
N'sikou: *Pachylobus büttneri* Engl.; *Burserac.;* Gab.
Nsing: *Irvingia gabonensis* Baill.; *Simarubac.;* Nig.
Nsingha: *Musanga smithi* R. Br.; *Morac.;* Gab.
? Nsiré: *Detarium sp.;* *Caesalpiniac.;* Gab.
Nsiré: *Strephonema sp.;* *Combretac.;* Gab.
Nsoko: *Garcinia manni* Oliv.; *Guttifer.;* Go.
Nsoko-dua: *Garcinia manni* Oliv.; *Guttifer.;* Go.
Nsoma: *Strombosiopsis rigida* Engl.; *Olacac.;* Gab.
Nsone-essoue: *Lavalleopsis densivenia* Engl.; *Olacac.;* Gab.
„ -sidé: *Heisteria trillesiana* Pierre; *Olacac.;* Gab.
„ -so: *Lavalleopsis densivenia* Engl.; *Olacac.;* Gab.
N'sonesou: *Grewia coriacea* Mast.; *Tiliac.;* Kam.
N'soro: *Scyphocephalium ochocoa* Warb., *S. grandifolium* Pierre; *Myristicac.;* Gab.
Nsossongo: *Rauwolfia macrophylla* Stapf; *Apocynac.;* Gab.
Nsotnjoc: *Kigelia africana* Bth.; *Bignoniac.;* Kam.
N'soultou: *Scyphocephalium ochocoa* Warb.; *Myristicac.;* Gab.

Nsu: *Daniella oblonga* Oliv.; *Caesalpiniac.*; Kam.
Ntaba: *Cola cordifolia* R. Br. var. ?; *Sterculiac.*; n. Elf.
Ntaguaya: *Mimusops lacera* Baker; *Sapotac.*; Elf.
Ntagué hié: *Rhizophora racemosa* G. F. W. Meyer; *Rhizophorac.*; Elf.
Ntakorowa, small: *Talbotiella genti* Hutch. et Greenway; *Caesalpiniac.*; Go.
Ntaleyerunga: *Zanthoxylum sp.*; *Rutac.*; br. O.-Af. (Uganda)
Ntan: *Rhizophora racemosa* G. F. W. Meyer; *Rhizophorac.*; Gab.
Ntand: *Rhizophora racemosa* G. F. W. Meyer; *Rhizophorac.*; Gab.
Ntanda: *Rhizophora racemosa* G. F. W. Meyer; *Rhizophorac.*; Gab.
Ntandi: *Kigelia aethiopica* Dcne.; *Bignoniac.*; D.-O.-Af.
Ntane: *Rhizophora racemosa* G. F. W. Meyer; *Rhizophorac.*; Gab.
Ntanga-tanga: *Albizzia sp.*; *Mimosac.*; s. C.-Af.
Ntangani: *Cocos nucifera* L.; *Palmac.*; Gab.
N'tébraïa: *Diospyros sanza-minika* Chev.; *Ebenac.*; Elf.
Ntèna: *Copaifera ? salikounda* Heckel; *Caesalpiniac.*; Gab.
Nténé: *Ficus vogeliana* Miq.; *Morac.*: Gab.
Ntenga: *Desbordesia insignis* Pierre; *Simarubac.*; Gab.
Ntèta: *Klainedoxa gabonensis* Pierre; *Simarubac.*; Gab.
Ntèva: *Desbordesia insignis* Pierre; *Simarubac.*; Gab.
N'ti bioume: *Dialium aff. connaroides* Harms; *Caesalpiniac.*; Gab. (Loango)
N'tiogo: *Spathodea campanulata* Beauv.; *Bignoniac.*; Gab.
Ntiombi: *Uapaca heudeloti* Baill.; *Euphorbiac.*; Gab.
Ntioubou: *Newtonia insignis* Baill.; *Caesalpiniac.*; Gab.
Ntioumbo: *Eugenia sp., Syzygium sp.*; *Myrtac.*; Gab.
Ntogolo: *Pseudospondias gigantea* Chev.; *Anacardiac.*; Gab.
Ntolokya: *Juniperus procera* Hochst.; *Conifer.*; br. O.-Af. (Uganda)
Ntom: *Pachypodanthium confine* Engl. et Diels; *Anonac.*; Kam., Gab
N'toma: *Sarcocephalus trillesi* Pierre, *S. pobeguini* Pob.; *Rubiac.*; Kam., Gab.
N'tombo: *Pachylobus balsamifera* Guill.; *Burserac.*; Gab.
    „   : *Uapaca sp.*; *Euphorbiac.*; Gab.
Ntomboti: *Excoecaria africana* Muell. Arg.; *Euphorbiac.*; sö. S.-Af. (Swaziland)
N'tomi: *Tamarindus indica* L.; *Caesalpiniac.*; Sudan.
Ntouleng: *Xylopia rubescens* Oliv.; *Anonac.*; Kam.
Ntoum: *Piptadenia sp.*; *Mimosac.*; Gab.
Ntouma: *Pachypodanthium confine* Engl. et Diels; *Anonac.*; Kam.
Ntovo: *Mitragyne macrophylla* Hiern; *Rubiac.*; Gab.
N'towo: *Mitragyne macrophylla* Hiern; *Rubiac.*; Elf., Gab.
Ntukuza: *Memecylon myrianthum* Gilg; *Melastomatac.*; br. O.-Af. (Uganda)
Ntuli: *Ongokea klaineana* Pierre; *Olacac.*; B.-K.
N'tyombi: *Uapaca sp.*; *Euphorbiac.*; Gab.
Ntzungu: *Dialium yambataense* Verm.; *Caesalpiniac.*; B.-K.
Nuati kuruzu: *Achatocarpus bicornutus* Schinz et Autran; *Phytolaccac.*; Arg.
    „   morotí: *Adelia membranifolia* Pax et K. Hoffm.; *Euphorbiac.*; Arg.
Nublay: *Polyalthia cerasoides* Bth. et Hook. f.; *Anonac.*; O.-Ind. (Madras)
Nûc nâc: *Oroxylum indicum* Vent.; *Bignoniac.*; Annam.
Nu-eh-blay-chu: *Casearia dinklagei* Gilg; *Flacourtiac.*; Lib.
Nuez: *Juglans cinerea* L.; *Juglandac.*; P.-R.
    „   moscada: *Ocotea sp., Nectandra sp.*; *Laurac.*; P.-R.
    „   cimmarróna: *Ocotea sp., Nectandra sp.*; *Laurac.*; P.-R.
    „   del pais: *Ocotea sp., Nectandra sp.*; *Laurac.*; P.-R.
    „   de calatola: *Calatola mollis* Standl.; *Icacinac.*; Mex.
    „   „ Queensland: *Macadamia ternifolia* F. v. M.; *Proteac.*; Cuba.
Nujunge: *Faurea speciosa* Welw., *F. usambarensis* Engl.; *Proteac.*; D.-O.-Af.
Nukasen: *Ulmus spp.*; *Ulmac.*; ö. Sib., ? Jap.
Nukui: *Holarrhena africana* A. DC.; *Apocynac.*; W.-Af. (S.-L.)
Nul ne s'y frotte: *Xanthostemon rubrum* Nied.; *Myrtac.*; N.-Kal.
    „   „   „ frisé: *Pleurocalyptus deplanchei* Brongn. et Griseb.; *Myrtac.*; N.-Kal.
Numhpunkap: *Albizzia lucida* Bth.; *Mimosac.*; Burma.
Numraw: *Schima wallichi* Choisy; *Theac.*; Burma.
Numugu: *Mimusops djave* Engl.; *Sapotac.*; W.-Af.
Nuna: *Morinda tinctoria* Roxb.; *Rubiac.*; O.-Ind. (Madras)

Nuna-akalu: Canthium didymum Roxb.; Rubiac.; C.-O.-Ind.
Nune: Hura crepitans L.; Euphorbiac.; Pan.
Nung: Barringtonia pterocarpa Kurz; Lecythidac.; Ind.-Ch.
Nungka: Lophira alata Banks; Ochnac.; W.-Af. (S.-L.)
Nungo: Fagara sp.; Rutac.; Kam., Gab.
Nungu: Mimusops africana Lecomte; Sapotac.; Gab.
„   -tsende: Fagara macrophylla Engl.; Rutac.; B.-K.
Nunnera: Anogeissus acuminata Wall.; Combretac.; O.-Ind. (Madras)
Nur: Cocos nucifera L.; Palmac.; O.-Ind. (Mysore)
Nurkal: Buchanania latifolia Roxb.; Anacardiac.; w. O.-Ind. (Bombay-Deccan)
Nurude: Rhus javanica L.; Anacardiac.; Jap.
Nusanali: Olea chrysophylla Lam.; Oleac.; br. O.-Af. (Uganda)
Nuso: Cleistopholis patens Engl. et Diels; Anonac.; Togo.
Nuss, Bitter- (d): Carya amara Nutt.; Juglandac.; ö. N.-Am.
„ , Butter- (d): Juglans cinerea L.; Juglandac.; N.-Am
„ , Ferkel- (d): Carya porcina Nutt.; Juglandac.; ö. USA.
„ , Flügel-, kaukasische (d): Pterocarya fraxinifolia Spach, P. caucasica C. A. Mey.;
„ , Grau- (d): Juglans cinerea L.; Juglandac.; N.-Am.   [Juglandac.; Kauk.
„ , Satin- (d): Liquidambar styraciflua L.; Hamamelidac.; USA.
„ , Schwarz- (d): Juglans nigra L.; Juglandac.; N.-Am.
Nussangue: Khaya anthotheca C. DC.; Meliac.; W.-Af.
Nussbaum (d): Juglans spp.; Juglandac.; Mitt.- & S.-Eu., As., N.-Am.
„   „ : Juglans regia L.; Juglandac.; s. Eu., Kauk.
„ , afrikanisch (d): Lovoa klaineana Pierre; Meliac.; W.-Af.
„ , amerikanisch (d): Juglans nigra L.; Juglandac.; N.-Am.
„ , Brasil- (d): Bertholletia exelsa Bonpl., B. nobilis Miers; Lecythidac.; Bras., Gu.
„ , grauer (d): Juglans cinerea L.; Juglandac.; N.-Am.
„ , ostindischer (d): Albizzia lebbek Bth.; Mimosac.; As.
„ , Pinien- (d): Pinus edulis Engelm.; Conifer.; n. Mex.
„ , Satin- (d): Liquidambar styraciflua L.; Hamamelidac.; s. N.-Am., C.-Am.
„ , türkischer (d): Juglans regia L.; Juglandac.; Türkei.
Nut, australian: Macadamia ternifolia F. v. M.; Proteac.; ö. Au.
„ , Brazil-: Bertholletia excelsa Bonpl.; Lecythidac.; n. Bras. (Amaz.)
„ , Butter-: Caryocar tomentosum Willd., C. glabrum Pers.; Caryocarac.; Sur.,
„ ,   „ : Caryocar villosum Pers.; Carycarac.; Sur., br. Gu.   [br. Gu.
„ , Cashew-: Anacardium occidentale L.; Anacardiac.; S.-Am.
„ , Clearin-: Strychnos potatorum L. f.; Loganiac.; O.-Ind.
„ , Dika-: Irvingia gabonensis Baill.; Simarubac.; W.-Af.
„ , Gaboon-: Coula edulis Baill.; Olacac.; Nig.
„ , Oil-: Juglans cinerea L.; Juglandac.; N.-Am.
„ , Opossum-: Macadamia praealta F. M. Bailey; Proteac.; ö. Au.
„ , Queensland-: Macadamia sp.; Proteac.; ö. Au.
„ , Sapucaia-: Lecythis ollaria L.; Myrtac.; Malay.
„ , white: Heisteria parvifolia Smith; Olacac.; Nig.
Nutmeg, african: Pycnanthus kombo Warb.; Myristicac.; Nig.
„ , „ : Monodora tenuifolia Bth.; Anonac.; Nig.
„ , Calabash-: Monodora myristica Dunal; Anonac.; Lib., Go.
„ , California (u): Tumion californicum Greene; Conifer.; USA.
„ , false: Pycnanthus schweinfurthi Warb.; Myristicac.; br. O.-Af. (Uganda)
„ , wild: Coelocaryon aff. oxycarpum Stapf; Myristicac.; Lib.
„ , „ : ? Ocotea pentagona Mez; Laurac.; Pan.
„ , „ : Rheedia lateriflora L.; Guttifer.; Trin.
„ , yellow-flowering: Monodora brevipes Bth.; Anonac.; Nig., Lib
Nutree: Fusanus cygnorum Bth.; Santalac.; Au.
Ñutssu-ñutssu: Cola cordifolia R. Br.; Sterculiac.; Togo.
Nutwood: Dicorynia paraënsis Bth.; Caesalpiniac.; holl. Gu.
N'vana: Didelotia duparquetiana ?; Caesalpiniac.; Gab.
Nvedzo: Guarea thompsoni Sprague et Hutch.; Meliac.; Elf.
Nwa: Terminalia glaucescens Planch.; Combretac.; Go.
Nwuama: Ricinodendron africanum Muell. Arg.; Euphorbiàc.; Go.
Ny: Shorea vulgaris Pierre; Dipterocarpac.; Ind.-Ch

Nya: *Acacia catechu* Willd.; *Mimosac.;* Burma.
Nyabuliko: *Gymnosporia senegalensis* Loes.; *Celastrac.;* br. O.-Af. (Uganda)
Nyakatoma: *Morus lactea* Mildbr.; *Morac.;* br. O.-Af. (Uganda)
Nyakpekpe: *Kigelia africana* Bth.; *Bignoniac.;* Togo.
Nyamá: *Bauhinia reticulata* DC.; *Caesalpiniac.;* Togo.
Nyamelebaka: *Alstonia congensis* Engl.; *Apocynac.;* Go.
Nyan: *Quercus serrata* Thunb.; *Fagac.;* Burma.
Nyanda: *Albizzia pospischili* Harms; *Mimosac.;* D.-O.-Af. (Kilossa)
Nyandwe: *Erythrophloeum africanum* Harms; *Caesalpiniac.;* B.-K. (Katanga)
Nyangala: *Aucoumea klaineana* Pierre; *Burserac.;* Gab., sp. Guin.
Nyangon: *Cola proteiformis* Chev.; *Sterculiac.;* Elf.
Nyankom: *Tarrietia utilis* Sprague; *Sterculiac.;* Go.
Nyanwen: *Tarrietia utilis* Sprague; *Sterculiac.;* Go.
Nyanwune: *Tarrietia utilis* Sprague; *Sterculiac.;* Go.
Nyanya ez'omuti: *Cyphomandra hartwegi* Sendt.; *Solanac.;* br. O.-Af. (Uganda)
Nyatoh: *Ganua motleyana* Pierre, *Madhuca spp., Palaquium spp.; Sapotac.;* Malay.
„   : *Payena lucida* A. DC., *P. maingayi* Clarke; *Sapotac.;* Malay.
„ bukit: *Madhuca watsoni* Ridl.; *Sapotac.;* Malay (Kuantan)
„ jangkang: *Palaquium spp.; Sapotac.;* Malay.
Nyatto pisang: *Palaquium sp.; Sapotac.;* br. Borneo (Sarawak)
Nyaung-bawdi: *Ficus religiosa* L.; *Morac.;* Burma.
„ -thabye: *Ficus glomerata* Roxb., *F. indica* L.; *Morac.;* Burma.
Nyé: *Ceiba pentandra* Gaertn. var. *dehiscens* Ulb.; *Bombacac.;* Elf.
„ -fe-ohn: *? Chytranthus setosus* Radlk.; *Sapindac.;* Lib.
Nyege: *Canarium schweinfurthi* Engl.; *Burserac.;* br. O.-Af. (Uganda)
Nyegye: *Canarium schweinfurthi* Engl.; *Burserac.;* br. O.-Af. (Uganda)
Nyi: *Acrocarpus fraxinifolius* Wight et Arn.; *Caesalpiniac.;* Burma.
Nyie-avu: *Albizzia zygia* MacBride; *Mimosac.;* Nig.
Nyieri: *Carapa spp.; Meliac.;* Burma (s. Tenasserim)
Nyimo: *Sarcocephalus sambucinus* K. Schum.; *Rubiac.;* Togo.
Nyimu: *Erythrina senegalensis* DC.; *Papilionac.;* Togo.
Nyina: *Ceiba pentandra* Gaertn.; *Bombacac.;* Go.
Nyireh: *Xylocarpus granatum* Koenig; *Meliac.;* br. Borneo (Sarawak)
„ batu: *Xylocarpus granatum* Koenig; *Meliac.;* br. Borneo (Sarawak)
Nyon: *Pentadesma leptonema* Pierre; *Guttifer.;* Kam.
Nyonne: *Macrolobium aff. limba* Scott Elliott; *Caesalpiniac.;* Kam.
„   : *Allanblackia klainei* Pierre; *Guttifer.;* Kam.
Nyonzo: *Allanblackia klainei* Pierre; *Guttifer.;* B.-K.
Nyowé: *Staudtia gabonensis* Warb.; *Myristicac.;* Gab.
Nyssa sylvestre (f): *Nyssa silvatica* Marsh.; *Cornac.;* ö. Kan. (sw. Ontario)
Nzali: *Pterocarpus soyauxi* Taub.; *Papilionac.;* B.-K.
Nzan: *Odyendea gabunensis* Engl.; *Simarubac.;* Gab.
Nzingu: *Mitragyne macrophylla* Hiern; *Rubiac.;* br. O.-Af. (Uganda)
Nzo: *Teclea nobilis* Delile; *Rutac.;* br. O.-Af. (Uganda)
N'zongui: *Daniella sp.; Caesalpiniac.;* Gab.
N'zonogwe: *Triumfetta rhomboidea* Lindl.; *Tiliac.;* br. O.-Af. (Niassa)
N'zou aillé: *Ricinodendron africanum* Muell. Arg.; *Euphorbiac.;* Kam.
Nzuza: *Pseudospondias microcarpa* Engl.; *Anacardiac.;* B.-K.
Nzybavoum: *Uvariastrum sp.; Anonac.;* Gab.

O

O-aw: *Adina cordifolia* Hook. f.; *Rubiac.;* Burma.
O rô: *Acanthus volubilis* Wall.; *Acanthac.;* Ind.-Ch.
Oabé: *Mimusops pierreana* Engl.; *Sapotac.;* Gab.
Oa-bo: *Calocarpum mammosum* Pierre; *Sapotac.;* Pan.
Oak (e): *Quercus spp.; Fagac.;* n. Hem.
„   : *Quercus pedunculata* Ehrh., *Q. sessiliflora* Salisb.; *Fagac.;* Eu.
„ : *Quercus corrugata* Hook.; *Fagac.;* br. Hond.
„   : *Picralima umbellata* Stapf; *Apocynac.;* Nig.
„   : *Quercus spp., Pasania spp.; Fagac.;* Malay.

Oak, african (e, u): *Chlorophora excelsa* Bth. et Hook.; *Morac.;* W.- & O.-Af.
   „   ,   „   : *Lophira alata* Banks; *Ochnac.;* Lib.
   „   ,   „   : *Oldfieldia africana* Bth. et Hook. f.; *Euphorbiac.;* Lib.
   „   , australian: *Casuarina equisetifolia* Forst.; *Casuarinac.;* Au.
   „   ,   „   : *Eucalyptus regnans* F. v. M.; *Myrtac.;* sö. Au. (Victoria), Tasm.
   „   , Barren-: *Quercus nigra* L.; *Fagac.;* N.-Am.
   „   , Basket-: *Quercus michauxi* Nutt.; *Fagac.;* N.-Am.
   „   , Bear-: *Quercus ilicifolia* Wang.; *Fagac.;* ö. USA.
   „   , black: *Quercus velutina* Lam., *Q. rubra* L., *Q. nigrescens* Rafin.; *Fagac.;* N.-Am.
   „   ,   „   : *Casuarina suberosa* Otto et Dietr.; *Casuarinac.;* ö. Au.
   „   , blue: *Quercus bicolor* Willd.; *Fagac.;* ö. N.-Am.
   „   , Bog-: *Quercus robur* L.; *Fagac.;* Gross-Britannien (Irland)
   „   , Box-, white: *Quercus stellata* Wang.; *Fagac.;* ö. USA.
   „   , Brash-, white: *Quercus alba* L.; *Fagac.;* N.-Am.
   „   , british (e): *Quercus pedunculata* Ehrh., *Q. sessiliflora* Salisb.; *Fagac.;* Gross-
   „   , British Columbia-: *Quercus garryana* Dougl.; *Fagac.;* w. Kan.   [Britan.
   „   , brown: *Quercus pedunculata* Ehrh., *Q. sessiliflora* Salisb.; *Fagac.;* England.
   „   ,   „   : *Tarrietia argyrodendron* Bth.; *Sterculiac.;* ö. Au.
   „   ,   „   : *Darlingia spectatissima* F. v. M.; *Proteac.;* ö. Au. (N.-S.-W.)
   „   ,   „   : *Musgravea stenostachya* F. v. M.; *Proteac.;* ö. Au. (N.-S.-W.)
   „   ,   „   , himalayan: *Quercus semecarpifolia* Smith; *Fagac.;* Afg., Himal.
   „   , Bull-: *Casuarina luehmanni* R. T. Baker; *Casuarinac.;* ö. Au. (Victoria)
   „   ,   „   : *Casuarina suberosa* Otto et Dietr.; *Casuarinac.;* Tasm.
   „   ,   „   : *Cardwellia sublimis* F. v. M.; *Proteac.;* Au.
   „   , Bur-: *Quercus macrocarpa* Michx.; *Fagac.;* ö. N.-Am.
   „   , Burr-: *Quercus pedunculata* Ehrh., *Q. sessiliflora* Salisb.; *Fagac.;* Gross-Brit.
   „   , Ceylon-: *Schleichera trijuga* Willd.; *Sapindac.;* tr. As.
   „   , Chestnut-: *Quercus montana* Willd., *Q. prinus* L.; *Fagac.;* N.-Am.
   „   ,   „   : *Quercus muehlenbergi* Engelm.; *Fagac.;* sö. Kan.
   „   ,   „   , Rock-: *Quercus prinus* L.; *Fagac.;* ö. N.-Am.
   „   ,   „   , Swamp-: *Quercus prinus* L.; *Fagac.;* USA.
   „   ,   „   , Upland-: *Quercus castanea* Née; *Fagac.;* N.-Am.
   „   , Chinquapin-: *Quercus muehlenbergi* Engelm., *Q. prinoides* Willd.; *Fagac.;* USA.
   „   ,   „   , Dwarf-: *Quercus muehlenbergi* Engelm., *Q. prinoides* Willd.; *Fagac.;*
   „   , common: *Quercus penduculata* Ehrh.; *Fagac.;* Gross-Britannien.   [N.-Am.
   „   , Durmast- (e): *Quercus sessiliflora* Salisb.; *Fagac.;* C.- & S.-Eu.
   „   , English: *Quercus pedunculata* Ehrh., *Q. sessiliflora* Salisb.; *Fagac.;* Engl.
   „   , European: *Quercus pedunculata* Ehrh., *Q. sessiliflora* Salisb.; Eu.
   „   , Evergreen-: *Quercus ilex* L.; *Fagac.;* M.-M.
   „   , Fiddleback-: *Eucalyptus sp.; Myrtac.;* Au.
   „   , Forest-: *Casuarina torulosa* Ait.; *Casuarinac.;* ö. Au.
   „   , Formosan: *Quercus pseudo-myrsinaefolia* Hay., *Q. sp.; Fagac.;* Form.
   „   , Garry-: *Quercus garryana* Dougl.; *Fagac.;* w. Kan.
   „   , Himalayan: *Quercus spicata* Smith; *Fagac.;* O.-Ind.
   „   , Holly-: *Quercus ilex* L.; *Fagac.;* N.-Af.
   „   ,   „   : *Quercus mori* Hayata; *Fagac.;* Form.
   „   , Holm-: *Quercus ilex* L.; *Fagac.;* M.-M.
   „   , home-grown (e): *Quercus pedunculata* Ehrh., *Q. sessiliflora* Salisb.; *Fagac.;*
   „   , Indian: *Tectona grandis* L. f.; *Verbenac.;* O.-Ind., Burma, Siam.   [Gross-Brit.
   „   ,   „   : *Barringtonia acutangula* Gaertn.; *Lecythidac.;* tr. As.
   „   , Iron-: *Quercus stellata* Wang.; *Fagac.;* ö. USA.
   „   , Japanese: *Quercus grosseserrata* Blume, *Q. crispula* Blume; *Fagac.;* Jap.
   „   ,   „   : *Quercus glandulifera* Blume, *Q. dentata* Thunb.; *Fagac.;* Jap.
   „   , Kellogg-: *Quercus californica* Cooper; *Fagac.;* N.-Am.
   „   , Live- (u): *Quercus virginiana* Mill.; *Fagac.;* USA.
   „   ,   „   : *Quercus chrysolepsis* Liebm., *Pasania densiflora* Oerst,; *Fagac.;* USA.
   „   ,   „   , Canyon-: *Quercus chrysolepis* Liebm.; *Fagac.;* ö. USA.
   „   , Maul-: *Quercus chrysolepis* Liebm.; *Fagac.;* w. N.-Am.
   „   , Mossy-: *Quercus cerris* L.; *Fagac.;* s. Eu.
   „   ,   „   : *Quercus macrocarpa* Michx.; *Fagac.;* sö. Kan.
   „   , Mossy-cup-: *Quercus macrocarpa* Michx.; *Fagac.;* ö. N.-Am.

379

Oak, Mountain-: *Quercus prinus* L.; *Fagac.*; N.-Am.
„ , „ : *Casuarina torulosa* Ait.; *Casuarinac.*; ö. Au.
„ , „ , white: *Quercus douglasi* Hook. et Arn.; *Fagac.*; N.-Am.
„ , Native-: *Avicennia nitida* Jacq.; *Verbenac.*; Jam.
„ , Nut-: *Macadamia ternifolia* F. v. M.; *Proteac.*; ö. Au.
„ , Oregon-: *Quercus garryana* Dougl.; *Fagac.*; w. N.-Am.
„ , „ , white: *Quercus garryana* Dougl.; *Fagac.*; N.-Am.
„ , Overcup-: *Quercus lyrata* Walt., *Q. macrocarpa* Michx.; *Fagac.*; ö. N.-Am.
„ , Pacific-, white: *Quercus garryana* Dougl.; *Fagac.*; w. N.-Am.
„ , Peach-: *Quercus phellos* L., *Pasania densiflora* Oerst.; *Fagac.*; N.-Am.
„ , Pear-: *Xylomelum pyriforme* Knight; *Proteac.*; ö. Au. (Queensl.)
„ , Pedunculate: *Quercus pedunculata* Ehrh.; *Fagac.*; Gross-Britannien.
„ , Pin-: *Quercus palustris* Muench.; *Fagac.*; ö. USA.
„ , Pollard-: *Quercus pedunculata* Ehrh., *Q. sessiliflora* Salisb.; *Fagac.*; Gross-Brit.
„ , Possum-: *Quercus aquatica* Walt.; *Fagac.*; USA.
„ , Post-: *Quercus stellata* Wang.; *Fagac.*; ö. N.-Am.
„ , „ -, pacific: *Quercus garryana* Dougl.; *Fagac.*; w. N.-Am.
„ , Punk-: *Quercus aquatica* Walt.; *Fagac.*; USA.
„ , Quebec-: *Quercus alba* L.; *Fagac.*; N.-Am.
„ , Quercitron-: *Quercus velutina* Lam.; *Fagac.*; N.-Am.
„ , red (u): *Quercus borealis-maxima* Ashe; *Fagac.*; USA.
„ , „ : *Quercus rubra* L., *Q. falcata* Michx., *Q. texana* Buckl.; *Fagac.*; N.-Am.
„ , „ : *Lophira procera* Chev.; *Ochnac.*; W.-Af.
„ , „ : *Berlinia spp.*; *Caesalpiniac.*; Lib.
„ , „ : *Carnarvonia araliaefolia* F. v. M.; *Proteac.*; ö. Au. (N.-S.-W.)
„ , „ , Atherton: *Carnarvonia araliaefolia* F. v. M.; *Proteac.*; ö. Au. (N.-S.-W.)
„ , „ , northern: *Quercus borealis-maxima* Ashe; *Fagac.*; USA.
„ , „ , southern: *Quercus rubra* L.; *Fagac.*; USA.
„ , „ , Swamp-: *Quercus rubra-pagodaefolia* Ashe; *Fagac.*; USA.
„ , „ , Texas-: *Quercus texana* Buckley; *Fagac.*; USA.
„ , River-: *Casuarina cunninghamiana* Miq.; *Casuarinac.*; ö. Au. (N.-S.-W.)
„ , Rock-: *Quercus prinus* L., *Q. muehlenbergi* Engelm.; *Fagac.*; N.-Am.
„ , Rose-: *Grevillea gibbosa* R. Br.; *Proteac.*; ö. Au. (N.-S.-W.)
„ , Russian: *Quercus pedunculata* Ehrh., *Q. sessiliflora* Salisb.; *Fagac.*; Russland
„ , Scarlet-: *Quercus coccinea* Muench.; *Fagac.*; USA.
„ , Scrub-: *Quercus macrocarpa* Michx., *Q. prinoides* Willd.; *Fagac.*; ö. N.-Am.
„ , „ , black: *Quercus ilicifolia* Wang.; *Fagac.*; ö. USA.
„ , „ , Chestnut-: *Quercus prinoides* Willd.; *Fagac.*; USA.
„ , „ , Sessile-: *Quercus sessiliflora* Salisb.; *Fagac.*; C.- & S.-Eu.
„ , She-: *Casuarina spp.*; *Casuarinac.*; ö. Au. (N.-S.-W.)
„ , „ (e): *Casuarina fraseriana* Miq.; *Casuarinac.*; Au.
„ , „ : *Casuarina equisetifolia* Forst.; *Casuarinac.*; Au., Tr.
„ , „ : *Casuarina stricta* Ait., *C. suberosa* Otto et Dietr.; *Casuarinac.*; sö. Au.
„ , „ , australian: *Casuarina equisetifolia* Forst.; *Casuarinac.*; Au., Tr.
„ , „ , Forest-: *Casuarina torulosa* Ait.; *Casuarinac.*; ö. Au. (N.-S.-W.)
„ , Shea-: *Casuarina fraseriana* Miq.; *Casuarinac.*; w. Au.
„ , Silky: *Carnarvonia araliaefolia* F. v. M.; *Proteac.*; ö. Au. (N.-S.-W.)
„ , „ : *Embothrium wickhami* F. v. M.; *Proteac.*; ö. Au. (N.-S.-W.)
„ , „ : *Grevillea gibbosa* R. Br., *G. hilliana* F. v. M.; *Proteac.*; N.-S.-W.
„ , „ : *Darlingia spectatissima* F. v. M.; *Proteac.*; N.-S.-W.
„ , „ : *Musgravea stenostachya* F. v. M.; *Proteac.*; N.-S.-W.
„ , „ : *Stenocarpus sinuatus* Endl., *S. salignus* R. Br.; *Proteac.*; ö. Au.
„ , „ : *Cardwellia sublimis* F. v. M.; *Proteac.*; ö. Au. (n. Queensl.)
„ , „ : *Orites excelsa* R. Br.; *Proteac.*; ö. Au. (s. Queensl., N.-S.-W.)
„ , „ : *Grevillea robusta* A. Cunn.; *Proteac.*; ö. Au. (s. Queensl., N.-S.-W.)
„ , „ , australian (e): *Cardwellia sublimis* F. v. M.; *Proteac.*; n. Queensl.
„ , „ „ „ : *Grevillea robusta* A. Cunn., *Orites excelsa* R. Br.; *Proteac.*;
„ , „ , Bull-: *Cardwellia sublimis* F. v. M.; *Proteac.*; Queensl. [ö. Au.
„ , „ , goldsprinkled: *Cardwellia sublimis* F. v. M.; *Proteac.*; N.-S.-W.
„ , „ , grey: *Stenocarpus sinuatus* Endl.; *Proteac.*; ö. Au., Tasm.
„ , „ , northern (e): *Orites excelsa* R. Br.; *Proteac.*; ö. Au.

Oak, Silky, pink: *Orites excelsa* R. Br.; *Proteac.;* ö. Au.
„ „ , prickly-leaved: *Orites excelsa* R. Br.; *Proteac.;* N.-S.-W.
„ „ , Queensland-: *Cardwellia sublimis* F. v. M.; *Proteac.;* n. Queensl.
„ „ „ : *Grevillea robusta* A. Cunn.; *Proteac.;* ö. Au.
„ „ „ : *Orites excelsa* R. Br.; *Proteac.;* ö. Au.
„ „ , red: *Stenocarpus salignus* R. Br.; *Proteac.;* ö. Au. (N.-S.-W.)
„ „ „ : *Grevillea hilliana* F. v. M.; *Proteac.;* ö. Au. (N.-S.-W.)
„ „ „ : *Embothrium wickhami* F. v. M.; *Proteac.;* N.-S.-W.
„ „ , satin: *Embothrium wickhami* F. v. M.; *Proteac.;* ö. Au.
„ „ , southern: *Grevillea robusta* A. Cunn.; *Proteac.;* ö. Au.
„ „ , white: *Stenocarpus sinuatus* Endl.; *Proteac.;* N.-S.-W.
„ , Stave-: *Quercus alba* L.; *Fagac.;* sö. Kan.
„ , Swamp-: *Quercus palustris* Muench.; *Fagac.;* ö. N.-Am.
„ , „ : *Casuarina sp.; Casuarinac.;* w. Au.
„ , „ : *Casuarina glauca* Sieber; *Casuarinac.;* ö. Au. (Victoria)
„ „ , curly: *Quercus bicolor* Willd.; *Fagac.;* ö. USA.
„ „ , white: *Quercus bicolor* Willd.; *Fagac.;* ö. USA.
„ , tasmanian: *Eucalyptus gigantea* Hook. f., *E. regnans* F. v. M.; *Myrtac.;* sö. Au.,
„ , thready-bark: *Casuarina inophloia* F. v. M. et Bailey; *Casuarinac.;* ö. Au. [Tasm.
„ , Tulip-: *Tarrietia actinophylla* F. M. Bailey, *T. argyrodendron* Bth.; *Sterculiac.;*
„ „ , blush: *Tarrietia actinophylla* F. M. Bailey; *Sterculiac.;* ö. Au. [ö. Au.
„ , Turkey-: *Quercus cinerea* Michx.; *Fagac.;* N.-Am.
„ , Victoria-: *Eucalyptus regnans* F. v. M., *E. gigantea* Hook. f.; *Myrtac.;* sö. Au.,
„ , Water-: *Quercus aquatica* Walt., *Q. laurifolia* Michx.; *Fagac.;* N.-Am. [Tasm.
„ , „ : *Quercus nigra* L., *Q. palustris* Muench.; *Fagac.;* N.-Am.
„ , white: *Quercus alba* L.; *Fagac.;* ö. N.-Am.
„ , „ : *Stenocarpus sinuatus* Endl.; *Proteac.;* ö. Au.
„ , Willow-: *Quercus phellos* L.; *Fagac.;* USA.
„ , yellow: *Quercus muehlenbergi* Engelm., *Q. velutina* Lam.; *Fagac.;* ö. USA.
„ , yellow-barked: *Quercus velutina* Lam.; *Fagac.;* ö. N.-Am.
Oamab: *Albizzia fastigiata* Oliv.; *Mimosac.;* B.-K.
Oanani: *Symphonia globulifera* L. f.; *Guttifer.;* Bras.
Oanka: *Commiphora africana* Engl.; *Burserac.;* n. O.-Af. (Abessinien)
Oassacú: *Hura crepitans* L.; *Euphorbiac.;* Bras.
Ob assi: *Entandrophragma aff. rederi* Harms; *Meliac.;* Kam.
Oba: *Bombax buonopozense* Beauv.; *Bombacac.;* Elf.
„ : *Ceiba pentandra* Gaertn.; *Bombacac.;* Elf.
„ : *Irvingia gabonensis* Baill.; *Simarubac.;* Nig., Kam., Gab.
„ : *Eugenia sp.; Myrtac.;* Borneo.
Oba-akô (jap.): *Ficus stipulosa* Miq.; *Morac.;* tr. As.
„ -harulan-inubiwa (jap.): *Ficus harlandi* Bth. var. *kotoënsis* Sata; *Morac.;* Form.
„ -suluk: *Shorea polysperma* Merr.; *Dipterocarpac.;* Borneo.
„ w'akowa: *? Irvingia klainei* Pierre; *Simarubac.;* Gab.
„ w'alombo: *? Irvingia klainei* Pierre; *Simarubac.;* Gab.
Obaa: *Pentaclethra macrophylla* Bth.; *Mimosac.;* Gab.
Obachi: *Triplochiton scleroxylon* K. Schum.; *Sterculiac.;* W.-Af.
Obachou: *? Sterculia sp.; Sterculiac.;* Gab.
Obad-sagekroe: *Garcinia picrorrhiza* Miq.; *Guttifer.;* Nied.-Ind.
Obada: *Pentaclethra macrophylla* Bth.; *Mimosac.;* Gab.
Obadan: *Ficus vogeli* Miq.; *Morac.;* Nig.
Obaeton: *Conopharyngia penduliflora* Stapf; *Apocynac.;* Kam.
Obaku: *Evodia glauca* Miq.; *Rutac.;* Jap.
Obala: *Pycnanthus kombo* Warb.; *Myristicac.;* Togo.
„ : *Pentaclethra macrophylla* Bth.; *Mimosac.;* Gab.
Obamba: *Croton oligandrum* Pierre; *Euphorbiac.;* Gab.
Obar suluk: *Shorea polysperma* Merr.; *Dipterocarpac.;* br. Nord-Borneo.
Obashinanoki: *Tilia maximoviczi* Shiras.; *Tiliac.;* Jap.
Obbo: *Cussonia barteri* Seem.; *Araliac.;* Togo.
Obé: *Canarium schweinfurthi* Engl.; *Burserac.;* Gab.
Obeche (e): *Triplochiton scleroxylon* K. Schum., *T. nigericum* Spr.; *Sterculiac.;* Nig.
„ ovbekhe (e): *Triplochiton scleroxylon* K. Schum.; *Sterculiac.;* Nig., Go.

Obechi: *Triplochiton scleroxylon* K. Schum., *T. nigericum* Spr.; *Sterculiac.;* Nig.
Obega: *Canarium schweinfurthi* Engl.; *Burserac.;* W.-Af. (? Gab.)
Obélé: *Canarium velutinum* Guill.; *Burserac.;* Gab.
Obero: *Astronium graveolens* Jacq.; *Anacardiac.;* Guat.
Obéro: *Picralima klaineana* Pierre; *Apocynac.;* Kam., Gab.
„ : *Picralima nitida* Th. et H. Dur., *P. umbellata* Stapf; *Apocynac.;* Gab. (Mayombe)
Obi: *Cola acuminata* Schott et Endl.; *Sterculiac.;* Togo, Go.
„ : *Klainedoxa gabonensis* Pierre; *Simarubac.;* Kam.
Obianguan: *Rhaptopetalum soyauxi* Oliv.; *Scytopetalac.;* Gab.
Obo: *Erythrophloeum guineense* G. Don; *Caesalpiniac.;* Nig.
„ de zopilote: *Spondias sp.; Anacardiac.;* Mex.
Oboba: *Ficus exasperata* Vahl; *Morac.;* Nig.
Obobonekhui: *Guarea thompsoni* Sprague et Hutch.; *Meliac.;* Nig., Go.
Obobonikiwi: *Guarea thompsoni* Sprague et Hutch.; *Meliac.;* Nig. (Benin)
Obolo: *Berlinia bracteosa* Bth.; *Caesalpiniac.;* Gab.
Obombé: *Elaeis guineensis* Jacq. var. *virescens* Chev.; *Palmac.;* Gab.
Obonawa: *Conopharyngia cumminsi* Stapf; *Apocynac.;* Go.
Oboountchoa: *Dactylopetalum barteri* Hook. f.; *Rhizophorac.;* Gab.
Obota: *Ochrocarpus africanus* Oliv.; *Guttifer.;* Elf.
Oboto: *Ochrocarpus africanus* Oliv.; *Guttifer.;* Elf.
„ : *Albizzia sp.; Mimosac.;* Elf.
„ : *Parinarium robustum* Oliv.; *Rosac.;* Gab.
„ : *Pynaertia occidentalis* Chev.; *Meliac.;* Gab.
Obotto: *Ochrocarpus africanus* Oliv.; *Guttifer.;* Elf.
Oboun m'lon: *Eremospatha macrocarpa* Wendl.; *Palmac.;* Kam.
Obrí: *Ormosia laxiflora* Bth.; *Papilionac.;* Togo.
Obungu: *Mimusops africana* Lecomte; *Sapotac.;* Gab.
Obwipe: *Podocarpus milanjianus* Rendle; *Conifer.;* br. O.-Af. (Uganda)
Ocatillo: *Cordia sp.; Borraginac.;* Mex.
Ochoco: *Scyphocephalium ochocoa* Warb.; *Myristicac.;* Gab.
Ochohó: *Hura crepitans* L.; *Euphorbiac.;* Bol.
Ochoko: *Scyphocephalium ochocoa* Warb.; *Myristicac.;* Gab.
Ochongo: *Anthostema aubryanum* Baill.; *Euphorbiac.;* Gab.
Ocobo: *Tabebuia pentaphylla* Hemsl.; *Bignoniac.;* Kol.
Ocote: *Pinus caribaea* Morelet, *P. oocarpa* Schiede; *Conifer.;* Guat., Hond.
„ blanco: *Pinus ayacahuite* Ehrenb., *P. montezumae* Lamb.; *Conifer.;* Mex.
„ „ : *Pinus leiophylla* Schlecht. et Cham.; *Conifer.;* Mex.
„ chino: *Pinus leiophylla* Schlecht. et Cham.; *Conifer.;* Mex.
„ hembro: *Pinus montezumae* Lamb.; *Conifer.;* Mex.
„ macho: *Pinus montezumae* Lamb., *P. oocarpa* Schiede; *Conifer.;* Mex.
Ocotillo: *Amyris elemifera* L.; *Rutac.;* Salv.
Ocotl: *Pinus spp.; Conifer.;* Mex.
Ocotzotl: *Liquidambar styraciflua* L.; *Hamamelidac.;* Mex.
Ocoula: *Coula edulis* Baill.; *Olacac.;* Gab.
Ocozol: *Liquidambar styraciflua* L.; *Hamamelidac.;* Mex.
Ocozote: *Liquidambar styraciflua* L.; *Hamamelidac.;* Mex.
Ocuje: *Calophyllum calaba* Jacq.; *Guttifer.;* Cuba.
Odahoma: *Piptadenia africana* Hook. f.; *Mimosac.;* Go.
Odahuma: *Piptadenia africana* Hook. f.; *Mimosac.;* Go.
Odal: *Sterculia villosa* Roxb.; *Sterculiac.;* n. O.-Ind. (United Prov.)
Odala: *Sterculia villosa* Roxb.; *Sterculiac.;* n. O.-Ind. (United Prov.)
„ : *Khaya grandifoliola* C. DC.; *Meliac.;* Nig.
Odani: *Piptadenia africana* Hook. f.; *Mimosac.;* Go.
Odanwoma: *Acacia adansoni* Guill. et Perr.; *Mimosac.;* Go.
O'dba: *Tamarix articulata* Vahl; *Tamaricac.;* Arab.
Odela: *Sterculia villosa* Roxb.; *Sterculiac.;* Assam.
Odenya: *Cylicodiscus gabunensis* Harms; *Papilionac.;* Go.
? Odenya: *Pentaclethra macrophylla* Bth.; *Mimosac.;* Go.
Odiau: *Pterocarpus blancoi* Merr., *P. echinatus* Perr.; *Papilionac.;* Phil. (Pangasinan)
„ : *Pterocarpus indicus* Willd.; *Papilionac.;* Phil. (Pangasinan)
Odiendlé: *Odyendea gabunensis* Engl.; *Simarubac.;* Gab.

Odienejé: *Odyendea gabunensis* Engl.; *Simarubac.;* Kam., Gab.
Odika: *Irvingia gabonensis* Baill.; *Simarubac.;* Gab.
Odiobozam: *Xylopia rubescens* Oliv.; *Anonac.;* Kam.
Odjakouna: *Scytopetalum sp.; Scytopetalac.;* Gab.
Odjengé: *Odyendea gabunensis* Engl.; *Simarubac.;* Gab.
Odjobi: *Xylopia rubescens* Oliv.; *Anonac.;* Kam.
Odjoué: *Xylopia africana* Oliv.; *Anonac.;* Kam.
Odjungé: *Mimusops africana* Lecomte; *Sapotac.;* Gab.
Odla: *Sterculia urens* Roxb.; *Sterculiac.;* Assam.
Odo: *Afzelia bipindensis* Harms; *Caesalpiniac.;* Nig.
Odoko: *Scottellia coriacea* Chev.; *Flacourtiac.;* Nig.
Odoliyan: *Afzelia bipindensis* Harms; *Caesalpiniac.;* Nig.
Odonomokyuku: *Santiriopsis klaineana* Pierre; *Burserac.;* Nig.
Odou: *Pachylobus edulis* G. Don; *Burserac.;* Gab.
Odoum: *Chlorophora excelsa* Bth. et Hook.; *Morac.;* Elf.
„ : *Ceiba pentandra* Gaertn.; *Bombacac.;* Kam., Gab.
Odu: *Celtis zenkeri* Engl.; *Ulmac.;* Kam.
Odubēn: *Haronga paniculata* Lodd.; *Guttifer.;* Go.
Odubin: *Khaya grandis* Stapf; *Meliac.;* Go.
Oduduku: *Sterculia tomentosa* Guill. et Perr.; *Sterculiac.;* Togo.
Odui: *Klainedoxa gabonensis* Pierre; *Simarubac.;* Kam.
Oduk: *Combretum splendens* Engl.; *Combretac.;* br. O.-Af. (Uganda)
Oduka: *Mimusops africana* Lecomte; *Sapotac.;* Gab.
Odum: *Chlorophora excelsa* Bth. et Hook.; *Morac.;* Go., Togo.
Odupon: *Khaya ivorensis* Chev.; *Meliac.;* Nig., Go.
Odwen: *Baphia nitida* Lodd., *B. pubescens* Hook. f.; *Papilionac.;* Go.
Odwene: *Baphia nitida* Lodd.; *Papilionac.;* Go.
Odwindwera: *Lecaniodiscus cupanioides* Planch.; *Sapindac.;* Go.
Odwindwina: *Lecaniodiscus cupanioides* Planch.; *Sapindac.;* Go.
Odwumā: *Musanga smithi* R. Br.; *Morac.;* Go.
Odyendié: *Odyendea gabunensis* Engl.; *Simarubac.;* tr. W.-Af.
Odyonni: *Phoenix spinosa* Schum. et Thonn.; *Palmac.;* Togo.
Odza kouna: *Scytopetalum sp.; Scytopetalac.;* Gab.
Odzenzé: *Odyendea gabunensis* Engl.; *Simarubac.;* Gab.
Odzikouna: *Scytopetalum sp.; Scytopetalac.;* Gab.
Odzobi: *Xylopia rubescens* Oliv.; *Anonac.;* Kam.
Odzori: *Baphia nitida* Lodd.; *Papilionac.;* Go.
Odzuma: *Musanga smithi* R. Br.; *Morac.;* Go.
Oeba sagehroe toeni: *Garcinia picrorrhiza* Miq.; *Guttifer.;* Nied.-Ind.
Oebi djindral: *Manihot utilissima* Pohl; *Euphorbiac.;* Javą.
„ kajoe: *Manihot utilissima* Pohl; *Euphorbiac.;* Java.
Oedoe lada: *Dalechampia bidentata* Blume; *Euphorbiac.;* Java.
Oehre: *Acer pseudoplatanus* L.; *Acerac.;* Mitt.- & S.-Eu.
Oeirana: *Salix humboldtiana* Willd.; *Salicac.;* Bras.
Oeja-oejahan: *Ficus copiosa* Steud., *F. quercifolia* Roxb., *F. ramentacea* Roxb.; *Morac.;*
„ „ bromo: *Ficus leptorrhyncha* Val.; *Morac.;* Java (Telemaja) [Java.
Oejah-oejahan: *Ficus rostrata* Lam. var. *urophylla* Wall.; *Morac.;* Java (Telemaja)
Oèko: *Poga oleosa* Pierre; *Rhizophorac.;* Gab.
Oelawabeta: *Swartzia tomentosa* DC.; *Caesalpiniac.;* Sur.
Oelé malie: *Couratari guianensis* Aubl., *C. sp.; Lecythidac.;* Sur.
„ mérie: *Couratari guianensis* Aubl., *C. sp.; Lecythidac.;* Sur.
Oelimani: *Lecythis ollaria* L.; *Lecythidac.;* Sur.
Oelin: *Eusideroxylon zwageri* Teijsm. et Binn.; *Laurac.;* ö. Borneo.
Oeloeloe-oe: *Parkia nitida* Miq., *P. microcephala* Kleinh.; *Mimosac.;* Sur.
Oemanbarklak: *Eschweilera corrugata* Miers, *E. sp.; Lecythidac.;* Sur.
Oemanbarkraki: *Eschweilera corrugata* Miers; *Lecythidac.;* Sur.
Oenbatapo: *Piratinera guianensis* Aubl., *P. spp.; Morac.;* Sur.
Oepas: *Antiaris toxicaria* Lesch.; *Morac.;* Java.
Oer: *Cocos nucifera* L.; *Palmac.;* n. Sum.
Oerang-oerangan: *Debregeasia longifolia* Wedd.; *Urticac.;* Java.
Oerebina: *Eschweilera corrugata* Miers; *Lecythidac.;* Sur.

Oerémérie: *Couratari guianensis* Aubl.; *Lecythidac.;* Sur.
Oéréré: *Mimusops djave* Engl., *M. pierreana* Engl.; *Sapotac.;* Kam., Gab.
Oeris-oerisan: *Glochidion obscurum* Hook. f.; *Euphorbiac.;* Java (Rogodjampi)
Oeroematta-hororadi-kore: *Clathrotropis brachypetalum* Kleinh.; *Papilionac.;* Sur.
Oeroeroe-oe: *Parkia nitida* Miq., *P. microcephala* Kleinh.; *Mimosac.;* Sur.
Oeroewasa: *Pithecolobium gonggrijpi* Kleinh., *P. kegeli* Meissn.; *Mimosac.;* Sur.
  „ : *Pithecolobium trapezifolium* Bth.; *Mimosac.;* Sur.
Oetjah-oetjahan: *Boehmeria clidemioides* Miq.; *Urticac.;* Java (Tosari)
Oévëi: *Elaeis guineensis* Jacq. var. ?; *Palmac.;* Gab.
Oewebina: *Eschweilera corrugata* Miers; *Lecythidac.;* Sur.
Ofa: *Triplochiton scleroxylon* K. Schum.; *Sterculiac.;* Elf.
Ofeo: *Pittosporum undulatum* Vent.; *Pittosporac.;* Südsee (Tahiti)
Offram: *Terminalia superba* Engl. et Diels; *Combretac.;* Nig., Go.
Offun: *Trema africana* Blume; *Ulmac.;* Nig.
Ofõ: *Antiaris africana* Engl.; *Morac.;* Go.
Ofõabiti: *Microdesmis puberula* Hook. f.; *Euphorbiac.;* Go.
Ofol: *? Isolona klaineana* Pierre; *Anonac.;* Gab.
Ofon criollo: *Uitex divaricata* Swartz; *Uerbenac.;* Cuba.
Oforto: *Glyphaea grewioides* Hook. f.; *Tiliac.;* Go.
Ofoss: *Pseudospondias longifolia* Engl.; *Anacardiac.;* Gab.
Ofowa: *Carpolobia lutea* G. Don; *Polygalac.;* Go.
Ofowafuo: *Carpolobia lutea* G. Don; *Polygalac.;* Go.
Ofrafraha: *Piptadenia africana* Hook. f.; *Mimosac.;* Go.
Ofram: *Terminalia superba* Engl. et Diels; *Combretac.;* Go.
Ofruntum: *Funtumia elastica* Stapf; *Apocynac.;* Go.
Ofun: *Mansonia altissima* Chev.; *Sterculiac.;* Go., Nig.
Ofuntum: *Funtumia elastica* Stapf; *Apocynac.;* Elf.
Oga: *Anogeissus leiocarpus* Guill. et Perr.; *Combretac.;* Togo.
Ogabezzi: *Hymenostegia afzeli* Harms; *Caesalpiniac.;* Nig.
Ogagoume: *Tetrapleura thonningi* Bth.; *Mimosac.;* Gab.
Ogal: *Bauhinia thonningi* K. Schum., *B. tomentosa* L.; *Caesalpiniac.;* br. O.-Af. (Uganda)
Ogali: *Bauhinia thonningi* K. Schum., *B. tomentosa* L.; *Caesalpiniac.;* br. O.-Af.
Ogamba: *Heisteria trillesiana* Pierre; *Olacac.;* Gab.   [(Uganda)
Ogambia-gambia: *Plectronia sp.;* *Rubiac.;* Gab.
Ogamignia: *Distemonanthus benthamianus* Baill.; *Caesalpiniac.;* tr. W.-Af.
Ogaminia: *Distemonanthus benthamianus* Baill.; *Caesalpiniac.;* Gab.
Ogana: *Xylopia aethiopica* A. Rich.; *Anonac.;* Gab.
Oganedjo: *Mitragyne macrophylla* Hiern; *Rubiac.;* Gab.
Oganga: *Copaifera sp.;* *Caesalpiniac.;* Gab.
Oganganhédo: *Antrocaryon klaineanum* Pierre; *Anacardiac.;* Gab.
Oganwe: *Khaya senegalensis* A. Juss., *K. grandis* Stapf; *Meliac.;* W.-Af.
Oganwo: *Khaya ivorensis* Chev.; *Meliac.;* Nig., Go.
Ogarwe: *Bridelia micrantha* Muell. Arg.; *Euphorbiac.;* br. O.-Af. (Uganda)
Ogatama-no-ki: *Michelia compressa* Maxim.; *Magnoliac.;* Form.
Ogba: *Parkia filicoidea* Welw., *Pentaclethra macrophylla* Bth..; *Mimosac.;* Nig.
Ogboin: *Irvingia gabonensis* Baill.; *Simarubac.;* Nig.
Ogbonélé: *Didelotia africana* Pierre; *Caesalpiniac.;* Kam.
Ogbon eli: *Copaifera aff. arnoldiana* de Wild. et Th. Dur.; *Caesalpiniac.;* Kam.
Ogéminia: *Distemonanthus benthamianus* Baill.; *Caesalpiniac.;* W.-Af.
Oggo: *Borassus flabellifer* L. var. *aethiopum* Warb.; *Palmac.;* Togo.
Oghávan: *Shorea palosapis* Merr.; *Dipterocarpac.;* Phil. (Samar)
Oghidi: *Scorodophloeus sp.;* *Caesalpiniac.;* B.-K.
Oghounghou: *Pachylobus balsamifera* Guill.; *Burserac.;* Gab.
Ogia: *Daniella ogea* Rolfe; *Caesalpiniac.;* Nig.
Ogikiomi-upakeka: *Hymenostegia afzeli* Harms; *Caesalpiniac.;* Nig. (Benin)
? Ogina: *Haronga madagascariensis* Choisy; *Guttifer.;* Gab.
Ogipogo: *Entandrophragma utile* Sprague; *Meliac.;* W.-Af.
Ogirili-okpi: *Parkia filicoidea* Welw.; *Mimosac.;* Nig.
Ogobé: *Staudtia gabonensis* Warb.; *Myristicac.;* Gab.
Ogoha: *? Isolona klaineana* Pierre; *Anonac.;* Gab.
Ogoma: *Klainedoxa gabonensis* Pierre; *Simarubac.;* Gab.

Ogonendjogou: *Mimusops africana* Lecomte; *Sapotac.;* Kam., Gab.
Ogoré: *Ongokea klaineana* Pierre; *Olacac.;* Gab.
Ogoué: *Tarrietia sp.; Sterculiac.;* Gab.
Ogoufé: *Ceiba pentandra* Gaertn. *var. dehiscens* Ulb.; *Bombacac.;* Dah.
Ogoula: *Coula edulis* Baill.; *Olacac.;* Kam., Gab.
Ogouma: *Ceiba pentandra* Gaertn.; *Bombacac.;* Kam., Gab.
Ogoumalanga: ? *Paivaeusa gabonensis* Chev.; *Euphorbiac.;* Gab.
Ogoumbvikoua: *Uvariastrum sp.; Anonac.;* Gab.
Ogounou: *Acolea missionis* Pierre; *Sapotac.;* Gab.
Ogourra: *Plagiostyles africana* Pierre; *Euphorbiac.;* Gab.
Ogouwa: *Mitragyne macrophylla* Hiern; *Rubiac.;* Elf.
Ogowéni: *Staudtia gabonensis* Warb.; *Myristicac.;* Gab.
Ogrisi: *Newbouldia laevis* Seem.; *Bignoniac.;* Nig.
Ogu: *Ceiba pentandra* Gaertn.; *Bombacac.;* Togo.
Oguanco: *Khaya grandis* Stapf; *Meliac.;* s. Nig.
Oguda: *Coula edulis* Baill.; *Olacac.;* Gab.
Oguembé: *Lecomtedoxa klaineana* Pierre; *Sapotac.;* Gab.
Oguéminia: *Distemonanthus benthamianus* Baill.; *Caesalpiniac.;* Gab.
Ogui: *Irvingia gabonensis* Baill.; *Simarubac.;* Nig. (Benin)
Ogüito: *Ixora rauwolfioides* Standl.; *Rubiac.;* Pan.
Ogula: *Coula edulis* Baill.; *Olacac.;* Gab.
Oguon: *Myrianthus arboreus* Beauv.; *Morac.;* Elf.
Ogwangu: *Khaya grandis* Stapf; *Meliac.;* W.-Af.
Ogwanwo: *Khaya ivorensis* Chev.; *Meliac.;* Nig., Go.
Ogwe: *Irvingia gabonensis* Baill.; *Simarubac.;* Nig. (Benin)
Ogwi: *Irvingia gabonensis* Baill.; *Simarubac.;* Nig. (Benin)
Ogyamba: *Musanga smithi* R. Br.; *Morac.;* Go.
Ohba-kurikashi: *Castanopsis kawakami* Hayata; *Fagac.;* Form.
„ -tabu: *Machilus kusanoi* Hayata; *Laurac.;* Form.
Ohe: *Piptadenia africana* Hook. f.; *Mimosac.;* Nig.
Ohez: *Cinnamomum zeylanicum* Breyn; *Laurac.;* sw. O.-Ind. (Burma)
Ohi: *Albizzia stipulata* Boivin; *Mimosac.;* nw. O.-Ind. (Punjab-Kangra)
Ohia: *Eugenia malaccensis* L.; *Myrtac.;* Pan.
„ : *Piptadenia africana* Hook. f.; *Mimosac.;* Nig.
„ lehua: *Metrosideros polymorpha* Gaudich.; *Myrtac.;* ö. Au., N.-Seel., Südsee
Ohinego: *Pterocarpus soyauxi* Taub.; *Papilionac.;* Gab.   [(Hawaii)
Ohnara: *Quercus grosseserrata* Blume, *Q. crispula* Blume; *Fagac.;* Jap.
Oho: *Ricinodendron africanum* Muell. Arg.; *Euphorbiac.;* Elf.
„ : *Staudtia gabonensis* Warb.; *Myristicac.;* Kam.
Oh-oh: *Mitragyne stipulosa* O. Ktze.; *Rubiac.;* br. O.-Af. (Uganda)
Ohombé: *Mitragyne macrophylla* Hiern; *Rubiac.;* Kam.
Ohura: *Albizzia zygia* MacBride; *Mimosac.;* Go.
Ohyonire: *Ulmus montana* Sm. *var. laciniata* Trautv.; *Ulmac.;* Jap.
Oi mo: *Psidium guiava* Raddi; *Myrtac.;* Ind.-Ch.
„ rung: *Cratoxylon polyanthum* Korth.; *Guttifer.;* Ind.-Ch.
„ tau: *Cratoxylon polyanthum* Korth. *var. pumilum* ?; *Guttifer.;* Ind.-Ch.
Oiakiá: *Pometia pinnata* Forst.; *Hippocastanac.;* Phil. (Mindoro)
Oil-bean tree: *Pentaclethra macrophylla* Bth.; *Mimosac.;* Lib.
„ -nut (u): *Juglans cinerea* L.; *Juglandac.;* ö. N.-Am.
„ -tree, meni: *Lophira alata* Banks; *Ochnac.;* Go.
Oïngo: *Scytopetalum brevipes* Pierre; *Scytopetalac.;* Gab.
Oi-ói: *Diospyros philippinensis* A. DC.; *Ebenac.;* Phil. (Bataan)
O-itabi (jap.): *Ficus pumila* L.; *Morac.;* ö. As.
Oiti: *Clarisia racemosa* Ruiz et Pavon; *Morac.;* Bras.
„ -mirim-ayra: *Piratinera sp.; Morac.;* Bras.
Oiticica: *Clarisia racemosa* Ruiz et Pavon; *Morac.;* Bras. (Bahia)
„ : *Couepia grandiflora* Bth.; *Rosac.;* nw. Bras.
„ : *Licania rigida* Bth., *L. sclerophylla* Mart.; *Rosac.;* Bras.
„ amarella: *Clarisia racemosa* Ruiz et Pavon; *Morac.;* Bras.
Oitiseiro: *Licania utilis* Fritsch; *Rosac.;* Bras.
Oity: *Moquilea tomentosa* Bth.; *Rosac.;* Bras. (Sao P)

Oity bravo: *Ficus maximiliana* Mart.; *Morac.;* Bras.
„ -coroya: *Moquilea tomentosa* Bth.; *Rosac.;* Bras.
„ da praia: *Moquilea tomentosa* Bth.; *Rosac.;* Bras.
„ de sertão: *Couepia sp.; Rosac.;* Bras.
Oiyo: *Albizzia coriaria* Welw.; *Mimosac.;* br. O.-Af. (Uganda)
Oja: *Parkia nitida* Miq., *P. microcephala* Kleinh.; *Mimosac.;* Sur.
Ojankuij: *Xylopia cf. thomsoni* Oliv.; *Anonac.;* Kam.
Ojé: *Ficus sp.; Morac.;* Peru.
Oji: *Chlorophora excelsa* Bth. et Hook.; *Morac.;* Nig.
„ karisi: *Newbouldia laevis* Seem.; *Bignoniac.;* Nig.
Ojite: *Trophis americana* L., *Ficus sp.; Morac.;* Mex.
Ojo de cangrejo: *Pithecolobium arboreum* Urb.; *Mimosac.;* Cuba.
„ „ muñeca: *Trichilia sp.; Meliac.;* Salv.
„ „ perdiz: *Brya ebenus* DC.; *Papilionac.;* Cuba.
„ „ venado: *Caesalpinia crista* L.; *Caesalpiniac.;* Mex. (Sinaloa)
Ojoche: *Brosimum alicastrum* Swartz; *Morac.;* Mex.
„ : *Brosimum terrabanum* Pittier; *Morac.;* Nic.
„ macho: *Trophis racemosa* Urban; *Morac.;* Pan.
Ojod gimeran: *Phyllanthus dalbergioides* Wall.; *Euphorbiac.;* Java (Soebah)
Ojot lawean: *Ficus foveolata* Wall.; *Morac.;* Java (Tangkil)
„ santenan: *Ficus callicarpa* Miq.; *Morac.;* Java.
Ojuste: *Brosimum sp.; Morac.;* Hond.
Okadi: *Elaeis guineensis* Jacq. var. *nigrescens* Chev.; *Palmac.;* Gab.
„ a mavasa: *Elaeis guineensis* Jacq. var. ?; *Palmac.;* Gab.
Okahimi: *Diospyros xanthochlamys* Gürke; *Ebenac.;* Nig.
Okaka: *Aucoumea klaineana* Pierre; *Burserac.;* Gab.
Okala: *Xylopia aethiopica* A. Rich.; *Anonac.;* Gab., B.-K.
„ -ehilé: *Sterculia tragacantha* Lindl.; *Sterculiac.;* Elf.
Okam: *Pseudospondias gigantea* Chev.; *Anacardiac.;* Gab.
Okambo: *Macrolobium palisoti* Bth.; *Caesalpiniac.;* Gab.
Okan: *Combretum micranthum* G. Don; *Combretac.;* Nig.
„ : *Cylicodiscus gabunensis* Harms; *Papilionac.;* Nig.
Okango: *Hymenocardia acida* Tul.; *Euphorbiac.;* br. O.-Af. (Uganda)
Okao: *Lophira procera* Chev.; *Ochnac.;* Gab.
„ koodo: *Ricinodendron africanum* Muell. Arg.; *Euphorbiac.;* Go.
Okèka: *Ongokea klaineana* Pierre; *Olacac.;* Gab.
Okène: *Dialium aff. connaroides* Harms; *Caesalpiniac.;* Gab.
Okengbo: *Ricinodendron africanum* Muell. Arg.; *Euphorbiac.;* Nig.
Okepea: *Lophira alata* Banks; *Ochnac.;* Nig.
Okerli: *Irvingia gabonensis* Baill.; *Simarubac.;* Nig. (Benin)
Okess: *Uapaca sp.; Euphorbiac.;* Gab.
Okha: *Ceiba pentandra* Gaertn.; *Bombacac.;* Nig.
Okhar: *Juglans regia* L.; *Juglandac.;* n. O.-Ind. (United Prov.)
Okhuen: *Ricinodendron africanum* Muell. Arg.; *Euphorbiac.;* Nig. (Benin)
Okikopom: *Lophira procera* Chev.; *Ochnac.;* Nig.
Okinawa-tsuge: *Buxus liukiuensis* Makino; *Buxac.;* Jap.
Okip: *Klainedoxa gabonensis* Pierre; *Simarubac.;* Gab.
Okisibiri: *Diospyros mespiliformis* Hochst.; *Ebenac.;* Go.
Okkwen nebo: *Ricinodendron africanum* Muell. Arg.; *Euphorbiac.;* Nig. (Benin)
Okoa: *Lophira procera* Chev.; *Ochnac.;* Kam.
Okoga: *Lophira procera* Chev.; *Ochnac.;* Kam.
Okoka: *Lophira procera* Chev.; *Ochnac.;* Kam., Gab.
Okola: *Xylopia aethiopica* A. Rich.; *Anonac.;* Kam.
„ : *Mimusops africana* Lecomte; *Sapotac.;* Kam., Gab.
„ angoume: *? Lecomtedoxa klaineana* Pierre; *Sapotac.;* Gab.
Okom: *Elaeis guineensis* Jacq. var. *virescens* Chev.; *Palmac.;* Gab.
Okómbokuij: *Spondianthus preussi* Engl.; *Euphorbiac.;* Kam.
Okomokoma: *Myrianthus arboreus* Beauv.; *Morac.;* Gab.
Okon: *Triumfetta rhomboidea* Jacq.; *Tiliac.;* Kam.
Okong: *Grewia coriacea* Mast.; *Tiliac.;* Gab.
Okongea: *Grewia coriacea* Mast.; *Tiliac.;* Gab.

Okongi: *Grewia coriacea* Mast.; *Tiliac.;* Gab.
Okóni: *Ehretia cymosa* Thonn.; *Borraginac.;* Togo.
Okot: *Lophira procera* Chev.; *Ochnac.;* Nig.
Okoua: *Lophira procera* Chev.; *Ochnac.;* Kam.
Okoué: *Baphia nitida* Afzel., *B. polygalacea* Baker; *Papilionac.;* Elf.
Okouem: *Hasskarlia didymostemon* Baill.; *Euphorbiac.;* Gab.
Okouka: *Alstonia congensis* Engl.; *Apocynac.;* Kam.
Okoumé: *Aucoumea klaineana* Pierre; *Burserac.;* Gab., Mitt.-Kongo.
   „   : *Canarium schweinfurthi* Engl.; *Burserac.;* Elf.
   „   cambogala: *Aucoumea klaineana* Pierre; *Burserac.;* Gab.
   „   combo-combo: *Aucoumea klaineana* Pierre; *Burserac.;* Gab.
   „   nyangala: *Aucoumea klaineana* Pierre; *Burserac.;* Gab.
   „   zouga: *Aucoumea klaineana* Pierre; *Burserac.;* Gab.
Okpe: *Parinarium kerstingi* Engl.; *Rosac.;* Togo.
Okpé: *Elaeis guineensis* Jacq.; *Palmac.;* Togo.
   „   : *Distemonanthus benthamianus* Baill.; *Caesalpiniac.;* Togo.
Okpoha: *Terminalia ivorensis* Chev.; *Combretac.;* Nig.
Okpunum: *Ricinodendron africanum* Muell. Arg.; *Euphorbiac.;* Nig.
Okrokro: *Hymenostegia afzeli* Harms; *Caesalpiniac.;* Nig.
Okromfẽ: *Dialium guineense* Willd.; *Caesalpiniac.;* Go.
Okslit: *Aegle marmelos* Correa; *Rutac.;* Burma.
Okué: *Strombosia pustulata* Oliv.; *Olacac.;* Elf.
Okue: *Ricinodendron africanum* Muell. Arg.; *Euphorbiac.;* Nig.
Okukduru: *Ricinodendron africanum* Muell. Arg.; *Euphorbiac.;* Go.
Okulaholz (d): *Mimusops heckeli* Lecomte; *Sapotac.;* W.-Af.
Okuma: *Klainedoxa gabonensis* Pierre; *Simarubac.;* Gab.
Okume (d): *Aucoumea klaineana* Pierre; *Burserac.;* Gab.
? Okuo: *Fagara macrophylla* Engl.; *Rutac.;* Go.
Okura: *Albizzia sp.; Mimosac.;* Go.
Okuraba: *Hymenostegia afzeli* Harms; *Caesalpiniac.;* Go.
Okuri: *Irvingia sp.; Simarubac.;* Go.
Okurimai: *Newbouldia laevis* Seem.; *Bignoniac.;* Nig.
Okuro: *Albizzia browni* Walp., *A. zygia* MacBride; *Mimosac.;* Go.
   „   -sante: *Millettia rhodantha* Baill.; *Papilionac.;* Go.
Okut: *Lophira procera* Chev.; *Ochnac.;* Nig.
   „   : *Ochrocarpus africanus* Oliv.; *Guttifer.;* Nig.
Okuturyang: *Acacia arabica* Willd.; *Mimosac.;* br. O.-Af. (Uganda)
Okwa: *Treculia africana* Dcne.; *Morac.;* Nig.
   „   -beke: *Artocarpus incisa* L.; *Morac.;* Nig.
Okwe: *Ricinodendron africanum* Muell. Arg.; *Euphorbiac.;* Nig. (Benin)
Okwein: *Ricinodendron sp.; Euphorbiac.;* Nig.
   „   : *Daniella sp., Brachystegia eurycoma* Harms; *Caesalpiniac.;* Nig.
Okwen: *Ricinodendron africanum* Muell. Arg.; *Euphorbiac.;* Nig. (Benin)
Okwenseva: *Ricinodendron africanum* Muell. Arg.; *Euphorbiac.;* Nig. (Benin)
Olaho: *Dichrostachys nutans* Bth.; *Mimosac.;* Togo.
Oláian: *Quercus sp.; Fagac.;* Phil.
Olandim: *Calophyllum brasiliense* Camb.; *Guttifer.;* Bras. (Sao P)
Olandy: *Calophyllum brasiliense* Camb.; *Guttifer.;* Bras.
Olangi: *Staudtia gabonensis* Warb.; *Myristicac.;* Gab.
Old William: *Vismia ferruginea* H. B. K.; *Guttifer.;* br. Hond.
   „   woman's walking stick: *Eupatorium hebebotryum* Hemsl.; *Composit.;* br. Hond.
Olèko: *Poga oleosa* Pierre; *Rhizophorac.;* Gab.
Olelang: *Chrysophyllum sp.; Sapotac.;* Kam.
Olélé: *Pycnanthus kombo* Warb.; *Myristicac.;* Elf.
Olemarie: *Couratari guianensis* Aubl.; *Lecythidac.;* Sur.
Olémellie: *Couratari guianensis* Aubl.; *Lecythidac.;* Sur.
Olene: *Irvingia grandifolia* Engl.; *Simarubac.;* s. Kam., n. Gab.
Olengué: *Canarium velutinum* Guill.; *Burserac.;* Gab.
Oleni: *Canarium schweinfurthi* Engl.; *Burserac.;* s. Kam., n. Gab.
Olensago: *Agathis alba* Foxw.; *Conifer.;* Phil. (Benguet)
Oleo: *Myroxylon peruiferum* L. f.; *Papilionac.;* Bras. (Sao P)

Oleo: *Copaifera sp.; Caesalpiniac.;* Bras. (Bahia)
„ branco: *Pterogyne nitens* Tul.; *Caesalpiniac.;* Bras. (Sao P)
„ jatahy: *Hymenaea stigonocarpa* Mart.; *Caesalpiniac.;* Bras. (Sao P)
„ pardo: *Myrocarpus frondosus* Allem.; *Papilionac.;* Bras. (Sao P)
„ vermelho: *Myroxylon peruiferum* L. f., *M. toluiferum* H. B. K.; *Papilionac.;* Bras.
„ de Cabórehyba: *Myrocarpus fastigiatus* Allem.; *Papilionac.;* Bras. (Sao P)
„ „ Copahuva: *Copaifera langsdorffi* Desf.; *Caesalpiniac.;* Bras. (Sao P)
„ „ Copahyba: *Copaifera langsdorffi* Desf.; *Caesalpiniac.;* Bras. (Sao P)
„ „ Jatahy: *Hymenaea stilbocarpa* Hayne; *Caesalpiniac.;* Bras. (Sao P)
„ „ „ : *Hymenaea stigonocarpa* Mart.; *Caesalpiniac.;* Bras. (Sao P)
„ „ Jatobá: *Hymenaea stigonocarpa* Mart.; *Caesalpiniac.;* Bras. (Sao P)
Oli: *Uapaca heudeloti* Baill.; *Euphorbiac.;* Togo.
Olibanum: *Boswellia serrata* Roxb.; *Burserac.;* O.-Ind.
Oliehout (h): *? Tapirira guianensis* Aubl.; *Anacardiac.;* fr. Gu.
Olie wapa (h): *Eperua falcata* Aubl.; *Caesalpiniac.;* fr. Gu.
Olijf, inheemsche: *Bontia daphnoïdes* Aubl.; *Myoporac.;* fr. Gu.
„ , wilde: *Bontia daphnoïdes* Aubl.; *Myoporac.;* fr. Gu.
Olinda: *Desbordesia insignis* Pierre; *Simarubac.;* Gab.
Olingué: *Daniella klainei* Pierre; *Caesalpiniac.;* Gab.
Oliva: *Capparis sp.; Capparidac.;* Curaçao.
„ macha: *Capparis sp.; Capparidac.;* Curaçao.
Olive (d): *Olea europaea* L.; *Oleac.;* s. Eu.
„ : *Olea hochstetteri* Baker; *Oleac.;* br. O.-Af. (Uganda)
„ , australian: *Olea paniculata* R. Br.; *Oleac.;* Au.
„ , brown: *Olea chrysophylla* Lam.; *Oleac.;* br. O.-Af. (Uganda)
„ , east african (e): *Olea hochstetteri* Baker; *Oleac.;* br. O.-Af. (Kenya)
„ , Kenya-: *Olea hochstetteri* Baker; *Oleac.;* br. O.-Af. (Kenya)
„ , Mock-: *Prunus caroliniana* Ait.; *Rosac.;* N.-Am.
„ , Native-: *Osmanthus americanus* Bth. et Hook.; *Oleac.;* N.-Am.
„ , „ : *Olea paniculata* R. Br.; *Oleac.;* Au.
„ , „ : *Notelaea ligustrina* Vent.; *Oleac.;* Tasm.
„ tree, black: *Bucida buceras* L.; *Combretac.;* N.-Am., Jam.
„ , white: *Halleria lucida* L.; *Scrophulariac.;* S.-Af. (Kap.)
„ , wilde: *Olea chrysophylla* Lam.; *Oleac.;* D.-O.-Af.
„ wood: *Olea verrucosa* Link; *Oleac.;* S.-Af. (Kap)
„ „ : *Elaeodendron australe* Vent.; *Celastrac.;* Au.
„ „ , indian: *Olea ferruginea* Royle; *Oleac.;* O.-Ind. (Punjab)
„ „ , Seychelles-: *Olea lancea* Lam.; *Oleac.;* Maur.
Olivenholz (d): *Olea europaea* L.; *Oleac.;* s. Eu., Orient.
Olivenhout: *Olea capensis* L.; *Oleac.;* S.-Af. (Kap)
Olivera: *Daphne mezereum* L.; *Tymelaeac.;* Sp.
Olivetier jaune: *Elaeodendron croceum* DC.; *Celastrac.;* S.-Af. (Kap)
Olivier: *? Terminalia obovata* Steud., *Bucida buceras* L.; *Combretac.;* Trin., Tob.
„ : *Olea laperrini* Batt. et Trab.; *Oleac.;* Sahara.
„ : *Elaeodendron curtipendulum* Endl.; *Celastrac.;* N.-Kal.
„ montagne: *Cyrilla racemiflora* L.; *Cyrillac.;* Guad.
„ sauvage: *Bontia daphnoïdes* Aubl.; *Myoporac.;* fr. Gu.
„ , white: *Terminalia obovata* Steud.; *Combretac.;* br. Hond., Trin.
„ , yellow: *Buchenavia capitata* Eichl.; *Combretac.;* Trin., Tob.
„ de sable: *Dodonaea viscosa* Jacq. *var. angustifolia* Bth.; *Sapindac.;* Jam.
„ „ la Grande Terre: *Byrsonima lucida* L. Cl. Rich.; *Malpighiac.;* Guad.
„ du pays: *Bontia daphnoïdes* Aubl.; *Myoporac.;* fr. Gu.
Olivilla: *Teucrium fruticans* L.; *Labiat.;* Sp.
Olivillo: *Kageneckia angustifolia* D. Don; *Rosac.;* Chile.
Olivo: *Capparis sp.; Capparidac.;* Hond.
„ : *Simaruba glauca* DC.; *Simarubac.;* Pan.
„ : *Sapium jamaicense* Sw.; *Euphorbiac.;* Pan.
„ : *Capparis odoratissima* Jacq.; *Capparidac.;* Ven., Kol.
„ : *Capparis breynia* Jacq., *C. indica* Fawc. et Rendle; *Capparidac.;* n. Kol.
„ macho: *Sapium thelocarpum* Schum. et Pitt.; *Euphorbiac.;* Pan.
„ -rá: *Chrysophyllum cainito* L.; *Sapotac.;* Arg.

Olla de mono: *Lecythis dubia* H. B. K., *L. elliptica* H. B. K.; *Lecythidac.;* n. Kol.
Olleto: *Lecythis melliana* Pittier, *L. elliptica* H. B. K.; *Lecythidac.;* Pan., n. Kol.
Olli: *Acacia stenocarpa* Hochst.; *Mimosac.;* br. O.-Af. (Uganda)
Ollito: *Eschweilera calyculata* Pittier; *Lecythidac.;* Pan.
Ollo: *Protium spp.; Burserac.;* Sur.
Olmo: *Ulmus campestris* L., *U. montana* With.; *Ulmac.;* Sp.
  „ alamo: *Ulmus campestris* L.; *Ulmac.;* Sp.
  „ montano: *Ulmus montana* With., *U. effusa* Willd.; *Ulmac.;* It.
  „ siberiano: *Ulmus pumila* L.; *Ulmac.;* It.
  „ de país: *Alnus sp.; Betulac.;* Mex.
Olo: *Protium spp.; Burserac.;* Sur.
  „ -karaubandi-korro: *Protium spp.; Burserac.;* Sur.
Ologbomodu: *Ochrocarpus africanus* Oliv.; *Guttifer.;* Nig.
Ologun sheshe: *Erythrina senegalensis* DC.; *Papilionac.;* Nig.
Oloko: *Chlorophora excelsa* Bth. et Hook.; *Morac.;* Nig.
Olokokima: *Afzelia africana* Smith; *Caesalpiniac.;* Nig.
Ol-ól: *Pinus insularis* Endl.; *Conifer.;* Phil. (Benguet)
Olom: *Scorodophloeus zenkeri* Harms; *Caesalpiniac.;* Kam.
  „ : *Afrostyrax kamerunensis* Perk.; *Styracac.;* Kam.
Olombi: *Afzelia africana* Smith; *Caesalpiniac.;* B.-K.
Olon (f): *Fagara macrophylla* Engl.; *Rutac.;* Kam., Gab.
Olonvogo: *Fagara sp.; Rutac.;* Kam., Gab.
  „ du Cameroun: *Sorindeia ochracea* Engl.; *Anacardiac.;* Kam.
Oloo: *Protium sp.; Burserac.;* br. Gu.
Oloque: *Syzygium guineense* DC.; *Myrtac.;* br. O.-Af. (Uganda)
Olosapo: *Couepia sp.; Rosac.;* C.-R.
Olos-ólos: *Litsea sp.; Laurac.;* Phil. (Pangasinan)
Oloumboli-loumboi: *Hernandia beninensis* Welw.; *Hernandiac.;* Gab.
Oloumi: *Lannea zenkeri* Engl. et Krause; *Anacardiac.;* Gab.
Oloun: *Erythrophloeum guineense* G. Don; *Caesalpiniac.;* Kam.
Oltarakwa: *Juniperus procera* Hochst.; *Conifer.;* br. O.-Af. (Kenya)
Olu: *Protium sp.; Burserac.;* br. Gu.
Olubu: *Vernonia amygdalina* Delile; *Composit.;* Nig.
Olumi: *Copaifera ? salikounda* Heckel; *Caesalpiniac.;* Gab.
  „ faux (f): *Detarium sp.; Caesalpiniac.;* Gab.
  „ na ngèndja: *Copaifera sp.; Caesalpiniac.;* Gab.
  „ noir (f): *Detarium sp.; Caesalpiniac.;* Gab.
  „ rouge (f): *Copaifera ? salikounda* Heckel; *Caesalpiniac.;* Gab.
  „ wa tenatena: *Copaifera ? salikounda* Heckel; *Caesalpiniac.;* Gab.
  „ „ vyovyo: *Detarium sp.; Caesalpiniac.;* Gab.
Olungholz (d): *Alstonia scholaris* R. Br.; *Apocynac.;* Tr.
Om beng thngé: *Pterospermum diversifolium* Blume; *Sterculiac.;* Ind.-Ch.
Omak: *Sterculia villosa* Roxb.; *Sterculiac.;* Assam.
Omamio: *Dracontomelum cumingianum* Baill.; *Anacardiac.;* Phil.
Oman: *Irvingia oblonga* Chev.; *Simarubac.;* Kam.
Omang: *Desbordesia glaucescens* Pierre; *Simarubac.;* Kam.
Omangatanga: *Cocos nucifera* L.; *Palmac.;* Gab.
Ombala: *Pentaclethra macrophylla* Bth.; *Mimosac.;* Gab.
Ombatapo: *Helicostylis poeppigiana* Tréc.; *Morac.;* Sur.
Ombatappo: *Helicostylis poeppigiana* Tréc.; *Morac.;* Sur.
Ombéga: *Khaya ivorensis* Chev., *Lovoa klaineana* Pierre; *Meliac.;* Gab., Mitt.-Kongo
  „ fiote: *Lovoa klaineana* Pierre; *Meliac.;* Gab.
  „ ouapoupou: *Khaya sp.; Meliac.;* Gab.
Ombéné: *Cola acuminata* Schott et Endl.; *Sterculiac.;* Gab.
  „ -obéni: *Cola vallaya* Cornu; *Sterculiac.;* Gab.
Ombïli: *Canarium schweinfurthi* Engl.; *Burserac.;* Gab. (Franceville)
Ombolowbolo: *Lovoa klaineana* Pierre; *Meliac.;* Gab.
Ombu: *Phytolacca dioica* L.; *Phytolaccac.;* Arg.
Ombù rá: *Gilibertia cuneifolia* Marchal; *Araliac.;* Arg.
Ome: *Miliusa velutina* Hook. f. et Thoms.; *Anonac.;* w. Beng. (Bihar)
Omebaba: *Didelotia africana* Pierre; *Caesalpiniac.;* Kam.

Omevah: *Combretum paniculatum* Vent., *Combretac.*; Go.
Omi taproepa: *Gustavia angusta* L., *G. pterocarpa* Poit.; *Lecythidac.*; Sur.
Omiry: *Humiria sp.*; *Humiriac.*; fr. Gu.
Omodon: *Ricinodendron africanum* Muell. Arg.; *Euphorbiac.*; Nig.
Omoé: *Irvingia sp.*; *Simarubac.*; Elf.
Omon: *Cordia milleni* Baker; *Borraginac.*; Nig.
Omorikaholz: *Picea omorika* Pančič; *Conifer.*; sö. Eu.
Omoué: *Irvingia sp.*; *Simarubac.*; Elf.
Omoungou: *Mimusops africana* Lecomte; *Sapotac.*; Kam., Gab.
Ompégwé: *Rauwolfia macrophylla* Stapf; *Apocynac.*; Gab.
Omphalier (f): *Omphalea diandra* L.; *Euphorbiac.*; fr. Gu.
Omponédé: *Corynanthe gabonensis* Chev.; *Rubiac.*; Gab.
Ompow: *Afzelia quanzensis* Welw.; *Caesalpiniac.*; port. O.-Af. (Moz.)
Omu: *Entandrophragma candollei* Harms; *Meliac.*; Nig. (Benin)
Omukaru: *Zizyphus mucronata* Willd.; *Rhamnac.*; tr. & S.-Af.
Omumbonde: *Acacia giraffae* Willd.; *Mimosac.*; S.-Af.
Omumu: *Coula edulis* Baill.; *Olacac.*; Nig.
Omveyala: *Xylopia striata* Engl.; *Anonac.*; Kam.
Omwali: *Garcinia klaineana* Pierre; *Guttifer.*; Gab.
On: *Persea americana* Mill.; *Laurac.*; br. Hond.
Onabou: *Entandrophragma ferruginea* Chev.; *Meliac.*; Elf.
Onara: *Quercus grosseserrata* Blume, *Q. crispula* Blume; *Fagac.*; Jap.
Onaslúm: *Cyclostemon sp.*; *Euphorbiac.*; Phil. (Negrosinsel)
Ondôn: *Litsea sebifera* Pers., *L. polyantha* Juss.; *Laurac.*; Burma.
„ -laukya: *Litsea polyantha* Juss.; *Laurac.*; Burma.
Ondulu: *Entada abyssinica* Steud.; *Mimosac.*; Togo.
Ong bâu: *Cordia bantamensis* Blume; *Borraginac.*; Tonkin.
„ can: *Cassia siamea* Lamk.; *Caesalpiniac.*; Ind.-Ch.
„ col: *Garcinia lanessani* Pierre; *Guttifer.*; Ind.-Ch.
Ongëï: *Ongokea klaineana* Pierre; *Olacac.*; Gab.
Ongék: *Ongokea klaineana* Pierre; *Olacac.*; Gab.
Ongeké: *Ongokea klaineana* Pierre; *Olacac.*; Gab.
Ongèko: *Ongokea klaineana* Pierre; *Olacac.*; Gab.
Onghé: *Alstonia congensis* Engl.; *Apocynac.*; Elf.
Ongle de chat: *Bignonia unguis-cati* L.; *Bignoniac.*; fr. Gu.
Ongo ayem: *Albizzia brownei* Oliv., *A. angolensis* Welw., *A. welwitschi* Oliv.;
Ongoo: *Terminalia glaucescens* Planch.; *Combretac.*; Go.   [*Mimosac.*; Kam.
Ongoumi: *Aucoumea klaineana* Pierre; *Burserac.*; Gab.
Ongueko: *Ongokea klaineana* Pierre; *Olacac.*; Gab., B.-K.
Onguent plan: *? Jacaranda copaia* G. Don; *Bignoniac.*; fr. Gu.
Onguié: *Alstonia congensis* Engl.; *Apocynac.*; Elf.
Ongumba: *Coula edulis* Baill.; *Olacac.*; Gab.
Onhidjo: *Terminalia ivorensis* Chev.; *Combretac.*; Elf.
Onhon: *Sarcocephalus esculentus* Afzel.; *Rubiac.*; Elf.
Oni-gasi (jap): *Lithocarpus castanopsifolia* Hayata; *Fagac.*; Form.
Onié: *Garcinia kola* Heckel, *G. manni* Oliv.; *Guttifer.*; Kam.
Oni-gurumi: *Juglans sieboldiana* Maxim.; *Juglandac.*; Jap.
Onionwood of New South Wales: *Owenia cepiodora* F. v. M.; *Meliac.*; Au.
Onisen: *Ulmus spp.*; *Ulmac.*; ? O.-Sib., ? Jap.
Onitoto: *Piptadenia africana* Hook. f.; *Mimosac.*; Nig.
Onja: *Symphonia gabonensis* Pierre; *Guttifer.*; Kam.
Onjam: *Antidesma ghaesembilla* Gaertn.; *Euphorbiac.*; Java.
Onko: *Taxus cuspidata* S. et Z.; *Conifer.*; Jap.
Onkola: *Mimusops africana* Lecomte; *Sapotac.*; Kam., Gab.
Onkolla: *Mimusops africana* Lecomte; *Sapotac.*; s. Kam., n. Gab.
Onoja-èrè-palli: *Mouriria acutiflora* Naud.; *Melastomatac.*; Sur.
Onöng: *Carpolobia alba* G. Don; *Polygalac.*; Kam.
Ononometiqué m'lon: *Ancistrophyllum opacum* Drude; *Palmac.*; Kam.
Ono-ore: *Betula bhojpattra* Wall., *B. schmidti* Regel; *Betulac.*; Jap., Korea.
Onoore-kanba: *Betula schmidti* Regel; *Betulac.*; Jap.
Onotello: *Calophyllum calaba* Jacq.; *Guttifer.*; Ven.

Onotillo: *Vismia ferruginea* H. B. K.; *Guttifer.;* Ven.
Onoto: *Bixa orellana* L.; *Bixac.;* Kol., Ven.
Onoungou: *Mimusops sp.; Sapotac.;* Gab.
Onsi: *Cocos nucifera* L.; *Palmac.;* Burma.
Onsongongo: *Antrocaryon klaineanum* Pierre; *Anacardiac.;* Gab.
Ontano bianco: *Alnus incana* Willd.; *Betulac.;* It.
  „ napolitano: *Alnus cordata* Desf.; *Betulac.;* It.
  „ nero: *Alnus glutinosa* Gaertn.; *Betulac.;* It.
Ontina: *Artemisia herba-alba* Asso; *Composit.;* Sp.
Onumu: *Canarium schweinfurthi* Engl.; *Burserac.;* Nig.
Onungu: *Mimusops africana* Lecomte; *Sapotac.;* Gab.
Onvong: *Dialium guineense* Willd.; *Caesalpiniac.;* Kam., Gab.
Onwama: *Ricinodendron africanum* Muell. Arg.; *Euphorbiac.;* Go.
Onwana: *Ricinodendron africanum* Muell. Arg.; *Euphorbiac.;* Go.
Onwong: *Dialium guineense* Willd.; *Caesalpiniac.;* Gab.
Onyã: *Ceiba pentandra* Gaertn.; *Bombacac.;* Go.
  „ -kene: *Ceiba pentandra* Gaertn. var. ?; *Bombacac.;* Go.
Onyamedua: *Alstonia congensis* Engl.; *Apocynac.;* Go.
Onyangba: *Pseudospondias microcarpa* Engl.; *Anacardiac.;* Togo.
Onyina: *Ceiba pentandra* Gaertn.; *Bombacac.;* Go.
Onyong: *Dialium guineense* Willd.; *Caesalpiniac.;* Gab.
Onzabili (f): *Antrocaryon klaineanum* Pierre; *Anacardiac.;* Kam.
Onzakon: *Antrocaryon klaineanum* Pierre; *Anacardiac.;* Gab.
Onzang: *Odyendea gabunensis* Engl.; *Simarubac.;* Gab.
Onzèng: *Odyendea gabunensis* Engl.; *Simarubac.;* Gab.
Onzon: *Odyendea gabunensis* Engl.; *Simarubac.;* Gab.
Ooba-tabu (jap.): *Machilus kusanoi* Hayata; *Laurac.;* Form.
Oodlu: *Bassia latifolia* Roxb.; *Sapotac.;* w. O.-Ind. (sw. Deccan)
Oo-kurikasi (jap.): *Castanopsis kawakami* Hayata; *Fagac.;* Form.
Ookwe: *Ricinodendron africanum* Muell. Arg.; *Euphorbiac.;* Nig. (Benin)
Oolemeriballi: *Brosimum sp.; Morac.;* Gu., Sur.
Ooline, pink: *Emmenospermum alphitonioides* F. v. M.; *Rhamnac.;* ö. Au. (Queensl.)
Oolu: *Icica schomburgkiana* Klotzsch; *Burserac.;* br. Gu.
Oop: *Anona purpurea* Moç. et Sessé; *Anonac.;* br. Hond.
Oori: *Melia japonica* G. Don; *Meliac.;* Jap.
Oouronzi: *Alstonia congensis* Engl.; *Apocynac.;* Elf.
Opapao: *Afzelia africana* Smith; *Caesalpiniac.;* Go.
Opapea: *Mimusops ? djave* Engl.; *Sapotac.;* Go.
Opapeh· *Afzelia africana* Smith; *Caesalpiniac.;* Go.
Opäti: *Terminalia baumanni* Engl. et Diels; *Combretac.;* Togo.
Opavess: *Desbordesia insignis* Pierre; *Simarubac.;* Gab.
Opay: *Bourreria mollis* Standl.; *Borraginac.;* br. Hond.
Opepe (e): *Sarcocephalus trillesi* Pierre, *S. esculentus* Afzel.; *Rubiac.;* Nig.
  „ : *Terminalia sp.; Combretac.;* Nig.
  „ , false: *Mitragyne macrophylla* Hiern; *Rubiac.:* W.-Af.
Opepea: *Phyllanthus discoideus* Muell. Arg.; *Euphorbiac.;* Go.
Opio-nuenue: *Carpolobia lutea* G. Don; *Polygalac.;* Go.
Oplái: *Alstonia scholaris* R. Br.; *Apocynac.;* Phil. (Cagayan)
Opo: *Ficus infectoria* Roxb.; *Morac.;* Java (Soebah)
Opok: *Ficus superba* Miq.; *Morac.;* Java (Soebah)
Opong-ópong: *Palaquium sp.; Sapotac.;* Phil. (Camarines)
  „ „ : *Sterculia sp.; Sterculiac.;* Phil. (Camarines)
Opopa: *Irvingia gabonensis* Baill.; *Simarubac.;* Nig.
Oppio: *Acer italum* Lauth; *Acerac.;* It.
Opputtu: *Antiaris sp.; Morac.;* Nig. (Benin)
Opuk: *Terminalia spekei* Rolfe; *Combretac.;* O.-Af. (Uganda)
Orabo: *Uapaca bingervillensis* Beille; *Euphorbiac.;* Elf.
Oralli: *Vismia macrophylla* H. B. K.; *Guttifer.;* br. Gu.
Orandamomi: *Cunninghamia sinensis* R. Br.; *Conifer.;* s. Ch.
Orang-aring: *Maoutia diversifolia* Wedd.; *Urticac.;* Java (Semarang, Kedoe)
  „ -orangan: *Debregeasia longifolia* Wedd.; *Urticac.;* Java (Ngebel)

Orange (e): *Citrus aurantium* L.; *Rutac.;* O.-Ind.
„   bark: *Celastrus dispermus* F. v. M.; *Celastrac.;* Au.
„   , Mock-: *Prunus caroliniana* Ait.; *Rosac.;* USA.
„   „   : *Bumelia lycioides* Gaertn.; *Sapotac.;* tr. Am.
„   , Native-: *Endiandra glauca* R. Br.; *Laurac.;* Au.
„   , Osage- (u): *Toxylon pomiferum* Raf.; *Morac.;* USA.
„   , sour: *Citrus aurantium* L.; *Rutac.;* Hond.
„   , sweet: *Citrus sinensis* Osbeck; *Rutac.;* Hond.
„   , wild: *Fagara flava* Krug et Urban; *Rutac.;* Jam.
„   , „   : *Prunus caroliniana* Ait.; *Rosac.;* USA.
„   , „   : *Tabernaemontana arborea* Rose, *T. grandiflora* Jacq.; *Apocynac.;* Pan.
„   wood: *Citrus aurantium* L.; *Rutac.;* Sur.
Orangenholz (d): *Citrus aurantium* L.; *Rutac.;* O.-Ind., Malay.
Oranger amer: *Citrus aurantium* L. *var. amara* L.; *Rutac.;* O.-Ind.
„   doux: *Citrus aurantium* L. *var. dulcis* L.; *Rutac.;* Tr.
„   montagne: *Ximenia americana* L.; *Olacac.;* fr. Gu.
„   de Malabar: *Aegle marmelos* Correa; *Rutac.;* Co.-Ch.
„   des falaises: *Ximenia americana* L.; *Olacac.;* Guad.
Orás: *Schizostachyum lumampao* Merr.; *Gramin.;* Phil. (Albay)
Orátan: *Celtis sp.; Ulmac.;* Phil. (Ilocos Norte, Abra)
Ordeal tree: *Erythrophloeum guineense* G. Don; *Caesalpiniac.;* Nig.
Oreganillo: *Eupatorium villosum* Sw.; *Composit.;* P.-R.
Oregon: *Pseudotsuga taxifolia* Britt.; *Conifer.;* USA.
Oreja de burro: *Clusia salvini* Donn. Smith; *Guttifer.;* Hond.
„   „   judio: *Enterolobium cyclocarpum* Griseb.; *Mimosac.;* Cuba.
„   „   negro: *Enterolobium timbouva* Mart.; *Mimosac.;* Arg.
Orejón: *Enterolobium cyclocarpum* Griseb.; *Mimosac.;* Mex.
Orelha, Brasilian: *Enterolobium sp.; Mimosac.;* Bras.
„   de negro: *Enterolobium timbouva* Mart.; *Mimosac.;* Bras. (Sao P)
„   „, Onça: *Lonchocarpus albiflorus* G. Don; *Papilionac.;* Bras. (Sao P)
Orendé rouge: *Cola ballayi* Cornu; *Sterculiac.;* Gab.
Orendeuva: *Astronium urundeuva* Engl.; *Anacardiac.;* Bras.
Oréré (f): *Mimusops djave* Engl., *M. pierreana* Engl.; *Sapotac.;* Kam., Gab.
Oreva: *Desbordesia insignis* Pierre; *Simarubac.;* Gab.
Orey wood: *Campnosperma panamensis* Standl.; *Anacardiac.;* Pan.
Orhamwood: *Ulmus americana* L.; *Ulmac.;* ö. Kan.
Ori: *Campnosperma panamensis* Standl.; *Anacardiac.;* Pan.
Oriental wood (u): *Endiandra palmerstoni* C. T. White; *Laurac.;* Au.
Oriera: *Enterolobium cyclocarpum* Griseb.; *Mimosac.;* Kol.
Orindeuva: *Astronium urundeuva* Engl.; *Anacardiac.;* Bras. (Sao P)
„   : ? *Schinus terebinthifolius* Raddi; *Anacardiac.;* Bras. (Sao P)
Orinello: *Fraxinus ornus* L.; *Oleac.;* It.
Orion (u): *Shorea guiso* Blume; *Dipterocarpac.;* Phil.
Oris: *Aporosa arborea* Muell. Arg., *A. microcalyx* Hassk.; *Euphorbiac.;* Java (Depok)
„   : *Koordersiodendron pinnatum* Merr.; *Anacardiac.;* Phil. (Ilocos Norte)
„   -nga-púrau: *Santiria nitida* Merr.; *Burserac.;* Phil. (Ilocos Norte)
Orli: *Vitex cienkowskyi* Kotschy et Peyr.; *Verbenac.;* Togo.
Orme bâtard: *Celtis occidentalis* L.; *Ulmac.;* ö. Kan.
„   blanc: *Ulmus americana* L.; *Ulmac.;* sö. Kan.
„   fauve: *Ulmus fulva* Michx.; *Ulmac.;* sö. Kan.
„   gras: *Ulmus fulva* Michx.; *Ulmac.;* sö. Kan.
„   liège: *Ulmus racemosa* Thomas; *Ulmac.;* sö. Kan.
„   maigre: *Ulmus americana* L.; *Ulmac.;* N.-Am.
„   pyramidal: *Guazuma ulmifolia* Lamk.; *Sterculiac.;* Ant.
„   rouge: *Ulmus fulva* Michx.; *Ulmac.;* sö. Kan.
„   d'Amérique: *Ulmus americana* L.; *Ulmac.;* sö. Kan.
„   „   (f): *Guazuma ulmifolia* Lamk.; *Sterculiac.;* tr. Am.
„   des Antilles: *Guazuma ulmifolia* Lamk.; *Sterculiac.;* Réunion.
„   „   bois: *Guazuma ulmifolia* Lamk.; *Sterculiac.;* Ant.
Ormigo: *Platymiscium polystachyum* Bth.; *Papilionac.;* Guat., Hond.
Oró: *Acer opulifolium* Vill.; *Acerac.;* Sp.

Oro: *Irvingia barteri* Hook. f., *I. gabonensis* Baill.; *Simarubac.*; Nig.
„ : *Antiaris africana* Engl.; *Morac.*; Nig.
„ : *Cistanthera papaverifera* Chev.; *Tiliac.*; Nig.
„ aiyo: *Antiaris toxicaria* Lesch.; *Morac.*; Nig.
Oroba: *Parinarium robustum* Oliv.; *Rosac.*; Elf.
Orobo: *Uapaca bingervillensis* Beille; *Euphorbiac.*; Elf.
Orokaike: *Protium spp.*; *Burserac.*; Sur.
Oroko: *Chlorophora excelsa* Bth. et Hook.; *Morac.*; tr. Af.
Orokpo: *Daniella thurifera* J. J. Benn.; *Caesalpiniac.*; Togo.
Oronapeta: *Swartzia tomentosa* DC.; *Caesalpiniac.*; Sur.
Oropa: *Irvingia gabonensis* Baill.; *Simarubac.*; Nig.
Orore: *Inga sp.*; *Mimosac.*; Ven.
Ororo: *Anthocleista nobilis* G. Don; *Loganiac.*; Gab.
Orotoguaje: *Acacia paniculata* Willd.; *Mimosac.*; Guat.
Orozuz: *Avicennia nitida* Jacq.; *Uerbenac.*; Mex.
Ortegon: *Coccoloba rugosa* Desf.; *Polygonac.*; P.-R.
Ortiga: *Jatropha urens* L.; *Euphorbiac.*; C.-Am.
Orumo: *Myrcia cucullata* Berg; *Myrtac.*; Ven.
Orumu: *Pachylobus edulis* G. Don; *Burserac.*; Nig.
Orungo: *Uernonia cinerea* Less.; *Composit.*; Nig.
Oruru: *Spathodea campanulata* Beauv.; *Bignoniac.*; Nig.
Orweiso: *Bridelia scleroneuroides* Pax; *Euphorbiac.*; br. O.-Af. (Uganda)
Orzech wloski (Polen): *Juglans regia* L.; *Juglandac.*; s. Eu., w. As.
Osa: *Pachylobus edulis* G. Don; *Burserac.*; Gab.
Osaga: *Piptadenia africana* Hook. f.; *Mimosac.*; Nig.
Osagedorn, orangenfarbiger: *Maclura aurantiaca* Nutt.; *Morac.*; N.-Am.
Osagho: *Pachylobus edulis* G. Don; *Burserac.*; Gab.
Osai: *Uitex peduncularis* Wall.; *Uerbenac.*; Assam.
Osegou: *Pachylobus edulis* G. Don; *Burserac.*; Gab.
Osena fufuo: *Dialium guineense* Willd.; *Caesalpiniac.*; Go.
Oséndjé: *Odyendea gabunensis* Engl.; *Simarubac.*; Gab.
Osēsēa: *Trema guineensis* Ficalho; *Ulmac.*; Go.
Osh: *Brosimum alicastrum* Swartz; *Morac.*; br. Hond.
Oshe: *Adansonia digitata* L.; *Bombacac.*; Nig.
Oshere: *Flacourtia ramontchi* L'Hérit.; *Flacourtiac.*; O.-Af. (Zambesi), Sey.
Oshima-zakura: *Prunus donarium* Siebold *subsp. speciosa* Koidz.; *Rosac.*; Jap.
Oshongo: *Anthostema aubryanum* Baill.; *Euphorbiac.*; Gab.
Osi: *Cussonia arborea* Hochst.; *Araliac.*; br. O.-Af. (Uganda)
Osier, golden: *Salix alba* L. *var. vitellina* Koch; *Salicac.*; ö. N.-Am.
„ red: *Cornus florida* L.; *Cornac.*; ö. Kan.
Osigho: *Pachylobus büttneri* Engl.; *Burserac.*; Gab.
Osighou: *Pachylobus büttneri* Engl.; *Burserac.*; Gab.
Osisiriw: *Spathodea campanulata* Beauv.; *Bignoniac.*; Go.
Osodou: *Symphonia globulifera* L., *S. gabonensis* Pierre; *Guttifer.*; Gab.
Osofoa: *Sterculia barteri* Mast., *S. tragacantha* Lindl.; *Sterculiac.*; Go.
Osok: *Ricinodendron africanum* Muell. Arg.; *Euphorbiac.*; Kam.
Osoko: *Scyphocephalium ochocoa* Warb.; *Myristicac.*; Gab.
Osokoru: *Carapa guianensis* Aubl.; *Meliac.*; Go.
Osol: *Symphonia gabonensis* Pierre; *Guttifer.*; Gab.
Osombé: *Elaeis guineensis* Jacq. *var.* ?; *Palmac.*; Gab.
Osongho: *Symphonia gabonensis* Pierre, *S. globulifera* L. f.; *Guttifer.*; Gab.
Osongobady: *Pseudospondias longifolia* Engl.; *Anacardiac.*; Gab.
Osongongo: *Antrocaryon klaineanum* Pierre; *Anacardiac.*; Gab.
Osongosongo: *Ricinodendron africanum* Muell. Arg.; *Euphorbiac.*; Gab.
Ossa: *Pachylobus edulis* G. Don; *Burserac.*; Gab.
Ossan weguia: *Pachylobus büttneri* Engl.; *Burserac..* Gab.
Ossang: *Diospyros gracilescens* Gürke; *Ebenac.*; Kam.
Ossanga: *Diospyros polystemon* Gürke; *Ebenac.*; Kam.
Ossé: *Markhamia tomentosa* K. Schum.; *Bignoniac.*; Kam.
Osségou: *Pachylobus edulis* G. Don; *Burserac.*; Gab.
? Osseh: *Maprounea membranacea* Pax et Hoffm.; *Euphorbiac.*; Gab.

Ossésaindé: *Bridelia speciosa* Muell. Arg.; *Euphorbiac.*; Gab.
Ossi: *Pachylobus edulis* G. Don *var. silvestris* Chev.; *Burserac.*; Gab.
Ossimiala: *Piptadenia sp.*; *Mimosac.*; Gab.
Ossimiale: *Piptadenia sp.*; *Mimosac.*; Gab.
Ossindio: *Anthocleista nobilis* G. Don; *Loganiac.*; Gab.
Ossoko: *Scyphocephalium ochocoa* Warb.; *Myristicac.*; Kam., Gab.
Ossol: *Moronobea coccinea* Aubl.; *Guttifer.*; Kam., Gab.
Ossongho: *Anthostema aubryanum* Baill.; *Euphorbiac.*; Gab.
Ossongo: *Anthostema aubryanum* Baill.; *Euphorbiac.*; Gab.
Ossoupou: *Mitragyne macrophylla* Hiern; *Rubiac.*; Gab.
Ostokrzen: *Ilex aquifolium* L.; *Aquifoliac.*; Polen.
Ostrolistnik: *Ilex aquifolium* L.; *Aquifoliac.*; Russland.
Ostrya de Virginie: *Ostrya virginiana* C. Koch; *Betulac.*; ö. Kan.
Ostryer de Virginie: *Ostrya virginiana* C. Koch; *Betulac.*; ö. Kan.
Osuabise: *Carapa guianensis* Aubl.; *Meliac.*; Go.
Osuobise: *Carapa procera* DC.; *Meliac.*; Go.
Osupawa: *Sarcocephalus esculentus* Afzel.; *Rubiac.*; Go.
Otangani: *Daniella sp.*; *Caesalpiniac.*; Gab.
Otélimo: *Pentadesma leucanthum* Chev.; *Guttifer.*; Elf.
Otènda: *Garcinia klaineana* Pierre; *Guttifer.*; Gab.
Otenga: *Dillenia indica* L.; *Dilleniac.*; Assam.
Otengah: *Dillenia indica* L.; *Dilleniac.*; Assam.
Otéva: *Desbordesia insignis* Pierre; *Simarubac.*; Gab.
Oti: *Pycnanthus kombo* Warb.; *Myristicac.*; Go.
„ -mirim-ayra: *Brosimum discolor* Schott; *Morac.*; Bras.
Otie: *Pycnanthus kombo* Warb.; *Myristicac.*; Go.
Otien-ogioro: *Ochrocarpus africanus* Oliv.; *Guttifer.*; Nig. (Benin)
„ -wuare: *Ochrocarpus africanus* Oliv.; *Guttifer.*; Nig. (Benin)
Otindia: *Bridelia speciosa* Muell. Arg.; *Euphorbiac.*; Kam., Gab.
„ : *Chrysophyllum sp.*; *Sapotac.*; Gab.
Otingah: *Dillenia indica* L.; *Dilleniac.*; Assam.
Otiturubá: *Lucuma sp.*; *Sapotac.*; Bras.
Otlug: *Lophira alata* Banks; *Ochnac.*; br. O.-Af. (Uganda)
Oto: *Pycnanthus kombo* Warb.; *Myristicac.*; W.-Af.
Otoba: *Dialyanthera otoba* Warb., *Virola sp.*; *Myristicac.*; Kol., Ven.
Otobaye: *Cordia batesi* Wernham; *Borraginac.*; Kam.
Otókotáka: *Bauhinia reticulata* DC.; *Caesalpiniac.*; Togo.
Otolbé: *Hernandia beninensis* Welw.; *Hernandiac.*; Gab.
Otómbi: *Copaifera ? salikounda* Heckel; *Caesalpiniac.*; Gab.
Otouan: *Xylopia quintasi* Engl. et Diels; *Anonac.*; Kam.
Otouma: *Ficus punctata* Lamk.; *Morac.*; Gab.
Otsagui: *Erismadelphus baudoni* Chev. et Russ.; *Vochysiac.*; fr. Kongo.
Otsoko: *Scyphocephalium ochocoa* Warb.; *Myristicac.*; Gab.
Otsuku: *Scyphocephalium ochocoa* Warb.; *Myristicac.*; Gab.
Ottindia: *Chrysophyllum sp.*; *Sapotac.*; Gab.
Ottugba: *Lophira alata* Banks; *Ochnac.*; Togo.
Otu: *Canarium schweinfurthi* Engl.; *Burserac.*; Kam.
Otua-nzam: *Discoglypremna caloneura* Prain; *Euphorbiac.*; Gab.
Otúgba: *Lophira alata* Banks; *Ochnac.*; Togo.
Otugua: *Peucedanum araliaceum* B. et H. *var. fraxinifolium* Engl.; *Umbellifer.*: br.
Otun: *Ochrocarpus africanus* Oliv.; *Guttifer.*; Nig.   [O.-Af. (Ug.)
Otuna: *? Cynometra dewevrei ?*; *Caesalpiniac.*; B.-K. (Kurubu)
Otunga: *? Polyalthia sp.*; *Anonac.*; Kam.
Otungui: *Xylopia quintasi* Engl. et Diels; *Anonac.*; Kam.
Oturu-itabi (jap.): *Ficus megacarpa* Merr.; *Morac.*; Form., Phil.
Otut: *Cistanthera papaverifera* Chev.; *Tiliac.*; Nig. (Lagos)
Otutu: *Cistanthera papaverifera* Chev.; *Tiliac.*; Go., Nig. (Lagos)
„ : *Mansonia altissima* Chev.; *Sterculiac.*; Nig.
Otwetwerede: *Ricinodendron africanum* Muell. Arg.; *Euphorbiac.*; Go.
Oua-oua: *Triplochiton scleroxylon* K. Schum.; *Sterculiac.*; Elf.
Ouabé: *Omphalea diandra* L.; *Euphorbiac.*; fr. Gu.

Ouala: *Funtumia africana* Stapf; *Apocynac.;* Elf.
„ : *Pentaclethra macrophylla* Bth.; *Mimosac.;* Gab.
Oualélé: *Pycnanthus kombo* Warb.; *Myristicac.;* Elf.
Oualio: *Schrebera arborea* Chev.; *Oleac.;* Elf.
Ouán tsé kò: *Hovenia dulcis* Thunb.; *Rhamnac.;* Ch.
Ouando: *Panda oleosa* Pierre; *Pandac.;* Kam., Gab.
Ouara: *Cola cordifolia* R. Br.; *Sterculiac.;* Elf.
Ouarouchi: *Virola surinamensis* Warb.; *Myristicac.;* Sur.
Ouatier (f): *Ceiba pentandra* Gaertn.; *Bombacac.;* Tr.
„ „ : *Bombax malabaricum* DC.; *Bombacac.;* O.-Ind.
Ouatta-pana: *Caesalpinia coriaria* Willd.; *Caesalpiniac.;* S.-Am. (? Ven.)
Ouaye: *Geonoma baculifera* Kunth; *Palmac.;* Sur.
Ouesse: *Triplochiton scleroxylon* K. Schum.; *Sterculiac.;* Kam.
Ouey: *Garcinia manni* Oliv.; *Guttifer.;* Kam.
„ : *Hexalobus crispiflorus* A. Rich.; *Anonac.;* Kam.
Oufo: *Ximenia americana* L.; *Olacac.;* s. Kam., n. Gab.
Ougéminia: *Distemonanthus benthamianus* Baill.; *Caesalpiniac.;* W.-Af.
Ougounou: *Acolea missionis* Pierre; *Sapotac.;* Gab.
Ou han: *Sarcocephalus esculentus* Afzel.; *Rubiac.;* Elf.
Ouhidji: *Terminalia ivorensis* Chev.; *Combretac.;* Elf.
Ouhom: *Sarcocephalus esculentus* Afzel.; *Rubiac.;* Elf.
Oulongua-palou: *Sloanea sp.; Elaeocarpac.;* Sur.
Oungou: *Parinarium glabrum* Oliv.; *Rosac.;* Gab.
Ouochi: *Albizzia zygia* MacBride; *Mimosac.;* Elf.
Ouofa: *Triplochiton scleroxylon* K. Schum.; *Sterculiac.;* Elf.
Ouogueso: *Tetrapleura thonningi* Bth.; *Mimosac.;* Gab.
Ouokissé: *Pterocarpus santalinoïdes* L'Hérit.; *Papilionac.;* Elf.
Ouolo: *Terminalia macroptera* Guill. et Perr.; *Combretac.;* Senl.
Ouologbaoué: *Tetrorchidium didymostemon* Pax et K. Hoffm.; *Euphorbiac.;* Elf.
Ouombé: *Haronga paniculata* Sprague; *Guttifer.;* Elf.
Ouo-oué: *Lophira procera* Chev.; *Ochnac.;* Elf.
Ouoré-ouoré: *Cleistanthus sp.; Euphorbiac.;* Elf.
Ouoro: *Terminalia macroptera* Guill. et Perr.; *Combretac.;* fr. Guin.
Ouossoupâlié: *Erythrina altissima* Chev., *E. sp.; Papilionac.;* Elf.
Ouotéra: *Allanblackia parviflora* Chev.; *Guttifer.;* Elf.
Ouregou: *Oxandra lanceolata* Baill.; *Anonac.;* Bras.
Ouricuri: *Cocos coronata* Mart.; *Palmac.;* Bras.
Ouroria: *Anthocleista vogeli* G. Don; *Loganiac.;* Gab.
Outeniqua: *Podocarpus elongata* L'Hérit.; *Conifer.;* S.-Af. (Kap.)
Outin: *Sterculia sp.; Sterculiac.;* Gab.
Ovaga: *Panda oleosa* Pierre; *Pandac.;* Gab.
Ovala: *Pentaclethra macrophylla* Bth.; *Mimosac.;* Elf., Kam., Gab.
Ovan: *Tylostemon membranaceus* Stapf; *Laurac.;* Kam.
Ovanda: *Panda oleosa* Pierre; *Pandac.;* Gab.
Ovang: *Didelotia africana* Pierre; *Caesalpiniac.;* Gab.
Ovao: *? Irvingia klainei* Pierre; *Simarubac.;* Gab.
Ovayo: *? Irvingia klainei* Pierre; *Simarubac.;* Gab.
Ovbevbe: *Mitragyne macrophylla* Hiern; *Rubiac.;* Nig.
Ovbiake: *Sarcocephalus trillesi* Pierre; *Rubiac.;* Nig., Go.
Ovèbe: *Schumanniophyton klaineanum* Pierre; *Rubiac.;* Gab.
Ovèi: *Elaeis guineensis* Jacq. var. ?; *Palmac.;* Gab.
Ovéng: *Copaifera demeusei* Harms, *C. tessmanni* Harms; *Caesalpiniac.;* Kam., Gab.
Ovèng'kol: *Sindora klaineana* Pierre; *Caesalpiniac.;* Gab.
Ovep: *Ochrocarpus africanus* Oliv.; *Guttifer.;* Nig.
Ovibe: *Schumanniophyton klaineanum* Pierre; *Rubiac.;* Gab.
Ovillo: *Hura polyandra* Baill.; *Euphorbiac.;* Mex.
Ovinezi: *Diospyros aggregata* Gürke; *Ebenac.;* Gab.
Ovoga: *Poga oleosa* Pierre; *Rhizophorac.;* Kam., Gab.
Ovogn: *Didelotia africana* Pierre; *Caesalpiniac.;* Kam.
Ovogo: *Poga oleosa* Pierre; *Rhizophorac.;* Gab.
Ovonda: *Hexalobus crispiflorus* A. Rich.; *Anonac.;* Gab.

Ovong: *Dialium guineense* Willd.; *Caesalpiniac.*; Gab.
Ovos: *Staudtia gabonensis* Warb., *S. kamerunensis* Warb.; *Myristicac.*; Kam.
Ovounda: *Hexalobus crispiflorus* A. Rich.; *Anonac.*; Gab.
Ovovo: *Ricinodendron africanum* Muell. Arg.; *Euphorbiac.*; Nig.
Ovu: *Antiaris sp.*; *Morac.*; Nig.
Owa zungéa: *Grewia canescens* A. Rich.; *Tiliac.*; n. Af. (Abessinien)
Owadè: *Garcinia klaineana* Pierre; *Guttifer.*; Gab.
Owala: *Pentaclethra macrophylla* Bth.; *Mimosac.*; Gab.
Owalè: *Garcinia klaineana* Pierre; *Guttifer.*; Gab.
Owama: *Ricinodendron africanum* Muell. Arg.; *Euphorbiac.*; Go.
Owang: *Didelotia africana* Pierre; *Caesalpiniac.*; Gab.
Owanghan: *Piptadenia africana* Hook. f.; *Mimosac.*; Nig.
Owati: *Parkia africana* R. Br.; *Mimosac.*; Togo.
Owawa: *Triplochiton scleroxylon* K. Schum.; *Sterculiac.*; Go.
Owebala: *Terminalia superba* Engl. et Diels; *Combretac.*; Nig.
Oweive: *Terminalia ivorensis* Chev.; *Combretac.*; Nig.
Owélé: *Canarium velutinum* Guill.; *Burserac.*; Gab.
Owimon: *Mimusops lacera* Baker; *Sapotac.*; Gab.
Owindefam: *Diospyros suaveolens* Gürke; *Ebenac.*; Kam.
Owingué: *Distemonanthus benthamianus* Baill.; *Caesalpiniac.*; Gab.
Owingue: *Klainedoxa latifolia* Pierre; *Simarubac.*; Kam., Gab.
Owiwi: *Baphia nitida* Lodd.; *Papilionac.*; Nig.
Owli: *Mimusops elengi* Roxb.; *Sapotac.*; sw. O.-Ind. (Bombay)
Owoë: *Hexalobus grandiflorus* Bth.; *Anonac.*; Kam.
Owöng: *Copaifera tessmanni* Harms; *Caesalpiniac.*; Kam.
Owoté limon: *Allanblackia parviflora* Chev.; *Guttifer.*; Elf.
Owounda: *Hexalobus crispiflorus* A. Rich.; *Anonac.*; Gab.
Owowa: *Triplochiton scleroxylon* K. Schum.; *Sterculiac.*; Go.
Owui: *Hexalobus crispiflorus* A. Rich.; *Anonac.*; Gab.
Owuru: *Entandrophragma cylindricum* Sprague; *Meliac.*; Nig.
Oxelbeere: *Pirus scandica* Fries; *Rosac.*; Norwegen, Schweden, Finland
Oxelbirn: *Sorbus intermedia* Pers.; *Rosac.*; Schweden.
Oyago: *Kigelia aethiopica* Dcne.; *Bignoniac.*; br. O.-Af. (Uganda)
Oyalo: *Vitex cienkowskyi* Kotschy et Peyr.; *Verbenac.*; br. O.-Af. (Uganda)
Oyamba: *Pentadesma butyraceum* Don; *Guttifer.*; Gab.
Oyamel: *Abies religiosa* Lindl.; *Conifer.*; Mex., Guat.
Oyankwren: *Ficus exasperata* Vahl; *Morac.*; Go.
Oyebé: *Randia acuminata* Bth.; *Rubiac.*; Kam.
Oyemvom: *Strombosiopsis tetrandra* Engl.; *Olacac.*; Kam.
Oyemvome: *Strombosiopsis tetrandra* Engl.; *Olacac.*; Kam.
Oyife: *Canarium sp.*; *Burserac.*; Nig.
Oyila: *Elaeis guineensis* Jacq. var. *nigrescens* Chev.; *Palmac.*; Gab.
? „ wi ndjógu: *Elaeis guineensis* Jacq. var. ?; *Palmac.*; Gab.
Oyo: *Albizzia coriaria* Welw.; *Mimosac.*; br. O.-Af. (Uganda)
Oyora: *Crataeva adansoni* DC.; *Capparidac.*; br. O.-Af. (Uganda)
Oyou: *Salvadora persica* L.; *Salvadorac.*; Sahara.
Oyoye: *Lovoa sp.*; *Meliac.*; B.-K. (Beni, Semliki)
Oyster wood, Cuban (u): *Gymnanthes lucida* Swartz; *Euphorbiac.*; Bah., W.-Ind.
Ozainegué: *Hymenostegia sp.*; *Caesalpiniac.*; Gab.
Ozakome: *Antrocaryon klaineanum* Pierre; *Anacardiac.*; Gab.
Ozakoum: *Antrocaryon klaineanum* Pierre; *Anacardiac.*; Gab.
Ozaneguilia: *Ricinodendron africanum* Muell. Arg.; *Euphorbiac.*; Gab.
Ozangilya: *Ricinodendron africanum* Muell. Arg.; *Euphorbiac.*; Gab.
Ozéndjé: *Odyendea gabunensis* Engl.; *Simarubac.*; Gab.
Ozenejé: *Guarea sp.*; *Meliac.*; Gab.
   „ : *Odyendea gabunensis* Engl.; *Simarubac.*; Gab.
Ozigo (f): *Pachylobus büttneri* Engl.; *Burserac.*; Gab.
Ozocote: *Liquidambar styraciflua* L.; *Hamamelidac.*; Mex.
Ozocotl: *Liquidambar styraciflua* L.; *Hamamelidac.*; Mex.
Ozogoué: *Dracaena fragrans* Ker-Gawl.; *Liliac.*; Gab.
Ozoli: *Symphonia globulifera* L.; *Guttifer.*; Gab.

Ozombi: *Uapaca sp.; Euphorbiac.;* Gab.
Ozonga: *Klainedoxa latifolia* Pierre; *Simarubac.;* Kam., Gab.
Ozongo: *Saccoglottis gabonensis* Urban; *Humiriac.;* Kam.
Ozouga (f): *Saccoglottis gabonensis* Urban; *Humiriac.;* Kam., Gab.
„ : *Klainedoxa latifolia* Pierre; *Simarubac.;* Kam., Gab.
„ : *Copaifera sp.; Caesalpiniac.;* Sur.
Ozougo: *Saccoglottis gabonensis* Urban; *Humiriac.;* Gab.
Ozové: *Ongokea klaineana* Pierre; *Olacac.;* Gab.
Ozozo: *Pseudospondias longifolia* Engl.; *Anacardiac.;* Gab.
Ozumi: *Pirus zumi* Matsum.; *Rosac.;* Jap.

p

Pá dông: *Pentacme siamea* Kurz; *Dipterocarpac.;* Ind.-Ch.
„ kshmuk: *Prunus capuli* Cav.; *Rosac.;* Mex.
„ nol: *Acronychia laurifolia* Blume; *Rutac.;* Ind.-Ch.
„ pai: *Trewia nudiflora* L.; *Euphorbiac.;* Ind.-Ch.
„ -raik: *Bombax insignis* Wall.; *Bombacac.;* Burma.
„ de canella: *Aydendron canella* Meissn.; *Laurac.;* Bras.
Paagáhan: *Adenanthera intermedia* Merr.; *Mimosac.;* Phil. (Laguna)
Paaloe boontsji: *Erythrina sp.; Papilionac.;* Curaçao.
Páang-balínis: *Terminalia oocarpa* Merr.; *Combretac.;* Phil. (Batangas)
„ -darága: *Alangium meyeri* Merr.; *Cornac.;* Phil. (Camarines)
Paardenfleesch: *Calocarpum mammosum* Pierre; *Sapotac.;* holl. Gu.
„ , oostindisch: *Schoutenia ovata* Korth.; *Tiliac.;* Java.
Paardenvlush: *Mimusops balata* L.; *Sapotac.;* Gu.
Paardepis: *Hippobromus alatus* Eckl. et Zeyh.; *Sapindac.;* S.-Af. (Kap)
Paardepram: *Zanthoxylum capense* Harv.; *Rutac.;* S.-Af. (Kap)
Paardevleesh-hout: *Mimusops surinamensis* Miq.; *Sapotac.;* Sur.
„ „ : *Mimusops balata* Gaertn., *M. globosa* Gaertn.; *Sapotac.;* Sur.
Pabba: *Chukrasia tabularis* Adr. Juss.; *Meliac.;* sw. O.-Ind. (Bombay)
Pabda: *Lagerstroemia hypoleuca* Kurz; *Lythrac.;* And.
Pacai: *Inga sp.; Mimosac.;* Peru, Bol.
„ de mono: *Inga sp.; Mimosac.;* Peru, Bol.
Pacanier: *Carya olivaeformis* Nutt.; *Juglandac.;* USA.
Pacapeua: *Swartzia sp.; Caesalpiniac.;* Bras.
Pacará: *Enterolobium contortisiliquum* Morong; *Mimosac.;* Arg.
„ negra: *Enterolobium sp.; Mimosac.;* Arg.
„ ploma: *Enterolobium sp.; Mimosac.;* Arg.
„ timbo: *Enterolobium contortisiliquum* Morong; *Mimosac.;* W.-Ind.
Pacay: *Inga sp.; Mimosac.;* Bras.
„ : *Inga feuillei* DC.; *Mimosac.;* Peru.
„ : *Andira inermis* H. B. K.; *Papilionac.;* Mex.
Pacaya: *Chamaedorea spp.; Palmac.;* Hond.
Paccai: *Inga feuillei* DC.; *Mimosac.;* Peru.
Pachunda: *Capparis grandis* L. f.; *Capparidac.;* O.-Ind.
Pachut: *Nephelium lappaceum* L.; *Sapindac.;* Malak.
Pacito: *Muntingia calabura* L.; *Elaeocarpac.;* Pan.
Paco-bala: *Zanthoxylum horridum* Welw., *Z. melanocanthum* Planch.; *Rutac.;* port. Af
Pacora de macaco: *Swartzia sp.; Caesalpiniac.;* Bras.
Pacouri: *Platonia insignis* Mart.; *Guttifer.;* fr. Gu.
Pacóva de Macado: *Swartzia langsdorffi* Raddi; *Caesalpiniac.;* Bras. (R.-J.)
Pacuca: *Geonoma binervia* Oerst.; *Palmac.;* Hond.
Pacún: *Sapindus saponaria* L.; *Sapindac.;* Salv.
Pacuquilla: *Geonoma spp.; Palmac.;* Hond.
Pacurero: *Torrubia pacurero* Standl.; *Nyctaginac.;* Ven.
Pacuri: *Rheedia brasiliensis* Pl. et Tr., *Platonia insignis* Mart.; *Guttifer.;* Arg.
Pacury guazu: *Platonia insignis* Mart.; *Guttifer.;* Par.
„ -mi: *Platonia sp.; Guttifer.;* Par.
Padâc: *Polyalthia cerasoïdes* Bth. et Hook. f.; *Anonac.;* Kamb.
Padal: *Stereospermum chelonioides* DC., *S. suaveolens* DC.; *Bignoniac.;* sw. & C.-O.-

Padal: *Stereospermum xylocarpum* Bth. et Hook. f.; *Bignoniac.;* C.-O.-Ind.　　[Ind.
Padam: *Juniperus macropoda* Boiss.; *Conifer.;* nw. O.-Ind. (Punjab)
Padang: *Mallotus moluccensis* Muell. Arg.; *Euphorbiac.;* Java (Besoeki)
Padar: *Stereospermum chelonoides* DC., *S. suaveolens* DC.; *Bignoniac.;* C.-O.-Ind.
　　„　: *Stereospermum xylocarpum* Bth. et Hook. f.; *Bignoniac.;* C.-O.-Ind.
Padaria: *Stereospermum suaveolens* DC.; *Bignoniac.;* n. O.-Ind. (United Prov.)
Padauk: *Pterocarpus dalbergioides* Roxb.; *Papilionac.;* And.
　　„　: *Pterocarpus macrocarpus* Kurz; *Papilionac.;* Burma.
　　„　, african: *Pterocarpus soyauxi* Taub., *P. santalinoides* L'Hérit.; *Papilionac.;* W.-Af.
　　„　　„　: *Pterocarpus angolensis* DC.; *Papilionac.;* W.-Af.
　　„　, andaman: *Pterocarpus dalbergioides* Roxb.; *Papilionac.;* And., O.-Ind.
　　„　, brown: *Pterocarpus macrocarpus* Kurz; *Papilionac.;* O.-Ind.
　　„　, Burma-: *Pterocarpus macrocarpus* Kurz; *Papilionac.;* Burma.
　　„　, Inland-: *Pterocarpus macrocarpus* Kurz; *Papilionac.;* O.-Ind., Burma.
　　„　, mixed-colored: *Pterocarpus macrocarpus* Kurz; *Papilionac.;* O.-Ind.
　　„　-ni: *Pterocarpus macrocarpus* Kurz; *Papilionac.;* O.-Ind.
　　„　-nyo: *Pterocarpus macrocarpus* Kurz; *Papilionac.;* Burma.
　　„　, Philippine-: *Pterocarpus indicus* Willd.; *Papilionac.;* Phil.
　　„　-pyo: *Pterocarpus macrocarpus* Kurz; *Papilionac.;* Burma.
　　„　, red: *Pterocarpus macrocarpus* Kurz; *Papilionac.;* Burma.
　　„　-sal: *Pterocarpus macrocarpus* Kurz; *Papilionac.;* Burma.
　　„　-sat: *Pterocarpus macrocarpus* Kurz; *Papilionac.;* Burma.
　　„　-wa: *Pterocarpus macrocarpus* Kurz; *Papilionac.;* Burma.
　　„　, white: *Pterocarpus macrocarpus* Kurz; *Papilionac.;* O.-Ind.
　　„　, yellow: *Pterocarpus macrocarpus* Kurz; *Papilionac.;* O.-Ind.
Paddam: *Prunus puddum* Roxb.; *Rosac.;* n. O.-Ind., (United Prov.)
Paddehout: *? Swartzia sp.; Caesalpiniac.;* fr. Gu.
Paddle wood: *Aspidosperma nitidum* Bth., *A. exelsum* Bth., *? A. oblongum* DC.;
　　„　　„　: *Uvaria büsgeni* Unwin; *Anonac.;* Nig.　　　　[*Apocynac.;* Gu.
　　„　„　, black: *Swartzia acuminata* Willd., *S. polyphylla* DC.; *Caesalpiniac.;* Sur.
　　„　„　,　„　: *Swartzia triphylla* Willd.; *Caesalpiniac.;* Sur.
Pader: *Stereospermum suaveolens* DC., *S. chelonioides* DC.; *Bignoniac.;* w. Beng.
Padiala: *Stereospermum suaveolens* DC.; *Bignoniac.;* n. O.-Ind. (United Prov.)
Padiwin: *Uvaria chamae* Beauv.; *Anonac.;* Togo.
Padjang: *Mallotus moluccanus* Muell. Arg.; *Euphorbiac.;* Java (Besoeki)
Padji: *Dryobalanops aromatica* Gaertn.; *Dipterocarpac.;* Borneo, Sum.
Padoli: *Stereospermum chelonioides* DC.; *Bignoniac.;* sw. O.-Ind. (Bombay)
Padoo: *Pterocarpus macrocarpus* Kurz; *Papilionac.;* O.-Ind., Burma.
Padouk: *Pterocarpus spp.; Papilionac.;* As., Af.
　　„　: *Pterocarpus macrocarpus* Kurz, *P. indicus* Willd.; *Papilionac.;* Burma.
　　„　: *Pterocarpus dalbergioides* Roxb.; *Papilionac.;* And.
　　„ (f): *Pterocarpus soyauxi* Taub.; *Papilionac.;* Kam., Gab.
　　„　: *Pterocarpus santalinoides* L'Hérit.; *Papilionac.;* tr. W.-Af.
　　„　: *Pterocarpus sp.; Papilionac.;* Nig.
　　„　, afrikanisches (d): *Pterocarpus soyauxi* Taub.; *Papilionac.;* W.-Af.
　　„　, Andaman: *Pterocarpus dalbergioides* Roxb.; *Papilionac.;* O.-Ind., Cey.
　　„　, brown: *Pterocarpus macrocarpus* Kurz; *Papilionac.;* O.-Ind.
　　„　, Burma-: *Pterocarpus macrocarpus* Kurz; *Papilionac.;* Burma.
　　„　, echtes: *Pterocarpus macrocarpus* Kurz; *Papilionac.;* Burma.
　　„　, figured: *Pterocarpus macrocarpus* Kurz; *Papilionac.;* O.-Ind.
　　„　, Manila-: *Pterocarpus indicus* Willd.; *Papilionac.;* Phil.
Padri (H): *Stereospermum chelonioides* DC.; *Bignoniac.;* C.-O.-Ind. (Gondwana)
　　„　　„　: *Stereospermum suaveolens* DC.; *Bignoniac.;* O.-Ind. (Madras)
Padu: *Pterocarpus macrocarpus* Kurz; *Papilionac.;* O.-Ind., Burma.
Pagaai-hout: *? Swartzia sp.; Caesalpiniac.;* fr. Gu.
Pagaksón: *Hopea philippinensis* Dyer; *Dipterocarpac.;* Phil. (Camarines)
Pagar anak: *Ixonanthes icosandra* Jack; *Linac.;* Malay.
　　„　„　: *Gordonia excelsa* Blume; *Theac.;* H.-Ind., Malay.
　　„　„　jantan: *Gordonia excelsa* Blume; *Theac.;* br. O.-Ind., Malay.
Pagatpat: *Sonneratia caseolaris* Engl.; *Sonneratiac.;* Phil.
Paghúbo: *Terminalia comintana* Merr.; *Combretac.;* Phil. (Rizal)

Paglumbóien: *Mangifera monandra* Merr.; *Anacardiac.;* Phil. (Pangasinan)
Pagoda tree: *Plumiera acutifolia* Poir.; *Apocynac.;* O.-Ind.
„ „ , japan (e): *Sophora japonica* L.; *Papilionac.;* ö. As.
Pagow ank: *Ixonanthes icosandra* Jack; *Linac.;* Malak.
Pagpágan: *Mimusops parvifolia* R. Br.; *Sapotac.;* Phil. (Cagayan)
Pagsaguáu: *Mangifera monandra* Merr.; *Anacardiac.;* Phil. (Leyte)
Pagsáhing: *Canarium villosum* F. Vill.; *Burserac.;* Phil. (Batangas)
Pagsahíngin: *Canarium villosum* F. Vill., *C. luzonicum* A. Gray; *Burserac.;* Phil.
„ : *Dipterocarpus grandiflorus* Blco.; *Dipterocarpac.;* Phil. (Laguna) [(Bataan)
„ : *Dipterocarpus hasselti* Blume; *Dipterocarpac.;* Phil. (Laguna)
„ : *Dipterocarpus vernicifluus* Blco.; *Dipterocarpac.;* Phil. (Laguna)
„ : *Vatica mangachapoi* Blco.; *Dipterocarpac.;* Phil. (Laguna)
Pagulíngin: *Cratoxylon sp.;* *Guttifer.;* Phil. (Rizal)
Pagulíngon: *Cratoxylon sp.;* *Guttifer.;* Phil. (Negrosinsel)
Pagun: *Bombax malabaricum* DC.; *Bombacac.;* Beng.
Paguríngon: *Cratoxylon sp.;* *Guttifer.;* Phil. (Camarines, Albay, Masbate)
Pah: *Canarium sikkimense* King; *Burserac.;* nö. O.-Ind. (Bhutan)
Paha: *Lophira alata* Banks; *Ochnac.;* Nig.
Pahad: *Stereospermum suaveolens* DC.; *Bignoniac.;* sw. O.-Ind. (Bombay)
Pahari: *Stereospermum suaveolens* DC.; *Bignoniac.;* nö. O.-Ind. (Nepal), sw. O.-Ind.
„ : *Populus ciliata* Wall.; *Salicac.;* n. O.-Ind. (United Prov.) [(Malabar)
„ pipal: *Populus ciliata* Wall.; *Salicac.;* n. O.-Ind. (United Prov.)
Pahautea: *Libocedrus bidwilli* Hook. f.; *Conifer.;* N.-Seel.
Pahé: *Fagara macrophylla* Engl.; *Rutac.;* Elf.
Pahi: *Tectona grandis* L. f.; *Verbenac.;* Burma.
Pahn: *Diospyros linderi* Hutch. et Dalz.; *Ebenac.;* Lib.
Páho': *Mangifera altissima* Blco.; *Anacardiac.;* Phil.
Pahu': *Mangifera altissima* Blco.; *Anacardiac.;* Phil.
Pahuhútan: *Mangifera altissima* Blco.; *Anacardiac.;* Phil.
Pahútan: *Mangifera altissima* Blco.; *Anacardiac.;* Phil.
Pái: *Litchi chinensis* Radlk.; *Sapindac.;* Ind.-Ch.
Pai: *Terminalia altissima* Chev.; *Combretac.;* Elf.
Paida: *Helicostylis poeppigiana* Tréc.; *Morac.;* Gu.
„ : *Piratinera guianensis* Aubl.; *Morac.;* Sur.
„ , Toekoesie-: *Helicostylis poeppigiana* Tréc.; *Morac.;* Gu.
„ , Wekere-: *Piratinera guianensis* Aubl.; *Morac.;* Sur.
Paidi: *Ficus glomerata* Roxb.; *Morac.;* O.-Ind. (Madras)
Paihápi: *Anisoptera thurifera* Blume; *Dipterocarpac.;* Phil. (Zambales)
Pailari: *Wrightia tomentosa* Roem. et Schultes; *Apocynac.;* C.-O.-Ind. (Gondwana)
Paildua: *Erythrina suberosa* Roxb.; *Papilionac.;* w. Beng. (Orissa)
Paillo: *Detarium sp.;* *Caesalpiniac.;* Gab.
Pain de Dika: *Irvingia gabonensis* Baill.; *Simarubac.;* Gab.
„ „ singe: *Adansonia digitata* L.; *Bombacac.;* Ant.
Painá: *Chorisia speciosa* St. Hil.; *Bombacac.;* Arg.
„ : *Hopea philippinensis* Dyer; *Dipterocarpac.;* Phil. (Tayabas)
„ : *Kingiodendron alternifolium* Merr.; *Papilionac.;* Phil.
„ : *Sindora supa* Merr.; *Caesalpiniac.;* Phil. (Camarines, Albay, Sorsogon)
„ banana: *Amphistelma angulatum* Fourn.; *Asclepiadac.;* Bras.
„ limpa: *Ceiba pentandra* Gaertn.; *Bombacac.;* Bras.
„ melao: *Fischeria maritima* Dcne.; *Asclepiadac.;* Bras.
„ ovo: *Chthamalia purpurea* Dcne.; *Asclepiadac.;* Bras.
„ pedra amarella: *Bombax campestris* K. Schum.; *Bombacac.;* Bras.
„ sinha: *Peckoltia pedalis* Fourn.; *Asclepiadac.;* Bras.
„ yvoty: *Lippia urticoides* Cham.; *Verbenac.;* Bras.
„ de arbusto: *Bombax marginatum* K. Schum.; *Bombacac.;* Bras.
„ „ Cuba: *Bombax aquaticum* K. Schum.; *Bombacac.;* Bras.
„ „ pennas: *Echites peltata* Vell.; *Asclepiadac.;* Bras.
„ „ Samohù: *Chorisia sp.;* *Bombacac.;* Arg.
„ „ sapo: *Asclepias curassavica* L.; *Asclepiadac.;* Bras.
„ „ seda: *Ceiba pentandra* Gaertn., *Chorisia speciosa* St. Hil.; *Bombacac.;* Bras.
„ „ „ : *Asclepias curassavica* L., *Araujia sericifera* Brot.; *Asclepiadac.;* Bras.

Painá de seda: *Gomphocarpus brasiliensis* Fourn.; *Asclepiadac.;* Bras.
„ „ „ do campo: *Oxypetalum campestre* Dcne.; *Asclepiadac.;* Bras.
„ do campo: *Araujia sericifera* Brot.; *Asclepiadac.;* Bras.
? Paineira: *Blepharodon linearis* Dcne.; *Asclepiadac.;* Bras.
„ : *Echites peltata* Vell.; *Apocynac.;* Bras.
„ : *Chorisia speciosa* St. Hil.; *Bombacac.;* Bras.
„ : *Ceiba sp.; Bombacac.;* Arg.
„ amarella das pedras: *Bombax campestris* K. Schum.; *Bombacac.;* Bras.
„ arbusto: *Bombax marginatum* K. Schum.; *Bombacac.;* Bras.
„ barriguda: *Chorisia insignis* H. B. K., *C. crispifolia* H. B. K.; *Bombacac.;* Bras.
„ branca: *Ceiba erianthos* K. Schum.; *Bombacac.;* Bras.
„ „ das pedras: *Bombax pentaphyllum* Vell.; *Bombacac.;* Bras.
„ cheirosa: *Ceiba jasminodora* K. Schum.; *Bombacac.;* Bras.
„ femea: *Chorisia speciosa* St. Hil., *C. crispifolia* H. B. K.; *Bombacac.;* Bras.
„ loura: *Echites peltigera* Stadelm.; *Apocynac.;* Bras.
„ macho: *Chorisia sp.; Bombacac.;* Bras.
„ mata pao: *Ceiba riviera* K. Schum.; *Bombacac.;* Bras.
„ a tôa: *Bombax endecaphyllum* Vell.; *Bombacac.;* Bras.
„ da praia: *Bombax endecaphyllum* Vell.; *Bombacac.;* Bras.
„ de castanha: *Bombax insigne* K. Schum.; *Bombacac.;* Bras.
„ „ Cuba: *Bombax aquaticum* K. Schum.; *Bombacac.;* Bras. (R.-J.)
„ „ harpoador: *Bombax cyathophorum* K. Schum.; *Bombacac.;* Bras.
„ „ seda: *Chorisia speciosa* St. Hil., *Ceiba pentandra* Gaertn.; *Bombacac.;* Bras.
„ do campo: *Bombax crenulatum* K. Schum., *B. spp.; Bombacac.;* Bras.
Paini mara: *Vateria indica* L.; *Dipterocarpac.;* sw. O.-Ind. (Malabar)
Pain-in-back: *Trema micrantha* Blume; *Ulmac.;* Bah.
Paira: *Piratinera guianensis* Aubl.; *Morac.;* br. Gu.
Pairawa: *Inga spp.; Mimosac.;* Sur.
Pairjawa: *Inga spp.; Mimosac.;* Sur.
Pairowa waikey: *Inga spp.; Mimosac.;* Gu.
Paít: *Eugenia sp.; Myrtac.;* Phil.
Paitán: *Eugenia costulata* C. B. Rob.; *Myrtac.;* Phil. (Pangasinan)
„ : *Eugenia similis* Merr.; *Myrtac.;* Phil. (Zambales, Pampanga)
„ : *Dehaasia triandra* Merr.; *Laurac.;* Phil. (Laguna)
Paja: *Prunus puddum* Roxb.; *Rosac.;* nw. O.-Ind. (Punjab)
Pajasan: *Ailanthus glandulosa* Desf.; *Simarubac.;* Jap., Ch.
Pajoeli: *Iryanthera sagotiana* Warb., *I. hostmanni* Warb.; *Myristicac.;* Sur.
Pajoelidan: *Iryanthera sagotiana* Warb., *I. hostmanni* Warb.; *Myristicac.;* Sur.
Pajoerian: *Xylopia frutescens* Aubl., *X. salicifolia* H. B. K.; *Anonac.;* Sur.
„ : *Xylopia grandiflora* St. Hil.; *Anonac.;* Sur.
Pajoerilan: *Iryanthera sagotiana* Warb., *I. hostmanni* Warb.; *Myristicac.;* Sur.
Pajui: *Bumelia buxifolia* Willd., *B. sp.; Sapotac.;* Ven.
Pajuil: *Anacardium occidentale* L.; *Anacardiac.;* P.-R.
Pákak: *Artocarpus communis* Forst., *A. treculiana* Elmer; *Morac.;* Phil.
„ : *Semecarpus sp.; Anacardiac.;* Phil.
Pakarán: *Palaquium sp.; Sapotac.;* Phil. (Pangasinan)
Pakié pakié: *Conopharyngia crassa* Stapf; *Apocynac.;* Elf.
? Pakilan: *Wetria macrophylla* J. J. Smith; *Euphorbiac.;* Java (Ngarengan)
Pakiria-sipioli: *Tetragastris balsamifera* O. Ktze., *Protium spp.; Burserac.;* Sur.
Pakoeli: *Platonia insignis* Mart., *P. sp.; Guttifer.;* Gu.
Pakoeri: *Platonia insignis* Mart., *P. sp.; Guttifer.;* Sur.
Pakoori: *Platonia insignis* Mart.; *Guttifer.;* br. Gu.
Pakoorie: *Platonia insignis* Mart.; *Guttifer.;* br. Gu.
Pakooru: *Platonia insignis* Mart.; *Guttifer.;* br. Gu.
Pakouli: *Ocotea rodioei* Mez; *Laurac.;* Sur.
Pakpa: *Afzelia sp.; Caesalpiniac.;* Dah.
Pakuri: *Platonia insignis* Mart., *P. sp.; Guttifer.;* Sur., br. Gu.
Pakyisie: *Crossopteryx kotschyana* Fenzl; *Rubiac.;* Go.
Pal-erouki-patté: *Chukrasia tabularis* A. Juss.; *Meliac.;* w. O.-Ind. (Deccan)
„ kurwan: *Wrightia tomentosa* Roem. et Schultes; *Apocynac.;* w. Beng. (Orissa)
„ malata: *Lacistema aggregatum* Rusby; *Lacistemac.;* br. Hond.

Pala: *Palaquium ellipticum* Bth., *Mimusops hexandra* Roxb.; *Sapotac.;* sw. O.-Ind.
„ : *Alstonia scholaris* R. Br., *Wrightia tomentosa* R. et Sch.; *Apocynac.;* O.-Ind.
„ : *Holarrhena antidysenterica* Wall.; *Apocynac.;* O.-Ind. (Madras) [(Madras)
Palaäta: *Mimusops surinamensis* Miq., *M. balata* Gaertn.; *Sapotac.;* Sur.
„ : *Mimusops globosa* Gaertn.; *Sapotac.;* Sur.
Palach: *Populus ciliata* Wall.; *Salicac.;* nw. O.-Ind. (Punjab)
Palacio: *Zuelania roussoviae* Pittier; *Flacourtiac.;* Hond.
Palaglar: *Dipterocarpus gracilis* Blume; *Dipterocarpac.;* Java.
„ : *Dipterocarpus trinervis* Blume, *D. hasselti* Blume; *Dipterocarpac.;* Sunda.
„ mienjak: *Dipterocarpus trinervis* Blume; *Dipterocarpac.;* Java, Sum.
„ „ : *Dipterocarpus retusus* Blume; *Dipterocarpac.;* Malay.
Palah: *Butea frondosa* Roxb.; *Papilionac.;* nw. O.-Ind. (Punjab)
Palahlar: *Dipterocarpus trinervis* Blume; *Dipterocarpac.;* Java.
Palai: *Mimusops hexandra* Roxb.; *Sapotac.;* O.-Ind.
„ : *Litsea zeylanica* Nees; *Laurac.;* O.-Ind., Cey.
Paláian: *Quercus sp.; Fagac.;* Phil. (Abra, Bontok, Benguet)
Palakoea: *Dimorphandra mora* Bth. et Hook.; *Caesalpiniac.;* Sur.
Palakpálak: *Palaquium sp.; Sapotac.;* Phil.
Palalaglar: *Dipterocarpus trinervis* Blume; *Dipterocarpac.;* Java, Sum.
Palalan: *Sonneratia caseolaris* Engl.; *Sonneratiac.;* Phil. (Cotabato)
Paláli: *Dillenia sp.; Dilleniac.;* Phil. (n. & C.-Luzon)
Palaloea: *Dimorphandra mora* Bth. et Hook.; *Caesalpiniac.;* Sur.
Palamkat: *Carallia lucida* Roxb.; *Rhizophorac.;* nö. O.-Ind. (Nepal)
Palan: *Mallotus philippinensis* Muell. Arg.; *Euphorbiac.;* Java (Kedoengdjati)
Palán-palán: *Nicotiana glauca* Grah.; *Solanac.;* Arg.
Palanco: *Anaxagorea guatemalensis* Standl.; *Anonac.;* br. Hond., Guat.
„ : *Sapranthus campechianus* Standl., *Xylopia frutescens* Aubl.; *Anonac.;* Hond.
Palang: *Heterophragma roxburghi* DC.; *Bignoniac.;* C.-O.-Ind.
Pálang: *Terminalia comintana* Merr.; *Combretac.;* Phil. (Rizal)
„ -batú: *Aphanomyxis cumingiana* Harms; *Meliac.;* Phil. (Laguna)
Palanggoengan: *Acalypha caturus* Blume; *Euphorbiac.;* Java (Pantjoer Idjen)
„ : *Villebrunea rubescens* Blume; *Urticac.;* Java.
Palangkomóg: *Buchanania sp.; Anacardiac.;* Phil. (Cagayan)
Palanto: *Semecarpus sp.; Anacardiac.;* Phil. (Cagayan)
Palara: *Soymida febrifuga* Adr. Juss.; *Meliac.;* sw. O.-Ind. (Bombay)
Palas: *Butea frondosa* Roxb.; *Papilionac.;* O.-Ind.
Palasin samatha: *Butea frondosa* Roxb.; *Papilionac.;* sw. O.-Ind. (Malabar)
Palata: *Mimusops surinamensis* Miq.; *Sapotac.;* Sur.
„ : *Mimusops balata* Gaertn., *M. globosa* Gaertn.; *Sapotac.;* Sur.
„ : *Calocarpum mammosum* Pierre; *Sapotac.;* Bras.
Palatángan: *Aglaia harmsiana* Perk.; *Meliac.;* Phil. (Cagayan, Ilocos Sur)
„ : *Albizzia procera* Bth.; *Mimosac.;* Phil. (Ilocos Sur)
Palatangen: *Aphanomyxis cumingiana* Harms; *Meliac.;* Phil.
Palatináu: *Maba buxifolia* Pers.; *Ebenac.;* Phil. (Pangasinan)
Palatpát: *Sonneratia caseolaris* Engl.; *Sonneratiac.;* Phil. (Bataan)
Palatu: *Hernandia peltata* Meissn.; *Hernandiac.;* s. O.-Ind., Cey.
Pálau: *Dipterocarpus vernicifluus* Blco.; *Dipterocarpac.;* Phil. (Rizal, Zamb.)
Palawa: *Garcinia speciosa* Wall.; *Guttifer.;* s. Burma (Tenasserim)
Palawan: *Tristania obovata* Benn.; *Myrtac.;* Bangka, Sumatra, Borneo.
Palawelanga: *Phoenix spinosa* Schum. et Thonn.; *Palmac.;* Togo.
Paldatam: *Ehretia laevis* Roxb.; *Borraginac.;* O.-Ind. (Madras)
Paldatan: *Ehretia laevis* Roxb.; *Borraginac.;* Pers., O.-Ind., Ch.
Palè-ie: *Eperua jenmani* Oliv.; *Caesalpiniac.;* Sur.
Palei: *Sideroxylon tomentosum* Roxb.; *Sapotac.;* O.-Ind. (Madras)
Paleng: *Buchanania sp.; Anacardiac.;* Phil. (Cagayan)
Paleta: *Dialium divaricatum* Vahl; *Caesalpiniac.;* Hond., br. Hond., Guat.
Paleto de blanco: *Pithecolobium aff. cognatum* Bth.; *Mimosac.;* Guat., Hond.
Palétuvier (f): *Rhizophora mangle* L.; *Rhizophorac.;* tr. Am., tr. Af.
„ „ : *Rhizophora racemosa* G. F. W. Meyer, *R. spp.; Rhizophorac.;* tr. Am.,
„ : *Conocarpus erecta* L.; *Combretac.;* tr. Am., tr. Af. [tr. Af.
„ : *Pterocarpus sp.; Papilionac.;* fr. Gu.

Palétuvier: *Inga spp.; Mimosac.;* Sur.
„ : *Bruguiera eriopetala* Wight et Arn.; *Rhizophorac.;* N.-Kal.
„ : *Rhizophora mucronata* Lam.; *Rhizophorac.;* N.-Kal.
„ : *Sonneratia alba* Smith; *Sonneratiac.;* N.-Kal.
„ blanc: *Avicennia nitida* Jacq.; *Uerbenac.;* fr. Gu.
„ „ : *Moronobea sp.; Guttifer.;* fr. Gu.
„ „ : *? Ilex casiquiarensis* Loes.; *Aquifoliac.;* fr. Gu.
„ „ : *Avicennia africana* Beauv.; *Uerbenac.;* Gab.
„ „ : *Avicennia officinalis* L.; *Uerbenac.;* N.-Kal.
„ commun: *Rhizophora racemosa* G. F. W. Meyer; *Rhizophorac.;* fr. Gu.
„ femelle: *Ceriops candolleana* Arn.; *Rhizophorac.;* Mad.
„ incarnat: *Rhizophora mucronata* Lam.; *Rhizophorac.;* port. O.-Af. (Moz., Quel.)
„ „ : *Bruguiera gymnorrhiza* Lamk.; *Rhizophorac.;* port. O.-Af. (M., Q.)
„ „ : *Ceriops candolleana* Arn.; *Rhizophorac.;* port. O.-Af. (Moz., Quel.)
„ jaune: *Moronobea coccinea* Aubl.; *Guttifer.;* Guad.
„ mâle: *Rhizophora mucronata* Lam.; *Rhizophorac.;* Mad.
„ mauve: *? Rhizophora mangle* L.; *Rhizophorac.;* port. W.-Af. (Angola-Zaire)
„ montagne: *Tabebuia dentata* Miers; *Bignoniac.;* fr. Gu.
„ „ : *Clusia sp.; Guttifer.;* Guad.
„ rouge: *Rhizophora mangle* L.; *Rhizophorac.;* fr. Gu.
„ „ : *Avicennia africana* Beauv.; *Uerbenac.;* fr. Gu.
„ „ : *Rhizophora racemosa* G. F. W. Meyer; *Rhizophorac.;* Gab.
„ „ : *Bruguiera gymnorrhiza* Lamk.; *Rhizophorac.;* port. O.-Af. (Moz., Quel.)
„ „ : *Ceriops candolleana* Arn.; *Rhizophorac.;* port. O.-Af. (Moz., Quel.)
„ „ : *Lumnitzera racemosa* Willd.; *Combretac.;* N.-Kal.
„ d'Afrique: *Rhizophora mangle* L., *R. racemosa* G. F. W. Meyer; *Rhizophorac.;*
„ de montagne: *Clusia venosa* Jacq.; *Guttifer.;* Ant. [W.-Af.
„ „ „ : *Inga bourgoni* DC.; *Mimosac.;* fr. Gu.
„ „ la plage: *? Rhizophora mangle* L.; *Rhizophorac.;* port. W.-Af. (Angola)
Pali: *Palaquium ellipticum* Baill.; *Sapotac.;* O.-Ind.
„ : *Erythrophloeum densiflorum* Merr.; *Caesalpiniac.;* Phil. (Lanao)
„ leche: *Bumelia sp.; Sapotac.;* Curaçao.
Palillo: *Esenbeckia hartmanni* Robins. et Fern.; *Rutac.;* Mex.
Palinā': *Kingiodendron alternifolium* Merr.; *Papilionac.;* Phil.
Páling: *Barringtonia spp.; Lecythidac.;* Phil. (Cagayan)
Palinlín: *Buchanania sp.; Anacardiac.;* Phil. (Zambales)
Palisander (d, e): *Dalbergia nigra* Allem.; *Papilionac.;* ö. Bras.
„ : *Dalbergia spp.; Papilionac.;* ö. Mad.
„ : *Acacia morondavensis* Drake, *Albizzia boinensis* R. Vig.; *Mimosac.;* ö. Mad.
„ : *Dalbergia latifolia* Roxb.; *Papilionac.;* O.-Ind.
Palissanderhout: *Machaerium scleroxylon* Tul.; *Papilionac.;* Sur.
„ : *Dalbergia nigra* Allem.; *Papilionac.;* Bras.
„ : *Dalbergia latifolia* Roxb.; *Papilionac.;* O.-Ind.
„ , onecht: *Jacaranda sp.; Bignoniac.;* tr. Am.
Palissandre (f): *Dalbergia spp.; Papilionac.;* Tr.
„ : *Dalbergia nigra* Allem.; *Papilionac.;* Bras.
„ : *Dalbergia ikopensis* Jum. et Perr.; *Papilionac.;* Mad. (Sakalave)
„ : *Dalbergia perrieri* Drake; *Papilionac.;* Mad. (Boina)
„ : *Dalbergia baroni* Baker; *Papilionac.;* Mad.
„ , brun: *Dalbergia baroni* Baker; *Papilionac.;* Mad.
„ , faux: *Jacaranda sp.; Bignoniac.;* tr. Am.
„ poutre de Cayenne: *Jacaranda sp.; Bignoniac.;* tr. Am.
„ violet: *Dalbergia greveana* Baill.; *Papilionac.;* Mad.
„ de Cayenne: *Jacaranda brasiliana* Pers., *J. filicifolia* D. Don; *Bignoniac.;* fr. Gu.
„ „ „ : *Machaerium scleroxylon* Tul.; *Bignoniac.;* fr. Gu.
„ „ Madagascar: *Dalbergia greveana* Baill.; *Papilionac.;* Mad. (Majunga)
„ „ „ : *Dalbergia sp.; Papilionac.;* Mad. (Tamatave)
„ „ Majunga: *Dalbergia greveana* Baill.; *Papilionac.;* Mad.
„ „ Tamatave: *Dalbergia sp.; Papilionac.;* Mad.
„ „ l'Afrique (f): *Macrolobium palisoti* Bth.; *Caesalpiniac.;* Kam., Gab.
„ „ l'Inde (f): *Dalbergia latifolia* Roxb.; *Papilionac.;* O.-Ind., Ind.-Ch.

Palissandre du Sénégal: *Pterocarpus erinaceus* Lam.; *Papilionac.*; Sudan.
Palissandro (i): *Dalbergia nigra* Allem.; *Papilionac.*; Bras.
Palla: *Mimusops hexandra* Roxb.; *Sapotac.*; O.-Ind. (Madras)
Pallewie: *Eperua jenmani* Oliv.; *Caesalpiniac.*; Sur.
Palm: *Mauritia flexuosa* L. f.; *Palmac.*; br. Gu.
  „ : *Cocos sp.*; *Palmac.*; Bras.
  „ : *Lodoicea sechellarum* Labill.; *Palmac.*; Sey.
  „ : *Oncosperma sp.*; *Palmac.*; Borneo.
  „ , Arac-: *Borassus flabellifer* L. var. *aethiopum* Warb.; *Palmac.*; Nig.
  „ , Awarra-: *Astrocaryum ? jauari* Mart.; *Palmac.*; br. Gu.
  „ . Bamboo-: *Raphia vinifera* Beauv.; *Palmac.*; Nig.
  „ , Bangalow-: *Archontophoenix cunninghami* H. Wendl. et Drude; *Palmac.*; Au.
  „ , Bertam-: *Eugeissonia tristis* Griff.; *Palmac.*; Malay.
  „ , Betelnut-: *Areca catechu* L.; *Palmac.*; Phil.
  „ , black: *Astrocaryum sp.*; *Palmac.*; Pan.
  „ , Botan-: *Sabal excelsa* Morris; *Palmac.*; br. Hond.
  „ , Buri-: *Corypha elata* Roxb.; *Palmac.*; Phil.
  „ , Cabbage-: *Oreodoxa oleracea* Morris; *Palmac.*; br. Hond.
  „ , Cabbage-tree-: *Livistona australis* Mart.; *Palmac.*; Au.
  „ , Carnaúba-: *Copernicia cerifera* Mart.; *Palmac.*; Bras.
  „ , Cock's tail-: *Geonoma sp.*; *Palmac.*; Pan.
  „ , Coco-nut-: *Cocos nucifera* L.; *Palmac.*; Tr.
  „ , Cohune-: *Attalea cohune* Mart.; *Palmac.*; C.-Am.
  „ , Cumare-: *Astrocaryum vulgare* Mart.; *Palmac.*; Kol.
  „ , Date-: *Phoenix dactylifera* L.; *Palmac.*; Hond.
  „ , „ -, wild: *Phoenix reclinata* Jacq.; *Palmac.*; br. O.-Af. (Niassa)
  „ , Dekzeil-: *Mauritia flexuosa* L.; *Palmac.*; fr. Gu.
  „ , Desert- (e): *Borassus flabellifer* L. var. *aethiopum* Warb.; *Palmac.*; Nig.
  „ -doh: *Chrysobalanus ellipticus* Soland.; *Rosac.*; Lib.
  „ , Doum- (e): *Hyphaene thebaica* Mart.; *Palmac.*; tr. Af.
  „ , Dum-: *Hyphaene thebaica* Mart.; *Palmac.*; n. Nig.
  „ , Dwerg-: *Geonoma baculifera* Kunth; *Palmac.*; Sur.
  „ , Egg-: *Guilielma speciosa* Mart.; *Palmac.*; Ek.
  „ , Fan-: *Mauritia carana* Wall.; *Palmac.*; Bras.
  „ , „ : *Borassus flabellifer* L. var. *aethiopum* Warb.; *Palmac.*; Go.
  „ , Ginger-bread-: *Hyphaene thebaica* Mart.; *Palmac.*; Nig.
  „ , Give and take-: *Acanthothrinax sp.*; *Palmac.*; br. Hond.
  „ , Gru-gru-: *Acrocomia sclerocarpa* Mart.; *Palmac.*; Jam., br. Hond.
  „ , Ita-: *Mauritia flexuosa* L.; *Palmac.*; Sur.
  „ , Jan-: *Borassus flabellifer* L. var. *aethiopum* Warb.; *Palmac.*; Nig.
  „ , Jupaty-: *Raphia taedigera* Mart.; *Palmac.*; nö. Bras. (Pará)
  „ , King oil-: *Elaeis guineensis* Jacq. var. *thompsoni* Chev.; *Palmac.*; Nig.
  „ , Kokerit-: *Maximiliana regia* Mart.; *Palmac.*; br. Gu.
  „ , Manicole-: *Euterpe edulis* Mart.; *Palmac.*; Ven., br. Gu.
  „ , Maripa-: *Maximiliana maripa* Mart.; *Palmac.*; Sur.
  „ , Mauritius-: *Mauritia flexuosa* L.; *Palmac.*; Sur.
  „ , Monkey tail-: *Euterpe edulis* Mart., *Chamaedorea sp.*; *Palmac.*; br. Hond.
  „ , Moriche-: *Mauritia flexuosa* L.; *Palmac.*; Ven.
  „ , Nibong-: *Oncosperma spp.*; *Palmac.*; Borneo.
  „ , Nipa-: *Nipa fruticans* Thunb.; *Palmac.*; ö. As.
  „ , Oil-: *Elaeis guineensis* Jacq.; *Palmac.*; W.-Af.
  „ , „ -, african: *Elaeis guineensis* Jacq.; *Palmac.*; br. O.-Af. (Niassa)
  „ , Palissaden-: *Euterpe oleracea* Mart.; *Palmac.*; Sur.
  „ , Palmyra-: *Borassus flabellifer* L. var. *aethiopum* Warb.; *Palmac.*; W.-Af.
  „ , Panama hat-: *Carludovica palmata* R. et P.; *Cylanthac.*; Kol., Ek.
  „ , Pina-: *Euterpe oleracea* Mart.; *Palmac.*; Sur.
  „ , Raphia-: *Raphia vinifera* Beauv.; *Palmac.*; W.-Af.
  „ , „ : *Raphia monbuttorum* Drude; *Palmac.*; br. O.-Af. (Uganda)
  „ , Rattam-: *Calamus deerratus* Mann et H. Wendl.; *Palmac.*; Nig.
  „ , Rower-: *Astrocaryum sp.*; *Palmac.*; Pan.
  „ , royal: *Oreodoxa regia* H. B. K.; *Palmac.*; tr. Am

Palm, Sago-: *Metroxylon rumphi* Mart.; *Palmac.;* Phil. (Mindanao, Bisaya)
„ , Silver thatch-: *Thrinax argentea* Lodd.; *Palmac.;* br. Hond.
„ , Sugar-: *Arenga pinnata* Merr.; *Palmac.;* Malay, Phil.
„ , Timiche-: *Manicaria saccifera* Gaertn.; *Palmac.;* Ven.
„ , Troeli-: *Manicaria saccifera* Gaertn.; *Palmac.;* Sur.
„ , Troolie-: *Manicaria saccifera* Gaertn.; *Palmac.;* br. Gu.
„ , Turu-: *Oenocarpus bacaba* Mart.; *Palmac.;* br. Gu.
„ , Waaier-: *Mauritia flexuosa* L.; *Palmac.;* fr. Gu.
„ , Wax-: *Ceroxylon sp.; Palmac.;* Kol., Ek.
„ , wild: *Quararibea pterocalyx* Hemsl.; *Bombacac.;* Pan.
„ , Wine-: *Acrocomia mexicana* Karw.; *Palmac.;* Hond.
„ , „ : *Raphia vinifera* Beauv., *R. hookeri* Mann et Wendl.; *Palmac.;* Nig.
Palma amarga: *Sabal mauritiaeformis* Griseb. et Wendl.; *Palmac.;* n. Kol.
„ araque: *Iriartea fusca* Drude; *Palmac.;* Ven.
„ bendita: *Ceroxylon klopstockiae* Mart.; *Palmac.;* Ven.
„ blanca: *Oenocarpus altissimus* Klotzsch; *Palmac.;* Ven.
„ brava: *Bactris sp.; Palmac.;* Pan.
„ „ : *Livistona rotundifolia* Mart., *L. merrilli* Becc.; *Palmac.;* Phil.
„ „ : *Livistona rotundifolia* Mart. var. *luzonensis* Becc.; *Palmac.;* Phil.
„ , Kavkazskaja-: *Buxus sempervirens* L.; *Buxac.;* Kauk.
„ mulata: *Zamia muricata* Willd.; *Cycadac.;* Ek.
„ negra: *Copernicia cerifera* Mart.; *Palmac.;* tr. Am.
„ „ : *Astrocaryum sp.; Palmac.;* Pan.
„ real: *Oreodoxa regia* H. B. K.; *Palmac.;* br. Hond., Guat.
„ „ : *Attalea gomphococca* Mart.; *Palmac.;* Pan.
„ „ : *Cocos butyracea* L. f.; *Palmac.;* Ek., Ven.
„ de cacho: *Iriartea altissima* Klotzsch; *Palmac.;* Ven.
„ „ cera: *Ceroxylon andicola* Humb. et Bonpl.; *Palmac.;* Ek., Ven.
„ „ coco: *Cocos nucifera* L.; *Palmac.;* tr. Am.
„ „ coroza: *Attalea cohune* Morris; *Palmac.;* Guat.
„ „ manaca: *Attalea cohune* Morris; *Palmac.;* Guat.
„ „ vino: *Scheelea butyracea* Karst.; *Palmac.;* n. Kol.
„ „ „ : *Attalea gomphococca* Mart.; *Palmac.;* Kol.
Palmaé: *Carapa sp.; Meliac.;* N.-Kal.
Palmani: *Lophopetalum wightianum* Arn.; *Celastrac.;* sw. O.-Ind. (Coorg)
Palme, Ago- (d): *Borassus flabellifer* L.; *Palmac.;* tr. As., tr. Af.
„ , Buri-: *Corypha umbraculifera* L.; *Palmac.;* Phil.
„ , Dattel-: *Phoenix dactylifera* L.; *Palmac.;* sw. As., Kanar.
„ , Deleb-: *Borassus flabellifer* L. var. *aethiopum* Warb.; *Palmac.;* tr. Af.
„ , Dum-: *Hyphaene thebaica* Mart., *H. spp.; Palmac.;* O.-Af.
„ , Gebang-: *Corypha gebanga* Blume; *Palmac.;* Nied.-Ind.
„ , Kohl-: *Oreodoxa oleracea* Mart.; *Palmac.;* W.-Ind.
„ , Kokos- (d): *Cocos nucifera* L.; *Palmac.;* Tr.
„ , Larga-: *Mauritia sp.; Palmac.;* br. W.-Ind.
„ , Manta-: *Phytelephas macrocarpa* R. et P.; *Palmac.;* S.-Am.
„ , Mia-: *Hyphaene coriacea* Gaertn.; *Palmac.;* O.-Af.
„ , Mützen-: *Manicaria saccifera* Gaertn.; *Palmac.;* P.-R.
„ , Palmyra-: *Borassus flabellifer* L.; *Palmac.;* s. As.
„ , Piassave-: *Attalea funifera* Mart.; *Palmac.;* tr. Am.
„ , Steinnuss-: *Phytelephas macrocarpa* R. et P.; *Palmac.;* Ek.
„ , „ : *Phytelephas microcarpa* R. et P.; *Palmac.;* Peru.
„ , Tagua-: *Phytelephas macrocarpa* R. et P.; *Palmac.;* Kol., Ek.
„ , Talipot- (d): *Corypha umbraculifera* L.; *Palmac.;* O.-Ind.
„ , Wachs-: *Copernicia cerifera* Mart.; *Palmac.;* tr. Am.
Palmeira leque: *Mauritia carana* Wall.; *Palmac.;* Bras. (Rio Negro)
Palmerina: *Thymelea hirsuta* Endl.; *Thymelaeac.;* Sp.
Palmetto: *Sabal palmetto* Lodd.; *Palmac.;* tr. Am.
„ : *Chamaerops tomentosa* Fulchiron; *Palmac.;* br. Hond.
Palmhout: *Buxus sempervirens* L.; *Buxac.;* As.
Palmiche: *Elaeis melanococca* Gaertn.; *Palmac.;* C.-R.
„ : *Iriartea durissima* Oerst.; *Palmac.;* Hond.

Palmiche: *Copernicia sanctae-marthae* Becc.; *Palmac.;* n. Kol.
„ : *Euterpe purpurea* Engl.; *Palmac.;* Ven. (Táchira)
„ morado: *Oenocarpus altissima* Klotzsch; *Palmac.;* Ven.
Palmier bâche: *Mauritia flexuosa* L.; *Palmac.;* fr. Gu.
„ bambou: *Raphia vinifera* Beauv.; *Palmac.;* Elf., Gab.
„ ban: *Raphia sp.; Palmac.;* Elf.
„ nain: *Phoenix sp.; Palmac.;* Elf.
„ nipa: *Nipa fruticans* Thunb.; *Palmac.;* Malay.
„ rotin: *Calamus spp.; Palmac.;* Kam.
„ à huile (f): *Elaeis guineensis* Jacq.; *Palmac.;* W.-Af.
Palmilla: *Lomatia ferruginea* R. Br.; *Proteac.;* Chile.
Palmistes des bois: *Acanthophoenix spp.; Palmac.;* Réunion.
Palmita: *Geonoma undata* Klotzsch; *Palmac.;* Ven. (Trujillo)
„ yeyih: *Euterpe edulis* Mart.; *Palmac.;* Arg.
Palmito: *Chamaerops humilis* L.; *Palmac.;* Sp.
Palmyre: *Borassus flabellifer* L.; *Palmac.;* O.-Ind., Siam, Co.-Ch.
Palo: *Prosopis oblonga* Bth.; *Mimosac.;* Togo.
„ amargo: *Pterogyne nitens* Tul.; *Caesalpiniac.;* Arg.
„ „ : *Picraena sp.; Simarubac.;* Arg.
„ amarillo: *Lonchocarpus sericeus* H. B. K.; *Papilionac.;* w. C.-Am.
„ „ : *Terminalia sp.; Combretac.;* Pan.
„ „ : *Phyllostylon rhamnoides* Taub., *P. brasiliensis* Cap.; *Ulmac.;* Arg.
„ „ : *Chlorophora tinctoria* Gaudich.; *Morac.;* Salv.
„ „ : *Diphysa carthaginensis* Jacq.; *Papilionac.;* Guat.
„ „ : *Terminalia australis* Cambess.; *Combretac.;* Arg.
„ anis: *Trichilia sp.; Meliac.;* Arg.
„ asta: *Cordia sp.; Borraginac.;* C.-R.
„ azul (s): *Haematoxylon campechianum* L.; *Caesalpiniac.;* tr. Am.
„ balsamo: *Bulnesia sarmienti* Lorentz; *Zygophyllac.;* Bol., Par., Arg.
„ barranco: *Carpinus caroliniana* Walt.; *Betulac.;* Mex.
„ barroso: *Acacia sp.; Mimosac.;* tr. Am.
„ „ : *Ruprechtia sp.; Polygonac.;* Arg.
„ blanco: *Simaruba glauca* DC.; *Simarubac.;* Cuba.
„ „ : *Drypetes glauca* Vahl; *Euphorbiac.;* P.-R.
„ „ : *Godmania aesculifolia* Standl.; *Bignoniac.;* Guat.
„ „ : *Vernonia patens* H. B. K.; *Composit.;* Pan.
„ „ : *Casearia sp.; Flacourtiac.;* P.-R., Kol.
„ „ : *Diplokeleba floribunda* N. E. Brown; *Sapindac.;* Arg.
„ „ : *Celtis mississippiensis* Bosc.; *Ulmac.;* n. Mex.
„ „ : *Phyllostylon brasiliensis* Capanema; *Ulmac.;* Arg.
„ „ : *Ichthyomethia mollis* Blake; *Papilionac.;* Mex.
„ „ : *Bombax sp.; Bombacac.;* Par.
„ „ : *Tecoma pentaphylla* Juss.; *Bignoniac.;* Cuba.
„ „ : *Calycophyllum multiflorum* Griseb.; *Rubiac.;* Arg. (Jujuy, Salto, Formosa)
„ bobo: *Morisonia americana* L.; *Capparidac.;* Mex.
„ „ : *Anona glabra* L.; *Anonac.;* sp. Ant.
„ „ : *Coccoloba sp.; Polygonac.;* P.-R.
„ „ : *Pisonia sp.; Nyctaginac.;* P.-R., Arg.
„ „ : *Erythrina glauca* Willd.; *Papilionac.;* Pan.
„ borracho: *Ceiba sp., Chorisia spp.; Bombacac.;* Arg.
„ botella: *Chorisia sp.; Bombacac.;* Arg.
„ brasil: *Haematoxylon campechianum* L.; *Caesalpiniac.;* Hond.
„ „ : *Caesalpinia echinata* Lam.; *Caesalpiniac.;* Bras.
„ bronco: ? *Malpighia urens* L.; *Malpighiac.;* Mex.
„ calabaza: *Bernouillia flammea* Oliv.; *Bombacac.;* Mex. (Oaxaca)
„ camarón: *Calycophyllum candidissimum* DC.; *Rubiac.;* Mex.
„ cascarudo: *Pithecolobium scalare* Griseb.; *Mimosac.;* Arg.
„ chancho: ? *Hieronyma alchorneoides* Allem.; *Euphorbiac.;* Pan.
„ chino: *Erythroxylon sp.; Erythroxylac.;* Mex.
„ „ : *Malpighia umbellata* Rose; *Malpighiac.;* Mex. (Sinaloa)
„ „ : *Bursera simaruba* Sarg.; *Burserac.;* br. Hond., Hond., Guat.

Palo chumbo: *Ouratea castaneifolia* Engl.; *Ochnac.;* Arg.
„ cochin: *Tetragastris balsamifera* O. Ktze.; *Burserac.;* fr. W.-Ind., fr. Gu.
„ cochino: *Tetragastris balsamifera* O. Ktze.; *Burserac.;* fr. W.-Ind., fr. Gu.
„ colorado: *Ruprechtia occidentalis* Standl.; *Polygonac.;* Mex.
„ „ : *Caesalpinia platyloba* Wats.; *Caesalpiniac.;* Mex.
„ „ : *Quercus reticulata* H. et B.; *Fagac.;* Mex.
„ „ : *Beilschmiedia pendula* Bth. et Hook. f.; *Laurac.;* P.-R.
„ „ : *Ocotea veraguensis* Mez; *Laurac.;* Nic.
„ „ : *Bursera simaruba* Sarg.; *Burserac.;* Mex.
„ „ : *Calderonia salvadorensis* Standl.; *Rubiac.;* Salv., Guat.
„ „ : *Sickingia klugei* Standl.; *Rubiac.;* Pan.
„ copal: *Bursera bipinnata* Engl.; *Burserac.;* Mex.
„ corra: *Suriana maritima* L.; *Simarubac.;* Curaçao.
„ creollo (s): *Minquartia guianensis* Aubl.; *Olacac.;* Pan.
„ cruz: *Tabebuia nodosa* Griseb.; *Bignoniac.;* Arg.
„ cuate: *Eysenhardtia polystachya* Sarg.; *Papilionac.;* Mex. (Sinaloa)
„ cucharo: *Sickingia erythroxylon* Willd.; *Rubiac.;* Ven.
„ danta: *Astronium graveolens* Jacq.; *Anacardiac.;* Ven.
„ diablo: *Capparis sp.*; *Capparidac.;* Cuba.
„ „ : *Krugiodendron ferreum* Urban; *Rhamnac.;* W.-Ind., C.-Am.
„ dulce: *Eysenhardtia polystachya* Sarg.; *Papilionac.;* Mex.
„ „ blanco: *Eysenhardtia polystachya* Sarg.; *Papilionac.;* Mex
„ duro: *Quercus sp.*; *Fagac.;* Mex.
„ „ : *Celtis sp.;* *Ulmac.;* Mex.
„ estaca: *Ruprechtia sp.*; *Polygonac.;* Arg.
„ fernambuco: *Caesalpinia crista* L.; *Caesalpiniac.;* Cuba.
„ guitarra: *Citharexylum caudatum* L.; *Verbenac.;* Cuba.
„ hormiguero: *Triplaris caracasana* Cham.; *Polygonac.;* Pan.
„ jiote: *Bursera simaruba* Sarg.; *Burserac.;* br. Hond., Hond., Guat.
„ lanza: *Phyllostylon brasiliensis* Capanema; *Ulmac.;* Arg.
„ „ negro: *Phyllostylon brasiliensis* Capanema; *Ulmac.;* Arg.
„ mabi: *Colubrina reclinata* Brong.; *Rhamnac.;* P.-R.
„ maceta: *Lonchocarpus campestris* Mart.; *Papilionac.;* Arg.
„ machete: *Erythrina sp.*; *Papilionac.;* Nic.
, „ : *Eperua falcata* Aubl.; *Caesalpiniac.;* Ven.
„ marango: *Chlorophora tinctoria* Gaudich.; *Morac.;* Trin.
„ Maria: *Calophyllum calaba* Jacq.; *Guttifer.;* C.-Am.
„ „ : *Calophyllum sp.*; *Guttifer.;* Ek., Indo-Malay.
„ „ : *Calophyllum inophyllum* L.; *Guttifer.;* Phil.
„ „ : *Kayea paniculata* Merr.; *Guttifer.;* Phil. (Tayabas, Mindoro)
„ „ : *Kingiodendron alternifolium* Merr.; *Papilionac.;* Phil. (Zamboanga)
„ „ barrabas: *Triplaris caracasana* Cham.; *Polygonac.;* Ven.
„ „ de la playa: *Calophyllum inophyllum* L.; *Guttifer.;* Phil.
„ mato: *Flotovia diacanthoides* Less.; *Composit.;* Chile.
„ mino: *Ottoschulzia domingensis* Urban; *Icacinac.;* St.-D.
„ misanteco: *Miscantea capitata* C. et S.; *Laurac.;* Mex., (Oaxaca, Vera Cruz)
„ mora: *Chlorophora tinctoria* Gaudich.; *Morac.;* Kol.
„ morado: *Peltogyne sp.*; *Caesalpiniac.;* Mex. (Acapulco)
„ moral: *Chlorophora tinctoria* Gaudich.; *Morac.;* Mex.
„ mortero: *Pterogyne nitens* Tul.; *Caesalpiniac.;* Arg.
„ „ : *Machaerium pseudotipe* Griseb.; *Papilionac.;* Arg.
„ mulato: *Pentaclethra filamentosa* Bth.; *Mimosac.;* Trin.
„ „ : *Celtis reticulata* Torr.; *Ulmac.;* Mex.
„ „ : *Triplaris auriculata* Meissn.; *Polygonac.;* Mex.
„ „ : *Zanthoxylum fagara* Sarg., *Z. pentanome* DC.; *Rutac.;* Mex.
„ „ : *Bursera simaruba* Sarg.; *Burserac.;* Mex., Hond.
„ „ : *Astronium graveolens* Jacq.; *Anacardiac.;* br. Hond.
„ „ : *Christiana africana* DC.; *Tiliac.;* br. Hond.
„ „ : *Lacistema aggregatum* Rusby; *Lacistemac.;* br. Hond.
„ „ de Mazatlan: *Zanthoxylum fraxineum* Willd.; *Rutac.;* Mex.
„ „ „ Mexico: *Zanthoxylum pentanome* DC.; *Rutac.;* Mex.

Palo mulato hoja larga: *Zanthoxylum clava-herculis* L.; *Rutac.;* Mex.
,, mulatto: *Astronium conzatti* Blake; *Anacardiac.;* br. Hond.
,, narango: *Chlorophora tinctoria* Gaudich.; *Morac.;* br. Hond., Trin., Jam.
,, naranjo: *Chlorophora tinctoria* Gaudich.; *Morac.;* Trin.
,, negro: *Dalbergia cf. lineata* Pittier; *Papilionac.;* Hond.
,, ,, : *Hybanthus guanacastensis* Standl.; *Violac.;* Nic.
,, ,, : *Diospyros foveo-reticulata* Merr., *D. plicata* Merr.; *Ebenac.;* Phil.
,, obero: *Astronium conzatti* Blake; *Anacardiac.;* Guat.   [(Cotabato)
,, ,, : *Astronium graveolens* Jacq.; *Anacardiac.;* Hond.
,. ortigo: *Urera elata* Griseb.; *Urticac.;* Pan.
,, picante: *Drimys winteri* Forst.; *Magnoliac.;* Mex.
,, pollo: *Pterocarpus sp.; Papilionac.;* P.-R.
,, pólvora: *Trema micrantha* Blume; *Ulmac.;* Arg.
,, prieto: *Celaenodendron mexicanum* Standl.; *Euphorbiac.;* w. C.-Am.
,, ,, : *Terminalia sp.; Combretac.;* Kol.
,, ,, : *Erythrina umbrosa* H. B. K.; *Papilionac.;* Ek.
,, rajador: *Casearia sp.; Flacourtiac.;* Arg.
,, rojo: *Pterocarpus soyauxi* Taub.; *Papilionac.;* sp. Guin.
,, rosa: *Aspidosperma polyneuron* Muell. Arg.; *Apocynac.;* n. Arg., s. Bras.
,, ,, : *Pterogyne nitens* Tul.; *Caesalpiniac.;* Arg.
,, San Antonio: *Rapanea laetevirens* Mez; *Myrsinac.;* Arg.
,, sangre: *Virola merendonis* Pittier; *Myristicac.;* br. Hond.
,, sano: *Bulnesia arborea* Engl.; *Zygophyllac.;* Ven.
,, sant: *Sorbus torminalis* Crantz; *Rosac.;* Sp.
,, Santo: *Guaiacum officinale* L.; *Zygophyllac.;* W.-Ind.
,, ,, : *Guaiacum sanctum* L.; *Zygophyllac.;* W.-Ind., Mex.
,, ,, : *Guaiacum coulteri* A. Gray; *Zygophyllac.;* Mex. (Oaxaca, Jalisco)
,, ,, : *Triplaris sp.; Polygonac.;* Kol.
,, ,, : *Triplaris caracasana* Cham.; *Polygonac.;* Peru.
,, ,, : *Bulnesia sarmienti* Lorentz; *Zygophyllac.;* Bol., Par., Arg.
,, ,, : *Platymiscium trifoliolatum* Bth.; *Papilionac.;* Mex. (Sinaloa)
,, ,, : *Erythrina glauca* Willd., *E. panamensis* Standl.; *Papilionac.;* Pan.
,, ,, : *Porliera lorentzi* Engl.; *Zygophyllac.;* Chile.
,, ,, : *Weinmannia trichosperma* Cav.; *Cunoniac.;* Chile.
,, ,, blanco: *Bulnesia gancedoi* Rojas; *Zygophyllac.;* Arg.
,, ,, nutante: *Bulnesia sarmienti* Lorentz; *Zygophyllac.;* Arg.
,, silo: *Carpinus caroliniana* Walt.; *Betulac.;* Mex.
,, torcido: *Mouriria sp.; Melastomatac.;* Cuba.
,, trebol: *Torresea cearensis* Allem.; *Papilionac.;* Arg.
,, ,, : *Stryphnodendron rotundifolium* Mart.; *Mimosac.;* Arg.
,, vívora: *Tabernaemontana sp.; Apocynac.;* Arg.
,, yugo: *Tecoma pentaphylla* Juss.; *Bignoniac.;* Mex.
,, zopilote: *Swietenia sp.; Meliac.;* Mex.
,, de aceite: *Tetragastris balsamifera* O. Ktze.; *Burserac.;* P.-R.
,, ,, agi: *Drimys winteri* Forst.; *Magnoliac.;* Kol.
,, ,, agua: *Gilibertia stenocarpa* Donn. Smith; *Araliac.;* Hond.
,, ,, ,, : *Ficus spp.; Morac.;* C.-R.
,, ,, ,, : *Bravaisia integerrima* Standl.; *Acanthac.;* n. Kol.
,, ,, ,, : *Capparis tenuisiliqua* Jacq.; *Capparidac.;* Kol.
,, ,, aguila: *Juglans sp.; Juglandac.;* Mex.
,, ,, ,, : *Celtis monoica* Hemsl.; *Ulmac.;* Mex.
,, ,, ,, : *Clusia salvini* Donn. Smith; *Guttifer.;* Mex.
,, ,, anastasio: *Trichilia sp.; Meliac.;* P.-R.
,, ,, asta: *Cordia gerascanthus* L.; *Borraginac.;* Guat.
,, ,, balsa: *Ochroma peruviana* Johnston; *Bombacac.;* Peru.
,, ,, balsamo: *Myroxylon toluiferum* H. B. K.; *Papilionac.;* Salv.
,, ,, boracho: *Chorisia insignis* H. B. K.; *Bombacac.;* Arg.
,, ,, boyo: *Erythrina sp.; Papilionac.;* P.-R.
,, ,, brasil: *Haematoxylon brasiletto* Karst.; *Caesalpiniac.;* n. Mex.
,, ,, buey: *Bauhinia candicans* Bth.; *Caesalpiniac.;* Arg.
,, ,, burro prieto: *Capparis sp.; Capparidac.;* P.-R.

Palo de cabra: *Trema micrantha* Blume; *Ulmac.;* P.-R.
„ „ „ : *Symplocos polyantha* Krug et Urban; *Symplocac.;* P.-R.
„ „ cacca: *Theobroma cacao* L.; *Sterculiac.;* Trin.
„ „ caja: *Allophylus cominia* Swartz; *Sapindac.;* Cuba.
„ „ Calentura: *Lucuma hypoglauca* Standl.; *Sapotac.;* Mex. (Vera Cruz)
„ „ Campeche: *Haematoxylon campechianum* L.; *Caesalpiniac.;* tr. Am.
„ „ „ : *Haematoxylon brasiletto* Karst.; *Caesalpiniac.;* s. Mex.
„ „ carnero: *Coccoloba schiedeana* Lindau; *Polygonac.;* Mex.
„ „ carretas: *? Machaerium latifolium* Pittier; *Papilionac.;* Mex.
„ „ cera: *Astronium sp.; Anacardiac.;* Mex.
„ „ cerdo: *Tetragastris balsamifera* O. Ktze.; *Burserac.;* Kol.
„ „ Chile: *Drimys winteri* Forst.; *Magnoliac.;* Mex.
„ „ cochino: *Tetragastris balsamifera* O. Ktze.; *Burserac.;* Cuba.
„ „ coco: *Ficus sp.; Morac.;* Mex.
„ „ corcho: *Anona glabra* L.; *Anonac.;* Mex.
„ „ craba: *Trema micranthum* Blume; *Ulmac.;* P.-R,
„ „ cruz: *Brownea grandiceps* Jacq.; *Caesalpiniac.;* Ven.
„ „ cupey: *Clusia rosea* Jacq.; *Guttifer.;* P.-R.
„ „ Doncello: *Byrsonima lucida* DC.; *Malpighiac.;* P.-R.
„ „ fraile: *Licania sp.; Rosac.;* Mex.
„ „ gallina: *Alchorneopsis portoricensis* Urban; *Euphorbiac.;* P.-R.
„ „ granata: *Myrsine marginata* Hook. et Arn.; *Myrsinac.;* Ind.-Ch.
„ „ hierro: *Krugiodendron ferreum* Urban; *Rhamnac.;* St.-D.
„ „ „ : *Lophira procera* Chev.; *Ochnac.;* sp. Guin.
„ „ „ : *Xanthostemon verdugonianus* Naves; *Myrtac.;* Phil.
„ „ incienso: *Amyris sp.; Rutac.;* Cuba.
„ „ jaqueca: *Thespesia populnea* Correa; *Malvac.;* P.-R.
„ „ jiote: *Bursera simaruba* Sarg.; *Burserac.;* Salv.
„ „ lama: *Alnus sp.; Betulac.;* Guat.
„ „ lanza: *Oxandra lanceolata* Baill.; *Anonac.;* tr. Am.
„ „ „ : *Ruprechtia excelsa* Griseb.; *Polygonac.;* Arg.
„ „ leche: *Brosimum utile* Pittier; *Morac.;* C.-Am., n. S.-Am.
„ „ „ : *Sapium biglandulosum* Muell. Arg.; *Euphorbiac.;* Ek.
„ „ lija: *Cecropia adenopus* Mart.; *Morac.;* Arg.
„ „ loro: *Prunus lusitanica* L.; *Rosac.;* Sp.
„ „ mambo: *Drimys winteri* Forst.; *Magnoliac.;* Arg.
„ „ marfil: *Amyris sp.; Rutac.;* Kol.
„ „ Maria: *Calophyllum calaba* Jacq.; *Guttifer.;* C.-Am., P.-R.
„ „ „ : *Calophyllum rekoi* Standl.; *Guttifer.;* Hond.
„ „ marrano: *Tetragastris balsamifera* O. Ktze.; *Burserac.;* Kol.
„ „ masa: *Tetragastris balsamifera* O. Ktze.; *Burserac.;* P.-R.
„ „ melón: *Crescentia cujete* L.; *Bignoniac.;* Mex., C.-Am.
„ „ mora: *Chlorophora tinctoria* Gaudich.; *Morac.;* P.-R., C.-R., Ven.
? „ „ mozo: *Machaerium sp.; Papilionac.;* Ven.
„ „ muerto: *Gustavia fustis-mortui* Pittier; *Lecythidac.;* Ven.
„ „ mulato: *Spondias sp.; Anacardiac.;* Guat.
„ „ muñeca: *Cordia sp.; Borraginac.;* P.-R.
„ „ nierro: *Krugiodendron ferreum* Urban; *Rhamnac.;* P.-R., Virgin. Inseln.
„ „ nuez: *Juglans sp.; Juglandac.;* P.-R.
„ „ pan: *Artocarpus communis* Forst.; *Morac.;* Hond.
„ „ paraiso: *Couroupita odoratissima* Seem.; *Lecythidac.;* Pan.
„ „ péndula: *Vitex sp.; Verbenac.;* P.-R.
„ „ perdiz: *Bernouillia flammea* Oliv.; *Bombacac.;* Mex. (Oaxaca)
„ „ piedra: *Homalium sp.; Flacourtiac.;* Mex.
„ „ „ : *Minquartia guianensis* Aubl.; *Olacac.;* C.-R.
„ „ „ : *Crudia obliqua* Griseb.; *Caesalpiniac.;* n. Kol.
„ „ „ corteño: *Casearia laevis* Standl.; *Flacourtiac.;* Mex.
„ „ pollo: *Pterocarpus officinalis* Jacq.; *Papilionac.;* P.-R.
„ „ rayo: *Peltophorum adnatum* Griseb.; *Caesalpiniac.;* Cuba.
„ „ resina: *Amyris sp.; Rutac.;* Cuba.
„ „ roble: *Amyris balsamifera* L.; *Rutac.;* Cuba.

Palo de ropa: *Zanthoxylum procerum* Donn. Smith; *Rutac.;* Mex.
„   „   „   : *Ottoschulzia rhodoxylon* Urban; *Icacinac.;* P.-R.
„   „   „   : *Tecoma pentaphylla* Juss.; *Bignoniac.;* Mex.
„   „   „   : *Guaiacum officinale* L.; *Zygophyllac.;* Mex.
„   „   sal: *Avicennia nitida* Jacq.; *Verbenac.;* Hond.
„   „   San Antonio: *Pentapanax angelicifolium* Griseb.; *Araliac.;* tr. Am.
„   „   „   „   : *Myrsine floribunda* R. Br.; *Myrsinac.;* tr. S.-Am.
„   „   sangre: *Myristica panamensis* Hemsl.; *Myristicac.;* br. Hond.
„   „   „   : *Machaerium glabratum* Pittier; *Papilionac.;* n. Kol.
„   „   ., : *Brosimum paraënse* Huber; *Morac.;* Peru.
„   „   : *Haematoxylon campechianum* L.; *Caesalpiniac.;* C.-Am., n. S.-Am.
„   „   seda: *Bombacopsis sp.; Bombacac.;* Curaçao.
„   „   tabaco: *Gymnanthes lucida* Swartz; *Euphorbiac.;* St.-D.
„   „   tea: *Amyris sp.; Rutac.;* P.-R.
„   „   tinta: *Haematoxylon campechianum* L.; *Caesalpiniac.;* tr. Am.
„   „   „   : *Haematoxylon brasiletto* Karst.; *Caesalpiniac.;* s. Mex.
„   „   vaca: *Couma guatemalensis* Standl.; *Apocynac.;* Guat.
„   „   „   : *Brosimum utile* Pittier; *Morac.;* Pan., n. S.-Am.
„   „   zope: *Ichthyomethia sp.; Papilionac.;* Salv.
„   „   zorillo: *Roupala borealis* Hemsl.; *Proteac.;* Mex.
„   del muerto: *Morisonia americana* L.; *Capparidac.;* Mex.
„   „   „   : *Ipomoea murucoides* Roem. et Schultes; *Convolvulac.;* Mex. (Oaxaca)
„   de la flecha: *Sapium sp.; Euphorbiac.;* C.-Am.
Paloeloe ibjo: *Erisma uncinatum* Warm.; *Vochysiac.;* Sur.
Palogápig: *Heritiera litoralis* Dry.; *Sterculiac.;* Phil. (Cagayan, Pangasinan, Zambales)
„   : *Tarrietia silvatica* Merr.; *Sterculiac.;* Phil. (Ilocos Norte, Tarlac)
Palokpálok: *Palaquium sp.; Sapotac.;* Phil.
Paloma: *Gilibertia smithiana* Johnston; *Araliac.;* Pan.
Palomaríang-babáe: *Kayea paniculata* Merr.; *Guttifer.;* Phil. (Tayabas, Mindoro)
Palomero: *Musanga smithi* R. Br.; *Morac.;* sp. Guin.
Palomino (u): *Ottoschulzia domingensis* Urban; *Icacinac.;* St.-D.
Palonápoi: *Quercus sp.; Fagac.;* Phil. (Zambales)
„   : *Tarrietia silvatica* Merr.; *Sterculiac.;* Phil. (Pangasinan)
Pálong: *Myristica sp.; Myristicac.;* Phil. (Ilocos Sur)
Palongápoi: *Heritiera litoralis* Dry.; *Sterculiac.;* Phil. (Cagayan, Pangas., Zambales)
Palosa: *Acacia modesta* Wall.; *Mimosac.;* Afg.
Palosápis: *Anisoptera curtisi* Dyer, *A. thurifera* Blume; *Dipterocarpac.;* Phil.
„   : *Vatica mangachapoi* Blco.; *Dipterocarpac.;* Phil.
Palote negro: *Guazuma ulmifolia* Lam.; *Sterculiac.;* Mex.
Paloz: *Acacia modesta* Wall.; *Mimosac.;* O.-Ind. (Himal., Punjab)
Palsahíngin: *Dipterocarpus vernicifluus* Blco.; *Dipterocarpac.;* Phil. (Laguna)
Palsaíngin: *Canarium villosum* F. Vill.; *Burserac.;* Phil. (Marinduque)
„   : *Dipterocarpus grandiflorus* Blco.; *Dipterocarpac.;* Phil. (Laguna)
Palte madar: *Erythrina indica* Lam.; *Papilionac.;* O.-Ind., Malay.
Palu: *Mimusops hexandra* Roxb.; *Sapotac.;* Cey.
Paluahan: *Dysoxylum decandrum* Merr., *D. euphlebium* Merr.; *Meliac.;* Phil. (Negros)
Paludar: *Cedrus deodara* Loudon; *Conifer.;* nw. O.-Ind. (Hazara)
„   : *Abies pindrow* Spach; *Conifer.;* nw. O.-Ind. (Punjab)
Palúmut: *Calophyllum blancoi* Pl. et Triana; *Guttifer.;* Phil. (Palawan)
Pamalalíen: *Dillenia sp.; Dilleniac.;* Phil. (n. & C.-Luzon)
„   : *Dipterocarpus grandiflorus* Blco., *D. spp.; Dipterocarpac.;* Phil. (Cagayan)
Pamalauágon: *Lagerstroemia piriformis* Koehne; *Lythrac.;* Phil. (Leyte)
Pamalitién: *Litsea sp.; Laurac.;* Phil. (Pangasinan)
Pamanságan: *Shorea teysmanniana* Dyer; *Dipterocarpac.;* Phil. (Camarines)
Pamantúlen: *Dipterocarpus grandiflorus* Blco.; *Dipterocarpac.;* Phil. (Pangasinan)
Pamantuling: *Dipterocarpus vernicifluus* Blco.; *Dipterocarpac.;* Phil.
Pamarauágon: *Lagerstroemia piriformis* Koehne; *Lythrac.;* Phil. (Leyte)
Pamarnísen: *Dipterocarpus grandiflorus* Blco.; *Dipterocarpac.;* Phil. (Cagayan)
Pamatágen: *Dysoxylum decandrum* Merr.; *Meliac.;* Phil. (Cagayan)
Pamayabásen: *Litsea sp.; Laurac.;* Phil. (Cagayan)
Pamayauásen: *Shorea balangeran* Dyer, *S. guiso* Blume; *Dipterocarpac.;* Phil. (Pangas.)

Pambané: *Coula edulis* Baill.; *Olacac.;* Elf.
Pambara: *Trewia nudiflora* L.; *Euphorbiac.;* sw. O.-Ind. (Malabar)
Pambuku: *Cardiogyne africana* Bureau; *Morac.;* O.-Af.
Pamelesian: *Polyalthia canangioides* Boerl.; *Anonac.;* Nied.-Ind.
Pamiásin: *Adenanthera intermedia* Merr.; *Mimosac.;* Phil. (Zambales)
Pamikláten: *Calophyllum blancoi* Pl. et Triana; *Guttifer.;* Phil. (Ilocos Sur)
Pamilín: *Buchanania sp.; Anacardiac.;* Phil. (Zambales)
Pamilingan: *Pygeum glandulosum* Merr., *P. presli* Merr.; *Rosac.;* Phil. (Ilocos N.)
Pamiltáogen: *Calophyllum cumingi* Pl. et Triana; *Guttifer.;* Phil. (Zambales)
Pamitaógen: *Calophyllum whitfordi* Merr., *C. spp.; Guttifer.;* Phil.
Pamitaóien: *Calophyllum blancoi* Pl. et Triana; *Guttifer.;* Phil. (Pangasinan)
Pamitlaten: *Calophyllum blancoi* Pl. et Triana; *Guttifer.;* Phil. (Ilocos Sur)
Pamito: *Euphoria cinerea* Radlk.; *Sapindac.;* Phil.
Pamittaógen: *Calophyllum cumingi* Pl. et Triana; *Guttifer.;* Phil. (Cagayan)
Pamparemo-caspi: *Duroia longifolia* Ducke; *Rubiac.;* Peru.
Pampe: *Albizzia sassa* MacBride; *Mimosac.;* Go.
Pampelmose (e): *Citrus decumana* Willd.; *Rutac.;* tr. As., tr. Am.
Pampena: *Albizzia sassa* MacBride; *Mimosac.,* Go.
Pamphunia: *Stereospermum chelonioides* DC.; *Bignoniac.;* w. Beng. (Orissa)
Pampi: *Swartzia madagascariensis* Desv.; *Caesalpiniac.;* B.-K.
Pamplemoussier (f): *Citrus decumana* Willd.; *Rutac.;* tr. As., tr. Am.
Pamplo: *Oxytenanthera ? abyssinica* Munro; *Gramin.;* Go.
Pam-prama: *Corynanthe paniculata* Welw.; *Rubiac.;* Go.
Pampru: *Oxytenanthera ? abyssinica* Munro; *Gramin.;* Go.
Pan del año: *Ficus sp.; Morac.;* Trin.
„ „ la vida: *Lucuma hypoglauca* Standl.; *Sapotac.;* C.-Am.
Pana: *Gilibertia arborea* E. March.; *Araliac.;* P.-R.
„ cimarróna: *Didymopanax morototoni* Dcne. et Pl.; *Araliac.;* P.-R.
Panacoco: *Swartzia spp.; Caesalpiniac.;* fr. Gu., Sur.
„ : *Bocoa provacensis* Aubl., *Ormosia spp.; Papilionac.;* fr. Gu., Sur.
„ , gran: *Swartzia tomentosa* DC.; *Caesalpiniac.;* fr. Gu.
„ noir: *Bocoa provacensis* Aubl.; *Papilionac.;* fr. Gu.
? „ petit: *Ormosia coccinea* Jacks., *O. spp.; Papilionac.;* Sur.
„ „ de Cayenne: *Ormosia coccinea* Jacks.; *Papilionac.;* Gu., Bras.
„ rouge: *Swartzia tomentosa* DC.; *Caesalpiniac.;* fr. Gu.
Panaga boonga: *Mesua ferrea* L.; *Guttifer.;* Malak.
Panagitmón: *Diospyros currani* Merr.; *Ebenac.;* Phil. (Camarines)
Panagulingon: *Cratoxylon sp.; Guttifer.;* Phil. (Zambales)
Panaka: *Pterocarpus santalinus* L. f.; *Papilionac.;* O.-Ind.
Panakitin: *Sterculia sp.; Sterculiac.;* Phil. (Mindoro)
Panalan: *Albizzia brownei* Oliv.; *Mimosac.;* Togo.
Panalsálan: *Dipterocarpus speciosus* Brandis; *Dipterocarpac.;* Phil. (Camarines)
Panamá: *Sterculia sp.; Sterculiac.;* C.-Am.
„ wood: *Sterculia apetala* Karst.; *Sterculiac.;* Pan.
Panan: *Ougeinia dalbergioides* Bth.; *Papilionac.;* w. Beng. (Monghyr)
Pananaäm: *Talauma villosa* Miq.; *Magnoliac.;* Nied.-Ind.
Panao: *Dipterocarpus vernicifluus* Blco., *D. grandiflorus* Blco.; *Dipterocarpac.;* Phil.
Panapary: *Clusia sp.; Guttifer.;* Bras.
Panasi: *Carallia lucida* Roxb.; *Rhizophorac.;* sw. O.-Ind. (Bombay)
Panási: *Sideroxylon ferrugineum* Hook. et Arn.; *Sapotac.;* Phil. (Camarines)
Pánau: *Dipterocarpus grandiflorus* Blco., *D. hasselti* Blume; *Dipterocarpac.;* Phil.
„ : *Hopea basilanica* Foxw.; *Dipterocarpac.;* Phil. (Basilan)
Panawar beas: *Rinorea javanica* O. Ktze.; *Uiolac.;* Java.
Panchman: *Anogeissus acuminata* Wall.; *Combretac.;* O.-Ind. (Madras)
Panchonta: *Palaquium ellipticum* Bth.; *Sapotac.;* sw. O.-Ind. (Bombay)
Panchu: *Bombax malabaricum* DC.; *Bombacac.;* Assam (Garo)
Panda: *Tabebuia sp.; Bignoniac.;* Sur.
„ : *Panda oleosa* Pierre; *Pandac.;* Kam.
Pandaru: *Canthium didymum* Roxb.; *Rubiac.;* O.-Ind.
Pandhra-khair: *Acacia ferruginea* DC.; *Mimosac.;* sw. O.-Ind. (Bombay)
Pandiki: *Kydia calycina* Roxb.; *Malvac.;* O.-Ind. (Madras)

Pandrai: *Albizzia procera* Bth.; *Mimosac.*; w. Beng. (Orissa)
Pandri: *Stereospermum suaveolens* DC.; *Bignoniac.*; C.-O.-Ind. (Gondwana)
Pandru: *Gardenia latifolia* Aiton; *Rubiac.*; sw. O.-Ind. (Bombay)
Pandry: *Murraya exotica* L.; *Rutac.*; w. O.-Ind. (Bombay)
Pandu kura: *Wrightia tomentosa* Roem. et Schultes; *Apocynac.*; C.-O.-Ind.
Pandur: *Anthocephalus cadamba* Miq.; *Rubiac.*; O.-Ind.
Paneira: *Chorisia ventricosa* Nees; *Bombacac.*; Bras.
Panen: *Daphniphyllum glaucescens* Blume; *Euphorbiac.*; Java (Takoka)
Paneque: *Zanthoxylum ochroxylum* P. DC.; *Rutac.*; Ven.
Pàng: *Citrus decumana* L.; *Rutac.*; Ind.-Ch.
Panga: *Xylia xylocarpa* Taub.; *Mimosac.*; sw. O.-Ind. (Travancore Hills)
  „ : *Terminalia chebula* Retz.; *Combretac.*; Burma.
Pangahútan: *Mangifera altissima* Blco.; *Anacardiac.*; Phil. (Tayabas)
Pangal-lingáen: *Pterospermum sp.*; *Sterculiac.*; Phil. (Ilocos Sur)
Pangalan: *Albizzia brownei* Oliv.; *Mimosac.*; Togo.
Pangalang: *Albizzia brownei* Oliv.; *Mimosac.*; Togo.
Pangalisoklóen: *Alstonia macrophylla* Wall.; *Apocynac.*; Phil. (Pangasinan)
Pangalusíten: *Terminalia calamansanai* Rolfe; *Combretac.*; Phil. (Abra)
Panganán: *Quercus sp.*; *Fagac.*; Phil. (Bataan)
Pangapaktulen: *Cyclostemon sp.*; *Euphorbiac.*; Phil. (Cagayan)
Pangar (H): *Aesculus indica* Colebr.; *Hippocastanac.*; n. O.-Ind. (United Prov.)
Pangara: *Erythrina suberosa* Roxb.; *Papilionac.*; sw. O.-Ind. (Bombay)
Pangarángin: *Cyclostemon sp.*; *Euphorbiac.*; Phil. (Bataan)
Pangardisen: *Bombycidendron campylosiphon* Warb.; *Malvac.*; Phil. (Ilocos S., Cagayan)
  „ : *Cyclostemon sp.*; *Euphorbiac.*; Phil. (Pangasinan)
Pangbaïnga: *Ziziphus mucronatus* Willd.; *Rhamnac.*; Togo.
Panggang: *Ficus pilosa* Reinw.; *Morac.*; Java (Karangasem)
Pangi: *Prosopis oblonga* Bth.; *Mimosac.*; Togo.
  „ : *Pangium edule* Reinw.; *Flacourtiac.*; Malay.
  „ : *Gymnartocarpus woodi* Merr.; *Morac.*; Phil. (Zambales)
Pangia: *Prunus puddum* Roxb.; *Rosac.*; n. O.-Ind. (United Prov.)
  „ : *Erythrina suberosa* Roxb.; *Papilionac.*; O.-Ind. (Himal.-Hind.)
Pangie: *Pangium edule* Reinw.; *Flacourtiac.*; Java, Malay.
Panginaháuan: *Homalium bracteatum* Bth.; *Flacourtiac.*; Phil. (Samar)
Panglingáen: *Pterospermum sp.*; *Sterculiac.*; Phil. (Nueva Ecija)
Panglongbóien: *Eugenia spp.*; *Myrtac.*; Phil.
  „ : *Neonauclea sp.*; *Rubiac.*; Phil. (Ilocos Sur, Nueva Ecija)
Panglumbóien: *Eugenia calubcob* C. B. Rob.; *Myrtac.*; Phil. (Tarlac)
Pangmanggáen: *Mangifera altissima* Blco.; *Anacardiac.*; Phil. (Ilocos Sur)
Pangnán: *Quercus sp.*; *Fagac.*; Phil. (Bataan)
Pango: *Hylodendron gabunense* Taub.; *Caesalpiniac.*; Gab. (Loango)
Pangó: *Eugenia clavifera* Roxb., *E. sp.*; *Myrtac.*; Phil. (Cagayan)
Pangoi: *Acer oblongum* Wall.; *Acerac.*; nw. O.-Ind. (Punjab, Jaunsar)
Pangolaksién: *Alstonia macrophylla* Wall.; *Apocynac.*; Phil. (Cagayan)
Pangót: *Eugenia sp.*; *Myrtac.*; Phil. (Cagayan)
Pangra: *Erythrina indica* Lam., *E. suberosa* Roxb.; *Papilionac.*; O.-Ind.
Pangugók: *Eugenia sp.*; *Myrtac.*; Phil. (Cagayan)
Pangugót: *Eugenia sp.*; *Myrtac.*; Phil. (Cagayan)
Panharya: *Acacia leucophloea* Willd.; *Mimosac.*; sw. O.-Ind. (Bombay)
Pani jama: *Salix tetrasperma* Roxb.; *Salicac.*; Beng.
Paniabila: *Gardenia latifolia* Aiton; *Rubiac.*; C.-O.-Ind. (Gondwana)
Paniagua: *Capparis subbiloba* H. B. K.; *Capparidac.*; Ven.
Paniala: *Bischofia javanica* Blume; *Euphorbiac.*; n. O.-Ind. (Dehra Dun)
Paniggáian: *Hopea plagata* Vid.; *Dipterocarpac.*; Phil. (Iloko)
Paniggáyen: *Vatica mangachapoi* Blco.; *Dipterocarpac.*; Phil. (Ilocos Norte)
Panisaj: *Terminalia myriocarpa* Heurck et Muell. Arg.; *Combretac.*; nö. O.-Ind. (Nepal,
Panji: *Randia dumetorum* Lamk.; *Rubiac.*; Beng.   [Assam)
Panjino: *Elaeagnus angustifolia* L.; *Elaeagnac.*; Sp.
Panjon: *Polyalthia cerasoides* Bth. et Hook. f.; *Anonac.*; w. Beng. (Chota Nagpur)
Pankakro: *Talauma hodgsoni* Hook. f. et Thoms.; *Magnoliac.*; Assam (Garo)
Panlag: *Heterophragma roxburghi* DC.; *Bignoniac.*; sw. O.-Ind. (Bombay)

Panna: *Prunus puddum* Roxb.; *Rosac.*; Burma.
Panni: *Prunus puddum* Roxb.; *Rosac.*; Burma.
Panniari: *Barringtonia acutangula* Gaertn.; *Lecythidac.*; n. O.-Ind. (United Pr.)
Panoh-deh: *Artocarpus chaplasha* Roxb.; *Morac.*; Burma.
Panpui: *Dillenia indica* L.; *Dilleniac.*; Assam (Garo)
Pansáket: *Terminalia nitens* Presl; *Combretac.*; Phil. (Tagalog)
Pansi: *Anogeissus acuminata* Wall.; *Combretac.*; O.-Ind. (Madras)
Panta: *Tabebuia sp.*; *Bignoniac.*; Sur.
Pantaga: *Calophyllum spectabile* Willd.; *Guttifer.*; s. Burma.
Pantano: *Hieronyma alchorneoides* Allem.; *Euphorbiac.*; Pan.
Pantappayan: *Canarium strictum* Roxb.; *Burserac.*; sw. O.-Ind. (Travancore)
Pantáuon: *Neonauclea sp.*; *Rubiac.*; Phil. (Camarines)
Panting-pantíng: *Lumnitzera littorea* Voigt; *Combretac.*; Phil. (Basilan)
Pantjal: *Phyllanthus indicus* Muell. Arg.; *Euphorbiac.*; Java (Tjemara)
„ kidang: *Phyllanthus indicus* Muell. Arg.; *Euphorbiac.*; Java (Tjemara, Poeger)
„ „ : *Cleistanthus sumatranus* Muell. Arg.; *Euphorbiac.*; Java (Rogodjampi)
Pantjar: *Antiaris toxicaria* Lesch.; *Morac.*; Java.
Pantou: *Cryptosepalum tetraphyllum* Bth.; *Caesalpiniac.*; Elf.
Panu: *Knema attenuata* Warb.; *Myristicac.*; sw. O.-Ind. (Travancore)
Panungsúngan: *Shorea polysperma* Merr.; *Dipterocarpac.*; Phil. (Camarines)
? Panwe: *Picralima klaineana* Pierre; *Apocynac.*; Go.
Panyero: *Vitex cienkowskyi* Kotschy et Peyr.; *Verbenac.*; Togo.
„ buda: *Vitex camporum* Buettn.; *Verbenac.*; Togo.
Panza: *Polylepis sp.*; *Rosac.*; Ek.
„ : *Pentaclethra macrophylla* Bth.; *Mimosac.*; Kongo.
Pao: *Croton joufra* Roxb.; *Euphorbiac.*; Ind.-Ch.
„ : *Aglaia korthalsi* Pellegr.; *Meliac.*; Ind.-Ch.
Páo: *Buchanania sp.*; *Anacardiac.*; Phil. (Pangasinan)
„ amarello: *Chlorophora tinctoria* Gaudich.; *Morac.*; Bras.
„ „ : *Euxylophora paraënsis* Huber; *Rutac.*; n. Bras.
„ „ : *Raputia magnifica* Engl.; *Rutac.*; Bras. (Sao P)
„ branco: *Phyllostylon brasiliensis* Capanema; *Ulmac.*; Bras.
„ „ : *Auxemma gardneriana* Miers; *Borraginac.*; nö. Bras. (Ceará)
„ brazil: *Caesalpinia echinata* Lam.; *Caesalpiniac.*; Bras. (Bahia)
„ „ claro: *Caesalpinia sp.*; *Caesalpiniac.*; Bras. (Sao P)
„ „ oscuro: *Caesalpinia sp.*; *Caesalpiniac.*; Bras. (Sao P)
„ cachorro: *Cordia sp.*; *Borraginac.*; Bras.
„ candeia: *Lychnophora sp.*; *Composit.*; Bras. (Sao P)
„ canudo: *Carpotroche brasiliensis* Endl.; *Flacourtiac.*; Bras. (Sao P)
„ choca: *Pentaclethra macroloba* Ktze.; *Mimosac.*; Bras. (Sao P)
„ concha: ? *Roupala brasiliensis* Klotzsch; *Proteac.*; Bras.
„ „ claro: ? *Roupala brasiliensis* Klotzsch; *Proteac.*; Bras.
„ „ rôxo: ? *Roupala brasiliensis* Klotzsch; *Proteac.*; Bras.
„ „ „ escura: ? *Roupala brasiliensis* Klotzsch; *Proteac.*; Bras.
„ cravo: *Dicypellium caryophyllatum* Nees; *Laurac.*; Bras.
„ „ : *Physocalymma floridum* Pohl; *Lythrac.*; Bras.
„ daguila: *Aquilaria agallocha* Roxb.; *Thymelaeac.*; Burma.
„ doce: *Vochysia sp.*; *Vochysiac.*; Bras.
„ „ : *Pradosia lactescens* Radkl.; *Bras.*; *Sapotac.*; Bras.
„ duro: *Esenbeckia leiocarpa* Engl.; *Rutac.*; s. Bras.
„ fedorento: *Gustavia spp.*; *Lecythidac.*; Bras.
„ ferro: *Casuarina stricta* Ait.; *Casuarinac.*; Gu., n. Bras.
„ „ : *Swartzia tomentosa* DC.; *Caesalpiniac.*; Gu., n. Bras.
„ „ : *Apuleia ferrea* Mart.; *Caesalpiniac.*; Bras. (Sao P)
„ „ : *Caesalpinia granadillo* Pittier, *C. ferrea* Mart.; *Caesalpiniac.*; Bras.
„ „ : *Machaerium scleroxylon* Tul.; *Papilionac.*; Bras.
„ „ claro: ? *Astronium sp.*; *Anacardiac.*; Bras. (Minas Geraes)
„ gamelo: *Clusia sp.*; *Guttifer.*; Bras.
„ -hua: *Machilus thunbergi* S. et Z.; *Laurac.*; Ch.
„ imbira: *Bombax sp.*; *Bombacac.*; Bras. (Sao P)
„ inglez: *Calophyllum brasiliana* Camb.; *Guttifer.*; Bras.

Páo jangada: *Apeiba tibourbou* Aubl.; *Tiliac.; Gu., Bras.
,, jantar: *Lonchocarpus sp.; Papilionac.; Bras.* (Sao P)
,, liso: *Balfourodendron riedelianum* Engl.; *Rutac.; Bras.* (Sao P)
,, macaco: *Lecythis sp.; Lecythidac.; Bras.*
,, mandioca: *Manihot sp.; Euphorbiac.; Bras.*
,, marfim: *Balfourodendron riedelianum* Engl.; *Rutac.; Bras.* (Sao P)
,, mulato: *Calycophyllum spruceanum* Bth.; *Rubiac.;* nö. Bras. (Pará)
,, mulatto: *Calycophyllum spruceanum* Bth.; *Rubiac.; Bras.*
,, muneunza: *Piptadenia africana* Hook. f.; *Mimosac.;* Ang. (Golungo alto)
,, musance: *Piptadenia africana* Hook. f.; *Mimosac.;* Ang. (Golungo alto)
,, muzungo: *Piptadenia africana* Hook. f.; *Mimosac.;* Ang. (Golungo alto)
,, oléo: *Pachylobus balsamifera* Guill.; *Burserac.;* Gab.
,, olho: *Copaifera sp.; Caesalpiniac.; Bras.*
,, parahyba: *Simaruba versicolor* St. Hil., *S. sp.; Simarubac.; Bras.* (Bahia)
,, ,, : *Tabebuia leucoxyla* P. DC.; *Bignoniac.; Bras.* (Sao P)
,, paratudo: *Drimys winteri* Forst.; *Magnoliac.; Bras.*
,, ,, : *Aspidosperma sp.; Apocynac.; Bras.* (Espirito Santo)
,, pente: *Platycyamus regnelli* Bth.; *Papilionac.; Bras.* (Sao P)
,, pereira: *Platycyamus regnelli* Bth.; *Papilionac.;* s. Bras.
,, ,, : *Aspidosperma macrocarpum* Mart.; *Apocynac.;* s. Bras.
,, Pernambuco: *Caesalpinia echinata* Lamk.; *Caesalpiniac.; Bras.* (Sao P)
,, peroba: *Cassia apoucouita* Aubl.; *Caesalpiniac.; Bras.*
,, pomba: *Simaruba amara* Aubl.; *Simarubac.;* n. Bras., fr. Gu., W.-Ind.
,, pombo: *Talauma sp.; Magnoliac.; Bras.*
,, ,, : *Tapirira guianensis* Aubl.; *Anacardiac.;* n. & ö. Bras.
,, rainha de listras: *Piratinera ? guianensis* Aubl.; *Morac.; Bras.*
,, rei: *Sterculia sp.; Sterculiac.; Bras.*
,, rojo: *Peltogyne spp.; Caesalpiniac.; Bras.*
,, rosa: *Dalbergia nigra* Allem.; *Papilionac.; Bras.*
,, ,, : *Aniba spp., Ocotea costulata* Mez; *Laurac.;* nö. Bras.
,, ,, : *Nectandra elaiophora* Barb. Rodr.; *Laurac.;* n. Bras. (Manaos)
,, ,, : *Aydendron riparium* Nees; *Laurac.; Bras.*
,, ,, branca: *Trichilia alta* Blake; *Meliac.; Bras.* (Bahia)
,, ,, da serra: *Aniba panurensis* Mez; *Laurac.; Bras.*
,, rosado: *Caesalpinia echinata* Lamk.; *Caesalpiniac.; Bras.* (Sao P)
,, roxo: *Peltogyne spp., Hymenaea sp.; Caesalpiniac.; Bras.*
,, sangue: *Pterocarpus sp.; Papilionac.; Bras.* (Bahia)
,, sanguinho: *Haematoxylon campechianum* L.; *Caesalpiniac.;* tr. Am.
,, santo: *Porliera hygrometrica* Ruiz et Pav.; *Zygophyllac.;* Mex., w. S.-Am.
,, ,, : *Zollernia paraënsis* Huber; *Caesalpiniac.;* nö. Bras. (Amaz.)
,, setim: *Euxylophora paraënsis* Huber; *Rutac.; Bras.*
,, ,, : *Aspidosperma eburneum* Allem.; *Apocynac.;* s. Bras. (Sao P, Victoria)
,, sha: *Abies beissneriana* Rehd. et Wilson; *Conifer.;* Ch.
,, tartarugo: *Piratinera ? guianensis* Aubl.; *Morac.; Bras.*
,, terra: *Qualea spp.; Vochysiac.; Bras.*
,, ,, : *Lafoensia glyptocarpa* Koehne; *Lythrac.; Bras.* (Sao P)
,, vermelho: *Piratinera ? guianensis* Aubl.; *Morac.; Bras.*
,, -yeh: *Machilus thunbergi* S. et Z.; *Laurac.;* Ch.
,, d'alho: *Gallesia scorododendrum* Casaretto; *Phytolaccac.; Bras.* (Sao P, Bahia)
,, ,, : *Crataeva tapia* L.; *Capparidac.; Bras.*
,, ,, do campo: *Agonandra brasiliensis* Bth.; *Olacac.; Bras.*
,, d'Angola: *Piper aff. acutifolium* Poepp.; *Piperac.;* nö. Bras. (Para)
,, d'aquilla: *Excoecaria africana* Muell. Arg.; *Euphorbiac.;* O.-Af., Sans.
,, d'arco: *Tecoma spp.; Bignoniac.; Bras.*
,, ,, amarello: *Tecoma spp.; Bignoniac.;* n. Bras. (Amaz.)
,, ,, rôxo: *Tecoma sp.; Bignoniac.;* n. Bras. (Amaz.)
,, d'oleo: *Copaifera spp.; Caesalpiniac.; Bras.*
,, de balsa: *Ochroma lagopus* Swartz; *Bombacac.; Bras.* (Amaz.)
,, ,, camerâo: *Clytostoma noterophilum* Bureau et K. Schum.; *Bignoniac.; Bras.*
,, ,, cheiro: *Eupatorium spp.; Composit.; Bras.*
,, ,, cinza: *Vochysia sp.; Vochysiac.; Bras.*

Páo de colher: *Tabernaemontana sp.; Apocynac.;* Bras.
  „ „   cortume: *Byrsonima sp.; Malpighiac.;* Bras.
  „ „   cotia: *Esenbeckia grandiflora* Mart.; *Rutac.;* s. Bras.
  „ „   Guine: *Anona sp.; Anonac.;* Bras.
  „ „   jangada: *Apeiba tibourbou* Aubl.; *Tiliac.;* Bras.
  „ „   lacre: *Vismia spp.; Guttifer.;* Bras.
  „ „   leite: *Sapium biglandulosum* Muell. Arg.; *Euphorbiac.;* Bras. (Sao P)
  „ „   macaco: *Eschweilera sp., Jugastrum sp.; Lecythidac.;* Bras.
  „ „   oleo: *Copaifera langsdorffi* Desf., *C. spp.; Caesalpiniac.;* Bras. (Sao P)
  „ „   polvora: *Trema micrantha* Blume; *Ulmac.;* Bras.
  „ „   rainha: *Centrolobium paraënse* Tul.; *Papilionac.;* n. Bras. (Amaz.)
  „ „   remo: *Swartzia tomentosa* DC.; *Caesalpiniac.;* Bras.
  „ „   rosa: *Aniba parviflora* Mez; *Laurac.;* Bras. (Pará)
  „ „   sabao: *Sapindus divaricatus* St. Hil.; *Sapindac.;* Bras.
  „ „   „   : *Guarea trichilioides* L.; *Meliac.;* Arg.
  „ „   San José: *Killmeyera coriacea* Mart.; *Guttifer.;* ö. & s. Bras.
  „ „   sangue: *Brosimum conduru* Allem.; *Morac.;* n. Bras. (Amaz.)
  „ „   santo: *Cabralea cangerana* Mart.; *Meliac.;* Bras. (Sao P)
  „ „   tucano: *Vochysia sp.; Vochysiac.;* Bras.
  „ „   vinho: *Vochysia sp.; Vochysiac.;* Bras.
  „ „   „ preto: *Vochysia sp.; Vochysiac.;* Bras.
  „ „   viola: *Citharexylum cinereum* L.; *Verbenac.;* s. Bras.
  „ „   vitem: *Schizolobium parahybum* Blake; *Caesalpiniac.;* Bras. (Bahia)
  „   do formige: *Cordia sp.; Borraginac.;* Bras.
Paong: *Chisocheton cochinchinensis* Pierre; *Meliac.;* Kamb.
? Pāpā: *Albizzia zygia* MacBride, *A. sassa* MacBride; *Mimosac.;* Go.
Papa: *Cistanthera papaverifera* Chev.; *Tiliac.;* Elf., Go.
  „   : *Gardenia latifolia* Aiton; *Rubiac.;* w. Beng. (Bihar)
Papache: *Randia armata* DC.; *Rubiac.;* Mex. (Sinaloa)
Papágan: *Mimusops parvifolia* R. Br.; *Sapotac.;* Phil. (Cagayan)
Papaguá-naranja: *Crataeva tapia* L.; *Capparidac.;* Arg.
Papaguayon: *Crataeva tapia* L.; *Capparidac.;* Arg.
Papaja, Bosch-: *Cecropia peltata* L.; *Morac.;* Sur.
  „ hoedoe: *Omphalea diandra* L.; *Euphorbiac.;* ? Sur.
Papallón: *Gustavia integrifolia* Standl.; *Lecythidac.;* Nic.
Papalón: *Coccoloba sp.; Polygonac.;* Salv.
Papalote: *Ulmus mexicana* Planch.; *Ulmac.;* Mex.
Papamiel: *Hamelia rovirosae* Wernh.; *Rubiac.;* Nic.
Papantie-ie: *Nectandra sp., Ocotea sp.; Laurac.;* Sur.
Papao: *Afzelia africana* Smith; *Caesalpiniac.;* Go., Togo.
Papar: *Gardenia latifolia* Aiton; *Rubiac.;* C.-O.-Ind.
  „   : *Pongamia glabra* Vent.; *Papilionac.;* n. Ö.-Ind. (United Prov.)
  „   : *Buxus wallichiana* Baill.; *Buxac.;* O.-Ind. (Punjab)
Papara: *Holoptelea integrifolia* Planch.; *Ulmac.;* sw. O.-Ind. (Bombay)
  „ -poulia-marom: *Adansonia digitata* L.; *Bombacac.;* w. O.-Ind. (Deccan)
Paparrón: *Coccoloba sp.; Polygonac.;* Salv.
Papasan: *Mallotus philippinensis* Muell. Arg.; *Euphorbiac.;* Java (Besoeki)
Papasco wood: *Acer saccharinum* L.; *Acerac.;* USA.
Papásil: *Lumnitzera littorea* Voigt; *Combretac.;* Phil. (Tayabas)
Papaturro: *Coccoloba uvifera* Jacq., *C. tuerckheimi* D. Sm.; *Polygonac.;* C.-Am.
  „ blanco: *Coccoloba caracasana* Meissn.; *Polygonac.;* Pan., C.-R.
  „ extranjero: *Coccoloba uvifera* Jacq.; *Polygonac.;* Hond.
Pápau: *Afzelia africana* Smith; *Caesalpiniac.;* Togo.
Papaw: *Asimina triloba* Dunal; *Anonac.;* sö. Kan.
Papawu: *Afzelia africana* Smith; *Caesalpiniac.;* Go.
Papaya: *Carica papaya* L., *C. dolicaula* D. Sm.; *Caricac.;* Hond., Pan.
Papayo: *Metopium toxiferum* Krug et Urb.; *Anacardiac.;* P.-R.
  „   : *Carica papaya* L.; *Caricac.;* Hond.
Papayote: *Cochlospermum vitifolium* Spreng.; *Cochlospermac.;* n. Kol.
Papayu: *Pygeum glandulosum* Merr., *P. presli* Merr.; *Rosac.;* Phil. (Bontok)
Papelillo: *Bursera sp.; Burserac.;* Mex.

Papenhoed: *Evonymus europaea* L.; *Celastrac.*; Holl.
Papenhout: *Evonymus europaea* L.; *Celastrac.*; Holl.
Paperbark (e): *Melaleuca leucadendron* L.; *Myrtac.*; Malay., Au.
Papirin-balé-ie: *Eperua schomburgkiana* Bth.; *Caesalpiniac.*; Sur.
Papita (H): *Sterculia campanulata* Wall.; *Sterculiac.*; And.
Pãplo: *Oxytenanthera* ? *abyssinica* Munro; *Gramin.*; Go.
Papo de zamba: *Lonchocarpus sanctae-marthae* Pittier; *Papilionac.*; n. Kol.
Papoa: *Afzelia africana* Smith; *Caesalpiniac.*; Go.
Papororoca: *Drimys winteri* Forst.; *Magnoliac.*; Bras.
Pappágan: *Buchanania sp.*; *Anacardiac.*; Phil. (Cagayan)
   „ : *Mimusops parvifolia* R. Br.; *Sapotac.*; Phil.
Pappel (d): *Populus spp.*; *Salicac.*; Eu., N.-Am.
   „ , Balsam-: *Populus balsamifera* L.; *Salicac.*; N.-Am.
   „ , Grau-: *Populus canescens* Smith; *Salicac.*; Eu.
   „ , grosszähnige: *Populus grandidentata* Michx.; *Salicac.*; N.-Am.
   „ , haarfrüchtige: *Populus trichocarpa* Torr. et Gray; *Salicac.*; N.-Am.
   „ , italienische: *Populus pyramidalis* Rozier; *Salicac.*; Pers.
   „ , kanadische: *Populus canadensis* Moench; *Salicac.*; N.-Am.
   „ , Pyramiden-: *Populus pyramidalis* Rozier; *Salicac.*; Pers.
   „ , Schwarz-: *Populus nigra* L.; *Salicac.*; Eu.
   „ , Silber-: *Populus alba* L.; *Salicac.*; Eu.
   „ , Weiss-: *Populus alba* L.; *Salicac.*; Eu
   „ , Zitter-: *Populus tremula* L.; *Salicac.*; Eu.
Papra: *Gardenia latifolia* Aiton; *Rubiac.*; C.-O.-Ind.
   „ : *Holoptelea integrifolia* Planch.; *Ulmac.*; C.-O.-Ind.
Paprang: *Buxus wallichiana* Baill.; *Buxac.*; n. O.-Ind. (Punjab)
Papri: *Buxus wallichiana* Baill.; *Buxac.*; n. O.-Ind. (United Prov.)
   „ : *Holoptelea integrifolia* Planch.; *Ulmac.*; n. O.-Ind. (Punjab, United Prov.)
Papur: *Gardenia latifolia* Aiton; *Rubiac.*; sw. O.-Ind. (Bombay)
Paqueretê: *Tabernaemontana macrophylla* Muell. Arg.; *Apocynac.*; Bras.
Para makka: *Bactris sp.*, *Astrocaryum paramaca* Mart.; *Palmac.*; Sur.
   „   „ : *Astrocaryum aculeatum* G. F. W. Meyer; *Palmac.*; Sur.
   „ markoesia: *Passiflora laurifolia* L.; *Passiflorac.*; Sur.
   „ -noot: *Bertholletia excelsa* H. B. K.; *Lecythidac.*; Sur.
Parabukúng: *Lagerstroemia speciosa* Pers.; *Lythrac.*; Phil. (Misamis)
Paracutáca: *Swartzia sp.*; *Caesalpiniac.*; Bras.
Paradakué: *Placodiscus pseudostipularis* Radlk.; *Sapindac.*; Elf.
Paradise tree: *Simaruba glauca* DC.; *Simarubac.*; tr. Am.
   „   „ : *Melia azedarach* L.; *Meliac.*; br. Hond.
   „ wood: *Aquilaria agallocha* Roxb.; *Thymelaeac.*; Burma.
Paradisier: *Melia azedarach* L.; *Meliac.*; Tr.
Paraguatá: *Sickingia erythroxylon* Widd.; *Rubiac.*; Ven.
Paraguatán: *Sickingia spp.*; *Rubiac.*; Ven.
Parahyba: *Simaruba versicolor* St. Hil.; *Simarubac.*; Bras.
Paraïba: *Simaruba sp.*; *Simarubac.*; Bras.
Paraiso: *Melia azedarach* L.; *Meliac.*; tr. Am.
   „ morado: *Melia azedarach* L.; *Meliac.*; Mex.
Parak: *Rhizophora candelaria* DC.; *Rhizophorac.*; Phil. (Culion)
Parakusan: *Swartzia acuminata* Willd., *S. spp.*; *Caesalpiniac.*; Sur.
Paral: *Stereospermum chelonioides* DC.; *Bignoniac.*; sw. O.-Ind. (Bombay)
   „ : *Stereospermum suaveolens* DC.; *Bignoniac.*; C.-O.-Ind.
   „ : *Stereospermum xylocarpum* Bth. et Hook. f.; *Bignoniac.*; C.-O.-Ind.
Parandédi: *Leptaulus daphnoïdes* Bth.; *Icacinac.*; Elf.
Parang dewo: *Mallotus philippinensis* Muell. Arg.; *Euphorbiac.*; Java (Soebah)
Paranga: *Acer oblongum* Wall.; *Acerac.*; n. O.-Ind. (United Prov.)
Parangi-shambirani: *Boswellia serrata* Roxb.; *Burserac.*; w. O.-Ind. (Decan)
Paranji: *Boswellia serrata* Roxb.; *Burserac.*; O.-Ind. (Madras)
Paránussbaum (d): *Bertholletia excelsa* Bonpl., *B. nobilis* Miers; *Lecythidac.*; Gu.,
Parapará: *Jacaranda copaia* D. Don; *Bignoniac.*; n. Bras. (Amaz.)   [n. Bras.
   „ : *Cordia sp.*; *Borraginac.*; Bras.
   „ : *Didymopanax morototoni* Dcne. et Pl.; *Araliac.*; Bras.

Parapará: *Sapindus saponaria* L.; *Sapindac.*; Ven.
Parapára: *Lophira alata* Banks; *Ochnac.*; Togo.
Parari: *Stereospermum chelonioides* DC.; *Bignoniac.*; nö. O.-Ind. (Nepal)
Paras: *Prunus padus* L.; *Rosac.*; nw. O.-Ind. (Punjab)
Parasablút: *Litsea sp.*; *Laurac.*; Phil. (Zambales)
Parasolier (f): *Musanga smithi* R. Br.; *Morac.*; W.-Af.
Paratodo: *? Tecoma sp.*, *? Tabebuia sp.*; *Bignoniac.*; sw. Bras.
Paratudo: *Drimys winteri* Forst.; *Magnoliac.*; Bras.
Parawaholz: *Avicennia nitida* Jacq.; *Verbenac.*; Sur.
Pará wood (e): *Caesalpinia echinata* Lam.; *Caesalpiniac.*; Bras.
Parcouri: *Platonia insignis* Mart., *Moronobea coccinea* Aubl.; *Guttifer.*; Gu.
    „   : *Clusia insignis* Mart.; *Guttifer.*; fr. Gu.
  „ -goupi: *Clusia sp.*; *Guttifer.*; fr. Gu.
  „ jaune (f): *Platonia insignis* Mart., *P. sp.*; *Guttifer.*; Gu.
  „ -manil: *Moronobea coccinea* Aubl.; *Guttifer.*; Sur.
  „ soufré (f): *Platonia insignis* Mart.; *Guttifer.*; fr. Gu.
Parcoury: *Platonia insignis* Mart.; *Guttifer.*; fr. Gu.
Pardeflesh: *Mimusops spp.*; *Sapotac.*; Sur.
Pardenvleesh: *Mimusops spp.*; *Sapotac.*; Sur.
Pardepis: *Hippobromus alatus* Eckl. et Zeyh.; *Sapindac.*; S.-Af. (Kap)
Pardillo: *Cordia gerascanthus* L.; *Borraginac.*; Ven., Trin.
    „   : *Cordia alliodora* Cham.; *Borraginac.*; Ven.
  „ blanco: *Cordia sp.*; *Borraginac.*; Ven.
  „ del monte: *Cordia gerascanthus* H. B. K.; *Borraginac.*; Ven.
Pardon: *Ilex aquifolium* L.; *Aquifoliac.*; Eu., n. Af.
Parè-èh: *Eperua falcata* Oliv., *E. schomburgkiana* Bth.; *Caesalpiniac.*; Sur.
Pareeya: *Ailanthus malabarica* A. DC.; *Simarubac.*; O.-Ind.
Parekha: *Dalbergia lanceolaria* L.; *Papilionac.*; C.-O.-Ind.
Parelhout: *Aspidosperma nitidum* Bth., *A. spp.*; *Apocynac.*; Sur.
  „ , witte: *Aspidosperma nitidum* Bth., *A. spp.*; *Apocynac.*; Sur.
  „ , zwart: *Swartzia acuminata* Willd., *S. sp.*; *Caesalpiniac.*; Sur.
  „ , „ : *Aspidosperma oblongum* A. DC.; *Apocynac.*; Sur.
Parengpeng: *Croton argyratus* Blume, *Macaranga javanica* Muell. Arg.; *Euphorbiac.*;
Parépou: *Guilielma speciosa* Mart.; *Palmac.*; fr. Gu.   [Java.
Paresh-pippul: *Thespesia populnea* Correa; *Malvac.*; Beng.
Pareya-auwal: *Stereospermum chelonioides* DC.; *Bignoniac.*; Assam (Cachar)
Pari: *Glochidion obscurum* Hook. f., *G. molle* Blume, *G. sp.*; *Euphorbiac.*: Java.
Paria: *Gardenia latifolia* Aiton; *Rubiac.*; C.-O.-Ind.
Pariá: *Myristica sp.*; *Myristicac.*; Phil. (Bataan)
Pariara: *Erythrina suberosa* Roxb.; *Papilionac.*; nw. O.-Ind. (Punjab)
Paricá: *Curatella americana* L.; *Dilleniac.*; fr. Gu.
    „   : *Piptadenia peregrina* Bth.; *Mimosac.*; Bras.
Parihoedoe: *Aspidosperma nitidum* Bth., *A. oblongum* A. DC., *A. sp.*; *Apocynac.*; Sur.
Pariná': *Kingiodendron alternifolium* Merr.; *Papilionac.*; Phil.
    „   : *Pinus insularis* Endl.; *Conifer.*; Phil. (Ilocos Sur)
    „   : *Sindora supa* Merr.; *Caesalpiniac.*; Phil. (Camarines, Albay, Sorsogon)
Parinari: *Parinarium campestre* Aubl., *? P. brachystachyum* Bth.; *Rosac.*; Sur.
    „   : *Couepia sp.*; *Rosac.*; Bras.
  „ , petit: *Parinarium campestre* Aubl.; *Rosac.*; fr. Gu.
  „ des Garipons, petit: *Parinarium campestre* Aubl., *? P. sp.*; *Rosac.*; Sur.
Parishia: *Parishia insignis* Hook. f.; *Anacardiac.*; O.-Ind.
Pariva: *Avicennia nitida* Jacq.; *Verbenac.*; Sur.
Parkia: *Parkia roxburghi* G. Don; *Mimosac.*; Sey.
Parmi: *Lannea grandis* Engl.; *Anacardiac.*; w. Beng. (Chota Nagpur)
Paroengan: *Macaranga javanica* Muell. Arg.; *Euphorbiac.*; Java (Pandanaroem)
Paroeroe pio: *Erisma uncinatum* Warm.; *Vochysiac.*; Sur.
Paroli: *Stereospermum chelonioides* DC.; *Bignoniac.*; Assam.
    „   : *Stereospermum suaveolens* DC.; *Bignoniac.*; sw. O.-Ind. (Malabar)
Paronápin: *Heritiera litoralis* Dry., *Tarrietia silvatica* Merr.; *Sterculiac.*; Phil.
Paronápoi: *Heritiera litoralis* Dry.; *Sterculiac.*; Phil. (Cagayan, Pangasinan, Zambales)
    „   : *Tarrietia silvatica* Merr.; *Sterculiac.*; Phil. (Pangasinan)

Paronot: *Planchonia spectabilis* Merr.; *Lecythidac.;* Phil. (Ilocos Norte)
Parota: *Enterolobium cyclocarpum* Griseb.; *Mimosac.;* Mex.
Parpat: *Acer oblongum* Wall.; *Acerac.;* nw. O.-Ind. (Kumaon)
Parra sılvestre: *Vitis vinifera* L.; *Vitac.;* Sp.
Parrewè: *Eperua jenmanni* Oliv.; *Caesalpiniac.;* Sur.
Parroria: *Parrotia jacquemontiana* Dcne.; *Hamamelidac.;* n. O.-Ind. (Himal.)
Pars: *Anogeissus acuminata* Wall.; *Combretac.;* C.-O.-Ind. (Gondwana)
Parsia: *Anogeissus acuminata* Wall.; *Combretac.;* C.-O.-Ind. (Gondwana)
Parsipu: *Thespesia populnea* Correa; *Malvac.;* O.-Ind. (Hind.)
Partridge berry: *Vaccinium vitis-idaea* L.; *Ericac.;* N.-Fundl.
Partridgecane: *Licuala acutifida* Mart.; *Palmac.;* Malay.
Partridgeholz (d): *Vouacapoua americana* Aubl.; *Caesalpiniac.;* Ven.
Partridge pea (e): *Heisteria coccinea* Jacq.; *Olacac.;* Ant.
Partridge tree: *Heisteria coccinea* Jacq.; *Olacac.;* Hond.
Partridge wood (e, u): *Vouacapoua americana* Aubl.; *Caesalpiniac.;* Sur.
   „   „ : *Caesalpinia melanocarpa* Griseb.; *Caesalpiniac.;* S.-Am. (Gran
   „   „ : *Caesalpinia granadillo* Pittier; *Caesalpiniac.;* Ven.   [Chaco)
   „   „ : *Heisteria coccinea* Jacq.; *Olacac.;* Ant.
Parug-án: *Knema glomerata* Merr.; *Myristicac.;* Phil. (Rizal)
Parul: *Stereospermum suaveolens* DC.; *Bignoniac.;* Beng., sw. O.-Ind. (Bombay)
Pa-ru-po: *Sterculia maingayi* Mast.; *Sterculiac.;* Malak.
Parur: *Stereospermum suaveolens* DC.; *Bignoniac.;* n. O.-Ind. (United Prov.)
Paruru: *Saccoglottis guianensis* Aubl.; *Humiriac.;* nö. Bras. (Amaz.)
Parutti: *Bombax malabaricum* DC.; *Bombacac.;* O.-Ind. (Madras)
Parwa: *Avicennia nitida* Jacq.; *Verbenac.;* Sur.
Pásak: *Bassia betis* Merr.; *Sapotac.;* Phil. (Manila)
  „ : *Mimusops parvifolia* R. Br.; *Sapotac.;* Phil.
  „ : *Parinarium sp.; Rosac.;* Phil. (Nueva Ecija, Manila)
  „ lenga: *Ilex macrophylla* Wall.; *Aquifoliac.;* Malak.
Pasaklák: *Cyclostemon sp.; Euphorbiac.;* Phil. (Union Prov.)
Pasang: *Quercus sp.; Fagac.;* Java.
Pasania: *Quercus junghuhni* Miq.; *Fagac.;* Form.
Pasau: *Crossopteryx africana* K. Schum.; *Rubiac.;* Togo.
Paschaholz (d): *Triplochiton scleroxylon* K. Schum.; *Sterculiac.;* W.-Af.
Pascua: *Euphorbia pulcherrima* Willd.; *Euphorbiac.;* Hond.
Pasēe: *Ochrocarpus africanus* Oliv.; *Guttifer.;* Go.
Pasfaré: *Sclerolobium denudatum* Vogel; *Caesalpiniac.;* Bras. (Sao P)
Pashaco: *Pithecolobium racemiflorum* Ducke; *Mimosac.;* Bras. (Amaz.)
  „ : *Entada sp.; Mimosac.;* Peru.
Pasi: *Anogeissus acuminata* Wall.; *Combretac.;* w. Beng. (Orissa)
  „ : *Heisteria trillesiana* Pierre; *Olacac.;* Gab.
Pasíngan: *Bambusa spinosa* Roxb.; *Gramin.;* Phil. (Cagayan)
Pasinia: *Quercus junghuhni* Miq.; *Fagac.;* Form.
Pasita: *Bumelia obovata* A. DC.; *Sapotac.;* n. Kol.
Paskapasan: *Croton argyratus* Blume; *Euphorbiac.;* Java (Tjoeramanis)
  „ : *Mallotus philippinensis* Muell. Arg.; *Euphorbiac.;* Java (Besoeki)
Pasnít: *Alstonia scholaris* R. Br.; *Apocynac.;* Phil. (Pangasinan)
Pasocho: *Berlinia sp.; Cǫesalpiniac.;* Togo.
Pasoemoeti: *Copaifera guianensis* Desf.; *Caesalpiniac.;* Sur.
Pasóso: *Eugenia mananquil* Blco.; *Myrtac.;* Phil. (Rizal)
Paspu: *Adina cordifolia* Hook. f.; *Rubiac.;* C.-O.-Ind. (Gondwana)
Passi: *Dalbergia lanceolaria* L.; *Papilionac.;* sw. O.-Ind. (Ajmer-Merwara)
Passiemoetie: *Copaifera guianensis* Desf.; *Caesalpiniac.;* Sur.
Passuaré: *Sclerolobium denudatum* Vogel; *Caesalpiniac.;* Bras. (Sao P)
Pasta: *Hirtella sp.; Rosac.;* Hond.
Pastelillo: *Coccoloba acuminata* H. B. K.; *Polygonac.;* Kol.
Pastuwanni: *Grewia oppositifolia* Roxb.; *Tiliac.;* nw. Himal.
Pata: *Acer pictum* Thunb.; *Acerac.;* nw. O.-Ind. (Kumaon)
  „ : *Shorea polysperma* Merr.; *Dipterocarpac.;* Phil. (Pangasinan)
  „ : *Ximenia americana* L.; *Olacac.;* Arg.
  „ fouanga: *Parkia agboënsis* Chev.; *Mimosac.;* Elf.

Pata de buey: *Bauhinia candicans* Bth.; *Caesalpiniac.*; Arg.
„ „ cabro: *Cynometra retusa* Britt. et Rose; *Caesalpiniac.*; br. Hond., Guat.
„ „ gallina: *Cordia macrostachya* Roem. et Schult.; *Borraginac.*; n. Kol.
„ „ pauji: *Mouriria pseudo-geminata* Pittier; *Melastomatac.*; Ven.
„ „ pava: *Comocladia sp.*; *Anacardiac.*; Mex.
„ „ vaca: *Bauhinia divaricata* L.; *Caesalpiniac.*; br. Hond.
„ „ „ : *Bauhinia mollifolia* Pittier, *B. spp.*; *Caesalpiniac.*; Kol.
„ „ „ : *Pithecolobium lanceolatum* Bth.; *Mimosac.*; Kol.
„ „ „ : *Bauhinia candicans* Bth.; *Caesalpiniac.*; Arg.
Pataban: *Laguncularia racemosa* Gaertn. f.; *Combretac.*; Kuba.
Pataboa: *Triplochiton scleroxylon* K. Schum.; *Sterculiac.*; Go.
Patagua: *Crinodendron patagua* Molina; *Elaeocarpac.*; Chile.
Patakli: *Celtis prantli* Priemer; *Ulmac.*; Togo.
Patala: *Stereospermum suaveolens* DC.; *Bignoniac.*; sw. O.-Ind. (Bombay)
Patandëu: *Lannea barteri* Engl.; *Anacardiac.*; Togo.
Patangawood: *Caesalpinia sappan* L.; *Caesalpiniac.*; Cey.
Pataoua: *Oenocarpus bataua* Mart.; *Palmac.*; fr. Gu.
Patapara: *Hymenostegia aubrevillei* Pellegr.; *Caesalpiniac.*; Elf.
„ : *Cynometra megalophylla* Harms; *Caesalpiniac.*; Elf.
Patapra: *Hymenostegia aubrevillei* Pellegr.; *Caesalpiniac.*; Elf.
Pataste: *Luehea sp.*; *Tiliac.*; Salv.
Patawa: *Oenocarpus bacaba* Mart., *O. bataua* Mart.; *Palmac.*; Sur., fr. Gu.
Patawnig: *Trewia nudiflora* L.; *Euphorbiac.*; Burma.
Pataxte: *Theobroma bicolor* Humb. et Bonpl.; *Sterculiac.*; Mex.
„ : *Luehea speciosa* Willd., *L. candida* Mart.; *Tiliac.*; Mex.
Pataxtillo: *Luehea speciosa* Willd.; *Tiliac.*; Mex.
Patazte: *Luehea candida* Mart.; *Tiliac.*; Mex.
Pate: *Panda oleosa* Pierre; *Pandac.*; Kam., Gab.
Paternillo: *Inga sp.*; *Mimosac.*; Salv.
Paterno: *Swartzia panamensis* Bth.; *Caesalpiniac.*; Hond.
„ : *Inga paterno* Harms; *Mimosac.*; Guat., Salv.
Patha: *Kydia calycina* Roxb.; *Malvac.*; n. O.-Ind. (United Prov.)
Pathiri: *Stereospermum xylocarpum* Bth. et Hook. f.; *Bignoniac.*; O.-Ind.-(Madras)
Pathor: *Bridelia retusa* Spreng.; *Euphorbiac.*; nw. O.-Ind. (Punjab)
Patiara sipioro: *Tetragastris balsamifera* O. Ktze., *Protium spp.*; *Burserac.*; Sur.
Patin dehoun: *Ceiba pentandra* Gaertn. *var. clausa* Ulb.; *Bombacac.*; Dah.
Patir: *Aegle marmelos* Correa; *Rutac.*; O.-Ind. (Madras)
Pativa: *Randia dumetorum* Lamk.; *Rubiac.*; w. Beng. (Orissa)
Patjahan: *Dalechampia bidentata* Blume; *Euphorbiac.*; Java.
Patjar: *Aglaia odorata* Lour.; *Meliac.*; Java (Pangentjongan)
„ goenoeng: *Vaccinium varingifolium* Miq.; *Ericac.*; Java (Idjenplateau)
Patle Katus: *Quercus lanceaefolia* Roxb.; *Fagac.*; n. O.-Ind.
Patobi: *Parinarium tenuifolium* Chev.; *Rosac.*; Elf.
Patoe: *Calophyllum inophyllum* L.; *Guttifer.*; Java.
Patol: *Erythrina leptorrhiza* DC.; *Papilionac.*; Mex.
„ colorín: *Erythrina sp.*; *Papilionac.*; Mex.
Patón: *Dendrocalamus latiflorus* Munro; *Gramin.*; Phil. (Sorsogon, Surigao)
Patotoo: *Adinandra dumosa* Jack; *Theac.*; Malak.
Patpát: *Sonneratia caseolaris* Engl.; *Sonneratiac.*; Phil. (Agusan)
Patpatta: *Talauma hodgsoni* Hook. f. et Thoms.; *Magnoliac.*; nö. O.-Ind. (Nepal)
Patphunnas: *Artocarpus hirsuta* Lamk.; *Morac.*; sw. O.-Ind. (Bombay)
Patru kurwa: *Holarrhena antidysenterica* Wall.; *Apocynac.*; w. Beng. (Orissa)
Patsahíngin: *Dipterocarpus vernicifluus* Blco.; *Dipterocarpac.*; Phil. (Pol.)
Patsaru: *Dalbergia lanceolaria* L.; *Papilionac.*; O.-Ind. (Madras)
Patte poule piquante: *Toddalia aculeata* Pers.; *Rutac.*; Maur.
„ de lièvre: *Ochroma lagopus* Swartz; *Bombacac.*; Ant., Sur.
„ „ poule: *Toddalia aculeata* Pers.; *Rutac.*; Mask.
Pattern wood: *Alstonia congensis* Engl.; *Apocynac.*; tr. W.-Af.
Pattewar: *Diospyros montana* Roxb.; *Ebenac.*; O.-Ind., Burma.
Pattia: *Albizzia stipulata* Boivin; *Mimosac.*; n. O.-Ind. (United Prov.)
Patuba: *Strombosia pustulata* Oliv.; *Olacac.*; Elf.

Pau = Pao
„ avangolja: *Licania heteromorpha* Bth., *L. spp.; Rosac.;* Sur.
„ Brasil: *Caesalpinia echinata* Lam.; *Caesalpiniac.;* Bras.
„ cadeiro: *Funtumia africana* Stapf; *Apocynac.;* W.-Af. (St. Thomé)
„ doce: *Glycoxylon praealtum* Ducke; *Sapotac.;* n. Bras. (Pará)
„ duro: *Esenbeckia leiocarpa* Engl.; *Rutac.;* Bras. (Sao P)
„ -fa-Holz: *Machilus thunbergi* S. et Z.; *Laurac.;* Jap., Ch.
„ hoi: *Machilus spp.; Laurac.;* Ch.
„ „ : *Phoebe nanmu* Gamble, *P. macrophylla* Gamble; *Laurac.;* s. Ch.
„ „ : *Firmiana simplex* F. N. Meyer; *Sterculiac.;* Central-Ch.
„ „ : *Ulmus pumila* L., *U. japonica* Sarg.; *Ulmac.;* N.-Ch.
„ jacaré: *Piptadenia communis* Bth.; *Mimosac.;* Bras. (Sao P)
„ -lai: *Oldfieldia africana* Bth. et Hook. f.; *Euphorbiac.;* Lib.
„ liso: *Balfourodendron riedelianum* Engl.; *Rutac.;* Bras. (Sao P)
„ marfim: *Balfourodendron riedelianum* Engl.; *Rutac.;* Bras. (Sao P)
„ paratudo: *Simaba cedron* Planch.; *Simarubac.;* n. Bras. (Pará)
„ pereira: *Platycyamus regnelli* Bth.; *Papilionac.;* Bras. (Sao P)
„ pipú: *Piptadenia incuriale* Bth.; *Mimosac.;* Bras. (Sao P)
„ -saandri: *Licania heteromorpha* Bth., *L. spp.; Rosac.;* Sur.
„ visco: *Funtumia africana* Stapf; *Apocynac.;* W.-Af. (St. Thomé)
„ de moquem: *Uernonia scabra* Pers.; *Composit.;* Bras.
„ „ rosa: *Machaerium firmum* Bth.; *Papilionac.;* Bras.
„ „ santo: *Cabralea cangerana* Sald.; *Meliac.;* Bras. (Sao P)
Pauh kijang: *Irvingia malayana* Oliv.; *Simarubac.;* Malay.
Pauji: *Mouriria pseudo-geminata* Pitt.; *Melastomatac.;* Ven.
Pauk: *Butea frondosa* Roxb.; *Papilionac.;* Burma.
Paukarmul: *Stereospermum suaveolens* DC.; *Bignoniac.;* C.-O.-Ind.
Paukatea: *Libocedrus doniana* Endl.; *Conifer.;* N.-Seel., Tasm.
„ : *Libocedrus bidwilli* Hook. f.; *Conifer.;* N.-Seel.
Paukkyan: *Dolichandrone stipulata* Bth.; *Bignoniac.;* Burma.
Paulati: *Oldfieldia africana* Bth.; *Euphorbiac.;* tr. W.-Af.
Paulétoe: *Piratinera guianensis* Aubl.; *Morac.;* Sur.
„ aledin: *Helicostylis poeppigiana* Tréc.; *Morac.;* Sur.
„ basra: *Helicostylis poeppigiana* Tréc.; *Morac.;* Sur.
Paumiprwah: *Stephegyne parvifolia* Korth.; *Rubiac.;* Burma.
Paungnyit: *Calophyllum inophyllum* L.; *Guttifer.;* Burma (s. Tenasserim)
Pauwebloem: *Caesalpinia pulcherrima* Swartz; *Caesalpiniac.;* Sur.
Pava: *Didymopanax morototoni* Dcne. et Pl.; *Araliac.;* C.-R.
Pavier blanc: *Ilex casiquiarensis* Loes.; *Aquifoliac.;* fr. Gu.
„ de California: *Pavia californica* Torr.; *Hippocastanac.;* w. N.-Am.
Pavito: *Jacaranda sp.; Bignoniac.;* Kol.
Pawleng: *Sterculia villosa* Roxb.; *Sterculiac.;* Burma.
Pawpan: *Butea frondosa* Roxb.; *Papilionac.;* Burma.
Pawpaw: *Asimina triloba* Dunal; *Anonac.;* USA.
„ : *Carica papaya* L.; *Caricac.;* ö. Bras.
Payan: *Prunus puddum* Roxb.; *Rosac.;* nw. O.-Ind. (Kumaon)
„ -utis: *Betula alnoides* Ham.; *Betulac.;* nw. O.-Ind. (Kumaon)
Payana: *Boswellia serrata* Roxb.; *Burserac.;* O.-Ind. (Malabar)
Payandi: *Pithecolobium sp.; Mimosac.;* Kol.
Payani: *Uateria indica* L.; *Dipterocarpac.;* sw .O.-Ind. (Malabar)
Pay-ápa: *Celtis sp.; Ulmac.;* Phil. (Bataan)
Payiná: *Uatica mangachapoi* Blco.; *Dipterocarpac.;* Phil. (Tayabas)
„ : *Kingiodendron alternifolium* Merr.; *Papilionac.;* Phil.
„ : *Sindora supa* Merr.; *Caesalpiniac.;* Phil. (Camarines, Albay, Sorsogon)
Payomko: *Tetrameles nudiflora* R. Br.; *Datiscac.;* nö. O.-Ind. (Nepal, Sikkim, Bhut.)
Payon: *Rhizophora mucronata* Lam.; *Rhizophorac.;* Burma (s. Tenasserim)
„ -ama: *Rhizophora mucronata* Lamk.; *Rhizophorac.;* Burma.
Pazinznyo: *Uitex peduncularis* Wall.; *Uerbenac.;* Burma.
P'dada: *Sonneratia acida* L. f.; *Sonneratiac.;* Malay.
Pdick: *Anisoptera cochinchinensis* Pierre; *Dipterocarpac.;* Ind.-Ch.
Pé: *Funtumia africana* Stapf; *Apocynac.;* Elf.

Pé làn: *Canarium commune* L.; *Burserac.;* Annam.
„ lin: *Funtumia africana* Stapf; *Apocynac.;* Elf.
„ mou: *Toona sinensis* Roem.; *Meliac.;* Ch.
„ „ : *Fokienia hodginsi* Henry et Thomas, *F. kawaî* Hay.; *Conifer.;* Ind.-Ch.
„ -ohn: *Polyalthia oliveri* Engl. et Diels; *Anonac.;* Lib.
„ -wan: *Berrya ammonilla* Roxb.; *Tiliac.;* Burma.
„ de boi: *Bauhinia macrostachya* Bth.; *Caesalpiniac.;* Bras.
„ „ perdiz: *Simaruba versicolor* St. Hil.; *Simarubac.;* Bras.
Pea, Congo- (e): *Cajanus indicus* Spreng.; *Papilionac.;* Tr.
„ , Pigeon-: *Cajanus indicus* Spreng.; *Papilionac.;* Tr.
„ tree: *Caragana arborescens* Lam.; *Papilionac.;* Sib., Mandschurei
Peach: *Prunus persica* S. et Z.; *Rosac.;* Sub-Tr.
„ , Native-: *Fusanus acuminatus* R. Br.; *Santalac.;* sö. Au. (Victoria)
„ , Negro-: *Sarcocephalus esculentus* Afzel.; *Rubiac.;* W.-Af.
„ , Sierra Leone-: *Sarcocephalus esculentus* Afzel.; *Rubiac.;* Lib.
„ wood: *Haematoxylon campechianum* L., *H. brasiletto* Karst.; *Caesalpiniac.;*
„ „ : *Caesalpinia echinata* Lam.; *Caesalpiniac.;* Bras. [tr. Am.
Peacock tree: *Adenanthera pavonina* L.; *Mimosac.;* Go.
Pear: *Pirus communis* L.; *Rosac.;* Eu.
„ : *Persea gratissima* Gaertn.; *Laurac.;* Jam.
„ , Alligator- (e): *Persea americana* Mill.; *Laurac.;* tr. Am.
„ , Avocado- (e): *Persea americana* Mill.; *Laurac.;* tr. Am.
„ , „ : *Persea americana* Mill. *var. drymifolia* Mez; *Laurac.;* Mex.
„ , Butter-: *Persea americana* Mill.; *Laurac.;* tr. Am.
„ , Garlic-: *Crataeva tapia* L.; *Capparidac.;* Trin.
„ , hard: *Olinia capensis* Klotzsch, *O. sp.; Oliniac.;* S.-Af. (Kap)
„ , „ : *Strychnos henningsi* Gilg; *Loganiac.;* S.-Af.
„ , indian, red: *Bursera serrata* Colebr.; *Burserac.;* O.-Ind.
„ , Native-: *Pachylobus edulis* G. Don; *Burserac.;* Nig.
„ , „ : *Xylomelum occidentale* R. Br.; *Proteac.;* w. Au.
„ , red: *Scolopia eckloni* Warb., *S. mundti* Warb.; *Flacourtiac.;* S.-Af.
„ , Thorn-: *Scolopia zeyheri* Szyszylowicz; *Flacourtiac.;* S.-Af. (Natal)
„ , white: *Apodytes dimidiata* E. Meyer; *Icacinac.;* S.-Af.
„ , wild: *Pirus communis* L.; *Rosac.;* Eu.
„ , „ : *Amelanchier canadensis* Medic.; *Rosac.;* ö. Kan.
„ , „ : *Persea sp.; Laurac.;* Jam.
„ , „ : *Persea schiedeana* Nees; *Laurac.;* br. Hond.
„ , „ : *Licania platypus* Fritsch; *Rosac.;* Pan.
„ , Wolf-: *Scolopia zeyheri* Szyszylowicz; *Flacourtiac.;* S.-Af. (Kap)
„ wood, white: *Pterocelastrus rostratus* Walp.; *Celastrac.;* S.-Af.
Pearlwood: *Nyssa spp.; Cornac.;* USA.
„ : *Pirus communis* L.; *Rosac.;* N.-Am., Hond.
„ : *Cassia apoucouita* Aubl.; *Caesalpiniac.;* Bras.
„ , african: *Mimusops lacera* Baker; *Sapotac.;* Kam., Gab.
„ , brown: *Lucuma amorphosperma* F. M. Bailey; *Sapotac.;* ö. Au.
Peau de letras: *Helicostylis poeppigiana* Tréc.; *Morac.;* Bras.
Pébé: *Monodora myristica* Dunal; *Anonac.;* Kam.
Pebedum: *Lovoa klaineana* Pierre; *Meliac.;* Go.
Pebôk: *Callicarpa arborea* Roxb.; *Verbenac.;* n. Burma (Ruby Mines)
Pec: *Pinus merkusi* Jungh. et de Vriese; *Conifer.;* Ind.-Ch.
Pecan nut: *Carya olivaeformis* Nutt.; *Juglandac.;* USA.
Peccary wood: *Caesalpinia vesicaria* L.; *Caesalpiniac.;* br. Hond.
Pecego: *Chytranthus manni* Hook. f.; *Sapindac.;* W.-Af. (St. Thomé)
Pecegueiro: *Chytrantus manni* Hook. f.; *Sapindac.;* W.-Af. (St. Thomé)
„ bravo: *Byrsonima spicata* Rich.; *Malpighiac.;* Bras.
„ do Matto: *Myrcianthes edulis* Berg; *Myrtac.;* s. Bras.
Pecha-da: *Diospyros kurzi* Hiern, *D. marmorata* Parker; *Ebenac.;* And.
Pêche des Nègres: *Sarcocephalus esculentus* Afzel.; *Rubiac.;* Elf.
Pèchi: *Funtumia elastica* Stapf; *Apocynac.;* Elf.
Pechiche: *Vitex gigantea* H. B. K.; *Verbenac.;* Ek.
Peckan nut: *Carya olivaeformis* Nutt.; *Juglandac.;* USA.

Pedada: *Sonneratia acida* L. f.; *Sonneratiac.;* Malay.
„ : *Sonneratia caseolaris* Engl.; *Sonneratiac.;* Phil. (Cotabato)
Pedaro: *Urandra sp.; Icacinac.;* br. Borneo (Sarawak)
Pedda: *Kydia calycina* Roxb.; *Malvac.;* O.-Ind. (Madras)
„ : *Ailanthus excelsa* Roxb.; *Simarubac.;* O.-Ind. (Madras)
„ -are: *Bauhinia purpurea* L.; *Caesalpiniac.;* O.-Ind. (Madras)
„ batava: *Cordia macleodi* Hook. f. et Thoms.; *Borraginac.;* O.-Ind. (Madras)
„ bikki: *Gardenia latifolia* Aiton; *Rubiac.;* O.-Ind. (Madras)
„ gomra: *Gmelina arborea* ·L.; *Verbenac.;* O.-Ind.
„ kalinga: *Dillenia indica* L.; *Dilleniac.;* sw. O.-Ind.
„ morali: *Buchanania angustifolia* Roxb.; *Anacardiac.;* s. O.-Ind. (Madras)
Peddachilka duduga: *Miliusa velutina* Hook. f. et Thoms.; *Anonac.;* O.-Ind. (Madras)
Peddah gomra: *Gmelina arborea* Roxb.; *Verbenac.;* O.-Ind., Malay.
Peddakalaguri: *Stereospermum suaveolens* DC.; *Bignoniac.;* O.-Ind. (Madras)
Peddaman: *Ailanthus excelsa* Roxb.; *Simarubac.;* O.-Ind. (Madras)
Peddanowli-eragu: *Holoptelea integrifolia* Planch.; *Ulmac.;* O.-Ind. (Madras)
Pedegi: *Pterocarpus marsupium* Roxb.; *Papilionac.;* O.-Ind. (Madras)
Pédis: *Garcinia venulosa* Choisy; *Guttifer.;* Phil. (Pampanga)
Pedra-hume-caá: *Myrcia sp.; Myrtac.;* n. Bras. (Amaz.)
Pedralejo: *Byrsonima crassifolia* H. B. K.; *Malpighiac.;* ö. C.-Am., n. S.-Am.
„ : *Curatella americana* L.; *Dilleniac.;* C.-Am., n. S.-Am.
Pedro Hernández: *Rhus juglandifolia* H. B. K.; *Anacardiac.;* Kol.
Pedu: *Ailanthus excelsa* Roxb.; *Simarubac.;* O.-Ind. (Madras)
Pee: *Ailanthus excelsa* Roxb.; *Simarubac.;* O.-Ind. (Madras)
Pee-doh: *Heinsia pulchella* K. Schum.; *Rubiac.;* Lib.
Peem: *Ceiba pentandra* Gaertn.; *Bombacac.;* Mex.
Peepul: *Ficus religiosa* L.; *Morac.;* O.-Ind.
Pèèr: *Ficus rostrata* Lamk. var. *urophylla* Wall.; *Morac.;* Java (Tjibodas)
Peer, Guyana-: *Couma guianensis* Aubl.; *Apocynac.;* fr. Gu.
Peetch-kitam: *Randia aculeata* L.; *Rubiac.;* br. Hond.
Pega-pega: *Befaria ledifolia* H. B. K.; *Ericac.;* Kol.
Pegamosco: *Befaria ledifolia* H. B. K.; *Ericac.;* Kol.
Pegblagei: *Heinsia pulchella* K. Schum.; *Rubiac.;* Lib.
Pegrecou: *Xylopia frutescens* Aubl.; *Anonac.;* Gu.
Pegrekoe: *Xylopia frutescens* Aubl., *X. salicifolia* H. B. K., *X. sp.; Anonac.;* Sur.
Péguéninga: *Acacia arabica* Willd.; *Mimosac.;* Sudan.
Pegui: *Caryocar tomentosum* Willd., *C. glabrum* Pers., *C. sp.; Caryocarac.;* Sur.
Pegya: *Pentadesma butyracea* Sabine; *Guttifer.;* Go.
Pehimbia-gass: *Filicium decipiens* Thw.; *Sapindac.;* O.-Ind. Cey.
Pehn: *Piper guineense* Schum. et Thonn.; *Piperac.;* Lib.
Pehuen: *Araucaria araucana* K. Koch; *Conifer.;* Chile.
Peigne macaque: *Apeiba ? aspera* Aubl.; *Tiliac.;* Bras., fr. Gu.
Peine: *Zanthoxylum pittieri* P. Wilson; *Rutac.;* Pan.
„ de mico: *Apeiba tibourbou* Aubl.; *Tiliac.;* Ven., Gu., Bras.
Peinecillo: *Apeiba tibourbou* Aubl.; *Tiliac.;* Pan.
Peini marum: *Vateria indica* L.; *Dipterocarpac.;* w. O.-Ind. (Deccan)
Peirah: *Buchanania latifolia* Roxb.; *Anacardiac.;* n. O.-Ind. (Oudh)
Peito de pomba: *Guarea sp.; Meliac.;* Bras.
Pejata: *Artocarpus hirsuta* Lamk.; *Morac.;* sw. O.-Ind. (Madras)
Pejiran: *Conceveiba guianensis* Aubl., *C. hostmanni* Bth.; *Euphorbiac.;* Sur.
Pejri padrai: *Melia azedarach* L.; *Meliac.;* O.-Ind. (Bombay)
Pek: *Pinus merkusi* Jungh. et de Vriese; *Conifer.;* Ind.-Ch.
Pekarakai: *Terminalia paniculata* Roth; *Combretac.;* sw. O.-Ind. (Madras)
Péké: *Irvingia gabonensis* Baill.; *Simarubac.;* W.-Af.
Pekea: *Caryocar tomentosum* Willd., *C. glabrum* Pers., *C. spp.; Caryocarac.;* Gu.
Pekeya: *Caryocar tomentosum* Willd.; *Caryocarac.;* Gu.
Péki (f): *Caryocar tomentosum* Willd., *C. butyrosum* Willd.; *Caryocarac.;* Gu.
Pekia: *Caryocar sp.; Caryocarac.;* fr. Gu.
Pekire: *Parinarium subcordatum* Oliv.; *Rosac.;* Togo.
Pekola batu: *Mimusops elengi* L.; *Sapotac.;* Malay.
Pela nariz: *Croton sp.; Euphorbiac.;* Hond.

Pelandjau: *Pentaspadon motleyi* Hook. f.; *Anacardiac.;* Bor., Sum., Neu-Guinea
Pelanggoengan: *Leucosyke capitellata* Wedd.; *Urticac.;* Java (Besoeki)
Pelas: *Ficus melinocarpa* Blume, *F. ampelos* Burm.; *Morac.;* Java.
„ -aloes: *Ficus pisifera* Wall.; *Morac.;* Java (Zuid-Bagelen)
„ -kebo: *Ficus parietalis* Blume; *Morac.;* Java (Soebah)
„ -krisik: *Ficus gibbosa* Blume; *Morac.;* Java.
„ -rambat: *Ficus subulata* Blume; *Morac.;* Java (Gadoengan)
Pelawan: *Tristania spp.; Myrtac.;* Malay.
Pélé: *Securidaca longipedunculata* Fresen.; *Polygalac.;* Gab. (Chari-Logoue)
Pele: *Streblus asper* Lour.; *Morac.;* Java.
„ labad: *Ficus punctata* Thunb., *F. punctata* Thunb. *var. falcata* King; *Morac.:*
Peleman: *Daphniphyllum glaucescens* Blume; *Euphorbiac.;* Java (Takoka)   [Java.
? Pelika patiara sipioro: *Tetragastris balsamifera* O. Ktze.; *Burserac.;* Sur.
„   „   „   „ : *Protium spp.; Burserac.;* Sur.
Pelleth: *Kleinhovia hospita* L.; *Sterculiac.;* Nied.-Ind.
Pellin: *Nothofagus obliqua* Mirb.; *Fagac.;* Chile.
Pelo rebi: *Eugenia perorebi* Parodi; *Myrtac.;* Arg.
„ de Indio: *Hirtella americana* L.; *Rosac.;* Nic.
Peloes: *Harpullia imbricata* Thw.; *Sapindac.;* Java.
Pelonquahuitl: *Schinus molle* L.; *Anacardiac.;* Mex.
Pelos: *Harpullia imbricata* Thw.; *Sapindac.;* Java.
Peloto: *Casearia spiralis* Johnston; *Flacourtiac.;* Kol.
Pelou: *Dillenia ovata* Wall.; *Dilleniac.;* Kamb.
„ phnom: *Dillenia elata* Pierre; *Dilleniac.;* Ind.-Ch.
„ pnom: *Dillenia elata* Pierre; *Dilleniac.;* Kamb.
Pelù: *Sophora tetraptera* J. Mill.; *Papilionac.;* Chile.
Peludilla: *Guettarda uruguensis* Cham. et Schl.; *Rubiac.;* s. Bras.
Pem: *Millettia rhodantha* Baill.; *Papilionac.;* Go.
Pemba: *Parinarium excelsum* Sabine; *Rosac.;* Gab.
Pemou: *Fokienia hodginsi* Henry et Thomas; *Conifer.;* Ind.-Ch.
Pempri: *Gardenia latifolia* Aiton; *Rubiac.;* sw. O.-Ind. (Malabar)
Pen: *Mallotus albus* Muell. Arg.; *Euphorbiac.;* Ind.-Ch.
Penaga: *Mesua ferrea* L.; *Guttifer.;* Malay.
„ : *Calophyllum inophyllum* L.; *Guttifer.;* Sum.
„ ayer: *Calophyllum spp.; Guttifer.;* Malay.
„ laut: *Calophyllum inophyllum* L.; *Guttifer.;* Malay.
Penak: *Balanocarpus heimi* King, *B. maximum* King; *Dipterocarpac.;* Malay.
„ : *Hopea sp.; Dipterocarpac.;* Malay (Temerloh, Pekan)
„ bunga: *Balanocarpus heimi* King; *Dipterocarpac.;* Malay.
„ sabut: *Balanocarpus heimi* King; *Dipterocarpac.;* Malay.
„ tembaga: *Balanocarpus heimi* King; *Dipterocarpac.;* Malay.
Penang lawyer: *Licuala acutifida* Mart.; *Palmac.;* Malay.
Penanga: *Calophyllum inophyllum* L.; *Guttifer.;* Celebes.
Penarahan: *Myristica maingayi* Hook. f., *M. spp.; Myristicac.;* Malay.
Penawar-beas: *Cyclostemon mucronatus* Blume; *Euphorbiac.;* Java.
„   „ : *Taxotropis javanica* Blume; *Morac.;* Java.
Pencil wood: *Juniperus occidentalis* Hook.; *Conifer.;* nw. N.-Am.
„   „ : *Juniperus barbadensis* L.; *Conifer.;* sö. USA., W.-Ind.
Penda: *Xanthostemon oppositifolius* F. M. Bailey; *Myrtac.;* sö. Au.
Pendejo: *Bupleurum spinosum* L.; *Umbellifer.;* Sp.
Pendjalinan: *Cupania helferi* Hiern., *C. lessertiana* Cambess.; *Sapindac.;* Java.
„   : *Arytera littoralis* Blume, *A. spp.; Sapindac.;* Java (Pringamba)
„   : *Harpullia cupanioides* Roxb.; *Sapindac.;* Java (Soebah)
„   : *Mallotus philippinensis* Muell. Arg.; *Euphorbiac.;* Java (Soebah)
„   : *Gironniera cuspidata* Kurz; *Ulmac.;* Java (Djenggolo)
„   : *Celtis wighti* Planch.; *Ulmac.;* Nied.-Ind.
Pendo: *Memecylon lopezianum* Chev.; *Melastomatac.;* Gab.
Péndola: *Citharexylum spinosum* L.; *Uerbenac.;* P.-R.
Pendoorn: *Celastrus rhombifolius* Eckl. et Zeyh.; *Celastrac.;* S.-Af. (Kap)
Pendra: *Randia dumetorum* Lamk.; *Rubiac.;* C.-O.-Ind.
„ , Safed: *Gardenia turgida* Roxb.; *Rubiac.;* C.-O.-Ind. (Central Prov.)

Péndula: *Vitex sp.; Verbenac.;* P.-R.
„ blanca: *Vitex divaricata* Sw.; *Verbenac.;* P.-R.
Pène: *Xylopia quintasi* Engl. et Diels; *Anonac.;* Kam.
Pengarawan: *Hopea spp.; Dipterocarpac.;* Malay.
„ penak: *Hopea spp.; Dipterocarpac.;* Malay.
Pengji: *Bridelia retusa* Spreng.; *Euphorbiac.;* nö. O.-Ind. (Nepal, Sikkim, Bhutan)
Pengoeng: *Ficus obtusa* Hassk.; *Morac.;* Java (Loemadjan)
Pengoh: *Laportea peltata* Gaudich.; *Urticac.;* Java (Madioen)
Pening pening: *Pasania spp., Quercus spp.; Fagac.;* Malay.
Peniow: *Shorea grandiflora* Brandis; *Dipterocarpac.;* br. Borneo (Sarawak)
Penipeniche: *Hippomane mancinella* L.; *Euphorbiac.;* Cuba.
Penitente: *Petraea glandulosa* Pittier; *Verbenac.;* Ven.
Penjan: *Excoecaria virgata* Zoll. et Mor.; *Euphorbiac.;* Java (Kedoengdjati)
Penku: *Harrisonia occidentalis* Engl.; *Simarubac.;* Go.
Penkwa: *Entandrophragma macrophyllum* Harms, *E. septentrionale* Chev.; *Meliac.;*
„ : *Lovoa klaineana* Pierre; *Meliac.;* Go.   [Go.
Pennagah: *Mesua ferrea* L.; *Guttifer.;* O.-Ind.
„ : *Calophyllum inophyllum* L.; *Guttifer.;* Malay.
Pennan kottai: *Sapindus laurifolius* Vahl; *Sapindac.;* w. O.-Ind. (Deccan)
Penoga: *Calophyllum sp.; Guttifer.;* Sur.
Penre: *Michelia excelsa* Blume; *Magnoliac.;* nö. O.-Ind. (Nepal, Sikkim, Bhutan)
Pentamón: *Taxodium mucronatum* Tenore; *Conifer.;* Mex.
Pentamu: *Taxodium mucronatum* Tenore; *Conifer.;* Mex.
Pente de macaco: *Apeiba spp.; Tiliac.;* Bras. (Amaz.)
„ „ „ preto: *Apeiba petoumo* Aubl.; *Tiliac.;* Bras.
Penticosia: *Hypericum lanceolatum* Lamk.; *Guttifer.;* Réunion.
Penyiau: *Shorea grandiflora* Brandis; *Dipterocarpac.;* br. Borneo (Sarawak)
Penza: *Dialium yambataense* Verm.; *Caesalpiniac.;* B.-K.
„ ndombe: *Parinarium glabrum* Oliv.; *Rosac.;* B.-K.
Peo: *Engelhardtia chrysolepis* Hance; *Juglandac.;* Ind.-Ch.
„ : *Tobagoa maleolens* Urban; *Rubiac.;* Ven.
Peon: *Calophyllum angustifolium* Roxb.; *Guttifer.;* Malay (Penang)
Peonía: *Erythrina sp.; Papilionac.;* Mex.
„ : *Ormosia fastigiata* Tul.; *Papilionac.;* Ven.
Pepa: *Morinda roioc* L.; *Rubiac.;* Ven.
Pepapati: *Detarium microcarpum* Guill. et Perr.; *Caesalpiniac.;* Togo.
Pepe: *Glochidion cyrtostylum* Miq.; *Euphorbiac.;* Java (Boekanegara)
„ : *Afzelia africana* Smith; *Caesalpiniac.;* Go.
„ cacáo: *Luehea speciosa* Willd.; *Tiliac.;* Mex.
„ nance: *Ximenia americana* L.; *Proteac.;* Salv.
? „ „ „ : *Zygia recordi* Britt. et Rose; *Mimosac.;* Guat.
Pépé sia: ? *Phyllanthus discoideus* Muell. Arg.; *Euphorbiac.;* Elf.
Pepedom: *Entandrophragma spp., Lovoa klaineana* Pierre; *Meliac.;* Go
Pepedoum: *Lovoa klaineana* Pierre; *Meliac.;* Go.
Pépépéré votomité: *Clusia alba* Jacq.; *Guttifer.;* fr. Gu., Sur.
Pépésia: *Phyllanthus discoideus* Muell. Arg.; *Euphorbiac.;* Elf.
Pepete: *Inga sp.; Mimosac.;* Salv.
Pepetillo: *Inga sp.; Mimosac.;* Salv.
Pepeto: *Inga sp.; Mimosac.;* Salv.
Pepita de San Ignacio: *Hura polyandra* Baill.; *Euphorbiac.;* Mex.
Pepito: *Inga sp.; Mimosac.;* Salv.
Pepper bark: ? *Drypetes ovata* Hutch.; *Euphorbiac.;* Lib.
„ berry tree: *Cryptocarya obovata* R. Br.; *Laurac.;* Au.
„ „ , North Queensland-: *Cryptocarya obovata* R. Br. var. *hypospodia* F. v. M.;
„ , Bush-: *Xylopia staudti* Engl.; *Anonac.;* Lib.   [*Laurac.;* Au.
„ , Guinea-: *Xylopia aethiopica* A. Rich.; *Anonac.;* Lib.
„ , Negro-: *Xylopia aethiopica* A. Rich.; *Anonac.;* Nig.
Pepperidge (u): *Nyssa silvatica* Marsh.; *Cornac.;* ö. USA.
Peppermint: *Eucalyptus amygdalina* Labill., *E. spp.; Myrtac.;* sö. Au., Tasm.
„ , black: *Eucalyptus amygdalina* Labill.; *Myrtac.;* Tasm.
„ , blue: *Eucalyptus risdoni* Hook. var. *elata* Dehnh.; *Myrtac.;* Tasm.

Peppermint, broad-leaved: *Eucalyptus dives* Schauer; *Myrtac.;* ö. Au.
„ , brown: *Eucalyptus amygdalina* Labill.; *Myrtac.;* Au.
„ , Mountain-: *Eucalyptus coccifera* Hook. f.; *Myrtac.;* Tasm.
„ , narrow-leaved: *Eucalyptus radiata* Sieber, *E. spp.; Myrtac.;* ö. Au., Tasm.
„ , Sydney-: *Eucalyptus piperita* Smith; *Myrtac.;* ö. Au. (N.-S.-W.)
„ , white: *Eucalyptus amygdalina* Labill., *E. linearis* Dehnh.; *Myrtac.;* Au., Tasm.
„ tree: *Schinus molle* L.; *Anacardiac.;* tr. Am., Nig.
„ „ : *Drimys axillaris* Forst.: *Magnoliac.;* N.-Seel.
„ wood: *Umbellularia californica* Nutt.; *Laurac.;* w. N.-Am.
., „ : *Zanthoxylum clava-herculis* L., *Z. sp.; Rutac.;* N.-Am., W.-Ind.
„ „ : *Licaria guianensis* Aubl.; *Laurac.;* fr. Gu.
Pepuli: *Zanthoxylum rhetsa* DC.; *Rutac.;* s. O.-Ind.
Pequi: *Caryocar tomentosum* Willd., *C. spp.; Caryocarac.;* Gu., Bras.
Pequiá: *Caryocar villosum* Pers.; *Caryocarac.;* n. Bras. (Amaz.)
„ : *Aspidosperma tomentosum* Mart.; *Apocynac.;* Bras. (Paraná)
„ amarella: *Aspidosperma tomentosum* Mart.; *Apocynac.;* Bras.
„ brava: *Caryocar brasiliense* St. Hil.; *Caryocarac.;* s. Bras.
„ marfim: *Balfourodendron riedelianum* Engl.; *Rutac.;* Bras. (Sao P)
„ „ : *Aspidosperma eburneum* Allem.; *Apocynac.;* s. Bras.
Pequy: *Caryocar sp.; Caryocarac.;* fr. Gu.
Pera: *Pirus communis* L.; *Rosac.;* Hond.
„ : *Couma guianensis* Aubl.; *Apocynac.;* Sur.
Perabisi: *Maprounea guianensis* Aubl.; *Euphorbiac.;* Sur.
Perah: *Elateriospermum tapos* Blume; *Euphorbiac.;* Malay.
Peral silvestre: *Pirus communis* L.; *Rosac.;* Sp.
Peralejo: *Byrsonima crassifolia* H. B. K.; *Malpighiac.;* Cuba, Kol.
„ : *Curatella americana* L.; *Dilleniac.;* Kol.
„ macho: *Byrsonima sp.; Malpighiac.;* Cuba.
„ del monte: *Byrsonima sp.; Malpighiac.;* Cuba.
Peralu: *Ficus bengalensis* L.; *Morac.;* sw. O.-Ind. (Malabar)
Peramán: *Moronobea coccinea* Aubl.; *Guttifer.;* Ven.
Peranjoli: *Hymenodictyon excelsum* Wall.; *Rubiac.;* O.-Ind. (Madras)
Perdrix: *Cassia siamea* Lam.; *Caesalpiniac.;* Ind.-Ch.
„ : *Vouacapoua americana* Aubl.; *Caesalpiniac.;* fr. Gu
„ : *Swartzia tomentosa* DC.; *Caesalpiniac.;* fr. Gu.
Péré: *Avicennia nitida* Jacq.; *Verbenac.;* Sur.
„ -péré: *Clusia alba* L.; *Guttifer.;* Sur.
Pereira: *Platycyamus regnelli* Bth.; *Papilionac.;* Bras. (Sao P)
„ amarella: *Platycyamus regnelli* Bth.; *Papilionac.;* Bras. (Sao P)
„ vermelha: *Platycyamus regnelli* Bth.; *Papilionac.;* Bras.
Pereiro: *Platycyamus regnelli* Bth.; *Papilionac.;* Bras.
Pereko: *Lovoa klaineana* Pierre; *Meliac.;* Nig.
Péréng: *Ficus subulata* Blume, *F. pisifera* Wall., *F. spp.; Morac.;* Java.
Peréng: *Uvaria chamae* Beauv.; *Anonac.;* Togo.
Perepat: *Sonneratia apetala* Buch.-Ham.; *Sonneratiac.;* br. Malay.
Pérépéré: *Clusia sp.; Guttifer.;* Gu.
Pereten: *Lovoa klaineana* Pierre; *Meliac.;* Nig.
Peri: *Mesua ferrea* L.; *Guttifer.;* sw. O.-Ind. (Travancore)
Perian: *Artocarpus rigida* Blume; *Morac.;* Malay.
Perico: *Thouinidium decandrum* Blake; *Sapindac.;* Mex.
Pericoa: *Erythrina sp.; Papilionac.;* Ven.
Pericoco: *Erythrina sp.; Papilionac.;* Ven.
Pericopsis wood: *Pericopsis mooniana* Thw.; *Papilionac.;* Borneo.
Perileo: *Myrcia splendens* DC.; *Myrtac.,* Kol.
Perilla: *Acnistus australis* Gr.; *Solanac.;* tr. Am.
Periquiteira: *Cochlospermum insigne* St. Hil., *C. spp.; Bixac.;* n. Bras. (Amaz.)
Péris: *Garcinia venulosa* Choisy; *Guttifer.;* Phil. (Batangas, Rizal)
Perkoso: *Croton argyratus* Blume; *Euphorbiac.;* Java (Telemojo, Ardjoeno)
Pernambucco (i): *Caesalpinia echinata* Lam.; *Caesalpiniac.;* ö. Bras.
Pernambuco wood (u): *Caesalpinia echinata* Lam.; *Caesalpiniac.;* ö. Bras.
Pernambukholz (d): *Caesalpinia echinata* Lam.; *Caesalpiniac.;* Bras.

Pernilla de casa: *Erythrina rubrinervis* H. B. K.; *Papilionac.*; Pan.
Pero: *Borassus flabellifer* L.; *Palmac.*; Togo.
„ rebi: *Eugenia perorebi* Parodi; *Myrtac.*; Arg.
Peroba: *Paratecoma peroba* Kuhlmann; *Bignoniac.*; Mitt.- & S.-Bras.
„ : *Sapota gonocarpa* Mart. et Eichl.; *Sapotac.*; Bras.
„ : *Aspidosperma gomezianum* A. DC., *A. polyneuron* Muell. Arg., *A. spp.; Apo-*
„ amarella: *Paratecoma peroba* Kuhlmann; *Bignoniac.*; Bras.   *[cynac.*; Bras.
„ „ : *Aspidosperma tomentosum* Mart.; *Apocynac.*; Bras. (Paraná)
„ assú: *Aspidosperma peroba* Allem.; *Apocynac.*; Bras. (Sao P)
„ branca: *Paratecoma peroba* Kuhlmann; *Bignoniac.*; Bras.
„ graúda: *Aspidosperma peroba* Allem., *A. polyneuron* Muell. Arg.; *Apocynac.*;
„ „ amarella: *Aspidosperma peroba* Allem.; *Apocynac.*; Bras.   [Bras. (Sao P)
„ „ parda: *Aspidosperma peroba* Allem., *A. gomezianum* A. DC.; *Apocynac.*;
„ „ preta: *Aspidosperma peroba* Allem.; *Apocynac.*; Bras. (Sao P) [Bras. (Sao P)
„ „ rosa: *Aspidosperma polyneuron* Muell. Arg.; *Apocynac.*; Bras. (Sao P)
„ „ vermelha: *Aspidosperma peroba* Allem.; *Apocynac.*; Bras. (Sao P)
„ mirim: *Sweetia sp.; Papilionac.*; Bras.
„ miuda: *Aspidosperma sp.; Apocynac.*; Bras. (Sao P)
„ poca: *Aspidosperma sp.; Apocynac.*; Bras. (Sao P)
„ preta: *Aspidosperma sp.; Apocynac.*; Bras. (Sao P)
„ rajada: *Aspidosperma polyneuron* Muell. Arg.; *Apocynac.*; Bras. (Sao P)
„ reseca: *Paratecoma peroba* Kuhlmann; *Bignoniac.*; Bras.
„ reversa: *Aspidosperma sp.; Apocynac.*; Bras. (Sao P)
„ revessa: *Aspidosperma sp.; Apocynac.*; Bras. (Sao P)
„ rosa: *Aspidosperma peroba* Allem.; *Apocynac.*; Bras.
„ „ : *Aspidosperma polyneuron* Muell. Arg.; *Apocynac.*; Bras. (Sao P)
„ „ : *Aspidosperma dasycarpon* A. DC.; *Apocynac.*; Bras. (Pará)
„ tremida: *Aspidosperma sp.; Apocynac.*; Bras. (Sao P)
„ verdadeira: *Paratecoma peroba* Kuhlmann; *Bignoniac.*; Bras.
„ vermelho: *Aspidosperma sp.; Apocynac.*; Bras. (Pará)
„ de Campos: *Paratecoma peroba* Kuhlmann; *Bignoniac.*; Bras.
„ do Bahia: *Sweetia sp.; Papilionac.*; Bras. (Bahia)
„ „ campo: *Aspidosperma sp.; Apocynac.*; Bras. (Sao P)
„ „ „ : *Paratecoma peroba* Kuhlmann; *Bignoniac.*; Bras. (R.-J., Bahia)
Perobinha: *Sweetia elegans* Bth.; *Papilionac.*; s. Bras. (Sao P)
„ tigre: *Sweetia elegans* Bth.; *Papilionac.*; Bras. (Sao P)
Perokoti: *Helicostylis poeppigiana* Tréc.; *Morac.*; Sur.
Peronilla: *Erythrina rubrinervia* H. B. K.; *Papilionac.*; n. Kol.
Peronío: *Pithecolobium unguis-cati* Mart.; *Mimosac.*; n. Kol.
Peropong: *Adina rubescens* Hemsl.; *Rubiac.*; Malay.
Perovaúina: *Melanoxylon brauna* Schott; *Caesalpiniac.*; Bras. (Sao P)
Perovaúna: *Melanoxylon brauna* Schott; *Caesalpiniac.*; Bras. (Sao P)
Perquetano: ? *Moquilea sp.; Rosac.*; Kol.
Perroquet à petites feuilles rouges: *Olax wightiana* Wall.; *Olacac.*; Maur., Réu.
Persicon: *Juglans regia* L.; *Juglandac.*; Griech.
Persil indien: *Piper sp.; Piperac.*; fr. Gu.
Persimmon (d, u): *Diospyros virginiana* L.; *Ebenac.*; ö. N.-Am.
„ , black: *Diospyros texana* Scheele; *Ebenac.*; s. USA., Mex.
„ , common: *Diospyros virginiana* L.; *Ebenac.*; USA.
„ , echtes: *Diospyros virginiana* L.; *Ebenac.*; ö. N.-Am.
„ , grey: *Diospyros pentamera* F. v. M.; *Ebenac.*; Au.
„ , japanese: *Diospyros kaki* L. f.; *Ebenac.*; Jap.
„ , mexikanisches: *Diospyros texana* Scheele; *Ebenac.*; s. USA., Mex.
„ , schwarzes: *Diospyros texana* Scheele; *Ebenac.*; s. USA., Mex.
Persimon (d): *Diospyros virginiana* L.; *Ebenac.*; ö. N.-Am.
„ (e): *Diospyros spp., Maba spp.; Ebenac.*; Tr.
Perta: *Aegialitis rotundifolia* Roxb.; *Plumbaginac.*; Burma (s. Tenasserim)
Peru: *Ailanthus excelsa* Roxb.; *Simarubac.*; O.-Ind. (Madras)
„ : *Schinus molle* L.; *Anacardiac.*; Mex.
Perua: *Buchanania latifolia* Roxb.; *Anacardiac.*; n. O.-Ind. (Oudh)
Perulu: *Randia dumetorum* Lamk.; *Rubiac.*; sw. O.-Ind. (Bombay)

Perum piney: *Vateria indica* L.; *Dipterocarpac.*; sw. O.-Ind. (Malabar)
Perunkuruntu: *Atalantia monophylla* Correa; *Rutac.*; O.-Ind. (Madras)
Pérusse: *Tsuga canadensis* Carr.; *Conifer.*; ö. Kan.
Pervinca: *Vinca minor* L.; *Apocynac.*; Sp.
Perzha: *Careya arborea* Roxb.; *Myrtac.*; sw. O.-Ind. (Malabar)
Pesang: *Polyalthia jenkinsi* Bth. et Hook. f.; *Anonac.*; Malak.
Pesigueiro bravo: *Prunus sphaerocarpa* Sw.; *Rosac.*; Arg. (Misiones)
Pésin: *Funtumia africana* Stapf; *Apocynac.*; Elf.
Peso: *Blighia sapida* Koenig; *Sapindac.*; Togo.
Pesse: *Picea excelsa* Link; *Conifer.*; Eu.
Pestano mula: *Heliocarpus appendiculatus* Turcz.; *Tiliac.*; Nic.
Peste à poux: *Picrasma excelsa* Planch.; *Simarubac.*; Guad.
Peta: *Myrceugenia planipes* Berg; *Myrtac.*; Chile.
Petai: *Parkia speciosa* Hassk.; *Mimosac.*; Malay, Sum., Java.
Petakara: *Chrysophyllum roxburghi* G. Don; *Sapotac.*; O.-Ind., Sunda.
Petalang: *Dryobalanops oblongifolium* Dyer; *Dipterocarpac.*; Sum.
Petalieng: *Strombosia sp.*; *Olacac.*; Nied.-Ind.
Petaling: *Ochanostachys amentacea* Mast., *O. bancana* Valeton; *Olacac.*; Malay, Sunda.
   „ ayer: *Homalium frutescens* King; *Flacourtiac.*; Malay.
   „ gajah: *Strombosia rotundifolia* King; *Olacac.*; Malay.
Petaquilla: *Tabernaemontana sp.*; *Apocynac.*; Ek.
Petari: *Trewia nudiflora* L.; *Euphorbiac.*; n. O.-Ind. (United Prov.)
Petarkura: *Gynocardia odorata* R. Br.; *Flacourtiac.*; Beng.
Pétembé: *Coula edulis* Baill.; *Olacac.*; Elf.
Pétépré: *Calpocalyx brevibracteatus* Harms; *Mimosac.*; Elf.
Peteprebi: *Gardenia ternifolia* Schum. et Thonn.; *Rubiac.*; Go.
Peterebí: *Cordia alliodora* Cham.; *Borraginac.*; sw. Bras.
   „ : *Cordia sp.*; *Borraginac.*; Arg. (Misiones)
   „ hú: *Cordia alliodora* Cham.; *Borraginac.*; Arg.
Petereby: *Cordia sp.*; *Borraginac.*; Arg. (Misiones)
Peteribi moroti: *Sida densiflora* Hook. et Arn.; *Malvac.*; Arg.
Petir: *Sindora intermedia* Baker; *Caesalpiniac.*; Borneo.
Petit bouis: *Bumelia sp.*; *Sapotac.*; fr. W.-Ind.
Petites feuilles: *Eugenia crassilana* ?; *Myrtac.*; Mart.
Petitia: *Petitia domingensis* Jacq.; *Verbenac.*; Jam.
Petlé-zin: *Vitex peduncularis* Wall.; *Verbenac.*; Burma.
Peto: *Dimorphandra mora* Bth. et Hook.; *Caesalpiniac.*; Sur.
Petoumo: *Apeiba sp.*; *Tiliac.*; Gu. (Galibis)
Petpuria: *Adina cordifolia* Hook. f.; *Rubiac.*; Beng.
Petra: *Myrceugenia pitra* Berg; *Myrtac.*; Chile.
Petshat: *Kydia calycina* Roxb.; *Malvac.*; Burma.
Petsut: *Engelhardtia spicata* Blume; *Juglandac.*; nw. Himal.
Petthan: *Dolichandrone stipulata* Bth.; *Bignoniac.*; Burma.
Petty-whin: *Ilex aquifolium* L.; *Aquifoliac.*; Eu., n. Af.
Petwoon: *Berrya ammonilla* Roxb.; *Tiliac.*; Burma.
Petwun: *Berrya ammonilla* Roxb.; *Tiliac.*; Burma, And.
Peu mou: *Fokienia hodginsi* Henry et Thomas; *Conifer.*; Ind.-Ch.
   „ nol: *Chisocheton globosus* Pierre; *Meliac.*; Ind.-Ch.
Peuh: *Hibiscus tiliaceus* L.; *Malvac.*; N.-Kal.
Peumo: *Cryptocarya peumus* Nees; *Laurac.*; Chile.
Peuplier: *Populus tremuloides* Michx.; *Betulac.*; Kan.
   „ baumier: *Populus balsamifera* L.; *Betulac.*; Kan.
   „ monalifère: *Populus deltoides* Marsh.; *Betulac.*; s. Kan.
   „ tremble: *Populus tremuloides* Michx.; *Betulac.*; Kan.
   „ à grandes dents: *Populus grandidentata* Michx.; *Betulac.*; sö. Kan.
   „ d'Amérique: *Coccoloba uvifera* Jacq.; *Polygonac.*; fr. Gu., fr. W.-Ind.
   „ d'Athènes: *Populus tremuloides* Michx.; *Betulac.*; Kan.
   „ de l'Ouest: *Populus trichocarpa* Torr. et Gray; *Betulac.*; w. Kan.
Peuris: *Aporosa microcalyx* Hassk.; *Euphorbiac.*; w. Java.
Peusar: *Artocarpus rigida* Blume; *Morac.*; Java.
Peuva: *Bignonia longiflora* Vell.; *Bignoniac.*; Bras. (Sao P)

Peuva: *Tecoma lapacho* K. Schum.; *Bignoniac.;* Bras. (Sao P)
Pey: *Ailanthus excelsa* Roxb.; *Simarubac.;* O.-Ind. (Madras)
Peypila: *Artocarpus hirsuta* Lamk.; *Morac.;* sw. O.-Ind. (Madras)
Pezzo: *Picea excelsa* Link; *Conifer.;* It.
Pfaffenhütchen: *Evonymus europaea* L.; *Celastrac.;* Eu.
Pfefferholz, brasilianisches: *Phoebe porosa* Mez; *Laurac.;* Bras.
Pfefferrohr: *Phyllostachys bambusoides* S. et Z.; *Gramin.;* Himal.
  „ , schwarzes: *Phyllostachys bambusoides* S. et Z.; *Gramin.;* Himal.
Pferdefleischholz (d): *Swartzia tomentosa* DC.; *Caesalpiniac.;* Gu.
  „   „ : *Mimusops balata* Gaertn., *M. spp.;* *Sapotac.;* Gu., n. Bras., Ven.
  „   : *Rhizophora mangle* L.; *Rhizophorac.;* tr. Am.
  „   : *Dolichandrone longissima* K. Schum.; *Bignoniac.;* O.-Ind.
Pferdepeitsche: *Luehea divaricata* Mart.; *Tiliac.;* s. Bras., n. Arg.
Pfirsichbaumholz: *Prunus persica* S. et Z.; *Rosac.;* Eu., As.
Pflaume: *Prunus domestica* L.; *Rosac.;* Eu., n. Pers.
  „ , Batoko-: *Flacourtia ramontchi* L'Hérit.; *Flacourtiac.;* O.-Af. (Zambesi)
  „ , Florida-: *Drypetes keyensis* Urban; *Euphorbiac.;* s. Fl., Bah.
  „ , Madagaskar-: *Flacourtia ramontchi* L'Hérit.; *Flacourtiac.;* Sey.
  „ , Maron-: *Flacourtia ramontchi* L'Hérit.; *Flacourtiac.;* Sey.
  „ , Mombin-, gelbe: *Spondias lutea* L.; *Anacardiac.;* S.-Am.
  „ , Schlehen-: *Prunus insititia* L.; *Rosac.;* ? Kauk.
Pfouey: *Alstonia congensis* Engl.; *Apocynac.;* Kam.
Pfriemengeisklee, weissblühender (d): *Cytisus multiflorus* Sweet-Bailey; *Papilionac.;*
Pfui: *? Fagara angolensis* Engl.; *Rutac.;* Lib.   [M.-M
Pha: *Randia exaltata* Pierre; *Rubiac.;* Ind.-Ch.
  „ may: *Aglaia gigantea* Pellegr.; *Meliac.;* Ind.-Ch.
Phaenila: *Sapindus laurifolius* Vahl; *Sapindac.;* O.-Ind.
Phai: *Duabanga sonneratioides* Buch.-Ham.; *Sonneratiac.;* Ind.-Ch.
Phaja: *Prunus puddum* Roxb.; *Rosac.;* n. O.-Ind.
Phalani: *Eugenia jambolana* Lamk.; *Myrtac.;* n. O.-Ind. (United Prov.)
Phaldu: *Stephegyne parvifolia* Korth.; *Rubiac.;* n. O.-Ind. (Garhwal)
  „ : *Hymenodictyon excelsum* Wall.; *Rubiac.;* n. O.-Ind. (United Prov.)
Phalinda: *Eugenia jambolana* Lamk.; *Myrtac.;* n. O.-Ind. (United Prov.)
Phals: *Populus ciliata* Wall.; *Salicac.;* nw. O.-Ind. (Punjab)
Phalsa: *Grewia tiliaefolia* Vahl; *Tiliac.;* n. O.-Ind. (United Prov.)
  „ : *Grewia elastica* Royle, *G. asiatica* L.; *Tiliac.;* C.-O.-Ind.
Phalwa: *Grewia elastica* Royle; *Tiliac.;* nw. O.-Ind. (Punjab)
Phamlet: *Machilus gammieana* King; *Laurac.;* nö. O.-Ind. (Nepal, Sikkim, Bhutan)
Phan: *Eugenia resinosa* Gagnep.; *Myrtac.;* Ind.-Ch.
  „ pim: *Paranephelium spirei* Lecomte; *Sapindac.;* Ind.-Ch.
  „ sat: *Payena elliptica* Pierre; *Sapotac.;* Ind.-Ch.
Phanat: *Quercus incana* Roxb.; *Fagac.;* n. O.-Ind. (Garhwal)
Phang: *Poinciana pulcherrima* L.; *Caesalpiniac.;* Ind.-Ch.
Phaong: *Calophyllum saigonense* Pierre; *Guttifer.;* Kamb.
Phara: *Grewia tiliaefolia* Vahl; *Tiliac.;* O.-Ind., Cey.
Pharas: *Butea frondosa* Roxb.; *Papilionac.;* C.-O.-Ind.
Pharat-singhali: *Quercus lamellosa* Smith; *Fagac.;* Beng.
Pharengala: *Acer oblongum* Wall.; *Acerac.;* Himal.
Pharsa: *Grewia tiliaefolia* Vahl; *Tiliac.;* n. O.-Ind. (United Prov.)
Pharsanyi: *Grewia elastica* Royle; *Tiliac.;* n. O.-Ind. (Garhwal)
Pharsawon: *Grewia elastica* Royle; *Tiliac.;* n. O.-Ind. (Garhwal)
Pharsia: *Grewia tiliaefolia* Vahl; *Tiliac.;* nw. O.-Ind. (Kumaon)
  „ : *Grewia elastica* Royle; *Tiliac.;* n. O.-Ind. (United Prov.)
Pharsuli: *Grewia elastica* Royle; *Tiliac.;* nw. O.-Ind. (Kumaon)
Pharwan: *Tamarix articulata* Vahl; *Tamaricac.;* nw. O.-Ind. (Punjab)
Phát: *Bischofia javanica* Blume; *Euphorbiac.;* Ind.-Ch.
Phatterphodi: *Bridelia retusa* Spreng.; *Euphorbiac.;* w. O.-Ind. (Bombay)
Phay: *Duabanga sonneratioides* Buch.-Ham.; *Sonneratiac.;* Ind.-Ch.
  „ : *Baccaurea cauliflora* L.; *Euphorbiac.;* Ind.-Ch.
  „ sùng: *Fokienia kawai* Hay., *F. hodginsi* Henry et Thomas; *Conifer.;* Ind.-Ch.
  „   „ : *Duabanga sonneratioides* Buch.-Ham.; *Sonneratiac.;* Ind.-Ch.

Phay vang môc: *Duabanga sonneratioides* Buch.-Ham.; *Sonneratiac.;* Ind.-Ch.
„ „ mop: *Duabanga sonneratioides* Buch.-Ham.; *Sonneratiac.;* Annam.
„ vi: *Sarcocephalus orientalis* Merr.; *Rubiac.;* Ind.-Ch.
„ „ : *Anthocephalus indicus* A. Rich.; *Rubiac.;* Ind.-Ch.
„ „ : *Duabanga sonneratioides* Buch.-Ham.; *Sonneratiac.;* Ind.-Ch.
Phaya: *Prunus puddum* Roxb.; *Rosac.;* n. O.-Ind. (United Prov.)
Phcai pruk: *Afzelia sp.; Caesalpiniac.;* Ind.-Ch.
Phcheck: *Shorea obtusa* Wall.; *Dipterocarpac.;* Ind.-Ch.
Phdiec: *Anisoptera cochinchinensis* Pierre; *Dipterocarpac.;* Ind.-Ch. (Kamb.)
„ so: *Anisoptera glabra* Kurz; *Dipterocarpac.;* Ind.-Ch. (Kamb.)
Phdieck: *Shorea hypochra* Hance; *Dipterocarpac.;* Ind.-Ch.
Pheasant wood (e): *Vouacapoua americana* Aubl.; *Caesalpiniac.;* Sur.
Phèn: *Schleichera trijuga* Willd.; *Sapindac.;* Ind.-Ch.
„ den: *Eugenia mékongensis* Gagnep.; *Myrtac.;* Ind.-Ch.
Phetra: *Gardenia turgida* Roxb.; *Rubiac.;* C.-O.-Ind.
Phetrak: *Gardenia turgida* Roxb.; *Rubiac.;* O.-Ind.
Pheu: *Protium serratum* Engl.; *Burserac.;* Ind.-Ch.
Phi: *Baccaurea sapida* Muell. Arg.; *Euphorbiac.;* Ind.-Ch.
Phkai pruk: *Afzelia sp.; Caesalpiniac.;* Ind.-Ch.
Phlong: *Ternstroemia penangiana* Choisy; *Theac.;* Ind.-Ch.
Phlu: *Dillenia aurea* Smith; *Dilleniac.;* Ind.-Ch. (Kamb.)
Phnel: *Vitex pubescens* Vahl; *Verbenac.;* Ind.-Ch.
Pho: *Ficus religiosa* L.; *Morac.;* Ind.-Ch.
Phòong: *Dillenia heterosepala* Finet et Gagnep.; *Dilleniac.;* Ind.-Ch.
Phootee: *Jacaranda copaia* D. Don; *Bignoniac.;* br. Gu.
Phú huong: *Litsea sp.; Laurac.;* Ind.-Ch.
„ luu: *Liquidambar formosana* Hance; *Hamamelidac.;* Ind.-Ch.
Phuj-pattra: *Betula alnoides* Buch.-Ham. *Betulac.;* n. O.-Ind. (United Prov.)
Phulai: *Acacia modesta* Wall.; *Mimosac.;* nw. O.-Ind. (Punjab)
Phulas: *Butea frondosa* Roxb.; *Papilionac.;* sw. O.-Ind. (Bombay)
Phullas: *Butea frondosa* Roxb.; *Papilionac.;* w. O.-Ind. (Gujarat)
Phum phát: *Bischofia javanica* Blume; *Euphorbiac.;* Ind.-Ch.
Phungali: *Excoecaria agallocha* L.; *Euphorbiac.;* sw. O.-Ind. (Bombay)
Phuòng: *Averrhoa carambola* L.; *Oxalidac.;* Ind.-Ch.
Phurkhona: *Adina cordifolia* Hook. f.; *Rubiac.;* Assam.
Physic, indian: *Magnolia fraseri* Walt.; *Magnoliac.;* USA.
„ nut, wild: *Jatropha gaumeri* Greenm.; *Euphorbiac.;* br. Hond.
Pi: *Ceiba pentandra* Gaertn. var. *dehiscens* Ulb.; *Bombacac.;* Sudan.
Piadák: *Xylocarpus muloccensis* M. Roem.; *Meliac.;* Phil. (Palawan)
Piagao: *Xylocarpus granatum* Koenig, *X. moluccensis* M. Roem.; *Meliac.;* Phil.
Piagau: *Xylocarpus granatum* Koenig, *X. moluccensis* M. Roem.; *Meliac.;* Phil.
Piago: *Xylocarpus moluccensis* M. Roem.; *Meliac.;* Phil.
Piak: *Alnus nepalensis* D. Don; *Betulac.;* O.-Ind.
Pial: *Buchanania latifolia* Roxb.; *Anacardiac.;* n. O.-Ind. (United Prov.)
Piaman: *Eugenia operculata* Roxb.; *Myrtac.;* n. O.-Ind. (United Prov.)
Pian pian: *Albizzia fastigiata* Oliv.; *Mimosac.;* Elf.
Piánga: *Bassia betis* Merr.; *Sapotac.;* Phil. (Cagayan, Isa.)
Piángan: *Bassia betis* Merr.; *Sapotac.;* Phil. (Cagayan, Isa.)
Pianggu: *Connariopsis spp.; Oxalidac.;* Malay, Borneo.
Pianro: *Duboscia macrocarpa* Bocq.; *Tiliac.;* Elf.
Piápi: *Avicennia alba* Blume; *Verbenac.;* Phil. (Misamis)
„ : *Avicennia officinalis* L.; *Verbenac.;* Phil. (Ilo-Ilo)
Piasal: *Pterocarpus marsupium* Roxb.; *Papilionac.;* w. Beng. (Orissa)
„ : *Terminalia tomentosa* Wight et Arn.; *Combretac.;* Beng.
Piassaba: *Attalea funifera* Mart.; *Palmac.;* n. Bras.
Picazu-rembiu: *Chrysophyllum sp.; Sapotac.;* Arg.
Picha picha: *Myrceugenia planipes* Berg; *Myrtac.;* Chile.
Piche: *Enterolobium cyclocarpum* Griseb.; *Mimosac.;* Mex.
Piché aboko: *Pentadesma leucantha* Chev.; *Guttifer.;* Elf.
Pichi: *Psidium guajava* L.; *Myrtac.;* br. Hond.
Pichipin: *Pinus sp.; Conifer.;* W.-Ind. (St.-D.)

Pichirina: *Vismia sp.; Guttifer.;* Peru.
Pichoco: *Erythrina sp.; Papilionac.;* Mex.
Pichuna: *Coccoloba sp.; Polygonac.;* Bras.
Pichwood (u): *Enterolobium cyclocarpum* Griseb.; *Mimosac.;* Mex., C.-Am.
Pichy pang: *Colubrina panamensis* Standl.; *Rhamnac.;* Pan.
Pidjietan: *Lansium domesticum* Jack; *Meliac.;* Sunda.
? Pie de venado: *Rinorea ulmifolia* Ktze.; *Violac.;* Kol.
Pied de poule: *Toddalia aculeata* Pers.; *Rutac.;* Mask.
Piegba: *Conopharyngia durissima* Stapf; *Apocynac.;* Elf.
Piengkoe: *Dysoxylum ramiflorum* Miq.; *Meliac.;* Nied.-Ind.
Piengo makka: *Guilielma speciosa* Mart.; *Palmac.;* Sur.
Pientrie: *Virola mycetis* Pulle; *Myristicac.;* Sur.
Pientro botrie: *Pouteria guianensis* Aubl.; *Sapotac.;* Sur.
Piesito de paloma: *Nectandra sanguinea* Rottb.; *Laurac.;* Mex.
Pieterselie, indiaansche: *Piper sp.; Piperac.;* fr. Gu.
Pietjoeng: *Pangium edule* Reinw.; *Flacourtiac.;* Sunda.
Pigeon berry: *Byrsonima sp.; Malpighiac.;* br. W.-Ind.
„ wood: *Coccoloba uvifera* L.; *Polygonac.;* Fl., W.-Ind.
„ „ : *Pisonia obtusata* Jacq.; *Nyctaginac.;* Fl., W.-Ind.
Pignut: *Carya glabra* Spach; *Juglandac.;* USA.
„ , little: *Carya microcarpa* Nutt.; *Juglandac.;* Kan.
Pigue pajaro: *Cassia occidentalis* L.; *Caesalpiniac.;* Nic.
Piim: *Ceiba pentandra* Gaertn., *C. aesculifolia* Britt. et Bak.; *Bombacac.;* Mex.
Pi-kopian: *Geniostoma oblongifolium* Koord. et Valet.; *Loganiac.;* Java (Pantjoer)
Pikpik-uág: *Artocarpus superba* Becc.; *Morac.;* Phil. (Basilan)
? „ uák: *Aglaia multifoliola* Merr.; *Meliac.;* Phil. (Basilan)
Piksík: *Avicennia officinalis* L.; *Verbenac.;* Phil. (Mindoro)
Pila: *Artocarpus hirsuta* Lamk.; *Morac.;* sw. O.-Ind. (Madras)
Pilachampa: *Michelia nilagirica* Zenker; *Magnoliac.;* sw. O.-Ind. (Bombay)
Pilahi: *Elaeocarpus tuberculatus* Roxb.; *Elaeocarpac.;* sw. O.-Ind. (Malabar)
Pilang: *Acacia leucophloea* Willd.; *Mimosac.;* Malay.
Piláuai: *Canarium luzonicum* A. Gray; *Burserac.;* Phil. (Tayabas)
Piláui: *Canarium luzonicum* A. Gray; *Burserac.;* Phil. (Tayabas)
Píldis: *Garcinia sp.; Guttifer.;* Phil. (Pampanga)
Píli: *Canarium luzonicum* A. Gray, *C. commune* L.; *Burserac.;* Phil.
„ -pili: *Aglaia harmsiana* Perk.; *Meliac.;* Phil. (Camarines)
„ vagei: *Albizzia stipulata* Boivin; *Mimosac.;* sw. O.-Ind. (Madras)
Pilíg: *Livistona rotundifolia* Mart.; *Palmac.;* Phil. (Pol.)
„ : *Livistona rotundifolia* Mart. var. *luzonensis* Becc.; *Palmac.;* Phil.
Pilinsuchil: *Bombax sp.; Bombacac.;* Salv.
Pilla murda wood: *Terminalia chebula* Retz.; *Combretac.;* O.-Ind.
Pillai maruthu: *Terminalia paniculata* Roth; *Combretac.;* sw. O.-Ind. (Malabar)
Pillarwood: *Weihea eiki* Engler, *W. africana* Bth.; *Rhizophorac.;* D.-O.-Af.
Pillo-pillo: *Ovidia pillopillo* Meissn.; *Thymelaeac.;* s. Arg., s. Chile.
Pilo: *Sophora tetraptera* J. Mill.; *Papilionac.;* Chile.
Pilón: *Hieronyma alchorneoides* Allem.; *Euphorbiac.;* Pan.
„ : *Vouacapoua americana* Aubl.; *Caesalpiniac.;* Sur., Ven.
Pilúkau: *Aglaia multifoliola* Merr.; *Meliac.;* Phil. (Leyte)
Pimenta gorda: *Pimenta officinalis* Lindl.; *Myrtac.;* Hond.
Pimenteira: *Trichilia alta* Blake; *Meliac.;* Bras. (Bahia)
Pimentero: *Schinus molle* L.; *Anacardiac.;* Chile.
Pimento: *Amomis caryophyllata* Krug et Urb.; *Myrtac.;* Trin.
„ : *Pimenta officinalis* Lindl.; *Myrtac.;* W.-Ind.
„ : *Myrcia oerstedtiana* Berg; *Myrtac.;* Pan.
„ : *Ocotea veraguensis* Mez; *Laurac.;* Salv.
„ de California: *Schinus molle* L.; *Anacardiac.;* C.-R.
Pimienta: *Amomis caryophyllata* Krug et Urb.; *Myrtac.;* Jam.
„ : *Amyris elemifera* L.; *Rutac.;* Hond.
„ de America: *Schinus molle* L.; *Anacardiac.;* Mex.
„ del diablo: *Schinus molle* L.; *Anacardiac.;* Arg.
Pimientillo: *Trichilia sp.; Meliac.;* Salv.

Pimientillo: *Moronobea coccinea* Aubl.; *Guttifer.;* Guat.
Pimiento: *Pimenta officinalis* Lindl.; *Myrtac.;* W.-Ind.
„ : *Schinus molle* L.; *Anacardiac.;* Kol., Arg.
„ -ché: *Colubrina ferruginosa* Brong.; *Rhamnac.;* Mex.
Pimpernuss: *Staphylea pinnata* L.; *Staphyleac.;* Mitt.-Eu., Kl.-As., Kauk.
Pin: *Cordia vestita* Hook. f. et Thoms.; *Borraginac.;* n. O-.Ind. (United Prov.)
„ : *Pinus khasya* Royle, *P. merkusi* Jungh. et de Vriese; *Conifer.;* Tonk., Annam.
„ argenté: *Pinus monticola* Dougl.; *Conifer.;* sw. Kan.
„ baliveau: *Pinus strobus* L.; *Conifer.;* sö. Kan.
„ blanc: *Pinus halepensis* Miller; *Conifer.;* M.-M.
„ „ : *Pinus strobus* L.; *Conifer.;* sö. Kan.
„ „ de l'Ouest: *Pinus flexilis* James; *Conifer.;* sw. Kan.
„ bleu: *Pinus excelsa* Wall.; *Conifer.;* O.-Ind. (Kaschmir)
„ bush: *Hakea sp.; Proteac.;* Au.
„ cembro: *Pinus cembra* L.; *Conifer.;* Mitt.-Eu.
„ chétif: *Pinus banksiana* Lamb.; *Conifer.;* sö. Kan.
„ colonnaire: *Araucaria cooki* R. Br.; *Conifer.;* N.-Kal.
„ gris: *Pinus banksiana* Lamb.; *Conifer.;* s. Kan.
„ jaune: *Pinus strobus* L.; *Conifer.;* sö. Kan.
„ kouèn: *Ceiba pentandra* Gaertn.; *Bombacac.;* Annam.
„ mân kiûn: *Flacourtia cataphracta* Roxb.; *Flacourtiac.;* Annam.
„ maritime: *Pinus pinaster* Sol.; *Conifer.;* Algier, Réunion.
„ noir: *Pinus pachyphylla* Christ.; *Conifer.;* Mitt.-Eu.
„ parasol: *Pinus pinea* L.; *Conifer.;* M.-M.
„ pleureux: *Pinus excelsa* Wall.; *Conifer.;* Himal.
„ potiron: *Pinus strobus* L.; *Conifer.;* sö. Kan.
„ résineux: *Pinus resinosa* Aiton; *Conifer.;* sö. Kan.
„ rouge: *Pinus resinosa* Aiton; *Conifer.;* sö. Kan.
„ suffin: *Pinus montana* Miller; *Conifer.;* Mitt.- & S.-Eu.
„ thong: *Pinus khasya* Royle, *P. merkusi* Jungh. et de Vriese; *Conifer.;* Ind.-Ch.
„ Weymouth: *Pinus strobus* L.; *Conifer.;* sö. Kan.
„ à blanche écorce: *Pinus albicaulis* Engelm.; *Conifer.;* sw. Kan.
„ „ bois lourd: *Pinus ponderosa* Dougl.; *Conifer.;* sw. Kan.
„ „ crochets: *Pinus uncinata* Willk.; *Conifer.;* Alpen, Pyr.
„ „ feuilles courtes: *Pinus echinata* Mill.; *Conifer.;* sö. N.-Am.
„ „ „ rigides: *Pinus rigida* Mill.; *Conifer.;* ö. Kan.
„ „ 3 aiguilles: *Pinus khasya* Royle; *Conifer.;* Ind.-Ch. (Kamb.)
„ de Banks: *Pinus banksiana* Lamb.; *Conifer.;* sö. Kan.
„ „ Douglas: *Pseudotsuga taxifolia* Britt.; *Conifer.;* N.-Am.
„ „ Murray: *Pinus murrayana* Balf.; *Conifer.;* w. Kan.
„ „ Norvège: *Pinus resinosa* Ait.; *Conifer.;* ö. USA.
„ „ l'Orégon: *Pseudotsuga taxifolia* Britt.; *Conifer.;* w. USA.
„ des rochers: *Pinus banksiana* Lamb.; *Conifer.;* s. Kan.
„ du Langbian: *Pinus khasya* Royle, *P. merkusi* Jungh. et de Vr.; *Conifer.;* Ind.-Ch.
„ „ lord: *Pinus strobus* L.; *Conifer.;* sö. Kan.
„ „ Paraná: *Araucaria angustifolia* O. Ktze.; *Conifer.;* Bras.
„ d'Alep: *Pinus halepensis* Mill.; *Conifer.;* Algier.
Pina blanca *Podocarpus sp.; Conifer.;* Pan.
Pinabete: *Pinus ponderosa* Dougl., *Cupressus arizonica* Greene; *Conifer.;* Mex.
„ : *Pseudotsuga taxifolia* Britt.; *Conifer.;* Mex.
Pinang: *Areca catechu* L.; *Palmac.;* Malay.
„ baik: *Vatica cinerea* King, *Shorea bracteolata* Dyer; *Dipterocarpac.;* Malay.
„ pinang goendie: *Adinandra dumosa* Jack; *Theac.;* Sum.
Pinari: *Sterculia foetida* L.; *Sterculiac.;* O.-Ind.
Pinavete: *Pinus spp.; Conifer.;* Hond.
Pincapincahan: *Oroxylum indicum* Bth.; *Bignoniac.;* Phil.
Pinchiricóto: *Cordia sp.; Borraginac.;* Bras.
Pinckai: *Calophyllum inophyllum* L.; *Guttifer.;* w. O.-Ind. (Bombay)
Pindahiba: *Duguetia lanceolata* St. Hil.; *Anonac.;* Bras. (Sao P)
Pindahuva: *Duguetia lanceolata* St. Hil.; *Anonac.;* Bras. (Sao P)
„ : *Xylopia emarginata* Mart.; *Anonac.;* Bras. (Sao P)

Pindahyba: *Oxandra lanceolata* Baill., *Anona sp.; Anonac.;* Bras.
  „   : *Xylopia emarginata* Mart., *X. sp.; Anonac.;* Bras. (Sao P)
  „ preta: *Xylopia emarginata* Mart.; *Anonac.;* Bras. (Sao P)
Pindahyva: *Xylopia emarginata* Mart., *X. spp., Anona sp.; Anonac.;* Bras. (Sao P)
Pindaiba: *Guatteria sp.; Anonac.;* Bras.
  „ branca: *Xylopia frutescens* Aubl.; *Anonac.;* Bras.
Pindalu: *Randia dumetorum* Lamk.; *Rubiac.;* C.-O.-Ind.
Pindaúba: *Anona sp., Xylopia spp.; Anonac.;* Bras.
Pindaúva: *Anona sp., Xylopia spp.; Anonac.;* Bras.
Pindi: *Myristica canarica* Bedd.; *Myristicac.;* sw. O.-Ind. (Madras)
Pindikai: *Myristica canarica* Bedd.; *Myristicac.;* sw. O.-Ind. (Madras)
Pindja: *Hylodendron gabunense* Taub.; *Caesalpiniac.;* Gab.
Pindó: *Cocos romanzoffiana* Cham.; *Palmac.;* Arg.
Pindra: *Randia dumetorum* Lamk.; *Rubiac.;* sw. O.-Ind. (Bombay)
Pindri: *Stereospermum chelonioides* DC.; *Bignoniac.;* C.-O.-Ind. (Gondwana)
Pine (e): *Pinus spp.; Conifer.;* Eu., N.-Am.
  „   : *Araucaria excelsa* R. Br.; *Conifer.;* Südsee (Norfolk-Inseln)
  „   : *Picea canadensis* B. S. P.; *Conifer.;* Kan.
  „   : *Pinus palustris* Mill.; *Conifer.;* Hond.
  „   : *Pinus caribaea* Morelet; *Conifer.;* Bah., br. Hond., Nic.
  „ , Alaska-: *Tsuga heterophylla* Sarg.; *Conifer.;* w. Kan.
  „ , Aleppo-: *Pinus halepensis* Mill.; *Conifer.;* M.-M.
  „ , australian: *Casuarina equisetifolia* Forst.; *Casuarinac.;* Au.
  „ , austrian: *Pinus nigra* Arn. *var. austriaca* Höss.; *Conifer.;* Eu.
  „ , Baltic-: *Picea excelsa* Link; *Conifer.;* Au. (kult.)
  „ , „ -, red: *Pinus silvestris* L., *Picea excelsa* Link; *Conifer.;* ö. Eu.
  „ , banksian: *Pinus banksiana* Lamb.; *Conifer.;* s. Kan.
  „ , Bark-, white: *Pinus bungeana* Zucc.; *Conifer.;* Ch.
  „ , Bastard-, black: *Podocarpus ferruginea* G. Benn.; *Conifer.;* N.-Seel.
  „ , Benguet-: *Pinus insularis* Endl.; *Conifer.;* Phil.
  „ , big: *Pinus ponderosa* Dougl.; *Conifer.;* w. Kan.
  „ , black: *Pinus nigra* Arn. *var. austriaca* Höss.; *Conifer.;* ö. USA.
  „ , „ : *Pinus murrayana* Balf.; *Conifer.;* w. Kan.
  „ , „ : *Pinus contorta* Dougl. *var. latifolia* Engelm.; *Conifer.;* w. Kan.
  „ , „ : *Pinus thunbergi* Parl.; *Conifer.;* Jap.
  „ , „ : *Agathis microstachya* J. F. Bailey et C. T. White; *Conifer.;* Au.
  „ , „ : *Callitris calcarata* R. Br.; *Conifer.;* Au.
  „ , „ : *Podocarpus spicata* R. Br., *P. ferruginea* G. Benn.; *Conifer.;* N.-Seel.
  „ , blue: *Pinus excelsa* Wall.; *Conifer.;* n. O.-Ind.
  „ , British Columbia-: *Pseudotsuga taxifolia* Britt.; *Conifer.;* w. Kan.
  „ , „ „ -, soft: *Pinus ponderosa* Dougl.; *Conifer.;* sw. Kan.
  „ , British Honduras-: *Pinus caribaea* Morelet; *Conifer.;* br. Hond.
  „ , brown: *Podocarpus elata* R. Br.; *Conifer.;* ö. Au.
  „ , Bull-: *Pinus ponderosa* Dougl., *P banksiana* Lamb.; *Conifer.;* w. Kan.
  „ , Bunya-: *Araucaria bidwilli* W. J. Hooker; *Conifer.;* ö. Au.
  „ , Bunya Bunya-: *Araucaria bidwilli* W. J. Hooker; *Conifer.;* Au.
  „ , Burma Hill-: *Pinus khasya* Royle; *Conifer.;* Burma.
  „ , California-, white: *Pinus ponderosa* Dougl.; *Conifer.;* w. USA.
  „ , „ -, yellow: *Pinus ponderosa* Dougl.; *Conifer.;* USA.
  „ , canadian, red: *Pinus resinosa* Aiton; *Conifer.;* ö. Kan.;
  „ , „ , white: *Pinus strobus* L.; *Conifer.;* ö. Kan.
  „ , „ , yellow: *Pinus strobus* L.; *Conifer.;* ö. Kan.
  „ , Canary Island-: *Pinus canariensis* C. Smith; *Conifer.;* Kanar.
  „ , carribean: *Pinus caribaea* Morelet; *Conifer.;* br. Hond.
  „ , Celery-: *Phyllocladus trichomanoides* D. Don; *Conifer.;* N.-Seel.
  „ , celery-leaved: *Phyllocladus trichomanoides* D. Don; *Conifer.;* N.-Seel.
  „ , celery top-: *Phyllocladus rhomboidalis* Rich.; *Conifer.;* Tasm.
  „ , celery-topped: *Phyllocladus trichomanoides* D. Don; *Conifer.;* N.-Seel.
  „ , Chilgoza-: *Pinus gerardiana* Wall.; *Conifer.;* O.-Ind. (Himal.)
  „ , Chili-: *Araucaria araucana* K. Koch; *Conifer.;* S.-Am.
  „ , chinese: *Pinus massoniana* D. Don *var. sinensis* Lamb.; *Conifer.;* Ch.

Pine, chinese, white: *Pinus armandi* Franch.; *Conifer.*; Ch.
„ , Chir-: *Pinus longifolia* Roxb.; *Conifer.*; O.-Ind.
„ , Cluster-: *Pinus pinaster* Sol.; *Conifer.*; Eu.
„ , Colonial-: *Araucaria cunninghami* Ait.; *Conifer.*; ö. Au.
„ , columbian: *Pseudotsuga taxifolia* Britton; *Conifer.*; USA.
„ , Cook's-: *Araucaria cooki* R. Br.; *Conifer.*; Malay.
„ , Cork-: *Pinus strobus* L.; *Conifer.*; ö. N.-Am.
„ , corsican: *Pinus nigra* Arn. var. *calabrica* Schneid.; *Conifer.*; Eu.
„ , Cowdie-: *Agathis australis* Salisb.; *Conifer.*; N.-Seel.
„ , cuban: *Pinus caribaea* Morelet; *Conifer.*; Guat., Hond., br. Hond.
„ , Cypress-: *Callitris robusta* R. Br., *C. spp.*; *Conifer.*; Au.
„ , „ -, black: *Callitris calcarata* R. Br.; *Conifer.*; N.-Seel.
„ , Danzig-: *Pinus silvestris* L.; *Conifer.*; n. Eu.
„ , Douglas-: *Pseudotsuga taxifolia* Britton; *Conifer.*; w. Kan.
„ , Dundathu-: *Agathis robusta* F. v. M.; *Conifer.*; ö. Au. (Queensl.)
„ , eastern, white: *Pinus strobus* L.; *Conifer.*; sö. Kan.
„ , fat: *Pinus caribaea* Morelet; *Conifer.*; sö. USA., C.-Am.
„ , Finger cone-: *Pinus monticola* Dougl.; *Conifer.*; w. Kan.
„ , Foxtail-: *Pinus edulis* Engelm.; *Conifer.*; USA.
„ , gray: *Pinus banksiana* Lamb.; *Conifer.*; ö. USA.
„ , hard: *Pinus rigida* Mill.; *Conifer.*; ö. USA.
„ , Hazel- (eu): *Liquidambar styraciflua* L.; *Hamamelidac.*; ö. USA., C.-Am.
„ , Hokkaido-: *Picea ajanensis* Fisch., *Abies sachalinensis* Mast.; *Conifer.*; Jap.
„ , Hoop-: *Araucaria cunninghami* Ait.; *Conifer.*; ö. Au.
„ , Huon-: *Dacrydium franklini* Hook. f.; *Conifer.*; Tasm.
„ , Idaho-, white: *Pinus monticola* Dougl.; *Conifer.*; N.-Am.
„ , Insignis-: *Pinus insignis* Dougl.; *Conifer.*; N.-Am.
„ , Jack-: *Pinus banksiana* Lamb., *P. rigida* Mill.; *Conifer.*; ö. N.-Am.
„ , „ -, eastern: *Pinus banksiana* Lamb.; *Conifer.*; Kan.
„ , „ -, western: *Pinus murrayana* Balf.; *Conifer.*; w. Kan.
„ , japanese, red: *Pinus densiflora* S. et Z.; *Conifer.*; Jap.
„ , Jeffrey-: *Pinus jeffreyi* A. Murr.; *Conifer.*;.w. USA.
„ , Kauri-: *Agathis robusta* F. M. Bailey, *A. spp.*; *Conifer.*; Au.
„ , „ -, New Zealand: *Agathis australis* Steud.; *Conifer.*; N.-Seel.
„ , „ -, South Queensland: *Agathis robusta* F. M. Bailey; *Conifer.*; ö. Au.
„ , King William-: *Arthrotaxis selaginoides* D. Don; *Conifer.*; Tasm.
„ , „ „ -: *Arthrotaxis cupressoides* D. Don; *Conifer.*; Tasm.
„ , Limber- (u): *Pinus flexilis* James; *Conifer.*; w. USA.
„ , Loblolly-: *Pinus taeda* L.; *Conifer.*; sö. USA.
„ , Lodgepole-: *Pinus murrayana* Balf., *P. contorta* Dougl.; *Conifer.*; w. N.-Am.
„ , Longleaf-: *Pinus palustris* Miller; *Conifer.*; s. & sö. USA.
„ , long-leaved: *Pinus longifolia* Roxb.; *Conifer.*; n. O.-Ind. (Himal.)
„ , Male-: *Zanthoxylum sp.*; *Rutac.*; W.-Ind. (St.-D.)
„ , maritime: *Pinus pinaster* Sol.; *Conifer.*; M.-M.
„ , mexican, white (u): *Pinus strobiformis* Engelm.; *Conifer.*; USA.
„ , Mindoro-: *Pinus merkusi* Jungh. et de Vriese; *Conifer.*; Phil.
„ , Monterey-: *Pinus insignis* Dougl.; *Conifer.*; w. USA.
„ , Moreton Bay-: *Araucaria cunninghami* Ait.; *Conifer.*; ö. Au. (Queensl.)
„ , Mountain-: *Pinus monticola* Dougl.; *Conifer.*; w. Kan.
„ , „ -: *Dacrydium sp.*; *Conifer.*; N.-Seel.
„ , New England-: *Pinus strobus* L.; *Conifer.*; ö. Kan.
„ , New Zealand-: *Agathis australis* Steud.; *Conifer.*; N.-Seel.
„ , „ „ -, black: *Podocarpus spicata* R. Br.; *Conifer.*; N.-Seel.
„ , „ „ -, red: *Dacrydium cupressinum* Soland.; *Conifer.*; N.-Seel.
„ , „ „ -, white: *Podocarpus dacrydioides* A. Rich.; *Conifer.*; N.-Seel.
„ , Norfolk Island-: *Araucaria excelsa* R. Br.; *Conifer.*; Südsee (Norfolk-Ins.) [Au.
„ , North Carolina-: *Pinus taeda* L., *P. echinata* Mill.; *Conifer.*; USA.
„ , northern: *Pinus silvestris* L.; *Conifer.*; Eu.
„ , „ : *Pinus strobus* L.; *Conifer.*; ö. Kan.
„ , „ , white: *Pinus strobus* L.; *Conifer.*; USA.
„ , Norway-: *Pinus resinosa* Ait.; *Conifer.*; ö. N.-Am.

Pine, Nut-: *Pinus edulis* Engelm.; *Conifer.;* USA.
„ , Old-: *Pinus taeda* L.; *Conifer.;* N.-Am.
„ , Old field-: *Pinus taeda* L.; *Conifer.;* USA.
„ , Oregon-: *Pseudotsuga taxifolia* Britt.; *Conifer.;* USA.
„ , Organ Mountains-: *Araucaria angustifolia* O. Ktze.; *Conifer.;* Bras.
„ , Ottawa-: *Pinus strobus* L.; *Conifer.;* ö. Kan.
„ , „ -, red: *Pinus resinosa* Ait.; *Conifer.;* ö. Kan.
„ , „ -, white: *Pinus strobus* L.; *Conifer.;* ö. Kan.
„ , Paraná-: *Araucaria angustifolia* O. Ktze.; *Conifer.;* Bras. (Paraná)
„ , Pattern-: *Pinus strobus* L.; *Conifer.;* sö. Kan.
„ , Peacock-: *Cryptomeria japonica* D. Don; *Conifer.;* Ch.
„ , Pitch-: *Pinus rigida* Mill.; *Conifer.;* ö. N.-Am.
„ , „ -, moirée: *Pinus palustris* Mill.; *Conifer.;* N.-Am.
„ , Pond-: *Pinus rigida* Mill. var. serotina Michx.; *Conifer.;* sö. N.-Am.
„ , Ponderosa-: *Pinus ponderosa* Dougl.; *Conifer.;* w. USA.
„ , Pondosa-: *Pinus ponderosa* Dougl.; *Conifer.;* nw. USA.
„ , poor: *Pinus echinata* Mill.; *Conifer.;* USA.
„ , Princess-: *Pinus banksiana* Lamb.; *Conifer.;* s. Kan.
„ , Puget Sound-: *Pseudotsuga taxifolia* Britt.; *Conifer.;* w. Kan.
„ , Pumpkin-: *Pinus strobus* L.; *Conifer.;* sö. Kan.
„ , Quebec-: *Pinus resinosa* Ait., *P. strobus* L.; *Conifer.;* sö. Kan.
„ , Queensland-: *Araucaria cunninghami* Ait., *A. bidwilli* Hook.; *Araucariac.;*
„ , red: *Pinus densiflora* S. et Z.; *Conifer.;* Jap.   [Au.
„ , „ (e): *Pinus resinosa* Ait.; *Conifer.;* ö. Kan.
„ , „ : *Pinus palustris* Mill.; *Conifer.;* sö. N.-Am.
„ , „ : *Pseudotsuga taxifolia* Britt.; *Conifer.;* w. Kan.
„ , „ : *Oxystigma stapfiana* Chev.; *Caesalpiniac.;* Lib.
„ , „ : *Dacrydium cupressinum* Soland.; *Conifer.;* N.-Seel.
„ , Riga-: *Pinus silvestris* L.; *Conifer.;* ö. N.-Eu.
„ , Rigid: *Pinus rigida* Mill.; *Conifer.;* sö. Kan.
„ , Rocky Mountains-: *Pinus flexilis* James; *Conifer.;* sw. Kan.
„ , Sand-: *Pinus clausa* Vasey; *Conifer.;* sö. USA.
„ , Sapling-: *Pinus strobus* L.; *Conifer.;* sö. Kan.
„ , scotch (u): *Pinus silvestris* L.; *Conifer.;* N.-Eu., ö. USA.
„ , Scrub-: *Pinus albicaulis* Engelm., *P. banksiana* Lamb., *P. sp.;* *Conifer.;* Kan.
„ , „ -, northern: *Pinus banksiana* Lamb.; *Conifer.;* ö. USA.
„ , „ -, western: *Pinus contorta* Dougl.; *Conifer.;* w. Kan.
„ , She-: *Podocarpus elata* R. Br.; *Conifer.;* ö. Au.
„ , Shore-: *Pinus murrayana* Balf.; *Conifer.;* w. Kan.
„ , Short leaf-: *Pinus echinata* Mill.; *Conifer.;* sö. N.-Am.
„ , Silver-: *Abies balsamea* Mill.; *Conifer.;* ö. Kan.
„ . „ -: *Pinus monticola* Dougl.; *Conifer.;* sw. Kan.
., , „ -: *Dacrydium westlandicum* T. Kirk, *D. colensoi* Hook.; *Conifer.;* N.-Seel.
„ , „ -, white: *Dacrydium westlandicum* T. Kirk; *Conifer.;* N.-Seel.
„ , Slash-: *Pinus caribaea* Morelet; *Conifer.;* s. USA., Guat.
„ , soft: *Pinus strobus* L.; *Conifer.;* ö. N.-Am.
„ , „ : *Pinus monticola* Dougl.; *Conifer.;* w. Kan.
„ , „ , Arkansas: *Pinus echinata* Mill.; *Conifer.;* USA.
„ , „ , British Columbia: *Pinus ponderosa* Dougl.; *Conifer.;* w. Kan.
„ , „ , white (u): *Pinus strobus* L.; *Conifer.;* USA.
„ , Spruce-: *Pinus strobus* L., *P. banksiana* Lamb.; *Conifer.;* ö. N.-Am.
„ , „ -: *Tsuga canadensis* Carr., *Picea mariana* B. S. P.; *Conifer.;* ö. Kan.
„ , „ -, Lowland: *Pinus glabra* Walt.; *Conifer.;* sö. N.-Am.
„ , „ -, Upland: *Pinus clausa* Vasey; *Conifer.;* N.-Am.
„ , Stone-: *Pinus pinea* L.; *Conifer.;* M.-M,
„ , Sugar- (u): *Pinus lambertiana* Dougl.; *Conifer.;* w. USA.
„ , · „ -, little: *Pinus monticola* Dougl.; *Conifer.;* w. Kan.
,. , Swiss-: *Pinus cembra* L., *Abies alba* Miller; *Conifer.;* Eu.
„ , Tea-: *Pinus canariensis* C. Smith; *Conifer.;* Kanar.
„ , „ -: *Pinus palustris* Mill.; *Conifer.;* sö. N.-Am.
„ , Tonawada-: *Pinus strobus* L.; *Conifer.;* ö. Kan.

Pine, Totara-: *Podocarpus totara* G. Benn.; *Conifer.;* N.-Seel.
„ , Virginia-: *Pinus virginiana* Mill.; *Conifer.;* N.-Am.
„ , Waika-: *Amyris elemifera* L.; *Rutac.;* br. Hond.
„ , Water-: *Glyptostrobus pensilis* C. Koch; *Conifer.;* Ch.
„ , „ -, chinese: *Glyptostrobus pensilis* C. Koch; *Conifer.;* Ch.
„ , Westland-: *Dacrydium westlandicum* T. Kirk; *Conifer.;* N.-Seel.
„ , Weymouth-: *Pinus strobus* L.; *Conifer.;* ö. N.-Am.
„ , „ -, Mountain: *Pinus monticola* Dougl.; *Conifer.;* w. Kan.
„ , whistling: *Casuarina equisetifolia* Forst.; *Casuarinac.;* Go.
„ , white: *Pinus strobus* L.; *Conifer.;* ö. USA.
„ , „ : *Pinus ponderosa* Dougl., *P. monticola* Dougl.; *Conifer.;* w. N.-Am.
„ , „ : *Callitris glauca* R. Br.; *Conifer.;* Au.
„ , „ : *Podocarpus dacrydioides* A. Rich.; *Conifer.;* N.-Seel.
„ , „ , american: *Pinus strobus* L.; *Conifer.;* N.-Am.
„ , „ , Arizona: *Pinus ponderosa* Dougl.; *Conifer.;* w. USA.
„ , „ , brazilian: *Simaruba amara* Aubl.; *Simarubac.;* nö. Bras. (Pará)
„ , „ , northern: *Pinus strobus* L.; *Conifer.;* nö. N.-Am.
„ , „ , western: *Pinus monticola* Dougl.; *Conifer.;* w. N.-Am.
„ , „ , „ , British Columbia: *Pinus monticola* Dougl.; *Conifer.;* w. Kan.
„ , whitebark- (u): *Pinus albicaulis* Engelm.; *Conifer.;* w. USA.
„ , white-stemmed: *Pinus albicaulis* Engelm.; *Conifer.;* sw. Kan.
„ , yellow: *Pinus strobus* L., *P. albicaulis* Engelm.; *Conifer.;* ö. N.-Am.
„ , „ : *Pinus resinosa* Ait., *P. rigida* Mill.; *Conifer.;* ö. N.-Am.
„ , „ : *Larix gmelini* Ledob.; *Conifer.;* Ch.
„ , „ : *Podocarpus elata* R. Br.; *Conifer.;* Au.
„ , „ , longleaved: *Pinus palustris* Mill.; *Conifer.;* sö. N.-Am.
„ , „ , Quebec: *Pinus strobus* L.; *Conifer.;* ö. Kan.
„ , „ , shortleaved: *Pinus echinata* Mill.; *Conifer.;* sö. N.-Am.
„ , „ , southern: *Pinus palustris* Mill., *P. echinata* Mill.; *Conifer.;* USA.
„ , „ , „ : *Pinus taeda* L., *P. caribaea* Morelet; *Conifer.;* USA.
„ , „ , „ , western: *Pinus ponderosa* Dougl.; *Conifer.;* w. N.-Am.
Pine barks, australian: *Callitris sp.; Conifer.;* Au.
Piney: *Hardwickia pinnata* Roxb.; *Caesalpiniac.;* O.-Ind.
„ maram: *Vateria indica* L.; *Dipterocarpac.;* sw. O.-Ind.
„ varnish tree: *Vateria indica* L.; *Dipterocarpac.;* sw. O.-Ind.
„ du Malabar: *Vateria indica* L.; *Dipterocarpac.;* sw. O.-Ind.
Ping: *Cynometra polyandra* Roxb.; *Caesalpiniac.;* Assam (Cachar)
Pinggapinggáhan: *Radermachera sp.; Bignoniac.;* Phil. (Laguna)
Pingkoe: *Dysoxylum ramiflorum* Miq.; *Meliac.;* Java.
Pingyat: *Crataegus oxyacantha* L.; *Rosac.;* O.-Ind.
Pinha: *Anona squamosa* L.; *Anonac.;* W.-Ind.
„ : *Anona sp.; Anonac.;* Bras.
Pinhão: *Oxandra lanceolata* Baill.; *Anonac.;* Bras.
Pinheira: *Duguetia lanceolata* St. Hil.; *Anonac.;* Bras. (Sao P)
Pinheirinho: *Podocarpus sellowi* Klotzsch, *P. lamberti* Klotzsch; *Conifer.;* s. Bras.
Pinheiro: *Araucaria angustifolia* O. Ktze.; *Conifer.;* Bras. (Paraná)
„ bravo: *Podocarpus sellowi* Klotzsch, *P. lamberti* Klotzsch; *Conifer.;* s. Bras.
„ do Brasil: *Araucaria angustifolia* O. Ktze.; *Conifer.;* Par.
„ „ brejo: *Talauma sp.; Magnoliac.;* Bras.
„ „ Paraná: *Araucaria angustifolia* O. Ktze.; *Conifer.;* Bras.
Pinho: *Araucaria angustifolia* O. Ktze.; *Conifer.;* Bras. (Sao P)
„ branco: *Araucaria angustifolia* O. Ktze.; *Conifer.;* Bras.
„ nacional: *Araucaria angustifolia* O. Ktze.; *Conifer.;* Bras. (Sao P)
„ vermelho: *Araucaria angustifolia* O. Ktze.; *Conifer.;* Bras. (Sao P)
„ de las Misiones: *Araucaria angustifolia* O. Ktze.; *Conifer.;* S.-Am.
„ „ brejo: *Talauma ovata* St. Hil.; *Magnoliac.;* Bras.
„ do Parana: *Araucaria angustifolia* O. Ktze.; *Conifer.;* Bras.
Pini jaune: *Zanthoxylum sp.; Rutac.;* fr. W.-Ind.
Pinie: *Pinus pinea* L.; *Conifer.;* M.-M.
Pinimo: *Burkea africana* Hook.; *Caesalpiniac.;* Go.
Piñipiñi: *Elaeodendron xylocarpum* DC.; *Celastrac.;* Cuba.

Pinipiniche: *Hippomane mancinella* L.; *Celastrac.;* Cuba.
   ,,   : *Elaeodendron xylocarpum* DC.; *Celastrac.;* Cuba.
Piñique: *Sapium aucuparium* Jacq., *S. hippomane* G.F.W.Meyer; *Euphorbiac.;* n.Kol
? Pinito: *Colubrina heteroneura* Standl.; *Rhamnac.;* n. Kol.
Pinjal: *Shorea robusta* Gaertn. f.; *Dipterocarpac.;* C.-O.-Ind. (Gondwana)
Pinkapínka: *Oroxylum indicum* Vent.; *Bignoniac.;* Phil. (Tagalog)
Pinkapinkáhan: *Oroxylum indicum* Vent.; *Bignoniac.;* Phil. (Tagalog)
Pinkheart: *Medicosma cunninghami* Hook. f.; *Rutac.;* Au.
Pinkoe: *Dysoxylum ramiflorum* Miq.; *Meliac.;* Nied.-Ind.
Pinkosknolle: *Araucaria angustifolia* O. Ktze.; *Conifer.;* Bras.
   ,,   : *Araucaria bidwilli* Hook.; *Conifer.;* ö. Au.
Pinkwood: *Beyeria viscosa* Miq.; *Euphorbiac.;* Tasm.
Pinlè-kabwè: *Casuarina equisetifolia* Forst.; *Casuarinac.;* Burma, And.
   ,, -kanazo: *Heritiera littoralis* Dry., *H. fomes* Buch.-Ham.; *Sterculiac.;* Burma
   ,, -mohwa: *Mimusops littoralis* Kurz; *Sapotac.;* And.
   ,, -on: *Carapa moluccensis* Lam., *C. obovata* Blume; *Meliac.;* Burma.
Pinnai: *Dillenia pentagyna* Roxb.; *Dilleniac.;* sw. O.-Ind.
Pinnay: *Calophyllum inophyllum* L.; *Guttifer.;* sw. O.-Ind. (Madras)
Pinniha: *Barringtonia acutangula* Gaertn.; *Lecythidac.;* n. O.-Ind. (Oudh)
Pino: *Pinus caribaea* Morelet; *Conifer.;* Guat., br. Hond.
   ,,   : *Abies religiosa* Schlecht. et Cham.; *Conifer.;* Hond.
   ,,   : *Casuarina equisetifolia* Forst.; *Casuarinac.;* Hond.
   ,,   : *Podocarpus sp.; Conifer.;* Kol., Ven., Arg.
   ,,   : *Araucaria angustifolia* O. Ktze.; *Conifer.;* Bras., Arg.
   ,, acahuite: *Pinus ayacahuite* K. Ehrenb.; *Conifer.;* Mex.
   ,, albar: *Pinus silvestris* L.; *Conifer.;* Sp.
   ,, barbón: *Pinus sp.; Conifer.;* Mex.
   ,, bianco: *Pinus silvestris* L.; *Conifer.;* It.
   ,, blanco: *Pinus oocarpa* Schiede, *P. spp.; Conifer.;* Hond.
   ,,   ,,   : *Pinus montezumae* Lamb.; *Conifer.;* Mex.
   ,,   ,,   : *Araucaria angustifolia* O. Ktze.; *Conifer.;* Par.
   ,,   ,,   : *Podocarpus angustifolia* Parl.; *Conifer.;* Arg.
   ,, bravo: *Pinus pinaster* Sol.; *Conifer.;* Sp.
   ,, cahuite: *Pinus ayacahuite* K. Ehrenb.; *Conifer.;* Mex.
   ,, calabrese: *Pinus brutia* Tenore; *Conifer.;* It.
   ,, carrasco: *Pinus halepensis* Mill.; *Conifer.;* Sp.
   ,, cembro: *Pinus cembra* L.; *Conifer.;* It.
   ,, colorado: *Araucaria angustifolia* O. Ktze.; *Conifer.;* Par.
   ,, domestico: *Pinus pinea* L.; *Conifer.;* It.
   ,, dondel: *Pinus pinea* L.; *Conifer.;* Sp.
   ,, kosso: *Pinus nigra* Arn.; *Conifer.;* s. Eu.
   ,, laricio: *Pinus nigra* Arn.; *Conifer.;* It.
   ,, macho: *Zanthoxylum flavum* Vahl; *Rutac.;* W.-Ind. (St.-D.)
   ,, manso: *Pinus pinea* L.; *Conifer.;* Sp.
   ,, maritimo: *Pinus pinaster* Sol.; *Conifer.;* It.
   ,, montano: *Pinus montana* Miller; *Conifer.;* It.
   ,, Montezuma: *Pinus montezumae* Lamb.; *Conifer.;* Mex.
   ,, mugo: *Pinus montana* Miller; *Conifer.;* It.
   ,, negral: *Pinus pinaster* Sol.; *Conifer.;* Sp.
   ,, nero: *Pinus nigra* Arn. *var. austriaca* Höss.; *Conifer.;* It.
   ,, ocote: *Pinus caribaea* Morelet, *P. oocarpa* Schiede; *Conifer.;* Hond.
   ,, Oregon (Chile): *Pseudotsuga douglasi* Carr.; *Conifer.;* USA.
   ,, piñomero: *Pinus pinea* L.; *Conifer.;* Sp.
   ,, piñón: *Araucaria imbricata* Pav.; *Conifer.;* Chile.
   ,,   ,,   : *Pinus cembroides* Zucc., *P. quadrifolia* Parry; *Conifer.;* Mex.
   ,, prieto: *Pinus sp.; Conifer.;* Mex.
   ,, pudio: *Pinus nigra* Arn.; *Conifer.;* Sp.
   ,, real: *Pinus pinea* L.; *Conifer.;* Sp.
   ,,   ,,   : *Pinus ayacahuite* K. Ehrenb., *P. teocote* Schl. et Cham.; *Conifer.;* Mex.
   ,,   ,,   : *Pinus ponderosa* Dougl., *P. oocarpa* Schiede; *Conifer.;* Mex.
   ,,   ,,   : *Pseudotsuga taxifolia* Britt.; *Conifer.;* Mex.

Pino rodeno: *Pinus pinaster* Sol.; *Conifer.;* Sp.
„ royo: *Pinus silvestris* L.; *Conifer.;* Sp.
„ salgareño: *Pinus nigra* Arn.; *Conifer.;* Sp.
„ salvatico: *Pinus pinaster* Sol.; *Conifer.;* It.
„ silvestre: *Pinus silvestris* L.; *Conifer.;* Sp., It.
„ tea: *Pinus canariensis* C. Smith; *Conifer.;* Kanar.
„ „ : *Pinus palustris* Mill.; *Conifer.;* sö. USA.
„ „ roja: *Pinus palustris* Mill.; *Conifer.;* sö. N.-Am.
„ triste: *Pinus lumholtzi* Robins. et Fern.; *Conifer.;* Mex.
„ veta: *Pinus caribaea* Morelet, *P. sp.; Conifer.;* Hond.
„ veti: *Podocarpus macrostachyus* Parl.; *Conifer.;* Ven.
„ d'Aleppo: *Pinus halepensis* Mill.; *Conifer.;* It.
„ da pinocchi: *Pinus pinea* L.; *Conifer.;* It.
„ de Alepo: *Pinus halepensis* Mill.; *Conifer.;* Sp.
„ „ azúcar: *Pinus sp.; Conifer.;* Mex.
„ „ corcho: *Pseudotsuga taxifolia* Britt.; *Conifer.;* Mex.
„ „ Gerusalemme: *Pinus halepensis* Mill.; *Conifer.;* It.
„ „ pacho: *Podocarpus sp.; Conifer.;* Kol.
„ del Chaco: *Pentapanax angelicifolium* Griseb.; *Araliac.;* Arg.
Piñon: *Pinus edulis* Engelm., *P. cembroides* Zucc.; *Conifer.;* s. USA., Mex.
„ : *Pinus quadrifolia* Parry; *Conifer.;* s. USA., Mex.
„ : *Enterolobium cyclocarpum* Griseb.; *Mimosac.;* Kol.
„ : *Jatropha gaumeri* Greenm., *J. curcas* L.; *Euphorbiac.;* br. Hond., Hond.
„ : *Sterculia apetala* Karst.; *Sterculiac.;* Kol.
„ : *Araucaria araucana* K. Koch; *Conifer.;* Chile.
? „ : *Gyrocarpus americanus* Jacq.; *Hernandiac.;* Kol.
„ botijo: *Erythrina grisebachi* Urban; *Papilionac.;* Cuba.
„ espinoso: *Erythrina corallodendron* L.; *Papilionac.;* Cuba.
„ florido: *Erythrina sp.; Papilionac.;* Cuba.
„ milagroso: *Erythrina sp.; Papilionac.;* Cuba.
„ real· *Erythrina grisebachi* Urban; *Papilionac.;* Cuba.
„ de costa: *Erythrina sp.; Papilionac.;* Cuba.
„ „ pito: *Erythrina sp.; Papilionac.;* Cuba.
„ „ sierra: *Erythrina sp.; Papilionac.;* Cuba.
Piñonero: *Pinus pinea* L.; *Conifer.;* Urug.
Pinot: *Euterpe oleracea* Mart.; *Palmac.;* fr. Gu.
Pinsell: *Hypericum ericoides* L.; *Guttifer.;* Sp.
Pinshi-caspi: *Aspidosperma sp.; Apocynac.;* Peru.
Pintabollo: *Arrabidaea sanctae-marthae* Sprague; *Bignoniac.;* n. Kol.
Pintadinha: *Licania sp.; Rosac.;* Bras.
Pintaiva: *Xylopia sericea* St. Hil.; *Anonac.;* Bras.
Pintri: *Virola mycetis* Pulle; *Myristicac.;* Sur.
Pinzan: *Pithecolobium dulce* Bth.; *Mimosac.;* Mex.
Pió: *Sterculia tragacantha* Lindl.; *Sterculiac.;* Kam.
„ : *Calophyllum inophyllum* L., *C. montanum* Vieillard; *Guttifer.;* N.-Kal.
Piocha: *Bursera simaruba* Sarg.; *Burserac.;* Mex.
„ : *Melia azedarach* L.; *Meliac.;* Mex.
Piojillo: *Karwinskia humboldtiana* Zucc.; *Rhamnac.;* Mex. (Sinaloa)
Pioppo: *Populus alba* L., *P. pyramidalis* Salisb.; *Salicac.;* It.
„ albero: *Populus nigra* L.; *Salicac.;* It.
Piorno: *Genista florida* L., *Cytisus purgans* Willk., *C. albus* Link; *Papilionac.;* Sp.
„ azul: *Erinacea pungens* Boiss.; *Papilionac.;* Sp.
Pipal: *Ficus religiosa* L.; *Morac.;* O.-Ind.
Piper: *Piper elongatum* Vahl; *Piperac.;* Pan.
Piphar: *Gardenia latifolia* Aiton; *Rubiac.;* C.-O.-Ind.
Pipí: *Albizzia saponaria* Blume; *Mimosac.;* Phil. (Negrosinsel)
Pipicho: *Saurauia sp.; Dilleniac.;* Mex.
Pipigbalé: *Parkia filicoidea* Welw.; *Mimosac.;* Elf.
Pipigpalé: *Parkia filicoidea* Welw.; *Mimosac.;* Elf.
Pipisíg: *Avicennia officinalis* L.; *Verbenac.;* Phil. (Tayabas, Mindoro, Camarines)
Pipisík: *Avicennia officinalis* L.; *Verbenac.;* Phil. (Tayabas, Mindoro, Camar.)

28*

Pipli (H): *Bucklandia populnea* R. Br.; *Hamamelidac.;* Beng.
„ : *Ficus religiosa* L.; *Morac.;* n. O.-Ind. (United Prov.)
Pipri: *Ficus religiosa* L.; *Morac.;* C.-O.-Ind.
Pipro: *Ficus religiosa* L.; *Morac.;* O.-Ind. (Bombay)
Piquiá: *Caryocar tomentosum* Willd., *C. glabrum* Pers.; *Caryocarac.;* Bras.
„ : *Caryocar villosum* Pers., *C. brasiliense* St. Hil.; *Caryocarac.;* Bras.
„ : *Aspidosperma sp.; Apocynac.;* Bras. (Bahia)
„ -eté: *Caryocar villosum* Pers.; *Caryocarac.;* n. Bras. (Amaz.)
„ -rana: *Caryocar glabrum* Pers.; *Caryocarac.;* nö. Bras. (Pará)
Piquillin: *Condalia lineata* A. Gray; *Rhamnac.;* Arg.
Piranheira: *Piranhea trifoliata* Baill.; *Euphorbiac.;* n. Bras. (Amaz.)
Pirara: *Sonneratia caseolaris* Engl.; *Sonneratiac.;* Phil. (Cotabato)
Piratiner: *Piratinera guianensis* Aubl.; *Morac.;* fr. Gu.
Piratinier: *Piratinera guianensis* Aubl., *P. spp.; Morac.;* Sur.
Piratininere: *Piratinera guianensis* Aubl.; *Morac.;* fr. Gu.
Piri: *Erythrina suberosa* Roxb.; *Papilionac.;* w. Beng. (Bihar, Orissa)
„ -piri: *? Urena lobata* L.; *Malvac.;* fr. Gu.
Pirigaramepé: *Gustavia sp.; Lecythidac.;* Gu.
Piris: *Garcinia vidali* Merr., *G. sp.; Guttifer.;* Phil.
Piro: *Dialium sp.; Caesalpiniac.;* Mad.
Pirokálau: *Zizyphus sp.; Rhamnac.;* Phil. (Camarines)
Piruetano: *Pirus communis* L.; *Rosac.;* Sp.
Pirul: *Schinus molle* L.; *Anacardiac.;* Mex.
Písa: *Canarium villosum* F. Vill.; *Burserac.;* Phil. (Abra)
Písak: *Hopea pierrei* Hance; *Dipterocarpac.;* Phil. (Cagayan)
„ : *Shorea guiso* Blume; *Dipterocarpac.;* Phil. (Ilocos Norte, Ilocos Sur)
Pisamo: *Erythrina sp.; Papilionac.;* Kol.
Piscala: *Caesalpinia gilliesi* Wall.; *Caesalpiniac.;* Arg.
Pisch-tong: *Phyllanthus glaucescens* H. B. K.; *Euphorbiac.;* br. Hond.
Písek: *Shorea guiso* Blume; *Dipterocarpac.;* Phil. (Ilocos Norte, Ilocos Sur)
Pisek: *Aglaia minahassae* Koord., *A. dulcis* T. et Binn.; *Meliac.;* Nied.-Ind.
Pisi: *Nectandra pisi* Miq., *Ocotea sp.; Laurac.;* Sur.
„ , echte: *Nectandra sp., Ocotea sp.; Laurac.;* Sur.
„ , harde: *Nectandra sp., Ocotea sp.; Laurac.;* Sur.
„ , Kaneel-, gewone: *Nectandra sp., Ocotea sp.; Laurac.;* Sur.
„ , Pira-: *Maprounea guianensis* Aubl.; *Euphorbiac.;* Sur.
? „ -wane: *Nectandra pisi* Miq.; *Laurac.;* Sur.
„ , witte: *Nectandra sp., Ocotea sp.; Laurac.;* Sur.
„ , zachte: *Nectandra sp., Ocotea sp.; Laurac.;* Sur.
„ , zwarte: *Nectandra sp., Ocotea sp.; Laurac.;* Sur.
Pisie: *Nectandra sp., Ocotea sp.; Laurac.;* Sur.
„ , brown: *Nectandra globosa* Mez; *Laurac.;* Sur.
„ , Pira-: *Maprounea guianensis* Aubl.; *Euphorbiac.;* Sur.
„ , Sabanna-: *Nectandra sp., Ocotea sp.; Laurac.;* Sur.
„ , Saniki-: *Qualea dinizi* Ducke, *Q. sp.; Vochysiac.;* Sur.
Pisigallo: *Aphelandra tetragona* Nees; *Acanthac.;* n. Kol.
Pisonai: *Erythrina crista-galli* L.; *Papilionac.;* Peru.
Pissaléboko: *Pentadesma leucantha* Chev., *P. butyraceum* Don; *Guttifer.;* Elf.
Pissietan: *Lansium domesticum* Jack; *Meliac.;* Sunda.
Pistachio wood: *Pistacia spp.; Anacardiac.;* O.-Ind.
Pistolet: *Guarea trichilioides* L.; *Meliac.;* tr. Am.
Pístula: *Cassia javanica* L.; *Caesalpiniac.;* Phil.
Pisul: *Stereospermum chelonioides* DC.; *Bignoniac.;* sw. O.-Ind. (Madras)
Pita jam: *Eugenia operculata* Roxb.; *Myrtac.;* w. Beng. (Orissa)
„ korwa: *Holarrhena antidysenterica* Wall.; *Apocynac.;* w. Beng. (Orissa)
Pitagoria: *Pongamia glabra* Vent.; *Papilionac.;* Beng.
Pitahaya: *Cereus spp.; Cactac.;* Mex.
„ dulce: *Cereus thurberi* Engelm.; *Cactac.;* Mex.
Pitaïca: *Swartzia sp.; Caesalpiniac.;* Bras.
Pitali: *Trewia nudiflora* L.; *Euphorbiac.;* Beng.
Pitamo: *Pedilanthus fendleri* Boiss.; *Euphorbiac.;* n. Kol.

Pitamo real: *Pedilanthus fendleri* Boiss.; *Euphorbiac.;* n. Kol.
Pitangi: *Eugenia uniflora* L.; *Myrtac.;* Malay.
Pitch tree, New Zealand-: *Agathis australis* Steud.; *Conifer.;* N.-Seel.
Pitinga: ? *Chusquea uruguaensis* Arech.; *Gramin.;* Arg.
Pitiréré: *Macrolobium macrophyllum* MacBride; *Caesalpiniac.;* Elf.
? Pitjis: *Distylium stellare* O. Ktze.; *Hamamelidac.;* Java (Dieng)
Pitjisan: *Vaccinium lucidum* Miq.; *Ericac.;* Java (Telemojo)
Pito: *Erythrina rubrinervia* H. B. K.; *Papilionac.;* Guat., br. Hond., Hond.
Pitombeira: *Pseudima frutescens* Radlk.; *Sapindac.;* Bras.
   „   de Marajo: *Simaba guianensis* Engl.; *Simarubac.;* Bras.
Pitón: *Ceiba pentandra* Gaertn.; *Bombacac.;* Mex.
   „   : *Erythrina rubrinervia* H. B. K.; *Papilionac.;* Guat.
Pitra: *Myrtus multiflora* Juss., *Myrceugenia pitra* Berg; *Myrtac.;* Chile.
Pitt: *Calophyllum inophyllum* L.; *Guttifer.;* N.-Kal.
Pitteraj: *Amoora rohituka* Wight et Arn.; *Meliac.;* Beng.
Pittospore ondulé: *Pittosporum undulatum* Vent.; *Pittosporac.;* Au.
Pittosporum, weeping: *Pittosporum phillyraeoïdes* DC.; *Pittosporac.;* Au.
Pitts: *Calophyllum inophyllum* L.; *Guttifer.;* N.-Kal.
Pitumba: *Casearia parvifolia* Willd.; *Flacourtiac.;* Bras.
Piu: *Sterculia tragacantha* Lindl.; *Sterculiac.;* Kam.
Piúva: *Tecoma chrysotricha* Mart., *T. lapacho* Schum.; *Bignoniac.;* Bras. (Sao P)
   „   : *Bignonia longiflora* Vell.; *Bignoniac.;* Bras. (Sao P)
Pivijay: *Ficus prinoides* H. et B., *F. nitida* Thunb.; *Morac.;* Kol.
Piwar: *Barringtonia acutangula* Gaertn.; *Lecythidac.;* sw. & C.-O.-Ind.
Pixoi: *Guazuma ulmifolia* Lam.; *Sterculiac.;* Mex., br. Hond.
Piyal: *Buchanania latifolia* Roxb.; *Anacardiac.;* n. O.-Ind. (United Prov.)
Pkhay: *Xylia dolabriformis* Bth.; *Mimosac.;* Burma.
Plai: *Alstonia scholaris* R. Br.; *Apocynac.;* tr. As., tr. Au.
Plaine bâtarde: *Acer spicatum* Lam.; *Acerac.;* sö. Kan.
   „   blanche: *Acer saccharinum* L.; *Acerac.;* sö. Kan.
   „   bleue: *Acer spicatum* Lam.; *Acerac.;* sö. Kan.
   „   rouge: *Acer rubrum* L.; *Acerac.;* sö. Kan.
   „   à Guigère: *Acer negundo* L.; *Acerac.;* s. Kan.
Plakonie: *Inga alba* Willd., *I. sertulifera* DC.; *Mimosac.;* Sur.
Plalan: *Mallotus philippinensis* Muell. Arg.; *Euphorbiac.;* Java (Kedoengdjati)
Plalar: *Shorea javanica* Koord. et Valet.; *Dipterocarpac.;* Java.
   „   : *Dipterocarpus trinervis* Blume; *Dipterocarpac.;* Sunda.
Plalavlenga: *Dipterocarpus trinervis* Blume; *Dipterocarpac.;* Sum.
Plampang poetih: *Gordonia sp.; Theac.;* Bangka.
Plan: *Mallotus cochinchinensis* Lour.; *Euphorbiac.;* Ind.-Ch.
   „   tree: *Acer pseudoplatanus* L.; *Acerac.;* Schottland.
Plane: *Platanus orientalis* L., *P. acerifolia* Willd.; *Platanac.;* Eu.
   „   (e): *Acer pseudoplatanus* L.; *Acerac.;* Eu.
   „   : *Platanus occidentalis* L.; *Platanac.;* USA.
   „   , Cape-: *Ochna arborea* Burch.; *Ochnac.;* S.-Af. (Kap)
   „   , eastern: *Platanus orientalis* L.; *Platanac.;* O.-Ind.
   „   , european: *Platanus orientalis* L.; *Platanac.;* Eu.
   „   , London- (e): *Platanus acerifolia* Willd.; *Platanac.;* Eu.
   „   , oriental: *Platanus orientalis* L.; *Platanac.;* Cypern.
   „   , Sycamore-: *Acer pseudoplatanus* L.; *Acerac.;* Eu.
Plank tree: *Heritiera fomes* Buch.-Ham.; *Sterculiac.;* O.-Ind., Burma, Borneo.
Plao: *Engelhardtia chrysolepis* Hance; *Juglandac.;* Ind.-Ch.
Plasa: *Butea frondosa* Roxb.; *Papilionac.;* Java.
Platane (d): *Platanus orientalis* L.; *Platanac.;* Kl.-As., Himal.
   „   „   : *Platanus occidentalis* L.; *Platanac.;* sö. Kan.
   „   , abendländische: *Platanus occidentalis* L.; *Platanac.;* s. Eu.
   „   , falsche: *Acer pseudoplatanus* L.; *Acerac.;* Mitt.- & S.-Eu.
   „   , fausse: *Acer pseudoplatanus* L.; *Acerac.;* Frankreich.
   „   , kalifornische: *Platanus racemosa* Nutt., *P. occidentalis* L.; *Platanac.;* w. USA.
   „   , morgenländische: *Platanus orientalis* L.; *Platanac.;* Orient.
Platanito: *Cassia bicapsularis* L., *C. bacillaris* L.; *Caesalpiniac.;* Kol.

Platano: *Minquartia guianensis* Aubl.; *Olacac.;* Nic.
„ : *Chimarrhis parviflora* Standl.; *Rubiac.;* Pan.
Platanus, false: *Acer pseudoplatanus* L.; *Acerac.;* Eu.
Plateado: *Croton niveus* Jacq.; *Euphorbiac.;* n. Kol.
Platiado: *Croton niveus* Jacq.; *Euphorbiac.;* Kol.
Plawan: *Tristania spp.; Myrtac.;* Borneo.
Play-je-nee: *Aulacocalyx jasminiflorus* Hook. f.; *Rubiac.;* Lib.
„ „ „ : *Olax manni* Oliv.; *Olacac.;* Lib.
Ple-chu: *Gardenia abbeokutae* Hiern; *Rubiac.;* Lib.
Pleehn: *Carpoloba spp.; Polygalac.;* Lib.
Pleh-ju-eh: *Homalium smythei* Hutch. et Dalz.; *Flacourtiac.;* Lib.
Plokonie: *Pithecolobium pedicellare* Bth.; *Mimosac.;* Sur.
Plong: *Eugenia mekongensis* Gagnep.; *Myrtac.;* Ind.-Ch. (Kamb.)
„ : *Memecylon edule* Roxb.; *Melastomatac.;* Ind.-Ch.
„ phmeas: *Memecylon edule* Roxb.; *Melastomatac.;* Ind.-Ch.
„ phua: *Memecylon edule* Roxb.; *Melastomatac.;* Ind.-Ch.
„ tuk: *Eugenia mekongensis* Gagnep.; *Myrtac.;* Ind.-Ch.
Plor-plor: *Ptychopetalum anceps* Oliv.; *Olacac.;* Lib.
„ „ : *Tetrorchidium spp.; Euphorbiac.;* Lib.
Plossobaum: *Butea frondosa* Roxb.; *Papilionac.;* O.-Ind., Malay.
Ploum: *Heteropanax fragrans* Seem.; *Araliac.;* Ind.-Ch.
Plum: *Prunus domestica* L.; *Rosac.;* Eu.
„ : *Prunus spp.; Rosac.;* USA.
„ : *Spondias spp.; Anacardiac.;* Hond.
„ : *Spondias mombin* L.; *Anacardiac.;* Pan.
„ , american: *Prunus americana* Marsh.; *Rosac.;* USA.
„ , black: *Eugenia jambolana* Lam.; *Myrtac.;* O.-Ind.
„ , Blood-: *Haematostaphis barteri* Hook. f.; *Anacardiac.;* W.-Af. (S.-L.)
„ , Burdekin-: *Pleiogynium solandri* Engl.; *Anacardiac.;* Au.
„ , Canada-: *Prunus nigra* Ait.; *Rosac.;* N.-Am.
„ , Coco-: *Chrysobalanus icaco* L.; *Rosac.;* tr. Am.
„ , „ -, wild: *Hirtella americana* L.; *Rosac.;* br. Hond.
„ , „ -, „ : *Linociera oblanceolata* Robinson; *Oleac.;* br. Hond.
„ , Cocoa-: *Chrysobalanus icaco* L.; *Rosac.;* tr. Am.
„ , Coffee-: *Doryalis zizyphoides* E. Meyer; *Flacourtiac.;* S.-Af.
„ , Courbaril (e): *Hymenaea courbaril* L.; *Caesalpiniac.;* tr. Am.
„ , Dog-: *Ekebergia capensis* Sparrm.; *Meliac.;* S.-Af.
„ , Downward-: *Bumelia angustifolia* Nutt.; *Sapotac.;* Fl., Bah., Cuba.
„ , european: *Prunus domestica* L.; *Rosac.;* Eu., USA.
„ , Florida-: *Drypetes keyensis* Urban; *Euphorbiac.;* s. Fl., Bah.
„ , Ginger bread-: *Parinarium macrophyllum* Sabine; *Rosac.;* n. Nig.
„ , Gopher-: *Chrysobalanus icaco* L.; *Rosac.;* tr. Am.
„ , grey: *Diospyros pentamera* F. v. M.; *Ebenac.;* ö. Au. (Queensl.)
„ , Guazuma-: *Guazuma ulmifolia* Lam.; *Sterculiac.;* br. W.-Ind.
„ , Hog-: *Prunus umbellata* Ell.; *Rosac.;* N.-Am.
„ , „ -: *Ximenia americana* L.; *Olacac.;* Fl., br. W.-Ind.
„ , „ -: *Metopium toxiferum* Krug et Urb.; *Anacardiac.;* Fl., br. W.-Ind.
„ , „ -: *Spondias mombin* L.; *Anacardiac.;* W.-Ind., C.-Am., Sur.
„ , „ -: *Spondias purpurea* L.; *Anacardiac.;* br. Hond.
„ , „ -: *Spondias dulcis* Forst.; *Anacardiac.;* Sur.
„ , Horse-: *Prunus nigra* Ait.; *Rosac.;* ö. N.-Am.
„ , japanese: *Prunus triflora* Roxb.; *Rosac.;* Jap., USA.
„ , Kaffir-: *Harpephyllum caffrum* Bernh.; *Anacardiac.;* S.-Af.
„ , Madagaskar-: *Flacourtia ramontchi* L'Hérit.; *Flacourtiac.;* O.-Ind., Burma.
„ , Maiden-: *Comocladia pinnatifolia* L.; *Anacardiac.;* Jam.
„ , Mountain-: *Ximenia americana* L.; *Olacac.;* Fl., br. W.-Ind.
„ , Pigeon-: *Coccoloba laurifolia* Jacq.; *Polygonac.;* s. Fl., W.-Ind., n. S.-Am.
„ , „ -: *Chrysobalanus icaco* L.; *Rosac.;* tr. Am., W.-Af.
„ , „ -: *Licania hypoleuca* Bth.; *Rosac.;* br. Hond.
„ , „ -, wild: *Hirtella triandra* Swartz; *Rosac.;* br. Hond.
„ , „ -, „ : *Ouratea wrighti* Riley; *Ochnac.;* Pan.

Plum, red: *Prunus nigra* Ait.; *Rosac.;* ö. N.-Am.
„ , rough-leaf: *Parinarium excelsum* Sabine; *Rosac.;* Lib.
„ , Saffron-: *Bumelia angustifolia* Nutt.; *Sapotac.;* Fl., Bah., Cuba.
„ , Seaside-: *Ximenia americana* L.; *Olacac.;* Trin.
„ , „ -: *Coccoloba uvifera* L.; *Polygonac.;* Fl., br. W.-Ind.
„ , Sebestan-: *Cordia myxa* L.; *Borraginac.;* Aegypten.
„ , sour: *Owenia venosa* F. v. M.; *Meliac.;* ö. Au. (Queensl.)
„ , spanish: *Ximenia americana* L.; *Olacac.;* Fl., br. W.-Ind.
„ , „ : *Spondias purpurea* L.; *Anacardiac.;* Hond.
„ , Sugar-: *Uapaca guineensis* Muell. Arg.; *Euphorbiac.;* Lib.
„ , sweet: *Owenia cerasifolia* F. v. M.; *Meliac.;* Au.
„ , Waika-: *Rheedia edulis* Tr. et Planch.; *Guttifer.;* br. Hond.
„ , wild: *Prunus nigra* Ait.; *Rosac.;* ö. N.-Am.
„ , „ : *Phyllanthus acidus* Skeels; *Euphorbiac.;* br. Hond.
„ , „ : *Spondias purpurea* L.; *Anacardiac.;* Pan.
„ , „ : *Pappea capensis* Eckl. et Zeyh.; *Sapindac.;* S.-Af. (Kap)
„ , „ : *Sideroxylon australe* Bth. et Hook.; *Sapotac.;* Au.
Plumajillo: *Schizolobium kellermani* Pitt., *S. parahybum* Blake; *Caesalpiniac.;* C.-Am.
Plumerillo: *Calliandra parvifolia* Speg., *C. tweediei* Bth.; *Mimosac.;* Arg.
Plumieria: *Plumiera spp.; Apocynac.;* br. W.-Ind.
Plumpang hitam: *Adinandra stylosa* Miq.; *Theac.;* Bangka.
„ poetie: *Adinandra glabra* Miq.; *Theac.;* Sum.
Plumwood, Tulip-: *Pleiogynium solandri* Engl.; *Anacardiac.;* ö. Au. (Queensl.)
Plywood, Queensland-: *Araucaria cunninghami* Ait.; *Conifer.;* ö. Au. (Queensl.)
Po: *Aleurites fordi* Hemsl.; *Euphorbiac.;* Ind.-Ch.
„ : *Hibiscus praeclaris* Gagnep.; *Malvac.;* Ind.-Ch.
„ : *Funtumia africana* Stapf; *Apocynac.;* Elf.
„ hsa: *Cupressus funebris* Endl.; *Conifer.;* Ch.
Poachwood (e): *Haematoxylon campechianum* L.; *Caesalpiniac.;* tr. Am.
Pobéroi: *Pynaertia occidentalis* Chev.; *Meliac.;* Elf.
Pobo: *Poga oleosa* Pierre; *Rhizophorac.;* Kam.
Poboi: *Mitragyne macrophylla* Hiern; *Rubiac.;* W.-Af.
Pobreroï: *Pynaertia occidentalis* Chev.; *Meliac.;* Elf.
Pochitoquillo: *Casearia javitensis* H. B. K.; *Flacourtiac.;* Mex.
Pochote: *Ceiba pentandra* Gaertn.; *Bombacac.;* Mex.
„ : *Cochlospermum vitifolium* Spreng.; *Cochlospermac.;* br. Hond.
Pochotl: *Ceiba pentandra* Gaertn., *C. sp.; Bombacac.;* Mex.
„ yaxche: *Ceiba pentandra* Gaertn.; *Bombacac.;* Mex.
Pochotle: *Ceiba pentandra* Gaertn., *C. sp.; Bombacac.;* Mex.
Pochte: *Bombax sp.; Bombacac.;* Pan.
Pockholz (d): *Guaiacum officinale* L.; *Zygophyllac.;* W.-Ind., n. S.-Am.
„ : *Dipteryx odorata* Willd.; *Papilionac.;* fr. Gu.
Pockhout: *Guaiacum officinale* L.; *Zygophyllac.;* Bras.
Pocouli: *Berlinia bracteosa* Bth.; *Caesalpiniac.;* Elf.
Podo: *Podocarpus gracilor* Pilger, *P. milanjianus* Rendle; *Conifer.;* O.-Af.
„ : *Podocarpus usambarensis* Pilger; *Conifer.;* D.-O.-Af. (Tanganyika)
Podocarpus: *Podocarpus gracilior* Pilger, *P. spp.; Conifer.;* D.-O.-Af.
Poé: *Strombosia pustulata* Oliv.; *Olacac.;* Elf.
Poehoen sikat: *Alchornea rugosa* Muell. Arg.; *Euphorbiac.;* Java (Depok)
Poelasan: *Nephelium mutabile* Blume; *Sapindac.;* Sunda.
Poelasari: *Vaccinium lucidum* Miq.; *Ericac.;* Java (Dieng)
Poelassan: *Nephelium mutabile* Blume; *Sapindac.;* Sunda.
Poeloe moto: *Iryanthera sagotiana* Warb., *I. hostmanni* Warb.; *Myristicac.;* Sur.
Poeloes: *Cleidion javanicum* Blume; *Euphorbiac.;* Java (Rogodjampi)
„ : *Laportea stimulans* Miq.; *Urticac.;* w. Java.
„ areuj: *Cnesmone javanica* Blume; *Euphorbiac.;* Java (Buitenzorg)
„ djalantong: *Cnesmone javanica* Blume; *Euphorbiac.;* Java (Buitenzorg)
„ djalatrong: *Laportea stimulans* Miq.; *Urtiac.;* w. Java.
Poenaga: *Calophyllum inophyllum* L.; *Guttifer.;* Celebes.
Poenay: *Erythroxylon retusum* Bauer; *Erythroxylac.;* Nied.-Ind.
Poendoeng: *Baccaurea javanica* Muell. Arg.; *Euphorbiac.;* Java (Ngebel)

Poendoengan: *Baccaurea javanica* Muell. Arg.; *Euphorbiac.;* Java (Tangkil)
Poe-neh-day: *Heisteria parvifolia* Smith; *Olacac.;* Lib.
Poeng: *Dichrostachys cinerea* Wight et Arn.; *Mimosac.;* Java.
Poering: *Codiaeum variegatum* Blume; *Euphorbiac.;* Java.
Poeroet: *Gymnartocarpus venenosa* Boerl.; *Morac.;* Java.
Poerprehati: *Peltogyne spp.; Caesalpiniac.;* Sur.
Poerwa djamboe: *Gaultheria fragrantissima* Wall. *var. punctata* J. J. Sm.; *Ericac.;*
„ geni: *Rhododendron javanicum* Benn., *R. loerzingi* J. J. Smith; *Ericac.;* [Java
? Poerwoko: *Cyclostemon macrophyllus* Blume; *Euphorbiac.;* Java (Pringombo) [Java
Poerworoko: *Gaultheria leucocarpa* Blume; *Ericac.;* Java (Dieng, Soembing)
Poerwosada: *Gaultheria leucocarpa* Blume; *Ericac.;* Java.
Poesoe: *Bridelia stipularis* Blume; *Euphorbiac.;* Java (Poeger)
Poesoetan: *Celtis wighti* Planch.; *Ulmac.;* Java (Ngebel)
Poespa: *Schima noronhae* Reinw., *S. wallichi* Choisy; *Theac.;* Nied.-Ind.
„ : *Gordonia excelsa* Blume; *Theac.;* Java.
Poes-poesi tère: *Amaranthus caudatus* L.; *Amarantac.;* Sur.
Poetat: *Planchonia valida* Blume; *Lecythidac.;* Java.
Poetijan: *Cleidion javanicum* Blume; *Euphorbiac.;* Java (Badjoelmati)
Poeting: *Payena leeri* Kurz; *Sapotac.;* Malay.
Poetis: *Cupania sp.; Sapindac.;* Nied.-Ind. (Bangka)
Poetjoeng: *Pangium edule* Reinw.; *Flacourtiac.;* Java, Malay.
Poevinga: *Piratinera guianensis* Aubl., *P. spp.; Morac.;* Sur.
„ : *Helicostylis poeppigiana* Trec.; *Morac.;* Sur.
Pogada: *Mimusops elengi* Roxb.; *Sapotac.;* s. O.-Ind. (Madras)
Pogaungsa: *Bischofia javanica* Blume; *Euphorbiac.;* Burma.
Poh shu: *Thuja orientalis* L.; *Conifer.;* Ch.
Pohn: *Deinbollia grandiflora* Hook. f.; *Sapindac.;* Lib.
„ : *Pseudospondias microcarpa* Engl, *Sorindeia deliviosa* Chev.; *Anacardiac.;* Lib.
Pohon: *Calophyllum saigonense* Pierre; *Guttifer.;* Ind.-Ch. (Kamb.)
„ -pedang: *Oroxylum indicum* Bth.; *Bignoniac.;* Malay.
„ -sapi: *Otophora amoena* Blume; *Sapindac.;* Java (Dépok)
„ -toewa: *Leea sambucina* Willd.; *Ampelidac.;* Java (Dépok)
Pohora: *Soymida febrifuga* Adr. Juss.; *Meliac.;* C.-O.-Ind. (Berar)
Pohutikawa: *Metrosideros tomentosa* A. Cunn.; *Myrtac.;* N.-Seel.
Pohutukawa: *Metrosideros tomentosa* A. Cunn.; *Myrtac.;* N.-Seel.
Poïé: *Strombosia pustulata* Oliv.; *Olacac.;* Elf.
Poinciana: *Poinciana regia* Bojer; *Caesalpiniac.;* Hond.
Poincillade: *Caesalpinia pulcherrima* Sw.; *Caesalpiniac.;* fr. Gu.
Poinsettia: *Euphorbia pulcherrima* Willd.; *Euphorbiac.;* Hond.
Poire petite: *Amelanchier canadensis* Medic.; *Rosac.;* ö. Kan.
„ à poudre des Pahouins: *Feronia gabonensis* M. Cornu; *Rutac.;* Gab.
Poirier: *Tecoma pentaphylla* Juss.; *Bignoniac.;* W.-Ind.
„ : *Terminalia sp.; Combretac.;* Trin.
„ : *Enterolobium schomburgki* Bth.; *Mimosac.;* fr. Gu. (Tollinche)
„ : *Couma guianensis* Aubl.; *Apocynac.;* Sur.
„ rouge: *Tecoma pentaphylla* Juss.; *Bignoniac.;* W.-Ind.
„ de la Guyane: *Couma guianensis* Aubl.; *Apocynac.;* fr. Gu.
„ „ „ Martinique: *Tecoma pentaphylla* Juss.; *Bignoniac.;* W.-Ind.
„ des Antilles: *Tecoma pentaphylla* Juss.; *Bignoniac.;* W.-Ind.
„ du pays: *Tecoma pentaphylla* Juss.; *Bignoniac.;* Mart.
Pois doux: *Inga laurina* Willd., *I. ingoides* Willd., *I. venosa* Gris.; *Mimosac.;* W.-Ind.
„ „ blanc: *Inga sp.; Mimosac.;* Guad., Mart.
„ perdrix: *Heisteria coccinea* Jacq.; *Olacac.;* Guad.
„ pigeon: *Cajanus indicus* Spreng.; *Papilionac.;* Tr.
„ sabre: *Eperua sp.; Caesalpiniac.;* fr. Gu.
„ sucré: *Inga dulcis* Mart., *I. spp.; Mimosac.;* fr. Gu., Sur.
„ „ rouge: *Inga sp.; Mimosac.;* fr. Gu.
„ d'Angola: *Cajanus indicus* Spreng.; *Papilionac.;* Tr.
„ de Congo: *Cajanus indicus* Spreng.; *Papilionac.;* Tr.
Poison, africain: *Mimusops djave* Engl.; *Sapotac.;* Gab.
„ , Fish-: *Paullinia pinnata* L.; *Sapindac.;* br. Hond.

Poison, tree (e): *Hura crepitans* L ; *Euphorbiac.;* tr. Am.
  „ „ : *Metopium toxiferum* Krug et Urb.; *Anacardiac.;* Fl., br. W.-Ind.
  „ „ , Scrub-: *Excoecaria dallachyana* Baill.; *Euphorbiac.;* Au.
  „ , Savanna-, white: *Cameraria belizensis* Standl.; *Apocynac.;* br. Hond.
  „ wood: *Gymnanthes lucida* Swartz; *Euphorbiac.;* Fl.
  „ „ : *Spondias sp.; Anacardiac.;* tr. Am.
  „ „ : *Metopium toxiferum* Krug et Urb.; *Anacardiac.;* Fl., W.-Ind.
  „ „ : *Metopium brownei* Urban; *Anacardiac.;* br. Hond.
  „ „ , black: *Metopium brownei* Urban, *? Mauria sp.; Anacardiac.;* br. Hond., Guat.
  „ „ , Ridge-, white: *Excoecaria aff. bicolor* Hassk.; *Euphorbiac.;* br. Hond.
  „ „ , Savannah-, white: *Cameraria belizensis* Standl.; *Apocynac.;* br. Hond.
  „ „ , white: *Cameraria belizensis* Standl.; *Apocynac.;* br. Hond.
Poivre indien: *Xylopia frutescens* Aubl.; *Anonac.;* tr. S.-Am.
  „ d'Ethiopie: *Xylopia aethiopica* A. Rich.; *Anonac.;* Gab.
  „ de singe: *Xylopia aethiopica* A. Rich.; *Anonac.;* W.-Af.
Poivrier: *Schinus terebinthifolius* Raddi; *Anacardiac.;* N.-Kal.
  „ catafaye: *Zanthoxylum aubertium* DC.; *Rutac.;* Réunion.
Pojoh: *Litsea polyantha* Juss.; *Laurac.;* w. Beng. (Chota Nagpur)
Poka: *Solanum torvum* Swartz; *Solanac.;* Java (Kedoengdjati)
Pokaka: *Elaeocarpus hookerianus* E. Raoul; *Elaeocarpac.;* N.-Seel.
Pokako: *Elaeocarpus hookerianus* E. Raoul; *Elaeocarpac.;* N.-Seel.
Pokenoboy: *Bactris horrida* Oerst.; *Palmac.;* br. Hond.
Pokhout (h): *Guaiacum officinale* L.; *Zygophyllac.;* Cuba, P.-R., St.-D., n. S.-Am.
  „ : *Guaiacum sanctum* L.; *Zygophyllac.;* P.-R.
  „ , Cayenne-: *Dipteryx odorata* Willd.; *Papilionac.;* Sur.
Pokka: *Bassia latifolia* Roxb.; *Sapotac.;* w. O.-Ind. (Deccan)
Pokki: *Gmelina arborea* L.; *Verbenac.;* sw. O.-Ind. (Malabar)
Poko kelinting: *Campnosperma griffithi* March.; *Anacardiac.;* Malay.
  „ tanjong: *Mimusops elengi* L.; *Sapotac.;* Malay.
Pokok ipoh: *Antiaris toxicaria* Lesch.; *Morac.;* Malay.
? Pokori: *Clusia insignis* Mart.; *Guttifer.;* Sur.
Pola: *Kydia calycina* Roxb.; *Malvac.;* n. O.-Ind. (United Prov.)
Polach: *Ochroma spp.; Bombacac.;* Nic.
Polack, Bastard-: *Belotia campbelli* Sprague; *Tiliac.;* br. Hond.
Polái: *Alstonia scholaris* R. Br.; *Apocynac.;* Phil. (Pangasinan)
Polak: *Ochroma bicolor* Rowlee; *Bombacac.;* br. Hond., Guat., Nic.
Polasan: *Nephelium mutabile* Blume; *Sapindac.;* Sunda.
Polewood: *Xylopia frutescens* Aubl.; *Anonac.;* br. Hond.
Poli: *Neolitsea vidali* Merr.; *Laurac.;* Phil. (Cagayan)
Polilla: *Caesalpinia palmeri* Watson; *Caesalpiniac.;* Mex. (Sinaloa)
Polillo: *Erythroxylon mexicanum* H. B. K.; *Erythroxylac.;* w. C.-Am.
Polisandro: *Stahlia monosperma* Urban; *Caesalpiniac.;* P.-R.
Polki: *Sterculia urens* Roxb.; *Sterculiac.;* O.-Ind. (Madras)
Polo blanco: *Myrsine grisebachi* Hieron.; *Myrsinac.;* Arg.
  „ piojo: *Willardia mexicana* Rose; *Papilionac.;* w. Mex.
Polok: *Ochroma limonensis* Rowlee; *Bombacac.;* Nic.
Polvillo: *Tabebuia chrysantha* Nicholson; *Bignoniac.;* n. Kol.
Pom: *Protium heptaphyllum* March., *P. copal* Engl.; *Burserac.;* Mex.
  „ : *Protium sp.; Burserac.;* br. Hond.
Poma: *Toona ciliata* M. Roem.; *Meliac.;* Assam.
  „ rosa: *Eugenia jambos* L.; *Myrtac.;* P.-R.
Pombathiri: *Stereospermum chelonioides* DC.; *Bignoniac.;* sw. O.-Ind. (Madras)
Pombeira: *Citharexylum cinereum* L.; *Verbenac.;* s. Bras.
Pomegranate: *Punica granatum* L.; *Punicac.;* Hond.
  „ , native: *Capparis nobilis* F. v. M.; *Capparidac.;* Au.
Pomelo: *Citrus decumana* L.; *Rutac.;* Malay.
Pomeloe: *Citrus decumana* L.; *Rutac.;* Ind.-Ch.
Pomleathiri: *Stereospermum chelonioides* DC.; *Bignoniac.;* sw. O.-Ind. (Madras)
Pomme: *Eugenia cymosa* Lam.; *Myrtac.;* Maur.
  „ -canelle: *Anona squamosa* L.; *Anonac.;* fr. Ant., Guat.
  „ cythère: *Spondias dulcis* Forst.; *Anacardiac.;* fr. Gu.

Pomme rose: *Eugenia jambos* L.; *Myrtac.;* Guad.
„ sûrette: *Chrysophyllum ? cainito* L.; *Sapotac.;* fr. W.-Ind.
„ d'acajou: *Anacardium occidentale* L.; *Anacardiac.;* Trin.
„ d'éléphant: *Feronia elephantum* Correa; *Rutac.;* Ind.-Ch.
„ de bois: *Feronia elephantum* Correa; *Rutac.;* Ind.-Ch.
„ „ cannelle: *Anona sp.; Anonac.;* Guad., Mart.
„ „ liane: *Passiflora laurifolia* L.; *Passiflorac.;* fr. Gu.
„ „ pin: *Lucuma sp.; Sapotac.;* fr. Gu.
„ des prés: *Vaccinium oxycoccus* L.; *Ericac.;* fr. W.-Ind.
Pommier (f): *Pirus malus* L.; *Rosac.;* Eu.
„ cajou: *Anacardium occidentale* L.; *Anacardiac.;* Guat.
„ canaque: *Jambosa pseudo-malaccensis* Vieill.; *Myrtac.;* N.-Kal.
„ -cannelle: *Anona sp.; Anonac.;* fr. Gu.
„ du Gabon: *Copaifera sp.; Caesalpiniac.;* Gab.
Pomo de Venus: *Spondias dulcis* Forst.; *Anacardiac.;* Bras.
Pompatharai: *Stereospermum chelonioides* DC.; *Bignoniac.;* sw. O.-Ind. (Madras)
Pompathorai: *Stereospermum chelonioides* DC.; *Bignoniac.;* sw. O.-Ind. (Madras)
Pompoléon: *Citrus decumana* L.; *Rutac.;* Ind.-Ch.
Pompoqua: *Spondias lutea* L.; *Anacardiac.;* Mex.
Pon liese: *Memecylon edule* Roxb.; *Melastomatac.;* Ind.-Ch.
Ponchia mara: *Heterophragma roxburghi* DC.; *Bignoniac.;* C.-O.-Ind. (Gondwana)
Pondei: *Randia malleifera* Bth. et Hook. f.; *Rubiac.;* Lib.
Pondosa: *Pinus ponderosa* Dougl.; *Conifer.;* w. USA.
Ponéné: *Parinarium robustum* Oliv.; *Rosac.;* Elf.
Pong kom: *Aglaia cambodiana* Pierre; *Meliac.;* Ind.-Ch.
Ponga: *Pongamia glabra* Vent.; *Papilionac.;* O.-Ind. (Madras)
Pongoi-hini: *Anthocleista nobilis* G. Don; *Loganiac.;* Lib.
Pongro: *Schleichera trijuga* Willd.; *Sapindac.;* Ind.-Ch. (Kamb.)
Pongu: *Calophyllum tomentosum* Wight; *Guttifer.;* sw. O.-Ind. (Madras)
Ponhon: *Lophira alata* Banks; *Ochnac.;* Nig.
Ponnagam: *Mallotus philippinensis* Muell. Arg.; *Euphorbiac.;* sw. O.-Ind. (Malabar)
Ponnan kottai: *Sapindus laurifolius* Vahl; *Sapindac.;* w. O.-Ind. (Deccan)
Ponnyet: *Calophyllum inophyllum* L.; *Guttifer.;* s. Burma (s. Tenasserim)
Ponponla: *Bombax buonopozense* Beauv.; *Bombacac.;* tr. Af.
Ponsigué: *Ziziphus mauritiana* Lam.; *Rhamnac.;* Ven.
Poo'd-mai: *Gardenia sp.; Rubiac.;* Siam.
Poon (H): *Calophyllum inophyllum* L., *C. tomentosum* Wight; *Guttifer.;* O.-Ind.
Poonam: *Bassia latifolia* Roxb.; *Sapotac.;* sw. O.-Ind. (Malabar)
Poonang: *Calophyllum inophyllum* L.; *Guttifer.;* w. Beng. (Orissa)
Poon-spar tree (e): *Calophyllum tomentosum* Wight; *Guttifer.;* O.-Ind.
Poor man's candle: *Carpolobia spp.; Polygalac.;* Lib.
Poose: *Sterculia tragacantha* Lindl.; *Sterculiac.;* W.-Af.
Poot: *Eucalyptus longicornis* F. v. M.; *Myrtac.;* w. Au.
Pooteli: *Litsea umbrosa* Nees; *Laurac.;* nö. O.-Ind. (Nepal)
Poöti: *Shorea sp., Hopea sp.; Dipterocarpac.;* Nied.-Ind.
Poovandiecottay: *Sapindus laurifolius* Vahl; *Sapindac.;* w. O.-Ind. (Deccan)
Popa: *Dialium guineense* Willd.; *Caesalpiniac.;* Gab.
Popal: *Areca catechu* L.; *Palmac.;* O.-Ind., Sunda.
Popél: *Shorea cochinchinensis* Pierre, *S. harmandi* Pierre; *Dipterocarpac.;* Kamb.
„ : *Vitex pubescens* Vahl; *Verbenac.;* Ind.-Ch.
„ massau: *Shorea cochinchinensis* Pierre; *Dipterocarpac.;* Kamb.
Popelma sau: *Shorea cochinchinensis* Pierre; *Dipterocarpac.;* Ind.-Ch.
Poplar (e): *Populus spp.; Salicac.;* n. Hemis.
„ : *Populus grandidentata* Michx., *P. balsamifera* L.; *Salicac.;* N.-Am.
„ : *Liriodendron tulipifera* L.; *Magnoliac.;* ö. N.-Am.
„ : *Mitragyne stipulosa* O. Ktze.; *Rubiac.;* Lib.
„ , Aspen-: *Populus tremuloides* Michx.; *Salicac.;* Kan.
„ , „ -, large-toothed: *Populus grandidentata* Michx.; *Salicac.;* sö. Kan.
„ , „ -, trembling: *Populus tremuloides* Michx.; *Salicac.;* Kan.
„ , Bahan-: *Populus euphratica* Oliv.; *Salicac.;* O.-Ind.
„ , Balm-: *Populus balsamifera* L.; *Salicac.;* Kan.

443

Poplar, Balm of Gilead-: *Populus balsamifera* L.; *Salicac.*; Kan.
„ , Balsam- (c, u): *Populus bulsumifera* L.; *Salicac.*; o. N.-Am.
„ , Bay-: *Nyssa aquatica* March.; *Cornac.*; USA.
„ , big-toothed: *Populus grandidentata* Michx.; *Salicac.*; sö. Kan.
„ , black: *Populus nigra* L., *P. serotina* Hartig; *Salicac.*; Eu.
„ , „ : *Populus balsamifera* L.; *Salicac.*; Kan.
„ , Carolina (u): *Populus deltoides* Marsh.; *Salicac.*; ö. USA.
„ , Downy-: *Populus heterophylla* L.; *Salicac.*; ö. N.-Am.
„ , english: *Populus alba* L., *P. nigra* L.; *Salicac.*; England.
„ , grey (e): *Populus canescens* Smith; *Salicac.*; Eu.
„ , himalayan: *Populus ciliata* Wall.; *Salicac.*; O.-Ind.
„ , indian: *Populus euphratica* Oliv.; *Salicac.*; O.-Ind.
„ , italian, black: *Populus serotina* Hartig; *Salicac.*; Eu.
„ , large (u): *Populus grandidentata* Michx.; *Salicac.*; ö. N.-Am.
„ , large-toothed (u): *Populus grandidentata* Michx.; *Salicac.*; ö. N.-Am.
„ , Lombardy-: *Populus nigra* L. var. *italica* Du Roi; *Salicac.*; Eu., USA.
„ , Necklace-: *Populus deltoides* Marsh.; *Salicac.*; ö. N.-Am.
„ , rough-barked: *Populus balsamifera* L.; *Salicac.*; Kan.
„ , Silver-: *Populus alba* L.; *Salicac.*; Eu., ö. USA.
„ , silver leaf-: *Populus alba* L.; *Salicac.*; Eu., ö. USA.
„ , smooth-barked: *Populus tremuloides* Michx.; *Salicac.*; Kan.
„ , white (e): *Populus alba* L., *P. tremuloides* Michx.; *Salicac.*; Eu.
„ , yellow (u): *Liriodendron tulipifera* L.; *Magnoliac.*; ö. USA.
Poplea: *Dillenia ? ovata* Hook. f. et Thoms.; *Dilleniac.*; Ind.-Ch. (Kamb.)
„ , pras: *Pterospermum grewiaefolium* Pierre; *Sterculiac.*; Ind.-Ch.
„ , „ : *Lagerstroemia loudoni* Teijsm. et Binn.; *Lythrac.*; Ind.-Ch.
„ , thom: *Grewia paniculata* Roxb.; *Tiliac.*; Ind.-Ch.
Poplear: *Grewia paniculata* Roxb.; *Tiliac.*; Ind.-Ch.
Popo: *Mitragyne macrophylla* Hiern; *Rubiac.*; fr. Guin.
„ : *Sterculia tragacantha* Lindl.; *Sterculiac.*; port. W.-Af. (Angola)
Popoaqua: *Spondias lutea* L., *Cyrtocarpa procera* H. B. K.; *Anacardiac.*; Mex.
Popoatti: *Peltogyne paniculata* Bth., *P. spp.*; *Caesalpiniac.*; Sur.
Popohatti: *Peltogyne paniculata* Bth., *P. spp.*; *Caesalpiniac.*; Sur.
Poposi: *Ricinodendron africanum* Muell. Arg.; *Euphorbiac.*; Elf.
Popossi: *Ricinodendron africanum* Muell. Arg.; *Euphorbiac.*; Elf.
Popoul achsat: *Vitex pubescens* Vahl; *Verbenac.*; Ind.-Ch.
„ thmar: *Vitex pubescens* Vahl; *Verbenac.*; Ind.-Ch.
Popple (u): *Populus tremuloides* Michx., *P. grandidentata* Michx.; *Salicac.*; ö. N.-Am
Popp tree: *Thespesia populnea* Correa; *Malvac.*; tr. As., tr. Af.
Popro: *Gardenia latifolia* Aiton; *Rubiac.*; w. Beng. (Chota Nagpur)
Popti: *Schrebera swietenioides* Roxb.; *Oleac.*; w. O.-Ind. (n. Bombay)
Popul: *Vitex pubescens* Vahl; *Verbenac.*; Ind.-Ch.
Porasu: *Butea frondosa* Roxb.; *Papilionac.*; w. Beng. (Orissa)
Porcupineholz: *Cocos nucifera* L.; *Palmac.*; Tr.
Porcupinewood: *Cocos nucifera* L.; *Palmac.*; Tr.
„ : *? Centrolobium paraense* Tul.; *Papilionac.*; n. Bras. (Amaz.)
„ : *Pandanus spp.*; *Pandanac.*; Indo-Malay.
Poré-poré: *Sterculia tragacantha* Lindl.; *Sterculiac.*; Elf.
Poris: *Thespesia populnea* Correa; *Malvac.*; O.-Ind.
„ : *Aporosa arborea* Muell. Arg.; *Euphorbiac.*; Java (Depok)
„ kebo: *Aporosa arborea* Muell. Arg.; *Euphorbiac.*; Java (Depok)
„ ketjil: *Antidesma montana* Blume; *Euphorbiac.*; Java (Depok)
Pork- and dough boy: *Bactris horrida* Oerst.; *Palmac.*; br. Hond.
„ fat apple: *Chrysobalanus icaco* L.; *Rosac.*; tr. Am., W.-Af.
„ wood: *Pisonia obtusata* Jacq.; *Nyctaginac.*; Fl., br. W.-Ind.
Poró: *Erythrina sp.*; *Papilionac.*; C.-R.
Pororoka: *Rapanea sp.*; *Myrsinac.*; Arg.
Porotillo: *Sesbania marginata* Bth.; *Papilionac.*; Arg.
Porponda: *Cordia macleodi* Hook. f. et Thoms.; *Borraginac.*; w. Beng.
Porpratti: *Peltogyne paniculata* Bth., *P. spp.*; *Caesalpiniac.*; Sur.
Portia tree: *Thespesia populnea* Correa; *Malvac.*; O.-Ind.

Poruwamara: *Diospyros insignis* L. f.; *Ebenac.;* s. O.-Ind.
Posa: *Morus alba* L., *M. laevigata* Wall.; *Morac.;* Burma.
Posh: *Anona cherimolia* Mill.; *Anonac.;* br. Hond.
Poshur: *Carapa moluccensis* Lam.; *Meliac.;* s. Beng.
Possentree: *Hura crepitans* L.; *Euphorbiac.;* Sur.
Possentrie: *Hura crepitans* L., *H. polyandra* Baill.; *Euphorbiac.;* Sur.
Possi: *Albizzia fastigiata* Oliv.; *Mimosac.;* Elf.
Possum tree (u): *Hura crepitans* L.; *Euphorbiac.;* Sur.
„ wood (u): *Hura crepitans* L., *H. polyandra* Baill.; *Euphorbiac.;* tr. Am.
Postentrie: *Hura crepitans* L , *H. polyandra* Baill.; *Euphorbiac.;* Sur.
Pot-bellied wood: *? Cavanillesia arborea* K. Schum.; *Bombacac.;* Bras.
Potab: *Randia dumetorum* Lamk.; *Rubiac.;* O.-Ind.
Potai: *Acer oblongatum* Wall.; *Acerac.;* nw. O.-Ind. (Kumaon)
Potari: *Kydia calycina* Roxb.; *Malvac.;* n. O.-Ind. (United Prov.)
Potean: *Millingtonia hortensis* L. f.; *Bignoniac.;* Burma (Tenasserim)
Potepote: *Parinarium curatellifolium* Planch.; *Rosac.;* Togo.
Pothidin: *Mallotus philippinensis* Muell. Arg.; *Euphorbiac.;* Burma.
Potli: *Acer pictum* Thunb.; *Acerac.;* n. O.-Ind. (Garhwal)
Potopoin: *Lannea acidissima* Chev.; *Anacardiac.;* Elf.
Potopoto: *Strephonema pseudocola* Chev.; *Combretac.;* Elf.
„ : *Ricinodendron africanum* Muell. Arg.; *Euphorbiac.;* Nig.
Potótan: *Bruguiera sexangula* Poir., *B. conjugata* Merr.; *Rhizophorac.;* Phil.
„ : *Bruguiera cylindrica* Blume; *Rhizophorac.;* Phil.
„ -laláki: *Bruguiera cylindrica* Blume; *Rhizophorac.;* Phil. (Tayabas, Mindoro)
„ -laláki: *Bruguiera parviflora* Wight et Arn.; *Rhizophorac.;* Phil.
Potowa: *Randia dumetorum* Lamk.; *Rubiac.;* w. Beng.
Potri: *Kydia calycina* Roxb.; *Malvac.;* O.-Ind. (Madras)
Potrodom: *Erythrophloeum micranthum* Harms; *Caesalpiniac.;* Go.
Potta vaga: *Albizzia stipulata* Boivin; *Mimosac.;* sw. O.-Ind. (Malabar)
Potur: *Hymenodictyon excelsum* Wall.; *Rubiac.;* C.-O.-Ind.
Poué: *Elaeocarpus ovigerus* Brong. et Griseb.; *Elaeocarpac.;* N.-Kal.
Pougro: *Schleichera trijuga* Willd.; *Sapindac.;* Ind.-Ch. (Kamb.)
Poui: *Tabebuia serratifolia* Nicholson; *Bignoniac.;* Trin., Tob.
„ , black: *Tabebuia glomerata* Urb., *T. serratifolia* Nich.; *Bignoniac.;* Trin., Tob.
„ , yellow: *Tabebuia glomerata* Urb., *T. serratifolia* Nich.; *Bignoniac.;* Trin.,
Poulitcha-maron: *Averrhoa bilimbi* L.; *Oxalidac.;* w. O.-Ind. (Deccan)   [Tob.
Poumboul: *Grewia coriacea* Mast.; *Tiliac.;* Kam.
Poun sva: *Spondias lakonensis* Pierre; *Anacardiac.;* Ind.-Ch.
Pouo: *Funtumia africana* Stapf; *Apocynac.;* Elf.
„ : *Strombosia pustulata* Oliv.; *Olacac.;* Elf.
Poupouia: *Vernonia sp.; Composit.;* Elf.
Pouriots: *Vaccinium myrtillus* L.; *Ericac.;* Frankreich.
Pourouma: *Lucuma sp.; Sapotac.;* fr. Gu.
Pouvandicotté: *Sapindus laurifolius* Vahl; *Sapindac.;* w. O.-Ind. (Deccan)
Pouverne (f): *Rhamnus frangula* L.; *Rhamnac.;* Eu.
Povo: *Poga oleosa* Pierre; *Rhizophorac.;* Kam.
Powassa-djoemoe: *Couratari guianensis* Aubl., *C. sp.; Lecythidac.;* Sur.
Powder bark: *Eucalyptus lane-poolei* Maiden; *Myrtac.;* w. Au.
Powhiwhi: *Plagianthus betulinus* A. Cunn., *Hoheria populnea* A. Cunn.; *Malvac.;*
Powo: *Mitragyne macrophylla* Hiern; *Rubiac.;* W.-Af.   [N.-Seel.
Poxot: *Ceiba pentandra* Gaertn.; *Bombacac.;* Nic.
Poyi: *Dalbergia melanoxylon* Guill. et Perr.; *Papilionac.;* br. O.-Af. (Uganda)
Poyn: *Careya arborea* Roxb.; *Myrtac.;* sw. O.-Ind. (Travancore Hills)
Poyu dua: *Funtumia elastica* Stapf; *Apocynac.;* Elf.
Pra kao: *Terminalia nigrovenulosa* Pierre; *Combretac.;* Ind.-Ch.
Pracachy: *Pentaclethra filamentosa* Bth.; *Mimosac.;* Bras.
Prah: *Elateriospermum tapos* Blume; *Euphorbiac.;* Malay.
Prahanaindi: *Aeglopsis chevalieri* Swingle; *Rutac.;* Elf.
Prahut tuk: *Fagraea racemosa* Jack; *Loganiac.;* Ind.-Ch.
Praing: *Xylia dolabriformis* Bth.; *Mimosac.;* Burma (Arakan)
Prak: *Machilus odoratissima* Nees; *Laurac.;* Ind.-Ch.

Prakao: *Terminalia nigrovenulosa* Pierre; *Combretac.*; Ind.-Ch.
Prakowa: *Dimorphandra mora* Bth. et Hook.; *Caesalpiniac.*; Sur.
Prampru: *Oxytenanthera ? abyssinica* Munro; *Gramin.*; Go.
Pran: *Xylia dolabriformis* Bth.; *Mimosac.*; Burma.
„ pran: *Albizzia fastigiata* Oliv.; *Mimosac.*; Elf.
Prapa: *Iriartea praemorsa* Klotzsch; *Palmac.*; Ven.
Prapat: *Sonneratia alba* Smith; *Sonneratiac* Malay.
Praphat: *Bombax malabaricum* DC.; *Bombacac.*; Ind.-Ch.
Prawan: *Tristania obovata* Benn.; *Myrtac.*; Sum., Bangka, Borneo.
Preecha: *Ctenolophon parvifolius* Oliv.; *Olacac.*; Malak.
Préfontaine: *Enterolobium schomburgki* Bth.; *Mimosac.*; fr. Gu. (Oyapock)
Prèh: *Ficus glabella* Blume, *F. pisifera* Wall., *F. spp.*; *Morac.*; Java.
Preisselbeere (d): *Vaccinium vitis-idaea* L.; *Ericac.*; Eu.
Prekese: *Tetrapleura thonningi* Bth.; *Mimosac.*; Togo.
Prembiloetan: *Flueggea virosa* Baill.; *Euphorbiac.*; Java (Semarang)
Prempeng herbo: *Leucosyke capitellata* Wedd.; *Urticac.*; Java (Pekalongan)
„ „ : *Maoutia diversifolia* Wedd.; *Urticac.*; Java (Banjoemas, Pekalongan)
Prengpreng: *Leucosyke capitellata* Wedd.; *Urticac.*; Java (Noesa Kambangan)
Prensilan: *Ilex triflora* Blume var. *acutata* Loes.; *Aquifoliac.*; Java.
Pri: *Funtumia africana* Stapf; *Apocynac.*; Elf.
Prickle, Main-, southern: *? Pithecolobium albicaulis* Br. et Rose; *Mimosac.*; br. Hond.
Prickle wood: *Pithecolobium macrandrium* Donn. Smith; *Mimosac.*; br. Hond.
„ „ : *Guettarda elliptica* Swartz; *Rubiac.*; br. Hond.
Prickly bark: *Eucalyptus todtiana* F. v. M.; *Myrtac.*; w. Au.
„ yellow: *Zanthoxylum sp.*; *Rutac.*; Jam.
„ „ : *Zanthoxylum kellermani* P. Wilson; *Rutac.*; br. Hond.
„ „ : *Acacia glomerosa* Bth.; *Mimosac.*; br. Hond.
„ „ , Bastard-: *Acacia glomerosa* Bth.; *Mimosac.*; br. Hond.
Prickwood (e): *Evonymus europaea* L.; *Celastrac.*; Eu.
Pride of India: *Melia azedarach* L.; *Meliac.*; O.-Ind., br. O.-Af. (Niassa)
? Prietjarie: *Zanthoxylum sp.*; *Rutac.*; Sur.
Prieto: *Dalbergia retusa* Hemsl., *D. hypoleuca* Pitt., *D. sp.*; *Papilionac.*; Pan.
Prih: *Ficus infectoria* Roxb. var. *lambertiana* King; *Morac.*; Java (Simpol)
„ : *Ficus gibbosa* Blume; *Morac.*; Java (Rogodjampi, Poeger)
Prima vera: *Tabebuia donnell-smithi* Rose; *Bignoniac.*; W.-Ind., C.-Am., n. S.-Am.
Primavera: *Wistaria sinensis* Sweet; *Papilionac.*; Arg.
„ , african: *Triplochiton scleroxylon* K. Schum.; *Sterculiac.*; W.-Af.
Prince wood: *Exostemma caribaeum* Roem. et Sch.; *Rubiac.*; W.-Ind., C.-Am.
„ „ : *Cordia gerascanthus* L.; *Borraginac.*; br. W.-Ind.
Pring: *Eugenia tinctoria* Gagnep.; *Myrtac.*; Ind.-Ch.
„ bai: *Eugenia brachiata* Roxb.; *Myrtac.*; Ind.-Ch.
„ chanh: *Eugenia zeylanica* Wight; *Myrtac.*; Ind.-Ch.
„ chanlos: *Eugenia chanlos* Gagnep.; *Myrtac.*; Ind.-Ch.
„ chom: *Eugenia longiflora* F. Villar; *Myrtac.*; Ind.-Ch.
„ das krebey: *Eugenia jambolana* Lam.; *Myrtac.*; Ind.-Ch.
„ lies: *Eugenia zeylanica* Wight; *Myrtac.*; Ind.-Ch.
„ phnom: *Eugenia longiflora* F. Villar; *Myrtac.*; Ind.-Ch.
„ sambac salp: *Eugenia longiflora* F. Villar; *Myrtac.*; Ind.-Ch.
„ thasar theas: *Eugenia tinctoria* Gagnep.; *Myrtac.*; Ind.-Ch.
„ thmar: *Eugenia tinctoria* Gagnep.; *Myrtac.*; Ind.-Ch.
„ thom: *Eugenia tinctoria* Gagnep.; *Myrtac.*; Ind.-Ch. (Kamb.)
Prinzholz: *Exostemma caribaeum* Roem. et Sch.; *Rubiac.*; W.-Ind., C.-Am.
Priti-jari: *Duguetia guianensis* DC.; *Anonac.*; Sur.
Pritjarie, witte: *Duguetia guianensis* DC.; *Anonac.*; Sur.
Pritjiarie: *Duguetia guianensis* DC.; *Anonac.*; Sur.
Privet, egyptian: *Lawsonia inermis* L.; *Lythrac.*; Nig.
Proekoeni: *Inga alba* Willd., *I. sertulifera* DC.; *Mimosac.*; Sur.
Proekoenie: *Pithecolobium pedicellare* Bth.; *Mimosac.*; Sur.
Prohut: *Garcinia vilersiana* Pierre; *Guttifer.*; Ind.-Ch.
„ phnôm: *Garcinia andersoni* Hook. f.; *Guttifer.*; Ind.-Ch. (Kamb.)
Prokoni: *Pithecolobium pedicellare* Bth.; *Mimosac.*; Sur.
„ : *Inga alba* Willd., *I. sertulifera* DC.; *Mimosac.*; Sur.

Prokonie: *Pithecolobium pedicellare* Bth.; *Mimosac.;* Sur.
„ : *Inga alba* Willd., *I. sertulifera* DC.; *Mimosac.;* Sur.
Promann: *Excoecaria oppositifolia* Muell. Arg.; *Euphorbiac.;* Ind.-Ch.
Proméné: *Parinarium robustum* Oliv.; *Rosac.;* Elf.
Prontolivia: *Guarea sp.; Meliac.;* Nic.
Propeal pros: *Pterospermum grewiaefolium* Pierre; *Sterculiac.;* Kamb.
Prosopis de Mexique: *Prosopis juliflora* DC.; *Mimosac.;* Senl.
Provision tree: *Pachira aquatica* Aubl., *P. macrocarpa* Walp.; *Bombacac.;* br. Hond.
Pruche: *Tsuga canadensis* Carr.; *Conifer.;* ö. Kan.
Prugnolo: *Prunus spinosa* L.; *Rosac.;* Eu., n. Af., Orient.
Pruim, witte: *Spondias lutea* L.; *Anacardiac.;* Sur.
Prune coton: *Chrysobalanus icaco* L.; *Rosac.;* fr. Gu.
„ monbin: *Spondias lutea* L.; *Anacardiac.;* fr. Gu.
„ d'Amérique: *Spondias sp.; Anacardiac.;* Guad.
„ d'anse: *Chrysobalanus icaco* L.; *Rosac.;* fr. Gu.
„ de Florida: *Drypetes keyensis* Urban; *Euphorbiac.;* s. Fl., Bah.
Prung: *Sarcocephalus cordatus* Miq.; *Rubiac.;* Burma.
Prunier canadien: *Prunus nigra* Ait.; *Rosac.;* ö. Kan.
„ épineux: *Ximenia americana* L.; *Olacac.;* Guad.
„ femelle: *Zizyphus jujuba* Lamk.; *Rhamnac.;* Réunion.
„ japonais: *Prunus triflora* Roxb.; *Rosac.;* Ch., Jap.
„ jaune d'oeuf: *Lucuma sp.; Sapotac.;* fr. Gu.
„ mâle: *Flacourtia ramontchi* L'Hérit.; *Flacourtiac.;* Réunion.
„ Myrobolan: *Prunus cerasifera* Ehrh.; *Rosac.;* As., N.-Af.
„ d'Espagne: *Spondias sp.; Anacardiac.;* tr. Am.
„ de Damas: *Prunus domestica* L. var. *damascena* Ehrh.; *Rosac.;* Algier.
„ „ Madagascar: *Flacourtia ramontchi* L'Hérit.; *Flacourtiac.;* Réunion.
„ du Gabon: *Chrysophyllum pruniformis* Pierre; *Sapotac.;* Gab.
? Pruno: *Cordia irvingi* Baker; *Borraginac.;* Go.
Prus: *Garcinia benthami* Pierre, *G. ferrea* Pierre, *G. sp.; Guttifer.;* Ind.-Ch.
„ pnóm: *Garcinia ferrea* Pierre; *Guttifer.;* Ind.-Ch. (Kamb.)
Prway: *Xylia dolabriformis* Bth.; *Mimosac.;* Burma.
Pseudo-senga: *Khaya canaliculata* de Wild.; *Meliac.;* B.-K.
Pu: *Schleichera trijuga* Willd.; *Sapindac.;* O.-Ind. (Madras)
Pu-bu-kuk: *Ocotea veraguensis* Mez; *Laurac.;* Guat.
Pu-marum-enney: *Schleichera trijuga* Willd.; *Sapindac.;* w. O.-Ind. (Deccan)
Pu maruthu: *Lagerstroemia flos-reginae* Retz.; *Lythrac.;* sw. O.-Ind. (Madras)
., prai: *Tritaxis gaudichaudi* Baill.; *Euphorbiac.;* Ind.-Ch.
Pua-veoveo: *Crataeva religiosa* Forst.; *Capparidac.;* Südsee (Tahiti)
Puant: *Foetidia mauritiana* Lam.; *Lecythidac.;* Maur.
Puarata: *Metrosideros polymorpha* Forst.; *Myrtac.;* Südsee (Samoa)
Puc: *Citrus decumana* L.; *Rutac.;* Ind.-Ch.
Puca-quiro: *Sickingia sp.; Rubiac.;* Peru.
Pucté: *Terminalia buceras* Bailey; *Combretac.;* Mex.
Puddum: *Mallotus philippinensis* Muell. Arg.; *Euphorbiac.;* Assam.
Pudek: *Calophyllum inophyllum* L.; *Guttifer.;* Malay.
Puerquito: *Escallonia floribunda* H. B. K.; *Saxifragac.;* Ven.
Pugáui-itim: *Diospyros aherni* Merr.; *Ebenac.;* Phil. (Laguna)
Pugúran: *Cyclostemon sp.; Euphorbiac.;* Phil. (Mindoro)
Pugusa: *Flacourtia ramontchi* L'Hérit.; *Flacourtiac.;* D.-O.-Af. (Tabora)
Puhatea: *Laurelia novae-zeelandiae* A. Cunn.; *Monimiac.;* N.-Seel.
Pui: *Tecoma serratifolia* G. Don; *Bignoniac.;* Trin.
Pujín: *Hesperomeles latifolia* M. Roem.; *Rosac.;* Ek.
Pujong: *Ganua motleyana* Pierre, *Madhuca spp.; Sapotac.;* Malay.
„ : *Palaquium spp., Payena spp.; Sapotac.;* Malay.
Puk-nam: *Trewia nudiflora* L.; *Euphorbiac.;* Burma
Pukan jantan: *Aphania paucijuga* Radlk.; *Sapindac.;* Malay.
Pukatea: *Laurelia novae-zeelandiae* A. Cunn.; *Monimiac.;* N.-Seel.
Pukté: *Terminalia buceras* Bailey; *Combretac.;* Mex.
Pula: *Kydia calycina* Roxb.; *Malvac.;* O.-Ind. (United Prov., Punjab)
„ : *Bombax malabaricum* DC.; *Bombacac.;* O.-Ind (Madras)

Puladhi: *Schleichera trijuga* Willd.; *Sapindac.;* O.-Ind. (Madras)
Pulángi: *Garcinia sp.; Guttifer.;* Phil. (Samar)
Pulanthi: *Elaeocarpus tuberculatus* Roxb.; *Tiliac.;* sw. O.-Ind. (Madras)
Pulasan: *Nephelium mutabile* Blume; *Sapindac.;* Malay.
Puli: *Kydia calycina* Roxb.; *Malvac.;* n. O.-Ind. (United Prov.)
„ vaga: *Albizzia odoratissima* Bth.; *Mimosac.;* sw. O.-Ind. (Malabar)
Pulian: *Kydia calycina* Roxb.; *Malvac.;* nw. O.-Ind. (Punjab)
Pulidh-dhakkay: *Averrhoa bilimbi* L.; *Oxalidac.;* w. O.-Ind. (Deccan)
Pulit: *Xylocarpus granatum* Koenig; *Meliac.;* Phil. (Basilan)
Pulla-maruthu: *Terminalia paniculata* Roth; *Combretac.;* sw. O.-Ind. (Madras)
Pullewood: *Dialium guineense* Willd.; *Caesalpiniac.;* W.-Af.
Pulu: *Kydia calycina* Roxb.; *Malvac.;* n. O.-Ind. (United Prov)
Pulusu-kayalu: *Averrhoa bilimbi* L.; *Oxalidac.;* w. O.-Ind. (n. Bombay)
Pulverholz: *Rhamnus frangula* L.; *Rhamnac.;* Eu., As., n. Af.
Pumacua: *Bixa orellana* L.; *Bixac.;* Mex.
Puma-maqui: *Oreopanax sp.; Araliac.;* Ek.
Pumboyapake: *Landolphia sp.; Apocynac.;* D.-O.-Af.
Pumpumjuche: *Pachira macrocarpa* Walp., *P. aquatica* Aubl.; *Bombacac.;* Salv.. Guat.
Pumpwood: *Cecropia sp.; Morac.;* br. Gu.
Puna: *Calophyllum inophyllum* L.; *Guttifer.;* .O.-Ind. (Madras)
Punag: *Mallotus philippinensis* Muell. Arg.; *Euphorbiac.;* Beng.
Punah: *Tetramerista glabra* Miq.; *Ochnac.;* Malay.
Punán: *Iryanthera tricornis* Ducke; *Myristicac.;* Bras. (Amaz.)
Punas: *Calophyllum inophyllum* L.; *Guttifer.;* O.-Ind. (Madras)
Pundúng: *Avicennia alba* Blume; *Verbenac.;* Phil. (Cotabato)
Punga: *Chlorophora excelsa* Bth. et Hook.; *Morac.;* B.-K. (Beni, Semliki)
Pungai: *Coelostegia griffithi* Bth.; *Bombacac.;* Malay.
Punggai: *Coelostegia griffithi* Bth.; *Bombacac.;* Malay.
Pungó': *Hopea philippinensis* Dyer; *Dipterocarpac.;* Phil. (Samar)
Puni: *Pithecolobium jupunba* Urb.; *Mimosac.;* Trin., Tob.
Punkwa: *Entandrophragma cylindricum* Sprague; *Meliac.;* W.-Af.
Punnai: *Calophyllum inophyllum* L.; *Guttifer.;* sw. O.-Ind. (Madras)
Punnapay: *Calophyllum inophyllum* L.; *Guttifer.;* sw. O.-Ind. (Malabar)
Punsdhi: *Carallia lucida* Roxb.; *Rhizophorac.;* sw. O.-Ind. (Bombay)
Punta de lanza: *Vismia guianensis* Pers., *V. aff. ferruginea* H. B. K.; *Guttifer.;* Kol.
„ „ sarvia: *Miconia impetiolaris* D. Don; *Melastomatac.;* Kol.
Puoi: *Lagerstroemia angustifolia* Pierre; *Lythrac.;* Ind.-Ch.
„ dông: *Lagerstroemia siamea* Gagnep.; *Lythrac.;* Ind.-Ch.
„ khao: *Lagerstroemia balansae* Koehne; *Lythrac.;* Ind.-Ch.
„ nam: *Terminalia chebula* Retz.; *Combretac.;* Ind.-Ch.
„ smoule: *Cleidion javanicum* Blume; *Euphorbiac.;* Ind.-Ch.
Pupalasu: *Butea frondosa* Roxb.; *Papilionac.;* sw. O.-Ind. (Travancore)
Pupúut: *Dysoxylum turczaninowi* C. DC.; *Meliac.;* Phil. (Ilocos Sur)
Purá': *Shorea palosapis* Merr.; *Dipterocarpac.;* Phil. (Camarines, Albay, Sorsogon)
Purapunna: *Calophyllum wightianum* Wall.; *Guttifer.;* sw. O.-Ind. (Malabar)
Purau: *Hibiscus tiliaceus* L.; *Malvac.;* Südsee (Tahiti)
Purbo: *Alstonia scholaris* R. Brown; *Apocynac.;* nö. O.-Ind. (Nepal, Sikkim, Bhutan)
Pureau: *Hibiscus tiliaceus* L.; *Malvac.;* Südsee (Tahiti)
Purendhequa: *Erythrina sp.; Papilionac.;* Mex.
Pureque: *Erythrina sp.; Papilionac.;* Mex.
Purga de gentio: *Joannesia princeps* Vell.; *Euphorbiac.;* Bras. (Sao P)
Purgación: *Machaerium arboreum* Vogel; *Papilionac.;* n. Kol.
Purgo: *Mimusops sp.; Sapotac.;* Ven.
Purguo: *Mimusops balata* Gaertn., *M. spp.; Sapotac.;* Ven.
Purio: *Oxandra lanceolata* Baill.; *Anonac.;* P.-R.
Puriri: *Vitex littoralis* A. Cunn., *V. lucens* T. Kirk; *Verbenac.;* N.-Seel.
Puroa: *Mallotus philippinensis* Muell. Arg.; *Euphorbiac.;* nö. O.-Ind. (Nepal, Sikkim)
Purõnöö: *Mansonia altissima* Chev.; *Sterculiac.;* Go.
Purpel law: *Condalia obovata* Hook.; *Rhamnac.;* s. USA.
Purperhart: *Peltogyne paniculata* Bth., *P. spp.; Caesalpiniac.;* Sur.
„ : *Copaifera bracteata* Bth.; *Caesalpiniac.;* Sur.
„ , echte (h): *Peltogyne paniculata* Bth.; *Caesalpiniac.;* Sur.

Purperhart, fijne: *Peltogyne paniculata* Bth.; *Caesalpiniac.; Sur.
„ , glanzende: *Peltogyne densiflora* Spruce; *Caesalpiniac.; Sur.
„ , grove: *Peltogyne confertiflora* Bth.; *Caesalpiniac.; Sur.
„ , roode: *Peltogyne sp.; Caesalpiniac.; Sur.
„ , witte: *Martiusia excelsa* Bth., *M. parvifolia* Bth.; *Caesalpiniac.; Sur.
Purperhout: *Peltogyne spp.; Caesalpiniac.; Sur.
Purpleheart: *Peltogyne porphyrocardia* Griseb.; *Caesalpiniac.; Trin., Tob.
„ „ : *Copaifera marti* Hayne; *Caesalpiniac.; Trin.
„ „ : *Peltogyne purpurea* Pittier; *Caesalpiniac.; Pan.
„ „ : *Peltogyne pubescens* Bth., *P. paniculata* Bth., *P. spp.; Caesalpiniac.;
„ „ : *Copaifera bracteata* Bth.; *Caesalpiniac.; Gu., Bras.   [br. Gu.
„ „ : *Hymenaea sp.; Caesalpiniac.; Gu., Bras.   •
„ wood (e): *Peltogyne spp.; Caesalpiniac.; C.-Am., S.-Am.
Purpurholz: *Copaifera bracteata* Bth.; *Caesalpiniac.; Gu., n. S.-Am.
Purpurhout: *Dalbergia nigra* Allem.; *Caesalpiniac.; W.-Ind., Bras., Arg.
Purput: *Randia dumetorum* Lamk.; *Rubiac.; C.-O.-Ind.
Purpuurheart (h): *Peltogyne spp.; Caesalpiniac.; C.-Am., S.-Am.
Purush: *Chloroxylon swietenia* DC.; *Rutac.; O.-Ind. (Madras)
Pusi pan: *Ehretia laevis* Roxb.; *Borraginac.; w. Beng. (Chota Nagpur)
Puska: *Schleichera trijuga* Willd.; *Sapindac.; O.-Ind. (Madras)
Pusku: *Schleichera trijuga* Willd.; *Sapindac.; O.-Ind. (Madras)
Pusopúso: *Neolitsea vidali* Merr., *Litsea sp.; Laurac.; Phil.
„ : *Buchanania sp.; Anacardiac.; Phil. (Mindoro)
„ : *Quercus sp.; Fagac.; Phil. (Bataan, Marinduque)
Pussur: *Carapa moluccensis* Lamk.; *Meliac.; Beng.
Pusum: *Malacantha warneckeana* Engl.; *Sapotac.; Togo.
Put: *Carica papaya* L.; *Caricac.; br. Hond.
Putah: *Psidium guajava* L.; *Myrtac.; br. Hond.
Putali: *Sterculia urens* Roxb.; *Sterculiac.; O.-Ind. (Madras)
Puta locus: *Pithecolobium racemiflorum* Ducke; *Mimosac.; Sur.
„ „ : *Stryphnodendron flammatum* Kleinh.; *Mimosac.; Sur.
Putat: *Barringtonia speciosa* Forst.; *Lecythidac.; Malay.
„ : *Barringtonia acutangula* Gaertn., *B. racemosa* Blume; *Lecythidac.; Phil.
„ : *Planchonia spectabilis* Merr.; *Lecythidac.; Phil. (Cagayan)
„ bukit: *Madhuca watsoni* Ridl.; *Sapotac.; Malay (Kuantan)
„ tepi: *Helicia petiolaris* Benn.; *Proteac.; Malay.
Puthang bolli: *Poeciloneuron indicum* Bedd.; *Guttifer.; sw. O.-Ind. (Madras)
Putían: *Alangium meyeri* Merr., *Beilschmiedia sp.; Laurac.; Phil.
„ : *Cyclostemon sp.; Euphorbiac.; Phil. (Mindoro)
„ : *Dehaasia triandra* Merr.; *Laurac.; Phil. (Negrosinsel)
„ : *Sideroxylon macranthum* Merr.; *Sapotac.; Phil. (Manila)
„ : *Vatica mangachapoi* Blco.; *Dipterocarpac.; Phil. (Pangasinan)
Putíg: *Cyclostemon sp.; Euphorbiac.; Phil. (Zamboanga)
Putik-putík: *Eugenia brevistylis* C. B. Rob.; *Myrtac.; Phil. (Zamboanga)
Putjung: *Pangium edule* Reinw.; *Flacourtiac.; Malay.
Putli: *Acer oblongum* Wall.; *Acerac.; nw. O.-Ind. (Kumaon)
Putumuju: *Centrolobium robustum* Mart., *C. spp.; Papilionac.; Bras.
„ vermelha: *Centrolobium sp.; Papilionac.; Bras. (Bahia)
Putunín: *Eupatorium albicaule* Schultz Bip.; *Composit.; Hond.
Putuputu: *Ricinodendron africanum* Muell. Arg.; *Euphorbiac.; Nig.
Putútan: *Bruguiera sexangula* Poir., *B. conjugata* Merr.; *Rhizophorac.; Phil.
Putzmucuy: *Bumelia retusa* Swartz; *Sapotac.; Mex. (Yucatan, Maya)
Puva: *Schleichera trijuga* Willd.; *Sapindac.; O.-Ind. (Madras)
Puvan: *Schleichera trijuga* Willd.; *Sapindac.; O.-Ind. (Madras)
Puwak: *Areca catechu* L.; *Palmac.; O.-Ind., Sunda.
Puxari: *Buxus sempervirens* L.; *Buxac.; Griechenland.
Puxos: *Buxus sempervirens* L.; *Buxac.; Griechenland.
Puyúgau: *Xylocarpus moluccensis* M. Roem.; *Meliac.; Phil. (Ticao)
Puzala: *Betula alnoides* Ham.; *Betulac.; n. O.-Ind. (Jaunsar)
Pwabet: *Cassia fistula* L.; *Caesalpiniac.; Burma.
Pyaukseik: *Holoptelea integrifolia* Planch.; *Ulmac.; Burma.
Pycnantha wattle: *Acacia pycnantha* Bth.; *Mimosac.; br. O.-Af. (Niassa)

Pyek: *Pinus khasya* Royle, *P. merkusi* Jungh.; *Conifer.;* Burma.
Pyi-nyaung: *Ficus bengalensis* L.; *Morac.;* Burma.
Pyin: *Xylia dolabriformis* Bth.; *Mimosac.;* Burma.
Pyingado: *Xylia dolabrifoimis* Bth.; *Mimosac.;* O.-Ind., Burma.
Pyingadu: *Xylia dolabriformis* Bth.; *Mimosac.;* O.-Ind., Burma.
Pyingog: *Xylia dolabriformis* Bth.; *Mimosac.;* O.-Ind., Burma.
Pyinkado (H): *Xylia dolabriformis* Bth.; *Mimosac.;* Burma.
„ , abo: *Xylia dolabriformis* Bth.; *Mimosac.;* Burma.
„ , ama: *Xylia dolabriformis* Bth.; *Mimosac.;* Burma.
„ , burmese: *Xylia dolabriformis* Bth.; *Mimosac.;* Burma.
Pyinma: *Lagerstroemia flos-reginae* Retz.; *Lythrac.;* Burma.
„ : *Lagerstroemia hypoleuca* Kurz; *Lythrac.;* And.
, , andaman: *Lagerstroemia hypoleuca* Kurz; *Lythrac.;* And.
„ -byu: *Lagerstroemia floribunda* Jack; *Lythrac.;* Burma.
„ -pyu: *Lagerstroemia tomentosa* Presl; *Lythrac.;* s. Burma (s. Tenasserim)
Pyitá: *Psidium guajava* L.; *Myrtac.;* S.-Am.
Pyu: *Rhizophora mucronata* Lamk.; *Rhizophorac.;* s. Burma (s. Tenasserim)

Q

Qua vai: *Nephelium litchi* Camb.; *Sapindac.;* Ind.-Ch. (Tonkin)
Quabajohuatli: *Hura polyandra* Baill.; *Euphorbiac.;* Mex.
Quachi: *Picraena excelsa* Lindl.; *Simarubac.;* fr. W.-Ind.
Quamwood: *Schizolobium kellermani* Pittier; *Caesalpiniac.;* br. Hond.
„ : *Schizolobium parahybum* Balke; *Caesalpiniac.;* br. Hond., Pan.
Quan nui: *Chaetocarpus castanocarpus* Thw.; *Euphorbiac.;* Ind.-Ch.
Quandong: *Elaeocarpus grandis* F. v. M.; *Elaeocarpac.;* Au.
„ , blue: *Elaeocarpus grandis* F. v. M.; *Elaeocarpac.;* Au.
„ , white: *Elaeocarpus kirtoni* F. v. M.; *Elaeocarpac.;* Au.
Quang-nam: *Strychnos sp.; Loganiac.;* Annam.
Quânh quack: *Dalbergia paniculata* Roxb.; *Papilionac.;* Ind.-Ch.
Quano: *Ficus dendrocida* H. B. K.; *Morac.;* Kol.
Quao: *Catalpa longissima* Sims; *Bignoniac.;* Ind.-Ch.
Quapalier à gros fruits: *Sloanea massoni* Swartz; *Tiliac.;* Ant.
Quapinole: *Hymenaea courbaril* L.; *Caesalpiniac.;* tr. Am.
„ jutahy: *Hymenaea courbaril* L.; *Caesalpiniac.;* P.-R.
Quapolier: *Sloanea sp.; Elaeocarpac.;* fr. Gu.
Quarabu: *Astronium faxinifolium* Schott; *Anacardiac.;* ö. Bras.
Quaresma: *Tibouchina mutabilis* Cogn.; *Melastomatac.;* Bras.
Quaresmeira: *Tibouchina granulosa* Cogn., *T. sp.; Melastomatac.;* Bras.
Quaruba: *Vochysia guatemalensis* J. D. Smith, *V. obidensis* Ducke; *Vochysiac.;* n. Bras.
„ de flores roxas: *Erisma uncinatum* Warm.; *Vochysiac.;* n. Bras. (Amaz.)
Quarush: *Quercus ilex* L.; *Fagac.;* n. Af.
Quasha: *Caesalpinia crista* L.; *Caesalpiniac.;* W.-Ind. (Grenada)
Quassia (u): *Quassia amara* L.; *Simarubac.;* n. S.-Am.
„ , Jamaica-: *Picranea excelsa* Lindl.; *Simarubac.;* W.-Ind.
„ jaune: *Picraena excelsa* Lindl.; *Simarubac.;* Ant.
„ , Surinam- (u): *Quassia amara* L.; *Simarubac.;* n. S.-Am.
„ verum: *Quassia amara* L.; *Simarubac.;* fr. W.-Ind.
Quassiaholz, echtes (d): *Quassia amara* L.; *Simarubac.;* n. S.-Am.
„ , Jamaica- (d): *Picraena excelsa* Lindl.; *Simarubac.;* W.-Ind.
Quassier amer: *Quassia amara* L.; *Simarubac.;* Gu.
„ de la Jamaïque: *Picraena exelsa* Lindl.; *Simarubac.;* Ant.
Quat: *Citrus japonica* Thunb.; *Rutac.;* Ind.-Ch.
„ hông bi: *Clausena lunsium* Skeels; *Rutac.;* Ind.-Ch.
Quatélé: *Lecythis sp.; Lecythidac.;* fr. Gu.
Quauhalagua: *Heliocarpus reticulatus* Rose; *Tiliac.;* Mex.
Quauhtlatlatzin: *Hura polyandra* Baill.; *Euphorbiac.;* Mex.
Que: *Cinnamomum parthenoxylon* Meissn.; *Laurac.;* Ind.-Ch.
Quebrachillo: *Acanthosyris spinescens* Griseb., *A. flacata* Gris.; *Santalac.;* S.-Am.
Quebracho: *Schinopsis lorentzi* Engl., *S. balansae* Engl.; *Anacardiac.;* Arg.
„ (u): *Sloanea sp.; Elaeocarpac.;* Jam.

Quebracho (u): *Tecoma sp.; Bignoniac.;* Hond.
„ : *Krugiodendron ferreum* Urban; *Rhamnac.;* br. Hond.
„ : *Lysiloma acapulcense* Bth.; *Mimosac.;* Hond.
„ : *Tabebuia chrysantha* Nicholson; *Bignoniac.;* Hond.
„ : *? Pithecolobium arboreum* Urban; *Mimosac.;* Guat., Hond.
„ : *Lonchocarpus michelianus* Pittier; *Papilionac.;* Salv.
„ : *Piptadenia constricta* MacBride; *Mimosac.;* Salv.
„ : *Lysiloma divaricata* Steud.; *Mimosac.;* Salv.
„ : *Astronium fraxinifolium* Schott; *Anacardiac.;* n. Kol.
„ blanco: *Poeppigia procera* Presl; *Caesalpiniac.;* Salv.
„ „ : *Aspidosperma sp.; Apocynac.;* Sur.
„ „ : *Aspidosperma quebracho-blanco* Schlecht.; *Apocynac.;* Par., Arg.
„ „ colorado: *Aspidosperma quebracho-blanco* Schlecht.; *Apocynac.;* Par., Arg.
„ „ lloron: *Aspidosperma quebracho-blanco* Schlecht.; *Apocynac.;* Par., Arg.
„ chaqueño: *Schinopsis balansae* Engl.; *Anacardiac.;* Arg.
„ colorado: *Schinopsis balansae* Engl., *S. lorentzi* Gris.; *Anacardiac.;* Arg.
„ crespo: *Schinopsis sp.; Anacardiac.;* Arg.
„ flojo: *Iodina rhombifolia* Hook. et Arn.; *Santalac.;* S.-Am.
„ macho: *Schinopsis sp.; Anacardiac.;* Arg.
„ moro: *Schinopsis sp.; Anacardiac.;* Arg.
„ negro: *Schinopsis sp.; Anacardiac.;* Arg.
„ , red: *Schinopsis lorentzi* Engl.; *Anacardiac.;* S.-Am.
„ rubio: *Schinopsis balansae* Engl., *S. lorentzi* Engl.; *Anacardiac.;* S.-Am.
„ santiagueño: *Schinopsis lorentzi* Engl.; *Anacardiac.;* Arg.
„ , white: *Aspidosperma quebracho-blanco* Schl.; *Apocynac.;* Arg., Urug., Par.
„ de cerro: *Diphysa robinoides* Bth.; *Papilionac.;* Hond.
Quebrachoholz, rotes: *Schinopsis balansae* Engl., *S. lorentzi* Engl.; *Anacardiac.;* S.-Am.
„ „ : *Aspidosperma quebracho-colorado* Schl.; *Apocynac.;* S.-Am.
„ , weisses (d): *Aspidosperma quebracho-blanco* Schl.; *Apocynac.;* Arg., Par., sw.
Quebracho bianco (i): *Aspidosperma quebracho-blanco* Schl.; *Apocynac.;* Arg.; [Bras.
Quebrahacha: *Astronium fraxinifolium* Jacq.; *Anacardiac.;* Kol. [Par., Bras.
Queensland Nut: *Macadamia ternifolia* F. v. M.; *Proteac.;* Au.
„ „ , small fruited: *Macadamia minor* F. v. M.; *Proteac.;* Au.
Queenwood: *Ocotea rodioei* Mez; *Laurac.;* br. Gu.
„ : *Piptadenia cebil* Griseb.; *Mimosac.;* Bras.
„ (u): *Endiandra palmerstoni* C. T. White; *Laurac.;* Au.
Queh-mahr: *Albizzia zygia* MacBride; *Mimosac.;* Lib.
Queiroca: *Erica cinerea* L.; *Ericac.;* Sp.
Quejigo: *Quercus lusitanica* Lam.; *Fagac.;* Sp.
Quélipé: *Ochrocarpus africanus* Oliv.; *Guttifer.;* Elf.
Quelipekelipe: *Ochrocarpus africanus* Oliv.; *Guttifer.;* Elf.
Quelo: *Parinarium anamense* Hance; *Rosac.;* Ind.-Ch. (Kamb.)
Quên: *Flacourtia cataphracta* Roxb.; *Flacourtiac.;* Ind.-Ch.
Quenepa: *Melicocca bijuga* L.; *Sapindac.;* P.-R.
Quenepe: *Melicocca bijuga* L.; *Sapindac.;* Trin.
Quenette: *Melicocca bijuga* L.; *Sapindac.;* tr. Am.
Quenua: *Polylepis racemosa* Ruiz et Pav.; *Rosac.;* Peru.
Quéo: *Mangifera duperreana* Pierre, *M. foetida* Lour.; *Anacardiac.;* Ind.-Ch.
Quercia: *Quercus pedunculata* Ehrh.; *Fagac.;* It.
Quercitron: *Quercus velutina* Lam.; *Fagac.;* ö. USA.
Queriche: *Quercus ilex* L.; *Fagac.;* M.-M.
Quesillo: *Malvaviscus grandiflorus* H. B. K.; *Malvac.;* Hond.
Quiarapaíba: *Tecoma sp.; Bignoniac.;* Bras.
Quibaba da Mussengue: *Khaya anthoteca* C. DC.; *Meliac.;* port. W.-Af. (Angola)
„ „ Queta: *Entandrophragma angolense* C. DC.; *Meliac.;* port. W.-Af. (Angola)
Quiebrahacha: *Allophylus occidentalis* Radlk.; *Sapindac.;* P.-R.
„ : *Thouinia striata* Radlk., *T. portoricensis* Radlk.; *Sapindac.;* P.-R.
„ : *Eugenia pseudopsidium* Jacq. *var. portoricensis* K. et U.; *Myrtac.;* P.-R.
„ : *? Copaifera hymenaefolia* Moric.; *Caesalpiniac.;* Cuba.
„ : *Schinopsis spp.; Anacardiac.;* S.-Am.
„ : *Bulnesia arborea* Engl.; *Zygophyllac.;* Mex.

Quiebrahacha: *Pseudocopaiva hymenaefolia* Britt. et Wilson; *Caesalpiniac.;* Cuba.
„ : *Caesalpinia punctata* Willd.; *Caesalpiniac.;* Ven.
Quiébra-hácha: *Hopea plagata* Vidal; *Dipterocarpac.;* Phil. (Zamboanga)
„ -muelas: *Drimys winteri* Forst.; *Magnoliac.;* C.-R.
Quigua: *Amyris balsamifera* L.; *Rutac.;* Ven.
Quilin: *Prosopis sp.; Mimosac.;* Arg.
Quilite: *Erythrina sp.; Papilionac.;* Salv.
Quillai: *Quillaja saponaria* Molina; *Rosac.;* Chile.
Quillay: *Gleditschia amorphoides* Taub.; *Caesalpiniac.;* Arg.
Quina: *Cinchona spp.; Rubiac.;* nw. S.-Am.
„ : *Cinchona pubescens* Vahl; *Rubiac.;* Ven.
„ : *Pogonopus febrifugus* Bth. et Hook.; *Rubiac.;* tr. Am.
„ : *Antirrhoea coriacea* Urb.; *Rubiac.;* P.-R.
„ : *Hasseltia mexicana* Standl.; *Flacourtiac.;* Guat.
„ : *Anacardium rhinocarpus* DC.; *Anacardiac.;* C.-R.
„ : ? *Macrocnemum roseum* Wedd.; *Rubiac.;* Kol.
„ : *Erythroxylum sp.; Erythroxylac.;* Bras. (Sao P)
„ : *Myroxylon toluiferum* H. B. K.; *Papilionac.;* Arg.
„ : *Quassia amara* L.; *Simarubac.;* Bras.
„ : *Aspidosperma sp.; Apocynac.;* Bras. (Bahia)
„ : *Coutarea hexandra* K. Schum.; *Rubiac.;* Arg.
„ brava: *Erythroxylum sp.; Erythroxylac.;* Bras. (Sao P)
„ „ : *Picraena excelsa* Lindl.; *Simarubac.;* Arg.
„ falsa: *Exostema caribaeum* Roem. et Schult.; *Rubiac.;* s. Fl., Mex.
„ de Cayenne: *Quassia amara* L.; *Simarubac.;* Gu.
„ do matto: *Esenbeckia febrifuga* Mart.; *Rutac.;* Bras.
„ -quina: *Myroxylon toluiferum* H. B. K.; *Papilionac.;* Bol., Arg.
„ „ : *Ladenbergia undata* Klotzsch; *Rubiac.;* Ven.
Quince: *Cydonia vulgaris* Pers.; *Rosac.;* s. Eu., USA.
„ : *Cydonia oblonga* Mill.; *Rosac.;* Hond.
„ , Bengal-: *Aegle marmelos* Correa; *Rutac.;* O.-Ind., Burma.
Quinguió: *Aptandra spruceana* Miers; *Olacac.;* Bras. (Amaz.)
Quinilla: *Warzcewiczia coccinea* Klotzsch; *Rubiac.;* Peru.
Quinon: *Drimys winteri* Forst.; *Magnoliac.;* Kol.
Quinquina africain: *Sarcocephalus esculentus* Afzel.; *Rubiac.;* Elf.
„ caraibe (f): *Exostemma caribaeum* Roem. et Sch.; *Rubiac.;* W.-Ind., C.-Am.
„ „ Cayenne: *Quassia amara* L.; *Simarubac.;* fr. W.-Ind.
„ „ Virginie: *Magnolia glauca* L.; *Magnoliac.;* ö. USA.
Quinua: *Polylepis sp.; Rosac.;* Ek.
Quiote: *Bursera simaruba* Sarg.; *Simarubac.;* Mex.
Quipito hediondo: *Randia spinosa* Karst.; *Rubiac.;* Ven.
Quipo: *Cavanillesia platanifolia* H. B. K.; *Bombacac.;* Pan.
Quira: *Platymiscium polystachyum* Bth., *P. sp.; Papilionac.;* Pan.
„ : *Andira inermis* H. B. K.; *Papilionac.;* Pan.
Quirindal: *Licania arborea* Seem.; *Rosac.;* Mex.
„ cacahuananche: *Licania arborea* Seem.; *Rosac.;* Mex.
Quiripyranga: *Apuleia ferrea* Mart.; *Caesalpiniac.;* Bras. (Sao P)
Quirola: *Erica umbellata* L.; *Ericac.;* Sp.
Quisanda: *Coccoloba pittieri* R. Knuth; *Polygonac.;* Ven.
Quishuar: *Gynoxis sp.; Composit.;* Ek.
„ : *Buddleia longifolia* H. B. K.; *Loganiac.;* Peru.
Quisjoche: ? *Brosimum costaricanum* Liebm., ? *B. terrebanum* Pitt.; *Morac.;* Mex.
Quìt: *Atalantia disticha* Merr.; *Rutac.;* Ind.-Ch.
„ trang: *Atalantia disticha* Merr.; *Rutac.;* Ind.-Ch.
Quitarán: *Colubrina ferruginosa* Brong.; *Rhamnac.;* P.-R.
Quitasol: *Escallonia tortuosa* H. B. K.; *Saxifragac.;* Ek., Ven.
Quito sol: *Licania arborea* Seem.; *Rosac.;* Kol.
Quizarrá: *Phoebe betazensis* Mez; *Laurac.;* C.-R.
Quon: *Schizolobium parahybum* Blake; *Caesalpiniac.;* Nic.
Quy: *Cinnamomum parthenoxylon* Meissn.; *Laurac.;* Ind.-Ch.

R

Ra giong: *Parkia streptocarpa* Hance; *Mimosac.;* Ind.-Ch.
„ huong: *Cinnamomum camphora* Nees et Eberm.; *Laurac.;* Ind.-Ch.
„ ran: *Holoptelea integrifolia* Planch.; *Ulmac.;* Ind.-Ch.
Rabai: *Cassia abbreviata* Oliv.; *Caesalpiniac.;* n. O.-Af. (it. Somali)
Rabba-rabba: *Inga urnifera* Kleinh.; *Mimosac.;* Sur.
Rabo: *Lonchocarpus albiflorus* Hassl.; *Caesalpiniac.;* Arg.
„ junco: *Casearia arborea* Urban; *Flacourtiac.;* P.-R.
„ macao: *Lonchocarpus nitidus* Bth.; *Papilionac.;* Arg.
„ molle: *Lonchocarpus costatus* Bth.; *Papilionac.;* Arg.
„ de arara: *Vochysia sp.; Vochysiac.;* Bras.
„ „ „ : *Warzcewiczia coccinea* Klotzsch; *Rubiac.;* Bras.
„ „ león: *Coccoloba acuminata* H. B. K.; *Polygonac.;* Hond.
„ „ macaco: *Lonchocarpus albiflorus* Hassl., *L. spp.; Papilionac.;* Arg.
„ „ tucano: *Vochysia oppugnata* Warm.; *Vochysiac.;* Bras.
Rac: *Melanorrhoea usitata* Wall.; *Anacardiac.;* Ind.-Ch.
Rach: *Sapium sebiferum* Roxb.; *Euphorbiac.;* Annam.
Racine: *Humiria sp.; Humiriac.;* fr. Gu.
„ de bruyère: *Erica arborea* L.; *Ericac.;* M.-M.
Racoube de la Guyane; *Homalium sp.; Flacourtiac.;* fr. Gu.
Racta-panga: *Curatella americana* L.; *Dilleniac.;* Peru.
Radal: *Lomatia obliqua* R. Br.; *Proteac.;* s. Chile, s. Arg.
Radbar: *Tecoma undulata* G. Don; *Bignoniac.;* Arab., w. O.-Ind.
Radja boenga: *Adenanthera tamarindifolia* Roxb.; *Mimosac.;* Java.
Ragat: *Soymida febrifuga* A. Juss.; *Meliac.;* C.-O.-Ind.
Ragatbera: *Pterocarpus marsupium* Roxb.; *Papilionac.;* C.-O.-Ind.
Ragha: *Abies pindrow* Spach; *Conifer.;* nw. O.-Ind. (Kumaon)
Ragi: *Ficus religiosa* L.; *Morac.;* O.-Ind.
Ragwood: *Pseudomorus brunoniana* Bur.; *Morac.;* Au.
Rahaina: *Kigelia aethiopica* Dcne. var. *bornuensis* Sprague; *Bignoniac.;* n. Nig.
Rahan: *Litsea sebifera* Pers.; *Laurac.;* nw. O.-Ind. (Punjab)
Rahira: *Tecoma undulata* G. Don; *Bignoniac.;* Arab., w. O.-Ind.
Rai: *Dillenia pentagyna* Roxb.; *Dilleniac.;* sw. O.-Ind.
„ : *Ficus religiosa* L.; *Morac.;* O.-Ind. (Gondwana, Madras)
„ : *Picea morinda* Link; *Conifer.;* n. O.-Ind. (Punjab, Jaunsar)
„ : *Abies pindrow* Spach; *Conifer.;* nw. O.-Ind. (Punjab)
Raiang: *Picea morinda* Link; *Conifer.;* nw. O.-Ind. (Kumaon)
Raicita: *Relbunium hypocarpium* Hemsl.; *Rubiac.;* Ven.
Raiga: *Ficus religiosa* L.; *Morac.;* O.-Ind. (Madras)
Raini: *Mimusops hexandra* Roxb.; *Sapotac.;* w. O.-Ind. (Bombay)
„ : *Mallotus philippinensis* Muell. Arg.; *Euphorbiac.;* n. O.-Ind. (Dehra Dun)
Rain tree (e): *Enterolobium saman* Prain; *Mimosac.;* C.-Am., n. S.-Am.
Raisella: *Pinus excelsa* Wall.; *Conifer.;* nw. O.-Ind. (Kumaon)
Raisin rouge: *Staphysora klaineana* Pierre; *Euphorbiac.;* Gab.
„ vert: *Staphysora albida* Pierre; *Euphorbiac.;* Gab.
„ d'ours: *Vaccinium myrtillus* L.; *Ericac.;* Frankreich.
„ de Corinthe: *Vitis vinifera* L. var. *apyrena* L.; *Vitac.;* Griechenland.
„ des bois: *Vaccinium myrtillus* L.; *Ericac.;* Frankreich.
„ du pahouin: *Emiliamarcelia sp.; Anacardiac.;* Gab.
Raisinier à feuilles de laurier: *Coccoloba laurifolia* Jacq.; *Polygonac.;* fr. W.-Ind.
„ à fruits: *Coccoloba uvifera* Jacq.; *Polygonac.;* fr. Gu.
„ „ grandes feuilles: *Coccoloba grandifolia* Jacq.; *Polygonac.;* fr. W.-Ind.
„ „ grappes: *Coccoloba uvifera* Jacq.; *Polygonac.;* fr. Gu.
„ de mer: *Coccoloba uvifera* Jacq.; *Polygonac.;* fr. Ant.
„ „ plage: *Coccoloba uvifera* Jacq.; *Polygonac.;* fr. W.-Ind.
„ des bords de la mer: *Coccoloba uvifera* Jacq.; *Polygonac.;* Ant.
„ du bord de mer: *Coccoloba uvifera* Jacq.; *Polygonac.;* fr. W.-Ind.
Raisin tree, japanese (e): *Hovenia dulcis* Thunb.; *Rhamnac.;* n. O.-Ind., Ch., Jap.
Raiuni: *Mallotus philippinensis* Muell. Arg.; *Euphorbiac.;* nw. O.-Ind. (Punjab)
Raj-birij: *Cassia fistula* L.; *Caesalpiniac.;* C.-O.-Ind. (Bopal)
„ -briksh: *Cassia fistula* L.; *Caesalpiniac.;* nw. O.-Ind. (Kumaon)

Rajador: *Casearia sp.; Flacourtiac.;* Arg.
Rajain: *Holoptelea integrifolia* Planch.; *Ulmac.;* nw. O.-Ind. (Punjab)
Rájate bien: *Vitex sp.; Verbenac.;* Guat., Hond.
„ „ : *Vernonia spp.; Composit.;* Salv.
„ luego: *Vernonia deppeana* Less., *V. triflosculosa* H. B. K.; *Composit.;* Salv.
Rajghur: *Camellia thea* Link; *Theac.;* Assam.
Raket-berar: *Murraya exotica* L.; *Rutac.;* C.-O.-Ind. (Gondwana)
Rakhal: *Taxus baccata* L.; *Conifer.;* nw. O.-Ind. (Punjab)
Rakta chandan: *Pterocarpus santalinus* L. f.; *Papilionac.;* sö. O.-Ind. (Madras)
„ -kanchan: *Bauhinia purpurea* L.; *Caesalpiniac.;* Beng.
Rakuda (u): *Hura crepitans* L., *H. polyandra* Baill.; *Euphorbiac.;* tr. Am.
Raldhup: *Canarium strictum* Roxb.; *Burserac.;* w. O.-Ind. (Bombay)
Ram: *Anogeissus rivularis* Gagnep., *A. acuminata* Wall.; *Combretac.;* Ind.-Ch.
Rama negra: *Ruprechtia sp.; Polygonac.;* Arg.
„ „ : *Cassia corymbosa* L.; *Caesalpiniac.;* Arg.
Ramaj: *Boehmeria nivea* Gaudich.; *Urticac.;* Java.
Rambai: *Baccaurea motleyana* Muell. Arg.; *Euphorbiac.;* Malay.
Rambeh daun: *Shorea acuminata* Dyer; *Dipterocarpac.;* Malay.
Ramboetan: *Nephelium lappaceum* L.; *Sapindac.;* Java.
„ atjeh: *Nephelium lappaceum* L.; *Sapindac.;* Malay, Sunda.
„ monjet: *Podadenia javanica* J. J. Sm.; *Euphorbiac.;* Java.
„ oetan: *Xerospermum noronhianum* Blume; *Sapindac.;* Sumatra, Java.
Ramboutan: *Nephelium lappaceum* L.; *Sapindac.;* Java, Sumatra.
Rambutan: *Nephelium lappaceum* L.; *Sapindac.;* Malay.
„ jantan: *Nephelium lappaceum* L. var. ?; *Sapindac.;* Malak.
„ pachat: *Xerospermum muricatum* Radlk.; *Sapindac.;* Malay.
„ pachut: *Nephelium lappaceum* L.; *Sapindac.;* Malak.
„ passeh: *Nephelium costatum* Hiern; *Sapindac.;* Malak.
Ramdala: *Duabanga sonneratioides* Buch.-Ham.; *Sonneratiac.;* Assam.
Rami: *Boehmeria nivea* Gaudich.; *Urticac.;* Java.
Ramkanta: *Acacia sp.; Mimosac.;* nw. O.-Ind.
Ramo: *Cupania vernalis* Cambess.; *Sapindac.;* Arg.
„ camboata: *Cupania uraguensis* Hook. et Arn.; *Sapindac.;* S.-Am.
„ colorado: *Cupania vernalis* Cambess.; *Sapindac.;* Arg.
Ramón: *Brosimum alicastrum* Swartz; *Morac.;* Mex., br. Hond.
„ : *Trophis racemosa* Urban; *Morac.;* Mex., Cuba, St.-D., Jam.
„ : *Cercocarpus fothergilloides* H. B. K.; *Rosac.;* Mex.
„ , white: *Trophis racemosa* Urban; *Morac.;* br. Hond.
Ramoncillo: *Coursetia arborea* Griseb.; *Papilionac.;* n. Kol.
„ : *Trophis racemosa* Urban; *Morac.;* P.-R.
„ : *Trichilia pallida* Sw.; *Meliac.;* P.-R.
Ramontchi: *Flacourtia ramontchi* L'Hérit.; *Flacourtic.;* Mad., Réu., Java.
Ramoon: *Trophis racemosa* Urban; *Morac.;* Jam.
„ : *Trophis macrostachya* Donn. Smith; *Morac.;* Pan.
Rampacho: *Clusia sp.; Guttifer.;* Kol.
Ramphal mechiaphal: *Dillenia indica* L.; *Dilleniac.;* nö. O.-Ind. (Nepal)
Ramy: *Canarium boivini* Engl.; *Burserac.;* Mad.
„ rouge: *Canarium boivini* Engl.; *Burserac.;* Mad.
Ran: *Artocarpus hirsuta* Lamk.; *Morac.;* sw. O.-Ind. (Bombay)
„ limbu: *Atalantia monophylla* Correa; *Rutac.;* O.-Ind., Cey.
Randai-gasi (jap.): *Lithocarpus randaiensis* Hayata; *Fagac.;* Form.
„ sugi (jap.): *Cunninghamia konishii* Hayata; *Conifer.;* Form.
? Randa-kawal: *Hebecoccus ferrugineus* Radlk.; *Sapindac.;* Java (Kaliporè)
Randegonde: *Catalpa longissima* Jacq.; *Bignoniac.;* Guad.
Randoe: *Ceiba pentandra* Gaertn.; *Bombacac.;* Java.
„ allas: *Bombax malabaricum* DC.; *Bombacac.;* Nied.-Ind.
Ráng: *Pentacme suavis* A. DC.; *Dipterocarpac.;* Siam.
„ : *Broussonetia papyrifera* Vent.; *Morac.;* Annam.
„ cua: *Excoecaria oppositifolia* Muell. Arg.; *Euphorbiac.;* Ind.-Ch.
„ „ : *Croton poilanei* Gagnep.; *Euphorbiac.;* Ind.-Ch.
„ ràng: *Spatholobus orientalis* Hassk.; *Papilionac.;* Ind.-Ch.
„ „ mât: *Spatholobus orientalis* Hassk.; *Papilionac.;* Ind.-Ch.

Ráng ràng vàng: *Spatholobus orientalis* Hassk.; *Papilionac.;* Ind.-Ch.
Rángas: *Buchanania sp.; Anacardiac.;* Phil. (Pangasinan)
Rangi: *Ficus religiosa* L.; *Morac.;* O.-Ind. (Madras)
Rangong: *Nephelium eriopetalum* Miq.; *Sapindac.;* Bangka.
Ranh: *Paracleistus subgracilis* Gagnep.; *Euphorbiac.;* Ind.-Ch.
Ranie: *Alseodaphne semecarpifolia* Nees; *Laurac.;* Cey.
Ranjal: *Mimusops elengi* Roxb.; *Sapotac.;* sw. O.-Ind. (Bombay)
Ranjana: *Mimusops hexandra* Roxb.; *Sapotac.;* O.-Ind. (Bombay)
Rankan-sidé (jap.): *Carpinus rankanensis* Hayata; *Betulac.;* Form.
Ranket: *Ficus bengalensis* L.; *Morac.;* Assam. (Garo).
Ranlimbu: *Atalantia monophylla* Correa; *Rutac.;* sw. O.-Ind. (Bombay)
Ranoe: *Adinandra glabra* Miq.; *Theac.;* Sumatra.
Ransinboku: *Pistacia chinensis* Bunge; *Anacardiac.;* Form.
Ransla: *Abies pindrow* Spach; *Conifer.;* nw. O.-Ind. (Kumaon)
Rao: *Picea morinda* Link; *Conifer.;* n. O.-Ind. (Garhwal)
„ ragha: *Abies pindrow* Spach; *Conifer.;* nw. O.-Ind. (Kumaon)
Raoe: *Dracontomelum mangiferum* Blume; *Anacardiac.;* Java.
Rapadura: *Licania utilis* Fritsch; *Rosac.;* Bras.
Rappa-rappa: *Hevea guianensis* Aubl.; *Euphorbiac.;* Sur.
Raque: *Vallea stipularis* L. f.; *Elaeocarpac.;* Kol.
Rara: *Randia dumetorum* Lamk.; *Rubiac.;* nw. O.-Ind. (Punjab)
Raral: *Lomatia obliqua* R. Br.; *Proteac.;* Arg.
Rarum: *Capparis nobilis* F. v. M.; *Capparidac.;* ö. Au. (Queensl.)
Rasalla: *Cordia myxa* L.; *Borraginac.;* C.-O.-Ind.
Rasamala: *Ficus glabella* Blume; *Morac.;* Nied.-Ind. (Madoera-Sampang)
„ : *Altingia excelsa* Noronha; *Hamamelidac.;* Nied.-Ind.
Ras-berasan: *Putranjiva roxburghi* Wall.; *Euphorbiac.;* Java (Soemberwaroe)
Rascavieja: *Ononis speciosa* Lagasca, *Adenocarpus grandiflorus* Boiss.; *Papilionac.;*
Raspa-guacal: *Curatella americana* L.; *Dilleniac.;* C.-R.   [Sp.
„ lengua: *Casearia sp.; Flacourtiac.;* Cuba, C.-R.
„ viejo: *Curatella americana* L.; *Dilleniac.;* Mex.
„ del árbor: *Curatella americana* L.; *Dilleniac.;* C.-R.
Raspberry jamwood: *Acacia acuminata* Bth.; *Mimosac.;* w. Au.
Rasp pod: *Flindersia australis* R. Br.; *Rutac.;* ö. Au. (Queensl.)
Rassak: *Vatica rassak* Blume; *Dipterocarpac.;* Java, Simalu.
„ : *Shorea barbata* Brandis; *Dipterocarpac.;* br. Malay.
„ : *Cotylelobium spp.; Dipterocarpac.;* tr. As.
„ durian: *Cotylelobium flavum* Pierre; *Dipterocarpac.;* br. Borneo (Sarawak)
Rasula: *Cupressus torulosa* D. Don; *Conifer.;* n. O.-Ind. (Garhwal)
Rata: *Amoora sp.; Meliac.;* Assam.
„ , northern: *Metrosideros robusta* A. Cunn.; *Myrtac.;* N.-Seel.
„ , southern: *Metrosideros lucida* A. Rich.; *Myrtac.,* N.-Seel.
R'ata-r'ata: *Triumfetta acuminata* H. B. K.; *Tiliac.;* Peru.
Ratah: *Sapindus laurifolius* Vahl; *Sapindac.;* Pers.
Ratmanti: *Litsea polyantha* Juss.; *Laurac.;* nö. O.-Ind. (Nepal)
Raton: *Colubrina ferruginosa* Brong.; *Rhamnac.;* P.-R.
Rattan cane: *Calamus rotang* L.; *Palmac.;* O.-Ind., Cey.
Rau: *Morinda citrifolia* L.; *Rubiac.;* Ind.-Ch.
„ bung: *Barringtonia acutangula* Gaertn.; *Lecythidac.;* Ind.-Ch.
Rauli: *Nothofagus procera* Oerst.; *Fagac.;* Chile.
Rauminiya: *Bouea burmanica* Griff.; *Anacardiac.;* Burma, Malay.
Rauni: *Mallotus philippinensis* Muell. Arg.; *Euphorbiac.;* C.- & n. O.-Ind. (Garhwal)
Raunj: *Acacia leucophloea* Willd.; *Mimosac.;* n. O.-Ind. (United Prov.)
Rauschbeere: *Vaccinium uliginosum* L.; *Ericac.;* Deutschland.
Ravi: *Ficus religiosa* L.; *Morac.;* O.-Ind. (Madras)
Rawadam: *Dillenia pentagyna* Roxb.; *Dilleniac.;* w. O.-Ind. (Deccan)
Rawan: *Excoecaria virgata* Zoll. et Mor.; *Euphorbiac.;* Java (Kedoengdjati)
Rawandan: *Dillenia pentagyna* Roxb.; *Dilleniac.;* sw. O.-Ind.
Rawuya: *Kigelia aethiopica* Dcne. *var. bornuensis* Sprague; *Bignoniac.;* n. Nig.
Rayan: *Mimusops hexandra* Roxb.; *Sapotac.;* w. O.-Ind. (n. Bombay)
Rayo del sol: *Ottoschulzia cubensis* Urban; *Icacinac.;* ö. Cuba.
Razo: *Heritiera fomes* Buch.-Ham.; *Sterculiac.;* Burma (Arakan)

Ré: *Picea morinda* Link; *Conifer.;* n. O.-Ind. (Garhwal, Kaschmir)
Rè: *Cinnamomum spp.; Laurac.;* Ind.-Ch.
 „ bâu: *Cinnamomum sp.; Laurac.;* Ind.-Ch.
 „ do: *Cinnamomum tetragonum* Chev.; *Laurac.;* Ind.-Ch.
 „ gung: *Litsea sp.; Laurac.;* Ind.-Ch.
 „ huong: *Cinnamomum parthenoxylon* Meissn.; *Laurac.;* Ind.-Ch.
 „ mât: *Cinnamomum sp.; Laurac.;* Ind.-Ch.
 „ mit: *Actinodaphne sp.; Laurac.;* Ind.-Ch.
 „ vàng: *Machilus odoratissima* Nees; *Laurac.;* Ind.-Ch.
Réang: *Pentacme suavis* A. DC.; *Dipterocarpac.;* Ind.-Ch. (Kamb.)
 „ phack: *Terminalia tomentosa* Wight et Arn.; *Combretac.;* Ind.-Ch.
 „ phnom: *Pentacme suavis* A. DC.; *Dipterocarpac.;* Ind.-Ch.
 „ „ : *Careya arborea* Roxb.; *Lecythidac.;* Ind.-Ch.
 „ tuk: *Oroxylon sp.; Bignoniac.;* Ind.-Ch. (Kamb.)
Réba: *Mangifera reba* Pierre; *Anacardiac.;* Ind.-Ch.
Rebhuhnholz (d): *Vouacapoua americana* Aubl.; *Caesalpiniac.;* Sur.
Reda paladante: *Ehretia laevis* Roxb.; *Borraginac.;* O.-Ind. (Madras)
Red berry: *Hamelia nodosa* Mart. et Gal.; *Rubiac.;* Pan.
Redbud: *Cercis canadensis* L.; *Caesalpiniac.;* ö. N.-Am.
Red fowl: *Hamelia erecta* Jacq.; *Rubiac.;* br. Hond.
 „ „ : *Pithecolobium brownei* Standl., *P. winzerlingi* Br. et Rose; *Mimosac.;* br.
Redheart: *Dissiliaria baloghioides* F. v. M.; *Euphorbiac.;* Au.   [Hond.
 „ : *Dysoxylum spectabile* Hook. f.; *Meliac.;* N.-Seel.
Redi-djedoe: *Sclerolobium sp.; Caesalpiniac.;* Sur.
 „ -tjabesi: *Andira coriacea* Pulle; *Papilionac.;* Sur.
Redie-tjabisi: *Andira coriacea* Pulle; *Papilionac.;* Sur.
Redjasan: *Baccaurea javanica* Muell. Arg.; *Euphorbiac.;* Java (Tangkil)
Red narrow: *Pterocarpus indicus* Willd.; *Papilionac.;* Phil.
Redwood (d, e, u): *Sequoia sempervirens* Endl.; *Conifer.;* w. USA.
 „ (u): *Pinus silvestris* L.; *Conifer.;* Eu.
 „ : *Adenanthera pavonina* L.; *Mimosac.;* tr. As.
 „ : *Soymida febrifuga* A. Juss.; *Meliac.;* O.-Ind. (Coromandel)
 „ : *Brownlowia tabularis* Pierre; *Tiliac.;* Co.-Ch.
 „ : *Guarea glabra* Vahl; *Meliac.;* Trin., Tob.
 „ : *Mosquitoxylum jamaicense* Krug et Urban; *Anacardiac.;* br. Hond.
 „ : *Erythroxylon affine* A. Rich.; *Erythroxylac.;* br. Hond.
 „ : *Calderonia salvadorensis* Standl.; *Rubiac.;* br. Hond.
 „ : *Cistanthera papaverifera* Chev.; *Tiliac.;* W.-Af.
 „ : *Baikiaea plurijuga* Harms; *Caesalpiniac.;* n. Rhod.
 „ : *Ochna arborea* Burch.; *Ochnac.;* S.-Af. (Kap, Natal)
 „ : *Eucalyptus transcontinentalis* Maiden; *Myrtac.;* w. Au.
 „ , andaman: *Pterocarpus dalbergioides* Roxb.; *Papilionac.;* And.
 „ , baltic: *Pinus silvestris* L.; *Conifer.;* ö. N.-Eu.
 „ , Brazil-: *Brosimum paraense* Huber; *Morac.;* n. Bras. (Amaz.)
 „ , giant (u): *Sequoia gigantea* Endl.; *Conifer.;* w. USA. (Kalif.)
 „ , indian: *Caesalpinia sappan* L.; *Caesalpiniac.;* O.-Ind.
 „ , „ : *Soymida febrifuga* A. Juss.; *Meliac.;* O.-Ind., Cey.
 „ , „ : *Chukrasia tabularis* A. Juss.; *Meliac.;* tr. As.
 „ , John Crow-: *Calderonia salvadorensis* Standl.; *Rubiac.;* br. Hond.
 „ , Panama- (e, u): *Platymiscium polystachyum* Bth.; *Papilionac.;* C.-Am., n. S.-Am.
 „ , rhodesian: *Baikiaea plurijuga* Harms; *Caesalpiniac.;* Rhod.
 „ , Ridge-: *Mosquitoxylum jamaicense* Krug et Urb.; *Anacardiac.;* br. Hond.
 „ , Scrub-: *Weinmannia lachnocarpa* F. v. M.; *Cunoniac.;* ö. Au.
 „ , Swamp-: *Erythroxylon obovatum* MacFad., *E. affine* A. Rich.; *Erythroxylac.;*
 „ de Coromandel: *Soymida febrifuga* A. Juss.; *Meliac.;* O.-Ind., Cey.   [br. Hond.
 „ of the Mountains (u): *Sequoia gigantea* Endl.; *Conifer.;* w. USA. (Kalif.)
Reejoeloe: *Hymenolobium flavum* Kleinh.; *Papilionac.;* Sur.
 „ erejoeroerian: *Hymenolobium flavum* Kleinh.; *Papilionac.;* Sur.
Reel wood: *Polyalthia cerasoides* Bth. et Hook. f.; *Anonac.;* O.-Ind.
Reetha: *Sapindus mukorossi* Gaertn., *S. laurifolius* Vahl; *Sapindac.;* O.-Ind. (Hind.)
Regi: *Zizyphus jujuba* Lamk.; *Rhamnac.;* O.-Ind. (Madras)
Regnans: *Eucalyptus regnans* F. v. M.; *Myrtac.;* Tasm.

Rehen: *Glochidion arborescens* Blume; *Euphorbiac.;* Java (Tjemara)
Reiné-ala: *Adansonia madagascariensis* Baill.; *Bombacac.;* Mad.
Reinette: *Dodonaea salicifolia* DC.; *Sapindac.;* O.-Ind., Réunion.
Reisstauholz: *Dipterocarpus grandiflorus* Blco., *D. spp.; Dipterocarpac.;* Malay.
? Rejo: *Sweetia panamensis* Bth.; *Papilionac.;* Kol.
Reko zhiko: *Chlorophora excelsa* Bth. et Hook.; *Morac.;* Nig.
Rela: *Cassia fistula* L.; *Caesalpiniac.;* O.-Ind. (Gondi)
Remang: *Cratoxylon hornschuchi* Blume; *Guttifer.;* Java.
Rembangoen: *Evodia sp.; Rutac.;* Bangka.
Rememe: *Glochidion philippicum* Robinson; *Euphorbiac.;* Java (Noesa Kambangan)
„ : *Glochidion kollmannianum* J. J. Sm.; *Euphorbiac.;* Java (G. Kapal)
Remenya: *Bouea macrophylla* Griff.; *Anacardiac.;* Malay.
Remgasz: *Gluta renghas* L.; *Anacardiac.;* Malay.
Remo-caspi: *Tocoyena sp.; Rubiac.;* Peru.
Remoe-epo: *Sideroxylon guianensis* A. DC.; *Sapotac.;* Sur.
Rempelas: *Ficus ampelas* Burm., *F. melinocarpa* Blume; *Morac.;* C.- & O.-Java.
„ -bawang: *Ficus melinocarpa* Blume; *Morac.;* Java (Waderan)
Remplas: *Ficus ampelas* Burm.; *Morac.;* C.- & O.-Java.
„ -badak: *Ficus glandulifera* Wall.; *Morac.;* Java (Sanggrawa, Palaboehanratoe)
Renaco: *Ficus sp.; Morac.;* Peru.
Renga: *Zizyphus jujuba* Lam.; *Rhamnac.;* O.-Ind.
Rengas: *Gluta spp., Melanorrhoea spp., Parishia spp.; Anacardiac.;* Malay.
„ : *Melanochyla spp., Swintonia spp.; Anacardiac.;* Malay.
„ : *Glutea renghas* L.; *Anacardiac.;* Java.
„ manau: *Melanorrhoea maingayi* Hook. f.; *Anacardiac.;* Malay.
Rengha: *Zizyphus jujuba* Lam.; *Rhamnac.;* O.-Ind. (Madras)
Renghas: *Gluta renghas* L.; *Anacardiac.;* Malay., Nied.-Ind.
Reni: *Soymida febrifuga* A. Juss.; *Meliac.;* O.-Ind.
Renjoeng: *Aporosa microcalyx* Hassk.; *Euphorbiac.;* Java (Tjampea)
Renkong: *Anisoptera thurifera* Blume, *A. spp.; Dipterocarpac.;* Malay (Penang)
Repala: *Wrightia tinctoria* P. Brown; *Apocynac.;* O.-Ind. (Madras)
Rere: *Cassia occidentalis* L.; *Caesalpiniac.;* Nig.
Réré: *Macrolobium heudelotianum* Baill.; *Caesalpiniac.;* Elf.
„ èrèparè: *Andira coriacea* Pulle, ? *A. retusa* H. B. K.; *Papilionac.;* Sur.
„ de Sassandra: *Macrolobium sp.; Caesalpiniac.;* Elf.
Rerek: *Sapindus rarak* DC.; *Sapindac.;* Java.
Rerey: *Canarium commune* L.; *Burserac.;* Celebes, Menado.
Reriang: *Fagraea gigantea* Ridl., *F. fragrans* Roxb.; *Loganiac.;* Malay.
Reru: *Acacia leucophloea* Willd.; *Mimosac.;* n. O.-Ind. (United Prov.)
Resak: *Cotylelobium spp.; Dipterocarpac.;* tr. As.
„ : *Pasania spp., Quercus spp.; Fagac.;* Malay.
„ : *Shorea barbata* Brandis; *Dipterocarpac.;* Malay.
„ : *Cotylelobium melanoxylon* Pierre; *Dipterocarpac.;* Malay.
„ : *Vatica spp.; Dipterocarpac.;* Borneo, Sumatra.
„ : *Shorea spp., Hopea spp., Isoptera spp.; Dipterocarpac.;* Malay.
„ batu: *Vatica spp., Cotylelobium spp.; Dipterocarpac.;* tr. As.
„ bukit: *Cotylelobium flavum* Pierre; *Dipterocarpac.;* Malay (Pekan, Singapore)
„ hitam: *Shorea ciliata* King, *S. costata* King; *Dipterocarpac.;* Malay.
„ laru: *Pachynocarpus stapfianus* King, *Vatica wallichi* Dyer; *Dipterocarpac.;*
[Malay.
„ „ : *Vatica ridleyana* Brandis, *V. ruminata* Brandis; *Dipterocarpac.;* Malay.
„ meniyur: *Shorea glauca* King; *Dipterocarpac.;* Malak.
„ merah: *Shorea ciliata* King, *S. costata* King; *Dipterocarpac.;* Malay.
„ minyak: *Shorea sp., Hopea sp., Isoptera sp.; Dipterocarpac.;* Malay.
„ padi: *Pachynocarpus stapfianus* King, *Vatica wallichi* Dyer; *Dipterocarpac.;*
[Malay.
„ „ : *Vatica ridleyana* Brandis, *V. ruminata* Brandis; *Dipterocarpac.;* Malay.
„ pasir: *Pachynocarpus stapfianus* King, *Vatica wallichi* Dyer; *Dipterocarpac.;*
[Malay.
„ „ : *Vatica ridleyana* Brandis, *V. ruminata* Brandis; *Dipterocarpac.;* Malay.
„ paya: *Pachynocarpus stapfianus* King, *Vatica wallichi* Dyer; *Dipterocarpac.;*
[Malay.

Resak paya: *Vatica ridleyana* Brandis, *V. ruminata* Brandis; *Dipterocarpac.;* Malay.
„ tempurong: *Shorea ciliata* King, *S. costata* King; *Dipterocarpac.;* Malay.
Resbalo mono: *Bursera simaruba* Sarg.; *Burserac.;* n. Kol.
Resep: *Antidesma tetrandrum* Blume; *Euphorbiac.;* Java (Pantjoer Idjen)
„ : *Celtis wighti* Planch.; *Ulmac.;* Java (Pantjoer Idjen)
Résolu: *Chimarrhis cymosa* Jacq.; *Rubiac.;* Guad.
Retam: *Retama raetam* Webb.; *Papilionac.;* Sahara.
Retama: *Retama sphaerocarpa* Boiss.; *Papilionac.;* Sp.
„ : *Parkinsonia aculeata* L.; *Caesalpiniac.;* s. USA., Bah., W.-Ind.
„ : *Spartium junceum* L.; *Papilionac.;* Arg.
„ : *Bulnesia retamo* Griseb.; *Zygophyllac.;* Arg.
„ blanca: *Genista florida* L., *G. monosperma* Lam.; *Papilionac.;* Sp.
„ de escobas (s): *Cytisus scoparius* Link; *Papilionac.;* Eu.
„ „ España: *Spartium junceum* L.; *Papilionac.;* M.-M., Kanar.
„ „ olor (s): *Spartium junceum* L.; *Papilionac.;* M.-M., Kanar.
„ „ las Canarias: *Cytisus canariensis* Kuntze; *Papilionac.;* Kanar.
Retamo: *Trichilia hirta* L.; *Meliac.;* P.-R.
„ : *Bulnesia retamo* Griseb.; *Zygophyllac.;* Arg.
„ : *Retanilla ephedra* Vent.-Brongn.; *Rhamnac.;* Chile.
Retorton: *Prosopis sp.; Mimosac.;* Arg.
Reunj: *Acacia leucophloea* Willd.; *Mimosac.;* C.-O.-Ind.
Revu: *Acacia leucophloea* Willd.; *Mimosac.;* nw. O.-Ind. (Punjab)
Rewar: *Abies pindrow* Spach; *Conifer.;* nw. O.-Ind. (Punjab)
Rewarewa: *Knightia excelsa* R. Br.; *Proteac.;* N.-Seel.
Rewari: *Abies pindrow* Spach; *Conifer.;* nw. O.-Ind. (Punjab)
Reyghoo: *Zizyphus jujuba* Lam.; *Rhamnac.;* w. O.-Ind. (n. Bombay)
Reylu: *Cassia fistula* L.; *Caesalpiniac.;* O.-Ind. (Madras)
Rézogué: *Tarrietia sp.; Sterculiac.;* Gab.
Rhai: *Picea morinda* Link; *Conifer.;* n. O.-Ind. (Garhwal)
Rhame: *Canarium commune* L.; *Burserac.;* Nossi-Bé.
Rhatt: *Combretum glutinosum* Guill. et Perr.; *Combretac.;* Senl.
Rhetsa: *Zanthoxylum rhetsa* DC.; *Rutac.;* sw. O.-Ind. (Madras)
Rheunja: *Acacia leucophloea* Willd.; *Mimosac.;* C.-O.-Ind.
Rhin: *Quercus incana* Roxb.; *Fagac.;* Himal.
Rhodiumholz: *Liquidambar orientale* Mill.; *Hamamelidac.;* Kl.-As.
Rian: *Mimusops hexandra* Roxb.; *Sapotac.;* O.-Ind. (n. Bombay)
Riar: *Picea morinda* Link; *Conifer.;* nw. O.-Ind. (Kaschmir)
Riariadan hororodikoro: *Hymenolobium flavum* Kleinh.; *Papilionac.;* Sur.
Ribbon tree: *Eucalyptus celastroides* Turcz.; *Myrtac.;* w. Au.
Ribbonwood: *Euroschinus falcatus* Hook. f.; *Anacardiac.;* Au.
„ : *Plagianthus betulinus* A. Cunn.; *Malvac.;* N.-Seel.
„ : *Hoheria populnea* A. Cunn.; *Malvac.;* N.-Seel. (Wellington)
Ribojo: *Leea angulata* Miq.; *Ampelidac.;* Java (Soebah)
Ridan: *Allophylus sp.; Sapindac.;* Bangka.
Riemhout: *Sideroxylon guianensis* A. DC.; *Sapotac.;* Sur.
Rifari: *Anisomeris paniculata* Standl.; *Rubiac.;* Peru.
„ : *Terminalia oblonga* Steud.; *Combretac.;* Peru.
Rikio (f): *Uapaca benguelensis* Muell. Arg., *U. bingervillensis* Beille; *Euphorbiac.;* Elf.
„ „ : *Uapaca staudti* Pax; *Euphorbiac.;* Kam., Gab.
„ des marais: *Uapaca sp.; Euphorbiac.;* Elf.
„ „ montagnes: *Uapaca togoensis* Pax; *Euphorbiac.;* Elf.
„ „ rivières: *Uapaca heudeloti* Baill.; *Euphorbiac.;* Elf.
Rímas: *Artocarpus communis* Forst.; *Morac.;* Phil.
Rimi: *Ceiba pentandra* Gaertn.; *Bombacac.;* n. Nig. (Kontagara)
Rimu: *Dacrydium cupressinum* Soland.; *Conifer.;* N.-Seel.
Rin: *Quercus incana* Roxb.; *Fagac.;* nw. O.-Ind. (Punjab-Hazara)
Rinboku: *Prunus spinulosa* S. et Z.; *Rosac.;* Jap.
Ringa: *Zizyphus jujuba* Lam.; *Rhamnac.;* C.-O.-Ind. (Gondwana)
„ : *Lophira procera* Chev.; *Ochnac.;* W.-Af. (S.-L.)
Ringas: *Melanorrhoea spp.; Anacardiac.;* Malay.
„ : *Buchanania sp.; Anacardiac.;* Phil. (Ilocos Norte)
Ringhala: *Cussonia sp.; Araliac.;* Elf.

Ringworm-hout: *Vatairia guianensis* Aubl.; *Papilionac.;* fr. Gu.
Rini: *Ceiba pentandra* Gaertn.; *Bombacac.;* Nig.
Rinj: *Quercus incana* Roxb.; *Fagac.;* nw. O.-Ind. (Punjab-Hazara)
Rinjal: *Shorea robusta* Gaertn. f.; *Dipterocarpac.;* C.-O.-Ind. (Gondwana)
Rinjra: *Acacia leucophloea* Willd.; *Mimosac.;* C.-O.-Ind.
Rinkaren: *Calophyllum inophyllum* L.; *Guttifer.;* Celebes.
Rintô (jap.): *Pandanus tectorius* Solander; *Pandanac.;* Form.
Riobu: *Clethra barbinervis* S. et Z.; *Clethrac.;* Jap.
Ripi: *Antiaris africana* Engl.; *Morac.;* br. O.-Af. (Uganda)
Riquesa: *Cola acuminata* R. Br.; *Sterculiac.;* port. W.-Af.
Riráu: *Sideroxylon duclitan* Blco.; *Sapotac.;* Phil. (Camarines)
Ri-ri-do: *Erismanthus indochinensis* Gagnep.; *Euphorbiac.;* Ind.-Ch.
Rishta: *Sapindus laurifolius* Vahl, *S. mukorossi* Gaertn.; *Sapindac.;* O.-Ind.
Rita: *Sapindus laurifolius* Vahl, *S. mukorossi* Gaertn.; *Sapindac.;* O.-Ind.
Ritha: *Sapindus laurifolius* Vahl, *S. mukorossi* Gaertn.; *Sapindac.;* O.-Ind.
Riukiumatzu: *Pinus luchuensis* Mayr; *Conifer.;* Jap.
Riu-yan: *Nephelium longana* Cambess.; *Sapindac.;* Jap.
Riverwood: *Pithecolobium vahlianum* DC.; *Mimosac.;* Pan.
Riviere: *Homalium paniculatum* O. Hoffm.; *Flacourtiac.;* Maur.
Riz de veau végétal: *Blighia sapida* Koenig; *Sapindac.;* Trin.
Rmà: *Schima noronha* Blume; *Theac.;* Ind.-Ch. (Kamb.)
Ro hoa: *Rhamnoneuron balansae* Gilg; *Thymelaeac.;* Annam.
„ ka: *Bombax albidum* Gagnep.; *Bombacac.;* Ind.-Ch.
„ ko: *Ceiba pentandra* Gaertn.; *Bombacac.;* Ind.-Ch. (Kamb.)
„ moc: *Wikstroemia indica* C. A. Meyer; *Thymelaeac.;* Annam.
Roan: *Cinnamomum iners* Reinw.; *Laurac.;* Ind.-Ch.
Roatanga: *Schleichera trijuga* Willd.; *Sapindac.;* O.-Ind. (Madras)
Roatangha-nune: *Schleichera trijuga* Willd.; *Sapindac.;* w. O.-Ind. (n. Deccan)
Robby: *Eugenia moorei* F. v. M.; *Myrtac.;* Au.
Robinia (i): *Robinia pseudacacia* L.; *Papilionac.;* N.-Am.
„ , clammy-barked: *Robinia viscosa* Vent.; *Papilionac.;* N.-Am.
„ , Hispid-: *Robinia hispida* L.; *Papilionac.;* N.-Am.
Robinie, gemeine (d): *Robinia pseudacacia* L.; *Papilionac.;* N.-Am.
Robinier faux-acacia (f): *Robinia pseudacacia* L.; *Papilionac.;* N.-Am.
„ hispide: *Robinia hispida* L.; *Papilionac.;* N.-Am.
„ visqueux: *Robinia viscosa* Vent.; *Papilionac.;* N.-Am.
Roble: *Tabebuia pentaphylla* Hemsl.; *Bignoniac.;* tr. Am.
„ : *Tabebuia rigida* Urban; *Bignoniac.;* P.-R.
„ (e): *Platymiscium trinitensis* Bth.; *Papilionac.;* Trin.
„ : *Platymiscium polystachyum* Bth.; *Papilionac.;* Trin., Tob., Ven.
„ : *Catalpa longissima* Sims; *Bignoniac.;* St.-D., Haiti, P.-R.
„ : *Tabebuia donnell-smithi* Rose; *Bignoniac.;* Mex.
? „ : *Arbutus xalapensis* H. B. K.; *Ericac.;* Mex. (Sinaloa)
„ : *Bourreria mollis* Standl., *B. oxyphylla* Standl.; *Borraginac.;* br. Hond.
„ : *Quercus spp.; Fagac.;* C.-Am.
„ : *Licania sp.; Rosac.;* Salv.
„ : *Dialyanthera otoba* Warb.; *Myristicac.;* w. Pan.
„ : *Tabebuia chrysea* Blake; *Bignoniac.;* n. Kol.
„ : *Stryphnodendron rotundifolium* Mart.; *Mimosac.;* Arg.
„ : *Torresea cearensis* Allem.; *Papilionac.;* Arg.
„ : *Nothofagus obliqua* Oerst.; *Fagac.;* Chile.
„ : *Nothofagus antarctica* Oerst.; *Fagac.;* Chile.
„ albar: *Quercus pedunculata* Ehrh., *Q. sessiliflora* Griseb.; *Fagac.;* Sp.
„ amarillo: *Citharexylum caudatum* L.; *Verbenac.;* Cuba.
„ „ : *Tabebuia chrysea* Blake; *Bignoniac.;* n. Kol.
„ antárctico: *Nothofagus antarctica* Oerst.; *Fagac.;* Chile.
„ blanco: *Tabebuia pentaphylla* Hemsl.; *Bignoniac.;* P.-R., Cuba, Mex.
„ „ : *Platymiscium polystachyum* Bth.; *Papilionac.;* Ven.
? „ „ : *Tabebuia sp.; Bignoniac.;* Sur.
„ cienego: *Nothofagus sp.; Fagac.;* Arg.
„ colorado: *Platymiscium polystachyum* Bth.; *Papilionac.;* Ven.
„ „ : *Tabebuia pentaphylla* Hemsl.; *Bignoniac.;* Ven.

Roble cortes: *Tabebuia chrysantha* Nicholson; *Bignoniac.;* s. Mex.
„ coyan: *Nothofagus obliqua* Oerst.; *Fagac.;* Chile.
„ enciniego: *Quercus lusitanica* Webb.; *Fagac.;* Sp.
„ encino: *Quercus sp.; Fagac.;* sp. Am.
„ guayo: *Petitia domingensis* Jacq., *Citharexylum caudatum* L.; *Verbenac.;* Cuba.
„ hualle: *Nothofagus obliqua* Oerst.; *Fagac.;* Chile.
„ morado: *Tabebuia pentaphylla* Hemsl.; *Bignoniac.;* n. Kol.
„ negro: *Quercus toza* Gillet; *Fagac.;* Sp.
„ „ : *Quercus sp.; Fagac.;* Salv.
„ pellin: *Nothofagus obliqua* Oerst.; *Fagac.;* s. Chile.
„ rebollo: *Quercus toza* Gillet.; *Fagac.;* Sp.
„ tozio: *Quercus toza* Gillet; *Fagac.;* Sp.
„ de Chiloè: *Nothofagus nitida* Reiche; *Fagac.;* Chile.
„ „ Colchagua: *Nothofagus macrocarpa* ?; *Fagac.;* Chile.
„ „ Magellanes: *Nothofagus betuloides* Oerst.; *Fagac.;* Chile.
„ „ Magellanos: *Nothofagus antarctica* Oerst.; *Fagac.;* s. Chile.
„ „ la Montaña: *Nothofagus antarctica* Oerst.; *Fagac.;* Chile.
„ „ olor: *Catalpa punctata* Griseb.; *Bignoniac.;* Cuba.
„ „ río: *Tabebuia pentaphylla* Hemsl.; *Bignoniac.;* n. Kol.
„ „ savana: *Tabebuia pentaphylla* Hemsl.; *Bignoniac.;* Pan.
„ „ yugo: *Tabebuia pentaphylla* Hemsl.; *Bignoniac.;* Cuba.
„ del pais: *Torresea cearensis* Allem.; *Papilionac.;* Arg.
Roblecito: *Quercus oleoides* Cham. et Schl., *Q. spp.; Fagac.;* Mex., Guat.
Robo raton: *Casearia sp.; Flacourtiac.;* P.-R.
Robojunco: *Casearia arborea* Urban; *Flacourtiac.;* P.-R.
Rocco: *Chlorophora excelsa* Bth. et Hook.; *Morac.;* Dah.
Roco: *Chlorophora excelsa* Bth. et Hook.; *Morac.;* Dah.
Roekum: *Flacourtia rukam* Zoll. et Moritzi; *Flacourtiac.;* Malay.
„ sepat: *Flacourtia cataphracta* Roxb.; *Flacourtiac.;* Java.
Roghu: *Anthocephalus cadamba* Miq., *Adina cordifolia* Hook. f.; *Rubiac.;* Assam.
Rohan: *Soymida febrifuga* A. Juss.; *Meliac.;* C.-O.-Ind. (Berar)
Rohin: *Soymida febrifuga* A. Juss.; *Meliac.;* w. O.-Ind. (Gujarat)
Rohina: *Soymida febrifuga* A. Juss.; *Meliac.;* Beng.
Rohini: *Soymida febrifuga* A. Juss.; *Meliac.;* C.-O.-Ind. (Berar)
Rohitoka: *Aphanomixis rohituka* Pierre; *Meliac.;* O.-Ind.
Rohituka: *Aphanomixis rohituka* Pierre; *Meliac.;* O.-Ind.
Rohn: *Soymida febrifuga* A. Juss.; *Meliac.;* Beng.
Rohni: *Mallotus philippinensis* Muell. Arg.; *Euphorbiac.;* n. O.-Ind. (Garhwal)
Rohun: *Soymida febrifuga* A. Juss.; *Meliac.;* Beng.
Rohuna: *Soymida febrifuga* A. Juss.; *Meliac.;* O.-Ind. (Hind.)
Rohuni: *Soymida febrifuga* A. Juss.; *Meliac.;* O.-Ind.
Roi: *Picea morinda* Link; *Conifer.;* n. O.-Ind. (Jaunsar)
„ : *Eugenia jambos* L.; *Myrtac.;* Ind.-Ch.
Rôi: *Garcinia benthami* Pierre, *G. thoreli* Pierre; *Guttifer.;* Ind.-Ch.
„ : *Garcinia schefferi* Pierre; *Guttifer.;* Ind.-Ch.
„ mât: *Garcinia ferrea* Pierre; *Guttifer.;* Annam.
Roibokida: *Acacia morondavensis* Drake; *Mimosac.;* Mad.
Roibontsika: *Acacia morondavensis* Drake; *Mimosac.;* Mad.
Roimbitsika: *Acacia morondavensis* Drake; *Mimosac.;* Mad.
Roini: *Mallotus philippinensis* Muell. Arg.; *Euphorbiac.;* n. O.-Ind. (Garhwal)
Rokam: *Flacourtia cataphracta* Roxb., *F. rukam* Zoll. et Moritzi; *Flacourtiac.;* Malay.
Rokko: *Chlorophora excelsa* Bth. et Hook.; *Morac.;* Dah.
„ -rokko: *Macoubea guianensis* Aubl.; *Apocynac.;* Sur.
Roko: *Chlorophora excelsa* Bth. et Hook.; *Morac.;* Elf.
Rol: *Terminalia chebula* Retz.; *Combretac.;* w. Beng. (Chota Nagpur)
Rola: *Terminalia chebula* Retz.; *Combretac.;* w. Beng. (Chota Nagpur)
Roldán: *Amyris elemifera* L.; *Rutac.;* Salv.
Roléai thom: *Nauclea sessilifolia* Roxb.; *Rubiac.;* Ind.-Ch.
Roli: *Mallotus philippinensis* Muell. Arg.; *Euphorbiac.;* nw. O.-Ind.
Rolón: *Pithecolobium unguis-cati* Bth.; *Mimosac.;* P.-R.
Rom deng: *Elaeocarpus madopetalus* Pierre; *Elaeocarpac.;* Ind.-Ch.
Romadizo: *Siparuna sp.; Monimiac.;* Kol.

Romero: *Rosmarinus officinalis* L.; *Labiat.*; Sp.
Rompe gibao: *Bumelia sp.*; *Sapotac.*; Bras.
Rompehueso: *Casearia sp.*; *Flacourtiac.*; Cuba.
Ronces: *Rubus spp.*; *Rosac.*; Guat.
Ronde: *Erythroxylon laurifolium* Lamk.; *Erythroxylac.*; Réunion.
Roné: *Testulea gabonensis* Pellegr.; *Ochnac.*; Kam. (Campo)
Rong: *Garcinia hanburyi* Hook. f.; *Guttifer.*; Ind.-Ch.
Rongle à grandes feuilles: *Erythroxylon laurifolium* Lamk.; *Erythoxylac.*; Réu.
Rônier: *Borassus flabellifer* L. var. *aethiopum* Warb.; *Palmac.*; Sudan, Elf.
Ronrón: *Astronium graveolens* Jacq., *A. conzatti* Blake; *Anacardiac.*; C.-Am.
Rooder, Bush-, Mountain-: *Zanthoxylum aff. procerum* G. Don; *Rutac.*; br. Hond.
  „   , wild: *Zanthoxylum procerum* Donn. Smith; *Rutac.*; br. Hond.
Roodhout: *Ochna arborea* Burch.; *Ochnac.*; S.-Af. (Kap, Natal)
Rookra: *Diospyros gabunensis* Gürke; *Ebenac.*; Lib.
Róp: *Nephelium longana* Cambess.; *Sapindac.*; Ind.-Ch.
Rope tree: *Firmiana barteri* K. Schum.; *Sterculiac.*; W.-Af.
Rora: *Mallotus philippinensis* Muell. Arg.; *Euphorbiac.*; w. Beng. (Chota Nagpur)
  „ -ek: *Ocotea rodioei* Mez; *Laurac.*; br. Gu.
Rori: *Mallotus philippinensis* Muell. Arg.; *Euphorbiac.*; C.-O.-Ind.
Rosadillo: *Swietenia sp.*; *Meliac.*; Mex.
  „   : *Cordia alliodora* Cham.; *Borraginac.*; Mex.
Rosal perruno: *Rosa canina* L.; *Rosac.*; Sp.
  „   silvestre: *Rosa canina* L.; *Rosac.*; Sp.
Rosamay (u): *Dysoxylum fraseranum* Bth.; *Meliac.*; ö. Au.
Rosawa: *Vitex vitiensis* Seem.; *Verbenac.*; Fid.
Rose femelle: *Acrodiclidium sp.*; *Laurac.*; Guad.
  „   , Mountain-: *Brownea latifolia* Jacq.; *Caesalpiniac.*; Trin., Tob.
  „   , wilde (d): *Rosa canina* L.; *Rosac.*; Eu.
Roselle: *Hibiscus sabdariffa* L.; *Malvac.*; Malay.
Rosenholz (d): *Hibiscus tiliaceus* L.; *Malvac.*; Tr.
  „   „   : *Amyris balsamifera* L.; *Rutac.*; W.-Ind., nw. S.-Am.
  „   „   : *Cordia gerascanthus* L.; *Borraginac.*; W.-Ind.
  „   „   : *Drypetes ilicifolia* Krug et Urb.; *Euphorbiac.*; Jam.
  „   „   : *Aniba rosaeodora* Ducke, *A. spp.*; *Laurac.*; nö. Bras. (Pará)
  „   „   : *Ocotea costulata* Mez, *Nectandra elaiophora* Barb. Rodr.; *Laurac.*:
  „   „   : *Dalbergia nigra* Allem.; *Papilionac.*; nö. Bras.   [nö. Bras.
  „   „   : *Dysoxylum fraseranum* Bth.; *Meliac.*; Au.
  „   „   : *Acacia excelsa* Bth.; *Mimosac.*; ö. Au.
  „   , afrikanisches: *Pterocarpus erinaceus* Poir.; *Papilionac.*; tr. Af.
  „   , brasilianisches (d): *Dalbergia nigra* Allem.; *Papilionac.*; nö. Bras.
  „   , Cayenne-: *Licaria guianensis* Aubl.; *Laurac.*; Bras.
  „   , Dominica-: *Cordia alliodora* Cham.; *Borraginac.*; tr. Am.
  „   , ostindisches (d): *Dalbergia latifolia* Roxb., *D. sissoo* Roxb.; *Papilionac.*; O.-Ind.
  „   , Queensland-: *Acacia sp.*; *Mimosac.*; ö. Au. (Queensl.)
  „   , Westafrika-: *Copaifera demeusi* Harms, *C. tessamanni* Harms; *Caesalpiniac.*;
Rosewood: *Dalbergia latifolia* Roxb.; *Papilionac.*; O.-Ind.   [Gab.
  „   : *Dalbergia cochinchinensis* Pierre; *Papilionac.*; Ind.-Ch.
  „   : *Zanthoxylum rhodoxylon* P. Wilson, *Amyris balsamifera* L.; *Rutac.*; Jam.
  „   : *Drypetes ilicifolia* Krug et Urban; *Euphorbiac.*; Jam.
  „   : *Aniba panurensis* Mez; *Laurac.*; Gu.
  „   : *Dalbergia retusa* Hemsl.; *Papilionac.*; C.-R., Nic., Pan.
  „   : *Dalbergia nigra* Allem.; *Papilionac.*; Bras.
  „   : *Dalbergia sp.*; *Papilionac.*; Mex.
  „   : *Dalbergia stevensoni* Standl., *D. laevigata* Standl.; *Papilionac.*; br. Hond.,
  „   : *Dalbergia cubilquitzensis* Pittier; *Papilionac.*; Hond.   [Hond.
  „   : *Dalbergia hypoleuca* Pittier; *Papilionac.*; Nic.
  „   : *Dalbergia granadillo* Pittier; *Papilionac.*; Nic.
  „   : *Pterocarpus erinaceus* Poir.; *Papilionac.*; Sudan, Gambia.
  „   : *Dalbergia spp.*; *Papilionac.*; Mad.
  „   : *Synoum glandulosum* A. Juss.; *Meliac.*; Au.
  „   : *Dysoxylum fraseranum* Bth.; *Meliac.*; ö. Au., N.-Seel.
  „   : *Dysoxylum lessertianum* Bth.; *Meliac.*; ö. Au. (N.-S.-W.)

Rosewood: *Acacia rhodoxylon* Maiden; *Mimosac.;* ö. Au. (Queensl.)
„   : *Acacia excelsa* Bth.; *Mimosac.;* ö. Au.
„   : *Beyeria viscosa* Miq.; *Euphorbiac.;* Tasm.
„   , african: *Pterocarpus erinaceus* Poir.; *Papilionac.;* Go., Nig.
„   , australian: *Dysoxylum fraseranum* Bth.; *Meliac.;* Au.
„   , Bahia-: *Dalbergia nigra* Allem.; *Papilionac.;* ö. Bras.
„   , black: *Dalbergia latifolia* Roxb.; *Papilionac.;* O.-Ind.
„   , Bombay-: *Dalbergia latifolia* Roxb.; *Papilionac.;* O.-Ind.
„   , Borneo-: *Melanorrhoea spp.;* *Anacardiac.;* Malay.
„   , brasilian (u): *Dalbergia nigra* Allem.; *Papilionac.;* Bras.
„   , British Honduras-: *Dalbergia stevensoni* Standl.; *Papilionac.;* W.-Ind.
„   , burmesse: *Pterocarpus macrocarpus* Kurz; *Papilionac.;* Burma.
„   , Ceylon-: *Albizzia odoratissima* Bth.; *Mimosac.;* O.-Ind.
„   , Dominica-: *Cordia gerascanthus* L.; *Borraginac.;* br. W-.Ind.
„   , east indian: *Dalbergia latifolia* Roxb.; *Papilionac.;* O.-Ind.
„   , faux: *Dalbergia greveana* Baill.; *Papilionac.;* Mad. (Majunga)
„   , „   : *Dalbergia sp.;* *Papilionac.;* Mad. (Tamatave)
„   , french: *Dalbergia greveana* Baill.; *Papilionac.;* Mad. (Majunga)
„   , „   : *Dalbergia sp.;* *Papilionac.;* Mad. (Tamatave)
„   , gambian: *Pterocarpus erinaceus* Poir.; *Papilionac.;* O.-Ind.
„   , Honduras-: *Dalbergia stevensoni* Standl.; *Papilionac.;* br. Hond.
„   , indian: *Dalbergia latifolia* Roxb.; *Papilionac.;* s. O.-Ind.
„   , Madagascar-: *Dalbergia greveana* Baill.; *Papilionac.;* Mad. (Majunga)
„   , „   : *Dalbergia sp.;* *Papilionac.;* Mad. (Tamatave)
„   , „   : *Dalbergia obtusa* Lecomte; *Papilionac.;* Mad.
„   , Majunga-: *Dalbergia greveana* Baill.; *Papilionac.;* Mad.
„   , Malabar-: *Dalbergia latifolia* Roxb.; *Papilionac.;* O.-Ind.
„   , northern-: *? Caesalpinia sp.;* *Caesalpiniac.;* br. Hond.
„   , Rio-: *Dalbergia nigra* Allem.; *Papilionac.;* ö. Bras.
„   , Rosetta-: *Dalbergia latifolia* Roxb.; *Papilionac.;* O.-Ind.
„   , scentless: *Synoum glandulosum* A. Juss.; *Meliac.;* Au.
„   , Seychelles- (e): *Thespesia populnea* Correa; *Malvac.;* Sey.
„   , south indian: *Dalbergia latifolia* Roxb.; *Papilionac.;* O.-Ind.
„   , Tamatave-: *Dalbergia sp.;* *Papilionac.;* Mad.
„   , white: *Dalbergia nigra* Allem.; *Papilionac.;* Bras.
„   , Yama- (u): *Platymiscium trifoliatum* Bth.; *Papilionac.;* Nic. (Corinto)
Rosier du Japon: *Camellia japonica* L.; *Theac.;* Jap.
Rosilla: *Eysenhardtia polystachya* Sarg.; *Papilionac.;* Mex. (Sinaloa)
Roso: *Vallea stipularis* Mutis; *Tiliac.;* Ven.
Rosul: *Dalbergia spp.;* *Papilionac.;* Hond.
Rot: *Ougeinia dalbergioides* Bth.; *Papilionac.;* w. Beng. (Chota Nagpur)
Rota-kaekuna: *Canarium commune* L.; *Burserac.;* s. O.-Ind.
Rotéang: *Hopea ferrea* Pierre; *Dipterocarpac.;* Ind.-Ch.
Rotholz, afrikanisch (d): *Pterocarpus soyauxi* Taub.; *Papilionac.;* W.-Af.
„   , Altata- (d): *Caesalpinia sp.;* *Caesalpiniac.;* ? Mex.
„   , Andamanen- (d): *Pterocarpus dalbergioides* Roxb.; *Papilionac.;* And.
„   , ostindisches: *Caesalpinia sappan* L.; *Caesalpiniac.;* O.-Ind.
Rotra: *Eugenia cuneifolia* Baker; *Myrtac.;* Mad.
„   mena: *Eugenia condensata* Baker; *Myrtac.;* Mad.
„   , small-leaf-: *Eugenia sp.;* *Myrtac.;* Mad.
Roucou: *Bixa orellana* L.; *Bixac.;* tr. Am.
Rouge blanc à grandes feuilles: *Elaeodendron orientale* Jacq.; *Celastrac.;* Réu.
„   oriental: *Elaeodendron orientale* Jacq.; *Celastrac.;* Réu.
„   à petites feuilles: *Elaeodendron orientale* Jacq.; *Celastrac.;* Réu.
Rough-leaf tree (e): *Curatella americana* L.; *Dilleniac.;* tr. Am.
Rougo: *Haronga paniculata* Lodd.; *Guttifer.;* Mad.
Rougou diguen: *Borassus flabellifer* L. var. *aethiopum* Warb.; *Palmac.;* Senl.
„   gor: *Borassus flabellifer* L. var. *aethiopum* Warb.; *Palmac.;* Senl.
Roura: *Quercus sessiliflora* Salisb.; *Fagac.;* Sp.
Routra: *Eugenia parkeri* Baker; *Myrtac.;* Mad.
Rovere: *Quercus sessiliflora* Salisb.; *Fagac.;* It.
Rowan tree: *Pirus americana* DC.; *Rosac.;* ö. N.-Am.

Rowan, Bog-: *Pirus aucuparia* Ehrh.; *Rosac.;* England (Wales)
Rowanra: *Erythrina suberosa* Roxb.; *Papilionac.;* n. O.-Ind. (United Prov.)
Roxinho: *Peltogyne sp.; Caesalpiniac.;* Bras.
Rozehout (h): *Licaria guianensis* Aubl.; *Laurac.;* fr. Gu.
„ , Mannetjes-: *Licaria sp.; Laurac.;* fr. Gu.
„ , Wijfjes-: *Licaria guianensis* Aubl.; *Laurac.;* fr. Gu.
Rozenhout (h): *? Aniba panurensis* Mez; *Laurac.;* Sur.
„ : *Dalbergia nigra* L.; *Papilionac.;* Bras.
„ , echt: *? Aniba panurensis* Mez; *Laurac.;* Sur.
„ , „ : *Aniba rosaeodora* Ducke; *Laurac.;* Gu., n. Bras.
R'pa: *Lagerstroemia angustifolia* Pierre; *Lythrac.;* Ind.-Ch.
Ru: *Aleurites cordata* R. Br.; *Euphorbiac.;* Annam.
„ : *Casuarina equisetifolia* Forst.; *Casuarinac.;* Malay, Borneo.
„ ri: *Calophyllum tonkinensis* Pitard; *Guttifer.;* Ind.-Ch.
„ ronan: *Casuarina sumatrana* Miq.; *Casuarinac.;* Malay.
„ ru: *Podocarpus chinensis* Wall.; *Conifer.;* Ind.-Ch.
Rubber tree: *Ficus populnea* Willd. *var. brevifolia* Warb.; *Morac.;* tr. Am.
„ „ : *Castilloa elastica* Cerv.; *Morac.;* Mex., C.-Am.
„ „ : *Castilloa ulei* Warb.; *Morac.;* n. Bras. (Amaz.)
„ „ : *Castilloa fallax* Cook; *Morac.;* Pan.
„ „ : *Funtumia africana* Stapf; *Apocynac.;* Lib.
„ „ , assam: *Ficus elastica* Roxb.; *Morac.;* O.-Ind.
„ „ , colombian: *Sapium biglandulosum* Muell. Arg.; *Euphorbiac.;* Kol.
„ „ , false: *Holarrhena wulfsbergi* Stapf; *Apocynac.;* W.-Af.
„ „ , Lagos-: *Ficus vogeli* Miq.; *Morac.;* W.-Af. (Lagos)
„ „ , male: *Castilloa fallax* Cook; *Morac.;* C.-Am.
„ „ , „ : *Holarrhena wulfsbergi* Stapf; *Apocynac.;* W.-Af.
„ „ , mexican: *Castilloa elastica* Cerv., *C. fallax* Cook; *Morac.;* Mex.
„ „ , Panama-: *Castilloa sp.; Morac.;* C.-Am.
„ „ , Pará-: *Hevea brasiliensis* Muell. Arg.; *Euphorbiac.;* nö. Bras.
„ „ , sterile: *Castilloa fallax* Cook; *Morac.;* C.-Am.
„ „ , wild: *Ficus aurea* Nutt.; *Morac.;* s. Fl., Bah., Cuba.
„ „ , „ : *Funtumia elastica* Stapf; *Apocynac.;* br. O.-Af. (Uganda)
„ „ , „ : *Funtumia latifolia* Stapf; *Apocynac.;* br. O.-Af. (Uganda)
Rubían: *Terminalia comintana* Merr.; *Combretac.;* Phil. (Laguna)
Rubiana peluda: *Cytisus triflorus* L'Hérit.; *Papilionac.;* Sp.
Ruby wood: *Pterocarpus santalinus* L. f.; *Papilionac.;* O.-Ind., Cey., Phil.
Rucuma: *Lucuma obovata* H. B. K.; *Sapotac.;* Peru.
Ruda: *Helichrysum serotinum* Boiss.; *Composit.;* Sp.
„ : *Zanthoxylum sp.; Rutac.;* Pan.
Rudrakshachettu: *Guazuma tomentosa* H. B. K.; *Sterculiac.;* w. O.-Ind. (n. Bombay)
Rudrakadapu: *Adina cordifolia* Hook. f.; *Rubiac.;* O.-Ind. (Madras)
Rudrakcham: *Elaeocarpus tuberculatus* Roxb.; *Tiliac.;* sw. O.-Ind. (Madras)
Rudruk-shamba: *Anthocephalus cadamba* Miq.; *Rubiac.;* O.-Ind. (Madras)
Rufindi: *Inga punctata* Willd.; *Mimosac.;* Peru.
Rugando: *Acacia stenocarpa* Hochst.; *Mimosac.;* br. O.-Af. (Uganda)
Rugían: *Bambusa spinosa* Roxb.; *Gramin.;* Phil. (Bikol)
Ruhen: *Soymida febrifuga* A. Juss.; *Meliac.;* w. Beng. (Bihar, Orissa)
Ruhin: *Mallotus philippinensis* Muell. Arg.; *Euphorbiac.;* sw. O.-Ind. (Bombay)
Ruinsh: *Cotoneaster bacillaris* Wall.; *Rosac.;* O.-Ind. (Himal.)
Ruivinha: *Relbunium hypocarpium* Hemsl.; *Rubiac.;* Ven.
Ruk: *Dillenia pentagyna* Roxb.; *Dilleniac.;* Ind.-Ch.
Rukaraigo: *Combretum abbreviatum* Engl.; *Combretac.;* br. O.-Af. (Uganda)
„ : *Combretum paniculatum* Vent.; *Combretac.;* br. O.-Af. (Uganda)
Rukh: *Tamarix articulata* Vahl; *Tamaricac.;* O.-Ind.
Rukt-mara: *Myristica attenuata* Wall.; *Myristicac.;* w. O.-Ind. (s. Bombay)
Rukum: *Flacourtia cataphracta* Roxb.; *Flacourtiac.;* Malay.
Rule wood: *Lecythis sp.; Lecythidac.;* Pan.
Rumadi: *Ficus glomerata* Roxb.; *Morac.;* w. O.-Ind. (s. Bombay)
Rumbal: *Ficus glomerata* Roxb.; *Morac.;* nw. O.-Ind. (Punjab)
Rumfu: *Cassia goratensis* Fresen.; *Caesalpiniac.;* n. Nig.
Rúngon: *Ceriops tagal* C. B. Rob.; *Rhizophorac.;* Phil. (Pangasinan)

Rungra: *Erythrina suberosa* Roxb.; *Papilionac.;* n. O.-Ind. (United Prov.)
Rúngus: *Buchanania sp.; Anacardiac.;* Phil. (Cagayan)
Runhu: *Cassia goratensis* Fresen.; *Caesalpiniac.;* n. Nig.
Ruoi: *Streblus asper* Lour.; *Morac.;* Ind.-Ch.
Rura: *Parinarium curatellaefolium* Planch.; *Rosac.;* n. Nig.
Ruri: *Mammea americana* L.; *Guttifer.;* Nic.
„ : *Calophyllum sp.; Guttifer.;* Ind.-Ch.
Ruriang: *Archytaea vahli* Choisy; *Theac.;* Malak.
Ruşam: *Schleichera trijuga* Willd.; *Sapindac.;* w. Beng. (Orissa)
Rusamala: *Altingia excelsa* Noronha; *Hamamelidac.;* tr. O.-As.
Rusche (d): *Ulmus campestris* L.; *Ulmac.;* Eu., As., N.-Af.
Rüster, Bast- (d): *Ulmus effusa* Willd.; *Ulmac.;* Mitt.-Eu.
„ , Berg- (d): *Ulmus montana* Withering; *Ulmac.;* Eu.
„ , Feld- (d): *Ulmus campestris* L.; *Ulmac.;* Eu.
„ , Flatter-: *Ulmus effusa* Willd.; *Ulmac.;* Mitt.-Eu.
„ , Rot- (d): *Ulmus campestris* L.; *Ulmac.;* Eu.
„ , Weiss- (d): *Ulmus montana* Withering; *Ulmac.;* Eu.
Ruta: *Ougeinia dalbergioides* Bth.; *Papilionac.;* O.-Ind. (w. Beng.)
Ruté: *Dalbergia latifolia* Roxb.; *Papilionac.;* O.-Ind. (? w. Beng.)
Ruthee chinta mola: *Gomphia ? sumatrana* Jack; *Ochnac.;* Malak.
Ruttliracham: *Elaeocarpus tuberculatus* Roxb.; *Tiliac.;* sw. O.-Ind. (Madras)
Ruvya: *Dillenia indica* L.; *Dilleniac.;* O.-Ind.

S

Sa: *Stephegyne parvifolia* Korth.; *Rubiac.;* O.-Ind. (? Burma)
„ : *Daniella thurifera* Bth.; *Caesalpiniac.;* Togo.
Saandoe: *Hymenolobium flavum* Kleinh., *H. sp.; Papilionac.;* Sur.
„ : *Parkia pendula* Bth.; *Mimosac.;* Sur.
„ awha: *Hymenolobium flavum* Kleinh., *H. sp.; Papilionac.;* Sur.
Saba: *Dialyanthera acuminata* Standl., *D. otoba* Warb.; *Myristicac.;* Pan.
„ : *Inga sp.; Mimosac.;* Pan.
Sabak-ché: *Exostemma indutum* Standl.; *Rubiac.;* br. Hond.
Sabanero: *Phyllostylon brasiliensis* Capanema; *Ulmac.;* n. Kol.
Sabanna-tranga-hoedoe: *Matayba surinamensis* Warb., *M. spp.; Sapindac.;* Sur.
„ „ „ : *Cupania scrobiculata* L. C. Rich.; *Sapindac.;* Sur.
Sabara: *Guiera senegalensis* Lam.; *Combretac.;* Nig.
Sabbir: *Hura crepitans* L.; *Euphorbiac.;* fr. W.-Ind.
Sabica: *Swietenia mahagoni* L.; *Meliac.;* Bah.
Sabicú: *Lysiloma sabicu* Bth.; *Mimosac.;* w. Cuba.
„ : *Peltophorum adnatum* Griseb.; *Caesalpiniac.;* Cuba.
„ : *Pithecolobium arboreum* Urb.; *Mimosac.;* Cuba.
„ , Bahama- (u): *Pithecolobium arboreum* Urb.; *Mimosac.;* Ant., Trin., C.-Am.
„ , Cuban- (u): *Lysiloma sabicu* Bth.; *Mimosac.;* Cuba.
„ moruro: *Peltophorum adnatum* Griseb.; *Mimosac.;* Cuba.
Sabicuholz (d): *Lysiloma sabicu* Bth.; *Mimosac.;* Cuba.
Sabidug: *Terminalia catappa* L.; *Combretac.;* Phil. (Batanes Gruppe)
Sabina: *Juniperus sabina* L., *J. phoenicea* L.; *Conifer.;* Sp.
„ : *Juniperus licayana* Britton; *Conifer.;* Cuba.
„ : *Juniperus sp.; Conifer.;* St.-D.
„ cimarrona: *Podocarpus angustifolia* Gris., *P. aristulatus* Parl.; *Conifer.;* Cuba.
„ meridonal: *Juniperus barbadensis* L.; *Conifer.;* sö. USA., W.-Ind.
„ rastrera: *Juniperus sabina* L.; *Conifer.;* Sp.
„ suave: *Juniperus phoenicea* L.; *Conifer.;* Sp.
„ de costa: *Juniperus lucayana* Britton; *Conifer.;* Cuba.
Sabino: *Juniperus mexicana* Spreng., *Taxodium mucronatum* Ten.; *Conifer.;* Mex.
„ : *Magnolia splendens* Urb.; *Magnoliac.;* P.-R.
Sablier (f): *Hura crepitans* L.; *Euphorbiac.;* tr. Am.
„ blanc: *Hura crepitans* L.; *Euphorbiac.;* Trin.
„ jaune: *Hura crepitans* L.; *Euphorbiac.;* Trin.
Sablót: *Litsea sp.; Laurac.;* Phil. (Ilocos Sur.)
Saboeiro: *Sapindus divaricatus* St. Hil.; *Sapindac.;* Bras.

Saboneteiro: *Sapindus saponaria* L.; *Sapindac.;* Bras. (Amaz.)
Sábong-kaág: *Parinarium sp.; Rosac.;* Phil. (Ilocos Sur)
Sabto: *Cordia alba* Roem. et Schultes; *Borraginac.;* Pan.
Sabuco: *Sambucus nigra* L.; *Caprifoliac.;* Sp.
Sacalacahuite: *Pinus ayacahuite* K. Ehrenb.; *Conifer.;* Mex.
Sacambú: *Platypodium floribundum* Vogel; *Papilionac.;* Bras. (Sao P)
Sácat: *Terminalia nitens* Presl, *T. spp.; Combretac.;* Phil.
Sac-bay-eck: *Aegiphila pauciflora* Standl.; *Borraginac.;* br. Hond.
Sacbey-eck: *Bourreria oxyphylla* Standl.; *Borraginac.;* br. Hond.
Sac-chacal: *Gilibertia concinna* Standl.; *Araliac.;* br. Hond.
Sac-pah: *Byrsonima crassifolia* DC.; *Malpighiac.;* br. Hond.
Sac-pom: *Cupania guatemalensis* Radlk.; *Sapindac.;* br. Hond.
Sacha cebil: *Piptadenia excelsa* Lillo; *Mimosac.;* Arg.
  „ -chope: *Gustavia sp.; Lecythidac.;* Bras.
  „ limón: *Zanthoxylum sp.; Rutac.;* Arg.
  „ manzana: *Ruprechtia sp.; Polygonac.;* Arg.
  „ membrillo: *Capparis jamaicensis* Jacq.; *Capparidac.;* Arg.
? „ paraiso: *Pentapanax angelicifolius* Griseb.; *Araliac.;* Arg.
  „ -peral: *Vallea stipularis* L. f.; *Tiliac.;* Ek.
? „ -piral: *Acnistus arborescens* Schlecht.; *Solanac.;* tr. S.-Am.
Sachavaya: *Gustavia longifolia* Poepp.; *Lecythidac.;* Bras., Peru.
Sacka: *Peltogyne pubescens* Bth.; *Caesalpiniac.;* br. Gu.
Sackwi-mbauwi: *Adansonia digitata* L.; *Bombacac.;* Lib.
Sada: *Ziziphus lotus* Lamk.; *Rhamnac.;* N.-Af.
Sadai: *Pometia pinnata* Forst.; *Sapindac.;* Phil. (Babuyan Gruppe)
Sadar: *Terminalia tomentosa* Wight et Arn.; *Combretac.;* C.-O.-Ind.
Sadâu: *Dysoxylum bailloni* Pierre; *Meliac.;* Annam.
Sadda-veppu: *Walsura piscidia* Roxb.; *Meliac.;* Cey.
Sadde-tree: *Nothofagus cunninghami* Oerst.; *Fagac.;* s. Au., Tasm.
Saddletree: *Liriodendron tulipifera* L.; *Magnoliac.;* USA.
Sadora: *Terminalia tomentosa* Wight et Arn.; *Combretac.;* C.-O.-Ind.
Sadra: *Terminalia arjuna* Bedd.; *Combretac.;* O.-Ind. (? s. Bombay)
  „ : *Terminalia tomentosa* Wight et Arn.; *Combretac.;* O.-Ind. (Central Prov.)
Sadri: *Terminalia tomentosa* Wight et Arn.; *Combretac.;* O.-Ind. (? s. Bombay)
Sadura: *Terminalia arjuna* Bedd.; *Combretac.;* O.-Ind. (Bombay)
Safan: *Talauma hodgsoni* Hook. f. et Thoms.; *Magnoliac.;* nö. O.-Ind. (ö. Nepal)
Safeda farast: *Populus ciliata* Wall.; *Salicac.;* nw. O.-Ind. (Kaschmir)
Saffeka: *Guarea gomma* Pulle; *Meliac.;* Sur.
Safforan wood: *Elaeodendron croceum* DC.; *Celastrac.;* S.-Af. (Natal)
Saffraanhout (h): *Elaeodendron croceum* DC.; *Celastrac.;* S.-Af. (Kap. Natal)
Saffronheart: *Halfordia drupifera* F. v. M.; *Rutac.;* Au.
Saffron tree: *Chrysophyllum ?* cainito L.; *Sapotac.;* br. W.-Ind.
  „ wood (e): *Elaeodendron croceum* DC.; *Celastrac.;* S.-Af. (Kap, Natal)
Safir: *Rhamnus alaternus* L.; *Rhamnac.;* Arab.
Safoukala: *Pachylobus saphu* Engl.; *Burserac.;* B.-K. (Mayombe)
Safoutier: *Pachylobus saphu* Engl.; *Burserac.;* Gab.
Safu kala: *Pachylobus pubescens* Verm.; *Burserac.;* B.-K. (Mayombe)
Sag: *Terminalia tomentosa* Bedd.; *Combretac.;* O.-Ind.
  „ : *Tectona grandis* L. f.; *Verbenac.;* sw. O.-Ind. (Bombay)
Saga: *Michelia champaca* L.; *Magnoliac.;* Burma.
  „ : *Adenanthera pavonina* L.; *Mimosac.;* Malay.
Sagada: *Schleichera trijuga* Willd.; *Sapindac.;* O.-Ind. (Bombay)
Sagapu: *Hymenodictyon excelsum* Wall.; *Rubiac.;* O.-Ind. (Madras)
Saga'sá': *Lumnitzera littorea* Voigt; *Combretac.;* Phil. (Dingatinsel)
Sagat: *Securidaca longipedunculata* Fresen.; *Polygalac.;* Arab.
Sagát: *Pterocarpus sp.; Papilionac.;* Phil. (Cagayan)
  „ : *Vitex parviflora* Juss.; *Verbenac.;* Phil.
  „ -kunkyan: *Quercus semiserrata* Roxb.; *Fagac.;* O.-Ind. (Burma)
Sagawa: *Michelia champaca* L.; *Magnoliac.;* Burma.
Sagawoita: *Zanthoxylum aff. macrophyllum* Oliv.; *Rutac.;* br. O.-Af. (Kenya)
Sage, black: *Cordia cylindrostachya* R. et S., *C. sebestena* L.; *Borraginac.;* W.-Ind.
  „ , wild: *Casearia silvestris* Sw.; *Flacourtiac.;* br. Hond.

Ságet: *Terminalia calamansanai* Rolfe; *Combretac.;* Phil. (Benguet)
Sagimsím: *Eugenia brevistylis* C. B. Rob.; *Myrtac.;* Phil. (Cotabato)
Sagna: *Corynanthe paniculata* Welw.; *Rubiac.;* B.-K. (Mayombe)
Sagoe-weerhout: *Garcinia picrorrhiza* Miq.; *Guttifer.;* Nied.-Ind.
Sagon: *Tectona grandis* L. f.; *Verbenac.;* C.-O.-Ind.
Sagouer: *Garcinia picrorrhiza* Miq.; *Guttifer.;* Moluk.
Saguaragy: *Colubrina rufa* Reiss.; *Rhamnac.;* Bras. (Sao P)
Saguaro: *Carnegica gigantea* Britt. et Rose; *Cactac.;* s. USA., Mex.
Sagué: *Monopetalanthus emarginatus* Hutch. et Dalz.; *Caesalpiniac.;* Elf.
Sagwan: *Tectona grandis* L. f.; *Verbenac.;* sw. & C.-O.-Ind.
Sahaju: *Terminalia tomentosa* Wight et Arn.; *Combretac.;* nö. O.-Ind. (Orissa)
Sahar: *Dillenia pentagyna* Roxb.; *Dilleniac.;* ö. Beng.
Sáhing: *Pinus insularis* Endl.; *Conifer.;* Phil. (Nueva Ecija)
  „ : *Canarium luzonicum* A. Gray; *Burserac.;* Phil. (Bataan Prov.)
 · „ : *Canarium villosum* F. Vill.; *Burserac.;* Phil. (Palawan)
Sahm: *Artocarpus chaplasha* Roxb.; *Morac.;* O.-Ind. (Assam)
Sah-sah: *Vitex micrantha* Gürke; *Verbenac.;* Lib.
Sahuara: *Carnegica gigantea* Britt. et Rose; *Cactac.;* Mex.
Sahuco: *Aralia laetevirens* C. Gay; *Araliac.;* Chile.
Sahuinto: *Psidium pyriferum* L.; *Myrtac.;* Peru.
Sai: *Broussonetia papyrifera* Vent.; *Morac.;* Java.
Saikachi: *Gleditschia horrida* Mak., *G. japonica* Miq.; *Caesalpiniac.;* Jap.
Saikaki: *Eurya japonica* Thunb.; *Theac.;* Jap.
Sain: *Terminalia tomentosa* Wight et Arn.; *Combretac.;* O.-Ind., Burma.
Sáingja: *Daniella thurifera* J. J. Benn.; *Caesalpiniac.;* Togo.
Saint-Jean: *Coccoloba sp.; Polygonac.;* fr. Gu., fr. W.-Ind.
  „ „ , rouge: *Virola surinamensis* Warb., *? V. sebifera* Aubl.; *Myristicac.;* fr. Gu.
  „ John's bread: *Ceratonia siliqua* L.; *Caesalpiniac.;* Eu.
  „ Martin: *Andira spp.; Papilionac.;* Sur.
  „ „ gris: *Dipteryx oppositifolia* Willd.; *Papilionac.;* fr. Gu.
  „ „ rouge: *Andira inermis* H. B. K.; *Papilionac.;* fr. Gu.
  „ „   „ : *Andira coriacea* Pulle; *Papilionac.;* Sur.
  „ „   „ : *Andira wachenheimi* R. Benoist; *Papilionac.;* fr. Gu.
Sainyia: *Securidaca longipedunculata* Fresen.; *Polygalac.;* n. Nig. (Kontagora)
Saisa: *Dalbergia latifolia* Roxb.; *Papilionac.;* n. O.-Ind. (United Prov.)
Saj: *Terminalia tomentosa* Wight et Arn.; *Combretac.;* C.-O.-Ind.
Sajadito: *Randia spinosa* Karst.; *Rubiac.;* Ven.
Saka: *Peltogyne ? paniculata* Bth., *P. sp.; Caesalpiniac.;* Sur., br. Gu.
  „ -balli: *Peltogyne sp.; Caesalpiniac.;* br. Gu.
Sakagnya: *Bosquiea welwitschi* Engl.; *Morac.;* B.-K. (Mayombe)
Sakaki: *Eurya japonica* Thunb., *E. ochnacea* Szysz.; *Theac.;* Jap.
Sakalang: *Elaeocarpus lanceaefolius* Roxb.; *Tiliac.;* O.-Ind., Assam.
Sakar: *Ehretia laevis* Roxb.; *Borraginac.;* nw. O.-Ind. (Punjab)
Sákat: *Terminalia nitens* Presl, *T. spp.; Combretac.;* Phil.
Sáket: *Terminalia calamansanai* Rolfe; *Combretac.;* Phil. (Benguet)
  „ : *Terminalia pellucida* Presl; *Combretac.;* Phil. (Pangasinan, Zambales, Laguna)
Saj: *Terminalia tomentosa* Bedd.; *Combretac.;* O.-Ind.
Sakhui: *Shorea robusta* Gaertn. f.; *Dipterocarpac.;* n. O.-Ind. (United Prov.)
Sakoa: *Sclerocarya caffra* Sond.; *Anacardiac.;* Mad.
Sakrum: *Payena elliptica* Pierre; *Sapotac.;* O.-Ind. (Kamb.)
Sakura: *Prunus pseudocerasus* Lindl.; *Rosac.;* Jap.
Sakwa: *Shorea robusta* Gaertn. f.; *Dipterocarpac.;* nö. O.-Ind. (Nepal)
Sal (f): *Shorea robusta* Gaertn. f.; *Dipterocarpac.;* n. O.-Ind. (Unit.-Prov., Assam)
  „ : *Schleichera trijuga* Willd.; *Sapindac.;* C.-O.-Ind. (Bihar)
  „ : *Terminalia tomentosa* Bedd.; *Combretac.;* C.-O.-Ind. (Bihar)
  „ , Burma- (e): *Shorea obtusa* Wall., *Pentacme suavis* A. DC.; *Dipterocarpac.;* O.-Ind.
  „ de Inchias: *Maytenus boaria* Molina; *Celastrac.;* Arg.
Sala: *Shorea robusta* Gaertn. f.; *Dipterocarpac.;* n. O.-Ind. (United Prov.)
  „ dhup: *Pinus longifolia* Roxb.; *Conifer.;* nö. O.-Ind. (Nepal)
Salabsálab: *Hopea pierrei* Hance; *Dipterocarpac.;* Phil. (Mindoro)
Salade soldat: *Peperomia procumbens* C. DC.; *Piperac.;* fr. Gu.
Salaee: *Boswellia serrata* Roxb.; *Burserac.;* O.-Ind. (Hind.)

Salah: *Eurya nitida* Korth.; *Theac.*; s. Sum.
Salaha: *Acacia arabica* Willd.; *Mimosac.*; Sudan.
Salai: *Boswellia serrata* Roxb.; *Burserac.*; O.-Ind.
Salaia-guggar: *Boswellia serrata* Roxb.; *Burserac.*; nw. O.-Ind. (United Prov.)
Salaisái: *Terminalia catappa* L.; *Combretac.*; Phil. (Benguet)
Salakín: *Aphanomyxis cumingiana* Harms; *Meliac.*; Phil. (Laguna)
Salakíng-pulá: *Aglaia diffusa* Merr.; *Meliac.*; Phil. (Laguna)
Salalángin: *Kingiodendron alternifolium* Merr.; *Papilionac.*; Phil. (Albay, Sorsogon)
Salam andking: *Cleistanthus myrianthus* Kurz; *Euphorbiac.*; Java (Tjemara)
Salamági: *Tamarindus indica* L.; *Caesalpiniac.*; Phil. (n. Luzon, Bisaya)
Salamíngai: *Aglaia bicolor* Merr.; *Meliac.*; Phil. (Rizal)
Sálamo: *Calycophyllum candidissimum* DC.; *Rubiac.*; C.-Am.
Salamoeli: *Cordia subcordata* Lam.; *Borraginac.*; Java.
Salamíngai: *Aglaia harmsiana* Perk.; *Meliac.*; Phil. (Batangas)
Salamúngi: *Aglaia turczaninowi* C. DC.; *Meliac.*; Phil. (Cav., Batangas)
Sálang: *Agathis alba* Foxw.; *Conifer.*; Phil. (Cagayan)
Salanggígí: *Albizzia saponaria* Blume; *Mimosac.*; Phil. (Mindoro)
Salangkúgi: *Albizzia saponaria* Blume; *Mimosac.*; Phil.
Salar: *Boswellia serrata* Roxb.; *Burserac.*; n. O.-Ind. (United Prov.)
Sala'sá: *Lumnitzera littorea* Voigt; *Combretac.*; Phil. (w. Negrosprovinz)
Saldaná: *Aglaia clarki* Merr.; *Meliac.*; Phil. (Cebu)
Saldawar: *Sterculia urens* Roxb.; *Sterculiac.*; O.-Ind. (Bombay)
Saldorija: *Satureja cuneifolia* Ten.; *Labiat.*; Sp.
Salee: *Acacia falcata* Willd.; *Mimosac.*; Au.
Saleh: *Boswellia serrata* Roxb.; *Burserac.*; C.-O.-Ind.
Salèlèbé: *Macrolobium chrysostachyum* Bth., *M. bifolium* Pers.; *Caesalpiniac.*; Sur.
Sáleng: *Pinus insularis* Endl.; *Conifer.*; Phil. (Ilocos Sur, Bontok, Benguet)
„ : *Agathis alba* Foxw.; *Conifer.*; Phil. (Cagayan)
Saler: *Boswellia serrata* Roxb.; *Burserac.*; O.-Ind. (Jeypore)
Salga: *Boswellia serrata* Roxb.; *Burserac.*; n. O.-Ind. (United Prov.), ö. Beng.
Salgueiro: *Salix aurita* L.; *Salicac.*; Sp.
„ : *Salix sp.*; *Salicac.*; Bras.
„ do matto: *Salix humboldtiana* Willd.; *Salicac.*; Bras.
„ „ „ : *Belangera tomentosa* Camb.; *Cunoniac.*; Bras.
Salhe: *Boswellia serrata* Roxb.; *Burserac.*; n. O.-Ind. (United Prov.)
Sali: *Chloroxylon swietenia* DC.; *Rutac.*; O.-Ind. (Bihar, Orissa)
„ : *Boswellia serrata* Roxb.; *Burserac.*; n. O.-Ind.
Salica: *Salix caprea* L.; *Salicac.*; It.
Salicone: *Salix caprea* L.; *Salicac.*; It.
Salie: *Tetragastris sp.*; *Burserac.*; Sur.
„ , red: *Protium sp.*; *Burserac.*; Sur.
„ , wilde: *Cordia graveolens* H. B. K.; *Borraginac.*; Sur.
„ , witte: *Hedwigia hostmanni* Engl., *H. balsamifera* Swartz; *Burserac.*; Sur.
„ , „ : *Heliotropium sp.*; *Borraginac.*; Sur.
Salieman: *Albizzia fastigiata* Oliv.; *Mimosac.*; Kam.
„ : *Afzelia africana* Smith; *Caesalpiniac.*; Kam.
Salifúngan: *Parinarium sp.*; *Rosac.*; Phil. (Cagayan)
Salikúgi: *Albizzia saponaria* Blume; *Mimosac.*; Phil. (Samar)
Salikút: *Palaquium ahernianum* Merr.; *Sapotac.*; Phil. (Agusan)
Salima: *Acer caesium* Wall.; *Acerac.*; nw. O.-Ind. (Punjab)
Salimbubú: *Sterculia sp.*; *Sterculiac.*; Phil. (Mindoro)
Salindángat: *Zizyphus sp.*; *Rhamnac.*; Phil. (Negrosinsel)
Salinggógon: *Cratoxylon blancoi* Blume, *C. spp.*; *Guttifer.*; Phil. (Camar., Albay)
Salingkápa: *Vitex parviflora* Juss.; *Uerbenac.*; Phil. (Guimaras)
Salingkúgi: *Albizzia saponaria* Blume; *Mimosac.*; Phil.
Salingkúki: *Albizzia saponaria* Blume; *Mimosac.*; Phil. (Bataan, Mindoro)
Salísai: *Terminalia catappa* L.; *Combretac.*; Phil. (Zambales, Bataan)
Sálit: *Pinus insularis* Endl.; *Conifer.*; Phil. (Zambales)
Salitrero: *Zanthoxylum sp.*; *Rutac.*; Salv.
Salla: *Boswellia serrata* Roxb.; *Burserac.*; C.-O.-Ind. (Gondwana)
„ : *Pinus longifolia* Roxb.; *Conifer.*; n. O.-Ind. (Garhwal)
Sallaci: *Boswellia serrata* Roxb.; *Burserac.*; O.-Ind.

Sallapugúd: *Aglaia harmsiana* Perk.; *Meliac.;* Phil. (Ilocos Sur)
\   „   :   *Hopea plagata* Vidal; *Dipterocarpac.;* Phil.
Sally, black: *Eucalyptus coriacea* A. Cunn.; *Myrtac.;* sö. Au.
   „   , white: *Eucalyptus coriacea* A. Cunn., *E. stellulata* Sieb.; *Myrtac.;* ö. Au., Tasm
Salmon wood: *Eriolaena candollei* Wall.; *Sterculiac.;* O.-Ind.
   „   „   :   *Cordia alliodora* Cham.; *Borraginac.;* C.-Am., Trin.
Salngán: *Vatica mangachapoi* Blco.; *Dipterocarpac.;* Phil. (Samar, Leyte)
Salngén: *Aglaia laevigata* Merr.; *Meliac.;* Phil. (Ilocos Sur, Pangasinan)
   „   :   *Vatica mangachapoi* Blco.; *Dipterocarpac.;* Phil. (Pangasinan)
Salo: *Adansonia digitata* L.; *Bombacac.;* Go.
Salom: *Lysiloma bahamense* Bth.; *Mimosac.;* br. Hond.
Sálong: *Agathis alba* Foxw.; *Conifer.;* Phil.
Salóngan: *Vatica mangachapoi* Blco.; *Dipterocarpac.;* Phil. (Union Prov.)
Salongsálong: *Vatica mangachapoi* Blco.; *Dipterocarpac.;* Phil. (Albay, Sorsogon)
Salongsalóngan: *Vatica mangachapoi* Blco.; *Dipterocarpac.;* Phil. (Agusan)
Salóyong: *Cordia myxa* L.; *Borraginac.;* Phil. (Tagalog)
Salphullie: *Boswellia serrata* Roxb.; *Burserac.;* O.-Ind. (Bombay)
Salsál: *Erythrophloeum densiflorum* Merr.; *Caesalpiniac.;* Phil. (Cagayan)
Salseparaille: *Smilax anceps* Willd.; *Liliac.;* Réunion.
Salt tree: *Halimodendron halodendron* Voss; *Papilionac.;* As.
   „   and oil tree: *Cleistopholis patens* Bth.; *Anonac.;* Go.
Salta cavallo: *Luehea divaricata* Mart., *L. paniculata* Mart.; *Tiliac.;* Bras. (Sao P)
Sal-tul: *Calocarpum mammosum* Pierre; *Sapotac.;* Guat.
Salua: *Sterculia villosa* Roxb.; *Sterculiac.;* Assam.
Salulúng: *Lagerstroemia piriformis* Koehne; *Lythrac.;* Phil. (Cagayan)
Salungkúgi: *Albizzia saponaria* Blume; *Mimosac.;* Phil.
Salutin: *Parinarium sp.;* *Rosac.;* Phil. (Cagayan)
Salvadera: *Hura crepitans* L.; *Euphorbiac.;* Cuba, Peru.
Salvia: *Buddleia wrighti* Rob., *B. spp.;* *Loganiac.;* Mex. (Sinaloa)
   „   real del Moncayo: *Salvia officinalis* L.; *Labiat.;* Sp.
   „   de la Alcarria: *Salvia lavandulaefolia* L.; *Labiat.;* Sp.
   „   del monte: *Acalypha villosa* Jacq.; *Euphorbiac.;* Kol.
Salwa: *Shorea robusta* Gaertn. f.; *Dipterocarpac.;* O.-Ind. (Orissa)
Salzstrauch, silberblättriger (d):   *Halimodendron halodendron* Voss; *Papilionac.;* As.
Sam (f):   *Triplochiton scleroxylon* K. Schum.; *Sterculiac.;* Elf.
   „   :   *Uapaca staudti* Pax; *Euphorbiac.;* Kam.
   „   :   *Lagerstroemia parviflora* Roxb.; *Lythrac.;* O.-Ind. (? Burma)
   „   :   *Artocarpus chaplasha* Roxb.; *Morac.;* Assam.
Sama (f):   *Triplochiton scleroxylon* K. Schum.; *Sterculiac.;* Elf.
   „   jawa: *Gordonia excelsa* Blume; *Theac.;* Malak.
   „   rupa: *Shorea sp.;* *Dipterocarpac.;* Malay.
   „   „   chengal: *Shorea bracteolata* Dyer; *Dipterocarpac.;* Malay.
   „   „   „   :   *Pachychlamys hemsleyana* Ridl.; *Dipterocarpac.;* Malay.
   „   „   „   :   *Shorea sp., Hopea sp., Isoptera sp.;* *Dipterocarpac.;* Malay.
Samadara gass: *Samandera indica* L.; *Simarubac.;* O.-Ind.
Samak: *Eugenia spp.;* *Myrtac.;* Malay.
Samal: *Rapanea andina* Mez; *Myrsinac.;* Ek.
Samalomba: *? Simaruba amara* Aubl.; *Simarubac.;* Sur.
Samán: *Enterolobium saman* Prain; *Mimosac.;* P.-R., Pan., Ven.
Samanibir: *Vitex cienkowskyi* Kotschy et Peyr.; *Verbenac.;* Go.
Samanigua: *Samanea samanigua* Pittier; *Mimosac.;* Ven.
Samanta: *Xylia evansi* Hutch.; *Mimosac.;* Elf.
   „   :   *Calpocalyx brevibracteata* Harms; *Mimosac.;* Go.
   „   :   *Bussea occidentalis* Hutch.; *Caesalpiniac.;* Go.
? Samaria wood: *Icica sp.;* *Burserac.;* tr. As.
Samarie-japo: *Cedrela odorata* L.; *Meliac.;* Sur.
Samarupa: *Jacaranda copaia* D. Don; *Bignoniac.;* Gu.
Samba (f):   *Triplochiton scleroxylon* K. Schum.; *Sterculiac.;* Elf.
Sambág: *Tamarindus indica* L.; *Caesalpiniac.;* Phil. (s. Luzon, Bisaya).
Sambahiba: *Curatella americana* L.; *Dilleniac.;* Bras.
Sambaibinha: *Curatella americana* L.; *Dilleniac.;* Bras.
Sambák: *Tamarindus indica* L.; *Caesalpiniac.;* Phil. (s. Luzon, Bisaya)

Sambalagán: *Koordersiodendron pinnatum* Merr.; *Anacardiac.*; Phil. (Bisaya)
Sambar: *Erythrina suberosa* Roxb.; *Papilionac.*; C.-O.-Ind. (Vindhja)
Sambaw: *Schima wallichi* Choisy; *Theac.*; w. Burma.
Sambi: *Schleichera trijuga* Willd.; *Sapindac.*; Burma, Java (Klangoen).
Sambia: *Anisophyllea sp.*; *Rhizophorac.*; D.-O.-Af.
Sambie: *Allanblackia stuhlmanni* Engl.; *Guttifer.*; D.-O.-Af.
Sambirodjo: *Gelonium glomerulatum* Hassk.; *Euphorbiac.*; Java (Poeger)
Samboera: *Simaruba amara* Aubl., *S. versicolor* St. Hil.; *Simarubac.*; Sur.
Sambogum: *Symphonia globulifera* L. f.; *Guttifer.*; Pan., C.-R.
Samborano: *Calophyllum sp.*; *Guttifer.*; Mad.
Sambos, wild: *Eugenia zeyheri* Harv.; *Myrtac.*; S.-Af. (Kap).
Sambrit: *Buchanania sp.*; *Anacardiac.*; Phil. (Cagayan).
Sambulauán: *Eugenia sp.*; *Myrtac.*; Phil. (s. Luzon).
      „   : *Koordersiodendron pinnatum* Merr.; *Anacardiac.*; Phil. (Bisaya)
      „   : *Santiria nitida* Merr.; *Burserac.*; Phil. (w. Negrosprovinz)
Sambuor meas le: *Leucaena glauca* Bth.; *Mimosac.*; Ind.-Ch.
Samburágat: *Terminalia calamansanai* Rolfe; *Combretac.*; Phil. (Palawan)
Samfi: *Uapaca guineensis* Muell. Arg.; *Euphorbiac.*; B.-K. (Ganda-Sundi)
Samfona: *Chrysophyllum aff. africanum* A. DC.; *Sapotac.*; Go.
Samgawngmaeot: *Careya arborea* Roxb.; *Myrtac.*; O.-Ind. (? Burma)
Samhkalam: *Lagerstroemia parviflora* Roxb.; *Lythrac.*; O.-Ind. (? Burma)
Sámia: *Tamarindus indica* L.; *Caesalpiniac.*; Togo.
Samíling: *Cinnamomum mercadoi* Vidal; *Laurac.*; Phil. (Bataan).
Saminoe-wéwé: *Paypayrola guianensis* Aubl.; *Violac.*; Sur.
Samkamba: *Triplochiton scleroxylon* K. Schum.; *Sterculiac.*; Elf.
Samkesar: *Albizzia lebbek* Bth.; *Mimosac.*; C.-O.-Ind. (Central Prov.)
Samo: *Pentaclethra eetveldeena* de Wild. et Th. Dur.; *Mimosac.*; Kongo (Haut-Ogoué)
Samohú: *Chorisia speciosa* St. Hil., *Ceiba sp.*; *Bombacac.*; Arg.
Samoka: *Holarrhena antidysenterica* Wall.; *Apocynac.*; C.-O.-Ind. (Gondwana)
Samokon: *Lonchocarpus sericeus* H. B. K.; *Papilionac.*; Elf.
Samondó: *Terminalia nitens* Presl; *Combretac.*; Phil. (Palawan)
Samou: *Cunninghamia lanceolata* Hook.; *Conifer.*; Ind.-Ch.
Sampálok: *Tamarindus indica* L.; *Caesalpiniac.*; Phil. (C.-Luzon).
Sampaughi: *Michelia champaca* L.; *Magnoliac.*; O.-Ind. (Madras)
Sampéjan: *Ficus pubinervis* Blume; *Morac.*; Java (Ragadjampi)
Sampeu: *Manihot utilissima* Pohl; *Euphorbiac.*; Java.
Sampige: *Michelia champaca* L.; *Magnoliac.*; sw. O.-Ind. (Madras)
Sampirodjo: *Elattostachys verrucosa* Radlk.; *Sapindac.*; Java.
Sampor: *Cudrania sp.*; *Morac.*; Ind.-Ch. (Kamb.)
Sampora: *Columbia javanica* Blume; *Tiliac.*; Nied.-Ind.
Samsheet (russisch): *Buxus sempervirens* L.; *Buxac.*; M.-M., ö. Eu., As.
Samuhú: *Chorisia speciosa* St. Hil., *Ceiba sp.*; *Bombacac.*; Arg.
  „ -pyta: *Chorisia insignis* H. B. K.; *Bombacac.*; Par.
Samulu porott: *Chorisia ventricosa* Nees; *Bombacac.*; Par.
Samundar: *Barringtonia acutangula* Gaertn.; *Lecythidac.*; O.-Ind. (Beng.)
Samundarphul: *Barringtonia acutangula* Gaertn.; *Lecythidac.*; n. O.-Ind. (United Prov.)
Samwood: *Cordia sp.*; *Borraginac.*; Pan.
San: *Melanorrhoea usitata* Wall.; *Anacardiac.*; Burma.
  „ : *Eugenia resinosa* Gagnep.; *Myrtac.*; Ind.-Ch. (Annam)
  „ : *Mimusops kauki* L.; *Sapotac.*; Malay.
  „ : *Albizzia fastigiata* Oliv.; *Mimosac.*; Elf.
  „ dân: *Sandoricum indicum* Cav.; *Meliac.*; Annam.
  „ „ : *Melia azedarach* L., *M. dubia* Cav.; *Meliac.*; Annam.
  „ où yé: *Aglaia odorata* Lour.; *Meliac.*; Ch.
Sana: *Elaeocarpus sp.*; *Elaeocarpac.*; Mad.
Sanaguare: *Enterolobium saman* Prain; *Mimosac.*; n. Kol.
Sanaí: *Canarium villosum* F. Vill.; *Burserac.*; Phil. (Negrosinsel)
Sanai: *Anisoptera thurifera* Blume; *Dipterocarpac.*; Malay.
Sanalego: *Vernonia patens* H. B. K.; *Composit.*; Pan .
Sanalinga: *Cinnamomum zeylanicum* Breyn; *Laurac.*; O.-Ind.
Sanapka: *Limonia acidissima* L.; *Rutac.*; Burma.
San Bartholomé: *Cordia sebestena* L.; *Borraginac.*; P.-R.

Tambor: *Schizolobium parahybum* Blake; *Caesalpiniac.;* Hond.
„ : *Enterolobium timbouva* Mart.; *Mimosac.;* Bras. (Sao P)
Tamboril: *Enterolobium timbouva* Mart.; *Mimosac.;* Bras. (Sao P)
Tambory: *Enterolobium timbouva* Mart.; *Mimosac.;* Bras. (Sao P)
Tamboryl: *Enterolobium timbouva* Mart.; *Mimosac.;* Bras. (Sao P)
Tamboti: *Excoecaria africana* Muell. Arg.; *Euphorbiac.;* S.-Af. (Transval)
Tambotie: *Excoecaria africana* Muell. Arg.; *Euphorbiac.;* S.-Af. (Transval)
Tambugai: *Shorea tumbuggaia* Roxb.; *Dipterocarpac.;* O.-Ind. (Madras)
Tambukh: *Croton macrostachys* Hochst.; *Euphorbiac.;* n. O.-Af. (Abessinien)
Tambulauán: *Eugenia sp.; Myrtac.;* Phil. (s. Luzon)
Tambulían: *Eusideroxylon zwageri* Teijsm. et Binn.; *Laurac.;* Phil. (Tawi-Tawi)
Tambuti: *Excoecaria africana* Muell. Arg.; *Euphorbiac.;* S.-W.-Af.
Tambuva: *Enterolobium sp.; Mimosac.;* Bras.
Tamè: *Eugenia jambolana* Lamk.; *Myrtac.;* n. Burma.
Tameague: *Eupatorium laeve* DC.; *Composit.;* Salv.
Tamenasi: *Bauhinia reticulata* DC.; *Caesalpiniac.;* Togo.
Tami: *Ochroma sp.; Bombacac.;* Bol.
Tamil: *Diospyros plicata* Merr.; *Ebenac.;* Phil. (Cotabato)
„ -lalaki: *Diospyros mindanaensis* Merr.; *Ebenac.;* Phil. (Lanao)
Taming-táming: *Dysoxylum decandrum* Merr.; *Meliac.;* Phil. (Basilan)
„ „ : *Hopea philippinensis* Dyer; *Dipterocarpac.;* Phil. (Zamboanga)
„ „ : *Cryptocarya zamboangensis* Merr.; *Laurac.;* Phil. (Sulu)
„ „ : *Diospyros plicata* Merr.; *Ebenac.;* Phil. (Sulu)
Tamma: *Acacia arabica* Willd.; *Mimosac.;* O.-Ind. (Madras)
Tamo: *Fraxinus sieboldiana* Blume *var. serrata* Nakai; *Oleac.;* Jap.
„ : *Fraxinus mandchurica* Rupr.; *Oleac.;* Jap.
Tamoene: *Eperua schomburgkiana* Bth., ? *E. jenmani* Oliv.; *Caesalpiniac.;* Sur.
„ -apoekoetja: *Aspidosperma nitidum* Bth., *A. spp.; Apocynac.;* Sur.
„ araurana: *Sclerolobium paniculatum* Vogel; *Caesalpiniac.;* Sur.
„ -kereoe: *Dicorynia paraënsis* Bth.; *Caesalpiniac.;* Sur.
„ koerahara: *Calophyllum sp.; Guttifer.;* Sur.
„ kwatéré: *Eschweilera corrugata* Miers, *E. sp.; Lecythidac.;* Sur.
Tamoenewawé: *Nectandra sp., Ocotea sp.; Laurac.;* Sur.
Tamoeni-tonorebjo· *Matayba spp., Cupania scrobiculata* L. C. Rich.; *Sapindac.;* Sur.
Tamoenin kwatéré: *Eschweilera spp.; Lecythidac.;* Sur.
Tamoiya: *Ehretia laevis* Roxb.; *Borraginac.;* C.-O.-Ind.
Tamók: *Pterospermum niveum* Vidal; *Sterculiac.;* Phil. (Bataan)
„ : *Shorea teysmanniana* Dyer, *S. polysperma* Merr.; *Dipterocarpac.;* Phil. (Ca-
Tampacho: *Clusia sp.; Guttifer.;* Kol. [gayan)
Tampines: *Sloetia sideroxylon* Teijsm. et Binn.; *Morac.;* Malay.
Tampinis: *Sloetia sideroxylon* Teijsm. et Binn.; *Morac.;* br. Borneo (Sarawak)
Tampoedau: *Dipterocarpus tampurau* Korth.; *Dipterocarpac.;* s. & ö. Borneo.
Tampoedow: *Dipterocarpus tampurau* Korth.; *Dipterocarpac.;* ö. Borneo.
Tampoerouw: *Dipterocarpus tampurau* Korth.; *Dipterocarpac.;* ö. Borneo.
Tampúi: *Eugenia calubcob* C. B. Rob.; *Myrtac.;* Phil. (Bulacan, Tayabas)
„ : *Eugenia xanthophylla* C. B. Rob.; *Myrtac.;* Phil. (Mindoro)
Tampunai: *Artocarpus rigida* Blume; *Morac.;* Malay.
Tamrisi: *Tamarindus indica* L.; *Caesalpiniac.;* Go.
Tamruj: *Diospyros melanoxylon* Roxb.; *Ebenac.;* O.-Ind.
? Tamsi: *Thespesia populnea* Correa; *Malvac.;* Go.
Tamujo: *Securinega buxifolia* Muell. Arg.; *Euphorbiac.;* Sp.
Tamuláuan: *Xanthostemon verdugonianus* Naves; *Myrtac.;* Phil. (Leyte)
Tamulero: *Coccoloba schiedeana* Lindau; *Polygonac.;* Mex.
Tan Georges: *Cupania laevis* Pers.; *Sapindac.;* Réunion.
„ mot: *Stereospermum chelonioides* DC.; *Bignoniac.;* Ind.-Ch. (Annam)
„ rouge: *Weinmannia macrostachya* DC.; *Cunoniac.;* Réunion.
Tana: *Durio sp.; Bombacac.;* Malay.
„ : *Panda oleosa* Pierre; *Pandac.;* Go.
Tanábog: *Pometia pinnata* Forst.; *Sapindac.;* Phil. (Palawan)
Tan'ág laláki: *Aleurites trisperma* Blco.; *Euphorbiac.;* Phil. (Rizal)
Tanam: *Ochna afzeli* R. Br.; *Ochnac.;* Togo.
Tananeo: *Peltogyne sp.; Caesalpiniac.;* n. Kol.

Tanaung: *Acacia leucophlcea* Willd.; *Mimosac.;* Burma.
Tanbor: *Ochroma sp.; Bombacac.;* Guat., Hond.
Tanda: *Rhizophora mangle* L.; *Rhizophorac.;* Kam.
Tandang-isók: *Ormosia calavensis* Azaola; *Papilionac.;* Phil. (Masbate)
Tandasi: *Pterocarpus erinaceus* Poir.; *Papilionac.;* Togo.
Tandjang: *Bruguiera gymnorrhiza* Lam., *B. eriopetala* W. et A.; *Rhizophorac.;* Java.
Tandjoeng: *Mimusops elengi* L.; *Sapotac.;* Malay.
Tandu: ? *Kayea sp.; Guttifer.;* Phil. (Sulu)
Tandúl: *Carallia lucida* Roxb.; *Rhizophorac.;* Phil. (Sulu)
Tanekaha: *Phyllocladus trichomanoides* D. Don; *Conifer.;* N.-Seel.
Tangá: *Pygeum glandulosum* Merr., *P. presli* Merr.; *Rosac.;* Phil. (Ilocos N., Bukidnon)
Tangál: *Ceriops tagal* C. B. Rob.; *Rhizophorac.;* Phil.
„ : *Ceriops roxburghiana* Arn.; *Rhizophorac.;* Phil. (Tayabas, Camarines)
„ : *Terminalia edulis* Blco.; *Combretac.;* Phil. (Camarines)
Tangalín: *Adenanthera intermedia* Merr.; *Mimosac.;* Phil. (Laguna)
Tangarana: *Triplaris sp.; Polygonac.;* Peru.
Tangás: *Semecarpus sp.; Anacardiac.;* Phil. (Negrosinsel)
Tangeao: *Litsea calicaris* Kirk; *Laurac.;* N.-Seel.
Tanghál: *Ceriops tagal* C. B. Rob.; *Rhizophorac.;* Phil. (Mindoro)
Tanghás: *Dolichandrone spathacea* K. Schum.; *Bignoniac.;* Phil. (Mindoro)
„ : *Wrightia sp.; Apocynac.;* Phil. (Laguna, Tayabas, Mindoro)
Tangíd: *Canangium odoratum* Baill.; *Anonac.;* Phil. (Camarines, Albay, Sorsogon)
Tangile: *Shorea negrosensis* Foxw.; *Dipterocarpac.;* Phil. (Manila)
„ : *Shorea polysperma* Merr.; *Dipterocarpac.;* Phil.
Tangíli: *Bassia ramiflora* Merr.; *Sapotac.;* Phil. (Bulacan, Tayabas, Mindoro)
„ : *Palaquium sp.; Sapotac.;* Phil. (Bulacan)
„ : *Shorea polysperma* Merr.; *Dipterocarpac.;* Phil. (Bataan)
Tangíling-bangóhan: *Aglaia harmsiana* Perk.; *Meliac.;* Phil. (Bulacan)
„ -kompól: *Palaquium cuneifolium* Merr.; *Sapotac.;* Phil. (Bulacan)
„ -palokpálok: *Palaquium sp.; Sapotac.;* Phil. (Bulacan)
Tangísan: *Terminalia comintana* Merr.; *Combretac.;* Phil. (Pampanga, Masbate)
„ -bagió: *Cyclostemon maquilingensis* Merr.; *Euphorbiac.;* Phil. (Tayabas)
Tangít: *Canangium odoratum* Baill.; *Anonac.;* Phil. (Camarines, Albay, Sorsogon)
Tangítang: *Alstonia scholaris* R. Br.; *Apocynac.;* Phil. (Bisaya)
„ : *Paralstonia clusiacea* Baill.; *Apocynac.;* Phil. (Negrosinsel)
Tangkelie: *Kleinhovia hospita* L.; *Sterculiac.;* Java.
Tangkil: *Gnetum gnemon* L.; *Gnetac.;* Java.
„ : *Ficus punctata* Thunb.; *Morac.;* Java (Pantjoer)
Tangkoeloe: *Aegle marmelos* Correa; *Rutac.;* Malay.
Tangkoiloh: *Kleinhovia hospita* L.; *Sterculiac.;* Java.
Tangkor: *Diospyros sp.; Ebenac.;* Kamb.
Tanglefoot: *Nothofagus gunni* Oerst.; *Fagac.;* Tasm.
Tanglin: *Adenanthera intermedia* Merr.; *Mimosac.;* Phil. (Bataan, Pampanga)
Tanglón: *Adenanthera intermedia* Merr.; *Mimosac.;* Phil. (Pampanga)
Tango: *Zollernia tango* Standl.; *Caesalpiniac.;* br. Hond., Hond.
„ : *Butyrospermum parki* Kotschy; *Sapotac.;* Elf.
Tangoeloeng: *Protium javanicum* Burm.; *Burserac.;* Nied.-Ind.
Tangolo: *Caloncoba glauca* Gilg; *Bixac.;* Kam.
Tanguile: *Shorca polysperma* Merr.; *Dipterocarpac.;* Phil.
Tani: *Terminalia belerica* Roxb., *T. tomentosa* Wight et Arn.; *Combretac.;* O.-Ind.
Tanibouca: *Terminalia tanibouca* Smith; *Combretac.;* n. Bras. (Amaz.)
Tanítan: *Alstonia scholaris* R. Br.; *Apocynac.;* Phil. (Bisaya)
Taniwayanagi: *Salix gracilistyla* Miq.; *Salicac.;* Jap.
Tankolo: *Kleinhovia hospita* L.; *Sterculiac.;* Java.
Tannaneo: *Peltogyne sp.; Caesalpiniac.;* Kol.
Tanne (d): *Abies spp.; Conifer.;* n. Hem.
„ , Balsam- (d): *Abies balsamea* Mill.; *Conifer.;* n. N.-Am.
„ , Chile-: *Araucaria araucana* K. Koch; *Conifer.;* Chile.
„ , Douglas-: *Pseudotsuga taxifolia* Britt.; *Conifer.;* N.-Am.
„ , Edel-: *Abies alba* Miller; *Conifer.;* Eu.
„ , gleichfarbige: *Abies concolor* Lindl. et Gord.; *Conifer.;* USA.
„ , große: *Abies grandis* Lindl.; *Conifer.;* w. N.-Am.

Tanne, Nordmann-: *Abies nordmanniana* Spach; *Conifer.*; Kauk.
„ , Norfolk-: *Araucaria excelsa* R. Br.; *Conifer.*; Südsee (Norfolkinsel)
„ , Pindrau-: *Abies pindrow* Spach; *Conifer.*; O.-Ind. (Himal.)
„ , Pindrow-: *Abies pindrow* Spach; *Conifer.*; O.-Ind. (Himal.)
„ , Rot-: *Picea excelsa* Link; *Conifer.*; Eu.
„ , Sachalin-: *Abies sachalinensis* Mast.; *Conifer.*; Jap.
„ , Silber-: *Abies alba* Mill.; *Conifer.*; Eu.
„ , Weiß-: *Abies alba* Mill.; *Conifer.*; Eu.
Tanóbong: *Pometia pinnata* Forst.; *Sapindac.*; Phil. (Palawan)
Tantai: *Albizzia lebbek* Bth.; *Mimosac.*; O.-Ind.
Tantalo: *Cedrela fissilis* Vell.; *Meliac.*; Arg.
Tanua: *Trichilia candollei* Chev.; *Meliac.*; Elf.
Tanuba: *Trichilia candollei* Chev.; *Meliac.*; Elf.
Tanuku: *Sterculia urens* Roxb.; *Sterculiac.*; n. O.-Ind. (United Prov.)
Tanwood: *Schinopsis spp.*; *Anacardiac.*; S.-Am.
Taô: *Quercus lanuginosa* D. Don; *Fagac.*; Ind.-Ch. (Annam)
Taóto: *Pterocymbium tinctorium* Merr.; *Sterculiac.*; Phil. (Bataan)
Tapabutija: *? Apeiba aspera* Aubl.; *Tiliac.*; Nic.
Tapaculo: *Guazuma ulmifolia* Lam.; *Sterculiac.*; Guat.
Tapana: *Hieronyma caribaea* Urban; *Euphorbiac.*; Trin., Tob.
„ : *Hieronyma alchorneoides* Allem.; *Euphorbiac.*; Trin.
Tapanare: *Hieronyma alchorneoides* Allem.; *Euphorbiac.*; Trin.
Tapang: *Koompassia excelsa* Taub.; *Caesalpiniac.*; Borneo.
Tapanhúma: *Cassia marçanahyba* Allem.; *Caesalpiniac.*; Bras. (Sao P)
„ : *Ocotea sp., Nectandra sp.*; *Laurac.*; s. Bras.
Taparo: *Crescentia cujete* L.; *Bignoniac.*; Ven. (Llanos)
Tapasi: *Holoptelea integrifolia* Planch.; *Ulmac.*; O.-Ind. (Madras)
Tapasquit: *Luehea seemanni* Tr. et Planch.; *Tiliac.*; br. Hond., Guat.
Tapatamal: *Coccoloba acuminata* H. B. K.; *Polygonac.*; Hond.
Tap-ché: *Rhizophora mangle* L.; *Rhizophorac.*; Mex.
Taperebá: *Spondias lutea* L.; *Anacardiac.*; Bras.
Tapé-riguá: *Trichilia sp.*; *Meliac.*; Arg.
Tapgás: *Parinarium corymbosum* Miq.; *Rosac.*; Phil. (Guimarao)
Tapia: *Crataeva tapia* L.; *Capparidac.*; Bras.
Tapiá: *Alchornea iricurana* Casar.; *Euphorbiac.*; Bras. (Sao P)
Tapiagosú: *Crataeva tapia* L.; *Capparidac.*; Bras.
Tapinág: *Sterculia crassiramea* Merr.; *Sterculiac.*; Phil.
„ , Mountain-: *Sterculia montana* Merr.; *Sterculiac.*; Phil.
Tapinhoan: *Silvia navalium* Allem.; *Laurac.*; s. Bras.
Tapirin-aurarama: *Scerolobium sp.*; *Caesalpiniac.*; Sur.
„ kwatéré: *Eschweilera corrugata* Merr., *E. sp.*; *Lecythidac.*; Sur.
„ -wonoe: *? Nectandra sp., ? Ocotea sp.*; *Laurac.*; Sur.
„ „ : *Terminalia buceras* Hook.; *Combretac.*; Sur.
Tapiriri: *Tapirira guianensis* Aubl.; *Anacardiac.*; Gu. (Galibis)
Tapir wood (e): *Maytenus obtusifolia* Mart.; *Celastrac.*; Bras. (Bahia)
Tapis: *Goniothalamus tapis* Miq.; *Anonac.*; Malay.
Tapoekin-mani: *Moronobea coccinea* Aubl.; *Guttifer.*; Sur.
Tapoekjen-ajawa: *Protium sagotianum* March., *P. spp.*; *Burserac.*; Sur.
Tapoeloe-kwatelie: *Eschweilera longipes* Miers, *E. subglandulosa* Miers; *Lecythidac.*;
Tapoeripa: *Gustavia spp.*; *Lecythidac.*; Sur.   [Sur.
„ : *Genipa americana* L., *? G. caruto* H. B. K.; *Rubiac.*; Sur.
„ , Hoogland-: *Gustavia pterocarpa* Poit.; *Lecythidac.*; Sur.
„ , Laagland-: *Gustavia angusta* L.; *Lecythidac.*; Sur.
Tapoeropo: *Genipa americana* L., *? G. caruto* H. B. K.; *Rubiac.*; Sur.
Taporo: *Citrus medica* Gall.; *Rutac.*; Südsee (Tahiti)
Tapria-siris: *Albizzia lucida* Bth.; *Mimosac.*; nö. O.-Ind. (Nepal)
Taproepa: *Genipa americana* L., *? G. caruto* H. B. K.; *Rubiac.*; Sur.
Tapsi: *Holoptelea integrifolia* Planch.; *Ulmac.*; O.-Ind. (Bombay)
Tapúlau: *Pinus merkusi* Jungh. et de Vr.; *Conifer.*; Phil. (Zambales)
Tapúrau: *Vatica mangachapoi* Blco.; *Dipterocarpac.*; Phil. (Albay)
Taquarizeiro: *Mabea spp.*; *Euphorbiac.*; n. Bras. (Amaz.)
Taquary: *Mabea paniculata* Bth.; *Euphorbiac.*; Bras.

Tara: *Oyedaea verbesinoides* DC.; *Composit.*; C.-Ven. (& Caracas)
„ : *Caesalpinia spinosa* Ktze., *C. pectinata* Cav.; *Caesalpiniac.*; tr. S.-Am.
„ blanca: *Montagnaea excelsa* Ernst; *Composit.*; Kol., Ven. (Caracas)
Tarachapa: *Adina cordifolia* Hook. f.; *Rubiac.*; O.-Ind. (Assam)
Taradák: *Cyclostemon sp.*; *Euphorbiac.*; Phil. (Cotabato)
Tarago: *Ilex latifolia* Thunb.; *Aquifoliac.*; Jap.
Tarahuilca: *Piptadenia sp.*; *Mimosac.*; Peru.
Taraia: *Ilex latifolia* Thunb.; *Aquifoliac.*; Jap.
Taraire: *Beilschmiedia tarairi* Kirk; *Laurac.*; N.-Seel.
Tarais: *Salix taxifolia* H. B. K.; *Salicac.*; Mex.
Taraka vepa: *Melia azedarach* L.; *Meliac.*; O.-Ind. (Madras)
Tarakit: *Juniperus procera* Hochst.; *Conifer.*; br. O.-Af. (Kenya)
Tarakuet: *Juniperus procera* Hochst.; *Conifer.*; br. O.-Af. (Kenya)
Tarala: *Dipteryx sp.*; *Papilionac.*; fr. Gu.
Tarami: *Beilschmiedia tarairi* Kirk; *Laurac.*; N.-Seel.
Taranoki: *Aralia chinensis* L. *var. glabrescens* Fr. et Sav.; *Araliac.*; Jap.
Tarantang: *Campnosperma auriculatum* Hook. f.; *Anacardiac.*; Sum.
Tarata: *Pittosporum eugenioides* A. Cunn.; *Pittosporac.*; N.-Seel.
Taratára: *Myristica sp.*; *Myristicac.*; Phil. (Tayabas)
Taráu: *Livistona rotundifolia* Mart., *L. sp.*; *Palmac.*; Phil. (Cagayan)
Taray: *Tamarix gallica* L.; *Tamaricac.*; Sp.
„ : *Salix taxifolia* H. B. K.; *Salicac.*; Mex.
„ : *Caesalpinia crista* L.; *Caesalpiniac.*; Mex.
„ : *Eysenhardtia polystachya* Sarg.; *Papilionac.*; Mex.
„ : *Amyris elemifera* L.; *Rutac.*; Hond.
„ del río: *Salix taxifolia* H. B. K.; *Salicac.*; Mex.
Tarayo: *Ilex latifolia* Thunb.; *Aquifoliac.*; Jap.
Tarco: *Jacaranda acutifolia* Humb. et Bonpl.; *Bignoniac.*; Arg.
„ : *Thouinia weinmannifolia* Griseb.; *Sapindac.*; Arg.
Taré: *Terminalia belerica* Roxb.; *Combretac.*; O.-Ind. (Bombay)
Tarecuen: *Quercus sp.*; *Fagac.*; Mex.
Taree: *Terminalia catappa* L.; *Combretac.*; O.-Ind. (Madras)
Tarfa: *Tamarix gallica* L.; *Tamaricac.*; Sahara, n. Sudan.
Tarimo-mèrèkè: *Conceveiba guianensis* Aubl., *C. hostmanni* Bth.; *Euphorbiac.*; Sur.
Tarisi: *Albizzia lebbek* Bth.; *Mimosac.*; Java.
Tariti: *Parinarium griffithianum* Bth.; *Rosac.*; Java.
Tarkhana: *Acer pictum* Thunb.; *Acerac.*; nw. O.-Ind. (Punjab)
Tarobam: *Albizzia coriaria* Welw.; *Mimosac.*; br. O.-Af. (Uganda)
Taroentoeng: *Lumnitzera coccinea* Wight et Arn.; *Combretac.*; Java (Moeara)
Taroko-gasi: *Quercus tarokoensis* Hayata; *Fagac.*; Form.
„ -sidé: *Carpinus hebestroma* Yamamoto; *Betulac.*; Form.
Tarongatingan: *Pterospermum obliquum* Blco.; *Sterculiac.*; Phil. (Camarines, Burias)
Tarop: *Buchanania latifolia* Roxb.; *Anacardiac.*; ö. Beng.
Tarout: *Cupressus dupreziana* A. Camus; *Conifer.*; Sahara.
Tarpandi: *Pterocarpus hypostictus* Miq.; *Papilionac.*; Sumatra.
Tarquí: *Hedyosmum scabrum* Solms; *Chloranthac.*; Ek.
Tarrafe: *Avicennia africana* Beauv.; *Verbenac.*; port. Guin.
„ : *Tamarix gallica* L.; *Tamaricac.*; Cap Verde-Inseln.
Tarrietie: *Tarrietia javanica* Blume; *Sterculiac.*; Nied.-Ind.
Tarro: *Bambusa aculeata* Hitchcock; *Gramin.*; br. Hond.
„ grande: *Bambusa aculeata* Hitchcock; *Gramin.*; Guat.
Tarroema: *Hieronyma laxiflora* Muell. Arg.; *Euphorbiac.*; Sur.
Tartago: *Jatropha curcas* L.; *Euphorbiac.*; P.-R.
Tartar tree: *Adansonia gregori* F. v. M.; *Bombacac.*; n. Au.
Tartboom, Cremor-: *Adansonia digitata* L.; *Bombacac.*; sw. Af.
Taruka: *Azadirachta indica* A. Juss.; *Meliac.*; O.-Ind. (Madras)
Taruma: *Vitex montevidensis* Cham.; *Verbenac.*; Bras.
Tarumá: *Citharexylum spp.*; *Verbenac.*; Bras. (Amaz.)
„ con espinas: *Citharexylum barbinerve* Cham.; *Verbenac.*; n. Arg.
„ dura: *Vitex sp.*; *Verbenac.*; Arg.
„ espinudo: *Citharexylum barbinerve* Cham.; *Verbenac.*; n. Arg.
„ frondosa: *Vitex orinocensis var. amazonica* Hub.; *Verbenac.*; Bras.

Tarumá guazu: *Vitex cymosa* Bert.; *Verbenac.*; Arg.
„ tuira: *Vitex sp.; Verbenac.*; Bras.
Taruman: *Vitex montevidensis* Cham., *V. sellowiana* Cham.; *Verbenac.*; Bras. (Sao P)
„ amarello: *Vitex montevidensis* Cham.; *Verbenac.*; Bras. (Sao P)
„ do Norte: *Vitex montevidensis* Cham.; *Verbenac.*; Bras. (Sao P)
„ pardo: *Vitex montevidensis* Cham.; *Verbenac.*; Bras. (Sao P)
„ vermelho: *Vitex montevidensis* Cham.; *Verbenac.*; Bras (Sao P)
Taruntum: *Lumnitzera littorea* Voigt; *Combretac.*; Malay.
Tarwa: *Tamarix articulata* Vahl; *Tamaricac.*; O.-Ind.
Tas: *Geonoma baculifera* Kunth; *Palmac.*; fr. Gu.
Tascate: *Juniperus pachyphloea* Torr., *J. mexicana* Spreng.; *Conifer.*; Mex.
Tasha: *Phyllanthus emblica* L.; *Euphorbiac.*; Burma.
Tasie: *Triplaris surinamensis* Cham.; *Polygonac.*; Sur.
Tassel plant: *Suriana maritima* L.; *Simarubac.*; Bah.
Tassie: *Triplaris surinamensis* Cham.; *Polygonac.*; Sur.
Tassièt: *Pentace burmanica* Kurz; *Tiliac.*; Kamb.
Tata: *Psychotria monticola* Hiern; *Rubiac.*; Kam.
„ blanca: *Celtis sellowiana* Miq.; *Ulmac.*; Bras.
„ iro: *Alchornea sp.; Euphorbiac.*; Elf.
„ yirá: *Chlorophora tinctoria* Gaudich.; *Morac.*; Arg.
Tataaro-sierwaballi: *Nectandra sp., Ocotea sp.; Laurac.*; Sur.
Tataboe: *? Andira retusa* H. B. K.; *Papilionac.*; Sur.
„ : *Diplotropis guianensis* Bth., *D. leptophylla* Kleinh.; *Papilionac.*; Sur.
Tataboo-hout: *Diplotropis guianensis* Bth., *D. sp.; Papilionac.*; Sur.
Tatabu: *Diplotropis guianensis* Bth., *D. leptophylla* Kleinh.; *Papilionac.*; Sur
Tatajiba: *Chlorophora tinctoria* Gaudich.; *Morac.*; Bras.
Tatajuba: *Bagassa guianensis* Aubl.; *Morac.*; Bras. (Pará)
„ : *Maclura affinis* Miq.; *Morac.*; Bras. (Sao P)
„ de tinta: *Chlorophora tinctoria* Gaudich.; *Morac.*; Bras.
Tatamaco: *Bursera tomentosa* Tr. et Planch.; *Burserac.*; Ant. (Neu-Granada)
Tatamaka: *Calophyllum inophyllum* L., *C. tacamahaca* Willd.; *Guttifer.*; Maur., Réu.
Tatané: *Pithecolobium vinhatico* Record, *P. tortum* Mart.; *Mimosac.*; Arg.
„ : *Pithecolobium scalare* Griseb.; *Mimosac.*; Arg.
„ blanco: *Pithecolobium vinhatico* Record; *Mimosac.*; Arg.
Tataré: *Pithecolobium vinhatico* Record; *Mimosac.*; Arg.
Tatascame: *Lippia myriocephala* Schl. et Cham.; *Verbenac.*; Salv.
Tatatían: *Comocladia repanda* Blake; *Anacardiac.*; Mex.
Tatatil: *Comocladia repanda* Blake; *Anacardiac.*; Mex.
Tatayibá: *Chlorophora tinctoria* Gaudich.; *Morac.*; Arg.
„ -pyitá: *Chlorophora tinctoria* Gaudich.; *Morac.*; Arg.
„ -saiyú: *Chlorophora tinctoria* Gaudich.; *Morac.*; Arg.
Tatayouba: *Caryocar tomentosum* Willd., *C. spp.; Caryocarac.*; Sur.
Tateyyoo: *Eurya acuminata* DC.; *Theac.*; Malak.
Tatô-gasi (jap.): *Lithocarpus taitoensis* Hayata; *Fagac.*; Form.
Tatrac: *Fagraea fragrans* Roxb.; *Loganiac.*; Kamb.
Tatri: *Dillenia pentagyna* Roxb.; *Dilleniac.*; nö. O.-Ind. (Nepal)
Tattan: *Strychnos potatorum* L. f.; *Loganiac.*; O.-Ind. (Madras)
Tattele: *Sterculia alata* Roxb.; *Sterculiac.*; sw. O.-Ind. (Bombay)
Tatú: *Eugenia axillaris* Vell.; *Myrtac.*: Bras. (Sao P)
„ graúdo: *Eugenia axillaris* Vell.; *Myrtac.*; Bras. (Sao P)
„ pequeño: *Eugenia axillaris* Vell.; *Myrtac.*; Bras. (Sao P)
Tatzungo: *Ichthyomethia americana* Blake; *Papilionac.*; Mex.
Taú: *Guarea sp.; Meliac.*; Bras.
Tau: *Vatica tonkinensis* Chev., *V. spp.; Dipterocarpac.*; Ind.-Ch.
„ : *Berlinia kerstingi* Harms; *Caesalpiniac.*; Togo.
Taua: *Beilschmiedia tawa* T. Kirk; *Laurac.*; N.-Seel.
Tauari: *Couratari tauari* Berg; *Lecythidac.*; nö. Bras. (Pará)
Taúba: *Maclura affinis* Miq.; *Morac.*; Bras. (Sao P)
Tauekaha: *Phyllocladus trichomanoides* D. Don; *Conifer.*; N.-Seel.
Taukkyan: *Terminalia tomentosa* Wight et Arn.; *Combretac.*; Burma.
Taung-letpan: *Bombax insigne* Wall.; *Bombacac.*; Burma.
„ peing: *Artocarpus chaplasha* Roxb.; *Morac.*; O.-Ind., And.

Taung peinnè: *Artocarpus chaplasha* Roxb.; *Morac.;* And.
„ -petwun: *Pterospermum acerifolium* Willd.; *Sterculiac.;* Burma.
„ -thayet: *Swintonia floribunda* Griff.; *Anacardiac.;* Burma.
„ -yemané: *Lophopetalum fimbriatum* Wight; *Celastrac.;* Burma.
Taungdama: *Toona serrata* M. Roem.; *Meliac.;* Burma.
„ : *Acrocarpus fraxinifolius* Wight; *Caesalpiniac.;* Burma.
Taungkaukno: *Litsea polyantha* Juss.; *Laurac.;* nö. Burma (Ruby Mines)
Taungmagyi: *Albizzia odoratissima* Bth.; *Mimosac.;* nö. Burma (Ruby Mines)
Taungméok: *Alstonia scholaris* R. Brown; *Apocynac.;* s. Burma (s. Tenasserim)
Taungnangyi: *Gmelina arborea* L.; *Uerbenac.;* Burma.
Taungsagaing: *Uatica scaphula* Dyer; *Dipterocarpac.;* w. Burma (Arrakan)
Taungsalet: *Wrightia tomentosa* Roem. et Schultes; *Apocynac.;* Burma.
Taungthayet: *Swintonia floribunda* Griff.; *Anacardiac.;* Burma.
Tauroniro: *Humiria sp.; Humiriac.;* br. Gu.
„ : *Humiria floribunda* Mart., *H. balsamifera* Aubl.; *Humiriac.;* Sur.
Tautu: *Pterocymbium tinctorium* Merr.; *Sterculiac.;* Phil. (Bataan)
Taúva: *Guarea tuberculata* Vell.; *Meliac.;* Bras.
„ : *Maclura affinis* Miq., *Chlorophora tinctoria* Gaudich.; *Morac.;* Bras.
Tavola: *Terminalia catappa* L.; *Combretac.;* Südsee (Fid.)
Tavolo: *Ravensara sp.; Laurac.;* Mad.
Tavoy wood (e): *Parashorea stellata* Kurz; *Dipterocarpac.;* Burma.
Taw: *Polyalthia simiarum* Bth. et Hook. f.; *Anonac.;* s. Burma (s. Tenasserim)
„ -thayet: *Mangifera silvatica* Roxb.; *Anacardiac.;* And.
„ -thidin: *Mallotus philippinensis* Muell. Arg.; *Euphorbiac.;* And.
Tawa: *Beilschmiedia tawa* T. Kirk; *Laurac.;* N.-Seel.
„ , indian: *Beilschmiedia sikkimensis* King; *Laurac.;* O.-Ind.
Tawai: *Nothofagus solanderi* Oerst., *N. fusca* Oerst.; *Fagac.;* N.-Seel.
Tawanonero: *Humiria floribunda* Mart., *H. balsamifera* Aubl.; *Humiriac.;* Sur.
Tawaranoe: *Humiria floribunda* Mart., *H. balsamifera* Aubl.; *Humiriac.;* Sur.
Tawaronero: *Humiria floribunda* Mart.; *Humiriac.;* Gu.
Tawatsa: *Entada abyssinica* Steud.; *Mimosac.;* Kam.
Tawhai: *Nothofagus fusca* Oerst., *N. solanderi* Oerst.; *Fagac.;* N.-Seel.
„ -rau-nui: *Nothofagus fusca* Oerst.; *Fagac.;* N.-Seel.
„ „ -riki: *Nothofagus solanderi* Oerst.; *Fagac.;* N.-Seel.
Tawhen: *Cedrela sp.; Meliac.;* Amboina.
Tawhero: *Weinmannia racemosa* L. f.; *Cunoniac.;* N.-Seel. (Auckland)
Tawitho: *Terminalia belerica* Roxb.; *Combretac.;* n. Burma.
Tawmagyi: *Elaeocarpus robustus* Roxb.; *Elaeocarpac.;* Burma.
Tawposa: *Morus laevigata* Wall.; *Morac.;* Burma.
Tawpwèsa: *Morus laevigata* Wall.; *Morac.;* Burma.
Tawsagasein: *Polyalthia simiarum* Bth. et Hook. f.; *Anonac.;* Burma.
Tawshauk: *Atalantia monophylla* Correa; *Rutac.;* Burma.
Tawthanbya thee: *Atalantia monophylla* Correa; *Rutac.;* And.
Tawthayet: *Nephelium longana* Camb.; *Sapindac.;* O.-Ind.
Taxate: *Juniperus mexicana* Spreng.; *Conifer.;* Mex.
Taxixte: *Zexmenia frutescens* Blake; *Composit.;* Guat.
Taya: *Phyllanthus emblica* L.; *Euphorbiac.;* n. Burma.
Tayanca: *Baccharis odorata* H. B. K.; *Composit.;* Peru.
Tayatáya: *Terminalia edulis* Blco.; *Combretac.;* Phil. (Guimaras)
Tayavevuia: *Tabebuia leucoxyla* DC.; *Bignoniac.;* Bras. (Sao P)
Tayaw: *Excoecaria agallocha* L.; *Euphorbiac.;* Burma.
„ -nyo: *Grewia laevigata* Vahl; *Tiliac.;* Burma.
„ -ywetwaing: *Eriolaena candollei* Wall.; *Sterculiac.;* Burma.
Tayé: *Tecoma sp.; Bignoniac.;* Arg.
Tayí: *Tecoma sp.; Bignoniac.;* Arg.
„ -hú: *Tabebuia avellanedae* Lorentz; *Bignoniac.;* Arg.
? Tay-ninh: *Terminalia bialata* Wall.; *Combretac.;* Co.-Ch.
Tayok-thè: *Bischofia javanica* Blume; *Euphorbiac.;* Burma.
Taytyoof: *Eurya acuminata* DC.; *Theac.;* Malak.
Tayum-táyum: *Cyclostemon littoralis* C. B. Rob.; *Euphorbiac.;* Phil. (Bataan)
Tayuva: *Chlorophora tinctoria* Gaudich.; *Morac.;* Bras. (Sao P)
Tcan-lol: *Ouratea nitida* Engl.; *Ochnac.;* br. Hond.

Tcan-sick: *Jacquinia aurantiaca* Ait.; *Theophrastac.;* br. Hond.
„ - „ : *Hyperbaena winzerlingi* Standl.; *Menispermac.;* br. Hond.
Tcha hoa: *Camellia sasanqua* Thunb.; *Theac.;* Ch.
„ yeou: *Camellia sasanqua* Thunb.; *Theac.;* Ch.
Tchaïchot: *Balanites aegyptiaca* Delile; *Zygophyllac.;* Sudan.
Tchampâ: *Michelia champaca* L.; *Magnoliac.;* O.-Ind. (Hind.)
Tchanchin: *Toona sinensis* M. Roem.; *Meliac.;* Jap.
Tchang lào: *Dryobalanops aromatica* Gaertn. f.; *Dipterocarpac.;* Annam.
Tchau-tchoun: *Ailanthus glandulosa* Desf.; *Simarubac.;* Ch.
Tchè kiú tsè: *Hovenia dulcis* Thunb.; *Rhamnac.;* Ch.
Tcheiray: *Taxus baccata* L.; *Conifer.;* nö. O.-Ind. (Nepal)
Tcheou tchoun chou: *Ailanthus glandulosa* Desf.; *Simarubac.;* Ch.
Tchiaba i kwa: *Camellia sasanqua* Thunb.; *Theac.;* Jap.
Tchiat kattsé: *Grumilea venosa* Bth.; *Rubiac.;* Elf.
Tchiatchia: *Albizzia lebbek* Bth.; *Mimosac.;* Haiti.
Tchiébuessain: *Xylia evansi* Hutch.; *Mimosac.;* Elf.
Tchickboué: *Tetrapleura thonningi* Bth.; *Mimosac.;* Elf.
Tchiko: *Lannea acidissima* Chev.; *Anacardiac.;* Elf.
Tchikoué: *Bridelia speciosa* Muell. Arg.; *Euphorbiac.;* Elf.
Tchikué: *Bridelia speciosa* Muell. Arg.; *Euphorbiac.;* Elf.
Tchio: *Lannea acidissima* Chev.; *Anacardiac.;* Elf.
Tchioukoué: *Bridelia speciosa* Muell. Arg.; *Euphorbiac.;* Elf.
Tchissafoukala: *Canarium velutinum* Guill.; *Burserac.;* Gab. (Loango)
Tchogo: *Spathodea campanulata* Beauv.; *Bignoniac.;* Gab.
Tchombi: *Uapaca sp.; Euphorbiac.;* Gab.
Tchomboko: *Bridelia speciosa* Muell. Arg.; *Euphorbiac.;* Gab.
Tchondoba: *Desbordesia sp.; Simarubac.;* Gab.
Tchontsi: *Erythrophloeum sp.; Caesalpiniac.;* Gab.
Tchoumbou: *Piptadenia sp.; Mimosac.;* Gab.
Tciao chang: *Machilus thunbergi* S. et Z.; *Laurac.;* Ch.
Té: *Terminalia altissima* Chev.; *Combretac.;* Elf.
Te: *Diospyros burmanica* Kurz; *Ebenac.;* Burma.
„ ake: *Alectryon dodonaeifolia* ?; *Sapindac.;* N.-Seel.
Tea (e): *Camellia thea* Link; *Theac.;* sö. As.
Téa: *Amyris balsamifera* L., *A. maritima* Jacq.; *Rutac.;* P.-R.
., : *Amyris silvatica* Jacq.; *Rutac.;* P.-R.
Tea bark: *Myrica mexicana* Willd.; *Myricac.;* br. Hond.
„ box: *Myrica mexicana* Willd.; *Myricac.;* br. Hond.
„ , New Jersey-: *Ceanothus americanus* L.; *Rhamnac.;* USA.
„ tree: *Leptospermum ericoides* A. Rich.; *Myrtac.;* N.-Seel.
„ „ : *Leptospermum lanigerum* Sm.; *Myrtac.;* Tasm.
„ „ , lemon-scented: *Leptospermum citratum* Challinor; *Myrtac.;* Au.
„ „ , white: *Leptospermum scoparium* Forst.; *Myrtac.;* N.-Seel.
Teak (e): *Tectona grandis* L. f.; *Verbenac.;* tr. As.
„ : *Hymenaea courbaril* L.; *Caesalpiniac.;* Trin., Tob.
., : *Flindersia australis* R. Br.; *Rutac.;* Au.
„ , african (e, u): *Chlorophora excelsa* Bth. et Hook.; *Morac.;* W.- & O.-Af.
„ , „ : *Copaifera coleosperma* Bth.; *Caesalpiniac.;* Rhod.
„ , Bajam-: *Afzelia bijuga* A. Gray; *Caesalpiniac.;* Borneo, Celebes.
„ , Borneo-: *Hopea spp., Shorea spp., Isoptera spp.; Dipterocarpac.;* n. Borneo.
„ , brasilianisches: *Andira excelsa* H. B. K.; *Caesalpiniac.;* Bras.
„ , Burma (e): *Tectona grandis* L. f.; *Verbenac.;* O.-Ind., Burma.
„ , Cape-: *Strychnos atherstonei* Harv.; *Loganiac.;* S.-Af.
., , east african (e): *Chlorophora excelsa* Bth. et Hook.; *Morac.;* br. O.-Af. (Uganda)
„ , falsches: *Andira excelsa* H. B. K.; *Caesalpiniac.;* Bras.
., , figured: *Tectona grandis* L. f.; *Verbenac.;* Burma.
„ , Goomar-: *Gmelina arborea* Roxb.; *Verbenac.;* O.-Ind., Burma, Cey.
„ , grey: *Gmelina leichhardti* F. v. M.; *Verbenac.;* Au.
., , Guiana-: *Dicorynia paraënsis* Bth.; *Caesalpiniac.;* Gu.
„ , Johore-: *Parinarium oblongifolium* Hook. f.; *Rosac.;* Malak. (Pahang)
„ , Mahoborn-: *Dryobalanops aromatica* Gaertn. f.; *Dipterocarpac.;* Borneo.
„ , Moule- (e): *Chlorophora excelsa* Bth. et Hook.; *Morac.;* O.-Af.

Teak, Moulmein- (e): *Tectona grandis* L. f.; *Verbenac.;* O.-Ind., Burma.
„ , native: *Flindersia australis* R. Br., *F. bennettiana* F. v. M.; *Rutac.;* ö. Au.
„ , New Zealand-: *Vitex littoralis* A. Cunn., *V. lucens* T. Kirk; *Verbenac.;* N.-Seel.
„ , nigerian (e): *Chlorophora excelsa* Bth. et Hook.; *Morac.;* W.-Af.
„ , Philippine-: *Lagerstroemia piriformis* Koehne; *Lythrac.;* Phil.
„ , Rangoon- (e): *Tectona grandis* L. f.; *Verbenac.;* O.-Ind., Burma.
„ , red: *Dysoxylum fraseranum* Bth.; *Meliac.;* ö. Au.
„ , rhodesian (e): *Baikiaea plurijuga* Harms; *Caesalpiniac.;* Rhod.
„ , „ : *Copaifera coleosperma* Bth.; *Caesalpiniac.;* Rhod.
„ , „ : *Pterocarpus angolensis* DC.; *Papilionac.;* Rhod.
„ , Seacoast-: *Guettarda speciosa* L.; *Rubiac.;* tr. As.
„ , Surinam-: *Dicorynia paraënsis* Bth.; *Caesalpiniac.;* Gu.
„ , Yang-: *Dipterocarpus tuberculatus* Roxb., *D. spp.; Dipterocarpac.;* tr. As.
Teakhout, surinaamsch (h): *Hymenaea courbaril* L.; *Caesalpiniac.;* Sur.
„ , westindisch (h): *Hymenaea courbaril* L.; *Caesalpiniac.;* Sur.
Tear-blanket: *Zanthoxylum fraxineum* Willd.; *Rutac.;* s. Fl.
Teatlale: *Cupressus benthami* Endl.; *Conifer.;* Mex.
Te-ay-chu: *Memecylon polyanthemum* Hook. f.; *Melastomatac.;* Lib.
Teba: *Allophylus africanus* Beauv.; *Sapindac.;* B.-K. (Mayombe)
Téba: *Irvingia ? oblonga* Chev.; *Simarubac.;* Gab.
Tèbè: *Desbordesia insignis* Pierre; *Simarubac.;* Gab.
Tebeldi: *Adansonia digitata* L.; *Bombacac.;* Arab.
Teben: *Crossopteryx kotschyana* Engl.; *Rubiac.;* Go.
Tébi: *Malacantha robusta* Chev.; *Sapotac.;* Elf.
Teborak: *Balanites aegyptiaca* Delile; *Zygophyllac.;* Sahara.
Teca: *Fagraea fragrans* Roxb.; *Loganiac.;* Phil. (Palawan, Mindoro)
Teck (f): *Tectona grandis* L. f.; *Verbenac.;* Ind.-Ch.
„ rouge: *Dipterocarpus spp.; Dipterocarpac.;* Ind.-Ch.
„ „ : *Manglietia glauca* Blume, *M. fordiania* Oliv.; *Magnoliac.;* Ind.-Ch.
„ d'Afrique: *Chlorophora excelsa* Bth. et Hook., *C. regia* Chev.; *Morac.;* W.-Af.
„ de Guyane: *Dicorynia paraënsis* Bth.; *Caesalpiniac.;* fr. Gu.
„ „ Madagascar: *Intsia bijuga* O. Ktȝe.; *Caesalpiniac.;* Mad.
Teclatilla: *Comocladia engleriana* Loes.; *Anacardiac.;* Mex.
Tecomaca: *Elaphrium sp.; Burserac.;* Mex.
Tecomate: *Crescentia cujete* L., *C. alata* H. B. K.; *Bignoniac.;* Mex.
Tecomatillo: *Cochlospermum vitifolium* Spreng.; *Cochlospermac.;* Guat.
Tedjan: *Cocculus laurifolius* DC.; *Menispermac.;* Java (Soebah)
Tefarshia: *Sarcocephalus russegeri* Kotschy; *Rubiac.;* n. Nig. (Kontagora)
Tegám: *Neonauclea sp.; Rubiac.;* Phil. (Abra, Bontok, Benguet)
Tegina: *Tectona grandis* L. f.; *Verbenac.;* O.-Ind. (Bombay)
Teh: *Camellia thea* Link; *Theac.;* Malay.
Tehoq: *Salvadora persica* L.; *Salvadorac.;* Sahara.
Tei-hatti: *Maprounea guianensis* Aubl.: *Euphorbiac.;* Sur.
Teil-tree (e): *Tilia parvifolia* Ehrh.; *Tiliac.;* Eu.
Teimadi: *Garuga pinnata* Roxb.; *Burserac.;* nw. O.-Ind. (Kumaon)
Tein: *Stephegyne parvifolia* Korth.; *Rubiac.;* Burma.
Teinkula: *Nauclea sessilifolia* Roxb.; *Rubiac.;* Burma.
Teinthe: *Stephegyne parvifolia* Korth.; *Rubiac.;* Burma.
Teja: *Cinnamomum spp.; Laurac.;* Malay.
Tejo: *Taxus baccata* L.; *Conifer.;* Sp.
„ : *Cassia sp.; Caesalpiniac.;* Nic.
Tek: *Tectona grandis* L. f.; *Verbenac.;* O.-Ind. (Bombay)
Teka: *Tectona grandis* L. f.; *Verbenac.;* C.-O.-Ind. (Gondwana)
Tekam: *Shorea guiso* Blume; *Dipterocarpac.;* br. N.-Borneo (Sarawak)
Tekbé: *Canthium sp.; Rubiac.;* Elf.
Téké: *Pycnanthus kombo* Warb.; *Myristicac.;* Elf.
Tekik: *Albizzia lebbek* Bth.; *Mimosac.;* Java.
Tekoe: *Adina cordifolia* Hook. f.; *Rubiac.;* n. O.-Ind. (United Prov.)
Teku: *Tectona grandis* L. f.; *Verbenac.;* O.-Ind. (Madras)
Tekúm: *Neonauclea vidali* Merr.; *Rubiac.;* Phil. (Abra, Bontok, Benguet)
Telambu: *Sterculia foetida* L.; *Sterculiac.;* s. O.-Ind., Cey.
Teledo wood (u): *Eschweilera corrugata* Miers; *Lecythidac.;* Sur.

Telelka: *Gardenia turgida* Roxb.; *Rubiac.;* O.-Ind. (Madras)
Téléloema: *? Nectandra sp., ? Ocotea sp.; Laurac.;* Sur.
„　　: *? Terminalia buceras* Hook.; *Combretac.;* Sur.
Telemboo: *Sterculia foetida* L.; *Sterculiac.;* Cey.
Teli: *Erythrophloeum guineense* G. Don; *Caesalpiniac.;* Elf.
Telia: *Acacia arabica* Willd.; *Mimosac.;* nw. O.-Ind.
„　: *Acacia scorpioides* W. F. Wight; *Mimosac.;* Senl., Sudan.
„　garjan: *Dipterocarpus alatus* Roxb.; *Dipterocarpac.;* Beng.
„　gurjum: *Dipterocarpus alatus* Roxb.; *Dipterocarpac.;* Beng.
Teliasag: *Cordia macleodi* Hook. f. et Thoms.; *Borraginac.;* C.-O.-Ind. (Centr. Pr.)
Tella chinduga: *Albizzia procera* Bth.; *Mimosac.;* O.-Ind. (Madras)
„　motku: *Ougeinia dalbergioides* Bth.; *Papilionac.;* O.-Ind.
„　pala: *Wrightia tinctoria* R. Br., *W. tomentosa* R. et Sch.; *Apocynac.;* O.-Ind.
„　sopara: *Albizzia procera* Bth.; *Mimosac.;* O.-Ind. (Madras)　　[(Madras)
„　tuma: *Acacia leucophloea* Willd.; *Mimosac.;* O.-Ind. (Madras)
„　turna: *Acacia ferruginea* DC.; *Mimosac.;* O.-Ind. (Madras)
Teloulout: *Capparis spinosa* L.; *Capparidac.;* Sahara.
Telu: *Stereospermum xylocarpum* Bth. et Hook. f.; *Bignoniac.;* C.-O.-Ind. (Centr. Pr.)
Telus: *Ougeinia dalbergioides* Bth.; *Papilionac.;* O.-Ind. (Bombay)
Teluto: *Pterocymbium tinctorium* Merr.; *Sterculiac.;* Phil.
Tema: *Chlorophora excelsa* Bth. et Hook.; *Morac.;* W.-Af. (S.-L.)
Temak: *Shorea macroptera* Dyer; *Dipterocarpac.;* Malay.
Temante: *Crescentia cujete* L.; *Bignoniac.;* Mex., C.-Am.
Temanu: *Calophyllum inophyllum* L.; *Guttifer.;* Südsee (Tahiti, Marquesas)
Tembaloe: *Dysoxylum acutangulum* Miq.; *Meliac.;* Nied.-Ind.
Temberary blanco: *? Zanthoxylum sp.; Rutac.;* Arg.
Tembetari: *Fagara hiemalis* Engl.; *Rutac.;* Arg.
Tembetarí: *Sesbania marginata* Bth.; *Papilionac.;* Arg.
„　guazu: *Fagara acutifolia* Sw.; *Rutac.;* Arg. (Misiones)
„　puitá: *Fagara hiemalis* Engl.; *Rutac.;* Arg.
Tembetarú: *Zanthoxylum sp.; Rutac.;* Bras.
Tembetary: *Zanthoxylum sp.; Rutac.;* Bras., Arg.
„　-hu: *Fagara hiemalis* Engl.; *Rutac.;* Par
„　hù: *Fagara naranjillo* Engl.; *Rutac.;* Arg.
„　moroti: *Fagara cujabensis* Engl.; *Rutac.;* Arg.
„　　„　: *Zanthoxylum petiolare* ?; *Rutac.;* Par.
„　-pyta: *Zanthoxylum rugosum* St. Hil. et Tul.; *Rutac.;* Par.
„　-sayyù: *Fagara riedeliana* Engl.; *Rutac.;* Arg.
Tembhruni: *Diospyros melanoxylon* Roxb.; *Ebenac.;* C.-O.-Ind. (Centr. Prov.)
Tembi: *Calamus barteri* Becc.; *Palmac.;* Nig.
Temblon: *Populus tremula* L.; *Salicac.;* Sp.
Tembru: *Diospyros melanoxylon* Roxb.; *Ebenac.;* O.-Ind. (Bombay)
Tembusa: *Fagraea gigantea* Ridl.; *Loganiac.;* Malay.
Tembusu: *Fagraea fragrans* Roxb., *F. gigantea* Ridl.; *Loganiac.;* Malay.
Tembusu paya: *Fagraea speciosa* Blume; *Loganiac.;* Malay.
„　talang: *Fagraea speciosa* Blume; *Loganiac.;* Malay.
„　tembaga: *Fagraea speciosa* Blume; *Loganiac.;* Malay.
Tembwe: *Cynometra alexandri* C. H. Wright; *Caesalpiniac.;* B.-K. (Beni, Semliki)
Temesu: *Fagraea fragrans* Roxb., *F. gigantea* Ridl.; *Loganiac.;* Malay.
Temigi: *Gaultheria leucocarpa* Blume; *Ericac.;* Java (Tosari)
„　: *Vaccinium varingifolium* Miq.; *Ericac.;* Java (Penandjaan)
„　kasar: *Gaultheria leucocarpa* Blume; *Ericac.;* Java (Tosari)
Teminini: *Guarea thompsoni* Sprague et Hutch.; *Meliac.;* Go.
Tempinis: *Sloetia sideroxylon* Teijsm. et Binn.; *Morac.;* Malay.
Tempisco: *Sideroxylon capiri* Pitt.; *Sapotac.;* Mex.
Tempisque: *Karwinskia humboldtiana* Zucc.; *Rhamnac.;* w. C.-Mex.
„　: *Sideroxylon capiri* Pittier; *Sapotac.;* Mex.
„　: *Sideroxylon tempisque* Pittier; *Sapotac.;* Guat., Salv.
„　: *Dipholis minutiflora* Pittier; *Sapotac.;* C.-R.
Tempixque: *Sideroxylon tempisque* Pittier; *Sapotac.;* Guat., Salv.
Tempixquillo: *Sideroxylon sp.; Sapotac.;* Guat., Salv.
Temple tree: *Plumiera acutifolia* Poir.; *Apocynac.;* O.-Ind.

Temporana: *Suriana maritima* L.; *Simarubac.*; P.-R.
Temria: *Gardenia turgida* Roxb.; *Rubiac.*; C.-O.-Ind. (Centr. Pr.)
Temru: *Diospyros tomentosa* Roxb.; *Ebenac.*; O.-Ind. (United Prov.)
  „ : *Diospyros melanoxylon* Roxb.; *Ebenac.*; C.-O.-Ind. (Central Prov.)
Temu: *Eugenia divaricata* Berg; *Myrtac.*; Chile.
Ten: *Pycnanthus kombo* Warb.; *Myristicac.*; Kam.
Tenaza: *Pithecolobium brevifolium* Bth.; *Mimosac.*; Mex.
Tenbeiro: *Ormosia excelsa* Spruce; *Papilionac.*; Bras.
Tenchigurumi: *Juglans regia* L. var. *sinensis* DC.; *Juglandac.*; Jap.
Tendalake: *Canarium strictum* Roxb.; *Burserac.*; sw. O.-Ind. (Coorg)
Tendru: *Trichilia heudeloti* Planch.; *Meliac.*; Go.
Tendu: *Diospyros ebenum* Koenig; *Ebenac.*; sw. O.-Ind. (Bombay)
  „ : *Diospyros tomentosa* Roxb.; *Ebenac.*; C.-O.-Ind
  „ : *Diospyros melanoxylon* Roxb.; *Ebenac.*; C.-O.-Ind., sw. O.-Ind. (Bombay)
Tenduri: *Trichilia heudeloti* Planch.; *Meliac.*; Go.
Teneo: *Weinmannia trichosperma* Cav.; *Cunoniac.*; s. Chile.
Teng: *Pycnanthus kombo* Warb.; *Myristicac.*; Kam.
Tengah: *Ceriops roxburghiana* Arn.; *Rhizophorac.*; Malay.
  „ puti: *Ceriops roxburghiana* Arn.; *Rhizophorac.*; Malay.
Tengjawang-blongseng: *Shorea balangeran* Burck; *Dipterocarpac.*; Malay.
Tengkawang: *Vatica spp., Shorea spp., Hopea spp., Isoptera spp.; Dipterocarpac.;*
  „ : *Isoptera borneensis* Scheff.; *Dipterocarpac.*; Malay.   [Malay.
  „ -lajar: *Shorea martiniana* Scheff.; *Dipterocarpac.*; Borneo.
  „ -seloengsoeng: *Shorea balangeran* Burck; *Dipterocarpac.*; Malay.
Tengoelan: *Protium javanicum* Burm.; *Burserac.*; Nied.-Ind.
Tèngsèk: *Dodonaea viscosa* Jacq.; *Sapindac.*; Java.
Tengue: *Pithecolobium arboreum* Urban; *Mimosac.*; Cuba.
Tenia: *Melicytus ramiflorus* Forst.; *Violac.*; Südsee (Tahiti)
Teninino: *Guarea thompsoni* Spr. et Hutch.; *Meliac.*; Go.
Tenio: *Weinmannia trichosperma* Cav.; *Cunoniac.*; s. Chile.
Teniu: *Weinmannia trichosperma* Cav.; *Cunoniac.*; Chile.
Ténouba: *Phyllanthus discoideus* Muell. Arg.; *Euphorbiac.*; Elf.
Tenoura: *Phyllanthus discoideus* Muell. Arg.; *Euphorbiac.*; Elf.
Tentout: *Strephonema pseudocola* Chev.; *Combretac.*; Lib.
Ténuba: *Trichilia candollei* Chev.; *Meliac.*; Elf.
Tenuro: *Trichilia heudcloti* Planch.; *Meliac.*; Go.
Te-ohn-way-doo: *Chlamydocarya capitata* Baill.; *Icacinac.*; Lib.
Tepalcahuite: *Coccoloba schiedeana* Lindau; *Polygonac.*; Mex.
Tepe: *Ficus religiosa* L.; *Morac.*; n. O.-Ind.
Tepecaulote: *Luehea sp.*; *Tiliac.*; Salv.
Tepehuexote: *Salix sp.*; *Salicac.*; Mex.
Tepemezquite: *Lysiloma divaricatum* Mac Bride; *Mimosac.*; Mex. (Sinaloa)
?   „ : *Prosopis juliflora* DC.; *Mimosac.*; Mex.
Tepemiste: *Poeppigia procera* Presl; *Caesalpiniac.*; Salv.
Tepezapote: *Clethra lanata* Mart. et Gal.; *Clethrac.*; Salv.
Teponastli: *Swietenia mahagoni* Jacq.; *Meliac.*; Mex.
Teponaxtle: *Bursera spp.*; *Burserac.*; Mex.
Teponaxtli: *Bursera spp.*; *Burserac.*; Mex.
Teponaztli: *Bursera spp.*; *Burserac.*; Mex.
Tepozán· *Buddleia wrighti* Robins.; *Loganiac.*; Mex. (Sinaloa)
Tepù: *Tepualia stipularis* Griseb.; *Myrtac.*; Chile.
Teque: *Aextoxicum punctatum* Ruiz et Pav.; *Euphorbiac.*; Chile.
Teraling: *Tarrietia simplicifolia* Mast.; *Sterculiac.*; Malay.
Terap: *Artocarpus kunstleri* King; *Morac.*; Malay.
Teraso-itabi (jap.): *Ficus megacarpa* Merr.; *Morac.*; Form., Phil.
Terblans: *Faurea macnaughtoni* Phill.; *Proteac.*; S.-Af.
Terblanz: *Faurea macnaughtoni* Phill.; *Proteac.*; S.-Af.
Terciopelo: *Clethra lanata* Mart. et Gal.; *Clethrac.*; Salv.
  „ : *Sloanea sp.*; *Elaeocarpac.*; Salv.
  „ : *Luehea sp.*; *Tiliac.*; Salv.
Terebinto: *Schinus molle* L.; *Anacardiac.*; Arg.
Terentang: *Campnosperma wallichi* King, *C auriculata* Hook. f.; *Anacardiac.*; Malay.

Tereté: *Sarcocephalus esculentus* Afzel.; *Rubiac.;* Elf.
Tereup-areuj: *Ficus pruniformis* Blume; *Morac.;* Java (Salak)
Teriha-inubiwa (jap.): *Ficus harlandi* Bth.; *Morac.;* tr. As.
Terjuchinta: *Albizzia odoratissima* Bth.; *Mimosac.;* C.-O.-Ind. (Gondwana)
Terjuhitta: *Albizzia odoratissima* Bth.; *Mimosac.;* C.-O.-Ind. (Gondwana)
Terminalia, black-barked: *Terminalia ivorensis* Chev.; *Combretac.;* W.-Af.
„ , yellow: *Terminalia ivorensis* Chev.; *Combretac.;* W.-Af.
Tèrong: *Solanum spp.; Solanac.;* Java.
„ -blanda: *Cyphomandra betacea* Sendtn.; *Solanac.;* Java.
„ , hollandsche: *Cyphomandra betacea* Sendtn.; *Solanac.;* Java.
Tèrongan: *Solanum torvum* Swartz; *Solanac.;* Java (Kaliparé, Tangkil)
Terpai: *Durio sp.; Bombacac.;* Malay.
Terrakat: *? Grewia populifolia* Vahl; *Tiliac.;* Sahara.
Terúkan: *Beilschmiedia glomerata* Merr.: *Laurac.;* Phil. (Ilocos Sur. Laguna, Bataan)
Terungtang: *Campnosperma wallichi* King; *Anacardiac.;* Malay.
Testucchio: *Acer campestre* L.; *Acerac.;* It.
Teta de burro: *Hirtella triandra* Swartz, *H. rugosa* Thuill.; *Rosac.;* P.-R.
„ „ „ : *Chrysophyllum olivaeforme* Lamk.; *Sapotac.;* P.-R.
„ „ „ cimarrón: *Hirtella sp.; Rosac.;* P.-R.
„ „ yegua: *Hirtella sp.; Rosac.;* Cuba.
Tété-hoedoe: *Eschweilera sp., ? Lecythis sp.; Lecythidac.;* Sur.
Tetekogyambra: *Macrolobium limba* Scott Ell.; *Caesalpiniac.;* Go.
Teter: *Solanum verbascifolium* L.; *Solanac.;* Java (Tjibodas)
Tétéré: *Sarcocephalus esculentus* Afzel.; *Rubiac.;* Elf.
Tetereta: *Hura polyandra* Baill.; *Euphorbiac.;* Guat.
Tétéroema: *? Nectandra sp., ? Ocotea sp.; Laurac.;* Sur.
„ : *? Terminalia buceras* Hook.; *Combretac.;* Sur.
Tétéroma: *? Nectandra sp., ? Ocotea sp.; Laurac.;* Sur.
„ : *? Terminalia buceras* Hook.; *Combretac.;* Sur.
Teterow sierwaballi: *Nectandra sp., Ocotea sp.; Laurac.;* Sur.
Tetiwon: *Macrolobium limba* Scott Ell.; *Caesalpiniac.;* Go.
Tetlate: *Elaphrium bipinnatum* Schlecht.; *Burserac.;* Mex.
„ : *Comocladia repanda* Blake; *Anacardiac.;* Mex.
Tetlatía: *Elaphrium bipinnatum* Schlecht.; *Burserac.;* Mex.
Tetlatián: *Elaphrium bipinnatum* Schlecht.; *Burserac.;* Mex.
Tetruma: *Nectandra sp.; Laurac.;* br. Gu.
„ : *? Nectandra sp., ? Ocotea sp.; Laurac.;* Sur.
„ : *? Terminalia buceras* Hook.; *Combretac.;* Sur.
Tetura: *Albizzia odoratissima* Bth.; *Mimosac.;* Beng.
Teul: *Ficus vogeliana* Miq.; *Morac.;* Gab.
Teureup: *Artocarpus elastica* Reinw.; *Morac.;* Java.
„ -areuj: *Ficus globosa* Blume; *Morac.;* Java (Palaboehan)
Tèva: *Desbordesia insignis* Pierre; *Simarubac.;* Gab.
Tevadarum: *Erythroxylon areolatum* L.; *Erythroxylac.;* O.-Ind. (s. Deccan)
Tèvè: *Desbordesia insignis* Pierre; *Simarubac.;* Gab.
Tewanák: *Bambusa vulgaris* Schrad.; *Gramin.;* Phil. (Laguna)
Tewas: *Ougeinia dalbergioides* Bth.; *Papilionac.;* O.-Ind. (Bombay)
Tewí: *Dolichandrone spathacea* K. Schum.; *Bignoniac.;* Phil. (Agusan)
Tewsa: *Ougeinia dalbergioides* Bth.; *Papilionac.;* C.-O.-Ind. (Vindhja)
Texcalama: *Ficus petiolaris* H. B. K.; *Morac.;* Mex.
Texmole: *Quercus fusiformis* Small; *Fagac.;* Mex.
Texo: *Taxus baccata* L.; *Conifer.;* Sp.
Teyei: *Mimusops clitandrifolium* Chev.; *Sapotac.;* Lib.
Tezonzapote: *Calocarpum mammosum* Pierre; *Sapotac.;* Mex.
Tgeloware: *Derris stuhlmanni* Harms; *Papilionac.;* Togo.
Thab: *Erythrina suberosa* Roxb.; *Papilionac.;* nw. O.-Ind. (Punjab)
Thabeik: *Quercus semiserrata* Roxb.; *Fagac.;* s. Burma (s. Tenasserim)
Thabut: *Polyalthia simiarum* Bth. et Hook. f.; *Anonac.;* s. Burma (s. Tenasserim)
„ -gyi: *Miliusa velutina* Hook. f. et Thoms.; *Anonac.;* Burma.
„ -kyi: *Miliusa velutina* Hook. f. et Thoms.; *Anonac.;* Burma.
„ -thein: *Polyalthia cerasoides* Bth. et Hook. f.; *Anonac.;* Burma.
Thabye: *Eugenia jambolana* Lamk.; *Myrtac.;* And.

Thabyebyu: *Eugenia jambolana* Lamk.; *Myrtac.*; Burma.
Thabyegyin: *Eugenia operculata* Roxb.; *Myrtac.*; Burma.
Thabyu: *Dillenia indica* L.; *Dilleniac.*; Burma.
Thach luc: *Xanthophyllum excelsum* Blume; *Polygalac.*; Ind.-Ch. (Annam)
Thadi: *Bursera serrata* Colebr.; *Burserac.*; Burma.
Thadisalu: *Grewia tiliaefolia* Vahl; *Tiliac.*; s. O.-Ind. (Coorg)
Thadsal: *Grewia tiliaefolia* Vahl; *Tiliac.*; O.-Ind. (Bombay)
Thain puchie pattai: *Guazuma tomentosa* H. B. K.; *Sterculiac.*; O.-Ind. (s. Deccan)
Thakabti: *Schleichera trijuga* Willd.; *Sapindac.*; n. Burma.
Thakutpo: *Stereospermum chelonioides* DC., *S. suaveolens* DC.; *Bignoniac.*; Burma.
Thalé: *Ulmus lancifolia* Roxb.; *Ulmac.*; Burma.
Thalè: *Holoptelea integrifolia* Planch.; *Ulmac.*; Burma.
Thalli: *Canarium strictum* Roxb.; *Burserac.*; sw. O.-Ind.
Thamba: *Shorea tumbuggaia* Roxb.; *Dipterocarpac.*; sö. O.-Ind. (Madras)
Thambagam: *Hopea parviflora* Bedd.; *Dipterocarpac.*; sw. O.-Ind.
Thamè: *Avicennia officinalis* L.; *Verbenac.*; s. Burma (s. Tenasserim)
Thaminza: *Randia dumetorum* Lamk.; *Rubiac.*; Burma.
Thaminzani: *Gardenia turgida* Roxb.; *Rubiac.*; Burma.
Than: *Juglans regia* L.; *Juglandac.*; O.-Ind.
„ : *Terminalia oliveri* Brandis; *Combretac.*; Burma.
Thanat: *Cordia myxa* L.; *Borraginac.*; Burma.
Thanatka: *Murraya exotica* L., *Limonia acidissima* L.; *Rutac.*; Burma.
Thandé: *Stereospermum chelonioides* DC., *S. suaveolens* DC.; *Bignoniac.*; Burma.
Thanela: *Gardenia turgida* Roxb.; *Rubiac.*; n. O.-Ind. (United Prov.)
Thanera: *Gardenia turgida* Roxb.; *Rubiac.*; n. O.-Ind. (Garhwal)
Thanh mai: *Myrica integrifolia* Roxb.; *Myricac.*; Ind.-Ch. (Annam)
Thani: *Terminalia belerica* Roxb.; *Combretac.*; sw. O.-Ind.
Thanmela: *Gardenia turgida* Roxb.; *Rubiac.*; Beng. (Bihar, Orissa)
Thanthat: *Albizzia lucida* Bth.; *Mimosac.*; Burma.
Thapan: *Ficus glomerata* Roxb.; *Morac.*; Burma.
Thapsi: *Holoptelea integrifolia* Planch.; *Ulmac.*; sw. O.-Ind. (Mysore)
Thara: *Terminalia belerica* Roxb.; *Combretac.*; O.-Ind. (nö. Madras)
Tharapi: *Calophyllum spectabile* Willd.; *Guttifer.*; Burma.
Tharfa: *Tamarix gallica* L.; *Tamaricac.*; Arab.
Tharra: *Grewia tiliaefolia* Vahl; *Tiliac.*; O.-Ind. (Madras)
Thaur: *Bauhinia retusa* Ham.; *Caesalpiniac.*; C.-O.-Ind. (Gondwana)
Thayet: *Mangifera indica* L.; *Anacardiac.*; Burma.
Thayetkan: *Swintonia floribunda* Griff.; *Anacardiac.*; Burma.
Thayet-kin: *Swintonia floribunda* Griff.; *Anacardiac.*; Burma.
Thayetlé: *Swintonia floribunda* Griff.; *Anacardiac.*; Burma.
Thayetsan: *Swintonia floribunda* Griff.; *Anacardiac.*; Burma.
Thayet-thitsi: *Gluta tavoyana* Hook. f.; *Anacardiac.*; Burma.
Thazzougart: *Zizyphus lotus* L.; *Rhamnac.*; n. Af. (Tunis, Algier)
Thbeng: *Dipterocarpus obtusifolius* Teysm.; *Dipterocarpac.*; Ind.-Ch. (Kamb.)
Thé, Camellia-: *Camellia sasanqua* Thunb.; *Theac.*; Jap., Ch.
„ des apalaches: *Ilex vomitoria* Ait.; *Aquifoliac.*; sö. USA.
Thebla: *Gmelina arborea* L.; *Verbenac.*; Burma.
Thekku: *Tectona grandis* L. f.; *Verbenac.*; w. & sw. O.-Ind.
Thekratenga kujitekra: *Carallia lucida* Roxb.; *Rhizophorac.*; Assam.
Thelawaw: *Careya arborea* Roxb.; *Myrtac.*; n. Burma.
Thembavoo: *Terminalia tomentosa* Wight et Arn.; *Combretac.*; sw. O.-Ind. (Travanc.)
Thèrbè: *Dalbergia cultrata* Graham; *Papilionac.*; n. Burma.
Thetlet: *Pentace burmanica* Kurz; *Tiliac.*; n. Burma.
Thevatharam: *Toona ciliata* M. Roem.; *Meliac.*; O.-Ind. (Madras)
Thewalaw: *Homalium tomentosum* Bth.; *Flacourtiac.*; n. Burma
Thi: *Feronia elephantum* Correa; *Rutac.*; Burma.
„ : *Diospyros rubra* Lecomte; *Ebenac.*; Ind.-Ch.
„ deù: *Diospyros nitidula* Lecomte; *Ebenac.*; Ind.-Ch. (Annam-Nhatrang)
„ dui nui: *Diospyros nitidula* Lecomte; *Ebenac.*; Ind.-Ch. (Annam-Nhatrang)
Thibet tree: *Albizzia lebbek* Bth.; *Mimosac.*; P.-R.
Thiêt dinh: *Dolichandrone stipulata* Bth.; *Bignoniac.*; Ind.-Ch. (Tonkin)
Thilla: *Excoecaria agallocha* L.; *Euphorbiac.*; sw. O.-Ind. (Madras)

Thingadu: *Parashorea stellata* Kurz; *Dipterocarpac.;* Burma.
Thingan (e): *Hopea odorata* Roxb.; *Dipterocarpac.;* Burma, And.
„ byu: *Hopea sp.; Dipterocarpac.;* Burma.
„ net: *Hopea sp.; Dipterocarpac.;* Burma.
„ , safed: *Hopea odorata* Roxb.; *Dipterocarpac.;* And.
„ wa: *Hopea sp.; Dipterocarpac.;* Burma.
Thinkadu: *Anisoptera glabra* Kurz; *Dipterocarpac.;* Burma.
Thinwin: *Millettia pendula* Bth.; *Papilionac.;* Burma.
„ : *Pongamia glabra* Vent.; *Papilionac.;* Burma, And.
Thiripu: *Bischofia javanica* Blume; *Euphorbiac.;* sw. O.-Ind. (Travancore)
Thirukkanamallay: *Berrya ammonilla* Roxb.; *Tiliac.;* O.-Ind. (Madras)
Thitcha: *Juglans regia* L.; *Juglandac.;* Burma.
„ : *Quercus semiserrata* Roxb.; *Fagac.;* Burma.
Thitcho: *Sideroxylon tomentosum* Roxb.; *Sapotac.;* Burma.
Thitè: *Quercus serrata* Thunb., *Q. semiserrata* Roxb., *Q. lamellosa* Smith.; *Fagac.*,
„ : *Castanopsis indica* A. DC., *Castanea indica* Roxb.; *Fagac.;* Burma. [Burma.
Thitègyin: *Castanopsis indica* A. DC., *Castanea indica* Roxb.; *Fagac.;* nö. Burma.
Thitka: *Pentace burmanica* Kurz; *Tiliac.;* Burma.
Thitkado: *Toona ciliata* M. Roem., *T. australis* Harms; *Meliac.;* O.-Ind., Burma.
Thitkaukhnyin: *Ulmus lancifolia* Roxb.; *Ulmac.;* Burma.
Thitkazaw: *Duabanga sonneratioides* Ham.; *Sonneratiac.;* Burma.
Thitkya: *Diospyros marmorata* Parker; *Ebenac.;* Burma.
Thitkyabo: *Cinnamomum zeylanicum* Breyn; *Laurac.;* Burma.
Thitmagyi: *Albizzia odoratissima* Bth.; *Mimosac.;* Burma.
Thitmin: *Podocarpus neriifolia* D. Don; *Conifer.;* And.
„ -po: *Podocarpus neriifolia* D. Don; *Conifer.;* Burma.
Thitni: *Amoora rohituka* Wight et Arn.; *Meliac.;* Burma.
Thitpagan: *Dalbergia lanceolaria* L.; *Papilionac.;* Burma.
Thitpok: *Tetrameles nudiflora* R. Br.; *Datiscac.;* Burma, And.
„ : *Diospyros ehretioides* Wall.; *Ebenac.;* s. Burma (s. Tenasserim)
Thitpyn: *Albizzia procera* Bth.; *Mimosac.;* O.-Ind.
Thitpyu: *Albizzia odoratissima* Bth., *A. procera* Bth.; *Mimosac.;* Burma.
Thitsein: *Terminalia belerica* Roxb.; *Combretac.;* Burma.
Thitsho: *Pentace griffithi* King; *Tiliac.;* O.-Ind. (? Burma)
Thitsi: *Melanorrhoea usitata* Wall.; *Anacardiac.;* Burma.
Thitsibo: *Buchanania latifolia* Roxb.; *Anacardiac.;* n. Burma.
Thitswèl-wé: *Schrebera swietenioides* Roxb.; *Oleac.;* Burma.
Thitto: *Sandoricum indicum* Cav.; *Meliac.;* Burma.
Thitya: *Shorea obtusa* Wall.; *Dipterocarpac.;* Burma.
„ -ingyin: *Pentacme suavis* A. DC.; *Dipterocarpac.;* Burma.
Thityah: *Schima wallichi* Choisy; *Theac.;* n. Burma.
Thlok: *Parinarium annamense* Hance; *Rosac.;* Ind.-Ch. (Kamb.)
Thmir: *Artocarpus sp.; Morac.;* Ind.-Ch. (Kamb.)
Thnong: *Sageraea elliptica* Hook. f. et Thoms.; *Anonac.;* Ind.-Ch. (Kamb.)
„ : *Pterocarpus pedatus* Pierre, *P. cambodianus* Pierre; *Papilionac.;* Kamb.
Tho boo: *Mangifera indica* L.; *Anacardiac.;* Ind.-Ch.
Thodappai: *Gluta travancorica* Bedd.; *Anacardiac.;* sw. O.-Ind. (Madras)
Thôi chanh: *Alangium chinense* Rehder, *Marlea begoniaefolia* Roxb.; *Cornac.;* Annam.
Thom-shui: *Cananga odorata* Hook. f. et Thoms.; *Anonac.;* Co -Ch.
„ „ : *Unona brandisana* Pierre; *Anonac.;* Annam.
Thonda pala: *Wrightia tinctoria* R. Br.; *Apocynac.;* sw. O.-Ind. (Travancore)
Thondo: *Pterocarpus angolensis* DC.; *Papilionac.;* port. O.-Af. (Mozambique)
Thong: *Pinus khasya* Royle, *P. merkusi* Jungh. et de Vr.; *Conifer.;* Ind.-Ch.
„ lau: *Cunninghamia sinensis* R. Br.; *Conifer.;* Ind.-Ch. (Tonkin, Laos)
Thoraro diamaroe: *Erisma uncinatum* Warm.; *Vochysiac.;* Sur.
Thorn, Ant-: *Acacia cooki* Safford; *Mimosac.;* br. Hond.
„ , Buck-, Cascara: *Rhamnus purshiana* DC.; *Rhamnac.;* Kan.
„ , „ -, common: *Rhamnus cathartica* L.; *Rhamnac.;* Eu.
„ , „ -, european: *Rhamnus cathartica* L.; *Rhamnac.;* Eu.
„ , „ -, island: *Rhamnus insularis* Greene; *Rhamnac.;* w. USA. (Kalif.)
„ , „ -, woolly: *Bumelia lanuginosa* Pers.; *Sapotac.;* sw. USA.
„ , Cat-: *Scutia capensis* Eckl. et Zeyh.; *Rhamnac.;* S.-Af. (Kap)

Thorn, Cockspur-: *Crataegus crus-galli* L.; *Rosac.;* USA.
„ , dotted-fruited: *Crataegus punctata* Jacq.; *Rosac.;* ö. N.-Am.
„ , green: *Ceanothus sorediatus* Hook. et Arn.; *Rhamnac.;* w. USA. (Kalif.)
„ , Haw-: *Crataegus arborescens* Ell., *C. spp.; Rosac.;* N.-Am.
„ , Knob-: *Zanthoxylum capense* Harv.; *Rutac.;* S.-Af. (Natal)
„ , „ -: *Acacia pallens* Rolfe; *Mimosac.;* S.-Af. (Transval)
„ , knobby: *Acacia pallens* Rolfe; *Mimosac.;* n. Rhod.
„ , Madras-: *Pithecolobium dulce* Bth.; *Mimosac.;* O.-Ind.
„ , Newcastle-: *Crataegus crus-galli* L.; *Rosac.;* N.-Am.
„ , red: *Crataegus coccinea* L.; *Rosac.;* N.-Am.
„ , Scarlet-: *Crataegus coccinea* L.; *Rosac.;* N.-Am.
„ , sweet: *Acacia horrida* Willd.; *Mimosac.;* S.-Af.
„ tree: *Gleditschia triacantha* L.; *Caesalpiniac.;* USA.
„ „ : *Acacia horrida* Willd.; *Mimosac.;* S.-Af. (Kap)
„ , Way- (u): *Rhamnus cathartica* L.; *Rhamnac.;* Eu.
„ , western: *Crataegus douglasi* Lindl.; *Rosac.;* nw. N.-Am.
„ , white: *Crataegus sp.; Rosac.;* ö. N.-Am.
„ „ : *Acacia sp.; Mimosac.;* n. Nig. (Kontagora)
„ , Wolfs-: *Phoberos zeyheri* Presl; *Flacourtiac.;* S.-Af. (Kap)
Thousand-jacket: *Hoheria populnea* A. Cunn.; *Malvac.;* N.-Seel. (Auckland)
Thriphal: *Zanthoxylum rhetsa* DC.; *Rutac.;* sw. O.-Ind. (Bombay)
Thuddu ponna: *Bruguiera gymnorrhiza* Lamk.; *Rhizophorac.;* w. O.-Ind. (Bombay)
Thui: *Parkia dongnaiensis* Pierre; *Mimosac.;* Ind.-Ch.
Thuja: *Thuja occidentalis* L.; *Conifer.;* sö. Kan.
„ : *Tetraclinus articulata* Mast.; *Conifer.;* n. Af.
Thujamaser, afrikanisch (d): *Tetraclinis articulata* Mast.; *Conifer.;* Algier, Marokko.
Thula: *Bombax insigne* Wall. *var. andamanica* Prain, *B. wighti* Prain; *Bombacac.;*
Thuna: *Taxus baccata* L.; *Conifer.;* nw. O.-Ind. (Punjab) [Beng.
Thuner: *Taxus baccata* L.; *Conifer.;* n. O.-Ind. (Jaunsar)
Thung: *Commersonia echinata* Forst. *var. platyphylla* Andr.; *Sterculiac.;* Ind.-Ch.
„ luc: *Randia pycnantha* Drake; *Rubiac.;* Ind.-Ch. (Annam)
„ mûc: *Wrightia balansana* Pierre; *Apocynac.;* Ind.-Ch. (Annam)
Thuniaru: *Taxus baccata* L.; *Conifer.;* n. O.-Ind. (Jaunsar)
Thunla: *Gardenia turgida* Roxb.; *Rubiac.;* nw. O.-Ind. (Kumaon)
Thura: *? Ceiba sp., ? Pachira sp., ? Bombax sp.; Bombacac.;* Mex.
Thuti: *Eugenia operculata* Roxb.; *Myrtac.;* n. O.-Ind. (Dehra Dun)
Thuya: *Tetraclinis articulata* Mast.; *Conifer.;* n. Af. (Algier, Marokko)
„ burl (u): *Tetraclinis articulata* Mast.; *Conifer.;* n. Af. (Algier)
„ géant: *Thuja gigantea* Nutt.; *Conifer.;* w. Kan.
Thwa-mu: *Lagerstroemia flos-reginae* Retz.; *Lythrac.;* n. Burma.
„ -u: *Lagerstroemia flos-reginae* Retz.; *Lythrac.;* n. Burma.
Thyaga: *Tectona grandis* L. f.; *Verbenac.;* sw. O.-Ind. (Coorg)
Tia koubé bia: *Pentadesma leucantha* Chev.; *Guttifer.;* Elf.
Tiaca: *Caldcluvia paniculata* D. Don; *Cunoniac.;* s. Chile.
Tiakdóg: *Quercus sp.; Fagac.;* Phil. (Benguet)
Tiama (f): *Entandrophragma septentrionale* Chev., *E. rufa* Chev.; *Meliac.;* Elf.
„ : *Entandrophragma macrophyllum* Chev.; *Meliac.;* Elf, Kam.
„ : *Entandrophragma ferrugineum* Chev.; *Meliac.;* Elf.
„ -tiama: *Khaya grandifolia* C. DC.; *Meliac.;* Go.
„ „ : *Entandrophragma ferrugineum* Chev.; *Meliac.;* Elf.
Tian: *Acer pictum* Thunb.; *Acerac.;* nw. O.-Ind. (Punjab)
„ : *Pycnanthus kombo* Warb.; *Myristicac.;* Kam.
Tianalin wéwé: *Piratinera ? guianensis* Aubl.; *Morac.;* Sur.
Tianúg: *Eugenia costulata* C. B. Rob.; *Myrtac.;* Phil. (Bataan)
Tiáong: *Shorea teysmanniana* Dyer; *Dipterocarpac.;* Phil. (Laguna, Leyte)
„ : *Pentacme contorta* Merr. et Rolfe; *Dipterocarpac.;* Phil. (Pol.)
Tiapotano: *Qualea coerulea* Aubl., *Q. albiflora* Warm.; *Vochysiac.;* Sur.
Tiassuba: *Duguetia sp.; Anonac.;* Bras.
Tibicusi: *Piratinera ? guianensis* Aubl.; *Morac.;* br. Gu.
Tibigaro: *Astronium graveolens* Jacq.; *Anacardiac.;* Kol., w. Ven.
Tibígi: *Xylocarpus moluccensis* M. Roem.; *Meliac.;* Phil. (Mindoro)
Tibikushi: *Piratinera ? guianensis* Aubl.; *Morac.;* br. Gu.

Tibisiri: *Mauritia flexuosa* L.; *Palmac.;* fr. Gu.
Tibourbou: *Apeiba tibourbou* Aubl.; *Tiliac.;* Gu.
Tibri: *Omphalocarpum anocentrum* Pierre; *Sapotac.;* Elf.
Tibúngau: *Aglaia harmsiana* Perk.; *Meliac.;* Phil. (Cagayan)
Tido: *Khaya grandifoliola* C. DC.; *Meliac.;* br. O.-Af. (Uganda)
Tiengi monni: *Protium sagotianum* March., *P. spp.; Burserac.;* Sur.
Tienjie monnie: *Protium sp.; Burserac.;* Sur.
Tiên lum: *Garcinia mangostana* L.; *Guttifer.;* Annam.
Tienndebelé: *Chytranthus edulis* Pierre; *Sapindac.;* Gab.
Tie-tie: *Anemopaegma belizeanum* Blake, *Bignonia dasyonix* Blake; *Bignoniac.;* br.
„   „ : *Banisteria laurifolia* L.; *Malpighiac.;* br. Hond.   [Hond.
„   „ : *Calopogonium coeruleum* Bth.; *Papilionac.;* br. Hond.
„   „ : *Cissus biformifolia* Standl.; *Vitac.;* br. Hond.
„   „ : *Connarus loncbotus* Blake; *Connarac.;* br. Hond.
„   „ : *Forsteronia viridescens* Blake; *Apocynac.;* br. Hond.
„   „ : *Paragonia pyramidata* Bur.; *Bignoniac.;* br. Hond.
„   „ , Barracouta-: *Liabum polyanthum* Klatt; *Composit.;* br. Hond.
„   „ , Water-: *Vitis sp.; Vitac.;* br. Hond.
„   -tongue: *Coccoloba sp.; Polygonac.;* br. W.-Ind.
Tíga: *Tristania decorticata* Merr., *T. micrantha* Merr.; *Myrtac.;* Phil.
„. : *Xanthostemon verdugonianus* Naves; *Myrtac.;* Phil. (Sibuyan)
Tigáui: *Pometia pinnata* Forst.; *Sapindac.;* Phil. (Camarines, Masbate, Ticao)
Tigerholz (d): *Piratinera guianensis* Aubl.; *Morac.;* Gu.
Tigerwood: *Erythrina rubrinervia* H. B. K.; *Papilionac.;* br. Hond.
„   : *? Machaerium schomburgki* Bth.; *Papilionac.;* Gu.
„   : *Astronium fraxinifolium* Schott; *Anacardiac.;* Bras.
„   (u): *Lovoa klaineana* Pierre; *Meliac.;* W.-Af.
Tiggèc: *Terminalia brownei* Fresen.; *Combretac.;* O.-Af. (Somali)
Tigían: *Albizzia saponaria* Blume; *Mimosac.;* Phil. (Guimaras)
Tiglio: *Tilia parvifolia* Ehrh.; *Tiliac.;* It.
„ a foglie grandi: *Tilia grandifolia* Ehrh.; *Tiliac.;* It.
„   „   „   piccole: *Tilia parvifolia* Ehrh.; *Tiliac.;* It.
Tigre: *Guarea sp.; Meliac.;* Kol.
Tigua: *Amyris balsamifera* L.; *Rutac.;* Ven.
Tigüilote: *Cordia alba* R. et S., *C. sp.; Borraginac.;* Hond., Salv.
„   : *Cordia diversifolia* Pavón; *Borraginac.;* Pan.
Tigulate: *Crescentia cujete* L.; *Bignoniac.;* Mex., C.-Am.
Tihiri: *Albizzia procera* Bth.; *Mimosac.;* sw. O.-Ind. (Bombay)
Tika: *Pycnanthus kombo* Warb.; *Myristicac.;* Go.
Tikal: *Livistona rotundifolia* Mart. var. *luzonensis* Becc.; *Palmac.;* Phil.
Tikalód: *Quercus soleriana* Vidal; *Fagac.;* Phil. (Bataan)
Tikám: *Litsea sp.; Laurac.;* Phil. (Bataan)
„   : *Adinandra montana* Merr.; *Theac.;* Phil. (Jg.)
Tikia: *Quercus incana* Roxb.; *Fagac.;* n. O.-Ind. (Dotial)
Tikím: *Neonauclea reticulata* Merr.; *Rubiac.;* Phil. (Abra, Bontok, Benguet)
„   : *Neonauclea vidalis* Merr.; *Rubiac.;* Phil. (Abra, Bontok, Benguet)
Tíkis: *Livistona rotundifolia* Mart.; *Palmac.;* Phil. (Zambales)
Tiklík: *Quercus obliquinervia* Merr., *Q. woodi* Hance; *Fagac.;* Phil. (Benguet)
Tikoki: *Alectryon excelsum* Gaertn.; *Sapindac.;* N.-Seel.
Tikóko: *Vitex philippinensis* Merr.; *Verbenac.;* Phil. (Leyte)
Tikoor: *Garcinia pedunculata* Roxb.; *Guttifer.;* ö. Beng.
Tikta: *Acer pictum* Thunb.; *Acerac.;* nw. O.-Ind. (Kumaon)
„. -raj: *Amoora rohituka* Wright et Arn.; *Meliac.;* Beng.
Tikul: *Garcinia pedunculata* Roxb.; *Guttifer.;* s. Assam (Sylhat), ö. Beng.
Tilai: *Excoecaria agallocha* L.; *Euphorbiac.;* O.-Ind. (n. Madras)
Tilajoe: *Erioglossum edule* Blume; *Sapindac.;* Java (Tjemara)
Tilani: *Acer pictum* Thunb.; *Acerac.;* n. O.-Ind.
Tilaunju: *Populus ciliata* Wall.; *Salicac.;* n. O.-Ind. (United Prov.)
Tíli: *Albizzia acle* Merr.; *Mimosac.;* Phil. (Zambales)
Tílis: *Albizzia acle* Merr.; *Mimosac.;* Phil. (Zambales)
Tiliya gurjun: *Dipterocarpus turbinatus* Gaertn. f.; *Dipterocarpac.;* Beng.
Tilleul noir: *Tilia americana* L.; *Tiliac.;* ö. N.-Am.

Tilleul sauvage: *Tilia parvifolia* Ehrh.; *Tiliac.;* Eu., Orient.
„ à grandes feuilles: *Tilia grandifolia* Ehrh.; *Tiliac.;* Mitt.- & s. Eu.
„ à pctites feuilles: *Tilia parvifolia* Ehrh.; *Tiliac.;* Eu., Orient.
„ d'Afrique: *Mitragyne macrophylla* Hiern; *Rubiac.;* W.-Af.
„ d'Amérique: *Tilia americana* L.; *Tiliac.;* ö. N.-Am.
„ de Hollande: *Tilia grandifolia* Ehrh.; *Tiliac.;* Mitt.- & s. Eu.
„ des bois: *Tilia parvifolia* Ehrh.; *Tiliac.;* Eu., Orient.
„ du Canada: *Tilia americana* L.; *Tiliac.;* ö. N.-Am.
Tilo: *Tilia grandifolia* Ehrh., *T. parvifolia* Ehrh., *T. intermedia* DC.; *Tiliac.;* Sp.
„ : *Tilia houghi* Rose; *Tiliac.;* Mex.
Tiloko-enoeroe: *Hieronyma laxiflora* Muell. Arg.; *Euphorbiac.;* Sur.
Tilong: *Quercus dilatata* Lindley; *Fagac.;* n. O.-Ind. (Kumaon, Garhwal)
Tilpattar: *Acer caesium* Wall.; *Acerac.;* nw. O.-Ind. (Punjab)
Tim: *Pterocarpus erinaceus* Poir.; *Papilionac.;* Togo.
Tima-tima: *Khaya grandifolia* C. DC.; *Meliac.;* Go.
Timák: *Dinochloa scandens* O. Kㅌe.; *Gramin.;* Phil. (Mindoro)
Timba: *Entandrophragma leplaei* Verm.; *Meliac.;* Kam.
„ eyidi: *Fagara macrophylla* Engl.; *Rutac.;* Kam.
„ moundi: *Sorindeia ochracea* Engl.; *Anacardiac.;* Kam.
Timbahy: *Enterolobium sp.;* *Mimosac.;* Par.
Timbania: *Scotellia kamerunensis* Gilg; *Flacourtiac.;* Elf.
Timbatau: *Lannea barteri* Engl.; *Anacardiac.;* Togo.
Timber sweet: *Phoebe helicterifolia* Mez, *Nectandra glabrescens* Bth.; *Laurac.* br.
Timbó: *Enterolobium timbouva* Mart.; *Mimosac.;* s. Bras., Arg.   [Hond.
„ atá: *Enterolobium sp.;* *Mimosac.;* Arg.
„ blanco: *Pithecolobium multiflorum* Bth.; *Mimosac.;* Bol.
? „ „ : *Enterolobium sp.;* *Mimosac.;* Arg.
„ cedro: *Enterolobium ? timbouva* Mart.; *Mimosac.;* Arg.
„ colorado: *Enterolobium timbouva* Mart.; *Mimosac.;* Arg.
„ negro: *Enterolobium sp.;* *Mimosac.;* Arg.
„ -páo: *Clathrotropis macrocarpa* Ducke; *Papilionac.;* n. Bras. (Amaz.)
„ pyitá: *Enterolobium sp.;* *Mimosac.;* Arg.
„ -rana: *Clathrotropis macrocarpa* Ducke; *Papilionac.;* n. Bras. (Amaz.)
„ urucú: *Lonchocarpus urucu* Killip et Smith; *Papilionac.;* n. Bras. (Amaz.)
„ y atá peró: *Enterolobium timbouva* Mart.; *Mimosac.;* Arg.
„ del campo: *Paullinia sp.;* *Sapindac.;* S.-Am.
Timboa: *Corynanthe johimbe* K. Schum.; *Rubiac.;* Kam.
„ mounganga: *Ricinodendron africanum* Muell. Arg.; *Euphorbiac.;* Kam.
Timboel: *Artocarpus incisa* L. f.; *Morac.;* Java.
Timboúba: *Enterolobium timbouva* Mart.; *Mimosac.;* Par., s. Bras.
„ : *Enterolobium sp.;* *Mimosac.;* Arg.
Timbourbou: *Apeiba timbourbou* Aubl.; *Tiliac.;* Gu. (Galibis)
Timboúva: *Enterolobium timbouva* Mart.; *Mimosac.;* Arg.
Timbre: *Inga spuria* Humb. et Bonpl.; *Mimosac.;* Mex.
Timbruni: *Diospyros melanoxylon* Roxb.; *Ebenac.;* sw. O.-Ind. (Bombay)
Timbuva: *Enterolobium sp.;* *Mimosac.;* s. Bras. (Sao P)
Timiouin: *Acacia arabica* Willd.; *Mimosac.;* Sudan.
Timoho: *Kleinhovia hospita* L.; *Sterculiac.;* Java.
Ti mono: *Morinda tinctoria* Roxb.; *Rubiac.;* Au.
Timsa: *Ougeinia dalbergioides* Bth.; *Papilionac.;* n. O.-Ind. (United Prov.)
Tináan: *Eugenia claviflora* Roxb.; *Myrtac.;* Phil. (Camarines)
„ : *Lagerstroemia piriformis* Koehne; *Lythrac.;* Phil. (Camar., Albay, Sorsogon)
„ -pantai: *Cyclostemon bordeni* Merr.; *Euphorbiac.;* Phil.
Tinadán: *Tristania decorticata* Merr.; *Myrtac.;* Phil. (Abra)
Tinagí: *Albizzia saponaria* Blume; *Mimosac.;* Phil. (Surigao)
Tinámalu: *Mauritia carana* Wall.; *Palmac.;* Bras., Ven.
Tinas: *Ougeinia dalbergioides* Bth.; *Papilionac.;* C.-O.-Ind. (Central Prov.)
Tindaló: *Pahudia rhomboidea* Prain; *Caesalpiniac.;* Phil.
„ : *Cassia javanica* L.; *Caesalpiniac.;* Phil. (Pangasinan)
„ : *Intsia acuminata* Merr.; *Caesalpiniac.;* Phil. (Tayabas)
Tinduc-tinducan: *Aegiceras corniculatum* Blco.; *Myrsinac.;* Phil.
Tiñe anzuelo: *Eupatorium pittieri* Klatt; *Composit.;* Hond.

Tiñe cordel: *Eupatorium albicaule* Schulq Bip.; *Composit.;* Hond.
Tinel: *Weinmannia trichosperma* Cav.; *Cunoniac.;* s. Chile.
Tinga: *Ceriops candolleana* Arn.; *Rhizophorac.;* Burma (s. Tenasserim)
Tingalo: *Diospyros mespiliformis* Hochst.; *Ebenac.;* Togo.
Tingantingan: *Pterospermum niveum* Vidal, *P. obliquum* Blco.; *Sterculiac.;* Phil.
Tingbatau: *Lannea barteri* Engl.; *Anacardiac.;* Togo.   [(Tayabas)
Tingee burong: *Evodia roxburghiana* Bth.; *Rutac.;* Malak.
Tinggiran poemay: *Erythroxylon retusum* Bauer; *Erythroxylac.;* Nied.-Ind.
Tingi hoedoe: *Lonchocarpus violaceus* Kunth; *Papilionac.;* fr. Gu.
   „   „   : *Piscidia erythrina* L.; *Papilionac.;* fr. Gu.
   „   -money: *Protium guianensis* March.; *Burserac.;* Gu.
Tingie-monnie: *Protium sagotianum* March., *P. spp.; Burserac.;* Sur.
Tingkál: *Cyclostemon sp.; Euphorbiac.;* Phil. (Zamboanga)
Tingkayád: *Palaquium sp.; Sapotac.;* Phil. (Rizal)
Tingle-tingle, red: *Eucalyptus jacksoni* Maiden; *Myrtac.;* w. Au.
   „   „   , yellow: *Eucalyptus guilfoylei* Maiden; *Myrtac.;* w. Au.
Tingma: *Vernonia senegalensis* Less.; *Composit.;* Togo.
Tingschi: *Taxus baccata* L.; *Conifer.;* Beng. (Bhutan)
Tinguaciba: *Zanthoxylum tinguassiba* St. Hil.; *Rutac.;* Bras.
Tinia: *Albizzia lebbek* Bth.; *Mimosac.;* sw. Beng. (Orissa)
Tinnas: *Ougeinia dalbergioides* Bth.; *Papilionac.;* n. O.-Ind. (United Prov.)
Tinsa: *Ougeinia dalbergioides* Bth.; *Papilionac.;* C.-O.-Ind. (Central Prov.)
Tinshu: *Pinus khasya* Royle, *P. merkusi* Jungh. et de Vr.; *Conifer.;* Burma.
Tinta: *Solanum nudum* H. B. K.; *Solanac.;* Guat.
   „   : *Haematoxylon campechianum* L.; *Caesalpiniac.;* Hond.
Tintatico: *Prosopis sp.; Mimosac.;* Arg.
Tinteira: *Coccoloba excelsa* Bth.; *Polygonac.;* Bras.
   „   : *Laguncularia racemosa* Gaertn.; *Combretac.;* Bras.
Tintenholz (d): *Exothea paniculata* Radlk.; *Sapindac.;* s. Fl., Bah., Ant.
Tinto: *Haematoxylon campechianum* L.; *Caesalpiniac.;* s. Mex., C.-Am., n. S.-Am.
Tinya: *Doona congestiflora* Thw.; *Dipterocarpac.,* Ceylon.
Tinyu: *Casuarina equisetifolia* Forst.; *Casuarinac.;* Burma, And.
   „   : *Pinus khasya* Royle, *P. merkusi* Jungh. et de Vr.; *Conifer.;* Burma.
Tiogo: *Spathodea campanulata* Beauv.; *Bignoniac.;* Gab.
Tiók: *Buchanania sp.; Anacardiac.;* Phil. (Tayabas)
Tiou: *Ceiba pentandra* Gaertn. *var. dehiscens* Ulb.; *Bombacac.;* Elf.
Tipa: *Tipuana tipu* Kunqe; *Papilionac.;* Arg.
? „ amarella: *Tipuana tipu* Kunqe; *Papilionac.;* Arg.
   „ blanca: *Machaerium pseudotipa* Griseb.; *Papilionac.;* tr. S.-Am.
   „   „   : *Tipuana tipu* Kunqe; *Papilionac.;* Arg.
   „ colorada: *Pterogyne nitens* Tul.; *Caesalpiniac.;* Arg.
Tipapop: *Terminalia catappa* L.; *Combretac.;* Südsee (Ponape, Carolinen)
Tipólo: *Artocarpus communis* Forst., *A. ovatifolia* Merr.; *Morac.;* Phil. (Camarines)
   „   : *Artocarpus treculiana* Elmer, *A. blancoi* Merr.; *Morac.;* Phil. (Camarines)
   „   : *Citrus limetta* Kramer; *Rutac.;* Südsee (Samoa)
   „ patupatu: *Citrus medica* L.; *Rutac.;* Südsee (Samoa)
Tipop: *Terminalia catappa* L.; *Combretac.;* Südsee (Ponape, Carolinen)
Tipuana: *Tipuana tipu* Kunqe; *Papilionac.;* Bras.
Tipúho: *Artocarpus treculiana* Elmer; *Morac.;* Phil. (Batanes)
Tipurús: *Palaquium polyandrum* C. B. Rob.; *Sapotac.;* Phil. (Cotabato)
Tique: *Aextoxicum punctatum* Ruiz et Pav.; *Euphorbiac.;* Chile.
Tira: *Melia azedarach* L.; *Meliac.;* Südsee (Tahiti)
Tiraco: *Pithecolobium dulce* Bth., *P. lanceolatum* Willd.; *Mimosac.;* n. Kol.
   „   : *Pithecolobium ligustrinum* Klotzsch, *P. oblongum* Bth.; *Mimosac.;* n. Kol.
Tiranglaí: *Vatica mangachapoi* Blco.; *Dipterocarpac.;* Phil. (Pangasinan)
Tirawa: *Xylia xylocarpa* Taub.; *Mimosac.;* sw. O.-Ind. (Bombay)
Tircana-malé-marom: *Berrya ammonilla* Roxb.; *Tiliac.;* O.-Ind. (s. Deccan)
Tiril: *Diospyros tomentosa* Roxb.; *Ebenac.;* n. O.-Ind.
Tirkha: *Randia dumetorum* Lamk.; *Rubiac.;* C.-O.-Ind. (Central Prov.)
Tirman: *Anogeissus latifolia* Wall.; *Combretac.;* O.-Ind. (Madras)
Tirorón: *Neonauclea philippinensis* Merr., *N. sp.; Rubiac.;* Phil.
   „   : *Terminalia comintana* Merr.; *Combretac.;* Phil. (Camarines)

Tirphal: *Zanthoxylum rhetsa* DC.; *Rutac.; sw. O.-Ind. (Bombay)
Tirpu: *Hopea parviflora* Bedd.; *Dipterocarpac.; sw. O.-Ind. (s. Bombay)
Tisate: *Perymenium grande* Hemsl.; *Composit.; Salv.
Tisemu: *Ficus umbrosa* Warb.; *Morac.; Togo.
Tisèze: *Pterocarpus soyauxi* Taub.; *Caesalpiniac.; Gab.
Tisoek: *Hibiscus vulpinus* Reinw.; *Malvac.; Java.
Tisomu: *Ficus platyphylla* Delile; *Morac.; Togo.
Tissaholz (d): *Rhus penthaphylla* Desf.; *Anacardiac.; n. Af.
Tissalala: *Piptadenia sp.; Mimosac.; Gab. (Loango)
Tissèze: *Pterocarpus soyauxi* Taub.; *Caesalpiniac.; Gab.
Tissoek: *Hibiscus vulpinus* Reinw.; *Malvac.; Java.
Tisswood (u): *Tetraclinis articulata* Mast.; *Conifer.; n. Af. (Algier)
Tisul: *Zanthoxylum rhetsa* DC.; *Rutac.; sw. O.-Ind. (Bombay)
Tita sopa: *Michelia champaca* L.; *Magnoliac.; Assam.
Títau: *Agathis alba* Foxw.; *Conifer.; Phil. (Abra)
Tité barkraki: *Eschweilera corrugata* Miers, *E. sp.; Lecythidac.; Sur.
„ merie: *Eschweilera corrugata* Miers, *E. sp.; Lecythidac.; Sur.
Titi: *Cliftonia ligustrina* Banks; *Cyrillac.; ö. & sö. USA.
„ , black: *Cliftonia ligustrina* Banks; *Cyrillac.; tr. Am.
„ , red: *Cyrilla racemiflora* L.; *Cyrillac.; tr. Am.
Titirillo: *Gyrocarpus americanus* Jacq.; *Hernandiac.; Guat.
Titoki: *Alectryon excelsum* Gaertn.; *Sapindac.; N.-Seel.
Tiupé: *Cynometra vogeli* Hook. f.; *Caesalpiniac.; Elf.
Tivar: *Barringtonia acutangula* Gaertn.; *Lecythidac.; C.-O.-Ind. (Central Prov.)
Tivila: *Diospyros evila* Pierre; *Ebenac.; Gab. (Loango)
Tiwarang: *Barringtonia acutangula* Gaertn.; *Lecythidac.; C.-O.-Ind. (Central Prov.)
Tiwas: *Ougeinia dalbergioides* Bth.; *Papilionac.; C.-O.-Ind. (Central Prov.)
Tiwí: *Dolichandrone spathodea* K. Schum.; *Bignoniac.; Phil. (Camarines)
Tiwoe landoe: *Artocarpus glauca* Blume; *Morac.; Java.
Tizerah: *Rhus pentaphylla* Desf.; *Anacardiac.; n. Af.
Tizra: *Rhus pentaphylla* Desf.; *Anacardiac.; w. N.-Af.
Tizrah: *Rhus pentaphylla* Desf.; *Anacardiac.; w. N.-Af.
Tjabisi: *Andira coriacea* Pulle; *Papilionac.; Sur.
Tjabisihoedoe: *Diplotropis guianensis* Bth., *D. leptophylla* Kleinh.; *Papilionac.; Sur.
Tjajoer: *Pterospermum blumeanum* Korth.; *Sterculiac.; Nied.-Ind. (Madura)
Tjalikangin: *Acer niveum* Blume; *Acerac.; Java (Tjibòdas)
Tjalingtjing: *Averrhoa bilimbi* L.; *Oxalidac.; w. Java.
Tjampaca: *Michelia champaca* L.; *Magnoliac.; Nied.-Ind.
Tjampaka boeloe: *Manglietia glauca* Blume; *Magnoliac.; Nied.-Ind.
„ oetan: *Talauma villosa* Miq.; *Magnoliac.; Nied.-Ind.
Tjampakka: *Michelia champaca* L.; *Magnoliac.; Nied.-Ind.
Tjampedak: *Artocarpus polyphema* Pers.; *Morac.; Java.
Tjanaren wakapoe: *Vouacapoua americana* Aubl.; *Papilionac.; Sur.
Tjangkoedoe: *Morinda citrifolia* L.; *Rubiac.; Java.
„ -badak: *Fagraea morindaefolia* Blume; *Loganiac.; Java (Tjemara)
„ -goenoeng: *Fagraea morindaefolia* Blume; *Loganiac.; Java (Palaboehan)
Tjangkolho: *Kleinhovia hospita* L.; *Sterculiac.; Java.
Tjantigi: *Vaccinium coriaceum* Miq., *V. spp.; Ericac.; Java.
„   : *Rhododendron retusum* Benn.; *Ericac.; Java (Gede)
„   : *Gaultheria fragrantissima* Wall. *var. punctata* J. J. Sm.; *Ericac.; Java (Gede)
„   : *Diplycosia heteriphylla* Blume; *Ericac.; Java.
„   : *Dodonaea viscosa* Jacq.; *Sapindac.; Java (Palaboeharantoe)
„ areuj: *Vaccinium lucidum* Miq., *V. ellipticum* Miq., *V. sp.; Ericac.; Java.
„ besar: *Rhododendron javanicum* Benn.; *Ericac.; Java (Tjikorai)
„ beureum: *Rhododendron retusum* Benn.; *Ericac.; Java (Gede)
„ bodas: *Gaultheria leucocarpa* Blume; *Ericac.; Java (Gede)
„ gede: *Vaccinium laurifolium* Miq.; *Ericac.; Java (Gede)
„ koneng: *Rhododendron album* Blume; *Ericac.; Java.
„ leutik: *Vaccinium lucidum* Miq.; *Ericac.; Java (Soendalandeu)
„ wangi: *Gaultheria leucocarpa* Blume; *Ericac.; Java (Gede)
„   „ : *Gaultheria fragrantissima* Wall. *var. punctata* J. J. Sm.; *Ericàc.; Java.
„ woengoe: *Vaccinium varingifolium* Miq.; *Ericac.; Java (Gede)

Tjaringin: *Ficus benjamina* L.; *Morac.;* w. Java.
Tjatjaboetja: *Vouacapoua americana* Aubl.; *Papilionac.;* Sur.
Tjatjankoedoean: *Fagraea morindaefolia* Blume; *Loganiac.;* Java (Palaboehan)
Tjavoeng: *Garcinia cornea* L.; *Guttifer.;* Java.
Tjawene sore: *Rhododendron javanicum* Benn.; *Ericac.;* Java (Gede)
Tjelemgan: *Crypteronia paniculata* Blume; *Sonneratiac.;* Malay.
Tjeliling: *Ficus edelfeldti* King; *Morac.;* Java (Pringamba)
Tjempaca kuning: *Michelia champaca* L.; *Magnoliac.;* Malay.
Tjemploeng: *Calophyllum inophyllum* L.; *Guttifer.;* Borneo.
Tjengal: *Hopea fagifolia* Miq.; *Dipterocarpac,;* Java, Bangka.
Tjengkarang: *Gnetum gnemon* L.; *Gnetac.;* Java.
Tjepoka: *Solanum torvum* Swartz; *Solanac.;* Java (Sepakoeng, Kledoeng)
Tjeralgal-monjeth: *Xerospermum noronhianum* Blume; *Sapindac.;* Nied.-Ind.
Tjerlang: *Pterospermum diversifolium* Blume; *Sterculiac.;* Java.
Tjermè-badak: *Meliosma ferruginea* Blume, *M. nervosa* Koord. et Val.; *Sapindac.:*
„ -beureum: *Meliosma nervosa* Koord. et Valet.; *Sapindac.;* Java. [Java.
Tjindoel: *Stelechocarpus burahol* Hook. f. et Thoms.; *Anonac.;* Java (Soebah)
Tjinkang: *Hopea mengarawan* Miq.; *Dipterocarpac.;* Sumatra, Bangka.
Tjoekilan: *Allophylus cobbe* Blume; *Sapindac.;* Mitt.-Java.
Tjoekilang: *Allophylus cobbe* Blume; *Sapindac.;* Mitt.-Java.
Tjoekoenda: *Jacaranda filicifolia* D. Don; *Bignoniac.;* Sur.
Tjoeng-beloet: *Solanum torvum* Swartz; *Solanac.;* Java (Ngebel)
Tjokla: *Theobroma cacao* L.; *Sterculiac.;* Java.
Tjulang: *Aglaia odorata* Lour.; *Meliac.;* Malay.
Tlachicón: *Curatella americana* L.; *Dilleniac.;* Mex.
Tlaia: *Tamarix articulata* Vahl; *Tamaricac.;* Sahara.
Tlapahoaxpatli: *Eysenhardtia polystachya* Sarg.; *Papilionac.;* Mex.
Tlapalezpatli: *Eysenhardtia polystachya* Sarg.; *Papilionac.;* Mex.
Tlascal: *Cupressus benthami* Endl.; *Conifer.;* Mex.
Tlascale: *Cupressus benthami* Endl.; *Conifer.;* Mex.
Tlatzcán: *Cupressus benthami* Endl.; *Conifer.;* Mex.
Tlazzcán: *Cupressus benthami* Endl.; *Conifer.;* Mex.
Tlitih: *Ficus pisifera* Wall.; *Morac.;* Java (Zuid-Pasoeroehan)
Tlone: *Sterculia hypochra* Pierre; *Sterculiac.;* Kamb., w. Annam.
Tnundaya: *Prunus capuli* Cav.; *Rosac.;* Mex.
Tnu-yáa: *Quercus reticulata* Humb. et Bonpl.; *Fagac.;* Mex.
„ -yaha: *Quercus reticulata* Humb. et Bonpl.; *Fagac.;* Mex.
Tnuyucu: *Taxodium mucronatum* Ten.; *Conifer.;* Mex.
To: *Randia uliginosa* DC.; *Rubiac.;* Ind.-Ch. (w. Annam)
Toatoa: *Phyllocladus trichomanoides* D. Don; *Conifer.;* N.-Seel.
Tobacco: *Pentagonia magnifica* Krause; *Rubiac.;* Pan.
„ , Tree-: *Nicotiana glauca* Graham; *Solanac.;* w. USA. (Kalif.)
„ , wild: *Nicotiana glauca* Graham; *Solanac.;* w. USA. (Kalif.)
„ „ : *Acnistus arborescens* Schlecht.; *Solanac.;* Trin., Tob.
Tobera: *Pittosporum tobira* Ait.; *Pittosporac.;* Jap.
Tobero: *Pittosporum tobira* Ait.; *Pittosporac.;* Jap.
Tobira riba: *Pittosporum tobira* Ait.; *Pittosporac.;* Jap.
Tobo: *Mitragyne macrophylla* Hiern; *Rubiac.;* Gab.
Toborochi: *Ceiba sp.; Bombacac.;* Bol.
Tobou: *Mitragyne macrophylla* Hiern; *Rubiac.;* Gab.
Tochi: *Aesculus turbinata* Blume; *Hippocastanac.;* Jap.
„ -noki: *Aesculus turbinata* Blume; *Hippocastanac.;* Java.
Toco: *Crataeva tapia* L.; *Capparidac.;* Trin.
„ : *Crataeva gynandra* L.; *Capparidac.;* Ven.
„ prieto: *Coccoloba humboldti* Meissn.; *Polygonac.;* Mex.
Tocque: *Crataeva tapia* L.; *Capparidac.;* Trin.
Tocuma: *Dipholis salicifolia* A. DC.; *Sapotac.;* sp. W.-Ind.
„ amarillo: *Sideroxylon mastichodendron* Jacq.; *Sapotac.;* sp. W.-Ind.
Todabuena: *Hypericum androsaemum* L.; *Hypericac.;* Sp.
Toddalie: *Toddalia asiatica* Baill.; *Rutac.;* O.-Ind.
Toddy-marom: *Adansonia digitata* L.; *Bombacac.;* O.-Ind. (s. Deccan)
Todinga: *Manilkara costata* Pierre; *Sapotac.;* Mad.

524

Todo: ? *Ficus lepicarpa* Blume; *Morac.;* Java (Pringamba, Telemojo)
„ : *Staudtia gabonensis* Warb.; *Myristicac.;* Gab.
Todomatsu: *Abies sachalinensis* Mast., *A. mariesi* Mast.; *Conifer.;* Jap.
Todouk: *Spathodea campanulata* Beauv.; *Bignoniac.;* Kam.
Todu: *Toona ciliata* M. Roem.; *Meliac.;* O.-Ind. (Bombay)
Toë: *Dialium guineense* Willd.; *Caesalpiniac.;* Togo.
Toelaroe fieroberoe: *Erisma uncinatum* Warm.; *Vochysiac.;* Sur.
Toelimani: *Lecythis ollaria* L.; *Lecythidac.;* fr. Gu.
Toemboes: *Campnosperma auriculata* Hook. f.; *Anacardiac.;* Sumatra (Loemet)
Toenba-lobbi: *Cordia tetrandra* Aubl.; *Borraginac.;* Sur.
Toendoen: *Xerospermum noronhianum* Blume; *Sapindac.;* Java (Tjilatjap)
Toengkei alie: *Eurycoma longifolia* Jacq.; *Simarubac.;* Nied.-Ind.
„ mempaloet: *Eurycoma longifolia* Jacq.; *Simarubac.;* Nied.-Ind.
Toentoen: *Nephelium lappaceum* L.; *Sapindac.;* Java, Sumatra.
Toepoeloe-koesali-epo: ? *Chaetocarpus schomburgkiana* Pax et Hoffm.; *Euphorbiac.;*
Toepoera apoekoetja: *Swartzia acuminata* Willd., *S. spp.; Caesalpiniac.;* Sur. [Sur.
Toepoeroe-apoekoetjarang: ? *Swartzia triphylla* Willd.; *Caesalpiniac..;* Sur.
„ -kwatéré: *Eschweilera longipes* Miers, *E. subglandulosa* Miers; *Lecythidac.;* Sur.
„ tonorebjo: *Matayba spp., Cupania scrobiculata* L. C. Rich.; *Sapindac.;* Sur.
„ wamé: *Nectandra sp., Ocotea sp.; Laurac.;* Sur.
Toeraroe-fieroeberoe: *Erisma uncinatum* Warm.; *Vochysiac.;* Sur.
Toerelie: *Inga heterophylla* Willd., *I. lateriflora* Miq.; *Mimosac.;* Sur.
Toeririe: *Inga heterophylla* Willd., *I. lateriflora* Miq.; *Mimosac.;* Sur.
Tofé: *Macaranga spp.; Euphorbiac.;* Elf.
Toga: *Tsuga sieboldi* Carr.; *Conifer.;* Jap.
Togar mogi: *Morinda tinctoria* Roxb.; *Rubiac.;* O.-Ind. (Madras)
Togari: *Morinda tinctoria* Roxb.; *Rubiac.;* O.-Ind. (Madras)
Togariba-maki: *Podocarpus nakai* Hayata; *Conifer.;* Form.
Togasawara: *Pseudotsuga japonica* Shir., *P. japonica* Shir.; *Conifer.;* Jap.
Togba: *Mitragyne macrophylla* Hiern; *Rubiac.;* Togo.
Tog-beli: *Millettia lane-poolei* Dunn; *Papilionac.;* Lib.
Togolo: *Pseudospondias gigantea* Chev.; *Anacardiac.;* Gab.
Tohi: *Picea hondoensis* Mayr, *P. obovata* Ledek.; *Conifer.;* Jap.
Tohn-we-eh: *Cola nitida* Chev.; *Sterculiac.;* Lib.
Tohog: *Petersianthus quadrialatus* Merr.; *Lecythidac.;* Phil. (Agusan)
Tói: *Miliusa velutina* Hook. f. et Thoms.; *Anonac.;* w. Annam.
Toï: *Alphitonia sizyphoides* Reiss.; *Celastrac.;* Südsee (Tahiti)
Toi: *Nothofagus sp.; Fagac.;* N.-Seel.
Tojo: *Ulex europaeus* L.; *Papilionac.;* Arg.
Tokei: *Coula edulis* Baill.; *Olacac.;* Lib.
Tokèwe: *Nectandra sp., Ocotea sp.; Laurac.;* Sur.
Tokiwa-inubiwa (jap.): *Ficus septica* Burm. f.; *Morac.;* tr. As.
Toko: *Eschweilera corrugata* Miers; *Lecythidac.;* Sur.
Tókod-langit: *Semecarpus sp.; Anacardiac.;* Phil. (Rizal)
Tokolo apolikiri: *Helicostylis poeppigiana* Tréc.; *Morac.;* Sur.
Tokoro: *Piratinera guianensis* Aubl., *P. spp.; Morac.;* Sur.
„ apolinore: *Helicostylis poeppigiana* Tréc.; *Morac.;* Sur.
„ apolli merie: *Piratinera guianensis* Aubl.; *Morac.;* Sur.
Tokun-yanagi (jap.): *Salix eriostroma* Hayata; *Salicac.;* Form.
Tol: *Ficus vogeliana* Miq.; *Morac.;* Kam., Gab.
Tola: *Colletia cruciata* Gill. et Hook., *C. ferox* Gill.; *Rhamnac.;* Arg.
Tolabitoum: *Piptadenia sp.; Mimosac.;* Gab.
Tolabu: *Grewia gigantiflora* K. Schum.; *Tiliac.;* Togo.
Tolelie: *Inga heterophylla* Willd., *I. lateriflora* Miq.; *Mimosac.;* Sur.
Toll: *Ficus vogeliana* Miq.; *Morac.;* Gab.
Tollon: *Heteromeles arbutifolia* Roem.; *Rosac.;* w. USA. (Kalif.)
Tolokyo: *Juniperus procera* Hochst.; *Conifer.* br. O.-Af. (Uganda)
Tolol: *Trichadenia zeylanica* Thw.; *Flacourtiac.;* Ceylon.
Tolondrón: *Coccoloba browniana* Standl.; *Polygonac.;* Hond.
Tolozin: *Heliocarpus sp.; Tiliac.;* Mex.
Tolú: *Bombacopsis sp.; Bombacac.;* Kol.
„ : *Myroxylon toluiferum* H. B. K.; *Papilionac.;* Kol.

Tolui: *Cola acuminata* Schott et Endl.; *Sterculiac.;* Lib.
Tom: *Pachylobus balsamifera* Guill.; *Burserac.;* Gab.
„ : *Sterculia pexa* Pierre; *Sterculiac.;* Ind.-Ch. (Annam, Kamb.)
„ xói: *Miliusa velutina* Hook. f. et Thoms.; *Anonac.;* w. Annam.
Toman: *Pseudocedrela kotschyi* Harms; *Meliac.;* Nig.
Tombi: *Tamarindus indica* L.; *Caesalpiniac.;* Sudan.
„ : *Parkia klainei* Pierre; *Mimosac.;* Kam.
Tombo: *Pachylobus balsamifera* Guill.; *Burserac.;* fr. Aeq.-Af.
„ : *Sarcocephalus trillesi* Pierre; *Rubiac.;* Gab.
Tomboti: *Excoecaria africana* Muell. Arg.; *Euphorbiac.;* O.-Af.
Tome: *Parkia klainei* Pierre; *Mimosac.;* Kam.
Tomeguin: *Zanthoxylum sp.; Rutac.;* Cuba.
Tomi: *Tamarindus indica* L.; *Caesalpiniac.;* fr. Guin.
Tomillo: *Thymus hirtus* Willd.; *Labiat.;* Sp.
„ aceitumero: *Thymus zygis* L.; *Labiat.;* Sp.
„ blanco: *Cistus salviaefolius* L.; *Cistac.;* Sp.
„ salsero: *Thymus vulgaris* L.; *Labiat.;* Sp.
Tomm: *Pachylobus klaineana* Guill.; *Burserac.;* Gab.
Tomoené-ira-kopie: *Qualea coerulea* Aubl., *Q. rosea* Aubl., *Q. sp.; Vochysiac.;* Sur.
Tomopio: *Minquartia guianensis* Aubl.; *Olacac.;* Sur.
Tonam: *Pseudocedrela kotschyi* Harms; *Meliac.;* Nig.
Tonas: *Pseudocedrela kotschyi* Harms; *Meliac.;* Nig.
Tonca bean: *Dipteryx odorata* Willd.; *Papilionac.;* n. S.-Am.
Tondo: *Cynometra vogeli* Hook. f.; *Caesalpiniac.;* Sudan.
Tondonbaka: *Copaifera demeusei* Harms; *Caesalpiniac.;* B.-K.
Tondsha: *Limonia acidissima* L.; *Rutac.;* O.-Ind. (Bombay)
Toneriko: *Fraxinus bungeana* A. DC.; *Oleac.;* Jap.
Tonga bean: *Dipteryx odorata* Willd.; *Papilionac.;* n. S.-Am.
Tongo: *Ceiba pentandra* Gaertn.; *Bombacac.;* Elf.
„ : *Celtis integrifolia* Lamk.; *Ulmac.;* Elf.
„ : *Dracaena fragrans* Ker-Gawl.; *Liliac.;* Kam.
Tongóg: *Ceriops tagal* C. B. Rob.; *Rhizophorac.;* Phil. (Masbate, Capiz, Negros, Tawi-
Tongrong: *Spondias mangifera* Pers.; *Anacardiac.;* O.-Ind. [tawi)
Tongwa: *Cissus adenocarpa* Gilg et Brandt; *Vitac.;* D.-O.-Af. (Tabora)
Tonka: *Dipteryx odorata* Willd.; *Papilionac.;* Sur.
„ bean: *Dipteryx odorata* Willd.; *Papilionac.;* n. S.-Am.
„ boon: *Dipteryx odorata* Willd.; *Papilionac.;* Gu.
„ fevrier: *Dipteryx odorata* Willd.; *Papilionac.;* fr. Gu.
„ -hout (h): *Dipteryx odorata* Willd.; *Papilionac.;* Sur.
Tonkin bean: *Dipteryx odorata* Willd.; *Papilionac.;* n. S.-Am.
Tonko: *Ceiba pentandra* Gaertn.; *Bombacac.;* Elf.
Tonorebjo: *Matayba spp., Cupania scrobiculata* L. C. Rich.; *Sapindac.;* Sur.
„ , zwarte: *Matayba spp., Cupania sp.; Sapindac.;* Sur.
Tonquin bean: *Dipteryx odorata* Willd.; *Papilionac.;* n. S.-Am.
Tonquine wood (u): *Chloroxylon swietenia* DC.; *Meliac.;* Ceylon.
Tontoe awha: *Parkia nitida* Miq., *P. microcephala* Kleinh.; *Mimosac.;* Sur.
Tooart: *Eucalyptus gomphocephala* A. DC.; *Myrtac.;* w. Au.
Tóob: *Bischofia javanica* Blume; *Euphorbiac.;* Phil. (Tayabas)
Tood: *Toona ciliata* M. Roem.; *Meliac.;* O.-Ind. (Madras, s. Deccan)
Tóog: *Petersianthus quadrialatus* Merr.; *Lecythidac.;* Phil.
„ : *Bischofia javanica* Blume; *Euphorbiac.;* Phil.
Toola: *Sterculia alata* Roxb.; *Sterculiac.;* O.-Ind.
Toom: *Garuga pinnata* Roxb.; *Burserac.;* Beng.
Toon: *Toona ciliata* M. Roem., *T. serrata* M. Roem.; *Meliac.;* O.-Ind., Burma.
„ : *Toona australis* Harms; *Meliac.;* Au.
„ : *Toona calantas* Merr. et Rolfe; *Meliac.;* Phil.
Toona: *Toona ciliata* M. Roem.; *Meliac.;* O.-Ind.
Toon-kor-doh: *Carapa procera* DC.; *Meliac.;* Lib.
Toonu: *Castilla fallax* Cook, *C. sp.; Morac.;* C.-Am.
Toor (e): *Cajanus indicus* Spreng.; *Papilionac.;* Tr.
Tooroo: *Oenocarpus bacaba* Mart.; *Palmac.;* fr. Gu.
Toothache tree: *Zanthoxylum americana* Mill.; *Rutac.;* nö. USA.

Toothache tree: *Zanthoxylum clava-herculis* L.; *Rutac.;* sö. USA.
Topa: *Ochroma sp.; Bombacac.;* Peru.
„ piaman: *Eugenia operculata* Roxb.; *Myrtac.;* O.-Ind. (Bihar, Orissa)
Topie: *Pouteria guianensis* Aubl.; *Sapotac.;* Sur.
„ : *Mouriria anomala* Pulle, *M. plaschaerti* Pulle; *Melastomatac.;* Sur.
„ -a-baäkawan: *Pouteria sp.; Sapotac.;* Sur.
Tópo: *Semecarpus gigantifolia* Vidal; *Anacardiac.;* Phil. (Camarines)
Topoeroe: *Sclerolobium sp.; Caesalpiniac.;* Sur.
Topol: *Populus nigra* L.; *Salicac.;* Sib.
Toporite: *Hernandia sonora* L.; *Hernandiac.;* Trin., Tob.
Toppie: *Mouriria anomala* Pulle, *M. plaschaerti* Pulle; *Melastomatac.;* Sur.
Toquilla: *Carludovica palmata* R. et P.; *Cyclanthac.;* Kol., Ek.
Toratti: *Hydnocarpus wightiana* Blume; *Flacourtiac.;* sw. O.-Ind. (Bombay)
Torch, Bastard-: *Ocotea sp., Nectandra sp.; Laurac.;* Bah.
„ , black: *Ocotea sp., Nectandra sp.; Laurac.;* Bah.
„ , white: *Amyris sp.; Rutac.;* br. W.-Ind.
Torchwood: *Amyris balsamifera* L., *A. sp.; Rutac.;* Jam., br. W.-Ind.
„ : *Ocotea rodioei* Mez; *Laurac.;* br. Gu.
„ , sweet: *Ocotea sp., Nectandra sp.; Laurac.;* Bah.
Torcido: *Mouriria sp.; Melastomatac.;* Cuba.
Torelaga: *Limonia acidissima* L.; *Rutac.;* O.-Ind. (Madras)
Torelie: *Inga heterophylla* Willd., *I. lateriflora* Miq.; *Mimosac.;* Sur.
Toreta: *Anona sp.; Anonac.;* Pan.
Toreto: *Anona purpurea* Moç. et Sessé, *Rollinia jimenzi* Safford; *Anonac.;* Pan.
Torillo: *Randia armata* DC.; *Rubiac.;* Guat.
Tornillo: *Prosopis pubescens* Bth.; *Mimosac.;* Mex.
Toro: *Morisonia americana* L.; *Capparidac.;* Kol.
Toromiro: *Podocarpus ferrugineus* Don; *Conifer.;* N.-Seel.
Toronja: *Citrus maxima* Merr.; *Rutac.;* Salv., Hond.
Toronya: *Citrus limonum* Risso; *Rutac.;* Par.
Toro pé: *Bauhinia candicans* Bth.; *Caesalpiniac.;* Arg.
„ pó: *Bauhinia candicans* Bth.; *Caesalpiniac.;* Arg.
„ -ratay: *Tabebuia nodosa* Griseb.; *Bignoniac.;* Arg.
Torote: *Elaphrium microphyllum* Rose, *E. odoratum* Rose; *Burserac.;* Mex
„ : *Elaphrium tecomaca* Standl.; *Burserac.;* Mex.
„ blanco: *Elaphrium microphyllum* Rose; *Burserac.;* Mex.
„ „ : *Pachycormus discolor* Coville; *Anacardiac.;* Mex.
„ prieto: *Elaphrium laxiflorum* Rose; *Burserac.;* Mex.
Torotoro: *Gluta turtur* March.; *Anacardiac.;* Mad.
Torreya, yew-leaved: *Torreya taxifolia* Arn.; *Conifer.;* sö. USA.
Tortoise-shell wood: *Cordia ? angiocarpa* A. Rich.; *Borraginac.;* Cuba.
„ „ „ : *Piratinera guianensis* Aubl.; *Morac.;* Gu.
Tortugo: *Sideroxylon mastichodendron* Jacq.; *Sapotac.;* P.-R.
„ : *Crataeva tapia* L., *Capparidac.;* Guat.
„ amarillo: *Sideroxylon mastichodendron* Jacq.; *Sapotac.;* P.-R.
„ „ : *Sideroxylon foetidissinum* Jacq.; *Sapotac.;* P.-R.
„ prieto: *Ravenia urbani* Engl.; *Rutac.;* P.-R.
„ „ : *Sideroxylon mastichodendron* Jacq.; *Sapotac.;* P.-R.
Tortuguillo: *Antirrhoea obtusifolia* Urban; *Rubiac.;* P.-R.
Torvisco: *Daphne gnidium* L.; *Thymelaeac.;* Sp.
Tos: *Picea morinda* Link, *Abies pindrow* Spach; *Conifer.;* nw. O.-Ind. (Punjab)
Toshirabe: *Abies nephrolepis* Maxim.; *Conifer.;* n. Korea.
Tostado: *Homalium racemosum* Jacq.; *Flacourtiac.;* P.-R.
„ prieto: *Homalium racemosum* Jacq.; *Flacourtiac.;* P.-R.
Tostón: *Boerhaavia paniculata* L. C. Rich.; *Nyctaginac.;* P.-R.
Totara: *Podocarpus totara* D. Don, *P. halli* T. Kirk; *Conifer.;* N.-Seel.
Toto amoeté: *Eperua schomburgkiana* Bth., *? E. jenmani* Oliv.; *Caesalpiniac.;* Sur.
Totoca: *? Bertholletia sp.; Lecythidac.;* Sur.
Totohoté: *Pterocarpus esculentus* Schum. et Thonn.; *Papilionac.;* Elf.
Totonai: *Heritiera littoralis* Dry.; *Sterculiac.;* Phil.
Totopostle: *Licania arborea* Seem.; Mex.
Totoro: *Lonchocarpus sericeus* H. B. K.; *Papilionac.;* Go.

Totoҕapotl: *Sideroxylon capiri* Pittier; *Sapotac.;* Mex.
Totui: *Jacaranda copaia* D. Don; *Bignoniac.;* Sur.
Totuma: *Crescentia cujete* L.; *Bignoniac.;* Kol., Ven.
Totҕi: *Aesculus turbinata* Blume; *Hippocastanac.;* Jap.
Tou: *Berlinia tomentosa* Harms; *Caesalpiniac.;* Togo.
Touamba: *Canarium schweinfurthi* Engl.; *Burserac.;* Kam.
Touambo: *Piptadenia africana* Hook. f.; *Mimosac.;* Kam.
Toubaki-marom: *Guazuma tomentosa* H. B. K.; *Sterculiac.;* O.-Ind. (Deccan)
Toubaouaté: *Monopetalanthus sp.; Caesalpiniac.;* Elf.
Toubesso: *Baphia pubescens* Hook. f.; *Papilionac.;* Elf.
Touca: *Bertholletia excelsa* H. B. K.; *Lecythidac.;* fr. Gu.
Touch me and you get it: *Ficus vogeli* Miq.; *Morac.;* n. Nig. (Kontagora)
Touckpong: *Sapium biglandulosum* Muell. Arg. var. ?; *Euphorbiac.;* br. Gu.
Touendo: *Vitex grandifolia* Baker; *Verbenac.;* Kam.
Touey: *Conopharyngia brachyantha* Stapf; *Apocynac.;* Kam.
Tougbi: *Bridelia atroviridis* Muell. Arg., *B. speciosa* Muell. Arg.; *Euphorbiac.;* Elf.
Tougbibi: *Bridelia micrantha* Baill.; *Euphorbiac.;* Elf.
Toul: *Piptadenia africana* Hook. f.; *Mimosac.;* Kam.
Toulici: *Toulicia guianensis* Aubl.; *Sapindac.;* Gu.
Touloucouna: *Carapa procera* DC.; *Meliac.;* Senl.
Touloul: *Glossolepis klainei* Pierre; *Sapindac.;* Gab.
Toum: *Piptadenia sp.; Mimosac.;* Gab.
Toumatou kourou: *Discaria toumatsou* Raoul; *Rhamnac.;* N.-Seel.
Toumbo: *Sarcocephalus trillesi* Pierre; *Rubiac.;* Kam.
Toumidio: *Malacantha alnifolia* Pierre; *Sapotac.;* Elf.
Toungkalamet: *Cordia fragrantissima* Kurz; *Borraginac.;* Burma.
Tounou: *Swartҕia sp.; Caesalpiniac.;* fr. Gu.
Touonne: *Anthocleista zenkeri* Gilg; *Loganiac.;* Kam.
Tourameira: *Humiria sp.; Humiriac.;* br. Gu.
Tourha: *Calotropis procera* Ait.; *Asclepiadac.;* Sahara.
Tourlouri: *Manicaria saccifera* Gaertn.; *Palmac.;* fr. Gu.
Touroutier: *Sterculia pruriens* K. Schum.; *Sterculiac.;* fr. Gu.
Tourou-tourou: *Sterculia sp.; Sterculiac.;* Gu. (Galibis)
Tourtour: *Gluta turtur* March.; *Anacardiac.;* Mad.
Tous fruitier à noyaux: *Prunus sp.; Rosac.;* Eu.
„   „   „ pépins: *Pirus sp.; Rosac.;* Eu.
Touté: *Discoglypremna caloneura* Prain; *Euphorbiac.;* Gab.
Toutou: *Parinarium curatellifolium* Planch.; *Rosac.;* Elf.
Towai: *Weinmannia racemosa* L. f.; *Cunoniac.;* N.-Seel.
Towanero: *Humiria sp.; Humiriac.;* br. Gu.
Towei: *Nothofagus sp.; Fagac.;* N.-Seel.
Toweroenierou: *Humiria sp.; Humiriac.;* fr. Gu.
Toweroeniroe: *Humiria floribunda* Mart., *H. balsamifera* Aubl.; *Humiriac.;* Sur.
Towhai: *Weinmannia racemosa* L. f.; *Cunoniac.;* N.-Seel.
Towoi: *Nothofagus sp.; Fagac.;* N.-Seel.
Toya: *Ficus glomerata* Roxb.; *Morac.;* C.-O.-Ind. (Gondwana)
To-ye: ? *Drypetes afzeli* Hutch.; *Euphorbiac.;* Lib.
Toyon: *Heteromeles arbutifolia* Roem.; *Rosac.;* w. USA. (Kalif.)
Tôzan-yanagi: *Salix transarisanensis* Hayata; *Salicac.;* Form.
Tra lac: *Vatica astrotricha* Hance; *Dipterocarpac.;* Ind.-Ch. (Kamb.)
„ meng: *Garcinia delpyana* Pierre; *Guttifer.;* Ind.-Ch. (Kamb.)
Trabeck-sralao: *Lagerstroemia angustifolia* Pierre; *Lythrac.;* Ind.-Ch.
„   „ : *Lagerstroemia flos-reginae* Retҕ.; *Lythrac.;* Ind.-Ch.
Trac: *Dalbergia cochinchinensis* Pierre, *D. latifolia* Roxb.; *Papilionac.;* Ind.-Ch.
„ lai: *Dalbergia cochinchinensis* Pierre; *Papilionac.;* Ind.-Ch.
Trach: *Dipterocarpus intricatus* Dyer; *Dipterocarpac.;* Kamb.
„ quach: *Adenanthera pavonina* L.; *Mimosac.;* Ind.-Ch.
Tragacanth, african (e): *Sterculia tragacantha* Lindl.; *Sterculiac.;* Nig., Go.
Trai: *Garcinia fagraeoides* Chev.; *Guttifer.;* Ind.-Ch. (Tonkin)
„ : *Adina polycephala* Bth.; *Rubiac.;* s. Ind.-Ch.
„ -cà: *Nephelium litchi* Camb.; *Sapindac.;* Annam.
Trakoukébia: *Pentadesma leucantha* Chev.; *Guttifer.;* Elf.

Trâlâk: ? *Spondias sp.; Anacardiac.;* Kamb.
Tralat: ? *Spondias sp.; Anacardiac.;* Kamb.
Traling: *Tarrietia simplicifolia* Mast.; *Sterculiac.;* Malak.
Tram: *Canarium spp.; Burserac.;* Ind.-Ch.
„ : *Melaleuca sp.; Myrtac.;* Co.-Ch.
„ den: *Canarium nigrum* Engl.; *Burserac.;* Ind.-Ch. (Phan-Luong)
„ hoang: *Kayea eugeniaefolia* Pierre; *Guttifer.;* Annam.
„ -hông: *Canarium nigrum* Engl.; *Burserac.;* Ind.-Ch. (Phan-Luong)
„ kang: *Peltophorum dasyrachis* Kurz; *Caesalpiniac.;* Kamb.
„ lanh: *Tristania burmanica* Griff.; *Myrtac.;* Ind.-Ch. (Annam)
„ lau: *Barringtonia sp.; Lecythidac.;* Ind.-Ch.
„ nên: *Canarium tonkinense* Engl.; *Burserac.;* Ind.-Ch. (Annam)
„ sung: *Eugenia tinctoria* Gagnep.; *Myrtac.;* Ind.-Ch. (Annam)
„ trâng: *Canarium tonkinense* Engl.; *Burserac.;* Ind.-Ch. (Phan-Luong)
Trameng: *Garcinia sp.; Guttifer.;* Ind.-Ch.
„ : *Carallia lucida* Roxb.; *Rhizophorac.;* Kamb.
Tramkan: *Peltophorum dasyrachis* Kurz; *Caesalpiniac.;* Kamb.
Tras: *Securinega flexuosa* Muell. Arg.; *Euphorbiac.;* Phil. (Cotabato)
Trasec: *Peltophorum dasyrachis* Kurz; *Caesalpiniac.;* Kamb.
Trâu: *Aleurites montana* Pierre; *Euphorbiac.;* Annam.
„ trau: *Ochrocarpus siamensis* T. Anders.; *Guttifer.;* Ind.-Ch.
Traumen: *Aralia laetevirens* C. Gay; *Araliac.;* Chile.
Trâyang: *Diospyros sp.; Ebenac.;* Kamb.
Tré: *Baphia nitida* Lodd.; *Papilionac.;* Elf.
Trebo: *Platymiscium polystachyum* Bth.; *Papilionac.;* Kol.
Trébol: *Platymiscium polystachyum* Bth.; *Papilionac.;* n. Kol.
Trebol: *Torresea cearensis* Allem.; *Papilionac.;* Bras.
Tree of Heaven (u): *Ailanthus glandulosa* Desf.; *Simarubac.;* n. Ch.
„ tomato: *Cyphomandra betacea* Sendtn.; *Solanac.;* br. O.-Af. (Uganda)
Trekhan: *Acer pictum* Thunb., *A. caesium* Wall.; *Acerac.;* nw. O.-Ind. (Punjab)
Tremble: *Populus tremuloides* Michx.; *Salicac.;* Kan.
„ , faux: *Populus tremuloides* Michx.; *Salicac.;* Kan.
„ , grand: *Populus grandidentata* Michx.; *Salicac.;* ö. Kan.
Trementina: *Schinus dependens* Ortega; *Anacardiac.;* Arg.
Trementino: *Guarea sp.; Meliac.;* Mex.
Tremilo: *Populus tremula* L.; *Salicac.;* It.
Trendj: *Citrus medica* Gall.; *Rutac.;* n. Af. (Tunis)
Trengaloen: *Protium javanicum* Burm.; *Burserac.;* Java.
Trenggoeli: *Cassia javanica* L.; *Caesalpiniac.;* Malay.
Treo: *Flotowia diacanthoides* Less.; *Composit.;* Chile.
Tres folhas vermelhas: *Esenbeckia febrifuga* Mart.; *Rutac.;* Bras.
„ Marías: *Forchammeria trifoliata* Radlk.; *Capparidac.;* br. Hond.
Tre-tre: *Calpocalyx brevibracteatus* Harms; *Mimosac.;* Go.
Triaca: *Caldcluvia paniculata* D. Don; *Cunoniac.;* Chile.
Triane: *Humiria sp.; Humiriac.;* fr. Gu.
Trincomalee wood (e): *Berrya ammonilla* Roxb.; *Tiliac.;* O.-Ind.
Trincomali wood: *Berrya ammonilla* Roxb.; *Tiliac.;* O.-Ind.
Tritih-berasan: *Ficus glaberrima* Blume; *Morac.;* Java (Tangkil)
Triwa: *Pontya excelsa* Chev.; *Urticac.;* Elf.
Tro: *Dipterocarpus tonkinensis* Chev.; *Dipterocarpac.;* Ind.-Ch. (Tonk., Laos, n. An-
Troeh: *Ficus edelfelti* King; *Morac.;* Java (Gadoengan)   [nam)
Troehn-dru-plohn: *Pausinystalia lane-poolei* Hutch.; *Rubiac.;* Lib.
Troène d'Amérique: *Duranta plumieri* Jacq.; *Uerbenac.;* N.-Kal.
Troi moi: *Antidesma sp.; Euphorbiac.;* Ind.-Ch.
Trom: *Sterculia hypochra* Pierre, *S. pexa* Pierre; *Sterculiac.;* Annam, Kamb.
Trompeto: *Guarea trichilioides* L.; *Meliac.;* Kol.
Trompette: *Cecropia sp.; Morac.;* Guad.
Trompillo: *Cordia boissieri* DC.; *Borraginac.;* Mex.
„ : *Rosa montezumae* Humb. et Bonpl.; *Rosac.;* Mex.
„ : *Euphorbia cotinifolia* L.; *Euphorbiac.;* Mex.
„ : *Lafoensia punicaefolia* DC.; *Lythrac.;* Mex.
„ : *Colubrina greggi* S. Wats., *C. macrocarpa* Don; *Rhamnac.;* Mex.

Trompillo: *Nectandra sinuata* Mez; *Laurac.;* Guat.
„ : *Guarea trichilioides* L.; *Meliac.;* Kol., Ven.
„ : *Laetia hirtella* H. B. K., *L. guazumaefolia* H. B. K.; *Bixac.;* Ven.
Trompito: *Jacquinia sp.; Theophrastac.;* Curacao.
„ : *Guarea trichilioides* L., *G. sp.; Meliac.;* Ven.
Tronador: *Hura crepitans* L.; *Euphorbiac.;* Pan.
Trul: *Xylia xylocarpa* Taub.; *Mimosac.;* sw. O.-Ind. (Travancore)
Trumalla: *Xylia xylocarpa* Taub.; *Mimosac.;* sw. O.-Ind.
Trumpet: *Cecropia mexicana* Hemsl.; *Morac.;* br. Hond.
„ tree: *Cecropia arachnoides* Pitt., *C. mexicana* Hemsl., *C. sp.; Morac.;* Pan.
„ „ : *Cecropia peltata* L.; *Morac.;* Trin., Tob.
Trunkelbeere (d): *Vaccinium uliginosum* L.; *Ericac.;* Eu.
Truntum: *Aegiceras corniculatum* Blco.; *Myrsinac.;* Malay.
Truong: *Pomelia pinnata* Forst.; *Sapindac.;* Ind.-Ch.
Truông: *Euphoria longana* Lamk., *? Nephelium sp.; Sapindac.;* Annam.
Trupi: *Acacia tortuosa* Willd.; *Mimosac.;* Kol.
Trupillo: *Prosopis chilensis* Stuntʒ; *Mimosac.;* Kol.
Trysil: *Pentaclethra filamentosa* Bth., *P. macroloba* Ktʒe.; *Mimosac.;* Gu.
Tsa: *Taxus baccata* L.; *Conifer.;* Beng. (Bhutan)
„ : *Ormocarpum bibracteatum* Baker; *Papilionac.;* n. Nig.
„ : *Erythrophloeum micranthum* Harms; *Caesalpiniac.;* Go.
„ : *Erythrophloeum guineense* G. Don; *Caesalpiniac.;* Togo.
Tsain: *Ricinodendron africanum* Muell. Arg.; *Euphorbiac.;* Elf.
T'sainfi: *Stenanthera hamata* Engl. et Diels; *Anonac.;* Elf.
Tsaklo: *Pentaclethra macrophylla* Bth.; *Mimosac.;* Go.
Tsamia: *Tamarindus indica* L.; *Caesalpiniac.;* n. Nig. (Kontagora)
Tsamiya: *Tamarindus indica* L.; *Caesalpiniac.;* Nig.
Tsangoti: *Alstonia gilleti* de Willd.; *Apocynac.;* B.-K. (Mayumbe)
Tsanum: *Agathis alba* Foxw.; *Conifer.;* Malay.
Tsanya: *Corynanthe paniculata* Welw.; *Rubiac.;* B.-K. (Mayumbe)
Tsapas sabani: *Calocarpum mammosum* Pierre; *Sapotac.;* Mex.
Tsatsagi: *Bauhinia rufescens* Lam.; *Caesalpiniac.;* n. Nig.
Tschabola buanda: *Hexalobus monepetalus* Engl. et Diels; *Anonac.;* Togo.
„ bunda: *Xylopia parviflora* Engl. et Diels; *Anonac.;* Togo.
Tschalo: *Daniella thurifera* Bth.; *Caesalpiniac.;* Togo.
Tschaloware: *Derris stuhlmanni* Harms; *Papilionac.;* Togo.
Tschamanu: *Cassia sieberiana* DC.; *Caesalpiniac.;* Togo.
Tschapea: *Syzygium guineense* DC.; *Myrtac.;* Togo.
Tschcan-an: *Hamelia erecta* Jacq.; *Rubiac.;* br. Hond.
Tschedia: *Ficus sp.; Morac.;* n. Nig. (Kontagora)
Tschengunga: *Crataeva religiosa* Forst.; *Capparidac.;* Togo.
Tschenyenga: *Hymenocardia acida* Tul.; *Euphorbiac.;* Togo.
Tschifulu-fulu: *Ongokea klaineana* Pierre; *Oleac.;* B.-K. (Kasai)
Tschingbelika: *Firmiana barteri* K. Schum.; *Sterculiac.;* Togo.
Tschinli: *Burkea africana* Hook.; *Caesalpiniac.;* Togo.
Tschipulu-pulu: *Ongokea klaineana* Pierre; *Oleac.;* B.-K. (Kasai)
Tschisili: *Burkea africana* Hook.; *Caesalpiniac.;* Togo.
Tschitschiñe: *Markhamia tomentosa* K. Schum.; *Bignoniac.;* Togo.
Tschufou: *Alchornea cordifolia* Muell. Arg.; *Euphorbiac.;* Togo.
Tsege: *Amblygonocarpus schweinfurthi* Harms; *Mimosac.;* n. Nig.
Tsetse: *Anogeissus leiocarpus* Guill. et Perr.; *Combretac.;* Togo.
Tshaloware: *Derris stuhlmanni* Harms; *Papilionac.;* Togo.
Tshibafubafu: *? Hua gaboni* Pierre; *Sterculiac.;* B.-K. (Kasai)
Tshibamba: *Pentaclethra macrophylla* Bth.; *Mimosac.;* B.-K. (Kasai)
Tshibambabamba: *Pentaclethra macrophylla* Bth.; *Mimosac.;* B.-K. (Kasai)
Tshilembelembe: *Anonidium manni* Engl. et Diels; *Anonac.;* B.-K. (Kasai)
Tshiloampusu: *Conopharyngia sp.; Apocynac.;* B.-K. (Kasai)
Tshinshiba: *Klainedoxa ovalifolia* Verm.; *Simarubac.;* B.-K. (Kasai)
Tshipamupamu: *Scorodophloeus sp.; Caesalpiniac.;* B.-K. (Kasai)
Tshipanda: *Spondianthus preussi* Engl.; *Euphorbiac.;* B.-K. (Katanga)
Tshisiei: *Burkea africana* Hook.; *Caesalpiniac.;* Togo.
Tshunkundu: *? Entandrophragma sp.; Meliac.;* B.-K.

Tsiandalana: *Dalbergia tricolor* Drake; *Papilionac.;* Mad.
„ : *Acacia bellula* Drake; *Mimosac.;* Mad.
Tsikibi: *Illicium anisatum* L.; *Magnoliac.;* Jap.
Tsilolona: *Bruguiera gymnorrhiza* Lam.; *Rhizophorac.;* Mad.
Tsim: *Pentaclethra eetveldeana* de Wild. et Th. Dur.; *Mimosac.;* Gab.
„ py: *Citrus limonum* Risso; *Rutac.;* Annam.
Tsisombo: *Scyphocephalium ochocoa* Warb.; *Myristicac.;* Gab.
Tsitohizanolomalaina: *Albizzia boinensis* R. Vig.; *Mimosac.;* Mad.
Tsitolona: *Bruguiera gymnorrhiza* Lam.; *Rhizophorac.;* Mad.
Tsitolony: *Bruguiera gymnorrhiza* Lam.; *Rhizophorac.;* Mad.
Tsjampa: *Cananga odorata* Hook. f. et Thoms.; *Anonac.;* Java.
Tso: *Xylopia aethiopica* A. Rich.; *Anonac.;* Togo.
Tsogo: *Scyphocephalium ochocoa* Warb.; *Myristicac.;* Gab.
Tsongati: *Alstonia gilleti* de Wild.; *Apocynac.;* B.-K. (Mayumbe)
Tsongoti: *Alstonia gilleti* de Wild.; *Apocynac.;* B.-K. (Mayumbe)
Tsou: *Entada abyssinica* Steud.; *Mimosac.;* Kam.
Tsoukoby: *Scotellia kamerunensis* Gilg; *Flacourtiac.;* Elf.
Tsoumbou: *Parkia klainei* Pierre; *Mimosac.;* Kam., Gab.
„ : *Piptadenia sp.; Mimosac.;* Gab.
Tsubaki: *Thea japonica* Nois.; *Theac.;* Jap.
„ , Hime-: *Thea japonica* Nois. *var. spontanea* Mak.; *Theac.;* Jap.
„ , „ -: *Schima confertiflora* Merr.; *Theac.;* Form.
Tsuga: *Tsuga sieboldi* Carr.; *Conifer.;* Jap.
„ de l'Ouest: *Tsuga heterophylla* Sarg.; *Conifer.;* w. Kan.
„ „ Patton: *Tsuga mertensiana* Carr.; *Conifer.;* w. Kan.
„ du Canada: *Tsuga canadensis* Carr.; *Conifer.;* ö. Kan.
Tsuge: *Buxus japonica* Muell. Arg.; *Buxac.;* Jap.
Tsugu: *Arenga saccharifera* Labill.; *Palmac.;* Jap.
Tsukubanegashi: *Quercus sessilifolia* Blume; *Fagac.;* Jap.
Tsung deuk: *Phyllostachys nigra* Munro; *Gramin.;* Ch. (Fukien)
Tsuribana: *Evonymus oxyphylla* Miq.; *Celastrac.;* Jap.
Tsuru-umemodoki: *Celastrus articulatus* Thunb.; *Celastrac.;* Jap.
Tsy yé choú: *Aesculus hippocastanum* L.; *Hippocastanac.;* Ch.
Tsze poh: *Juniperus chinensis* L.; *Conifer.;* Ch.
Ttâhî: *Prosopis juliflora* DC.; *Mimosac.;* Mex.
Ttzacthûmi: *Schinus molle* L.; *Anacardiac.;* Mex.
Ttzacthunni: *Schinus molle* L.; *Anacardiac.;* Mex.
Tuá: *Bischofia javanica* Blume; *Euphorbiac.;* Phil. (Rizal, Laguna)
Tuáan: *Kingiodendron alternifolium* Merr.; *Papilionac.;* Phil. (Misamis)
Tuai: *Bischofia javanica* Blume; *Euphorbiac.;* Phil. (Bataan, Albay, Mindoro)
Tuak-tuak: *Sideroxylon ferrugineum* Hook.; *Sapotac.;* Malay.
Tualang: *Koompassia excelsa* Taub., *K. parviflora* Prain; *Caesalpiniac.;* Malay.
Tuáling-bakúlau: *Cassia javanica* L.; *Caesalpiniac.;* Phil. (Zambales)
Tuanga: *? Maesobotrya stapfiana* Beille; *Euphorbiac.;* Elf.
Tuar: *Wrightia tomentosa* Roem. et Schultes; *Apocynac.;* n. O.-Ind.
Tuart (e): *Eucalyptus gomphocephala* A. DC.; *Myrtac.;* w. Au.
Tuatuka: *Stereospermum chelonioides* DC.; *Bignoniac.;* sw. O.-Ind. (Bombay)
Tuba: *Pterygopodium balsamiferum* Verm.; *Papilionac.;* B.-K. (Mayumbe)
„ merah: *Derris malaccensis* Prain; *Papilionac.;* Malay.
„ puteh: *Derris elliptica* Bth.; *Papilionac.;* Malay.
„ rabut: *Derris malaccensis* Prain *var. sarawakensis* ?; *Papilionac.;* Malay.
„ root: *Derris elliptica* Bth., *D. polyantha* Miq., *D. sp.; Papilionac.;* Malay.
Tubhás: *Litsea glutinosa* C. B. Rob.; *Laurac.;* Phil. (Batanes)
Tubroos: *Enterolobium cyclocarpum* Griseb.; *Mimosac.;* br. Hond.
Tuca: *Bertholletia excelsa* Bonpl., *B. nobilis* Miers; *Lecythidac.;* Bras.
Tucan-calao: *Aglaia clarki* Merr.; *Meliac.;* Phil.
Tucáng-cálao: *Aglaia clarki* Merr.; *Meliac.;* Phil.
Tucano: *Vochysia tucanorum* Mart.; *Vochysiac.;* Bras.
Tucib: *Anona cherimola* Mill.; *Anonac.;* br. Hond.
Tucum: *Astrocaryum aculeatum* G. F. W. Meyer; *Palmac.;* Bras.
Tucuma: *Astrocaryum aculeatum* G. F. W. Meyer; *Palmac.;* br. Gu.
Tud-wan: *Ixora laxiflora* Smith; *Rubiac.;* Lib.

Tu-eh: *Ostryocarpus riparius* Hook. f.; *Papilionac.;* Lib.
Tuél: *Bischofia javanica* Blume; *Euphorbiac.;* Phil. (Abra)
Tuenga: *Maesobotrya stapfiana* Beille; *Euphorbiac.;* Elf.
Tugás: *Vitex parviflora* Juss., *V. turczaninowi* Merr.; *Verbenac.;* Phil.
Tugáui: *Pometia pinnata* Forst.; *Sapindac.;* Phil. (Camarines, Masbate, Ticao)
„ : *Shorea polysperma* Merr.; *Dipterocarpac.;* Phil. (Camarines)
Tugge: *Aquilaria agallocha* Roxb.; *Thymelaeac.;* Burma.
Tugher: *Acacia adansoni* Guill. et Perr.; *Mimosac.;* n. O.-Af. (it. Somali)
Tugli: *Albizzia amara* Boivin; *Mimosac.;* O.-Ind. (Bombay)
Tugui: *Ocotea rodioei* Mez; *Laurac.;* br. Gu.
Tugul: *Ocotea rodioei* Mez; *Laurac.;* br. Gu.
Tuhidja: *Terminalia ivorensis* Chev.; *Combretac.;* Elf.
Tui: *Dolichandrone spathacea* K. Schum.; *Bignoniac.;* Phil.
Tuibesso: *Baphia pubescens* Hook. f., *B. sp.; Papilionac.;* Elf.
Tuinynelli: *Chrysophyllum obovatum* G. Don; *Sapotac.;* Lib.
Tukang-kalau: *Aglaia clarki* Merr.; *Meliac.;* Phil.
Tuki: *Diospyros ebenum* Koenig; *Ebenac.;* O.-Ind. (Madras)
Tukla: *Mallotus philippinensis* Muell. Arg.; *Euphorbiac.;* n. O.-Ind. (ö. Nepal)
Tukubane-gasi: *Quercus paucidentata* Franch.; *Fagac.;* Form.
Túkud-lángit: *Polyscias nodosa* Seem.; *Araliac.;* Phil. (Bataan)
Tukusi-byakusin: *Juniperus chinensis* L. *var. tsukushiensis* Masamune; *Conifer.;* Form.
Tula: *Sterculia alata* Roxb.; *Sterculiac.;* Beng.
Tulang daing: *Millettia atropurpurea* Bth.; *Papilionac.;* Malay.
„ putih: *Aphania paucijuga* Radlk.; *Sapindac.;* Malay.;
Tulé: *Glossolepis klainei* Pierre; *Bignoniac.;* Gab.
Tulipán: *Malvaviscus grandiflorus* H. B. K.; *Malvac.;* br. Hond.
Tulip flower: *Stenocarpus sinuatus* Endl.; *Proteac.;* Au.
„ tree: *Thespesia populnea* Correa; *Malvac.;* O.-Ind.
„ „ : *Liriodendron tulipifera* L.; *Magnoliac.;* ö. N.-Am.
„ „ : *Hibiscus tiliaceus* L.; *Malvac.;* Ant.
„ „ : *Spathodea campanulata* Beauv.; *Bignoniac.;* Nig., Go.
„ „ , african: *Spathodea campanulata* Beauv.; *Bignoniac.;* Nig., Go.
„ wood: *Liriodendron chinense* Sarg.; *Magnoliac.;* Ch.
„ „ : *Dalbergia nigra* Allem.; *Papilionac.;* nö. Bras.
„ „ : *Harpullia pendula* Planch.; *Sapindac.;* ö. Au.
„ „ , Burma-: *Dalbergia oliveri* Gamble; *Papilionac.;* Burma.
„ , yellow: *Hemicyclia australasica* Muell. Arg.; *Euphorbiac.;* Au.
Tulipero de Virginia: *Liriodendron tulipifera* L.; *Magnoliac.;* ö. N.-Am.
Tulipier: *Liriodendron tulipifera* L.; *Magnoliac.;* ö. N.-Am.
„ : *Spathodea campanulata* Beauv.; *Bignoniac.;* Kam., Gab.
„ , Laurier-: *Magnolia grandiflora* L.; *Magnoliac.;* USA.
„ d'Afrique: *Spathodea campanulata* Beauv.; *Bignoniac.;* Gab.
„ d'Australie: *Harpullia pendula* Planch., *H. hilli* F. v. M.; *Sapindac.;* Au.
„ de Virginie: *Liriodendron tulipifera* L.; *Magnoliac.;* ö. USA.
„ du Gabon (f): *Spathodea campanulata* Beauv.; *Bignoniac.;* Kam.
Tulla: *Tetrameles nudiflora* R. Br.; *Datiscac.;* O.-Ind. (Assam)
Tulpenbaum: *Liriodendron tulipifera* L., *Magnolia spp.; Magnoliac.;* ö. N.-Am.
„ : *Spathodea campanulata* Beauv.; *Bignoniac.;* Togo.
Tulpenboom, amer. Populier-: *Liriodendron tulipifera* L.; *Magnoliac.;* Holl.
Tulpenboomhout: *Liriodendron tulipifera* L.; *Magnoliac.;* N.-Am.
Tulpenholz: *Dalbergia sp.; Papilionac.;* Bras.
Tul-ul: *Calocarpum mammosum* Pierre; *Sapotac.;* Guat.
Tulus: *Anisoptera thurifera* Blume; *Dipterocarpac.;* Phil. (Rizal)
Tuma: *Acacia arabica* Willd.; *Mimosac.;* O.-Ind. (Madras)
Tumatakuru: *Discaria toumatou* Raoul; *Rhamnac.;* N.-Seel.
Tumbalaka: *Hexalobus monopetalus* Engl. et Diels; *Anonac.;* Togo.
Tumbaol: *Apeiba sp.; Tiliac.;* Trin.
Tumbaríla: *Garuga littoralis* Merr. *var. paucijuga* Merr.; *Burserac.;* Phil. (Palawan)
Tumbi: *Diospyros ebenum* Koenig; *Ebenac.;* O.-Ind. (Madras)
Tumbile: *? Ceiba sp., ? Pachira sp., ? Bombax sp.; Bombacac.;* Mex.
Tumbong-uák: *Cyclostemon sp.; Euphorbiac.;* Phil. (Palawan)
Tumbrung: *Nyssa sessiliflora* Hook. f.; *Cornac.;* n. O.-Ind. (ö. Nepal)

Tumi: *Diospyros melanoxylon* Roxb.; *Ebenac.*; O.-Ind. (Madras)
Tumki: *Diospyros tomentosa* Roxb.; *Ebenac.*; O.-Ind. (Central Prov., Gondwana)
„ : *Diospyros melanoxylon* Roxb.; *Ebenac.*; O.-Ind. (Madras)
Tummer: *Diospyros tomentosa* Roxb.; *Ebenac.*; C.-O.-Ind. (Gondwana)
Tumri: *Diospyros tomentosa* Roxb., *D. melanoxylon* Roxb.; *Ebenac.*; O.-Ind. (Gond⸗
„ : *Trewia nudiflora* L.; *Euphorbiac.*; O.-Ind. (Bombay)   [wana⸜
Tumu: *Bruguiera gymnorrhiza* Lam., *Kandelia rheedi* W. et A.; *Rhizophorac.*; Malay.
Tumulubo: *Artocarpus rubrovenia* Warb.; *Morac.*; Phil. (Isa.)
Tumutugáui: *Shorea polysperma* Merr.; *Dipterocarpac.*; Phil. (Camarines)
Tun: *Toona ciliata* M. Roem.; *Meliac.*; Beng.
„ -paw-man: *Cordia myxa* L.; *Borraginac.*; O.-Ind. (Burma)
„ -sa-se: *Acacia catechu* Willd.; *Mimosac.*; Burma.
Tuna: *Opuntia eliator* Mill.; *Cactac.*; Kol.
„ : *Cereus sp.*; *Cactac.*; Arg.
„ : *Pseudocedrela kotschyi* Harms; *Meliac.*; Nig.
Tuncunaré merecá: *Mouriria sp.*; *Melastomatac.*; Bras.
Tunda: *Toona ciliata* M. Roem.; *Meliac.*; Kanar.
Tundu: *Toona ciliata* M. Roem.; *Meliac.*; O.-Ind. (Bombay)
Tung: *Mallotus philippinensis* Muell. Arg.; *Euphorbiac.*; Beng.
„ : *Taxus baccata* L., *Abies pindrow* Spach; *Conifer.*; O.-Ind. (Kaschmir)
„ -ju: *Aleurites cordata* R. Br.; *Euphorbiac.*; subtr. As.
„ -oil tree: *Aleurites cordata* R. Br., *A. spp.*; *Euphorbiac.*; Jap.
„ yu: *Aleurites fordi* Hemsl.; *Euphorbiac.*; Ch.
Tungfiam: *Trewia nudiflora* L.; *Euphorbiac.*; n. O.-Ind. (ö. Nepal)
Tungu: *Pistacia integerrima* Stewart; *Anacardiac.*; nw. O.-Ind. (Punjab)
Tungúd: *Ceriops tagal* C. B. Rob.; *Rhizophorac.*; Phil.
Tungúg: *Ceriops tagal* C. B. Rob., *C. roxburghiana* Arn.; *Rhizophorac.*; Phil.
Tuni: *Toona ciliata* M. Roem.; *Meliac.*; Beng.
Tunki: *Diospyros melanoxylon* Roxb.; *Ebenac.*; O.-Ind. (Madras)
Tunna: *Toona ciliata* M. Roem.; *Meliac.*; O.-Ind.
Tuno: *Castilla fallax* Cook; *Morac.*; Nic.
Tunsi: *Taxus baccata* L.; *Conifer.*; n. O.-Ind. (ö. Nepal)
Túnt: *Morus alba* L., *M. serrata* Roxb.; *Morac.*; n. O.-Ind. (Dehra Dun).
Tuntri: *Morus alba* L.; *Morac.*; n. O.-Ind. (United Prov.)
Tunu: *Castilla fallax* Cook, *C. sp.*; *Morac.*; C.-Am.
Tunua: *Acacia leucophloea* Willd.; *Mimosac.*; O.-Ind. (Madras)
Tuo: *Hymenodictyon floribunda* Robinson; *Rubiac.*; Elf.
Túog: *Petersianthus quadrialatus* Merr.; *Lecythidac.*; Phil.
Túol: *Bischofia javanica* Blume; *Euphorbiac.*; Phil. (Ifug.)
Tupa: *Canthium didymum* Roxb.; *Rubiac.*; w. O.-Ind. (Bombay)
Tupelo (u): *Nyssa silvatica* Marsh., *N. aquatica* L., *N. biflora* Walt.; *Cornac.*; N.-Am.
„ , sour: *Nyssa ogeche* Marsh.; *Cornac.*; USA.
Tupkela: *Randia dumetorum* Lamk.; *Rubiac.*; C.-O.-Ind. (Central Prov.)
Tura: *Ficus djurensis* Warb.; *Morac.*; Togo.
Turamán: *Vitex sp.*; *Verbenac.*; Arg.
„ do alagado: *Vitex sp.*; *Verbenac.*; Bras.
Turamira: *Humiria sp.*; *Humiriac.*; Bras.
Turcre: *Citrus medica* L.; *Rutac.*; Pers.
Turi: *Humiria sp.*; *Humiriac.*; fr. Gu.
Turingi: *Albizzia amara* Boivin; *Mimosac.*; O.-Ind. (Madras)
Turka vepa: *Melia azedarach* L.; *Meliac.*; O.-Ind. (s. Deccan)
Turkey gill: *Pithecolobium arboreum* Urban; *Mimosac.*; br. Hond.
„ victuals: *Morinda panamensis* Seem.; *Rubiac.*; br. Hond.
Turma de perro: *Tabernaemontana sp.*; *Apocynac.*; Bras.
Turnipwood: *Akania hilli* Hook. f.; *Sapindac.*; Au.
Turpentine (e): *Syncarpia laurifolia* Tenore; *Myrtac.*; Au.
„ , Fraser Island-: *Syncarpia hilli* Bailey; *Myrtac.*; ö. Au. (Fraserinsel)
„ , Scrub-: *Rhodamnia trinervis* Blume; *Myrtac.*; Au.
Turtle bone: *Croton sp.*; *Euphorbiac.*; br. Hond.
„ „ : *Inga stevensoni* Standl.; *Mimosac.*; br. Hond.
„ „ : *Zygia latifolia* St. Hil., *Z. pecki* Britt. et Rose; *Mimosac.*; br. Hond.
.. „ . Bastard-: *Calyptranthes belizensis* Standl.; *Myrtac.*; br. Hond.

Turtle bough, green: *Laguncularia racemosa* Gaertn.; *Combretac.;* Bah.
Turu: *Oenocarpus bacaba* Mart.; *Palmac.;* br. Gu.
„ dogo: *Albizzia gigantea* Chev.; *Mimosac.;* Elf.
„ „ bossolo: *Albizzia gigantea* Chev.; *Mimosac.;* Elf.
Turucasa: *Porliera hygrometrica* Ruiz et Pav.; *Zygophyllac.;* Peru.
Turukop bumi: *Cassia nodosa* Ham.; *Caesalpiniac.;* Malay.
Tururibi: *Lasiosiphon sp.; Thymelaeac.;* n. Nig. (Kontagora)
Tusca: *Acacia farnesiana* Willd., *A. lutea* N. L. Britton; *Mimosac.;* Arg.
„ : *Acacia macrantha* Humb. et Bonpl. *var. moniliformis* Hieron.; *Mimosac.;* Arg.
Tusimá: *Vernonia senegalensis* Less.; *Composit.;* Togo.
Tút: *Morus alba* L.; *Morac.;* nw. O.-Ind. (Punjab)
„ : *Morus serrata* Roxb.; *Morac.;* n. O.-Ind. (Dehra Dun)
„ : *Morus laevigata* Wall.; *Morac.;* n. O.-Ind. (United Prov.)
Tútri: *Morus alba* L.; *Morac.;* n. O.-Ind. (United Prov.)
Tuʒ: *Attalea cohune* Mart.; *Palmac.;* br. Hond.
Tuwe: *Bischofia javanica* Blume; *Euphorbiac.;* Phil. (Lepanto)
Tuwi: *Dolichandrone spathacea* K. Schum.; *Bignoniac.;* Phil.
Twafoyeden: *Harrisonia occidentalis* Engl.; *Simarubac.;* Go.
Twar: *Bauhinia retusa* Ham.; *Caesalpiniac.;* sw. Beng. (Orissa)
Tweappia: *Garcinia kola* Heckel; *Guttifer.;* Go.
Tweeabiri: *? Diospyros sp.; Ebenac.;* Go.
Tweneduro: *Cordia irvingi* Baker; *Borraginac.;* Go.
Twi: *Koordersiodendron pinnatum* Merr.; *Anacardiac.;* Phil. (Negrosinsel)
„ : *Funtumia africana* Stapf, *F. elastica* Stapf; *Apocynac.;* Elf.
Tyanká: *Limonia warneckei* Engl.; *Rutac.;* Togo.
Tyentyém: *Celtis integrifolia* Lam.; *Ulmac.;* Togo.
Tyenyeolo: *Crossopteryx africana* K. Schum.; *Rubiac.;* Togo.
Tyetyebu: *Lannea acida* A. Rich.; *Anacardiac.;* Togo.
Tyttebar: *Vaccinium vitis-idaea* L.; *Ericac.;* Norwegen, Dänemark.
Tzagwodo: *Sterculia cinerea* A. Rich.; *Sterculiac.;* br. O.-Af. (Uganda)
Tzapotl: *Calocarpum mammosum* Pierre; *Sapotac.;* Mex.
Tzicozapotl: *Achras sapota* L.; *Sapotac.;* Mex.
Tzim-ché: *Pithecolobium unguis-cati* Bth.; *Mimosac.;* Mex.
Tzinacanquahuitl: *Erythrina sp.; Papilionac.;* Mex.
Tzin-ché: *Pithecolobium unguis-cati* Bth.; *Mimosac.;* Mex.
Tziritsequa: *Prosopis juliflora* DC.; *Mimosac.;* Mex.
Tzogar: *Morinda tinctoria* Roxb.; *Rubiac.;* O.-Ind. (Madras)
Tzompantle: *Erythrina sp.; Papilionac.;* Mex.
Tzompantli: *Erythrina sp.; Papilionac.;* Mex.
Tzopilotl: *Swietenia sp.; Meliac.;* Mex.
Tzopilotlzontecomatl: *Swietenia sp.; Meliac.;* Mex.

U

U deuk: *Phyllostachys nigra* Munro; *Gramin.;* Ch. (Fukien)
Uacá: *Chrysophyllum cainito* L.; *Sapotac.;* Bras.
Uachuá: *Saccoglottis guianensis* Bth.; *Humiriac.;* n. Bras. (Amaz.)
Uacima da praia: *Hibiscus tiliaceus* L.; *Malvac.;* Bras.
„ do campo: *Luehea sp.; Tiliac.;* Bras.
Uakátan: *Rhizophora conjugata* L.; *Rhizophorac.;* Phil. (Mindoro)
„ : *Alphitonia zizyphoides* A. Gray; *Rhamnac.;* Phil.
Ua-nassu: *Orbignia martiana* Barb. Rodr.; *Palmac.;* Bras.
Uaó: *Celtis luzonica* Warb., *C. philippinensis* Blco.; *Ulmac.;* Phil. (Surigao)
Uapiranga: *Mouriria apiranga* Spruce; *Melastomatac.;* Bras. (Tapajoz)
Uapuim-assú: *Ficus anthelmintica* Mart.; *Morac.;* Bras. (Amaz.)
Uarié: *Dialium guineense* Willd.; *Caesalpiniac.;* Elf.
Uas-uása: *Parinarium corymbosum* Blume; *Rosac.;* Phil. (Cagayan)
Uaspu mundi: *Adina cordifolia* Hook. f.; *Rubiac.;* C.-O.-Ind. (Gondwana)
Uassacú: *Hura crepitans* L.; *Euphorbiac.;* Bras.
Uayam-ché: *Hirtella americana* L.; *Rosac.;* br. Hond.
Uba coupari: *Platonia insignis* Mart.; *Guttifer.;* Bras.

Ubajai ? cerezo: *Myrcianthes edulis* Berg; *Myrtac.;* tr. Am.
Ubajay: *Myrcianthes edulis* Berg.; *Myrtac.;* S.-Am.
Ubam: *Albizzia zygia* MacBride, *Piptadenia africana* Hook. f.; *Mimosac.;* Nig.
Ubamegashi: *Quercus ilex* L. var. *phillyraeoides* Franch.; *Fagac.;* Jap., Form.
Ubamugashi: *Quercus ilex* L. var. *phillyraeoides* Franch.; *Fagac.;* Jap.
Uban: *Planchonia spectabilis* Merr.; *Lecythidac.;* Phil. (Tayabas)
Ubánan: *Shorea palosapis* Merr.; *Dipterocarpac.;* Phil. (Agusan)
Ubara: *Baphia nitida* Lodd., *B. pubescens* Hook.; *Papilionac.;* Nig.
Ubatan: *Astronium fraxinifolium* Schott; *Anacardiac.;* Bras.
„ amarello: *Astronium sp.; Anacardiac.;* Bras.
Ubatão: *Astronium gracile* Engl., *A. sp.; Anacardiac.;* Bras. (Sao P)
? „ : *Vantanea paniculata* Urb. var. *parvifolia* Hoehne; *Humiriac.;* Bras. (Sao P)
Ubatinga: *Luehea divaricata* Mart., *L. paniculata* Mart.; *Tiliac.;* Bras. (Sao P)
„ : *Eugenia sp.; Myrtac.;* Bras. (Sao P)
Ubero: ? *Coccoloba caracasana* Meissn.; *Polygonac.;* Pan.
Ubién: *Artocarpus cumingiana* Tréc.; *Morac.;* Phil. (Ilocano)
„ : *Artocarpus lamellosa* Blco., *A. rubrovenia* Warb.; *Morac.;* Phil. (Pangasinan)
Ubilesa: *Entandrophragma cylindricum* Sprague; *Meliac.;* Nig., Go.
Ubiri: *Terminalia ivorensis* Chev.; *Combretac.;* Nig.
Ubolu-mili: *Mitragyne macrophylla* Hiern; *Rubiac.;* Nig.
Uburu: *Mitragyne macrophylla* Hiern; *Rubiac.;* Nig.
Ucar: *Terminalia buceras* Bailey; *Combretac.;* P.-R.
Uche: *Fagara xanthoxyloides* Lam.; *Rutac.;* Togo.
Uchi: *Saccoglottis verrucosa* Ducke; *Humiriac.;* Bras.
„ -curua: *Saccoglottis cuspidata* Urban; *Humiriac.;* n. Bras. (Amaz.)
„ -pucú: *Saccoglottis uchi* Huber; *Humiriac.;* n. Bras. (Amaz.)
„ -rana: *Couepia sp.; Rosac.;* Bras.
„ „ : *Andira retusa* H. B. K.; *Papilionac.;* Bras.
„ „ : *Saccoglottis amazonica* Mart.; *Humiriac.;* n. Bras. (Amaz.)
„ „ : *Vantanea cupularis* Huber, *V. guianensis* Aubl.; *Humiriac.;* Bras. (Amaz.)
Uchpa-caspi: *Faranea candelabrum* ?; *Rubiac.;* Peru.
Uchypucú: *Saccoglottis uchi* Huber; *Humiriac.;* nö. Bras. (Pará)
Uchyrana: *Couepia glaucescens* Spruce; *Rosac.;* Bras.
Ucle: *Cereus sp.; Cactac.;* Arg.
Ucuhubá: *Virola sebifera* Aubl.; *Myristicac.;* Bras.
Ucuiba: *Virola sp.; Myristicac.;* Bras.
Ucuquirana: *Ecclinusa balata* Ducke, *E. spuria* Ducke; *Sapotac.;* n. Bras. (Amaz.)
Ucuúba: *Virola surinamensis* Warb.; *Myristicac.;* Bras.
„ branca: *Virola sp.; Myristicac.;* Bras.
„ vermelha: *Virola sp.; Myristicac.;* Bras.
Udal: *Sterculia villosa* Roxb.; *Sterculiac.;* n. O.-Ind. (United Prov.)
Udar: *Sterculia villosa* Roxb.; *Sterculiac.;* n. O.-Ind. (United Prov.)
Udayú: *Celtis philippensis* Blco.; *Ulmac.;* Phil. (Surigao)
Udé: *Stereospermum xylocarpum* Bth. et Hook. f.; *Bignoniac.;* sw. O.-Ind. (Coorg)
Udi: *Stereospermum xylocarpum* Bth. et Hook. f.; *Bignoniac.;* sw. O.-Ind. (Coorg)
Udika: *Irvingia gabonensis* Baill.; *Simarubac.;* Gab.
Udis: *Alnus nepalensis* D. Don; *Betulac.;* n. O.-Ind. (Nepal)
Udivi-nima: *Atalantia monophylla* Correa; *Rutac.;* O.-Ind. (Madras)
Uella: *Abutilon vitifolium* Presl; *Malvac.;* Chile.
Ufi: *Piptadenia africana* Hook. f.; *Mimosac.;* Nig.
Ufie: *Baphia nitida* Lodd.; *Papilionac.;* Nig.
Ufili: *Scorodophloeus zenkeri* Harms; *Caesalpiniac.;* B.-K.
Ufiri: *Scorodophloeus sp.; Caesalpiniac.;* B.-K. (Stanleyville)
Ugad: *Tetrameles nudiflora* R. Br.; *Datiscac.;* w. O.-Ind. (Bombay)
Ugar: *Aquilaria agallocha* Roxb.; *Thymelaeac.;* Beng.
Ugáu: *Myristica nitida* Merr.; *Myristicac.;* Phil. (Camarines)
„ : *Garcinia lateriflora* Blume; *Guttifer.;* Phil.
„ : *Diospyros mindanaënsis* Merr.; *Ebenac.;* Phil.
Ugbeberi: *Lophira alata* Banks; *Ochnac.;* Nig.
Uging: *Cratoxylon cochinchinense* Blume; *Guttifer.;* Phil. (Cagayan)
Ugíngen: *Cratoxylon cochinchinense* Blume; *Guttifer.;* Phil. (Cagayan)
Ugob: *Artocarpus communis* Forst.; *Morac.;* Phil. (Bicol)

Uguru: *Excoecaria agallocha* L.; *Euphorbiac.*; Beng.
Uhee-tee: *Dialium divaricatum* Vahl; *Caesalpiniac.*; br. Hond.
Uhie: *Baphia nitida* Lodd.; *Papilionac.*; Nig.
Uiakiá: *Pometia pinnata* Forst.; *Sapindac.*; Phil. (Mindoro)
Uique: *Calocarpum mammosum* Pierre; *Sapotac.*; Bras.
Uísak: *Neonauclea spp.*; *Rubiac.*; Phil. (Rizal, Laguna)
Ujio: *Irvingia gabonensis* Baill.; *Simarubac.*; Nig.
Ujiri: *Irvingia gabonensis* Baill.; *Simarubac.*; Nig.
Ujuju: *Vernonia conferta* Bth.; *Composit.*; Nig.
Ukan: *Tamarix articulata* Vahl; *Tamaricac.*; nw. O.-Ind. (Punjab)
Ukana: *Pentaclethra macrophylla* Bth.; *Mimosac.*; Nig.
Uke: *Acacia pennaza* Willd.; *Mimosac.*; Nig.
Ukhan: *Tamarix articulata* Vahl; *Tamaricac.*; O.-Ind.
Ukim: *Ceiba pentandra* Gaertn., *Bombax buonopozense* Beauv.; *Bombacac.*; Nig.
Ukloba: *Chlorophora excelsa* Bth. et Hook.; *Morac.*; Togo.
Ukola: *Dumoria africana* Dubard; *Sapotac.*; sp. Guin.
Ukolo: *Mimusops africana* Lecomte; *Sapotac.*; Gab.
Ukonk: *Ficus asperifolia* Miq.; *Morac.*; Nig.
Ukpa: *Baphia nitida* Lodd.; *Papilionac.*; Nig.
Ukpakeka: *Hymenostegia afzeli* Harms; *Caesalpiniac.*; Nig.
Ukudo: *Coula edulis* Baill.; *Olacac.*; Gab.
Ukwok: *Ficus asperifolia* Miq.; *Morac.*; Nig.
Ul: *Mangifera indica* L.; *Anacardiac.*; ö. Beng.
Ulaga: *Genista scorpius* DC.; *Papilionac.*; Sp.
Uláian: *Quercus llanosi* A. DC., *Q. sp.*; *Fagac.*; Phil.
Ulaló: *Celtis luzonica* Warb., *C. philippensis* Blco.; *Ulmac.*; Phil. (Capiz)
Ulanda: *Amyris sp.*; *Rutac.*; Kol.
Ularóg: *Celtis luzonica* Warb., *C. philippensis* Blco.; *Ulmac.*; Phil. (Catanduanes)
Ulazo: *Holoptelea grandis* Mildbr.; *Ulmac.*; Nig. (Benin)
Ule: *Castilla elastica* Cerv.; *Morac.*; Guat., br. Hond., Hond.
„ : *Castilla fallax* Cook, *C. panamensis* Cook; *Morac.*; Pan.
„ de charro: *Castilla spp.*; *Morac.*; Guat.
„ „ lagrimo: *Castilla spp.*; *Morac.*; Guat.
„ -ule: *Castilla sp.*; *Morac.*; C.-Am.
Ulezapote: *Couepia sp.*; *Rosac.*; Salv.
Uli: *Mangifera indica* L.; *Anacardiac.*; n. O.-Ind.
„ : *Agathis alba* Foxw.; *Conifer.*; Phil. (Zambales)
Ulin: *Eusideroxylon zwageri* Teijsm. et Binn.; *Laurac.*; ö. Borneo.
Uling: *Cratoxylon blancoi* Blume, *C. celebicum* Blume; *Guttifer.*; Phil. (Abra)
„ : *Fagraea cochinchinensis* Chev.; *Loganiac.*; Phil.
Ulingon: *Cratoxylon celebicum* Blume; *Guttifer.*; Phil. (Surigao, Agusan)
Ulíuan: *Cinnamomum mercadoi* Vidal; *Laurac.*; Phil. (Cagayan)
Ulme (d): *Ulmus spp.*; *Ulmac.*; Eu.
„ , Berg-: *Ulmus montana* With.; *Ulmac.*; Eu.
„ , Feld-: *Ulmus campestris* Spach; *Ulmac.*; Eu.
„ , Flatter-: *Ulmus effusa* Willd.; *Ulmac.*; Mitt.-Eu.
„ , Hasel-: *Ulmus montana* With.; *Ulmac.*; Mitt.-Eu.
„ , Trauben-: *Ulmus racemosa* Thomas; *Ulmac.*; N.-Am.
Ulmo: *Eucryphia cordifolia* Cav.; *Eucryphiac.*; s. Chile.
Ulozapote: *Lucuma palmeri* Fernald; *Sapotac.*; Salv.
„ : *Couepia sp.*; *Rosac.*; Salv.
Uluhivu: *Raphia monbuttorum* Drude; *Palmac.*; br. O.-Af. (Uganda)
Ulula: *Acacia sp.*; *Mimosac.*; D.-O.-Af. (Tabora)
Ulundu: *Chlorophora excelsa* Bth. et Hook.; *Morac.*; B.-K. (Katanga)
Uluzapote: *Couepia sp.*; *Rosac.*; Salv.
Ulvi: *Terminalia paniculata* Roth; *Combretac.*; O.-Ind. (Bombay)
Umar: *Ficus glomerata* Roxb.; *Morac.*; C.-O.-Ind. (Central Prov.)
Umary: *Paraqueiba paraënsis* Ducke; *Icacinac.*; Par., Arg.
Umaryrana: *Couepia subcordata* Hook. f.; *Rosac.*; Bras.
Umbaba: *Calodendron capense* Thunb.; *Rutac.*; S.-Af.
Umbar: *Ficus glomerata* Roxb.; *Morac.*; C.-O.-Ind. (Central Prov.)
Umbomvo: *Calodendron capense* Thunb.; *Rutac.*; S.-Af.

Umbovani: *Elaeodendron croceum* DC.; *Celastrac.;* S.-Af.
Umbrella tree (e): *Thespesia popuinea* Correa; *Malvac.;* Tr.
    „   „   :   *Magnolia tripetala* L.; *Magnoliac.;* ö. USA.
    „   „   :   *Musanga smithi* R. Br.; *Morac.;* W.-Af.
    „   „   :   *Panax murrayi* F. v. M.; *Araliac.;* Au.
    „   „   , large-leaved: *Magnolia macrophylla* Michx.; *Magnoliac.:* USA.
Umbri: *Ficus glomerata* Roxb.; *Morac.;* C.-O.-Ind. (Central Prov.)
Umbu: *Phytolacca dioica* L.; *Phytolaccac.;* Arg.
Um bukhesa: *Crataeva religiosa* Forst.; *Capparidac.;* s. Sudan.
Umburana: *Torresea cearensis* Allem.; *Papilionac.;* Bras.
   „ cheira: *Torresea cearensis* Allem.; *Papilionac.;* Bras.
Umdagana: *Apodytes dimidiata* E. Meyer; *Olacac.;* S.-Af.
Umdus: *Parkia filicoidea* Bth.; *Mimosac.;* s. Sudan.
Uıngesisa: *Cussonia umbellifera* Sond.; *Araliac.;* S.-Af.
Umhlagela: *Drypetes gerardi* Hutch.; *Euphorbiac.;* S.-Af.
Umiri: *Humiria floribunda* Mart., *H. balsamifera* Aubl.; *Humiriac.;* Sur.
   „   :   *Humiria floribunda* Mart.; *Humiriac.;* Bras.
   „ balsamo: *Humiria* ? *balsamifera* Aubl.; *Humiriac.;* fr. Gu.
Umirirana: *Qualea retusa* Warm., *Q. wittrocki* Malme; *Uochysiac.;* Bras.
Umiry: *Humiria floribunda* Mart.; *Humiriac.;* Bras., Gu.
   „ de casca cheiro: *Humiria sp.; Humiriac.;* Bras.
   „   „ cheiro: *Humiria sp.; Humiriac.;* Bras.
Umkamba: *Acacia spp.; Mimosac.;* O.-Af.
   „   :   *Afzelia quanzensis* Welw.; *Caesalpiniac.;* s. Rhod.
Umkulu: *Elaeodendron croceum* DC.; *Celastrac.;* S.-Af.
Umnoiso: *Curtisia faginea* Ait.; *Cornac.;* S.-Af. (Natal)
Umnunquazi: *Zanthoxylum capense* Harv.; *Rutac.;* S.-Af.
Umowenek: *Lophira procera* Chev.; *Ochnac.;* Nig.
Umpanzite: *Cryptocarya latifolia* Sond.; *Laurac.;* S.-Af.
Umpapa: *Baikiaea plurijuga* Harms; *Caesalpiniac.;* port. W.-Af. (Angola)
Umpenek: *Lophira procera* Chev.; *Ochnac.;* Nig.
Umpeque: *Ximenia americana* L.; *Olacac.;* port. W.-Af. (Angola-Sambesi)
Umra: *Ficus glomerata* Roxb.; *Morac.;* nw. O.-Ind. (Kumaon)
Um shutter: *Kigelia aethiopica* Dcne.; *Bignoniac.;* s. Sudan.
Um sisi: *Ficus sp.; Morac.;* Sudan (Kordofan)
Umtata: *Ptaeroxylon utile* Eck. et Zeyh.; *Sapindac.;* S.-Af.
Umtiza: *Umtiza listeriana* Th. Sim; *Papilionac.;* S.-Af.
Umtobotie: *Spirostachys africana* Sond.; *Euphorbiac.;* S.-Af. (Natal)
Umtomboti: *Excoecaria africana* Muell. Arg.; *Euphorbiac.;* O.-Af.
Umukengeri: *Rhus villosa* L. f.; *Anacardiac.;* br. O.-Af. (Uganda)
Umukindo: *Phoenix reclinata* Jacq.; *Palmac.;* br. O.-Af. (Uganda)
Umusambya: *Markhamia platycalyx* Sprague; *Bignoniac.;* br. O.-Af. (Uganda)
Umusogasoga: *Croton macrostachys* A. Rich.; *Euphorbiac.;* br. O.-Af. (Uganda)
Umuvuma: *Ficus eryobotryoides* Kunth et Bouché; *Morac.;* br. O.-Af. (Uganda)
Umuyonza: *Carissa edulis* Vahl; *Apocynac.;* br. O.-Af. (Uganda)
Umvagaz: *Pterocarpus angolensis* DC.; *Papilionac.;* s. Rhod.
Umvangazi: *Pterocarpus angolensis* DC.; *Papilionac.;* s. Rhod.
Umwonron: *Terminalia superba* Engl. et Diels; *Combretac.;* Nig.
Umyeamate: *Ekebergia capensis* Sparrm.; *Meliac.;* S.-Af.
Umzani: *Toddalia lanceolata* Lamk.; *Rutac.;* S.-Af. (Kap, Natal)
Umzimbeet: *Millettia caffra* Meissn.; *Papilionac.;* S.-Af.
Uña de gato: *Genista anglica* L.; *Papilionac.;* Sp.
   „   „   „   :   *Celtis sp.; Ulmac.;* Cuba.
   „   „   „   :   *Pithecolobium unguis-cati* Bth.; *Mimosac.;* Nic.
   „   „   „   :   *Zanthoxylum fagara* Sarg.; *Rutac.;* Kol.
   „   „   „   :   *Solanum lanceifolium* Jacq.; *Solanac.;* Nic.
   „   „   „   :   *Acacia bonariensis* Gill.; *Mimosac.;* Arg., Urug.
   „   „ tigre: *Pithecolobium ligustrinum* Klotzsch; *Mimosac.;* Kol.
Unab: *Zizyphus jujuba* Lamk.; *Rhamnac.;* O.-Ind. (Hind.)
Uñagato: *Zanthoxylon fagara* Sarg.; *Rutac.;* Mex.
Undi: *Calophyllum inophyllum* L.; *Guttifer.;* w. O.-Ind. (s. Bombay)
   „   :   *Sonneratia apetala* Ham.; *Sonneratiac.;* w. O.-Ind. (s. Bombay)

Undipanu: *Myristica attenuata* Wall.; *Myristicac.*; sw. O.-Ind. (Travancore)
Ung: *Dipterocarpus tuberculatus* Roxb.; *Dipterocarpac.*; Burma.
Ungubu: *Staudtia gabonensis* Warb.; *Myristicac.*; Gab.
Ungwe: *Ceiba pentandra* Gaertn.; *Bombacac.*; Lib.
Unha d'anta: *Sterculia sp.*; *Sterculiac.*; Bras.
„   „ : *Sweetia dasycarpa* Bth.; *Papilionac.*; Bras. (Goyaz)
Uník: *Albizzia marginata* Merr.; *Mimosac.*; Phil. (Bataan, Pampanga, Tarlac)
Uníp: *Albizzia marginata* Merr.; *Mimosac.*; Phil. (Bataan, Pampanga, Tarlac)
Unnué bollina: *Xylopia africana* Oliv.; *Anonac.*; W.-Af. (port. Inseln, Principé)
Uno: *Musanga smithi* R. Br.; *Morac.*; Nig.
Unu: *Grewia tiliaefolia* Vahl; *Tiliac.*; O.-Ind. (Madras)
Uolí: *Vitex cienkowskyi* Kotschy et Peyr.; *Verbenac.*; Togo.
Uong deuk: *Phyllostachys nigra* Munro; *Gramin.*; Ch. (Fukien)
Uóos: *Sterculia oblongata* R. Br.; *Sterculiac.*; Phil. (Camarines, Bíkol)
Uós: *Sterculia oblongata* R. Br.; *Sterculiac.*; Phil. (Camarines, Bíkol)
Upa: *Afzelia africana* Smith, *Hymenostegia afzeli* Harms; *Caesalpiniac.*; Nig.
„ : *Irvingia gabonensis* Baill.; *Simarubac.*; Kam.
Upak: *Sterculia cuneata* R. Br.; *Sterculiac.*; Phil. (Pampanga)
Upando: *Panda oleosa* Pierre; *Pandac.*; Gab.
Upas tree: *Antiaris toxicaria* Lesch.; *Morac.*; Malay.
„   „ : *Antiaris africana* Engl.; *Morac.*; Nig.
Upay: *Cordia alba* Roem. et Sch., *C. diversifolia* Pav.; *Borraginac.*; Guat.
Upoo-poma: *Rhizophora mucronata* Lamk.; *Rhizophorac.*; sw. O.-Ind. (Madras)
Upung-upung: *Terminalia pellucida* Presl; *Combretac.*; Phil. (Samar)
Ur: *Minquartia guianensis* Aubl.; *Olacac.*; Pan.
Urai-gasi: *Lithocarpus uraiana* Hayata; *Fagac.*; Form.
Urajirogashi: *Quercus myrsinaefolia* Blume; *Fagac.*; Jap.
Urajiro-inugaya (jap.): *Amentotaxus argotaenia* Pilger; *Conifer.*; Ch., Form.
Urajiroutsugi: *Deutzia hypoleuca* Maxim.; *Saxifragac.*; Jap.
Uram: *Hardwickia pinnata* Roxb.: *Caesalpiniac.*; sw. O.-Ind. (Travancore)
Urape: *Bauhinia multinervia* DC.; *Caesalpiniac.*; Ven.
Urari: *Minquartia sp.*; *Olacac.*; Pan.
Uraróg: *Celtis philippensis* Blco.; *Ulmac.*; Phil. (Camarines)
Urat mata: *Parashorea plicata* Brandis; *Dipterocarpac.*; br. N.-Borneo.
„   „ : *Pentacme contorta* Merr. et Rolfe; *Dipterocarpac.*; br. N.-Borneo.
Urátan: *Celtis philippensis* Blco.; *Ulmac.*; Phil. (Ilocos Norte, Abra)
„ : *Chaetospermum glutinosum* Swingle; *Rutac.*; Phil. (Ilóko)
„ : *Gonocaryum calleryanum* Becc.; *Icacinac.*; Phil. (Ilóko)
Uravu: *Machilus macrantha* Nees; *Laurac.*; sw. O.-Ind.
Uraziro-enoki: *Trema orientalis* Blume; *Ulmac.*; Jap.
Uroe: *Erica arborea* L.; *Ericac.*; Sp.
Urhearike: *Lovoa klaineana* Pierre; *Meliac.*; Nig.
Urhur (e): *Cajanus indicus* Spreng.; *Papilionac.*; Tr.
Uriam: *Bischofia javanica* Blume; *Euphorbiac.*; Assam.
Uricuru: *Scheelea martiana* Burret; *Palmac.*; Bras.
Urihadakaede: *Acer rufinerve* S. et Z.; *Acerac.*; Jap.
Urindeuva: *Astronium urundeuva* Engl.; *Anacardiac.*; Bras. (Sao P)
Uríngon: *Cratoxylon spp.*; *Guttifer.*; Phil. (Masbate)
Uripapa: *Holocalyx balansae* Micheli; *Caesalpiniac.*; Par.
Uripu: *Hopea parviflora* Bedd.; *Dipterocarpac.*; sw. O.-Ind. (Madras)
Urirana: *Hieronyma alchorneoides* Allem.; *Euphorbiac.*; Bras.
Uris: *Koordersiodendron pinnatum* Merr.; *Anacardiac.*; Phil. (Ilocos Norte)
Urísan: *Koordersiodendron pinnatum* Merr.; *Anacardiac.*; Phil. (Cagayan)
Urmè: *Cassia abbreviata* Oliv.; *Caesalpiniac.*; n. O.-Af. (Somali)
Urmu: *Bursera serrata* Colebr.; *Burserac.*; ö. Beng.
Urodibe: *Minquartia guianensis* Aubl.; *Olacac.*; Pan.
Urraco: *Licania platypus* Fritsch; *Rosac.*; Hond.
Urrucay: *Protium sp.*; *Burserac.*; Kol.
Uruazeiro: *Cordia alliodora* Cham.; *Borraginac.*; Bras.
Urubasu: *Funtumia elastica* Stapf; *Apocynac.*; Elf.
Uruca: *Trichilia havanensis* Jacq.; *Meliac.*; C.-R.
Urucu: *Bixa orellana* L.; *Bixac.*; Mex.

Urucuca: *Vochysia oppugnata* Warming; *Vochysiac.;* Bras.
Urucurana: *Hieronyma alchorneoides* Allem.; *Euphorbiac.;* Bras.
     „   :   *Sloanea sp.; Elaeocarpac.;* Bras.
„   mirim: *Hieronyma alchorneoides* Allem.; *Euphorbiac.;* Bras. (Sao P)
„   de. leite: *Hieronyma alchorneoides* Allem.; *Euphorbiac.;* Bras. (Sao P)
„   „ prégo: *Hieronyma alchorneoides* Allem.; *Euphorbiac.;* Bras. (Sao P)
Urucuri: *Scheelea martiana* Burret, *S. huebneri* Burret; *Palmac.;* Bras.
Urundai: *Astronium balansae* Engl., *A. candollei* Engl.; *Anacardiac.;* Par., Arg.
Urunday: *Astronium urundeuva* Engl.; *Anacardiac.;* Arg.
     „   :   *Astronium balansae* Engl., *A. candollei* Engl.; *Anacardiac.;* Par., Arg.
„   blanco: *Diplokeleba floribunda* N. E. Brown; *Sapindac.;* Arg.
„   crespo: *Astronium balansae* Engl., *A. candollei* Engl.; *Anacardiac.;* Par., Arg.
„   -mi: *Astronium balansae* Engl., *A. candollei* Engl.; *Anacardiac.;* Par., Arg.
„   -nra: *Diplokeleba floribunda* N. E. Brown; *Sapindac.;* Arg.
„   -pardó: *Astronium urundeuva* Engl.; *Anacardiac.;* Arg.
„   - „   :   *Astronium balansae* Engl., *A. candollei* Engl.; *Anacardiac.;* Par., Arg.
„   -pichái: *Astronium balansae* Engl., *A. candollei* Engl.; *Anacardiac.;* Par., Arg.
„   -rá: *Nematopogon sp.; Bignoniac.;* Arg.
Urundel: *Astronium urundeuva* Engl.; *Anacardiac.;* Arg.
     „   :   *Astronium balansae* Engl., *A. candollei* Engl.; *Anacardiac.;* Par., Arg.
Urundeuva: *Astronium urundeuva* Engl.; *Anacardiac.;* Bras. (Sao P)
Urundey: *Astronium balansae* Engl., *A. candollei* Engl.; *Anacardiac.;* Par., Arg.
„   crespo: *Astronium urundeuva* Engl.; *Anacardiac.;* S.-Am.
Urung: *Fagraea fragrans* Roxb.; *Loganiac.;* Phil. (Palawan)
Urupagua: *Aveledoa nucifera* Pittier; *Olacac.;* Ven. (Coro)
Urupariba: *Tecoma sp.; Bignoniac.;* Bras.
Urury: *Mouriria pseudo-geminata* Pittier; *Melastomatac.;* Bras.
Urushinoki: *Rhus vernicifera* DC.; *Anacardiac.;* Jap.
Urza: *Eysenhardtia polystachya* Sarg.; *Papilionac.;* Mex. (Otomi)
Usa: *Mimusops pierreana* Engl.; *Sapotac.;* Gab.
Usan: *Terminalia tomentosa* Wight et Arn.; *Combretac.;* Beng.
Ushikoroshi: *Photinia villosa* DC.; *Rosac.;* Jap.
Usi: *Mimusops pierreana* Engl.; *Sapotac.;* Gab.
     „   :   *Cussonia arborea* Hochst.; *Araliac.;* br. O.-Af. (Uganda)
Usiere: *Erythrina senegalensis* DC.; *Papilionac.;* Nig.
Usiu: *Dinochloa scandens* O. Kɣe., *Schizostachyum dielsianum* Merr.; *Gramin.;* Phil.
Usombo: *Elaeis guineensis* Jacq. var.?; *Palmac.;* Gab.
Usoro: *Scyphocephalium ochocoa* Warb.; *Myristicac.;* Gab.
Usozo: *Scyphocephalium ochocoa* Warb.; *Myristicac.;* Gab.
Usto: *Ficus religiosa* L.; *Morac.;* s. Beng.
Ustumri: *Strychnos potatorum* L. f.; *Loganiac.;* C.-O.-Ind. (Gondwana)
Usuba-ké-inubiwa: *Ficus bucheyana* Hook. et Arn. *forma tenuifolium* Sata; *Morac.;*
Usuba-kusunoki: *Machilus acuminatissimum* Kanehira; *Laurac.;* Form.   [Form.
Usuba-tabu: *Machilus acuminatissimum* Kanehira; *Laurac.;* Form.
Utevo: *Desbordesia insignis* Pierre; *Simarubac.;* Gab.
Uthi: *Lannea grandis* Engl.; *Anacardiac.;* O.-Ind. (Madras)
Utis: *Betula alnoides* Ham.; *Betulac.;* O.-Ind. (Garhwal)
     „   :   *Alnus nepalensis* D. Don, *A. nitida* Endl.; *Betulac.;* O.-Ind. (United Prov.)
Utrej: *Citrus medica* Gall.; *Rutac.;* Arab.
Utsugi: *Deutɣia scabra* Thunb.; *Saxifragac.;* Jap.
Utuenyi: *Khaya grandifoliola* C. DC.; *Meliac.;* Nig.
Uu-ti: *Cola cordifolia* R. Br.; *Sterculiac.;* Togo.
Uva: *Dillenia indica* L.; *Dilleniac.;* sw. O.-Ind.
     „   :   *Coccoloba uvifera* Jacq.; *Polygonac.;* W.-Ind., C.-Am.
     „   :   *Conostegia xalapensis* D. Don; *Melastomatac.;* Hond.
„   caleta: *Coccoloba uvifera* L.; *Polygonac.;* Cuba.
„   gomosa: *Cordia alba* Roem. et Schult.; *Borraginac.;* Cuba.
„   de montaña: *Ardisia amplifolia* Standl.; *Myrsinac.;* Hond.
„   „ la playa: *Coccoloba uvifera* Jacq.; *Polygonac.;* Hond.
„   del mar: *Coccoloba uvifera* Jacq.; *Polygonac.;* tr. Am.
Uvando: *Panda oleosa* Pierre; *Pandac.;* Gab.
Uvëï: *Elaeis guineensis* Jacq. var.?; *Palmac.;* Gab.

Uverillo: *Coccoloba laurifolia* Jacq.; *Polygonac.;* sp. W.-Ind., n. S.-Am.
  „ : *Coccoloba obtusifolia* Jacq.; *Polygonac.;* P.-R.
Uvero: *Coccoloba uvifera* Jacq.; *Polygonac.;* Mex., C.-Am., Kol., Cuba.
  „ : *Coccoloba caracasana* Meissn.; *Polygonac.;* n. Kol., Ven. (Guárico)
  „ : *Cordia alba* Roem. et Schult.; *Borraginac.;* Pan.
  „ macho: *Coccoloba caracasana* Meissn.; *Polygonac.;* n. Kol., Ven. (Caracas)
  „ de playa: *Coccoloba uvifera* Jacq.; *Polygonac.;* C.-Am., Ven.
  „ del monte: *Coccoloba sp.; Polygonac.;* Trin.
Uvifero: *Coccoloba uvifera* Jacq.; *Polygonac.;* Mex., C.-Am., Kol., Cuba.
Uvillo: *Coccoloba laurifolia* Jacq.; *Polygonac.;* sp. W.-Ind., n. S.-Am.
  „ : *Coccoloba uvifera* Jacq.; *Polygonac.;* St.-D.
  „ : *Coccoloba obtusifolia* Jacq.; *Polygonac.;* P.-R.
  „ : *Cordia alba* Roem. et Schult.; *Borraginac.;* Pan.
Uvinya: *Klainedoxa gabonensis* Pierre; *Simarubac.;* Gab.
Uvita: *Ardisia amplifolia* Standl.; *Myrsinac.;* Nic.
Uvito: *Icacorea revoluta* Standl.; *Myrsinac.;* Pan.
  „ : *Cordia alba* Roem. et Schultes; *Borraginac.;* n. Kol.
  „ de murciélago: *Cordia alba* Roem. et Schultes; *Borraginac.;* n. Kol.
Uvovo: *Poga oleosa* Pierre; *Rhizophorac.;* Gab.
Uvre: *Oncoba laurina* Warb.; *Flacourtiac.;* Pan.
Uwamizu-sakura: *Prunus grayana* Maxim.; *Rosac.;* Jap.
Uwar yara: *Ficus capensis* Thunb.; *Morac.;* n. Nig.
Uwen: *Mitragyne macrophylla* Hiern; *Rubiac.;* Nig.
Uwowo: *Poga oleosa* Pierre; *Rhizophorac.;* Gab.
Uyók: *Buchanania arborescens* Blume; *Anacardiac.;* Phil. (Benguet)
Uyu: *Monodora myristica* Dunal; *Anonac.;* Togo.
Uzi: *Cylicodiscus gabunensis* Harms; *Mimosac.;* Nig.
Uzu: *Citrus limonum* L.; *Rutac.,* Jap.

V

Vaa: *Macrolobium sp.; Caesalpiniac.;* Elf.
Vaalboom (h): *Terminalia sericea* Burch.; *Combretac.;* S.-Af.
Vaalbos (h): *Tarchonanthus camphoratus* L.; *Composit.;* S.-Af.
Vaal bosch (h): *Terminalia sericea* Burch.; *Combretac.;* s. C.-Af.
Vabbina: *Polyalthia cerasoides* Bth. et Hook. f.; *Anonac.;* O.-Ind. (Bombay)
Vaca-buey: *Curatella americana* L.; *Dilleniac.;* Cuba.
  „ hosca: *Couma sapida* Pittier; *Apocynac.;* Ven.
Vacapou: *Vouacapoua americana* Aubl.; *Caesalpiniac.;* Bras.
Vaciet: *Vaccinium myrtillus* L.; *Ericac.;* Frankreich.
Vácima: *Guazuma ulmifolia* Lam.; *Sterculiac.;* Mex.
Vaco: *Brosimum utile* Pittier; *Morac.;* Ven.
Vacoas marrons: *Pandanus spp.; Pandanac.;* Réunion.
Vaconné-cheddi: *Grewia laevigata* Vahl; *Tiliac.;* O.-Ind. (s. Deccan)
Vacu: *Schmidelia edulis* St. Hil.; *Sapindac.;* Arg.
Vadaduri: *Lecythis grandiflora* Aubl.; *Lecythidac.;* Sur.
Vadari: *Zizyphus jujuba* Lamk.; *Rhamnac.;* O.-Ind.
Vadencarni: *Stereospermum xylocarpum* Bth. et Hook. f.; *Bignoniac.;* sw. O.-Ind.
Vadi: *Ficus bengalensis* L.; *Morac.;* O.-Ind. (Bombay)
Vaga: *Albizzia lebbek* Bth.; *Mimosac.;* sw. O.-Ind.
Vagai: *Albizzia lebbek* Bth.; *Mimosac.;* O.-Ind.
Vagamani: *Tapirira myriantha* Tr. et Pl.; *Anacardiac.;* Pan. (Colon)
Vagei: *Albizzia lebbek* Bth.; *Mimosac.;* O.-Ind. (Madras)
Vahn-chu: *Amphimas pterocarpoides* Harms; *Papilionac.;* Lib.
  „ -mleh-doo: *Duboscia sp.; Tiliac.;* Lib.
  „ -te-ohn: *Octoknema borealis* Hutch. et Dalz.; *Olacac.;* Lib.
  „ -vchn: *Myrianthus serratus* Bth. et Hook. f.; *Morac.;* Lib.
Vai thiêu: *Nephelium lappaceum* L.; *Sapindac.;* Ind.-Ch.
Vaina de espada: *Eugenia axillaris* Willd.; *Myrtac.;* br. Hond.
Vainillo: *Inga leptoloba* Schlecht., *I. xalapensis* Bth.; *Mimosac.;* Mex.
  „ : *Inga leiocarpa* Bth.; *Mimosac.;* Mex.
  „ : *Cassia occidentalis* L.; *Caesalpiniac.;* Mex.

Vainillo: *Cassia emarginata* L.; *Caesalpiniac.;* Guat., Nic.
Vaivai: *Serianthes myriadenia* Planch.; *Mimosac.;* Südsee (Fid.)
Vaiya: *Poeciloneuron indicum* Bedd.; *Guttifer.;* sw. O.-Ind.
Vakenar: *Sterculia villosa* Roxb.; *Sterculiac.;* O.-Ind. (Madras)
Valaiváian: *Radermachera fenicis* Merr.; *Bignoniac.;* Phil. (Batanes)
Vallay kungiliam: *Vateria indica* L.; *Dipterocarpac.;* sw. O.-Ind. (Madras)
Valmurichha: *Filicium decipiens* Thw.; *Sapindac.;* sw. O.-Ind. (Travancore)
Valnod: *Juglans spp.;* *Juglandac.;* Dänemark, Norwegen.
Van rao: *Schleichera trijuga* Willd.; *Sapindac.;* Ind.-Ch. (Annam)
„ wijk's hout (h): *Bolusanthus speciosus* Harms; *Papilionac.;* S.-Af.
Vanamani: *Tapirira myriantha* Tr. et Planch.; *Anacardiac.;* Pan. (Colon)
Vandagué: *Millettia sp.;* *Papilionac.;* Elf.
Vandrika: *Craspidospermum verticillatum* Boyer; *Apocynac.;* Mad.
Vang: *Mallotus spp.;* *Euphorbiac.;* fr. Gu.
Vâng: *Mallotus sp.;* *Euphorbiac.;* Ind.-Ch.
„ : *Barringtonia sp.;* *Lecythidac.;* Ind.-Ch.
„ anh: *Saraca dives* Pierre; *Caesalpiniac.;* Ind.-Ch. (Laos)
„ chung: *Endospermum chinense* Bth.; *Euphorbiac.;* Ind.-Ch.
„ lan: *Albizzia lucida* Bth.; *Mimosac.;* Ind.-Ch.
„ nhúa: *Garcinia vilersiana* Pierre; *Guttifer.;* Ind.-Ch. (Annam)
„ tam: *Manglietia glauca* Blume, *M. fordiana* Oliv.; *Magnoliac.;* Ind.-Ch.
„ „ dât: *Fagraea obovata* Wall.; *Loganiac.;* Ind.-Ch. (Tonkin-Chapa)
Vangh nghe: *Garcinia sp.;* *Guttifer.;* Ind.-Ch.
Vanillier faux: *Duranta plumieri* Jacq.; *Verbenac.;* N.-Kal.
Vap: *Mesua ferrea* L.; *Guttifer.;* Ind.-Ch.
„ cat: *Gordonia sp.;* *Theac.;* Ind.-Ch. (Annam)
Vaporiti: *Eugenia micheli* Lam.; *Myrtac.;* S.-Am.
Var: *Abies balsamea* Miller; *Conifer.;* Kan.
Vara blanca: *Aegiphila monstrosa* Moldenke; *Verbenac.;* C.-Am.
„ „ : *Capparis incana* H. B. K.; *Capparidac.;* Mex.
„ „ : *Trema micrantha* Blume; *Ulmac.;* Mex., C.-R.
„ „ : *Croton niveus* Jacq.; *Euphorbiac.;* Mex.
„ „ : *Casearia nitida* Jacq.; *Flacourtiac.;* Guat., Hond., n. Kol.
„ „ : *Aegiphila fasciculata* Donn. Smith; *Verbenac.;* Hond.
„ „ : *Sorocea columbiana* Standl.; *Morac.;* Kol.
„ „ : *Protium insigne* Engl.; *Burserac.;* Kol.
„ colorada: *Acacia rosei* Standl.; *Mimosac.;* Mex. (Sinaloa)
„ dulce: *Eysenhardtia polystachya* Sarg.; *Papilionac.;* Mex. (Durango)
„ santa: *Triplaris caracasana* Cham.; *Polygonac.;* Pan.
„ „ : *Triplaris americana* L.; *Polygonac.;* Kol.
„ „ overa: *Triplaris laxa* Blake; *Polygonac.;* Kol.
„ de piedra: *Casearia tremula* Griseb.; *Flacourtiac.;* n. Kol.
Varacchi: *Albizzia amara* Boivin; *Mimosac.;* sw. O.-Ind.
Varaduz: *Eysenhardtia polystachya* Sarg.; *Papilionac.;* Mex. (Durango)
Varanga: *Carallia lucida* Roxb.; *Rhizophorac.;* sw. O.-Ind.
Varía: *Luehea sp.;* *Tiliac.;* Cuba.
„ : *Cordia gerascanthoides* H. B. K., *C. gerascanthus* L.; *Borraginac.;* Cuba.
„ blanca: *Cordia sp.;* *Borraginac.;* Cuba.
Varilla: *Calophyllum calaba* Jacq.; *Guttifer.;* St.-D.
„ negra: *Cordia sp.;* *Borraginac.;* C.-R.
Varillo: *Calophyllum rekoi* Standley; *Guttifer.;* Salv.
Varital: *Drypetes keyensis* Urban; *Euphorbiac.;* P.-R.
„ : *Bauhinia kappleri* Sagot; *Mimosac.;* P.-R.
„ : *Dipholis montana* Griseb.; *Sapotac.;* P.-R.
Varnish tree, black: *Melanorrhoea usitata* Wall.; *Anacardiac.;* O.-Ind.
„ „ , japanese: *Ailanthus glandulosa* Desf.; *Simarubac.;* Jap.
Varo: *Hibiscus tiliaceus* L.; *Malvac.;* Mad.
Varongy: *Ocotea trichophlebia* Baker; *Laurac.;* Mad. (Farafangana)
„ mainty: *Mespilodaphne tapack* Danguy; *Laurac.;* Mad.
„ mavo: *Mespilodaphne racemosa* Danguy; *Laurac.;* Mad.
„ noir: *Ocotea trichophlebia* Baker; *Laurac.;* Mad.
Varul: *Ailanthus excelsa* Roxb.; *Simarubac.;* O.-Ind. (Bombay)

Vassourinha: *Dodonaea viscosa* Jacq.; *Sapindac.;* Bras. (Sao P)
Vasurinha: *Chrysophyllum cainito* L.; *Sapotac.;* Arg.
Vatchkuran: *Canthium didymum* Roxb.; *Rubiac.;* O.-Ind.
Vatehuli: *Artocarpus lakoocha* Roxb.; *Morac.;* sw. O.-Ind. (Madras)
Vatinga: *Eugenia sp.; Myrtac.;* Bras. (Sao P)
Vauvau: *Laurelia aromatica* Juss.; *Monimiac.;* s. S.-Am.
Vavangu: *Dipterocarpus indicus* Bedd.; *Dipterocarpac.;* sw. O.-Ind. (Travancore)
Vavli: *Holoptelea integrifolia* Planch.; *Ulmac.;* sw. O.-Ind. (Bombay)
Vay: *Litchi chinensis* Sonn.; *Sapindac.;* Ind.-Ch. (Annam)
„ ôc: *Calophyllum pulcherrimum* Wall.; *Guttifer.;* Ind.-Ch. (Annam)
„ „ : *Diospyros filipendula* Pierre; *Ebenac.;* Co.-Ch.
Vayana: *Cinnamomum zeylanicum* Breyn; *Laurac.;* sw. O.-Ind.
Vayila: *Poeciloneuron indicum* Bedd.; *Guttifer.;* sw. O.-Ind.
Ve-ay-dweh: *Parinarium sp.; Rosac.;* Lib.
Vébou: *Macrolobium sp.; Caesalpiniac.;* Gab.
Vedam: *Terminalia catappa* L.; *Combretac.;* O.-Ind. (Madras)
Vedangkomai: *Stereospermum cyclocarpum* Bth. et Hook. f.; *Bignoniac.;* sw. O.-Ind.
Vedi-babul: *Acacia arabica* Willd.; *Mimosac.;* O.-Ind. (Bombay)
„ vembu: *Toona ciliata* M. Roem.; *Meliac.;* sw. O.-Ind. (Travancore)
Vedupla: *Cullenia excelsa* Wight; *Malvac.;* sw. O.-Ind. (Madras)
Vekka: *Sterculia villosa* Roxb.; *Sterculiac.;* sw. O.-Ind.
Vela: *Feronia elephantum* Correa; *Rutac.;* O.-Ind. (Madras)
Velaga: *Feronia elephantum* Correa; *Rutac.;* O.-Ind. (Madras)
Velame do campo: *Vitex sp.; Verbenac.;* Arg.
Velayani: *Dipterocarpus indicus* Bedd.; *Dipterocarpac.;* sw. O.-Ind. (Travancore)
Velayil: *Holoptelea integrifolia* Planch.; *Ulmac.;* O.-Ind. (Madras)
Velenge: *Pterospermum suberifolium* Willd.; *Sterculiac.;* Ceylon.
Velita: *Lippia hemisphaerica* Cham.; *Verbenac.;* n. Kol.
Vella agil: *Dysoxylum malabaricum* Bedd.; *Meliac.;* sw. O.-Ind. (Madras)
„ cadamba: *Anthocephalus cadamba* Miq.; *Rubiac.;* O.-Ind. (Madras)
„ kadamba: *Hymenodictyon excelsum* Wall.; *Rubiac.;* sw. O.-Ind.
„ kondrikam: *Vateria indica* L.; *Dipterocarpac.;* sw. O.-Ind.
„ marda: *Terminalia arjuna* Bedd.; *Combretac.;* O.-Ind. (Madras)
Vellaikonju: *Hopea parviflora* Bedd.; *Dipterocarpac.;* sw. O.-Ind. (Madras)
Vellam: *Feronia elephantum* Correa; *Rutac.;* O.-Ind. (Madras)
Vellanga: *Feronia elephantum* Correa; *Rutac.;* O.-Ind. (Madras)
Vellapiney: *Vateria indica* L.; *Dipterocarpac.;* O.-Ind.
Vellay: *Sterculia urens* Roxb.; *Sterculiac.;* O.-Ind. (Madras)
„ naga: *Anogeissus latifolia* Wall.; *Combretac.;* O.-Ind. (Madras)
Velvelam: *Acacia leucophloea* Willd., *A. ferruginea* DC.; *Mimosac.;* O.-Ind. (Madras)
Vem-maruthu: *Terminalia paniculata* Roth; *Combretac.;* O.-Ind. (Madras)
Vembou: *Azadirachta indica* A. Juss.; *Meliac.;* O.-Ind. (Madras)
Vembu: *Azadirachta indica* A. Juss.; *Meliac.;* O.-Ind. (Madras)
Vempu: *Azadirachta indica* A. Juss.; *Meliac.;* O.-Ind. (Madras)
Vên: *Pterocarpus erinaceus* Poir.; *Papilionac.;* W.-Af. (Dah.)
Ven-ven: *Anisoptera cochinchinensis* Pierre; *Dipterocarpac.;* Ind.-Ch.
„ „ : *Anisoptera robusta* Pierre, *A. glabra* Kurz; *Dipterocarpac.;* Ind.-Ch.
„ „ : *Shorea hypochra* Hance; *Dipterocarpac.;* Ind.-Ch. (Annam)
„ „ trang: *Anisoptera glabra* Kurz; *Dipterocarpac.;* Ind.-Ch. (Annam)
„ „ „ : *Anisoptera cochinchinensis* Pierre; *Dipterocarpac.;* Annam.
„ „ xanh: *Anisoptera costata* Korth., *A. glabra* Kurz; *Dipterocarpac.;* Annam.
Venaco: *Trema micrantha* Blume; *Ulmac.;* Kol.
Venadillo: *Swietenia cirrhata* Blake; *Meliac.;* Mex.
Vencola (u): *Platymiscium polystachyum* Bth.; *Papilionac.;* Ven.
Venda: *Lagerstroemia lanceolata* Wall.; *Lythrac.;* sw. O.-Ind. (Travancore)
Véne: *Pterocarpus erinaceus* Poir.; *Papilionac.;* fr. Gu., W.-Af.
Vénékouer: *Lonchocarpus sericeus* H. B. K.; *Papilionac.;* Gab.
Venenito: *Rauwolfia canescens* L.; *Apocynac.;* n. Kol.
Venesia (u): *Bulnesia arborea* Engl.; *Zygophyllac.;* Ven.
Venga: *Pterocarpus marsupium* Roxb.; *Papilionac.;* sw. O.-Ind.
Vengai: *Pterocarpus marsupium* Roxb.; *Papilionac.;* O.-Ind. (Madras)
Vengalam: *Lagerstroemia lanceolata* Wall.; *Lythrac.;* sw. O.-Ind. (Travancore)

Vengalkattai: *Lophopetalum wightianum* Arn.; *Celastrac.;* sw. O.-Ind. (Madras)
Vengur: *Pterocarpus marsupium* Roxb.; *Papilionac.;* C.-O.-Ind. (Gondwana)
Venkottei: *Lophopetalum wightianum* Arn.; *Celastrac.;* sw. O.-Ind. (Madras)
Ventaku: *Lagerstroemia lanceolata* Wall.; *Lythrac.;* sw. O.-Ind. (Madras)
Venteak: *Lagerstroemia lanceolata* Wall.; *Lythrac.;* O.-Ind.
Venthekku: *Lagerstroemia lanceolata* Wall.; *Lythrac.;* sw. O.-Ind.
Ventura: *Ichthyomethia piscipula* Hitch.; *Papilionac.;* P.-R.
Vepa: *Azadirachta indica* A. Juss.; *Meliac.;* O.-Ind. (Madras)
Vepali: *Holarrhena antidysenterica* Wall.; *Apocynac.;* w. O.-Ind. (Deccan)
Veppam: *Azadirachta indica* A. Juss.; *Meliac.;* O.-Ind. (Madras)
Vera: *Guaiacum sanctum* L.; *Zygophyllac.;* Haiti.
„   : *Bulnesia arborea* Engl.; *Zygophyllac.;* Kol., Ven.
„   aceituna: *Bulnesia arborea* Engl.; *Zygophyllac.;* Ven., Kol.
„   amarilla: *Guaiacum sanctum* L.; *Zygophyllac.;* Cuba.
„   azul: *Bulnesia arborea* Engl ; *Zygophyllac.;* Ven.
„   blanca: *Bulnesia arborea* Engl.; *Zygophyllac.;* Kol., Ven.
„   prieta: *Guaiacum sanctum* L.; *Zygophyllac.;* Cuba.
Veraco: *Tabernaemontana psychotrifolia* H. B. K.; *Apocynac.;* Ven.
Verbena tree, sweet: *Backhousia citriodora* F. v. M.; *Myrtac.;* Au.
Verdecillo: *Tecoma chrysantha* DC.; *Bignoniac.;* Mex.
Verde lucero: *? Pithecolobium sp.;* *Mimosac.;* br. Hond.
Verescled: *Evonymus europaea* L.; *Celastrac.;* Rußland.
Vería negra: *Cordia gerascanthoides* H. B. K.; *Borraginac.;* Cuba.
„   prieta: *Cordia gerascanthoides* H. B. K.; *Borraginac.;* Cuba.
Vermoui: *Distostigma corymbosa* Panch. et Sebert; *Guttifer.;* N.-Kal.
Vern: *Tamarix articulata* Vahl; *Tamaricac.;* O.-Ind.
Vernis de la Chine: *Ailanthus glandulosa* Desf.; *Simarubac.;* Ch.
„   du Japon: *Ailanthus glandulosa* Desf.; *Simarubac.;* Jap.
Verno: *Ailanthus glandulosa* Desf.; *Simarubac.;* s. Frankreich.
Veroiti: *? Quararibea sp.;* *Bombacac.;* Bras.
Verwermorebessen (h): *Chlorophora tinctoria* Gaudich.; *Morac.;* tr. Am.
Verzino (i): *Caesalpinia echinata* Lam.; *Caesalpiniac.;* ö. Bras.
Vescicaria (i): *Colutea arborescens* L.; *Papilionac.;* M.-M.
Veseet (u): *Acacia glomerosa* Bth.; *Mimosac.;* br. Hond.
Vesi: *Intsia bijuga* O. Kče.; *Mimosac.;* Südsee (Fid.)
Vevala: *Lagerstroemia lanceolata* Wall.; *Lythrac.;* w. O.-Ind. (Deccan)
Vi: *Haematostaphis barteri* Hook. f.; *Anacardiac.;* Elf.
Vianpou: *Craibia atlantica* Dunn; *Papilionac.;* Elf.
Vias: *Heisteria trillesiana* Pierre; *Olacac.;* s. Kam., n. Gab.
Viass: *Heisteria trillesiana* Pierre; *Olacac.;* Gab.
Viazogeld: *Ilex aquifolium* L.; *Aquifoliac.;* Russland.
? Vidana: *Clematis vitalba* L.; *Ranunculac.;* Sp.
Vidi: *Cordia myxa* L.; *Borraginac.;* w. O.-Ind. (Deccan)
Vidraria de hojas estrechas: *Clematis flammula* L.; *Ranunculac.;* Sp.
Viêt: *Payena elliptica* Pierre; *Sapotac.;* Ind.-Ch. (Annam)
Vijaguillo: *Muntingia calabura* L.; *Elaeocarpac.;* n. Kol.
Vikèko: *Klainedoxa gabonensis* Pierre; *Simarubac.;* Gab.
Vila: *Feronia elephantum* Correa; *Rutac.;* O.-Ind. (Madras)
Vilam-pazham: *Feronia elephantum* Correa; *Rutac.;* O.-Ind. (Madras)
Vilatti: *Feronia elephantum* Correa; *Rutac.;* O.-Ind. (Madras)
Vilayata imli: *Pithecolobium dulce* Bth.; *Mimosac.;* O.-Ind.
Villa de mar: *Caesalpinia crista* L.; *Caesalpiniac.;* Mex.
Villangay: *Feronia elephantum* Correa; *Rutac.;* O.-Ind. (Madras)
Vilva: *Aegle marmelos* Correa; *Rutac.;* O.-Ind. (Madras)
Vin vin: *Anisoptera glabra* Kurz, *A. cochinchinensis* Pierre; *Dipterocarpac.;* Annam.
„   „ : *Shorea hypochra* Hance; *Dipterocarpac.;* Co.-Ch.
„   „ nghe: *Shorea hypochra* Hance; *Dipterocarpac.;* Ind.-Ch. (Annam)
„   „ trang: *Shorea hypochra* Hance; *Dipterocarpac.;* Annam.
„   „ xanh: *Shorea hypochra* Hance; *Dipterocarpac.;* Ind.-Ch. (Annam)
Vinaigrier: *Rhus typhina* L.; *Anacardiac.;* sö. Kan.
Vinal: *Prosopis ruscifolia* Griseb.; *Mimosac.;* Bol. (Chaco), Arg.
Vindou: *Vitex pachyphylla* Baker; *Verbenac.;* W.-Af.

Vine, Witch-: *Landolphia leonensis* Stapf; *Apocynac.*; Lib.
Vinette: *Erythroxylon areolatum* L.; *Erythroxylac.*; Guad.
„ , grosse: *Erythroxylon squamatum* Vahl; *Erythroxylac.*; Guad.
Vinhatico: *Plathymenia reticulata* Bth.; *Mimosac.*; Bras.
„ : *Pithecolobium vinhatico* Record; *Mimosac.*; Bras.
„ : *Enterolobium ellipticum* Bth.; *Mimosac.*; Bras. (Sao P)
„ : *Echirospermum balthazari* Allem.; *Mimosac.*; Bras.
„ amarello: *Plathymenia reticulata* Bth.; *Mimosac.*; ö. Bras.
„ „ : *Echirospermum balthazari* Allem.; *Mimosac.*; ö. Bras.
„ cabelleira: *Plathymenia reticulata* Bth.; *Mimosac.*; Bras.
„ castanho: *Plathymenia reticulata* Bth.; *Mimosac.*; Bras. (Bahia)
„ espinho: *Pithecolobium vinhatico* Record; *Mimosac.*; Bras.
„ flor de algodão: *Plathymenia reticulata* Bth.; *Mimosac.*; Bras.
„ orelha de macaco: *Plathymenia reticulata* Bth.; *Mimosac.*; Bras. (Sao P)
„ pé de boi: *Plathymenia reticulata* Bth.; *Mimosac.*; Bras.
„ testa de boi: *Plathymenia reticulata* Bth.; *Mimosac.*; Bras. (Sao P)
„ de algodão: *Plathymenia reticulata* Bth.; *Mimosac.*; Bras. (Sao P)
„ „ boi: *Plathymenia reticulata* Bth.; *Mimosac.*; Bras. (Sao P)
„ „ espinho: *Pithecolobium vinhatico* Record; *Mimosac.*; Bras. (Bahia)
„ „ „ : *Pithecolobium tortum* Mart.; *Mimosac.*; Bras.
„ „ macaco: *Pithecolobium vinhatico* Record; *Mimosac.*; Bras.
„ do campo: *Plathymenia reticulata* Bth.; *Mimosac.*; Bras. (Sao P)
„ „ espinho: *Pithecolobium tortum* Mart.; *Mimosac.*; Bras.
„ „ matto: *Acacia maleolens* Mart.; *Mimosac.*; Bras.
Vinheiro do matto: *Vochysia tucanorum* Mart. var. *elongata* Schl.; *Vochysiac.*: Bras.
Vintanina: *Calophyllum parviflorum* Bojer; *Guttifer.*; Mad.
Violet: *Peltogyne sp.*; *Caesalpiniac.*; fr. Gu.
„ de montagne: *Peltogyne sp.*; *Caesalpiniac.*; fr. Gu.
„ „ savane: *Peltogyne sp.*; *Caesalpiniac.*; fr. Gu.
Violete: *Dalbergia cearensis* Ducke; *Papilionac.*; n. Bras. (Amaz.)
Violettholz (d): *Peltogyne paniculata* Bth., *P. spp.*; *Caesalpiniac.*; Sur.
Violet wood: *Peltogyne sp.*; *Caesalpiniac.*; Gu.
Vioss: *Heisteria trillesiana* Pierre; *Olacac.*; Gab.
Virapitá: *Peltophorum vogelianum* Bth.; *Caesalpiniac.*; Arg.
Viraró: *Ruprechtia viraru* Griseb., *R. polystachya* Griseb.; *Polygonac.*; Arg.
„ : *Pterogyne nitens* Tul.; *Caesalpiniac.*; Arg.
Viraru: *Ruprechtia polystachya* Griseb.; *Polygonac.*; Bras., Arg.
„ colorado: *Ruprechtia triflora* Griseb.; *Polygonac.*; Arg.
Virgilia: *Cladrastis lutea* Koch; *Papilionac.*; ö. & sö. USA.
Viri: *Calophyllum tomentosum* Wight; *Guttifer.*; sw. O.-Ind. (Travancore)
Virki: *Cordia myxa* L.; *Borraginac.*; C.-O.-Ind. (Gondwana)
Viroity: *Quararibea sp.*; *Bombacac.*; Bras. (Bahia)
Virola: *Virola sp.*; *Myristicac.*; Gu.
Virusham: *Cordia myxa* L.; *Borraginac.*; sw. O.-Ind.
Viruviru: *Ocotea rodioei* Mez; *Laurac.*; Ven.
Visco: *Acacia visite* Griseb.; *Mimosac.*; Arg.
„ -arca: *Acacia visco* Lorentz; *Mimosac.*; S.-Am.
Viscote: *Acacia visite* Griseb.; *Mimosac.*; Arg.
Visiebia: *Paypayrola guianensis* Aubl.; *Violac.*; Sur.
Visnal: *Prosopis ruscifolia* Griseb.; *Mimosac.*; Arg.
Visong'u: ? *Irvingia klainei* Pierre; *Simarubac.*; Gab.
Vitali: *Pterocarpus blancoi* Merr., *P. spp.*; *Papilionac.*; Phil. (Zamboanga)
Vitanina: *Calophyllum sp.*; *Guttifer.*; Mad.
Vivaona: *Dilobeia sp.*; *Proteac.*; Mad.
Vivaro: *Ruprechtia sp.*; *Polygonac.*; Urug.
„ : *Pterogyne nitens* Tul.; *Caesalpiniac.*; Urug.
Vivaruru: *Ruprechtia viraru* Griseb.; *Polygonac.*; W.-Ind.
Vivaseca: *Cathorsium mangense* Dugand; *Papilionac.*; n. Kol.
„ : *Diphysa carthaginensis* Jacq.; *Papilionac.*; Kol.
Vive eperou: ? *Perebea sp.*; *Morac.*; Gu.
Vleeschhout (h): *Schoutenia ovata* Korth.; *Tiliac.*; Nied.-Ind.
Vó bo: *Sterculia balanghas* L.; *Sterculiac.*; Ind.-Ch. (Annam)

Vó dang: *Ilex sp.; Aquifoliac.;* Ind.-Ch. (Annam)
Voakotry: *Calophyllum inophyllum* L.; *Guttifer.;* Mad.
Voamboana: *Dalbergia baroni* Baker; *Papilionac.;* Mad.
„ mainty: *Dalbergia baroni* Baker; *Papilionac.;* Mad.
„ mavo: *Dalbergia baroni* Baker; *Papilionac.;* Mad.
„ mena: *Dalbergia baroni* Baker; *Papilionac.;* Mad.
Voanana: *Elaeocarpus rhodanthus* Baker; *Elaeocarpac.;* Mad.
Voandrano: *Rhizophora mucronata* Lam.; *Rhizophorac.;* Mad.
Voapaka: *Uapaca thouarsi* Baill.; *Euphorbiac.;* Mad.
„ : *Homalium scleroxylon* Baill.; *Flacourtiac.;* Mad.
Vogelbeerbaum (d): *Sorbus aucuparia* L.; *Rosac.;* Eu.
Vogelkop: *Sideroxylon guianensis* A. DC.; *Sapotac.;* Sur.
Vohilahitra: *Cinnamosma fragrans* Baill.; *Canellac.;* Mad.
Voi mat: *Garcinia sp.; Guttifer.;* Ind.-Ch.
Voirouchi: *Virola sp.; Myristicac.;* Gu.
Volador: *Gyrocarpus americanus* Jacq.; *Hernandiac.;* Kol., Ven.
„ : *Ruprechtia ramiflora* C. A. Meyer; *Polygonac.;* Kol., Ven.
Volomborana: *Albizzia sassa* Bruce; *Mimosac.;* Mad. (Betsimisaraka)
„ : *Albizzia fastigiata* Oliv.; *Mimosac.;* Mad.
Vomital colorado: *Cordia sebestena* L.; *Borraginac.;* P.-R.
Vonkhout: *Licania pachystachya* Kleinh., *L. micrantha* Miq.; *Rosac.;* Sur.
„ : *Parinarium campestre* Aubl., *? P. brachystachyum* Bth.; *Rosac.;* Sur.
„ : *Hirtella caudata* Kleinh., *H. americana* Aubl., *H. sp.; Rosac.;* Sur.
Voo-fohn: *Voacanga bracteata* Stapf; *Apocynac.;* Lib.
Votomite: *Clusia sp.; Guttifer.;* Gu.
Vouacapouholz (d): *Vouacapoua americana* Aubl.; *Caesalpiniac.;* Gu., nö. Bras.
Vouacapú: *Vouacapoua americana* Aubl.; *Caesalpiniac.;* Bras.
Voukou: *Mitragyne macrophylla* Hiern; *Rubiac.;* B.-K.
Vounda: *Chrysophyllum subnudum* Baker; *Sapotac.;* Gab.
Vreemoesoehoedoe: *Andira coriacea* Pulle, *? A. retusa* H. B. K.; *Papilionac.;* Sur.
Vu bó: *Unona mesnyi* Pierre; *Anonac.;* Ind.-Ch. (Annam)
Vuku: *Mitragyne macrophylla* Hiern; *Rubiac.;* B.-K. (Boma, Malela)
Vullam-marum: *Feronia elephantum* Roxb.; *Rutac.;* O.-Ind. (Madras)
Vum-maray: *Chloroxylon swietenia* DC.; *Rutac.;* O.-Ind. (Madras)
Vuma: *Calophyllum inophyllum* L.; *Guttifer.;* sw. O.-Ind. (Bombay)
Vung: *Careya sphaerica* Roxb.; *Lecythidac.;* Ind.-Ch. (Annam)
Vunga: *Nelitris vitiensis* A. Gray; *Myrtac.;* Südsee (Fid.)
Vutálau: *Calophyllum inophyllum* L.; *Guttifer.;* Phil. (Batanes)
Vya: *Heisteria trillesiana* Pierre; *Olacac.;* Gab.
Vyas: *Heisteria trillesiana* Pierre; *Olacac.;* Gab.

W

Wa: *Cassia siamea* Lam.; *Caesalpiniac.;* Ceylon.
„ : *Ficus sp.; Morac.;* n. Nig. (Kontagora)
„ lélé: *Pycnanthus kombo* Warb.; *Myristicac.;* Elf.
„ -pasang: *Eugenia jambolana* Lamk.; *Myrtac.;* O.-Ind.
Waäta alaan-kopie: *Erisma uncinatum* Warm.; *Vochysiac.;* Sur.
„ gwégwé: *Pterocarpus draco* L.; *Papilionac.;* Sur.
Wabé: *Mimusops pierreana* Engl.; *Sapotac.;* Gab.
Wacapoehout (h): *Vouacapoua americana* Aubl.; *Caesalpiniac.;* tr. S.-Am.
Wacapou: *Vouacapoua americana* Aubl.; *Caesalpiniac.;* Gu.
Wacholder: *Juniperus communis* L., *J. sabina* L.; *Conifer.;* n. Hem.
„ , gemeiner: *Juniperus communis* L.; *Conifer.;* Eu.
„ , südlicher: *Juniperus barbadensis* L.; *Conifer.;* sö. USA., W.-Ind.
Wad: *Ficus bengalensis* L.; *Morac.;* O.-Ind. (Bombay)
Wadaduri: *Lecythis sp.; Lecythidac.;* br. Gu.
Wadala: *Couratari guianensis* Aubl., *C. sp.; Lecythidac.;* Sur.
Wadang: *Pterospermum diversifolium* Blume, *P. javanicum* Jungh.; *Sterculiac.;* Java.
„ : *Pterospermum suberifolium* Lam.; *Sterculiac.;* Java.
„ oerang: *Pterospermum suberifolium* Lam.; *Sterculiac.;* Java.

Wadara: *Couratari guianensis* Aubl., *C. sp.; Lecythidac.;* Sur.
„ kakaralli: *Couratari guianensis* Aubl., *C. sp.; Lecythidac.;* Sur.
Waddywood: *Pseudomorus brunoniana* Bureau; *Morac.;* Au.
Wadile-hoeroewassa: *Sweetia nitens* Bth.; *Papilionac.;* Sur.
„ -lanaballi-diamaro: *Eschweilera sp.; Lecythidac.;* Sur.
Wadilie-kakarallie: *Eschweilera spp.; Lecythidac.;* Sur.
Wadodorie: *Eschweilera longipes* Miers; *Lecythidac.;* Sur.
„ lanaballi diamaro: *Eschweilera sp.; Lecythidac.;* Sur.
Wadoedoelie: *Couratari guianensis* Aubl.; *Lecythidac.;* Sur.
„ : *Eschweilera sp.; Lecythidac.;* Sur.
Wagaaka: *Nectandra sp., Ocotea sp.; Laurac.;* Sur.
Wagbebe: *Strychnos laxa* Solereder; *Loganiac.;* Togo.
Wagen-boom (h): *Protea grandiflora* Thunb.; *Proteac.;* S.-Af. (Kap)
Wah: *Protomegabaria stapfiana* Hutch.; *Euphorbiac.;* Lib.
Wahoo: *Ulmus alata* Michx.; *Ulmac.;* s. C.-USA.
„ : *Tilia americana* L.; *Tiliac.;* N.-Am.
„ : *Euonymus atropurpureus* Jacq.; *Celastrac.;* USA.
Waiakarra: *Nectandra sp., Ocotea sp.; Laurac.;* Sur.
Waibaima: *Nectandra sp.; Laurac.;* Gu.
Waika: *Moronobea coccinea* Aubl.; *Guttifer.;* br. Hond.
„ ribbon: *Acalypha sp.; Euphorbiac.;* br. Hond.
Waikara-pisi: *Nectandra sp., Ocotea sp.; Laurac.;* Sur.
Waikey: *Inga nobilis* Willd.; *Mimosac.;* br. Gu.
Waikie: *Inga urnifera* Kleinh.; *Mimosac.;* Sur.
„ ingaoe: *Inga spp.; Mimosac.;* Sur.
Wainop: *Ocotea rodioei* Mez; *Laurac.;* br. Gu.
Wait-a-bit: *Zanthoxylum fraxineum* Willd.; *Rutac.;* USA.
Waitjarra: *Nectandra sp., Ocotea sp.; Laurac.;* Sur.
Wajaaka: *Nectandra sp., Ocotea sp.; Laurac.;* Sur.
Waje: *Geonoma baculifera* Kunth; *Palmac.;* fr. Gu.
Wajewoe: *Swartzia minutiflora* Kleinh., *S. spp.; Caesalpiniac.;* Sur.
Wajoeli: *Xylopia spp.; Anonac.;* Sur.
Wak: *Pentacme suavis* A. DC.; *Dipterocarpac.;* Burma.
Waka: *Copaifera demeusei* Harms; *Caesalpiniac.;* B.-K.
Wakaboe: *Vouacapoua americana* Aubl.; *Caesalpiniac.;* Sur.
Wakabu: *Vouacapoua americana* Aubl.; *Caesalpiniac.;* Sur.
Wak-a-my: *Warscewiczia coccinea* Klotzsch; *Rubiac.;* Trin., Tob.
Wakapo: *Vouacapoua americana* Aubl.; *Caesalpiniac.;* Sur.
Wakbau: *Pentacme suavis* A. DC.; *Dipterocarpac.;* Burma.
Wakere: *Chrysophyllum balansae* Baill.; *Sapotac.;* N.-Kal.
Wakie: *Inga spp.; Mimosac.;* Sur.
Wakoekwa-tokon: *Cordia tetrandra* Aubl.; *Borraginac.;* Sur.
Wakradani: *Miconia guianensis* Cogn.; *Melastomatac.;* br. Gu.
Wala: *Funtumia africana* Stapf; *Apocynac.;* Elf.
Walaba: *Eperua falcata* Aubl.; *Caesalpiniac.;* Sur.
„ hariroe: *Eperua schomburgkiana* Bth., *? E. jenmani* Oliv.; *Caesalpiniac.;* Sur.
„ , roode: *Eperua sp.; Caesalpiniac.;* Sur.
„ , soft: *Eperua falcata* Aubl.; *Caesalpiniac.;* Sur.
Walang: *Pterospermum diversifolium* Blume; *Sterculiac.;* Java.
Walddistel: *Ilex decidua* Walt.; *Aquifoliac.;* s. & sö. USA.
Waldgeissblatt: *Lonicera periclymenum* L.; *Caprifoliac.;* Eu.
Waldrebe: *Clematis vitalba* L.; *Ranunculac.;* Eu.
„ , gemeine: *Clematis vitalba* L.; *Ranunculac.;* Eu.
Walèn: *Ficus ribes* Reinw.; *Morac.;* w. Java.
Wali: *Garcinia klaineana* Pierre; *Guttifer.;* Gab.
Waliklar: *Acer niveum* Blume; *Acerac.;* Java (Dieng)
Walikoekoen: *Schoutenia ovata* Korth.; *Tiliac.;* Java.
Walikoekon: *Schoutenia ovata* Korth.; *Tiliac.;* Nied.-Ind.
Walikoekoon: *Schoutenia ovata* Korth.; *Tiliac.;* Nied.-Ind.
Walimia: *Tapirira aff. guianensis* Aubl.; *Anacardiac.;* Sur.
Walk-naked wood: *Calyptranthes millspaughi* Urban; *Myrtac.;* br. Hond.
Wallaba: *Eperua rubiginosa* Miq., *E. spp.; Caesalpiniac.;* br. Gu.

Wallaba, Bimiti-: *Eperua schomburgkiana* Bth.; *Caesalpiniac.;* br. Gu.
„ , Clump-: *Dicymbe corymbosa* Bth.; *Caesalpiniac.;* br. Gu.
„ , Itoori-: *Eperua sp.; Caesalpiniac.;* br. Gu.
„ , Ituri-: *Eperua jenmani* Oliv.; *Caesalpiniac.;* br. Gu.
„ , soft-: *Eperua falcata* Aubl.; *Caesalpiniac.;* br. Gu.
„ , Water-: *Macrolobium chrysostachyum* Bth., *M. bifolium* Pers.; *Caesalpiniac.;*
„ , „ -:*Eperua schomburgkiana* Bth.; *Caesalpiniac.;* br. Gu. [Sur.
„ , witte: *Eperua schomburgkiana* Bth., *? E. jenmani* Oliv.; *Caesalpiniac.;* Sur.
„ , Yoboka-: *Eperua sp.; Caesalpiniac.;* br. Gu.
Walnuss (d): *Juglans spp.; Juglandac.;* n. Hem.
„ , afrikanisch: *Lovoa klaineana* Pierre; *Meliac.;* W.-Af.
„ , gemeine: *Juglans regia* L.; *Juglandac.;* ö. & s. Eu., w. As.
„ , kaukasisch: *Juglans regia* L.; *Juglandac.;* ö. Eu.
„ , schwarze: *Juglans nigra* L.; *Juglandac.;* ö. N.-Am.
Walnut (e): *Juglans spp.; Juglandac.;* n. Hem.
„ : *Carya ovata* K. Koch; *Juglandac.;* ö. N.-Am.
„ : *Juglans peruviana* Dode; *Juglandac.;* Peru.
„ : *Endiandra palmerstoni* C. T. White; *Laurac.;* Au.
„ , african (e, u): *Lovoa klaineana* Pierre; *Meliac.;* W.-Af.
„ , „ : *Coula edulis* Baill.; *Olacac.;* Lib.
„ , american (e, u): *Juglans nigra* L.; *Juglandac.;* ö. N.-Am.
„ , Ancona- (e): *Juglans regia* L.; *Juglandac.;* s. Eu.
„ , Arizona-: *Juglans rupestris* Engelm.; *Juglandac.;* s. USA., n. Mex.
„ , australian (e, u): *Endiandra palmerstoni* C. T. White; *Laurac.;* ö. Au.
„ , Auvergne-: *Juglans regia* L.; *Juglandac.;* s. Frankreich.
„ , Bean-: *Endiandra palmerstoni* C. T. White; *Laurac.;* Au.
„ , Benin-: *Lovoa klaineana* Pierre; *Meliac.;* Nig.
„ , black (e, u): *Juglans nigra* L.; *Juglandac.;* ö. N.-Am.
„ , „ : *Endiandra palmerstoni* C. T. White; *Laurac.;* Au.
„ , „ , american: *Juglans nigra* L.; *Juglandac.;* ö. N.-Am.
„ , „ , Queensland-: *Endiandra palmerstoni* C. T. White; *Laurac.;* Au.
„ , Black Sea- (e): *Juglans regia* L.; *Juglandac.;* ö. Eu.
„ , Blush-: *Beilschmiedia obtusifolia* Bth.; *Laurac.;* ö. Au.
„ , bolivian (u): *Juglans boliviana* Dode; *Juglandac.;* Bol.
„ , brazilian (u): *Phoebe porosa* Mez; *Laurac.;* s. Bras.
„ , „ , „ : *Cordia goeldiana* Huber; *Borraginac.;* Bras.
„ , british (e): *Juglans regia* L.; *Juglandac.;* Eu.
„ , California-: *Juglans californica* Watson; *Juglandac.;* w. USA. (Kalif.)
„ , canadian: *Juglans nigra* L.; *Juglandac.;* ö. Kan.
„ , caucasian (e, u): *Juglans regia* L.; *Juglandac.;* ö. Eu.
„ , central american (u): *Enterolobium cyclocarpum* Griseb.; *Papilionac.;* tr. Am.
„ , chinese: *Juglans cathayensis* Dode; *Juglandac.;* Ch.
„ , circassian (e, u): *Juglans regia* L.; *Juglandac.;* n. Hem.
„ , common: *Juglans regia* L.; *Juglandac.;* ö. & s. Eu.
„ , Congo-: *Terminalia superba* Engl. et Diels; *Combretac.;* B.-K.
„ , east indian (e, u): *Albizzia lebbek* Bth.; *Mimosac.;* O.-Ind., And.
„ , english: *Juglans regia* L.; *Juglandac.;* Eu.
„ , european: *Juglans regia* L.; *Juglandac.;* Eu., w. As.
„ , french (e, u): *Juglans regia* L.; *Juglandac.;* s. Frankreich.
„ , grey: *Juglans cinerea* L.; *Juglandac.;* N.-Am.
„ , „ : *Beilschmiedia elliptica* C. T. White; *Laurac.;* Au.
„ , Honduras-: *Metopium brownei* Urban; *Anacardiac.;* C.-Am., W.-Ind.
„ , indian: *Aleurites moluccana* Willd.; *Euphorbiac.;* Tr., Subtr.
„ , italian (e, u): *Juglans regia* L.; *Juglandac.;* It.
„ , Jamaica-: *Juglans sp.; Juglandac.;* br. W.-Ind.
„ , japanese (e, u): *Juglans sieboldiana* Maxim.; *Juglandac.;* Jap.
„ , Lagos-: *Lovoa klaineana* Pierre; *Meliac.;* W.-Af.
„ , manchurian (e): *Juglans mandschurica* Maxim.; *Juglandac.;* O.-As.
„ , Medang-: *Endiandra palmerstoni* C. T. White; *Laurac.;* Au.
„ , mexican: *Juglans rupestris* Engelm.; *Juglandac.;* s. USA., n. Mex.
„ , „ (u): *Enterolobium cyclocarpum* Griseb.; *Mimosac.;* Mex.
„ , nigerian: *Lovoa klaineana* Pierre; *Meliac.;* Nig.

Walnut, oriental (u): *Endiandra palmerstoni* C. T. White; *Laurac.;* Au.
„ , persian (u): *Juglans regia* L.; *Juglandac.;* Pers.
„ , peruvian: *Juglans neotropica* Diels, *J. peruviana* Dode; *Juglandac.;* Peru.
„ , Queensland- (e, u): *Endiandra palmerstoni* C. T. White; *Laurac.;* Au.
„ , rose (e, u): *Cryptocarya erythroxylon* Maiden et Betche; *Laurac.;* Queensl.
„ , royal (u): *Juglans regia* L.; *Juglandac.;* Eu., w. As.
„ , russian (u): *Juglans regia* L.; *Juglandac.;* Russland.
„ , Satin- (e): *Liquidambar styraciflua* L.; *Hamamelidac.;* s. N.-Am., C.-Am.
„ , south american (u): *Enterolobium cyclocarpum* Griseb.; *Mimosac.;* tr. Am.
„ , spanish (u): *Juglans regia* L.; *Juglandac.;* Sp.
„ , Swamp- (u): *Salix nigra* Marsh.; *Salicac.;* s. USA.
„ , Texas-: *Juglans rupestris* Engelm.; *Juglandac.;* s. USA.
„ , turkish (u): *Juglans regia* L.; *Juglandac.;* Türkei.
„ , west african (e): *Lovoa klaineana* Pierre; *Meliac.;* W.-Af.
„ , „ indian: *Juglans sp.; Juglandac.;* br. W.-Ind.
„ , white (e, u): *Juglans cinerea* L.; *Juglandac.;* ö. N.-Am.
„ , „ : *Cryptocarya obovata* R. Br.; *Laurac.;* Au.
„ , yellow: *Beilschmiedia bancroftei* C. T. White; *Laurac.;* Au.
Wal-sapu: *Michelia nilagirica* Zenker; *Magnoliac.;* Ceylon.
Walung: *Salix tetrasperma* Roxb.; *Salicac.;* C.-O.-Ind. (Central Prov.)
Walunj: *Salix tetrasperma* Roxb.; *Salicac.;* sw. O.-Ind. (Bombay)
Wama: *Ricinodendron africanum* Muell. Arg.; *Euphorbiac.;* Go.
Wamara: *Swartzia tomentosa* DC.; *Caesalpiniac.;* Gu.
Wamaru: *Swartzia tomentosa* DC.; *Caesalpiniac.;* br. Gu.
Wamba: *Ricinodendron africanum* Muell. Arg.; *Euphorbiac.;* Go.
Wambola-ya-ndumbili: *Heinsia pulchella* K. Schum.; *Rubiac.;* D.-O.-Af. (Tabora)
Wana: *Nectandra wana* ?, ? *N. pisi* Miq.; *Laurac.;* Sur.
„ : *Triplochiton scleroxylon* K. Schum.; *Sterculiac.;* Elf., Go.
Wanania: *Minquartia guianensis* Aubl.; *Olacac.;* Sur.
Wanasoro: *Cecropia sp.; Morac.;* br. Gu.
Wanatah: *Murraya sumatrana* Roxb.; *Rutac.;* Nied.-Ind. (Menado)
Wandoe: *Eucalyptus redunca* Schauer; *Myrtac.;* Au.
Wandoo: *Eucalyptus redunca* Schauer, *E. regnans* F. v. M.; *Myrtac.;* w. Au.
Wandoo, Powder bark-: *Eucalyptus accedens* W. V. Fitzg.; *Myrtac.;* w. Au.
„ , White gum-: *Eucalyptus redunca* Schauer var. *elata* Bth.; *Myrtac.;* w. Au.
Wandra: *Salix tetrasperma* Roxb.; *Salicac.;* C.-O.-Ind. (Central Prov.)
Wane: ? *Ocotea rubra* Mez, *Nectandra sp.; Laurac.;* Sur.
Wanebala: *Swartzia tomentosa* DC.; *Caesalpiniac.;* br. Gu.
Waneprokonie: *Inga urnifera* Kleinh.; *Mimosac.;* Sur.
Wangsa: *Canangium odoratum* Baill.; *Anonac.;* Java (Ragadjampi)
Wangu: *Terminalia glaucescens* Planch.; *Combretac.;* Go.
„ : *Baphia odorata* de Wild.; *Papilionac.;* B.-K. (Kasai)
Wangwata: *Sarcocephalus trillesi* Pierre; *Rubiac.;* B.-K. (Kurubu)
Wanosoro: *Cecropia sp.; Morac.;* br. Gu.
Wansawa: *Turraeanthus aff. africana* Pellegr.; *Meliac.;* Go.
Wansenwa: *Turraeanthus aff. africana* Pellegr.; *Meliac.;* Go.
Wap: *Mesua ferrea* L.; *Guttifer.;* Co.-Ch.
Wapa: *Eperua falcata* Aubl., *E. rubiginosa* Miq.; *Caesalpiniac.;* Gu.
„ : *Eperua kourouensis* R. Benoist; *Caesalpiniac.;* Gu.
„ blanc: *Eperua sp.; Caesalpiniac.;* fr. Gu.
? „ courbaril: *Eperua sp.; Caesalpiniac.;* fr. Gu.
„ gras: *Eperua falcata* Aubl.; *Caesalpiniac.;* fr. Gu.
„ gris: *Eperua falcata* Aubl.; *Caesalpiniac.;* fr. Gu.
„ huileux: *Eperua falcata* Aubl.; *Caesalpiniac.;* Gu.
„ patouvé: *Eperua falcata* Aubl.; *Caesalpiniac.;* Bras.
„ -sec: *Macrolobium chrysostachyum* Bth., *M. bifolium* Pers.; *Caesalpiniac.;* Sur.
Wapi: *Pouteria guianensis* Aubl.; *Sapotac.;* Gu.
Wapo: *Pouteria guianensis* Aubl., *Ecclinusa sanguinolenta* Pierre; *Sapotac.;* Gu.
War: *Ficus bengalensis* L.; *Morac.;* C.-O.-Ind. (Central Prov.)
Waracoori: *Tabebuia sp.;* Bignoniac.; Sur.
Warakuri: *Tabebuia longipes* Baker, *T. sp.; Bignoniac.;* Gu.
Waramia: *Tapirira guianensis* Aubl.; *Anacardiac.:* br. Gu.

Warang: *Kydia calycina* Roxb.; *Malvac.*; sw. O.-Ind. (Bombay)
Warapa kwarè: *Erisma uncinatum* Warm.; *Vochysiac.*; Sur.
„ -kwarèrè: *Erisma uncinatum* Warm.; *Vochysiac.*; Sur.
Waras: *Heterophragma roxburghi* DC.; *Bignoniac.*; sw. O.-Ind. (Bombay)
Waratah: *Telopea truncata* R. Br.; *Proteac.*; Tasm.
„ , Gippsland-: *Telopea oreades* F. v. M.; *Proteac.*; sö. Au. (Vict.)
Warau: *Pterocarpus sp.*; *Papilionac.*; br. Gu.
Warawaili: *Stereospermum xylocarpum* Bth. et Hook. f.; *Bignoniac.*; w. O.-Ind. (n.
War-bueh: *Ixora congesta* Stapf; *Rubiac.*; Lib.   [Bombay)
Waré: *Pterygota cordifolia* Chev.; *Sterculiac.*; Elf.
Warfine: *Diospyros aggregata* Gürke; *Ebenac.*; Gab.
Wariemia: *Tapirira aff. guianensis* Aubl.; *Anacardiac.*; Sur.
Warimiaballi: *Helicostylis poeppigiana* Tréc.; *Morac.*; Sur.
Warimiballi: *Brosimum paraense* Huber; *Morac.*; Sur.
Waringin: *Ficus benjamina* L.; *Morac.*; ö. Java.
Waroe: *Hibiscus tiliaceus* L., *H. similis* Blume; *Malvac.*; Nied.-Ind.
„ goeneng: *Hibiscus elatus* Swartz; *Malvac.*; Nied.-Ind.
„ gombong: *Hibiscus tiliaceus* L.; *Malvac.*; Nied.-Ind.
„ laoet: *Hibiscus tiliaceus* L.; *Malvac.*; Nied.-Ind.
„ laut: *Hibiscus tiliaceus* L.; *Malvac.*; Nied.-Ind.
„ linggies: *Hibiscus sp.*; *Malvac.*; Nied.-Ind.
„ -lôt: *Neesia altissima* Blume; *Bombacac.*; Java.
Waroekoeli: *Tabebuia sp.*; *Bignoniac.*; Sur.
Waroesi: *Virola surinamensis* Warb.; *Myristicac.*; Sur.
Waroesierian: *Virola mycetis* Pulle; *Myristicac.*; Sur.
Warokorie: *Tabebuia sp.*; *Bignoniac.*; Sur.
Warokotie: *Virola mycetis* Pulle; *Myristicac.*; Sur.
Warokotti: *Virola mycetis* Pulle; *Myristicac.*; Sur.
Warosirian: *Virola mycetis* Pulle; *Myristicac.*; Sur.
Warrec: *Caesalpinia yucatanensis* Greenm.; *Caesalpiniac.*; br. Hond.
Warri wood: *Flindersia grayleyana* F. v. M.; *Rutac.*; Au.
Warrie: *Caesalpinia yucatanensis* Greenm.; *Caesalpiniac.*; br. Hond.
Warsi: *Heterophragma roxburghi* DC.; *Bignoniac.*; C.-O.-Ind. (Centr. Prov.)
Warua: *Cassia leiandra* Bth.; *Caesalpiniac.*; br. Gu.
Was: *Crescentia cujete* L.; *Bignoniac.*; br. Hond.
Wasane: *Monotes kerstingi* Gilg; *Dipterocarpac.*; Nig.
Wasani: *Monotes kerstingi* Gilg; *Dipterocarpac.*; n. Nig.
Washiba: *Tecoma leucoxylon* Mart. *var. pentaphylla* Juss.; *Bignoniac.*; Sur.
„ : *Tecoma araliacea* P. DC.; *Bignoniac.*; Sur.
Wasieba: *Tecoma leucoxylon* Mart. *var. pentaphylla* Juss.; *Bignoniac.*; Sur.
„ : *Tecoma araliacea* P. DC.; *Bignoniac.*; Sur.
Wassé-poehoe: *Pouteria guianensis* Aubl.; *Sapotac.*; Sur.
Wassiba: *Tecoma leucoxylon* Mart. *var. pentaphylla* Juss.; *Bignoniac.*; Sur.
„ : *Tecoma araliacea* P. DC.; *Bignoniac.*; Sur.
„ -omera-kodikoro: *Sweetia nitens* Bth.; *Papilionac.*; Sur.
? Wassila: *Vouacapoua americana* Aubl.; *Caesalpiniac.*; Gu.
Wassi-wassi: *Vochysia guianensis* Aubl.; *Vochysiac.*; fr. Gu.
„ „ : *Nectandra pisi* Miq.; *Laurac.*; fr. Gu.
Wassie-wassie: *Qualea dinizi* Ducke, *Q. sp.*; *Vochysiac.*; Sur.
Watakpuo: *Cola cordifolia* R. Br.; *Sterculiac.*; Go.
Watapa: *Macrolobium chrysostachyum* Bth., *M. bifolium* Pers.; *Caesalpiniac.*; Sur.
Watepuo: *Cola cordifolia* R. Br.; *Sterculiac.*; Go.
Water-bébé: *Pterocarpus draco* L.; *Papilionac.*; Sur.
Water-djemoe: *Couratari guianensis* Aubl., *C. sp.*; *Lecythidac.*; Sur.
„ kopi: *Couepia guianensis* Aubl.; *Rosac.*; fr. Gu.
„ tree: *Tetracera leiocarpa* Stapf; *Dilleniac.*; Lib.
„ „ , red: *Erythrophloeum guineense* G. Don; *Caesalpiniac.*; Nig.
„ wiss: *Vitis sp.*; *Vitac.*; br. Hond.
„ wood: *Cassipourea elliptica* Poir.; *Rhizophorac.*; br. Hond.
„ „ , Bastard-: *Miconia pteropoda* Naud.; *Melastomatac.*; br. Hond.
„ care: *Chytroma idatimon* Miers; *Lecythidac.*; Trin.

Watjie: *Inga spp.; Mimosac.;* Sur.
Watrabiriehoedoe: *Macrolobium chrysostachyum* Bth.; *Caesalpiniac.;* Sur.
      „   : *Macrolobium bifolium* Pers.; *Caesalpiniac.;* Sur.
Watragrien: *Sweetia nitens* Bth.; *Papilionac.;* Sur.
Watra-kwali hoedoe: *Vochysia guianensis* Aubl., *V. coerulea* Aubl.; *Vochysiac.;* Sur.
Watramama-bobbi: *Gustavia angusta* L.; *Lecythidac.;* Sur.
Watrapa: *Macrolobium chrysostachyum* Bth., *M. bifolium* Pers.; *Caesalpiniac.;* Sur.
Watratamalin: *Pithecolobium kegeli* Meissn., *P. trapezifolium* Bth.; *Mimosac.;* Sur.
      „   : *Pithecolobium gonggrijpi* Kleinh.; *Mimosac.;* Sur.
Watta: *Osyris abyssinica* Hochst.; *Santalac.;* n. O.-Af. (Somali)
Wattle, black: *Acacia decurrens* Willd. *var. mollis* Lindl.; *Mimosac.;* sö. Au., Tasm.
  „   „ : *Acacia decurrens* Willd. *var. normalis* Bth.; *Mimosac.;* N.-S.-W.
  „   „ : *Acacia melanoxylon* R. Br.; *Mimosac.;* s. Au.
  „   „ : *Callicoma serratifolia* Andr.; *Cunoniac.;* ö. Au.
  „   „ , feather-leaved: *Acacia decurrens* Willd. *var. mollis* Lindl.; *Mimosac.;* sö.
  „ , Box-: *Notobuxus natalensis* Oliv.; *Buxac.;* S.-Af.   [Au., Tasm.
  „ , broad-leaved: *Acacia pycnantha* Bth.; *Mimosac.;* s. & sö. Au.
  „ . Cootamundra-: *Acacia baileyana* F. v. M.; *Mimosac.;* Au.
  „ , golden: *Acacia pycnantha* Bth.; *Mimosac.;* s. & sö. Au.
  „ , golden rain-: *Acacia prominens* A. Cunn.; *Mimosac.;* ö. & nö. Au.
  „ . green: *Acacia decurrens* Willd. *var. normalis* Bth.; *Mimosac.;* N.-S.-W.
  „   „ : *Acacia decurrens* Willd. *var. mollis* Lindl.; *Mimosac.;* sö. Au. (Vict.),
  „ , knife-leaved: *Acacia cultriformis* A. Cunn.; *Mimosac.;* sö. Au   [Tasm.
  „ , Mount Morgan-: *Acacia podalyriaefolia* A. Cunn.; *Mimosac.;* nö. Au.
  „ , Scrub-: *Acacia bakeri* J. H. Maiden; *Mimosac.;* Au.
  „ , Sidney-: *Acacia decurrens* Willd. *var. normalis* Bth.; *Mimosac.;* N.-S.-W.
  „ , Silver-: *Acacia decurrens* Willd. *var. dealbata* F. v. M.; *Mimosac.;* sö. Au., Tasm.
  „ ,   „ : *Acacia decurrens* Willd. *var. normalis* Bth.; *Mimosac.;* sö. Au., Tasm.
  „ ,   „ : *Acacia decurrens* Willd. *var. mollis* Lindl.; *Mimosac.;* Au., Tasm.
Wattoe moonba: *Virola mycetis* Pulle; *Myristicac.;* Sur.
Waua: *Dipterocarpus grandiflorus* Blco.; *Dipterocarpac.;* Java.
Wavuli: *Holoptelea integrifolia* Planch.; *Ulmac.;* C.-O.-Ind. (Central Prov.)
Wawa: *Triplochiton scleroxylon* K. Schum.; *Sterculiac.;* Elf., Go.
Wawabima: *Sterculia rhinopetala* K. Schum.; *Sterculiac.;* Go.
Wawahi: *Prosopis juliflora* DC.; *Mimosac.;* Curacao.
Wawampe: *Pterygota macrocarpa* K. Schum.; *Sterculiac.;* Go.
Wawé: *Nectandra sp., Ocotea sp.; Laurac.;* Sur.
Waye: *Pentadesma butyracea* Sab.; *Guttifer.;* Lib.
  „ -doo: *Landolphia leonensis* Stapf; *Apocynac.;* Lib.
  „ -kpay: *Pentadesma butyracea* Sab.; *Guttifer.;* Lib.
  „ -pu: *Symphonia gabonensis* Pierre *var. macrantha* Hutch. et Dalz.; *Guttifer.;* Lib.
Webb, billy: *Sweetia panamensis* Bth.; *Papilionac.;* br. Hond.
Wedi-quah-dro: *Napoleona leonensis* Hutch. et Dalz.; *Lecythidac.;* Lib.
  , -wa-doon: *Napoleona leonensis* Hutch. et Dalz.; *Lecythidac.;* Lib.
Wee-chu: *Lonchocarpus cyanescens* Bth.; *Papilionac.;* Lib.
Wegabaholz (d): *Vouacapoua americana* Aubl.; *Caesalpiniac.;* Gu., nö. Bras.
Wegil: *Aphania montana* Blume; *Sapindac.;* Java (Ngandang)
Weh-boh: *Conopharyngia durissima* Stapf; *Apocynac.;* Lib.
Wéhètè: *Tecoma leucoxylon* Mart. *var. pentaphylla* Juss.; *Bignoniac.;* Sur.
  „ : *Tecoma araliacea* P. DC.; *Bignoniac.;* Sur.
Wehn-flay: *Hippocratea loesneriana* Hutch. et Moss; *Hippocastanac.;* Lib.
Wehu: *Albizzia fastigiata* Oliv.; *Mimosac.;* B.-K. (Yangambi)
Weichsel (d): *Prunus mahaleb* L.; *Rosac.;* w. & s. Eu., Orient.
  „ , türkische: *Prunus mahaleb* L.; *Rosac.;* w. & s. Eu., Orient.
Weide, Asch- (d): *Salix cinerea* L.; *Salicac.;* Eu.
  „ , Bach-: *Salix purpurea* L.; *Salicac.;* s. & Mitt.-Eu.
  „ , Bein-: *Lonicera xylosteum* L.; *Caprifoliac.;* Eu.
  „ , Bruch-: *Salix fragilis* L.; *Salicac..;* Eu.
  „ , Dorf-: *Salix alba* L.; *Salicac.;* Eu.
  „ , Dotter-: *Salix alba* L.; *Salicac.;* Eu.
  „ , kaspische: *Salix acutifolia* Willd.; *Salicac.;* s. Eu.

Weide, Knack-: *Salix fragilis* L.; *Salicac.;* Eu.
„ , Korb-: *Salix viminalis* L.; *Salicac.;* Eu.
„ , Lavendel-: *Salix incana* Schrank; *Salicac.;* s. Eu.
„ , Lorbeer-: *Salix pentandra* L.; *Salicac.;* Mitt.- & n. Eu.
„ , Mandel-: *Salix amygdalina* L.; *Salicac.;* Eu.
„ , Moor-: *Salix repens* L.; *Salicac.;* Eu.
„ , Ohr-: *Salix aurita* L.; *Salicac.;* Eu.
„ , Öl-: *Elaeagnus angustifolia* L.; *Elaeagnac.;* w. As., s. Eu.
„ , Palm-: *Salix caprea* L.; *Salicac.;* Eu.
„ , Purpur-: *Salix purpurea* L.; *Salicac.;* Eu.
„ , Rain-: *Ligustrum vulgare* L.; *Oleac.;* n. Hem.
„ , „ -, gemeine: *Ligustrum vulgare* L.; *Oleac.;* n. Hem.
„ , Reif-: *Salix daphnoides* Vill.; *Salicac.;* Eu.
„ , Sal-: *Salix caprea* L.; *Salicac.;* Eu.
„ , Schimmel-: *Salix daphnoides* Vill.; *Salicac.;* Eu.
„ , schlesische: *Salix silesiaca* Willd.; *Salicac.;* n. Eu.
„ , Schwarz-: *Salix nigricans* Sm.; *Salicac.;* Eu.
„ , Silber-: *Salix alba* L.; *Salicac.;* Eu.
„ , Trauer-: *Salix babylonica* L.; *Salicac.;* Eu.
„ , Weiss-: *Salix alba* L.; *Salicac.;* Eu.
Weihrauch: *Boswellia serrata* Roxb.; *Burserac.;* O.-Ind.
Weissdorn, einweibiger: *Crataegus monogyna* Jacq.; *Rosac.;* Eu.
„ , stumpfblättriger: *Crataegus oxyacantha* L · *Rosac.;* Eu.
Wékéroe koepésinirian: *Licania micrantha* Miq.: *Rosac.:* Sur.
Wekie: *Inga spp.; Mimosac.;* Sur.
Wele: *Pterocarpus soyauxi* Taub.; *Papilionac.;* B.-K. (Stanleyville, Yangambi)
Welengéle: *Cleistopholis patens* Engl. et Diels; *Anonac.;* Togo
Wèli: *Garcinia klaineana* Pierre; *Guttifer.;* Gab
Wellingtonia: *Sequoia gigantea* Dcne.; *Conifer.;* w. N.-Am.
Welu: *Afzelia africana* Smith; *Caesalpiniac.;* Togo.
Wena: ? *Eschweilera corrugata* Miers; *Lecythidac.;* br. Gu.
Wengé: *Millettia laurenti* de Wild.; *Papilionac.;* fr. Kongo.
Wenzi: *Scorodophloeus sp.; Caesalpiniac.;* B.-K. (Mayumbe)
Wepetano tamoené: *Swartʒia spp.; Caesalpiniac.;* Sur.
„ -waroesie: *Iryanthera sagotiana* Warb., *I. hostmanni* Warb.; *Myristicac.;* Sur.
Werak: *Sapindus rarak* DC.; *Sapindac.;* Java (Gadoengan-Pare)
Wernham: *Cordia batesi* Wernh.; *Borraginac.;* Kam.
Weroe: *Albizzia procera* Bth.; *Mimosac.;* Java.
Wessépoeboe: *Pouteria guianensis* Aubl.; *Sapotac.;* Sur.
Weti: *Allophylus africanus* Beauv.; *Sapindac.;* Togo.
Wetie wan bie-ie hoedoe: *Eperua schomburgkiana* Bth.; *Caesalpiniac.;* Sur.
Wewarana: *Alseodaphne semecarpifolia* Nees; *Laurac.;* Cey.
Wha-la-koh: *Strombosia pustulata* Oliv.; *Olacac.;* Lib.
Whalebone tree: *Pseudomorus brunoniana* Bureau; *Morac.;* ö. Au.
Wharangipiro: ? *Melicope simplex* A. Cunn.; *Rutac.;* N.-Seel.
What-o'clock: *Jacaranda coerulea* Griseb.; *Bignoniac.;* Bah.
Whauai: *Plagianthus betulinus* A. Cunn.; *Malvac.;* N.-Seel.
Whauwhi: *Hoheria populnea* A. Cunn.; *Malvac.;* N.-Seel.
Wheel of fire: *Stenocarpus sinuatus* Endl.; *Proteac.;* Au.
Whin (e): *Ulex europaeus* L.; *Papilionac.;* Eu.
Whistlewood: *Acer pennsylvanicum* L.; *Acerac.;* ö. N.-Am.
Whitebark: *Cryptocarya erythroxylon* Maiden et Betche; *Laurac.;* Au.
„ : *Endiandra compressa* C. T. White; *Laurac.;* Au.
White leaf: *Alphitonia petriei* Braid et C. T. White; *Rhamnac.;* Au.
Whitewood: *Pinus silvestris* L.; *Conifer.;* Eu.
„ (u): *Picea excelsa* Link; *Conifer.;* n. Eu.
„ (e, u): *Liriodendron tulipifera* L.; *Magnoliac.;* ö. N.-Am.
„ : *Populus grandidentata* Michx.; *Salicac.;* ö. N.-Am.
„ : *Tilia americana* L.; *Tiliac.;* ö. USA.
„ : *Canella winterana* Gaertn.; *Canellac.;* Fl., Bah.
„ : *Ocotea sp., Nectandra sp.; Laurac.;* Jam.

Whitewood: *Drypetes lateriflora* Krug et Urban; *Euphorbiac.;* tr. Am.
„ : *Psychotria chiapensis* Standl.; *Rubiac.;* br. Hond.
„ : *Elaeocarpus kirtoni* F. v. M.; *Elaeocarpac.;* Au.
„ : *Atalaya hemiglauca* F. v. M.; *Sapindac.;* Au.
„ : *Schizomeria ovata* D. Don; *Cunoniac.;* ö. Au.
„ : *Pittosporum bicolor* Hook.; *Pittosporac.;* ö. Au., Tasm.
„ , african: *Triplochiton scleroxylon* K. Schum., *T. nigericum* Spr.; *Sterculiac.;*
„ , baltic: *Picea excelsa* Link; *Conifer.;* nö. Eu. [Nig., Go.
„ , Borneo-: *? Parashorea plicata* Brandis; *Dipterocarpac.;* Borneo.
„ , Canary-: *Liriodendron tulipifera* L.; *Magnoliac.;* ö. N.-Am.
„ , Florida-: *Drypetes keyensis* Urban; *Euphorbiac.;* s. Fl., Bah.
„ , Mowbullan-: *Elaeocarpus kirtoni* F. v. M.; *Elaeocarpac.;* Au.
„ , nigerian: *Triplochiton scleroxylon* K. Schum., *T. nigericum* Spr.; *Sterculiac.*
„ , northern: *Picea excelsa* Link; *Conifer.;* n. Eu. [Nig.
Whoe: *Daniella similis* Craib.; *Caesalpiniac.;* Lib.
Whykee: *Inga nobilis* Willd.; *Mimosac.;* br. Gu.
Wicke, Kronen-, strauchige: *Coronilla emerus* L.; *Papilionac.;* s. Eu.
Widget wood: *Erythrina indica* Lam.; *Papilionac.;* O.-Ind.
Wiedzo: *Borassus flabellifer* L. *var. aethiopum* Warb.; *Palmac.;* Go.
Wierookboom: *Protium guianense* March.; *Burserac.;* Gu.
Wijoejang: *Ficus annulata* Blume; *Morac.;* Java (Tjandiroeboe)
Wilada: *Ficus vrieseana* Miq.; *Morac.;* Java (Tjilatjap)
„ -banjoe: *Ficus vrieseana* Miq.; *Morac.;* Java (Pringamba)
„ toja: *Ficus vrieseana* Miq.; *Morac.;* Java.
Wiladan: *Ficus melinocarpa* Blume; *Morac.;* Java (Gadoengan)
Wilga, Scrub-: *Geijera salicifolia* Schott; *Rutac.;* Au.
Wiliwili: *Erythrina sp.;* *Papilionac.;* Südsee (Hawaii)
Willow (e, u): *Salix spp.;* *Salicac.;* n. Hem., Tr.
„ : *Salix gracilistyla* Miq.; *Salicac.;* Jap.
„ : *Salix nigra* Marsh.; *Salicac.;* s. USA.
„ : *Salix humboldtiana* Willd.; *Salicac.;* tr. Am.
„ : *Salix chilensis* Molina; *Salicac.;* br. Hond., Guat., Hond.
„ , almond-leaved: *Salix amygdaloides* Anders.: *Salicac.;* USA.
„ , Bat-: *Salix coerulea* Smith. *S. alba* L.; *Salicac.;* Eu.
„ , Bigelow-: *Salix lasiolepis* Bebb.; *Salicac.;* w. USA. (Kalif.)
„ , black: *Salix nigra* Marsh.; *Salicac.;* ö. USA., Mex.
„ „ : *Capparis sp.;* *Capparidac.;* Bah.
„ „ , western: *Salix lasiandra* Bth.; *Salicac.;* w. USA.
„ , Button-: *Cephalanthus occidentalis* L.; *Rubiac.;* USA.
„ , California-, black: *Salix laevigata* Bebb; *Salicac.;* w. USA. (Kalif.)
„ , „ -, white: *Salix lasiolepis* Bebb; *Salicac.;* w. USA. (Kalif.)
„ , Cape-: *Salix capensis* Thunb.; *Salicac.;* S.-Af. (Kap)
„ , Close bark-: *Salix alba* L. *var. coerulea* Koch; *Salicac.;* Eu.
„ , Cricket bat-: *Salix alba* L., *var. coerulea* Koch; *Salicac.;* Eu.
„ , Desert-: *Chilopsis linearis* P. DC.; *Bignoniac.;* s. USA., Mex.
„ , european: *Salix alba* L. *var. coerulea* Koch, *S. alba* L.; *Salicac.;* Eu.
„ , flowering: *Chilopsis saligna* D. Don; *Bignoniac.;* w. USA.
„ , indian: *Salix spp.;* *Salicuc.;* O.-Ind.
„ , japanese: *Salix urbaniana* Seem.; *Salicac.;* Jap.
„ , long-stalk: *Salix longipes* Anders.; *Salicac.;* ö. & sö. USA.
„ , Missouri-: *Salix missouriensis* Muehl.; *Salicac.;* ö. & sö. USA.
„ , Nuttall-: *Salix nuttalli* Sarg.; *Salicac.;* w. USA.
„ , Peach-: *Salix amygdaloides* Anders.; *Salicac.;* USA.
„ , red: *Cornus florida* L.; *Cornac.;* ö. Kan.
„ , Sandbar-: *Salix fluviatilis* Nutt.; *Salicac.;* N.-Am., n. Mex.
„ , silky: *Salix sitchensis* Sanson; *Salicac.;* w. N.-Am.
„ , Sitka-: *Salix sitchensis* Sanson; *Salicac.;* w. N.-Am.
„ , Ward-: *Salix longipes* Anders.; *Salicac.;* ö. & sö. USA.
„ , weeping: *Salix babylonica* L.; *Salicac.;* tr. Am.
„ , „ , european: *Salix babylonica* L.; *Salicac.;* USA.
„ , white, european: *Salix alba* L.; *Salicac.;* ö. USA.

Willow, wild: *Salix alba* L.; *Salicac.*; Eu.
„ , yellow: *Salix alba* L.; *Salicac.*; ö. USA.
Wilodo: *Ficus fistulosa* Reinw., *F. lepicarpa* Blume; *Morac.*; ö. Java.
„ -banjoe: *Ficus lepicarpa* Blume; *Morac.*; Java.
Wimon: *Platysepalum chevalieri* Harms; *Papilionac.*; B.-K. (Eala)
Winong: *Tetrameles nudiflora* R. Br.; *Datiscac.*; ö. Java.
Winterberry: *Ilex decidua* Walt.; *Aquifoliac.*; s. & sö. USA.
Winter bloom: *Hamamelis virginiana* L.; *Hamamelidac.*; sö. Kan.
Winter's-bark tree: *Drimys winteri* Forst.; *Magnoliac.*; tr. Am.
Wipa: *Irvingia gabonensis* Baill.; *Simarubac.*; Kam.
Wisa ne nkyene: *Cleistopholis patens* Bth.; *Anonac.*; Go.
Wisak: *Neonauclea spp.*; *Rubiac.*; Phil. (Rizal, Laguna)
Wish-more: *Tarrietia utilis* Sprague; *Sterculiac.*; Lib.
Wistaria, chinese: *Wistaria sinensis* Sweet; *Papilionac.*; Ch.
Wistarie, chinesische: *Wistaria sinensis* Sweet; *Papilionac.*; Ch.
Witchhazel: *Hamamelis virginiana* L.; *Hamamelidac.*; ö. USA.
Witie-apokita: *Aspidosperma nitidum* Bth., *A. excelsum* Bth.; *Apocynac.*; Sur.
„   „   : *Aspidosperma oblongum* A. DC.; *Apocynac.*; Sur.
„ -parihoedoe: *Aspidosperma nitidum* Bth., *A. excelsum* Bth.; *Apocynac.*; Sur.
„   -   „   : *Aspidosperma oblongum* A. DC.; *Apocynac.*; Sur.
Witpeer: *Pterocelastrus rostratus* Meissn.; *Celastrac.*; S.-Af. (Kap)
Witti-apakwie-ie: ? *Chaetocarpus schomburgkiana* Pax et Hoffm.; *Euphorbiac.*; Sur.
„ -hoedoe: *Erisma uncinatum* Warm.; *Vochysiac.*; Sur.
Wlo: *Borassus flabellifer* L. *var. aethiopum* Warb.; *Palmac.*; Go.
Wo: *Parkia africana* R. Br.; *Mimosac.*; Togo.
„ : *Entandrophragma utile* Sprague; *Meliac.*; Kam.
Woacrolie root: *Sarcocephalus esculentus* Afzel.; *Rubiac.*; Elf.
Woagbum: *Talisiopsis oliviformis* Radlk.; *Sapindac.*; Togo.
Woapa: *Eperua falcata* Aubl.; *Caesalpiniac.*; Sur.
Wobyebye: *Allophylus africanus* Beauv.; *Sapindac.*; B.-K. (Kundu)
Wodier: *Lannea grandis* Engl.; *Anacardiac.*; O.-Ind. (Madras)
? Woenen: *Ilex javanica* Valeton; *Aquifoliac.*; Java.
Woenoet: *Ficus glabrescens* Bl., *F. retusa* L., *F. spp.*; *Morac.*; Java.
„ -banjoe: *Ficus pilosa* Reinw., *F. glabella* Blume, *F. sp.*; *Morac.*; Java.
„ -bibis: *Ficus callophylla* Blume, *F. infectoria* Roxb.; *Morac.*; Java (Poeger)
„ -boeloe: *Ficus pilosa* Reinw.; *Morac.*; Java (Pantjoer Idjen)
Woeroe: *Fagraea obovata* Wall.; *Loganiac.*; Java (Pringamba)
„ -gesik: *Myrica longifolia* Teijsm. et Binn.; *Myricac.*; Java.
„ -dapoeng: *Acer niveum* Blume; *Acerac.*; Java (Sepakong)
„ -kembang: *Acer niveum* Blume; *Acerac.*; Java (Simpar)
? „ -ketingi: *Acer niveum* Blume; *Acerac.*; Java (Pringamba)
„ -poeti: *Acer niveum* Blume; *Acerac.*; Java (G. Karang)
„ -poetih: *Acer niveum* Blume; *Acerac.*; Java (Sepakoeng)
„ -tjangkok: *Myrica longifolia* Teijsm. et Binn.; *Myricac.*; Java.
Woitè: *Tecoma leucoxylon* Mart. *var. pentaphylla* Juss.; *Bignoniac.*; Sur.
„ : *Tecoma araliacea* P. DC.; *Bignoniac.*; Sur.
Woïwa: *Terminalia brownei* Fresen.; *Combretac.*; n. O.-Af. (Abessinien)
Wokomolo kotèlè: *Gustavia pterocarpa* Poit.; *Lecythidac.*; Sur.
Wokopopé: *Mouriria acutiflora* Naud.; *Melastomatac.*; Sur.
Woleh: *Aglaia ganggo* Miq.; *Meliac.*; Celebes.
Woloh: *Pterospermum acerifolium* Willd.; *Sterculiac.*; Celebes.
Woma: *Xylopia striata* Engl.; *Anonac.*; Kam.
Womagunguna: *Securidaca longipedunculata* Fresen.; *Polygalac.*; n. Nig. (Kontagora)
Woman's tongue tree: *Albizzia lebbek* Bth.; *Mimosac.*; W.-Ind.
Womara: *Swartʒia tomentosa* DC.; *Caesalpiniac.*; Sur.
Womba: *Xylopia striata* Engl.; *Anonac.*; Kam.
Wombolu: *Irvingia wombolu* Verm.; *Simarubac.*; B.-K.
Wonitan: *Saccopetalum horsfieldi* Benn.; *Anonac.*; Java (Besoeki, Ragadjampi)
Wonjangasanga: *Ricinodendron africanum* Muell. Arg.; *Euphorbiac.*; Kam.
Wonjasanga: *Ricinodendron africanum* Muell. Arg.; *Euphorbiac.*; Kam.
Wonoe: ? *Nectandra sp.*, ? *Ocotea sp.*; *Laurac.*; Sur.

Wonoe: *? Terminalia buceras* Hook.; *Combretac.;* Sur.
Wonta: *Artocarpus lakoocha* Roxb.; *Morac.;* sw. O.-Ind. (Madras)
Wontemara: *Artocarpus lakoocha* Roxb.; *Morac.;* w. O.-Ind. (Bombay)
Wonton: *Morus mesozygia* Stapf; *Morac.;* Go.
Wood of life: *Guaiacum officinale* L.; *Zygophyllac.;* W.-Ind., n. S.-Am.
Wooitano ajawa: *Protium spp.; Burserac.;* Sur.
  „ -koembetassi: *Hirtella caudata* Kleinh., *H. hirsuta* Lam.; *Rosac.;* Sur.
  „ -   „   : *Hirtella americana* Aubl.; *Rosac.;* Sur.
Woolly bark: *Eucalyptus obliqua* L'Hérit.; *Myrtac.;* ö. Au.
Woollybutt: *Eucalyptus gigantea* Hook. f., *E. obliqua* L'Hérit.; *Myrtac.;* sö. Au.
  „   : *Eucalyptus miniata* A. Cunn.; *Myrtac.;* w. Au.
  „ , Silvertop-: *Eucalyptus fastigiata* Deane et Maiden; *Myrtac.;* ö. Au.
  „ , white top-: *Eucalyptus fastigiata* Deane et Maiden; *Myrtac.;* ö. Au.
Wora: *Ficus bengalensis* L.; *Morac.;* C.-O.-Ind. (Central Prov.)
Worisjeporo apotopo: *Inga urnifera* Kleinh.; *Mimosac.;* Sur.
Wormbark: *? Andira inermis* H. B. K.; *Papilionac.;* tr. Am.
Wormbast: *Hymenolobium flavum* Kleinh., *H. sp.; Papilionac.;* Sur.
Wormboom: *Andira retusa* H. B. K.; *Papilionac.;* Sur.
  „   : *Hymenolobium flavum* Kleinh., *H. sp.; Papilionac.;* Sur.
Worr worr: *Acacia sp.; Mimosac.;* n. O.-Af. (Abessinien)
Wortelboom: *Conocarpus erecta* L.; *Combretac.;* Sur., fr. Gu.
  „   : *Rhizophora mangle* L.; *Rhizophorac.;* Sur.
  „ , Berg- (h): *Inga bourgoni* DC.; *Mimosac.;* Gu.
Wortelhout, amboneesch (h): *Pterocarpus indicus* Willd.; *Papilionac.;* O.-Ind.
Wosiono-sipio: *Protium spp.; Burserac.;* Sur.
Wosjie-wosjie: *Vochysia melinoni* Beckmann; *Vochysiac.;* Sur.
Wossie-jossie: *Qualea dinizi* Ducke, *Q. sp.; Vochysiac.;* Sur.
  „ -wossie: *Vochysia melinoni* Beckmann; *Vochysiac.;* Sur.
Wotó: *Rhizophora mangle* L.; *Rhizophorac.;* Togo.
  „ kwaleli: *Qualea coerulea* Aubl., *Q. rosea* Aubl.; *Vochysiac.;* Sur.
  „   „   : *Qualea albiflora* Warm.; *Vochysiac.;* Sur.
Woto kware: *Qualea coerulea* Aubl., *Q. rosea* Aubl.; *Vochysiac.;* Sur.
  „   „   : *Qualea albiflora* Warm.; *Vochysiac.;* Sur.
Wotobé: *Allanblackia parviflora* Chev.; *Guttifer.;* Elf.
Wotomba: *Artocarpus lakoocha* Roxb.; *Morac.;* sw. O.-Ind. (Bombay)
Wouapa: *Eperua falcata* Aubl.; *Caesalpiniac.;* Sur.
Woula: *Coula edulis* Baill.; *Olacac.;* Kam.
  „ bedino: *Strombosiopsis tetrandra* Engl.; *Olacac.;* Kam.
Wounian: *Myrianthus sp.; Morac.;* Elf.
  „ à feuilles entières: *Myrianthus serratus* Bth. et Hook.; *Morac.;* Elf.
Wouwoko: *Mitragyne macrophylla* Hiern; *Rubiac.;* B.-K. (Lulonga)
Wovali: *Mimusops elengi* Roxb.; *Sapotac.;* sw. O.-Ind. (Bombay)
Wowoli: *Holoptelea integrifolia* Planch.; *Ulmac.;* sw. O.-Ind. (Bombay)
Wowoué: *Lophira procera* Chev.; *Ochnac.;* Elf.
Wowovoko: *Mitragyne macrophylla* Hiern; *Rubiac.;* B.-K.
Wowowoko: *Mitragyne macrophylla* Hiern; *Rubiac.;* B.-K. (Lulonga)
Wrahn-beh: *Coelocaryon off. oxycarpum* Stapf; *Myristicac.;* Lib.
Wringin: *Ficus kurzi* King; *Morac.;* Java.
Wro-doo: *Canthium hispidum* Bth.; *Rubiac.;* Lib.
Wu: *Ceiba pentandra* Gaertn.; *Bombacac.;* Togo.
Wueti: *Mimusops lacera* Baker; *Sapotac.;* Togo.
Wula: *Coula edulis* Baill.; *Olacac.;* Kam.
? Wulfran: *Macrolobium limba* Scott Ell.; *Caesalpiniac.;* Go.
Wuma: *Calophyllum inophyllum* L.; *Guttifer.;* O.-Ind.
Wundi: *Ochrocarpus longifolius* Bth. et Hook.; *Guttifer.;* O.-Ind.
Wunga: *Piptadenia africana* Hook. f.; *Mimosac.;* Kam.
Wunja: *Albizzia amara* Boivin; *Mimosac.;* O.-Ind. (Madras)
Wur: *Ficus bengalensis* L.; *Morac.;* nw. O.-Ind. (Bombay)
Wurmrindenbaum: *? Andira inermis* H. B. K.; *Papilionac.;* tr. Am.
Wurus: *Heterophragma roxburghi* DC.; *Bignoniac.;* sw. O.-Ind.
Wuwoko: *Mitragyne macrophylla* Hiern; *Rubiac.;* B.-K.

Wuyan damo: *Combretum sokodense* Engl.; *Combretac.;* Nig.
Wycot: *Symphonia globulifera* L. f.; *Guttifer.;* br. Hond.
Wytil: *Sickingia maxoni* Standl.; *Rubiac.;* Pan.

X

Xa-ma: *Vatica dyeri* Pierre; *Dipterocarpac.;* Ind.-Ch. (Annam)
„ mou: *Cunninghamia sinensis* R. Br.; *Conifer.;* Ind.-Ch. (Tonkin, Laos)
„ muc: *Cunninghamia sinensis* R. Br.; *Conifer.;* Ind.-Ch. (Tonkin, Laos)
Xagua: *Genipa americana* L.; *Rubiac.;* Mex.
Xalócotl: *Pinus teocote* Schlecht., *Abies religiosa* Schlecht.; *Conifer.;* Mex.
Xàm ran: *Sterculia mastersi* Pierre; *Sterculiac.;* Ind.-Ch. (Kamb.)
Xanacol: *Bombax ellipticum* H. B. K.; *Bombacac.;* Mex.
Xang da: *Xanthophyllum colubrinum* Gagnep.; *Polygalac.;* Ind.-Ch. (Annam)
„ mói: *Miliusa bailloni* Pierre; *Anonac.;* Ind.-Ch. (Annam)
„ oi: *Terminalia corticosa* Pierre; *Combretac.;* Ind.-Ch.
Xao den: *Hopea odorata* Roxb.; *Dipterocarpac.;* Ind.-Ch. (Annam)
Xay: *Dialium cochinchinense* Pierre; *Caesalpiniac.;* Ind.-Ch. (Annam)
Xáza: *Schinus molle* L.; *Anacardiac.;* Mex.
Xcunchó: *Bombax ellipticum* H. B. K.; *Bombacac.;* Mex.
Xen: *Shorea cochinchinensis* Pierre; *Dipterocarpac.;* Ind.-Ch. (Annam)
„ do: *Shorea harmandi* Pierre; *Dipterocarpac.;* Ind.-Ch. (Annam)
„ hoqua: *Shorea henryana* Pierre; *Dipterocarpac.;* Ind.-Ch. (Annam)
Xengua: *Prunus capuli* Cav.; *Rosac.;* Mex.
Xeti: *Fagara xanthoxyloides* Lam.; *Rutac.;* Go.
Xetsi: *Fagara xanthoxyloides* Lam.; *Rutac.;* Go.
Xiaxek: *Lysiloma sabicu* Bth.; *Mimosac.;* Mex.
Xic-ché: *Zenthoxylum fagara* Sarg.; *Rutac.;* Mex.
Xihuicxan: *Bombax ellipticum* H. B. K.; *Bombacac.;* Mex.
Xiloxochitl: *Ceiba pentandra* Gaertn., *Bombax ellipticum* H. B. K.; *Bombacac.;* Mex.
Ximbo: *Enterolobium timbouva* Mart.; *Mimosac.;* Bras. (Sao P)
Xina: *Afzelia quanzensis* Welw.; *Caesalpiniac.;* port. O.-Af. (Mozambique)
Xioquauitl: *Bursera simaruba* Sarg.; *Simarubac.;* Mex.
Xkinin-bohó: *Spondias lutea* L.; *Anacardiac.;* Mex.
Xkuk-ché: *Ximenia americana* L.; *Olacac.;* Mex.
Xkulinsis: *Trichilia hirta* L.; *Meliac.;* Mex.
Xo mo ron: *Sterculia hypochra* Pierre; *Sterculiac.;* Ind.-Ch. (w. Annam)
Xoai: *Mangifera indica* L.; *Anacardiac.;* Ind.-Ch. (Annam)
Xoan: *Melia azedaràch* L.; *Meliac.;* Ind.-Ch.
„ dao: *Melia azadirachta* L.; *Meliac.;* Ind.-Ch.
„ „ : *Pygeum parviflorum* Teijsm. et Binn.; *Rosac.;* Ind.-Ch. (Tonkin)
„ môc: *Toona ciliata* M. Roem.; *Meliac.;* Ind.-Ch.
Xoay: *Dialium cochinchinense* Pierre; *Caesalpiniac.;* Ind.-Ch.
„ : *Evonymus hamiltoniana* Wall.; *Celastrac.;* Co.-Ch.
Xobo: *Spondias mombin* L.; *Anacardiac.;* Mex.
Xochicopal: *Elaphrium aloexylon* Schiede; *Burserac.;* Mex.
Xocot: *Spondias mombin* L.; *Anacardiac.;* Mex.
„ : *Malpighia glabra* L.; *Malpighiac.;* Mex.
Xoloẓin: *Heliocarpus sp.;* *Tiliac.;* Mex.
Xpaxakil: *Simaruba glauca* DC.; *Simarubac.;* Mex.
Xpu-ku-sikil: *Trichilia minutiflora* Standl.; *Meliac.;* br. Hond.
Xtolo: *Hibiscus tiliaceus* L.; *Malvac.;* Mex.
Xúa: *Albizzia lebbek* Bth.; *Mimosac.;* Ind.-Ch. (Kamb.)
Xun liên: *Melia azedarach* L.; *Meliac.;* Ind.-Ch. (Annam)
Xuong: *Zanthoxylum usitatum* Pierre; *Rutac.;* Co.-Ch.
„ ca: *Canthium didymum* Roxb., *C. didymum* Roxb. var. rostrata ?; *Rubiac.;* Ind.-
„ cô: *Carapa mekongensis* Pellegr.; *Meliac.;* Ind.-Ch.   [Ch.
„ môt: *Toona febrifuga* M. Roem.; *Meliac.;* Ind.-Ch.
Xylaloe: *Aquilaria agallocha* Roxb.; *Thymelaeac.;* Burma.

Y

Ya: *Achras sapota* L.; *Sapotac.;* Mex., br. Hond.
Yaa: *Butyrospermum parki* Kotschy; *Sapotac.;* br. O.-Af. (Uganda)
Yaba: *? Andira inermis* H. B. K.; *Papilionac.;* Cuba, Mex.
Yabe: *Mouriria sp.; Melastomatac.;* Bol.
Yabo: *? Andira inermis* H. B. K.; *Papilionac.;* Cuba, Mex.
Yabunikkei: *Cinnamomum pedunculatum* J. S. Presl; *Laurac.;* Jap., Form.
Yaca: *Tilia houghi* Rose; *Tiliac.;* Mex.
„ : *Byrsonima sp.; Malpighiac.;* Kol.
„ de cuero: *Trema micrantha* Blume; *Ulmac.;* Mex.
Yacal: *Shorea balangeran* Burck, *S. spp.; Dipterocarpac.;* Phil.
„ : *Hopea odorata* Roxb., *H. basilanica* Foxw., *H. plagata* Vid.; *Dipterocarpac.;*
„ blanco: *Vatica mangachapoi* Blco.; *Dipterocarpac.;* Phil. (Tayabas) [Phil.
„ , Borneo-: *Shorea balangeran* Burck, *S. spp.; Dipterocarpac.;* br. N.-Borneo.
„ , „ -: *Hopea odorata* Roxb., *H. spp.; Dipterocarpac.;* br. N.-Borneo.
Yacca: *Podocarpus coriaceus* L. C. Rich.; *Conifer.;* br. Hond.
Yachidamo: *Fraxinus mandschurica* Rupr.; *Oleac.;* Jap.
„ : *Fraxinus mandschurica* Rupr. var. *japonica* Maxim.; *Oleac.;* Jap.
Yaco granadillo: *Guazuma ulmifolia* Lam.; *Sterculiac.;* Mex.
„ de la costa: *Bombax ellipticum* H. B. K.; *Bombacac.;* Mex.
Yacure: *Pithecolobium dulce* Bth.; *Mimosac.;* Ven.
Yafo: *Parinarium curatellifolium* Planch.; *Rosac.;* Togo.
Yaga-biche: *Astronium conzattii* Blake; *Anacardiac.;* Mex.
„ -biguichi: *Pithecolobium dulce* Bth.; *Mimosac.;* Mex.
„ -bito: *Liquidambar styraciflua* L.; *Hamamelidac.;* Mex.
„ -bixihui: *Pithecolobium dulce* Bth.; *Mimosac.;* Mex.
„ -bizie: *Alnus sp.; Betulac.;* Mex.
„ -bizigni: *Liquidambar styraciflua* L.; *Hamamelidac.;* Mex.
„ -chichicino: *Taxodium mucronatum* Ten.; *Conifer.;* Mex.
„ -cino: *Quercus sp.; Fagac.;* Mex.
„ -gueza: *Salix sp.; Salicac.;* Mex.
„ -guichi-xiña: *Taxodium mucronatum* Ten.; *Conifer.;* Mex.
„ -guichi: *Heliocarpus sp.; Tiliac.;* Mex.
„ , huil: *Chlorophora tinctoria* Gaudich.; *Morac.;* Mex.
„ , na: *Guaiacum palmeri* Vail; *Zygophyllac.;* Mex.
„ -reche: *Quercus sp.; Fagac.;* Mex.
„ -xeni: *? Ceiba sp., ? Pachira sp., ? Bombax sp.; Bombacac.;* Mex.
„ -xoo: *Quercus sp.; Fagac.;* Mex.
„ -yoo: *Quercus sp.; Fagac.;* Mex.
Yagalache: *Rhus juglandifolia* H. B. K.; *Anacardiac.;* Mex.
Yagrume: *Didymopanax morototoni* Dcne. et Pl.; *Araliac.;* P.-R.
„ hembra: *Cecropia peltata* L.; *Morac.;* P.-R.
„ macho: *Didymopanax morototoni* Dcne. et Pl.; *Araliac.;* P.-R.
Yagrumo: *Cecropia sp.; Morac.;* Arg.
„ hembra: *Cecropia peltata* L.; *Morac.;* P.-R.
„ „ : *Cecropia obtusa* Tréc.; *Morac.;* Cuba.
Yagua: *Oreodoxa oleracea* Mart.; *Palmac.;* Hond.
„ -pinda: *Pisonia sp.; Nyctaginac.;* Arg.
Yaguero: *Roupala spp.; Proteac.;* Ven.
Yaha: *Curatella americana* L.; *Dilleniac.;* br. Hond.
Yahal: *Pourouma aspera* Tréc.; *Urticac.;* Nic.
Yaibi: *Gymnanthes lucida* Swartz; *Euphorbiac.;* Cuba, P.-R.
Yaití macho: *Gymnanthes lucida* Swartz; *Euphorbiac.;* Cuba.
Yaje: *Lysiloma sp., Leucaena sp.; Mimosac.;* Guat.
Yaka: *Dacrydium elatum* Wall.; *Conifer.;* Südsee (Fid.)
Yakal: *Isoptera borneensis* Scheff.; *Dipterocarpac.;* Phil. (Zamboanga)
„ : *Shorea balangeran* Dyer, *Hopea sp.; Dipterocarpac.;* Phil.
Yakalele: *Swartzia sp.; Caesalpiniac.;* fr. Gu.
Yakitá: *Brosimum gaudichaudi* Tréc.; *Morac.;* Arg.
Yaksaru: *Andira sp.; Papilionac.;* br. Gu.

Yaksauru: *Andira sp.; Papilionac.;* br. Gu.
Yakusugi: *Cryptomeria japonica* D. Don; *Conifer.;* Jap.
Yalam: *Cedrela sp.; Meliac.;* Nic.
Yali: *Acer campbelli* Hook. f. et Thoms.; *Acerac.;* nö. O.-Ind. (ö. Nepal)
Yalla: *Oxandra lanceolata* Baill.; *Anonac.;* Pan.
Yamaboshi: *Cornus kousa* Buerger; *Cornac.;* Jap., Korea.
Yamagourouma: *Trochodendron aralioides* S. et Z.; *Magnoliac.;* Jap.
? Yamagua: *Guarea sp.; Meliac.;* Cuba.
? „ colorada: *Guarea sp.; Meliac.;* Cuba.
Yamaguruma: *Trochodendron aralioides* S. et Z.; *Magnoliac.;* Jap.
Yamaguwa: *Morus alba* L. *var. stylosa* Bureau; *Morac.;* Jap.
Yamahannoki: *Alnus incana* Willd. *var. glauca* Ait.; *Betulac.;* Jap.
Yamamomizi: *Acer palmatum* Thunb. *subsp. matsumura* Koidz.; *Acerac.;* Jap.
Yamamomo: *Myrica rubra* S. et Z. *var. rubra* Mak.; *Myricac.;* Jap.
„ : *Myrica integrifolia* Roxb.; *Myricac.;* Form.
Yamanarashi: *Populus tremula* L. *var. villosa* Wesm.; *Salicac.;* Jap.
Yamane: *Gmelina arborea* L.; *Verbenac.;* O.-Ind., Burma.
Yamáo· *Guarea trichilioides* L.; *Meliac.;* Cuba.
Yamashiba: *Acer carpinifolium* S. et Z.; *Acerac.;* Jap.
Yamazakura: *Prunus pseudocerasus* Lindl. *var. spontanea* Maxim.; *Rosac.;* Jap.
Yamban: *Shorea balangeran* Dyer, *S. guiso* Blume; *Dipterocarpac.;* Phil.
„ -aromui: *Palaquium sp.; Sapotac.;* Phil. (Zambales)
Yana: *Ximenia americana* L.; *Olacac.;* Cuba.
Yanagiba-gasi: *Lithocarpus dodonaeifolia* Hayata; *Fagac.;* Form.
Yandiira: *Calophyllum brasiliense* Camb.; *Guttifer.;* Bras.
Yandira: *Minquartia guianensis* Aubl.; *Bignoniac.;* nö. Bras. (w. Pará)
Yandiroba: *Carapa guianensis* Aubl.; *Meliac.;* Bras.
Yang (H): *Dipterocarpus spp.; Dipterocarpac.;* Burma.
„ tâo hoà: *Averrhoa carambola* L.; *Oxalidac.;* Ind.-Ch. (Annam)
Yangon: *Cola proteiformis* Chev.; *Sterculiac.;* Elf.
Yangwood: *Dipterocarpus tuberculatus* Roxb.; *Dipterocarpac.;* Ind.-Ch.
Yanilla: *Allophylus cominia* Swartz; *Sapindac.;* Cuba.
Yankole: *Cleistopholis patens* Engl.; *Anonac.;* B.-K. (Bena-Dibele)
Yaó-si-y-guazú: *Phyllostylon brasiliensis* Capanema; *Ulmac.;* Arg.
Yaouia: *Spondias lakonensis* Pierre; *Anacardiac.;* Ind.-Ch. (Annam)
Yapa: *Azadirachta indica* A. Juss.; *Meliac.;* O.-Ind. (Madras)
„ : *Hardwickia binata* Roxb.; *Caesalpiniac.;* O.-Ind. (Madras)
Yapán: *Acacia visite* Griseb.; *Mimosac.;* Arg.
Yapi: *Terminalia ivorensis* Chev.; *Combretac.;* Elf.
Yaque: *Caesalpinia granadillo* Pittier; *Caesalpiniac.;* Ven.
„ blanco: *Prosopis juliflora* DC.; *Mimosac.;* Ven.
„ negro: *Prosopis juliflora* DC.; *Mimosac.;* Ven.
Yar yar: *Mitragyne macrophylla* Hiern; *Rubiac.;* Go.
Yara: *? Terminalia altissima* Chev.; *Combretac.;* Kam.
Yararu, Bastard-: *Swartzia sp.; Caesalpiniac.;* br. Gu.
Yarey: *Copernicia sp.; Palmac.;* Cuba.
Yareyón: *Copernicia gigas* Burret; *Palmac.;* Cuba.
Yari-yari: *Duguetia guianensis* DC.; *Anonac.;* br. Gu.
„ „ : *Oxandra lanceolata* Baill.; *Anonac.;* br. Gu.
Yarri-yarri: *Oxandra lanceolata* Baill.; *Anonac.;* Sur.
Yarúa: *Coccoloba retusa* Griseb.; *Polygonac.;* Cuba.
Yarumo: *Cecropia peltata* L.; *Morac.;* Kol.
„ de savana: *Didymopanax morototoni* Dcne. et Pl.; *Araliac.;* Ven.
Yaruri: *Aspidosperma excelsum* Bth., *Tabernaemontana sp.; Borraginac.;* fr. Gu.
Yaruru: *Aspidosperma nitidum* Bth., *A. oblongum* A. DC., *A. excelsum* Bth.; *Apocy-*
Yasche: *Ceiba pentandra* Gaertn.; *Bombacac.;* br. Hond. [*nac.;* Gu.
Yashabushi: *Alnus firma* S. et Z.; *Betulac.;* Jap.
Yash-can-an: *Psychotria chiapensis* Standl.; *Rubiac.;* br. Hond.
Yash-eck: *? Pithecolobium sp.; Mimosac.;* br. Hond.
„ -hulup: *Oreopanax meiocephalum* D. Sm.; *Araliac.;* br. Hond.
Yashi (jap.): *Cocos nucifera* L.; *Palmac.;* Tr.

Yashnik: *Vitex gaumeri* Greenm.; *Verbenac.*; br. Hond.
Yasi: *Santalum yasi* Seem.; *Santalac.*; Südsee (Fid.)
Yasiyasi: *Eugenia effusa* A. Gray; *Myrtac.*; Südsee (Fid.)
Yassa: ? *Terminalia altissima* Chev.; *Combretac.*; Kam.
Yat: *Carallia lucida* Roxb.; *Rhizophorac.*; Burma.
Yatayba: *Hymenaea courbaril* L.; *Caesalpiniac.*; Bras.
Yate: *Eucalyptus cornuta* Labill.; *Myrtac.*; w. Au.
„ , black: *Eucalyptus gracilis* F. v. M.; *Myrtac.*; w. Au.
„ , flat-topped-: *Eucalyptus occidentalis* Endl.; *Myrtac.*; w. Au.
„ , Swamp-: *Eucalyptus occidentalis* Endl.; *Myrtac.*; w. Au.
Yatevo: *Bambusa tvini* Nees; *Gramin.*; Arg.
Yatita: *Brosimum gaudichaudi* Tréc.; *Morac.*; Arg.
Yatli: *Acer campbelli* Hook. f. et Thoms.; *Acerac.*; nö. O.-Ind. (ö. Nepal)
Yatsude: *Fatsia japonica* Dcne.; *Araliac.*; Jap., Form.
Yatu: *Canarium occidentale* Chev.; *Burserac.*; Elf.
Yauaan: *Parashorea malaanonan* Merr.; *Dipterocarpac.*; Phil. (Surigao)
Yaupon: *Ilex vomitoria* Ait.; *Aquifoliac.*; sö. USA.
Yava: ? *Andira inermis* H. B. K.; *Papilionac.*; Cuba.
Yawhooballi: *Mollia sphaerocarpa* Gleason; *Tiliac.*; br. Gu.
Yawi: *Tarrietia utilis* Sprague; *Sterculiac.*; Lib.
Yawrukunan: *Tephrosia toxicaria* Pers.; *Papilionac.*; br. Gu.
Yaxnik: *Vitex gaumeri* Greenm.; *Verbenac.*; br. Hond.
Yax-puken: *Colubrina greggi* S. Wats.; *Rhamnac.*; Mex.
„ -pukim: *Colubrina greggi* S. Wats.; *Rhamnac.*; Mex.
Yaxche: *Ceiba pentandra* Gaertn., *C. aesculifolia* Britt. et Baker; *Bombacac.*; Mex.
Yaya: *Oxandra lanceolata* Baill., *Uvaria neglecta* A. Rich.; *Anonac.*; C.-Am., W.-Ind.
„ : *Curatella americana* L.; *Dilleniac.*; br. Hond.
„ : *Nectandra laurel* Kl. et Karst.; *Laurac.*; Pan.
„ : *Tabernaemontana utilis* Wight et Arn.; *Apocynac.*; fr. Gu.
„ : *Duguetia sp.*; *Anonac.*; Kol.
„ blanca: *Oxandra lanceolata* Baill.; *Anonac.*; P.-R.
„ „ : *Unonopsis pittieri* Safford; *Anonac.*; Pan.
„ cimmaróna: *Mouriria sp.*; *Melastomatac.*; Cuba.
„ comun: *Oxandra lanceolata* Baill.; *Anonac.*; Cuba.
„ mansa: *Mouriria sp.*; *Melastomatac.*; Cuba.
„ pino: *Duguetia vallicola* MacBride; *Anonac.*; Kol.
Ya-ya: ? *Protium sp.*; *Burserac.*; fr. Gu.
„ „ : *Mitragyne macrophylla* Hiern; *Rubiac.*; Go.
Yayabé: *Hannoa undulata* Planch.; *Simarubac.*; Togo.
Yayajabico: *Colubrina reclinata* Brong.; *Rhamnac.*; Cuba.
Yayamadoe: *Virola surinamensis* Warb.; *Myristicac.*; Sur.
Yayamadou: *Myristica sebifera* Sw., *M. sp.*, *Virola sp.*; *Myristicac.*; fr. Gu.
Yayo: *Trichilia sp.*; *Meliac.*; Kol.
„ colorado: *Trichilia sp.*; *Meliac.*; Kol.
Yba-pomo: *Melicocca bijuga* L.; *Sapindac.*; Par.
Ybaro: *Sapindus divaricatus* St. Hil.; *Sapindac.*; Par.
Ybatan: *Astronium sp.*; *Anacardiac.*; Bras.
Yberá-peré: *Apuleia praecox* Mart.; *Caesalpiniac.*; Arg.
Ybirá-pyitá: *Peltophorum vogelianum* Bth.; *Caesalpiniac.*; Arg.
„ - „ -guazú: *Peltophorum vogelianum* Bth.; *Caesalpiniac.*; Arg.
„ -rá. *Pterocarpus sp.*; *Papilionac.*; Arg.
„ -ré: *Piptadenia sp.*; *Mimosac.*; Arg.
„ -ro: *Pterogyne nitens* Tul.; *Caesalpiniac.*; Arg.
„ -ro-mi: *Pterogyne nitens* Tul.; *Caesalpiniac.*; Arg.
Ybopé-hú: *Prosopis sp.*; *Mimosac.*; Arg.
Ybura-pepe: *Holocalyx balansae* Micheli; *Caesalpiniac.*; Arg.
Ycy: *Protium icicariba* March.; *Burserac.*; Par.
Ydil: *Coccoloba schiedeana* Lindau; *Polygonac.*; br. Hond.
Ye-kanazo: *Heritiera fomes* Buch.; *Sterculiac.*; Burma.
Ye-ma-u: *Anthocephalus cadamba* Miq.; *Rubiac.*; Burma.
Ye-padauk: *Bischofia javanica* Blume; *Euphorbiac.*; Burma.

Ye-thapan: *Ficus glomerata* Roxb.; *Morac.;* Burma.
Yea: *Fagara xanthoxyloides* Lam., *F. macrophylla* Engl.; *Rutac.;* Go.
Yébaro: *Eperua purpurea* Bth.; *Caesalpiniac.;* n. Bras. (Amaz.)
Yedama: *Toona ciliata* M. Roem.; *Meliac.;* Burma.
Yedi: *Cestrum panamense* Standl.; *Solanac.;* Pan.
Yedra: *Hedera helix* L.; *Araliac.;* Sp.
Yegi: *Pterocarpus marsupium* Roxb.; *Papilionac.;* O.-Ind. (Madras)
Yegonoki: *Styrax japonica* S. et Z.; *Styracac.;* Jap., Korea.
Yehmyôt: *Trewia nudiflora* L.; *Euphorbiac.;* Burma.
Yela: *Terminalia belerica* Roxb.; *Combretac.;* O.-Ind. (Bombay)
Yelchi: *Zizyphus jujuba* Lamk.; *Rhamnac.;* O.-Ind. (Bombay)
Yelim-buriki: *Schleichera trijuga* Willd.; *Sapindac.;* O.-Ind. (Madras)
Yella maddi: *Anogeissus latifolia* Wall.; *Combretac.;* O.- Ind. (Madras)
Yellal: *Canthium didymum* Roxb.; *Rubiac.;* O.-Ind. (Bombay)
Yellande: *Zizyphus jujuba* Lamk.; *Rhamnac.;* O.-Ind. (Madras)
Yellanga: *Feronia elephantum* Correa; *Rutac.;* O.-Ind. (Madras)
Yellow-heart, Burma-: *Fagraea fragrans* Roxb.; *Loganiac.;* O.-Ind.
Yellow prickle: *Zanthoxylum fraxineum* Willd.; *Rutac.;* Jam.
    ,,   ,, : *Zanthoxylum kellermani* P. Wilson; *Rutac.;* br. Hond.
    ,,   ,, : *Zanthoxylum panamense* P. Wilson; *Rutac.;* Pan.
Yellow mixed wood: *? Enterolobium schomburgki* Bth.; *Mimosac.;* br. Hond.
Yellow wood: *Sarcocephalus cordatus* Miq.; *Rubiac.;* tr. As.
   ,,   ,, : *Tsuga sieboldi* Carr., *Podocarpus nagi* Pilger; *Conifer.;* Jap.
   ,,   ,, : *Cladrastis lutea* Koch; *Papilionac.;* ö. USA.
   ,,   ,, : *Schaefferia frutescens* Jacq.; *Celastrac.;* subtr. & tr. Am.
   ,,   ,, : *Chlorophora tinctoria* Gaudich.; *Morac.;* Trin., Jam., br. Hond.
   ,,   ,, : *Zanthoxylum flavum* Vahl; *Rutac.;* br. W.-Ind.
   ,,   ,, : *Enantia chlorantha* Oliv.; *Anonac.;* Lib.
   ,,   ,, : *Podocarpus gracilior* Pilger; *Conifer.;* br. O.-Af.
   ,,   ,, : *Podocarpus milanjianus* Rendle; *Conifer.;* br. O.-Af.
   ,,   ,, : *Terminalia sericea* Burch.; *Combretac.;* s. C.-Af.
   ,,   ,, : *Podocarpus elongatus* L'Hérit.; *Conifer.;* S.-Af.
   ,,   ,, : *Podocarpus spp.; Conifer.;* Au.
   ,,   ,, : *Flindersia oxleyana* F. v. M.; *Rutac.;* ö. Au. (Queensl.)
   ,,   ,, : *Coprosma linariifolia* Hook. f.; *Rubiac.;* N.-Seel.
   ,, ,, , african: *Enantia chlorantha* Oliv.; *Anonac.;* Nig.
   ,, ,, , australian (e): *Flindersia oxleyana* F. v. M.; *Meliac.;* ö. Au.
   ,, ,, , Bastard-: *Podocarpus elongatus* L'Hérit.; *Conifer.;* S.-Af.
   ,, ,, , brazilian (H): *Plathymenia reticulata* Bth.; *Mimosac.;* ö. Bras.
   ,, ,, , common: *Podocarpus elongatus* L'Hérit.; *Conifer.;* S.-Af.
   ,, ,, , cuban: *Zanthoxylum sp.; Rutac.;* Bah.
   ,, ,, , deep: *Rhodosphaera rhodanthema* Engl.; *Anacardiac.;* Au.
   ,, ,, , Falcate-: *Podocarpus falcata* R. Br.; *Conifer.;* S.-Af.
   ,, ,, , light: *Flindersia oxleyana* F. v. M.; *Rutac.;* ö. Au. (Queensl.)
   ,, ,, , Moreton Bay-: *Acronychia laevis* Forst.; *Rutac.;* Au.
   ,, ,, , Outeniqua-: *Podocarpus elongatus* L'Hérit.; *Conifer.;* S.-Af. (Kap)
   ,, ,, , real: *Podocarpus thunbergi* Hook., *P. latifolius* R. Br.; *Conifer.;* S.-Af.
   ,, ,, , thorny: *Zanthoxylum brachyacanthum* F. v. M.; *Rutac.;* Au.
   ,, ,, , upright: *Podocarpus thunbergi* Hook.; *Conifer.;* S.-Af. (Kap)
   ,, ,, de la Floride: *Schaefferia frutescens* Jacq.; *Celastrac.;* Fl.
   ,, ,, of South Africa: *Podocarpus milanjianus* Rendle; *Conifer.;* br. O.-Af. (Nias-
Yeludijamaram: *Adina cordifolia* Hook. f.; *Rubiac.;* sw. O.-Ind.   [sa)
Yema de huevo: *Cassia bacillaris* L.; *Caesalpiniac.;* Kol.
   ,,   ,,   ,, : *Morinda panamensis* Seem.; *Rubiac.;* Kol.
Yemané: *Gmelina arborea* L.; *Verbenac.;* Burma.
   ,, -apyu: *Lophopetalum fimbriatum* Wight; *Celastrac.;* Burma.
Yeme: *Hardwickia pinnata* Roxb.; *Caesalpiniac.;* sw. O.-Ind. (Bombay)
Yemeri: *Vochysia hondurensis* Sprague; *Vochysiac.;* br. Hond.
Yemoke: *Vochysia hondurensis* Sprague; *Vochysiac.;* br. Hond.
Yené: *Salix tetrasperma* Roxb.; *Salicac.;* Burma.
Yenghe: *Scorodophloeus sp.; Caesalpiniac.;* B.-K. (Uelé)
Yennemara: *Dipterocarpus indicus* Bedd.; *Dipterocarpac.;* sw. O.-Ind. (Coorg)

Yenoki: *Celtis sinensis* Pers.; *Ulmac.;* Jap.
Yeôu: *Citrus decumana* Willd.; *Rutac.;* Ind.-Ch. (Annam)
Yepa: *Azadirachta indica* A. Juss.; *Meliac.;* O.-Ind. (Madras)
Yepi: *Hardwickia binata* Roxb.; *Caesalpiniac.;* O.-Ind. (Madras)
Yerba doncella: *Vinca major* L., *V. media* Hoffm. et Link; *Apocynac.;* Sp.
„ de comadreja: *Capparis sp.; Capparidac.;* Arg.
„ „ lagarto: *Casearia sp.; Flacourtiac.;* Arg.
„ del toro: *Jacobinia mexicana* Seem.; *Acanthac.;* Mex.
Yerbhicky: *Gardenia gummifera* L.; *Rubiac.;* O.-Ind. (Bombay)
Yerindi: *Dysoxylum binectariferum* Hook. f.; *Meliac.;* sw. O.-Ind. (Bomb.)
Yerma: *Anogeissus latifolia* Wall.; *Combretac.;* C.-O.-Ind. (Gondwana)
Yermaddi: *Terminalia arjuna* Bedd.; *Combretac.;* O.-Ind. (Madras)
Yeromatsu: *Picea ajanensis* Fisch.; *Conifer.;* Jap.
Yerra: *Sterculia urens* Roxb.; *Sterculiac.;* O.-Ind. (Madras)
„ : *Dalbergia lanceolaria* L.; *Papilionac.;* O.-Ind. (Madras)
„ bikki: *Gardenia turgida* Roxb.; *Rubiac.;* O.-Ind. (Madras)
„ chandanam: *Pterocarpus santalinus* L. f.; *Papilionac.;* O.-Ind. (Madras)
„ munukudu: *Atalantia monophylla* Correa; *Rutac.;* O.-Ind. (Madras)
Yerri bikki: *Gardenia lucida* Roxb.; *Rubiac.;* w. O.-Ind. (n. Bombay)
Yertchuk: *Eugenia consideniana* Maiden; *Myrtac.;* ö. Au.
Yerugudu: *Dalbergia latifolia* Roxb.; *Papilionac.;* O.-Ind. (Madras)
Yerul: *Xylia xylocarpa* Taub.; *Mimosac.;* sw. O.-Ind.
Yesé: *Klainedoxa latifolia* Pierre; *Simarubac.;* Kam.
Yesomatsu: *Picea ajanensis* Fisch.; *Conifer.;* Jap.
Yetama: *Acrocarpus fraxinifolius* Wight; *Caesalpiniac.;* Burma.
Yetega: *Adina cordifolia* Hook. f., *Stephegyne parvifolia* Korth.; *Rubiac.;* w. O.-Ind.
Yethabye: *Eugenia operculata* Roxb.; *Myrtac.;* Burma.
„ : *Salix tetrasperma* Roxb.; *Salicac.;* Burma.
Yettama: *Anogeissus latifolia* Wall.; *Combretac.;* O.-Ind. (Madras)
Yetti: *Strychnos nux-vomica* L.; *Loganiac.;* w. O.-Ind. (Deccan)
Yeux crabes: *Blighia sapida* Koenig; *Sapindac.;* Mart.
„ de chat: *Caesalpinia crista* L.; *Caesalpiniac.;* Mart.
Yevo-gboma: *Eriocoelum kerstingi* Gilg; *Sapindac.;* Togo.
Yevune: *Cocos nucifera* L.; *Palmac.;* Togo.
Yevutsa: *Anacardium occidentale* L.; *Anacardiac.;* Go.
Yevutsitoe: *Tamarindus indica* L.; *Caesalpiniac.;* Go.
Yew (e, u): *Taxus baccata* L., *T. spp.; Conifer.;* n. Hem.
„ : *Taxus cuspidata* S. et Z.; *Conifer.;* ö. As.
„ , american: *Taxus canadensis* Willd.; *Conifer.;* ö. N.-Am.
„ , common: *Taxus baccata* L.; *Conifer.;* Eu., As.
„ , european (u): *Taxus baccata* L.; *Conifer.;* Eu., As.
„ , Florida- (u): *Taxus floridana* Chapm.; *Conifer.;* s. USA. (Fl.)
„ , New Zealand-: *Podocarpus totara* G. Benn.; *Conifer.;* N.-Seel.
„ , Pacific- (u): *Taxus brevifolia* Nutt.; *Conifer.;* w. USA.
„ , Tecoma- (u): *Andira inermis* H. B. K.; *Papilionac.;* W.-Ind.
„ , western: *Taxus brevifolia* Nutt.; *Conifer.;* sw. Kan.
Yezo matsu: *Picea ajanensis* Fisch.; *Conifer.;* Jap.
Yikwi: *Monodora myristica* Dunal; *Anonac.;* Go.
Yin-chien-mu: *Cunninghamia lanceolata* Hook.; *Conifer.;* Ch.
Yindaik: *Dalbergia cultrata* Graham; *Papilionac.;* Burma.
Yinma: *Chukrasia tabularis* A. Juss.; *Meliac.;* Burma.
Yinye: *Lumnitzera littorea* Voigt; *Combretac.;* Burma.
Yinzat: *Dalbergia fusca* Pierre; *Papilionac.;* Burma.
Yir: *Salix tetrasperma* Roxb.; *Salicac.;* nw. O.-Ind. (Kaschmir)
Yitpadi: *Bursera serrata* Colebr.; *Burserac.;* Burma.
Ymirá-itá: ? *Caesalpinia ferrea* Mart.; *Caesalpiniac.;* nö. Bras.
„ piranga: *Caesalpinia echinata* Lam.; *Caesalpiniac.;* Bras.
Yo: *Butyrospermum parki* Kotschy; *Sapotac.;* Togo.
Yoá blanco: *Lycium cestroides* Cham. et Schlt.; *Solanac.;* Arg.
„ -hú-y: *Seguieria paraguayensis* Morong; *Phytolaccac.;* Arg.
„ -moroti: *Lycium cestroides* Cham. et Schlt.; *Solanac.;* Arg.
„ -sỳ-y: *Celtis brasiliensis* Planch.; *Ulmac.;* Arg.

Yodo: *Desbordesia insignis* Pierre; *Simarubac.;* Gab.
Yogusominebari: *Betula ulmifolia* S. et Z.; *Betulac.;* Jap.
Yohimbé: *Corynanthe yohimbe* K. Schum.; *Rubiac.;* Kam.
Yoibi: *Protium icicariba* March.; *Burserac.;* Par.
Yoke (e): *Piptadenia peregrina* Bth.; *Mimosac.;* Trin.
„ : *Astronium obliquum* Griseb.; *Anacardiac.;* Trin., Tob.
„ , Savannah-: *Piptadenia peregrina* Bth.; *Mimosac.;* Trin., Tob.
„ wood: *Catalpa longissima* Sims; *Bignoniac.;* Trin.
Yokharaúgu: *Strychnos laxa* Solereder; *Loganiac.;* Togo.
Yolosuchil: *Talauma mexicana* Don; *Magnoliac.;* Mex.
Yoloxochitl: *Talauma mexicana* Don; *Magnoliac.;* Mex.
Yoma wood: *Vitex glabrata* R. Br.; *Verbenac.;* O.-Ind.
Yomate: *Astronium graveolens* Jacq.; *Anacardiac.;* Kol.
Yomi: *Scorodophloeus zenkeri* Harms; *Caesalpiniac.;* Kam.
Yòn: *Anogeissus acuminata* Wall.; *Combretac.;* Burma.
Yongmaw: *Adina cordifolia* Hook. f.; *Rubiac.;* O.-Ind.
Yorocónte: *Talauma sp.; Magnoliac.;* Guat., Hond.
Yorrell: *Eucalyptus gracilis* F. v. M.; *Myrtac.;* w. Au.
Yos: *Sapium sp.; Euphorbiac.;* C.-R.
Yotsa: *Butyrospermum parki* Kotschy; *Sapotac.;* Togo.
Youcamoney: *Protium guianense* March., *P. sp.; Burserac.;* Gu.
You rouoari: *Uncaria guianensis* Gmel.; *Rubiac.;* Ven.
Youro: *Securidaca longipedunculata* Fresen.; *Polygalac.;* Sudan.
Youzou: *Citrus limonum* Risso; *Rutac.;* Jap.
Yowı: *Mitragyne macrophylla* Hiern; *Rubiac.;* Togo.
Ypil: *Intsia bijuga* O. Ktze.; *Mimosac.;* ö. As.
Yrewo: *Erythrina senegalensis* DC.; *Papilionac.;* Togo.
Ytayucuine: *Bourreria huanita* Hemsl.; *Borraginac.;* Mex.
Ytú: *Apuleia ferrea* Mart.; *Caesalpiniac.;* Bras. (Sao P)
Yu-latsen: *Vitex peduncularis* Wall.; *Verbenac.;* O.-Ind.
„ mou: *Melia azedarach* L.; *Meliac.;* Ind.-Ch. (Annam)
Yuba wood: *Eucalyptus sp.; Myrtac.;* Au.
Yuca antígua: *Manihot pittieri* Pax et Hoffm.; *Euphorbiac.;* n. Kol.
Yuchan: *Chorisia insignis* H. B. K.; *Bombacac.;* Arg.
Yuco: *Byrsonima crassifolia* DC., *B. sp.; Malpighiac.;* Kol.
Yucu-ndatura: *Taxodium mucronatum* Ten.; *Conifer.;* Mex.
Yudzu: *Citrus limonum* Risso; *Rutac.;* Jap.
Yuguazú: *Gleditschia amorphoides* Taub.; *Caesalpiniac.;* Arg.
Yuguilla: *Zamia sp.; Cycadac.;* ö. Cuba.
Yühngo: *Treculia africana* Dcne.; *Morac.;* Elf.
Yukerí-ruzú-rá: *Pisonia praecox* O. Ktze.; *Nyctaginac.;* Arg.
„ -sy: *Pisonia praecox* O. Ktze.; *Nyctaginac.;* Arg.
Yuki-rá: *Maytenus boaria* Molina; *Celastrac.;* Arg.
Yukirú-ruzú: *Pisonia praecox* O. Ktze.; *Nyctaginac.;* Arg.
Yulu: *Swietenia macrophylla* King; *Meliac.;* Nic.
Yumbu: *Grewia gigantiflora* K. Schum.; *Tiliac.;* Togo.
Yung: *Anogeissus acuminata* Wall.; *Combretac.;* w. Burma (Arrecan)
Yuquerí-buzú: *Pisonia praecox* O. Ktze.; *Nyctaginac.;* Arg.
Yuquery guazú: *Acacia riparia* H. B. K.; *Mimosac.;* Arg.
Yuriballi: ? *Protium sp.; Burserac.;* br. Gu.
Yutahí: *Hymenaea courbaril* L.; *Caesalpiniac.;* Bras.
Yutahy: *Hymenaea courbaril* L.; *Caesalpiniac.;* Bras.
Yutnusatnu: *Pinus montezumae* Lamb.; *Conifer.;* Mex.
Yutnu-itne: *Juniperus sp.; Conifer.;* Mex.
„ -nuu: *Salix sp.; Salicac.;* Mex.
„ -tandaa: *Guaiacum palmeri* Vail; *Zygophyllac.;* Mex.
Yuvia: *Bertholletia excelsa* Bonpl.; *Lecythidac.;* Gu., n. Bras., Ven.
Yuy: *Crataeva tapia* L.; *Capparidac.;* br. Hond.
Yuyubo: *Ziziphus mauritiana* Lam ; *Rhamnac.;* Ven.
Yuzuriha: *Daphniphyllum macropodum* Miq.; *Euphorbiac.;* Jap.
? Yvá-si-y-guazú: *Phyllostylon brasiliensis* Capanema; *Ulmac.;* Arg.
Yvirá-pitá: *Peltophorum vogelianum* Bth.; *Caesalpiniac.;* Arg.

Ywegyi: *Adenanthera pavonina* L.; *Mimosac.;* O.-Ind.
Yzerhart: *Swarʒia sp.; Caesalpiniac.;* Sur.
„ , Bastard- (h): *Chlorophora tinctoria* Gaudich.; *Morac.;* C.-Am , S.-Am.
Yzer-hout: *Olea capensis* L.; *Oleac.;* S.-Af. (Kap)

Z

Za: *Adansonia madagascariensis* Baill.; *Bombacac.;* Mad.
Zaad-broodboom: *Artocarpus integrifolia* L.; *Morac.;* fr. Gu.
Zabel, doux: *Persea sp.; Laurac.;* Guad.
Zaboca: *Persea americana* Mill.; *Laurac.;* Mex.
Zabon: *Citrus decumana* Willd.; *Rutac.;* Jap.
Zaccone: *Balanites aegyptiaca* Delile; *Zygophyllac.;* Arab.
Zachum: *Balanites aegyptiaca* Delile; *Zygophyllac.;* Arab.
Zacoba: *Charia indeniensis* Chev.; *Meliac.;* Elf.
Zacon: *Balanites aegyptiaca* Delile; *Zygophyllac.;* Arab.
Zada bauakh: *Abutilon longicuspe* Hochst.; *Malvac.;* n. O.-Af. (Abessinien)
Zaehthout: *Hura crepitans* L.; *Euphorbiac.;* Sur.
Zagat: *Quercus semiserrata* Roxb.; *Fagac.;* Burma.
Zagrinette: *Bactris pectinata* Mart.; *Palmac.;* fr. Gu.
Zah: *Necepsia afzeli* Prain; *Euphorbiac.;* Lib.
Zaham borizano: *Eugenia jambosa* L.; *Myrtac.;* Mad.
Zahano: *Phyllarthron bojerianum* DC.; *Bignoniac.;* Mad.
Zahdi: *Juniperus procera* Hochst.; *Conifer.;* n. O.-Af. (Abessinien)
Zahn: *Pterocarpus santalinoides* L'Hérit.; *Papilionac.;* Lib.
Zahnbürstenholz (d): *Salvadora persica* L.; *Salvadorac.;* As., O.-Af.
Zahnpuʒholz (d): *Garcinia kola* Heckel; *Guttifer.;* Go.
Zaitun: *Olea ferruginea* Royle; *Oleac.;* O.-Ind. (Afg.)
Zaklu: *Detarium microcarpum* Guill. et Perr.; *Caesalpiniac.;* Togo.
Zakuériombi: *Linociera manni* Solereder; *Oleac.;* Elf.
Zalatni: *Rhododendron arboreum* Sm.; *Ericac.;* Burma.
Zamanguila: *Khaya sp.; Meliac.;* Gab.
? Zambèze: *Haronga madagascariensis* Choisy; *Guttifer.;* Gab.
Zamniguila: *Khaya ivorensis* Chev.; *Meliac.;* Gab.
Zamo baboso: *Heliocarpus attenuatus* S. Wats.; *Tiliac.;* Mex.
Zamourette: *Drepanocarpus lunatus* G. F. W. Meyer; *Papilionac.;* fr. Gu.
Zandkokerboom: *Hura crepitans* L., *H. polyandra* Baill.; *Euphorbiac.;* Sur.
Zante wood: *Chloroxylon swietenia* DC.; *Rutac.;* O.-Ind., Ceylon.
Zapallo: *Pisonia sp.; Nyctaginac.;* Arg.
„ -caspi: *Pisonia zaballo* Griseb.; *Nyctaginac.;* Arg.
Zapatero: *Peltophorum adnatum* Griseb.; *Caesalpiniac.;* Cuba.
„ : *Peltogyne sp.; Caesalpiniac.;* Trin.
„ : *Hieronyma alchorneoides* Allem.; *Papilionac.;* Pan.
„ : *Casearia praecox* Griseb.; *Flacourtiac.;* Ven.
„ : *Aspidosperma vargasi* A. DC.; *Apocynac.;* Ven.
„ : *Tecoma pentaphylla* Juss.; *Bignoniac.;* Ant.
„ de Vénézuela: *Casearia praecox* Griseb.; *Flacourtiac.;* Ven.
Zapatillo: *Clusia mexicana* Vesque; *Guttifer.;* Mex.
Zapatón: *Pachira macrocarpa* Walp.; *Bombacac.;* Guat.
Zapote: *Achras sapota* L.; *Sapotac.;* Cuba, Mex.
„ : *Calocarpum mammosum* Pierre; *Sapotac.;* Cuba, Mex., C.-Am., Ven.
„ : *Calocarpum viride* Pittier; *Sapotac.;* Nic.
„ : *Licania platypus* Fritsch; *Rosac.;* C.-R., Pan.
„ amarillo: *Licania platypus* Fritsch; *Rosac.;* Mex.
„ „ : *Lucuma salicifolia* H. B. K.; *Sapotac.;* Mex.
„ „ : *Couepia polyandra* Rose; *Rosac.;* Mex.
„ blanco: *Casimiroa edulis* Llave et Lexarza; *Rutac.;* Mex.
„ „ : *Lucuma hypoglauca* Standl.; *Sapotac.;* Mex. (Yucatan)
„ „ : *Achras sapota* L.; *Sapotac.;* C.-Am.
„ „ : *Calocarpum viride* Pittier; *Sapotac.;* C.-R.
„ bobo: *Pachira aquatica* Aubl.; *Bombacac.;* Mex.
„ bolo: *Couepia sp.; Rosac.;* Salv.

Zapote borracho: *Licania platypus* Fritsch; *Rosac.;* Mex.
„   „   : *Lucuma salicifolia* H. B. K.; *Sapotac.;* Mex.
„ cabello: *Licania platypus* Fritsch; *Rosac.;* Mex.
„ chico: *Achras sapota* L.; *Sapotac.;* Mex., Guat.
„ colorado: *Calocarpum mammosum* Pierre; *Sapotac.;* Mex.
„   „   : *Achras sapota* L.; *Sapotac.;* C.-Am.
„ Domingo: *Mammea americana* L.; *Guttifer.;* Mex.
„ faisán: *Sideroxylon amygdalinum* Standl.; *Sapotac.;* Guat.
„   „   : *Dipholis stevensoni* Standl.; *Sapotac.;* br. Hond.
„   „   : *Dipholis minutiflora* Pittier; *Sapotac.;* C.-R.
„ ingerto: *Calocarpum viride* Pittier, *C. mammosum* Pierre; *Sapotac.;* Guat., Salv.
„ mamey: *Mammea americana* L.; *Guttifer.;* Mex.
„   „   : *Calocarpum mammosum* Pierre; *Sapotac.;* Mex., Kol.
„ morado: *Achras sapota* L.; *Sapotac.;* C.-Am.
„ negro: *Diospyros ebenaster* Retz., *D. sp.; Ebenac.;* Mex.
„ niño: *Mammea americana* L.; *Guttifer.;* Mex.
„   „   : *Calocarpum parvum* Pierre; *Sapotac.;* Mex.
„ prieto: *Diospyros ebenaster* Retz.; *Ebenac.;* Mex.
„ reventador: *Pachira macrocarpa* Walp.; *Bombacac.;* Mex.
„ de agua: *Pachira aquatica* Aubl.; *Bombacac.;* Mex.
„ „ ave: *Sideroxylon capiri* Pittier; *Sapotac.;* Mex.
„ „ mono: *Couroupita darienensis* Pitt., *C. parviflora* Standl.; *Lecythidac.;* Pan.
„ „ niño: *Lucuma salicifolia* H. B. K.; *Sapotac.;* C.-Am.
„ „ Santo Domingo: *Mammea americana* L.; *Guttifer.;* Mex.
Zapotile: *Cordia sp.; Borraginac.;* C.-R.
Zapotillo: *Achras sapota* L.; *Sapotac.;* Mex.
„   : *Sideroxylon meyeri* Standl.; *Sapotac.;* Mex., (Campeche), br. Hond.
„   : *Lucuma durlandi* Standl., *L. salicifolia* H. B. K.; *Sapotac.;* C.-Am.
„   : *Achras chicle* Pittier; *Sapotac.;* Guat., Hond.
„   : *Couepia polyandra* Rose, *C. dodecandra* Hemsl.; *Rosac.;* Guat., Hond.
„   : *Calocarpum viride* Pittier; *Sapotac.;* Hond.
„   : *Saurauia villosa* DC.; *Dilleniac.;* Hond.
„   : *Labatia standleyana* Pittier; *Sapotac.;* w. Hond.
„   · *Chrysophyllum olivaeforme* L.; *Sapotac.;* Salv.
„   : *Couepia sp.; Rosac.;* Salv.
„   : *Crataeva tapia* L.; *Capparidac.;* Mex.
„ amarillo: *Couepia sp.; Rosac.;* Salv.
„ calenturiente: *Lucuma laeteviridis* Pittier; *Sapotac.;* Guat.
„ de montaña: *Clethra lanata* Mart. et Gal.; *Clethrac.;* Salv.
„ „ peña: *Bumelia leiogyna* Donn. Smith; *Sapotac.;* Salv.
Zapotón: *Pachira macrocarpa* Walp., *P. aquatica* Aubl.; *Bombacac.;* C.-Am.
Zapotte: *Calocarpum mammosum* Pierre; *Sapotac.;* fr. W.-Ind.
„ . große: *Calocarpum mammosum* Pierre; *Sapotac.;* fr. W.-Ind.
„ à crême: *Calocarpum mammosum* Pierre; *Sapotac.;* fr. W.-Ind.
Zaquenette: *Bactris pectinata* Mart.; *Palmac.;* fr. Gu.
Zaragatoa: *Evonymus europaea* L.; *Celastrac.;* Port.
Zar-deh-jay: *Gaertnera cooperi* Hutch. et Moss; *Loganiac.;* Lib.
Zarza: *Rubus discolor* Weihe et Nees; *Rosac.;* Sp.
„ : *Celtis sp., Ulmac.;* Cuba.
„ : *Mimosa polycarpa* Bth.; *Mimosac.;* Guat.
„ colorada: *Piptadenia communis* Bth.; *Mimosac.;* n. Kol.
„ escaramujera: *Rosa rubiginosa* L.; *Rosac.;* Sp.
Zarzamora: *Rubus amoenus* Port., *R. thyrsoideus* Wimm.; *Rosac.;* Sp.
Zar-zreh-wehn-ye: *Macaranga sp.; Euphorbiac.;* Lib.
Zatzumbo: *Ichthyomethia americana* Blake; *Papilionac.;* Mex.
Zauh: *Hannoa klaineana* Pierre et Engl.; *Simarubac.;* Lib.
Zaungbale: *Lagerstroemia parviflora* Roxb.; *Lythrac.;* Burma.
Zaungi: *Dalbergia cultrata* Graham; *Papilionac.;* Burma.
Zeb-blo: *Scytopetalum tieghemi* Hutch. et Dalz.; *Scytopetalac.;* Lib.
Zebraholz, afrikanisch: *Berlinia bracteosa* Bth., *Macrolobium sp.; Caesalpiniac.;* Kam.
Zebrahout (h): *Connarus guianensis* Lamb.; *Connarac.;* fr. Gu.
Zébrano: *Cynometra aff. lujaei* de Wild., *Macrolobium sp.; Caesalpiniac.;* Kam., Gab.

563

Zebra wood: *Diospyros kurzi* Hiern, *D. oocarpa* Thwaites; *Ebenac.;* O.-Ind.
„ „ : *Diospyros marmorata* Parker; *Ebenac.; And.
„ „ : *Connarus guianensis* Lamb.; *Connarac.;* br. Gu.
„ „ : *Pithecolobium racemiflorum* Ducke; *Mimosac.;* Gu., n. Bras.
„ „ : ? *Caesalpinia paraense* Ducke; *Caesalpiniac.;* Trin.
„ „ : *Astronium fraxinifolium* Schott; *Anacardiac.;* Bras.
„ „ , african: *Macrolobium sp.; Caesalpiniac.;* W.-Af.
„ „ , Mountain-: *Eugenia fragrans* Willd.; *Myrtac.;* Jam.
„ „ , red: ? *Diospyros oocarpa* Thwaites; *Ebenac.;* Burma.
Zeder (d): *Cedrus spp.; Conifer.;* Kl.-As., O.-Ind., n. Af.
„ : *Juniperus procera* Hochst.; *Conifer.;* D.-O.-Af.
„ , Alaska-: *Chamaecyparis nootkaënsis* Spach; *Conifer.;* w. N.-Am.
„ , Atlas-: *Cedrus atlantica* Manetti; *Conifer.;* n. Af.
„ , Bermudas-: *Juniperus bermudiana* L.; *Conifer.;* W.-Ind. (Bermudas)
„ , Bleistift- (d): *Juniperus virginiana* L.; *Conifer.;* ö. N.-Am.
„ , „ -: *Juniperus procera* Hochst.; *Conifer.;* O.-Af.
„ , „ -: *Dysoxylum fraserianum* Bth.; *Meliac.;* Au.
„ , Deodar-: *Cedrus deodara* Loud.; *Conifer.;* O.-Ind. (Himal.)
„ , Florida-: *Juniperus bermudiana* L.; *Conifer.;* W.-Ind. (Bermudas)
„ , Fluss-, kalifornische: *Libocedrus decurrens* Torr.; *Conifer.;* w. USA.
„ , Haiti-: *Juniperus gracilior* Pilger; *Conifer.;* Haiti.
„ , Libanon-: *Cedrus libani* Barr.; *Conifer.;* Kl.-As.
„ , Neuseeland-: *Libocedrus doniana* Endl.; *Conifer.;* N.-Seel., Tasm.
„ , Oregon-: *Chamaecyparis lawsoniana* Parl.; *Conifer.;* w. N.-Am.
„ , Ostafrika-: *Juniperus procera* Hochst.; *Conifer.;* O.-Af.
„ , Schumewald-: *Juniperus procera* Hochst.; *Conifer.;* O.-Af.
„ , virginische: *Juniperus virginiana* L.; *Conifer.;* ö. N.-Am.
„ , Weihrauch-: *Libocedrus decurrens* Torr.; *Conifer.;* w. USA.
Zedernholz (d): *Cedrela odorata* L., *C. spp.; Meliac.;* tr. Am.
Zedrelaholz (d): *Cedrela spp.; Meliac.;* tr. Am.
? Zeedrat: *Peltogyne sp.; Caesalpiniac.;* Sur.
Ze-eh: *Leea guineensis* G. Don; *Vitac.;* Lib.
Zee-druif: *Coccoloba uvifera* L.; *Polygonac.;* Sur.
Zeep-hout: *Caryocar tomentosum* Willd.; *Caryocarac.;* Sur.
„ „ : *Caryocar glabrum* Pers., *C. villosum* Pers.; *Caryocarac.;* Sur.
Zellim bauakh: *Dombeya bruceana* Rich.; *Sterculiac.;* n. O.-Af. (Abessinien)
Zellimo: *Gymnosporia luteola* Loes.; *Celastrac.;* n. O.-Af. (Abessinien)
Zembi: *Klainedoxa gabunensis* Pierre; *Simarubac.;* Kam.
Zembila: *Irvingia grandifolia* Engl.; *Simarubac.;* B.-K. (Mayumbe)
Zen-nieh: *Strombosia pustulata* Oliv.; *Olacac.;* Lib.
Zercotte (u): ? *Astronium conzatti* Blake; *Anacardiac.;* C.-Am.
Zerongo: *Sarcocephalus trillesi* Pierre, *S. pobeguini* Pob.; *Rubiac.;* Elf.
Zi: *Zizyphus jujuba* Lamk.; *Rhamnac.;* Burma.
Zibyu: *Phyllanthus emblica* L.; *Euphorbiac.;* Burma.
Zidaw: *Zizyphus jujuba* Lamk.; *Rhamnac.;* Burma.
Zidéno: *Albizzia amara* Boiv.; *Mimosac.;* n. O.-Af. (Abessinien)
Zigbli: *Dialium guineense* Willd.; *Caesalpiniac.;* Togo, Go.
Zimingelia: *Terminalia superba* Engl. et Diels; *Combretac.;* Kam.
Zimmt, weisser (d): *Canella winterana* Gaertn.; *Canellac.;* s. Fl., W.-Ind., Kol., Ven.
Zina: *Cereus schotti* Engelm.; *Cactac.;* Mex.
Zinbye-bo: *Lagerstroemia parviflora* Roxb.; *Lythrac.;* nö. Burma (Ruby Mines)
Zinbyun: *Dillenia pentagyna* Roxb.; *Dilleniac.;* Burma.
Zing: *Monodora myristica* Dunal; *Anonac.;* Gab.
Zingan: *Pseudocedrela kotschyi* Harms; *Meliac.;* W.-Af.
Zingana (f): *Cynometra aff. lujaei* de Wild.; *Caesalpiniac.;* Kam., Gab.
„ : *Macrolobium sp.; Caesalpiniac.;* Gab.
Zingili: *Prosopis oblonga* Bth.; *Mimosac.;* br. O.-Af (nw. Uganda)
Ziopou: *Tessmannia baikiaeoïdes* Hutch. et Dalz.; *Caesalpiniac.;* Elf.
Zipote: *Mimusops sp.; Sapotac.;* P.-R.
Zirbe (d): *Pinus cembra* L.; *Conifer.;* Mitt.-Eu.
Ziricote: *Cordia dodecandra* DC.; *Borraginac.;* s. Mex.
„ : *Cordia sebestena* L.; *Borraginac.;* br. Hond.

36*

Zitan: *Dipterocarpus vernicifluus* Blco.; *Dipterocarpac.*; Phil. (Cagayan)
Zitronenholz, afrikanisch (d): *Distemonanthus benthamianus* Baill.; *Caesalpiniac.*;
Zizako: *Cynometra sp.*; *Caesalpiniac.*; Kam. [W.-Af.
Zoe: *Macrolobium spp.*; *Caesalpiniac.*; Lib.
„ -kpoe: *Tylostemon manni* Stapf; *Laurac.*; Lib.
Zoélé: *Khaya ivorensis* Chev.; *Meliac.*; Kam.
Zogalagandi: *Moringa pterygosperma* Gaertn.; *Moringac.*; n. Nig.
Zompancle: *Erythrina sp.*; *Papilionac.*; Mex.
Zompantle: *Erythrina sp.*; *Papilionac.*; Mex.
Zompantli: *Erythrina sp.*; *Papilionac.*; Mex.
Zonderhart (h): *? Perebea sp.*; *Morac.*; Sur.
Zongolica: *Bursera simaruba* Sarg.; *Burserac.*; Mex.
„ : *Astronium conzattii* Blake; *Anacardiac.*; Mex.
Zoo: *Randia acuminata* Bth.; *Rubiac.*; Kam.
Zoofim: *? Diospyros spp.*, *Maba spp.*; *Ebenac.*; Kam.
Zope: *Ichthyomethia grandifolia* Blake; *Papilionac.*; Salv.
Zopilocuahuitl: *Swietenia sp.*; *Meliac.*; Mex.
Zopilote: *Swietenia sp.*; *Meliac.*; Mex.
„ : *Pithecolobium arboreum* Urban; *Mimosac.*; br. Hond.
„ : *Plumiera multiflora* Standl.; *Apocynac.*; br. Hond.
„ : *Cornutia grandifolia* Schauer; *Verbenac.*; Hond.
„ : *Solanum torvum* Sw.; *Solanac.*; Nic.
„ colorado: *Swietenia sp.*; *Meliac.*; Mex.
„ negro: *Swietenia sp.*; *Meliac.*; Mex.
Zopilotl: *Swietenia sp.*; *Meliac.*; Mex.
Zopilo-zontecomacuahuitl: *Swietenia sp.*; *Meliac.*; Mex.
Zorillo: *Zanthoxylum sp.*; *Rutac.*; Cuba.
„ : *Roupala sp.*; *Proteac.*; Salv.
Zorra: *Schizolobium parahybum* Blake, *S. kellermani* Pittier; *Caesalpiniac.*; Guat.,
„ : *Alvaradoa amorphoides* Liebm.; *Simarubac.*; Hond. [Hond.
„ : *Enterolobium saman* Prain; *Mimosac.*; Salv.
Zorrocloco: *Morisonia americana* L.; *Capparidac.*; Ven.
Zouga: *Aucoumea klaineana* Pierre; *Burserac.*; Gab.
Zoulé: *Tylostemon sp.*; *Laurac.*; Kam.
Zouocébé: *Loesenera kalantha* Harms; *Caesalpiniac.*; Elf.
Zoy: *Sideroxylon gaumeri* Pittier; *Sapotac.*; br. Hond.
Zree-chu: *Pachypodanthium staudti* Engl. et Diels; *Anonac.*; Lib.
Zrok: *Platymiscium polystachyum* Bth.; *Papilionac.*; Pan.
Zucker-bos: *Protea mellifera* Thunb.; *Proteac.*; S.-Af.
Zuckerkistenholz (d): *Cedrela odorata* L., *C. spp.*; *Meliac.*; tr. Am.
Zuinan: *Erythrina crista-galli* L.; *Papilionac.*; Arg.
Zum: *Prunus padus* L.; *Rosac.*; nw. O.-Ind. (Punjab)
Zumpantle: *Erythrina sp.*; *Papilionac.*; Mex.
Zungea: *Grewia ferruginea* Hochst.; *Tiliac.*; n. O.-Af. (Abessinien)
Zürgelbaum (d): *Celtis australis* L.; *Ulmac.*; M.-M.
„ , europäische: *Celtis australis* L.; *Ulmac.*; M.-M.
Zuurzak (h): *Anona muricata* L.; *Anonac.*; Gu.
Zuwo: *Celtis integrifolia* Lam.; *Ulmac.*; n. Nig.
Zwahn: *Parinarium kerstingi* Engl.; *Rosac.*; Lib.
Zwartbast: *Royena lucida* L.; *Ebenac.*; S.-Af. (Kap)
Zweer-hout: *? Jacaranda copaia* D. Don; *Bignoniac.*; Gu.
Zweren-zalf: *? Jacaranda copaia* D. Don; *Bignoniac.*; fr. Gu.
Zwetschenbaum (d): *Prunus domestica* L.; *Rosac.*; Eu.
Zybast (h): *Celastrus acuminatus* L.; *Celastrac.*; S.-Af. (Kap)
Zypresse (d): *Cupressus sempervirens* L.; *Conifer.*; It.
„ , Feuer-: *Chamaecyparis obtusa* S. et Z.; *Conifer.*; Jap.
„ , mexikanische: *Taxodium mucronatum* Ten.; *Conifer.*; Mex.
„ , Nutka-: *Chamaecyparis nutkaënsis* Spach; *Conifer.*; w. N.-Am.
„ , Sawara-: *Chamaecyparis pisifera* S. et Z.; *Conifer.*; Jap.
„ , Sitka-: *Chamaecyparis nutkaënsis* Spach; *Conifer.*; nw. N.-Am.
„ , Sumpf-: *Taxodium distichum* Rich.; *Conifer.*; sö. N.-Am.
„ , Zeder-: *Chamaecyparis thyoides* Britt.; *Conifer.*; ö. N.-Am.